2012 IBC®

CODE AND COMMENTARY

Volume 2

2012 International Building Code® Commentary

First Printing: September 2011

ISBN:978-1-60983-063-2 (soft-cover edition)

PRINTED IN THE U.S.A.

PREFACE

The principal purpose of the Commentary is to provide a basic volume of knowledge and facts relating to building construction as it pertains to the regulations set forth in the 2012 *International Building Code*. The person who is serious about effectively designing, constructing and regulating buildings and structures will find the Commentary to be a reliable data source and reference to almost all components of the built environment

As a follow-up to the International Building Code, we offer a companion document, the *International Building Code Commentary—Volume II.* Volume II covers Chapters 16 through 35 and the appendices of the 2012 *International Building Code*. The basic appeal of the Commentary is thus: it provides in a small package and at reasonable cost thorough coverage of many issues likely to be dealt with when using the *International Building Code* — and then supplements that coverage with historical and technical background. Reference lists, information sources and bibliographies are also included.

Throughout all of this, effort has been made to keep the vast quantity of material accessible and its method of presentation useful. With a comprehensive yet concise summary of each section, the Commentary provides a convenient reference for regulations applicable to the construction of buildings and structures. In the chapters that follow, discussions focus on the full meaning and implications of the code text. Guidelines suggest the most effective method of application, and the consequences of not adhering to the code text. Illustrations are provided to aid understanding; they do not necessarily illustrate the only methods of achieving code compliance.

The format of the Commentary includes the full text of each section, table and figure in the code, followed immediately by the commentary applicable to that text. At the time of printing, the Commentary reflects the most up-to-date text of the 2012 *International Building Code*. As stated in the preface to the *International Building Code*, the content of sections in the code which begin with a letter designation (i.e., Section [P]2903.1) are maintained by another code development committee. Each section's narrative includes a statement of its objective and intent, and usually includes a discussion about why the requirement commands the conditions set forth. Code text and commentary text are easily distinguished from each other. All code text is shown as it appears in the *International Building Code*, and all commentary is indented below the code text and begins with the symbol ❖.

Readers should note that the Commentary is to be used in conjunction with the *International Building Code* and not as a substitute for the code. The Commentary is advisory only; the code official alone possesses the authority and responsibility for interpreting the code.

Comments and recommendations are encouraged, for through your input, we can improve future editions. Please direct your comments to the Codes and Standards Development Department at the Chicago District Office.

The International Code Council would like to extend its thanks to the following individuals for their contributions to the technical contentof this commentary:

Chris Marion — Gregory Cahanin
Jeff Tubbs — David Cooper
Rebecca Quinn — Dave Collins
Joann Surmar — Vickie Lovell
James Milke — John Valiulis
Richard Walke — Marcelo Hirschler
Dave Adams — Edward Keith
Zeno Martin — Phillip Samblanet
Jason Thompson

TABLE OF CONTENTS

Chapter 16:
Structural Design

General Comments

This chapter contains the commentary for the following structural topics: definitions of structural terms, construction document requirements, load combinations, dead loads, live loads, snow loads, wind loads, soil lateral loads, rain loads, flood loads and earthquake loads. This chapter provides minimum design requirements so that all buildings and structures are proportioned to resist the loads and forces that are likely to be encountered. The loads specified herein have been established through research and service performance of buildings and structures. The application of these loads and adherence to the serviceability criteria will enhance the protection of life and property. The earthquake loads, wind loads and snow loads in this chapter are based on the 2010 edition of ASCE 7. The earthquake criteria and ASCE 7 load requirements are based on the National Earthquake Hazards Reduction Program's (NEHRP) *Recommended Provisions for Seismic Regulations for New Buildings and other Structures* (FEMA 450). The NEHRP provisions were prepared by the Building Seismic Safety Council (BSSC) for the Federal Emergency Management Agency (FEMA).

Purpose

The purpose of this chapter is to prescribe minimum structural loading requirements for use in the design and construction of buildings and structures with the intent to minimize hazard to life and improve the occupancy capability of essential facilities after a design level event or occurrence.

SECTION 1601
GENERAL

1601.1 Scope. The provisions of this chapter shall govern the structural design of buildings, structures and portions thereof regulated by this code.

❖ While a significant portion of Chapter 16 is dedicated to the determination of minimum design loads, it also includes other important criteria that impact the design of structures, such as the permitted design methodologies, as well as the combinations of design loads used to establish the required minimum strength of structural members. Unless stated otherwise, the criteria found in this chapter are applicable to all buildings and structures. See Chapter 34 for application of these requirements to alterations, additions or repairs to existing structures.

SECTION 1602
DEFINITIONS AND NOTATIONS

1602.1 Definitions. The following terms are defined in Chapter 2:

❖ Definitions facilitate the understanding of code provisions and minimize potential confusion. To that end, this section lists definitions of terms associated with structural design. Note that these definitions are found in Chapter 2. The use and application of defined terms, as well as undefined terms, are set forth in Section 201.

ALLOWABLE STRESS DESIGN.

DEAD LOADS.

DESIGN STRENGTH.

DIAPHRAGM.

Diaphragm, blocked.

Diaphragm boundary.

Diaphragm chord.

Diaphragm flexible.

Diaphragm, rigid.

DURATION OF LOAD.

ESSENTIAL FACILITIES.

FABRIC PARTITION.

FACTORED LOAD.

HELIPAD.

ICE-SENSITIVE STRUCTURE.

IMPACT LOAD.

LIMIT STATE.

LIVE LOAD.

LIVE LOAD (ROOF).

LOAD AND RESISTANCE FACTOR DESIGN (LRFD).

LOAD EFFECTS.

LOAD FACTOR.

LOADS.

NOMINAL LOADS.

OTHER STRUCTURES.

PANEL (PART OF A STRUCTURE).

RESISTANCE FACTOR.

RISK CATEGORY.

STRENGTH, NOMINAL.

STRENGTH, REQUIRED.

STRENGTH DESIGN.

SUSCEPTIBLE BAY.

VEHICLE BARRIER.

NOTATIONS.

D = Dead load.

D_i = Weight of ice in accordance with Chapter 10 of ASCE 7.

E = Combined effect of horizontal and vertical earthquake induced forces as defined in Section 12.4.2 of ASCE 7.

F = Load due to fluids with well-defined pressures and maximum heights.

F_a = Flood load in accordance with Chapter 5 of ASCE 7.

H = Load due to lateral earth pressures, ground water pressure or pressure of bulk materials.

L = Roof live load greater than 20 psf (0.96 kN/m^2) and floor live load.

L_r = Roof live load of 20 psf (0.96 kN/m^2) or less.

R = Rain load.

S = Snow load.

T = Self-straining load.

V_{asd} = Nominal design wind speed (3-second gust), miles per hour (mph) (km/hr) where applicable.

V_{ult} = Ultimate design wind speeds (3-second gust), miles per hour (mph) (km/hr) determined from Figures 1609A, 1609B, or 1609C or ASCE 7.

W = Load due to wind pressure.

W_i = Wind-on-ice in accordance with Chapter 10 of ASCE 7.

❖ These notations are used to refer to specific nominal loads that are determined in this chapter for use in the load combinations in Section 1605:

- *D* is the nominal dead load determined in Section 1606. Also see the definition of "Dead load."
- D_i is the weight of ice. See the ASCE 7 provisions referenced in Section 1614.
- Earthquake load effect, *E*, in Section 12.4.2 of ASCE 7 includes the effects of the horizontal load, E_h, as well as a vertical component, E, E_h is the product of the redundancy factor, ρ, and Q_E, the effects of horizontal earthquake forces. E„ accounts for vertical acceleration due to earthquake ground motion, taken as $0.2S_{DS}D$. Note that its magnitude is not intended to represent a total vertical response, since that is not likely to coincide with the maximum horizontal response. It is essentially a portion of the dead load, D, that is added in "additive" load combinations or subtracted in "counteractive" load combinations. The term S_{DS}, design spectral response acceleration at short periods, is explained in the commentary to Section 1613.3.4.

For example, when this expression is used in the alternative allowable stress design load combinations of Section 1605.3.2 that include earthquake load effects the resulting combinations are as follows:

Equation 16-21

$$D + L + S + E/1.4 = (1 + 0.143S_{DS})D + L + S + \rho Q_E/1.4$$

Equation 16-22

$$0.9D + E/1.4 = (0.9 - 0.143S_{DS})D + \rho Q_E/1.4$$

Earthquake design criteria is provided in Section 1613, which, in turn, references the relevant ASCE 7 provisions for computation of the earthquake load effects. While these loads are necessary for establishing the required strength, the computed forces approximate the expected deformations under the design earthquake ground motions and are not applied to a structure in an actual earthquake.

- *F* refers to the nominal load due to fluids having "well defined pressures and maximum heights." Unlike most other nominal loads, there is no code section governing the determination of fluid loads. Also note that *F* includes a vertical component (fluid weight), as well as a horizontal component (lateral pressure).
- F_a is used to refer to the flood load that is determined under Chapter 5 of ASCE 7. Note that F_a is not explicitly included under other loads listed for the alternative ASD combination in Section 1605.3.2.
- *H* is used to refer to the nominal load resulting from lateral soil pressure, lateral pressure of ground water or the lateral pressure of bulk materials. Section 1610 specifies minimum requirements for lateral soil loads. Note that there are not specific provisions for the determination of load resulting from the lateral pressure of bulk materials.
- *L* in the nominal live load determined in accordance with Section 1607 (also see the definition of "Live load"). In addition to floor live loads, it includes roof live loads that exceed the limit on L_r. L_r represents nominal roof live loads up to 20 psf (0.96 N/m^2).
- *R* is the nominal rain load determined in accordance with Section 1611.

- *S* is the nominal snow load determined in accordance with Section 1608.
- *T* is used to refer to self-straining forces resulting from contraction or expansion due to temperature change, shrinkage, moisture change or creep, as well as movement due to differential settlement. A thermal gradient at an exterior wall is an example of a structural element where these self-straining forces can affect the design. Unlike most other nominal loads, there is no code section governing the determination of self-straining forces. *T* is not included directly in the load combinations, but reference to it is found in Sections 1605.2.1 and 1605.3.1.2.
- V_{asd} is the term used to refer to nominal design wind speeds that are determined in Section 1609.3.1.
- V_{ult} is the term used to refer to the mapped wind speeds in order to differentiate them from the nominal design wind speeds.
- *W* is the strength-level wind load determined in accordance with Section 1609.
- W_i is the wind-on-ice loading. See the ASCE 7 provisions referenced in Section 1614.

SECTION 1603 CONSTRUCTION DOCUMENTS

1603.1 General. *Construction documents* shall show the size, section and relative locations of structural members with floor levels, column centers and offsets dimensioned. The design loads and other information pertinent to the structural design required by Sections 1603.1.1 through 1603.1.9 shall be indicated on the *construction documents*.

Exception: *Construction documents* for buildings constructed in accordance with the *conventional light-frame construction* provisions of Section 2308 shall indicate the following structural design information:

1. Floor and roof live loads.
2. Ground snow load, P_g.
3. Ultimate design wind speed, V_{ult}, (3-second gust), miles per hour (mph) (km/hr) and nominal design wind speed, V_{asd}, as determined in accordance with Section 1609.3.1 and wind exposure.
4. *Seismic design category* and *site class*.
5. Flood design data, if located in *flood hazard areas* established in Section 1612.3.
6. Design load-bearing values of soils.

❖ The term "construction documents" is defined in Chapter 2. It is commonly used to refer to calculations, drawings and specifications but it includes other data that is required to indicate compliance with the code as described in Section 107. The purpose of this section is to specifically require the design professional to provide the building official with the appropriate structural details, criteria and design load data for verifying compliance with the provisions of this chapter. Note that additional structural information and specific submittal documents may also be required to be incorporated by Chapters 17 through 23.

The construction documents are required to contain sufficient detail for the building official to perform plan review and field inspection, as well as for construction activity. Dimensions indicated on architectural drawings are not required to be duplicated on the structural drawings and vice versa. The design loads, to be indicated by the design professional on the construction documents, are to be consistent with the loads used in the structural calculations. Note that the loads are not required to be on the construction drawings but must be included within the construction documents in a manner such that the design loads are clear. The building official is to compare the loads on the construction documents with the applicable minimum required loads as specified by this chapter. The inclusion of the load design information is an indication that the structure has been designed for the loads required by the code. It should be emphasized that these requirements for construction documents are applicable regardless of the involvement of a registered design professional, which is regulated by the applicable state's licensing laws. The exception provides a less extensive list of structural data to be indicated for buildings constructed in accordance with the conventional wood-frame provisions of Section 2308. This is appropriate in view of the prescriptive nature of these requirements.

1603.1.1 Floor live load. The uniformly distributed, concentrated and impact floor live load used in the design shall be indicated for floor areas. Use of live load reduction in accordance with Section 1607.10 shall be indicated for each type of live load used in the design.

❖ The purpose of the requirement in this section is to provide information for the building official to facilitate the plan review process. The floor live loads, which are indicated on the construction documents by the design professional, are required to meet or exceed the loads in Section 1607. Any live load reductions taken are also to be indicated.

1603.1.2 Roof live load. The roof live load used in the design shall be indicated for roof areas (Section 1607.12).

❖ This section provides information allowing the building official to facilitate the plan review process. The roof live loads, indicated on the construction documents by the design professional, are required to meet or exceed the loads in Section 1607.12.

1603.1.3 Roof snow load data. The ground snow load, P_g, shall be indicated. In areas where the ground snow load, P_g, exceeds 10 pounds per square foot (psf) (0.479 kN/m^2), the following additional information shall also be provided, regardless of whether snow loads govern the design of the roof:

1. Flat-roof snow load, P_f.

2. Snow exposure factor, C_e.
3. Snow load importance factor, *I*.
4. Thermal factor, C_t.

❖ The roof snow load design basis, indicated on the construction documents (design drawings or specifications) by the design professional, provides information allowing the building official to facilitate the plan review process. The flat-roof snow load, snow exposure factor, snow load importance factor, *I*, and roof thermal factor are not to be less than the minimum requirements established by Section 1608.

1603.1.4 Wind design data. The following information related to wind loads shall be shown, regardless of whether wind loads govern the design of the lateral force-resisting system of the structure:

1. Ultimate design wind speed, V_{ult}, (3-second gust), miles per hour (km/hr) and nominal design wind speed, V_{asd}, as determined in accordance with Section 1609.3.1.
2. *Risk category.*
3. Wind exposure.Where more than one wind exposure is utilized, the wind exposure and applicable wind direction shall be indicated.
4. The applicable internal pressure coefficient.
5. Components and cladding. The design wind pressures in terms of psf (kN/m^2) to be used for the design of exterior component and cladding materials not specifically designed by the *registered design professional.*

❖ The wind load design basis, indicated on the construction documents (design drawings or specifications) by the design professional, provides information allowing the building official to facilitate the plan review process. All five of the indicated items are to be on the submitted construction documents. Each of the indicated items is an important parameter in the determination of the wind resistance that is required in the building framework. The building official should verify that the information is on the construction documents during the plan review process. The correctness of the listed items is the responsibility of the owner or the owner's design professional.

1603.1.5 Earthquake design data. The following information related to seismic loads shall be shown, regardless of whether seismic loads govern the design of the lateral force-resisting system of the structure:

1. *Risk category.*
2. Seismic importance factor, I_e.
3. Mapped spectral response acceleration parameters, S_S and S_1.
4. *Site class.*
5. Design spectral response acceleration parameters, S_{DS} and S_{D1}.
6. *Seismic design category.*
7. Basic seismic force-resisting system(s).
8. Design base shear(s).
9. Seismic response coefficient(s), C_S.
10. Response modification coefficient(s), *R*.
11. Analysis procedure used.

❖ The earthquake load design basis, indicated on the construction documents by the design professional, provides information that allows the building official to facilitate the plan review process. All buildings, except those indicated in the exceptions to Section 1613.1, are to be designed for earthquake effects. The earthquake design data for a specific building are required to meet or exceed the minimum requirements established by Section 1613.

1603.1.6 Geotechnical information. The design load-bearing values of soils shall be shown on the *construction documents*.

❖ Load-bearing values for soils must be documented so that the foundation design can be verified.

1603.1.7 Flood design data. For buildings located in whole or in part in *flood hazard areas* as established in Section 1612.3, the documentation pertaining to design, if required in Section 1612.5, shall be included and the following information, referenced to the datum on the community's Flood Insurance Rate Map (FIRM), shall be shown, regardless of whether flood loads govern the design of the building:

1. In *flood hazard areas* not subject to high-velocity wave action, the elevation of the proposed lowest floor, including the basement.
2. In *flood hazard areas* not subject to high-velocity wave action, the elevation to which any nonresidential building will be dry flood proofed.
3. In *flood hazard areas* subject to high-velocity wave action, the proposed elevation of the bottom of the lowest horizontal structural member of the lowest floor, including the basement.

❖ The flood hazard elevation information to be shown on the construction documents by the registered design professional provides information that allows the building official to facilitate the plan review process. By providing the design documentation required in Section 1612.5 and by citing the specified flood information, the registered design professional is indicating that the building was designed in accordance with the flood hazard requirements in Section 1612. If any portion of a building is in a flood hazard area, then the building must meet the corresponding flood requirements.

Depending on the nature of the designated flood hazard area, certain elevation requirements are to be met. In flood hazard areas not subject to high-velocity wave action (commonly called A/AE zones), the lowest floor of all buildings and structures, or the elevation to which nonresidential buildings are dry floodproofed, is to be located at or above the elevation specified in Section 1612.4 (which references ASCE 24). In flood hazard areas subject to high-velocity wave action (commonly called V or VE zones), the bottom of the lowest horizontal structural member is to be located at or above the elevation specified in Section 1612.4 (which references ASCE 24). These elevations are the main factor used in determining flood insurance premium rates. Constructing a building or structure with its lowest floor (or dry floodproofing) below the required elevation will result in significantly higher flood insurance premiums for the building owner.

1603.1.8 Special loads. Special loads that are applicable to the design of the building, structure or portions thereof shall be indicated along with the specified section of this code that addresses the special loading condition.

❖ Indication of special loads on the construction documents by the design professional provides information that allows the building official to facilitate the plan review process. The design professional is expected to identify any special loads that the occupancy will impose on the structure. These could include the operating weight of specialty equipment, for instance. There are also instances outside of Chapter 16 where the code specifies loading criteria that the structural design must address. For example, Section 415.8.3 requires that liquid petroleum gas facilities be in accordance with NFPA 58. In that document, a room housing a liquefied petroleum gas distribution facility must be separated from an adjacent use with ceilings and walls that are designed for a static pressure of 100 pounds per square foot (psf) (4788 Pa).

1603.1.9 Systems and components requiring special inspections for seismic resistance. *Construction documents* or specifications shall be prepared for those systems and components requiring *special inspection* for seismic resistance as specified in Section 1705.11 by the *registered design professional* responsible for their design and shall be submitted for approval in accordance with Section 107.1. Reference to seismic standards in lieu of detailed drawings is acceptable.

❖ This section provides for construction documents for the systems and components specified in Section 1705.11. Generally, the systems and components are the seismic-force-resisting systems, certain mechanical and electrical equipment and systems and architectural components for buildings where the seismic performance category is high. Construction documents are needed for these important items to verify that they comply with code requirements.

SECTION 1604 GENERAL DESIGN REQUIREMENTS

1604.1 General. Building, structures and parts thereof shall be designed and constructed in accordance with strength design, *load and resistance factor design*, *allowable stress design*, empirical design or conventional construction methods, as permitted by the applicable material chapters.

❖ This section identifies the various design methods that are permitted by the code and referenced design standards. The details for these design methods are either explicitly specified in the code or are located in the design standards that are referenced in the structural material design chapters for concrete, aluminum, masonry, steel and wood. For example, empirical design of masonry is addressed in Section 2109. The design of masonry using allowable stress design is required to be in accordance with TMS 402/ACI 530/ASCE 5 with the modifications in Section 2107.

1604.2 Strength. Buildings and other structures, and parts thereof, shall be designed and constructed to support safely the factored loads in load combinations defined in this code without exceeding the appropriate strength limit states for the materials of construction. Alternatively, buildings and other structures, and parts thereof, shall be designed and constructed to support safely the *nominal loads* in load combinations defined in this code without exceeding the appropriate specified allowable stresses for the materials of construction.

Loads and forces for occupancies or uses not covered in this chapter shall be subject to the approval of the *building official*.

❖ This section describes the strength and allowable stress design methods in the code. It also gives the building official approval authority for structural loads in buildings used for occupancies that are not specifically addressed in this chapter.

1604.3 Serviceability. Structural systems and members thereof shall be designed to have adequate stiffness to limit deflections and lateral drift. See Section 12.12.1 of ASCE 7 for drift limits applicable to earthquake loading.

❖ The stated objective in this section is to limit member deflections and system drifts by providing adequate stiffness. Note that the definition of "Limit state" describes a serviceability limit state as a condition beyond which a structure or member is no longer useful for its intended purpose. This would be in contrast to a strength limit state beyond which the structure or member is considered unsafe. The code does not directly address a serviceability limit state. Instead the code limits certain member deflections for specific nominal loads in addition to the story drift limits for earthquake load effects.

The deflection limits for structural members must be in accordance with the requirements in this and subsequent sections. This section also provides a reference to the story drift limitations in the earthquake provisions. Generally, deflection limits are needed for the comfort of the building occupants and so that the structural member's deflection does not cause damage to supported construction. Excessive deflection can also contribute to excessive vibration, which is discomforting to the occupants of the building.

TABLE 1604.3. See below.

❖ The deflection limits in this table apply when they are more restrictive than those in the structural design standards that are indicated in Sections 1604.3.2 through 1604.3.5. Note that the deflection limits for exterior walls and interior partitions vary depending on the type of finish (i.e., flexible, plaster and stucco or other brittle finishes). A flexible finish is intended to be one that has been designed to accommodate the higher deflection indicated and remain serviceable. A brittle finish is any finish that has not been designed to accommodate the deflection allowed for a flexible finish. The more restrictive limit for plaster or stucco is based on ASTM C 926. The deflection limit for a roof member supporting a plaster ceiling is intended to apply only for a plaster ceiling. The limit for a roof member supporting a gypsum board ceiling is that listed in the table for "supporting a nonplaster ceiling."

The answers to the following frequently asked questions provide further guidance on applying the deflection limits of the code.

Frequently Asked Questions—Table 1604.3

Q1. For purposes of checking the deflection limits of Table 1604.3, should the calculated wind load be used directly or the combined loads from Section 1605.3.1 or 1605.3.2?

A1. In computing deflections to verify compliance with Table 1604.3 limits, the loads shown in the column headings of Table 1604.3 are the only loads that must be applied to the member. The load used to check the deflection in this case is W, the nominal wind load in accordance with Section 1609. It is not necessary to use the load combinations of Section 1605.3 for verifying that the deflection limits have been met.

Q2. Note f permits the wind load to be taken as 0.42 times the "component and cladding" loads for the purpose of determining deflection limits. Please explain the basis for this 0.42 adjustment.

A2. There are two aspects to this adjustment. The first is the conversion from a wind speed with a 50-year mean recurrence interval (MRI) to a 10-year MRI event at an allowable stress design level. The mapped wind speeds of the code have generally been based on an MRI of 50 years. Serviceability checks, such as deflection, have typically been based on a lower MRI (for example, 10 years). The ASCE 7 commentary to the 2005 edition provided factors to covert to wind speeds with MRIs other than 50 years in

TABLE 1604.3
DEFLECTION LIMITS[a, b, c, h, i]

CONSTRUCTION	L	S or W[f]	$D + L$[d, g]
Roof members:[e]			
Supporting plaster or stucco ceiling	l/360	l/360	l/240
Supporting nonplaster ceiling	l/240	l/240	l/180
Not supporting ceiling	l/180	l/180	l/120
Floor members	l/360	—	l/240
Exterior walls and interior partitions:			
With plaster or stucco finishes	—	l/360	—
With other brittle finishes	—	l/240	—
With flexible finishes	—	l/120	—
Farm buildings	—	—	l/180
Greenhouses	—	—	l/120

For SI: 1 foot = 304.8 mm.

a. For structural roofing and siding made of formed metal sheets, the total load deflection shall not exceed l/60. For secondary roof structural members supporting formed metal roofing, the live load deflection shall not exceed l/150. For secondary wall members supporting formed metal siding, the design wind load deflection shall not exceed l/90. For roofs, this exception only applies when the metal sheets have no roof covering.

b. Interior partitions not exceeding 6 feet in height and flexible, folding and portable partitions are not governed by the provisions of this section. The deflection criterion for interior partitions is based on the horizontal load defined in Section 1607.14.

c. See Section 2403 for glass supports.

d. For wood structural members having a moisture content of less than 16 percent at time of installation and used under dry conditions, the deflection resulting from $L + 0.5D$ is permitted to be substituted for the deflection resulting from $L + D$.

e. The above deflections do not ensure against ponding. Roofs that do not have sufficient slope or camber to assure adequate drainage shall be investigated for ponding. See Section 1611 for rain and ponding requirements and Section 1503.4 for roof drainage requirements.

f. The wind load is permitted to be taken as 0.42 times the "component and cladding" loads for the purpose of determining deflection limits herein.

g. For steel structural members, the dead load shall be taken as zero.

h. For aluminum structural members or aluminum panels used in skylights and sloped glazing framing, roofs or walls of sunroom additions or patio covers, not supporting edge of glass or aluminum sandwich panels, the total load deflection shall not exceed l/60. For continuous aluminum structural members supporting edge of glass, the total load deflection shall not exceed l/175 for each glass lite or l/60 for the entire length of the member, whichever is more stringent. For aluminum sandwich panels used in roofs or walls of sunroom additions or patio covers, the total load deflection shall not exceed l/120.

i. For cantilever members, l shall be taken as twice the length of the cantilever.

Table C6-7. For wind speeds between 85 mph and 100 mph, the factor for 10 years in Table C6-7 was 0.84. This conversion factor applies to the wind speed, *V*. Since design wind pressure is a function of V^2, the conversion factor must be squared before applying it to the design wind pressure. The factor 0.84 squared is 0.7056, which is rounded off to 0.7, which was the factor given in previous editions of the code. The second aspect is the conversion of the wind load from strength level to allowable stress design level which means the 0.7 factor is multiplied by 0.6, giving rise to the factor of 0.42 in the footnote.

Q3. In Table 1604.3, Note g states "dead load shall be taken as zero for structural steel members." Would this apply to the precomposition check of composite beam deflection limits under wet weight of concrete?

A3. No. The serviceability requirements of Section 1604.3 apply to the finished construction. The loading condition described would be a construction consideration, which is not directly regulated by the serviceability criteria.

1604.3.1 Deflections. The deflections of structural members shall not exceed the more restrictive of the limitations of Sections 1604.3.2 through 1604.3.5 or that permitted by Table 1604.3.

❖ The deflection of structural members is limited by the code, as well as certain material standards for damage control of supported construction and human comfort. Generally, the public equates visible deflection, or even detectable vibration, with a potentially unsafe condition (which in many cases is not true). The intent of this section is that deflection is not to exceed either the limitations in the applicable material design standard or the applicable specified requirements.

1604.3.2 Reinforced concrete. The deflection of reinforced concrete structural members shall not exceed that permitted by ACI 318.

❖ The deflection limitations in ACI 318 are not to be exceeded for reinforced (and prestressed) concrete (see ACI 318 for detailed deflection requirements).

1604.3.3 Steel. The deflection of steel structural members shall not exceed that permitted by AISC 360, AISI S100, ASCE 8, SJI CJ-1.0, SJI JG-1.1, SJI K-1.1 or SJI LH/DLH-1.1, as applicable.

❖ The design standard to be met depends on the type of steel structural member. For example, steel joists are to meet the deflection limitations in the Steel Joist Institute's (SJI) standard, which is applicable to the type of joist, and rolled steel members are to meet the deflection criteria in AISC 360.

1604.3.4 Masonry. The deflection of masonry structural members shall not exceed that permitted by TMS 402/ACI 530/ASCE 5.

❖ The deflection limits for masonry beams and lintels is specified in TMS 402/ACI 530/ASCE 5.

1604.3.5 Aluminum. The deflection of aluminum structural members shall not exceed that permitted by AA ADM1.

❖ The deflection of aluminum structural members is limited to that in AA-ADM1 for aluminum or that permitted by Table 1604.3 of the code.

1604.3.6 Limits. The deflection limits of Section 1604.3.1 shall be used unless more restrictive deflection limits are required by a referenced standard for the element or finish material.

❖ The limits specified in Table 1604.3 apply to the indicated members for any structural material. As indicated in Section 1604.3.1, the deflection limits in the structural material standards apply when they are more restrictive than indicated in the table.

The deflection limits that are applicable to a flat concrete roof member that does not support any nonstructural elements likely to be damaged by large deflections are summarized in Table 1604.3.6. As can be seen, the code deflection criteria is more stringent and would govern the design of this type of member.

Table 1604.3.6
CONCRETE ROOF MEMBER DEFLECTION CRITERIA

Construction	LOADS		
	L	*S* or *W*	*D* + *L*
IBC Table 1604.3			
Roof members	*l*/360	*l*/360	*l*/240
Concrete per ACI 318 See Table 9.5(b)			
Flat roofs	*l*/180	Note a	Note a

a. No deflection limit specified in ACI 318.

1604.4 Analysis. *Load effects* on structural members and their connections shall be determined by methods of structural analysis that take into account equilibrium, general stability, geometric compatibility and both short- and long-term material properties.

Members that tend to accumulate residual deformations under repeated service loads shall have included in their analysis the added eccentricities expected to occur during their service life.

Any system or method of construction to be used shall be based on a rational analysis in accordance with well-established principles of mechanics. Such analysis shall result in a system that provides a complete load path capable of transferring loads from their point of origin to the load-resisting elements.

The total lateral force shall be distributed to the various vertical elements of the lateral force-resisting system in proportion to their rigidities, considering the rigidity of the horizontal bracing system or diaphragm. Rigid elements assumed not to be a part of the lateral force-resisting system are permitted to be incorporated into buildings provided their effect on the action of the system is considered and provided for in the design. Except where diaphragms are flexible, or are permitted to be analyzed as flexible, provisions shall be made for the increased forces induced on resisting elements of the structural system resulting from torsion due to eccentricity between the center of application of the lateral forces and the center of rigidity of the lateral force-resisting system.

Every structure shall be designed to resist the overturning effects caused by the lateral forces specified in this chapter. See Section 1609 for wind loads, Section 1610 for lateral soil loads and Section 1613 for earthquake loads.

❖ This section includes the general requirements for structural analysis. The principles stated in this section are those commonly found in structural engineering textbooks. The requirement that structural analysis be capable of demonstrating a complete load path is essential to the adequate resistance of the structural system to wind loads or earthquake effects. The load path is to be capable of transferring all of the loads from their point of application onto the structure to the foundation. It is also important that nonstructural rigid elements be properly accounted for in the design. For example, a partial-height rigid masonry wall placed between steel columns in a steel frame will resist the horizontal shear load and cause bending in the column unless a flexible joint is provided between the wall and the column.

The definition of "Diaphragm" in Section 202.1 includes definitions that distinguish a rigid diaphragm from a flexible diaphragm. The distribution of forces in buildings with flexible diaphragms differs from those having rigid diaphragms. Where the diaphragm is determined to be flexible, the effect of diaphragm rigidity on the distribution of lateral forces is considered to be negligible and can therefore be neglected in the structural analysis. Otherwise, for structures having rigid diaphragms, this section requires the engineer to distribute lateral forces to the vertical supporting elements in proportion to their rigidities, and to include the effect of the increased forces induced on the vertical supporting elements resulting from torsion due to eccentricity between the center of mass and the center of rigidity.

1604.5 Risk category. Each building and structure shall be assigned a *risk category* in accordance with Table 1604.5. Where a referenced standard specifies an occupancy category, the *risk category* shall not be taken as lower than the occupancy category specified therein.

❖ This section requires classification of the risk category of any building in accordance with the nature of occupancy as described in Table 1604.5. The risk category serves as a threshold for a variety of code provisions related to earthquake, flood, snow and wind loads. Particularly noteworthy are the importance factors that are used in the calculation of design earthquake, snow and wind loads. The value of the importance factor generally increases with the importance of the facility. Structures assigned greater importance factors must be designed for larger forces. The result is a more robust structure that would be less likely to sustain damage under the same conditions than a structure with a lower importance factor. The intent is to enhance a structure's performance based upon its use or the need to remain in operation during and after a design event.

The impact of a higher risk category classification is not limited to increasing the design loads. Compared to Risk Category I, II or III, for instance, a Risk Category IV classification can lead to a higher seismic design category classification that can, in turn, require more stringent seismic detailing and limitations on the seismic-force-resisting system. This can also affect the seismic design requirements for architectural, mechanical and electrical components and systems.

TABLE 1604.5. See page 16-9.

❖ The risk category determined in this table generally increases with the importance of the facility, which relates to the availability of the facility after an emergency, and the consequence of a structural failure on human life. The categories range from Risk Category I, which represents the lowest hazard to life, through Risk Category IV, which encompasses essential facilities.

Risk Category IV: These are buildings that are considered to be essential in that their continuous use is needed, particularly in response to disasters. Fire, rescue and police stations, and emergency vehicle garages must remain operational during and after major events, such as earthquakes, floods or hurricanes. The phrase "designated as essential facilities" refers to designation by the building official that certain facilities are required for emergency response or disaster recovery. This provides jurisdictions the latitude to identify specific facilities that should be considered essential in responding to various types of emergencies. These could include structures that would not otherwise be included in this risk category. This designation would only be made with consideration of broader public policy, as well as emergency preparedness planning within the jurisdiction in question. The reasons for including facilities, such as hospitals, fire stations, police stations, emergency response operations centers, etc., should be self-evident. Some items warranting additional discussion are as follows:

- *Designated emergency shelters and designated emergency response facilities.* These items repeat the term "designated," which is referring to designation by the building official that the facilities have been identified as necessary for sheltering evacuees or responding to emergencies (see discussion of "designated-above). For example, an elementary school having an occupant load of 275 would typically be considered a Risk Category III facility. If that school is designated as an emergency shelter, then the school will be considered a Risk Category IV building.

- *Facilities supplying emergency backup power for Risk Category IV.* A power-generating station or other utility (such as a natural gas facility) is to be classified as Risk Category IV only if the facility serves an emergency backup function for a Risk Category IV building, such as a fire station or police station. Otherwise, the power-generating station or utility should be classified as Risk Category III.
- *Structures with quantities of highly toxic materials in excess of the quantities permitted for a control area in Table 307.1(2).* This applies only to "Highly toxic materials" (see definition in Section 307.2), which are covered in the second row of Table 307.1(2). That table lists the maximum allowable quantities per control area of materials posing a health hazard. Since the use of control areas is permitted by Table 307.1(2), it is reasonable to recognize the control area for the purpose of making this risk category determination. In other words, this applies to occupancies

TABLE 1604.5
RISK CATEGORY OF BUILDINGS AND OTHER STRUCTURES

RISK CATEGORY	NATURE OF OCCUPANCY
I	Buildings and other structures that represent a low hazard to human life in the event of failure, including but not limited to: • Agricultural facilities. • Certain temporary facilities. • Minor storage facilities.
II	Buildings and other structures except those listed in Risk Categories I, III and IV
III	Buildings and other structures that represent a substantial hazard to human life in the event of failure, including but not limited to: • Buildings and other structures whose primary occupancy is public assembly with an occupant load greater than 300. • Buildings and other structures containing elementary school, secondary school or day care facilities with an occupant load greater than 250. • Buildings and other structures containing adult education facilities, such as colleges and universities, with an occupant load greater than 500. • Group I-2 occupancies with an occupant load of 50 or more resident care recipients but not having surgery or emergency treatment facilities. • Group I-3 occupancies. • Any other occupancy with an occupant load greater than 5,000[a]. • Power-generating stations, water treatment facilities for potable water, waste water treatment facilities and other public utility facilities not included in Risk Category IV. • Buildings and other structures not included in Risk Category IV containing quantities of toxic or explosive materials that: Exceed maximum allowable quantities per control area as given in Table 307.1(1) or 307.1(2) or per outdoor control area in accordance with the *International Fire Code*; and Are sufficient to pose a threat to the public if released [b].
IV	Buildings and other structures designated as essential facilities, including but not limited to: • Group I-2 occupancies having surgery or emergency treatment facilities. • Fire, rescue, ambulance and police stations and emergency vehicle garages. • Designated earthquake, hurricane or other emergency shelters. • Designated emergency preparedness, communications and operations centers and other facilities required for emergency response. • Power-generating stations and other public utility facilities required as emergency backup facilities for Risk Category IV structures. • Buildings and other structures containing quantities of highly toxic materials that: Exceed maximum allowable quantities per control area as given in Table 307.1(2) or per outdoor control area in accordance with the *International Fire Code*; and Are sufficient to pose a threat to the public if released [b]. • Aviation control towers, air traffic control centers and emergency aircraft hangars. • Buildings and other structures having critical national defense functions. • Water storage facilities and pump structures required to maintain water pressure for fire suppression.

a. For purposes of occupant load calculation, occupancies required by Table 1004.1.2 to use gross floor area calculations shall be permitted to use net floor areas to determine the total occupant load.

b. Where approved by the building official, the classification of buildings and other structures as Risk Category III or IV based on their quantities of toxic, highly toxic or explosive materials is permitted to be reduced to Risk Category II, provided it can be demonstrated by a hazard assessment in accordance with Section 1.5.3 of ASCE 7 that a release of the toxic, highly toxic or explosive materials is not sufficient to pose a threat to the public.

that are classified as Group H-4 based on the quantities of highly toxic material exceeding the permitted quantity within a control area. However, recognizing control areas means that both the risk category as well as the occupancy classification could be lowered by adding either fire-resistance-rated walls or floor/ceiling assemblies in order to divide the building into a number of smaller control areas. The additional wording "....sufficient to pose a threat to the public if released" places a further qualification on the material quantity, but it is subjective since the threat to the public could be difficult to determine. Also note that a Group H-4 occupancy classification could be based on exceeding the quantities permitted for toxics or corrosives [see Table 307.1(2)], but the presence of those materials would not affect the assessment of the facility as Risk Category IV.

Risk Category III: Risk Category III buildings include those occupancies that have relatively large numbers of occupants because of the overall size of the building. They also include uses that pose an elevated life-safety hazard to the occupants, such as public assembly, schools or colleges. In addition, Risk Category III includes uses where the occupant's ability to respond to an emergency is either restricted, such as in jails, or otherwise impaired, such as in nursing homes housing patients that require skilled nursing care. A discussion of some of the specific table listings follows:

- *Buildings and other structures with a primary occupancy that is public assembly with an occupant load greater than 300.* Public assembly occupancies meeting this criterion will typically be classified as Group A in Chapter 3. The wording requires agreement on the determination that a building's "primary occupancy" is in fact public assembly. This could be as simple as verifying that the portion of the building housing the public assembly occupancy is more than 50 percent of the total building area.
- *Group 1-2 occupancies with an occupant load of 50 or more resident patients but not having surgery or emergency treatment facilities.* This category applies to health care facilities with at least 50 resident patients. The term "resident patient" is not defined or used elsewhere in the code, but would seem to refer to locations where those patients receive around-the-clock (24-hour) care as opposed to ambulatory surgery centers or outpatient units. This table entry covers facilities where patients have difficulty responding to an emergency or are incapable of self-preservation.
- *Buildings having an occupant load greater than 5,000.* Uses that pose elevated life-safety concerns, such as public assembly uses, schools and health care facilities, are covered elsewhere and have a much lower threshold based on the number of occupants. This table entry covers buildings that are large enough to have more than 5,000 occupants, providing added protection for the occupants of larger structures whatever the use happens to be. In order to determine occupant load, the methods outlined in Section 1004 are normally used. Chapter 10 sets forth standards that provide a reasonably conservative number of occupants for all spaces, and while actual loads are commonly less than the design amount, it is not unusual in the life of a space in a building to have periods when high actual occupant loads exist. Because there is no clear rationale that connects the occupant load used to calculate minimum means of egress requirements to the risks associated with structural design, Note a provides some reasonable adjustment to this determination by permitting the use of net floor area. It provides a more reasonable approach for occupancies, such as office, mercantile and residential, that are required to base occupant load on gross area—an area that includes corridors, stairways, elevators, closets, accessory areas, structural walls and columns, etc.
- *Power-generating stations, potable water treatment facilities, wastewater treatment and other public utilities not included in Risk Category IV.* A failure and subsequent shutdown of these types of facilities would not pose an imme-diate threat to life safety. These infrastructure items are considered Risk Category III because of the impact an extended disruption in service can have on the public.
- *Buildings not included in Risk Category IV containing quantities of toxic or explosive materials that exceed permitted quantities per control area in Table 307.1(1) or 307.1(2) and are sufficient to pose a threat to the public if released.* Buildings included under Risk Category IV would be those containing quantities of highly toxic materials that exceed the permitted quantity in Table 307.1(2) (see discussion under "Risk Category IV" above). This item addresses buildings with explosives or toxic substances, both of which are defined in Section 307.2.

Risk Category II: Risk Category II buildings represent a lesser hazard to life because of fewer building occupants and smaller building size compared to those that are considered Risk Category III. Since Risk Categories III and IV represent buildings with higher risk or essential facilities, on a relative scale Risk Category II can be thought of as a "standard occupancy" building as evidenced by importance factors for earthquake, snow and wind that are all equal to 1.

Risk category I: Risk Category I buildings exhibit the lowest hazard to life since they have little or no human occupants or, for those that are temporary, the exposure to the hazards of earthquakes, floods, snow and wind would be considerably less than that of a permanent structure. Note that this category includes "minor storage facilities," but the code does not provide an explanation of which storage facilities could be considered minor.

1604.5.1 Multiple occupancies. Where a building or structure is occupied by two or more occupancies not included in the same *risk category*, it shall be assigned the classification of the highest *risk category* corresponding to the various occupancies. Where buildings or structures have two or more portions that are structurally separated, each portion shall be separately classified. Where a separated portion of a building or structure provides required access to, required egress from or shares life safety components with another portion having a higher *risk category*, both portions shall be assigned to the higher *risk category*.

❖ Buildings are frequently occupied by a mixture of uses or occupancies. A single-use building is probably the exception rather than the rule. Where multiple occupancies are proposed in a building, the risk category of each one must be considered. In some cases, the proposed occupancies in a building will fall into more than one risk category and the requirements for multiple occupancies stated in Section 1604.5.1 must be satisfied. These requirements were previously part of the earthquake load provisions. As of the 2006 edition, they were relocated so that they now apply regardless of the type of load being considered. The code identifies two design options for mixed-use buildings. The entire structure can be designed as a single unit based on the requirements for the most stringent risk category for the building. Alternatively, the engineer can structurally separate portions of the structure containing distinct occupancy categories and design each portion accordingly based on its risk category.

This section also provides direction regarding access to and egress from adjacent structures that fall into different occupancy categories by making the requirements for the more stringent risk category applicable to both structures. This requirement is the result of lessons learned from events, such as earthquakes, in which essential functions have been rendered unusable because of a failure in an adjacent structure.

1604.6 In-situ load tests. The *building official* is authorized to require an engineering analysis or a load test, or both, of any construction whenever there is reason to question the safety of the construction for the intended occupancy. Engineering analysis and load tests shall be conducted in accordance with Section 1709.

❖ The building official has the option of requiring either a structural analysis, an on-site in-situ load test, or both, in accordance with Section 1714, on an existing structure, building or portion thereof if there is reasonable doubt as to structural integrity. The building official should document his or her reasons for the testing requirement. Whenever possible, the concern should be addressed by structural analysis since load testing a structure is very expensive. One example would be an analysis by a third-party engineering firm acceptable to both the building official and the owner. The structural integrity may be examined for items such as visible signs of excessive settlement or lateral deflection, such as cracks in concrete foundation walls or excessive vibration when the assembly is loaded. The procedure must simulate the actual load conditions to which the structure is subjected during normal use.

1604.7 Preconstruction load tests. Materials and methods of construction that are not capable of being designed by *approved* engineering analysis or that do not comply with the applicable referenced standards, or alternative test procedures in accordance with Section 1707, shall be load tested in accordance with Section 1710.

❖ The alternative test procedure described in Section 1707 and the preconstruction load test procedure described in Section 1710 are intended to apply to materials or an assembly of structural materials that do not have an accepted analysis technique; thus, they are approved for use by way of the alternative test procedure. The preconstruction test procedure in Section 1710 includes the determination of the allowable superimposed design load (see Section 1710 for details).

1604.8 Anchorage. Buildings and other structures, and portions thereof, shall be provided with anchorage in accordance with Sections 1604.8.1 through 1604.8.3, as applicable.

❖ Sections 1604.8.1 through 1606.8.3 contain general anchorage requirements as well as specific requirements for the anchorage of walls and decks.

1604.8.1 General. Anchorage of the roof to walls and columns, and of walls and columns to foundations, shall be provided to resist the uplift and sliding forces that result from the application of the prescribed loads.

❖ This section states a specific load path requirement for anchorage that is already required in more general terms by Section 1604.4. Required anchorage resistance to uplift and sliding is determined in accordance with the appropriate load combinations in Section 1605, which account for the effects of earthquakes, fluid pressures, lateral earth pressure or wind (also see Section 1604.9).

An example of the condition described in this section is roof uplift as a result of high winds. Table 2308.10.1 requires roof uplift connectors for wood-frame construction designed and installed in accordance with the prescriptive requirements for conventional light-frame construction. Note that net roof uplift occurs at all wind speeds indicated in the table; thus, uplift connectors are required for roof wood framing for all of the indicated wind speed locations.

1604.8.2 Structural walls. Walls that provide vertical load-bearing resistance or lateral shear resistance for a portion of the structure shall be anchored to the roof and to all floors and members that provide lateral support for the wall or that are supported by the wall. The connections shall be capable of resisting the horizontal forces specified in Section 1.4.4 of ASCE 7 for walls of structures assigned to Seismic Design Category A and to Section 12.11 of ASCE 7 for walls of structures assigned to all other seismic design categories. Required anchors in masonry walls of hollow units or cavity walls shall be embedded in a reinforced grouted structural element of the wall. See Sections 1609 for wind design requirements and 1613 for earthquake design requirements.

❖ This section refers to the ASCE 7 technical provisions for the minimum strength design load on anchorages of walls to the horizontal diaphragms that provide lateral support. The connection requirements apply to both bearing walls and walls that resist lateral loads. The wall anchorage must be a positive connection that does not rely on friction for load transfer. Earthquake detailing provisions may result in higher anchorage design forces based on the seismic design category of the building, as well as the mass of the wall.

1604.8.3 Decks. Where supported by attachment to an *exterior wall*, decks shall be positively anchored to the primary structure and designed for both vertical and lateral loads as applicable. Such attachment shall not be accomplished by the use of toenails or nails subject to withdrawal. Where positive connection to the primary building structure cannot be verified during inspection, decks shall be self-supporting. Connections of decks with cantilevered framing members to exterior walls or other framing members shall be designed for both of the following:

1. The reactions resulting from the dead load and live load specified in Table 1607.1, or the snow load specified in Section 1608, in accordance with Section 1605, acting on all portions of the deck.
2. The reactions resulting from the dead load and live load specified in Table 1607.1, or the snow load specified in Section 1608, in accordance with Section 1605, acting on the cantilevered portion of the deck, and no live load or snow load on the remaining portion of the deck.

❖ This requirement for the positive anchorage of decks is a result of failures that have occurred primarily on nonengineered residential decks. Toenail connections are very weak since many of the field connections split the wood framing member. Nails that are installed in line with applied tension forces pull out easily. Wood deck framing could be attached to a supporting ledger board by joist hangers that provide nails loaded in shear for the vertical and horizontal loads.

In order to make the loading considerations perfectly clear for decks with cantilevers, this section states the two loading conditions that are applicable. The second condition covers the case where uplift can occur at the point of connection between the wood framing and the exterior wall for cantilevered deck framing. This highlights the need for resistance to uplift by a positive connection at the exterior wall. These conditions also clarify that snow load should be considered.

1604.9 Counteracting structural actions. Structural members, systems, components and cladding shall be designed to resist forces due to earthquakes and wind, with consideration of overturning, sliding and uplift. Continuous load paths shall be provided for transmitting these forces to the foundation. Where sliding is used to isolate the elements, the effects of friction between sliding elements shall be included as a force.

❖ This section requires that all elements of the structure be designed to resist the required earthquake and wind effects. The required level of resistance will be based on the appropriate load combinations of Section 1605. The term "counteracting" used in the section title refers to so-called counteracting load combinations as opposed to those in which the load effects are additive. The effects of overturning, uplift and sliding are most pronounced in these counteractive load combinations. These would consist of the following depending on the design method chosen:

- Load and resistance factor design or strength design, Section 1605.2:

  $0.9D + 1.0W + 1.6H$ Equation 16-6

  $0.9(D + F) + 1.0E + 1.6H$ Equation 16-7

- Basic allowable stress design load combinations, Section 1605.3.1:

  $0.6D + 0.6W + H$ Equation 16-15

  $0.6(D + F) + 0.7E + H$ Equation 16-16

- Alternative allowable stress design load combinations, Section 1605.3.2:

  $0.9D + E/1.4$ Equation 16-22

  (Note the adjustment to the vertical component of earthquake load effect that is permitted for evaluating foundations.)

  $(^2/_3)D + (0.6 \times 1.3)W$ Equation 16-18

  (Equation 16-18 with $L = 0$ in accordance with Section 1605.1 requirement to investigate each load combination with one or more variable loads equal to zero, two-thirds of the dead load in accordance with requirement in Section 1605.3.2 and w set equal to 1.3 for 3-second gust wind speed.)

As can be seen, for the critical counteracting load combinations, resistance to overturning, uplift and sliding is largely provided by dead load. Note that Section 1605.3.2 requires the designer to consider only the dead load likely to be in place during a design wind event when using the alternative allowable stress load combinations. Regardless of the design methodology utilized, the designer should be cautious in estimating dead loads that resist overturning, sliding and uplift. In order to remain conservative, estimated dead load should be no more than the actual dead load in these cases (see design dead load in Section 1606.2). To keep a structure from sliding horizontally, the dead load must generate enough friction at the base of the structure to resist the horizontal base shear due to earthquake and wind. Otherwise, adequate anchor-age must be provided to resist the base shear.

1604.10 Wind and seismic detailing. Lateral force-resisting systems shall meet seismic detailing requirements and limitations prescribed in this code and ASCE 7, excluding Chapter 14 and Appendix 11A, even when wind *load effects* are greater than seismic *load effects*.

❖ This section is needed to clarify that the seismic detailing requirements of the code always apply, even where the wind load effects are higher than the seismic load effects. This is required because the calculated earthquake load is dependent upon an assumed level of ductility to be provided in the selected seismic-force-resisting system. If the designer does not follow through by providing the detailing required to achieve the ductility of the selected system, the assumption is invalid.

For example, consider the case where the earthquake load provisions apply in accordance with Section 1613.1, the seismic-force-resisting system is an ordinary concentrically braced steel frame and, based on the seismic design category, Section 2205 requires that the AISC 341 seismic provisions are to be met. For this case, the requirements in Section 14 of AISC 341 must be satisfied in the design of the steel frame, even if the wind load effects exceed the seismic load effects.

SECTION 1605 LOAD COMBINATIONS

1605.1 General. Buildings and other structures and portions thereof shall be designed to resist:

1. The load combinations specified in Section 1605.2, 1605.3.1 or 1605.3.2;
2. The load combinations specified in Chapters 18 through 23; and
3. The seismic load effects including overstrength factor in accordance with Section 12.4.3 of ASCE 7 where required by Section 12.2.5.2, 12.3.3.3 or 12.10.2.1 of ASCE 7. With the simplified procedure of ASCE 7 Section 12.14, the seismic load effects including overstrength factor in accordance with Section 12.14.3.2 of ASCE 7 shall be used.

Applicable loads shall be considered, including both earthquake and wind, in accordance with the specified load combinations. Each load combination shall also be investigated with one or more of the variable loads set to zero.

Where the load combinations with overstrength factor in Section 12.4.3.2 of ASCE 7 apply, they shall be used as follows:

1. The basic combinations for strength design with overstrength factor in lieu of Equations 16-5 and 16-7 in Section 1605.2.
2. The basic combinations for *allowable stress design* with overstrength factor in lieu of Equations 16-12, 16-14 and 16-16 in Section 1605.3.1.
3. The basic combinations for *allowable stress design* with overstrength factor in lieu of Equations 16-21 and 16-22 in Section 1605.3.2.

❖ Generally, there are two types of load combinations specified in the code: those to be used with a strength design or load and resistance factor design (LRFD) and those to be used with allowable stress design. Where additional load combinations are specified in the structural material chapters, they apply also. Note that this section also requires that for a given load combination engineers also consider additional load cases where one or more variable loads are not acting concurrently with other variable loads. It is necessary to explore such possibilities, since they can at times result in the most critical load effect for some elements of a structure. According to the definition of "Loads," loads that are not considered permanent are variable loads.

The load combinations with overstrength factor in Section 12.4.3.2 of ASCE 7 only apply where specifically required by the seismic provisions. They do not replace the applicable load combinations for strength design or allowable stress design. They constitute an additional requirement that must be considered in the design of specific structural elements.

The purpose of these load combinations with overstrength factor is to account for the maximum earthquake load effect, E_m, which considers "system overstrength." This system characteristic is accounted for by multiplying the effects of the lateral earthquake load by the overstrength factor, Ω_0, for the seismic-force-resiting system that is utilized. It is representative of the upper bound system strength for purposes of designing nonyielding elements for the maximum expected load. Under the design earthquake ground motions, the forces generated in the seismic-force-resisting system can be much greater than the prescribed seismic design forces. If not accounted for, the system overstrength effect can cause failures of structural elements that are subjected to these forces. Because system overstrength is unavoidable, design for the maximum earthquake force that can be developed is warranted for certain elements. The intent is to provide key elements with sufficient overstrength so that inelastic (ductile) response/behavior appropriately occurs within the vertical-resisting elements.

The ASCE 7 earthquake load provisions that require consideration of the load combinations with over-strength factor are stated in this section. Note that these load combinations are not general load cases, but are only to be applied where specified in the indicated earthquake load provisions (see above) or in the structural materials requirements (see Section 1810.3.11.2 for an example of the latter).

For example, Section 12.3.3.3 of ASCE 7 applies to structural elements that support discontinuous frames or shear wall systems where the discontinuity is severe enough to be deemed a structural irregularity. See Figure 1605.1(1) for examples of building configurations that include discontinuous vertical systems. Also see the commentary to Section 12.3.3.3 of ASCE 7.

Section 12.10.2.1 of ASCE 7 applies to collector elements in structures assigned to Seismic Design Category C or higher. The term "collector" describes an element used to transfer forces from a diaphragm to the supporting element(s) of the lateral-force-resisting system where the length of the vertical element is less than the diaphragm dimension at that location. Figure 1605.1(2) illustrates this relationship. Failures of collectors can result in loads not being delivered to resisting elements and separation of a portion of a building from its lateral-force-resisting system. Where a collector also supports gravity loads, such a failure would result in the loss of a portion of the gravity load-carrying system. The intent is to provide collector elements that have sufficient overstrength so that any inelastic behavior appropriately occurs within the seismic-force-resisting system rather than the collector elements.

Maximum earthquake load effect, E_m, in Section 12.4.3 of ASCE 7 includes effects of the horizontal load, E_{mh}, as well as a vertical component, E_v. Emh is the product of the overstrength factor, Ω_0, and Q_E, the effects of horizontal earthquake forces. E_v is the same vertical component used for the earthquake load effect, E (see commentary, Section 1602.1).

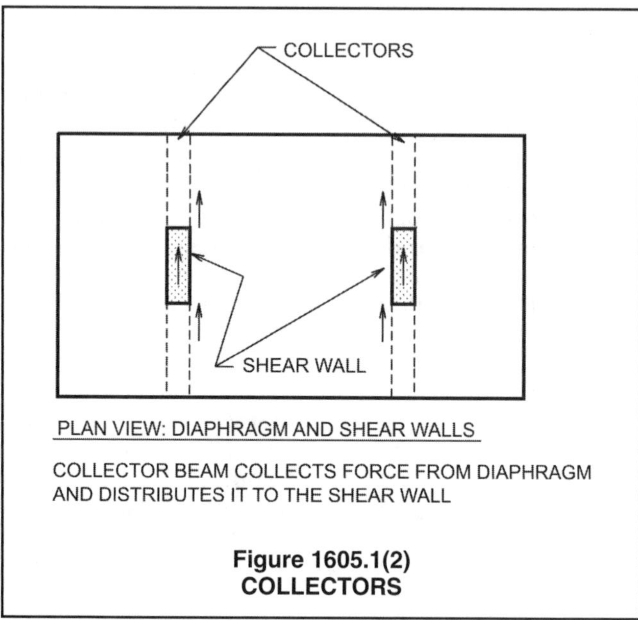

Figure 1605.1(2)
COLLECTORS

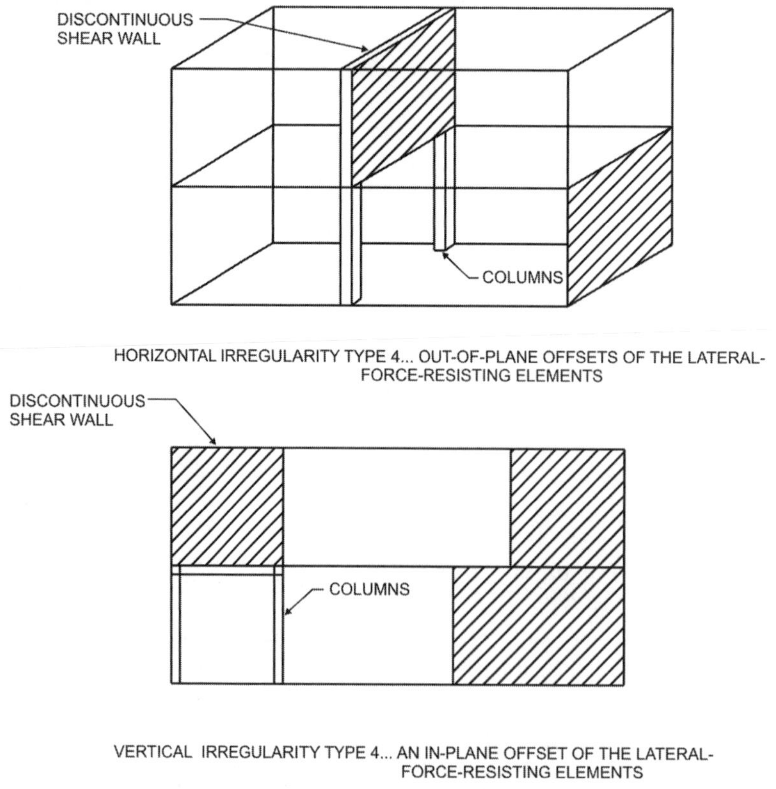

DISCONTINUOUS SHEAR WALL

COLUMNS

HORIZONTAL IRREGULARITY TYPE 4... OUT-OF-PLANE OFFSETS OF THE LATERAL-FORCE-RESISTING ELEMENTS

DISCONTINUOUS SHEAR WALL

COLUMNS

VERTICAL IRREGULARITY TYPE 4... AN IN-PLANE OFFSET OF THE LATERAL-FORCE-RESISTING ELEMENTS

Figure 1605.1(1)
EXAMPLES OF DISCONTINUOUS SHEAR WALLS

1605.1.1 Stability. Regardless of which load combinations are used to design for strength, where overall structure stability (such as stability against overturning, sliding, or buoyancy) is being verified, use of the load combinations specified in Section 1605.2 or 1605.3 shall be permitted. Where the load combinations specified in Section 1605.2 are used, strength reduction factors applicable to soil resistance shall be provided by a *registered design professional*. The stability of retaining walls shall be verified in accordance with Section 1807.2.3.

❖ This section permits soil resistance and strength reduction factors to be considered where strength design factored loads are used in foundation design.

1605.2 Load combinations using strength design or load and resistance factor design. Where strength design or load and resistance factor design is used, buildings and other structures, and portions thereof, shall be designed to resist the most critical effects resulting from the following combinations of factored loads:

$1.4(D + F)$ **(Equation 16-1)**

$1.2(D + F) + 1.6(L + H) + 0.5(L_r \text{ or } S \text{ or } R)$ **(Equation 16-2)**

$1.2(D + F) + 1.6(L_r \text{ or } S \text{ or } R) + 1.6H + (f_1L \text{ or } 0.5W)$ **(Equation 16-3)**

$1.2(D + F) + 1.0W + f_1L + 1.6H + 0.5(L_r \text{ or } S \text{ or } R)$ **(Equation 16-4)**

$1.2(D + F) + 1.0E + f_1L + 1.6H + f_2S$ **(Equation 16-5)**

$0.9D + 1.0W + 1.6H$ **(Equation 16-6)**

$0.9(D + F) + 1.0E + 1.6H$ **(Equation 16-7)**

where:

f_1 = 1 for places of public assembly live loads in excess of 100 pounds per square foot (4.79 kN/m^2), and parking garages; and 0.5 for other live loads.

f_2 = 0.7 for roof configurations (such as saw tooth) that do not shed snow off the structure, and 0.2 for other roof configurations.

Exceptions:

1. Where other factored load combinations are specifically required by other provisions of this code, such combinations shall take precedence.
2. Where the effect of H resists the primary variable load effect, a load factor of 0.9 shall be included with H where H is permanent and H shall be set to zero for all other conditions.

❖ This section lists the load combinations for strength design or load and resistance factor design methods. See the definitions in Section 1602 for an explanation of the load notations. The basis for these load combinations is Section 2.3.2 of ASCE 7. These combinations of factored loads are the agreed-upon strength limit states that establish the required strength to be provided in the structural component being designed. In spite of the precise appearance of the load and resistance factor design load combinations, one should keep in mind their probabilistic nature. The goal is to allow a wide variety of structures to be designed economically with an acceptably low probability that the strength of the structure will be exceeded. Doing so necessitates combining loads in scenarios that are likely to occur. Dead load is a permanent load and it appears in every combination. The load combinations are constructed by adding the dead load to one of the variable loads at its maximum value, which is typically indicated by the load factor of 1.6. In addition, other variable loads are included with load factors that are less than 1.0. Those so-called "companion loads" represent arbitrary point-in-time values for those loads. The exception is the maximum earthquake load that has a load factor of 1.0 (see the commentary to the definitions "Factored load" and "Nominal load" in Section 202).

1605.2.1 Other loads. Where flood loads, F_a, are to be considered in the design, the load combinations of Section 2.3.3 of ASCE 7 shall be used. Where self-straining loads, T, are considered in design, their structural effects in combination with other loads shall be determined in accordance with Section 2.3.5 of ASCE 7. Where an ice-sensitive structure is subjected to loads due to atmospheric icing, the load combinations of Section 2.3.4 of ASCE 7 shall be considered.

❖ Under a cooperative agreement with FEMA, American Society of Civil Engineers completed an extensive analysis of flood loads and flood load combinations. The results of this study were first included in the 1998 edition of ASCE 7. For all buildings and structures located in flood hazard areas, this section specifies that flood load and flood load combinations using the strength design method are to be determined in accordance with Section 2.3.3 of ASCE 7, which states:

> 2.3.3. Load Combinations Including Flood Load. When a structure is located in a flood zone (Section 5.3.1), the following load combinations shall be considered in addition to the basic combinations in Section 2.3.2:
>
> 1. In V Zones or Coastal A Zones, 1.0W in combinations (4) and (6) shall be replaced by 1.0W + 2.0$_a$.
> 2. In noncoastal A Zones, 1.0 W in combinations (4) and (6) shall be replaced with 0.5 W + 1.0F_a.

1605.3 Load combinations using allowable stress design.

1605.3.1 Basic load combinations. Where *allowable stress design* (working stress design), as permitted by this code, is used, structures and portions thereof shall resist the most critical effects resulting from the following combinations of loads:

$D + F$ **(Equation 16-8)**

$D + H + F + L$ **(Equation 16-9)**

$D + H + F + (L_r \text{ or } S \text{ or } R)$ **(Equation 16-10)**

$D + H + F + 0.75(L) + 0.75(L_r \text{ or } S \text{ or } R)$ **(Equation 16-11)**

$D + H + F + (0.6W \text{ or } 0.7E)$ **(Equation 16-12)**

$D + H + F + 0.75(0.6W) + 0.75L + 0.75(L_r \text{ or } S \text{ or } R)$ **(Equation 16-13)**

$D + H + F + 0.75\ (0.7\ E) + 0.75\ L + 0.75\ S$ **(Equation 16-14)**

$0.6D + 0.6W + H$ **(Equation 16-15)**

$0.6(D + F) + 0.7E + H$ **(Equation 16-16)**

Exceptions:

1. Crane hook loads need not be combined with roof live load or with more than three-fourths of the snow load or one-half of the wind load.
2. Flat roof snow loads of 30 psf (1.44 kN/m^2) or less and roof live loads of 30 psf (1.44 kN/m^2) or less need not be combined with seismic loads. Where flat roof snow loads exceed 30 psf (1.44 kN/m^2), 20 percent shall be combined with seismic loads.
3. Where the effect of *H* resists the primary variable load effect, a load factor of 0.6 shall be included with *H* where *H* is permanent and *H* shall be set to zero for all other conditions.
4. In Equation 16-15, the wind load, *W*, is permitted to be reduced in accordance with Exception 2 of Section 2.4.1 of ASCE 7.
5. In Equation 16-16, 0.6 *D* is permitted to be increased to 0.9 *D* for the design of special reinforced masonry shear walls complying with Chapter 21.

❖ See Section 1602 for an explanation of the notations. These basic load combinations for allowable stress design are based on Section 2.4.1 of ASCE 7. Note that a 0.75 factor is applied where these combinations include more than one variable load. This reduces the combined effect of these variable loads in recognition of the lower probability that two or more variable loads will reach their maximum values simultaneously.

Previous editions of the model codes specified that the overturning moment and sliding due to wind load could not exceed two-thirds of the dead load stabilizing moment; however, it was not typically applied to all elements in the building. In the code this limitation on dead load is accomplished through the load combinations. The applicable combination is $0.6D + 0.6W + H$. This load combination limits the dead load resisting wind loads to 60 percent ($^2/_3 = 0.67$; round down) but it applies to all elements. In this form, it is required that this safety factor on dead load applies to all actions where dead load is resisting wind loads.

The load combination, $0.6\ (D + F) + 0.7E + H$, applies throughout the structure and provides for overall stability of the structure similar to the load combination for wind. In determining seismic load effect, *E*, the definition of effective seismic weight requires inclusion of a portion of roof snow under certain conditions. Frequently the question of whether it is logical to include a component of snow load in determining *E*, but not to use that same weight of snow in the load case to resist the overturning or uplift due to *E* is asked. This apparent inconsistency has been built into the seismic requirements for some time. It introduces a bit more conservatism into the earthquake loading where heavier snow loads are concerned.

Exception 1 for crane hook loads has been a longstanding allowance in the alternative basic load combinations of Section 1605.3.1 under the legacy model codes and is now permitted in the code. It allows special consideration in combining crane loads with wind, snow and roof live loads due to a lower probability that these maximum loads occur simultaneously. As noted above, these basic load combinations are based on ASCE 7 allowable stress load combinations, but there is no corresponding exception for crane loads under ASCE 7. In the development of the 2000 edition of this code, this exception was added to the basic load combinations in an attempt to provide "parity" with the alternative basic load combinations. It is worth noting that these two sets of load combinations come from two different sources as described above and, from an overall perspective, they never have provided parity. Crane loads are considered live loads (see Section 1607.13). Exception 1 has no effect on dead load, other floor live loads, rain loads or earthquake loads; thus, where crane live loads are to be considered, the exception eliminates the roof live load, L_r, altogether and reduces the snow load to $0.75S$ and wind load to $0.5(0.6W)$. This exception does not negate the need to combine live loads other than the crane live loads in the usual manner with wind and snow loads. In other words, the load combinations in Equations 16-11 and 16-13 must be investigated without the crane live load and then modified versions of these combinations that include crane live load (use L_0 to denote crane live load) would be considered to make use of the exception as follows:

Equation 16-11
$D + H + F + 0.75L + 0.75\ (L_r \text{ or } S \text{ or } R)$

Equation 16-11c
$D + H + F + 0.75\ (L + L_0) + 0.75\ (0.75S \text{ or } R)$

Equation 16-13
$D + H + F + 0.75\ (0.6W) + 0.75L + 0.75\ (L_r \text{ or } S \text{ or } R)$

Equation 16-13
$D + H + F + 0.75\ (0.5 \times 0.6W) + 0.75[L + L_0] + 0.75\ (0.75S \text{ or } R)$

Equation 16-14
$D + H + F + 0.75\ (0.7E) + 0.75L + 0.75S$

Equation 16-14
$D + H + F + 0.75\ (0.7E) + 0.75[L + L_0] + 0.75\ (0.75S)$

Exception 2 states that "flat roof snow loads of 30 psf (1.44 kN/m^2) or less need not be combined with seismic loads..." This is an exception to the requirement to combine the effects of snow and earthquake load as would otherwise be required by Equation 16-14.

1605.3.1.1 Stress increases. Increases in allowable stresses specified in the appropriate material chapter or the referenced standards shall not be used with the load combinations of Section 1605.3.1, except that increases shall be permitted in accordance with Chapter 23.

❖ Some code sections and some referenced standards permit increases in the allowable stress where load combinations include either wind or earthquake loads. Such increases are expressly prohibited with the basic allowable stress load combinations. This prohibition is due to the 0.75 reduction applied to multiple variable loads in these load combinations, which has replaced the concept of allowable stress increases. Several adjustment factors for the design of wood construction allow increases to reference design values, such as the load duration factor, size factor, flat use factor, repetitive member factor, buckling stiffness factor and bearing area factor. These wood adjustment factors are considered to be material dependent. Since they are not related to the presence of particular design loads, this section recognizes their use.

1605.3.1.2 Other loads. Where flood loads, F_a, are to be considered in design, the load combinations of Section 2.4.2 of ASCE 7 shall be used. Where self-straining loads, T, are considered in design, their structural effects in combination with other loads shall be determined in accordance with Section 2.4.4 of ASCE 7. Where an ice-sensitive structure is subjected to loads due to atmospheric icing, the load combinations of Section 2.4.3 of ASCE 7 shall be considered.

❖ The study discussed in the commentary to Section 1605.2.1 determined flood load and flood load combination factors for use with the allowable stress method. This section specifies that flood load and flood load combinations using the allowable stress design method are to be determined in accordance with Section 2.4.2 of ASCE 7, which states:

2.4.2. Load Combinations Including Flood Load. When a structure is located in a flood zone, the following load combinations shall be considered in addition to the basic combinations in Section 2.4.1:

1. In V zones or Coastal A zones (see Section 5.3.1), 1.5F_a shall be added to other load combinations (5), (6) and (7), and *E* shall be set equal to zero in (5) and (6).
2. In noncoastal A zones, 0.75F_a, shall be added to the combinations (5), (6) and (7), and *E* shall be set at zero in (5) and (6).

1605.3.2 Alternative basic load combinations. In lieu of the basic load combinations specified in Section 1605.3.1, structures and portions thereof shall be permitted to be designed for the most critical effects resulting from the following combinations. When using these alternative basic load combinations that include wind or seismic loads, allowable stresses are permitted to be increased or load combinations reduced where permitted by the material chapter of this code or the referenced standards. For load combinations that include the counteracting effects of dead and wind loads, only two-thirds of the minimum dead load likely to be in place during a design wind event shall be used. When using allowable stresses which have been increased or load combinations which have been reduced as permitted by the material chapter of this code or the referenced standards, where wind loads are calculated in accordance with Chapters 26 through 31 of ASCE 7, the coefficient (ω) in the following equations shall be taken as 1.3. For other wind loads, (ω) shall be taken as 1. When allowable stresses have not been increased or load combinations have not been reduced as permitted by the material chapter of this code or the referenced standards, (ω) shall be taken as 1. When using these alternative load combinations to evaluate sliding, overturning and soil bearing at the soil-structure interface, the reduction of foundation overturning from Section 12.13.4 in ASCE 7 shall not be used. When using these alternative basic load combinations for proportioning foundations for loadings, which include seismic loads, the vertical seismic *load effect*, E_v, in Equation 12.4-4 of ASCE 7 is permitted to be taken equal to zero.

$D + L + (L_r \text{ or } S \text{ or } R)$ **(Equation 16-17)**

$D + L + 0.6\ \omega W$ **(Equation 16-18)**

$D + L + 0.6\ \omega W + S/2$ **(Equation 16-19)**

$D + L + S + 0.6\ \omega W/2$ **(Equation 16-20)**

$D + L + S + E/1.4$ **(Equation 16-21)**

$0.9D + E/1.4$ **(Equation 16-22)**

Exceptions:

1. Crane hook loads need not be combined with roof live loads or with more than three-fourths of the snow load or one-half of the wind load.
2. Flat roof snow loads of 30 psf (1.44 kN/m^2) or less and roof live loads of 30 psf (1.44 kN/m^2) or less need not be combined with seismic loads. Where flat roof snow loads exceed 30 psf (1.44 kN/m^2), 20 percent shall be combined with seismic loads.

❖ These are alternative load combinations to those in Section 1605.3.1 for use with allowable stress design. They are based on the allowable stress load combinations in the *Uniform Building Code*® (UBC™).

Note that the dead load is limited to two-thirds of the dead load likely to be in place during a design wind event. Section 1604.9 requires that all elements of a building be anchored to resist overturning, uplift and sliding. Often, a considerable portion of this resistance is provided by the dead load of a building, including the weight of foundations and any soil directly above them. This section requires the designer to give consideration to the dead load used to resist wind loads by stating that only the minimum dead load likely to be in place during a design wind event is permitted to be used. This, however, does not imply that certain parts or elements of a building are not designed to remain in place. This criteria simply cautions the designer against using dead loads that may not be installed, or in place, such as can occur where a design includes an allowance for a planned future expansion.

Exception 1 for crane hook loads has been a long-standing allowance under the legacy model codes that were also the source of these alternative basic load combinations. It allows special consideration in combining crane live loads with wind, snow and roof live loads due to a lower probability that these maximum loads occur simultaneously.

Exception 2 states that "flat roof snow loads of 30 psf (1.44 N/m^2) or less need not be combined with seismic loads...". This is an exception to the requirement to combine the effects of snow and earthquake load as would otherwise be required by Equation 16-21.

Also see the commentary to Section 1602.1 for discussion of earthquake load effect, *E*, as well as other loads required by these load combinations.

1605.3.2.1 Other loads. Where *F*, *H* or *T* are to be considered in the design, each applicable load shall be added to the combinations specified in Section 1605.3.2. Where self-straining loads, *T*, are considered in design, their structural effects in combination with other loads shall be determined in accordance with Section 2.4.4 of ASCE 7.

❖ See Section 1602 for an explanation of the notation used in this section. As indicated, the applicable loads are to be added to each of the load combinations in Section 1605.3.2.

SECTION 1606 DEAD LOADS

1606.1 General. Dead loads are those loads defined in Section 1602.1. Dead loads shall be considered permanent loads.

❖ The nominal dead load, *D*, is determined in accordance with Section 1606. Similar to the general section on live loads (see Section 1607.1), this section provides a reminder that the term "dead loads" is defined, establishing exactly what should be considered a dead load. It also states that dead loads are considered permanent. This affects how dead loads are classified (permanent versus variable), which is a necessary distinction to make when applying the provisions for load combinations.

1606.2 Design dead load. For purposes of design, the actual weights of materials of construction and fixed service equipment shall be used. In the absence of definite information, values used shall be subject to the approval of the *building official.*

❖ When determining the design dead loads, the actual weights of all materials, equipment, construction, etc., are to be used for the structural design. Often the exact type of materials and equipment to be installed has not yet been determined by the design team at the time the structural design is initiated. Where the actual weights are not known, it is common to use an estimate of the anticipated materials and equipment. While the actual dead load itself does not vary, the estimate of the dead load that is used in structural computations can vary. Such estimates of dead loads are typically greater than the actual dead loads so that the computations are conservative and a redesign of the structure will not be necessary when the actual weights become known. This section clarifies that the building official must approve these estimated dead loads.

Overestimating the actual dead load, so that the design computations are conservative, is acceptable when considering load combinations in Section 1605 that are additive. But the same cannot be said when using counteracting load combinations (see commentary, Section 1604.9). In determining the anchorage required to resist the overturning or uplift effects of wind or seismic loads, resistance is typically provided by the dead load. For example, uplift forces due to wind cause tension to develop in hold-down connections. If the dead load used to resist the wind uplift is overestimated, the result may be an unconservative design. Note that Section 1605.3.2 restricts the dead load used to counteract the effects of overturning and uplift to be the minimum dead load likely to be in place during a design wind event.

As a design guideline, see the unit weights of common construction materials and assemblies in Tables C3-1 and C3-2 of the commentary to ASCE 7. The unit dead loads listed in the tables for assembled elements are usually given in units of pounds per square foot (psf) of surface area (i.e., floor areas, wall areas, ceiling areas, etc.). Unit dead loads for materials used in construction are given in terms of density. The unit weights given in the tables are generally single values, even though a range of weights may actually exist. The average unit weights given are generally suitable for design purposes; however, where there is reason to believe that the actual weights of assembled elements or construction materials may substantially exceed the tabular values, then the situation should be investigated and the highest values used.

SECTION 1607 LIVE LOADS

1607.1 General. Live loads are those loads defined in Section 1602.1.

❖ Nominal live loads are determined in accordance with Section 1607. The live load requirements for the design of buildings and structures are based on the type of occupancy. Live loads are transient loads that vary with time. Generally, the design live load is that which is believed to be near the maximum transient load for a given occupancy.

TABLE 1607.1. See page 16-20.

❖ The design values of live loads for both uniform and concentrated loads are shown in the table as a function of occupancy. The values given are conservative and include both the sustained and variable portions of the live load. Section 1607.3 directs the designer to utilize the greater live loads produced by the intended occupancy, but not less than the minimum uniformly distributed live loads listed in Table 1607.1. It should also be noted that the "occupancy" category listed is not necessarily group specific. For example, an office building may be classified as Group B, but still contain incidental storage areas. Depending on the type of storage, the areas may warrant storage live loads of either 125 or 250 psf (5.98 or 11.9 kN/m^2) to be applied to the space in question.

Table 1607.1 specifies minimum uniform live load in residential attics for three distinct conditions: uninhabitable attics without storage; uninhabitable attics with limited storage; and habitable attics and sleeping areas. Commentary Table 1607.1(1) summarizes the uniform live loads that apply to uninhabitable attics based on the criteria contained in Notes i, j and k to Table 1607.1. The process for determining which load is applicable is summarized in the flow chart in Figure 1607.1.

One noteworthy distinction made by Note k is that any uninhabitable attic that is served by a stairway (other than a pull-down type) must be designed using the load of 30 psf (1.44 kN/m^2) that is applicable to habitable attics (see Note k). The implication is that the presence of a stairway is more conducive to using the attic for storage and therefore warrants a greater design live load. This recognizes that the stairway is likely to serve an attic with greater headroom, providing more storage capacity. For an attic that is accessed by other means, such as a framed opening or pull-down stairs, it is necessary to determine whether the attic storage load applies in accordance with the criteria contained in Notes i and j.

Historically, a minimum load of 10 psf (0.48 kN/m^2) has been viewed as appropriate where occasional access to the attic is anticipated for maintenance purposes, but significant storage is restricted by physical constraints, such as low clearance or the configuration of truss webs. It provides a minimum degree of structural integrity, allowing for occasional access to an attic space for maintenance purposes. Allowing the application of this load to be independent of other live loads is deemed appropriate, since it would be rare for this load and other maximum live loads to occur at once.

Note m clarifies that a live load reduction is not permitted unless specific exceptions of Section 1607.10 apply. The note appears at each specific use or occupancy in Table 1607.1 where a live load reduction is restricted, which serves to clarify the limitations on live load reduction. References appear in Sections 1607.10.1 and 1607.10.2 to correlate with the note.

Table 1607.1(1)
SUMMARY OF MINIMUM LIVE LOAD REQUIRED IN UNINHABITABLE ATTICS

Description	Uniform Load
Without storage	10 psf
With storage	20 psf
Served by a stairway	30 psf

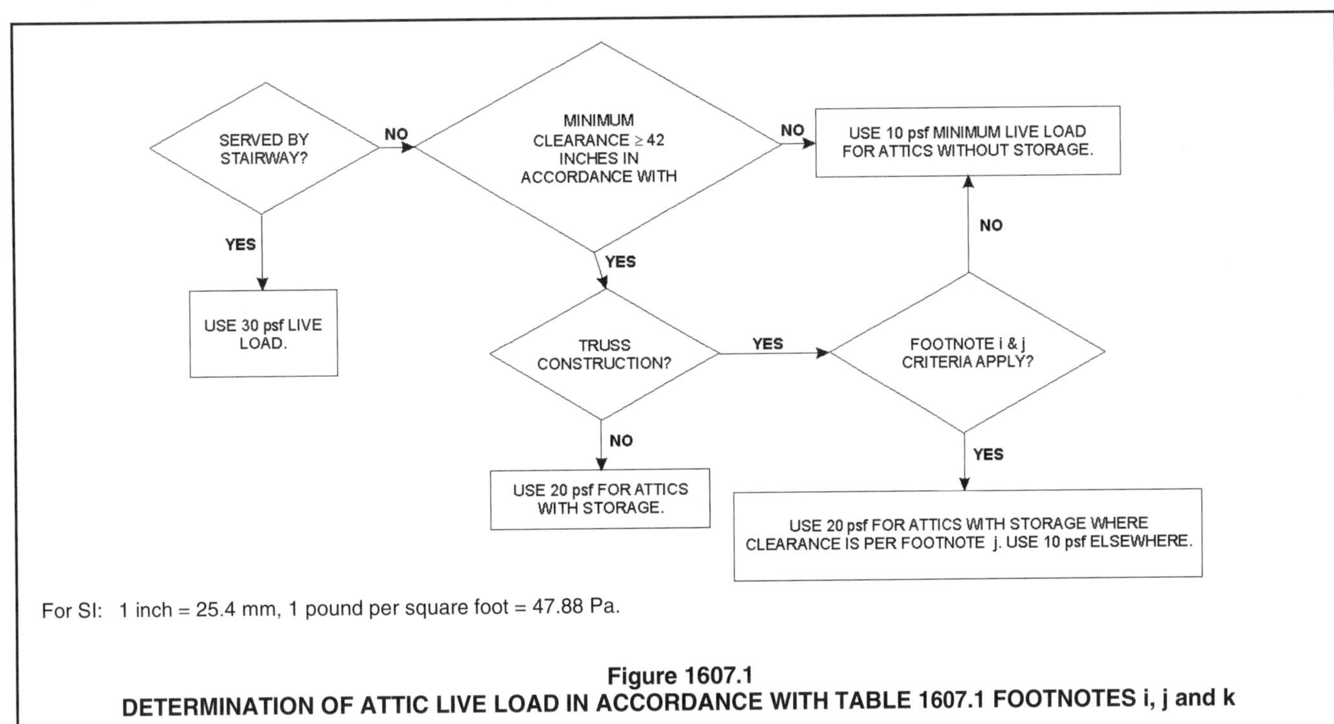

For SI: 1 inch = 25.4 mm, 1 pound per square foot = 47.88 Pa.

Figure 1607.1
DETERMINATION OF ATTIC LIVE LOAD IN ACCORDANCE WITH TABLE 1607.1 FOOTNOTES i, j and k

TABLE 1607.1 MINIMUM UNIFORMLY DISTRIBUTED LIVE LOADS, L_o, AND MINIMUM CONCENTRATED LIVE LOADS[g]

OCCUPANCY OR USE	UNIFORM (psf)	CONCENTRATED (lbs.)
1. Apartments (see residential)	—	—
2. Access floor systems Office use Computer use	 50 100	 2,000 2,000
3. Armories and drill rooms	150[m]	—
4. Assembly areas Fixed seats (fastened to floor) Follow spot, projections and control rooms Lobbies Movable seats Stage floors Platforms (assembly) Other assembly areas	 60[m] 50 100[m] 100[m] 150[m] 100[m] 100[m]	—
5. Balconies and decks[h]	Same as occupancy served	—
6. Catwalks	40	300
7. Cornices	60	—
8. Corridors First floor Other floors	 100 Same as occupancy served except as indicated	—
9. Dining rooms and restaurants	100[m]	—
10. Dwellings (see residential)	—	—
11. Elevator machine room grating (on area of 2 inches by 2 inches)	—	300
12. Finish light floor plate construction (on area of 1 inch by 1 inch)	—	200
13. Fire escapes On single-family dwellings only	100 40	—
14. Garages (passenger vehicles only) Trucks and buses	40[m] See Section 1607.7	Note a
15. Handrails, guards and grab bars	See Section 1607.8	
16. Helipads	See Section 1607.6	
17. Hospitals Corridors above first floor Operating rooms, laboratories Patient rooms	 80 60 40	 1,000 1,000 1,000
18. Hotels (see residential)	—	—
19. Libraries Corridors above first floor Reading rooms Stack rooms	 80 60 150[b, m]	 1,000 1,000 1,000
20. Manufacturing Heavy Light	 250[m] 125[m]	 3,000 2,000
21. Marquees	75	—
22. Office buildings Corridors above first floor File and computer rooms shall be designed for heavier loads based on anticipated occupancy Lobbies and first-floor corridors Offices	 80 — 100 50	 2,000 — 2,000 2,000

(continued)

TABLE 1607.1—continued MINIMUM UNIFORMLY DISTRIBUTED LIVE LOADS, L_o, AND MINIMUM CONCENTRATED LIVE LOADS[g]

OCCUPANCY OR USE	UNIFORM (psf)	CONCENTRATED (lbs.)
23. Penal institutions Cell blocks Corridors	 40 100	—
24. Recreational uses: Bowling alleys, poolrooms and similar uses Dance halls and ballrooms Gymnasiums Reviewing stands, grandstands and bleachers Stadiums and arenas with fixed seats (fastened to floor)	 75[m] 100[m] 100[m] 100[c, m] 60[c, m]	—
25. Residential One- and two-family dwellings Uninhabitable attics without storage[i] Uninhabitable attics with storage[i, j, k] Habitable attics and sleeping areas[k] All other areas Hotels and multifamily dwellings Private rooms and corridors serving them Public rooms[m] and corridors serving them	 10 20 30 40 40 100	—
26. Roofs All roof surfaces subject to maintenance workers Awnings and canopies: Fabric construction supported by a skeleton structure All other construction Ordinary flat, pitched, and curved roofs (that are not occupiable) Where primary roof members are exposed to a work floor, at single panel point of lower chord of roof trusses or any point along primary structural members supporting roofs: Over manufacturing, storage warehouses, and repair garages All other primary roof members Occupiable roofs: Roof gardens Assembly areas All other similar areas	 5 nonreducible 20 20 100 100[m] Note 1	 300 2,000 300 Note 1
27. Schools Classrooms Corridors above first floor First-floor corridors	 40 80 100	 1,000 1,000 1,000
28. Scuttles, skylight ribs and accessible ceilings	—	200
29. Sidewalks, vehicular drive ways and yards, subject to trucking	250[d, m]	8,000[e]

(continued)

TABLE 1607.1—continued
MINIMUM UNIFORMLY DISTRIBUTED LIVE LOADS, L_o, AND MINIMUM CONCENTRATED LIVE LOADS[g]

OCCUPANCY OR USE	UNIFORM (psf)	CONCENTRATED (lbs.)
30. Stairs and exits		
One- and two-family dwellings	40	300[f]
All other	100	300[f]
31. Storage warehouses (shall be designed for heavier loads if required for anticipated storage)		—
Heavy	250[m]	
Light	125[m]	
32. Stores		
Retail		
First floor	100	1,000
Upper floors	75	1,000
Wholesale, all floors	125[m]	1,000
33. Vehicle barriers	See Section 1607.8.3	
34. Walkways and elevated platforms (other than exitways)	60	—
35. Yards and terraces, pedestrians	100[m]	—

For SI: 1 inch = 25.4 mm, 1 square inch = 645.16 mm^2,
1 square foot = 0.0929 m^2,
1 pound per square foot = 0.0479 kN/m^2, 1 pound = 0.004448 kN,
1 pound per cubic foot = 16 kg/m^3.

a. Floors in garages or portions of buildings used for the storage of motor vehicles shall be designed for the uniformly distributed live loads of Table 1607.1 or the following concentrated loads: (1) for garages restricted to passenger vehicles accommodating not more than nine passengers, 3,000 pounds acting on an area of 4.5 inches by 4.5 inches; (2) for mechanical parking structures without slab or deck that are used for storing passenger vehicles only, 2,250 pounds per wheel.

b. The loading applies to stack room floors that support nonmobile, double-faced library book stacks, subject to the following limitations:
1. The nominal bookstack unit height shall not exceed 90 inches;
2. The nominal shelf depth shall not exceed 12 inches for each face; and
3. Parallel rows of double-faced book stacks shall be separated by aisles not less than 36 inches wide.

c. Design in accordance with ICC 300.

d. Other uniform loads in accordance with an approved method containing provisions for truck loadings shall also be considered where appropriate.

e. The concentrated wheel load shall be applied on an area of 4.5 inches by 4.5 inches.

f. The minimum concentrated load on stair treads shall be applied on an area of 2 inches by 2 inches. This load need not be assumed to act concurrently with the uniform load.

g. Where snow loads occur that are in excess of the design conditions, the structure shall be designed to support the loads due to the increased loads caused by drift buildup or a greater snow design determined by the building official (see Section 1608).

h. See Section 1604.8.3 for decks attached to exterior walls.

i. Uninhabitable attics without storage are those where the maximum clear height between the joists and rafters is less than 42 inches, or where there are not two or more adjacent trusses with web configurations capable of accommodating an assumed rectangle 42 inches in height by 24 inches in width, or greater, within the plane of the trusses. This live load need not be assumed to act concurrently with any other live load requirements.

(continued)

TABLE 1607.1—continued
MINIMUM UNIFORMLY DISTRIBUTED LIVE LOADS, L_o, AND MINIMUM CONCENTRATED LIVE LOADS[g]

j. Uninhabitable attics with storage are those where the maximum clear height between the joists and rafters is 42 inches or greater, or where there are two or more adjacent trusses with web configurations capable of accommodating an assumed rectangle 42 inches in height by 24 inches in width, or greater, within the plane of the trusses.

The live load need only be applied to those portions of the joists or truss bottom chords where both of the following conditions are met:

i. The attic area is accessible from an opening not less than 20 inches in width by 30 inches in length that is located where the clear height in the attic is a minimum of 30 inches; and

ii. The slopes of the joists or truss bottom chords are no greater than two units vertical in 12 units horizontal.

The remaining portions of the joists or truss bottom chords shall be designed for a uniformly distributed concurrent live load of not less than 10 lb./ft^2.

k. Attic spaces served by stairways other than the pull-down type shall be designed to support the minimum live load specified for habitable attics and sleeping rooms.

l. Areas of occupiable roofs, other than roof gardens and assembly areas, shall be designed for appropriate loads as approved by the building official. Unoccupied landscaped areas of roofs shall be designed in accordance with Section 1607.12.3.

m. Live load reduction is not permitted unless specific exceptions of Section 1607.10 apply.

1607.2 Loads not specified. For occupancies or uses not designated in Table 1607.1, the live load shall be determined in accordance with a method *approved* by the *building official*.

❖ Whenever an occupancy or use of a structure cannot be identified with the listing shown in Table 1607.1, then the live load values used for design are required to be determined by the design professional and subject to the approval of the building official. Aside from the obvious intent of this requirement, however, which is to prescribe a minimum design load value, some caution needs to be exercised by the design professional in determining the appropriate design live load value. For example, the table shows that heavy storage areas must be designed for a uniform live load of 250 psf (11.9 kN/m^2). This is a minimum value. Storage warehouses or storage areas within manufacturing facilities containing items, such as automobile parts, electrical goods, coiled steel, plumbing supplies and bulk building materials, generally have live loads ranging between 300 and 400 psf (14.4 and 19 kN/m^2). Similarly, storage facilities containing dry goods, paints, oil, groceries or liquor often have loadings that range between 200 to 300 psf (9.6 to 14.4 kN/m^2). Another example is a heavy manufacturing facility that makes generators for the electric power industry. Some of the production areas in this type of facility require structural floors that support loads of 1,000 psf (47.9 kN/m^2) or more, which is about seven times the live load specified in Table 1607.1.

1607.3 Uniform live loads. The live loads used in the design of buildings and other structures shall be the maximum loads expected by the intended use or occupancy but shall in no case be less than the minimum uniformly distributed live loads given in Table 1607.1.

❖ Studies have shown that building live loads consist of a sustained portion based on the day-to-day use of the facilities, and a variable portion created by unusual events, such as remodeling, temporary storage of materials, the extraordinary assemblage of people for an occasional business meeting or social function (i.e., holiday party) and similar events. The sustained portion of the live load will likely vary during the life of a building because of tenant changes, rearrangement of office space and furnishings, changes in the nature of the occupancy (i.e., number of people or type of business), traffic patterns and so on. In light of this variability of loadings that are apt to be imposed on a building, the code provisions simplify the design procedure by expressing the applicable load as either a uniformly distributed live load or a concentrated live load on the floor area. It should be pointed out that this section does not require the concurrent application of uniform live load and concentrated live load. In other words, this section requires that either the uniform load or the concentrated load be applied, so long as the type of load that produces the greater stress in the structural element under consideration is utilized.

1607.4 Concentrated live loads. Floors and other similar surfaces shall be designed to support the uniformly distributed live loads prescribed in Section 1607.3 or the concentrated live loads, in pounds (kiloNewtons), given in Table 1607.1, whichever produces the greater *load effects*. Unless otherwise specified, the indicated concentration shall be assumed to be uniformly distributed over an area of $2^1/_2$ feet by $2^1/_2$ feet (762 mm by 762 mm) and shall be located so as to produce the maximum *load effects* in the structural members.

❖ A building or portion thereof is subjected to concentrated floor loads commensurate with the use of the facility. For example, in Group B, a law office may have stacks of books and files that impose large concentrations of loads on the supporting structural elements. An industrial facility may have a tank full of liquid material on a mezzanine that feeds a machine on the floor below. The structural floor of a stockroom may support heavy bins containing metal parts and so on. The exact locations or nature of such concentrated loadings is not usually known at the time of the design of the building. Furthermore, new sources of concentrated loadings will be added during the life of the structure, while some or all of the existing sources will be relocated; therefore, because of the uncertainties of the sources of concentrated loads, as well as their weights and locations, the code provides typical loads to be used in the design of structural floors consistent with the type of use of the facility. The minimum concentrated loads to be used for design are contained in Table 1607.1. Concentrated loads are not required to be applied simultaneously, with the uniform live loads also specified in Table 1607.1. Concentrated loads are to be applied as an independent load condition at the location on the floor that produces the greatest stress in the structural members being designed. The single concentrated load is to be placed at any location on the floor. For example, in an office use area, the floor system is to be designed for either a 2,000-pound (8897 N) concentrated load (unless the anticipated actual concentrated load is higher) applied at any location in the office area, or the 50 psf (2.40 kN/m^2) live load specified in Table 1607.1, whichever results in the greater stress in the supporting structural member.

1607.5 Partition loads. In office buildings and in other buildings where partition locations are subject to change, provisions for partition weight shall be made, whether or not partitions are shown on the *construction documents*, unless the specified live load exceeds 80 psf (3.83 kN/m^2). The partition load shall not be less than a uniformly distributed live load of 15 psf (0.72 kN/m^2).

❖ Provision for the weight of partitions must be made in the structural design. The weight of any built-in partitions should be considered a dead load in accordance with the definition in Section 202. Buildings where partitions are readily relocated must include a live load of 15 psf (0.74 kN/m^2) if the uniform floor live load is 80 psf (3.83 kN/m^2) or less. This partition allowance is included under live loads because of its variable nature.

1607.6 Helipads. Helipads shall be designed for the following live loads:

1. A uniform live load, *L*, as specified below. This load shall not be reduced.
   - 1.1. 40 psf (1.92 kN/m^2) where the design basis helicopter has a maximum take-off weight of 3,000 pounds (13.35 kN) or less.
   - 1.2. 60 psf (2.87 kN/m^2) where the design basis helicopter has a maximum take-off weight greater than 3,000 pounds (13.35 kN).
2. A single concentrated live load, *L*, of 3,000 pounds (13.35 kN) applied over an area of 4.5 inches by 4.5 inches (114 mm by 114 mm) and located so as to produce the maximum load effects on the structural elements under consideration. The concentrated load is not required to act concurrently with other uniform or concentrated live loads.
3. Two single concentrated live loads, *L*, 8 feet (2438 mm) apart applied on the landing pad (representing the helicopter's two main landing gear, whether skid type or wheeled type), each having a magnitude of 0.75 times the maximum take-off weight of the helicopter, and located so as to produce the maximum load effects on the structural elements under consideration. The concentrated loads shall be applied over an area of 8 inches by 8 inches (203 mm by 203 mm) and are not required to act concurrently with other uniform or concentrated live loads.

Landing areas designed for a design basis helicopter with maximum take-off weight of 3,000 pounds (13.35 kN) shall be identified with a 3,000 pound (13.34 kN) weight limitation. The landing area weight limitation shall be indicated by the numeral "3" (kips) located in the bottom right corner of the landing area as viewed from the primary approach path. The indication for the landing area weight limitation shall be a minimum 5 feet (1524 mm) in height.

❖ Detailed requirements for helistops and heliports, including definitions, can be found in Section 412 of the code. For example Section 412.7.1 requires a minimum landing area size of 20 feet by 20 feet (6096 mm by 6096 mm) for helicopters weighing less than 3,500 pounds (15.57 kN). In addition, Section 2007 of the *International Fire Code*® (IFC®) regulates helistops and heliports. Section 2007.8 of the IFC requires Federal Aviation Administration (FAA) approval of these facilities.

The term "helipad" is used to describe a helicopter landing area, which is the subject of this provision. These structural design requirements establish the minimum live load criteria that are specific to the design of the landing area and the supporting structural elements.

Items 1 and 2 specify the uniform live load based on the threshold weight of 3,000 pounds (13.34 kN). The majority of helicopters used in general aviation have a gross weight of 3,000 pounds (13.34 kN) or less. With weights comparable to those of passenger vehicles and considering the size of the landing area, the equivalent uniform load on these landing areas is actually lower than the minimum uniform live load required for a passenger vehicle parking garage. Thus, a reduced design live load is permitted for the design of such helipads. Since marking the weight limitation of the landing area is standard practice, as well as an FAA recommendation, indicating the weight limitation on the landing area was included as a condition for using the uniform load associated with helicopters weighing up to 3,000 pounds (13.34 kN).

1607.7 Heavy vehicle loads. Floors and other surfaces that are intended to support vehicle loads greater than a 10,000 pound (4536 kg) gross vehicle weight rating shall comply with Sections 1607.7.1 through 1607.7.5.

❖ Sections 1607.7.1 through 1607.7.5 give criteria for addressing heavy vehicle loads, including fire trucks and forklifts.

1607.7.1 Loads. Where any structure does not restrict access for vehicles that exceed a 10,000-pound (4536 kg) gross vehicle weight rating, those portions of the structure subject to such loads shall be designed using the vehicular live loads, including consideration of impact and fatigue, in accordance with the codes and specifications required by the jurisdiction having authority for the design and construction of the roadways and bridges in the same location of the structure.

❖ Where heavy highway-type vehicles have access onto a structure, this section requires the structure to be designed using the same requirements that are applicable to roadways in that jurisdiction. This loading may in fact be the loading from American Association of State Highway and Transportation Officials (AASHTO), or the loading for other elements such as lids of large detention tanks or utility vaults. The registered design professional should consult with the jurisdiction for design loads for these special conditions.

1607.7.2 Fire truck and emergency vehicles. Where a structure or portions of a structure are accessed and loaded by fire department access vehicles and other similar emergency vehicles, the structure shall be designed for the greater of the following loads:

1. The actual operational loads, including outrigger reactions and contact areas of the vehicles as stipulated and approved by the building official; or
2. The live loading specified in Section 1607.7.1.

❖ This establishes two criteria for addressing heavy vehicle loads due to fire trucks and other similar emergency vehicles.

1607.7.3 Heavy vehicle garages. Garages designed to accommodate vehicles that exceed a 10,000 pound (4536 kg) gross vehicle weight rating, shall be designed using the live loading specified by Section 1607.7.1. For garages the design for impact and fatigue is not required.

Exception: The vehicular live loads and load placement are allowed to be determined using the actual vehicle weights for the vehicles allowed onto the garage floors, provided such loads and placement are based on rational engineering principles and are approved by the building official, but shall not be less than 50 psf (2.9 kN/m^2). This live load shall not be reduced.

❖ This section helps clarify that the passenger vehicle garage loads are not applicable to garages that accommodate heavier vehicles.

1607.7.4 Forklifts and movable equipment. Where a structure is intended to have forklifts or other movable equipment present, the structure shall be designed for the total vehicle or equipment load and the individual wheel loads for the anticipated vehicles as specified by the owner of the facility. These loads shall be posted per Section 1607.7.5.

❖ Heavy vehicle loads due to forklifts and other moveable equipment require that a structure be designed for the total vehicle or equipment load as well as the individual wheel loads. The owner of the facility needs to make the planned usage of such a vehicle clear to the design team. As a precaution these loads must be posted (see Section 1607.7.5).

1607.7.4.1 Impact and fatigue. Impact loads and fatigue loading shall be considered in the design of the supporting structure. For the purposes of design, the vehicle and wheel loads shall be increased by 30 percent to account for impact.

❖ This section clarifies that consideration of impact and fatigue loading is a design requirement for structures subjected to heavy moving equipment, such as forklifts.

1607.7.5 Posting. The maximum weight of the vehicles allowed into or on a garage or other structure shall be posted by the owner in accordance with Section 106.1.

❖ As a precaution against overloading a structure, the maximum weight of the vehicles that are anticipated and used in the design should be posted by the owner (see Section 106.1).

1607.8 Loads on handrails, guards, grab bars, seats and vehicle barriers. Handrails, *guards*, grab bars, accessible seats, accessible benches and vehicle barriers shall be designed and constructed to the structural loading conditions set forth in this section.

❖ The requirements of this section are intended to provide an adequate degree of structural strength and stability to handrails, guards, grab bars and vehicle barriers.

1607.8.1 Handrails and guards. Handrails and *guards* shall be designed to resist a linear load of 50 pounds per linear foot (plf) (0.73 kN/m) in accordance with Section 4.5.1 of ASCE 7. Glass handrail assemblies and *guards* shall also comply with Section 2407.

Exceptions:

1. For one- and two-family dwellings, only the single concentrated load required by Section 1607.8.1.1 shall be applied.
2. In Group I-3, F, H and S occupancies, for areas that are not accessible to the general public and that have an *occupant load* less than 50, the minimum load shall be 20 pounds per foot (0.29 kN/m).

❖ The loading in this section represents the maximum anticipated load on a handrail or guard due to a crowd of people on the adjacent walking surface. The exceptions allow lower loads for handrails in locations that are not typically open to the public. These loads, depicted in Figure 1607.8.1, are permitted to be applied independent of other loads.

1607.8.1.1 Concentrated load. Handrails and guards shall also be designed to resist a concentrated load of 200 pounds (0.89 kN) in accordance with Section 4.5.1 of ASCE 7.

❖ The concentrated loading in this section is not to be applied with any other design load; it is a separate load case. The load simulates the maximum anticipated load from a person grabbing or falling into the handrail or guard.

1607.8.1.2 Intermediate rails. Intermediate rails (all those except the handrail), balusters and panel fillers shall be designed to resist a concentrated load of 50 pounds (0.22 kN) in accordance with Section 4.5.1 of ASCE 7.

❖ This is a localized design load for the guard members and is not to be applied with any other loads. It is to be applied horizontally at a 90-degree (1.57 rad) angle with the guard members. The number of balusters that would resist this load are those within the 1 square foot (0.093 m^2) area in the plane of the guard as shown in Figure 1607.8.1.2.

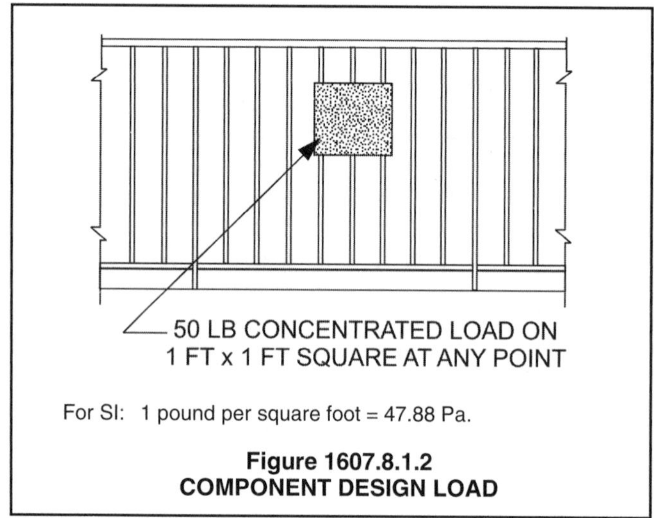

For SI: 1 pound per square foot = 47.88 Pa.

Figure 1607.8.1.2
COMPONENT DESIGN LOAD

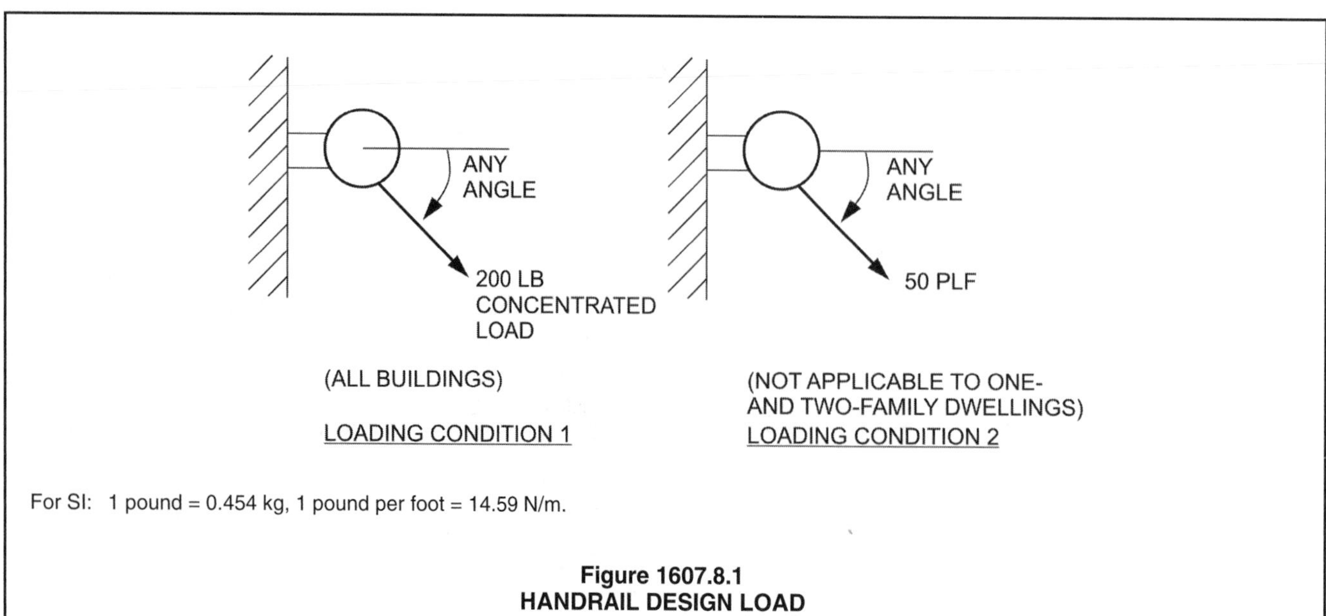

For SI: 1 pound = 0.454 kg, 1 pound per foot = 14.59 N/m.

Figure 1607.8.1
HANDRAIL DESIGN LOAD

1607.8.2 Grab bars, shower seats and dressing room bench seats. Grab bars, shower seats and dressing room bench seat systems shall be designed to resist a single concentrated load of 250 pounds (1.11 kN) applied in any direction at any point on the grab bar or seat so as to produce the maximum load effects.

❖ These live loads provide for the normal anticipated loads from the use of the grab bars, shower seats and dressing room bench seats. These structural requirements provide consistency with Americans with Disabilities Act (ADA) *Accessibility Guidelines* (ADAAG).

1607.8.3 Vehicle barriers. Vehicle barriers for passenger vehicles shall be designed to resist a concentrated load of 6,000 pounds (26.70 kN) in accordance with Section 4.5.3 of ASCE 7. Garages accommodating trucks and buses shall be designed in accordance with an *approved* method that contains provisions for traffic railings.

❖ Vehicle barriers provide a passive restraint system in locations where vehicles could fall to a lower level (see definition of "Vehicle barrier," Section 202). Figure 1607.8.3 depicts criteria for the design of passenger car and light truck vehicle barriers. The 6,000-pound (26.70 kN) load considers impact. The load is applied at a height, h, that is representative of vehicle bumper heights. Due to the variety of barrier configurations and anchorage methods, it is necessary to consider any height within the specified size in order to determine the most critical load effects for design of the barrier. For bus and heavy truck vehicle barrier design criteria, a state's Department of Transportation should be contacted (also see Section 1607.7).

1607.9 Impact loads. The live loads specified in Sections 1607.3 through 1607.8 shall be assumed to include adequate allowance for ordinary impact conditions. Provisions shall be made in the structural design for uses and loads that involve unusual vibration and impact forces.

❖ In cases where "unusual" live loads occur in a building that impose impact or vibratory forces on structural elements (i.e., elevators, machinery, craneways, etc.), additional stresses and deflections are imposed on the structural system. Where unusual vibration (dynamic) and impact loads are likely to occur, the code requires that the structural design take these effects into account. Typically, the dynamic effects are approximated through the application of an equivalent static load equal to the dynamic load effects. In most cases, an equivalent static load is sufficient. A dynamic analysis is usually not required.

1607.9.1 Elevators. Members, elements and components subject to dynamic loads from elevators shall be designed for impact loads and deflection limits prescribed by ASME A17.1.

❖ The static load of an elevator must be increased to account for the effect of the elevator's motion. For example, when an elevator comes to a stop, the load on the elevator's supports is significantly higher than the weight of the elevator and the occupants. This effect varies with the acceleration and deceleration rate of the elevator. This section clarifies that the impact load from elevators applies specifically to members, elements and components subject to dynamic loading from the elevator mechanism and directs the code user to the elevator standard to determine the increases.

1607.9.2 Machinery. For the purpose of design, the weight of machinery and moving loads shall be increased as follows to allow for impact: (1) light machinery, shaft- or motor-driven, 20 percent; and (2) reciprocating machinery or power-

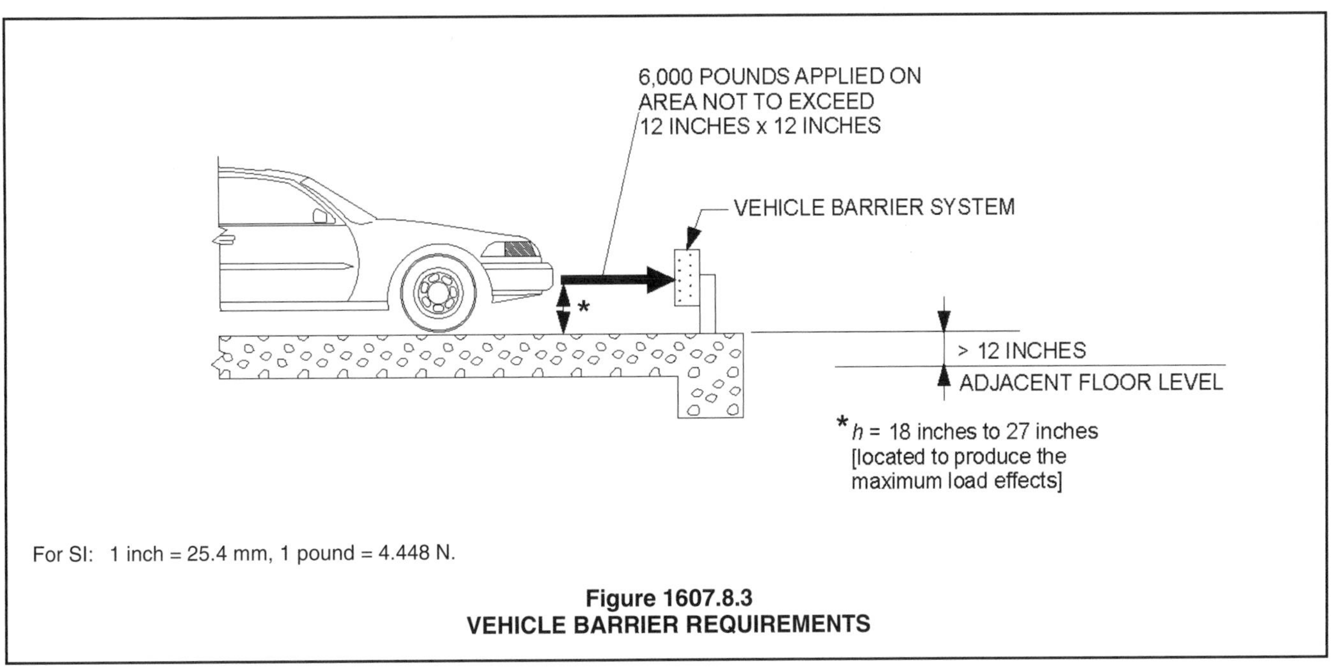

Figure 1607.8.3
VEHICLE BARRIER REQUIREMENTS

driven units, 50 percent. Percentages shall be increased where specified by the manufacturer.

❖ The specified increases for machinery loads include the vibration of the equipment, which increases the effective load. The load increase for reciprocating machinery versus rotating shaft-driven machinery is to account for the higher vibration.

1607.10 Reduction in uniform live loads. Except for uniform live loads at roofs, all other minimum uniformly distributed live loads, L_o, in Table 1607.1 are permitted to be reduced in accordance with Section 1607.10.1 or 1607.10.2. Uniform live loads at roofs are permitted to be reduced in accordance with Section 1607.12.2.

❖ Small floor areas are more likely to be subjected to the full uniform load than larger floor areas. Unloaded or lightly loaded areas tend to reduce the total load on the structural members supporting those floors. The specified uniformly distributed live loads from Table 1607.1 are permitted to be reduced, with some notable exceptions or limitations, in recognition that the larger the tributary area of a structural member, the lower the likelihood that the full live load will be realized. The basis for the live load reduction in Sections 1607.10.1 through 1607.10.1.3 is ASCE 7. An alternative method of live load reduction, retained from legacy model codes, is provided in Section 1607.10.2.

1607.10.1 Basic uniform live load reduction. Subject to the limitations of Sections 1607.10.1.1 through 1607.10.1.3 and Table 1607.1, members for which a value of $K_{LL}A_T$ is 400 square feet (37.16 m²) or more are permitted to be designed for a reduced uniformly distributed live load, L, in accordance with the following equation:

$$L = L_o\left(0.25 + \frac{15}{\sqrt{K_{LL}A_T}}\right) \qquad \textbf{(Equation 16-23)}$$

For SI: $L = L_o\left(0.25 + \frac{4.57}{\sqrt{K_{LL}A_T}}\right)$

where:

L = Reduced design live load per square foot (m²) of area supported by the member.

L_o = Unreduced design live load per square foot (m²) of area supported by the member (see Table 1607.1).

K_{LL} = Live load element factor (see Table 1607.10.1).

A_T = Tributary area, in square feet (m²).

L shall not be less than $0.50L_o$ for members supporting one floor and L shall not be less than $0.40L_o$ for members supporting two or more floors.

❖ This section provides a method of reducing uniform floor live loads that is based on the provisions of ASCE 7. The concept is that where the design live load is governed by the minimum live loads in Table 1607.1, the actual load on a large area of the floor is very likely to be less than the nominal live load in the table. Thus, the allowable reduction increases with the tributary area of the floor that is supported by a structural member; therefore, a girder that supports a large tributary area would be allowed to be designed for somewhat lower uniform live load than a floor beam that supports a smaller floor area.

The following example demonstrates the live load reduction calculation for the conditions shown in Figure 1607.10.1:

Solution:

For interior beam K_a = 2 (Table 1607.10.1)

K, A_T = (2)(750 sq. ft.) = 1500 sq. ft. (139 m²)

Using Equation 16-23

$$L = 50\left(0.25 + \frac{15}{\sqrt{1500}}\right) = 35\text{psf} \ (1.68 \text{ kN/m}^2)$$

$0.5\ L_o$ = 25 psf < 32 psf (1.53 kN/m²)

Use reduced live load, L = 32 psf (1.53 kN/m²)

TABLE 1607.10.1
LIVE LOAD ELEMENT FACTOR, K_{LL}

ELEMENT	K_{LL}
Interior columns Exterior columns without cantilever slabs	4 4
Edge columns with cantilever slabs	3
Corner columns with cantilever slabs Edge beams without cantilever slabs Interior beams	2 2 2
All other members not identified above including: Edge beams with cantilever slabs Cantilever beams One-way slabs Two-way slabs Members without provisions for continuous shear transfer normal to their span	1

❖ The purpose of this table is to provide tributary area adjustment factors, K_{LL}, for determining live load reductions in Section 1607.10.1. The factor converts the tributary area of the structural member to an "influence area." This "influence area" of a structural member is considered to be the adjacent floor area from which it derives any of its load. These adjustments to the tributary area range from 1 through 4 based on the type of structural element being designed and are meant to reflect the element's ability to share load with adjacent elements.

1607.10.1.1 One-way slabs. The tributary area, A_T, for use in Equation 16-23 for one-way slabs shall not exceed an area defined by the slab span times a width normal to the span of 1.5 times the slab span.

❖ This section limits the tributary area that can be utilized to determine a live load reduction for one-way slabs.

1607.10.1.2 Heavy live loads. Live loads that exceed 100 psf (4.79 kN/m^2) shall not be reduced.

Exceptions:

1. The live loads for members supporting two or more floors are permitted to be reduced by a maximum of 20 percent, but the live load shall not be less than L as calculated in Section 1607.10.1.
2. For uses other than storage, where *approved*, additional live load reductions shall be permitted where shown by the *registered design professional* that a rational approach has been used and that such reductions are warranted.

❖ The purpose of this section is to prohibit live load reductions where the live loads exceed 100 psf (4.79 kN/m^2). Such live loads are typically intended for storage or related purposes. It is more likely that the full live load will be realized at a given floor level for such occupancies. Thus, reduced live loads are not allowed for these conditions except as described in this section.

In Exception 1, the loads on structural members, such as columns and bearing walls that support two or more floors, are allowed to be reduced by 20 percent. Surveys have indicated that it is rare for the total live load on any story to exceed 80 percent of the tabulated uniform live loads. Conservatively, the full load should apply to beams and girders, but a member supporting multiple floors is allowed some live load reduction.

Recognizing that there are circumstances under which live loads exceed 100 psf (4.79 kN/m^2) in occupancies other than storage, Exception 2 allows the registered design professional to present a "rational" live load reduction proposal to be applied for members with larger tributary areas (e.g., girders, columns, foundations, etc.). Examples would be mechanical rooms, electrical rooms, process mezzanines in industrial buildings, etc. These types of areas may have very high localized uniform loads under the equipment footprints, for instance, but the live loads to members having larger tributary areas are much less, on average.

1607.10.1.3 Passenger vehicle garages. The live loads shall not be reduced in passenger vehicle garages.

Exception: The live loads for members supporting two or more floors are permitted to be reduced by a maximum of 20 percent, but the live load shall not be less than L as calculated in Section 1607.10.1.

❖ This section limits the live load reduction for passenger vehicle garages to only those members that support more than two floors. Thus, floor framing members that support only a part of one floor do not warrant a reduction to the live load of 40 psf (1.92 kN/m^2) that is specified in Table 1607.1. The rationale for allowing some live load reduction for members supporting multiple floors is similar to that given under Section 1607.10.1.2, Exception 1.

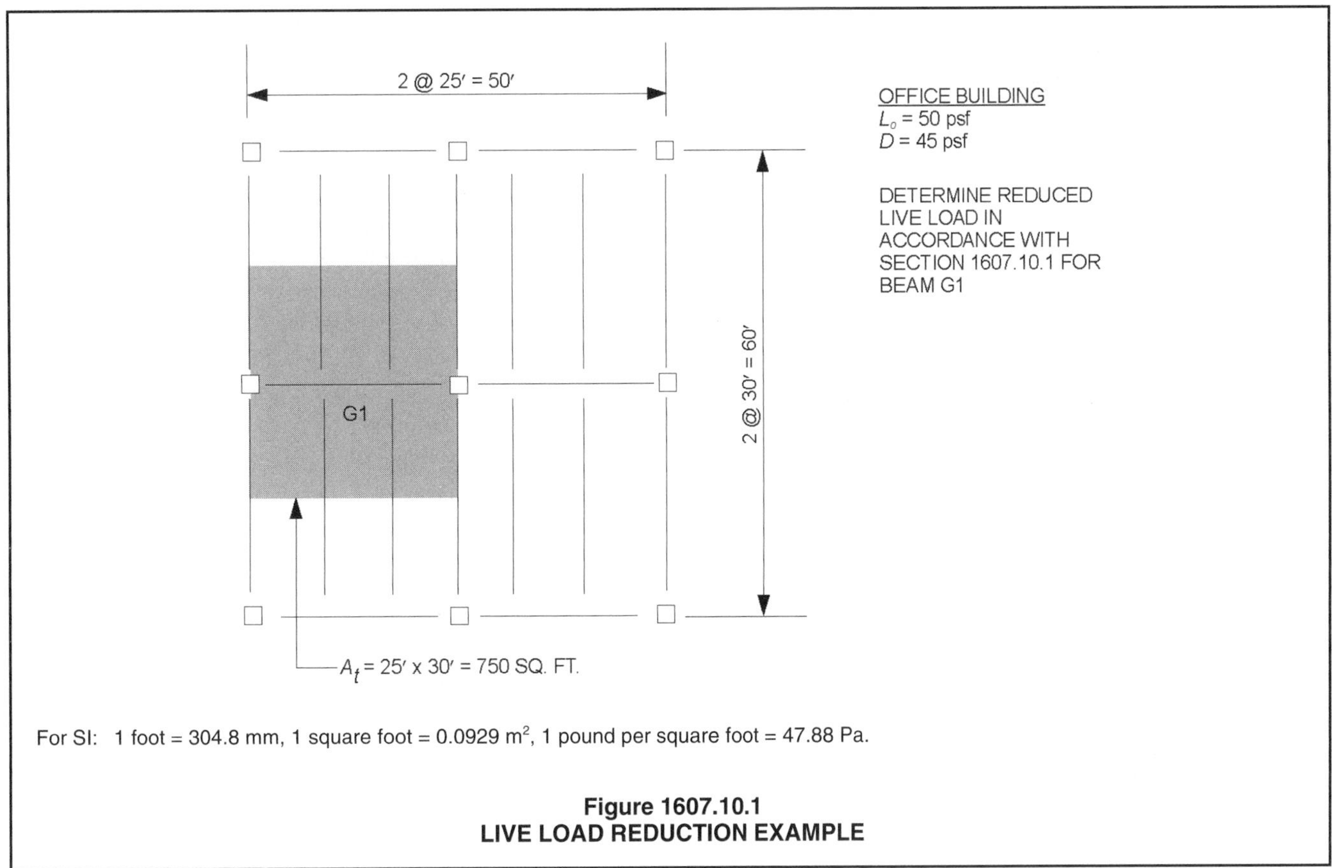

For SI: 1 foot = 304.8 mm, 1 square foot = 0.0929 m^2, 1 pound per square foot = 47.88 Pa.

Figure 1607.10.1
LIVE LOAD REDUCTION EXAMPLE

1607.10.2 Alternative uniform live load reduction. As an alternative to Section 1607.10.1 and subject to the limitations of Table 1607.1, uniformly distributed live loads are permitted to be reduced in accordance with the following provisions. Such reductions shall apply to slab systems, beams, girders, columns, piers, walls and foundations.

1. A reduction shall not be permitted where the live load exceeds 100 psf (4.79 kN/m^2) except that the design live load for members supporting two or more floors is permitted to be reduced by a maximum of 20 percent.

   Exception: For uses other than storage, where *approved*, additional live load reductions shall be permitted where shown by the *registered design professional* that a rational approach has been used and that such reductions are warranted.

2. A reduction shall not be permitted in passenger vehicle parking garages except that the live loads for members supporting two or more floors are permitted to be reduced by a maximum of 20 percent.
3. For live loads not exceeding 100 psf (4.79 kN/m^2), the design live load for any structural member supporting 150 square feet (13.94 m^2) or more is permitted to be reduced in accordance with Equation 16-24.
4. For one-way slabs, the area, A, for use in Equation 16-24 shall not exceed the product of the slab span and a width normal to the span of 0.5 times the slab span.

$$R = 0.08(A - 150) \qquad \textbf{(Equation 16-24)}$$

For SI: $R = 0.861(A - 13.94)$

Such reduction shall not exceed the smallest of:

1. 40 percent for horizontal members;
2. 60 percent for vertical members; or
3. R as determined by the following equation.

$$R = 23.1(1 + D/L_o) \qquad \textbf{(Equation 16-25)}$$

where:

A = Area of floor supported by the member, square feet (m^2).

D = Dead load per square foot (m^2) of area supported.

L_o = Unreduced live load per square foot (m^2) of area supported.

R = Reduction in percent.

❖ This section includes an alternative floor live load method that is permitted to be used instead of the method indicated in Sections 1607.10.1 through 1607.10.1.3. The basis for this section is the 1997 UBC.

Where reductions are permitted, they are allowed at a rate of 0.08 percent per square foot of area in excess of 150 square feet (14 m^2). This value cannot exceed 60 percent for vertical members, such as columns or bearing walls, or 40 percent for horizontal members, such as beams or girders. Additionally, the reduction cannot be more than the value determined by Equation 16-25.

Example of an alternate floor live load reduction:

For beam G1 given in Figure 1607.10.1, determine the reduced floor live load in accordance with Section 1607.10.2.

Solution: A_T = 750 square feet (69 m^2) > 150 square feet (14 m^2); therefore, a reduction is permitted.

Equation 16-24 $R = (0.08\%)(750 - 150) = 48\% > 40\%$

Equation 16-25

$$R = (23.1\%)\left(1 + \frac{45}{50}\right) = 43.7\% > 40\%$$

Use the maximum 40 percent reduction allowed for horizontal members.

Use $L = 50(1\text{-}0.4) = 30$ psf (144 kN/m^2).

1607.11 Distribution of floor loads. Where uniform floor live loads are involved in the design of structural members arranged so as to create continuity, the minimum applied loads shall be the full dead loads on all spans in combination with the floor live loads on spans selected to produce the greatest *load effect* at each location under consideration. Floor live loads are permitted to be reduced in accordance with Section 1607.10.

❖ For continuous floor members loaded such that the live loads of a building are distributed in some bays and not in others, some of the structural elements will be subjected to greater stresses because of partial loading conditions as compared to full loading on all spans. This code section requires the engineer to consider partial loadings that produce the greatest design forces for any location in the design of continuous floor elements.

For example, Figure 1607.11 shows a continuous multispan girder with partial loading. The Type I loading condition shows that only the alternate spans have uniform live loads, which produces:

- Maximum positive moments at the centers of the loaded spans (A-B, C-D, E-F) and
- Maximum negative moments at the centers of the unloaded spans (B-C, D-E, F-G).

The Type II live load distribution shows two loaded adjacent spans with alternate spans loaded beyond these, which produces:

- Maximum negative moment at Support D and
- Maximum girder shears.

To obtain the maximum total stresses imposed on the girder, the dead load moments and shears must be added to those produced by the partial live loadings.

1607.12 Roof loads. The structural supports of roofs and marquees shall be designed to resist wind and, where applicable, snow and earthquake loads, in addition to the dead load of construction and the appropriate live loads as prescribed in this section, or as set forth in Table 1607.1. The live loads acting on a sloping surface shall be assumed to act vertically on the horizontal projection of that surface.

❖ In addition to dead and live loads, the roof's structural system is to be designed and constructed to resist environmental loads caused by wind, snow and earthquakes. According to the definition of "Roof live loads" in Section 202, these are typically an allowance for maintenance of equipment as well as the roof system itself. Other roof live-loads must be considered where appropriate, such as "occupiable roofs" (see Section 1607.12.3) where an occupancy-related live load would be applicable.

1607.12.1 Distribution of roof loads. Where uniform roof live loads are reduced to less than 20 psf (0.96 kN/m^2) in accordance with Section 1607.12.2.1 and are applied to the design of structural members arranged so as to create continuity, the reduced roof live load shall be applied to adjacent spans or to alternate spans, whichever produces the most unfavorable *load effect*. See Section 1607.12.2 for reductions in minimum roof live loads and Section 7.5 of ASCE 7 for partial snow loading.

❖ For continuous roof construction, where live loads are reduced to less than 20 psf (0.96 kN/m^2) as permitted in Section 1607.12.2.1, partial loadings must be included in the design of structural elements to determine the governing loading situation. For example, Figure 1607.11 shows a continuous multispan girder with partial loading. The Type I loading condition shows that only the alternate spans have uniform live loads, which produces:

- Maximum positive moments at the centers of the loaded spans (A-B, C-D, E-F) and
- Maximum negative moments at the centers of the unloaded spans (B-C, D-E, F-G).

The Type II live load distribution shows two loaded adjacent spans with alternate spans loaded beyond these, which produces:

- Maximum negative moment at Support D and
- Maximum girder shears.

To obtain the maximum total stresses imposed on the girder, the dead load moments and shears must be added to those produced by the partial live loadings.

1607.12.2 General. The minimum uniformly distributed live loads of roofs and marquees, L_o, in Table 1607.1 are permitted to be reduced in accordance with Section 1607.12.2.1.

❖ The minimum roof live loads typically reflect loads that occur during roof maintenance, construction or repair. In addition to the standard roof live load of 20 psf (0.96 kN/m^2), Table 1607.1 includes roof live loads for special purpose roofs and fabric awnings. While this section seems to refer to reducing any of the tabulated uniformly distributed live loads, the actual reduction method provided in Section 1607.12.2.1 is limited to the 20 psf (0.96 kN/m^2) live load. This is made evident by the limits on Equation 16-26 of $12 \leq L_r \leq 20$.

1607.12.2.1 Ordinary roofs, awnings and canopies. Ordinary flat, pitched and curved roofs, and awnings and canopies other than of fabric construction supported by a skeleton structure, are permitted to be designed for a reduced uniformly distributed roof live load, L_r, as specified in the following equations or other controlling combinations of loads as specified in Section 1605, whichever produces the greater *load effect*.

In structures such as greenhouses, where special scaffolding is used as a work surface for workers and materials during maintenance and repair operations, a lower roof load than

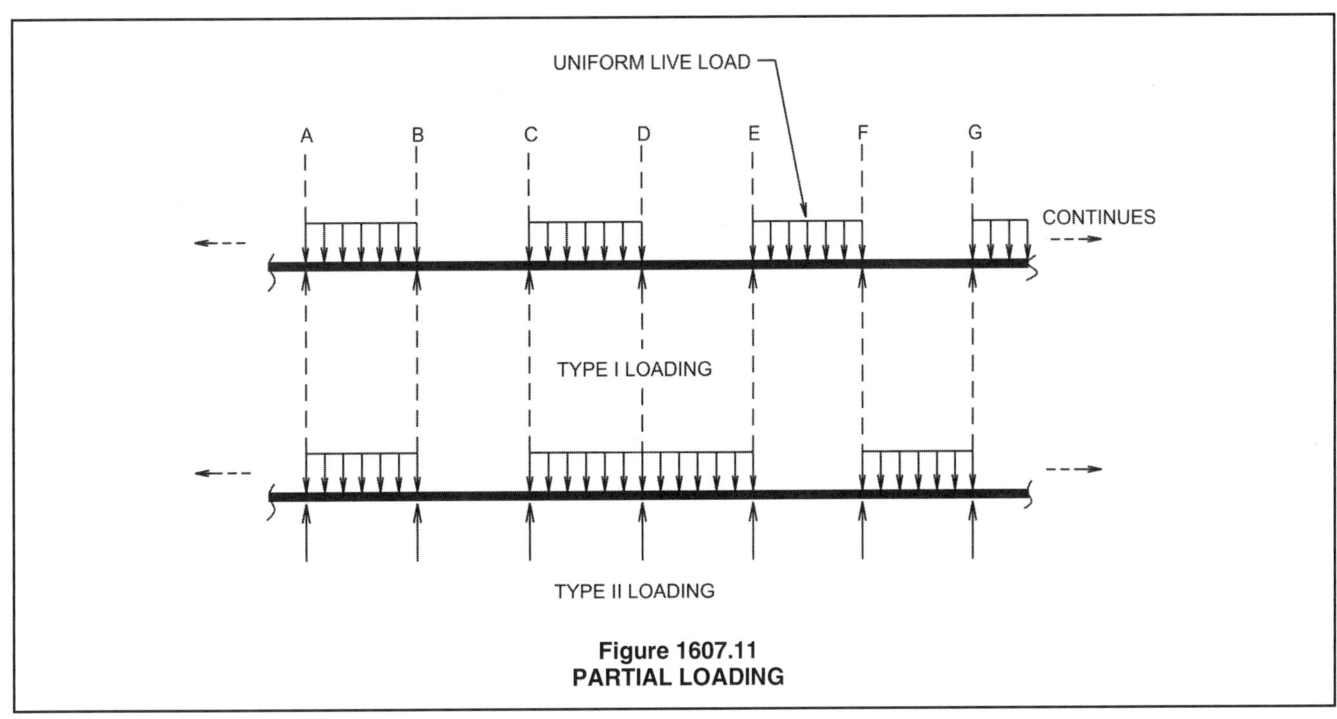

Figure 1607.11
PARTIAL LOADING

specified in the following equations shall not be used unless *approved* by the *building official*. Such structures shall be designed for a minimum roof live load of 12 psf (0.58 kN/m^2).

$L_r = L_o R_1 R_2$ **(Equation 16-26)**

where: $12 \leq L_r \leq 20$

For SI: $L_r = L_o R_1 R_2$

where: $0.58 \leq L_r \leq 0.96$

L_o = Unreduced roof live load per square foot (m^2) of horizontal projection supported by the member (see Table 1607.1).

L_r = Reduced roof live load per square foot (m^2) of horizontal projection supported by the member.

The reduction factors R_1 and R_2 shall be determined as follows:

$R_1 = 1$ for $A_t \leq 200$ square feet (18.58 m^2) **(Equation 16-27)**

$R_1 = 1.2 - 0.001A_t$ for 200 square feet $< A_t <$ 600 square feet **(Equation 16-28)**

For SI: $1.2 - 0.011A_t$ for 18.58 square meters $< A_t <$ 55.74 square meters

$R_1 = 0.6$ for $A_t \geq 600$ square feet (55.74 m^2) **(Equation 16-29)**

where:

A_t = Tributary area (span length multiplied by effective width) in square feet (m^2) supported by the member, and

$R_2 = 1$ for $F \leq 4$ **(Equation 16-30)**

$R_2 = 1.2 - 0.05\ F$ for $4 < F < 12$ **(Equation 16-31)**

$R_2 = 0.6$ for $F \geq 12$ **(Equation 16-32)**

where:

F = For a sloped roof, the number of inches of rise per foot (for SI: $F = 0.12 \times$ slope, with slope expressed as a percentage), or for an arch or dome, the rise-to-span ratio multiplied by 32.

❖ This section provides a formula for the determination of the live load for the design of flat, pitched or curved roofs. The live load from Table 1607.1 that applies is 20 psf (0.96 kN/m^2). Reduced roof live loads are based on the roof slope and the tributary area of the member being considered. The portion of the live load reduction based on tributary area does not apply to roof members that support small tributary areas of less than 200 square feet (18.58 m^2). The load can be reduced as the tributary area increases but never to less than 12 psf (0.58 kN/m^2). For roof slopes between 4:12 and 12:12, live load reductions based on slope apply. Figure 1607.12.2.1 shows the roof

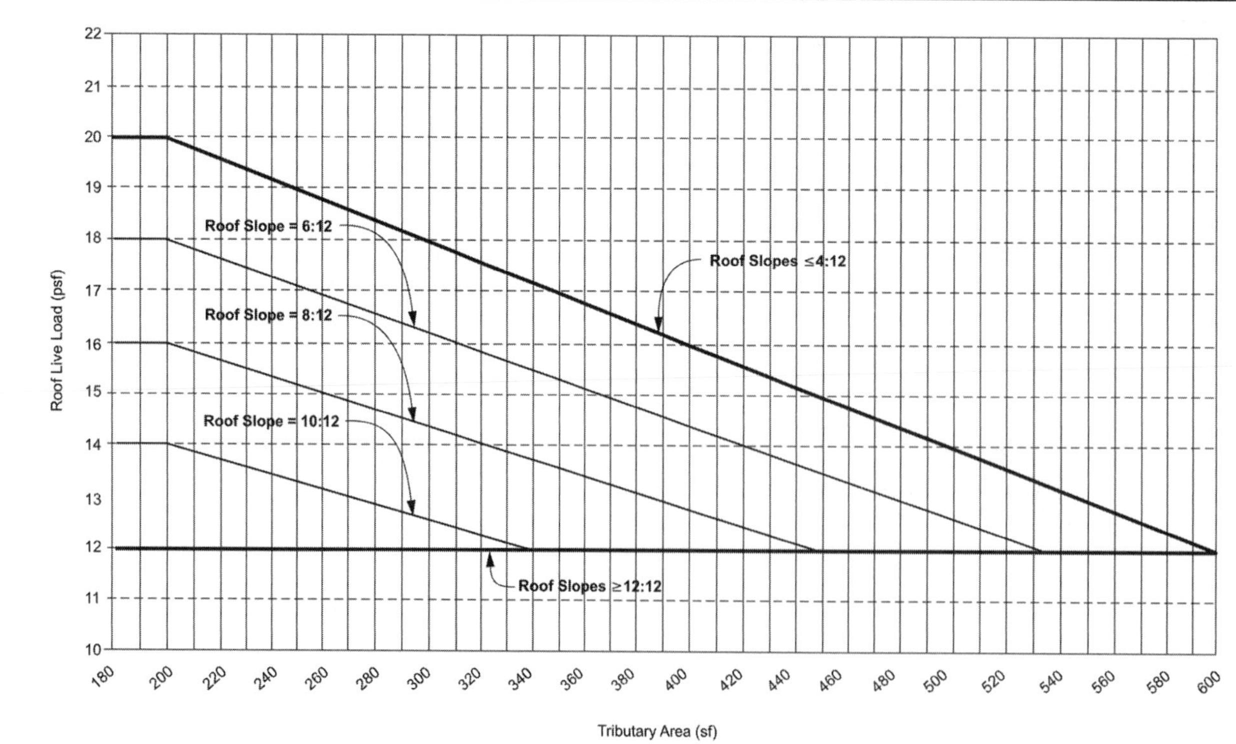

Figure 1607.12.2.1
MINUMUM ROOF LIVE LOAD IN ACCORDANCE WITH EQUATION 16-26

live load, L_r, determined by Equation 16-26 for increments of roof slope. Since the relationship of variables is linear, intermediate values can be interpolated from the figure.

This section also provides for a lower roof live load for a greenhouse, since it is not likely that loads from maintenance or repair will exceed the specified 12 psf (0.58 kN/m^2).

1607.12.3 Occupiable roofs. Areas of roofs that are occupiable, such as roof gardens, or for assembly or other similar purposes, and marquees are permitted to have their uniformly distributed live loads reduced in accordance with Section 1607.10.

❖ Roofs that are to be occupied during social events incidental to the principal use of the facility are to be designed for a minimum uniform live load of 60 psf (2.87 kN/m^2). The promenade deck of a residential penthouse located on the main roof of an apartment building is an example of this type of use. Where roofs are designed to be used as roof gardens or to support large gatherings of people as a function accompanying the educational or assembly uses of a facility, the roofs are required to be designed to a minimum live load of 100 psf (4.79 kN/m^2). The minimum live loads specified in Table 1607.1 are not required to be added to the design load requirements for occupiable roofs specified in this section.

1607.12.3.1 Landscaped roofs. The uniform design live load in unoccupied landscaped areas on roofs shall be 20 psf (0.958 kN/m^2). The weight of all landscaping materials shall be considered as dead load and shall be computed on the basis of saturation of the soil.

❖ Those areas of a roof that are to be landscaped are required to be designed for a minimum uniform live load of 20 psf (0.96 kN/m^2) to accommodate the occasional loads associated with the maintenance of plantings. The weight of landscaping materials and saturated soil is to be considered a dead load in the design of the roof structure, which is to be combined with the live load (see Section 1605 for applicable load combinations).

1607.12.4 Awnings and canopies. Awnings and canopies shall be designed for uniform live loads as required in Table 1607.1 as well as for snow loads and wind loads as specified in Sections 1608 and 1609.

❖ Awning structures are lightweight frames that are typically covered with fabric materials and are designed to sustain a live load of 5 psf (0.24 kN/m^2), as well as the specified snow and wind loads of Sections 1608 and 1609. The live load, snow load and wind load are to be combined according to Section 1605.

1607.13 Crane loads. The crane live load shall be the rated capacity of the crane. Design loads for the runway beams, including connections and support brackets, of moving bridge cranes and monorail cranes shall include the maximum wheel loads of the crane and the vertical impact, lateral and longitudinal forces induced by the moving crane.

❖ This section provides a general description of the crane loads that are required to be included in the design. The supporting structure for the crane is to be designed for a combination of the maximum wheel load, vertical impact and horizontal load as a simultaneous load combination. The typical arrangement for a top-running bridge crane is shown in Figure 1607.13.

1607.13.1 Maximum wheel load. The maximum wheel loads shall be the wheel loads produced by the weight of the bridge, as applicable, plus the sum of the rated capacity and the weight of the trolley with the trolley positioned on its runway at the location where the resulting load effect is maximum.

❖ The maximum vertical wheel load occurs when the trolley is moved as close as possible to the supporting beams under consideration. This results in the greatest portion of the crane weight, the design weight lifted load and the wheel vertical impact load on the supporting beams.

1607.13.2 Vertical impact force. The maximum wheel loads of the crane shall be increased by the percentages shown below to determine the induced vertical impact or vibration force:

Monorail cranes (powered) .... 25 percent

Cab-operated or remotely operated bridge cranes (powered) .... 25 percent

Pendant-operated bridge cranes (powered) .... 10 percent

Bridge cranes or monorail cranes with hand-geared bridge, trolley and hoist .... 0 percent

❖ A vertical impact force is necessary to account for the impact from the starting and stopping movement of the suspended weight from the crane. Vertical impact is also created by the movement of the crane along the rails.

1607.13.3 Lateral force. The lateral force on crane runway beams with electrically powered trolleys shall be calculated as 20 percent of the sum of the rated capacity of the crane and the weight of the hoist and trolley. The lateral force shall be assumed to act horizontally at the traction surface of a runway beam, in either direction perpendicular to the beam, and shall be distributed with due regard to the lateral stiffness of the runway beam and supporting structure.

❖ This section is necessary to define the design lateral force on the crane supports. Lateral force at the right angle to the crane rail is caused by the lateral movement of the lifted load and from the frame action of the crane.

1607.13.4 Longitudinal force. The longitudinal force on crane runway beams, except for bridge cranes with hand-geared bridges, shall be calculated as 10 percent of the maximum wheel loads of the crane. The longitudinal force shall be assumed to act horizontally at the traction surface of a runway beam, in either direction parallel to the beam.

❖ This section is needed to define the longitudinal force on the crane supports, which is caused from the longitudinal motion of the crane with the lifted load.

1607.14 Interior walls and partitions. Interior walls and partitions that exceed 6 feet (1829 mm) in height, including their finish materials, shall have adequate strength to resist the loads to which they are subjected but not less than a horizontal load of 5 psf (0.240 kN/m^2).

Exception: Fabric partitions complying with Section 1607.14.1 shall not be required to resist the minimum horizontal load of 5 psf (0.24 kN/m^2).

❖ The minimum lateral live load is intended to provide sufficient strength and durability of the wall framing and of the finished construction to provide a minimum level of resistance to nominal impact loads that commonly occur in the use of a facility, such as impacts from moving furniture or equipment, as well as to resist heating, ventilating and air-conditioning (HVAC) pressurization.

1607.14.1 Fabric partitions. Fabric partitions that exceed 6 feet (1829 mm) in height, including their finish materials, shall have adequate strength to resist the following load conditions:

1. A horizontal distributed load of 5 psf (0.24 kN/m^2) applied to the partition framing. The total area used to determine the distributed load shall be the area of the fabric face between the framing members to which the fabric is attached. The total distributed load shall be uniformly applied to such framing members in proportion to the length of each member.
2. A concentrated load of 40 pounds (0.176 kN) applied to an 8-inch diameter (203 mm) area [50.3 square inches (32 452 mm^2)] of the fabric face at a height of 54 inches (1372 mm) above the floor.

❖ This section provides criteria for fabric partitions (see definition, Section 202) as an alternative to Section 1607.14. The construction of these partitions is unique, which makes it difficult to meet the full requirements of Section 1607.14. Condition 1 requires the partition framing to be capable of resisting a minimum lateral load. Condition 2 approximates the load of a person leaning against the fabric using their hand as the point of contact. This criterion is based on test standards that are used to evaluate the tip-over resistance of office furniture panel systems that are often used to provide open plan offices.

SECTION 1608
SNOW LOADS

1608.1 General. Design snow loads shall be determined in accordance with Chapter 7 of ASCE 7, but the design roof load shall not be less than that determined by Section 1607.

❖ The determination of the nominal snow load, S, must be in accordance with this section. The intent of this section is that the code requirements are based on the technical requirements in Chapter 7 of ASCE 7. The snow load provisions in ASCE 7 are based on over 40 years of ground snow load data, and include consideration of thermal resistance of the roof structure, a rain-on-snow surcharge, partial loading on

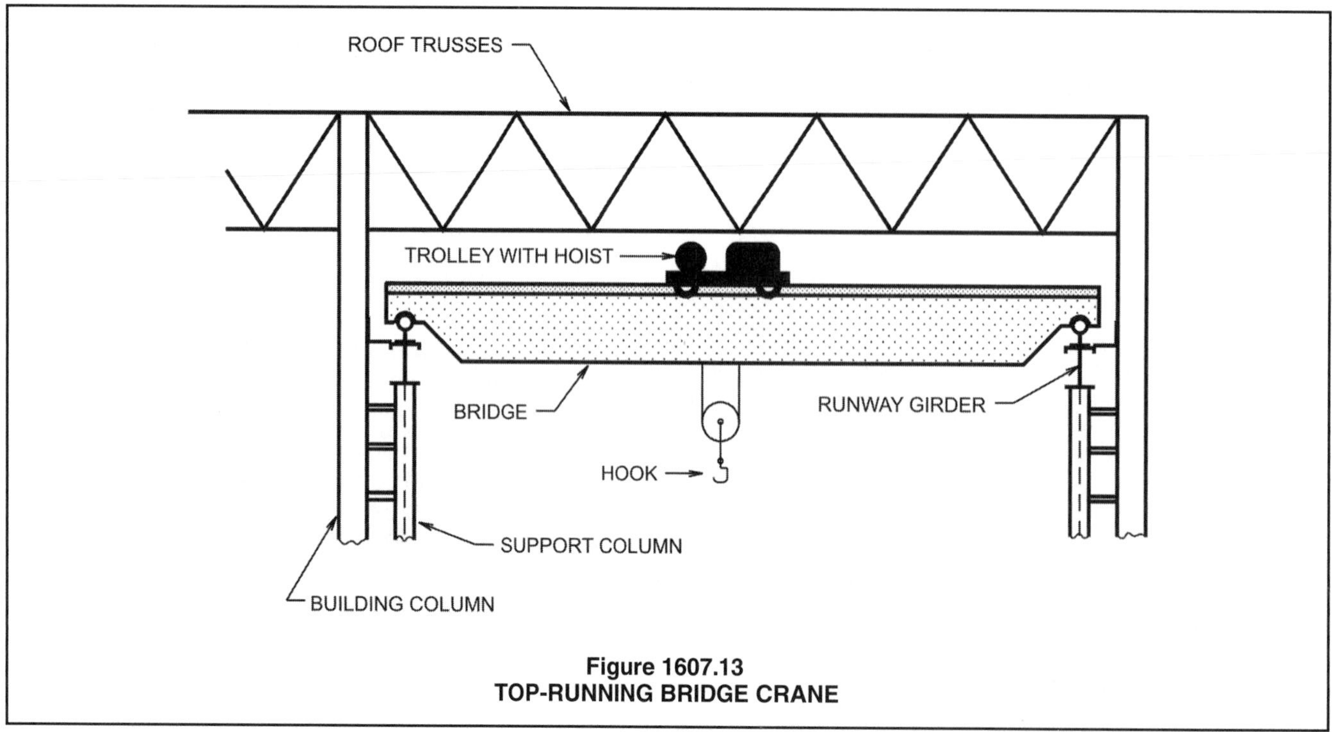

Figure 1607.13
TOP-RUNNING BRIDGE CRANE

continuous beam systems and ponding instability from melting snow or rain on snow. The variables that affect the determination of roof snow loads are:

- Ground snow load (p_g) – see Section 1608.2.
- Importance factor (I) – a factor determined in Section 7.3.3 of ASCE 7, ranging from 0.8 to 1.2, based on the risk category assigned in accordance with Section 1604.5.
- Exposure factor (C_e) and Thermal factor (C_t) – see Sections 7.3.1 and 7.3.2 of ASCE 7.

In addition to the above, the roof slope affects the snow load determination, as well as other design considerations. Roofs with low slopes are designed for a flat roof snow load, p_f determined in accordance with Section 7.3 of ASCE 7, using the previously mentioned criteria. Other considerations for low roof slopes include ponding instability and rain-on-snow surcharge loading. For sloped roofs, the sloped roof snow load, p_s in accordance with Section 7.4 of ASCE 7 applies. The sloped roof snow load essentially modifies the flat roof snow load by the slope factor (C_s), which varies from 1.0 at low a slope (i.e., no effect) to zero at a slope of 70 degrees (1.36 rad).

The flat roof snow load or sloped roof snow load applied uniformly to the entire roof is referred to as the balanced snow load condition. This loading is always a design consideration. Depending on factors such as the type of roof structural system, the geometry of the roof, etc., the following additional snow loadings may require evaluation:

- Partial loading (see Section 7.5, ASCE 7) is a pattern consisting of balanced snow load and one-half of balanced snow load arranged to produce the maximum effects on the structural member being considered.
- Unbalanced snow loads (see Section 7.6, ASCE 7) reflect an uneven loading pattern, such as can occur on a sawtooth roof [see Figure 1608.1(1)].
- Drifting (see Sections 7.7 and 7.8, ASCE 7) is a concern where adjacent roof surfaces are at different elevations [see Figures 1608.1(2) and (3)] or at projections above the roof level, such as at equipment or parapets.

1608.2 Ground snow loads. The ground snow loads to be used in determining the design snow loads for roofs shall be determined in accordance with ASCE 7 or Figure 1608.2 for the contiguous United States and Table 1608.2 for Alaska. Site-specific case studies shall be made in areas designated "CS" in Figure 1608.2. Ground snow loads for sites at elevations above the limits indicated in Figure 1608.2 and for all sites within the CS areas shall be *approved*. Ground snow load determination for such sites shall be based on an extreme value statistical analysis of data available in the vicinity of the site using a value with a 2-percent annual probability of being exceeded (50-year mean recurrence interval). Snow loads are zero for Hawaii, except in mountainous regions as *approved* by the *building official*.

❖ The ground snow loads on the maps in Figure 1608.2 of the code are generally based on over 40 years of snow depth records. The snow loads on the maps are those that have a 2-percent annual probability of being exceeded (a 50-year mean recurrence interval). The maps were generated from data through the winter of 1991-92, and from data through the winter of 1993-94 where the snows were heavy. The mapped snow loads are not increased much from a single snowy winter, since most reporting stations have more than 20 years of snow data. The map values indicate the ground snow load in pounds per square foot. In mountainous areas, the map also indi-

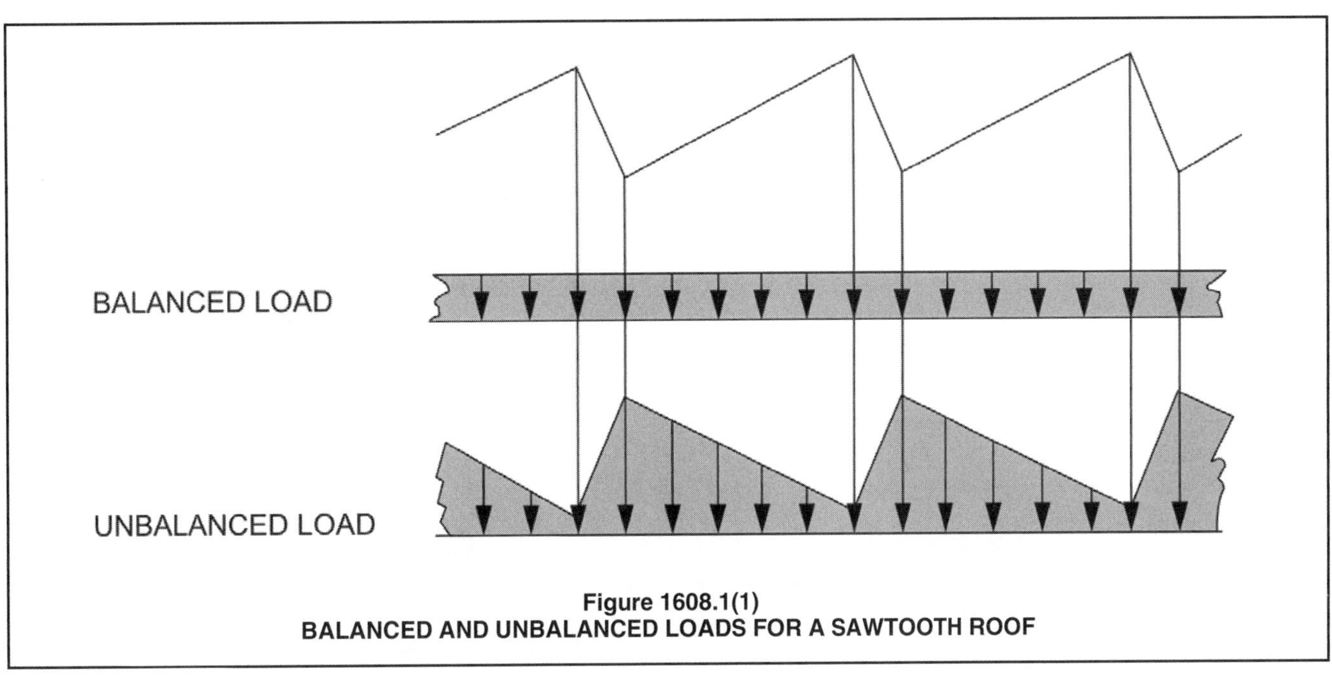

Figure 1608.1(1)
BALANCED AND UNBALANCED LOADS FOR A SAWTOOTH ROOF

cates the highest elevation that is appropriate for the use of the associated snow load. Where the elevation limit is exceeded, a site-specific case study is necessary to establish the appropriate ground snow load. In some areas the ground snow load is too variable to allow mapping. These areas are noted as "CS," which indicates a site-specific case study is necessary. Assistance in the determination of an appropriate ground snow load for these areas may be obtained from the U.S. Department of Army Cold Regions Research and Engineering Laboratory (CRREL) in Hanover, New Hampshire.

TABLE 1608.2. See page 16-35.

❖ Since Alaska in not shown on Figure 1608.2, this table provides the ground snow loads, Pg, for Alaskan locations. These values are needed to determine the appropriate roof snow loads for the indicated locations. The roof snow load is to be determined according to Section 7.3 of ASCE 7 for flat roofs and Section 7.4 of ASCE 7 for sloped roofs.

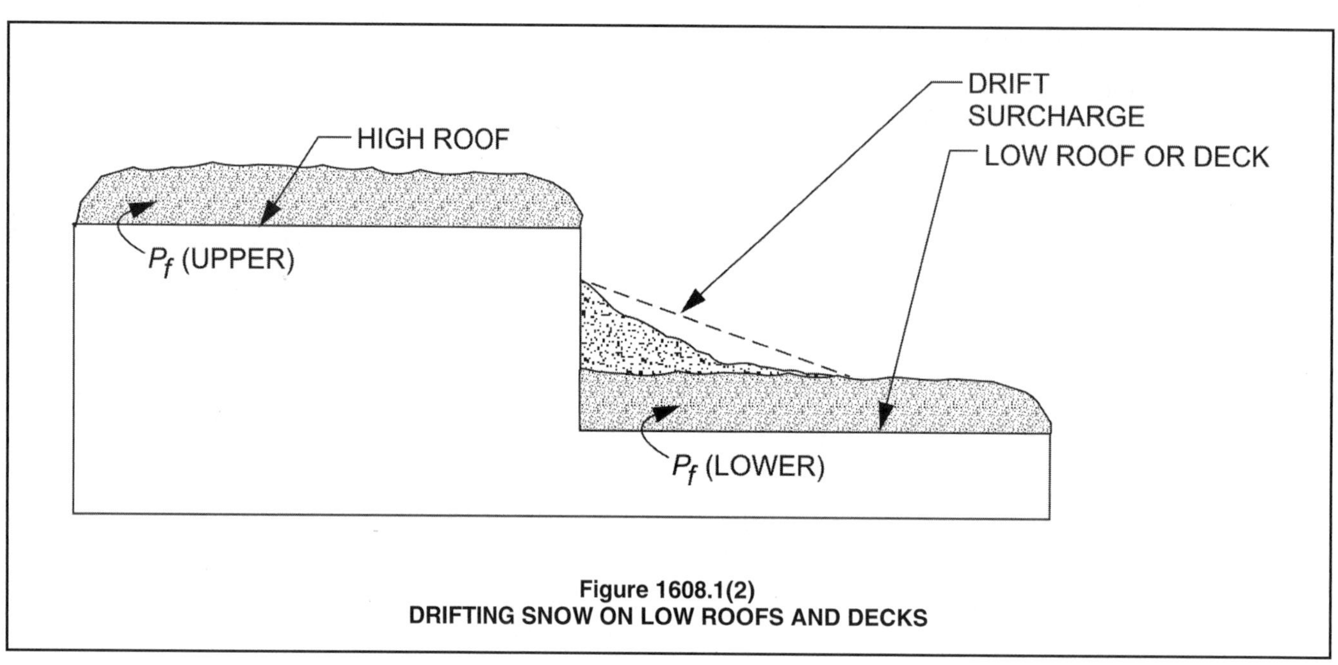

Figure 1608.1(2)
DRIFTING SNOW ON LOW ROOFS AND DECKS

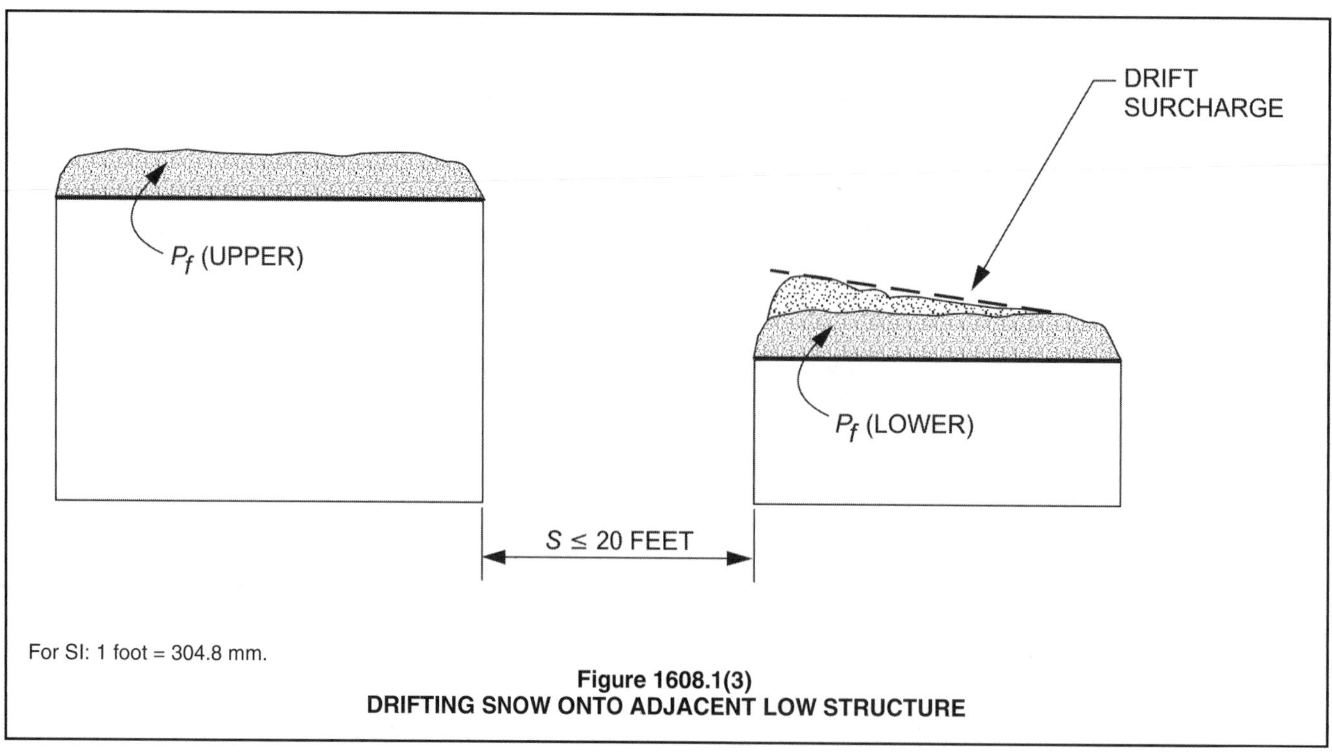

For SI: 1 foot = 304.8 mm.

Figure 1608.1(3)
DRIFTING SNOW ONTO ADJACENT LOW STRUCTURE

FIGURE 1608.2. See page 16-36.

❖ See the commentary to Section 1608.2 for an overview of the snow load map. See the commentary for Section 7.2 of ASCE 7 for a complete description of methodology used in developing the contour lines shown on the figure.

1608.3 Ponding instability. Susceptible bays of roofs shall be evaluated for ponding instability in accordance with Section 7.11 of ASCE 7.

❖ Susceptible bays of roofs are required to meet the technical provisions of ASCE 7 for consideration of progressive deflection. The term "Susceptible bay" is defined in Section 202 and provides a technical basis for determining which bays of a roof need to be investigated for ponding instability. Section 1611.2 also relies on the determination of which bays are susceptible bays.

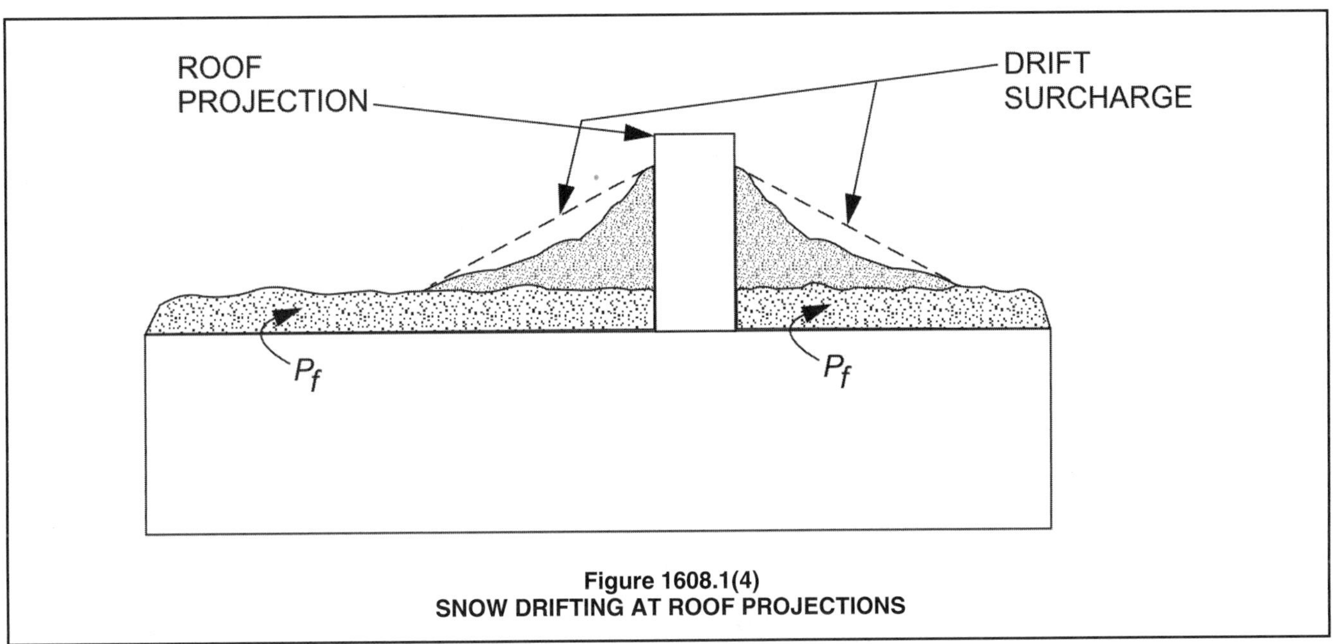

Figure 1608.1(4)
SNOW DRIFTING AT ROOF PROJECTIONS

TABLE 1608.2
GROUND SNOW LOADS, p_g, FOR ALASKAN LOCATIONS

LOCATION	POUNDS PER SQUARE FOOT	LOCATION	POUNDS PER SQUARE FOOT	LOCATION	POUNDS PER SQUARE FOOT
Adak	30	Galena	60	Petersburg	150
Anchorage	50	Gulkana	70	St. Paul Islands	40
Angoon	70	Homer	40	Seward	50
Barrow	25	Juneau	60	Shemya	25
Barter Island	35	Kenai	70	Sitka	50
Bethel	40	Kodiak	30	Talkeetna	120
Big Delta	50	Kotzebue	60	Unalakleet	50
Cold Bay	25	McGrath	70	Valdez	160
Cordova	100	Nenana	80	Whittier	300
Fairbanks	60	Nome	70	Wrangell	60
Fort Yukon	60	Palmer	50	Yakutat	150

For SI: 1 pound per square foot = 0.0479 kN/m^2.

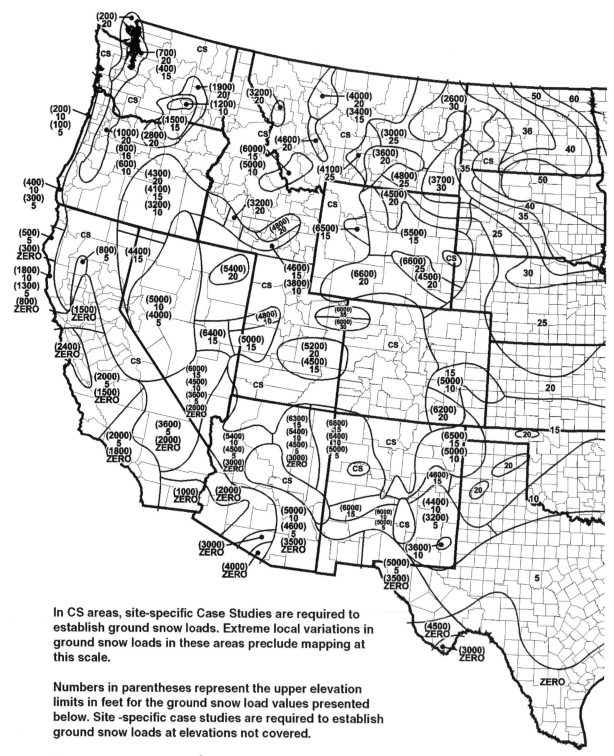

In CS areas, site-specific Case Studies are required to establish ground snow loads. Extreme local variations in ground snow loads in these areas preclude mapping at this scale.

Numbers in parentheses represent the upper elevation limits in feet for the ground snow load values presented below. Site -specific case studies are required to establish ground snow loads at elevations not covered.

To convert lb/sq ft to kNm^2, multiply by 0.0479.

To convert feet to meters, multiply by 0.3048.

FIGURE 1608.2—continued
GROUND SNOW LOADS, p_g, FOR THE UNITED STATES (psf)

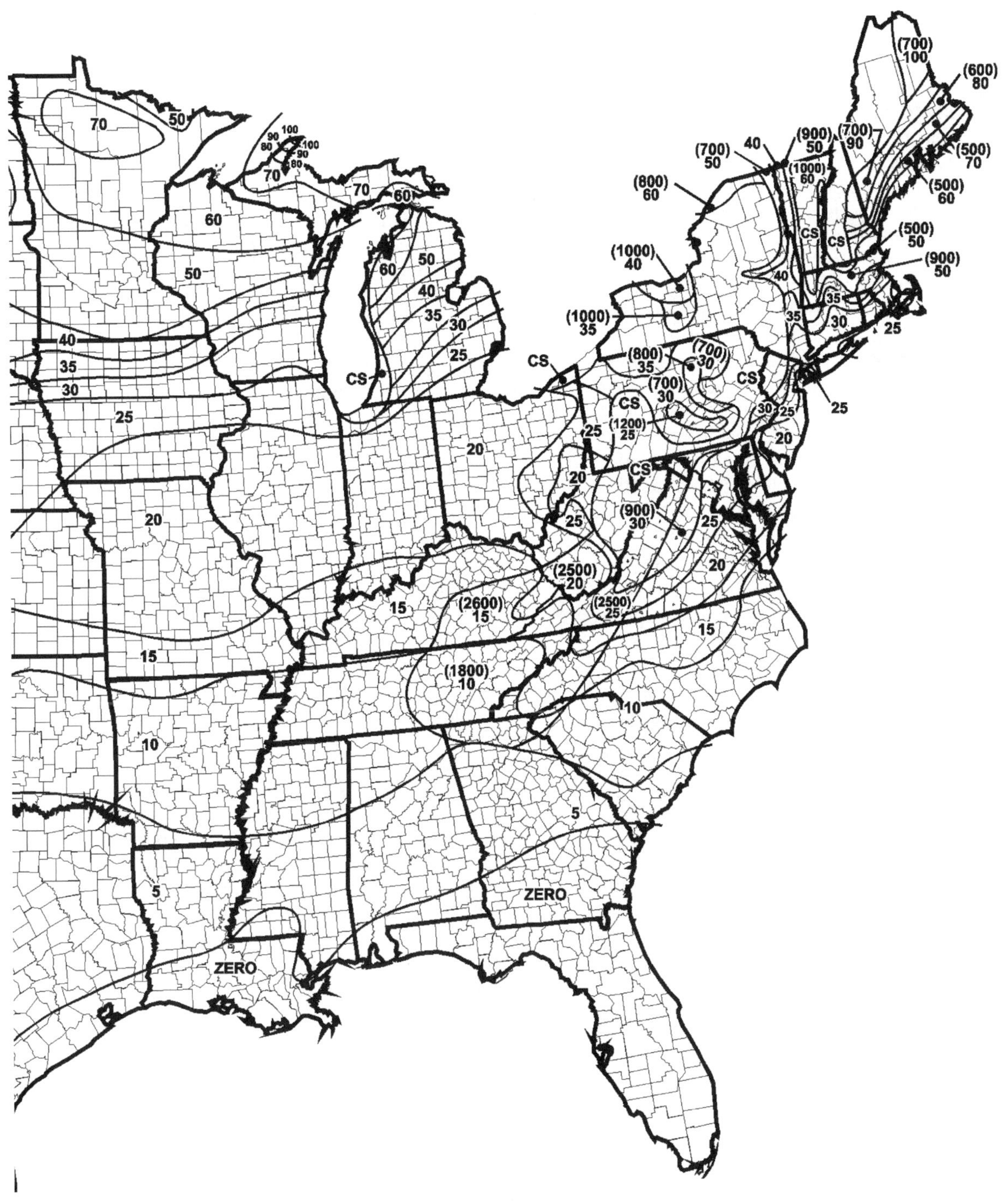

FIGURE 1608.2—continued
GROUND SNOW LOADS, p_g, FOR THE UNITED STATES (psf)

SECTION 1609 WIND LOADS

FIGURES 1609A, 1609B & 1609C. See pages 16-41 through 16-43.

❖ Over the past decade, new data and research have indicated that the mapped hurricane wind speeds have been overly conservative. Significantly more hurricane data has become available, which in turn allows for improvements in the hurricane simulation model that is used to develop wind speed maps. The new hurricane hazard model yields hurricane wind speeds that are lower than those given in previous editions of ASCE 7 and the code, even though the overall rate of intense storms (as defined by central pressure) produced by the new model is increased compared to those produced by the hurricane simulation model used to develop previous maps. In preparing new maps, it was decided to use multiple ultimate event or strength design maps in conjunction with a wind load factor of 1.0 for strength design (see Section 1605.2). For allowable stress design, the load factor has been reduced from 1.0 to 0.6 (see Section 1605.3). Several factors that are important to an accurate wind load standard led to this decision:

1. An ultimate event or strength design wind speed map makes the overall approach consistent with that used in seismic design in that they both map ultimate events and use a load factor of 1.0 for strength design.
2. Utilizing different maps for the different risk categories eliminates the problems associated with using "importance factors" that vary with category. The difference in the importance factors in hurricane-prone and nonhurricane-prone regions for Risk Category I structures, which prompted many questions, is gone.
3. The use of multiple maps eliminates the confusion associated with the recurrence interval associated with the previous wind speed map—the map was not a uniform 50-year return period map. This created a situation where the level of safety that was provided within the overall design was not consistent along the hurricane coast.

Utilizing the new wind speed maps and integrating their use into the code necessitated the introduction of the terms "V_{ult}" and "V_{asd}" to be associated with the "ultimate" design wind speed and the "nominal" design wind speed, respectively. Because of the number of different provisions that use the wind speed map to "trigger" different requirements, it was necessary to provide a conversion methodology in Section 1609.3.1 so that those provisions were not affected. The terms "ultimate design wind speed" and "nominal design wind speed" have been incorporated to clarify the different levels of wind speed.

Prior to 1998, earlier editions of ASCE 7 as well as the legacy model codes incorporated fastest-mile wind speed maps. "Fastest mile" is defined as the average speed of one mile of air that passes a specific reference point. It is important to recognize the differences in averaging times between fastest-mile and the 3-second-gust maps. The averaging time for a 90-mph fastest-mile wind speed is (t = 3,600/V) 40 seconds. Obviously, due to greater averaging time, for a given location the fastest-mile wind speed will be less than the 3-second-gust wind speed. The fact that the wind speed values are higher does not necessarily indicate higher wind loads. Buildings and structures resist wind loads, not wind speeds. Wind speed, although a significant contributor, is only one of several variables and factors that affect wind forces. Wind loads are affected by atmosphere and aerodynamics. Other elements that affect actual wind forces as wind flows across a bluff body include shape factors (C_a), gust effect factors (G) and the velocity pressure that is a function of wind speed, exposure and topography, among others.

The change from the fastest-mile wind speed map to a 3-second-gust map in the 2000 code (based on the 1998 edition of ASCE 7) was necessary for the following reasons. First, weather stations across the United States no longer collect fastest-mile wind speed data. Additionally, the perception of the general public will be more favorable where the code wind speeds are higher, although the design wind pressures were not changed significantly. The map includes a more complete analysis of hurricane wind speeds than previous maps, since more data was available for sites away from the coast. In western states, the 85- to 90-mph contour boundary follows along the Washington, Oregon and California eastern state lines. This is because inland wind data was such that there was no statistical basis to place them elsewhere. The reference to the 50-year MRI used in previous maps was removed, reflecting the fact that the MRI is greater than 50 years along the hurricane coastline. However, nonhurricane wind speeds are based on a 50-year MRI, as Section 1609.3 specifically requires when estimating basic wind speeds from local climatic data.

1609.1 Applications. Buildings, structures and parts thereof shall be designed to withstand the minimum wind loads prescribed herein. Decreases in wind loads shall not be made for the effect of shielding by other structures.

❖ The determination of the wind load, W, must be in accordance with Section 1609. The intent of this section is to provide minimum criteria for the design and construction of buildings and other structures to resist wind loads. These regulations serve to reduce the potential for damage to property caused by windstorms and to provide an acceptable level of protection to building occupants. The objective also includes the prevention of damage to adjacent properties because of the possible detachment of major building components (e.g., walls, roofs, etc.), structural collapse or flying debris and for the safety of people in the immediate vicinity.

The criteria for wind design given in this section of the code generally reflect the wind load provisions of ASCE 7. For a better understanding of the wind load provisions it is important to have a fundamental knowledge of the effects of high-velocity wind forces on buildings and other structures. Wind/structure interactions can be characterized as follows:

When wind encounters a stationary object, such as a building, the airflow changes direction and produces several effects on the building that are illustrated in Figure 1609.1(1). Exterior walls and other vertical surfaces facing the wind (windward side) and perpendicular to its path are subjected to inward (positive) pressures; however, wind does not stop on contact with a facing surface, but flows around and over the building surfaces. This airflow does not instantaneously change directions at surface discontinuities, such as corners of walls or eaves, or over ridges and roof corners. Instead it separates from the downwind surfaces due to high turbulence and localized pressures, resulting in outward (negative) pressures. This phenomena produces suction or outward pressures (negative) on the sidewalls, leeward wall and, depending on geometry, the roof.

Figure 1609.1(1) shows a flat roof and the resulting negative pressures caused by external wind; however, pressures may differ on sloping roofs in the direction perpendicular to the ridge. Roof surfaces (on the windward side) with shallow slopes are generally subjected to outward (negative) pressures—the same as flat roofs. Moderately sloping roofs [about 30 degrees (0.5235 rad)] may be subjected to either inward (positive) pressures, outward (negative) pressures or both—negative pressure in the lower part of the roof and positive pressure in the area of the ridge; however, the code does not require consideration of this scenario. High-sloping roofs (windward side) respond similar to walls and sustain positive wind pressures. The leeward side of a sloped roof is subjected to negative pressure, regardless of the angle. When the wind acts parallel to the ridge, the pressures on a sloped roof are similar to the pressures on a flat roof, meaning the roof is subject to negative pressures.

Openings in the building envelope can impact the magnitude of wind pressure on a structure by increasing the internal pressures that, in turn, will affect the net pressures on the structure. Figure 1609.1(2) illustrates the effects of openings in a building's exterior walls. An opening in the leeward wall causes negative pressure on the interior, increasing the total load on the windward wall. Openings on the windward wall cause positive internal pressure against all the walls from the interior of the building. As the figure shows, the result is an increase in the total load on the leeward wall. In extreme cases, fail-

WINDWARD WALL
SEPARATION BUBBLE
LEEWARD WALL
WIND FLOW
(+) PRESSURES (-)
(-)
(-)
WALL DEFLECTION
SIDEWALL
PLAN
SEPARATION BUBBLE
FLAT ROOF
(-)
WIND FLOW
(+) PRESSURES (-)
ELEVATION

Figure 1609.1(1)
WIND PRESSURES CAUSED BY EXTERNAL WIND FLOW

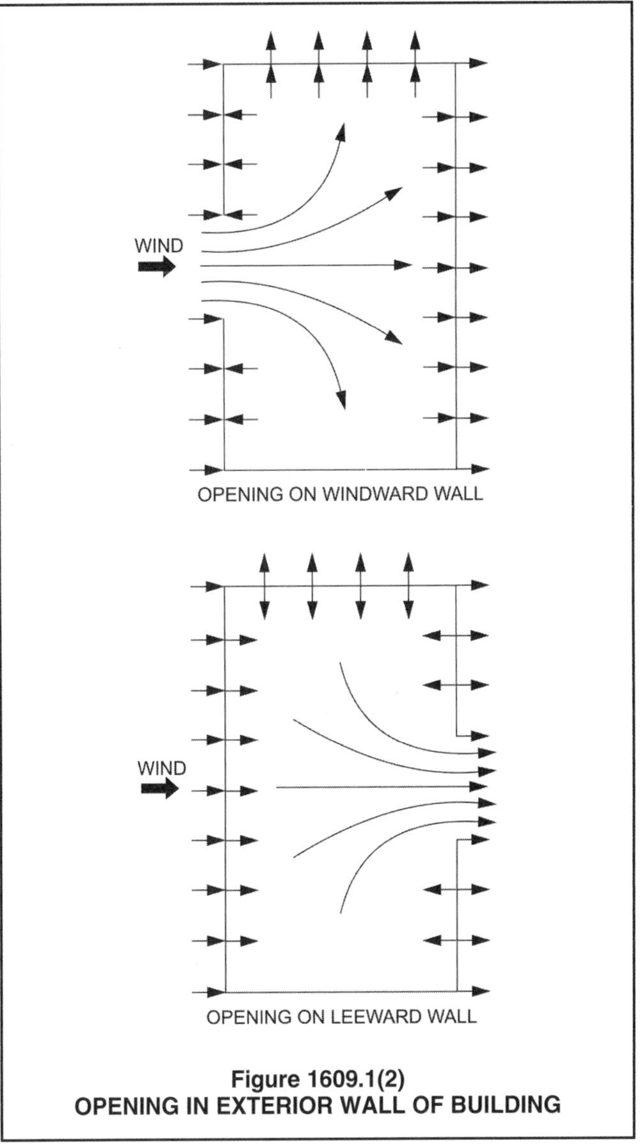

Figure 1609.1(2)
OPENING IN EXTERIOR WALL OF BUILDING

ures resulting from this type of wind flow appear as if the building has exploded.

The definition of "Openings" is worth noting, since in the nonstructural code provisions the term "openings" typically refers to doors, ducts or windows. The definition of "Openings" that is relevant to wind load is in Section 26.2 of ASCE 7. They are all described as "apertures or holes in the building envelope that allow air to flow through the building envelope and that are designed as 'open' during design wind loads." Such openings are then considered in classifying the building as enclosed, partially enclosed or open. The wind load considerations will differ based on that classification.

1609.1.1 Determination of wind loads. Wind loads on every building or structure shall be determined in accordance with Chapters 26 to 30 of ASCE 7 or provisions of the alternate all-heights method in Section 1609.6. The type of opening protection required, the ultimate design wind speed, V_{ult}, and the exposure category for a site is permitted to be determined in accordance with Section 1609 or ASCE 7. Wind shall be assumed to come from any horizontal direction and wind pressures shall be assumed to act normal to the surface considered.

Exceptions:

1. Subject to the limitations of Section 1609.1.1.1, the provisions of ICC 600 shall be permitted for applicable Group R-2 and R-3 buildings.
2. Subject to the limitations of Section 1609.1.1.1, residential structures using the provisions of AF&PA WFCM.
3. Subject to the limitations of Section 1609.1.1.1, residential structures using the provisions of AISI S230.
4. Designs using NAAMM FP 1001.
5. Designs using TIA-222 for antenna-supporting structures and antennas, provided the horizontal extent of Topographic Category 2 escarpments in Section 2.6.6.2 of TIA-222 shall be 16 times the height of the escarpment.
6. Wind tunnel tests in accordance with Chapter 31 of ASCE 7.

The wind speeds in Figures 1609A, 1609B and 1609C are ultimate design wind speeds, V_{ult}, and shall be converted in accordance with Section 1609.3.1 to nominal design wind speeds, V_{asd}, when the provisions of the standards referenced in Exceptions 1 through 5 are used.

❖ The intent of Section 1609 is to require that buildings and structures be designed and constructed to resist the wind loads quantified in Chapters 26 through 30 of ASCE 7.

There are six exceptions to using the provisions of ASCE 7 for the determination of wind loads.

Exception 1 provides for the use of ICC 600 for Group R-2 and R-3 buildings where they are located within Exposure B or C as defined in Section 1609.4 and not sited on the upper half of an isolated hill, escarpment or ridge with the characteristics described in Section 1609.1.1.1. ICC 600 has prescriptive construction requirements and required load capacity tables that replace the requirement for structural analysis, which is intended to provide improved design construction details to achieve greater structural performance for single- and multiple-family dwellings in a high-wind event (see ICC 600 for other detailed application limitations).

Exception 2 provides for the use of the AF&PA *Wood Frame Construction Manual for One- and Two-family Dwellings* where the building is sited within Exposure B or C as defined in Section 1609.4 and not on the upper half of an isolated hill, escarpment or ridge with the characteristics described in Section 1609.1.1.1. The AF&PA *Wood Frame Construction Manual for One- and Two-family Dwellings* has prescriptive construction requirements and required load-resistance tables that replace the requirement for structural analysis. The tabulated engineered and prescriptive design provisions apply to one- and two-family wood frame dwellings where the fastest-mile basic wind speed is between 90 and 120 mph. See Table 1609.3.1 for conversion between the 3-second-gust and fastest-mile wind speeds (see the AF&PA *Wood Frame Construction Manual for One- and Two-family Dwellings*, Chapter 1, for other detailed application limitations). Similarly, Exception 3 allows the cold-formed steel prescriptive standard for one- and two-family dwellings.

Exceptions 4 and 5 simply refer to national standards dealing specifically with the design of flag poles and telecommunication towers.

Exception 5 allows the use of TIA 222-G for antennas and their supporting structures. This standard provides a simplification of the topographic wind speed-up effect. The TIA 222-G standard allows a designer to use the full topographic wind speed-up method of ASCE 7 in order to avoid the conservatism of the simplified method. A code modification makes sure that the simplified method is safe in all cases.

TIA 222-G accounts for the worst-case wind speed-up at the crest for a steep slope, but overlooks the fact that lesser sloped escarpments create pressure increases that cannot be safely ignored beyond the "steep slope" influence. As written, the standard stops considering the topographic wind speed-up effect at "8 x height" from the crest. At this distance, a shallow slope can still increase wind pressure by more than 20 percent.

The modification changes "8 x height" in the standard to "16 x height." The intent is that the "Topographic category 2" definition in TIA 222-G should be applied as follows:

> Category 2. Structures located at or near the crest of an escarpment. Wind speed-up shall be considered to occur in all directions. Structures located vertically on the lower half of an escarpment or horizontally beyond 16 times the height of the escarpment from its crest, shall be permitted to be considered as Topographic Category 1.

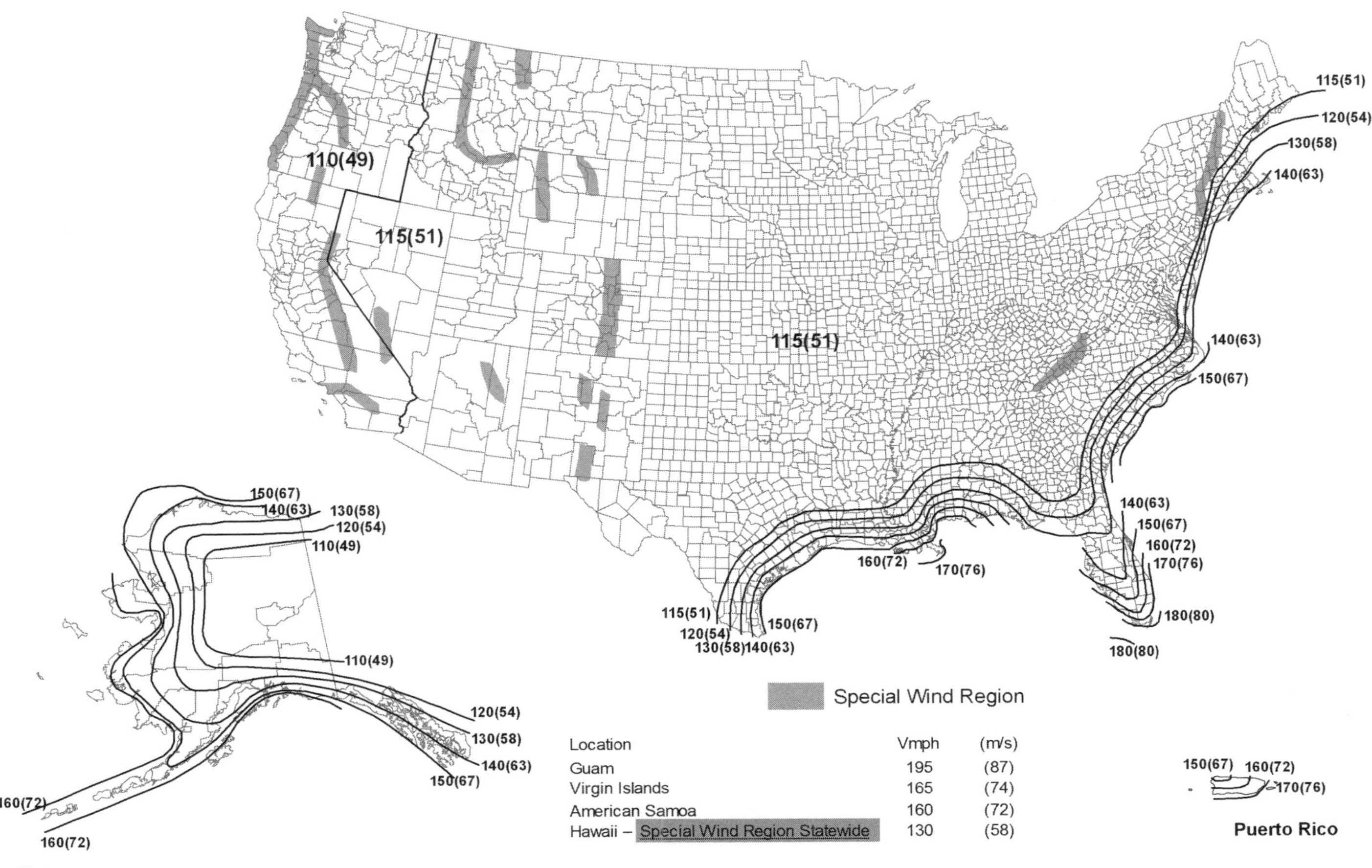

Notes:

1. Values are nominal design 3-second gust wind speeds in miles per hour (m/s) at 33 ft (10m) above ground for Exposure C category.
2. Linear interpolation between contours is permitted.
3. Islands and coastal areas outside the last contour shall use the last wind speed contour of the coastal area.
4. Mountainous terrain, gorges, ocean promontories, and special wind regions shall be examined for unusual wind conditions.
5. Wind speeds correspond to approximately a 7% probability of exceedance in 50 years (Annual Exceedance Probability = 0.00143, MRI = 700 Years).

FIGURE 1609A
ULTIMATE DESIGN WIND SPEEDS, V_{ULT}, FOR RISK CATEGORY II BUILDINGS AND OTHER STRUCTURES

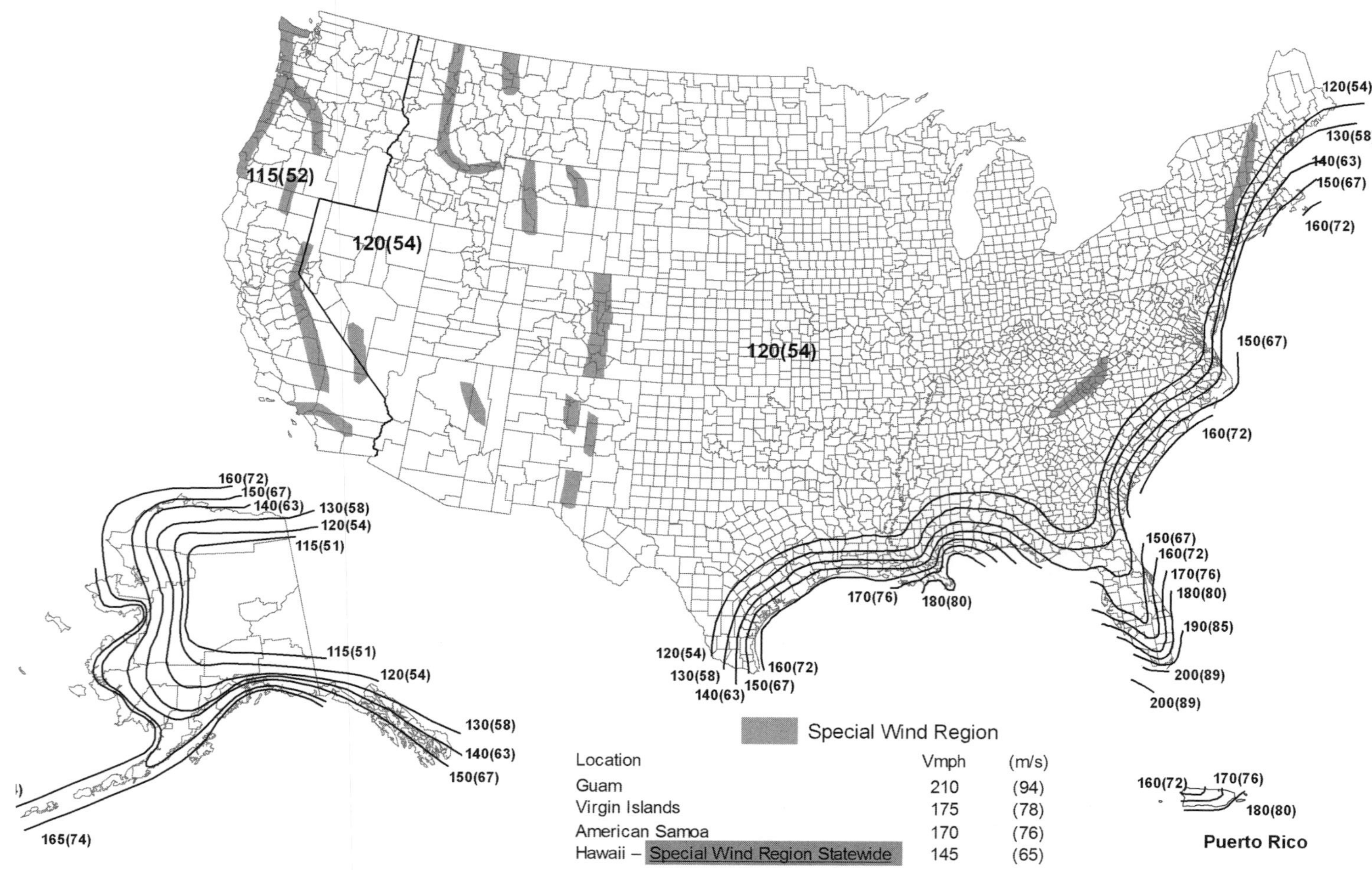

Location	Vmph	(m/s)
Guam	210	(94)
Virgin Islands	175	(78)
American Samoa	170	(76)
Hawaii – Special Wind Region Statewide	145	(65)

tes:

/alues are nominal design 3-second gust wind speeds in miles per hour (m/s) at 33 ft (10m) above ground for Exposure C category.

.inear interpolation between contours is permitted.

slands and coastal areas outside the last contour shall use the last wind speed contour of the coastal area.

/lountainous terrain, gorges, ocean promontories, and special wind regions shall be examined for unusual wind conditions.

Vind speeds correspond to approximately a 3% probability of exceedance in 50 years (Annual Exceedance Probability = 0.000588, MRI = 1700 Years).

FIGURE 1609B
ULTIMATE DESIGN WIND SPEEDS, V_{ULT}, FOR RISK CATEGORY III AND IV BUILDINGS AND OTHER STRUCTURES

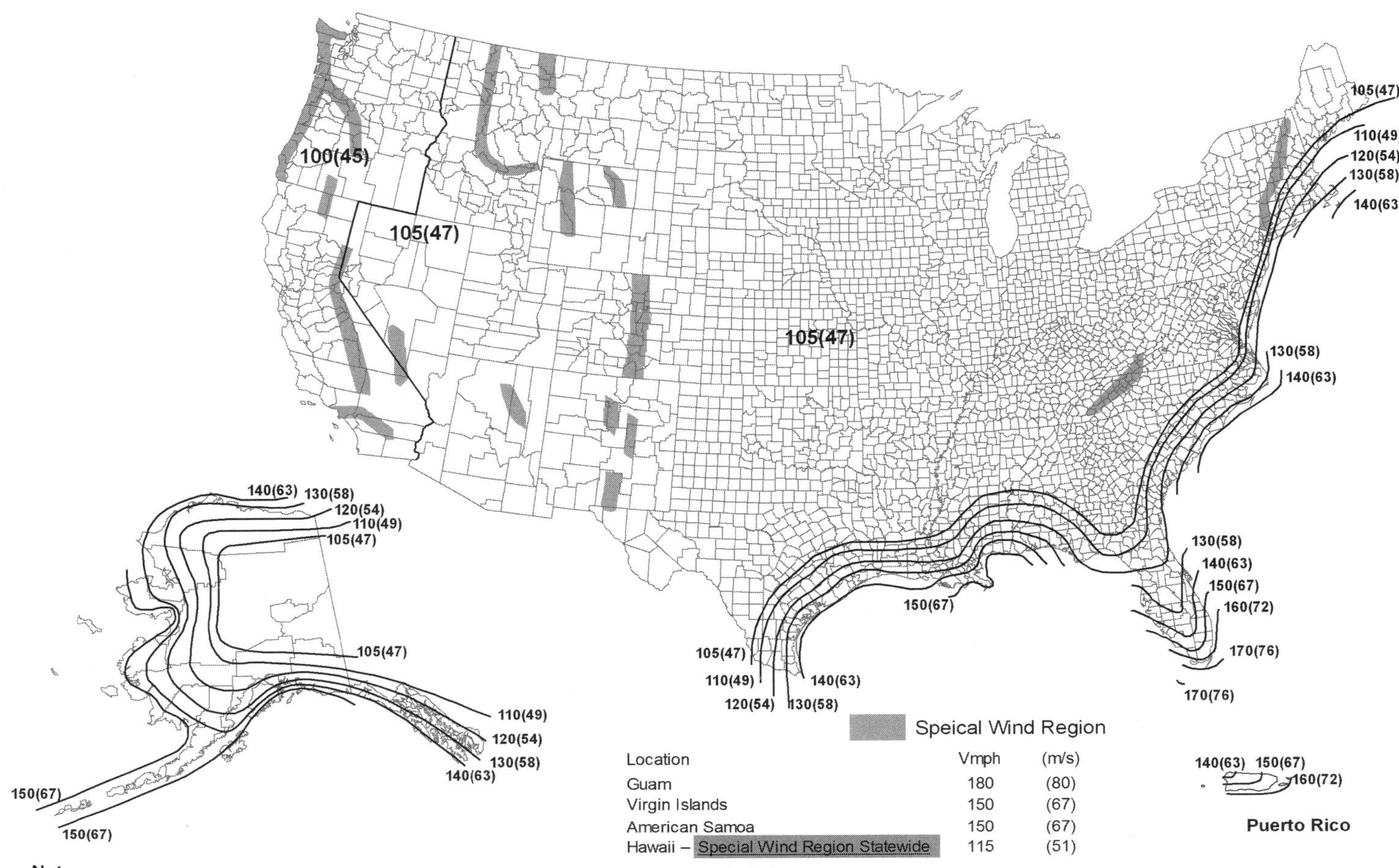

Notes:
1. Values are nominal design 3-second gust wind speeds in miles per hour (m/s) at 33 ft (10m) above ground for Exposure C category.
2. Linear interpolation between contours is permitted.
3. Islands and coastal areas outside the last contour shall use the last wind speed contour of the coastal area.
4. Mountainous terrain, gorges, ocean promontories, and special wind regions shall be examined for unusual wind conditions.
5. Wind speeds correspond to approximately a 15% probability of exceedance in 50 years (Annual Exceedance Probability = 0.00333, MRI = 300 Years).

FIGURE 1609C
ULTIMATE DESIGN WIND SPEEDS, V_{ULT}, FOR RISK CATEGORY I BUILDINGS AND OTHER STRUCTURES

1609.1.1.1 Applicability. The provisions of ICC 600 are applicable only to buildings located within Exposure B or C as defined in Section 1609.4. The provisions of ICC 600, AF&PA WFCM and AISI S230 shall not apply to buildings sited on the upper half of an isolated hill, ridge or escarpment meeting the following conditions:

1. The hill, ridge or escarpment is 60 feet (18 288 mm) or higher if located in Exposure B or 30 feet (9144 mm) or higher if located in Exposure C;
2. The maximum average slope of the hill exceeds 10 percent; and
3. The hill, ridge or escarpment is unobstructed upwind by other such topographic features for a distance from the high point of 50 times the height of the hill or 1 mile (1.61 km), whichever is greater.

❖ This section places limitations on the use of ICC 600 and the AF&PA *Wood Frame Construction Manual for One- and Two-family Dwellings*. Neither of these standards accounts for the effect of isolated hills, ridges or escarpments. As illustrated in Figure 1609.1.1.1, the wind speed increases when a mass of air passes over those terrain features. This phenomenon is referred to as the "wind speed-up" effect and requires the use of ASCE 7.

1609.1.2 Protection of openings. In *wind-borne debris regions*, glazing in buildings shall be impact resistant or protected with an impact-resistant covering meeting the requirements of an *approved* impact-resistant standard or ASTM E 1996 and ASTM E 1886 referenced herein as follows:

1. Glazed openings located within 30 feet (9144 mm) of grade shall meet the requirements of the large missile test of ASTM E 1996.
2. Glazed openings located more than 30 feet (9144 mm) above grade shall meet the provisions of the small missile test of ASTM E 1996.

Exceptions:

1. Wood structural panels with a minimum thickness of $^7/_{16}$ inch (11.1 mm) and maximum panel span of 8 feet (2438 mm) shall be permitted for opening protection in one- and two-story buildings classified as Group R-3 or R-4 occupancy. Panels shall be precut so that they shall be attached to the framing surrounding the opening containing the product with the glazed opening. Panels shall be predrilled as required for the anchorage method and shall be secured with the attachment hardware provided. Attachments shall be designed to resist the components and cladding loads determined in accordance with the provisions of ASCE 7, with corrosion-resistant attachment hardware provided and anchors permanently installed on the building. Attachment in accordance with Table 1609.1.2 with corrosion-resistant attachment hardware provided and anchors permanently installed on the building is permitted for buildings with a mean roof height of 45 feet (13 716 mm) or less where V_{asd} determined in accordance with Section 1609.3.1 does not exceed 140 mph (63 m/s).
2. Glazing in *Risk Category* I buildings as defined in Section 1604.5, including greenhouses that are occupied for growing plants on a production or research basis, without public access shall be permitted to be unprotected.
3. Glazing in *Risk Category* II, III or IV buildings located over 60 feet (18 288 mm) above the ground and over 30 feet (9144 mm) above aggregate surface roofs located within 1,500 feet (458 m) of the building shall be permitted to be unprotected.

❖ The purpose of this section is to address risks associated with wind-borne debris in high-wind areas. See the definitions of "Wind-borne debris region" and "Hurricane-prone regions" in Section 1609.2 for an explanation of where these provisions apply. This section requires protection of glazed openings in buildings in wind-borne debris regions. This can be provided in the form of a protective assembly that is impact tested or by using impact-resistant glazing.

During a hurricane, buildings are impacted from

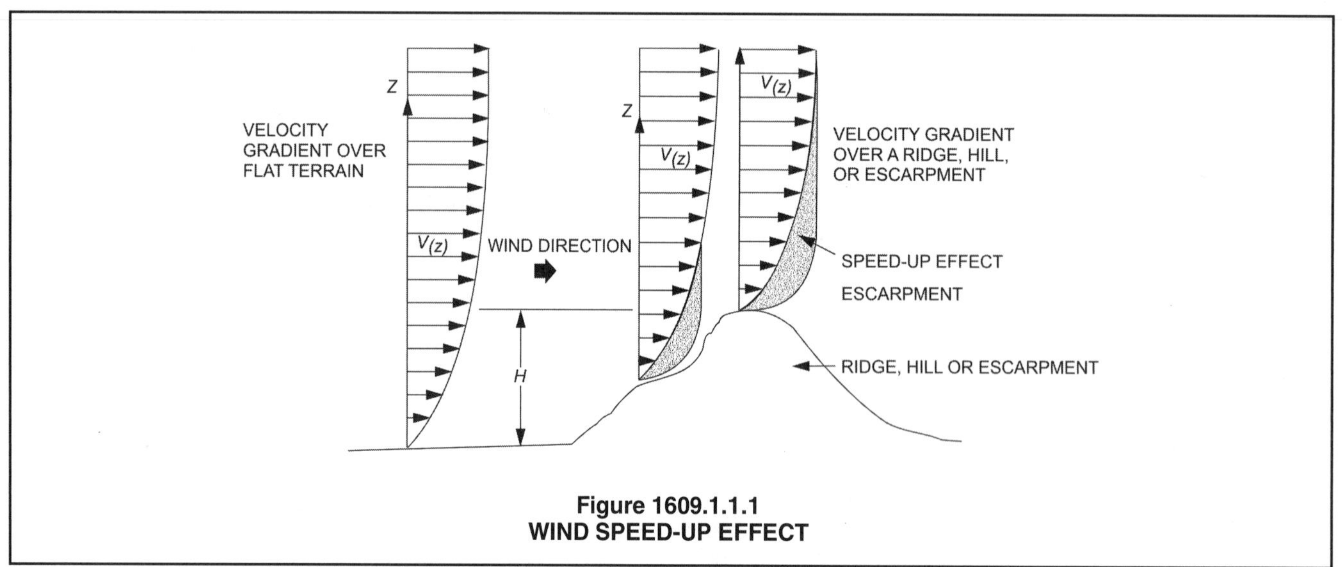

Figure 1609.1.1.1
WIND SPEED-UP EFFECT

wind-borne debris due to high-velocity winds. This debris can impact the glazing, causing breakage and creating an opening within the building envelope. The presence of openings in the building envelope can have a significant effect on the magnitude of the total wind pressure required to be resisted by each structural element of a building. Depending on the location of these openings with respect to wind direction and the amount of background porosity, external and internal pressures may act in the same direction to produce higher forces on some walls and the roof. An example of this is shown in Figure 1609.1.2. In this scenario, as the wind flows over the building, pressures are developed on the external surface, as shown. Introduction of an opening in the windward wall causes the wind to rush into the building, exerting internal pressures (positive) against all interior surfaces. This type of opening has the net effect of producing potentially high internal pressures that will act in the same direction as the external pressures on the roof, side and leeward walls. Considering the high probability of wind-borne debris during a hurricane and the effect of an unintended opening in the building envelope, the code requires glazing in designated regions to be protected.

Where wind-borne protection is provided, the section specifies two types of tests to demonstrate adequate resistance: the large missile test to simulate large debris up to 30 feet (9144 mm) above grade and the small missile test to simulate smaller debris up to 60 feet (18 288 mm) above grade, both of which are common during very high winds. An example of small debris is gravel from the surrounding area that becomes air borne.

Impact-resistant coverings or glazing must meet the test requirements of an approved impact standard or ASTM E 1886 and ASTM E 1996. Other impact standards the building official may consider are SBCCI SSTD 12 or Florida Building Code Test Protocol TAS 201, TAS 202 and TAS 203. These standards specify similar-type testing with a large missile test (2 by 4), small missile test (2 gram balls) and cyclic pressure loading test. ASTM E 1886 and ASTM E 1996 work together with the standard test method (E 1886) and the test specification (E 1996), including scoping, technical requirements and pass/fail criteria.

Exception 1 permits the use of $^{7}/_{16}$-inch (11.1 mm) wood structural panels with maximum spans of 8 feet (2438 mm) as an impact-resistant covering for one- and two-story buildings. Panel attachments have to be designed to resist the component and cladding pressure from ASCE 7 or attached in accordance with Table 1609.1.2. This protective system has been tested and meets the requirements of SBCCI SSTD 12. The intent is that precut panel coverings and attachment hardware are provided on site.

Exception 2 exempts low-hazard (Risk Category I) buildings from this requirement for protecting openings against wind-borne debris.

Exception 3 exempts openings in buildings that are classified in Risk Category II, III or IV where the opening locations meet the stated conditions.

TABLE 1609.1.2. See page 16-46.

❖ This table provides the connections for the wood structural panel impact-resistant covering that is described in the exception to Section 1609.1.2. The table lists the spacing of the screws around the perimeter of the panel for the indicated panel spans. Note that Table 1609.1.2 is only applicable to buildings with a mean roof height of 33 feet (10 058 mm) or less and located where the basic wind speed is 130 mph or less.

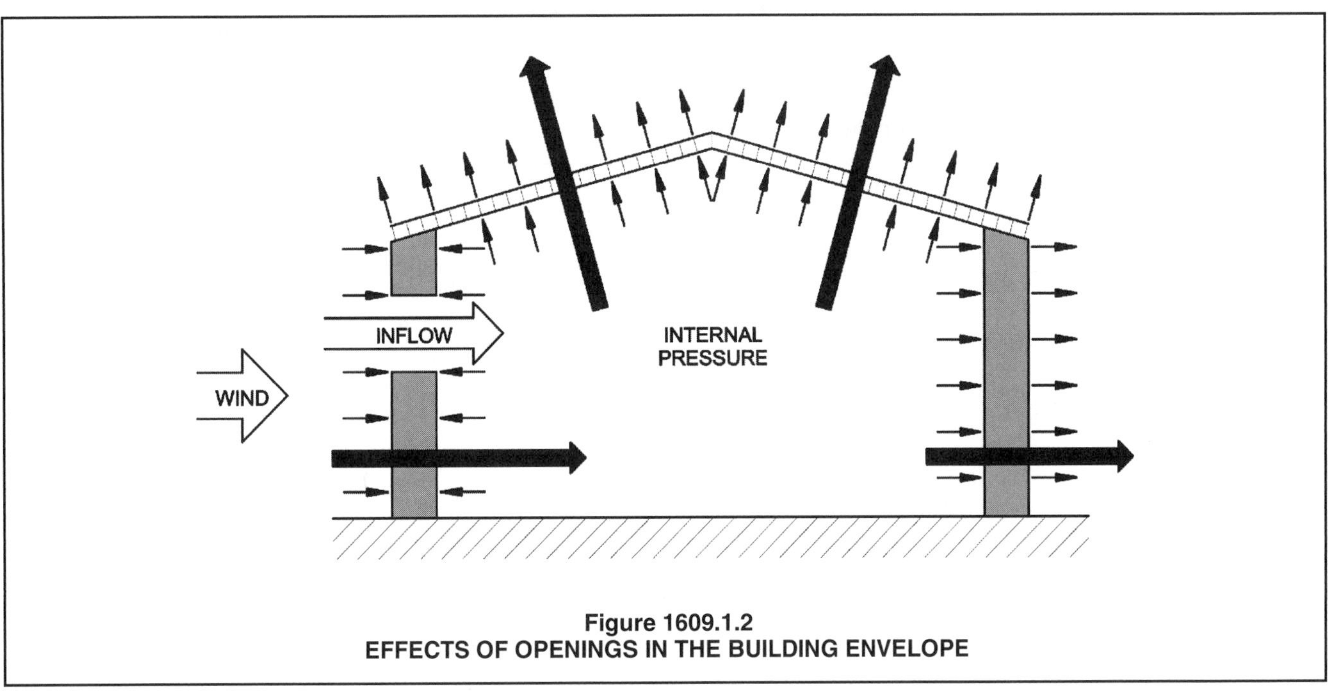

Figure 1609.1.2
EFFECTS OF OPENINGS IN THE BUILDING ENVELOPE

TABLE 1609.1.2
WIND-BORNE DEBRIS PROTECTION FASTENING SCHEDULE FOR WOOD STRUCTURAL PANELS[a, b, c, d]

FASTENER TYPE	FASTENER SPACING (inches)		
	Panel Span ≤ 4 feet	4 feet < Panel Span ≤ 6 feet	6 feet < Panel Span ≤ 8 feet
No. 8 wood-screw-based anchor with 2-inch embedment length	16	10	8
No. 10 wood-screw-based anchor with 2-inch embedment length	16	12	9
$^1/_4$-inch diameter lag-screw-based anchor with 2-inch embedment length	16	16	16

For SI: 1 inch = 25.4 mm, 1 foot = 304.8 mm, 1 pound = 4.448 N, 1 mile per hour = 0.447 m/s.

a. This table is based on 140 mph wind speeds and a 45-foot mean roof height.

b. Fasteners shall be installed at opposing ends of the wood structural panel. Fasteners shall be located a minimum of 1 inch from the edge of the panel.

c. Anchors shall penetrate through the exterior wall covering with an embedment length of 2 inches minimum into the building frame. Fasteners shall be located a minimum of $2^1/_2$ inches from the edge of concrete block or concrete.

d. Where panels are attached to masonry or masonry/stucco, they shall be attached using vibration-resistant anchors having a minimum ultimate withdrawal capacity of 1,500 pounds.

1609.1.2.1 Louvers. Louvers protecting intake and exhaust ventilation ducts not assumed to be open that are located within 30 feet (9144 mm) of grade shall meet the requirements of AMCA 54.

❖ This section provides direction for impact testing of louvers that cover intake or exhaust duct openings in wind-borne debris regions. Louvers often have blades or slats affixed to and covering an opening in the exterior envelope, making them similar to certain types of porous shutters used to protect glazed openings. When a louver in an exterior wall is damaged by wind-borne debris during a high-wind event, the air-leakage-rated damper inside the ventilation duct may also be exposed to damage.

The scope of ASTM E 1996 covers impact testing of exterior building features, such as windows, glazed curtain walls, doors and storm shutters, in buildings located in geographic regions that are prone to hurricanes, simulating impact by both large and small missiles. For glazed openings and nonporous shutters that protect the fenestration assembly, the impact testing is followed by a cyclic loading test. There is no specific provision in the standard for testing louvers that cover ventilation openings, either for impact or air pressure cycling. Nevertheless, in the absence of an appropriate specification and test method for louvers, some jurisdictions have based their product approvals on the performance criteria of the large missile test of ASTM E 1886 and ASTM E 1996. The Air Movement Control Association (AMCA) has developed a standard specification for louvers that provides a uniform set of guidelines and a consistent basis for evaluating the ability of the louver to maintain its integrity during the large missile test of ASTM E 1886 and ASTM E 1996.

1609.1.2.2 Application of ASTM E 1996. The text of Section 6.2.2 of ASTM E 1996 shall be substituted as follows:

6.2.2 Unless otherwise specified, select the wind zone based on the strength design wind speed, V_{ult}, as follows:

6.2.2.1 *Wind Zone 1*—130 mph ≤ ultimate design wind speed, V_{ult} < 140 mph.

6.2.2.2 *Wind Zone 2*—140 mph ≤ ultimate design wind speed, V_{ult} < 150 mph at greater than one mile (1.6 km) from the coastline. The coastline shall be measured from the mean high water mark.

6.2.2.3 *Wind Zone 3*—150 mph (58 m/s) ≤ ultimate design wind speed, V_{ult} ≤ 160 mph (63 m/s), or 140 mph (54 m/s) ≤ ultimate design wind speed, V_{ult} ≤ 160 mph (63 m/s) and within one mile(1.6 km) of the coastline. The coastline shall be measured from the mean high water mark.

6.2.2.4 *Wind Zone 4*— ultimate design wind speed, V_{ult} >160 mph (63 m/s).

❖ The purpose of this section is to correlate the wind zones of ASTM E 1996 with the new wind speed maps. It is needed to delineate the wind zones because ASTM E 1996 does not use V_{ult}.

1609.1.2.3 Garage doors. Garage door glazed opening protection for wind-borne debris shall meet the requirements of an *approved* impact-resisting standard or ANSI/DASMA 115.

❖ This provision references a standard for the wind-borne debris resistance testing of glazing installed in garage doors. Because ASTM E 1886 and ASTM E 1996 require interpretation regarding their application to garage doors, DASMA 115 is the primary standard referenced for this purpose.

1609.2 Definitions. For the purposes of Section 1609 and as used elsewhere in this code, the following terms are defined in Chapter 2.

❖ Definitions facilitate the understanding of code provisions and minimize potential confusion. To that end, this section lists definitions of terms associated with wind loads. Note that these definitions are found in Chapter 2. The use and application of defined terms, as well as undefined terms, are set forth in Section 201.

HURRICANE-PRONE REGIONS.

WIND-BORNE DEBRIS REGION.

WIND SPEED, V_{ult}.

WIND SPEED, V_{asd}.

1609.3 Basic wind speed. The ultimate design wind speed, V_{ult}, in mph, for the determination of the wind loads shall be determined by Figures 1609A, 1609B and 1609C. The ultimate design wind speed, V_{ult}, for use in the design of Risk Category II buildings and structures shall be obtained from Figure 1609A. The ultimate design wind speed, V_{ult}, for use in the design of Risk Category III and IV buildings and structures shall be obtained from Figure 1609B. The ultimate design wind speed, V_{ult}, for use in the design of Risk Category I buildings and structures shall be obtained from Figure 1609C. The ultimate design wind speed, V_{ult}, for the special wind regions indicated near mountainous terrain and near gorges shall be in accordance with local jurisdiction requirements. The ultimate design wind speeds, V_{ult}, determined by the local jurisdiction shall be in accordance with Section 26.5.1 of ASCE 7.

In nonhurricane-prone regions, when the ultimate design wind speed, V_{ult}, is estimated from regional climatic data, the ultimate design wind speed, V_{ult}, shall be determined in accordance with Section 26.5.3 of ASCE 7.

❖ This section establishes the wind speed that is to be used for design. The ultimate design wind speed maps identify special wind regions where the speeds vary substantially within a very short distance due to topographic effects, such as mountains and valleys. This section specifies appropriate recurrence interval criteria to be used for estimating the ultimate design wind speeds from regional climatic data in other than hurricane- prone regions. See Section 26.5.3 of ASCE 7 for issues to be addressed in the data analysis and the associated commentary for further details (also see commentary, Figures 1609A, 1609B and 1609C).

1609.3.1 Wind speed conversion. When required, the ultimate design wind speeds of Figures 1609A, 1609B and 1609C shall be converted to nominal design wind speeds, V_{asd}, using Table 1609.3.1 or Equation 16-33.

$$V_{asd} = V_{ult}\sqrt{0.6} \qquad \textbf{(Equation 16-33)}$$

where:

V_{asd} = nominal design wind speed applicable to methods specified in Exceptions 1 through 5 of Section 1609.1.1.

V_{ult} = ultimate design wind speeds determined from Figures 1609A, 1609B or 1609C.

❖ Because many code provisions are driven by the wind speed, it is necessary to include a mechanism that provides a comparable wind speed so that the provisions triggered by wind speed are not affected. The terms "ultimate design wind speed" and "nominal design wind speed" have been incorporated in numerous locations to help the code user distinguish between them. For example, in a case where the code previously imposed requirements where the basic wind speed exceeds 100 mph (45 m/s), it now imposes the requirements where V_{asd} exceeds 100 mph (45 m/s). A nominal design speed, V_{asd}, equal to 100 mph (45 m/s) corresponds to an ultimate design wind speed, V_{ult}, equal to 129 mph (58 m/s). The conversion between the two is accomplished using Equation 16-32 or Table 1609.3.1. This conversion equation is the result of the wind load being proportional to the square of the velocity pressure and the ASD wind load being 0.6 times the strength level wind load.

It should be noted that the term "basic wind speed" remains in ASCE 7, but it corresponds to the "ultimate design wind speed" in the code. For a comparison of ASCE 7-93 fastest-mile wind speeds and ASCE 7-05 3-second gust (ASD) wind speeds to ASCE 7-10 3-second gust wind speeds, refer to Table C26.5-6 in the ASCE 7-10.

TABLE 1609.3.1. See below.

❖ The table converts the mapped ultimate wind speed to nominal design wind speed for use where necessary, such as in the standards that are referenced in Section 1609.1.1.

1609.4 Exposure category. For each wind direction considered, an exposure category that adequately reflects the characteristics of ground surface irregularities shall be determined for the site at which the building or structure is to be constructed. Account shall be taken of variations in ground surface roughness that arise from natural topography and vegetation as well as from constructed features.

❖ The concept of exposure categories provides a means of accounting for the relative roughness of the boundary layer. The earth's surface exerts a horizontal drag force on wind due to ground obstructions that retard the flow of air close to the ground. The reduction in the flow of air is a function of height above ground and terrain roughness. Wind speeds increase with height above ground, and the relationship between height above ground and wind speed is exponential. The rate of increase in wind speeds with height is a function of the terrain features. The rougher the terrain (such as large city centers), the shallower the slope of the wind speed profile. The

TABLE 1609.3.1
WIND SPEED CONVERSIONS[a, b, c]

V_{ult}	100	110	120	130	140	150	160	170	180	190	200
V_{asd}	78	85	93	101	108	116	124	132	139	147	155

For SI: 1 mile per hour = 0.44 m/s.

a. Linear interpolation is permitted.

b. V_{asd} = nominal design wind speed applicable to methods specified in Exceptions 1 through 5 of Section 1609.1.1.

c. V_{ult} = ultimate design wind speeds determined from Figures 1609A, 1609B, or 1609C.

smoother the terrain (open water), the steeper the slope of the wind speed profile.

The definitions of exposure and roughness categories are used to account for this roughness in the boundary layer and are intended to provide an adequate assessment of surface roughness for most situations. Exposure B is considered the roughest boundary layer condition. Exposure D is considered the smoothest boundary layer condition. Accordingly, calculated wind loads are less for Exposure B, which has more surface obstructions, as compared to Exposure D, with all other variables the same (see ASCE 7 commentary for guidance on performing a more detailed analysis of surface roughness).

1609.4.1 Wind directions and sectors. For each selected wind direction at which the wind loads are to be evaluated, the exposure of the building or structure shall be determined for the two upwind sectors extending 45 degrees (0.79 rad) either side of the selected wind direction. The exposures in these two sectors shall be determined in accordance with Sections 1609.4.2 and 1609.4.3 and the exposure resulting in the highest wind loads shall be used to represent winds from that direction.

❖ An exposure category must be determined for each direction in which wind loading is being considered. This section describes the method for doing so, and requires use of the more restrictive exposure category where the two sectors in a given wind direction would be classified in different exposure categories.

1609.4.2 Surface roughness categories. A ground surface roughness within each 45-degree (0.79 rad) sector shall be determined for a distance upwind of the site as defined in Section 1609.4.3 from the categories defined below, for the purpose of assigning an exposure category as defined in Section 1609.4.3.

Surface Roughness B. Urban and suburban areas, wooded areas or other terrain with numerous closely spaced obstructions having the size of single-family dwellings or larger.

Surface Roughness C. Open terrain with scattered obstructions having heights generally less than 30 feet (9144 mm). This category includes flat open country, and grasslands.

Surface Roughness D. Flat, unobstructed areas and water surfaces. This category includes smooth mud flats, salt flats and unbroken ice.

❖ This section defines three surface roughness categories that are used in evaluating each sector and, subsequently, in determining an exposure category. These surface roughnesses were previously included in the corresponding exposure category definition. The required upwind distance that must be considered varies according to the exposure category definition in Section 1609.4.3.

A significant philosophical change in determination of exposure categories occurred with the inclusion of shorelines in hurricane-prone regions in the definition of Exposure Category C (Surface Roughness C). Exposure Category D (Surface Roughness D) had been used for wind flowing over open water until further research determined that wave action at the water's surface in a hurricane, due to the intensity of the turbulence, produced substantial surface obstructions and friction that reduces the wind profile values to be more in line with Surface Roughness C as opposed to Surface Roughness D. Surface Roughness D would still apply to inland waterways and shorelines that are not in the hurricane-prone regions, such as coastal California, Oregon, Washington and Alaska.

1609.4.3 Exposure categories. An exposure category shall be determined in accordance with the following:

Exposure B. For buildings with a mean roof height of less than or equal to 30 feet (9144 mm), Exposure B shall apply where the ground surface roughness, as defined by Surface Roughness B, prevails in the upwind direction for a distance of at least 1,500 feet (457 m). For buildings with a mean roof height greater than 30 feet (9144 mm), Exposure B shall apply where Surface Roughness B prevails in the upwind direction for a distance of at least 2,600 feet (792 m) or 20 times the height of the building, whichever is greater.

Exposure C. Exposure C shall apply for all cases where Exposures B or D do not apply.

Exposure D. Exposure D shall apply where the ground surface roughness, as defined by Surface Roughness D, prevails in the upwind direction for a distance of at least 5,000 feet (1524 m) or 20 times the height of the building, whichever is greater. Exposure D shall also apply where the ground surface roughness immediately upwind of the site is B or C, and the site is within a distance of 600 feet (183 m) or 20 times the building height, whichever is greater, from an exposure D condition as defined in the previous sentence.

❖ This section defines the three exposure categories that are used to determine various wind load parameters. Exposure categories are used along with basic wind speed elsewhere in the code as a threshold for wind requirements, such as establishing the need for structural observation in Section 1704.5.

Exposure B is the most common type of exposure category in the country. A study by the National Association of Home Builders (NAHB) indicated that perhaps up to 80 percent of all buildings were located in Exposure B.

1609.5 Roof systems. Roof systems shall be designed and constructed in accordance with Sections 1609.5.1 through 1609.5.3, as applicable.

❖ This section clarifies the design wind loads that are applied to roof decks and roof coverings.

1609.5.1 Roof deck. The roof deck shall be designed to withstand the wind pressures determined in accordance with ASCE 7.

❖ This section specifies the wind load criteria for the roof deck. The roof deck is a structural component of the building and must resist the applicable wind pressures of ASCE 7. This section is referenced by Section 1609.5.2 as the criteria for the wind design for roof coverings that are relatively impermeable.

1609.5.2 Roof coverings. Roof coverings shall comply with Section 1609.5.1.

Exception: Rigid tile roof coverings that are air permeable and installed over a roof deck complying with Section 1609.5.1 are permitted to be designed in accordance with Section 1609.5.3.

Asphalt shingles installed over a roof deck complying with Section 1609.5.1 shall comply with the wind-resistance requirements of Section 1507.2.7.1.

❖ This section establishes the wind design criteria for roof coverings. The exception references the use of Section 1609.5.3 for air-permeable rigid tile roof coverings. If the roof deck is relatively impermeable, wind pressures will act through it to the building frame system. The roof covering may or may not be subjected to the same wind pressures as the roof deck. If the roof covering is also relatively impermeable and fastened to the roof deck, the two components will react to and resist the same wind pressures. If the roof covering is not impermeable, the wind pressures will be able to develop on both the top of, and underneath, the roof covering. This "venting action" will negate some wind pressure on the roof covering.

1609.5.3 Rigid tile. Wind loads on rigid tile roof coverings shall be determined in accordance with the following equation:

$$M_a = q_h C_L b L L_a [1.0 - GC_p] \quad \textbf{(Equation 16-34)}$$

For SI: $$M_a = \frac{L(q_h C_L b) L_a [1.0 - GC_p]}{1,000}$$

where:

b = Exposed width, feet (mm) of the roof tile.

C_L = Lift coefficient. The lift coefficient for concrete and clay tile shall be 0.2 or shall be determined by test in accordance with Section 1711.2.

GC_p = Roof pressure coefficient for each applicable roof zone determined from Chapter 30 of ASCE 7. Roof coefficients shall not be adjusted for internal pressure.

L = Length, feet (mm) of the roof tile.

L_a = Moment arm, feet (mm) from the axis of rotation to the point of uplift on the roof tile. The point of uplift shall be taken at 0.76L from the head of the tile and the middle of the exposed width. For roof tiles with nails or screws (with or without a tail clip), the axis of rotation shall be taken as the head of the tile for direct deck application or as the top edge of the batten for battened applications. For roof tiles fastened only by a nail or screw along the side of the tile, the axis of rotation shall be determined by testing. For roof tiles installed with battens and fastened only by a clip near the tail of the tile, the moment arm shall be determined about the top edge of the batten with consideration given for the point of rotation of the tiles based on straight bond or broken bond and the tile profile.

M_a = Aerodynamic uplift moment, feet-pounds (N-mm) acting to raise the tail of the tile.

q_h = Wind velocity pressure, psf (kN/m^2) determined from Section 27.3.2 of ASCE 7.

Concrete and clay roof tiles complying with the following limitations shall be designed to withstand the aerodynamic uplift moment as determined by this section.

1. The roof tiles shall be either loose laid on battens, mechanically fastened, mortar set or adhesive set.
2. The roof tiles shall be installed on solid sheathing which has been designed as components and cladding.
3. An underlayment shall be installed in accordance with Chapter 15.
4. The tile shall be single lapped interlocking with a minimum head lap of not less than 2 inches (51 mm).
5. The length of the tile shall be between 1.0 and 1.75 feet (305 mm and 533 mm).
6. The exposed width of the tile shall be between 0.67 and 1.25 feet (204 mm and 381 mm).
7. The maximum thickness of the tail of the tile shall not exceed 1.3 inches (33 mm).
8. Roof tiles using mortar set or adhesive set systems shall have at least two-thirds of the tile's area free of mortar or adhesive contact.

❖ This section includes the wind design method for clay or concrete rigid tile roofs. The method consists of the calculation of the aerodynamic uplift moment from the wind that acts to raise the end of the tile. This section includes the characteristics and the type of installation of the concrete or clay roof tile that are required for the use of the design method.

In certain types of installations, the roof covering is not exposed to the same wind loads as the roof deck. Concrete and clay roof tiles are typical of this type of installation and are not subject to wind loads that would be obtained from current wind-loading criteria. This is due to the gaps at tile joints allowing some equalization of pressure between the inner and outer face of the tiles, leading to reduced loads. A procedure has been developed through research for determining the uplift moment on loose-laid and mechanically fastened roof tiles when laid over sheathing with an underlayment. The procedure is based on practical measurements on real tiles to determine the effect of air being able to penetrate the roof covering.

1609.6 Alternate all-heights method. The alternate wind design provisions in this section are simplifications of the ASCE 7 Directional Procedure.

❖ In response to concerns from design engineers on the complexity of wind design procedures, member organizations of the National Council of Standard Engineering Associations (NCSEA) assembled this alternative method for determining wind loads. The procedure has been developed to give results equal to or more conservative than the Directional Procedure that is found in ASCE 7. The intention is to reduce the effort required in determining wind forces for the main wind-force-resisting system, as well as for components and cladding (C & C).

1609.6.1 Scope. As an alternative to ASCE 7 Chapters 27 and 30, the following provisions are permitted to be used to determine the wind effects on regularly shaped buildings, or other structures that are regularly shaped, which meet all of the following conditions:

1. The building or other structure is less than or equal to 75 feet (22 860 mm) in height with a height-to-least-width ratio of 4 or less, or the building or other structure has a fundamental frequency greater than or equal to 1 hertz.
2. The building or other structure is not sensitive to dynamic effects.
3. The building or other structure is not located on a site for which channeling effects or buffeting in the wake of upwind obstructions warrant special consideration.
4. The building shall meet the requirements of a simple diaphragm building as defined in ASCE 7 Section 26.2, where wind loads are only transmitted to the main windforce-resisting system (MWFRS) at the diaphragms.
5. For open buildings, multispan gable roofs, stepped roofs, sawtooth roofs, domed roofs, roofs with slopes greater than 45 degrees (0.79 rad), solid free-standing walls and solid signs, and rooftop equipment, apply ASCE 7 provisions.

❖ While ASCE 7 already includes a simplified procedure, it necessarily includes numerous limitations. Similarly, this alternative design procedure is limited, though it was developed to apply to a broader range of buildings.

Note that Item 1 limits application to buildings with a frequency of at least 1 hertz. In other words, this can only be used for rigid structures (see definition of "Rigid buildings and other structures" in Section 26.2 of ASCE 7). Item 1 also allows any building up to 75 feet (22 860 mm) in height that has a height-to-least-width ratio of 4 or less to qualify without calculating the frequency. Buildings must also be regularly shaped, simple diaphragm buildings with envelopes classified as either enclosed or partially enclosed. Item 5 provides a partial list of structures that must be analyzed using the ASCE 7 provisions.

1609.6.1.1 Modifications. The following modifications shall be made to certain subsections in ASCE 7: in Section 1609.6.2, symbols and notations that are specific to this section are used in conjunction with the symbols and notations in ASCE 7 Section 26.3.

❖ These alternative provisions are an adaptation of the ASCE 7 analytical procedure. In providing the intended simplifications, this section alerts the code user that these are modifications to the ASCE 7 procedure.

1609.6.2 Symbols and notations. Coefficients and variables used in the alternative all-heights method equations are as follows:

C_{net} = Net-pressure coefficient based on K_d [(G) (C_p) - (GC_{pi})], in accordance with Table 1609.6.2.

G = Gust effect factor for rigid structures in accordance with ASCE 7 Section 26.9.1.

K_d = Wind directionality factor in accordance with ASCE 7 Table 26-6.

P_{net} = Design wind pressure to be used in determination of wind loads on buildings or other structures or their components and cladding, in psf (kN/m^2).

❖ The primary simplification is accomplished by generating a table of net pressure coefficients (C_{net}), combining a number of parameters in a simple, yet conservative manner. These are shown in the definition of notation C_{net}. Application of the tabulated net pressure coefficients reduces the number of steps required for performing a wind analysis, resulting in net wind forces that meet or exceed those calculated using the Directional Procedure of ASCE 7. A gust factor of 0.85 is used for the tabulated C_{net} values, as Section 26.9.2 of ASCE 7 permits for rigid structures.

TABLE 1609.6.2. See page 16-52.

❖ Net pressure coefficients are tabulated for both enclosed and partially enclosed structures (see Section 26.10 of ASCE 7).

For main windforce-resisting system roofs Table 1609.6.2 refers to Conditions 1 and 2 for "windward roof slopes," which correspond to the two values of C_p listed in Figure 27.4.1-1 of ASCE 7 (Note 3 at the bottom of that figure also refers to "both conditions").

The basis is $C_{net} = K_d [(G)(C_p) - (GC_{pi})]$ (see Section 1609.6.2) and the values obtained for C_p. When considering the wind loads on the "windward roof slope," see Figure 27.4.1-1 of ASCE 7. On the windward side of a sloping roof for "Gable, hip roof," there are two wind load conditions, one "upward" and the other "downward" as indicated by the two arrows. Two conditions on the windward roof slope provide two values of C_p, and thereby two values of C_{net} shown in Table 1609.6.2.

1609.6.3 Design equations. When using the alternative all-heights method, the MWFRS, and components and cladding of every structure shall be designed to resist the effects of wind pressures on the building envelope in accordance with Equation 16-35.

$$P_{net} = 0.00256V^2K_zC_{net}K_{zt} \quad \textbf{(Equation 16-35)}$$

Design wind forces for the MWFRS shall not be less than 16 psf (0.77 kN/m^2) multiplied by the area of the structure projected on a plane normal to the assumed wind direction (see ASCE 7 Section 27.4.7 for criteria). Design net wind pressure for components and cladding shall not be less than 16 psf (0.77 kN/m^2) acting in either direction normal to the surface.

❖ This section provides the formula that is used for the design wind pressure. It also incorporates the ASCE 7 minimum wind pressure for main wind-force-resisting systems and components and cladding.

1609.6.4 Design procedure. The MWFRS and the components and cladding of every building or other structure shall be designed for the pressures calculated using Equation 16-35.

❖ Using Equation 16-35, the design pressures can be calculated for the main wind-force-resisting system. Similarly, the components and cladding design pressures are calculated for the various portions of the building envelope.

1609.6.4.1 Main windforce-resisting systems. The MWFRS shall be investigated for the torsional effects identified in ASCE 7 Figure 27.4.6.

❖ A reference is made to the ASCE 7 figure that illustrates the design wind load cases that must be considered, specifically making mention of the torsional load cases.

1609.6.4.2 Determination of K_z and K_{zt}. Velocity pressure exposure coefficient, K_z, shall be determined in accordance with ASCE 7 Section 27.3.1 and the topographic factor, K_{zt}, shall be determined in accordance with ASCE 7 Section 26.8.

1. For the windward side of a structure, K_{zt} and K_z shall be based on height z.
2. For leeward and sidewalls, and for windward and leeward roofs, K_{zt} and K_z shall be based on mean roof height h.

❖ For the velocity pressure exposure coefficient, the user is referred to the corresponding ASCE 7 section. For the topographic factor, the user is referred to the corresponding ASCE 7 section. Note that these are evaluated only at the mean roof height for leeward walls, sidewalls and roofs.

1609.6.4.3 Determination of net pressure coefficients, C_{net}. For the design of the MWFRS and for components and cladding, the sum of the internal and external net pressure shall be based on the net pressure coefficient, C_{net}.

1. The pressure coefficient, $C_{net,}$ for walls and roofs shall be determined from Table 1609.6.2.
2. Where C_{net} has more than one value, the more severe wind load condition shall be used for design.

❖ The tabulated C_{net} values represent the sum of external and internal pressure coefficients as shown in the notation defined in Section 1609.6.2.

1609.6.4.4 Application of wind pressures. When using the alternative all-heights method, wind pressures shall be applied simultaneously on, and in a direction normal to, all building envelope wall and roof surfaces.

❖ This section clarifies how the wind pressures are applied relative to the surfaces of the building envelope.

1609.6.4.4.1 Components and cladding. Wind pressure for each component or cladding element is applied as follows using C_{net} values based on the effective wind area, A, contained within the zones in areas of discontinuity of width and/or length "a," "2a" or "4a" at: corners of roofs and walls; edge strips for ridges, rakes and eaves; or field areas on walls or roofs as indicated in figures in tables in ASCE 7 as referenced in Table 1609.6.2 in accordance with the following:

1. Calculated pressures at local discontinuities acting over specific edge strips or corner boundary areas.
2. Include "field" (Zone 1, 2 or 4, as applicable) pressures applied to areas beyond the boundaries of the areas of discontinuity.
3. Where applicable, the calculated pressures at discontinuities (Zone 2 or 3) shall be combined with design pressures that apply specifically on rakes or eave overhangs.

❖ The C_{net} values for C & C pressures are separately tabulated for roof and walls. These portions of the structure are further divided into those that are in areas of discontinuities and those that are not. Effective wind area is another factor in selecting the correct C & C value. While the lateral-load-resisting system design may be controlled by earthquake forces, it is necessary to consider these C & C wind pressures, which can control the design of the component and attachments.

TABLE 1609.6.2
NET PRESSURE COEFFICIENTS, C_{net}[a, b]

<table>
<tr><th>STRUCTURE OR PART THEREOF</th><th colspan="2">DESCRIPTION</th><th colspan="4">C_{net} FACTOR</th></tr>
<tr><td rowspan="40">1. Main windforce-resisting frames and systems</td><td colspan="2" rowspan="2">Walls:</td><td colspan="2">Enclosed</td><td colspan="2">Partially enclosed</td></tr>
<tr><td>+ Internal pressure</td><td>- Internal pressure</td><td>+ Internal pressure</td><td>- Internal pressure</td></tr>
<tr><td colspan="2">Windward wall</td><td>0.43</td><td>0.73</td><td>0.11</td><td>1.05</td></tr>
<tr><td colspan="2">Leeward wall</td><td>-0.51</td><td>-0.21</td><td>-0.83</td><td>0.11</td></tr>
<tr><td colspan="2">Sidewall</td><td>-0.66</td><td>-0.35</td><td>-0.97</td><td>-0.04</td></tr>
<tr><td rowspan="2">Parapet wall</td><td>Windward</td><td colspan="2">1.28</td><td colspan="2">1.28</td></tr>
<tr><td>Leeward</td><td colspan="2">-0.85</td><td colspan="2">-0.85</td></tr>
<tr><td colspan="2">Roofs:</td><td colspan="2">Enclosed</td><td colspan="2">Partially enclosed</td></tr>
<tr><td colspan="2">Wind perpendicular to ridge</td><td>+ Internal pressure</td><td>- Internal pressure</td><td>+ Internal pressure</td><td>- Internal pressure</td></tr>
<tr><td colspan="2">Leeward roof or flat roof</td><td>-0.66</td><td>-0.35</td><td>-0.97</td><td>-0.04</td></tr>
<tr><td colspan="6">Windward roof slopes:</td></tr>
<tr><td rowspan="2">Slope < 2:12 (10°)</td><td>Condition 1</td><td>-1.09</td><td>-0.79</td><td>-1.41</td><td>-0.47</td></tr>
<tr><td>Condition 2</td><td>-0.28</td><td>0.02</td><td>-0.60</td><td>0.34</td></tr>
<tr><td rowspan="2">Slope = 4:12 (18°)</td><td>Condition 1</td><td>-0.73</td><td>-0.42</td><td>-1.04</td><td>-0.11</td></tr>
<tr><td>Condition 2</td><td>-0.05</td><td>0.25</td><td>-0.37</td><td>0.57</td></tr>
<tr><td rowspan="2">Slope = 5:12 (23°)</td><td>Condition 1</td><td>-0.58</td><td>-0.28</td><td>-0.90</td><td>0.04</td></tr>
<tr><td>Condition 2</td><td>0.03</td><td>0.34</td><td>-0.29</td><td>0.65</td></tr>
<tr><td rowspan="2">Slope = 6:12 (27°)</td><td>Condition 1</td><td>-0.47</td><td>-0.16</td><td>-0.78</td><td>0.15</td></tr>
<tr><td>Condition 2</td><td>0.06</td><td>0.37</td><td>-0.25</td><td>0.68</td></tr>
<tr><td rowspan="2">Slope = 7:12 (30°)</td><td>Condition 1</td><td>-0.37</td><td>-0.06</td><td>-0.68</td><td>0.25</td></tr>
<tr><td>Condition 2</td><td>0.07</td><td>0.37</td><td>-0.25</td><td>0.69</td></tr>
<tr><td rowspan="2">Slope = 9:12 (37°)</td><td>Condition 1</td><td>-0.27</td><td>0.04</td><td>-0.58</td><td>0.35</td></tr>
<tr><td>Condition 2</td><td>0.14</td><td>0.44</td><td>-0.18</td><td>0.76</td></tr>
<tr><td colspan="2">Slope = 12:12 (45°)</td><td>0.14</td><td>0.44</td><td>-0.18</td><td>0.76</td></tr>
<tr><td colspan="2">Wind parallel to ridge and flat roofs</td><td>-1.09</td><td>-0.79</td><td>-1.41</td><td>-0.47</td></tr>
<tr><td colspan="6">Nonbuilding Structures: Chimneys, Tanks and Similar Structures:</td></tr>
<tr><td colspan="3" rowspan="2"></td><td colspan="3">h/D</td></tr>
<tr><td>1</td><td>7</td><td>25</td></tr>
<tr><td colspan="3">Square (Wind normal to face)</td><td>0.99</td><td>1.07</td><td>1.53</td></tr>
<tr><td colspan="3">Square (Wind on diagonal)</td><td>0.77</td><td>0.84</td><td>1.15</td></tr>
<tr><td colspan="3">Hexagonal or Octagonal</td><td>0.81</td><td>0.97</td><td>1.13</td></tr>
<tr><td colspan="3">Round</td><td>0.65</td><td>0.81</td><td>0.97</td></tr>
<tr><td colspan="3">Open signs and lattice frameworks</td><td colspan="3">Ratio of solid to gross area</td></tr>
<tr><td colspan="3"></td><td>< 0.1</td><td>0.1 to 0.29</td><td>0.3 to 0.7</td></tr>
<tr><td colspan="3">Flat</td><td>1.45</td><td>1.30</td><td>1.16</td></tr>
<tr><td colspan="3">Round</td><td>0.87</td><td>0.94</td><td>1.08</td></tr>
</table>

(continued)

TABLE 1609.6.2—continued
NET PRESSURE COEFFICIENTS, C_{net}[a, b]

STRUCTURE OR PART THEREOF	DESCRIPTION		C_{net} FACTOR	
2. Components and cladding not in areas of discontinuity—roofs and overhangs	**Roof elements and slopes**		**Enclosed**	**Partially enclosed**
	Gable of hipped configurations (Zone 1)			
	Flat < Slope < 6:12 (27°) See ASCE 7 Figure 30.4-2B Zone 1			
	Positive	10 square feet or less	0.58	0.89
		100 square feet or more	0.41	0.72
	Negative	10 square feet or less	-1.00	-1.32
		100 square feet or more	-0.92	-1.23
	Overhang: Flat < Slope < 6:12 (27°) See ASCE 7 Figure 30.4-2A Zone 1			
	Negative	10 square feet or less	-1.45	
		100 square feet or more	-1.36	
		500 square feet or more	-0.94	
	6:12 (27°) < Slope < 12:12 (45°) See ASCE 7 Figure 30.4-2C Zone 1			
	Positive	10 square feet or less	0.92	1.23
		100 square feet or more	0.83	1.15
	Negative	10 square feet or less	-1.00	-1.32
		100 square feet or more	-0.83	-1.15
	Monosloped configurations (Zone 1)		**Enclosed**	**Partially enclosed**
	Flat < Slope < 7:12 (30°) See ASCE 7 Figure 30.4-5B Zone 1			
	Positive	10 square feet or less	0.49	0.81
		100 square feet or more	0.41	0.72
	Negative	10 square feet or less	-1.26	-1.57
		100 square feet or more	-1.09	-1.40
	Tall flat-topped roofs h > 60 feet		**Enclosed**	**Partially enclosed**
	Flat < Slope < 2:12 (10°) (Zone 1) See ASCE 7 Figure 30.8-1 Zone 1			
	Negative	10 square feet or less	-1.34	-1.66
		500 square feet or more	-0.92	-1.23
3. Components and cladding in areas of discontinuities—roofs and overhangs (continued)	Gable or hipped configurations at ridges, eaves and rakes (Zone 2)			
	Flat < Slope < 6:12 (27°) See ASCE 7 Figure 30.4-2B Zone 2			
	Positive	10 square feet or less	0.58	0.89
		100 square feet or more	0.41	0.72
	Negative	10 square feet or less	-1.68	-2.00
		100 square feet or more	-1.17	-1.49
	Overhang for Slope Flat < Slope < 6:12 (27°) See ASCE 7 Figure 30.4-2B Zone 2			
	Negative	10 square feet or less	-1.87	
		100 square feet or more	-1.87	
	6:12 (27°) < Slope < 12:12 (45°) Figure 30.4-2C		**Enclosed**	**Partially enclosed**
	Positive	10 square feet or less	0.92	1.23
		100 square feet or more	0.83	1.15
	Negative	10 square feet or less	-1.17	-1.49
		100 square feet or more	-1.00	-1.32
	Overhang for 6:12 (27°) < Slope < 12:12 (45°) See ASCE 7 Figure 30.4-2C Zone 2			
	Negative	10 square feet or less	-1.70	
		500 square feet or more	-1.53	

(continued)

TABLE 1609.6.2—continued
NET PRESSURE COEFFICIENTS, C_{net}[a, b]

STRUCTURE OR PART THEREOF	DESCRIPTION		C_{net} FACTOR	
3. Components and cladding in areas of discontinuities—roofs and overhangs	**Roof elements and slopes**		**Enclosed**	**Partially enclosed**
	Monosloped configurations at ridges, eaves and rakes (Zone 2)			
	Flat < Slope < 7:12 (30°) See ASCE 7 Figure 30.4-5B Zone 2			
	Positive	10 square feet or less	0.49	0.81
		100 square feet or more	0.41	0.72
	Negative	10 square feet or less	-1.51	-1.83
		100 square feet or more	-1.43	-1.74
	Tall flat topped roofs *h* > 60 feet		**Enclosed**	**Partially enclosed**
	Flat < Slope < 2:12 (10°) (Zone 2) See ASCE 7 Figure 30.8-1 Zone 2			
	Negative	10 square feet or less	-2.11	-2.42
		500 square feet or more	-1.51	-1.83
	Gable or hipped configurations at corners (Zone 3) See ASCE 7 Figure 30.4-2B Zone 3			
	Flat < Slope < 6:12 (27°)		**Enclosed**	**Partially enclosed**
	Positive	10 square feet or less	0.58	0.89
		100 square feet or more	0.41	0.72
	Negative	10 square feet or less	-2.53	-2.85
		100 square feet or more	-1.85	-2.17
	Overhang for Slope Flat < Slope < 6:12 (27°) See ASCE 7 Figure 30.4-2B Zone 3			
	Negative	10 square feet or less	-3.15	
		100 square feet or more	-2.13	
	6:12 (27°) < 12:12 (45°) See ASCE 7 Figure 30.4-2C Zone 3			
	Positive	10 square feet or less	0.92	1.23
		100 square feet or more	0.83	1.15
	Negative	10 square feet or less	-1.17	-1.49
		100 square feet or more	-1.00	-1.32
	Overhang for 6:12 (27°) < Slope < 12:12 (45°)		**Enclosed**	**Partially enclosed**
	Negative	10 square feet or less	-1.70	
		100 square feet or more	-1.53	
	Monosloped Configurations at corners (Zone 3) See ASCE 7 Figure 30.4-5B Zone 3			
	Flat < Slope < 7:12 (30°)			
	Positive	10 square feet or less	0.49	0.81
		100 square feet or more	0.41	0.72
	Negative	10 square feet or less	-2.62	-2.93
		100 square feet or more	-1.85	-2.17
	Tall flat topped roofs *h* > 60 feet		**Enclosed**	**Partially enclosed**
	Flat < Slope < 2:12 (10°) (Zone 3) See ASCE 7 Figure 30.8-1 Zone 3			
	Negative	10 square feet or less	-2.87	-3.19
		500 square feet or more	-2.11	-2.42
4. Components and cladding not in areas of discontinuity—walls and parapets (continued)	Wall Elements: *h* = 60 feet (Zone 4) Figure 30.4-1		**Enclosed**	**Partially enclosed**
	Positive	10 square feet or less	1.00	1.32
		500 square feet or more	0.75	1.06
	Negative	10 square feet or less	-1.09	-1.40
		500 square feet or more	-0.83	-1.15
	Wall Elements: *h* > 60 feet (Zone 4) See ASCE 7 Figure 30.8-1 Zone 4			
	Positive	20 square feet or less	0.92	1.23
		500 square feet or more	0.66	0.98

(continued)

TABLE 1609.6.2—continued
NET PRESSURE COEFFICIENTS, C_{net} [a, b]

STRUCTURE OR PART THEREOF	DESCRIPTION		C_{net} FACTOR	
4. Components and cladding not in areas of discontinuity-walls and parapets	Negative	20 square feet or less	-0.92	-1.23
		500 square feet or more	-0.75	-1.06
	Parapet Walls			
	Positive		2.87	3.19
	Negative		-1.68	-2.00
5. Components and cladding in areas of discontinuity—walls and parapets	Wall elements: $h \leq 60$ feet (Zone 5) Figure 30.4-1		Enclosed	Partially enclosed
	Positive	10 square feet or less	1.00	1.32
		500 square feet or more	0.75	1.06
	Negative	10 square feet or less	-1.34	-1.66
		500 square feet or more	-0.83	-1.15
	Wall elements: $h > 60$ feet (Zone 5) See ASCE 7 Figure 30.8-1 Zone 4			
	Positive	20 square feet or less	0.92	1.23
		500 square feet or more	0.66	0.98
	Negative	20 square feet or less	-1.68	-2.00
		500 square feet or more	-1.00	-1.32
	Parapet walls			
	Positive		3.64	3.95
	Negative		-2.45	-2.76

For SI: 1 foot = 304.8 mm, 1 square foot = 0.0929m^2, 1 degree = 0.0175 rad.

a. Linear interpolation between values in the table is permitted.

b. Some C_{net} values have been grouped together. Less conservative results may be obtained by applying ASCE 7 provisions.

SECTION 1610 SOIL LATERAL LOADS

1610.1 General. Foundation walls and retaining walls shall be designed to resist lateral soil loads. Soil loads specified in Table 1610.1 shall be used as the minimum design lateral soil loads unless determined otherwise by a geotechnical investigation in accordance with Section 1803. Foundation walls and other walls in which horizontal movement is restricted at the top shall be designed for at-rest pressure. Retaining walls free to move and rotate at the top shall be permitted to be designed for active pressure. Design lateral pressure from surcharge loads shall be added to the lateral earth pressure load. Design lateral pressure shall be increased if soils at the site are expansive. Foundation walls shall be designed to support the weight of the full hydrostatic pressure of undrained backfill unless a drainage system is installed in accordance with Sections 1805.4.2 and 1805.4.3.

> **Exception:** Foundation walls extending not more than 8 feet (2438 mm) below grade and laterally supported at the top by flexible diaphragms shall be permitted to be designed for active pressure.

❖ Nominal loads attributable to lateral earth pressures are determined in accordance with this section. This section provides lateral loads for various soil types. This section requires that foundation and retaining walls be designed to be capable of resisting the lateral soil loads specified by Table 1610.1 where a specific soil investigation has not been performed. Consideration must be given to additional lateral soil pressures due to surcharge loads that result from sloping backfill, driveways or parking spaces that are close to a foundation wall, as well as the foundation of an adjacent structure.

TABLE 1610.1. See page 16-56.

❖ The table lists at-rest and active soil pressures for a number of different types of moist soils. The basis of the soil classification into the various types listed is ASTM D 2487. Soils identified by Note b in Table 1610.1 have unpredictable characteristics. These are called expansive soils. Because of their ability to absorb water, they shrink and swell to a higher degree than other soils. As expansive soils swell, they are capable of exerting large forces on soil-retaining structures; thus, these types of soils are not to be used as backfill.

SECTION 1611 RAIN LOADS

1611.1 Design rain loads. Each portion of a roof shall be designed to sustain the load of rainwater that will accumulate on it if the primary drainage system for that portion is blocked plus the uniform load caused by water that rises above the inlet of the secondary drainage system at its design flow. The design rainfall shall be based on the 100-year hourly rainfall rate indicated in Figure 1611.1 or on other rainfall rates determined from *approved* local weather data.

$R = 5.2(d_s + d_h)$ **(Equation 16-36)**

For SI: $R = 0.0098(d_s + d_h)$

where:

d_h = Additional depth of water on the undeflected roof above the inlet of secondary drainage system at its design flow (i.e., the hydraulic head), in inches (mm).

d_s = Depth of water on the undeflected roof up to the inlet of secondary drainage system when the primary drainage system is blocked (i.e., the static head), in inches (mm).

R = Rain load on the undeflected roof, in psf (kN/m_2). When the phrase "undeflected roof" is used, deflections from loads (including dead loads) shall not be considered when determining the amount of rain on the roof.

❖ The nominal rain load, *R*, is determined in accordance with this section. It represents the weight of accumulated rainwater, assuming a blockage of the primary roof drainage system. The design of the roof drainage systems must be in accordance with Chapter 11 of the *International Plumbing Code*® (IPC®). The primary roof drainage system can include roof drains, leaders, conductors and horizontal storm drains within the structure. Drainage system design is based on a specified design rainfall intensity, as well as the roof area it drains. The criteria for sizing the components of the drainage system are provided in Section 1106 of the IPC. Where the building is configured such that water will not collect on the roof there is no requirement for a secondary drainage system [see Figure 1611.1(1)]. Likewise, there would be no rain load required in the design of the roof.

It is not uncommon to find that roof drains have become blocked by debris, leading to ponding of rainwater where the roof construction is conducive to retaining water. While the objective of providing roof drainage is typically to prevent the accumulation of water, the code also recognizes controlled drainage systems that are engineered to retain rainwater (see Section 1611.3). The important point is that wherever the potential exists for the accumulation of rainwater on a roof, whether it is intentional or otherwise, the roof must be designed for this load. Furthermore, Section 1101.7 of the IPC requires the maximum depth of water to be determined, assuming all primary roof drainage to be blocked. The water will rise above the primary roof drain until it reaches the elevation of the roof edge, scuppers or another serviceable drain. At the design rainfall intensity, this depth will be based on the flow rate of the secondary drainage system. This depth, referred to as the hydraulic head, can be determined from Table 1611.1(2) for various types of drains and flow rates. Its use is illustrated in the example on page 16-65. Section 1108 of

TABLE 1610.1
LATERAL SOIL LOAD

DESCRIPTION OF BACKFILL MATERIAL[c]	UNIFIED SOIL CLASSIFICATION	DESIGN LATERAL SOIL LOAD[a] (pound per square foot per foot of depth)	
		Active pressure	At-rest pressure
Well-graded, clean gravels; gravel-sand mixes	GW	30	60
Poorly graded clean gravels; gravel-sand mixes	GP	30	60
Silty gravels, poorly graded gravel-sand mixes	GM	40	60
Clayey gravels, poorly graded gravel-and-clay mixes	GC	45	60
Well-graded, clean sands; gravelly sand mixes	SW	30	60
Poorly graded clean sands; sand-gravel mixes	SP	30	60
Silty sands, poorly graded sand-silt mixes	SM	45	60
Sand-silt clay mix with plastic fines	SM-SC	45	100
Clayey sands, poorly graded sand-clay mixes	SC	60	100
Inorganic silts and clayey silts	ML	45	100
Mixture of inorganic silt and clay	ML-CL	60	100
Inorganic clays of low to medium plasticity	CL	60	100
Organic silts and silt clays, low plasticity	OL	Note b	Note b
Inorganic clayey silts, elastic silts	MH	Note b	Note b
Inorganic clays of high plasticity	CH	Note b	Note b
Organic clays and silty clays	OH	Note b	Note b

For SI: 1 pound per square foot per foot of depth = 0.157 kPa/m, 1 foot = 304.8 mm.

a. Design lateral soil loads are given for moist conditions for the specified soils at their optimum densities. Actual field conditions shall govern. Submerged or saturated soil pressures shall include the weight of the buoyant soil plus the hydrostatic loads.

b. Unsuitable as backfill material.

c. The definition and classification of soil materials shall be in accordance with ASTM D 2487.

the IPC specifically requires a secondary roof drainage system where the building construction extends above the roof at the perimeter. This applies to parapet walls, stepped buildings or any other construction that would allow rainwater to pond on the roof. The sizing of a secondary drainage system is identical to the process used for the primary system. Instead of using a "piped" secondary system, designers may prefer to install scuppers to allow rainwater to overflow the roof. Examples of both types of secondary systems are shown in Figure 1611.1(3). Also note that the IPC requires a secondary system to be completely separate and to discharge above grade. Since the secondary system serves as an emergency backup, requiring it to discharge above grade provides a means of signaling that there is a blockage of the primary drainage system.

Some roof failures have been attributed to the increased loads from ponding water. This section requires the roof to be capable of resisting the maximum water depth that can occur if the primary means of roof drainage becomes blocked. Blockages are typically caused by debris at the inlet to the primary roof drains, but they can occur anywhere along the primary piping system, such as an under-slab pipe collapse. Computation of rain load, R, is in accordance with Equation 16-36. The coefficient of that equation is merely the conversion of the unit weight of water to an equivalent unit load per inch of water depth as Figure 1611.1(4) illustrates. Two variables are considered to determine rain load: the depth of the water on the undeflected roof, as measured from the low point elevation to the inlet elevation of the secondary drain; and the additional depth of water at the secondary drainage flow, respectively referred to as static head and hydraulic head. The sum of these depths is the design depth for computing rain load, R, as indicated in Equation 16-36. An example of the

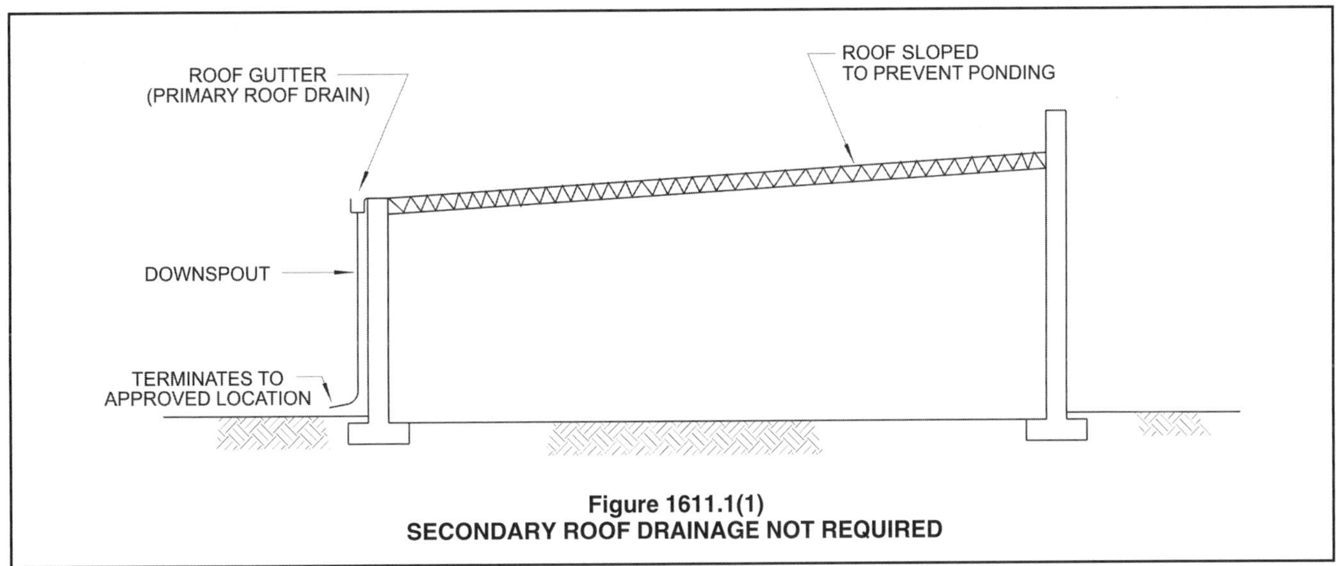

Figure 1611.1(1)
SECONDARY ROOF DRAINAGE NOT REQUIRED

DRAINAGE SYSTEM	FLOW RATE (gpm) Depth of water above drain inlet (hydraulic head) (inches)									
	1	2	2.5	3	3.5	4	4.5	5	7	8
4-inch-diameter drain	80	170	180							
6-inch-diameter drain	100	190	270	380	540					
8-inch-diameter drain	125	230	340	560	850	1,100	1,170			
6-inch-wide, open-top scupper	18	50	*	90	*	140	**	194	321	393
24-inch-wide, open-top scupper	72	200	*	360	*	560	*	776	1,284	1,572
6-inch-wide, 4-inch-high, closed-top scupper	18	50	*	90	*	140	*	177	231	253
24-inch-wide, 4-inch-high, closed-top scupper	72	200	*	360	*	560	*	708	924	1,012
6-inch-wide, 6-inch-high, closed-top scupper	18	50	*	90	*	140	*	194	303	343
24-inch-wide, 6-inch-high, closed-top scupper	72	200	*	360	*	560	*	776	1,212	1,372

For SI: 1 inch = 25.4 mm, 1 gallon per minute = 3.785 L/m.
Source: Factory Mutual Engineering Corp. Loss Prevention Data 1-54.

Table 1611.1(2)
FLOW RATE, IN GALLONS PER MINUTE, OF VARIOUS ROOF DRAINS AT VARIOUS WATER DEPTHS AT DRAIN INLETS (INCHES)

computation of rain load is provided in the following example [also see Figure 1611.1(5)].

EXAMPLE
Rain Load on Roof with Overflow Scuppers

Given:

Primary roof drain and overflow scupper shown in Figure 1611.1(5).

Static head, d_s = 7 inches (178 mm)

Tributary area, A = 5,400 square feet (502 m^2)

Rainfall rate, i = 2.5 inches/hour

= 0.208333 feet/hour (0.0635 m/hr)

Determine:

Hydraulic head, d_h

Rain load, R

Calculate required flow rate, Q, at scupper in gallons per minute (gpm).

$Q = A \times i$ = 5,400 square feet (502 m^2) × 0.2083 feet/hour

= 1125 cubic feet/hour (31.9 m^3/hr)

= 18.75 cubic feet/minute (0.531 m^3/min)

= 140.25 gpm (531 L/m)

Look up hydraulic head using Table 1611.1(2).

For 6 inches wide 6 inches (152 mm) high scupper, find 140 gpm. (530 L/m)

d_h = 4 inches (102 mm)

Determine rain load, R

Total head = $(d_s + d_h)$ = 11 inches (279 mm)

$R = 5.2\ (d_s + d_h)$

R = 57.2 psf (2.74 kN/m^2)

FIGURE 1611.1. See page 16-60.

❖ Including these IPC maps in the code provides the structural designer with design criteria that are essential for determining the rain load. Figure 1611.1 consists of five maps for various regions of the country. This figure provides the rainfall rates for a storm of 1-hour duration that has a 100-year return period. Rainfall rates indicate the maximum rate of rainfall within the given period of time occurring at the stated frequency. For example, the map indicates a rainfall rate of 2.1 inches (53 mm) per hour for Burlington, Vermont. Thus, it is predicted that it will rain 2.1 inches (53 mm) within 1 hour once every 100 years. The rainfall rates are calculated by a statistical analysis of weather records. Because the statistics are based on previous or historical weather conditions, it is conceivable to have two 100-year storms in one week's time. The probability of this occurring, however, is very low.

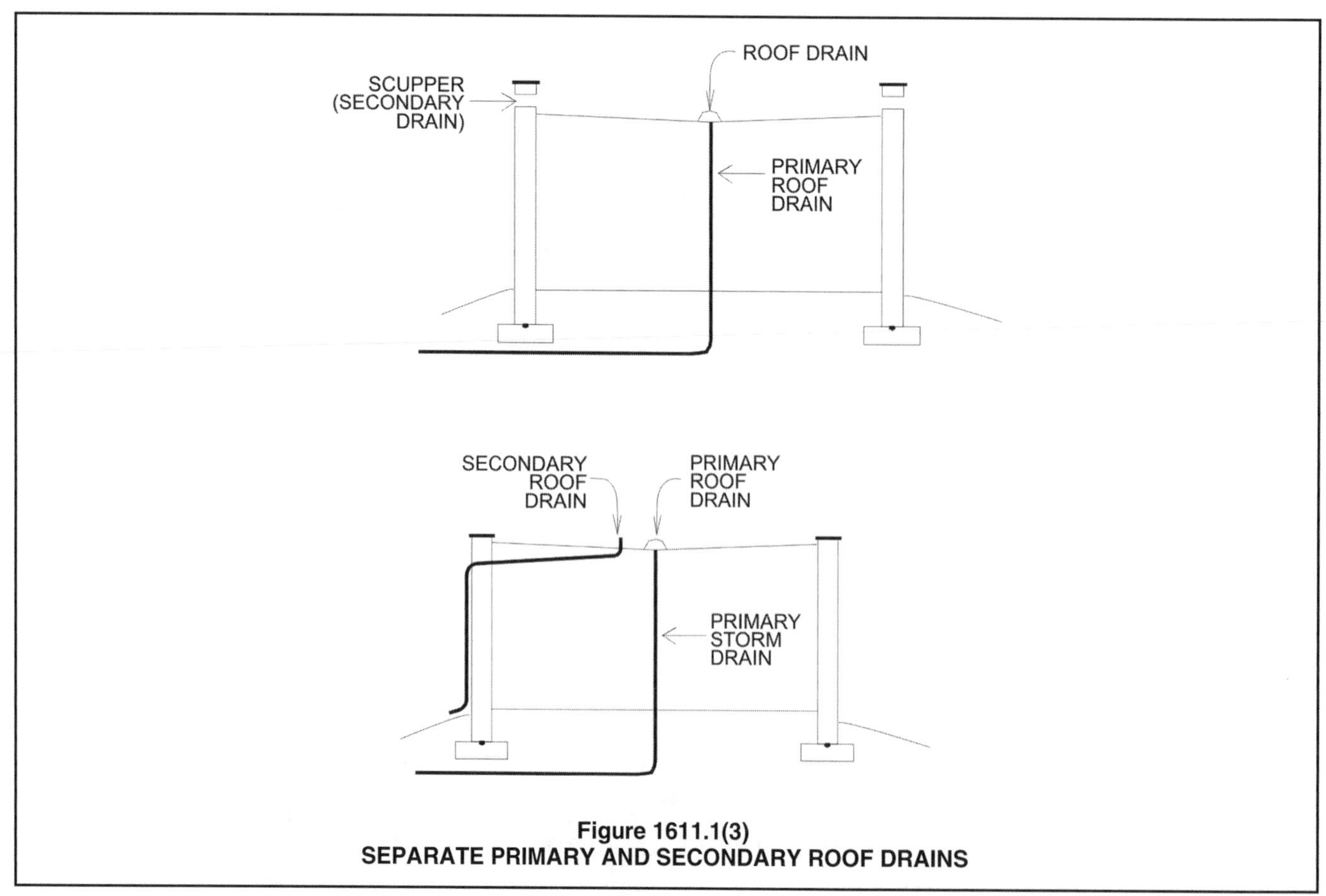

Figure 1611.1(3)
SEPARATE PRIMARY AND SECONDARY ROOF DRAINS

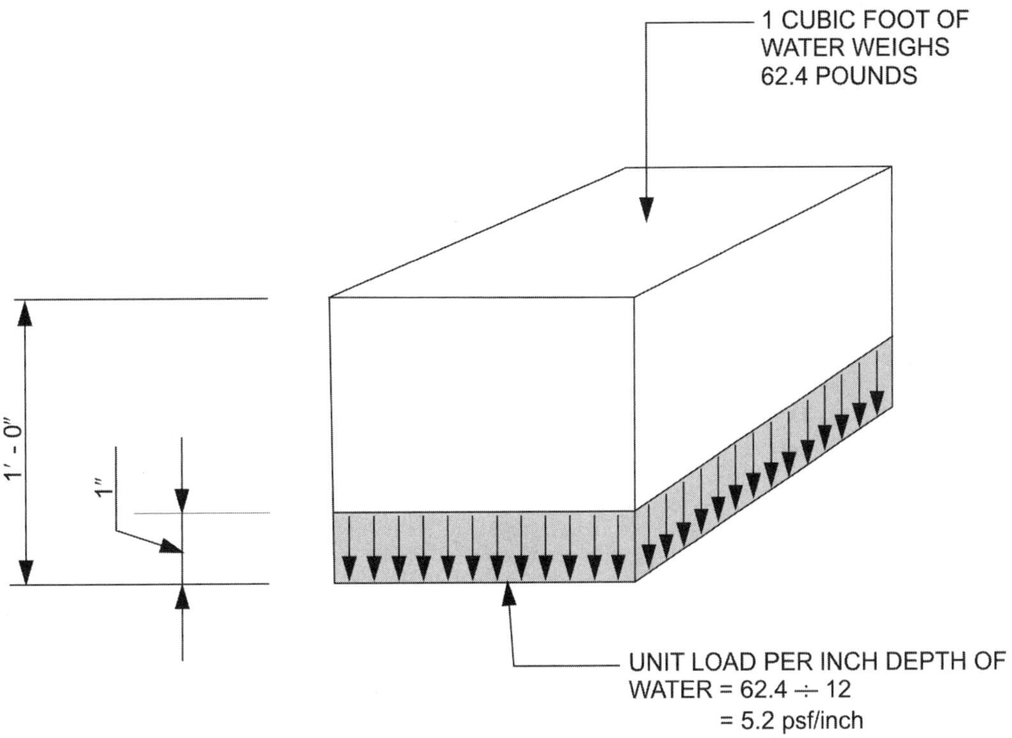

1 CUBIC FOOT OF WATER WEIGHS 62.4 POUNDS

1′ - 0″

1″

UNIT LOAD PER INCH DEPTH OF WATER = 62.4 ÷ 12 = 5.2 psf/inch

For SI: 1 inch = 25.4 mm, 1 foot = 304.8 mm, 1 cubic foot = 0.02832 m^2, 1 pound = 0.454 kg, 1 pound per square foot = 47.88 Pa.

Figure 1611.1(4)
EQUATION 16-35 COEFFICIENT

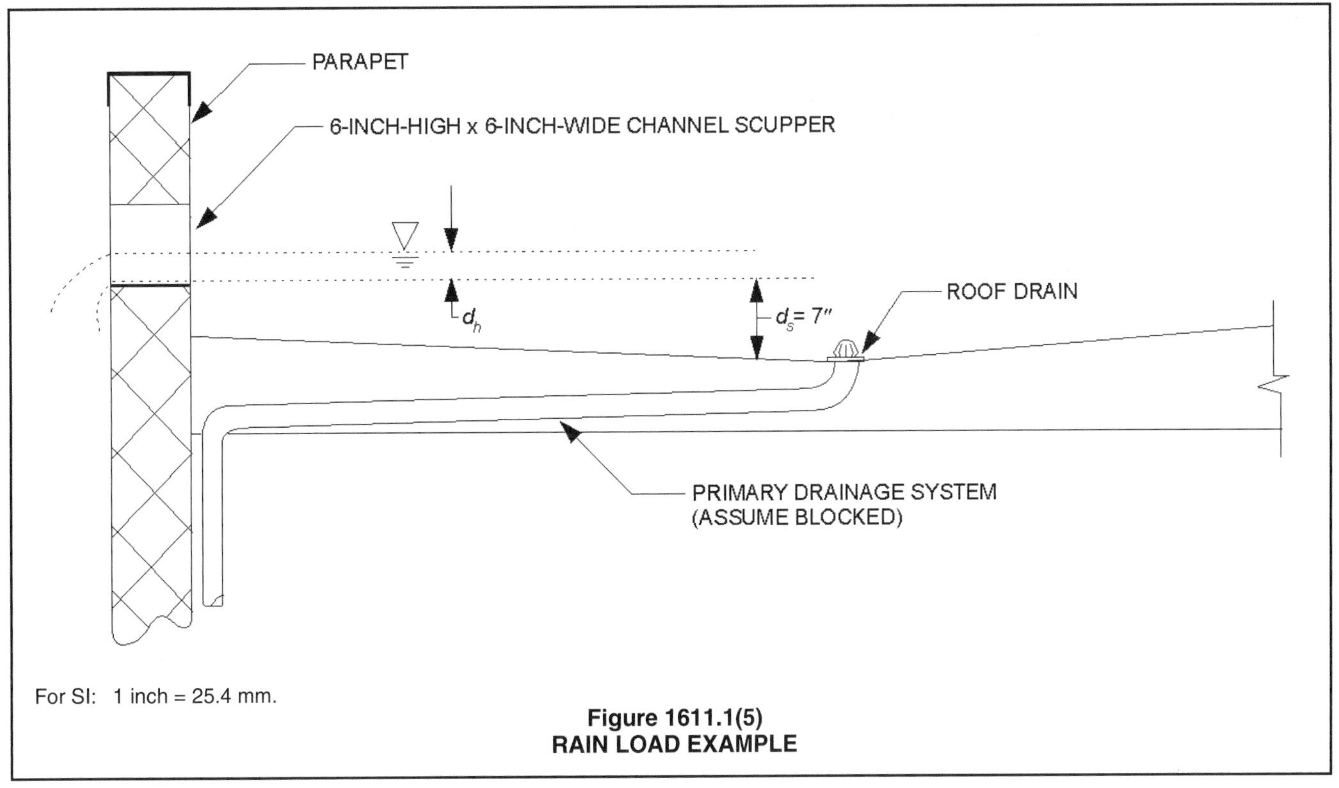

For SI: 1 inch = 25.4 mm.

Figure 1611.1(5)
RAIN LOAD EXAMPLE

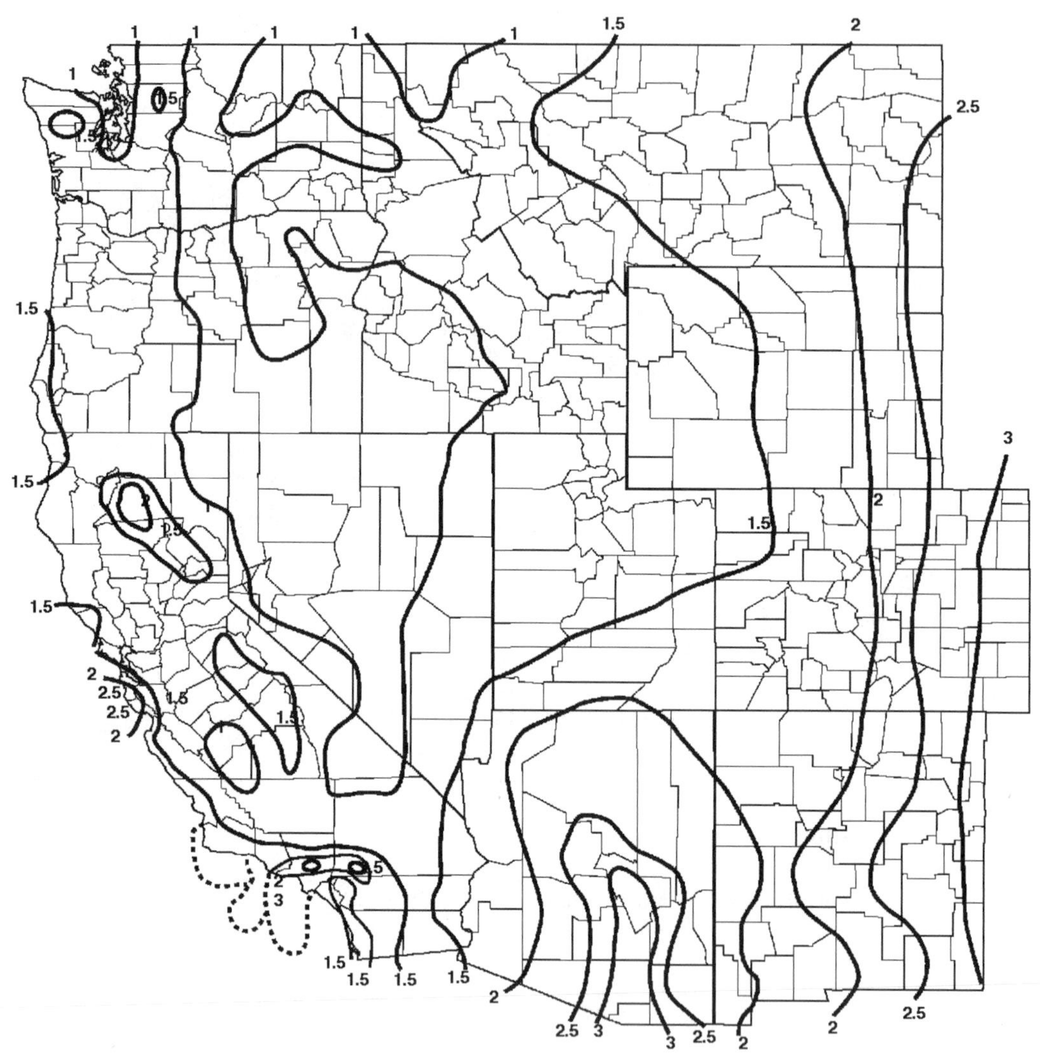

[P] FIGURE 1611.1
100-YEAR, 1-HOUR RAINFALL (INCHES) WESTERN UNITED STATES

For SI: 1 inch = 25.4 mm.
Source: National Weather Service, National Oceanic and Atmospheric Administration, Washington, DC.

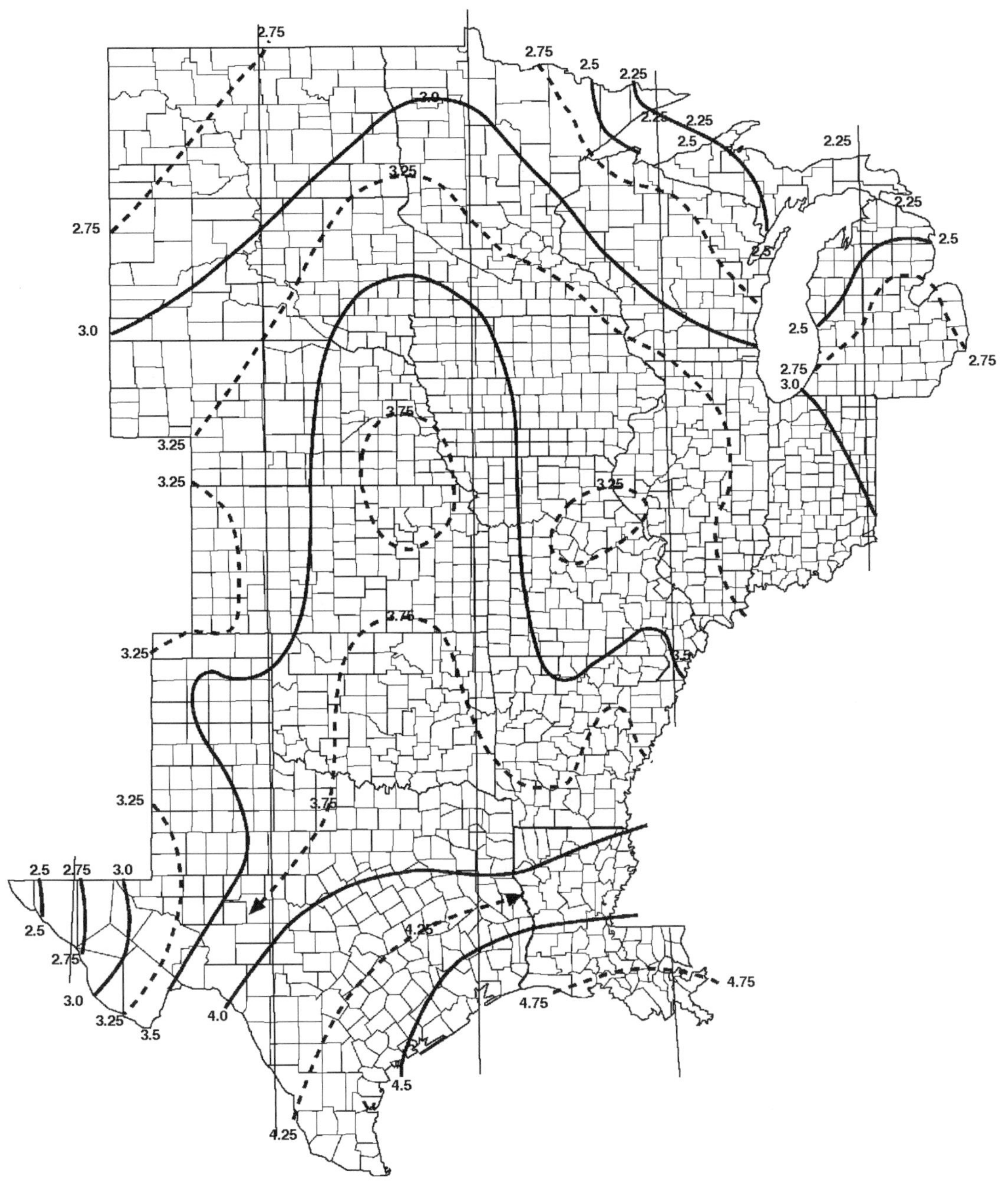

[P] FIGURE 1611.1—continued
100-YEAR, 1-HOUR RAINFALL (INCHES) CENTRAL UNITED STATES

For SI: 1 inch = 25.4 mm.
Source: National Weather Service, National Oceanic and Atmospheric Administration, Washington, DC.

[P] FIGURE 1611.1—continued
100-YEAR, 1-HOUR RAINFALL (INCHES) EASTERN UNITED STATES

For SI: 1 inch = 25.4 mm.
Source: National Weather Service, National Oceanic and Atmospheric Administration, Washington, DC.

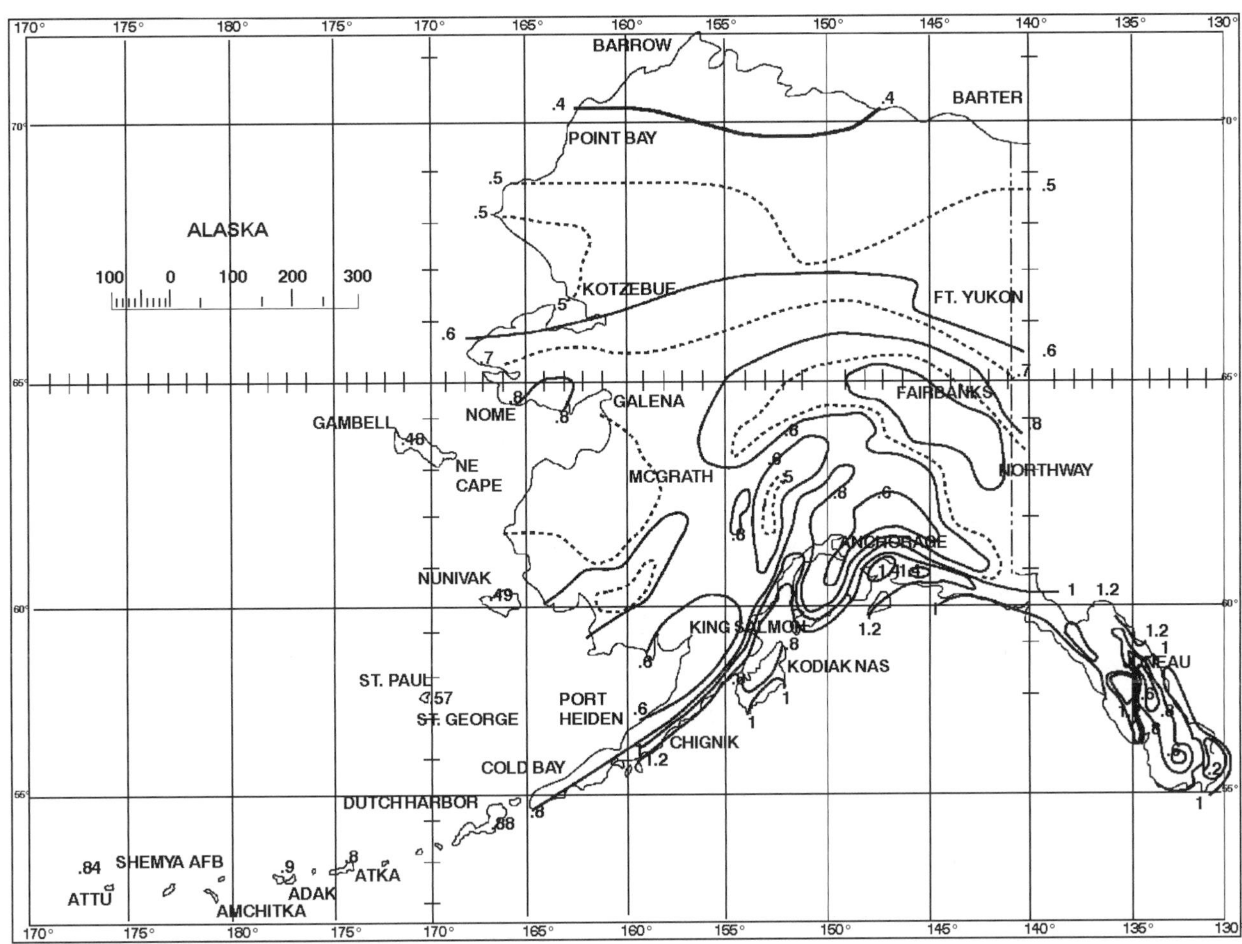

[P] FIGURE 1611.1—continued
100-YEAR, 1-HOUR RAINFALL (INCHES) ALASKA

For SI: 1 inch = 25.4 mm.
Source: National Weather Service, National Oceanic and Atmospheric Administration, Washington, DC.

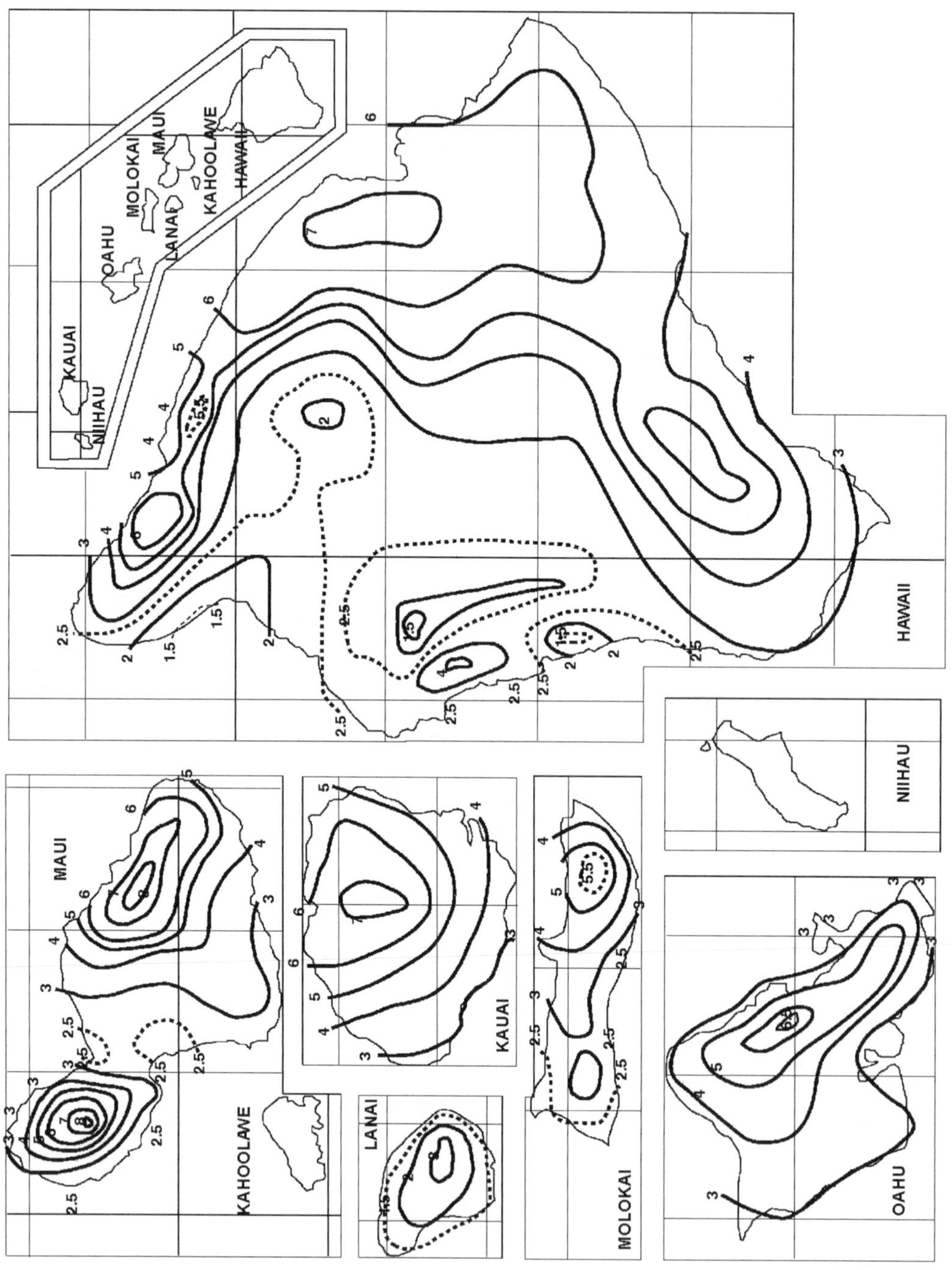

[P] FIGURE 1611.1—continued
100-YEAR, 1-HOUR RAINFALL (INCHES) HAWAII

For SI: 1 inch = 25.4 mm.
Source: National Weather Service, National Oceanic and Atmospheric Administration, Washington, DC.

1611.2 Ponding instability. Susceptible bays of roofs shall be evaluated for ponding instability in accordance with Section 8.4 of ASCE 7.

❖ In roofs lacking sufficient framing stiffness, a condition known as "ponding instability" can occur where increasingly larger deflections caused by the continued accumulation of rainwater are large enough to overload the structure and result in a roof collapse. This must be countered by providing adequate stiffness in order to prevent increasingly larger deflections due to the buildup of rainwater. Another means to minimize the accumulation of rainwater is to camber the roof framing. This section requires a check for ponding instability at susceptible bays (see definition in Section 202). A ponding instability check is to be made assuming the primary roof drains are blocked. The determination of ponding instability is typically done by an iterative structural analysis where the incremental deflection is determined and the resulting increased rain load from the deflection is added to the original rain load.

1611.3 Controlled drainage. Roofs equipped with hardware to control the rate of drainage shall be equipped with a secondary drainage system at a higher elevation that limits accumulation of water on the roof above that elevation. Such roofs shall be designed to sustain the load of rainwater that will accumulate on them to the elevation of the secondary drainage system plus the uniform load caused by water that rises above the inlet of the secondary drainage system at its design flow determined from Section 1611.1. Such roofs shall also be checked for ponding instability in accordance with Section 1611.2.

❖ Controlled drainage is the limitation of the drainage flow rate to a rate that is less than the rainfall rate such that the depth of the water intentionally builds up on the roof during a design rainfall. Controlled flow roof drain systems must be designed in accordance with Section 1111 of the IPC. A secondary roof drain system is needed to limit the buildup of water to a specific depth for roof design. The depth of water on the roof is also to include the depth of the water above the inlet of the secondary drain when the design flow rate is reached. Consideration of the effect of the accumulated rainwater is identical to assuming a blockage in the primary drainage system as discussed under Section 1611.1.

SECTION 1612 FLOOD LOADS

1612.1 General. Within *flood hazard areas* as established in Section 1612.3, all new construction of buildings, structures and portions of buildings and structures, including substantial improvement and restoration of substantial damage to buildings and structures, shall be designed and constructed to resist the effects of flood hazards and flood loads. For buildings that are located in more than one *flood hazard area*, the provisions associated with the most restrictive *flood hazard area* shall apply.

❖ This section addresses requirements for all buildings and structures in flood hazard areas. These areas are commonly referred to as "flood plains" and are shown on a community's Flood Insurance Rate Map (FIRM) prepared by FEMA or on another adopted flood hazard map. Code users should be aware that floods often affect areas outside the flood hazard area boundaries shown on FIRMs and can exceed the base flood elevation.

Through the adoption of the code, communities meet a significant portion of the flood plain management regulation requirements necessary to participate in the National Flood Insurance Program (NFIP). To participate in the NFIP, a jurisdiction must adopt regulations that at least meet the requirements of federal regulations in Section 60.3 of 44 CFR. This requirement can be satisfied by adopting the *International Building Code*® (IBC®) (including Appendix G) and the *International Residential Code*® (IRC®). [The *International Existing Building Code*® (IEBC®) also has provisions that are consistent with the NFIP.] If Appendix G (Flood-resistant Construction) is not enforced, its provisions must be captured in a companion flood plain management ordinance adopted by the community. The NFIP requires communities to regulate all development in flood hazard areas. Section 201.2 of Appendix G defines "Development" as "any man-made change to improved or unimproved real estate, including but not limited to buildings or other structures; temporary or permanent storage of materials; mining; dredging; filling; grading; paving; excavations; operations and other land-disturbing activities."

The NFIP was established to reduce flood losses, to better indemnify individuals from flood losses and to reduce federal expenditures for disaster assistance. A community that has been determined to have flood hazard areas elects to participate in the NFIP to protect health, safety and property, so that its citizens can purchase federally backed flood insurance. FEMA administers the NFIP, which includes monitoring community compliance with the flood plain management requirements of the NFIP.

New buildings and structures, and substantial improvements to existing buildings and structures, are to be designed and constructed to resist flood forces to minimize damage. Flood forces include flotation, lateral (hydrostatic) pressures, moving water (hydrodynamic) pressures, wave impact and debris impact. Flood-related hazards may include erosion and scour.

Many states and communities have elected to regulate flood plain development to a higher standard than the minimum required to participate in the NFIP. Communities considering using the code and other *International Codes*® to meet the flood plain management requirements of the NFIP are advised to consult with their state NFIP coordinator or the appropriate FEMA regional office.

If located in flood hazard areas, buildings and structures that are damaged by any cause are to be examined by the building official to determine if the damage constitutes substantial damage, in which the cost to repair or restore the building or structure to its predamaged condition equals or exceeds 50 percent of its market value before the damage occurred. All substantial improvements and repairs of buildings and structures that are substantially damaged must meet the flood-resistant provisions of the code. For additional guidance, see FEMA P-758, *Substantial Improvement/Substantial Damage Desk Reference*.

Some buildings are proposed to be located such that either they are in more than one flood zone or only a portion of the building is in a flood zone. Where this occurs, the entire building or structure is required to be designed and constructed according to the requirements of the more restrictive flood zone. For example, if a building is partially in a flood hazard area subject to high-velocity wave action (V zone), then the entire building must meet the requirements for that area. Similarly, if a building is partially in a flood hazard area and partially out of the mapped flood plain, then the entire building must be flood resistant.

For additional guidance on how to use the *International Codes®* to participate in the NFIP, see *Reducing Flood Losses through the International Codes: Meeting the Requirements of the National Flood Insurance Program*.

1612.2 Definitions. The following terms are defined in Chapter 2:

BASE FLOOD.

BASE FLOOD ELEVATION.

BASEMENT.

DESIGN FLOOD.

DESIGN FLOOD ELEVATION.

DRY FLOODPROOFING.

EXISTING CONSTRUCTION.

EXISTING STRUCTURE.

FLOOD or FLOODING.

FLOOD DAMAGE-RESISTANT MATERIALS.

FLOOD HAZARD AREA.

FLOOD HAZARD AREA SUBJECT TO HIGH-VELOCITY WAVE ACTION.

FLOOD INSURANCE RATE MAP (FIRM).

FLOOD INSURANCE STUDY.

FLOODWAY.

LOWEST FLOOR.

SPECIAL FLOOD HAZARD AREA.

START OF CONSTRUCTION.

SUBSTANTIAL DAMAGE.

SUBSTANTIAL IMPROVEMENT.

❖ Definitions facilitate the understanding of code provisions and minimize potential confusion. To that end, this section lists definitions of terms associated with flood requirements. Note that these definitions are found in Chapter 2. The use and application of defined terms, as well as undefined terms, are set forth in Section 201.

1612.3 Establishment of flood hazard areas. To establish *flood hazard areas*, the applicable governing authority shall adopt a flood hazard map and supporting data. The flood hazard map shall include, at a minimum, areas of special flood hazard as identified by the Federal Emergency Management Agency in an engineering report entitled "The Flood Insurance Study for **[INSERT NAME OF JURISDICTION]**," dated **[INSERT DATE OF ISSUANCE]**, as amended or revised with the accompanying Flood Insurance Rate Map (FIRM) and Flood Boundary and Floodway Map (FBFM) and related supporting data along with any revisions thereto. The adopted flood hazard map and supporting data are hereby adopted by reference and declared to be part of this section.

❖ Flood maps and studies are prepared by FEMA and are to be used as a community's official map, unless the community chooses to adopt a map that shows more extensive flood hazard areas. Most communities have multiple flood map panels, all of which should be listed in the adopting ordinance by panel number and date so that the appropriate effective map is used.

From time to time, FEMA's flood plain maps and studies may be revised and republished. In recent years, revised FIRMs have been produced in a digital format (referred to as Digital FIRMs or DFIRMs). Communities that prefer to cite the digital data should obtain a legal opinion. DFIRMs are registered to the primary coordinate system of the state or community. FEMA advises that the horizontal location of flood hazard areas relative to specific sites should be determined using the coordinate grid rather than planimetric base map features such as streets.

When maps are revised and flood hazard areas are changed, FEMA involves the community and provides a formal opportunity to review the documents. Once the revisions are finalized, FEMA requires the community to adopt the new maps. Communities may be able to minimize having to adopt each revision by referencing the date of the original map and study and all future revisions. This is a method by which subsequent revisions to flood maps and studies may be adopted administratively without requiring legislative action on the part of the community. Communities will need to determine whether this "adoption by reference" approach is allowed under their state's enabling authority and due process requirements. If not allowed, communities are to follow their state's requirements, which typically require public notices, hearings and specific adoption of revised maps by the community's legislative body.

1612.3.1 Design flood elevations. Where design flood elevations are not included in the *flood hazard areas* established in Section 1612.3, or where floodways are not designated, the *building official* is authorized to require the applicant to:

1. Obtain and reasonably utilize any design flood elevation and floodway data available from a federal, state or other source; or
2. Determine the design flood elevation and/or floodway in accordance with accepted hydrologic and hydraulic engineering practices used to define special flood hazard areas. Determinations shall be undertaken by a *registered design professional* who shall document that the technical methods used reflect currently accepted engineering practice.

❖ The purpose of this provision is to clarify how design flood elevations are to be determined for those flood hazard areas shown on community flood hazard maps that do not have the flood elevation already specified. Section 107.2 requires that the construction documents submitted are to be accompanied by a site plan, which includes flood hazard areas, floodways or design flood elevations, as applicable. While flood elevations are often available, a large percentage of areas that are mapped as special flood hazard areas by the NFIP do not have flood elevations or do not have floodway designations. This section clarifies the authority of the building official to require use of data, which may be obtained from other sources, or to require the applicant to develop flood hazard data, and is based on the NFIP regulation in 44 CFR §60.3(b)(4).

1612.3.2 Determination of impacts. In riverine *flood hazard areas* where design flood elevations are specified but floodways have not been designated, the applicant shall provide a floodway analysis that demonstrates that the proposed work will not increase the design flood elevation more than 1 foot (305 mm) at any point within the jurisdiction of the applicable governing authority.

❖ This section requires a floodway analysis to determine impacts. Development in riverine flood plains can increase flood levels and loads on other properties, especially if it occurs in areas known as "floodways" that must be reserved to convey flood flows. Commercial software for these analyses is readily available and FEMA provides software and technical guidance at http://www.fema.gov/plan/prevent/fhm/frm_soft.shtm.

This section provides consistency with the NFIP, which requires applicants to demonstrate whether their proposed work will increase flood levels in floodplains where the NFIP's FIRM shows base flood elevations, but floodways are not shown. A small percentage of flood plains where FEMA has specified base flood elevations do not have designated floodways.

1612.4 Design and construction. The design and construction of buildings and structures located in *flood hazard areas*, including flood hazard areas subject to high-velocity wave action, shall be in accordance with Chapter 5 of ASCE 7 and with ASCE 24.

❖ FEMA uses multiple designations for flood hazard areas shown on each FIRM, including A, AO, AH, A1-30, AE, A99, AR, AR/A1-30, AR/AE, AR/A0, AR/AH, AR/A, VO or V1-30, VE and V. Along many open coasts and lake shorelines where wind-driven waves are predicted, the flood hazard area is commonly referred to as the "V zone." Flood hazard areas that are inland of areas subject to high-velocity wave action and flood hazard areas along rivers and streams are commonly referred to as "A zones." Due to waves, the flood loads in areas subject to high-velocity wave action differ from those in other flood hazard areas. Some recent FIRMS in coastal communities show the "Limit of Moderate Wave Action," which is the inland extent of the 1.5-foot (457 mm) wave height. In ASCE 7 and ASCE 24, as well as in common usage, the area between the V zone and Limit of Moderate Wave Action is called "Coastal A Zone."

ASCE 24 outlines in detail the specific requirements that are to be applied to buildings and structures in all flood hazard areas. Communities that have AR zones and A99 zones shown on their FIRMs should consult with the appropriate state agency or FEMA regional office for guidance on the requirements that apply. AR zones are areas that result from the decertification of a previously accredited flood protection system (such as a levee) that is determined to be in the process of being restored to provide base flood protection. A99 zones are areas subject to inundation by the 1-percent-annual-chance flood event, but which will ultimately be protected upon completion of an under-construction federal flood protection system.

1612.5 Flood hazard documentation. The following documentation shall be prepared and sealed by a *registered design professional* and submitted to the *building official*:

1. For construction in *flood hazard areas* not subject to high-velocity wave action:
   1.1. The elevation of the lowest floor, including the basement, as required by the lowest floor elevation inspection in Section 110.3.3.
   1.2. For fully enclosed areas below the design flood elevation where provisions to allow for the automatic entry and exit of floodwaters do not meet the minimum requirements in Section 2.6.2.1 of ASCE 24, *construction documents* shall include a statement that the design will provide for equalization of hydrostatic flood forces in accordance with Section 2.6.2.2 of ASCE 24.

1.3. For dry floodproofed nonresidential buildings, *construction documents* shall include a statement that the dry floodproofing is designed in accordance with ASCE 24.

2. For construction in flood hazard areas subject to high-velocity wave action:

2.1. The elevation of the bottom of the lowest horizontal structural member as required by the lowest floor elevation inspection in Section 110.3.3.

2.2. *Construction documents* shall include a statement that the building is designed in accordance with ASCE 24, including that the pile or column foundation and building or structure to be attached thereto is designed to be anchored to resist flotation, collapse and lateral movement due to the effects of wind and flood loads acting simultaneously on all building components, and other load requirements of Chapter 16.

2.3. For breakaway walls designed to have a resistance of more than 20 psf (0.96 kN/m^2) determined using allowable stress design, *construction documents* shall include a statement that the breakaway wall is designed in accordance with ASCE 24.

❖ The NFIP requires that certain documentation be submitted in order to demonstrate compliance with provisions of the code that cannot be easily verified during a site inspection. The most common, described in Item 1.1, is the documentation of the lowest floor elevation. It provides documents that the lowest floor is at or above the required minimum elevation. Elevation is one of the most important aspects of flood-resistant construction and is a significant factor used to determine flood insurance premium rates. FEMA Form 81-31, *Elevation Certificate*, which includes illustrations and instructions, is recommended (download from www.fema.gov/business/nfip/elvinst.shtm). Building owners need elevation certificates to obtain NFIP flood insurance, and insurance agents use the certificates to compute the proper flood insurance premium rates.

The criteria for the minimum number and size of flood openings to allow free inflow and outflow of floodwaters under all types of flood conditions are set in ASCE 24. The statement described in Item 1.2 is required in the construction documents if the engineered openings are used (see Section 2.6.2.2 of ASCE 24); this statement is not required if nonengineered (prescriptive) openings meet the criteria of Section 2.6.2.1 of ASCE 24. For further guidance, refer to FEMA TB #1, *Openings in Foundation Walls and Walls of Enclosures Below Elevated Buildings in Special Flood Hazard Areas.*

The statement described in Item 1.3 is to be included in the construction documents for nonresidential buildings that are designed to be dry floodproofed. It is important to note that dry floodproofing is allowed only for nonresidential buildings and structures that are located in flood hazard areas not subject to high-velocity wave action. The registered design professional who seals the construction documents is indicating that, based upon development or review of the structural design, specifications and plans for construction, the design and methods of construction are in accordance with accepted standards of practice in ASCE 24 to meet the following provisions: (1) the structure, together with attendant utilities and sanitary facilities, is water tight to the floodproofed design elevation indicated with walls that are substantially impermeable to the passage of water and (2) all structural components are capable of resisting hydrostatic and hydrodynamic flood forces, including the effects of buoyancy and anticipated debris impact forces. The use of FEMA Form 81-65, *Floodproofing Certificate*, is recommended (download from http://www.fema.gov/plan/prevent/floodplain/nfipkeywordsfloodproofing_certificate.shtm). This certificate is used by insurance agents to determine NFIP flood insurance premium rates for dry floodproofed nonresidential buildings. For further guidance, refer to FEMA TB #3, *Nonresidential Floodproofing—Requirements and Certification for Buildings Located in Special Flood Hazard Areas.*

Certain documentation must be submitted in order to demonstrate compliance with provisions of the code that cannot be verified readily during a site inspection. The most common, in Item 2.1, provides evidence that the bottom of the lowest horizontal members of buildings constructed in flood hazard areas subject to high-velocity wave action (V zones) are elevated to or above the minimum required height. Buildings located in flood hazard areas subject to high-velocity wave action and winds are expected to experience significant flood and wind loads simultaneously. FEMA and coastal communities report significant damage to buildings that are not built to current code. The statement described in Item 2.2 is included in the construction documents to indicate that the design meets the flood load provisions of ASCE 24 and other loads required by this chapter.

The documentation described in Item 2.3 is used only for specific situations in which properly elevated buildings in flood hazard areas subject to high-velocity wave action have enclosures beneath them, and then only if the walls of the enclosures are designed to resist more than 20 psf (0.96 kN/m^2) determined using ASD. Because breakaway walls will fail under flood conditions, building materials can become water-borne debris that may damage adjacent buildings. Refer to FEMA TB #5, *Free-of-Obstruction Requirements for Buildings Located in Coastal High-hazard Areas,* and FEMA TB #9, *Design and Construction Guidance for Breakaway Walls Below Elevated Buildings in Coastal High Hazard Areas.*

SECTION 1613 EARTHQUAKE LOADS

1613.1 Scope. Every structure, and portion thereof, including nonstructural components that are permanently attached to structures and their supports and attachments, shall be designed and constructed to resist the effects of earthquake motions in accordance with ASCE 7, excluding Chapter 14 and Appendix 11A. The *seismic design category* for a structure is permitted to be determined in accordance with Section 1613 or ASCE 7.

Exceptions:

1. Detached one- and two-family dwellings, assigned to *Seismic Design Category* A, B or C, or located where the mapped short-period spectral response acceleration, S_S, is less than 0.4 g.
2. The seismic force-resisting system of wood-frame buildings that conform to the provisions of Section 2308 are not required to be analyzed as specified in this section.
3. Agricultural storage structures intended only for incidental human occupancy.
4. Structures that require special consideration of their response characteristics and environment that are not addressed by this code or ASCE 7 and for which other regulations provide seismic criteria, such as vehicular bridges, electrical transmission towers, hydraulic structures, buried utility lines and their appurtenances and nuclear reactors.

❖ These code provisions provide the requirements essential to determining a building's seismic design category. The balance of the earthquake load provisions are contained in the ASCE 7 load standard. The chapters that are noted as excluded from the ASCE 7 referenced standard are those that can create conflicts with Chapters 17 through 23. The balance of ASCE 7 earthquake load provisions are as follows:

Chapter	Subject
11	Seismic Design Criteria
12	Seismic Design Requirements for Building-Structures
13	Seismic Design Requirements for Nonstructural Components
15	Seismic Design Requirements for Non-building Structures
16	Seismic Response History Procedures
17	Seismic Design Requirements for Seismically Isolated Structures
18	Seismic Design Requirements for Structures with Damping Systems
19	Soil Structure Interaction for Seismic Design
20	Site Classification Procedure for Seismic Design
21	Site-specific Ground Motion Procedures for Seismic Design
22	Seismic Ground Motion and Long Period Transition Maps
23	Seismic Design Reference Documents

These seismic requirements are based, for the most part, on the NEHRP *Recommended Provisions for Seismic Regulations for New Buildings and Other Structures*. A new set of NEHRP recommended provisions has been prepared by the BSSC approximately every three years since the first edition in 1985. The code uses the NEHRP recommended provisions as the technical basis for seismic design requirements because of the nationwide input into the development of these design criteria. The NEHRP recommended provisions present up-to-date criteria for the design and construction of buildings subject to earthquake ground motions that are applicable anywhere in the nation. The requirements are intended to minimize the hazard to life for all buildings, increase the expected performance of higher occupancy buildings as compared to ordinary buildings and improve the capability of essential facilities to function during and after an earthquake. These minimum criteria are considered to be prudent and economically justified for the protection of life safety in buildings subject to earthquakes. Achieving the intended performance, however, depends on a number of factors, including the type of structural framing type, configuration, construction materials and as-built details of construction.

Detailed descriptions of changes made for the 2009 NEHRP *Recommended Provisions for Seismic Regulations for New Building and Other Structures* are available at www.bssconline.org under the explanation of changes made for the 2009 edition of the Provisions.

There are four exceptions to seismic design requirements included in this section. The following discussion addresses each of the exceptions.

Exception 1 exempts detached one- and two-family dwellings under two conditions. The first applies to structures assigned to Seismic Design Category A, B or C (meaning that both S_{ips} < 0.5g and S_{DI} < 0.2g). The second is for structures having a value for mapped short-period spectral response acceleration, S_s, less than 0.4g. The latter condition is derived from the ASCE 7 seismic provisions. Since it is based solely on the mapped value for short periods, it may allow structures to qualify more directly for this exception than will the first condition. In other words, the value of S_s should be checked first. If the structure qualifies based on S_s, then it is not necessary to check the seismic design category.

Exception 2 exempts conventional light-frame wood construction from the seismic requirements in this chapter. It should be noted that the limitations for conventional light-frame wood construction are included in Section 2308.2. There is no limitation on the use of the structure, except that Risk Category IV structures are not permitted to be constructed using the conventional light-frame wood construction provisions. Similarly, irregular portions of Seismic Design Category D and E structures are not permitted to be constructed using conventional light-frame wood construction (see commentary, Section 2308.12). The conventional light-frame wood construction provisions are deemed to provide equivalent seismic resistance as compared to construction designed in accordance with the requirements of this chapter based on the history of such conventional construction.

In Exception 3, agricultural buildings are exempt because they present a minimal life-safety hazard due to the low probability of human occupancy.

Exception 4 corresponds to an ASCE 7 exception regarding structures that are covered under other regulations, making it unnecessary to comply with the earthquake load requirements of the code.

1613.2 Definitions. The following terms are defined in Chapter 2:

❖ Definitions facilitate the understanding of code provisions and minimize potential confusion. To that end, this section lists definitions of terms associated with earthquake loads. Note that these definitions are found in Chapter 2. The use and application of defined terms, as well as undefined terms, are set forth in Section 201.

DESIGN EARTHQUAKE GROUND MOTION.

MECHANICAL SYSTEMS.

ORTHOGONAL.

RISK-TARGETED MAXIMUM CONSIDERED EARTHQUAKE (MCE_R) GROUND MOTION RESPONSE ACCELERATION.

SEISMIC DESIGN CATEGORY.

SEISMIC FORCE-RESISTING SYSTEM.

SITE CLASS.

SITE COEFFICIENTS.

1613.3 Seismic ground motion values. Seismic ground motion values shall be determined in accordance with this section.

❖ The design earthquake ground motion levels determined in this section may result in damage, both structural and nonstructural, from the high stresses that occur because of the dynamic nature of seismic events. For most structures, damage from the design earthquake ground motion would be repairable, but might be so costly as to make it economically undesirable. For essential facilities, it is expected that damage from the design earthquake ground motion would not be so severe as to prevent continued occupancy and function of the facility. For ground motions greater than the design levels, the intent is that there be a low likelihood of structural collapse.

1613.3.1 Mapped acceleration parameters. The parameters S_s and S_1 shall be determined from the 0.2 and 1-second spectral response accelerations shown on Figures 1613.3.1(1) through 1613.3.1(6). Where S_1 is less than or equal to 0.04 and S_s is less than or equal to 0.15, the structure is permitted to be assigned to *Seismic Design Category* A. The parameters S_S and S_1 shall be, respectively, 1.5 and 0.6 for Guam and 1.0 and 0.4 for American Samoa.

❖ The mapped maximum considered earthquake spectral response accelerations at 0.2-second period (S_s) and 1-second period (S_1) for a particular site are to be determined from Figures 1613.3.1(1) through (6). Where a site is between contours, as would usually be the case, straight-line interpolation or the value of the higher contour may be used. Areas that are considered to have a low seismic risk based solely on the mapped ground motions are placed directly into Seismic Design Category A.

The mapped maximum considered earthquake spectral response accelerations for a site may also be obtained using a seismic parameter program that has been developed by the United States Geological Survey (USGS) in cooperation with BSSC and FEMA. This program is available on the USGS earthquake hazards web site. The data are interpolated for a specific latitude-longitude or zip code, which the user enters. Output for an entry uses the built-in database to interpolate for the specific site. Caution should be used when using a zip code. In regions with highly variable mapped ground motions, the design parameters within a zip code may vary considerably from the value at the centroid of the zip code area. The code/NEHRP maximum considered earthquake (MCE) output for a site are the two spectral values required for design. The user may also use the program to calculate an MCE response spectrum, with or without site coefficients. Site coefficients can be calculated and included in calculations by simply selecting the site class; the program then calculates the site coefficient.

FIGURES 1613.3.1(1) through 1613.3.1(6). See page 16-74 through 16-81.

❖ These code figures provide the 5-percent damped spectral response accelerations at 0.2-second period (S_s), as well as at 1-second period (S_i) for Site Class B soil profiles. The code has incorporated updated earthquake ground motion maps that reflect the 2008 maps developed by the USGS National Seismic Hazard Mapping Project as well as technical changes adopted for the 2009 NEHRP *Recommended Seismic Provisions for New Buildings and Other Structures* (FEMA P750). In the NEHRP update process, the title for these maps was revised from "Maximum Considered Earthquake (MCE) Ground Motions" to "Risk-Targeted Earthquake (RTE) Ground Motions."

The seismic hazard maps incorporate new information on earthquake sources and ground motion prediction equations, including the new Next Generation Attenuation (NGA) relations. The ground motion maps further incorporate technical changes that reflect the use of: (1) risk-targeted ground motions; (2) maximum direction ground motions and (3) near-source 84th percentile ground motions.

Precise design values can be obtained from a USGS web site (http://earthquake.usgs.gov/research/hazmaps/design/index.php) using the longitude and latitude of the building site, obtained from GPS mapping programs or web sites.

1613.3.2 Site class definitions. Based on the site soil properties, the site shall be classified as *Site Class* A, B, C, D, E or F in accordance with Chapter 20 of ASCE 7. Where the soil properties are not known in sufficient detail to determine the site class, Site Class D shall be used unless the building official or geotechnical data determines Site Class E or F soils are present at the site.

❖ Each site is to be classified as one of six site classes (A through F), as defined in ASCE 7, based on one of three soil properties measured over the top 100 feet (30 480 mm) of the site. If the top 100 feet (30 480 mm) are not homogeneous, the ASCET provisions address how to determine average properties. Site Class A is hard rock typically found in the eastern United States. Site Class B is softer rock, typical of the western parts of the country. Site Class C, D or E indicates progressively softer soils. From an earthquake-resistance perspective, rock is the best material for most structures to be founded on. Site Class F indicates soil so poor that site-specific geotechnical investigation and dynamic site-response analysis are needed to determine appropriate site coefficients.

The three soil properties forming the basis of site classification are: shear wave velocity, standard penetration resistance or blow count (determined in accordance with ASTM D 1586) and undrained shear strength (determined in accordance with ASTM D 2166 or ASTM D 2850). Site Class A and B designations must be based on shear wave velocity measurements or estimates.

Where soil property measurements to a depth of 100 feet (30 480 mm) are not feasible, the registered design professional performing the geotechnical investigation may estimate appropriate soil properties based on known geologic conditions.

When soil properties are not known in sufficient detail to determine the site class, the default site class is D, unless the building official determines that Site Class E or F soil may exist at the site.

1613.3.3 Site coefficients and adjusted maximum considered earthquake spectral response acceleration parameters. The maximum considered earthquake spectral response acceleration for short periods, S_{MS}, and at 1-second period, S_{M1}, adjusted for *site class* effects shall be determined by Equations 16-37 and 16-38, respectively:

$$S_{MS} = F_a S_s \quad \textbf{(Equation 16-37)}$$

$$S_{M1} = F_v S_1 \quad \textbf{(Equation 16-38)}$$

where:

F_a = Site coefficient defined in Table 1613.3.3(1).

F_v = Site coefficient defined in Table 1613.3.3(2).

S_S = The mapped spectral accelerations for short periods as determined in Section 1613.3.1.

S_1 = The mapped spectral accelerations for a 1-second period as determined in Section 1613.3.1.

❖ Table 1613.3.3(1) defines an acceleration-related or short-period site coefficient, F_a, as a function of site class and the mapped spectral response acceleration at 0.2-second period, S_s. Table 1613.3.3(2) similarly defines a velocity-related or long-period site coefficient, F_v, as a function of site class and the mapped spectral response acceleration at 1-second period, S_1. The acceleration-related or short-period site coefficient F_a times S_s is S_{MS}, the 5-percent damped soil-modified maximum considered earthquake spectral response acceleration at short periods. The velocity-related or long-period site coefficient F_v times S_1 is S_{M1}, the 5-percent damped soil-modified maximum considered earthquake spectral response acceleration at 1-second period. Such modification by site coefficients is necessary because the mapped quantities are for Site Class B soils. Softer soils (Site Classes C through E) would typically amplify, and stiffer soils (Site Class A) would deamplify ground motion referenced to Site Class B.

As one would expect, both the short-period site coefficient, F_a, and the long-period site coefficient, F_v, are equal to unity for the benchmark Site Class B, irrespective of seismicity of the site. For Site Class A, both coefficients are smaller than unity, indicating reduction of benchmark Site Class B ground motion caused by the stiffer soils. For Site Classes C through E, both site coefficients, with the exception of F_a for Site Class E where $S_s \geq 1.00$, are larger than 1, indicating amplification of benchmark Site Class B ground motion on softer soils. For the same site class, each site coefficient is typically larger in areas of low seismicity than in areas of high seismicity. The basis of this lies in observations that low-magnitude subsurface rock motion is amplified to a larger extent by overlying softer soils than is high-magnitude rock motion. The site coefficients typically become larger for progressively softer soils. The only exception is provided by the short-period site coefficient in areas of high seismicity (S_s > 0.75), which remain unchanged or even decrease as the site class changes from D to E. The basis for this also lies in observations that very soft soils are not capable of amplifying the short-period components of subsurface rock motion; deamplification, in fact, takes place when the subsurface rock motion is high in magnitude.

TABLE 1613.3.3(1). See page 16-72.

❖ F_a, the acceleration-related or short-period site coefficient, is defined in this table as a function of site class and the seismicity at the site in the form of the mapped spectral response acceleration at 0.2-second period, S_s.

TABLE 1613.3.3(2). See below.

❖ F_v, the velocity-related or long-period site coefficient, is defined in this table as a function of site class and the seismicity at the site in the form of the mapped spectral response acceleration at 1-second period, S_1.

1613.3.4 Design spectral response acceleration parameters. Five-percent damped design spectral response acceleration at short periods, $S_{DS,}$ and at 1-second period, $S_{DS,}$ shall be determined from Equations 16-39 and 16-40, respectively:

$$S_{DS} = \frac{2}{3}S_{MS} \qquad \textbf{(Equation 16-39)}$$

$$S_{D1} = \frac{2}{3}S_{M1} \qquad \textbf{(Equation 16-40)}$$

where:

S_{MS}= The maximum considered earthquake spectral response accelerations for short period as determined in Section 1613.3.3.

S_{M1}= The maximum considered earthquake spectral response accelerations for 1-second period as determined in Section 1613.3.3.

❖ The design spectral response acceleration at 0.2-second period, S_{DS}, is two-thirds of the S_{MS} value calculated in accordance with Section 1613.3.3. The design spectral response acceleration at 1-second period, S_{D1}, is two-thirds of the S_{M1} value calculated in accordance with Section 1613.3.3. Two-thirds is the reciprocal of 1.5; thus, the design ground motion is 1/1.5 times the soil-modified maximum considered earthquake ground motion. This is in recognition of the inherent margin contained in the NEHRP provisions that would make collapse unlikely under 1.5 times the design-level ground motion. The idea is to avoid collapse when a structure is subjected to the soil-modified maximum considered earthquake ground motion.

1613.3.5 Determination of seismic design category. Structures classified as *Risk Category* I, II or III that are located where the mapped spectral response acceleration parameter at 1-second period, S_1, is greater than or equal to 0.75 shall be assigned to *Seismic Design Category* E. Structures classified as *Risk Category* IV that are located where the mapped spectral response acceleration parameter at 1-second period, S_1, is greater than or equal to 0.75 shall be assigned to *Seismic Design Category* F. All other structures shall be assigned to a *seismic design category* based on their *risk category* and the design spectral response acceleration parameters, S_{DS} and S_{D1}, determined in accordance with Section 1613.3.4 or the site-specific procedures of ASCE 7. Each building and structure shall be assigned to the more severe *seismic design category* in accordance with Table 1613.3.5(1) or 1613.3.5(2), irrespective of the fundamental period of vibration of the structure, *T*.

❖ The seismic design category classification provides a relative scale of earthquake risk to structures. The seismic design category considers not only the seismicity of the site in terms of the mapped spectral response accelerations, but also the site soil profile and the nature of the structure's risk category (see the step-by-step description of this process under the

TABLE 1613.3.3(1)
VALUES OF SITE COEFFICIENT F_a [a]

SITE CLASS	MAPPED SPECTRAL RESPONSE ACCELERATION AT SHORT PERIOD				
	$S_s \leq 0.25$	$S_s = 0.50$	$S_s = 0.75$	$S_s = 1.00$	$S_s \geq 1.25$
A	0.8	0.8	0.8	0.8	0.8
B	1.0	1.0	1.0	1.0	1.0
C	1.2	1.2	1.1	1.0	1.0
D	1.6	1.4	1.2	1.1	1.0
E	2.5	1.7	1.2	0.9	0.9
F	Note b	Note b	Note b	Note b	Note b

a. Use straight-line interpolation for intermediate values of mapped spectral response acceleration at short period, S_s.
b. Values shall be determined in accordance with Section 11.4.7 of ASCE 7.

TABLE 1613.3.3(2)
VALUES OF SITE COEFFICIENT F_V [a]

SITE CLASS	MAPPED SPECTRAL RESPONSE ACCELERATION AT 1-SECOND PERIOD				
	$S_1 \leq 0.1$	$S_1 = 0.2$	$S_1 = 0.3$	$S_1 = 0.4$	$S_1 \geq 0.5$
A	0.8	0.8	0.8	0.8	0.8
B	1.0	1.0	1.0	1.0	1.0
C	1.7	1.6	1.5	1.4	1.3
D	2.4	2.0	1.8	1.6	1.5
E	3.5	3.2	2.8	2.4	2.4
F	Note b	Note b	Note b	Note b	Note b

a. Use straight-line interpolation for intermediate values of mapped spectral response acceleration at 1-second period, S_1.
b. Values shall be determined in accordance with Section 11.4.7 of ASCE 7.

definition of "Seismic design category"). It is important to note that there are two tables that are used in order to establish the most restrictive seismic design category classification.

There are two conditions that allow the seismic design category to be determined based upon mapped spectral accelerations, making it unnecessary to go through the process described above. The first is Section 1613.3.1, which identifies areas that have a low seismic risk and are classified as Seismic Design Category A without the need to go through the usual steps. The second instance is an area that is close to a major active fault where S_1 is greater than or equal to 0.75g. These areas are deemed to be areas of considerable seismic risk, and the seismic design category is classified as E or F depending on the structure's risk category.

The seismic design category classification is a key criterion in using and understanding the seismic requirements because the analysis method, general design, structural material detailing and the structure's component and system design requirements are determined, at least in part, by the seismic design category. Some of the special inspection requirements in Section 1705.11 and structural observation requirements in Section 1704.5 are dependent on the seismic design category classification, as well.

TABLE 1613.3.5(1). See below.

❖ This table defines seismic design category as a function of risk category and the short-period (0.2 second) spectral response acceleration at the site of a structure. As the value of S_{DS} increases, the structure is assigned a higher seismic design category and the earthquake design requirements become more stringent.

TABLE 1613.3.5(2). See below.

❖ This table defines seismic design category as a function of risk category and the long-period (1-second) spectral response acceleration at the site of a structure. As the value of S_{M1} increases, the structure is assigned a higher seismic design category and the earthquake design requirements become more stringent.

1613.3.5.1 Alternative seismic design category determination. Where S_1 is less than 0.75, the *seismic design category* is permitted to be determined from Table 1613.3.5(1) alone when all of the following apply:

1. In each of the two orthogonal directions, the approximate fundamental period of the structure, Ta, in each of the two orthogonal directions determined in accordance with Section 12.8.2.1 of ASCE 7, is less than 0.8 T_s determined in accordance with Section 11.4.5 of ASCE 7.
2. In each of the two orthogonal directions, the fundamental period of the structure used to calculate the story drift is less than T_s.
3. Equation 12.8-2 of ASCE 7 is used to determine the seismic response coefficient, C_s.
4. The diaphragms are rigid as defined in Section 12.3.1 of ASCE 7 or, for diaphragms that are flexible, the distances between vertical elements of the seismic force-resisting system do not exceed 40 feet (12 192 mm).

❖ This section permits the seismic design category of structures meeting the four listed conditions to be based solely on the short-period design spectral coefficient, S_{DS}. This is similar to the approach taken in mapping seismic design categories for use in the IRC, as well as for the simplified earthquake analysis procedure.

TABLE 1613.3.5(1)
SEISMIC DESIGN CATEGORY BASED ON SHORT-PERIOD (0.2 second) RESPONSE ACCELERATIONS

VALUE OF S_{DS}	RISK CATEGORY		
	I or II	III	IV
$S_{DS} < 0.167g$	A	A	A
$0.167g \le S_{DS} < 0.33g$	B	B	C
$0.33g \le S_{DS} < 0.50g$	C	C	D
$0.50g \le S_{DS}$	D	D	D

TABLE 1613.3.5(2)
SEISMIC DESIGN CATEGORY BASED ON 1-SECOND PERIOD RESPONSE ACCELERATION

VALUE OF S_{D1}	RISK CATEGORY		
	I or II	III	IV
$S_{D1} < 0.067g$	A	A	A
$0.067g \le S_{D1} < 0.133g$	B	B	C
$0.133g \le S_{D1} < 0.20g$	C	C	D
$0.20g \le S_{D1}$	D	D	D

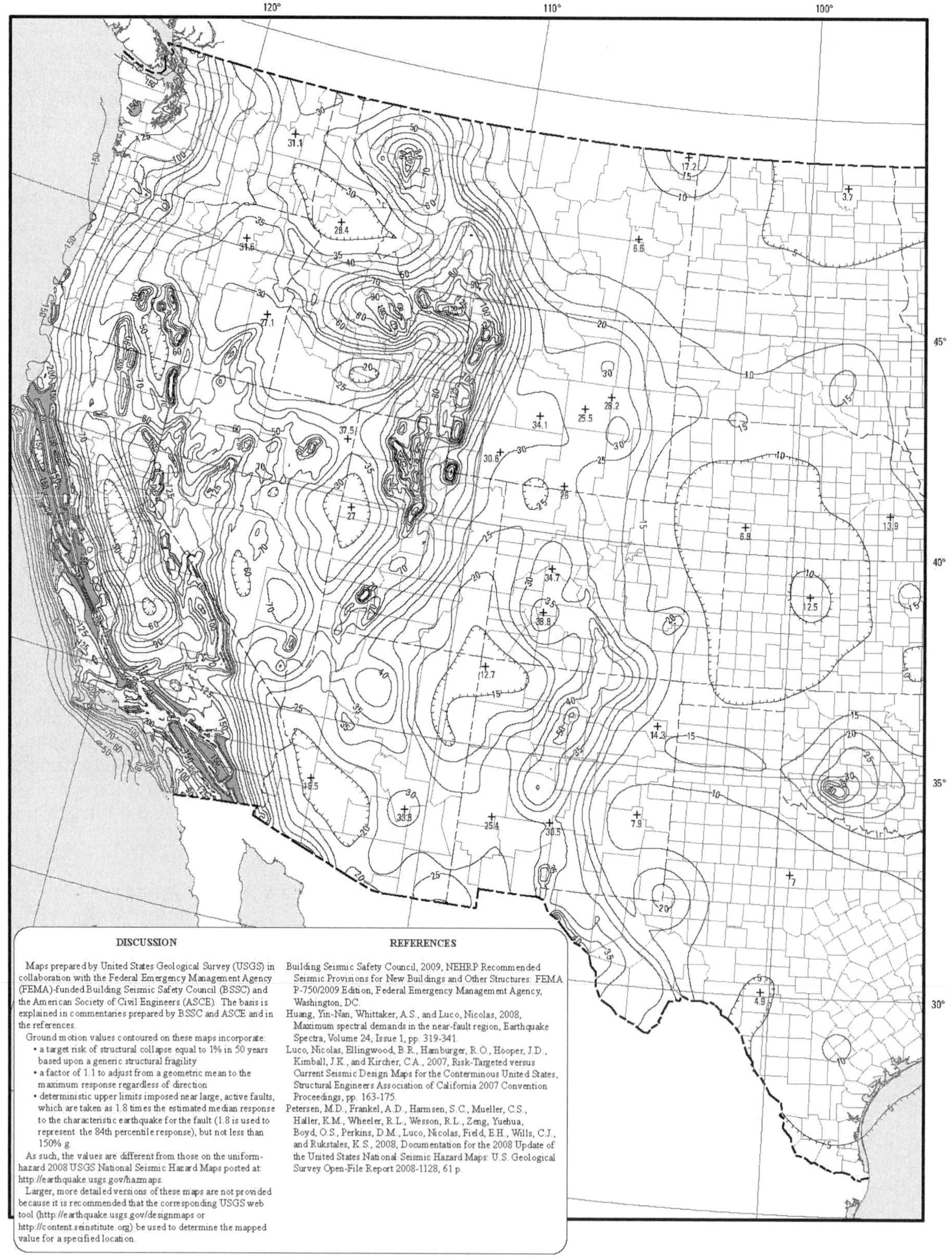

FIGURE 1613.3.1(1)
RISK-TARGETED MAXIMUM CONSIDERED EARTHQUAKE (MCE_R) GROUND MOTION RESPONSE ACCELERATIONS FOR THE CONTERMINOUS UNITED STATES OF 0.2-SECOND SPECTRAL RESPONSE ACCELERATION (5% OF CRITICAL DAMPING), SITE CLASS B

(continued)

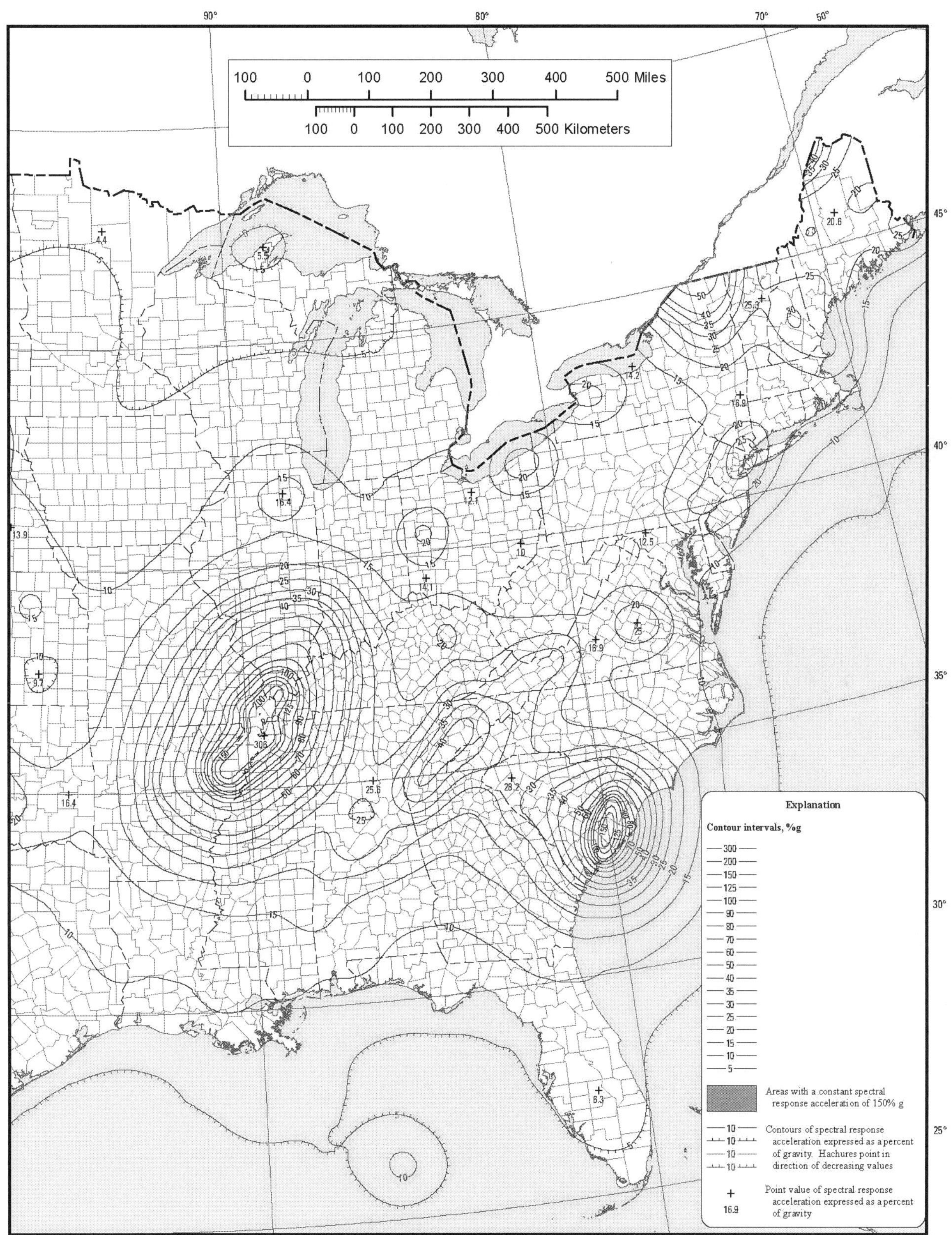

FIGURE 1613.3.1(1)—continued
RISK-TARGETED MAXIMUM CONSIDERED EARTHQUAKE (MCE_R) GROUND MOTION RESPONSE ACCELERATIONS FOR THE CONTERMINOUS UNITED STATES OF 0.2-SECOND SPECTRAL RESPONSE ACCELERATION (5% OF CRITICAL DAMPING), SITE CLASS B

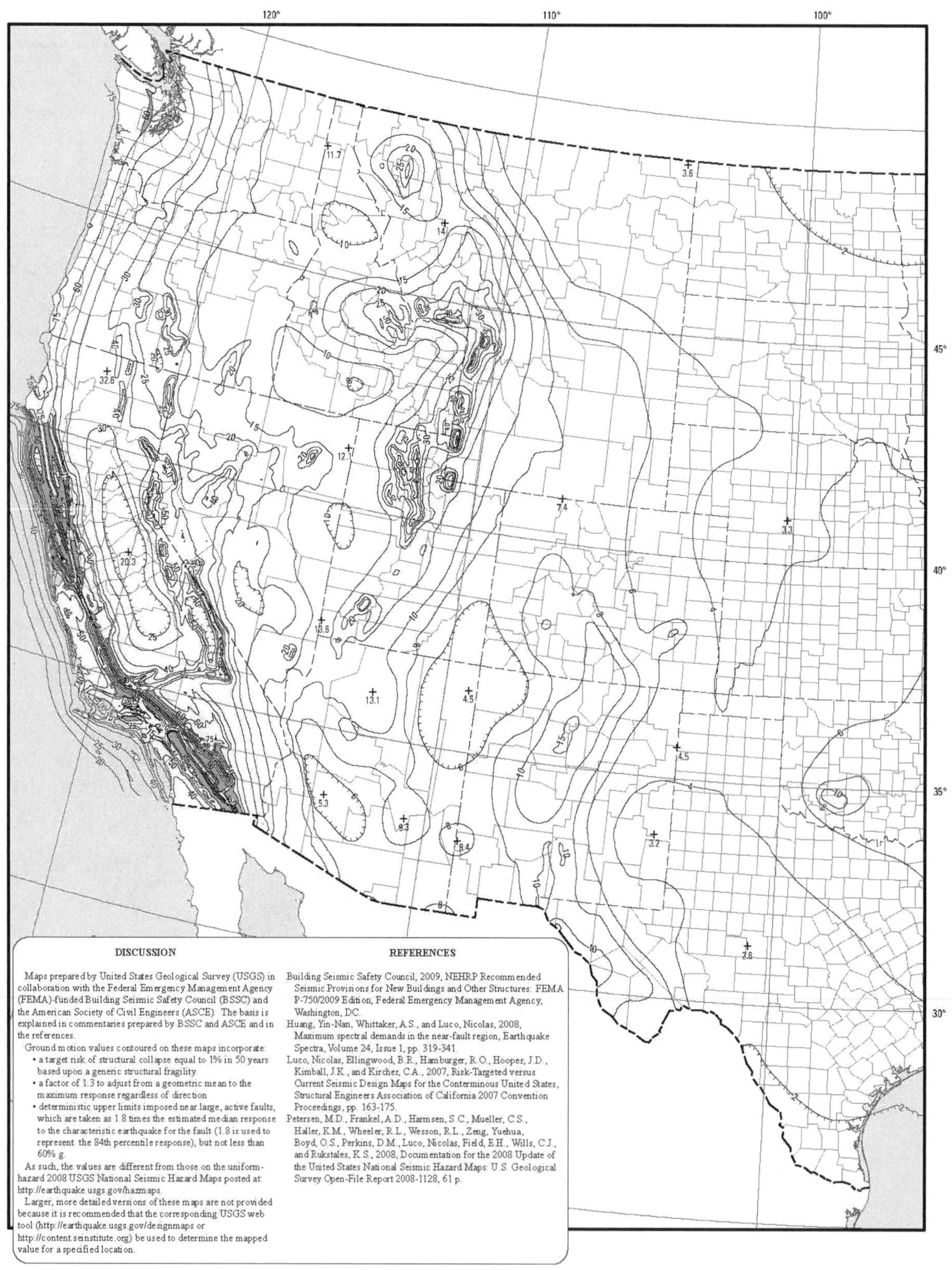

FIGURE 1613.3.1(2)
RISK-TARGETED MAXIMUM CONSIDERED EARTHQUAKE (MCE_R) GROUND MOTION RESPONSE ACCELERATIONS FOR THE CONTERMINOUS UNITED STATES OF 1-SECOND SPECTRAL RESPONSE ACCELERATION (5% OF CRITICAL DAMPING), SITE CLASS B

(continued)

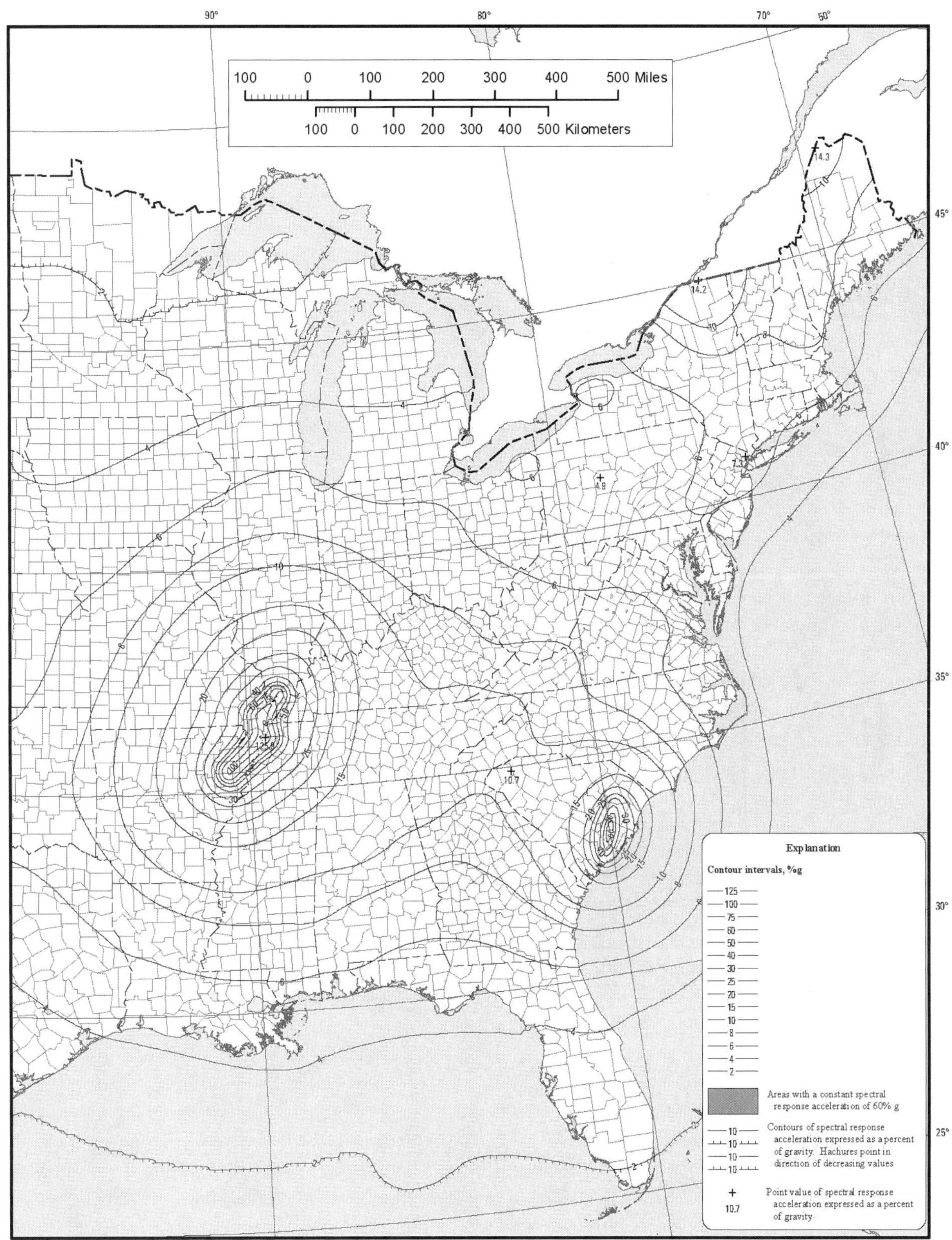

FIGURE 1613.3.1(2)—continued
RISK-TARGETED MAXIMUM CONSIDERED EARTHQUAKE (MCE_R) GROUND MOTION RESPONSE ACCELERATIONS FOR THE CONTERMINOUS UNITED STATES OF 1-SECOND SPECTRAL RESPONSE ACCELERATION (5% OF CRITICAL DAMPING), SITE CLASS B

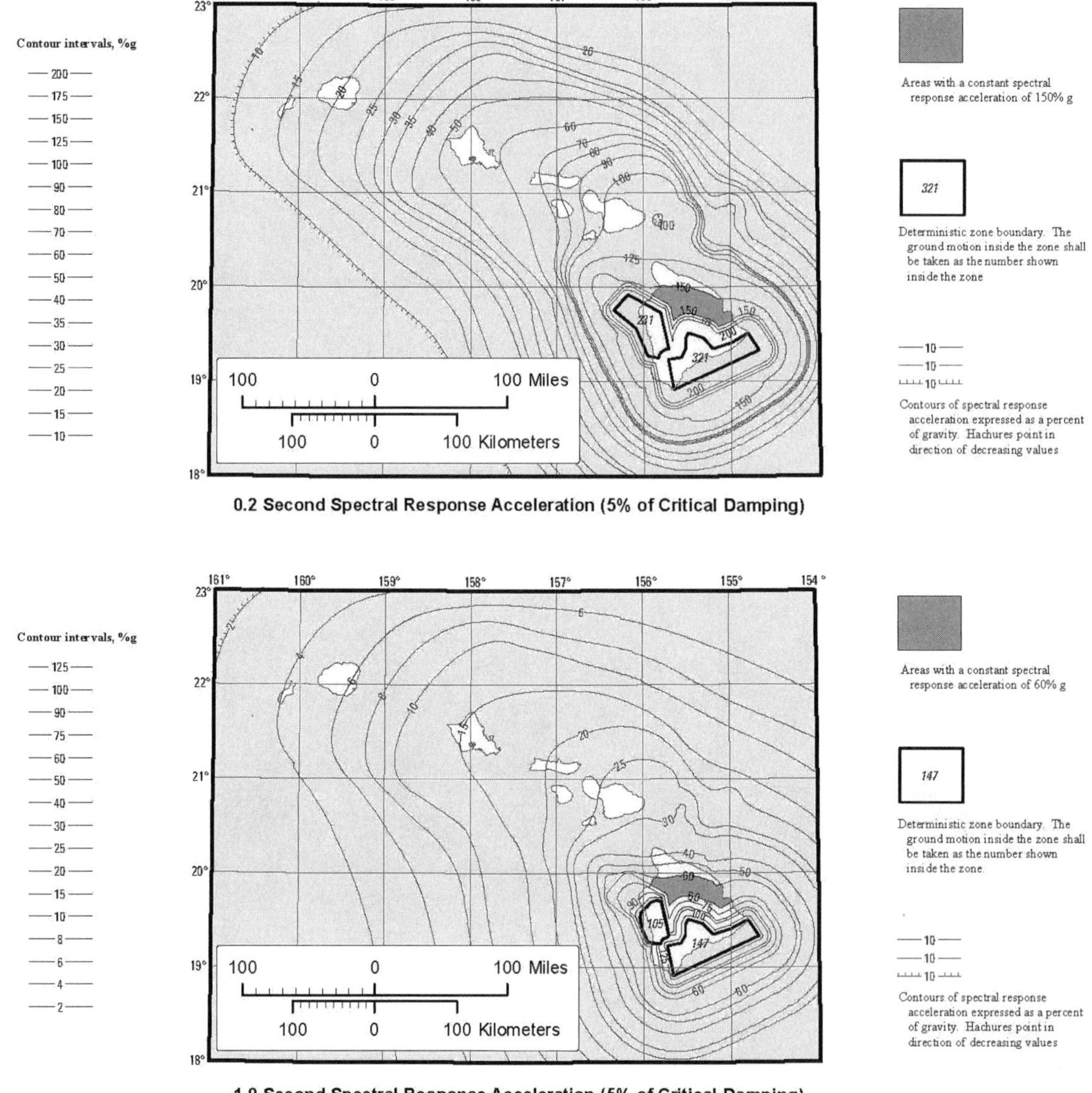

DISCUSSION

Maps prepared by United States Geological Survey (USGS) in collaboration with the Federal Emergency Management Agency (FEMA)-funded Building Seismic Safety Council (BSSC) and the American Society of Civil Engineers (ASCE). The basis is explained in commentaries prepared by BSSC and ASCE and in the references.

Ground motion values contoured on these maps incorporate:

- a target risk of structural collapse equal to 1% in 50 years based upon a generic structural fragility
- deterministic upper limits imposed near large, active faults, which are taken as 1.8 times the estimated median response to the characteristic earthquake for the fault (1.8 is used to represent the 84th percentile response), but not less than 150% and 60% g for 0.2 and 1.0 sec, respectively.

As such, the values are different from those on the uniform-hazard 1998 USGS National Seismic Hazard Maps for Hawaii posted at http://earthquake.usgs.gov/hazmaps.

Larger, more detailed versions of these maps are not provided because it is recommended that the corresponding USGS web tool (http://earthquake.usgs.gov/designmaps or http://content.seinstitute.org) be used to determine the mapped value for a specified location.

REFERENCES

Building Seismic Safety Council, 2009, NEHRP Recommended Seismic Provisions for New Buildings and Other Structures: FEMA P-750/2009 Edition, Federal Emergency Management Agency, Washington, DC.

Huang, Yin-Nan, Whittaker, A.S., and Luco, Nicolas, 2008, Maximum spectral demands in the near-fault region, Earthquake Spectra, Volume 24, Issue 1, pp. 319-341.

Klein, F., Frankel, A.D., Mueller, C.S., Wesson, R.L., and Okubo, P., 2001, Seismic hazard in Hawaii: high rate of large earthquakes and probabilistic ground-motion maps, Bulletin of the Seismological Society of America, Volume 91, pp. 479-498.

Luco, Nicolas, Ellingwood, B.R., Hamburger, R.O., Hooper, J.D., Kimball, J.K., and Kircher, C.A., 2007, Risk-Targeted versus Current Seismic Design Maps for the Conterminous United States, Structural Engineers Association of California 2007 Convention Proceedings, pp. 163-175.

FIGURE 1613.3.1(3)
RISK-TARGETED MAXIMUM CONSIDERED EARTHQUAKE (MCE_R) GROUND MOTION RESPONSE ACCELERATIONS FOR HAWAII OF 0.2- AND 1-SECOND SPECTRAL RESPONSE ACCELERATION (5% OF CRITICAL DAMPING), SITE CLASS B

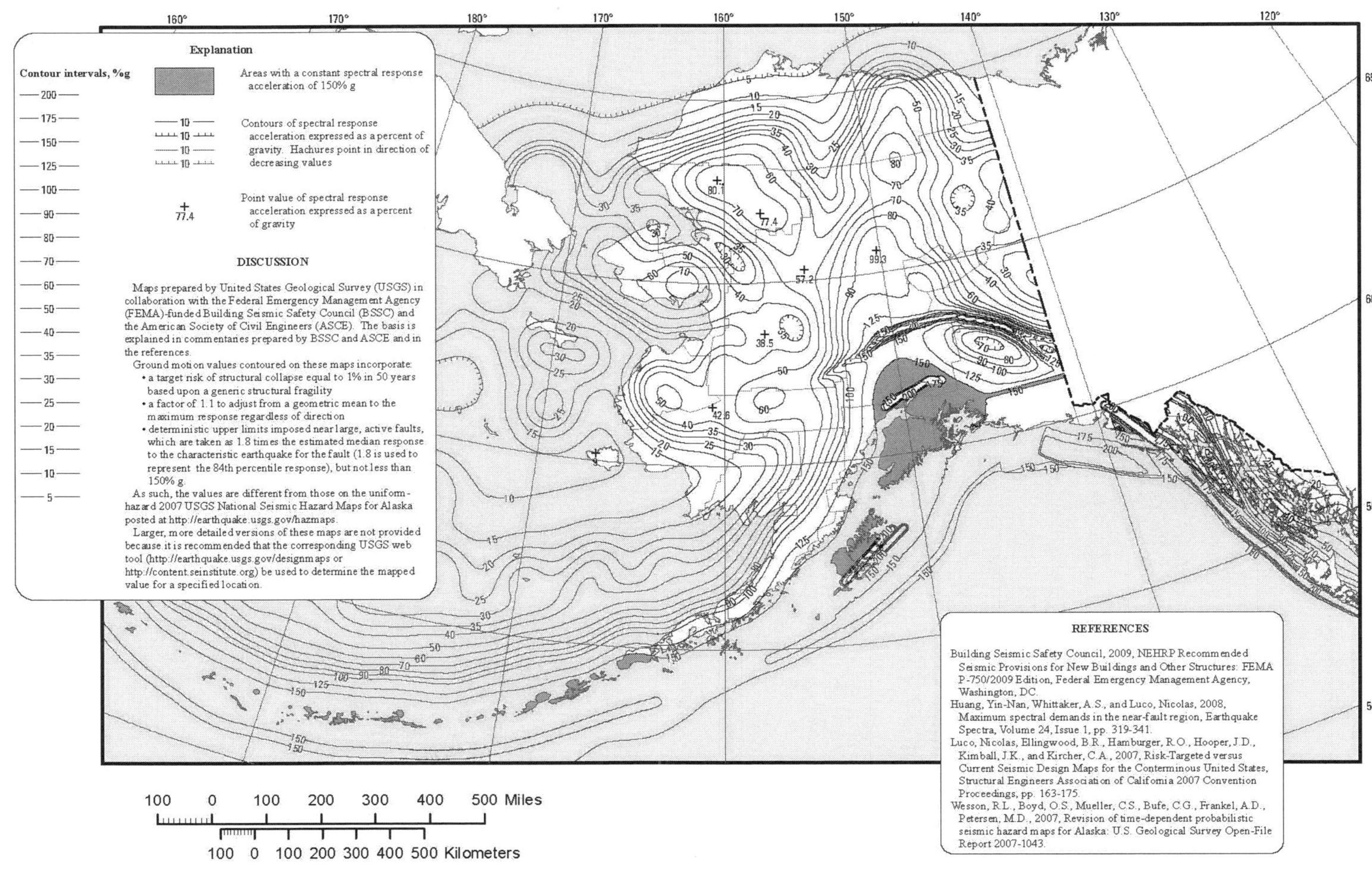

FIGURE 1613.3.1(4)
RISK-TARGETED MAXIMUM CONSIDERED EARTHQUAKE (MCE_R) GROUND MOTION RESPONSE ACCELERATIONS FOR ALASKA OF 0.2-SECOND SPECTRAL RESPONSE ACCELERATION (5% OF CRITICAL DAMPING), SITE CLASS B

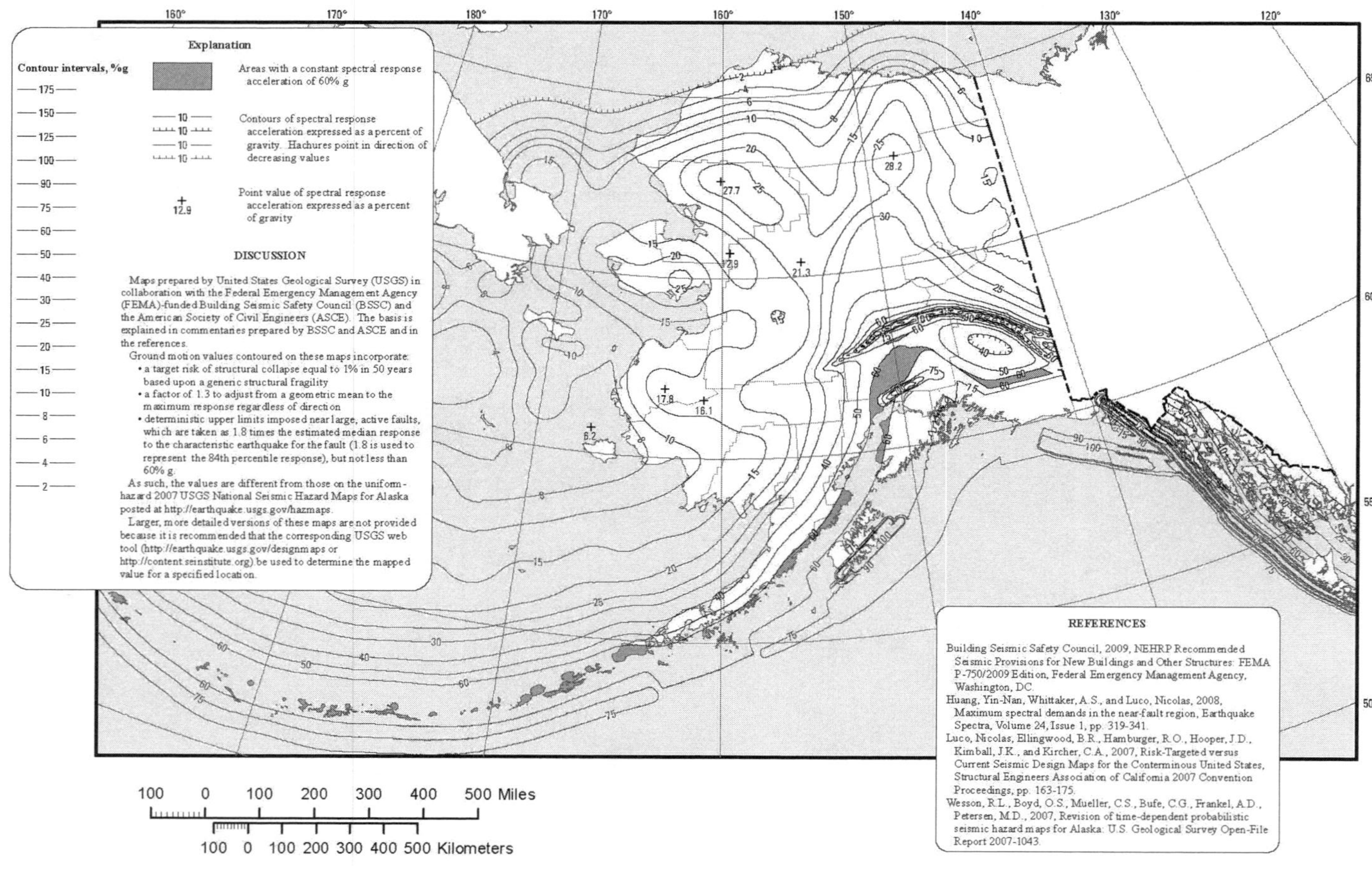

FIGURE 1613.3.1(5)
RISK-TARGETED MAXIMUM CONSIDERED EARTHQUAKE (MCE_R) GROUND MOTION RESPONSE ACCELERATIONS FOR ALASKA OF 1.0-SECOND SPECTRAL RESPONSE ACCELERATION (5% OF CRITICAL DAMPING), SITE CLASS B

Contour intervals, %g

200, 175, 150, 125, 100, 90, 80, 70, 60

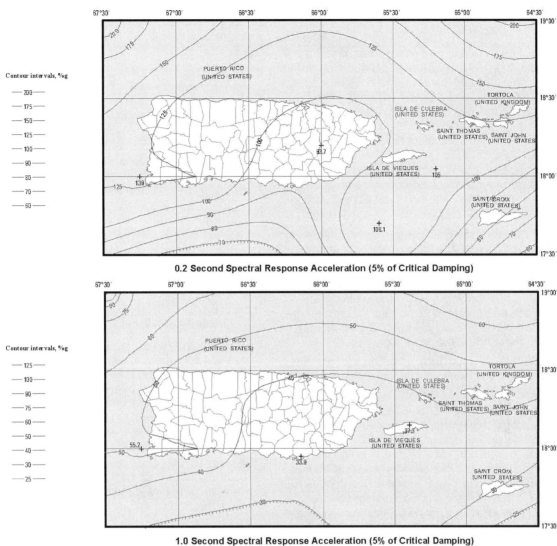

0.2 Second Spectral Response Acceleration (5% of Critical Damping)

Contour intervals, %g

125, 100, 90, 75, 60, 50, 40, 30, 25

1.0 Second Spectral Response Acceleration (5% of Critical Damping)

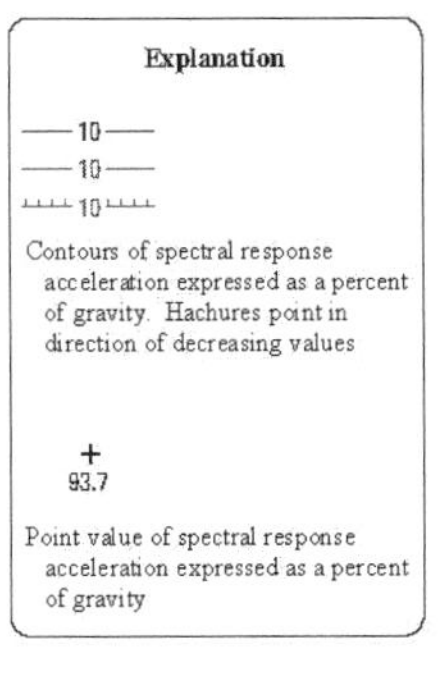

DISCUSSION

Maps prepared by United States Geological Survey (USGS) in collaboration with the Federal Emergency Management Agency (FEMA)-funded Building Seismic Safety Council (BSSC) and the American Society of Civil Engineers (ASCE). The basis is explained in commentaries prepared by BSSC and ASCE and in the references.

Ground motion values contoured on these maps incorporate:

- a target risk of structural collapse equal to 1% in 50 years based upon a generic structural fragility
- a factor of 1.1 and 1.3 for 0.2 and 1.0 sec, respectively, to adjust from a geometric mean to the maximum response regardless of direction
- deterministic upper limits imposed near large, active faults, which are taken as 1.8 times the estimated median response to the characteristic earthquake for the fault (1.8 is used to represent the 84th percentile response), but not less than 150% and 60% g for 0.2 and 1.0 sec, respectively.

As such, the values are different from those on the uniform-hazard 2003 USGS National Seismic Hazard Maps for Puerto Rico and the U.S. Virgin Islands posted at http://earthquake.usgs.gov/hazmaps.

Larger, more detailed versions of these maps are not provided because it is recommended that the corresponding USGS web tool (http://earthquake.usgs.gov/designmaps or http://content.seinstitute.org) be used to determine the mapped value for a specified location.

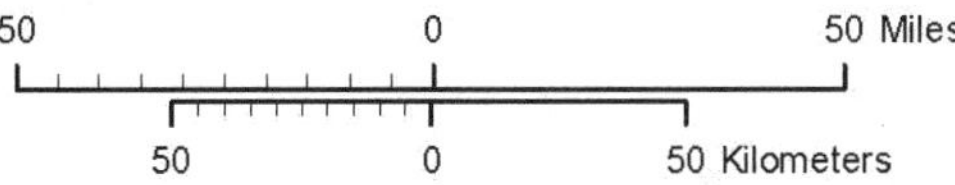

REFERENCES

Building Seismic Safety Council, 2009, NEHRP Recommended Seismic Provisions for New Buildings and Other Structures: FEMA P-750/2009 Edition, Federal Emergency Management Agency, Washington, DC.

Huang, Yin-Nan, Whittaker, A.S., and Luco, Nicolas, 2008, Maximum spectral demands in the near-fault region, Earthquake Spectra, Volume 24, Issue 1, pp. 319-341.

Luco, Nicolas, Ellingwood, B.R., Hamburger, R.O., Hooper, J.D., Kimball, J.K., and Kircher, C.A., 2007, Risk-Targeted versus Current Seismic Design Maps for the Conterminous United States, Structural Engineers Association of California 2007 Convention Proceedings, pp. 163-175.

Mueller, C.S., Frankel, A.D., Petersen, M.D., and Leyendecker, E.V., 2003, Documentation for the 2003 USGS Seismic Hazard Maps for Puerto Rico and the U.S. Virgin Islands: U.S. Geological Survey Open-File Report 03-379.

FIGURE 1613.3.1(6)
RISK-TARGETED MAXIMUM CONSIDERED EARTHQUAKE (MCE_R) GROUND MOTION RESPONSE ACCELERATIONS FOR PUERTO RICO AND THE UNITED STATES VIRGIN ISLANDS OF 0.2- AND 1-SECOND SPECTRAL RESPONSE ACCELERATION (5% OF CRITICAL DAMPING), SITE CLASS B

1613.3.5.2 Simplified design procedure. Where the alternate simplified design procedure of ASCE 7 is used, the *seismic design category* shall be determined in accordance with ASCE 7.

❖ A simplified earthquake procedure was first introduced under the 1997 UBC and continued on with minor changes through the first two editions of the code. The 2005 edition of ASCE 7 introduced the first stand- alone version of a simplified earthquake analysis for simple bearing wall or building frame systems that comply with the 12 limitations listed in Section 12.14.1.1 of ASCE 7. For structures that qualify to use this method of analysis, the simplified procedure permits the determination of the seismic design category to be based solely on the short-period design spectral coefficient, S_{DS}.

1613.4 Alternatives to ASCE 7. The provisions of Section 1613.4 shall be permitted as alternatives to the relevant provisions of ASCE 7.

❖ This section is intended to provide permitted alternatives to the corresponding ASCE 7 requirements. As such, they should be viewed as options that are available to the design professional in demonstrating compliance with the earthquake provisions.

1613.4.1 Additional seismic force-resisting systems for seismically isolated structures. Add the following exception to the end of Section 17.5.4.2 of ASCE 7:

> **Exception:** For isolated structures designed in accordance with this standard, the Structural System Limitations and the Building Height Limitations in Table 12.2-1 for ordinary steel concentrically braced frames (OCBFs) as defined in Chapter 11 and ordinary moment frames (OMFs) as defined in Chapter 11 are permitted to be taken as 160 feet (48 768 mm) for structures assigned to *Seismic Design Category* D, E or F, provided that the following conditions are satisfied:
>
> 1. The value of R_I as defined in Chapter 17 is taken as 1.
> 2. For OMFs and OCBFs, design is in accordance with AISC 341.

❖ The ASCE 7 provisions include requirements for structures that utilize seismic base isolation. It is worth noting that use of seismic isolation is never required by the code. This method is an optional method of design for earthquake loading that has been recognized under legacy model codes, such as the UBC. The modification adds an exception to the ASCE 7 requirements for structural systems above the isolation system. It corrects an oversight that unnecessarily changes prior practice by restricting the use of certain seismic-force-resisting systems that have been used successfully. As modified, it would allow ordinary concentric braced frames and ordinary moment frame systems with heights up to 160 feet (48 768 mm) when the stated conditions are met.

SECTION 1614
ATMOSPHERIC ICE LOADS

1614.1 General. Ice-sensitive structures shall be designed for atmospheric ice loads in accordance with Chapter 10 of ASCE 7.

❖ This section provides charging text in the code in conjunction with technical provisions in ASCE 7 for computing atmospheric ice loads. This section relies on the determination of which structures are ice-sensitive structures in order to determine the need to comply with the applicable provisions of ASCE 7. An "Ice-sensitive structure" is defined in Section 202 and provides a technical basis for determining which structures are ice-sensitive structures.

SECTION 1615
STRUCTURAL INTEGRITY

1615.1 General. High-rise buildings that are assigned to *Risk Category* III or IV shall comply with the requirements of this section. Frame structures shall comply with the requirements of Section 1615.3. Bearing wall structures shall comply with the requirements of Section 1615.4.

❖ The requirements in the codes and standards together with the common structural design and construction practices prevalent in the United States have provided the overwhelming majority of structures with adequate levels of reliability and safety. Many building simply do not have integrity issues. Low-rise buildings do not represent the same risk as taller, high-rise buildings. The structural integrity provisions reflect this relative risk. By limiting this requirement to Risk Category III and IV buildings that are also high-rise buildings, the provision does not affect buildings that are commonly built and do not exhibit integrity issues.

1615.2 Definitions. The following words and terms are defined in Chapter 2:

❖ Definitions facilitate the understanding of code provisions and minimize potential confusion. To that end, this section lists definitions of terms associated with these provisions for structural integrity. Note that these definitions are found in Chapter 2. The use and application of defined terms, as well as undefined terms, are set forth in Section 201.

BEARING WALL STRUCTURE.

FRAME STRUCTURE.

1615.3 Frame structures. Frame structures shall comply with the requirements of this section.

❖ These provisions enhance the overall structural integrity and resistance of frame structures by establishing minimum requirements for tying together the primary structural elements.

1615.3.1 Concrete frame structures. Frame structures constructed primarily of reinforced or prestressed concrete, either cast-in-place or precast, or a combination of these, shall conform to the requirements of ACI 318 Sections 7.13, 13.3.8.5, 13.3.8.6, 16.5, 18.12.6, 18.12.7 and 18.12.8 as applicable. Where ACI 318 requires that nonprestressed reinforcing or prestressing steel pass through the region bounded by the longitudinal column reinforcement, that reinforcing or prestressing steel shall have a minimum nominal tensile strength equal to two-thirds of the required one-way vertical strength of the connection of the floor or roof system to the column in each direction of beam or slab reinforcement passing through the column.

Exception: Where concrete slabs with continuous reinforcement having an area not less than 0.0015 times the concrete area in each of two orthogonal directions are present and are either monolithic with or equivalently bonded to beams, girders or columns, the longitudinal reinforcing or prestressing steel passing through the column reinforcement shall have a nominal tensile strength of one-third of the required one-way vertical strength of the connection of the floor or roof system to the column in each direction of beam or slab reinforcement passing through the column.

❖ This section refers to ACI 318 requirements for structural integrity that are applicable to concrete frames. These provisions establish minimum requirements for tying together the primary structural elements. These ACI 318 structural integrity requirements are already incorporated in the code by the general reference to ACI 318 in Section 1901.2 for the design and construction of structural concrete. The only additional requirement for concrete frame structures is the minimum nominal tensile strength that is established for reinforcement that must pass through the longitudinal column reinforcement (see Section 7.13.2.5 of ACI 318). The exception reduces this minimum nominal tensile strength for slab construction that meets certain conditions.

1615.3.2 Structural steel, open web steel joist or joist girder, or composite steel and concrete frame structures. Frame structures constructed with a structural steel frame or a frame composed of open web steel joists, joist girders with or without other structural steel elements or a frame composed of composite steel or composite steel joists and reinforced concrete elements shall conform to the requirements of this section.

❖ The provisions contained in this section have been adapted from requirements contained in the ACI 318 standard for many years. By adapting those requirements to structural steel, open web steel joist or joist girder, or composite steel and concrete frame structures, their overall structural integrity is enhanced.

1615.3.2.1 Columns. Each column splice shall have the minimum design strength in tension to transfer the design dead and live load tributary to the column between the splice and the splice or base immediately below.

❖ The additional requirement for the tensile strength of column splices enhances the column's performance in unforeseen events.

1615.3.2.2 Beams. End connections of all beams and girders shall have a minimum nominal axial tensile strength equal to the required vertical shear strength for *allowable stress design* (ASD) or two-thirds of the required shear strength for *load and resistance factor design* (LRFD) but not less than 10 kips (45 kN). For the purpose of this section, the shear force and the axial tensile force need not be considered to act simultaneously.

Exception: Where beams, girders, open web joist and joist girders support a concrete slab or concrete slab on metal deck that is attached to the beam or girder with not less than $^{3}/_{8}$-inch-diameter (9.5 mm) headed shear studs, at a spacing of not more than 12 inches (305 mm) on center, averaged over the length of the member, or other attachment having equivalent shear strength, and the slab contains continuous distributed reinforcement in each of two orthogonal directions with an area not less than 0.0015 times the concrete area, the nominal axial tension strength of the end connection shall be permitted to be taken as half the required vertical shear strength for ASD or one-third of the required shear strength for LRFD, but not less than 10 kips (45 kN).

❖ Providing the required tensile strength for all beam and girder connections provides some ability to carry, transfer and/or redistribute load in the event there is loss of support. The exception allows a reduced tensile strength in the beam and girder connection where a concrete slab is utilized.

1615.4 Bearing wall structures. Bearing wall structures shall have vertical ties in all load-bearing walls and longitudinal ties, transverse ties and perimeter ties at each floor level in accordance with this section and as shown in Figure 1615.4.

❖ These provisions enhance the overall structural integrity and resistance of bearing wall structures by establishing minimum requirements for tying together the primary structural elements.

FIGURE 1615.4. See page 16-84.

❖ This figure illustrates the ties that are required in Sections 1614.4.1 through 1614.4.2.4.

1615.4.1 Concrete wall structures. Precast bearing wall structures constructed solely of reinforced or prestressed concrete, or combinations of these shall conform to the requirements of Sections 7.13, 13.3.8.5 and 16.5 of ACI 318.

❖ This section refers to the ACI 318 requirements that are applicable to concrete wall structures. These provisions establish minimum requirements for tying together the primary structural elements. These ACI 318 structural integrity requirements are already incorporated in the code by the general reference to ACI 318 in Section 1901.2 for the design and construction of structural concrete.

1615.4.2 Other bearing wall structures. Ties in bearing wall structures other than those covered in Section 1615.4.1 shall conform to this section.

❖ The provisions contained in this section have been adapted from requirements contained within the ACI

318 standard for many years. By adapting those requirements to other bearing wall structures, these provisions enhance their overall structural integrity.

1615.4.2.1 Longitudinal ties. Longitudinal ties shall consist of continuous reinforcement in slabs; continuous or spliced decks or sheathing; continuous or spliced members framing to, within or across walls; or connections of continuous framing members to walls. Longitudinal ties shall extend across interior load-bearing walls and shall connect to exterior load-bearing walls and shall be spaced at not greater than 10 feet (3038 mm) on center. Ties shall have a minimum nominal tensile strength, T_T, given by Equation 16-41. For ASD the minimum nominal tensile strength shall be permitted to be taken as 1.5 times the allowable tensile stress times the area of the tie.

$$T_T = w\,LS \leq \alpha_T S \qquad \textbf{(Equation 16-41)}$$

where:

L = The span of the horizontal element in the direction of the tie, between bearing walls, feet (m).

w = The weight per unit area of the floor or roof in the span being tied to or across the wall, psf (N/m^2).

S = The spacing between ties, feet (m).

α_T = A coefficient with a value of 1,500 pounds per foot (2.25 kN/m) for masonry bearing wall structures and a value of 375 pounds per foot (0.6 kN/m) for structures with bearing walls of cold-formed steel light-frame construction.

❖ Requirements for tying together horizontal elements in the longitudinal direction (spanning between bearing walls—noted as "L" in Figure 1615.4) provides for transfer and/or redistribution of loads in the event there is loss of support.

1615.4.2.2 Transverse ties. Transverse ties shall consist of continuous reinforcement in slabs; continuous or spliced decks or sheathing; continuous or spliced members framing to, within or across walls; or connections of continuous framing members to walls. Transverse ties shall be placed no farther apart than the spacing of load-bearing walls. Transverse ties shall have minimum nominal tensile strength T_T, given by Equation 16-46. For ASD the minimum nominal tensile strength shall be permitted to be taken as 1.5 times the allowable tensile stress times the area of the tie.

❖ Requirements for ties in the transverse direction at bearing walls (noted as "T" in Figure 1615.4) provides for transfer and/or redistribution of loads in the event there is loss of support.

1615.4.2.3 Perimeter ties. Perimeter ties shall consist of continuous reinforcement in slabs; continuous or spliced decks or sheathing; continuous or spliced members framing to, within

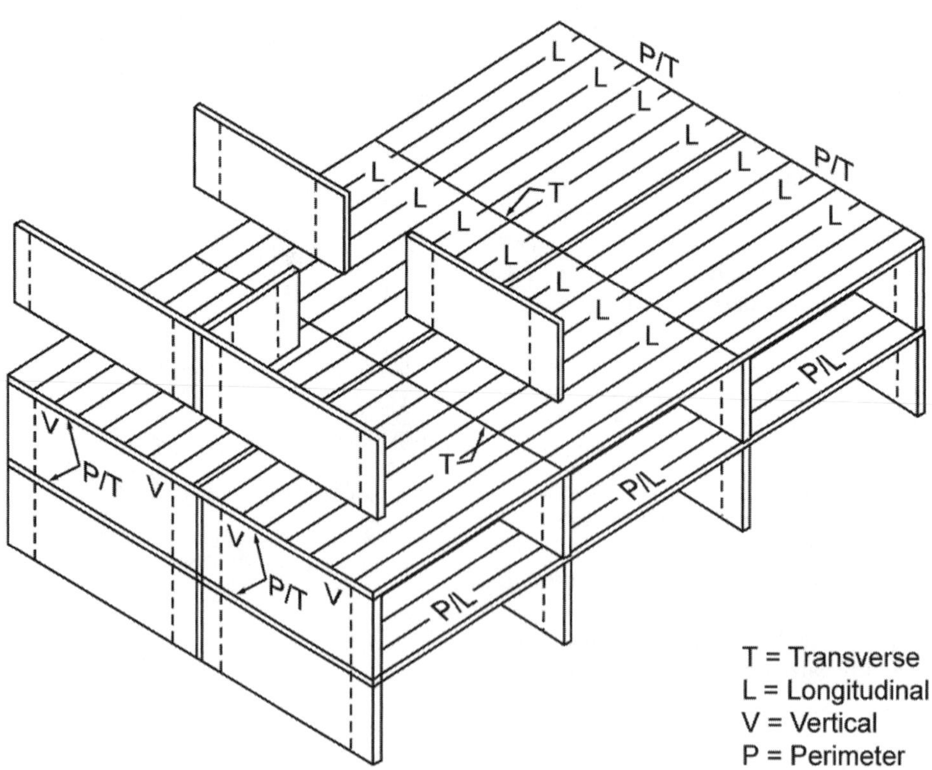

FIGURE 1615.4
LONGITUDINAL, PERIMETER, TRANSVERSE AND VERTICAL TIES

or across walls; or connections of continuous framing members to walls. Ties around the perimeter of each floor and roof shall be located within 4 feet (1219 mm) of the edge and shall provide a nominal strength in tension not less than T_p, given by Equation 16-42. For ASD the minimum nominal tensile strength shall be permitted to be taken as 1.5 times the allowable tensile stress times the area of the tie.

$T_p = 200w \leq \beta_T$ **(Equation 16-42)**

For SI: $T_p = 90.7w \leq \beta_T$

where:

w = As defined in Section 1615.4.2.1.

β_T = A coefficient with a value of 16,000 pounds (7200 kN) for structures with masonry bearing walls and a value of 4,000 pounds (1300 kN) for structures with bearing walls of cold-formed steel light-frame construction.

❖ Requirements for tying together horizontal elements at the perimeter in both the longitudinal and transverse directions (noted as "P/L" and "P/T" in Figure 1615.4) provides for transfer and/or redistribution of load in the event there is loss of support.

1615.4.2.4 Vertical ties. Vertical ties shall consist of continuous or spliced reinforcing, continuous or spliced members, wall sheathing or other engineered systems. Vertical tension ties shall be provided in bearing walls and shall be continuous over the height of the building. The minimum nominal tensile strength for vertical ties within a bearing wall shall be equal to the weight of the wall within that *story* plus the weight of the diaphragm tributary to the wall in the *story* below. No fewer than two ties shall be provided for each wall. The strength of each tie need not exceed 3,000 pounds per foot (450 kN/m) of wall tributary to the tie for walls of masonry construction or 750 pounds per foot (140 kN/m) of wall tributary to the tie for walls of cold-formed steel light-frame construction.

❖ The additional requirement for continuous vertical ties enhances the performance of bearing walls in unforeseen events (noted as "V" in Figure 1615.4).

Bibliography

The following resource materials are referenced in this chapter or are relevant to the subject matter addressed in this chapter.

AA ADM 1-10, *Aluminum Design Manual*. Washington, DC: Aluminum Association, 2010.

ACI 318-11, *Building Code Requirements for Structural Concrete*. Farmington Hills, MI: American Concrete Institute, 2011.

ADA, *Accessibility Guidelines for Building and Facilities*. 28 CFR, Part 36 revised, July 1, 1994.

AF&PA WFCM-12, *Wood Frame Construction Manual for One- and Two-family Dwellings*. Washington, DC: American Forest and Paper Association, 2012.

AISC 341-10, *Seismic Provisions for Structural Steel Buildings, including Supplement No. 1*. Chicago: American Institute of Steel Construction, Inc., 2010.

AISC 360-10, *Specification for Structural Steel Buildings*. Chicago, IL: American Institute of Steel Construction, Inc., 2010.

AISI 100-07, *North American Specification for Design of Cold-formed Steel Structural Members*. Washington, DC: American Iron and Steel Institute, 2007.

Algermissen, S.T. *An Introduction to the Seismicity of the United States*. Oakland, CA: Earthquake Engineering Research Institute, 1983.

Ambrose, James and Dimitry Vergun. *Design for Lateral Forces*. New York: John A. Wiley & Sons, 1987.

ASCE 7-05, *Minimum Design Loads for Buildings and Other Structures*. New York: American Society of Civil Engineers, 2005.

ASCE 7-10, *Minimum Design Loads for Buildings and Other Structures*. New York: American Society of Civil Engineers, 2010.

ASCE 7-88, *Minimum Design Loads for Buildings and Other Structures*. Reston, VA: American Society of Civil Engineers, 1988.

ASCE 7-95, *Minimum Design Loads for Buildings and Other Structures*. New York: American Society of Civil Engineers, 1995.

ASCE 8-02, *Standard Specification for Design of Cold-formed Stainless Steel Structural Members*. Reston, VA: American Society of Civil Engineers, 2002.

ASCE 24-05, *Flood-resistance Design and Construction Standard*. Reston, VA: American Society of Civil Engineers, 2005.

ASME A17.1-07, *Safety Code for Elevators and Escalators*. New York: American Society of Mechanical Engineers, 2007.

ASME B31.4-95, *Boilers and Pressure Vessels Code*. New York: American Society of Mechanical Engineers, 1995.

ASTM D 1586-08a, *Specifications for Penetration Test and Split-barrel Sampling of Soils*. West Conshohocken, PA: ASTM International, 2008.

ASTM D 2166-06, *Test Method for Unconfined Compressive Strength of Cohesive Soil*. West Conshohocken, PA: ASTM International, 2006.

ASTM D 2216-05, *Test Method for Laboratory Determination of Water (Moisture) Content of Soil and Rock by Mass*. West Conshohocken, PA: ASTM International, 2005.

ASTM D 2487-06e1, *Standard Classification of Soils for Engineering Purposes Unified Soil Classification*

System. West Conshohocken, PA: ASTM International, 2006.

ASTM D 2850-03a (2007), *Test Method for Unconsolidated, Undrained Triaxial Compression Test on Cohesive Soil*. West Conshohocken, PA: ASTM International, 2007.

ASTM D 4318-05, *Test Methods for Liquid Limit, Plastic Limit and Plasticity Index of Soils*. West Conshohocken, PA: ASTM International, 2005.

ASTM E 1886-05, *Test Method for Performance of Exterior Windows, Curtain Walls, Doors and Storm Shutters Impacted by Missiles and Exposed to Cyclic Pressure Differentials*. West Conshohocken, PA: ASTM International, 2005.

ASTM E 1996-09, *Specification for Performance of Exterior Windows, Curtain Walls, Doors and Storm Shutters Impacted by Wind-borne Debris in Hurricanes*. West Conshohocken, PA: ASTM International, 2009.

Becker, Roy, Farzad Naeim and Edward J. Teal. *Seismic Design Practice for Steel Buildings*. Los Angeles: The Steel Committee of California, Revised 1988.

Bolt, Bruce A. *Earthquakes*. New York: W.H. Freeman and Company, 1988.

Chen, Wai-Fah and Charles Scaw thorn. *Earthquake Engineering Handbook*. Boca Raton, FL: CRC Press LLC, 2003.

Choppra, Anil K. *Dynamics of Structures: A Primer*. Oakland, CA: Earthquake Engineering Research Institute, 1981.

Cobeen, Kelly, James Russell and J. Daniel Dolan. *CUREE W30a & 30b: Recommendations for Earthquake Resistance in the Design and Construction of Woodframe Buildings*. Consortium of Universities for Research in Earthquake Engineering (CUREE), 2004.

Cook, N.J. *The Designer's Guide to Wind Loading of Building Structures, Part I: Building Research Establishment*. Cambridge: The University Press, 1985.

DASMA 115-05, *Standard Method for Testing Sectional Garage Doors and Rolling Doors: Determination of Structural Performance Under Missile Impact and Cyclic Wind Pressure*. Cleveland, OH: Door and Access Systems Manufacturers Association International, 2005.

Diaphragms. Tacoma, WA: American Plywood Association, 1989.

Dowrick, David J. *Earthquake-resistant Design for Engineers and Architects, 2nd ed.* New York: John Wiley & Sons, 1987.

Fanella, David A., S.K. Ghosh and Xuemei Liang. *Seismic and Wind Design of Concrete Buildings (2003 IBC, ASCE 7-02, AC/ 318-02)*. Washington, DC: International Code Council, 2005.

FEMA 44 CFR, Parts 59-73, *National Flood Insurance Program (NFIP)*. Washington, DC: Federal Emergency Management Agency.

FEMA 55, *Coastal Construction Manual: Principles and Practices of Planning, Siting, Designing, Constructing, and Maintaining Residential Buildings in Coastal Areas*. Washington, DC: Federal Emergency Management Agency, 2000.

FEMA 259, *Engineering Principles and Practices for Retrofitting Flood-prone Residential Buildings*. Washington, DC: Federal Emergency Management Agency, 1995.

FEMA 265, *Managing Floodplain Development in Approximate Zone A Areas: A Guide for Obtaining and Developing Base (100-year) Flood Elevations*. Washington, DC: Federal Emergency Management Agency, 1995.

FEMA 302, NEHRP *Recommended Provisions for Seismic Regulations for New Buildings and Other Structures, Part 1*. Washington, DC: Prepared by the Building Seismic Safety Council for the Federal Emergency Management Agency, 1998.

FEMA 303, NEHRP *Recommended Provisions for Seismic Regulations for New Buildings and Other Structures Commentary, Part 2 Commentary*. Washington, DC: Prepared by the Building Seismic Safety Council for the Federal Emergency Management Agency, 1998.

FEMA 311, *Guidance on Estimating Substantial Damage Using the NFIP Substantial Damage Estimator*. Washington, DC: Federal Emergency Management Agency, 2009.

FEMA 348, *Protecting Building Utilities from Flood Damage: Principles and Practices for the Design and Construction of Flood-resistant Building Utility Systems*. Washington, DC: Federal Emergency Management Agency, 1999.

FEMA 450-1, NEHRP *Recommended Provisions for Seismic Regulations for New Buildings and Other Structures, Part 1*. Washington, DC: Prepared by the Building Seismic Safety Council for the Federal Emergency Management Agency, 2004.

FEMA 450-2, NEHRP *Recommended Provisions for Seismic Regulations for New Buildings and Other Structures, Part 2 Commentary*. Washington, DC: Prepared by the Building Seismic Safety Council for the Federal Emergency Management Agency, 2004.

FEMA (various dates). NFIP *Technical Bulletin Series*. Washington, DC: National Flood Insurance Program. Available at www.fema.gov,plan/prevent/floodplain/techbul.shtm.

FEMA Form 81-31, *Elevation Certificate* (2009). Available at www.fema.gov/business/nfip/elvinst.shtm.

FEMA Form 81-65, *Floodproofing Certificate* (2009). Available at www.fema.gov/plan/prevent/fhm/dl_fpc.shtm.

FEMA P-85, *Protecting Manufactured Homes from Floods and Other Hazards: A Multi-Hazard Foundation and Installation Guide.* Washington, DC: Federal Emergency Management Agency, 2009.

FEMA P-467-1, *Floodplain Management Bulletin on the Elevation Certificate.* Washington, DC: Federal Emergency Management Agency, 2004.

FEMA P-467-2, *Floodplain Management Bulletin on Historic Structures.* Washington, DC: Federal Emergency Management Agency, 2008.

FEMA P750-1, NEHRP *Recommended Provisions for Seismic Regulations for New Buildings and Other Structures, Part 1.* Washington, DC: Prepared by the Building Seismic Safety Council for the Federal Emergency Management Agency, 2009.

FEMA P750-2, NEHRP *Recommended Provisions for Seismic Regulations for New Buildings and Other Structures, Part 2 Commentary.* Washington, DC: Prepared by the Building Seismic Safety Council for the Federal Emergency Management Agency, 2009.

FEMA P-758, *Substantial Improvement/Substantial Damage Desk Reference.* Washington, DC: Federal Emergency Management Agency, 2009.

FEMA P-784, *Substantial Damage Estimator.* Washington, DC: Federal Emergency Management Agency, 2009.

FEMA TB #1, *Openings in Foundation Walls and Walls of Enclosures Below Elevated Buildings Located in Special Flood Hazard Areas.* Washington, DC: Federal Emergency Management Agency, 2008.

FEMA TB #2, *Flood Damage-Resistant Materials Requirements for Buildings Located in Special Flood Hazard Areas.* Washington, DC: Federal Emergency Management Agency, 2008.

FEMA TB #3, *Nonresidential Floodproofing Requirements and Certification for Buildings Located in Special Flood Hazard Areas.* Washington, DC: Federal Emergency Management Agency,1993.

FEMA TB #4, *Elevator Installation for Buildings Located in Special Flood Hazard Areas.* Washington, DC: Federal Emergency Management Agency, 2011.

FEMA TB #5, *Free-of-Obstruction Requirements for Buildings Located in Coastal High-hazard Areas.* Washington, DC: Federal Emergency Management Agency, 2008.

FEMA TB #6, *Below Grade Parking Requirements for Buildings Located in Special Flood Hazard Areas.* Washington, DC: Federal Emergency Management Agency, 1993.

FEMA TB #7, *Wet Floodproofing Requirements for Structures Located in Special Flood Hazard Areas.* Washington, DC: Federal Emergency Management Agency, 1993.

FEMA TB #8, *Corrosion Protection for Metal Connectors in Coastal Areas for Structures Located in Special Flood Hazard Areas.* Washington, DC: Federal Emergency Management Agency,1996.

FEMA TB #9, *Design and Construction Guidance for Breakaway Walls Below Elevated Coastal Buildings.* Washington, DC: Federal Emergency Management Agency, 2008.

FEMA TB #10, *Ensuring that Structures Built on Fill In or Near Special Flood Hazard Areas are Reasonably Safe from Flooding.* Washington, DC: Federal Emergency Management Agency, 2001.

FEMA TB #11, *Crawl Space Construction for Buildings Located in Special Flood Hazard Areas.* Washington, DC: Federal Emergency Management Agency, 2001.

Florida Building Code, Test Protocols for High-velocity Hurricane Zones. Tallahassee, FL: Florida Building Commission, 2007.

Fuller, Myron. *The New Madrid Earthquake.* Marion, IL: U.S. Department of the Interior/Geological Survey, 1912 (reprinted 1988).

Ghosh, S.K. *Seismic Design Using Structural Dynamics (2000 IBC).* Washington, DC: International Code Council, 2003.

Glen, Berg V. *Seismic Design Codes and Procedures.* Oakland, CA: Earthquake Engineering Research Institute, 1983.

Green, Norman. *Earthquake-resistant Building Design and Construction.* New York: Elsevier Science Publishing Company, 1987.

Guide for the Design and Construction of Mill Buildings (AISE Technical Report #13). Washington, DC: Association of Iron and Steel Engineers, August 1, 1979.

Hall, W.J. and N.M. Newmark. *Earthquake Spectra and Design.* Oakland, CA: Earthquake Engineering Research Institute, 1982.

ICC-300-12, *Standard for Bleachers, Folding and Telescopic Seating, and Grandstands.* Washington, DC: International Code Council, 2011.

ICC-600-08, *Standard for Residential Construction in High-wind Regions.* Washington, DC: International Code Council, 2008.

IFC-12, *International Fire Code.* Washington, DC: International Code Council, 2011.

IPC-12, *International Plumbing Code.* Washington, DC: International Code Council, 2011.

IRC-12, *International Residential Code.* Washington, DC: International Code Council, 2011.

Isyumor, N. and T. Tschanz. *Building Motion in Wind.* New York: American Society of Civil Engineers, 1986.

Liu, Henry. *Wind Engineering: A Handbook for Structural Engineers.* Englewood Cliffs, NJ: Prentice Hall, Inc., 1991.

Low-rise Building Systems Manual. Cleveland, OH: Metal Building Manufacturers Association, 1986.

Lubbock, Richard Marshall, Kishor C. Mehta and Dale Perry. *Guide to the Use of the Wind Load Provisions of ASCE 7-88 (formerly ANSI A58.1).* New York: American Society of Civil Engineers, 1991.

Mehta, Kishor C. and Joseph E. Minor. "*Wind Damage Observations and Implications.*" Journal of the Structural Division. New York: American Society of Civil Engineers, Vol. 105, November 1979.

NAAMM FP 1001-07, *Guide Specification for Design of Metal Flag Poles.* Chicago: National Association of Architectural Metal Manufacturers, 2007.

Naeim, Farzad. *The Seismic Design Handbook.* Norwell, MA: Kluwer Academic Publishers, 2001.

NCMA-TEK No. 109A, *Concrete Masonry Design for Seismic Forces.* Herndon, VA: National Concrete Masonry Association, Revised, 1989.

NFPA 13-10, *Installation of Sprinkler Systems.* Quincy, MA: National Fire ProtPection Association, 2010.

NFPA 58- 11, *Liquefied Petroleum Gas Code.* Quincy, MA: National Fire Protection Association, 2011.

Pakiser, Louis C. *Earthquakes.* Washington, DC: U.S. Department of the Interior/Geological Survey, 1988.

Proceedings of the 20th Joint Meeting of the U.S.—Japan Cooperative Program in Natural Resources, Wind and Seismic Effects. National Institute of Standards and Technology, NIST SP 760, January 1989.

Proceedings of the First International Conference on Snow Engineering. Santa Barbara, CA: U.S. Army Corps of Engineers—Cold Regions Research & Engineering Laboratory, Special Report 89-6, July 1988.

Recommendations for Direct-hung Acoustical Tile and Lay-in Panel Ceilings, Seismic Zones 0-2. St. Charles, IL: Ceiling and Interior Systems Construction Association, 2004.

Recommendations for Direct-hung Acoustical Tile and Lay-in Panel Ceilings, Seismic Zones 3-4. St. Charles, IL: Ceiling and Interior Systems Construction Association, 2004.

Recommended Lateral Force Requirements and Commentary (SEAOC Blue Book). Sacramento, CA: Structural Engineers Association of California, 1999.

Recommended Lateral Force Requirements and Tentative Commentary. San Francisco: Structural Engineers Association of California, 1988.

Reducing Flood Losses Through the International Code Series: Meeting the Requirements of the National Flood Insurance Program (3rd ed.). Falls Church, VA: International Code Council, 2008.

SEAW's *Handbook of a Rapid-solutions Methodology™ for Wind Design.* Redwood City, CA: Structural Engineers Association of Washington, Applied Technology Council, 2004.

SEAW/ATC 60, SEAW *Commentary on Wind Code Provisions, Volume 1.* Redwood City, CA: Applied Technology Council, 2004.

SEAW/ATC 60, SEAW *Commentary on Wind Code Provisions, Volume 2-Example Problems.* Redwood City, CA: Applied Technology Council, 2004.

SJI CJ-10, *Standard Specification for Composite Steel Joists, CJ-series.* Myrtle Beach, SC: Steel Joist Institute, 2010.

SJI JG-10, *Standard Specification for Joist Girders.* Myrtle Beach, SC: Steel Joist Institute, 2010.

SJI K-10, *Standard Specification for Open Web Steel Joists, K Series.* Myrtle Beach, SC: Steel Joist Institute, 2010.

SJI LH/DLH-10, *Standard Specification for Long- span Steel Joists, LH Series and Deep Longspan Steel Joists, DLH Series.* Myrtle Beach, SC: Steel Joist Institute, 2010.

Specifications for Electric Overhead Traveling Cranes (CMAA Specification #70). New York: Crane Manufacturers Association of America, Inc., 1988.

Taly, Narendra. *Loads and Load Paths in Buildings: Principles of Structural Design.* Washington, DC: International Code Council, 2003.

Taly, Narendra. *Loads and Load Paths in Buildings: Principles of Structural Design-Problems and Solutions.* Washington, DC: International Code Council, 2005.

Taranath, Bungale. *Wind- and Earthquake-resistant Buildings Structural Analysis and Design.* New York: Marcel Dekker, 2005.

The Loma Prieta Earthquake of October 17, 1989. Washington, DC: U.S. Department of the Interior/ Geological Survey, 1989.

The Severity of an Earthquake. Washington, DC: U.S Department of the Interior/Geological Survey, 1985.

TMS 402/ACI 530/ASCE 5-11, *Building Code Requirements for Masonry Structures.* Farmington Hills, MI: American Concrete Institute; New York: American Society of Civil Engineers; Boulder, CO: The Masonry Council, 2011.

UBC-97, *Uniform Building Code*. Whittier, CA: International Conference of Building Officials, 1997.

Williams, Alan. *Seismic- and Wind-forces Structural Design Examples, Second Edition*. Washington, DC: International Code Council, 2005.

Wind-loading and Wind-induced Structural Response. Committee on Wind Effects. New York: American Society of Civil Engineers, 1987.

Chapter 17: Special Inspections and Tests

General Comments

In this chapter, the code sets minimum quality standards for the acceptance of materials used in building construction. It also establishes requirements for special inspections, structural observations and load testing.

Section 1701 contains the scope statement and general statement for new and used materials.

The terms primarily related to this chapter are listed in Section 1702 and their definitions are provided in Chapter 2.

Section 1703 addresses the approval process and labeling by approved agencies. Special inspections, contractor responsibility and structural observation are specified in Section 1704.

Section 1704 also includes the detailed requirements pertaining to the statement of special inspections.

Section 1705 contains detailed special inspection and verification requirements for various building elements based on the type of construction involved. Included are special inspection and verification requirements for steel, concrete, masonry, wood, soils, deep foundations, wind, seismic, fire resistance, Exterior Insulation and Finish Systems (EIFS) and smoke control. Structural testing and qualification for seismic resistance is also addressed in Section 1705.

The general requirements for determining the design strengths of materials are in Section 1706.

Section 1707 provides for an alternative test procedure in the absence of approved standards.

Provisions for a test load are addressed in Section 1708.

Section 1709 includes requirements for field load testing of a structure.

Preconstruction load testing of materials and methods of construction that are not capable of being designed by an approved analysis is covered by Section 1710.

Section 1711 includes specific material and test standards for joist hangers and concrete and clay tile roof covering.

Chapter 17 provides information regarding the evaluation, inspection and approval process for any material or system proposed for use as a component of a structure. These are general requirements that expand on the requirements of Chapter 1 relating to the roles and responsibilities of the building official regarding approval of building components. Additionally, the chapter includes general requirements relating to the roles and responsibilities of the owner, contractor, special inspectors and architects or engineers.

Purpose

This chapter provides procedures and criteria for: testing materials or assemblies; labeling materials; systems and assemblies and special inspection and verification of structural assemblies.

SECTION 1701 GENERAL

1701.1 Scope. The provisions of this chapter shall govern the quality, workmanship and requirements for materials covered. Materials of construction and tests shall conform to the applicable standards listed in this code.

❖ This chapter gives provisions for quality, workmanship, testing and labeling of all materials covered within. In general, all construction materials and tests must conform to the standards, or portions thereof, that are referenced in the code. This chapter provides requirements for materials and tests when there are no applicable standards; specific tests and standards are referenced in other chapters of the code. Additionally, this chapter provides basic requirements for labeling construction materials and assemblies, and for special inspection and verification of structural systems and components.

1701.2 New materials. New building materials, equipment, appliances, systems or methods of construction not provided for in this code, and any material of questioned suitability proposed for use in the construction of a building or structure, shall be subjected to the tests prescribed in this chapter and in the *approved* rules to determine character, quality and limitations of use.

❖ Testing is required to be performed on materials that are not specifically provided for in the code. For example, suppose a manufacturer of a sandwich panel consisting of aluminum skins and a foam plastic core wishes to use this panel as an exterior weather covering. The material does not conform to any of the standards referenced in Chapter 14, so an appropriate test protocol must be developed. The same provision for acceptance of alternative materials is already given in Section 104.11. That section provides a strong, definitive statement for performance requirements for alternative materials, requiring the proposed alternative to be equivalent to that prescribed by the code in quality, strength, effectiveness, durability and safety. Section 1701.2 simply reasserts that alternative materials (new materials) may be used, as long as the performance characteristics and quality can be established.

1701.3 Used materials. The use of second-hand materials that meet the minimum requirements of this code for new materials shall be permitted.

❖ Materials and assemblies may be reused, provided that they meet the requirements of the code for new materials (see Section 104.9.1 of the code regarding reuse of materials and equipment). Caution should be exercised in approving a used material for reuse. The applicable material standards must be consulted to determine if certain reuses are prohibited and to determine the characteristics of the used material that must be carefully checked before reuse is approved.

One example is a high-strength structural steel bolt. Reuse of the bolt is restricted by Research Council on Structural Connections (RCSC), *Specification for Structural Joints Using ASTM A 325 or A 490 Bolts*. Even a piece of structural steel, such as a wide flange, would need to be carefully checked to determine that dimensional tolerances for a new piece of structural steel are met (see ASTM A 6 and A 36).

SECTION 1702 DEFINITIONS

1702.1 Definitions. The following terms are defined in Chapter 2:

❖ Definitions facilitate the understanding of code provisions and minimize potential confusion. To that end, this section lists definitions of terms associated with special inspections and testing. Note that these definitions are found in Chapter 2. The use and application of defined terms, as well as undefined terms, are set forth in Section 201.

APPROVED AGENCY.

APPROVED FABRICATOR.

CERTIFICATE OF COMPLIANCE.

DESIGNATED SEISMIC SYSTEM.

FABRICATED ITEM.

INSPECTION CERTIFICATE.

INTUMESCENT FIRE-RESISTANT COATINGS.

MAIN WINDFORCE-RESISTING SYSTEM.

MASTIC FIRE-RESISTANT COATINGS.

SPECIAL INSPECTION.

Continuous special inspection.

Periodic special inspection.

SPECIAL INSPECTOR.

SPRAYED FIRE-RESISTANT MATERIALS.

STRUCTURAL OBSERVATION.

SECTION 1703 APPROVALS

1703.1 Approved agency. An *approved agency* shall provide all information as necessary for the *building official* to determine that the agency meets the applicable requirements.

❖ This section specifies the information that an approved agency must provide to the building official to enable him or her to determine if the agency meets the applicable requirements.

1703.1.1 Independence. An *approved agency* shall be objective, competent and independent from the contractor responsible for the work being inspected. The agency shall also disclose possible conflicts of interest so that objectivity can be confirmed.

❖ As part of the basis for a building official's approval of a particular inspection agency, the agency must demonstrate its objectivity and competence. The judgement of objectivity is linked to the financial and fiduciary independence of the agency. The competence of the agency is judged by its experience and organization, and the experience of its personnel.

For example, suppose that ACME Agency is the inspection agency employed by Builder's, Inc. for factory-built fireplaces. During an investigation of the agency, it is discovered that ACME and Builder's are subsidiaries of the same parent company, Conglomerate, Inc. The inspection agency and manufacturer clearly have a relationship that is undesirable from the standpoint of independence.

1703.1.2 Equipment. An *approved agency* shall have adequate equipment to perform required tests. The equipment shall be periodically calibrated.

❖ As part of judging the ability of a testing or inspection agency, the building official should determine that the agency has the proper equipment to perform the required tests or inspections.

1703.1.3 Personnel. An *approved agency* shall employ experienced personnel educated in conducting, supervising and evaluating tests and/or inspections.

❖ The competence of an inspection or testing agency is also based on the experience and background of its personnel. For example, if 10 engineering graduates form an agency, the building official should question whether or not this newly formed agency is sufficiently experienced to perform the tests.

If the services being provided by the inspection or test agency come within the purview of the professional registration laws of the state in which the building is being constructed, the building official should request evidence that the personnel are qualified to perform the work in accordance with this professional registration law, as well.

1703.2 Written approval. Any material, appliance, equipment, system or method of construction meeting the requirements of this code shall be *approved* in writing after satisfactory completion of the required tests and submission of required test reports.

❖ In order to have a documented record of the approval and basis for it, including any conditions or limitations, materials and systems must be approved in writing by the building official. The code also requires the approval to be granted within a reasonable period of time, after all documentation has been satisfactorily developed and submitted, so as to avoid any unnecessary delay in completion of construction.

1703.3 Approved record. For any material, appliance, equipment, system or method of construction that has been *approved*, a record of such approval, including the conditions and limitations of the approval, shall be kept on file in the *building official's* office and shall be open to public inspection at appropriate times.

❖ Written approvals must be kept on file by the building official, and be available and open to the public. This provides reasonable access to the records on approvals of materials and systems should there be any subsequent investigation or further evaluation.

1703.4 Performance. Specific information consisting of test reports conducted by an *approved* testing agency in accordance with the appropriate referenced standards, or other such information as necessary, shall be provided for the *building official* to determine that the material meets the applicable code requirements.

❖ When conformance to the code is predicated on the performance and quality of materials, the building official must require the submittal of testing reports from an approved agency. In the absence of such reports, the building official must accept specific information and details that prove compliance with the intent of the applicable code requirements.

1703.4.1 Research and investigation. Sufficient technical data shall be submitted to the *building official* to substantiate the proposed use of any material or assembly. If it is determined that the evidence submitted is satisfactory proof of performance for the use intended, the *building official* shall approve the use of the material or assembly subject to the requirements of this code. The costs, reports and investigations required under these provisions shall be paid by the applicant.

❖ This section is usually used in conjunction with Section 104.11 when analysis of any construction material, such as new and innovative materials, is required to determine code compliance. The analysis is based entirely upon technical data. All costs of testing and investigations must be paid by the applicant.

1703.4.2 Research reports. Supporting data, where necessary to assist in the approval of materials or assemblies not specifically provided for in this code, shall consist of valid research reports from *approved* sources.

❖ Reports prepared by approved agencies, such as those published by organizations affiliated with model code groups, may be accepted as part of the information needed by the building official to evaluate proposed construction and form the basis for approval. Such reports can supplement the building department resources by eliminating the need for the building official to conduct a detailed analysis on each new product, material or system. It is important that such material be truly objective and credible, and not consist merely of the manufacturer's brochures or similar proprietary information. It is also important to note that when the building official is utilizing research reports in evaluating compliance with the code, such as those issued by organizations affiliated with model code groups, he or she is not mandated to approve these reports just because the code is the legally adopted building code in the jurisdiction. These reports are not code text; they are advisory only and intended for technical reference.

1703.5 Labeling. Where materials or assemblies are required by this code to be *labeled*, such materials and assemblies shall be *labeled* by an *approved agency* in accordance with Section 1703. Products and materials required to be labeled shall be labeled in accordance with the procedures set forth in Sections 1703.5.1 through 1703.5.4.

❖ This section provides requirements for third-party inspection of a manufacturer of a material or assembly when the code says that the material or assembly must be labeled. The materials or assemblies required to be labeled are given in other chapters of the code, the *International Mechanical Code*® (IMC®), the *International Fire Code*® (IFC®) and the *International Plumbing Code*® (IPC®). Labeling provides a readily available source of information that is useful for field inspection of installed products. The label identifies the product or material and provides other information that can be investigated further if there is any question as to its suitability for the specific installation.

Some examples are gas appliances, fire doors, prefabricated construction (when the building official does not inspect it), electrical appliances, glass, factory-built fireplaces, plywood and other wood members when used structurally, lumber and foam plastics.

1703.5.1 Testing. An *approved agency* shall test a representative sample of the product or material being *labeled* to the relevant standard or standards. The *approved agency* shall maintain a record of the tests performed. The record shall provide sufficient detail to verify compliance with the test standard.

❖ As a basis for the allowed use of an agency's label, the agency is required to perform testing on the material or product in accordance with the standard referenced by the code. For example, Section 903.1 of the IMC requires that factory-built fireplaces be tested in accordance with the referenced standard UL 127 and states that factory-built fireplaces are required to be listed and labeled by an approved agency.

1703.5.2 Inspection and identification. The *approved agency* shall periodically perform an inspection, which shall be in-plant if necessary, of the product or material that is to be *labeled*. The inspection shall verify that the labeled product or material is representative of the product or material tested.

❖ The approved agency whose label is to be applied to a product must perform periodic inspections. The primary objective of these inspections is to determine that the manufacturer is, indeed, making the same product that was tested. For example, using the factory-built fireplace discussed in the commentary to Section 1703.5.1, if the fire chamber wall in the test was $^{3}/_{8}$-inch-thick (9.5 mm) steel, the inspection agency must check to see that this thickness is being used. If the manufacturer has decided to use $^{1}/_{4}$-inch (6.4 mm) steel, then the inspection agency would be required to withdraw the use of its label and listing.

1703.5.3 Label information. The *label* shall contain the manufacturer's or distributor's identification, model number, serial number or definitive information describing the product or material's performance characteristics and *approved agency's* identification.

❖ This section lists the information that is required on a label (see Figure 1703.5.3). The purpose is to provide sufficient information for the inspector to verify that the installed product is consistent with what was approved during the plan review process.

Figure 1703.5.3
TYPICAL LABEL INFORMATION

1703.5.4 Method of labeling. Information required to be permanently identified on the product shall be acid etched, sand blasted, ceramic fired, laser etched, embossed or of a type that, once applied, cannot be removed without being destroyed.

❖ The section requires that permanent labeling be of a nature that cannot be removed and specifies acceptable methods of permanent labeling.

1703.6 Evaluation and follow-up inspection services. Where structural components or other items regulated by this code are not visible for inspection after completion of a prefabricated assembly, the applicant shall submit a report of each prefabricated assembly. The report shall indicate the complete details of the assembly, including a description of the assembly and its components, the basis upon which the assembly is being evaluated, test results and similar information and other data as necessary for the *building official* to determine conformance to this code. Such a report shall be *approved* by the *building official*.

❖ As an alternative to physical inspection by the building official in the plant or location where prefabricated components are manufactured, such as modular homes, trusses, etc., the building official has the option of accepting an evaluation report from an approved agency detailing such inspections.

1703.6.1 Follow-up inspection. The applicant shall provide for *special inspections* of fabricated items in accordance with Section 1704.2.5.

❖ The owner is required to provide special inspections of fabricated assemblies at the fabrication plant in accordance with Section 1704.2.5.

1703.6.2 Test and inspection records. Copies of necessary test and inspection records shall be filed with the *building official*.

❖ All testing and inspection records related to a fabricated assembly must be filed with the building official so as to maintain a complete and legal record of the assembly and erection of the building components.

SECTION 1704
SPECIAL INSPECTIONS, CONTRACTOR RESPONSIBILITY AND STRUCTURAL OBSERVATIONS

1704.1 General. This section provides minimum requirements for special inspections, the statement of special inspections, contractor responsibility and structural observations.

❖ Section 1704 provides minimum requirements for special inspections, what is required to be included in the statement of special inspections, contractor responsibility requirements related to construction of the lateral-force-resisting systems for wind and seismic loads and structural observations.

1704.2 Special inspections. Where application is made for construction as described in this section, the owner or the *registered design professional in responsible charge* acting as the owner's agent shall employ one or more *approved agen-*

cies to perform inspections during construction on the types of work listed under Section 1705. These inspections are in addition to the inspections identified in Section 110.

Exceptions:

1. *Special inspections* are not required for construction of a minor nature or as warranted by conditions in the jurisdiction as *approved* by the *building official.*
2. Unless otherwise required by the *building official, special inspections* are not required for Group U occupancies that are accessory to a residential occupancy including, but not limited to, those listed in Section 312.1.
3. Special inspections are not required for portions of structures designed and constructed in accordance with the cold-formed steel light-frame construction provisions of Section 2211.7 or the conventional light-frame construction provisions of Section 2308.

❖ Special inspections provide a means of quality assurance. Structural properties of the concrete or steel that is used in most structures are not usually discernable by a mere visual inspection. Typically, construction materials must be tested and their installation must be monitored in order to provide a finished structure that performs in accordance with the construction documents. Trained specialists that provide these inspections give the building official and engineer an indication that the required structural performance will be achieved. The permit applicant is responsible for hiring the special inspector and must incur all associated costs. According to Section 105.1, the permit applicant may be the owner or authorized agent in connection with the project (see Section 105.1 for further details).

Special inspections do not replace inspections performed by the jurisdiction. Rather, they are intended as an enhancement to those inspections.

Exceptions to the requirement for special inspections include minor work, Group U occupancies accessory to residential occupancies and prescriptive light-frame construction consisting of cold-formed steel or wood framing.

In addition to exempting minor work, Exception 1 refers to “conditions in the jurisdiction” as a possible exception. The primary “condition” envisioned is one in which the jurisdiction has the resources and skills to perform these inspection tasks. This exception should not be interpreted as one that can be invoked by the permit applicant. A local jurisdiction should not be obligated to invoke this exception. The purpose of this exception is merely to allow jurisdictions to perform these inspections to the extent they are capable of doing so, rather than requiring a special inspector.

Exception 2 eliminates the special inspection requirement for Group U occupancies that are accessory to a residential occupancy and where allowed by the building official. Because Group U occupancies could have elements or structural systems that require special inspection, the building official has the authority to require special inspection in such cases.

Exception 3 waives the requirement for special inspections for portions of structures designed and constructed in accordance with the prescriptive provisions of Section 2211.7 for cold-formed steel light-frame structures or of Section 2308 for conventional light-frame wood structures.

1704.2.1 Special inspector qualifications. The special inspector shall provide written documentation to the building official demonstrating his or her competence and relevant experience or training. Experience or training shall be considered relevant when the documented experience or training is related in complexity to the same type of *special inspection* activities for projects of similar complexity and material qualities. These qualifications are in addition to qualifications specified in other sections of this code.

The *registered design professional in responsible charge* and engineers of record involved in the design of the project are permitted to act as the *approved agency* and their personnel are permitted to act as the special inspector for the work designed by them, provided they qualify as special inspectors.

❖ This section attempts to standardize special inspector qualifications in an effort to provide for the availability of an adequate pool of qualified and knowledgeable special inspectors. The code requires the special inspector to demonstrate competence to the satisfaction of the building official. There are many certification and training programs for various facets of special inspections that may provide guidance to the building official in making this judgment. The provision allows the registered design professional in responsible charge and engineers of record that designed the project to act as an approved agency and perform special inspections for work designed by them, provided they meet the same qualification requirements for special inspectors.

1704.2.2 Access for special inspection. The construction or work for which special inspection is required shall remain accessible and exposed for special inspection purposes until completion of the required special inspections.

❖ This section is similar to Section 110.1 in requiring that work remain accessible to the special inspector, until any required inspections have been performed.

1704.2.3 Statement of special inspections. The applicant shall submit a statement of *special inspections* in accordance with Section 107.1 as a condition for permit issuance. This statement shall be in accordance with Section 1704.3.

Exception: A statement of *special inspections* is not required for portions of structures designed and constructed in accordance with the cold-formed steel light-frame construction provisions of Section 2211.7 or the conventional light-frame construction provisions of Section 2308.

❖ The applicant must submit for approval a statement of special inspections, in addition to other construction documents, before issuance of the building permit. This section refers to Section 1704.3 for specific details, while Section 1704.3.1 identifies the materials, components and work to be covered in the statement of special inspections.

The exception addresses where the statement of special inspections document is not needed, which is for portions of structures designed and constructed in accordance with the prescriptive provisions of Section 2211.7 for cold formed steel light-frame structures or of Section 2308 for conventional light-frame wood structures.

1704.2.4 Report requirement. Special inspectors shall keep records of inspections. The special inspector shall furnish inspection reports to the *building official*, and to the *registered design professional in responsible charge*. Reports shall indicate that work inspected was or was not completed in conformance to *approved construction documents*. Discrepancies shall be brought to the immediate attention of the contractor for correction. If they are not corrected, the discrepancies shall be brought to the attention of the *building official* and to the *registered design professional in responsible charge* prior to the completion of that phase of the work. A final report documenting required *special inspections* and correction of any discrepancies noted in the inspections shall be submitted at a point in time agreed upon prior to the start of work by the applicant and the *building official*.

❖ Records of each inspection must be submitted to the building official so as to compile a complete legal record of the project. These records must include all inspections made, violations and discrepancies. Before a certificate of occupancy is issued, a final report must be submitted indicating that all special inspections have been made and all discrepancies have been resolved or removed in order to show compliance with the applicable code requirements. It is the responsibility of the special inspector to document and submit inspection records to the building official and the registered design professional in responsible charge of the project.

1704.2.5 Inspection of fabricators. Where fabrication of structural load-bearing members and assemblies is being performed on the premises of a fabricator's shop, *special inspection* of the fabricated items shall be required by this section and as required elsewhere in this code.

❖ Inspection of in-plant fabrications and the requirements for special in-plant inspections are addressed herein. This section should be used in conjunction with Section 1703.6 relating to evaluation and follow-up inspections.

1704.2.5.1 Fabrication and implementation procedures. The special inspector shall verify that the fabricator maintains detailed fabrication and quality control procedures that provide a basis for inspection control of the workmanship and the fabricator's ability to conform to *approved construction documents* and referenced standards. The special inspector shall review the procedures for completeness and adequacy relative to the code requirements for the fabricator's scope of work.

Exception: *Special inspections* as required by Section 1704.2.5 shall not be required where the fabricator is *approved* in accordance with Section 1704.2.5.2.

❖ The special inspector is required to verify not only that the fabricator complies with the design details and in-house quality control procedures at the plant, but also the plant's ability to construct/fabricate to the approved drawings, standards and specifications. An example of this would be an inspection of proper placement and rolling of truss-plate connectors at a wood-truss manufacturing plant. Improper procedures could result in the connectors "popping out" or "peeling back" after the truss is concealed and loaded, thus causing structural failure. Special inspections are not necessary where an approved independent agency conducts in-house inspections during fabrication.

1704.2.5.2 Fabricator approval. *Special inspections* required by Section 1705 are not required where the work is done on the premises of a fabricator registered and *approved* to perform such work without *special inspection*. Approval shall be based upon review of the fabricator's written procedural and quality control manuals and periodic auditing of fabrication practices by an *approved special inspection* agency. At completion of fabrication, the *approved* fabricator shall submit a *certificate of compliance* to the *building official* stating that the work was performed in accordance with the *approved construction documents*.

❖ "Approved fabricator" is defined in Section 202 and this section provides the basis for such approval by the building official. If the fabricator is approved by the building official, then its internal quality control procedures are deemed to be sufficient without the need for a special inspection. A certificate of compliance provides the building official with evidence on each project that the work has been performed in accordance with the code and construction documents.

1704.3 Statement of special inspections. Where *special inspection* or testing is required by Section 1705, the *registered design professional in responsible charge* shall prepare a statement of special inspections in accordance with Section 1704.3.1 for submittal by the applicant in accordance with Section 1704.2.3.

Exception: The statement of *special inspections* is permitted to be prepared by a qualified person *approved* by the *building official* for construction not designed by a *registered design professional*.

❖ The statement of special inspections is required to be prepared by the registered design professional responsible for the building or structure. This is because the special inspections statement relates directly to the construction documents, which are the responsibility of the registered design professional. The exception states conditions under which the statement of special inspections may be prepared by someone other than the registered design professional. The section outlines the basic content required to be included in the statement of special inspections as well as the specific requirements for seismic and wind resistance, arranging them in logical order. The intent is to document the required inspections and testing in order to foster a systematic approach that helps to achieve the goal of a finished structure that meets or exceeds the minimum performance expectations.

1704.3.1 Content of statement of special inspections. The statement of special inspections shall identify the following:

1. The materials, systems, components and work required to have *special inspection* or testing by the *building official* or by the *registered design professional* responsible for each portion of the work.
2. The type and extent of each *special inspection*.
3. The type and extent of each test.
4. Additional requirements for *special inspection* or testing for seismic or wind resistance as specified in Sections 1705.10, 1705.11 and 1705.12.
5. For each type of *special inspection*, identification as to whether it will be continuous *special inspection* or periodic *special inspection*.

❖ This section details the areas to be addressed in the statement of special inspection. It requires a list of materials and work subject to special inspection, the type and frequency of inspections and whether the inspections are continuous or periodic.

1704.3.2 Seismic requirements in the statement of special inspections. Where Section 1705.11 or 1705.12 specifies special inspection, testing or qualification for seismic resistance, the statement of special inspections shall identify the designated seismic systems and seismic force- resisting systems that are subject to *special inspections*.

❖ Where special inspection or testing for seismic resistance is required, the statement of special inspections must identify the designated seismic systems and seismic force-resisting systems that are required to have special inspection.

1704.3.3 Wind requirements in the statement of special inspections. Where Section 1705.10 specifies special inspection for wind requirements, the statement of special inspections shall identify the main windforce-resisting systems and wind-resisting components subject to *special inspection*.

❖ Where special inspection for wind resistance is required, the statement of special inspections must identify the main wind-force-resisting systems and components and cladding that are required to have special inspection.

1704.4 Contractor responsibility. Each contractor responsible for the construction of a main wind- or seismic force-resisting system, designated seismic system or a wind- or seismic-resisting component listed in the statement of special inspections shall submit a written statement of responsibility to the *building official* and the owner prior to the commencement of work on the system or component. The contractor's statement of responsibility shall contain acknowledgement of awareness of the special requirements contained in the statement of *special inspection*.

❖ The statement of contractor responsibility is required wherever the statement of special inspections includes additional wind- or seismic-resistance inspections. This statement by the contractor is separate from the statement of special inspections. It is the contractor's acknowledgment of the special inspections or testing that are beyond what is typically required.

1704.5 Structural observations. Where required by the provisions of Section 1704.5.1 or 1704.5.2, the owner shall employ a *registered design professional* to perform structural observations as defined in Section 1702.

Prior to the commencement of observations, the structural observer shall submit to the *building official* a written statement identifying the frequency and extent of structural observations.

At the conclusion of the work included in the permit, the structural observer shall submit to the *building official* a written statement that the site visits have been made and identify any reported deficiencies which, to the best of the structural observer's knowledge, have not been resolved.

❖ This section requires that a registered design professional such as an engineer or architect be employed by the owner to provide on-site visits to observe compliance with the structural draw-ings when required by Section 1704.5.1 or 1704.5.2. Structural observations are required under certain conditions in high-seismic and high-wind locations (see Section 202 for the definition of "Structural observation"), providing an additional level of inspections beyond those provided by the building inspector or a special inspector. The intent of requiring structural observations by a registered design professional for these structures is to verify that the structural systems are constructed in general conformance with the approved construction documents.

This section provides guidance on the frequency of observations. It requires submittal of a written statement by the structural observer to the building official prior to the commencement of observations identifying the frequency and extent of structural observations. The determination of the frequency of structural observations should be by the structural observer and the owner in consultation with the local building official. The structural observer must also submit a written statement to the building official at the conclusion of the work included in the permit.

1704.5.1 Structural observations for seismic resistance. Structural observations shall be provided for those structures assigned to *Seismic Design Category* D, E or F where one or more of the following conditions exist:

1. The structure is classified as *Risk Category* III or IV in accordance with Table 1604.5.
2. The height of the structure is greater than 75 feet (22 860 mm) above the base.
3. The structure is assigned to *Seismic Design Category* E, is classified as *Risk Category* I or II in accordance with Table 1604.5, and is greater than two *stories above grade plane*.
4. When so designated by the *registered design professional* responsible for the structural design.
5. When such observation is specifically required by the *building official*.

❖ This section lists the thresholds of seismic risk that necessitate structural observation. Items 1 through 3 are specific conditions that require structural observation, while Items 4 and 5 give discretion to the registered design professional or building official to require structural observation in other instances (see commentary, Section 1704.5).

1704.5.2 Structural observations for wind requirements. Structural observations shall be provided for those structures sited where V_{asd} as determined in accordance with Section 1609.3.1 exceeds 110 mph (49 m/sec), where one or more of the following conditions exist:

1. The structure is classified as *Risk Category* III or IV in accordance with Table 1604.5.
2. The *building height* of the structure is greater than 75 feet (22 860 mm).
3. When so designated by the *registered design professional* responsible for the structural design.
4. When such observation is specifically required by the *building official*.

❖ This section lists the thresholds of high-wind exposure that necessitate structural observation. Items 1 and 2 are specific conditions that require structural observation, while Items 3 and 4 give discretion to the registered design professional or building official to require structural observation in other instances (see commentary, Section 1704.5).

SECTION 1705
REQUIRED VERIFICATION AND INSPECTION

1705.1 General. Verification and inspection of elements of buildings and structures shall be as required by this section.

❖ The specific elements that require verification by special inspection are covered in Section 1705.

1705.1.1 Special cases. *Special inspections* shall be required for proposed work that is, in the opinion of the *building official*, unusual in its nature, such as, but not limited to, the following examples:

1. Construction materials and systems that are alternatives to materials and systems prescribed by this code.
2. Unusual design applications of materials described in this code.
3. Materials and systems required to be installed in accordance with additional manufacturer's instructions that prescribe requirements not contained in this code or in standards referenced by this code.

❖ This section requires special inspections for proposed work that is unique and not specifically addressed in the code or in standards that are referenced by the code. For example, a designer chooses to utilize a new shear wall anchorage system that is not specifically covered by the code. Because the product is approved as an alternative under Section 104.11, the inspector must rely on installation requirements such as embedment, edge distances, etc., given in approved reports from the manufacturer, provided that the system is previously approved for installation by the building official. Code evaluation reports that require special inspection are based on the requirements in this section.

1705.2 Steel construction. The *special inspections* for steel elements of buildings and structures shall be as required in this section.

> **Exception:** *Special inspection* of the steel fabrication process shall not be required where the fabricator does not perform any welding, thermal cutting or heating operation of any kind as part of the fabrication process. In such cases, the fabricator shall be required to submit a detailed procedure for material control that demonstrates the fabricator's ability to maintain suitable records and procedures such that, at any time during the fabrication process, the material specification, and grade for the main stress-carrying elements are capable of being determined. Mill test reports shall be identifiable to the main stress-carrying elements when required by the approved construction documents.

❖ The requirements to be followed by the special inspector for the erection and fabrication of steel elements of building construction are covered in Section 1705.2.

The exception is allowed if the fabrication plant does not utilize any facilities or methods that may alter the physical characteristics or properties of the steel members or components, such as welding, thermal cutting or heating operations. The fabricator would, in any case, need to provide evidence that procedures are used that verify that the proper material specification and grade for the main stress-carrying elements are supplied in accordance with the job specifications and shop drawings.

1705.2.1 Structural steel. Special inspection for structural steel shall be in accordance with the quality assurance inspection requirements of AISC 360.

❖ The 2010 edition of AISC 360-10, *AISC Specification for Structural Steel Buildings* contains quality assurance and inspection requirements for structural steel buildings. This section references ASIC 360 for special inspection requirements related to structural steel.

1705.2.2 Steel construction other than structural steel. Special inspection for steel construction other than structural steel shall be in accordance with Table 1705.2.2 and this section.

❖ Special inspection requirements for steel elements other than structural steel, such as cold-formed steel deck and reinforcing steel, are given in Table 1705.2.2.

TABLE 1705.2.2. See below.

❖ The table gives the type of steel elements requiring verification by the special inspector, whether the inspection is continuous or periodic, and the applicable ASTM International (ASTM) material standard or American Welding Society (AWS) welding standard. Note a of the table references Section 1705.11 for any additional special inspections related to seismic resistance.

1705.2.2.1 Welding. Welding inspection and welding inspector qualification shall be in accordance with this section.

❖ This section provides the necessary references that provide guidance on welding inspection for cold-formed steel and reinforcing steel. Welding requirements for structural steel are covered in AISC 360.

1705.2.2.1.1 Cold-formed steel. Welding inspection and welding inspector qualification for cold-formed steel floor and roof decks shall be in accordance with AWS D1.3.

❖ This section provides a reference to the AWS D1.3 standard that applies to welding inspection and welder inspector qualifications for cold-formed steel floor and roof decks.

1705.2.2.1.2 Reinforcing steel. Welding inspection and welding inspector qualification for reinforcing steel shall be in accordance with AWS D1.4 and ACI 318.

❖ This section provides the necessary reference to AWS D1.4 and ACI 318 that provides guidance on welding inspection and welder inspector qualifications for reinforcing steel.

1705.2.2.2 Cold-formed steel trusses spanning 60 feet or greater. Where a cold-formed steel truss clear span is 60 feet (18 288 mm) or greater, the special inspector shall verify that the temporary installation restraint/bracing and the permanent individual truss member restraint/bracing are installed in accordance with the *approved* truss submittal package.

❖ Longer span trusses require additional care during handling and installation due to their size and weight. Special inspection of the bracing provides verification that these critical elements are properly installed in accordance with the approved truss submittal.

1705.3 Concrete construction. The *special inspections* and verifications for concrete construction shall be as required by this section and Table 1705.3.

Exception: *Special inspections* shall not be required for:

1. Isolated spread concrete footings of buildings three stories or less above *grade plane* that are fully supported on earth or rock.
2. Continuous concrete footings supporting walls of buildings three stories or less above *grade plane* that are fully supported on earth or rock where:
   - 2.1. The footings support walls of light-frame construction;
   - 2.2. The footings are designed in accordance with Table 1809.7; or

TABLE 1705.2.2
REQUIRED VERIFICATION AND INSPECTION OF STEEL CONSTRUCTION OTHER THAN STRUCTURAL STEEL

VERIFICATION AND INSPECTION	CONTINUOUS	PERIODIC	REFERENCED STANDARD[a]
1. Material verification of cold-formed steel deck:			
a. Identification markings to conform to ASTM standards specified in the approved construction documents.	—	X	Applicable ASTM material standards
b. Manufacturer's certified test reports.	—	X	
2. Inspection of welding:			
a. Cold-formed steel deck:			
1) Floor and roof deck welds.	—	X	AWS D1.3
b. Reinforcing steel:			
1) Verification of weldability of reinforcing steel other than ASTM A 706.	—	X	AWS D1.4 ACI 318: Section 3.5.2
2) Reinforcing steel resisting flexural and axial forces in intermediate and special moment frames, and boundary elements of special structural walls of concrete and shear reinforcement.	X	—	
3) Shear reinforcement.	X	—	
4) Other reinforcing steel.	—	X	

For SI: 1 inch = 25.4 mm.

a. Where applicable, see also Section 1705.11, Special inspections for seismic resistance.

2.3. The structural design of the footing is based on a specified compressive strength, f'_c, no greater than 2,500 pounds per square inch (psi) (17.2 MPa), regardless of the compressive strength specified in the *construction documents* or used in the footing construction.

3. Nonstructural concrete slabs supported directly on the ground, including prestressed slabs on grade, where the effective prestress in the concrete is less than 150 psi (1.03 MPa).

4. Concrete foundation walls constructed in accordance with Table 1807.1.6.2.

5. Concrete patios, driveways and sidewalks, on grade.

❖ This section establishes criteria for special inspections of elements of buildings and structures of concrete construction. Exceptions to the requirements of this section address concrete components that have little or no load-carrying requirements, such as nonstructural slabs on grade, driveways, patios, etc., or footings and foundations that require no reinforcement and carry relatively low loads.

TABLE 1705.3. See below.

❖ Required verifications and inspections during concrete construction operations are listed in Table

TABLE 1705.3
REQUIRED VERIFICATION AND INSPECTION OF CONCRETE CONSTRUCTION

VERIFICATION AND INSPECTION	CONTINUOUS	PERIODIC	REFERENCED STANDARD[a]	IBC REFERENCE
1. Inspection of reinforcing steel, including prestressing tendons, and placement.	—	X	ACI 318: 3.5, 7.1-7.7	1910.4
2. Inspection of reinforcing steel welding in accordance with Table 1705.2.2, Item 2b.	—	—	AWS D1.4 ACI 318: 3.5.2	—
3. Inspection of anchors cast in concrete where allowable loads have been increased or where strength design is used.	—	X	ACI 318: 8.1.3, 21.2.8	1908.5, 1909.1
4. Inspection of anchors post-installed in hardened concrete members[b].	—	X	ACI 318: 3.8.6, 8.1.3, 21.2.8	1909.1
5. Verifying use of required design mix.	—	X	ACI 318: Ch. 4, 5.2-5.4	1904.2, 1910.2, 1910.3
6. At the time fresh concrete is sampled to fabricate specimens for strength tests, perform slump and air content tests, and determine the temperature of the concrete.	X	—	ASTM C 172 ASTM C 31 ACI 318: 5.6, 5.8	1910.10
7. Inspection of concrete and shotcrete placement for proper application techniques.	X	—	ACI 318: 5.9, 5.10	1910.6, 1910.7, 1910.8
8. Inspection for maintenance of specified curing temperature and techniques.	—	X	ACI 318: 5.11-5.13	1910.9
9. Inspection of prestressed concrete: a. Application of prestressing forces. b. Grouting of bonded prestressing tendons in the seismic force-resisting system.	 X X	—	 ACI 318: 18.20 ACI 318: 18.18.4	—
10. Erection of precast concrete members.	—	X	ACI 318: Ch. 16	—
11. Verification of in-situ concrete strength, prior to stressing of tendons in post-tensioned concrete and prior to removal of shores and forms from beams and structural slabs.	—	X	ACI 318: 6.2	—
12. Inspect formwork for shape, location and dimensions of the concrete member being formed.	—	X	ACI 318: 6.1.1	—

For SI: 1 inch = 25.4 mm.

a. Where applicable, see also Section 1705.11, Special inspections for seismic resistance.

b. Specific requirements for special inspection shall be included in the research report for the anchor issued by an approved source in accordance with ACI 355.2 or other qualification procedures. Where specific requirements are not provided, special inspection requirements shall be specified by the registered design professional and shall be approved by the building official prior to the commencement of the work.

1705.3. This table provides a complete list of the types of inspections required, whether the inspection is continuous or periodic, and the applicable referenced standards for the placing, curing, prestressing and erection of concrete construction. Note a of the table references Section 1705.11 for any additional special inspections related to seismic resistance.

1705.3.1 Materials. In the absence of sufficient data or documentation providing evidence of conformance to quality standards for materials in Chapter 3 of ACI 318, the building official shall require testing of materials in accordance with the appropriate standards and criteria for the material in Chapter 3 of ACI 318. Weldability of reinforcement, except that which conforms to ASTM A 706, shall be determined in accordance with the requirements of Section 3.5.2 of ACI 318.

❖ Concrete materials, such as cement, aggregates, admixtures and water, must comply with the standards of Chapter 3 of ACI 318, which regulates materials and addresses specific standards. In the absence of sufficient data or documentation, the building official must require testing in accordance with the standards listed in Chapter 3 of ACI 318.

ASTM A 706 is the standard for reinforcing steel that is weldable, meaning that the chemical composition and manufacturing processes are such that the material is well suited for an acceptable quality of weld. Section 3.5.2 of ACI 318 states that any standard other than ASTM A 706 used for reinforcement material would need to be supplemented for weldability requirements. The intent of this provision is that, where welding of reinforcing steel is required, the steel specified and delivered must be checked for weldability.

1705.4 Masonry construction. Masonry construction shall be inspected and verified in accordance with TMS 402/ACI 530/ASCE 5 and TMS 602/ACI 530.1/ASCE 6 quality assurance program requirements.

Exception: *Special inspections* shall not be required for:

1. Empirically designed masonry, glass unit masonry or masonry veneer designed by Section 2109, 2110 or Chapter 14, respectively, where they are part of structures classified as *Risk Category* I, II or III in accordance with Section 1604.5.
2. Masonry foundation walls constructed in accordance with Table 1807.1.6.3(1), 1807.1.6.3(2), 1807.1.6.3(3) or 1807.1.6.3(4).
3. Masonry fireplaces, masonry heaters or masonry chimneys installed or constructed in accordance with Section 2111, 2112 or 2113, respectively.

❖ Because the 2011 edition of the Masonry Standards Joint Committee (MSJC) Code (TMS 402/ACI 530/ASCE 5) contains quality assurance and inspection requirements for masonry structures, this section references TMS 402/ACI 530/ASCE 5 and TMS 602/ACI 530.1/ASCE 6 MSJC Code and Specification for special inspection requirements related to masonry construction. The three exceptions exempt special inspection for: (1) empirically designed masonry, glass unit masonry or masonry veneer in structures classified as Risk Category I, II or III; (2) prescriptive masonry foundation walls constructed in accordance with the tables in Section 1807.1.6.3; and (3) masonry fireplaces, heaters and chimneys constructed in accordance with the requirements of Chapter 21.

1705.4.1 Empirically designed masonry, glass unit masonry and masonry veneer in Risk Category IV. The minimum *special inspection* program for empirically designed masonry, glass unit masonry or masonry veneer designed by Section 2109, 2110 or Chapter 14, respectively, in structures classified as *Risk Category* IV, in accordance with Section 1604.5, shall comply with TMS 402/ACI 530/ASCE 5 Level B Quality Assurance.

❖ This section defines the minimum level of special inspections required for empirically designed masonry, glass unit masonry and masonry veneer in Risk Category IV buildings. The section requires compliance with the Level B Quality Assurance provisions contained in TMS 402/ACI 530/ASCE 5.

1705.4.2 Vertical masonry foundation elements. *Special inspection* shall be performed in accordance with Section 1705.4 for vertical masonry foundation elements.

❖ Vertical masonry foundation elements such as masonry foundation walls must meet the special inspection requirements specified for other masonry elements. The section refers to the general provisions for special inspection of masonry structures in Section 1705.4. It should be noted that masonry foundation walls constructed in accordance with the prescriptive provisions and tables in Section 1807.1.6 are specifically exempted from special inspection.

1705.5 Wood construction. *Special inspections* of the fabrication process of prefabricated wood structural elements and assemblies shall be in accordance with Section 1704.2.5. *Special inspections* of site-built assemblies shall be in accordance with this section.

❖ The fabrication process of wood structural elements and assemblies (such as wood trusses) that is being performed on the premises of a fabricator's shop must have special inspection in accordance with Section 1704.2.5. As noted in Section 1704.2.5.2, special inspection is not required in an approved fabricator's shop.

1705.5.1 High-load diaphragms. High-load diaphragms designed in accordance with Section 2306.2 shall be installed with *special inspections* as indicated in Section 1704.2. The special inspector shall inspect the wood structural panel sheathing to ascertain whether it is of the grade and thickness shown on the *approved* building plans. Additionally, the special inspector must verify the nominal size of framing members at adjoining panel edges, the nail or staple diameter and length, the number of fastener lines and that the spacing between fasteners in each line and at edge margins agrees with the *approved* building plans.

❖ This section requires special inspection of specific portions of high-load diaphragms with multiple rows of fasteners that are designed in accordance with Section 2306.2 [see Table 2306.2(2)]. These "high-load" diaphragms have multiple rows of fasteners and are designed to carry higher wind or seismic loads, making it important to have a special inspection during installation.

1705.5.2 Metal-plate-connected wood trusses spanning 60 feet or greater. Where a truss clear span is 60 feet (18 288 mm) or greater, the special inspector shall verify that the temporary installation restraint/bracing and the permanent individual truss member restraint/bracing are installed in accordance with the *approved* truss submittal package.

❖ Longer span wood trusses require additional care during handling and installation due to their size and weight. Special inspection of the bracing provides verification that these critical elements are properly installed in accordance with the approved truss submittal.

1705.6 Soils. *Special inspections* for existing site soil conditions, fill placement and load-bearing requirements shall be as required by this section and Table 1705.6. The *approved* geotechnical report, and the *construction documents* prepared by the *registered design professionals* shall be used to determine compliance. During fill placement, the special inspector shall determine that proper materials and procedures are used in accordance with the provisions of the *approved* geotechnical report.

> **Exception:** Where Section 1803 does not require reporting of materials and procedures for fill placement, the special inspector shall verify that the in-place dry density of the compacted fill is not less than 90 percent of the maximum dry density at optimum moisture content determined in accordance with ASTM D 1557.

❖ The load-bearing capacity of the supporting soil has a significant impact on the structural integrity of any building. The amount of compaction and the methods vary depending on the particular design. Use of proper compaction, lift and density, however, is critical to achieving the desired bearing capacity (see Table 1705.6 for the specific inspections that are required). When required by Section 1803, a geotechnical report must be provided, which would list the soil criteria that must be verified. Section 1804.5 clarifies that compacted fill used to support foundations must be in accordance with the criteria of an approved geotechnical report, except for fill depths of 12 inches (305 mm) or less. The exception in this section provides minimum verification requirements of compacted fill when it is not necessary to provide a geotechnical report.

TABLE 1705.6. See below.

❖ Tabular formatting of soil inspections clearly conveys the intended requirements for testing and inspection of soils for controlled fill and determination of soil-bearing capacity. The verifications prior to fill placement include verifying that the site preparation meets specified requirements, including proper excavation depth, removal of all deleterious material and any other special requirements that the soils engineer deems necessary for the design (see Items 1, 2 and 5). Verifications of the fill material and the placement operation are treated in Items 3 and 4. Without observing and documenting that the proper material is used and the specified compaction techniques and lifts are employed, the specified load-bearing capacity may not be achieved. A major factor in the design of the fill is the in-place density. This evaluation is needed so that the compaction methods result in adequate soil-bearing capacity.

1705.7 Driven deep foundations. *Special inspections* shall be performed during installation and testing of driven deep foundation elements as required by Table 1705.7. The *approved instruction documents* prepared by the *registered design professionals*, shall be used to determine compliance.

❖ Table 1705.7 lists the special inspection verifications that are required for installation and testing of driven elements of deep foundations. A geotechnical investigation is required by Section 1803.5.5. The section establishes specific criteria that the geotechnical report should include, which would also serve as the basis for the field verifications by a special inspector.

TABLE 1705.7. See page 17-13.

❖ This table provides a concise list of requirements for inspection and verification of materials, testing and

TABLE 1705.6
REQUIRED VERIFICATION AND INSPECTION OF SOILS

VERIFICATION AND INSPECTION TASK	CONTINUOUS DURING TASK LISTED	PERIODICALLY DURING TASK LISTED
1. Verify materials below shallow foundations are adequate to achieve the design bearing capacity.	—	X
2. Verify excavations are extended to proper depth and have reached proper material.	—	X
3. Perform classification and testing of compacted fill materials.	—	X
4. Verify use of proper materials, densities and lift thicknesses during placement and compaction of compacted fill.	X	—
5. Prior to placement of compacted fill, observe subgrade and verify that site has been prepared properly.	—	X

installation of driven deep foundation elements and indicates whether the inspection is continuous or periodic.

1705.8 Cast-in-place deep foundations. *Special inspections* shall be performed during installation and testing of cast-in-place deep foundation elements as required by Table 1705.8. The *approved* geotechnical report, and the *construction documents* prepared by the *registered design professionals*, shall be used to determine compliance.

❖ Table 1705.8 lists the special inspections that are required for cast-in-place elements of deep foundations. A geotechnical investigation is required by Section 1803.5.5. This section establishes specific criteria that the geotechnical report should include, which would also serve as the basis for field verifications by a special inspector.

TABLE 1705.8. See below.

❖ This table provides a concise list of requirements for inspection and verification of materials, testing and installation of cast-in-place deep foundation elements and indicates whether the inspection is continuous or periodic.

1705.9 Helical pile foundations. *Special inspections* shall be performed continuously during installation of helical pile foundations. The information recorded shall include installation equipment used, pile dimensions, tip elevations, final depth, final installation torque and other pertinent installation data as required by the *registered design professional in responsible charge*. The *approved* geotechnical report and the *construction documents* prepared by the *registered design professional* shall be used to determine compliance.

❖ Helical piles are premanufactured and installed by rotating them into the ground (see definition of "Helical pile" in Section 202). As with other types of deep foundation elements, special inspection must be provided to verify that their installation conforms with the approved geotechnical report and construction documents.

TABLE 1705.7
REQUIRED VERIFICATION AND INSPECTION OF DRIVEN DEEP FOUNDATION ELEMENTS

VERIFICATION AND INSPECTION TASK	CONTINUOUS DURING TASK LISTED	PERIODICALLY DURING TASK LISTED
1. Verify element materials, sizes and lengths comply with the requirements.	X	—
2. Determine capacities of test elements and conduct additional load tests, as required.	X	—
3. Observe driving operations and maintain complete and accurate records for each element.	X	—
4. Verify placement locations and plumbness, confirm type and size of hammer, record number of blows per foot of penetration, determine required penetrations to achieve design capacity, record tip and butt elevations and document any damage to foundation element.	X	—
5. For steel elements, perform additional inspections in accordance with Section 1705.2.	—	—
6. For concrete elements and concrete-filled elements, perform additional inspections in accordance with Section 1705.3.	—	—
7. For specialty elements, perform additional inspections as determined by the registered design professional in responsible charge.	—	—

TABLE 1705.8
REQUIRED VERIFICATION AND INSPECTION OF CAST-IN-PLACE DEEP FOUNDATION ELEMENTS

VERIFICATION AND INSPECTION TASK	CONTINUOUS DURING TASK LISTED	PERIODICALLY DURING TASK LISTED
1. Observe drilling operations and maintain complete and accurate records for each element.	X	—
2. Verify placement locations and plumbness, confirm element diameters, bell diameters (if applicable), lengths, embedment into bedrock (if applicable) and adequate end-bearing strata capacity. Record concrete or grout volumes.	X	—
3. For concrete elements, perform additional inspections in accordance with Section 1705.3.	—	—

1705.10 Special inspections for wind resistance. *Special inspections* itemized in Sections 1705.10.1 through 1705.10.3, unless exempted by the exceptions to Section 1704.2, are required for buildings and structures constructed in the following areas:

1. In wind Exposure Category B, where V_{asd} as determined in accordance with Section 1609.3.1 is 120 miles per hour (52.8 m/sec) or greater.
2. In wind Exposure Category C or D, where V_{asd} as determined in accordance with Section 1609.3.1 is 110 mph (49 m/sec) or greater.

❖ This section contains the required additional special inspections in areas that experience higher wind forces. The list of items that need the additional special inspections is related to the general requirement in Section 1704.3.3. This section focuses on the specific areas of concern with respect to wind resistance as detailed in Sections 1705.10.1 for structural wood, Section 1705.10.2 for cold-formed steel and Section 1705.10.3 for components and cladding.

1705.10.1 Structural wood. Continuous special inspection is required during field gluing operations of elements of the main windforce-resisting system. Periodic special inspection is required for nailing, bolting, anchoring and other fastening of components within the main windforce-resisting system, including wood shear walls, wood diaphragms, drag struts, braces and hold-downs.

Exception: *Special inspection* is not required for wood shear walls, shear panels and diaphragms, including nailing, bolting, anchoring and other fastening to other components of the main windforce-resisting system, where the fastener spacing of the sheathing is more than 4 inches (102 mm) on center.

❖ The risk of damage in buildings having main wind-force-resisting systems that are constructed of wood warrants additional special inspections. The exception relaxes the special inspection requirement where demands are low, which is reflected by the sheathing fastener spacing of more than 4 inches (102 mm) on center.

1705.10.2 Cold-formed steel light-frame construction. Periodic special inspection is required during welding operations of elements of the main windforce-resisting system. Periodic special inspection is required for screw attachment, bolting, anchoring and other fastening of components within the main windforce-resisting system, including shear walls, braces, diaphragms, collectors (drag struts) and hold-downs.

Exception: *Special inspection* is not required for cold-formed steel light-frame shear walls, braces, diaphragms, collectors (drag struts) and hold-downs where either of the following apply:

1. The sheathing is gypsum board or fiberboard.
2. The sheathing is wood structural panel or steel sheets on only one side of the shear wall, shear panel or diaphragm assembly and the fastener spacing of the sheathing is more than 4 inches (102 mm) on center (o.c.).

❖ The risk of damage in buildings having main wind-force-resisting systems that are constructed of cold-formed steel warrants additional special inspections. The exception relaxes the special inspection requirement where demands are low, which is reflected by the sheathing fastener spacing of more than 4 inches (102 mm) on center, or where gypsum board or fiberboard sheathing is used.

1705.10.3 Wind-resisting components. Periodic special inspection is required for the following systems and components:

1. Roof cladding.
2. Wall cladding.

❖ The hazard addressed by this section is the cladding on buildings and structures in areas expected to expe-rience higher wind forces. Damage to buildings due to high wind forces often begins with a failure of the cladding system, which in turn exposes the main wind-force-resisting system to forces that this system is typically not designed to withstand, as well as the effects of wind-driven rain. Wind-driven rain damage occurs to building interiors when roof coverings are blown off, windows are broken or other parts of the structure fail. These are major causes of insurance claims. While such damage is not necessarily life-threatening to the occupants, it does result in costly repairs and renovations. Estimates from Hurricane Andrew, for example, indicate that water damage was responsible for roughly 60 percent of the insured losses.

1705.11 Special inspections for seismic resistance. *Special inspections* itemized in Sections 1705.11.1 through 1705.11.8, unless exempted by the exceptions of Section 1704.2, are required for the following:

1. The seismic force-resisting systems in structures assigned to *Seismic Design Category* C, D, E or F in accordance with Sections 1705.11.1 through 1705.11.3, as applicable.
2. Designated seismic systems in structures assigned to *Seismic Design Category* C, D, E or F in accordance with Section 1705.11.4.
3. Architectural, mechanical and electrical components in accordance with Sections 1705.11.5 and 1705.11.6.
4. Storage racks in structures assigned to *Seismic Design Category* D, E or F in accordance with Section 1705.11.7.
5. Seismic isolation systems in accordance with Section 1705.11.8.

Exception: Special inspections itemized in Sections 1705.11.1 through 1705.11.8 are not required for structures designed and constructed in accordance with one of the following:

1. The structure consists of light-frame construction; the design spectral response acceleration at short periods, S_{DS}, as determined in Section 1613.3.4, does not exceed 0.5; and the building height of the structure does not exceed 35 feet (10 668 mm).
2. The seismic force-resisting system of the structure consists of reinforced masonry or reinforced concrete; the design spectral response acceleration at short periods, S_{DS}, as determined in Section 1613.3.4, does not exceed 0.5; and the building height of the structure does not exceed 25 feet (7620 mm).
3. The structure is a detached one- or two-family dwelling not exceeding two *stories above grade plane* and does not have any of the following horizontal or vertical irregularities in accordance with Section 12.3 of ASCE 7:
   - 3.1. Torsional or extreme torsional irregularity.
   - 3.2. Nonparallel systems irregularity.
   - 3.3. Stiffness-soft story or stiffness-extreme soft story irregularity.
   - 3.4. Discontinuity in lateral strength-weak story irregularity.

❖ The added special inspection requirements for seismic resistance in this section are an important consideration in carrying out the intent of the seismic provisions of the code. A certain amount of inelastic behavior is inherent in a building's response to the design earthquake. Strong ground shaking caused by earthquakes tends to expose any underlying flaws in a building's construction or design. Thus, the code specifies additional special inspections of various structural and nonstructural components for seismic resistance in order to provide further verification that these portions of the finished structure are constructed in accordance with the construction documents. It is imperative that the registered design professional who designs systems that are critical to the earthquake performance also identifies them and specifies the necessary inspection and testing.

Item 1 states the seismic risk threshold for seismic-force-resisting systems base on seismic design category. The inspections of items described in Section 1705.11.1 through 1705.11.3 would apply to these structures. Item 2 covers designated seismic systems that are nonstructural components and systems that are assigned an importance factor greater than 1. The only inspections that apply specifically to designated seismic systems are given in Section 1705.11.4, but other requirements, such as those in Section 1705.11.6 for mechanical and electrical components in general, would also be applicable where the component is given an importance factor greater than 1. Item 3 covers special inspections for architectural components in Section 1705.11.5 and for electrical and mechanical components in Section 1705.11.6. Note that the applicable Seismic Design Categories (SDCs) are not specified in Item 3, so the specific SDCs for the various items covered in Sections 1705.11.5 and 1705.11.6 must be followed.

Periodic special inspection is required for storage racks 8 feet or greater in height in buildings in SDC D, E or F as required by Section 0405.11.7.

Section 1705.11.8 requires periodic special inspection during fabrication and installation of isolator units and energy dissipation devices for seismic isolation systems.

Exceptions 1 and 2 apply to light-frame structures with a height no greater than 35 feet and concrete or masonry structures not more than 25 feet in height that have a relatively low seismic risk (where S_{DS} does not exceed 0.50g).

Exception 3 applies to detached one- or two-family dwellings not exceeding two stories above grade plane without significant structural irregularities. The exception is limited to those structures that do not have any of the following irregularities: torsional irregularity; extreme torsional irregularity; nonparallel systems; stiffness irregularity (soft story); stiffness irregularity (extreme soft story) or discontinuity in capacity (weak story). It is important to emphasize that this exception is for the exemption from additional special inspections, not for the design of the structure in accordance with the structural the requirements of the code.

1705.11.1 Structural steel. *Special inspection* for structural steel shall be in accordance with the quality assurance requirements of AISC 341.

> **Exception:** *Special inspections* of structural steel in structures assigned to *Seismic Design Category* C that are not specifically detailed for seismic resistance, with a response modification coefficient, *R*, of 3 or less, excluding cantilever column systems.

❖ Section 1705.11 requires the special inspection of seismic-force-resisting systems of structures classified as Seismic Design Category C, D, E or F. This section specifically requires special inspection of the structural steel system in accordance with AISC 341 with the one exception. Also note that Section 2205.2.2 requires structural steel seismic-force-resisting systems to be designed in accordance with AISC 341 for structures in Seismic Design Category D, E or F.

Section 1705.11 makes a general reference to Seismic Design Category C and higher for seismic-force-resisting systems because many of the structural systems that are required utilize special seismic detailing; however, the general requirement does not reflect the allowance in Section 2205.2.1 for steel systems that are not specifically detailed for seismic resistance. The exception recognizes that buildings classified as Seismic Design Category C may use a seismic system that is designed using a response coefficient, $R = 3$. This recognizes the inherent ductility of these structures that are permitted to be constructed in accordance with AISC 360 (i.e., not detailed in accordance with the provisions of AISC 341). As these construction details and connections are the same as would be used in any steel buildings designed according to AISC 360, no additional inspection or testing is required beyond that required for typical steel buildings in Section 1705.2.1.

1705.11.2 Structural wood. Continuous special inspection is required during field gluing operations of elements of the seismic force-resisting system. Periodic special inspection is required for nailing, bolting, anchoring and other fastening of components within the seismic force-resisting system, including wood shear walls, wood diaphragms, drag struts, braces, shear panels and hold-downs.

Exception: *Special inspection* is not required for wood shear walls, shear panels and diaphragms, including nailing, bolting, anchoring and other fastening to other components of the seismic force-resisting system, where the fastener spacing of the sheathing is more than 4 inches (102 mm) on center (o.c.).

❖ Continuous special inspection of field gluing operations of structural wood is required, while special inspection of fastenings of components within the seismic-force-resisting system is required on a periodic basis. Since special inspection should only be necessary for complex, highly loaded systems, an exception waives the special inspection of fastening in shear walls and diaphragms that are lightly loaded. Lower design loads mean these elements will have less stringent requirements for bolting, anchoring and fastening. The building official who is already inspecting the sheathing will also inspect fastening in these instances. Rather than being based on the component's design shear, the exception is based on the sheathing fastener spacing as a more practical threshold. All things being equal, as the fastener spacing decreases the component's design load is higher and the tighter spacing increases the potential for splitting framing members. The latter concern is a quality issue associated primarily with nailing.

1705.11.3 Cold-formed steel light-frame construction. Periodic special inspection is required during welding operations of elements of the seismic force-resisting system. Periodic special inspection is required for screw attachment, bolting, anchoring and other fastening of components within the seismic force-resisting system, including shear walls, braces, diaphragms, collectors (drag struts) and hold-downs.

Exception: *Special inspection* is not required for cold-formed steel light-frame shear walls, braces, diaphragms, collectors (drag struts) and hold-downs where either of the following apply:

1. The sheathing is gypsum board or fiberboard.
2. The sheathing is wood structural panel or steel sheets on only one side of the shear wall, shear panel or diaphragm assembly and the fastener spacing of the sheathing is more than 4 inches (102 mm) o.c.

❖ Periodic special inspection is required for cold-formed steel framing and its fastening. The exception relaxes the special inspection requirement where demands are low, which is reflected by the sheathing fastener spacing of more than 4 inches (102 mm) on center, or where gypsum board or fiberboard sheathing is used.

1705.11.4 Designated seismic systems. The special inspector shall examine designated seismic systems requiring seismic qualification in accordance with Section 1705.12.3 and verify that the *label*, anchorage or mounting conforms to the *certificate of compliance*.

❖ For elements of the designated seismic system as defined in Section 11.2 of ASCE 7, the special inspector is required to verify that the component label, anchorage or mounting conforms to the certificate of compliance.

1705.11.5 Architectural components. Periodic *special inspection* is required during the erection and fastening of exterior cladding, interior and exterior nonbearing walls and interior and exterior veneer in structures assigned to *Seismic Design Category* D, E or F.

Exceptions:

1. *Special inspection* is not required for exterior cladding, interior and exterior nonbearing walls and interior and exterior veneer 30 feet (9144 mm) or less in height above grade or walking surface.
2. *Special inspection* is not required for exterior cladding and interior and exterior veneer weighing 5 psf (24.5 N/m^2) or less.
3. *Special inspection* is not required for interior nonbearing walls weighing 15 psf (73.5 N/m^2) or less.

❖ Although this section is titled "Architectural components," the requirements refer specifically to exterior cladding, interior and exterior nonbearing walls and interior and exterior veneer. It follows that architectural components other than exterior cladding, interior and exterior nonbearing walls and interior and exterior veneer are not subject to these inspections. It applies to nonbearing walls, which are considered nonstructural components as opposed to bearing walls, which are structural elements, and are therefore subject to the special inspections required for the material used to construct the wall. Periodic special inspection of the installation of the listed architectural components verifies that the intent of the design is carried out in the field. The listed components are often supported at different levels of a structure, meaning the component and connection design must consider relative movements. Where connections are designed to allow for relative displacements, verifying that the installation is in accordance with the construction documents is a prudent measure. The special inspector should verify that the method of anchoring or fastening and the number, spacing and types of fasteners actually used conform with the approved construction documents for the component installed.

Exception 1 exempts construction that is not more than 30 feet (9144 mm) above grade or above a walking surface. Exception 2 exempts cladding and veneer weighing 5 psf (24.5 N/m^2) or less from special inspection. Exception 3 exempts interior nonbearing walls, but not exterior nonbearing walls, weighing 15 psf (73.5 N/m^2) or less from special inspection. Restated, this section requires periodic special inspection for the erection and fastening of exterior nonbearing walls, cladding and veneer weighing more than 5 psf (24.5 N/m^2) and interior nonbearing walls weighing more than 15 psf (73.5 N/m^2) when any of these components are installed more than 30 feet (9144 mm) above grade or a walking surface.

1705.11.5.1 Access floors. Periodic *special inspection* is required for the anchorage of access floors in structures assigned to *Seismic Design Category* D, E or F.

❖ Access floors consist of a system of panels and supports that create a raised floor above the actual structural floor system. By raising the floor, a space is created in between the raised floor and the structural floor where various components like wiring for power, voice, and data can be routed. This space has also become increasingly valuable for heating, ventilation, and air-conditioning (HVAC) distribution either as a plenum space or with defined ductwork. Because failure of the floor system can pose a threat to the occupants in high seismic areas, anchorage of access floors require periodic special inspection in buildings in Seismic Design Category D, E and F.

1705.11.6 Mechanical and electrical components. *Special inspection* for mechanical and electrical components shall be as follows:

1. Periodic special inspection is required during the anchorage of electrical equipment for emergency or standby power systems in structures assigned to *Seismic Design Category* C, D, E or F;
2. Periodic special inspection is required during the anchorage of other electrical equipment in structures assigned to *Seismic Design Category* E or F;
3. Periodic special inspection is required during the installation and anchorage of piping systems designed to carry hazardous materials and their associated mechanical units in structures assigned to *Seismic Design Category* C, D, E or F;
4. Periodic special inspection is required during the installation and anchorage of ductwork designed to carry hazardous materials in structures assigned to *Seismic Design Category* C, D, E or F; and
5. Periodic special inspection is required during the installation and anchorage of vibration isolation systems in structures assigned to *Seismic Design Category* C, D, E or F where the *construction documents* require a nominal clearance of $^1/_4$ inch (6.4 mm) or less between the equipment support frame and restraint.

❖ It is anticipated that the minimum requirements for mechanical and electrical components will be complied with when the special inspector is satisfied that the method of anchoring or fastening, and the number, spacing and types of fasteners actually used, conforms to the approved construction documents for the components installed. It is noted that such special inspection requirements are for selected electrical, lighting, piping and ductwork components in Seismic Design Category C, D, E or F. Periodic special inspection is required during the anchorage of electrical equipment for emergency or standby power systems, installation and anchorage of piping systems designed to carry hazardous materials and their mechanical units, and installation and anchorage of ductwork designed to carry hazardous materials in structures assigned to Seismic Design Category C, D, E or F. Periodic special inspection is required during the anchorage for other electrical equipment in structures assigned to Seismic Design Category E or F.

Item 5 requires the inspection of vibration isolation systems where an optional lower clearance is specified by the design. Typically, vibration-isolated mechanical equipment requires an increased seismic design force. Note b of Table 13.6-1 of ASCE 7 provides an option to reduce this seismic design force on vibration-isolated components and systems having a clearance of $^1/_4$ inch (6.4 mm) or less between the equipment support frame and restraint. To confirm that the design intent is carried out in the field, isolated equipment installations that utilize a reduced clearance are subject to special inspection.

1705.11.7 Storage racks. Periodic *special inspection* is required during the anchorage of storage racks 8 feet (2438 mm) or greater in height in structures assigned to *Seismic Design Category* D, E or F.

❖ Tall storage racks such as those found in large building supply stores pose a threat to the public in high seismic areas. Because anchorage is the most critical element, anchorage of storage racks greater than or equal to 8 feet (2438 mm) in height in structures assigned to Seismic Design Category D, E or F require periodic special inspection.

1705.11.8 Seismic isolation systems. Periodic special inspection shall be provided for seismic isolation systems during the fabrication and installation of isolator units and energy dissipation devices.

❖ Seismic isolation units and energy dissipation devices are required to have periodic special inspection during fabrication and installation. The appropriate threshold for requiring such inspections would be that listed for seismic-force-resisting systems in Item 1 of Section 1705.11.

1705.12 Testing and qualification for seismic resistance. The testing and qualification specified in Sections 1705.12.1 through 1705.12.4, unless exempted from *special inspections* by the exceptions of Section 1704.2 are required as follows:

1. The seismic force-resisting systems in structures assigned to *Seismic Design Category* C, D, E or F shall meet the requirements of Sections 1705.12.1 and 1705.12.2, as applicable.
2. Designated seismic systems in structures assigned to *Seismic Design Category* C, D, E or F and subject to the certification requirements of ASCE 7 Section 13.2.2 shall comply with Section 1705.12.3.
3. Architectural, mechanical and electrical components in structures assigned to *Seismic Design Category* C, D, E or F and where the requirements of ASCE 7 Section 13.2.1 are met by submittal of manufacturer's certification, in accordance with Item 2 therein, shall comply with Section 1705.12.3.
4. The seismic isolation system in seismically isolated structures shall meet the testing requirements of Section 1705.12.4.

❖ This section specifies when material seismic-resistance tests for seismic-force-resisting systems and designated seismic systems are required. These requirements supplement the test requirements contained in the referenced standards given in other sections of the code. The seismic provisions of the code are frequently based on assumed material behavior. Material tests are key to verifying the quality of material that is used for certain seismic-resistant construction.

Some inelastic behavior can be anticipated in a building's response to the design earthquake. Strong ground shaking caused by earthquakes tends to expose any underlying flaws in a building's construction or design. Thus, the code specifies testing of some structural and nonstructural components as additional verification that the seismic resistance provided in the finished structure meets the intent of the design. It is imperative that the professional who designs systems that are critical to the earthquake performance also identifies them and specifies the necessary testing in the statement of special inspections.

Item 1 states the seismic risk threshold for seismic-force-resisting systems. The testing of items in Sections 1705.12.1 and 1705.12.2 would apply to these structures. Item 2 covers designated seismic systems that are nonstructural components and systems that are assigned an importance factor greater than 1. The testing applicable to designated seismic systems is given in Section 1705.12.3. Item 3 requires testing for architectural, electrical and mechanical components where the general design requirements of ASCE 7 Section 13.2.1, Item 2 for manufacturer's certification are satisfied by testing. Testing of seismic isolation systems is covered in Section 1705.12.4 which references Section 17.8 of ASCE 7.

1705.12.1 Concrete reinforcement. Where reinforcement complying with ASTM A 615 is used to resist earthquake-induced flexural and axial forces in special moment frames, special structural walls and coupling beams connecting special structural walls, in structures assigned to *Seismic Design Category* B, C, D, E or F, the reinforcement shall comply with Section 21.1.5.2 of ACI 318. Certified mill test reports shall be provided for each shipment of such reinforcement. Where reinforcement complying with ASTM A 615 is to be welded, chemical tests shall be performed to determine weldability in accordance with Section 3.5.2 of ACI 318.

❖ Certified material test reports are required for rebar. When ASTM A 615 is used in special moment frames and shear walls, the testing requirements of Section 21.1.5.2 of ACI 318 must be used. Where ASTM A 615 rebar is to be welded, a material properties report must be provided in accordance with ACI 318 Section 3.5.2 to determine that the weldability of the steel conforms to AWS D1.4.

1705.12.2 Structural steel. Testing for structural steel shall be in accordance with the quality assurance requirements of AISC 341.

> **Exception:** Testing for structural steel in structures assigned to *Seismic Design Category* C that are not specifically detailed for seismic resistance, with a response modification coefficient, *R*, of 3 or less, excluding cantilever column systems.

❖ Structural steel must be tested as required by AISC 341. Appendix Q of AISC 341 provides the code user with the minimum acceptable requirements for a quality assurance plan that applies to the construction of welded joints, bolted joints and other details in the seismic-force-resisting system. Where appropriate, the appendix references AWS D1.1 for specific acceptance criteria.

Section 1705.12 makes a general reference to Seismic Design Category C and higher for seismic-force-resisting systems because many of the structural systems that are required utilize special detailing; however, the general requirement does not reflect the allowance in Section 2205.2.1 for steel systems not specifically detailed for seismic resistance.

The exception recognizes that buildings classified as Seismic Design Category C may employ a seismic system that is designed using a response coefficient, $R = 3$. This recognizes the inherent ductility of these structures that are permitted to be constructed in accordance with AISC 360, and not be detailed in accordance with the provisions of AISC 341. As these construction details and connections are the same as would be used in any steel building designed according to AISC 360, no additional inspection or testing is required beyond that required for typical steel buildings in Section 1705.2.1.

1705.12.3 Seismic certification of nonstructural components. The *registered design professional* shall specify on the construction documents the requirements for certification by analysis, testing or experience data for nonstructural components and designated seismic systems in accordance with Section 13.2 of ASCE 7, where such certification is required by Section 1705.12.

❖ The registered design professional is required to specify the applicable seismic certification requirements on the construction documents for non structural components and designated seismic systems as required by Section 1705.12 Items 3 and 4. This is necessary for the manufacturer who is required to provide a certificate of compliance. Affected mechanical and electrical components are required to demonstrate seismic performance by testing or analytical methods. The intent is to verify the ability of critical equipment to not only survive an earthquake but also to remain functional. This is accomplished by the special certification of designated seismic systems.

All nonstructural components must comply with the general design provisions of Section 13.2.1 of ASCE 7, which permits justification of components by project-specific design or certification by the manufacturer (through analysis, testing or experience data). Item 2 is meant to clarify this intention.

A certificate of compliance for review and acceptance is also required. The registered design professional responsible must then review and accept the certificate of compliance, and the building official must approve it. ICC Evaluation Service's AC 156, *Acceptance Criteria for Seismic Qualification by Shake-table Testing of Nonstructural Components and Systems* provides testing criteria that can be used to comply with this provision.

1705.12.4 Seismic isolation systems. Seismic isolation systems shall be tested in accordance with Section 17.8 of ASCE 7.

❖ The referenced section of ASCE 7 contains detailed provisions for isolation system testing. This testing provides effective stiffness and damping values to be used in the design of a seismically isolated structure. A minimum of two full-size specimens must be tested for each proposed type and size of isolator. These prototypes are not to be used in the construction, unless approved by both the registered design professional and building official.

1705.13 Sprayed fire-resistant materials. *Special inspections* for sprayed fire-resistant materials applied to floor, roof and wall assemblies and structural members shall be in accordance with Sections 1705.13.1 through 1705.13.6. *Special inspections* shall be based on the fire-resistance design as designated in the *approved construction documents*. The tests set forth in this section shall be based on samplings from specific floor, roof and wall assemblies and structural members. *Special inspections* shall be performed after the rough installation of electrical, automatic sprinksler, mechanical and plumbing systems and suspension systems for ceilings, where applicable.

❖ Section 704.13 contains code requirements for the application of sprayed fire-resistant materials (SFRM). The application must be in accordance with the terms and conditions of the listing and the manufacturer's instructions. The special inspection of SFRM verifies that the requirements for thickness, density and bond strength that are specified in the design have been satisfied by the actual installation. These inspections are to be performed after the rough installation of mechanical, electrical, plumbing, automatic sprinkler systems and ceiling suspension systems.

1705.13.1 Physical and visual tests. The *special inspections* shall include the following tests and observations to demonstrate compliance with the listing and the fire-resistance rating:

1. Condition of substrates.
2. Thickness of application.
3. Density in pounds per cubic foot (kg/m^3).
4. Bond strength adhesion/cohesion.
5. Condition of finished application.

❖ To verify that an SFRM performs as intended, certain conditions are required to be met. The verifications include: substrate conditions; thickness and density of material, as well as the condition of the finished application. Section 704.13 states the substrate condition and finished condition requirements.

1705.13.2 Structural member surface conditions. The surfaces shall be prepared in accordance with the *approved* fire-resistance design and the written instructions of *approved* manufacturers. The prepared surface of structural members to be sprayed shall be inspected before the application of the sprayed fire-resistant material.

❖ The integrity of an SFRM system depends on the conditions of the surface of the steel member to which it is to be applied. The system must be fully adhered to the surface for proper performance, in accordance with design values. See Section 704.13.3.1 for requirements related to the substrate surface conditions.

1705.13.3 Application. The substrate shall have a minimum ambient temperature before and after application as specified in the written instructions of *approved* manufacturers. The area for application shall be ventilated during and after application as required by the written instructions of *approved* manufacturers.

❖ During application of SFRMs, and immediately thereafter during cure of the material, several items must be controlled, including the ambient temperature during the application and temperature of the substrate and the SFRMs. Temperature control is important to determine that the necessary chemical reactions needed to make a particular material bond to the steel surfaces and hold together do, in fact, happen. The minimum or maximum temperatures necessary for proper bond and cure depend on the specific type of material (see Section 704.13.4 for more information). The scope of the special inspection also includes verification of the proper ventilation during application, as well as for curing.

1705.13.4 Thickness. No more than 10 percent of the thickness measurements of the sprayed fire-resistant materials applied to floor, roof and wall assemblies and structural members shall be less than the thickness required by the *approved* fire-resistance design, but in no case less than the minimum allowable thickness required by Section 1705.13.4.1.

❖ For the system to provide the required design fire-resistance rating, it must be applied at the appropriate thickness. This section establishes an acceptable percentage of thickness readings that can fall below the specified design value, provided none of these readings are less than the minimum established in Section 1705.13.4.1. These limitations on the thickness measurements are intended to provide a high confidence level that the installed material meets, or exceeds, the design requirements.

1705.13.4.1 Minimum allowable thickness. For design thicknesses 1 inch (25 mm) or greater, the minimum allowable individual thickness shall be the design thickness minus $^{1}/_{4}$ inch (6.4 mm). For design thicknesses less than 1 inch (25 mm), the minimum allowable individual thickness shall be the design thickness minus 25 percent. Thickness shall be determined in accordance with ASTM E 605. Samples of the sprayed fire-resistant materials shall be selected in accordance with Sections 1705.13.4.2 and 1705.13.4.3.

❖ This requirement prevents the combination of very thin readings with thicker readings in order to show compliance. The required sampling provided for floor, roof and wall assemblies in Section 1705.13.4.2 and structural members in Section 1705.13.4.3 is based on ASTM E 605 with some modifications. This standard also provides testing methods that are commonly used by the industry.

1705.13.4.2 Floor, roof and wall assemblies. The thickness of the sprayed fire-resistant material applied to floor, roof and wall assemblies shall be determined in accordance with ASTM E 605, making not less than four measurements for each 1,000 square feet (93 m^2) of the sprayed area, or portion thereof, in each *story*.

❖ Sampling of an SFRM for membrane components (floors, roofs or walls) is based on the square footage of the components. The number of samples is increased in the code by using a sampling area of every 1,000 square feet (93 m^2) rather than 10,000 square feet (929 m^2) as is specified in ASTM E 605. This is intended to provide a higher level of confidence in the performance of the installed assembly.

1705.13.4.3 Cellular decks. Thickness measurements shall be selected from a square area, 12 inches by 12 inches (305 mm by 305 mm) in size. A minimum of four measurements shall be made, located symmetrically within the square area.

❖ This section is consistent with the procedure in ASTM E 605 for flat decks.

1705.13.4.4 Fluted decks. Thickness measurements shall be selected from a square area, 12 inches by 12 inches (305 mm by 305 mm) in size. A minimum of four measurements shall be made, located symmetrically within the square area, including one each of the following: valley, crest and sides. The average of the measurements shall be reported.

❖ This section is consistent with the procedure in ASTM E 605 for fluted decks.

1705.13.4.5 Structural members. The thickness of the sprayed fire-resistant material applied to structural members shall be determined in accordance with ASTM E 605. Thickness testing shall be performed on not less than 25 percent of the structural members on each floor.

❖ Sampling of the SFRM for structural elements is based on the square footage of each floor. Sample size and number of elements represented are based on ASTM E 605.

1705.13.4.6 Beams and girders. At beams and girders thickness measurements shall be made at nine locations around the beam or girder at each end of a 12-inch (305 mm) length.

❖ This section is consistent with the procedure in ASTM E 605 for beams.

1705.13.4.7 Joists and trusses. At joists and trusses, thickness measurements shall be made at seven locations around the joist or truss at each end of a 12-inch (305 mm) length.

❖ This section is consistent with the procedure in ASTM E 605 for joists.

1705.13.4.8 Wide-flanged columns. At wide- flanged columns, thickness measurements shall be made at 12 locations around the column at each end of a 12-inch (305 mm) length.

❖ This section is consistent with the procedure in ASTM E 605 for columns.

1705.13.4.9 Hollow structural section and pipe columns. At hollow structural section and pipe columns, thickness measurements shall be made at a minimum of four locations around the column at each end of a 12-inch (305 mm) length.

❖ This section provides guidance for the thickness sampling of hollow structural sections that is not provided in ASTM E 605.

1705.13.5 Density. The density of the sprayed fire-resistant material shall not be less than the density specified in the *approved* fire-resistance design. Density of the sprayed fire-resistant material shall be determined in accordance with ASTM E 605. The test samples for determining the density of the sprayed fire-resistant materials shall be selected as follows:

1. From each floor, roof and wall assembly at the rate of not less than one sample for every 2,500 square feet (232 m^2) or portion thereof of the sprayed area in each *story*.

2. From beams, girders, trusses and columns at the rate of not less than one sample for each type of structural

member for each 2,500 square feet (232 m^2) of floor area or portion thereof in each *story*.

❖ The density of an SFRM will have an impact on the fire-resistance rating of the system; therefore, it is important that the density of the material be measured to verify that the product is as designed. The sampling requirements differ from those for thickness measurements. The required method of determining density is provided in ASTM E 605, but the sample size is based on a sampling area of every 2,500 square feet (232 m^2) rather than 10,000 square feet (929 m^2), as is specified in ASTM E 605. This is intended to provide a higher level of confidence in the performance of the finished assembly.

1705.13.6 Bond strength. The cohesive/adhesive bond strength of the cured sprayed fire-resistant material applied to floor, roof and wall assemblies and structural members shall not be less than 150 pounds per square foot (psf) (7.18 kN/m^2). The cohesive/adhesive bond strength shall be determined in accordance with the field test specified in ASTM E 736 by testing in-place samples of the sprayed fire-resistant material selected in accordance with Sections 1705.13.6.1 through 1705.13.6.3.

❖ The adhesion of a sprayed-on material is critical to its performance. This is the key factor in minimizing the chances of the material becoming dislodged. A minimum cohesive/adhesive bond strength of 150 pounds per square foot (psf) (7.18 kN/m^2) is required in this section based on the American Institute of Architects (AIA) Master Specification and the recommendations of the General Services Administration (GSA) for durability and serviceability of the material. While this minimum bond strength may be generally suitable, note that Section 403.2.4 provides more stringent requirements that apply to high-rise buildings, where the consequences of dislodged materials can be much greater.

1705.13.6.1 Floor, roof and wall assemblies. The test samples for determining the cohesive/adhesive bond strength of the sprayed fire-resistant materials shall be selected from each floor, roof and wall assembly at the rate of not less than one sample for every 2,500 square feet (232 m^2) of the sprayed area, or portion thereof, in each *story*.

❖ The sampling rate for bond in this section matches the sampling rate for determining density.

1705.13.6.2 Structural members. The test samples for determining the cohesive/adhesive bond strength of the sprayed fire-resistant materials shall be selected from beams, girders, trusses, columns and other structural members at the rate of not less than one sample for each type of structural member for each 2,500 square feet (232 m^2) of floor area or portion thereof in each *story*.

❖ The bond strength sampling rate in this section is the same as that indicated in Section 1705.13.5 for determining density.

1705.13.6.3 Primer, paint and encapsulant bond tests. Bond tests to qualify a primer, paint or encapsulant shall be conducted when the sprayed fire-resistant material is applied to a primed, painted or encapsulated surface for which acceptable bond-strength performance between these coatings and the fire-resistant material has not been determined. A bonding agent *approved* by the SFRM manufacturer shall be applied to a primed, painted or encapsulated surface where the bond strengths are found to be less than required values.

❖ The in-place adhesion of SFRM can be reduced by a factor of 10 when applied over certain primers as opposed to the adhesion obtained by the rated material applied on bare, clean steel. Where the listing does not consider application over such materials, then its bond strength must be determined as described in this section. It is necessary to apply a bonding agent when the bond strength is less than required due to the effect of an encapsulated, painted or primed surface. Also see Section 704.13.3.2 for additional conditions on application of the SFRM.

1705.14 Mastic and intumescent fire-resistant coatings. *Special inspections* for mastic and intumescent fire-resistant coatings applied to structural elements and decks shall be in accordance with AWCI 12-B. *Special inspections* shall be based on the fire-resistance design as designated in the *approved construction documents*.

❖ Special inspection of mastic and intumescent fire-resistant coatings is justified because these products are often complex systems that require special expertise from applicators and quality assurance personnel. It is essential to confirm that their installation is in accordance with the manufacturer's instructions and the terms of their listing so that they will perform as expected.

1705.15 Exterior insulation and finish systems (EIFS). *Special inspections* shall be required for all EIFS applications.

Exceptions:

1. *Special inspections* shall not be required for EIFS applications installed over a *water-resistive barrier* with a means of draining moisture to the exterior.
2. *Special inspections* shall not be required for EIFS applications installed over masonry or concrete walls.

❖ Special inspections are required for all EIFS installations except for the two exceptions in this section. Exception 1 recognizes that EIFS, which are installed over a water-resistive barrier and incorporate flashings at penetrations and terminations, and a means of drainage to the exterior, afford a built-in redundancy to water penetration that makes the need for special inspections less critical.

Exception 2 recognizes that concrete and masonry substrates are relatively durable and the exposure to moisture in wall conditions does not necessarily have a detrimental effect on these materials.

1705.15.1 Water-resistive barrier coating. A *water-resistive barrier* coating complying with ASTM E 2570 requires *special inspection* of the *water-resistive barrier* coating when installed over a sheathing substrate.

❖ Where an EIFS is utilized in residential occupancies of Type V construction, an EIFS with drainage is required by Chapter 14. This specific type of system incorporates a means of drainage applied over a water-resistive barrier and Chapter 14 allows a water-resistive barrier coating complying with ASTM E 2570 as an option. This section requires special inspection of that barrier coating.

1705.16 Fire-resistant penetrations and joints. In high-rise buildings or in buildings assigned to *Risk Category* III or IV in accordance with Section 1604.5, special inspections for through-penetrations, membrane penetration firestops, fire-resistant joint systems, and perimeter fire barrier systems that are tested and listed in accordance with Sections 714.3.1.2, 714.4.1.2, 715.3 and 715.4 shall be in accordance with Section 1705.16.1 or 1705.16.2.

❖ Through-penetration and membrane-penetration firestop systems, as well as fire-resistant joint systems and perimeter fire barrier systems, are critical to maintaining the fire-resistive integrity of fire-resistance-rated construction, including fire walls, fire barriers, fire partitions, smoke barriers and horizontal assemblies. The proper selection and installation of such systems must be in compliance with the code and/or appropriate listing. With thousands of listed firestop and joint systems available, each with variations that multiplies possible systems for a building exponentially, the selection of the correct system is not a generic process. Where such systems are used in two types of buildings considered as "high risk," it is mandatory that they be included as a part of the special inspection process. Such "high risk" buildings are identified as:

- Buildings assigned to Risk Category III or IV in accordance with Section 1604.5 and
- High-rise buildings.

Although the proper application of firestop and joint system requirements is very important in all types and sizes of buildings, the requirement for special inspection is limited to specific building types that represent a substantial hazard to human life in the event of a system failure or that are considered to be essential facilities. Inspection to ASTM E 2174 for penetration firestop systems and ASTM E 2393 for fire-resistant joint systems brings an increased level of review to this important discipline.

1705.16.1 Penetration firestops. Inspections of penetration firestop systems that are tested and listed in accordance with Sections 714.3.1.2 and 714.4.1.2 shall be conducted by an approved inspection agency in accordance with ASTM E 2174.

❖ A primary method of addressing a penetration of a fire-resistance-rated wall assembly is through the use of an approved firestop system installed in accordance with ASTM E 814 or UL 1479. The system must have an F rating that is not less than the fire-resistance rating of the wall being penetrated. It is critical that the firestop system be appropriate for the penetration being protected. The choice of firestop systems varies based upon the size and material of the penetrating item, as well as the construction materials and fire-resistance rating of the wall being penetrated. Special inspection of the firestop system is intended to verify that the appropriate system has been specified and the installation is in conformance with its listing.

1705.16.2 Fire-resistant joint systems. Inspection of fire-resistant joint systems that are tested and listed in accordance with Sections 715.3 and 715.4 shall be conducted by an approved inspection agency in accordance with ASTM E 2393.

❖ A "Joint" is defined as a "linear opening in or between adjacent fire-resistance-rated assemblies that is designed to allow independent movement of the building in any plane caused by thermal, seismic, wind or any other loading." The joint creates an interruption of the fire-resistant integrity of the wall or floor system, requiring the use of an appropriate fire-resistant joint system. The code mandates general installation criteria for such systems and requires them to be tested in accordance with ASTM E 1966 or UL 2079. Much like the inspection of penetration firestop systems, the proper choice and installation of fire-resistant joint systems can be verified through a comprehensive special inspection process.

Although regulated under the provisions for fire-resistant joint systems, a second type of system is technically not a joint but rather an extension of protection afforded by a horizontal assembly. The void created at the intersection of an exterior curtain wall assembly and a fire-resistance-rated floor or floor/ceiling assembly must be filled in a manner that maintains the integrity of the horizontal assembly. The system utilized to fill the void must be in compliance with ASTM E 2307 and able to resist the passage of flame for a time period equal to that of the floor assembly. Special inspection is necessary to verify that the appropriate joint system is chosen and installed.

[F] 1705.17 Special inspection for smoke control. Smoke control systems shall be tested by a special inspector.

❖ Because smoke control systems are unique and complex life safety systems, this section requires that all smoke control systems be tested by special inspection.

[F] 1705.17.1 Testing scope. The test scope shall be as follows:

1. During erection of ductwork and prior to concealment for the purposes of leakage testing and recording of device location.

2. Prior to occupancy and after sufficient completion for the purposes of pressure difference testing, flow measurements and detection and control verification.

❖ Special inspections need to occur at two different stages during the installation of a smoke control system. The first round of special inspections occurs before concealment of the ductwork or fire protection elements. At this stage, the special inspector needs to verify that duct leakage is in accordance with Section 909.10.2. Additionally, the location of all fire protection devices needs to be verified and documented at this time. The second round of special inspections occurs just prior to occupancy in order to more closely replicate the conditions under which the system must operate. The inspections include the verification of pressure differences across smoke barriers as required in Sections 909.5.1 and 909.18.6, the verification of appropriate volumes of airflow as noted in the design, and finally the verification of the appropriate operation of the detection and control mechanisms as required in Sections 909.18.1 and 909.18.7 (see Section 909.18.8.3 for report requirements).

[F] 1705.17.2 Qualifications. *Special inspection* agencies for smoke control shall have expertise in fire protection engineering, mechanical engineering and certification as air balancers.

❖ This provision establishes a certain level of qualifications for the inspection of smoke control systems that would include the need for expertise in fire protection engineering, mechanical engineering and certification as air balancers.

SECTION 1706
DESIGN STRENGTHS OF MATERIALS

1706.1 Conformance to standards. The design strengths and permissible stresses of any structural material that are identified by a manufacturer's designation as to manufacture and grade by mill tests, or the strength and stress grade is otherwise confirmed to the satisfaction of the *building official*, shall conform to the specifications and methods of design of accepted engineering practice or the *approved* rules in the absence of applicable standards.

❖ Structural materials must conform to applicable design standards, approved rules and accepted methods of engineering practice. Conformance to these provisions and to the manufacturer's designations provides the building official with the information needed to verify that the materials will perform their intended function satisfactorily.

1706.2 New materials. For materials that are not specifically provided for in this code, the design strengths and permissible stresses shall be established by tests as provided for in Section 1707.

❖ Materials that are not explicitly covered by the code are allowed when subjected to the appropriate testing demonstrating adequate performance (see Section 1701.2).

SECTION 1707
ALTERNATIVE TEST PROCEDURE

1707.1 General. In the absence of *approved* rules or other *approved* standards, the *building official* shall make, or cause to be made, the necessary tests and investigations; or the *building official* shall accept duly authenticated reports from *approved agencies* in respect to the quality and manner of use of new materials or assemblies as provided for in Section 104.11. The cost of all tests and other investigations required under the provisions of this code shall be borne by the applicant.

❖ Test reports from approved agencies may be used as a basis for approval of materials that are not within the purview of any approved rules (i.e., "new materials" as mentioned in Section 1701.2). This section directly references Section 104.11. It is within the power of the building official to accept reports from an approved agency. In determining the approval, the building official should check that the agency is an independent third-party agency with no financial or fiduciary affiliations with the applicant or material supplier. The capability and competency of the agency must also be examined. It should be noted that this section assigns responsibility for the costs of testing to the applicant.

SECTION 1708
TEST SAFE LOAD

1708.1 Where required. Where proposed construction is not capable of being designed by *approved* engineering analysis, or where proposed construction design method does not comply with the applicable material design standard, the system of construction or the structural unit and the connections shall be subjected to the tests prescribed in Section 1710. The *building official* shall accept certified reports of such tests conducted by an *approved* testing agency, provided that such tests meet the requirements of this code and *approved* procedures.

❖ Testing to determine safe load is required when a structural component cannot be designed in accordance with approved engineering practices or where the construction design method does not fully comply with the respective material design standard listed in Chapter 35. If either of these situations exist, the structural components are required to be subjected to the prescriptive tests listed in Section 1710, which address loading and deflection criteria.

An example of a structural component that cannot be designed by approved engineering practice is a composite concrete and steel slab in which the shear connector is some type of new configuration called a "widget." The horizontal shear that can be developed is unknown; therefore, a complete analysis cannot be performed. This section also restates the building official's option of accepting data from an approved testing agency, as previously stated in Section 1703.4.

SECTION 1709
IN-SITU LOAD TESTS

1709.1 General. Whenever there is a reasonable doubt as to the stability or load-bearing capacity of a completed building, structure or portion thereof for the expected loads, an engineering assessment shall be required. The engineering assessment shall involve either a structural analysis or an in-situ load test, or both. The structural analysis shall be based on actual material properties and other as-built conditions that affect stability or load-bearing capacity, and shall be conducted in accordance with the applicable design standard. If the structural assessment determines that the load-bearing capacity is less than that required by the code, load tests shall be conducted in accordance with Section 1709.2. If the building, structure or portion thereof is found to have inadequate stability or load-bearing capacity for the expected loads, modifications to ensure structural adequacy or the removal of the inadequate construction shall be required.

❖ The intent of this section is to utilize an engineering analysis to verify the adequacy of the structure, if possible. The load test requirement should only be done if an engineering analysis does not verify structural adequacy. Load tests are last resort options, and the building official should document his or her reasons for any load testing requirement.

An example of the executed structural analysis would be an analysis by a third-party engineering firm acceptable to both the building official and owner. The structural integrity may be questioned for items such as visible signs of excessive settlement or lateral deflection, such as cracks in concrete foundation walls or excessive vibration when the assembly is loaded.

A load test procedure must simulate the actual load conditions to which the structure is subjected during normal use (see Section 1709.3 for details).

1709.2 Test standards. Structural components and assemblies shall be tested in accordance with the appropriate referenced standards. In the absence of a standard that contains an applicable load test procedure, the test procedure shall be developed by a *registered design professional* and *approved.* The test procedure shall simulate loads and conditions of application that the completed structure or portion thereof will be subjected to in normal use.

❖ When load test procedures for materials are given by the applicable referenced material standard, the test procedure outlined in that specific standard must be adhered to without variation. If a referenced standard lacks a load test procedure, or a material or assembly does not have a specific referenced standard, then such a test must be developed by a registered design professional and approved by the building official. The test procedure must be representative of, and simulate the actual loading conditions that, the completed structure or portion thereof will be subjected to during normal use.

1709.3 In-situ load tests. In-situ load tests shall be conducted in accordance with Section 1709.3.1 or 1709.3.2 and shall be supervised by a *registered design professional.* The test shall simulate the applicable loading conditions specified in Chapter 16 as necessary to address the concerns regarding structural stability of the building, structure or portion thereof.

❖ The criteria for in-situ load tests are set forth for two categories: procedures specified, which are regulated by Section 1709.3.1, and procedures not specified, which are regulated by Section 1709.3.2. This section further requires that the test be performed under the supervision of a registered design professional and that it simulates the actual loads and conditions of the completed structure or portion thereof.

1709.3.1 Load test procedure specified. Where a referenced standard contains an applicable load test procedure and acceptance criteria, the test procedure and acceptance criteria in the standard shall apply. In the absence of specific load factors or acceptance criteria, the load factors and acceptance criteria in Section 1709.3.2 shall apply.

❖ The load test must be in accordance with the applicable referenced standard. Section 1709.3.2 must only be utilized in the absence of either a specific standard or specific load factors and acceptance criteria from applicable referenced standards.

1709.3.2 Load test procedure not specified. In the absence of applicable load test procedures contained within a standard referenced by this code or acceptance criteria for a specific material or method of construction, such *existing structure* shall be subjected to a test procedure developed by a *registered design professional* that simulates applicable loading and deformation conditions. For components that are not a part of the seismic load-resisting system, the test load shall be equal to two times the unfactored design loads. The test load shall be left in place for a period of 24 hours. The structure shall be considered to have successfully met the test requirements where the following criteria are satisfied:

1. Under the design load, the deflection shall not exceed the limitations specified in Section 1604.3.
2. Within 24 hours after removal of the test load, the structure shall have recovered not less than 75 percent of the maximum deflection.
3. During and immediately after the test, the structure shall not show evidence of failure.

❖ If the applicable standards do not specify load factor or testing criteria acceptance methods, then the testing criteria listed in this section must be followed. Note that the design load includes design live load and all dead loads that are not yet in place, such as the dead load from tenant walls in a speculative office building.

SECTION 1710
PRECONSTRUCTION LOAD TESTS

1710.1 General. In evaluating the physical properties of materials and methods of construction that are not capable of being designed by *approved* engineering analysis or do not comply with the applicable referenced standards, the structural adequacy shall be predetermined based on the load test criteria established in this section.

❖ This section establishes requirements for load testing structural assemblies that are either incapable of being designed or those that, for one reason or another, do not comply with the applicable material design standards. This section does not govern the load testing of existing buildings, which is governed by Section 1709.

The different categories of preconstruction load tests are addressed herein. Specified load test procedures are regulated by Section 1710.2. Load test procedures that are not specified are regulated by Section 1710.3. Wall and partition assemblies are regulated by Section 1710.4. Exterior window and door assemblies are regulated by Section 1710.5, skylights and sloped glazing are regulated by Section 1710.6 and test specimens are regulated by Section 1710.7.

1710.2 Load test procedures specified. Where specific load test procedures, load factors and acceptance criteria are included in the applicable referenced standards, such test procedures, load factors and acceptance criteria shall apply. In the absence of specific test procedures, load factors or acceptance criteria, the corresponding provisions in Section 1710.3 shall apply.

❖ This section has priority over Section 1710.3, provided that load factors and acceptance criteria are established in the applicable design standards.

1710.3 Load test procedures not specified. Where load test procedures are not specified in the applicable referenced standards, the load-bearing and deformation capacity of structural components and assemblies shall be determined on the basis of a test procedure developed by a *registered design professional* that simulates applicable loading and deformation conditions. For components and assemblies that are not a part of the seismic force-resisting system, the test shall be as specified in Section 1710.3.1. Load tests shall simulate the applicable loading conditions specified in Chapter 16.

❖ In the absence of load factors and acceptance criteria in the applicable design standards and in accordance with Section 1710.2, this section is to be used by the building official to determine if conformance to the applicable code requirements has been achieved. Additionally, this section provides the design professional and building official with specific loading and pass/fail criteria.

1710.3.1 Test procedure. The test assembly shall be subjected to an increasing superimposed load equal to not less than two times the superimposed design load. The test load shall be left in place for a period of 24 hours. The tested assembly shall be considered to have successfully met the test requirements if the assembly recovers not less than 75 percent of the maximum deflection within 24 hours after the removal of the test load. The test assembly shall then be reloaded and subjected to an increasing superimposed load until either structural failure occurs or the superimposed load is equal to two and one-half times the load at which the deflection limitations specified in Section 1710.3.2 were reached, or the load is equal to two and one-half times the superimposed design load. In the case of structural components and assemblies for which deflection limitations are not specified in Section 1710.3.2, the test specimen shall be subjected to an increasing superimposed load until structural failure occurs or the load is equal to two and one-half times the desired superimposed design load. The allowable superimposed design load shall be taken as the lesser of:

1. The load at the deflection limitation given in Section 1710.3.2.
2. The failure load divided by 2.5.
3. The maximum load applied divided by 2.5.

❖ Load test criteria relating to superimposed design loads are established herein. These requirements are a compilation of commonly accepted engineering practices to adequately test against structural failure.

In the case of structural components and assemblies for which maximum deflection limitations are not addressed in Section 1710.3.2, the test assemblies must be subjected to increasing superimposed loads until failure occurs or the load is equal to two and one-half times the superimposed design load, whichever occurs first.

1710.3.2 Deflection. The deflection of structural members under the design load shall not exceed the limitations in Section 1604.3.

❖ Acceptance criteria for deflection of structural systems when subjected to the allowable design load are used to demonstrate adequate structural performance and are addressed in Section 1604.3.

1710.4 Wall and partition assemblies. *Load-bearing wall* and partition assemblies shall sustain the test load both with and without window framing. The test load shall include all design load components. Wall and partition assemblies shall be tested both with and without door and window framing.

❖ Load-bearing wall and partition assemblies must sustain loads with and without window framing. It is not appropriate to assume that a wall will sustain loads better if window framing is involved in the test. Each individual design must be evaluated separately based on the construction of that assembly. All design loads, such as vertical and lateral forces, must be included in the test.

1710.5 Exterior window and door assemblies. The design pressure rating of exterior windows and doors in buildings shall be determined in accordance with Section 1710.5.1 or 1710.5.2.

> **Exception:** Structural wind load design pressures for window units smaller than the size tested in accordance with Section 1710.5.1 or 1710.5.2 shall be permitted to be higher than the design value of the tested unit provided such higher pressures are determined by accepted engineering analysis. All components of the small unit shall be the same as the tested unit. Where such calculated design

pressures are used, they shall be validated by an additional test of the window unit having the highest allowable design pressure.

❖ This section allows two methods of load testing for exterior window and door assemblies. The first method, provided for in Section 1710.5.1, allows products to be tested and labeled as conforming to AAMA/WMDA/CSA101/I.S.2/A440. The second method allows products to be tested in accordance with ASTM E 330 and the glazing must comply with Section 2403. The exception allows window units smaller than the size tested to have higher design pressures, provided the higher pressures are determined by accepted engineering analysis, all components of the smaller unit are the same as the tested unit and an additional test of the smaller unit having the highest calculated design pressure is performed in accordance with Section 1710.5.1 or 1710.5.2.

1710.5.1 Exterior windows and doors. Exterior windows and sliding doors shall be tested and labeled as conforming to AAMA/WDMA/CSA101/I.S.2/A440. The *label* shall state the name of the manufacturer, the *approved* labeling agency and the product designation as specified in AAMA/ WDMA/ CSA101/I.S.2/A440. Exterior side-hinged doors shall be tested and *labeled* as conforming to AAMA/WDMA/ CSA101/I.S.2/A440 or comply with Section 1710.5.2. Products tested and labeled as conforming to AAMA/WDMA/ CSA 101/I.S.2/A440 shall not be subject to the requirements of Sections 2403.2 and 2403.3.

❖ This section requires exterior windows and doors complying with AAMA/WMDA/CSA 101/I.S.2/A440 to be labeled as such. Products so tested and labeled must not be required to meet the provisions of Sections 2403.2 and 2403.3. By requiring the product to be labeled, the building official does not have to interpret test results and determine load-carrying capacities, or accept the manufacturer's interpretation of tests.

1710.5.2 Exterior windows and door assemblies not provided for in Section 1710.5.1. Exterior window and door assemblies shall be tested in accordance with ASTM E 330. Structural performance of garage doors and rolling doors shall be determined in accordance with either ASTM E 330 or ANSI/DASMA 108, and shall meet the acceptance criteria of ANSI/DASMA 108. Exterior window and door assemblies containing glass shall comply with Section 2403. The design pressure for testing shall be calculated in accordance with Chapter 16. Each assembly shall be tested for 10 seconds at a load equal to $1^1/_2$ times the design pressure.

❖ This section allows an alternative to the provisions of Section 1710.5.1. This procedure is to be used to verify the integrity of door and window assemblies as a whole, and its results do not supersede, but rather complement, the requirements of Chapter 24. Glass thickness must be determined in accordance with the provisions of Chapter 24. The assemblies must comply with Section 2403 and the design pressure for testing is determined from Chapter 16. Each assembly is required to be tested for 10 seconds at a load equal to one and one-half times the design pressure. When testing a product line that has a variety of sizes, the most critical size can usually be tested and the results used to qualify other similar products within its family. In general, the larger size and most heavily loaded of each particular design, type, construction or configuration should be tested. The code does not specify how many specimens are to be tested. ASTM E 330 states that if only one sample is tested, it should be selected by the specifying authority. ANSI/DASMA 108 addresses the pressure testing of garage doors and includes garage door acceptance criteria, which is not provided by ASTM E 330.

1710.6 Skylights and sloped glazing. Unit skylights and tubular daylighting devices (TDDs) shall comply with the requirements of Section 2405. All other skylights and sloped glazing shall comply with the requirements of Chapter 24.

❖ A tubular daylighting device (TDD) is typically field-assembled from a manufactured kit, unlike a unit skylight which is typically shipped as a factory-assembled unit. The dome of a TDD is not necessarily constructed out of a single panel of glazing material. Thus, a separate definition of a TDD was added to Chapter 2 based on the definition in AAMA/WDMA A440. The section refers to Section 2405 for sloped unit skylight and TDD requirements.

1710.7 Test specimens. Test specimens and construction shall be representative of the materials, workmanship and details normally used in practice. The properties of the materials used to construct the test assembly shall be determined on the basis of tests on samples taken from the load assembly or on representative samples of the materials used to construct the load test assembly. Required tests shall be conducted or witnessed by an *approved agency*.

❖ The test specimen must resemble and simulate as much as possible, the design being tested using materials and workmanship that could be expected in the actual construction or fabrication. The test itself must be witnessed or conducted by an agency acceptable to and approved by the building official.

SECTION 1711
MATERIAL AND TEST STANDARDS

1711.1 Joist hangers. Testing of joist hangers shall be in accordance with Sections 1711.1.1 through 1711.1.3, as applicable.

❖ This section prescribes the test standard and criteria to be used for joist hangers and connectors. This criteria is meant to be used in establishing the load capacity of joist hangers and connectors used in wood construction for which there is no calculated procedure recognized by the code.

The referenced test standard, ASTM D 1761, requires the joist length to be twice the joist depth plus 10 inches (254 mm), but this can result in longer joists that fail before the hanger. In establishing a hanger's allowable load, the objective is to have the hanger fail in the test, rather than the joist. The exception sets a maximum joist length in order to facilitate the appropriate failure mode.

1711.1.1 General. The vertical load-bearing capacity, torsional moment capacity and deflection characteristics of joist hangers shall be determined in accordance with ASTM D 1761 using lumber having a specific gravity of 0.49 or greater, but not greater than 0.55, as determined in accordance with AF&PA NDS for the joist and headers.

Exception: The joist length shall not be required to exceed 24 inches (610 mm).

❖ The specified ASTM standard test method provides a procedure for evaluating the vertical load-carrying capacity, torsional moment capacity and deflection characteristics of joist hangers and similar devices used to connect wood joists to headers of wood or other materials. The lumber used for the test specimen must have a specific gravity equal to or greater than 0.49, but not greater than 0.55.

1711.1.2 Vertical load capacity for joist hangers. The vertical load-bearing capacity for the joist hanger shall be determined by testing a minimum of three joist hanger assemblies as specified in ASTM D 1761. If the ultimate vertical load for any one of the tests varies more than 20 percent from the average ultimate vertical load, at least three additional tests shall be conducted. The allowable vertical load-bearing of the joist hanger shall be the lowest value determined from the following:

1. The lowest ultimate vertical load for a single hanger from any test divided by three (where three tests are conducted and each ultimate vertical load does not vary more than 20 percent from the average ultimate vertical load).
2. The average ultimate vertical load for a single hanger from all tests divided by three (where six or more tests are conducted).
3. The average from all tests of the vertical loads that produce a vertical movement of the joist with respect to the header of $^1/_8$ inch (3.2 mm).
4. The sum of the allowable design loads for nails or other fasteners utilized to secure the joist hanger to the wood members and allowable bearing loads that contribute to the capacity of the hanger.
5. The allowable design load for the wood members forming the connection.

❖ The method prescribed establishes the allowable load for normal duration, as defined by the American Forest & Paper Association (AF&PA) *National Design Specification (NDS) for Wood Construction.* Additionally, allowable stresses cannot exceed those allowed by the code. For example, published allowable loads cannot contain nail loads higher than those allowed by AF&PA NDS, nor can tension in steel strapping exceed that allowed by the steel design standards noted in Chapter 22.

For loads of other than normal duration, the stresses or loads may be increased or must be decreased as noted by the appropriate design standard, but in no case can the load exceed that which will produce $^1/_8$-inch (3.2 mm) movement of the joist.

EXAMPLE:

Given:

A manufacturer's test results for a particular joist hanger are as follows:

Test 1 Ultimate load = 1,000 pounds with the $^1/_8$-inch deflection occurring at 400 pounds.

Test 2 Ultimate load = 1,100 pounds with the $^1/_8$-inch deflection occurring at 350 pounds.

Test 3 Ultimate load = 900 pounds with the $^1/_8$-inch deflection occurring 375 pounds.

For SI: 1 inch = 25.4 mm, 1 pound = 0.454 kg.

The manufacturer submitted structural calculations indicating that the allowable design load of the nails is 250 pounds (114 kg), and the allowable shear load in the wood joists framing to the hangers is 280 pounds (127 kg). Joist hanger geometry does not allow for any meaningful calculation of stresses in the steel sections of the hanger.

Find: The allowable load for the joist hanger.

Solution:

Average ultimate load = (1,000 + 1,100 + 900) ÷ 3.0 = 1,000 pounds (454 kg)

Since 20 percent of 1,000 is 200, the test scatter is within the allowable range of plus and minus 20 percent of the average ultimate load; therefore, three tests are sufficient to establish allowable load.

Thus, the allowable load is 250 pounds (114 kg) based on the lesser of:

Lowest ultimate load ÷ 3.0 = 300 pounds (137 kg)

$^1/_8$-inch deflection in any test = 350 pounds (159 kg)

Allowable nail load = 250 pounds (114 kg)

Allowable joist shear = 280 pounds (127 kg)

In this case, the allowable vertical load for a normal duration is limited by the calculated allowable design load of the fastener [250 pounds (113 kg)] in accordance with Item 4 of Section 1711.1.2. This value is permitted to be modified by a duration of loading factor; however, the modified value cannot exceed the lowest test value as determined in accordance with Items 1, 2 and 3 of Section 1711.1.2.

1711.1.2.1 Design value modifications for joist hangers. Allowable design values for joist hangers that are determined by Item 4 or 5 in Section 1711.1.2 shall be permitted to be modified by the appropriate load duration factors as specified in AF&PA NDS but shall not exceed the direct loads as determined by Item 1, 2 or 3 in Section 1711.1.2. Allowable design values determined by Item 1, 2 or 3 in Section 1711.1.2 shall not be modified by load duration factors.

❖ The calculated allowable design values, as determined by Item 4 or 5 of Section 1711.1.2 and modified by duration of loading factors, must not exceed the lowest test value as determined by Item 1, 2 or 3 of Section 1711.1.2.

1711.1.3 Torsional moment capacity for joist hangers. The torsional moment capacity for the joist hanger shall be determined by testing at least three joist hanger assemblies as specified in ASTM D 1761. The allowable torsional moment of the joist hanger shall be the average torsional moment at which the lateral movement of the top or bottom of the joist with respect to the original position of the joist is $^1/_8$ inch (3.2 mm).

❖ The allowable torsional moment capacity for a joist hanger is determined by testing in accordance with ASTM D 1761 with the limitation that rotational deflection of the top or bottom of the joist with respect to the header must not exceed 0.125 inch (3.2 mm).

1711.2 Concrete and clay roof tiles. Testing of concrete and clay roof tiles shall be in accordance with Sections 1711.2.1 and 1711.2.2, as applicable.

❖ This section prescribes the test standards and criteria to be used to determine the overturning resistance and wind characteristics of concrete and clay roof tiles.

1711.2.1 Overturning resistance. Concrete and clay roof tiles shall be tested to determine their resistance to overturning due to wind in accordance with SBCCI SSTD 11 and Chapter 15.

❖ Section 1711.2.1 requires concrete and clay tiles to be tested to determine their overturning resistance in accordance with SBCCI SSTD 11. SBCCI SSTD 11 prescribes methods for determining the allowable overturning moment for mechanically fastened, adhesive-set and mortar-set tiles. A test procedure is also prescribed for determining the allowable uplift loads on hip/ridge tiles.

1711.2.2 Wind tunnel testing. Where concrete and clay roof tiles do not satisfy the limitations in Chapter 16 for rigid tile, a wind tunnel test shall be used to determine the wind characteristics of the concrete or clay tile roof covering in accordance with SBCCI SSTD 11 and Chapter 15.

❖ The wind tunnel test procedures in SBCCI SSTD 11 must be used if the roof tiles do not meet the limitations of Chapter 16 for rigid tile.

Bibliography

The following resource materials are referenced in this chapter or are relevant to the subject matter addressed in this chapter.

AAMA/WMDA/CSA 101/1.S.2/A440-11, *Voluntary Specifications for Aluminum Vinyl (PVC) and Wood Windows and Glass Doors*. Schaumburg, IL: American Architectural Manufacturers Association, 2011.

ACI 318-11, *Building Code Requirements for Structural Concrete*. Farmington Hills, MI: American Concrete Institute, 2011.

AF&PA NDS-2012, *National Design Specification for Wood Construction*. Washington, DC: American Forest & Paper Association, 2012.

AISC 341-10, *Seismic Provisions for Structural Steel Buildings*. Chicago: American Institute of Steel Construction, 2010.

AISC 360-10, *Specification for Structural Steel Buildings*. Chicago: American Institute of Steel Construction, 2010.

ASCE 7-10, *Minimum Design Loads for Buildings and Other Structures*. New York: American Society of Civil Engineers, 2010.

ASTM A 6-07, *Specification for General Requirements for Rolled Structural Bars, Plates, Shapes and Sheet Piling*. West Conshohocken, PA: ASTM International, 2007.

ASTM A 36/A 36M-08, *Specification for Carbon Structural Steel*. West Conshohocken, PA: ASTM International, 2008.

ASTM A 325-04a, *Specification for Structural Bolts, Steel, Heat Treated, 120/105 ksi Minimum Tensile Strength*. West Conshohocken, PA: ASTM International, 2004.

ASTM A 435-90 (2001), *Specification for Straight Beam Ultrasound Examination of Steel Plates*. West Conshohocken, PA: ASTM International, 2001.

ASTM A 490-04a, *Specification for Structural Bolts, Alloy Steel, Heat Treated, 150 ksi Minimum Tensile Strength*. West Conshohocken, PA: ASTM International, 2004.

ASTM A 568-06a, *Specification for Steel Sheet, Carbon and High-strength, Low-alloy, Hot-rolled and Cold-rolled, General Requirements For*. West Conshohocken, PA: ASTM International, 2006.

ASTM A 615/A 615M-09, *Specification for Deformed and Plain Billet-steel Bars for Concrete Reinforcement*. West Conshohocken, PA: ASTM International, 2009.

ASTM A 706-09, *Specification for Low-alloy Steel Deformed and Plain Bars for Concrete Reinforcement*. West Conshohocken, PA: ASTM International, 2009.

ASTM D 1761-06, *Standard Test Method for Mechanical Fasteners in Wood*. West Conshohocken, PA: ASTM International, 2006.

ASTM E 330-02, *Standard Test Methods for Structural Performance of Exterior Windows, Curtain Walls, and Doors by Uniform Static Air Pressure Difference*. West Conshohocken, PA: ASTM International, 2002.

ASTM E 605-06, *Test Methods for Thickness and Density of Sprayed Fire-resistive Material Applied to Structural Members*. West Conshohocken, PA: ASTM International, 2006.

ASTM E 2570-07, *Standard Test Method for Evaluating Water-resistive Barrier (WRB) Coatings Used Under Exterior Industrial Finish Systems (EIFS) for EIFS with Drainage*. West Conshohocken, PA: ASTM International, 2007.

AWCI 12-A, *Technical Manual 12-A, Standard Practice for the Testing and Inspection of Field Applied Fire-resistive Materials; an Annoted Guide, Third Edition*. Falls Church, VA: Association of the Wall and Ceiling Industries International, 1997.

AWCI 12-B, *Technical Manual 12-B Standard Practice for the Testing and Inspection of Field Applied Thin-film Intumescent Fire-resistive Materials; an Annoted Guide, First Edition*. Falls Church, VA: The Association of the Wall and Ceiling Industries International, 1998.

AWS D1.1-04, *Structural Welding Code—Steel*. Miami, FL: American Welding Society, 2004.

AWS D1.3-98, *Structural Welding Code—Sheet Steel*. Miami, FL: American Welding Society, 1998.

AWS D1.4-98, *Structural Welding Code—Reinforced Steel*. Miami, FL: American Welding Society, 1998.

AWS QC1-88, *Standard and Guide for Qualification and Certification of Welding Inspectors*. Miami, FL: American Welding Society, 1988.

DASMA 108-05, *Standard Method for Testing Sectional Garage Doors and Rolling Doors Determination of Structural Performance Under Uniform Static Pressure*. Cleveland, OH: Door and Access Systems Manufacturers Association International, 2005.

FEMA 450, *NEHRP Recommended Provisions for Seismic Regulations for New Buildings and Other Structures*. Washington, DC: Federal Emergency Management Agency, 2004.

ICC-ES AC 156, *Acceptance Criteria for Seismic Qualification Testing of Nonstructural Components*. Whittier, CA: International Code Council Evaluation Service, 2000.

IFC-12, *International Fire Code*. Washington, DC: International Code Council, 2011.

IMC-12, *International Mechanical Code*. Washington, DC: International Code Council, 2011.

IPC-12, *International Plumbing Code*. Washington, DC: International Code Council, 2011.

IRC-12, *International Residential Code*. Washington, DC: International Code Council, 2011.

SBCCI SSTD 11-97, *Standard for Determining Wind Resistance of Concrete or Clay Roof Tiles*. Birmingham, AL: Southern Building Code Congress International, 1997.

Specification for Structural Joints Using A325 or A490 Bolts. Chicago: Research Council on Structural Connections (c/o American Institute of Steel Construction, Inc.), 2004.

TMS 402/ACI 530/ASCE 5-11, *Building Code Requirements for Masonry Structures*. Farmington Hills, MI: American Concrete Institute; New York: American Society of Civil Engineers; Boulder, CO: The Masonry Council, 2011.

TMS 602/ACI 530.1/ASCE 6-11, *Specifications for Masonry Structures*. Farmington Hills, MI: American Concrete Institute; New York: American Society of Civil Engineers; Boulder, CO: The Masonry Council, 2011.

UL 127-08, *Factory-built Fireplaces*. Northbrook, IL: Underwriters Laboratories Inc., 2008.

Chapter 18: Soils and Foundations

General Comments

Chapter 18 contains provisions regulating the design and construction of foundations for buildings and other structures and involves geotechnical and structural considerations in the selection and installation of adequate supports for the loads transferred from the structure above.

Section 1801 gives the general scope or purpose of the provisions in Chapter 18.

Section 1802 lists definitions that pertain to foundations.

Section 1803 provides the requirements for foundation and geotechnical investigations that are to be conducted at the site prior to design and construction.

Section 1804 includes excavation, grading and backfill provisions.

Section 1805 provides specifications for the dampproofing and waterproofing of floor slabs and below-grade walls.

Section 1806 establishes the allowable load-bearing values for soils where site soil tests do not verify that higher soil values are appropriate.

Section 1807 provides requirements for foundation walls, retaining walls and embedded posts.

Section 1808 gives general requirements for foundations.

Section 1809 provides the specifications for shallow foundations.

Section 1810 includes the provisions for deep foundations.

The proper design and construction of a foundation system is critical to the satisfactory performance of the entire building structure that it supports.

Foundation problems are not uncommon and vary greatly. They may be of a simple or complex nature and may be manageable or without practical remedy. The uncertainties of foundation construction make it extremely difficult to address every potential problem area within the text of the code. The provisions of the code are meant to set forth and regulate the minimum standards and conventional practices needed for the design and construction of foundation systems so as to provide adequate safety to life and property. Due care must be exercised in the planning and design of foundation systems based on obtaining sufficient soils information, the use of accepted engineering procedures, experience and good technical judgement.

Essentially, there are two parts to the foundation system: the substructure and the soil. The substructure consists of structural components that serve as the medium through which the building loads are transmitted to the supporting earth (soil or rock). The substructure components may consist of shallow foundations, such as basement walls, grade walls, beams, isolated spread footings or combinations of these components. Shallow foundations may also involve the use of mat or raft foundations. As may be required, the substructure can consist of deep foundations involving the use of piles; drilled shafts or piers; caissons or other similar deep foundations. The second part of the foundation system involves the use of soil (including rock) as a structural material to carry the load of the building or any other load transmitted through the substructure.

The substructure and the soil are interdependent elements of the foundation system and must be understood and dealt with as a composite engineering consideration. Indeed, the selection of the kind of substructure to be used, whether it employs any of the different types of shallow foundations commonly used or adopts the use of a deep foundation, is a direct function of the nature of the soil encountered at the project site. Chapter 18 broadly outlines the conventional systems of foundation construction. Although it does not specifically include special or patented systems, it does not preclude their use because most such construction will fall into the general categories of foundation systems prescribed in the provisions and, thus, meet the intent of the code.

In determining the load-bearing capacity and other values of the soil mass, the code provisions address such considerations in terms of "undisturbed" soil. Special provisions are included where prepared fill is to be utilized for foundation support.

Purpose

The provisions of this chapter set forth the minimum requirements for the design and construction of foundation systems for buildings and other structures.

SECTION 1801
GENERAL

1801.1 Scope. The provisions of this chapter shall apply to building and foundation systems.

❖ The provisions contained in this chapter for foundation design and construction apply to all structures.

1801.2 Design basis. Allowable bearing pressures, allowable stresses and design formulas provided in this chapter shall be used with the *allowable stress design* load combinations specified in Section 1605.3. The quality and design of materials used structurally in excavations and foundations shall comply with the requirements specified in Chapters 16, 19, 21, 22 and 23 of this code. Excavations and fills shall also comply with Chapter 33.

❖ Design requirements in Chapter 18 are generally based on an allowable stress design (ASD) approach. Allowable stresses and service loads should not be used with the load combinations for strength design, and vice versa. This section clarifies the applicable load combinations from Chapter 16 that are to be used for the design of foundations.

SECTION 1802
DEFINITIONS

1802.1 Definitions. The following words and terms are defined in Chapter 2:

DEEP FOUNDATION.

DRILLED SHAFT.

Socketed drilled shaft.

HELICAL PILE.

MICROPILE.

SHALLOW FOUNDATION.

❖ Definitions are intended to facilitate the understanding of code provisions and to minimize potential confusion. To that end, this section lists the definitions of terms associated with foundations that can be found in Chapter 2. The use and application of all defined terms, as well as undefined terms, are set forth in Section 201.

SECTION 1803
GEOTECHNICAL INVESTIGATIONS

1803.1 General. Geotechnical investigations shall be conducted in accordance with Section 1803.2 and reported in accordance with Section 1803.6. Where required by the *building official* or where geotechnical investigations involve in-situ testing, laboratory testing or engineering calculations, such investigations shall be conducted by a *registered design professional*.

❖ This section addresses the conditions that mandate a geotechnical investigation, as well as the information that must be included in the report. The investigation of soils is to be done by a registered design professional in recognition that the testing and calculations necessitate individuals with significant experience in soil and foundation analysis. The field of soil mechanics and foundation engineering is diverse and complicated, and since it is not an exact science, its application requires specialized knowledge and judgment based on experience. Where subsurface conditions are found or suspected to be of a critical nature, the building official is encouraged to seek the professional advice of experienced foundation engineers.

1803.2 Investigations required. Geotechnical investigations shall be conducted in accordance with Sections 1803.3 through 1803.5.

Exception: The *building official* shall be permitted to waive the requirement for a geotechnical investigation where satisfactory data from adjacent areas is available that demonstrates an investigation is not necessary for any of the conditions in Sections 1803.5.1 through 1803.5.6 and Sections 1803.5.10 and 1803.5.11.

❖ Soils investigations to determine subsurface conditions should be made prior to the design and construction of new buildings and other structures. Such investigations should also be conducted when additions to existing facilities are considered and are of such a scope that would significantly increase or change the distribution of foundation loads.

There are two main objectives for conducting a soils investigation. The first is of a confirmatory nature. Its purpose is to obtain information already known from adjacent structures, such as soil-boring records, field test results, laboratory test data and analyses and any other knowledge useful in the design of the foundation system. The second objective is of an exploratory nature. It is warranted where soils information does not exist or is insufficient or unsatisfactory for use in the design of the foundation system.

Regardless of the objective of the soils investigation, the information generally required includes one or more (or all) of the following items for determining subsurface conditions:

1. The depth, thickness and composition of each soil stratum;
2. For rock, the characteristics of the rock stratum (or strata), including the thickness of the rock to a reasonable depth;
3. The depth of ground water below the site surface; and
4. The engineering properties of the soil and rock strata that are pertinent for the proper design and performance of the foundation system.

For shallow foundations, the soils investigation should yield sufficient information to establish the character and load-bearing capacity of the soil (or rock) at depths that will receive the foundations.

Foundation problems are not uncommon and may vary greatly, ranging from very simple and manageable problems to very complex situations that may be either manageable or without practical remedy.

As indicated in the exception, where geotechnical data from adjacent areas are well known, the building official can accept the use of local engineering practices for the design of foundations.

1803.3 Basis of investigation. Soil classification shall be based on observation and any necessary tests of the materials disclosed by borings, test pits or other subsurface exploration made in appropriate locations. Additional studies shall be made as necessary to evaluate slope stability, soil strength, position and adequacy of load-bearing soils, the effect of moisture variation on soil-bearing capacity, compressibility, liquefaction and expansiveness.

❖ When soils are required to be classified, the classification must be based on observations and tests, such as borings or test pits. In addition to the situations specified that require a soils investigation and classification, the evaluation of slope stability, soil strength, position and adequacy of load-bearing soils, moisture effects, compressibility and liquefaction are required to be performed when deemed necessary by the building official or registered design professional.

1803.3.1 Scope of investigation. The scope of the geotechnical investigation including the number and types of borings or soundings, the equipment used to drill or sample, the in-situ testing equipment and the laboratory testing program shall be determined by a *registered design professional*.

❖ Whenever the load capacity of a soil is in doubt and a field investigation is necessary, exploratory borings are to be made to determine the load-bearing value of the soil. The investigation is to be performed by a registered design professional, which in most cases would be a geotechnical engineer. Exploratory borings and their associated tests should be conducted in each area of relatively dissimilar subsoil conditions.

1803.4 Qualified representative. The investigation procedure and apparatus shall be in accordance with generally accepted engineering practice. The *registered design professional* shall have a fully qualified representative on site during all boring or sampling operations.

❖ A qualified representative is required on site to verify that the information obtained from the investigation will be adequate, valid and acceptable to the building official.

1803.5 Investigated conditions. Geotechnical investigations shall be conducted as indicated in Sections 1803.5.1 through 1803.5.12.

❖ Sections 1803.5.1 through 1803.5.12 state conditions that necessitate a subsurface investigation.

1803.5.1 Classification. Soil materials shall be classified in accordance with ASTM D 2487.

❖ Where required, soils are to be classified in accordance with ASTM D 2487. This standard provides a system for classifying soils for engineering purposes based on laboratory determination of particle size characteristics, liquid limit and plasticity index. The classification system identifies three major soil divisions—course-grained, fine-grained and highly organic—which are separated further into 15 basic soil groups. ASTM D 2487 is the ASTM International version of the Unified Soil Classification System.

1803.5.2 Questionable soil. Where the classification, strength or compressibility of the soil is in doubt or where a load-bearing value superior to that specified in this code is claimed, the *building official* shall be permitted to require that a geotechnical investigation be conducted.

❖ Whenever relevant soil characteristics are in doubt, or where a design is based upon load-bearing values that are greater than those specified in the code, the building official may require investigation and testing of the soil.

One such method of investigation includes construction test pits for field load-bearing tests. Test pits are usually required to be at least 4 square feet (0.37 m^2) in area and be excavated down to the elevation of the proposed bearing surface. The typical apparatus for making such tests involves placing the test loads on a platform supported on a post through which the applied loads are transferred to a bearing plate of a specified size and, in turn, to the soil below. A typical setup for field load tests is shown in Figure 1803.5.2.

It is important that the load (weights on platform) is applied such that all of it will be transmitted to the soil as a static load without impact, fluctuation or eccentricity. The load should be applied incrementally, and continuous records of all settlements should be kept. Measurements are usually made by settlement

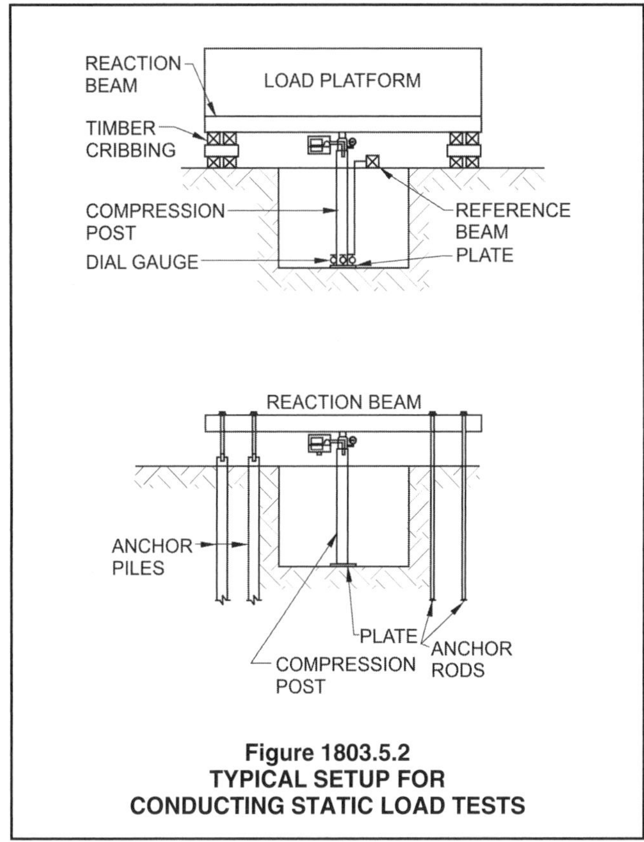

Figure 1803.5.2
TYPICAL SETUP FOR CONDUCTING STATIC LOAD TESTS

recording devices, such as dial gauges, capable of measuring the settlement of the test bearing plate to an accuracy of at least 0.01 inch (0.25 mm).

The test is continued until either the maximum test load is reached or the ratio of load increment to settlement increment reaches a minimum, steady magnitude sustained for a period of 48 hours. After the load is released, the elastic rebound of the soil is also measured for a period of time.

Load test results are normally presented in a load settlement diagram in which the applied test load measured in tons per square foot is plotted in relation to the settlement readings recorded in fractions of an inch. The bearing capacity of the soil can be computed from the test results.

There are some drawbacks to the use of field load tests for determining soil-bearing capacity. Test results can be misleading if the soil under the footing is not uniform for the full depth of load influence, which is equal to about twice the width of the footing. Also, since a load test is conducted for a short duration, settlements that occur due to the consolidation of the soil over a very long time cannot be predicted. Since this type of field load-bearing testing is relatively expensive, it is not widely used.

Other methods for determining the safe bearing capacity of soils may be more appropriate. Standard laboratory tests can usually produce sufficient proof of satisfactory bearing capacity and settlement information. However, there are conditions when standard laboratory tests may not produce reliable results, such as when clay materials contain a pattern of cracks or when stiff clays may have suffered differential movement or expansion. As stated in Section 1803.3.1, a registered design professional must establish an appropriate testing program.

1803.5.3 Expansive soil. In areas likely to have expansive soil, the *building official* shall require soil tests to determine where such soils do exist.

Soils meeting all four of the following provisions shall be considered expansive, except that tests to show compliance with Items 1, 2 and 3 shall not be required if the test prescribed in Item 4 is conducted:

1. Plasticity index (PI) of 15 or greater, determined in accordance with ASTM D 4318.
2. More than 10 percent of the soil particles pass a No.200 sieve (75 μm), determined in accordance with ASTM D 422.
3. More than 10 percent of the soil particles are less than 5 micrometers in size, determined in accordance with ASTM D 422.
4. Expansion index greater than 20, determined in accordance with ASTM D 4829.

❖ Expansive soils, often referred to as "swelling soils," contain montmorillonite minerals and have the characteristics of absorbing water and swelling, or shrinking and cracking when drying. Significant volume changes can cause serious damage to buildings and other structures as well as to pavements and sidewalks. Swelling soils are found throughout the nation, but are more prevalent in regions with dry or moderately arid climates.

There is a general relationship between the plasticity index (PI) of a soil as determined by the ASTM D 4318 standard test method and the potential for expansion, as shown in Figure 1803.5.3.

This section defines "Expansive soil" as any plastic material with a PI of 15 or greater (Item 1); with more than 10 percent of the soil particles passing a No. 200 sieve (Item 2); less than 5 micrometers in size (Item 3) and having an expansion index (EI) greater than 20 (Item 4). Alternatively, the EI in accordance with ASTM D 4829 can be used exclusively. The EI value is a measure of the swelling potential of the soil. A soil with an EI value of 20 or less has a very low potential for expansion.

The amount and depth of potential swelling that can occur in a clay material are, to some extent, functions of the cyclical moisture content in the soil. In dryer climates where the moisture content in the soil near the ground surface is low because of evaporation, there is a greater potential for extensive swelling than the same soil in wetter climates where the variations of moisture content are not as severe. Volume changes in highly expansive soils range between 7 and 10 percent, but experience has shown that occasionally, under abnormal conditions, they can reach as high as 25 percent.

SWELLING POTENTIAL	PLASTICITY INDEX
Low	0-15
Medium	10-35
High	20-55
Very high	35 and above

Figure 1803.5.3
SWELLING POTENTIAL OF SOILS AND PLASTICITY INDEX

Source: R.B. Peck, W.E. Hanson and T.H. Thornburn. *Foundation Engineering*, 2nd ed. (New York: John Wiley & Sons, 1974).

1803.5.4 Ground-water table. A subsurface soil investigation shall be performed to determine whether the existing ground-water table is above or within 5 feet (1524 mm) below the elevation of the lowest floor level where such floor is located below the finished ground level adjacent to the foundation.

Exception: A subsurface soil investigation to determine the location of the ground-water table shall not be required where waterproofing is provided in accordance with Section 1805.

❖ There are several reasons for conducting a subsurface investigation to determine the level of ground water at a construction site. If the ground-water table is above subsurface slabs (i.e., basement floors), then walls and floors need to be designed to resist hydrostatic pressures. Foundation walls and basement slabs may need to be dampproofed or waterproofed, depending on the location of the ground-water table. A subsurface investigation will also determine the type of drainage system needed as a permanent installation, whether there will be any major water problems that could affect the excavation operations and construction of the foundation system, and if it is necessary to provide a temporary drainage system of a type and size that will control ground-water seepage.

Ground-water levels can vary significantly over a year's time, as well as from year to year. While it would be ideal to make ground-water table observations that encompass a full annual cycle, the reality is that such an undertaking would not normally facilitate a design/construction program and would be impractical. Ground-water observations must be made in shorter time intervals; however, this situation poses some real problems. For example, measurements of ground-water levels in bore holes taken 24 hours after completion of the soil borings can provide an acceptable indication of the water level in permeable soils, such as sand, gravel or sand/gravel mixtures. Fine-grained soils of low permeability, such as clays, require the use of observation tubes (piezometers) and, depending on the specific properties of the soil, a time period of 10 weeks or longer to obtain acceptable readings.

Water levels established by either of the two methods described above are sufficient indication of the water conditions at the time of measurement, but do not necessarily represent the highest possible ground-water levels that can occur. For design purposes, the water levels established by field observations may need to be adjusted with the climatological and hydrological records of the region in order to establish the high and low points.

As indicated in the exception, a subsurface investigation is not required where floors, walls, joints and penetrations are waterproofed as required in Section 1805.3.

1803.5.5 Deep foundations. Where deep foundations will be used, a geotechnical investigation shall be conducted and shall include all of the following, unless sufficient data upon which to base the design and installation is otherwise available:

1. Recommended deep foundation types and installed capacities.
2. Recommended center-to-center spacing of deep foundation elements.
3. Driving criteria.
4. Installation procedures.
5. Field inspection and reporting procedures (to include procedures for verification of the installed bearing capacity where required).
6. Load test requirements.
7. Suitability of deep foundation materials for the intended environment.
8. Designation of bearing stratum or strata.
9. Reductions for group action, where necessary.

❖ A foundation investigation is required when deep foundations are proposed. Such investigations are needed to define as accurately as possible the subsurface conditions of soil and rock materials, establish the soil and rock profiles across the construction site and locate the ground-water table. Sometimes, it may also be necessary to determine specific soil properties, such as shear strength, relative density, compressibility and other such technical data required for analyzing subsurface conditions. Foundation investigations may also be used to render such valuable data as information on existing construction at the site or on neighboring properties (including boring and test records), the type and condition of the existing structures, their age, the type of foundations used and performance over the years. Other helpful information includes knowledge of existing deleterious substances in the soils that could affect the durability (as well as the performance) of the piles, data on geologic conditions at the site (including such information as the existence of mines, earth cavities, underground streams or other adverse water conditions), as well as a history of any seismic activity.

The types of information described above are usually obtained by means of soil and rock borings; laboratory and field tests and engineering analyses. Such information is used for determining design loads; types and lengths of piles; driving criteria and selection of equipment and probable durability of pile materials in relation to subsurface conditions.

1803.5.6 Rock strata. Where subsurface explorations at the project site indicate variations or doubtful characteristics in the structure of the rock upon which foundations are to be constructed, a sufficient number of borings shall be made to a depth of not less than 10 feet (3048 mm) below the level of the foundations to provide assurance of the soundness of the foundation bed and its load-bearing capacity.

❖ Rock may be found at levels near or at the earth's surface, upon which shallow foundations can be supported or will range downward to very low levels, upon which piles and other types of deep foundations can bear.

Most intact rock will have compressive strengths that far exceed the requirements for foundation support. It is most common, however, to find cracks, joints and other defects in rock formations that will increase the compressibility of the material. Depending on the nature and extent of the defects, settlement may become the governing factor in determining allowable load-bearing capacity rather than rock strength.

Where the condition of the rock is in doubt, borings must be made at least 10 feet (305 mm) into the rock stratum below the bottom of the footings to verify the soundness of the material and to determine its load-bearing capacity.

1803.5.7 Excavation near foundations. Where excavation will remove lateral support from any foundation, an investigation shall be conducted to assess the potential consequences and address mitigation measures.

❖ Section 1804.1 addresses soil stability when excavations are made adjacent to existing foundations. This provision makes it clear that a geotechnical investigation is necessary in these situations.

1803.5.8 Compacted fill material. Where shallow foundations will bear on compacted fill material more than 12 inches (305 mm) in depth, a geotechnical investigation shall be conducted and shall include all of the following:

1. Specifications for the preparation of the site prior to placement of compacted fill material.
2. Specifications for material to be used as compacted fill.
3. Test methods to be used to determine the maximum dry density and optimum moisture content of the material to be used as compacted fill.
4. Maximum allowable thickness of each lift of compacted fill material.
5. Field test method for determining the in-place dry density of the compacted fill.
6. Minimum acceptable in-place dry density expressed as a percentage of the maximum dry density determined in accordance with Item 3.
7. Number and frequency of field tests required to determine compliance with Item 6.

❖ The information on the prepared fill in the soils report is necessary to permit a reasonable prediction as to the load-bearing capacity of the fill material. Where prepared fill is to be utilized for foundation support, the geotechnical report is to contain detailed information for approval of the fill. The information on the prepared fill in the soils report is necessary to permit a reasonable prediction as to the load-bearing capacity of the fill material. Additionally, when prepared fill is to be utilized where special inspections are required, the fill operation itself must be performed under the scrutiny of a special inspector. For example, if one (or more) of the exceptions in Section 1704.1 applies, special inspection of the prepared fill operation is not required.

1803.5.9 Controlled low-strength material (CLSM). Where shallow foundations will bear on controlled low-strength material (CLSM), a geotechnical investigation shall be conducted and shall include all of the following:

1. Specifications for the preparation of the site prior to placement of the CLSM.
2. Specifications for the CLSM.
3. Laboratory or field test method(s) to be used to determine the compressive strength or bearing capacity of the CLSM.
4. Test methods for determining the acceptance of the CLSM in the field.
5. Number and frequency of field tests required to determine compliance with Item 4.

❖ As an alternative to compacted fill, the code permits the use of controlled low-strength material (CLSM) for the support of footings. CLSM must be placed in accordance with an approved report that includes requirements for the material strength and field verification.

In Chapter 2, "CLSM" is defined as "self-compacting cementitious materials." This class of material is commonly referred to by a variety of other names, which include flowable fill, controlled density fill, unshrinkable fill and soil-cement slurry. Guidance on the use of these materials can be found in ACI 229R. Additional documents that may be useful references for sampling and testing these materials include the following ASTM International standards:

ASTM D 4832, *Standard Test Method for Preparation and Testing of Controlled Low-strength Material (CLSM) Test Cylinders.*

ASTM D 5971, *Standard Practice for Sampling Freshly Mixed Controlled Low-strength Material.*

ASTM D 6023, *Standard Test Method for Unit Weight, Yield, Cement Content, and Air Content (Gravimetric) of Controlled Low-strength Material (CLSM).*

ASTM D 6024, *Standard Test Method for Ball Drop on Controlled Low-strength Material (CLSM) to Determine Suitability for Load Application.*

ASTM D 6103, *Standard Test Method for Flow Consistency of Controlled Low-strength Material (CLSM).*

1803.5.10 Alternate setback and clearance. Where setbacks or clearances other than those required in Section 1808.7 are desired, the *building official* shall be permitted to require a geotechnical investigation by a *registered design professional* to demonstrate that the intent of Section 1808.7 would be satisfied. Such an investigation shall include consideration of material, height of slope, slope gradient, load intensity and erosion characteristics of slope material.

❖ Section 1808.7 regulates the placement of foundations adjacent to slopes that are greater than one unit vertical to three units horizontal (33.3-percent slope). This section provides the building official the authority to approve alternative setbacks and clearances to those required in Section 1808.7. The building official has the authority to require an investigation by a registered design professional to show that the intent of the code has been met. This item also specifies the parameters that must be considered by the registered design professional in the investigation.

1803.5.11 Seismic Design Categories C through F. For structures assigned to *Seismic Design Category* C, D, E or F, a geotechnical investigation shall be conducted, and shall include an evaluation of all of the following potential geologic and seismic hazards:

1. Slope instability.
2. Liquefaction.
3. Total and differential settlement.
4. Surface displacement due to faulting or seismically induced lateral spreading or lateral flow.

❖ The potential for liquefaction, surface rupture or slope instability at a building site is greater in areas of moderate and high seismicity than in areas of low seismicity. Also, the consequences of damage resulting from such hazards are more severe for buildings in higher risk categories (such as essential facilities). Thus, this section requires an investigation report for building sites that are assigned to Seismic Design Category C or higher, that includes an evaluation of the specific earthquake hazards that are listed. The purpose of this section is to reduce the hazard of large ground movement and the damaging effects on the structure.

Liquefaction of saturated granular soils has been a major source of building damage during past earthquakes. For example, many structures in Niigata, Japan suffered major damage as a consequence of liquefaction during its 1964 earthquake. Loss of bearing strength, differential settlement and differential horizontal displacement due to lateral spread were the direct causes of damage. Many structures have been similarly damaged by differential ground displace-ments during U.S. earthquakes, such as the San Fernando Valley Juvenile Hall during the 1971 San Fernando, California earthquake and the Marine Sciences Laboratory at Moss Landing, California during the 1989 Loma Prieta event.

For more information regarding evaluation of slope instability, liquefaction and surface rupture due to faulting or lateral spreading, see Section 7.4 of the National Earthquake Hazards Reduction Program (NEHRP) Provisions commentary (FEMA 450).

1803.5.12 Seismic Design Categories D through F. For structures assigned to *Seismic Design Category* D, E or F, the geotechnical investigation required by Section 1803.5.11 shall also include all of the following as applicable:

1. The determination of dynamic seismic lateral earth pressures on foundation walls and retaining walls supporting more than 6 feet (1.83 m) of backfill height due to design earthquake ground motions.
2. The potential for liquefaction and soil strength loss evaluated for site peak ground acceleration, earthquake magnitude, and source characteristics consistent with the maximum considered earthquake ground motions. Peak ground acceleration shall be determined based on:
   - 2.1 A site-specific study in accordance with Section 21.5 of ASCE 7; or
   - 2.2 In accordance with Section 11.8.3 of ASCE 7.
3. An assessment of potential consequences of liquefaction and soil strength loss, including, but not limited to:
   - 3.1. Estimation of total and differential settlement;
   - 3.2. Lateral soil movement;
   - 3.3. Lateral soil loads on foundations;
   - 3.4. Reduction in foundation soil-bearing capacity and lateral soil reaction;
   - 3.5. Soil downdrag and reduction in axial and lateral soil reaction for pile foundations;
   - 3.6. Increases in soil lateral pressures on retaining walls; and
   - 3.7. Flotation of buried structures.
4. Discussion of mitigation measures such as, but not limited to:
   - 4.1. Selection of appropriate foundation type and depths;
   - 4.2. Selection of appropriate structural systems to accommodate anticipated displacements and forces;
   - 4.3. Ground stabilization; or
   - 4.4. Any combination of these measures and how they shall be considered in the design of the structure.

❖ This section includes additional requirements for the soil investigation report for sites with structures assigned to Seismic Design Categories D and higher. The investigation must determine lateral earth pressures on basement and retaining walls due to earthquake motions. Earthquake motions create increased lateral soil pressure on walls below the ground surface, especially in soft soils in areas of high seismicity. This requirement makes certain that the dynamic soil pressures are included in the design of basement and retaining walls. Because the requirement can be onerous for small structures and retaining walls, the applicability is limited to those walls that are higher than 6 feet (1.83 m). Section 7.5.1 of the 1997 NEHRP Provisions commentary includes a discussion about how earth-retaining structures have been designed for dynamic loads.

Additionally, a thorough assessment of potential consequences of any liquefaction and soil strength loss needs to be made and considered in the design of the structure. See the commentary to Section 1803.5.11 for a discussion of earthquake damage due to liquefaction. Design to mitigate damage due to liquefaction consists of three parts: evaluation of liquefaction hazard; evaluation of potential ground displacement and designing to resist ground displacement, reducing the potential for liquefaction or choosing an alternative site with a lower hazard. The assessment is required to be made for the site's peak ground accelerations that are consistent with the maximum considered earthquake (MCE) ground motions. In this way the potential for liquefaction as well as the effects of liquefaction during the MCE are considered in design. This is consistent with the risk-based targets for collapse prevention as a performance goal and other evaluations for the MCE.

1803.6 Reporting. Where geotechnical investigations are required, a written report of the investigations shall be submitted to the *building official* by the owner or authorized agent at the time of *permit* application. This geotechnical report shall include, but need not be limited to, the following information:

1. A plot showing the location of the soil investigations.
2. A complete record of the soil boring and penetration test logs and soil samples.
3. A record of the soil profile.
4. Elevation of the water table, if encountered.
5. Recommendations for foundation type and design criteria, including but not limited to: bearing capacity of natural or compacted soil; provisions to mitigate the effects of expansive soils; mitigation of the effects of liquefaction, differential settlement and varying soil strength; and the effects of adjacent loads.
6. Expected total and differential settlement.
7. Deep foundation information in accordance with Section 1803.5.5.
8. Special design and construction provisions for foundations of structures founded on expansive soils, as necessary.
9. Compacted fill material properties and testing in accordance with Section 1803.5.8.
10. Controlled low-strength material properties and testing in accordance with Section 1803.5.9.

❖ If a written report is required by the building official, it is required to include at a minimum the items listed in this section. These items will establish a retrievable and verifiable record of the soil conditions if problems are encountered in the future. These items also provide the minimum necessary information for compliance with the code and an adequate foundation system.

SECTION 1804 EXCAVATION, GRADING AND FILL

1804.1 Excavation near foundations. Excavation for any purpose shall not remove lateral support from any foundation without first underpinning or protecting the foundation against settlement or lateral translation.

❖ The purpose of this section is to provide for stability of adjacent foundations when excavations are made. The method to be used, whether it be shoring or underpinning, must be addressed by a geotechnical investigation as required in Section 1803.5.7. Due to their lack of shear strength, cohesionless soils, such as sand, will slide to the bottom of an excavation until a certain slope of the sides is reached. This slope is known as the angle of repose of natural slope and is independent of the depth of the excavation.

Cohesive (fine-grained) soils, such as clay, behave much differently compared to granular materials. For example, unsupported vertical cuts of 20 feet (6096 mm) or more can be made in stiff plastic clay materials. This is due to a firm bond between the particles of the cohesive soil. But the strength of this bond (cohesiveness) will vary based on the conditions of the soil, such as density, water content, plasticity and sensitivity (loss of shear strength upon disturbance).

In cohesive soils, when a certain critical depth of excavation is reached, the sides of the cut will fail and the soil mass will fall to the bottom. Unlike granular materials, such as sand, the steepest slope at which a cohesive soil will stand decreases as the depth of the excavation increases. Technically, in cohesive soils, the resistance against sliding is a function of the shearing resistance of the material and its corresponding angle of internal friction (frictional resistance between particles).

The use of the angle of internal friction (and other factors) to calculate slope stability is applicable not only to cohesive soils, but also to granular materials. For example, when the slope angle of an excavation in a bed of sand exceeds the angle of internal friction of the material, the sand will slide down the slope; therefore, the steepest slope that sand can attain is equal to the angle of internal friction. The angle of repose (previously discussed) will be approximately the same value as the angle of internal friction only when the sand is in a dry and loose condition (such as in a stockpile) or is fully immersed in water.

For simplicity, what we have been dealing with in this part of the commentary is slope stability as it relates to homogeneous soils, such as sand and clay. In nature, however, soils often occur as mixtures or layers (strata) of different materials, making the determination of slope stability a highly technical and complex subject. Normally for shallow excavations, determination of safe slopes is a matter of applying local experience. In cases of deep cuts, however, slope stability is best determined through tests and analytical methods performed by professionals experienced in foundation engineering.

Some texts dealing with soil mechanics contain tables that indicate the angles of slopes that can be expected for various soil materials commonly found throughout the country. While such tables provide useful information, they should be used only as a guide. They should not be used for design purposes, nor employed when a critical subsurface condition exists or when the type of soil (established by borings) does not closely fit those described in the tables.

1804.2 Placement of backfill. The excavation outside the foundation shall be backfilled with soil that is free of organic material, construction debris, cobbles and boulders or with a controlled low-strength material (CLSM). The backfill shall be placed in lifts and compacted in a manner that does not damage the foundation or the waterproofing or dampproofing material.

Exception: CLSM need not be compacted.

❖ This section requires that soils used for backfilling foundation excavations must be free of organic material, construction debris or large rocks. The type of soil used for backfill purposes becomes an important consideration in the design of foundation walls. For example, clean sand, gravel or a mixture of these two granular materials is considered the best kind of backfill to use because each is free draining and generally has frost-free properties. On the other hand, fine-grained soils, such as clays, tend to accumulate moisture and are susceptible to swelling and shrinking, as well as frost action. Such backfill materials, particularly at times when shrinkage cracks occur, can become loaded with rainwater, thus subjecting foundation walls and basement floors to hydrostatic pressures and possible structural damage.

Backfilling and related work should be performed in such a way as to prevent the movement of the earth of adjoining properties or the subsequent caving in of backfilled areas. Backfilling should not be done until retaining walls, foundation walls or other construction against which backfill is to be placed is in a suitable condition to resist lateral pressures. This section requires that backfill be free of organic materials, construction debris, cobbles and boulders.

In addition to carefully selecting the backfill material, the soil should be placed in lifts, usually 9 inches (229 mm) or less, and compacted to prevent significant subsidence due to consolidation under its own weight. While compaction is done by hand-operated tampers or other portable compaction equipment, care should be taken to prevent any possible damage to waterproofing or dampproofing installations and to avoid overcompaction of backfill since it may cause excessive earth pressure against foundation walls. As an alternative to compacted backfill, the code also permits CLSM (see commentary, Section 1804.6).

1804.3 Site grading. The ground immediately adjacent to the foundation shall be sloped away from the building at a slope of not less than one unit vertical in 20 units horizontal (5-percent slope) for a minimum distance of 10 feet (3048 mm) measured perpendicular to the face of the wall. If physical obstructions or lot lines prohibit 10 feet (3048 mm) of horizontal distance, a 5-percent slope shall be provided to an *approved* alternative method of diverting water away from the foundation. Swales used for this purpose shall be sloped a minimum of 2 percent where located within 10 feet (3048 mm) of the building foundation. Impervious surfaces within 10 feet (3048 mm) of the building foundation shall be sloped a minimum of 2 percent away from the building.

Exception: Where climatic or soil conditions warrant, the slope of the ground away from the building foundation shall be permitted to be reduced to not less than one unit vertical in 48 units horizontal (2-percent slope).

The procedure used to establish the final ground level adjacent to the foundation shall account for additional settlement of the backfill.

❖ This section requires that the ground immediately adjacent to the foundation be sloped away from the building. The intent is to facilitate water drainage and reduces the potential for water standing under and around the building.

Where a full 10 feet (3048 mm) of slope is not available on a building site, an alternative method of diverting water away from the foundation is permitted by providing a minimum 5-percent slope to an approved diversion structure. The use of swales to convey surface water is recognized, provided the minimum slope is provided where necessary.

The exception permits the slope to be reduced to a rate of 1 unit vertical in 48 units horizontal (2-percent slope) where climatic or soil conditions warrant. This exception would be applicable in arid areas or sites that are surrounded by free-draining soils, such as sand.

1804.4 Grading and fill in flood hazard areas. In *flood hazard areas* established in Section 1612.3, grading and/or fill shall not be *approved*:

1. Unless such fill is placed, compacted and sloped to minimize shifting, slumping and erosion during the rise and fall of flood water and, as applicable, wave action.
2. In floodways, unless it has been demonstrated through hydrologic and hydraulic analyses performed by a *registered design professional* in accordance with standard engineering practice that the proposed grading or fill, or both, will not result in any increase in flood levels during the occurrence of the *design flood*.
3. In flood hazard areas subject to high-velocity wave action, unless such fill is conducted and/or placed to avoid diversion of water and waves toward any building or structure.
4. Where design flood elevations are specified but floodways have not been designated, unless it has been demonstrated that the cumulative effect of the proposed *flood hazard area* encroachment, when combined with all other existing and anticipated *flood hazard area* encroachment, will not increase the design flood elevation more than 1 foot (305 mm) at any point.

❖ This section puts limitations on fill and grading in flood hazard areas. In many flood hazard areas it is common to use fill to achieve the appropriate elevation of the building's lowest floor. Item 1 intends to minimize the risk of fills becoming unstable in a flood. Fill materials that become saturated during conditions of flooding may become unstable and fill slopes may be exposed to erosive velocities and waves. When placed in a flood hazard area, fill and grading must remain stable as floodwaters rise and, in particular, as floodwaters fall and the saturated materials drain.

Item 2 permits grading or fill in a floodway only if it is demonstrated that it will not adversely affect surrounding areas by increasing the design flood elevation. As the definition indicates, a "Floodway" is that portion of flood hazard areas along rivers and streams that must be reserved for the discharge of the design flood event. The National Flood Insurance Program (NFIP) requires that the impact of development or encroachment into the floodway must be considered.

In flood hazard areas that are subject to high-velocity wave action, fill can divert the flow of floodwaters and increase flood risks on other properties. Item 3 provides coordination with ASCE 24 provisions relating to fill in flood hazard areas subject to high-velocity wave action. These areas are commonly referred to as "coastal high hazard areas" or "V zones," and typically are indicated on flood hazard maps of communities along the open coast. In these areas, changing the shape of the ground through grading or fill can divert erosive flows and increase wave energies that, in turn, increase the forces that affect a building and any adjacent structures. It is also notable that ASCE 24 specifies that fill may not be used for structural support of buildings in these areas.

Although the Federal Emergency Management Agency (FEMA) designates floodways on many rivers and streams shown on Flood Insurance Rate Maps (FIRMs), some riverine flood hazard areas have base flood elevations, but do not have delineated floodways. In these areas, the effect of flood-plain development on flood elevations has not been evaluated. Item 4 provides consistency with NFIP with respect to development in areas where a base flood elevation has been established, but no floodway has been designated. It allows a proposed development to be approved if it is demonstrated that the flood elevations will not be increased by more than 1 foot (305 mm). Development in riverine flood plains can increase flood levels and loads on other properties, particularly if it occurs in areas considered floodways that must be reserved to convey flood flows (see the definition of "Floodway").

FEMA TB #10, *Ensuring That Structures Built on Fill In or Near Special Flood Hazard Areas Are Reasonably Safe From Flooding*, offers guidance to communities to determine whether structures on fill are reasonably safe from flooding. This determination is required by FEMA if property owners submit documentation requesting that FEMA show that filled land is (or will be) no longer subject to flooding by the base flood. If the documentation meets FEMA's requirements, a Letter of Map Revision based on Fill (LOMR-F) is issued to revise the FIRM.

1804.5 Compacted fill material. Where shallow foundations will bear on compacted fill material, the compacted fill shall comply with the provisions of an *approved* geotechnical report, as set forth in Section 1803.

> **Exception:** Compacted fill material 12 inches (305 mm) in depth or less need not comply with an *approved* report, provided the in-place dry density is not less than 90 percent of the maximum dry density at optimum moisture content determined in accordance with ASTM D 1557. The compaction shall be verified by *special inspection* in accordance with Section 1705.6.

❖ Where prepared fill is to be utilized for foundation support, the geotechnical report is to contain detailed information for approval of the fill. The information on the prepared fill in the geotechnical report is necessary to permit a reasonable prediction as to the load-bearing capacity of the fill material. Additionally, when prepared fill is to be utilized where special inspections are required, the fill operation itself must be performed under the scrutiny of a special inspector. The exception permits a limited depth of fill to be placed in accordance with prescriptive criteria rather than requiring details of fill placement in a geotechnical report. Special inspection is a requirement.

1804.6 Controlled low-strength material (CLSM). Where shallow foundations will bear on controlled low-strength material (CLSM), the CLSM shall comply with the provisions of an *approved* geotechnical report, as set forth in Section 1803.

❖ As an alternative to compacted fill in accordance with Section 1804.5, this section permits the use of CLSM for the support of footings. CLSM must be placed in accordance with an approved report that includes requirements for the material strength and field verification. In Chapter 2, CLSM is defined as "self-compacting cementitious materials." This class of material is commonly referred to by a variety of other names, which include flowable fill, controlled density fill, unshrinkable fill and soil-cement slurry. Guidance on the use of these materials can be found in ACI 229R (see Section 1803.5.9).

SECTION 1805
DAMPPROOFING AND WATERPROOFING

1805.1 General. Walls or portions thereof that retain earth and enclose interior spaces and floors below grade shall be waterproofed and dampproofed in accordance with this section, with the exception of those spaces containing groups other than residential and institutional where such omission is not detrimental to the building or occupancy.

Ventilation for crawl spaces shall comply with Section 1203.4.

❖ Section 1805 covers the requirements for waterproofing and dampproofing those parts of substructure construction that need to be provided with moisture protection. It identifies the locations where moisture barriers are required and specifies the materials to be used and the methods of application. The provisions also deal with subsurface water conditions, drainage systems and other protection requirements.

The term "waterproofing" is at times used where dampproofing is the minimum requirement. Although both terms are intended to apply to the installation and use of moisture barriers, dampproofing does not furnish the same degree of protection against moisture.

Dampproofing generally refers to the application of one or more coatings of a compound or other materials that are impervious to water, which are used to prevent the passage of water vapor through walls or other building components, and which restrict the flow of water under slight hydrostatic pressure. Waterproofing, on the other hand, refers to the application of coatings and sealing materials to walls or other building components to prevent moisture from penetrating in either a vapor or liquid form, even under conditions of significant hydrostatic pressure. Hydrostatic pressure is created by the presence of water under pressure. This pressure can occur when the ground-water table rises above the bottom of the foundation wall, or the soil next to the foundation wall becomes saturated with water caused by uncontrolled storm water runoff.

Section 1805.1 is an overall requirement that waterproofing and dampproofing applications are to be made to horizontal and vertical surfaces of below-ground spaces where the occupancy would normally be affected by the intrusion of water or moisture. Moisture or water in a floor below grade can cause damage to structural members, such as columns, posts or load-bearing walls, as well as pose a health hazard by promoting the growth of bacteria or fungi and adversely affect any mechanical and electrical appliances that may be located at that level. It can also cause a great deal of damage to goods that may be located or stored in that lower level. These vertical and horizontal surfaces include foundation walls, retaining walls, underfloor spaces and floor slabs. Waterproofing and dampproofing are not required in locations other than residential and institutional occupancies where the omission of moisture barriers would not adversely affect the use of the spaces. An example of a location where waterproofing or dampproofing would not be required is in an open parking structure, as long as the structural components are individually protected against the effects of water. Waterproofing and dampproofing are not permitted to be omitted from residential and institutional occupancies where people may be sleeping or services are provided on the floor below grade. A person walking in a flooded basement may be in a very hazardous situation, particularly if the possibility of an electrical charge in the water exists that is caused by electrical service at that level.

Section 1805.1.1 addresses the type of problem faced when a portion of a story is above grade, while Section 1805.1.2 limits any infiltration of water into crawl spaces so as to protect this area from potential water damage and prevent ponding of water. Both of these sections reference other applicable sections of the code, as well as the exceptions.

1805.1.1 Story above grade plane. Where a basement is considered a *story above grade plane* and the finished ground level adjacent to the basement wall is below the basement floor elevation for 25 percent or more of the perimeter, the floor and walls shall be dampproofed in accordance with Section 1805.2 and a foundation drain shall be installed in accordance with Section 1805.4.2. The foundation drain shall be installed around the portion of the perimeter where the basement floor is below ground level. The provisions of Sections 1803.5.4, 1805.3 and 1805.4.1 shall not apply in this case.

❖ The provisions of this section, stated in another way, require that where a basement is deemed to be a story above grade plane (see definition, Section 202), the section of the basement floor that occurs below the exterior ground level and the walls that bound that part of the floor are to be dampproofed in accordance with the requirements of Section 1805.2.

The use of dampproofing, rather than waterproofing, is permitted here since hydrostatic pressure will not tend to develop against the walls if the basement is a story above grade plane and the ground level adjacent to the basement wall is below the basement floor elevation for no less than 25 percent of the basement perimeter.

Any water pressure that may occur against the walls below ground or under the basement floor would be relieved by the water drainage system required in this section. The drainage system would be installed at the base of the wall construction in accordance with Section 1805.4.2 for a minimum distance along those portions of the wall perimeter where the basement floor is below ground level. Because of the relationship of grade to the basement floor and the inclusion of foundation drains, the potential for hydrostatic pressure buildup is not significant; therefore, a ground-water table investigation, waterproofing and the basement floor gravel base course is not required.

The objective of Section 1805.4.1 is to prevent moisture migration in basement spaces. In story-above-grade construction that meets the requirements of this section, the basement floor would be only partly below ground level (sometimes a small part) and the need for moisture protection as required by Section 1805.4.1 would be unnecessary. Dampproofing of the floor slab would be required, however, in accordance with Section 1805.2.1.

1805.1.2 Under-floor space. The finished ground level of an under-floor space such as a crawl space shall not be located below the bottom of the footings. Where there is evidence that the ground-water table rises to within 6 inches (152 mm) of the ground level at the outside building perimeter, or that the surface water does not readily drain from the building site, the ground level of the under-floor space shall be as high as the outside finished ground level, unless an *approved* drainage system is provided. The provisions of Sections 1803.5.4, 1805.2, 1805.3 and 1805.4 shall not apply in this case.

❖ The requirements of this section are designed to prevent any ponding of water in underfloor spaces, such as crawl spaces. Crawl spaces are particularly susceptible to ponding of water, since they are usually uninhabitable spaces that are observed very infrequently. Water can build up in these spaces and remain for an extended period of time without being noticed by the building occupants. This type of stagnant water under a building, which can harbor disease, mold and disease-carrying insects, such as mosquitoes, can result in a serious health concern. Water buildup in a crawl space can also damage the structural integrity of the building. Wood exposed to water will deteriorate and rot, while concrete and masonry exposed to water will deteriorate with a loss of strength.

Steel exposed to water or high humidity can eventually rust to the extent that effective structural capability is jeopardized. Water buildup in a crawl space can also damage any mechanical or electrical appliances, which may be located in the space, by causing corrosion of electrical parts or metal skins and deterioration to insulation used to protect heating elements.

Where it is known that the water table can rise to within 6 inches (152 mm) of the outside ground level, or where there is evidence that surface water cannot readily drain from the site, then the finished ground surface in underfloor spaces is to be set at an elevation equal to the outside ground level around the perimeter of the building unless an approved drainage system is provided. In order for the drainage system to be approved, it must be demonstrated to be adequate to prevent the infiltration of water into the underfloor space. This is done by determining the maximum possible flow of water near the foundation wall and footing and designing the drainage system to remove that flow of water as it occurs, without permitting the buildup of water at the foundation wall.

To prevent the ponding of water in the underfloor space from a rise in the ground-water table, or from storm water runoff, the finished ground level of an underfloor space is not to be located below the bottom of the foundation footings.

Dampproofing (see Section 1805.2) the foundation walls, waterproofing (see Section 1805.3) and providing subsoil drainage (see Section 1805.4) is not necessary if the ground level of the underfloor space is as high as the ground level at the outside of the building perimeter, as the foundation walls do not enclose an interior space below grade. Compliance with Sections 1805.2, 1805.3 and 1805.4 would still be required where the finished ground surface of the underfloor space is below the outside ground level.

1805.1.2.1 Flood hazard areas. For buildings and structures in flood hazard areas as established in Section 1612.3, the finished ground level of an under-floor space such as a crawl space shall be equal to or higher than the outside finished ground level on at least one side.

Exception: Under-floor spaces of Group R-3 buildings that meet the requirements of FEMA/FIA-TB-11.

❖ The definition of "Basement" for buildings and structures located in flood hazard areas is "the portion of a building having its floor subgrade (below grade) on all sides." This definition pertains to enclosed areas below elevated buildings and structures whether or not there is enough clearance for such areas to be occupied, such as a crawl space. Neither the use of the enclosed space nor the clear height is the deciding factor as to whether an enclosed area below an elevated building is considered a basement under Section 1612.2. The controlling factor is whether the interior grade is below the exterior grade on all sides, even if the interior grade is established by the footing excavation that was not backfilled.

Enclosed areas below elevated buildings or structures must meet the requirements of Section 1612.4, which references ASCE 24. In the event that a building or structure is not built in accordance with these requirements, particularly if it is determined to have a basement that is subgrade on all sides, NFIP flood insurance premium rates may be significantly higher.

In recognition of common construction practices in some parts of the country, the exception permits crawl spaces to be as much as 2 feet (610 mm) below the lowest adjacent exterior grade in accordance with FEMA Technical Bulletin # 11, provided the additional requirements of that document are met [specifically, the maximum depth below grade is limited to 2 feet (210 mm) and maximum depth of floodwater above grade is limited to 2 feet (210 mm)]. Communities that choose to allow this practice must amend their ordinances to include these additional requirements.

1805.1.3 Ground-water control. Where the ground-water table is lowered and maintained at an elevation not less than 6 inches (152 mm) below the bottom of the lowest floor, the floor and walls shall be dampproofed in accordance with Section 1805.2. The design of the system to lower the ground-water table shall be based on accepted principles of engineering that shall consider, but not necessarily be limited to, permeability of the soil, rate at which water enters the drainage system, rated capacity of pumps, head against which pumps are to operate and the rated capacity of the disposal area of the system.

❖ After completion of building construction, it is often necessary to maintain the water table at a level that is at least 6 inches (152 mm) below the bottom of the lowest floor in order to prevent the flow or seepage of water into the basement. Where the site consists of well-draining soil and the highest point of the water table occurs naturally at or lower than the required level stated above, there is no need to provide a site drainage system specifically designated to control the ground-water level. Where the soil characteristics and site topography are such that the water table can rise to a level that will produce a hydrostatic pressure against the basement structure, then a site drainage system may be installed to reduce the water level if there is sufficient land area to accomplish the purpose. When ground-water control in accordance with this section is provided, waterproofing in accordance with Section 1805.3 is not required.

There are many types of site drainage systems that can be employed to control ground-water levels. The most commonly used systems may involve the installation of drainage ditches or trenches filled with pervious materials, sump pits and discharge pumps, well point systems, drainage wells with deep-well pumps, sand-drain installations, etc. This section requires that all such systems be designed and constructed using accepted engineering principles and practices based upon considerations that include the permeability of the soil, amount and rate at which water enters the system, pump capacity, capacity of the disposal area and other such factors that are necessary for the complete design of an operable drainage system.

1805.2 Dampproofing. Where hydrostatic pressure will not occur as determined by Section 1803.5.4, floors and walls for other than wood foundation systems shall be dampproofed in accordance with this section. Wood foundation systems shall be constructed in accordance with AF&PA PWF.

❖ For a general definition of "Dampproofing," see the commentary to Section 1805.1. Where a ground-water table investigation made in accordance with the requirements of Section 1803.5.4 (see commentary) has established that the high water table will occur at such a level that the building substructure will not be subjected to hydrostatic pressure, then dampproofing in accordance with this section and a subsoil drain in accordance with Section 1805.4 are sufficient to control moisture in the floor below grade. Since the wall will not be subject to water under pressure, the more restrictive provisions of waterproofing, as outlined in Section 1805.3, are not required. Wood foundation systems specified in Section 1807.1.4 (see commentary) are to be dampproofed as required by the American Forest & Paper Association (AF&PA) *Permanent Wood Foundation Design Specification* (PWF).

Dampproofing products having ICC Evaluation Service reports can be viewed at www.icc-es.org.

1805.2.1 Floors. Dampproofing materials for floors shall be installed between the floor and the base course required by Section 1805.4.1, except where a separate floor is provided above a concrete slab.

Where installed beneath the slab, dampproofing shall consist of not less than 6-mil (0.006 inch; 0.152 mm) polyethylene with joints lapped not less than 6 inches (152 mm), or other *approved* methods or materials. Where permitted to be installed on top of the slab, dampproofing shall consist of mopped-on bitumen, not less than 4-mil (0.004 inch; 0.102 mm) polyethylene, or other *approved* methods or materials. Joints in the membrane shall be lapped and sealed in accordance with the manufacturer's installation instructions.

❖ Floors requiring dampproofing in accordance with Section 1805.2 are to employ materials as specified in Section 1805.2.1. The dampproofing materials must be placed between the floor construction and the supporting gravel or stone base as shown in Figure 1805.4.2. Even if a floor base in accordance with Section 1805.4.1 is not required, dampproofing is still to be placed under the slab unless otherwise specified in Section 1805.2.1.

The installation is intended to provide a moisture barrier against the passage of water vapor or seepage into below-ground spaces.

The dampproofing material most commonly used for underslab installations consists of a polyethylene film no less than 6 mil [0.006 inch; (0.152 mm)] in thickness, which is applied over the gravel or stone base required in Section 1805.4.1. Care must be used in the installation of the material over the rough surface of the base and during the concreting operations so as not to puncture the polyethylene. Joints must be lapped at least 6 inches (152 mm). Other materials used in a similar way are made of neoprene or butyl rubber. Dampproofing materials can also be applied on top of the base concrete slab if a separate floor is provided above the base slab, since dampproofing is provided to prevent moisture infiltration of the interior space, not the concrete slab.

Materials commonly used for dampproofing floors are listed in Figure 1805.2.2.

MATERIAL	CONSTRUCTION
Asphalt	ASTM D 449
Asphalt primer	ASTM D 41
Coal-tar	ASTM D 450
Concrete and masonry oil primer (for coal-tar applications only)	ASTM D 43
Treated glass fabric	ASTM D 1668

Figure 1805.2.2
MATERIALS FOR WATERPROOFING AND DAMPROOFING INSTALLATIONS

1805.2.2 Walls. Dampproofing materials for walls shall be installed on the exterior surface of the wall, and shall extend from the top of the footing to above ground level.

Dampproofing shall consist of a bituminous material, 3 pounds per square *yard* (16 N/m^2) of acrylic modified cement, $^1/_8$ inch (3.2 mm) coat of surface-bonding mortar complying with ASTM C 887, any of the materials permitted for waterproofing by Section 1805.3.2 or other *approved* methods or materials.

❖ Walls requiring dampproofing in accordance with Section 1805.2 are first to be prepared as required in Section 1805.2.2.1 and then coated with a bituminous material, cement or mortar as specified in Section 1805.2.2 or other approved materials and methods of application. Approved materials are those that will prevent moisture from penetrating the foundation wall when water is present but not under pressure.

Coatings are applied to cover prepared exterior wall surfaces extending from the top of the wall footings to slightly above ground level so that the entire wall that contacts the ground is protected. Surfaces are usually primed to provide a bond coat and then dampproofed with a protective coat of asphalt or tar pitch. Emulsion-type coatings may be applied directly on "green" concrete or unit masonry walls; however, because they are water-soluble materials, their use is not generally recommended. Installation should comply with the manufacturer's instructions.

Any of the materials specified in Section 1805.3.2 for waterproofing are also allowed to be used for dampproofing. Figure 1805.2.2 provides a list of bituminous materials that can be used, including the applicable standards that may be used as the basis of acceptance of such materials. Included in Figure 1805.2.2 is ASTM D 1668 for glass fabric that is treated with asphalt (Type I), coal-tar pitch (Type II) or organic resin (Type III).

Surface-bonding mortar complying with ASTM C 887 may be utilized. This specification covers the materials, properties and packaging of dry, combined materials for use as surface-bonding mortar with concrete masonry units that have not been prefaced, coated or painted. Since this specification does not address design or application, manufacturers' recommendations should be followed. This standard covers proportioning, physical requirements, sampling and testing. The minimum thickness of the coating is $^1/_8$ inch (3.2 mm).

Acrylic-modified cement coatings may be utilized at the rate of 3 pounds per square yard (16 N/m^2). These types of materials have been used and performed successfully as dampproofing materials for foundation walls. Surface-bonding mortar and acrylic-modified cement are limited in use to dampproofing. The ability of these two types of products to bridge nonstructural cracks, as required in Section 1805.3.2 for waterproofing materials, is not known; therefore, their use is limited to dampproofing and they are not permitted to be used as waterproofing. Dampproofing may also include other materials and methods of installation acceptable to the building official.

1805.2.2.1 Surface preparation of walls. Prior to application of dampproofing materials on concrete walls, holes and recesses resulting from the removal of form ties shall be sealed with a bituminous material or other *approved* methods or materials. Unit masonry walls shall be parged on the exterior surface below ground level with not less than $^3/_8$ inch (9.5 mm) of Portland cement mortar. The parging shall be coved at the footing.

Exception: Parging of unit masonry walls is not required where a material is *approved* for direct application to the masonry.

❖ Before applying dampproofing materials, the concrete must be free of any holes or recesses that could affect the proper sealing of the wall surfaces. Air trapped beneath a dampproofing coating or membranes can cause blistering, while rocks and other sharp objects can puncture membranes. Irregular surfaces can also create uneven layering of coatings, which can result in vulnerable areas of dampproofing. Surface irregularities commonly associated with concrete wall construction can be sealed with bituminous materials or filled with portland cement grout or other approved methods.

Unit masonry walls are usually parged (plastered) with a $^1/_2$-inch-thick (12.7 mm) layer of portland cement and sand mix (1:2$^1/_2$ by volume) or with Type M mortar proportioned in accordance with the requirements of ASTM C 270, and applied in two $^1/_4$-inch-thick (6.4 mm) layers. In no case is parging to result in a final thickness of less than $^3/_8$ inch (9.5 mm). The parging is to be coved at the joint formed by the base of the wall and the top of the wall footing to prevent the accumulation of water at that location.

The moisture protection of unit masonry walls provided by the parging method may not be required where approved dampproofing materials, such as grout coatings, cement-based paints or bituminous coatings, can be applied directly to masonry surfaces.

1805.3 Waterproofing. Where the ground-water investigation required by Section 1803.5.4 indicates that a hydrostatic pressure condition exists, and the design does not include a ground-water control system as described in Section 1805.1.3, walls and floors shall be waterproofed in accordance with this section.

❖ For a general definition of "Waterproofing," and the distinction between dampproofing and waterproofing, see the commentary to Section 1805.1.

The significance of waterproofing installations is that they are intended to provide moisture barriers against water seepage that may be forced into below-ground spaces by hydrostatic pressure.

Where a ground-water table investigation made in accordance with the requirements of Section 1803.5.4 (see commentary) has established that the high water table will occur at such a level that the building substructure will be subjected to hydrostatic pressure, and where the water table is not lowered by a water control system, as prescribed in the commentary to Section 1805.1.3, all floors and walls below ground level are to be waterproofed in accordance with Sections 1805.3.1, 1805.3.2 and 1805.3.3. Waterproofing products having ICC Evaluation Service reports can be viewed at www.icc-es.org.

1805.3.1 Floors. Floors required to be waterproofed shall be of concrete and designed and constructed to withstand the hydrostatic pressures to which the floors will be subjected.

Waterproofing shall be accomplished by placing a membrane of rubberized asphalt, butyl rubber, fully adhered/fully bonded HDPE or polyolefin composite membrane or not less than 6-mil [0.006 inch (0.152 mm)] polyvinyl chloride with joints lapped not less than 6 inches (152 mm) or other *approved* materials under the slab. Joints in the membrane shall be lapped and sealed in accordance with the manufacturer's installation instructions.

❖ Since floors that are required to be waterproofed are subjected to hydrostatic uplift pressures, such floors must, for all practical purposes, be made of concrete and designed and constructed to resist the maximum hydrostatic pressures possible. It is particularly important that the floor slab be properly designed, since severe cracking or movement of the concrete would allow water seepage into below-ground spaces because the ability of the waterproofing materials to bridge small cracks would be exceeded. Concrete floor construction is to comply with the applicable provisions of Chapter 19.

Below-ground floors subjected to hydrostatic uplift pressures are to be waterproofed with membrane materials placed as underslab or split-slab installations, including such materials as rubberized asphalt, butyl rubber, fully bonded High Density Polyethylene (HDPE), fully bonded polyolefin and neoprene or with polyvinyl chloride (PVC) not less than 6 mil [0.006 inch; (0.152 mm)] in thickness, lapped at least 6 inches (152 mm). There are many proprietary membrane products available in the marketplace that are specifically made for waterproofing floors and walls (i.e., polyethylene sheets sandwiched between layers of asphalt), which may be used for that purpose when approved by the building official.

All membrane joints are to be lapped and sealed in accordance with the manufacturer's instructions to form a continuous, impermeable moisture barrier.

1805.3.2 Walls. Walls required to be waterproofed shall be of concrete or masonry and shall be designed and constructed to withstand the hydrostatic pressures and other lateral loads to which the walls will be subjected.

Waterproofing shall be applied from the bottom of the wall to not less than 12 inches (305 mm) above the maximum elevation of the ground-water table. The remainder of the wall shall be dampproofed in accordance with Section 1805.2.2. Waterproofing shall consist of two-ply hot-mopped felts, not less than 6-mil (0.006 inch; 0.152 mm) polyvinyl chloride, 40-mil (0.040 inch; 1.02 mm) polymer-modified asphalt, 6-mil (0.006 inch; 0.152 mm) polyethylene or other *approved* methods or materials capable of bridging nonstructural cracks. Joints in the membrane shall be lapped and sealed in accordance with the manufacturer's installation instructions.

❖ Walls that are required to be waterproofed in accordance with Section 1805.3 must first be prepared as required in Section 1805.3.2.1.

The walls must be designed to resist the hydrostatic pressure anticipated at the site, as well as any other lateral loads to which the wall will be subjected, such as soil pressures or seismic loads. As with floors required to be waterproofed, it is particularly important that walls required to be waterproofed be properly designed to resist all loads present, since cracking and other damage would allow water seepage into below-ground spaces. Water seepage can lead to deterioration of the foundation as wood rots, concrete and masonry erode and steel rusts. Such deterioration can cause structural failure of the foundation. More importantly, failure of the foundation wall can lead to structural failure of the building, since the foundation supports the building structure. Masonry or concrete construction must comply with the applicable provisions of Chapters 21 and 19, respectively.

Figures 1805.2.2 and 1805.3.2 list materials commonly used for the installation of moisture barriers in wall construction and the related standard that may be used as a basis for acceptance of such materials. Asphalt and coal-tar products are not compatible and should not be used together.

Waterproofing installations are to extend from the bottom of the wall to a height no less than 12 inches (305 mm) above the maximum elevation of the ground-water table determined in accordance with the requirements of Section 1803.5.4 (see commentary). The remainder of the wall below ground level (if the height is small) may be waterproofed as a continuation of the installation or must be dampproofed in accordance with the requirements of Section 1805.2.2. If the ground-water table investigation is

MATERIAL	CONSTRUCTION
Asphalt-saturated asbestos felt	ASTM D 250
Asphalt-saturated burlap fabric	ASTM D 1327
Asphalt-saturated cotton fabric	ASTM D 173
Asphalt-saturated organic felt	ASTM D 226
Coal-tar-saturated burlap fabric	ASTM D 1327
Coal-tar-saturated cotton fabric	ASTM D 173
Coal-tar-saturated organic felt	ASTM D 227

Figure 1805.3.2
MATERIALS FOR WATERPROOFING INSTALLATIONS

not conducted on the basis of the exception to Section 1803.5.4, then waterproofing should be provided from a point below the footing to above the ground level.

This section requires that waterproofing must consist of two-ply hot-mopped felts. The practice of the waterproofing industry is to select the number of plies of membrane material based on the hydrostatic head (height of water pressure against the wall). As a general practice, if the head of water is between 1 foot (305 mm) and 3 feet (914 mm), two plies of felt or fabric membrane are used; between 4 feet (1219 mm) and 10 feet (3048 mm), three-ply construction is needed and between 11 feet (3353 mm) and 25 feet (7620 mm), four-ply construction is necessary.

Waterproofing installations may also use polyvinyl chloride materials of no less than 6-mil [0.006 inch (0.152 mm)] thick, 40-mil [0.040 inch (1.02 mm)] polymer-modified asphalt or 6-mil [0.006 inch (0.152 mm)] polyethylene. These materials have been widely recognized for their effectiveness in bridging nonstructural cracks. Other approved materials and methods may be used provided that the same performance standards are met. All membrane joints must be lapped and sealed in accordance with the manufacturer's instructions.

1805.3.2.1 Surface preparation of walls. Prior to the application of waterproofing materials on concrete or masonry walls, the walls shall be prepared in accordance with Section 1805.2.2.1.

❖ Before applying waterproofing materials to concrete or masonry walls, the surfaces must be prepared in accordance with the requirements of Section 1805.2.2.1, which requires the sealing of all holes and recesses. Surfaces to be waterproofed must also be free of any projections that might puncture or tear membrane materials that are applied over the surfaces.

1805.3.3 Joints and penetrations. Joints in walls and floors, joints between the wall and floor and penetrations of the wall and floor shall be made water-tight utilizing *approved* methods and materials.

❖ This section requires that all joints occurring in floors and walls and at locations where floors and walls meet, as well as all penetrations in floors and walls, be made water tight by approved methods. Sealing joints and penetrations is of primary importance to ensuring the effectiveness of the waterproofing. If the joints or penetrations are not sealed properly, they can develop leaks, which become a passageway for water to enter the building. Since the remainder of the foundation is wrapped in waterproofing, moisture can actually become trapped in the foundation walls or floor slab, and serious damage to these structural components can occur. Such methods may involve the use of construction keys (e.g., between the base of the wall and the top of the footing) or, if there is hydrostatic pressure, floor and wall joints may require the use of manufactured waterstops made of metal, rubber, plastic or mastic materials. Figures 1805.3.3(1) and 1805.3.3(2) illustrate examples of joint treatment and penetration treatment of waterproofing. Floor edges along walls and floor expansion joints may employ the use of any number of preformed expansion joint materials, such as asphalt, polyurethane, sponge rubber, self-expanding cork, cellular fibers bonded with bituminous materials, etc., which all comply with applicable ASTM International (ASTM) or American Association of State Highway and Transportation Officials (AASHTO) standards or other federal specifications. A variety of sealants may be used together with the preformed joint materials. Gaskets made of neoprene and other materials are also available for use in concrete and masonry joints. The National Roofing Contractors Association's (NRCA's) *Roofing and Waterproofing Manual* pro-

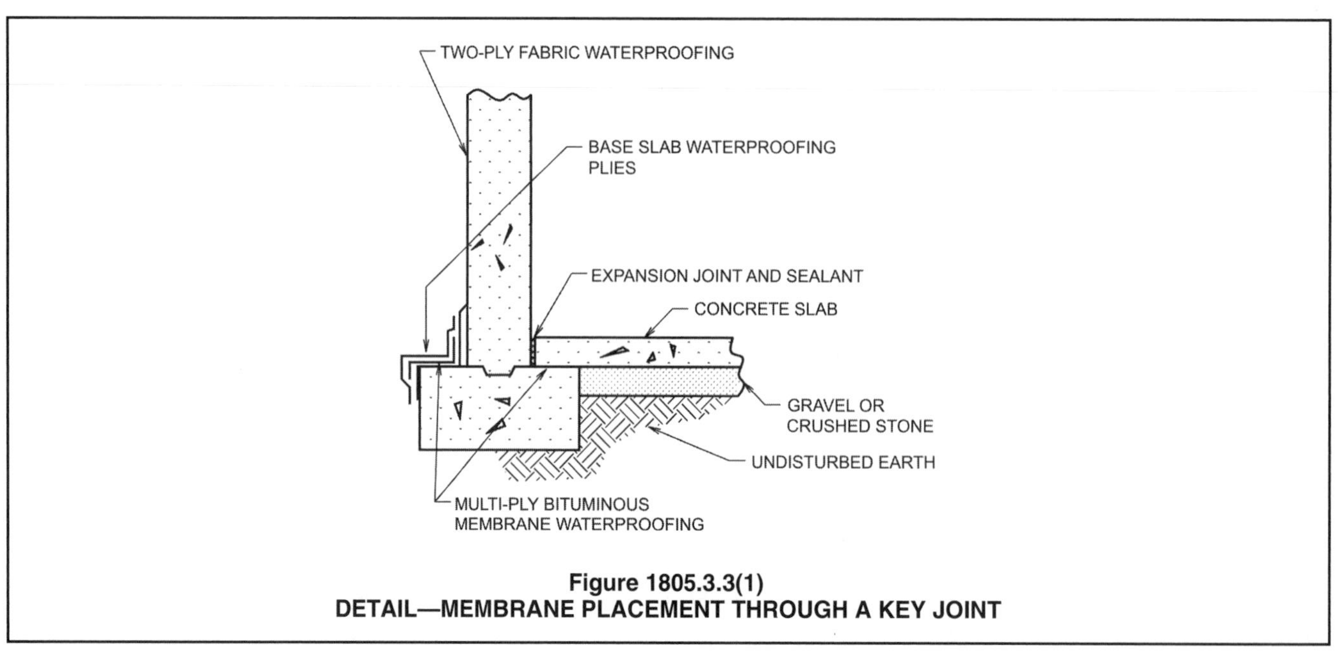

Figure 1805.3.3(1)
DETAIL—MEMBRANE PLACEMENT THROUGH A KEY JOINT

vides details for the reinforcement of membrane terminations, corners, intersections of slabs and walls, through-wall and slab penetrations and other locations.

Penetrations in walls and floors may be made water-tight with grout or manufactured fill materials and sealants made for that purpose.

1805.4 Subsoil drainage system. Where a hydrostatic pressure condition does not exist, dampproofing shall be provided and a base shall be installed under the floor and a drain installed around the foundation perimeter. A subsoil drainage system designed and constructed in accordance with Section 1805.1.3 shall be deemed adequate for lowering the groundwater table.

❖ This section covers subsoil drainage systems in conjunction with dampproofing (see Section 1805.2) to protect below-ground spaces from water seepage. Such systems are not used where basements or other below-ground spaces are subject to hydrostatic pressure because they would not be effective in disposing of the amount of water anticipated if a hydrostatic pressure condition exists. Ground-water tables may be reduced to acceptable levels by methods described in the commentary to Section 1805.1.3.

The details of subsoil drainage systems are covered in the requirements of Sections 1805.4.1 through 1805.4.3.

1805.4.1 Floor base course. Floors of basements, except as provided for in Section 1805.1.1, shall be placed over a floor base course not less than 4 inches (102 mm) in thickness that consists of gravel or crushed stone containing not more than 10 percent of material that passes through a No. 4 (4.75 mm) sieve.

> **Exception:** Where a site is located in well-drained gravel or sand/gravel mixture soils, a floor base course is not required.

❖ This section requires that basement floors, except for story-above-grade construction, must be placed on a gravel or stone base no less than 4 inches (102 mm) thick. Not more than 10 percent of the material is to pass a No. 4 sieve so as to provide a porous installation. Material that passes a No. 4 sieve would be fine silt or clay that would not permit the free movement of water through the floor base. This requirement serves three purposes. The first is to provide an adjustment to the irregularities of a compacted subgrade so as to produce a level surface upon which to cast a concrete slab. The second is to provide a capillary break so that moisture from the soil below will not rise to the underside of the floor. Finally, where required, the porous base can act as a drainage system to expel underslab water by means of gravity or the use of a sump pump or other approved methods.

The exception allows for the omission of the floor base when the natural soils beneath the floor slab consist of well-draining granular materials, such as sand, stone or mixtures of these materials. The exception is consistent with the requirements of Section 1805.4.3. Some caution, however, is justified in the use of this exception. If the granular soils contain an excessive percentage of fine materials, the porosity and the ability of the soil to provide a capillary break may be considerably diminished. The exception is only to be applied if the natural base is equivalent to the floor base otherwise required by this section.

1805.4.2 Foundation drain. A drain shall be placed around the perimeter of a foundation that consists of gravel or crushed stone containing not more than 10-percent material that passes through a No. 4 (4.75 mm) sieve. The drain shall extend a minimum of 12 inches (305 mm) beyond the outside edge of the footing. The thickness shall be such that the bottom of the drain is not higher than the bottom of the base under the floor, and that the top of the drain is not less than 6 inches (152 mm) above the top of the footing. The top of the drain shall be covered with an *approved* filter membrane material. Where a drain tile or perforated pipe is used, the

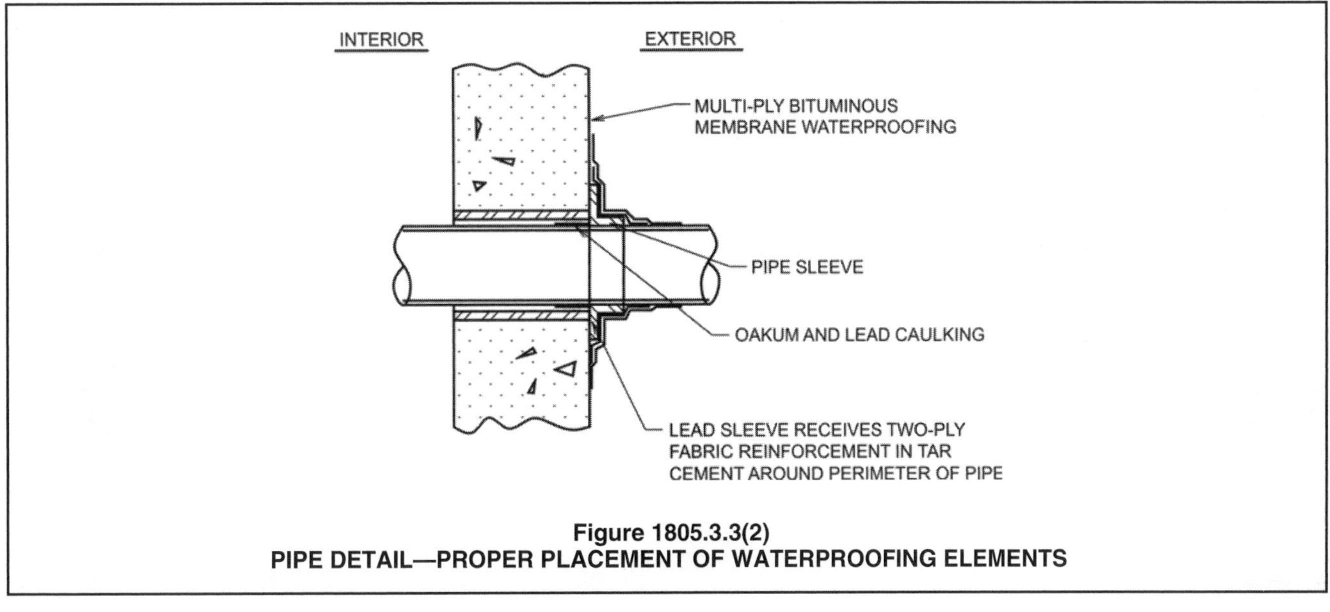

Figure 1805.3.3(2)
PIPE DETAIL—PROPER PLACEMENT OF WATERPROOFING ELEMENTS

invert of the pipe or tile shall not be higher than the floor elevation. The top of joints or the top of perforations shall be protected with an *approved* filter membrane material. The pipe or tile shall be placed on not less than 2 inches (51 mm) of gravel or crushed stone complying with Section 1805.4.1, and shall be covered with not less than 6 inches (152 mm) of the same material.

❖ This section describes the materials and features of construction required for the installation of foundation drain systems. Perimeter foundation drains provide a means to remove free ground water and prevent leakage into habitable below-grade spaces. The first part of this provision describes a foundation drain of gravel or crushed stone. The balance of the provision describes the installation of drain tile or perforated pipe where either of those is provided. The use of drain tiles is important in areas having moderate to heavy rainfall as well as in soils that have low percolation rates. The requirements of this section are illustrated in Figure 1805.4.2.

This type of drain system is suitable where the water table occurs at such elevation that there is no hydrostatic pressure exerted against the basement floor and walls, and where the amount of seepage from the surrounding soil is so small that the water can be readily discharged by gravity or mechanical means into sewers or ditches. The objective is to combine the protection afforded by the dampproofing of walls and floors (see Section 1805.2) and that given by perimeter drains so as to maintain below-ground spaces in a dry condition.

Gravel or crushed stone drain material is to be placed in the excavation so that it will extend out from the edge of the wall footing a distance of at least 12 inches (305 mm), with the bottom of the fill being no higher than the bottom of the base under the floor (see Section 1805.4.1) and the top of the filter material being no less than 6 inches (152 mm) above the top of the wall footing so that water will not collect along the top of the footing. Requiring the bottom of the gravel or crushed rock drain to be no higher than the bottom of the floor base is necessary so that if the water table rises into the floor base, it will also be able to rise unobstructed into the foundation drain. The foundation drain will then drain the water away from the building, as required by Section 1805.4.3. The top of the drain must be covered with an approved filter membrane to allow water to pass through without allowing, or at least greatly reducing, the possibility of fine soil materials entering the drainage system.

Where drain tile or perforated drain pipe is utilized,

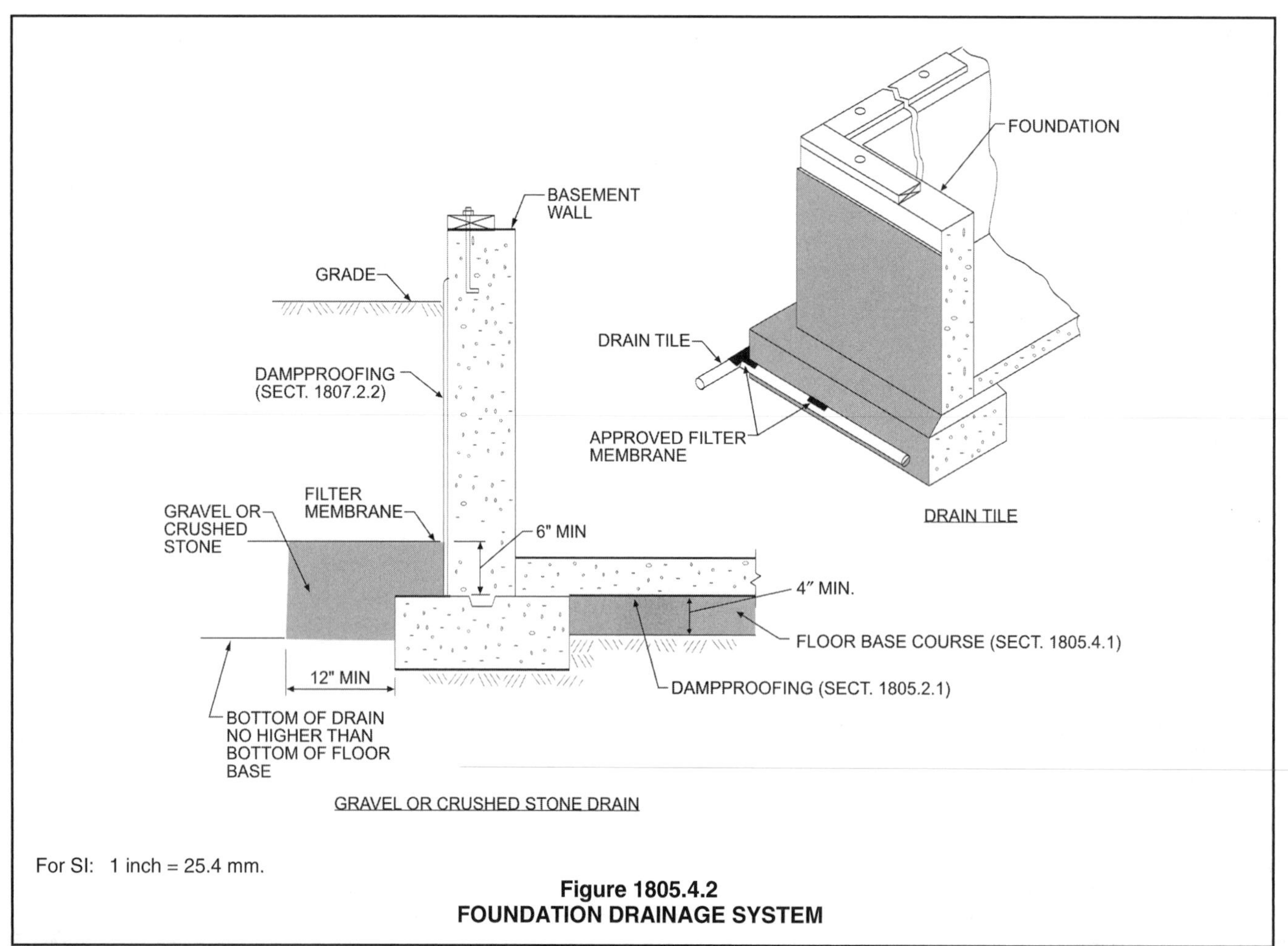

Figure 1805.4.2
FOUNDATION DRAINAGE SYSTEM

the drain system consists of clay or concrete drain tiles or corrugated metal or nonmetallic pipes installed in a filter bed with at least 2 inches (51 mm) of filter material and covered with at least 6 inches (152 mm) of filter material to maintain good water flow into the drain tile or pipe. The foundation drain is set adjacent to the wall footing and extends around the perimeter of the building. Drain tiles are placed end to end with open joints to permit water to enter the system. Metallic and nonmetallic drains are made with perforations at the invert (bottom) section of the pipe and are installed with connected ends. The drain pipe invert is not to be set higher than the basement floor line such that water conveyed by the drain does not seep into the filter material and create a hydrostatic pressure condition against the foundation wall and footing. The inverts should not be placed below the bottom of the adjacent wall footings so as to avoid carrying away fine soil particles that, in time, could undermine the footing and settlement of the foundation walls.

Tile joints or pipe perforations should be covered with an approved filter membrane material to prevent them from becoming clogged, preventing fine particles that may be contained in the surrounding soil from entering the system and being carried away by water. The filter material around the drain tiles or pipes (not to be confused with filter membrane material) is to consist of selected gravel and crushed stone containing no more than 10 percent of material that passes through a No. 4 sieve. The filter materials should be selected to prevent the movement of particles from the protected soil surrounding the drain installation into the drain.

1805.4.3 Drainage discharge. The floor base and foundation perimeter drain shall discharge by gravity or mechanical means into an *approved* drainage system that complies with the *International Plumbing Code.*

> **Exception:** Where a site is located in well-drained gravel or sand/gravel mixture soils, a dedicated drainage system is not required.

❖ This section references the *International Plumbing Code*® (IPC®) for the requirements of installing piping systems for the disposal of water from the floor base (see Section 1805.4.1) and the foundation drains (see Section 1805.4.2). Chapter 11 of the IPC deals with the piping materials, applicable standards and methods of installation of subsurface storm drains to facilitate water discharge either by gravity or mechanical means.

Where the soil at the site consists of well-drained granular materials (i.e., gravel or sand-gravel mixtures) to prevent the occurrence of hydrostatic pressure against the foundation walls and under the floor slab, the use of a dedicated drainage system as prescribed in the IPC is not required, since the site soils would permit natural drainage.

SECTION 1806 PRESUMPTIVE LOAD-BEARING VALUES OF SOILS

1806.1 Load combinations. The presumptive load-bearing values provided in Table 1806.2 shall be used with the *allowable stress design* load combinations specified in Section 1605.3. The values of vertical foundation pressure and lateral bearing pressure given in Table 1806.2 shall be permitted to be increased by one-third where used with the alternative basic load combinations of Section 1605.3.2 that include wind or earthquake loads.

❖ This section clarifies the use of the presumptive load-bearing values relating to foundations and footing design. Foundation design is based on the ASD approach. Since Chapter 16 includes provisions and load combinations for strength design [load and resistance factor design (LRFD)], it is necessary to clarify that these values are to be used with the ASD load combinations.

1806.2 Presumptive load-bearing values. The load-bearing values used in design for supporting soils near the surface shall not exceed the values specified in Table 1806.2 unless data to substantiate the use of higher values are submitted and *approved.* Where the *building official* has reason to doubt the classification, strength or compressibility of the soil, the requirements of Section 1803.5.2 shall be satisfied.

Presumptive load-bearing values shall apply to materials with similar physical characteristics and dispositions. Mud, organic silt, organic clays, peat or unprepared fill shall not be assumed to have a presumptive load-bearing capacity unless data to substantiate the use of such a value are submitted.

> **Exception:** A presumptive load-bearing capacity shall be permitted to be used where the *building official* deems the load-bearing capacity of mud, organic silt or unprepared fill is adequate for the support of lightweight or temporary structures.

❖ Where the load-bearing capacity of the soil has not been determined by borings, as specified in Section 1803, the presumptive load-bearing values listed in Table 1806.2 are intended to apply in the design of shallow foundation systems.

While fill material (unconsolidated), mud, muck, peat, organic silt, soft clay and other unprepared fill materials are considered to have no presumptive load-bearing value, soil tests may show that they do have some limited load-bearing capacity and, based on this type of evidence, the building official may approve the construction of lightweight structures upon such soils.

The presumption is that the building official possesses sufficient technical knowledge on the character and behavior of subsurface materials to render a valid judgment on the adequacy of the soil to support satisfactorily the lightweight or temporary structure, or he or she has sought and gained specific advice through consultation with professionals who are competent in the field of foundation engineering. It would be an unwise practice to authorize construction on exceptionally weak soils without the benefit of technical knowledge to make judgmental decisions.

TABLE 1806.2. See below.

❖ The values in Table 1806.2 are intended to be lower-bound allowable pressures to be used where the bearing value is not determined by borings as specified in Section 1803. The classes of soil and rock listed in Table 1806.2 are those materials most commonly found at construction sites around the country. The allowable foundation pressures expressed in terms of pounds per square foot (psf) for each class of subsurface materials listed in Table 1806.2 are based on experience with the behavior of these materials under loaded conditions.

Should local soil conditions be significantly different than those listed in Table 1806.2, then, with the approval of the building official, local experience can be employed in the design of foundations, particularly where actual load-bearing values are less than the allowable unit pressures given in the table.

Whether the allowable load-bearing values of Table 1806.2 are used or local conditions on soil capacity prevail, many precautions must be taken in the design of foundation systems that are not regulatory functions of the code, but rather are the professional considerations and design applications of those who are engaged in foundation engineering. The building official needs to pay particular attention to proper selection of the allowable load-bearing capacity of the soil because it is the source of many foundation failures. The problem arises when the load-bearing material directly under a foundation overlies a stratum (or strata) of weaker material having a smaller allowable load-bearing capacity. The selection of the load-bearing value to be used in the design of the foundation should take into account the load distribution at the weaker stratum (or strata) so that the pressure on the soil does not exceed its allowable load-bearing capacity. In this respect, it is important to have a soil profile showing the classes of material at the construction site or at properties in the near vicinity of or adjacent to the area of construction. Such information can be a part of the records or other data required by Section 1803.6 or the results of a soil investigation.

With this information, the building official can consult with the registered design professional in charge of the foundation design to ascertain that due care was exercised in adopting proper soil load-bearing values.

1806.3 Lateral load resistance. Where the presumptive values of Table 1806.2 are used to determine resistance to lateral loads, the calculations shall be in accordance with Sections 1806.3.1 through 1806.3.4.

❖ When the tabulated values for lateral load resistance of soils are utilized, certain limitations must be considered, and they are contained in the following subsections.

1806.3.1 Combined resistance. The total resistance to lateral loads shall be permitted to be determined by combining the values derived from the lateral bearing pressure and the lateral sliding resistance specified in Table 1806.2.

❖ This section permits a foundation's lateral resistance to be determined by combining the lateral bearing and lateral sliding values from Table 1806.2.

1806.3.2 Lateral sliding resistance limit. For clay, sandy clay, silty clay, clayey silt, silt and sandy silt, in no case shall the lateral sliding resistance exceed one-half the dead load.

❖ For cohesive soils, the tabulated lateral sliding value is set at 130 psf (6.23 kPa). However, this section also stipulates that for clay, sandy clay, silty clay and clayey silt, the lateral sliding resistance cannot exceed one-half the dead load.

**TABLE 1806.2
PRESUMPTIVE LOAD-BEARING VALUES**

CLASS OF MATERIALS	VERTICAL FOUNDATION PRESSURE (psf)	LATERAL BEARING PRESSURE (psf/ft below natural grade)	LATERAL SLIDING RESISTANCE	
			Coefficient of friction[a]	Cohesion (psf)[b]
1. Crystalline bedrock	12,000	1,200	0.70	—
2. Sedimentary and foliated rock	4,000	400	0.35	—
3. Sandy gravel and/or gravel (GW and GP)	3,000	200	0.35	—
4. Sand, silty sand, clayey sand, silty gravel and clayey gravel (SW, SP, SM, SC, GM and GC)	2,000	150	0.25	—
5. Clay, sandy clay, silty clay, clayey silt, silt and sandy silt (CL, ML, MH and CH)	1,500	100	—	130

For SI: 1 pound per square foot = 0.0479kPa, 1 pound per square foot per foot = 0.157 kPa/m.

a. Coefficient to be multiplied by the dead load.

b. Cohesion value to be multiplied by the contact area, as limited by Section 1806.3.2.

1806.3.3 Increase for depth. The lateral bearing pressures specified in Table 1806.2 shall be permitted to be increased by the tabular value for each additional foot (305 mm) of depth to a maximum of 15 times the tabular value.

❖ The lateral sliding resistance is calculated as the sum of the lateral bearing and lateral sliding values from Table 1806.2. The lateral bearing values are determined as the product of the tabular value and the depth below natural grade. This section essentially limits the foundation depth for which the lateral bearing values can be increased to 15 feet (4572 mm).

1806.3.4 Increase for poles. Isolated poles for uses such as flagpoles or signs and poles used to support buildings that are not adversely affected by a $^1/_2$ inch (12.7 mm) motion at the ground surface due to short-term lateral loads shall be permitted to be designed using lateral bearing pressures equal to two times the tabular values.

❖ For isolated poles used as supports for flagpoles or signs, and poles used to support buildings that are not adversely affected by a $^1/_2$-inch (12.7 mm) motion at the ground surface due to short-term lateral loads, the tabulated lateral bearing values are permitted to be doubled.

SECTION 1807
FOUNDATION WALLS, RETAINING WALLS AND EMBEDDED POSTS AND POLES

1807.1 Foundation walls. Foundation walls shall be designed and constructed in accordance with Sections 1807.1.1 through 1807.1.6. Foundation walls shall be supported by foundations designed in accordance with Section 1808.

❖ Foundation walls typically serve as the enclosure for a basement or crawl space as well as a below-grade load-bearing foundation component. Where applicable, they are designed to resist lateral soil pressures as well as dead, live, snow, wind, and seismic loads. This section contains the provisions applicable to the design and construction of foundation walls.

1807.1.1 Design lateral soil loads. Foundation walls shall be designed for the lateral soil loads set forth in Section 1610.

❖ This section requires that foundation and retaining walls be designed to resist the lateral soil loads in accordance with Section 1610. The code provides lateral soil loads in Table 1610.1 for use where a specific geotechnical investigation has not been performed. Consideration must be also be given to hydrostatic loads in addition to lateral pressures resulting from surcharge loads, such as a sloping backfill.

1807.1.2 Unbalanced backfill height. Unbalanced backfill height is the difference in height between the exterior finish ground level and the lower of the top of the concrete footing that supports the foundation wall or the interior finish ground level. Where an interior concrete slab on grade is provided and is in contact with the interior surface of the foundation wall, the unbalanced backfill height shall be permitted to be measured from the exterior finish ground level to the top of the interior concrete slab.

❖ This provision establishes unbalanced backfill height as the difference in height of the exterior and interior finished ground levels. The height of unbalanced backfill quantifies the magnitude of the lateral soil load for the different classifications of soils presented in the prescriptive foundation wall tables. For an illustration of unbalanced backfill height where an interior slab on grade is in contact with the foundation wall, see Figure 1807.1.2.

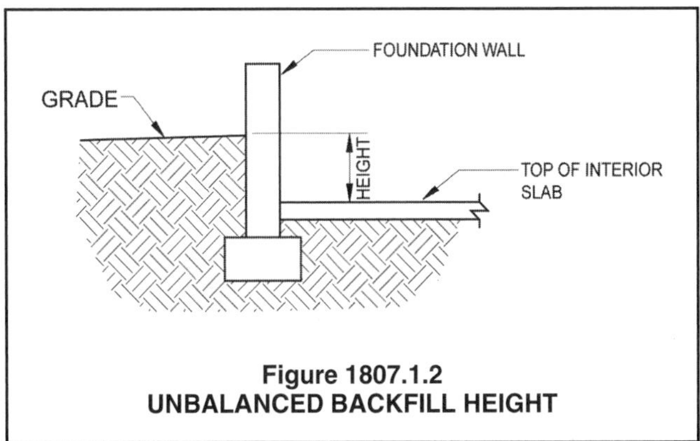

Figure 1807.1.2
UNBALANCED BACKFILL HEIGHT

1807.1.3 Rubble stone foundation walls. Foundation walls of rough or random rubble stone shall not be less than 16 inches (406 mm) thick. Rubble stone shall not be used for foundation walls of structures assigned to *Seismic Design Category* C, D, E or F.

❖ Foundation walls of rough or random rubble stone are limited to a thickness of 16 inches (406 mm). Because of the use of rough stones of irregular size and shape, a larger thickness is required (as compared to concrete or masonry walls) for adequate bonding of the stone and mortar to provide structural stability against soil pressures, as well as protection from water penetration.

1807.1.4 Permanent wood foundation systems. Permanent wood foundation systems shall be designed and installed in accordance with AF&PA PWF. Lumber and plywood shall be treated in accordance with AWPA U1 (Commodity Specification A, Use Category 4B and Section 5.2) and shall be identified in accordance with Section 2303.1.8.1.

❖ This section covers the design and installation of a multicomponent wood foundation system complying with the specifications of AF&PA PWF. The construction is essentially a below-grade, load-bearing, wood-frame system that serves as an enclosure for basements and crawl spaces as well as the structural foundation for the support of light-frame structures. According to AF&PA PWF, such systems are "engineered to support lateral soil pressures as well as dead, live, snow, wind, and seismic loads," where the foundation wall behavior is primary, and vertical load behavior is secondary.

This foundation system consists of several principal components:

1. Walls that consist of plywood panels attached to a wood stud framing system;
2. A composite footing consisting of a continuous wood plate set on a bed of granular materials, such as sand, gravel or crushed stone, which supports the foundation walls and transmits their loads to the bearing soil below;
3. Polyethylene film that serves as a vapor barrier and covers the exterior side of the plywood foundation walls from grade down to the footing plate;
4. Caulking compounds used for sealing joints in plywood walls as well as bonding agents for attaching the polyethylene film to the plywood and for sealing the film joints;
5. Metal fasteners made of silicon, bronze, copper or stainless steel or hot-dipped, zinc-coated steel nails or staples; and
6. Pressure-treated plywood and lumber to protect the foundation material against decay, termites and other insects.

A typical cross section of a wood foundation basement wall is shown in Figure 1807.1.4.

All plywood and lumber used in the foundation system are required to receive a preservative treatment to protect the material against decay, termites and other insects in accordance with American Wood Preserver's Association (AWPA U1), Specification A, Use Category 4B and Section 5.2. Specification A of AWPA U1 pertains to sawn wood products, while Use Category 4B refers to wood that is in contact with the ground. The additional reference to Section 5.2 incorporates requirements under Specification A that are unique to permanent wood foundations. This section also requires the treated wood to be identified in accordance with Section 2303.1.8.1. See the commentary to that section for a discussion of identification requirements.

Ordinary metal fasteners in contact with preservatives used in the treatment of wood products will corrode over time and can cause serious structural problems. For this reason, AF&PA PWF prescribes that only fasteners made of silicon bronze, copper or Type 304 or 316 stainless steel, as defined by the American Iron and Steel Institute (AISI), are permitted to be used in a wood foundation system. Hot-dipped, zinc-coated steel nails are permitted to be used in a

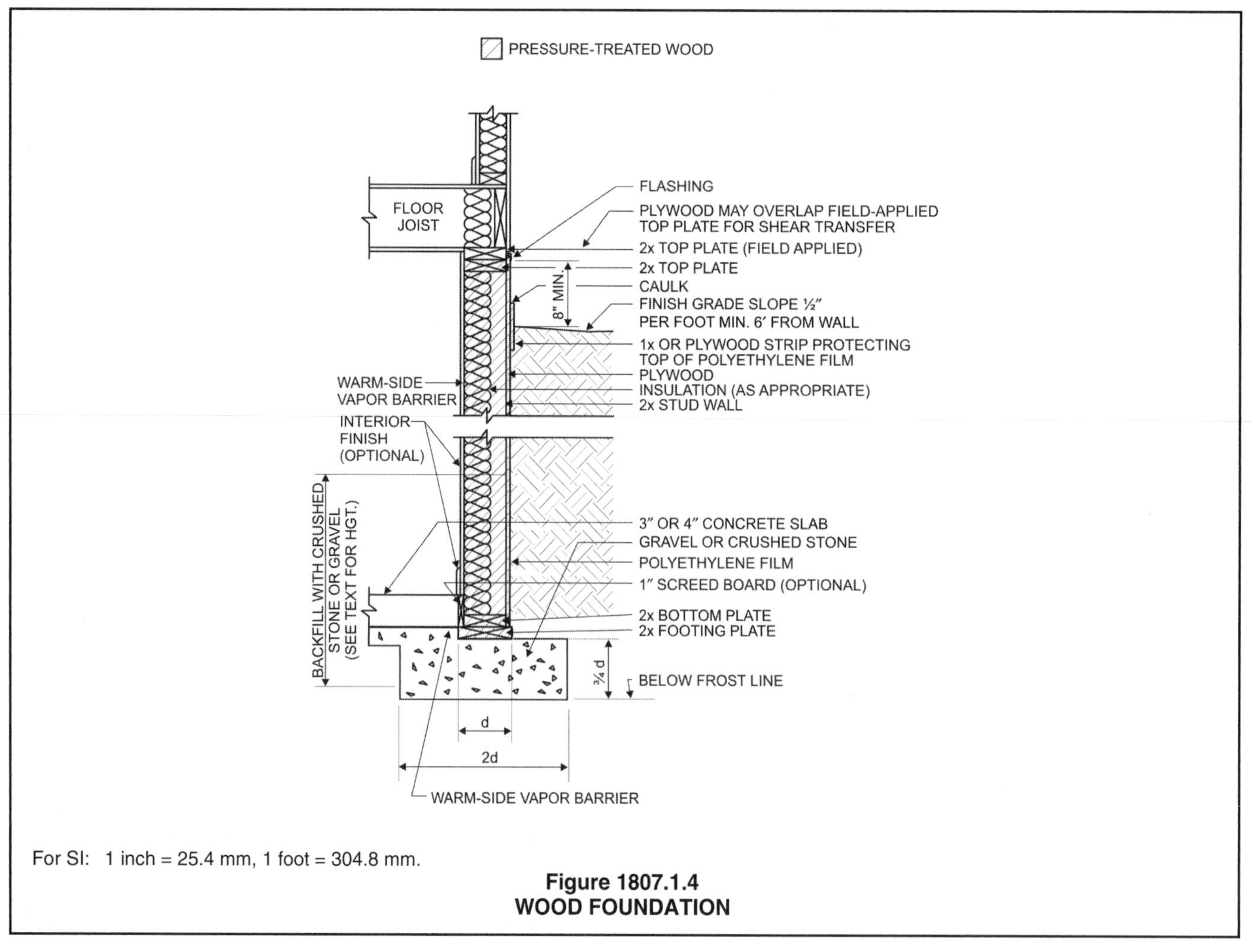

For SI: 1 inch = 25.4 mm, 1 foot = 304.8 mm.

Figure 1807.1.4
WOOD FOUNDATION

wood foundation system when installed in accordance with the specific conditions contained in AF&PA PWF for the surface treatment of nails and moisture protection of the foundation.

Certain cold-weather precautions should be taken when installing a wood foundation system. The composite footing consisting of a wood plate supported on a bed of stone or sand fill should not be placed on frozen ground. While the bottom of the wood plate footing would normally be placed below the frost line (i.e., basement construction), under certain drainage conditions, AF&PA PWF permits the wood plate to be set above the frost line level. The building official should ascertain that such an alternative design satisfies the intent of the code. Most important is the use of proper sealants during very cold weather. All manufacturers of sealants and bonding agents impose temperature restrictions on the use of their products. Only sealants and bonding agents specifically produced for cold-weather conditions should be used. This is an important consideration since sealants are a factor in verifying the water-tight integrity of the foundation installation.

The wood foundation system is an innovative design consisting of many parts and several kinds of materials. It can be expected to perform satisfactorily as a foundation for light construction only if the comprehensive provisions and recommendations of AF&PA PWF are strictly followed.

1807.1.5 Concrete and masonry foundation walls. Concrete and masonry foundation walls shall be designed in accordance with Chapter 19 or 21, as applicable.

Exception: Concrete and masonry foundation walls shall be permitted to be designed and constructed in accordance with Section 1807.1.6.

❖ Foundation walls are usually designed and constructed in accordance with the accepted engineering practices to carry the vertical loads from the structure above, resist wind and any lateral forces transmitted to the foundations and sustain earth pressures exerted against the walls, including any forces that may be imposed by frost action. For reinforced and plain concrete, the physical properties and design criteria are provided in Chapter 19. For masonry construction, the physical properties and design criteria are provided in Chapter 21. An exception permits the use of prescriptive foundation walls when the applicable limitations are met.

1807.1.6 Prescriptive design of concrete and masonry foundation walls. Concrete and masonry foundation walls that are laterally supported at the top and bottom shall be permitted to be designed and constructed in accordance with this section.

❖ This section allows prescriptive design of concrete or masonry foundation walls that are primarily intended for, but not necessarily limited to, basement construction in residential and light commercial buildings or other light structures. For the actual load limit on these prescriptive concrete and masonry foundation walls, see Section 1807.1.6.2, Item 7 and Section 1807.1.6.3, Item 8, respectively. Since the designs in the tables are only applicable to foundation walls that are laterally supported at the top and bottom, foundation walls that are not laterally supported at the top and bottom must be designed as described in Section 1807.1.5.

1807.1.6.1 Foundation wall thickness. The thickness of prescriptively designed foundation walls shall not be less than the thickness of the wall supported, except that foundation walls of at least 8-inch (203 mm) nominal width shall be permitted to support brick-veneered frame walls and 10-inch-wide (254 mm) cavity walls provided the requirements of Section 1807.1.6.2 or 1807.1.6.3 are met.

❖ The minimum thicknesses in this section are to facilitate the support of the wall above grade. The thickness requirements in this section are empirical and have been used successfully for many years.

1807.1.6.2 Concrete foundation walls. Concrete foundation walls shall comply with the following:

1. The thickness shall comply with the requirements of Table 1807.1.6.2.
2. The size and spacing of vertical reinforcement shown in Table 1807.1.6.2 is based on the use of reinforcement with a minimum yield strength of 60,000 pounds per square inch (psi) (414 MPa). Vertical reinforcement with a minimum yield strength of 40,000 psi (276 MPa) or 50,000 psi (345 MPa) shall be permitted, provided the same size bar is used and the spacing shown in the table is reduced by multiplying the spacing by 0.67 or 0.83, respectively.
3. Vertical reinforcement, when required, shall be placed nearest the inside face of the wall a distance, d, from the outside face (soil face) of the wall. The distance, d, is equal to the wall thickness, t, minus 1.25 inches (32 mm) plus one-half the bar diameter, d_b, [$d = t - (1.25 + d_b / 2)$]. The reinforcement shall be placed within a tolerance of ± $^3/_8$ inch (9.5 mm) where d is less than or equal to 8 inches (203 mm) or ± $^1/_2$ inch (12.7 mm) where d is greater than 8 inches (203 mm).
4. In lieu of the reinforcement shown in Table 1807.1.6.2, smaller reinforcing bar sizes with closer spacings that provide an equivalent cross-sectional area of reinforcement per unit length shall be permitted.
5. Concrete cover for reinforcement measured from the inside face of the wall shall not be less than $^3/_4$ inch (19.1 mm). Concrete cover for reinforcement measured from the outside face of the wall shall not be less than $1^1/_2$ inches (38 mm) for No. 5 bars and smaller, and not less than 2 inches (51 mm) for larger bars.
6. Concrete shall have a specified compressive strength, f'_c, of not less than 2,500 psi (17.2 MPa).

7. The unfactored axial load per linear foot of wall shall not exceed $1.2\,tf'_c$ where t is the specified wall thickness in inches.

❖ The section contains the limitations on construction details and material properties that are needed to utilize the prescriptive designs of Table 1807.1.6.2 for concrete foundation walls. The concrete wall thickness and reinforcing (if any) must be as shown in Table 1807.1.6.2, based on the wall's height and depth of any supported backfill as well as the applicable lateral soil pressure.

The minimum required concrete strength is given as well as the yield strength of reinforcement steel that was used as the basis for the tabulated vertical reinforcement. The use of other grades of reinforcing steel is permitted provided the appropriate adjustment is made to the spacing of the vertical bars.

TABLE 1807.1.6.2. See below.

❖ The wall thickness and, where required, the size and spacing of reinforcing bars in the table are based on the design requirements from ACI 318.

Design lateral soil pressures correspond to those for some, but not all, of the soils that are tabulated in Table 1610.1. Section 1610.1 requires that the Table 1610.1 lateral pressures be used unless other values are substantiated by a geotechnical investigation. Note d coordinates with requirements in Section 1610.1. Generally, foundation walls that are restrained at the top must be designed for at-rest lateral soil loads. An exception permits the use of active pressures for designing walls that are no more than 8 feet (2438 mm) in height and are supported at the top by a flexible diaphragm. Note that in Table 1610.1, the minimum at-rest design lateral load is 60 psf per foot (2873 Pa/mm) of depth. Depending upon the soil's classification (see Table 1610.1), the prescriptive foundation wall table does not apply to soils with design lateral pressures exceeding 60 psf per foot (2873 Pa/mm) of depth.

Footnote c explains that "PC" as used in this table means "plain concrete," but it actually indicates that

TABLE 1807.1.6.2
CONCRETE FOUNDATION WALLS[b, c]

MAXIMUM WALL HEIGHT (feet)	MAXIMUM UNBALANCED BACKFILL HEIGHT[e] (feet)	MINIMUM VERTICAL REINFORCEMENT-BAR SIZE AND SPACING (inches)								
		Design lateral soil load[a] (psf per foot of depth)								
		30[d]			45[d]			60		
		Minimum wall thickness (inches)								
		7.5	9.5	11.5	7.5	9.5	11.5	7.5	9.5	11.5
5	4	PC	PC	PC	PC	PC	PC	PC	PC	PC
	5	PC	PC	PC	PC	PC	PC	PC	PC	PC
6	4	PC	PC	PC	PC	PC	PC	PC	PC	PC
	5	PC	PC	PC	PC	PC	PC	PC	PC	PC
	6	PC	PC	PC	PC	PC	PC	PC	PC	PC
7	4	PC	PC	PC	PC	PC	PC	PC	PC	PC
	5	PC	PC	PC	PC	PC	PC	PC	PC	PC
	6	PC	PC	PC	PC	PC	PC	#5 at 48	PC	PC
	7	PC	PC	PC	#5 at 46	PC	PC	#6 at 48	PC	PC
8	4	PC	PC	PC	PC	PC	PC	PC	PC	PC
	5	PC	PC	PC	PC	PC	PC	PC	PC	PC
	6	PC	PC	PC	PC	PC	PC	#5 at 43	PC	PC
	7	PC	PC	PC	#5 at 41	PC	PC	#6 at 43	PC	PC
	8	#5 at 47	PC	PC	#6 at 43	PC	PC	#6 at 32	#6 at 44	PC
9	4	PC	PC	PC	PC	PC	PC	PC	PC	PC
	5	PC	PC	PC	PC	PC	PC	PC	PC	PC
	6	PC	PC	PC	PC	PC	PC	#5 at 39	PC	PC
	7	PC	PC	PC	#5 at 37	PC	PC	#6 at 38	#5 at 37	PC
	8	#5 at 41	PC	PC	#6 at 38	#5 at 37	PC	#7 at 39	#6 at 39	#4 at 48
	9[d]	#6 at 46	PC	PC	#7 at 41	#6 at 41	PC	#7 at 31	#7 at 41	#6 at 39
10	4	PC	PC	PC	PC	PC	PC	PC	PC	PC
	5	PC	PC	PC	PC	PC	PC	PC	PC	PC
	6	PC	PC	PC	PC	PC	PC	#5 at 37	PC	PC
	7	PC	PC	PC	#6 at 48	PC	PC	#6 at 35	#6 at 48	PC
	8	#5 at 38	PC	PC	#7 at 47	#6 at 47	PC	#7 at 35	#7 at 47	#6 at 45
	9[d]	#6 at 41	#4 at 48	PC	#7 at 37	#7 at 48	#4 at 48	#6 at 22	#7 at 37	#7 at 47
	10[d]	#7 at 45	#6 at 45	PC	#7 at 31	#7 at 40	#6 at 38	#6 at 22	#7 at 30	#7 at 38

For SI: 1 inch = 25.4 mm, 1 foot = 304.8 mm, 1 pound per square foot per foot = 0.157 kPa/m.

a. For design lateral soil loads, see Section 1610.

b. Provisions for this table are based on design and construction requirements specified in Section 1807.1.6.2.

c. "PC" means plain concrete.

d. Where unbalanced backfill height exceeds 8 feet and design lateral soil loads from Table 1610.1 are used, the requirements for 30 and 45 psf per foot of depth are not applicable (see Section 1610).

e. For height of unbalanced backfill, see Section 1807.1.2.

no vertical reinforcing is required. The term is not to be misconstrued with the more general use of the term "plain concrete" in ACI 318, Chapter 19, as well as Section 1807.1.6.2.1.

1807.1.6.2.1 Seismic requirements. Based on the *seismic design category* assigned to the structure in accordance with Section 1613, concrete foundation walls designed using Table 1807.1.6.2 shall be subject to the following limitations:

1. *Seismic Design Categories* A and B. Not less than one No. 5 bar shall be provided around window, door and similar sized openings. The bar shall be anchored to develop f_y in tension at the corners of openings.
2. *Seismic Design Categories* C, D, E and F. Tables shall not be used except as allowed for plain concrete members in Section 1905.1.8.

❖ For structures classified as either Seismic Design Category A or B, the prescriptive concrete foundation wall Table 1807.1.6.2 applies along with the provisions stated in Item 1 of this section. For other seismic design categories, the use of the prescriptive concrete foundation wall Table 1807.1.6.2 is limited by Section 1905.1.8. That section allows the use of plain concrete foundation walls (i.e., walls that do not comply with the ACI 318 definition of "reinforced concrete") in one- and two-family dwellings.

1807.1.6.3 Masonry foundation walls. Masonry foundation walls shall comply with the following:

1. The thickness shall comply with the requirements of Table 1807.1.6.3(1) for plain masonry walls or Table 1807.1.6.3(2), 1807.1.6.3(3) or 1807.1.6.3(4) for masonry walls with reinforcement.
2. Vertical reinforcement shall have a minimum yield strength of 60,000 psi (414 MPa).
3. The specified location of the reinforcement shall equal or exceed the effective depth distance, *d*, noted in Tables 1807.1.6.3(2), 1807.1.6.3(3) and 1807.1.6.3(4) and shall be measured from the face of the exterior (soil) side of the wall to the center of the vertical reinforcement. The reinforcement shall be placed within the tolerances specified in TMS 602/ ACI 530.1/ASCE 6, Article 3.4.B.8 of the specified location.
4. Grout shall comply with Section 2103.13.
5. Concrete masonry units shall comply with ASTM C 90.
6. Clay masonry units shall comply with ASTM C 652 for hollow brick, except compliance with ASTM C 62 or ASTM C 216 shall be permitted where solid masonry units are installed in accordance with Table 1807.1.6.3(1) for plain masonry.
7. Masonry units shall be laid in running bond and installed with Type M or S mortar in accordance with Section 2103.9.
8. The unfactored axial load per linear foot of wall shall not exceed 1.2 $t f'_m$ where t is the specified wall thickness in inches and f'_m is the specified compressive strength of masonry in pounds per square inch.
9. At least 4 inches (102 mm) of solid masonry shall be provided at girder supports at the top of hollow masonry unit foundation walls.
10. Corbeling of masonry shall be in accordance with Section 2104.2. Where an 8-inch (203 mm) wall is corbeled, the top corbel shall not extend higher than the bottom of the floor framing and shall be a full course of headers at least 6 inches (152 mm) in length or the top course bed joint shall be tied to the vertical wall projection. The tie shall be W2.8 (4.8 mm) and spaced at a maximum horizontal distance of 36 inches (914 mm). The hollow space behind the corbelled masonry shall be filled with mortar or grout.

❖ This section contains the limitations on construction details and material properties that are needed to utilize the prescriptive designs of Tables 1807.1.6.3(1) through 1807.1.6.3(4) for masonry foundation walls.

Item 3 provides specific placement requirements regarding the location of reinforcement in connection with Tables 1807.1.6.3(2) through 1807.1.6.3(4). This dimension is necessary so that the wall is capable of developing the required bending capacity to resist the lateral soil loads.

Where corbeling is utilized to match the width of a masonry cavity wall above, Item 10 directs the reader to the appropriate requirements in Chapter 21. A header course is provided at the corbel to transfer the load deeper into the supporting wall. An alternative to the header utilizes a filled collar joint and ties to accomplish this load transfer. This option is illustrated in Figure 1807.1.6.3.

TABLE 1807.1.6.3(1). See page 18-26.

❖ This table provides wall thickness designs for various wall heights and unbalanced backfill depths. The thicknesses in the table are proportional to the lateral soil loads for each classification of soils. The lateral soil loads in the table are consistent with some of the soil loads in Table 1610.1.

The thicknesses in the table for plain masonry are based on research by the National Concrete Masonry Association (NCMA) in the report *Research Evaluation of the Flexural Strength of Concrete Masonry*, Project No. 93-172, December 7, 1993.

Footnote f provides a caution related to the applicability of 30 and 45 psf (1.43 and 2.156 kPa) design lateral soil loads. Section 1610.1 requires that foundation walls be designed for at-rest pressures, except foundation walls up to 8 feet (2438 mm) in height that are supported at the top by a flexible diaphragm. Where lateral soil loads from Table 1610.1 are used, the values of 30 and 45 psf (1.43 and 2.156 kPa) occur only for active pressure. Therefore, where both of these conditions apply, the prescriptive design table entries for 30 and 45 psf (1.43 and 2.156 kPa) cannot be used.

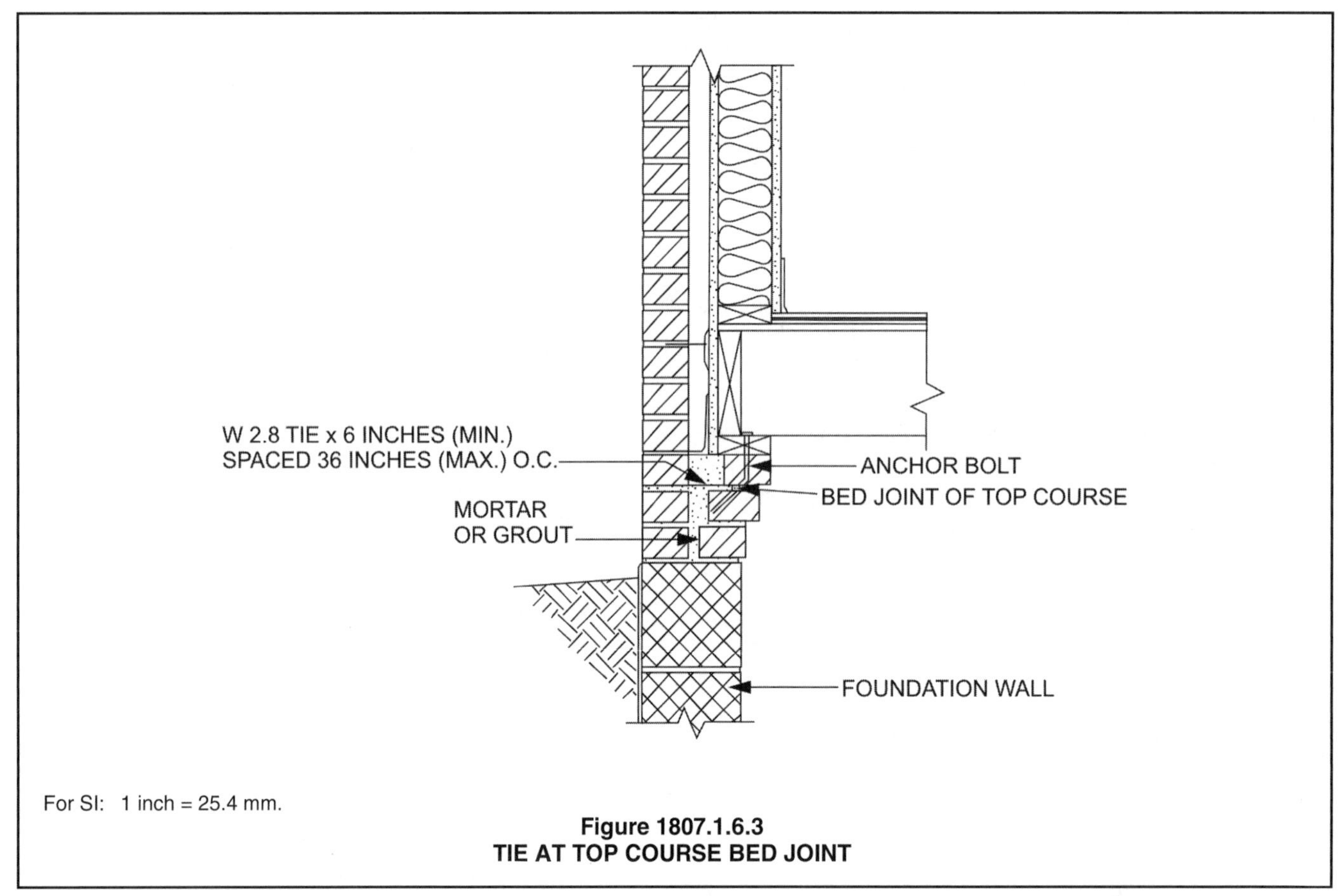

For SI: 1 inch = 25.4 mm.

Figure 1807.1.6.3
TIE AT TOP COURSE BED JOINT

TABLE 1807.1.6.3(1)
PLAIN MASONRY FOUNDATION WALLS[a, b, c]

MAXIMUM WALL HEIGHT (feet)	MAXIMUM UNBALANCED BACKFILL HEIGHT[e] (feet)	MINIMUM NOMINAL WALL THICKNESS (inches)		
		Design lateral soil load[a] (psf per foot of depth)		
		30[f]	45[f]	60
7	4 (or less)	8	8	8
	5	8	10	10
	6	10	12	10 (solid[c])
	7	12	10 (solid[c])	10 (solid[c])
8	4 (or less)	8	8	8
	5	8	10	12
	6	10	12	12 (solid[c])
	7	12	12 (solid[c])	Note d
	8	10 (solid[c])	12 (solid[c])	Note d
9	4 (or less)	8	8	8
	5	8	10	12
	6	12	12	12 (solid[c])
	7	12 (solid[c])	12 (solid[c])	Note d
	8	12 (solid[c])	Note d	Note d
	9[f]	Note d	Note d	Note d

For SI: 1 inch = 25.4 mm, 1 foot = 304.8 mm, 1 pound per square foot per foot = 0.157 kPa/m.

a. For design lateral soil loads, see Section 1610.

b. Provisions for this table are based on design and construction requirements specified in Section 1807.1.6.3.

c. Solid grouted hollow units or solid masonry units.

d. A design in compliance with Chapter 21 or reinforcement in accordance with Table 1807.1.6.3(2) is required.

e. For height of unbalanced backfill, see Section 1807.1.2.

f. Where unbalanced backfill height exceeds 8 feet and design lateral soil loads from Table 1610.1 are used, the requirements for 30 and 45 psf per foot of depth are not applicable (see Section 1610).

e. For height of unbalanced backfill, see Section 1807.1.2.

f. Where unbalanced backfill height exceeds 8 feet and design lateral soil loads from Table 1610.1 are used, the requirements for 30 and 45 psf per foot of depth are not applicable (see Section 1610).

TABLE 1807.1.6.3(2). See below.

❖ For general comments regarding this table, see the commentary to Table 1807.1.6.3(1). The reinforcement size and spacing in the table are based on the design requirements from TMS 402/ACI 530/ASCE 5.

Three different tables are provided for three distinct values of effective depth, d. For an explanation of the effective depth, see the commentary to Section 1807.1.6.3.

TABLE 1807.1.6.3(3). See page 18-28.

❖ See the commentary to Table 1807.1.6.3(2).

TABLE 1807.1.6.3(4). See page 18-29.

❖ See the commentary to Table 1807.1.6.3(2).

1807.1.6.3.1 Alternative foundation wall reinforcement. In lieu of the reinforcement provisions for masonry foundation walls in Table 1807.1.6.3(2), 1807.1.6.3(3) or 1807.1.6.3(4), alternative reinforcing bar sizes and spacings having an equivalent cross-sectional area of reinforcement per linear foot (mm) of wall shall be permitted to be used, provided the spacing of reinforcement does not exceed 72 inches (1829 mm) and reinforcing bar sizes do not exceed No. 11.

❖ The purpose of this section is to provide an alternative for the reinforcement requirements in Tables 1807.1.6.3(2), 1807.1.6.3(3) and 1807.1.6.3(4) that would result in the same capacity for lateral soil load resistance. For example, if No. 5 reinforcement bars were installed where the table indicates No. 4 at 48 inches (1219 mm) on center (o.c.), the spacing of the No. 5 bars for equivalent reinforcement area per foot of wall would be 72 inches (1829 mm) o.c. The increase in spacing would be allowed, since the cross-sectional area of a No. 5 bar is approximately 50 percent higher than a No. 4 reinforcement bar.

1807.1.6.3.2 Seismic requirements. Based on the *seismic design category* assigned to the structure in accordance with Section 1613, masonry foundation walls designed using Tables 1807.1.6.3(1) through 1807.1.6.3(4) shall be subject to the following limitations:

1. *Seismic Design Categories* A and B. No additional seismic requirements.

TABLE 1807.1.6.3(2)
8-INCH MASONRY FOUNDATION WALLS WITH REINFORCEMENT WHERE d ≥ 5 INCHES[a, b, c]

MAXIMUM WALL HEIGHT (feet-inches)	MAXIMUM UNBALANCED BACKFILL HEIGHT[d] (feet-inches)	MINIMUM VERTICAL REINFORCEMENT-BAR SIZE AND SPACING (inches) Design lateral soil load[a] (psf per foot of depth)		
		30[e]	45[e]	60
7-4	4-0 (or less)	#4 at 48	#4 at 48	#4 at 48
	5-0	#4 at 48	#4 at 48	#4 at 48
	6-0	#4 at 48	#5 at 48	#5 at 48
	7-4	#5 at 48	#6 at 48	#7 at 48
8-0	4-0 (or less)	#4 at 48	#4 at 48	#4 at 48
	5-0	#4 at 48	#4 at 48	#4 at 48
	6-0	#4 at 48	#5 at 48	#5 at 48
	7-0	#5 at 48	#6 at 48	#7 at 48
	8-0	#5 at 48	#6 at 48	#7 at 48
8-8	4-0 (or less)	#4 at 48	#4 at 48	#4 at 48
	5-0	#4 at 48	#4 at 48	#5 at 48
	6-0	#4 at 48	#5 at 48	#6 at 48
	7-0	#5 at 48	#6 at 48	#7 at 48
	8-8[e]	#6 at 48	#7 at 48	#8 at 48
9-4	4-0 (or less)	#4 at 48	#4 at 48	#4 at 48
	5-0	#4 at 48	#4 at 48	#5 at 48
	6-0	#4 at 48	#5 at 48	#6 at 48
	7-0	#5 at 48	#6 at 48	#7 at 48
	8-0	#6 at 48	#7 at 48	#8 at 48
	9-4[e]	#7 at 48	#8 at 48	#9 at 48
10-0	4-0 (or less)	#4 at 48	#4 at 48	#4 at 48
	5-0	#4 at 48	#4 at 48	#5 at 48
	6-0	#4 at 48	#5 at 48	#6 at 48
	7-0	#5 at 48	#6 at 48	#7 at 48
	8-0	#6 at 48	#7 at 48	#8 at 48
	9-0[e]	#7 at 48	#8 at 48	#9 at 48
	10-0[e]	#7 at 48	#9 at 48	#9 at 48

For SI: 1 inch = 25.4 mm, 1 foot = 304.8 mm, 1 pound per square foot per foot = 0.157 kPa/m.

a. For design lateral soil loads, see Section 1610.

b. Provisions for this table are based on design and construction requirements specified in Section 1807.1.6.3.

c. For alternative reinforcement, see Section 1807.1.6.3.1

d. For height of unbalanced backfill, see Section 1807.1.2

e. Where unbalanced backfill height exceeds 8 feet and design lateral soil loads from Table 1610.1 are used, the requirements for 30 and 45 psf per foot of depth are not applicable. See Section 1610.

2. *Seismic Design Category* C. A design using Tables 1807.1.6.3(1) through 1807.1.6.3(4) is subject to the seismic requirements of Section 1.18.4.3 of TMS 402/ ACI 530/ASCE 5.

3. *Seismic Design Category* D. A design using Tables 1807.1.6.3(2) through 1807.1.6.3(4) is subject to the seismic requirements of Section 1.18.4.4 of TMS 402/ ACI 530/ASCE 5.

4. *Seismic Design Categories* E and F. A design using Tables 1807.1.6.3(2) through 1807.1.6.3(4) is subject to the seismic requirements of Section 1.18.4.5 of TMS 402/ACI 530/ASCE 5.

❖ For structures classified as either Seismic Design Category A or B, the prescriptive masonry foundation wall tables apply without further requirements. For all other seismic design categories, the use of these prescriptive tables is limited by the referenced code sections. For structures classified as Seismic Design Categories D or higher, only the tables that require reinforcement can be used.

1807.2 Retaining walls. Retaining walls shall be designed in accordance with Sections 1807.2.1 through 1807.2.3.

❖ This section provides design considerations for a retaining wall.

1807.2.1 General. Retaining walls shall be designed to ensure stability against overturning, sliding, excessive foundation pressure and water uplift. Where a keyway is extended below the wall base with the intent to engage passive pressure and enhance sliding stability, lateral soil pressures on both sides of the keyway shall be considered in the sliding analysis.

❖ This provision for retaining wall analyses is intended to provide guidance on the considerations that are appropriate for retaining wall design. Designs for retaining walls with keyways should account for the driving pressures acting on the keyway. Since there are indications that these are at times overlooked, this provision clarifies that such forces must be considered.

TABLE 1807.1.6.3(3)
10-INCH MASONRY FOUNDATION WALLS WITH REINFORCEMENT WHERE d ≥ 6.75 INCHES [a, b, c]

MAXIMUM WALL HEIGHT (feet-inches)	MAXIMUM UNBALANCED BACKFILL HEIGHT[d] (feet-inches)	MINIMUM VERTICAL REINFORCEMENT-BAR SIZE AND SPACING (inches)		
		Design lateral soil load[a] (psf per foot of depth)		
		30[e]	45[e]	60
7-4	4-0 (or less)	#4 at 56	#4 at 56	#4 at 56
	5-0	#4 at 56	#4 at 56	#4 at 56
	6-0	#4 at 56	#4 at 56	#5 at 56
	7-4	#4 at 56	#5 at 56	#6 at 56
8-0	4-0 (or less)	#4 at 56	#4 at 56	#4 at 56
	5-0	#4 at 56	#4 at 56	#4 at 56
	6-0	#4 at 56	#4 at 56	#5 at 56
	7-0	#4 at 56	#5 at 56	#6 at 56
	8-0	#5 at 56	#6 at 56	#7 at 56
8-8	4-0 (or less)	#4 at 56	#4 at 56	#4 at 56
	5-0	#4 at 56	#4 at 56	#4 at 56
	6-0	#4 at 56	#4 at 56	#5 at 56
	7-0	#4 at 56	#5 at 56	#6 at 56
	8-8[e]	#5 at 56	#7 at 56	#8 at 56
9-4	4-0 (or less)	#4 at 56	#4 at 56	#4 at 56
	5-0	#4 at 56	#4 at 56	#4 at 56
	6-0	#4 at 56	#5 at 56	#5 at 56
	7-0	#4 at 56	#5 at 56	#6 at 56
	8-0	#5 at 56	#6 at 56	#7 at 56
	9-4[e]	#6 at 56	#7 at 56	#7 at 56
10-0	4-0 (or less)	#4 at 56	#4 at 56	#4 at 56
	5-0	#4 at 56	#4 at 56	#4 at 56
	6-0	#4 at 56	#5 at 56	#5 at 56
	7-0	#5 at 56	#6 at 56	#7 at 56
	8-0	#5 at 56	#7 at 56	#8 at 56
	9-0[e]	#6 at 56	#7 at 56	#9 at 56
	10-0[e]	#7 at 56	#8 at 56	#9 at 56

For SI: 1 inch = 25.4 mm, 1 foot = 304.8, 1 pound per square foot per foot = 1.157 kPa/m.

a. For design lateral soil loads, see Section 1610.

b. Provisions for this table are based on design and construction requirements specified in Section 1807.1.6.3

c. For alternative reinforcement, see Section 1807.1.6.3.1.

d. For height of unbalanced backfill, See Section 1807.1.2.

e. Where unbalanced backfill height exceeds 8 feet and design lateral soil loads from Table 1610.1 are used, the requirements for 30 and 45 psf per foot of depth are not applicable. See Section 1610.

1807.2.2 Design lateral soil loads. Retaining walls shall be designed for the lateral soil loads set forth in Section 1610.

❖ Section 1610.1 requires that the Table 1610.1 lateral pressures be used unless other values are substantiated by a geotechnical investigation. The lateral pressure of the soil against the retaining wall is greatly influenced by soil moisture. Backfill is usually kept from being saturated for an extended length of time by placing drains near the base of the retaining wall to remove the water in the soil behind it.

1807.2.3 Safety factor. Retaining walls shall be designed to resist the lateral action of soil to produce sliding and overturning with a minimum safety factor of 1.5 in each case. The load combinations of Section 1605 shall not apply to this requirement. Instead, design shall be based on 0.7 times nominal earthquake loads, 1.0 times other *nominal loads*, and investigation with one or more of the variable loads set to zero. The safety factor against lateral sliding shall be taken as the available soil resistance at the base of the retaining wall foundation divided by the net lateral force applied to the retaining wall.

Exception: Where earthquake loads are included, the minimum safety factor for retaining wall sliding and overturning shall be 1.1.

❖ The safety factor for stability of retaining walls predates the development of load combinations. This section clarifies that load combinations do not apply when evaluating the stability of retaining walls under sliding and overturning.

1807.3 Embedded posts and poles. Designs to resist both axial and lateral loads employing posts or poles as columns embedded in earth or in concrete footings in earth shall be in accordance with Sections 1807.3.1 through 1807.3.3.

❖ The criteria in this section apply to the lateral resistance of posts or poles embedded either directly in earth or in a concrete footing. They were originally developed to reduce the complexity of analyzing the

TABLE 1807.1.6.3(4)
12-INCH MASONRY FOUNDATION WALLS WITH REINFORCEMENT WHERE d ≥ 8.75 INCHES[a, b, c]

MAXIMUM WALL HEIGHT (feet-inches)	MAXIMUM UNBALANCED BACKFILL HEIGHT[d] (feet-inches)	MINIMUM VERTICAL REINFORCEMENT-BAR SIZE AND SPACING (inches) Design lateral soil load[a] (psf per foot of depth)		
		30[e]	45[e]	60
7-4	4 (or less)	#4 at 72	#4 at 72	#4 at 72
	5-0	#4 at 72	#4 at 72	#4 at 72
	6-0	#4 at 72	#4 at 72	#5 at 72
	7-4	#4 at 72	#5 at 72	#6 at 72
8-0	4 (or less)	#4 at 72	#4 at 72	#4 at 72
	5-0	#4 at 72	#4 at 72	#4 at 72
	6-0	#4 at 72	#4 at 72	#5 at 72
	7-0	#4 at 72	#5 at 72	#6 at 72
	8-0	#5 at 72	#6 at 72	#8 at 72
8-8	4 (or less)	#4 at 72	#4 at 72	#4 at 72
	5-0	#4 at 72	#4 at 72	#4 at 72
	6-0	#4 at 72	#4 at 72	#5 at 72
	7-0	#4 at 72	#5 at 72	#6 at 72
	8-8[e]	#5 at 72	#7 at 72	#8 at 72
9-4	4 (or less)	#4 at 72	#4 at 72	#4 at 72
	5-0	#4 at 72	#4 at 72	#4 at 72
	6-0	#4 at 72	#5 at 72	#5 at 72
	7-0	#4 at 72	#5 at 72	#6 at 72
	8-0	#5 at 72	#6 at 72	#7 at 72
	9-4[e]	#6 at 72	#7 at 72	#8 at 72
10-0	4 (or less)	#4 at 72	#4 at 72	#4 at 72
	5-0	#4 at 72	#4 at 72	#4 at 72
	6-0	#4 at 72	#5 at 72	#5 at 72
	7-0	#4 at 72	#6 at 72	#6 at 72
	8-0	#5 at 72	#6 at 72	#7 at 72
	9-0[e]	#6 at 72	#7 at 72	#8 at 72
	10-0[e]	#7 at 72	#8 at 72	#9 at 72

For SI: 1 inch = 25.4 mm, 1 foot = 304.8 mm, 1 pound per square foot per foot = 0.157 kPa/m.

a. For design lateral soil loads, see Section 1610.

b. Provisions for this table are based on design and construction requirements specified in Section 1807.1.6.3.

c. For alternative reinforcement, see Section 1807.1.6.3.1.

d. For height of unbalanced backfill, see Section 1807.1.2.

e Where unbalanced backfill height exceeds 8 feet and design lateral soil loads from Table 1610.1 are used, the requirements for 30 and 45 psf per foot of depth are not applicable. See Section 1610.

soil-structure interaction, thereby facilitating the design of cantilevered poles used to support outdoor advertising. While more accurate methods are available for determining the resistance of posts or poles to axial and lateral loads, the other methods are much more complex, and the results do not differ significantly from this method.

1807.3.1 Limitations. The design procedures outlined in this section are subject to the following limitations:

1. The frictional resistance for structural walls and slabs on silts and clays shall be limited to one-half of the normal force imposed on the soil by the weight of the footing or slab.
2. Posts embedded in earth shall not be used to provide lateral support for structural or nonstructural materials such as plaster, masonry or concrete unless bracing is provided that develops the limited deflection required.

Wood poles shall be treated in accordance with AWPA U1 for sawn timber posts (Commodity Specification A, Use Category 4B) and for round timber posts (Commodity Specification B, Use Category 4B).

❖ The limitations imposed by this section are intended to address both structural stability and serviceability. The limitation of the frictional resistance for structural walls and slabs on silts and clays is consistent with Section 1806.3 and Table 1806.2, which also limit the sliding resistance to one-half the dead load. The limitations on the types of construction materials that utilize lateral support of poles are based on the brittle nature of the materials. In order to prevent excessive distortions, which would produce cracking of these brittle materials, this section limits the use of poles unless some type of rigid cross bracing is provided to limit the deflections to those that can be tolerated by the materials.

1807.3.2 Design criteria. The depth to resist lateral loads shall be determined using the design criteria established in Sections 1807.3.2.1 through 1807.3.2.3, or by other methods *approved* by the *building official*.

❖ The design criteria for the use of poles embedded in the ground or in concrete footings in the ground were originally developed in the 1940s. The design criteria address conditions where constraint is provided at the ground surface, such as a rigid floor, and where no constraint is provided. The original design criteria established a $^1/_2$-inch (12.7 mm) lateral pole deformation at the surface of the ground. [Note that where $^1/_2$ inch (12.7 mm) movement can be tolerated, Section 1806.3.4 allows the lateral bearing pressure to be increased.] These criteria were also based on field tests conducted in a range of sandy and gravelly soils and of silts and clays.

1807.3.2.1 Nonconstrained. The following formula shall be used in determining the depth of embedment required to resist lateral loads where no lateral constraint is provided at the ground surface, such as by a rigid floor or rigid ground surface pavement, and where no lateral constraint is provided above the ground surface, such as by a structural diaphragm.

$$d = 0.5A\{1 + [1 + (4.36h/A)]^{1/2}\} \qquad \textbf{(Equation 18-1)}$$

where:

$A = 2.34P/(S_1 b)$

b = Diameter of round post or footing or diagonal dimension of square post or footing, feet (m).

d = Depth of embedment in earth in feet (m) but not over 12 feet (3.658 m) for purpose of computing lateral pressure.

h = Distance in feet (m) from ground surface to point of application of "P."

P = Applied lateral force in pounds (kN).

S_1 = Allowable lateral soil-bearing pressure as set forth in Section 1806.2 based on a depth of one-third the depth of embedment in pounds per square foot (psf) (kPa).

❖ The required embedment depth for posts or poles that are not restrained at or above the ground surface is determined by Equation 18-1. Nonconstrained conditions include flexible pavements, such as asphalt. The code further restricts the use of this equation to structures that have no lateral support, such as from a floor diaphragm, provided above the ground level. From an analysis standpoint, the embedded pole is free to rotate and translate at the point of load application. The pole rotates about a point below the soil surface, which is where the direction of the passive pressure on the embedded pole changes direction.

Determining the minimum embedment depth of a vertical member resisting lateral loads requires two steps:

1. Assuming an embedment depth, calculate the soil pressure that the embedded member imposes on the soil, and
2. Compare the calculated pressure to an allowable lateral soil bearing pressure.

This two-step process is expressed as a single trial and-error equation, which requires an iterative process to arrive at a satisfactory depth of embedment.

Note that the allowable lateral soil bearing pressure is in accordance with Section 1806.2. This includes increases permitted in Section 1806.3.4. In particular, the allowable lateral bearing values can be doubled for poles supporting structures that can tolerate 0.5 inches (12.7 mm) of movement at the ground surface—generally considered applicable to isolated poles.

1807.3.2.2 Constrained. The following formula shall be used to determine the depth of embedment required to resist lateral loads where lateral constraint is provided at the ground surface, such as by a rigid floor or pavement.

$$d = \sqrt{\frac{4.25Ph}{S_3 b}} \qquad \textbf{(Equation 18-2)}$$

or alternatively

$$d = \sqrt{\frac{4.25M_g}{S_3 b}} \qquad \textbf{(Equation 18-2)}$$

where:

M_g = Moment in the post at grade, in foot-pounds (kN-m).

S_3 = Allowable lateral soil-bearing pressure as set forth in Section 1806.2 based on a depth equal to the depth of embedment in pounds per square foot (kPa).

❖ The required embedment depth for posts or poles that are restrained at the ground surface is determined by Equation 18-2 or 18-3. A constrained post has stiff resistance at grade, compared to the stiffness of the soil. This point of support acts as a fulcrum for the lateral load. The pole is free to rotate about this point rather than being fixed.

1807.3.2.3 Vertical load. The resistance to vertical loads shall be determined using the vertical foundation pressure set forth in Table 1806.2.

❖ While no specific limit is provided with respect to vertical loading, this methodology was originally developed for cantilevered posts where the primary concern is lateral resistance (see commentary, Section 1807.3). The above determination of embedment length is independent of the vertical load (see the commentary, Section 1806.2 and Table 1806.2).

1807.3.3 Backfill. The backfill in the annular space around columns not embedded in poured footings shall be by one of the following methods:

1. Backfill shall be of concrete with a specified compressive strength of not less than 2,000 psi (13.8 MPa). The hole shall not be less than 4 inches (102 mm) larger than the diameter of the column at its bottom or 4 inches (102 mm) larger than the diagonal dimension of a square or rectangular column.
2. Backfill shall be of clean sand.The sand shall be thoroughly compacted by tamping in layers not more than 8 inches (203 mm) in depth.
3. Backfill shall be of controlled low-strength material (CLSM).

❖ In order for the post or pole to meet the conditions and limitations of research for which the above criteria was established, backfill in the annular space around a column not embedded in a concrete footing must be either 2,000 psi (13.8 MPa) concrete, CLSM (see commentary, Section 1804.6) or clean sand thoroughly compacted by tamping in layers not more than 8 inches (203 mm) in depth.

SECTION 1808 FOUNDATIONS

1808.1 General. Foundations shall be designed and constructed in accordance with Sections 1808.2 through 1808.9. Shallow foundations shall also satisfy the requirements of Section 1809. Deep foundations shall also satisfy the requirements of Section 1810.

❖ Section 1808 contains the provisions that apply to all foundation types. Then specific requirements for shallow and deep foundations are given in Sections 1809 and 1810, respectively.

1808.2 Design for capacity and settlement. Foundations shall be so designed that the allowable bearing capacity of the soil is not exceeded, and that differential settlement is minimized. Foundations in areas with expansive soils shall be designed in accordance with the provisions of Section 1808.6.

❖ Regardless of the type of shallow foundation used, the allowable bearing capacity of soil must not be exceeded. There are two premises by which allowable soil-bearing pressures are established. The first premise requires that the safety factor against ultimate shear failure of the soil be adequate. The second premise requires that settlements under allowable bearing pressures not exceed tolerable values. In most cases, settlement governs the value established for allowable soil-bearing capacity. Bearing capacity is usually determined from a soils investigation and engineering analysis.

When the soils profile of a construction site is established by a sufficient number of test borings, and it indicates that a nonuniform soil condition exists where the strata of suitable bearing materials occurs at varying thicknesses or different depths, the foundation design must be adjusted to the subsurface condition to provide for the proper and safe performance of the foundation system.

Under such circumstances, it becomes necessary in the design of shallow foundations to determine the different depths at which isolated or continuous stepped footings need to be placed in order to obtain equal bearing pressures and avoid serious structural damage caused by differential (unequal) settlement of the different parts of the foundation system.

Another design method for obtaining equal bearing pressures and keeping the footings at a common elevation is to size the footings in accordance with the allowable bearing capacity of the soil at each location, thus producing a balanced design of the foundation system and preventing differential settlement.

Section 1808.6 stipulates required methods and design of foundations on expansive soils.

1808.3 Design loads. Foundations shall be designed for the most unfavorable effects due to the combinations of loads specified in Section 1605.2 or 1605.3. The dead load is permitted to include the weight of foundations and overlying fill. Reduced live loads, as specified in Sections 1607.10 and 1607.12, shall be permitted to be used in the design of foundations.

❖ Foundations are permitted to be designed using either ASD or strength design. The appropriate load combinations are investigated to determine the most severe structural effects. Live load reductions permitted in Section 1607 apply equally to the foundation design loads. This section also clarifies what portions of the foundation construction are considered dead loads. See the commentary to Section 1606 for a discussion of dead load estimates.

1808.3.1 Seismic overturning. Where foundations are proportioned using the load combinations of Section 1605.2 or 1605.3.1, and the computation of seismic overturning effects is by equivalent lateral force analysis or modal analysis, the proportioning shall be in accordance with Section 12.13.4 of ASCE 7.

❖ This provision correlates with ASCE 7 earthquake load requirements that are based on the NEHRP *Recommended Provisions for Seismic Regulations for New Buildings* (FEMA 450). When using LRFD load combinations to size foundations, the seismic overturning computed by the equivalent lateral force method or the modal analysis method is permitted to be reduced.

ASCE 7 permits the reduction of seismic overturning for foundation design where either strength design or ASD load combinations are used. This provision refers to Sections 1605.2 and 1605.3.1, which correspond to the ASCE 7 load combinations. Because the load combinations in Section 1605.3.2 include 0.9D, where overturning is assessed (rather than 0.6D as in Section 1605.3.1), reduction of seismic overturning would be unconservative where those load combinations are used.

1808.4 Vibratory loads. Where machinery operations or other vibrations are transmitted through the foundation, consideration shall be given in the foundation design to prevent detrimental disturbances of the soil.

❖ Vibrations emanating from machinery operations and transmitted to the soil through the foundation may cause serious settlement to occur, particularly to foundations bearing on granular materials. While granular soils generally have a considerable volume of voids, the foundation pressure is usually carried and distributed by the bearing of grain on grain without detrimental deformations. Granular materials subjected to strong vibrations, however, may result in the particles readjusting and slipping into the void spaces. Essentially, the soil mass is consolidated and reduced in volume, causing vertical settlement.

Vibratory loads that cause a disturbance of the soil will flush water out of the material. In saturated granular soils, such as sand, the flushing action may cause a "quick" condition that allows sudden flow beneath the foundation, sometimes resulting in serious structural damage.

In soils that are rather impermeable, such as clays, vibration may also cause water to flush out, but it will take a very long time. Also, foundation settlement will occur over a prolonged period.

Care should be taken in the design of foundations to eliminate completely or minimize the transmission of vibratory loads to load-bearing soils.

1808.5 Shifting or moving soils. Where it is known that the shallow subsoils are of a shifting or moving character, foundations shall be carried to a sufficient depth to ensure stability.

❖ Shallow foundations placed on or within a mass of shifting or moving soils must be designed with great rigidity and strength in order to adequately resist soil movement and avoid damage. This section requires foundations to be carried to a sufficient depth for adequate stability. Adequate stability implies, among other things, the consideration of uplift and perpendicular forces exerted on the footings because of soil movement.

1808.6 Design for expansive soils. Foundations for buildings and structures founded on expansive soils shall be designed in accordance with Section 1808.6.1 or 1808.6.2.

Exception: Foundation design need not comply with Section 1808.6.1 or 1808.6.2 where one of the following conditions is satisfied:

1. The soil is removed in accordance with Section 1808.6.3; or
2. The *building official* approves stabilization of the soil in accordance with Section 1808.6.4.

❖ Expansive soils, often referred to as "swelling soils," contain montmorillonite minerals and have the characteristics of absorbing water and swelling, or shrinking and cracking when drying. Significant volume changes can cause serious damage to buildings and other structures as well as to pavements and sidewalks. Swelling soils are found throughout the nation, but are more prevalent in regions with dry or moderately arid climates. It is in areas where these soils experience large fluctuations in water content that the swelling is more likely to affect structures. Heavy rains will cause the soil to expand and push upward on a foundation. Movement is most noticeable at the perimeter of the foundation where the increase in water content is greatest.

There is a general relationship between the PI of a soil as determined by the ASTM D 4318 standard test method and the potential for expansion. This relationship is shown in Figure 1803.5.3.

Section 1803.5.3 defines "Expansive soil" as any plastic material with a plasticity index of 15 or greater with more than 10 percent of the soil particles passing a No. 200 sieve and less than 5 micrometers in size. As an alternative, tests in accordance with ASTM D 4829 can be used to determine if a soil is expansive. The expansion index is a measure of the swelling potential of the soil.

The amount and depth of potential swelling that can occur in a clay material are, to some extent, functions of the cyclical moisture content in the soil. In dryer climates where the moisture content in the soil near the ground surface is low because of evaporation, there is a greater potential for extensive swelling than the same soil in wetter climates where the variations of moisture content are not as severe.

Volume changes in highly expansive soils range between 7 and 10 percent, but experience has shown that occasionally, under abnormal conditions, they can reach as high as 25 percent.

The exception refers to the provisions that govern removal or stabilization of the expansive soil (see the commentary to Section 1803.5.3 for a discussion regarding expansive soils).

1808.6.1 Foundations. Foundations placed on or within the active zone of expansive soils shall be designed to resist differential volume changes and to prevent structural damage to the supported structure. Deflection and racking of the supported structure shall be limited to that which will not interfere with the usability and serviceability of the structure.

Foundations placed below where volume change occurs or below expansive soil shall comply with the following provisions:

1. Foundations extending into or penetrating expansive soils shall be designed to prevent uplift of the supported structure.
2. Foundations penetrating expansive soils shall be designed to resist forces exerted on the foundation due to soil volume changes or shall be isolated from the expansive soil.

❖ Shallow foundation systems placed on or within a mass of expansive soil must be designed with great rigidity and strength in order to adequately resist swelling pressures and avoid serious structural damage. Sometimes footings and piers, as well as foundation walls and grade beams, are isolated from the swelling soils by intervening fills or granular materials. While this type of insulation serves to cushion or diminish lateral pressures, it will not prevent structure heaving, except for grade beams constructed on collapsible forms.

One method of foundation construction in expansive soils is the use of drilled piers with belled footings extending below the zone of swelling activity, or at least to a soil stratum where the seasonal moisture content of the expansive soil will remain within a tolerable range. This type of construction is made possible in swelling soils because the material usually consists of stiff clays that do not contain free water, providing excellent conditions for drilling holes in the ground.

Piers with belled footings have been widely used for foundations in expansive soils, even in the construction of single-family dwellings.

The concrete shaft must be reinforced for its entire length because the swelling soil is apt to exert high uplift forces and subject the drilled pier construction to tensile stresses. This condition can be prevented or sufficiently mitigated by isolating the concrete from the swelling soil by surrounding the shaft with a vertical layer of granular soil or other suitable materials that possess little or no shearing strength (e.g., vermiculite).

1808.6.2 Slab-on-ground foundations. Moments, shears and deflections for use in designing slab-on-ground, mat or raft foundations on expansive soils shall be determined in accordance with *WRI/CRSI Design of Slab-on-Ground Foundations* or *PTI Standard Requirements for Analysis of Shallow Concrete Foundations on Expansive Soils.* Using the moments, shears and deflections determined above, nonprestressed slabs-on-ground, mat or raft foundations on expansive soils shall be designed in accordance with *WRI/CRSI Design of Slab-on-Ground Foundations* and post-tensioned slab-on-ground, mat or raft foundations on expansive soils shall be designed in accordance with *PTI Standard Requirements for Design of Shallow Post-Tensioned Concrete Foundations on Expansive Soils.* It shall be permitted to analyze and design such slabs by other methods that account for soil-structure interaction, the deformed shape of the soil support, the plate or stiffened plate action of the slab as well as both center lift and edge lift conditions. Such alternative methods shall be rational and the basis for all aspects and parameters of the method shall be available for peer review.

❖ It is not uncommon for concrete slabs on ground to be reinforced (prestressed) with post-tensioned strands so that the flexural stresses induced in the slab from uplift of the soil are reduced, thus avoiding cracking of the concrete at its top surface. All things being equal, a post-tensioned slab on ground can be more economical than a conventionally reinforced slab because the precompression reduces the flexural stresses, making it unnecessary to increase the thickness of the slab or use a higher strength concrete. It has also been a practice to cast the slab on cellular forms made of cardboard or other collapsible materials that will support wet concrete, but will yield when subjected to swelling pressures. Since there is no rebound of the form material after subsidence of the soil, floors must be designed as structural slabs.

This section recognizes the analytical procedures in Wire Reinforcing Institute/Concrete Reinforcing Steel Institute (WRI/CRSI) *Design of Slab-on-Ground Foundations* and Post-tensioning Institute (PTI) *Standard Requirements for Analysis of Shallow Concrete Foundations on Expansive Soils* for slab-on-ground foundations located on expansive soils. The WRI/CRSI document also provides design requirements for reinforced concrete slabs. The PTI *Standard Requirements for Design of Shallow Post-Tensioned Concrete Foundations on Expansive Soils* addresses design requirements for post-tensioned concrete slabs on expansive soils.

1808.6.3 Removal of expansive soil. Where expansive soil is removed in lieu of designing foundations in accordance with Section 1808.6.1 or 1808.6.2, the soil shall be removed to a depth sufficient to ensure a constant moisture content in the remaining soil. Fill material shall not contain expansive soils and shall comply with Section 1804.5 or 1804.6.

Exception: Expansive soil need not be removed to the depth of constant moisture, provided the confining pressure in the expansive soil created by the fill and supported structure exceeds the swell pressure.

❖ The best way to avoid the problems associated with expansive soils is to remove such soil from the construction site and, where necessary, replace it with suitable compacted fill material. Expansive soil shrinks when the water content decreases and swells when the water content increases. To minimize the chances of swelling and shrinking, the code requires removal to a depth that provides a constant moisture content in the soil that is left in place. The exception to removing to this depth alludes to an alternative method of stabilization that is discussed in the commentary to Section 1808.6.4.

1808.6.4 Stabilization. Where the active zone of expansive soils is stabilized in lieu of designing foundations in accordance with Section 1808.6.1 or 1808.6.2, the soil shall be stabilized by chemical, dewatering, presaturation or equivalent techniques.

❖ This section allows expansive soils to be stabilized by either chemical, presaturation or dewatering methods or by other equivalent techniques. These methods, however, have limited use.

Chemical stabilization may be effectively accomplished by the use of lime thoroughly mixed with the soil and compacted at approximately the optimum moisture content. The purpose of using lime is to reduce the plasticity of the soil, which will reduce the swelling potential (see commentary, Section 1808.6). Since this method requires a uniform mix of lime and soil, however, it is generally limited to compacted fills. There is a method of pressure injecting lime slurry into heavily fissured clay materials, but the method is not appropriate for all site conditions.

Presaturation by flooding the site is rarely a totally effective way to stabilize expansive soils, since it takes a very long time for the water to penetrate to any great depth. Dewatering of expansive soils that consist generally of dense clays without free water is not an effective way to control moisture content. One good method of stabilization is to place sufficient fill material on the site so that the downward pressures of the fill will, to the extent possible, balance the upward swelling pressures of the soil. The effectiveness of this balancing concept depends on the pressures developed by the expansive soil as a function of estimated volume change and the depth of compacted fill material necessary to counteract these pressures. The application of this stabilization method is usually economically feasible when the soil pressures are low, around 500 psf (23.9 kPa); however, pressures occasionally have reached as high as 20 tons per square foot (1915 kPa).

1808.7 Foundations on or adjacent to slopes. The placement of buildings and structures on or adjacent to slopes steeper than one unit vertical in three units horizontal (33.3-percent slope) shall comply with Sections 1808.7.1 through 1808.7.5.

❖ The provisions of this section apply to buildings placed on or adjacent to slopes steeper than 1 unit vertical in 3 units horizontal (33.3-percent slope).

1808.7.1 Building clearance from ascending slopes. In general, buildings below slopes shall be set a sufficient distance from the slope to provide protection from slope drainage, erosion and shallow failures. Except as provided in Section 1808.7.5 and Figure 1808.7.1, the following criteria will be assumed to provide this protection. Where the existing slope is steeper than one unit vertical in one unit horizontal (100-percent slope), the toe of the slope shall be assumed to be at the intersection of a horizontal plane drawn from the top of the foundation and a plane drawn tangent to the slope at an angle of 45 degrees (0.79 rad) to the horizontal. Where a retaining wall is constructed at the toe of the slope, the height of the slope shall be measured from the top of the wall to the top of the slope.

❖ Code Figure 1808.7.1 depicts criteria for locating buildings adjacent to slopes. The setback is intended to provide protection to the structure not only from shallow slope failures (sometimes referred to as "sloughing") but also from erosion and slope drainage. At ascending slopes, the setback also provides access around the building and helps create a light and open-air environment. Where an existing slope is steeper than 1:1, the toe of the slope is assumed to be at a distance determined by the intersection of a horizontal line at the top of the foundation and a line drawn at a 45-degree (0.79 rad) angle to the horizontal line, and terminating at the top of the slope. This determination of slope height and measurement of the setback is illustrated in Figure 1808.7.1(1).

FIGURE 1808.7.1. See page 18-35.

❖ This figure illustrates required setbacks at slopes. The setback is a function of the height of the slope (see commentary, Sections 1808.7.1 and 1808.7.2).

1808.7.2 Foundation setback from descending slope surface. Foundations on or adjacent to slope surfaces shall be founded in firm material with an embedment and set back from the slope surface sufficient to provide vertical and lateral support for the foundation without detrimental settlement. Except as provided for in Section 1808.7.5 and Figure 1808.7.1, the following setback is deemed adequate to meet the criteria. Where the slope is steeper than 1 unit vertical in 1 unit horizontal (100-percent slope), the required setback shall be measured from an imaginary plane 45 degrees (0.79 rad) to the horizontal, projected upward from the toe of the slope.

❖ The provisions of this section restrict the placement of footings at the top of slopes so that vertical and lateral support for the footing is provided. This setback is shown in code Figure 1808.7.1. When the slope is greater than one unit vertical in one unit horizontal (100-percent slope), the setback is measured from a top of the slope that is established as the intersection of the ground surface and an imaginary plane at 45 degrees (0.79 rad) to the horizontal, projected from the toe of the slope. This requirement is illustrated in Figure 1808.7.2. For most conditions, the required setbacks will provide adequate lateral support for the foundations.

1808.7.3 Pools. The setback between pools regulated by this code and slopes shall be equal to one-half the building footing setback distance required by this section. That portion of the pool wall within a horizontal distance of 7 feet (2134 mm) from the top of the slope shall be capable of supporting the water in the pool without soil support.

❖ This section specifies the required setback distances for pools located near ascending or descending slopes. The minimum setback distance is established as one-half the required building setback from Section 1808.7.2. The pool wall that is within 7 feet (2134 mm) of the top slope is required to be self-supporting without support from the soil, and is intended to provide additional safety measures should localized minor sliding and sloughing occur.

1808.7.4 Foundation elevation. On graded sites, the top of any exterior foundation shall extend above the elevation of the street gutter at point of discharge or the inlet of an *approved* drainage device a minimum of 12 inches (305 mm) plus 2 percent. Alternate elevations are permitted subject to the approval of the *building official*, provided it can be demonstrated that required drainage to the point of discharge and away from the structure is provided at all locations on the site.

❖ Figure 1808.7.4 depicts the requirements of this section regarding elevation for exterior foundations with respect to the street, gutter or point of inlet of a drainage device. The elevation of the street or gutter shown is that point at which drainage from the site reaches the street or gutter.

This requirement is intended to protect the building from water encroachment in case of heavy or unprecedented rain and may be modified on the approval of the building official if he or she finds that positive drainage slopes are provided to drain water away from the building and that the drainage pattern is not subject to temporary flooding due to landscaping or

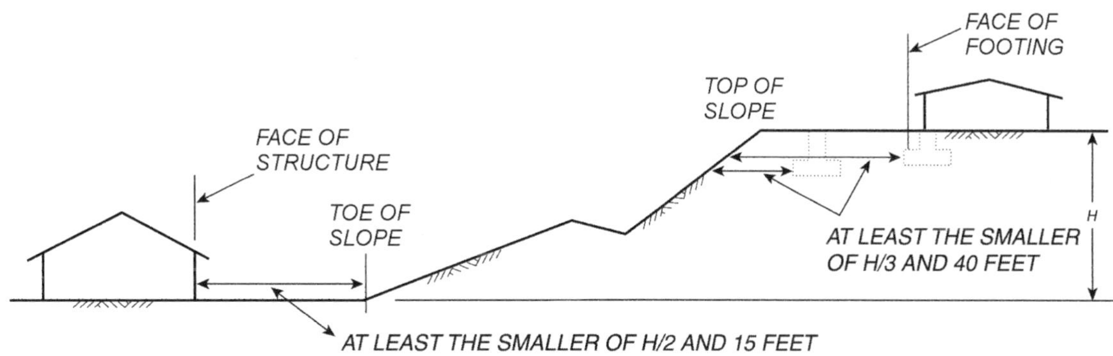

For SI: 1 foot = 304.8 mm.

FIGURE 1808.7.1
FOUNDATION CLEARANCES FROM SLOPES

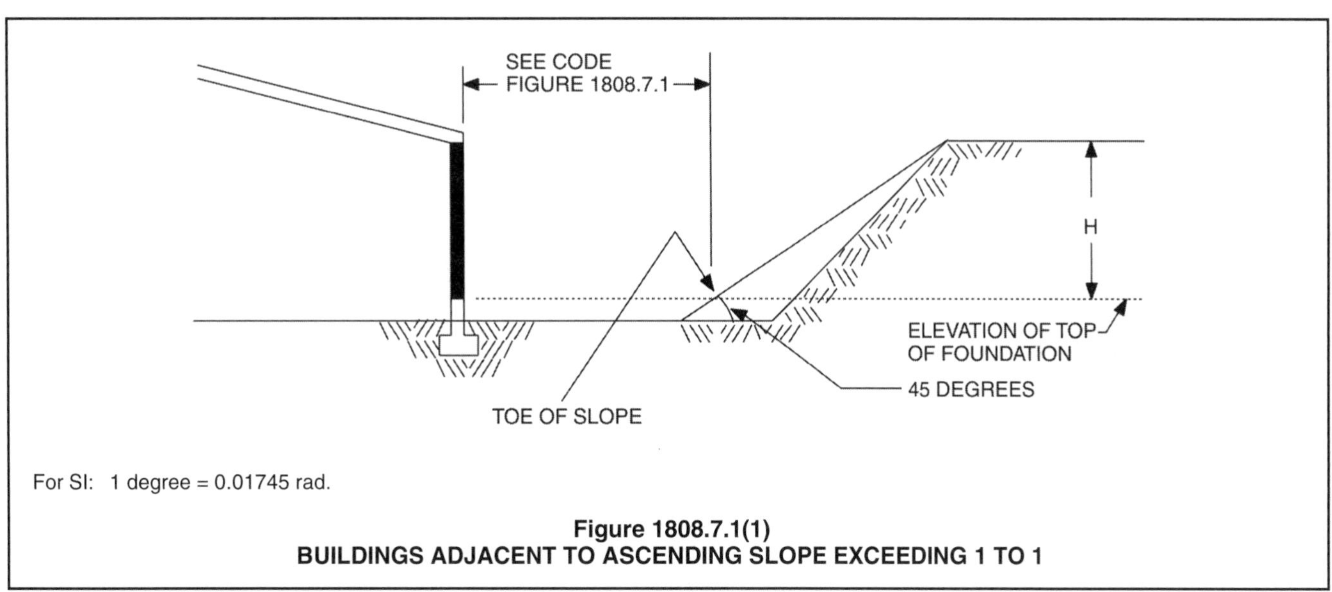

For SI: 1 degree = 0.01745 rad.

Figure 1808.7.1(1)
BUILDINGS ADJACENT TO ASCENDING SLOPE EXCEEDING 1 TO 1

other impediments to drainage. This section is related more directly to the provisions for site grading in Section 1804.3 than it is to Section 1808.7 requirements for footings adjacent to slopes.

1808.7.5 Alternate setback and clearance. Alternate setbacks and clearances are permitted, subject to the approval of the *building official*. The *building official* shall be permitted to require a geotechnical investigation as set forth in Section 1803.5.10.

❖ This section provides the building official the authority to approve alternative setbacks and clearances from slopes, provided he or she is satisfied that the intent of this section has been met. The building official has the authority to require a foundation investigation by a registered design professional to show that the intent of the code has been met. This item also specifies the parameters that must be considered by the registered design professional in the investigation.

1808.8 Concrete foundations. The design, materials and construction of concrete foundations shall comply with Sections 1808.8.1 through 1808.8.6 and the provisions of Chapter 19.

Exception: Where concrete footings supporting walls of light-frame construction are designed in accordance with Table 1809.7, a specific design in accordance with Chapter 19 is not required.

❖ The design, construction and materials used for concrete foundations are to comply with Chapter 19 which, of course, references ACI 318, and also the provisions of this section which are specific to concrete foundations.

The exception applies to concrete footings support-

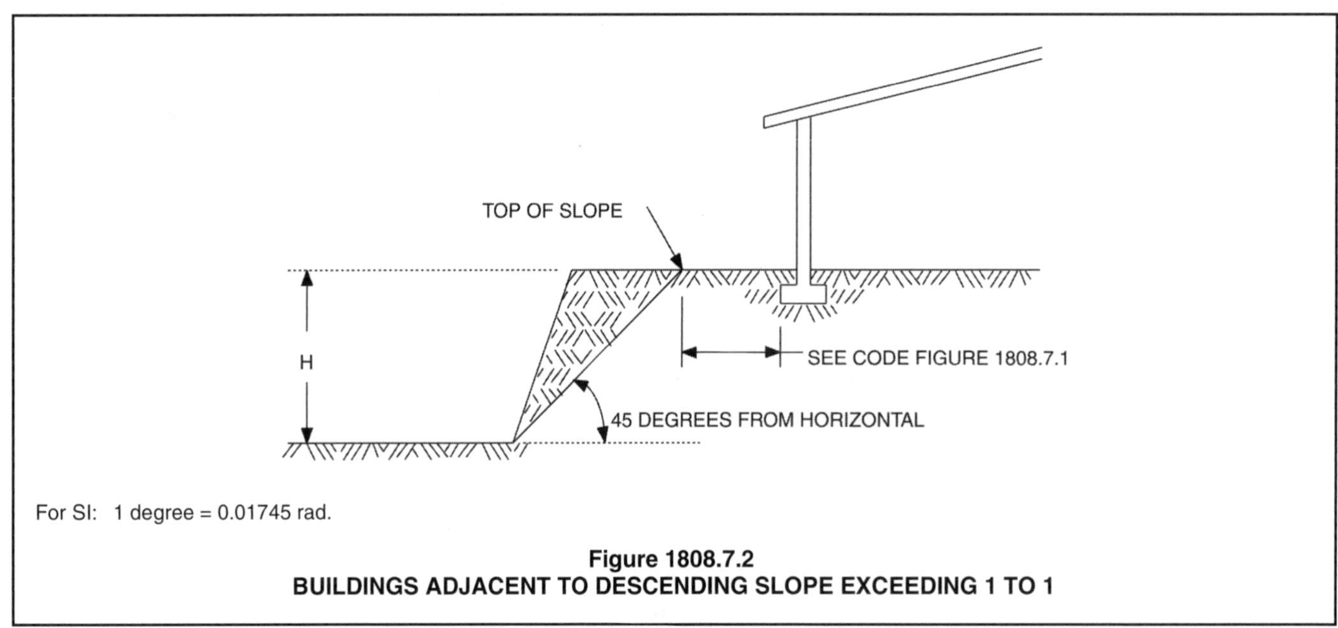

For SI: 1 degree = 0.01745 rad.

Figure 1808.7.2
BUILDINGS ADJACENT TO DESCENDING SLOPE EXCEEDING 1 TO 1

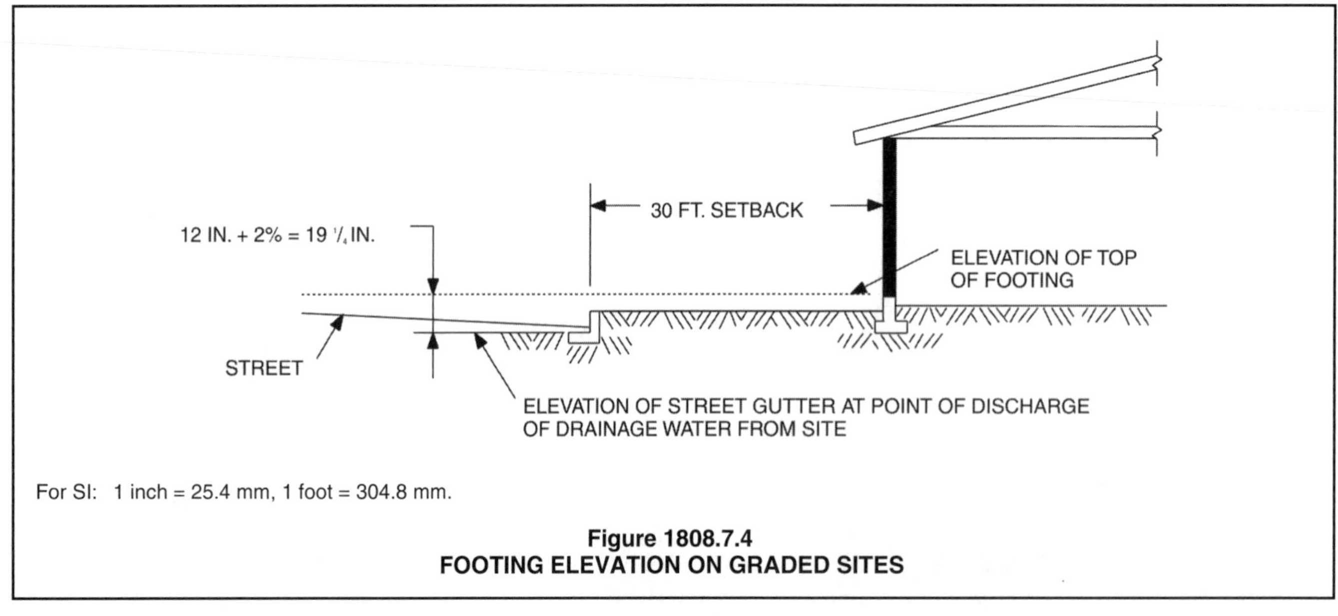

For SI: 1 inch = 25.4 mm, 1 foot = 304.8 mm.

Figure 1808.7.4
FOOTING ELEVATION ON GRADED SITES

ing walls of light-frame construction when a specific footing design is not provided. The purpose of Table 1809.7 is to specify footing sizes that can be used to safely support walls of light-frame construction. The table is based on anticipated loads on foundations due to wall, floor and roof systems.

1808.8.1 Concrete or grout strength and mix proportioning. Concrete or grout in foundations shall have a specified compressive strength (f'_c) not less than the largest applicable value indicated in Table 1808.8.1.

Where concrete is placed through a funnel hopper at the top of a deep foundation element, the concrete mix shall be designed and proportioned so as to produce a cohesive workable mix having a slump of not less than 4 inches (102 mm) and not more than 8 inches (204 mm). Where concrete or grout is to be pumped, the mix design including slump shall be adjusted to produce a pumpable mixture.

❖ This section of the code establishes a minimum compressive strength for concrete and grout that is used in foundations. The code's intent is that concrete or grout used as structural material for foundation construction will be of sufficient strength and durability to satisfy safety requirements.

This section also requires that the concrete mix for deep foundation elements be designed and proportioned to produce a cohesive workable mix with a consistency (slump) ranging between 4 inches (102 mm) and 8 inches (203 mm) where it is placed using a funnel hopper. This provision is within the bounds required for the proper placement of conventional structural concrete for cased cast-in-place elements. The slump requirements stated here, however, should not be considered so absolute that, with the use of engineering judgement, the consistency of the concrete cannot be adjusted to satisfy prevailing conditions of construction. For example, concrete placement conducted under difficult conditions—such as for elements containing heavy reinforcement, for very long cased elements, for element shells driven on steep batters or for other demanding situations—may require the employment of special concrete mixes using reduced quantities of coarse aggregates with corresponding increases in sand and cement content. Under such conditions, slumps of 4 inches (102 mm) with a tolerance of plus 2 inches (51 mm) or minus 1 inch (25 mm) are in order.

The consistency of concrete required for uncased cast-in-place concrete elements may be altogether different than the slumps specified and used for cased construction. For example, concrete placed in drilled holes should have a slump of at least 6 inches (152 mm) so that the concrete flows properly and full bond is attained with the reinforcement.

Furthermore, high slump concrete is used so that the complete volume of the hole is filled, including natural soil crevices and pockets and surface irregularities caused by the drilling operations. High slumps are also used where the concrete must displace drilling slurry.

Concrete slumps as high as 8 inches (203 mm) are sometimes needed where temporary casings are used to facilitate pile construction and are subsequently extracted as the concrete is placed.

Concrete is usually placed through a funnel hopper at the top of the element opening, but other approved methods, such as pumping or tremie, may also be used.

TABLE 1808.8.1. See below.

❖ This table lists the minimum compressive strength for concrete and grout based on the type of foundation.

1808.8.2 Concrete cover. The concrete cover provided for prestressed and nonprestressed reinforcement in foundations shall be no less than the largest applicable value specified in Table 1808.8.2. Longitudinal bars spaced less than $1^1/_2$ inches (38 mm) clear distance apart shall be considered bundled bars for which the concrete cover provided shall also be no less than that required by Section 7.7.4 of ACI 318. Concrete cover shall be measured from the concrete surface to the outermost surface of the steel to which the cover requirement applies. Where concrete is placed in a temporary or permanent casing or a mandrel, the inside face of the casing or mandrel shall be considered the concrete surface.

❖ A minimum thickness of concrete cover must be provided to protect reinforcement against corrosion from the moisture or severe environments, such as saltwater.

TABLE 1808.8.1
MINIMUM SPECIFIED COMPRESSIVE STRENGTH f'_c OF CONCRETE OR GROUT

FOUNDATION ELEMENT OR CONDITION	SPECIFIED COMPRESSIVE STRENGTH, f'_c
1. Foundations for structures assigned to Seismic Design Category A, B or C	2,500 psi
2a. Foundations for Group R or U occupancies of light-frame construction, two stories or less in height, assigned to Seismic Design Category D, E or F	2,500 psi
2b. Foundations for other structures assigned to Seismic Design Category D, E or F	3,000 psi
3. Precast nonprestressed driven piles	4,000 psi
4. Socketed drilled shafts	4,000 psi
5. Micropiles	4,000 psi
6. Precast prestressed driven piles	5,000 psi

For SI: 1 pound per square inch = 0.00689 MPa.

TABLE 1808.8.2. See below.

❖ This table lists the minimum concrete cover based on the type of foundation and the type of exposure.

1808.8.3 Placement of concrete. Concrete shall be placed in such a manner as to ensure the exclusion of any foreign matter and to secure a full-size foundation. Concrete shall not be placed through water unless a tremie or other method *approved* by the *building official* is used. Where placed under or in the presence of water, the concrete shall be deposited by *approved* means to ensure minimum segregation of the mix and negligible turbulence of the water. Where depositing concrete from the top of a deep foundation element, the concrete shall be chuted directly into smooth-sided pipes or tubes or placed in a rapid and continuous operation through a funnel hopper centered at the top of the element.

❖ Placing concrete under water should be avoided wherever possible. The risk of segregation of the concrete mixture is much greater when depositing under water as opposed to air; however, when concrete must be placed under water, it should be done by any one of several accepted methods used for such construction, including tremie. Due care must be exercised in the concreting operations so as to avoid or minimize segregation of the mix and turbulence of the water.

Generally, when concrete is to be placed under water, the mixture should be proportioned to provide a good plastic mix and high workability so that it will flow without segregation. The slump of the concrete should be 5 inches (127 mm) or greater. This desired consistency can be obtained by the use of rounded aggregates, a higher percentage of fines and entrained air. Cement content should be increased by 10 to 15 percent above the quantities required for similar mixtures placed in air to compensate for increases in water-cement ratios (see Chapter 19). In no case should the cement content be less than 600 pounds per cubic yard (355 kg/m^3) of concrete.

1808.8.4 Protection of concrete. Concrete foundations shall be protected from freezing during depositing and for a period of not less than five days thereafter. Water shall not be allowed to flow through the deposited concrete.

❖ Concrete for foundations should not be placed during rain, sleet or snow or in freezing weather unless adequate protection, as approved by the building official, is provided. Such protection, when required, is to be provided during the concreting operations and for a period of not less than five days thereafter. Rainwater or water from other sources must not be allowed to flow through freshly deposited concrete so as to increase the mixing water content or to damage the surface finish. For detailed information on materials, methods and procedures used in cold-weather operations, refer to ACI 306R, ACI 306.1 and Chapter 19.

1808.8.5 Forming of concrete. Concrete foundations are permitted to be cast against the earth where, in the opinion of the *building official*, soil conditions do not require formwork. Where formwork is required, it shall be in accordance with Chapter 6 of ACI 318.

❖ Where earth cuts are used as the concrete form in foundation construction, the soil must have sufficient stiffness to maintain the desired shape and dimensions before and during concreting operations. In the event that the soil is deemed to be unstable for such purpose, the building official is to require that formwork be built in accordance with the provisions of ACI 318.

1808.8.6 Seismic requirements. See Section 1908 for additional requirements for foundations of structures assigned to *Seismic Design Category* C, D, E or F.

For structures assigned to *Seismic Design Category* D, E or F, provisions of ACI 318, Sections 21.12.1 through

TABLE 1808.8.2
MINIMUM CONCRETE COVER

FOUNDATION ELEMENT OR CONDITION	MINIMUM COVER
1. Shallow foundations	In accordance with Section 7.7 of ACI 318
2. Precast nonprestressed deep foundation elements	
Exposed to seawater	3 inches
Not manufactured under plant conditions	2 inches
Manufactured under plant control conditions	In accordance with Section 7.7.3 of ACI 318
3. Precast prestressed deep foundation elements	
Exposed to seawater	2.5 inches
Other	In accordance with Section 7.7.3 of ACI 318
4. Cast-in-place deep foundation elements not enclosed by a steel pipe, tube or permanent casing	2.5 inches
5. Cast-in-place deep foundation elements enclosed by a steel pipe, tube or permanent casing	1 inch
6. Structural steel core within a steel pipe, tube or permanent casing	2 inches
7. Cast-in-place drilled shafts enclosed by a stable rock socket	1.5 inches

For SI: 1 inch = 25.4 mm.

21.12.4, shall apply where not in conflict with the provisions of Sections 1808 through 1810.

Exceptions:

1. Detached one- and two-family dwellings of light-frame construction and two stories or less above *grade plane* are not required to comply with the provisions of ACI 318, Sections 21.12.1 through 21.12.4.
2. Section 21.12.4.4(a) of ACI 318 shall not apply.

❖ This section first provides an important reference to Section 1908 for additional requirements concerning footings and foundations of buildings assigned to Seismic Design Category C, D, E or F. As specified in Section 1908.1.8, all footings and foundations in these structures require reinforcement except for footings supporting walls in detached one- and two-family dwellings three stories or less in height and constructed with stud-bearing walls. The concern is that plain concrete footings may be susceptible to damage, which can be reduced or avoided if reinforcement is provided. Foundation damage is obviously very expensive to repair and is better avoided.

The second paragraph of this section refers to provisions in ACI 318 concerning the design and construction of foundations of buildings assigned to high seismic design categories. Chapter 21 of ACI 318 is titled "Earthquake-resistant Structures," and Section 21.12 contains the requirements pertaining to foundations. Additionally, relatively small, detached dwellings classified as Seismic Design Category D or E are exempted from the referenced ACI 318 provisions. These structures would only need to comply with requirements for Seismic Design Category C.

1808.9 Vertical masonry foundation elements. Vertical masonry foundation elements that are not foundation piers as defined in Section 202 shall be designed as piers, walls or columns, as applicable, in accordance with TMS 402/ACI 530/ASCE 5.

❖ This provision directs the code user to the definition of "Masonry foundation pier," which states that it has a height less than or equal to four times its thickness. Furthermore, the code user is referred to TMS 402/ ACI 530/ASCE5 for the appropriate design requirements where the definition limits are exceeded. A vertical masonry element will be designed as a foundation pier, a pier, a wall or a column, depending on the dimensions.

SECTION 1809 SHALLOW FOUNDATIONS

1809.1 General. Shallow foundations shall be designed and constructed in accordance with Sections 1809.2 through 1809.13.

❖ Sections 1809.2 through 1809.13 address requirements related to proper design and installation of shallow foundations.

1809.2 Supporting soils. Shallow foundations shall be built on undisturbed soil, compacted fill material or controlled low-strength material (CLSM). Compacted fill material shall be placed in accordance with Section 1804.5. CLSM shall be placed in accordance with Section 1804.6.

❖ It is important that shallow foundations be built on undisturbed soil of known bearing value or properly compacted fill, with known bearing capacity. As an alternative to compacted fill, the code permits the use of CLSM (see commentary, Section 1804.6).

1809.3 Stepped footings. The top surface of footings shall be level. The bottom surface of footings shall be permitted to have a slope not exceeding one unit vertical in 10 units horizontal (10-percent slope). Footings shall be stepped where it is necessary to change the elevation of the top surface of the footing or where the surface of the ground slopes more than one unit vertical in 10 units horizontal (10-percent slope).

❖ The tops and bottoms of footings are required to be essentially level, with a slope of 1 unit vertical in 10 units horizontal (10-percent slope) permitted for the bottom of footings. Where the slope of the surface of the ground exceeds one unit vertical in 10 units horizontal (10-percent slope), footings are required to be stepped. Although not specifically mentioned, crack propagation at the joints should be considered when determining the overlapping and vertical dimensions of the steps.

1809.4 Depth and width of footings. The minimum depth of footings below the undisturbed ground surface shall be 12 inches (305 mm). Where applicable, the requirements of Section 1809.5 shall also be satisfied. The minimum width of footings shall be 12 inches (305 mm).

❖ Footings are required to extend below the ground surface a minimum of 12 inches (305 mm). This is considered a minimum depth to protect the footing from movement of the soil caused by freezing and thawing in mild climate areas (see Section 1809.5 for general frost protection requirements).

1809.5 Frost protection. Except where otherwise protected from frost, foundations and other permanent supports of buildings and structures shall be protected from frost by one or more of the following methods:

1. Extending below the frost line of the locality;
2. Constructing in accordance with ASCE 32; or
3. Erecting on solid rock.

Exception: Free-standing buildings meeting all of the following conditions shall not be required to be protected:

1. Assigned to *Risk Category* I, in accordance with Section 1604.5;
2. Area of 600 square feet (56 m^2) or less for light-frame construction or 400 square feet (37 m^2) or less for other than light-frame construction; and
3. Eave height of 10 feet (3048 mm) or less.

Shallow foundations shall not bear on frozen soil unless such frozen condition is of a permanent character.

❖ Shallow foundations must be placed on soil strata with adequate load-bearing capacity and at depths to which freezing cannot penetrate. In winter, frost action can raise the ground level (frost heave), whereas in springtime the same area will soften and settle back to its previous state. If foundations are built in soil strata that can freeze, then the heave or vertical movement of the ground, which is rarely uniform, can cause serious damage to buildings and other structures. Frost heave can become particularly aggravated in clay soils. Well-drained soils, such as sand and gravel, are not as susceptible to extensive movements.

Unless the exception applies, the foundation is to be protected from frost in accordance with this section. A common method of accomplishing this is by placing the footing bottom below the frost line. The "Frost line" is defined as the lowest level below the ground surface to which a temperature of 32°F (0°C) extends. The factors determining the depth of the frost line are air temperature and the length of time the temperature is below freezing [32°F (0°C)], as well as the ability of the soil to conduct heat and its level of thermal conductivity. Frost lines vary significantly throughout the country, ranging from 5 inches (127 mm) in the deep south to 100 inches (2540 mm) in the uppermost northern regions. The frost-free depth for shallow foundations is dependent on the frost line set for the particular locality of construction.

Another form of protection is the use of frost-protected shallow foundations (FPSF) in accordance with ASCE 32. This type of frost protection utilizes slab edge insulation to minimize heat loss at the slab edge. By retaining heat from the building in the ground, it has the effect of raising the frost line around the perimeter of the building.

Foundations are not to be placed on frozen soil because when the ground thaws, uneven settlement of the structure is apt to occur, thereby causing structural damage. This section does, however, permit footings to be constructed on permanently frozen soil. In permafrost areas, special precautions are necessary to prevent heat from the structure from thawing the soil beneath the foundation.

The exception to frost protection applies to low-risk structures, such as a detached garage. The permitted areas are intended to accommodate garage sizes that are commonly constructed. Note that the minimum footing depth of Section 1809.4 would apply where this exception is invoked.

1809.6 Location of footings. Footings on granular soil shall be so located that the line drawn between the lower edges of adjoining footings shall not have a slope steeper than 30 degrees (0.52 rad) with the horizontal, unless the material supporting the higher footing is braced or retained or otherwise laterally supported in an *approved* manner or a greater slope has been properly established by engineering analysis.

❖ The bottoms of adjacent footings bearing on granular soil are to be located so that a line drawn between their closest edges would not be steeper than 30 degrees (0.52 rad) from the horizontal. The exceptions are where the soil surrounding the higher footing is laterally braced or retained as approved by the building official or where engineering analysis shows that a greater slope can be tolerated. The purpose of this restriction is to provide conditions of safety against the possible influence of lateral and vertical soil pressures transmitted to the lower footing(s) by the loads of higher adjacent foundations (see Figure 1809.6).

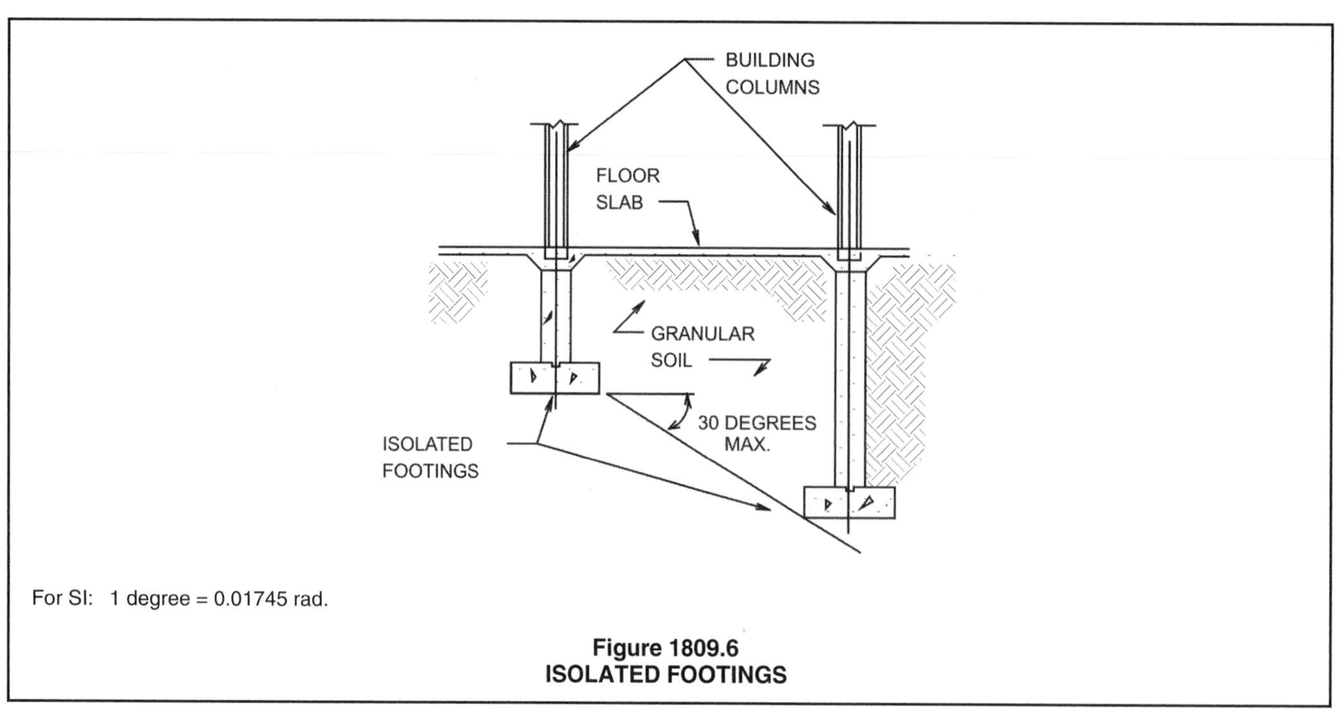

Figure 1809.6
ISOLATED FOOTINGS

1809.7 Prescriptive footings for light-frame construction. Where a specific design is not provided, concrete or masonry-unit footings supporting walls of light-frame construction shall be permitted to be designed in accordance with Table 1809.7.

❖ This section allows prescriptive designs for concrete and masonry-unit footings supporting walls of light-frame construction when a specific design is not provided.

TABLE 1809.7
PRESCRIPTIVE FOOTINGS SUPPORTING WALLS OF LIGHT-FRAME CONSTRUCTION[a, b, c, d, e]

NUMBER OF FLOORS SUPPORTED BY THE FOOTING[f]	WIDTH OF FOOTING (inches)	THICKNESS OF FOOTING (inches)
1	12	6
2	15	6
3	18	8[g]

For SI: 1 inch = 25.4 mm, 1 foot = 304.8 mm.

a. Depth of footings shall be in accordance with Section 1809.4.

b. The ground under the floor shall be permitted to be excavated to the elevation of the top of the footing.

c. Interior stud-bearing walls shall be permitted to be supported by isolated footings. The footing width and length shall be twice the width shown in this table, and footings shall be spaced not more than 6 feet on center.

d. See Section 1905 for additional requirements for concrete footings of structures assigned to Seismic Design Category C, D, E or F.

e. For thickness of foundation walls, see Section 1807.1.6.

f. Footings shall be permitted to support a roof in addition to the stipulated number of floors. Footings supporting roof only shall be as required for supporting one floor.

g. Plain concrete footings for Group R-3 occupancies shall be permitted to be 6 inches thick.

❖ Table 1809.7 provides prescriptive designs for footings supporting walls of light-frame construction when a specific design is not provided. The first column of the table is specifically intended to apply to the number of floors supported by the footing, not necessarily the number of stories of a building. This table provides for the minimum width and thickness of footings.

1809.8 Plain concrete footings. The edge thickness of plain concrete footings supporting walls of other than light-frame construction shall not be less than 8 inches (203 mm) where placed on soil or rock.

> **Exception:** For plain concrete footings supporting Group R-3 occupancies, the edge thickness is permitted to be 6 inches (152 mm), provided that the footing does not extend beyond a distance greater than the thickness of the footing on either side of the supported wall.

❖ An isolated spread footing of square or rectangular shape is the most common type of shallow foundation used in building construction. Basically, the footing is a slab of concrete supporting a column or other concentrated load. The footing thickness or depth is generally a function of shear strength and is established by the more severe of two structural conditions. The thickness of a footing determined by one-way shear action may be compared to the shear in a beam where the vertical shear plane extends across the entire structural section. Thickness established by two-way shear action (punching shear) is based on the consideration that the column (or column pedestal) tends to punch through the footing and failure occurs as a fracture in the form of a truncated pyramid concentrated under the load (see Figure 1809.8). Foundation designs will normally result in concrete footing thicknesses that are greater than the minimum thicknesses prescribed in the code. The minimum thickness requirements for plain concrete footings are intended to provide adequate construction based on experience.

In compliance with the requirements of ACI 318, the edge thickness for plain concrete footings is not to be less than 8 inches (203 mm). In computing footing stresses (flexure, combined flexure, axial load and shear), however, the overall depth is to be 2 inches (51 mm) less than the actual thickness of the footing for all foundations on soil (see ACI 318 for design of plain concrete members) to allow for the irregularities of excavated earth surfaces and for possible contamination of the concrete by the soil in contact with the construction.

The exception permits the edge thickness of plain concrete footings to be reduced to 6 inches (152 mm) for occupancies in Group R-3 of light-frame construction where the footing does not extend beyond a distance greater than the thickness of the footing on either side of the supported wall. For lightweight construction with the dimensional limitations given, shear stresses in the concrete will not usually govern the design thicknesses of such foundations. Plain concrete is not to be used for footings on piles.

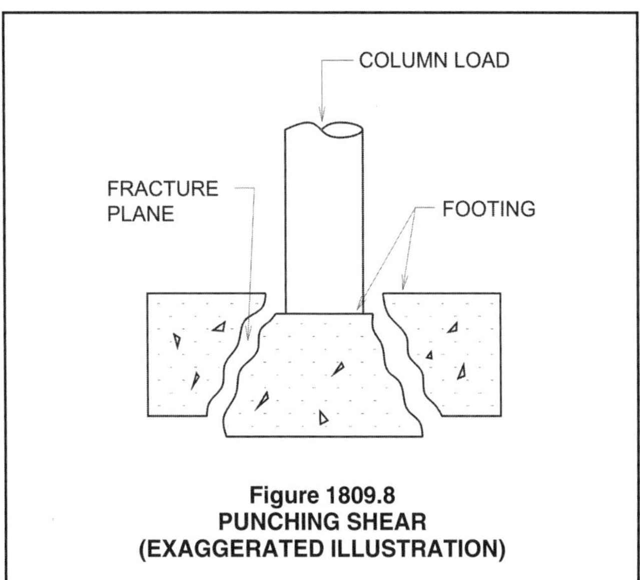

Figure 1809.8
PUNCHING SHEAR
(EXAGGERATED ILLUSTRATION)

1809.9 Masonry-unit footings. The design, materials and construction of masonry-unit footings shall comply with Sections 1809.9.1 and 1809.9.2, and the provisions of Chapter 21.

> **Exception:** Where a specific design is not provided, masonry-unit footings supporting walls of light-frame

construction shall be permitted to be designed in accordance with Table 1809.7.

❖ Sections 1809.9.1 and 1809.9.2 provide general requirements for the construction of solid masonry footings. This type of foundation was widely used until the middle of the last century when it was replaced by steel or wood grillage and eventually by more economical plain and reinforced concrete foundations. At that time, masonry foundations were built of either stone cut to specific sizes or of rubble stone of random sizes bonded together with mortar. Masonry footings under columns were usually constructed as high piers in the shape of truncated pyramids. While stone footings may still exist in very old structures, they are extinct in modern construction. Today, although scarcely used, masonry footings may be constructed of hard-burned brick in cement mortar to support the walls of lightweight construction, such as for low residential buildings. Brick footings are usually set on a full bed of mortar spread upon the earth or a grout sill. Stepped footing courses are recommended to be built with the brick units laid on edge, which are capable of resisting a greater transverse stress than flat courses.

The exception applies to footings supporting walls of light-frame construction when a specific footing design is not provided. The purpose of Table 1809.7 is to specify footing sizes and depths that can be used to safely support walls of light-frame construction. The table is based on anticipated loads on foundations due to wall, floor and roof systems.

1809.9.1 Dimensions. Masonry-unit footings shall be laid in Type M or S mortar complying with Section 2103.9 and the depth shall not be less than twice the projection beyond the wall, pier or column. The width shall not be less than 8 inches (203 mm) wider than the wall supported thereon.

❖ Footings made of masonry units are to be laid in Type M or S mortar complying with the requirements of Section 2103.8. Type M is a high-strength mortar suitable for general use and, in particular, where maximum masonry compressive strength is required. It is also suitable for unreinforced masonry below grade and in construction that is in contact with earth. Type S is a general-use mortar where high lateral strength of masonry is required. It is specifically recommended for use in reinforced masonry.

Projections of footings beyond a wall, pier or column base are not to exceed dimensions by more than one-half the depth of the foundation. For example, a footing supporting a wall and consisting of three courses of brick with a depth of about 8 inches (203 mm) is not to project more than 4 inches (102 mm) ($^1/_2$ by 8 inches) beyond the base of the wall. This is to keep the shearing stresses in the footing within a safe limitation.

Masonry footings are not to be less than 8 inches (203 mm) wider than the thickness of the supported wall so as to provide adequate distribution of the wall load to the load-bearing soil.

1809.9.2 Offsets. The maximum offset of each course in brick foundation walls stepped up from the footings shall be $1^1/_2$ inches (38 mm) where laid in single courses, and 3 inches (76 mm) where laid in double courses.

❖ Foundation walls that entail stepping back successive courses of brick (called racking) are not to have horizontal offsets exceeding $1^1/_2$ inches (38 mm) from the face of the course below. If the steps are made in double course increments, then the offsets must not exceed 3 inches (76 mm).

Sometimes, wide footings are required so as not to exceed the safe load-bearing capacity of the supporting soil. Since footing projections beyond a wall are limited in dimension (see Section 1809.9.1), the wall must inevitably be stepped out to provide the proper base on the footing (see Figure 1809.9.2).

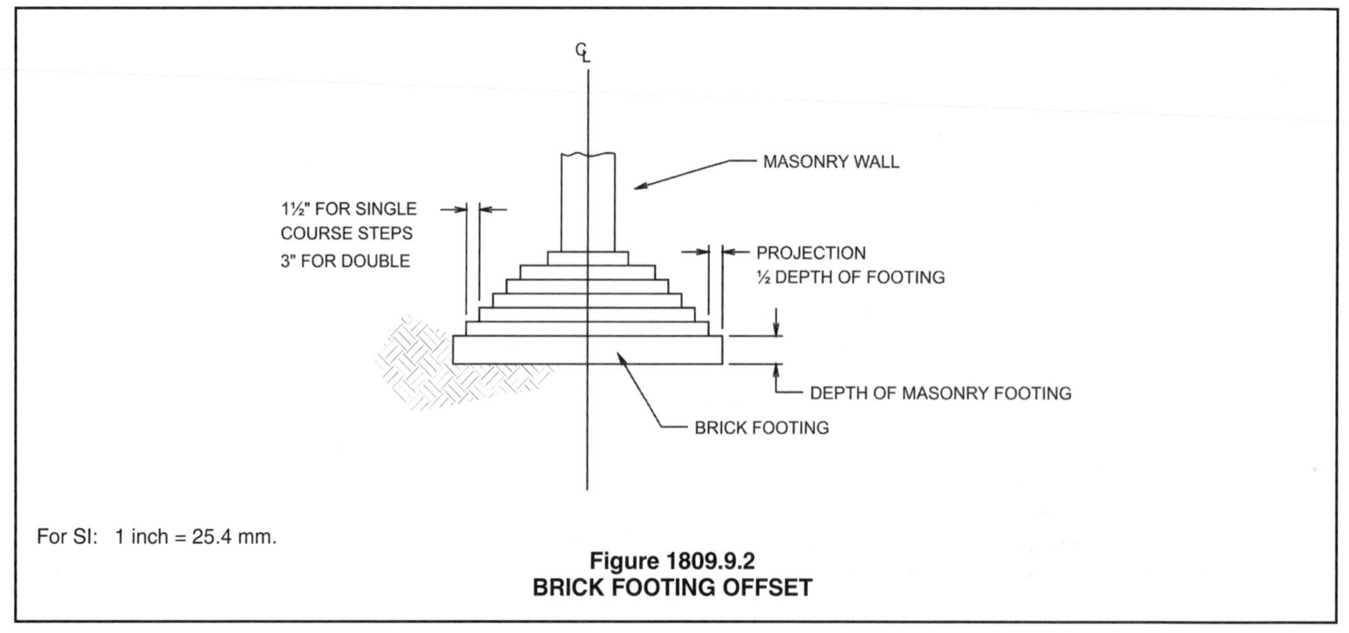

Figure 1809.9.2
BRICK FOOTING OFFSET

1809.10 Pier and curtain wall foundations. Except in *Seismic Design Categories* D, E and F, pier and curtain wall foundations shall be permitted to be used to support light-frame construction not more than two *stories above grade plane*, provided the following requirements are met:

1. All load-bearing walls shall be placed on continuous concrete footings bonded integrally with the *exterior wall* footings.
2. The minimum actual thickness of a load-bearing masonry wall shall not be less than 4 inches (102 mm) nominal or $3^5/_8$ inches (92 mm) actual thickness, and shall be bonded integrally with piers spaced 6 feet (1829 mm) on center (o.c.).
3. Piers shall be constructed in accordance with Chapter 21 and the following:
   - 3.1. The unsupported height of the masonry piers shall not exceed 10 times their least dimension.
   - 3.2. Where structural clay tile or hollow concrete masonry units are used for piers supporting beams and girders, the cellular spaces shall be filled solidly with concrete or Type M or S mortar.

     Exception: Unfilled hollow piers shall be permitted where the unsupported height of the pier is not more than four times its least dimension.
   - 3.3. Hollow piers shall be capped with 4 inches (102 mm) of solid masonry or concrete or the cavities of the top course shall be filled with concrete or grout.
4. The maximum height of a 4-inch (102 mm) load-bearing masonry foundation wall supporting wood frame walls and floors shall not be more than 4 feet (1219 mm) in height.
5. The unbalanced fill for 4-inch (102 mm) foundation walls shall not exceed 24 inches (610 mm) for solid masonry, nor 12 inches (305 mm) for hollow masonry.

❖ Pier and curtain wall foundation systems are now and have been a very popular method of construction for many years, particularly throughout the eastern United States. The origin of the design is seen in many precolonial wood-frame buildings. Pier and curtain wall foundations are only permitted to support structures of light-frame construction (wood or light-gage steel framing members) not more than two stories in height and assigned to Seismic Design Category A, B or C. Seismic detailing requirements for higher seismic design categories have not yet been developed. The term “pier” used in this application means “pilaster” rather than a type of foundation. The term “curtain wall” refers to minimum 4-inch-thick (102 mm) masonry load-bearing walls. The provisions apply to simple wood-frame buildings where the combined loads are minimal.

This type of foundation system is also addressed in Section R404.1.5.3 of the *International Residential Code*® (IRC®), which includes an accompanying figure illustrating the code requirements. This IRC figure is reproduced here as Figure 1809.10.

1809.11 Steel grillage footings. Grillage footings of structural steel shapes shall be separated with *approved* steel spacers and be entirely encased in concrete with at least 6 inches (152 mm) on the bottom and at least 4 inches (102 mm) at all other points. The spaces between the shapes shall be completely filled with concrete or cement grout.

❖ Steel grillage footings were extensively used during the latter part of the last century, but the development and use of reinforced concrete foundations have made this type of construction all but obsolete. They are, however, still used for underpinning purposes. There are many steel grillage footings in existence in old buildings.

A typical grillage footing consists of two or more tiers of steel beams (usually I-sections) with each tier placed at right angles to the one below it. The beams in each tier are usually held together by a system of bolts and pipe spacers. The beams should be clean and unpainted, and the whole system completely filled and encased in concrete with at least 6 inches (152 mm) of cover on the bottom and 4 inches (102 mm) at all other points. In lieu of concrete (other than the encasement), the spaces between the steel beams may be filled with cement grout.

1809.12 Timber footings. Timber footings shall be permitted for buildings of Type V construction and as otherwise *approved* by the *building official*. Such footings shall be treated in accordance with AWPA U1 (Commodity Specification A, Use Category 4B). Treated timbers are not required where placed entirely below permanent water level, or where used as capping for wood piles that project above the water level over submerged or marsh lands. The compressive stresses perpendicular to grain in untreated timber footings supported upon treated piles shall not exceed 70 percent of the allowable stresses for the species and grade of timber as specified in the AF&PA NDS.

❖ The use of timber footings is allowed only in Type V construction, or as otherwise approved by the building official. Such footings are commonly built as grillages of heavy timbers, using large sections, such as railroad ties.

If the footings are to be constructed at depths above the water table, they are to be given a preservative treatment by pressure processes to protect the materials from decay, fungi and harmful insects. Pressure treatment is to be in accordance with the requirements of the U1 standard of the American Wood Protection Association (AWPA). AWPA U1, Specification A, Use Category 4B, refers to sawn wood products that are used in contact with the ground. Except for use on cut surfaces, brush or spray applications of preservatives are not acceptable methods of treatment.

Preservative treatment by the pressure process within the limitations specified in AWPA U1 should not significantly affect the strength of the wood. Part of the process, however, involves the conditioning of timbers before treatment. Timbers conditioned by steaming or boiling under vacuum can suffer significant reductions in strength. This condition is recognized in the American Forest and Paper Association *National Design Specification*® (AF&PA NDS) by the use of the untreated factor C_u.

The design values for treated round timber piles given in this specification are adjusted to compensate for strength reductions because of conditioning prior to treatment. It is recommended that the values given for the several species of wood be used in the design of treated timber footings. Design values given in other tables contained in the AF&PA referenced standard are for general structural purposes using untreated lumber. While the section of this specification for pressure-preservative treatment stipulates that the design values apply to wood products that are pressure impregnated by an approved process, it is not apparent that reductions have been made to allow for possible strength loss in the wood because of the treatment process.

Untreated timber may be used when the footings are completely embedded in soil below the water level. Experience has shown that timber foundations permanently confined in water will stay sound and durable indefinitely. Wood submerged in fresh water cannot decay because the necessary air is excluded. It is not uncommon, however, to have changing ground-water levels because of changes in adjacent drainage systems and a variety of other subterranean

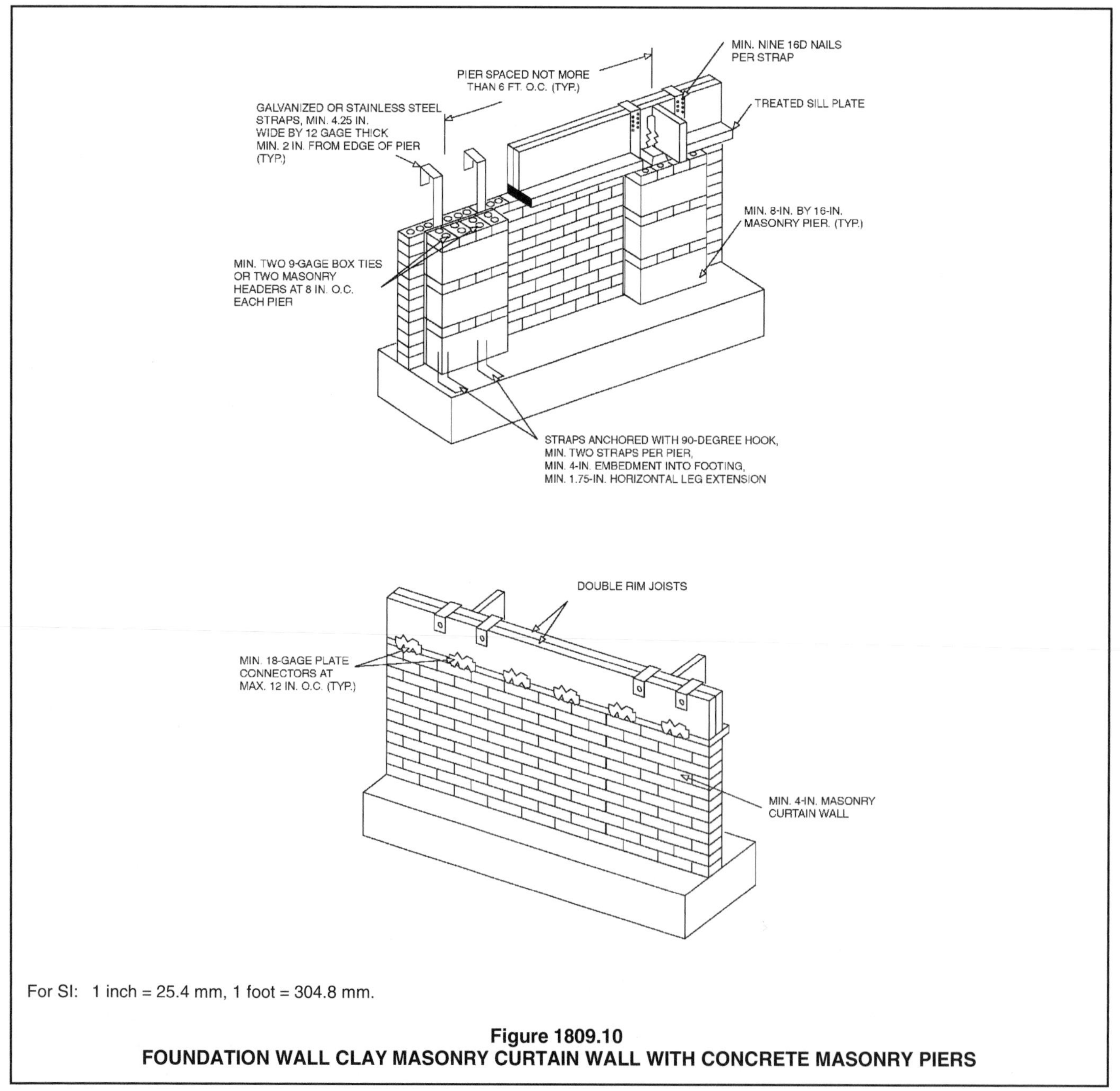

For SI: 1 inch = 25.4 mm, 1 foot = 304.8 mm.

Figure 1809.10
FOUNDATION WALL CLAY MASONRY CURTAIN WALL WITH CONCRETE MASONRY PIERS

conditions. Precautions should be taken so that untreated timber footings are placed at depths sufficiently below the groundwater level such that small drops in the water level will not expose the footings to air. Otherwise, decay will set in, causing settlements and possible structural failures.

The code also prescribes that compressive stresses perpendicular to grain in untreated timber footings supported upon piles are not to exceed 70 percent of the allowable stresses specified for the species and grade of timber given in the AF&PA referenced standard.

1809.13 Footing seismic ties. Where a structure is assigned to *Seismic Design Category* D, E or F, individual spread footings founded on soil defined in Section 1613.3.2 as *Site Class* E or F shall be interconnected by ties. Unless it is demonstrated that equivalent restraint is provided by reinforced concrete beams within slabs on grade or reinforced concrete slabs on grade, ties shall be capable of carrying, in tension or compression, a force equal to the lesser of the product of the larger footing design gravity load times the seismic coefficient, S_{DS}, divided by 10 and 25 percent of the smaller footing design gravity load.

❖ This section requires that spread footings on soft soil profiles be interconnected by ties when they support structures assigned to Seismic Design Categories D and higher. The purpose of this section is to preclude excessive movement of one column or wall with respect to another. One of the prerequisites of adequate structural performance during an earthquake is that the foundation of the structure acts as a unit. This is typically accomplished by tying together the individual footings with ties capable of carrying, in tension or compression, the smaller of 10 percent of the larger footing gravity load multiplied by S_{DS} or 25 percent of the smaller footing design gravity load. S_{DS} is the design spectral response acceleration at short periods, as determined in Section 1613.3.4. This tie can be provided by concrete floor slabs or tie beams. The differential movement of the foundation should not exceed that included in the design of the seismic-force-resisting system.

SECTION 1810
DEEP FOUNDATIONS

1810.1 General. Deep foundations shall be analyzed, designed, detailed and installed in accordance with Sections 1810.1 through 1810.4.

❖ This section sets forth the general rules for analyzing, designing, detailing and installing deep foundations.

1810.1.1 Geotechnical investigation. Deep foundations shall be designed and installed on the basis of a geotechnical investigation as set forth in Section 1803.

❖ A foundations investigation is mandatory when deep foundations are used. Such investigations are needed to define as accurately as possible the subsurface conditions of soil and rock materials, establish the soil and rock profiles across the construction site and locate the ground-water table. Sometimes it may also be necessary to determine specific soil properties, such as shear strength, relative density, compressibility and other such technical data required for analyzing subsurface conditions. Foundation investigations may also be used to render such valuable data as information on existing construction at the site or on neighboring properties (including boring and test records), the type and condition of the existing structures, their ages, the type of foundations used and performance over the years. Knowledge of existing deleterious substances in the soils that could affect the durability (as well as the performance) of foundation elements, data on geologic conditions at the site as well as a history of seismic activity are also important.

1810.1.2 Use of existing deep foundation elements. Deep foundation elements left in place where a structure has been demolished shall not be used for the support of new construction unless satisfactory evidence is submitted to the *building official*, which indicates that the elements are sound and meet the requirements of this code. Such elements shall be load tested or redriven to verify their capacities. The design load applied to such elements shall be the lowest allowable load as determined by tests or redriving data.

❖ After the demolition of an existing building, any deep foundation elements remaining in place cannot be reused to support a new structure unless sufficient and reliable information is provided to the building official showing that the new loads to be imposed will be adequately supported by the existing deep foundation elements. This requirement is necessary because of the lack of adequate soil data and technical information on the material used and the unavailability of pile-driving records made during the construction of the existing deep foundation. The current condition of the deep foundation element is not known, since it may have deteriorated over time, possibly reducing its load capacity. Deep foundation capacities may be determined by load tests, or the elements may be retracted and redriven to verify their load capacities.

1810.1.3 Deep foundation elements classified as columns. Deep foundation elements standing unbraced in air, water or fluid soils shall be classified as columns and designed as such in accordance with the provisions of this code from their top down to the point where adequate lateral support is provided in accordance with Section 1810.2.1.

> **Exception:** Where the unsupported height to least horizontal dimension of a cast-in-place deep foundation element does not exceed three, it shall be permitted to design and construct such an element as a pedestal in accordance with ACI 318.

❖ The section addresses the condition where deep foundation elements are not laterally supported by soil and, therefore, must be designed as columns. This design condition applies from the top of the element to a point at which the soil can be assumed to provide lateral support as described in Section 1810.2.1. The exception addresses concrete foundation elements with a height no greater than three times the least horizontal dimension, recognizing that these elements can be designed as pedestals under ACI 318.

1810.1.4 Special types of deep foundations. The use of types of deep foundation elements not specifically mentioned herein is permitted, subject to the approval of the *building official*, upon the submission of acceptable test data, calculations and other information relating to the structural properties and load capacity of such elements. The allowable stresses for materials shall not in any case exceed the limitations specified herein.

❖ Deep foundations are generally identified according to the materials used (concrete, steel or wood) or the methods of construction or installation. While the most commonly used types of deep foundations are addressed in this section, there are many variations of deep foundation types used in construction, including some special or proprietary types that are beyond the scope of the code.

However, while it is not the intent to preclude the use of such special or proprietary types of deep foundations, it is necessary to substantiate their structural performance by submitting test data, calculations, information on structural properties and load capacity and installation procedures to the building official for approval.

1810.2 Analysis. The analysis of deep foundations for design shall be in accordance with Sections 1810.2.1 through 1810.2.5.

❖ This section provides requirements applicable to the analysis of deep foundations, such as lateral support, stability, settlement, lateral loads and group effects.

1810.2.1 Lateral support. Any soil other than fluid soil shall be deemed to afford sufficient lateral support to prevent buckling of deep foundation elements and to permit the design of the elements in accordance with accepted engineering practice and the applicable provisions of this code.

Where deep foundation elements stand unbraced in air, water or fluid soils, it shall be permitted to consider them laterally supported at a point 5 feet (1524 mm) into stiff soil or 10 feet (3048 mm) into soft soil unless otherwise *approved* by the *building official* on the basis of a geotechnical investigation by a *registered design professional.*

❖ The primary concern that is addressed in this section is the consideration of slenderness effects in the design of a compression element. Experience with deep foundation performance under loaded conditions has shown that elements embedded in earth, including even relatively soft and compressible clays, exhibit lateral restraint that is sufficient to prevent buckling.

Deep foundation elements that are driven into fluid soils, such as saturated silts, as well as portions of deep foundation elements that project above the supporting soil are not laterally supported and, therefore, may be susceptible to buckling. Under such conditions, these elements must be designed as columns in accordance with the applicable provisions of the code.

This section allows the assumption of lateral support at a prescribed embedment without requiring confirmation by a soils investigation. The embedment required is the distance into either stiff soil or soft soil (not necessarily the distance below the ground surface). Note that, while the terms "stiff soil" and "soft soil" are not defined, they are generally consistent with the terms used in Section 1613.3.2 for Site Classes D and E. This provision essentially provides a starting point for analyzing the fixity of the deep foundation element.

1810.2.2 Stability. Deep foundation elements shall be braced to provide lateral stability in all directions. Three or more elements connected by a rigid cap shall be considered braced, provided that the elements are located in radial directions from the centroid of the group not less than 60 degrees (1 rad) apart. A two-element group in a rigid cap shall be considered to be braced along the axis connecting the two elements. Methods used to brace deep foundation elements shall be subject to the approval of the *building official.*

Deep foundation elements supporting walls shall be placed alternately in lines spaced at least 1 foot (305 mm) apart and located symmetrically under the center of gravity of the wall load carried, unless effective measures are taken to provide for eccentricity and lateral forces, or the foundation elements are adequately braced to provide for lateral stability.

Exceptions:

1. Isolated cast-in-place deep foundation elements without lateral bracing shall be permitted where the least horizontal dimension is no less than 2 feet (610 mm), adequate lateral support in accordance with Section 1810.2.1 is provided for the entire height and the height does not exceed 12 times the least horizontal dimension.

2. A single row of deep foundation elements without lateral bracing is permitted for one- and two-family dwellings and lightweight construction not exceeding two *stories above grade plane* or 35 feet (10 668 mm) in *building height*, provided the centers of the elements are located within the width of the supported wall.

❖ A group of deep foundation elements designed to support a common load and, as may be required, to resist horizontal forces, must be braced or rigidly tied together to act as a single structural unit that will provide lateral stability in all directions. Deep foundation elements that are connected by a rigid, reinforced concrete pile cap are deemed to be braced construction that serves the intent of this provision.

Three or more deep foundation elements are generally used to support a building column load or other isolated concentrated load. In a three-pile group, lateral stability is provided by requiring that the elements are located such that they will not be less than 60 degrees (1.0 rad) apart as measured from the centroid of the group in a radial direction. For stability of deep foundation elements supporting a wall structure, the elements are braced by a continuous rigid footing and are alternately staggered in two lines that are at least 1 foot (305 mm) apart and symmetrically located on each side of the center of gravity of the wall. Other approved deep foundation arrangements may be used to support walls, provided that the elements are adequately braced and the lateral stability of the foundation construction is ensured.

Exception 1 addresses the use of isolated elements without lateral bracing. To qualify elements must have widths (or diameters) of at least 2 feet (610 mm) and be proportioned so that the height (length) of the element is no more than 12 times the least lateral dimension. This is an empirical requirement that is intended to offset the concerns that typically require consideration in more slender elements.

For one- and two-family dwellings as well as lightweight construction that does not exceed two stories above grade plane or 35 feet (10 668 mm) in height, Exception 2 permits a single row of elements, rather than the minimum 1 foot (305 mm) offset, provided the deep foundation elements are located within the width of the supported wall. In this case, stability of the group is theoretically afforded only in the direction of the line of the deep foundation elements. However, in this limited group of relatively lighter buildings, the cross walls, floor slabs and other structural components are assumed to provide some degree of lateral stability to the deep foundation system.

1810.2.3 Settlement. The settlement of a single deep foundation element or group thereof shall be estimated based on *approved* methods of analysis. The predicted settlement shall cause neither harmful distortion of, nor instability in, the structure, nor cause any element to be loaded beyond its capacity.

❖ A settlement analysis is performed to design a deep foundation system that will maintain the stability and structural integrity of the supported building or structure. Foundation systems that suffer serious settlements, particularly differential settlements, can cause structural damage to the supported structure as well as to the foundation itself.

The settlement analysis of an individual element is a complex procedure. In comparison, the analysis of a group of elements is even more complex because of the overlapping soil stresses caused by closely spaced piles. Analytical procedures vary with the type of elements and especially with the type of soil.

Settlements are of two basic types: immediate settlements are those that occur as soon as the load is applied and usually take place within a period of less than seven days; consolidation settlements are time dependent and take place over a long period of time. All cohesionless soils, such as granular materials consisting of sand, gravel or a mixture of both, which have a large coefficient of permeability (rapid draining properties), undergo immediate settlements. All fine-grained, saturated, cohesive soils, such as clays, undergo time-dependent consolidation settlements. Settlement analysis would generally include cases involving piers and end-bearing piles on rock or hard soils as well as friction-type piles in both granular and cohesive soils. Load tests are often performed to provide data that will aid the settlement analysis.

The code places no specific limitation on the amount of settlement; instead, it states that the effects of any settlement, particularly differential settlements, should not be harmful to the structure. Thus, the tolerable settlement is largely dependent on the type of structure being considered. Guidance on tolerable settlements can be found in engineering textbooks, such as *Soil Mechanics in Engineering Practice* (Terzaghi and Peck), *Foundation Analysis and Design* (Bowles) and *Foundation Engineering* (Hanson, Peck and Thornburn). As a rule, these references indicate that a total settlement of 1 inch (25 mm) is acceptable for the majority of structures, while other structures can tolerate even greater settlements without distress. On the other hand, a more restrictive settlement criterion can be necessary based on the needs of a specific structure. Current engineering practice is often based on an allowable total settlement of 1 inch (25 mm) with the objective of controlling the differential settlements to $^3/_4$ inch (19 mm) or less. The effects of differential settlements need to be considered in the structural design as a self-straining force, *T*, when designing for the load combinations of Section 1605.

1810.2.4 Lateral loads. The moments, shears and lateral deflections used for design of deep foundation elements shall be established considering the nonlinear interaction of the shaft and soil, as determined by a *registered design professional*. Where the ratio of the depth of embedment of the element to its least horizontal dimension is less than or equal to six, it shall be permitted to assume the element is rigid.

❖ This section addresses miscellaneous issues unique to the seismic design of deep foundations. If the length is less than or equal to six times the least horizontal dimension of the element, it can be assumed to be rigid. Then moments, shears and lateral deflections can be calculated accordingly. Where the length exceeds six times the least horizontal dimension, the nonlinear interaction of the element and soil effects are to be included in the analysis. The effect of abrupt changes in soil deposits, such as changes from soft to firm or loose to dense soils, should be included in the analysis.

1810.2.4.1 Seismic Design Categories D through F. For structures assigned to *Seismic Design Category* D, E or F, deep foundation elements on *Site Class* E or F sites, as determined in Section 1613.3.2, shall be designed and constructed to withstand maximum imposed curvatures from earthquake ground motions and structure response. Curvatures shall include free-field soil strains modified for soil-foundation-structure interaction coupled with foundation element deformations associated with earthquake loads imparted to the foundation by the structure.

Exception: Deep foundation elements that satisfy the following additional detailing requirements shall be deemed to comply with the curvature capacity requirements of this section.

1. Precast prestressed concrete piles detailed in accordance with Section 1810.3.8.3.3.
2. Cast-in-place deep foundation elements with a minimum longitudinal reinforcement ratio of 0.005 extending the full length of the element and detailed in accordance with Sections 21.6.4.2, 21.6.4.3 and 21.6.4.4 of ACI 318 as required by Section 1810.3.9.4.2.2.

❖ The first paragraph of this section requires special consideration of flexural loads on deep foundation elements due to earthquake motions. The following discussion taken from the NEHRP Provisions commentary, Section 7.5.4, provides justification for these requirements:

Special consideration is required in the design of concrete piles subject to significant bending during earthquake shaking. Bending can become crucial to element design where portions of the deep foundation elements are supported in soils, such as loose granular materials or soft soils that are susceptible to large deformations or strength degradation. Severe pile bending problems can result from various combinations of soil conditions during strong ground shaking. For example:

- Soil settlement at the pile-cap interface either from consolidation of soft soil prior to the earthquake or from soil compaction during the earthquake can create a free-standing short column adjacent to the pile cap.
- Large deformations, a reduction in strength or both, resulting from liquefaction of loose, granular materials, can cause bending or conditions of free-standing columns.
- Large deformations in soft soils can cause varying degrees of element bending. The degree of bending will depend upon thickness and strength of the soft soil layer(s) and the properties of the soft/stiff soil interface(s).

The designer needs to consider the variation in soil conditions and driven element lengths in providing for pile ductility at potential high curvature interfaces. Interaction between the geotechnical and structural engineers is essential.

It is prudent to design deep foundation elements to remain functional during and following earthquakes in view of the fact that it is difficult to repair foundation damage. The desired foundation performance can be accomplished by proper selection and detailing of the pile foundation system. Such design should accommodate bending from both reaction to the building's inertial loads and those induced by the motions of the soils themselves.

1810.2.5 Group effects. The analysis shall include group effects on lateral behavior where the center-to-center spacing of deep foundation elements in the direction of lateral force is less than eight times the least horizontal dimension of an element. The analysis shall include group effects on axial behavior where the center-to-center spacing of deep foundation elements is less than three times the least horizontal dimension of an element.

❖ This section also prescribes conditions under which group effects on the nominal pile strength, lateral as well as vertical, need to be considered in the analysis.

1810.3 Design and detailing. Deep foundations shall be designed and detailed in accordance with Sections 1810.3.1 through 1810.3.12.

❖ This section provides requirements and guidance for the design and detailing of deep foundation elements. This includes material-specific criteria, allowable loads, dimensions, splices, pile caps, grade beams and seismic ties.

1810.3.1 Design conditions. Design of deep foundations shall include the design conditions specified in Sections 1810.3.1.1 through 1810.3.1.6, as applicable.

❖ This section covers design methods for concrete elements, guidance for composite elements, effects of mislocation, driven piles, helical piles and casings.

1810.3.1.1 Design methods for concrete elements. Where concrete deep foundations are laterally supported in accordance with Section 1810.2.1 for the entire height and applied forces cause bending moments no greater than those resulting from accidental eccentricities, structural design of the element using the load combinations of Section 1605.3 and the allowable stresses specified in this chapter shall be permitted. Otherwise, the structural design of concrete deep foundation elements shall use the load combinations of Section 1605.2 and *approved* strength design methods.

❖ This provision states what is typical in current practice for foundation design. For decades structural concrete design has primarily been based on the strength design method. However, there is also a long tradition of using simple allowable stress design approaches for the proportioning of deep foundation elements (for both soil-foundation behavior and structural design). This section recognizes allowable stress design for elements that are concentrically loaded and laterally supported.

1810.3.1.2 Composite elements. Where a single deep foundation element comprises two or more sections of different materials or different types spliced together, each section of the composite assembly shall satisfy the applicable requirements of this code, and the maximum allowable load in each section shall be limited by the structural capacity of that section.

❖ Composite elements are made of two or more sections of different materials or of different types that are spliced together to form a single deep foundation element. These elements are typically used when significant lengths are needed. Due to economic considerations and problems with splicing, composite elements are now used less frequently. An illustration of typical composite elements is shown in Figure 1810.3.1.2.

This section gives the general requirement that each section of a composite element must comply with the applicable provisions for that element type as well as the material comprising that section. Furthermore the element's overall load capacity must be based on the most restrictive permitted value for all of the sections.

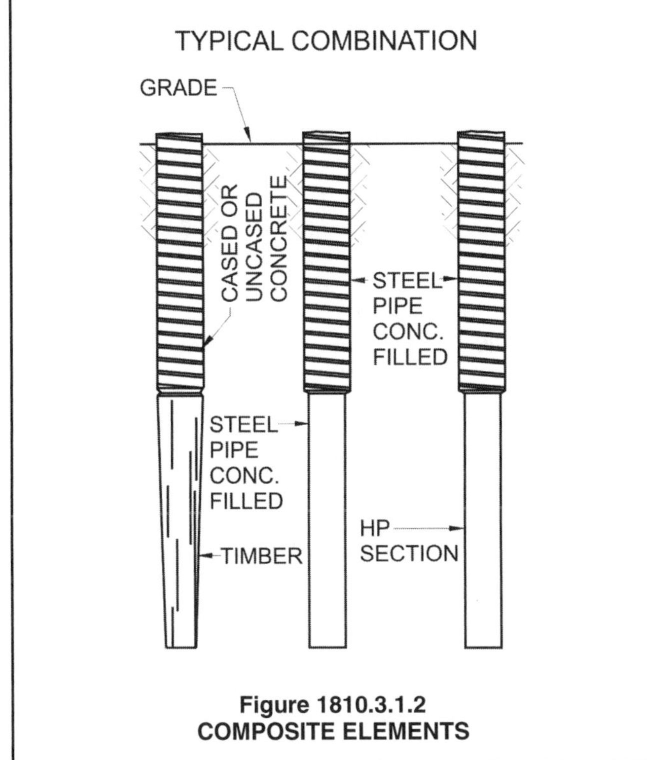

Figure 1810.3.1.2
COMPOSITE ELEMENTS

1810.3.1.3 Mislocation. The foundation or superstructure shall be designed to resist the effects of the mislocation of any deep foundation element by no less than 3 inches (76 mm). To resist the effects of mislocation, compressive overload of deep foundation elements to 110 percent of the allowable design load shall be permitted.

❖ Because of subsurface obstructions or other reasons, it is sometimes necessary to offset deep foundation elements a small distance from their intended locations or they may be driven out of position. In such cases, the load distribution in a group of elements may be changed from the design requirements and cause some of the elements to be overloaded. This section requires that the maximum compressive load on any deep foundation element caused by mislocation should not exceed 110 percent of the allowable design load. Elements exceeding this limitation must be extracted and installed in the proper location or other approved remedies must be applied, such as installing additional elements to balance the group.

1810.3.1.4 Driven piles. Driven piles shall be designed and manufactured in accordance with accepted engineering practice to resist all stresses induced by handling, driving and service loads.

❖ Precast elements must be properly designed to resist the stresses induced both by handling and driving operations, and later imposed by service loads. Care must be given during handling and installation to minimize or avoid possible damage to these elements, such as cracking, crushing or spalling.

1810.3.1.5 Helical piles. Helical piles shall be designed and manufactured in accordance with accepted engineering practice to resist all stresses induced by installation into the ground and service loads.

❖ See the definition of "Helical pile" in Section 1802.1. This section gives only general guidance to design in accordance with accepted engineering practice. Helical pile systems having ICC Evaluation Service reports can be viewed at www.icc-es.org (also see ICC Evaluation Service Acceptance Criteria 358).

1810.3.1.6 Casings. Temporary and permanent casings shall be of steel and shall be sufficiently strong to resist collapse and sufficiently water tight to exclude any foreign materials during the placing of concrete. Where a permanent casing is considered reinforcing steel, the steel shall be protected under the conditions specified in Section 1810.3.2.5. Horizontal joints in the casing shall be spliced in accordance with Section 1810.3.6.

❖ Because a casing is exposed to earth and water pressures, this section requires that casings are of adequate strength to resist serious damage or collapse and to maintain sufficient water tightness so as to prevent any foreign materials from entering during concrete placement (see Figure 1810.3.1.6 for an illustration of steel casings). Steel pipe casings driven with a mandrel can be torn or otherwise damaged because of underground obstructions, such as rock crevices. If this happens, the water tightness and structural integrity of the deep foundation element may be affected. Sometimes the buildup of ground pressures during driving may cause pile casings to squeeze or even collapse after the mandrel has been withdrawn.

1810.3.2 Materials. The materials used in deep foundation elements shall satisfy the requirements of Sections 1810.3.2.1 through 1810.3.2.8, as applicable.

❖ This section specifies minimum requirements for concrete, prestressing steel, structural steel timber, etc., for use in deep foundations.

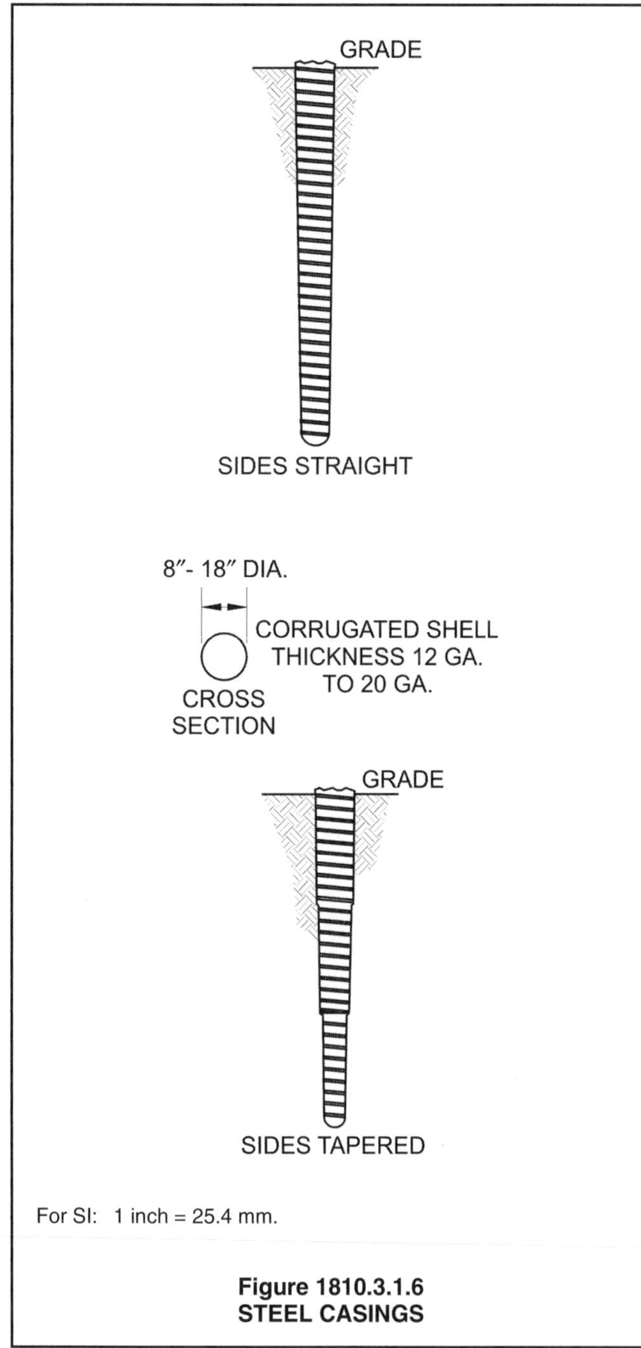

Figure 1810.3.1.6
STEEL CASINGS

1810.3.2.1 Concrete. Where concrete is cast in a steel pipe or where an enlarged base is formed by compacting concrete, the maximum size for coarse aggregate shall be $^3/_4$ inch (19.1 mm). Concrete to be compacted shall have a zero slump.

❖ This section limits the aggregate size in cased elements and enlarged base elements (see Section 1810.4.7). Concrete materials generally are to comply with the applicable requirements of Chapter 19. Coarse aggregate materials used in the concrete mix are not to exceed $^3/_4$ inch (19.1 mm) in size. Concrete that will be compacted must have a zero slump in order to provide a stiff mix capable of being compacted by a heavy drop weight.

1810.3.2.1.1 Seismic hooks. For structures assigned to *Seismic Design Category* C, D, E or F, the ends of hoops, spirals and ties used in concrete deep foundation elements shall be terminated with seismic hooks, as defined in ACI 318, and shall be turned into the confined concrete core.

❖ In structures that have moderate or high seismic risk, transverse reinforcement confines the concrete core of deep foundation elements and provides lateral supports for longitudinal bars. The requirement to terminate hoops, spirals and ties with seismic hooks parallels requirements in Chapter 21 of ACI 318.

1810.3.2.1.2 ACI 318 Equation (10-5). Where this chapter requires detailing of concrete deep foundation elements in accordance with Section 21.6.4.4 of ACI 318, compliance with Equation (10-5) of ACI 318 shall not be required.

❖ In deep foundation elements the axial compression is limited by the capacity of the soil-foundation interface. This is also reflected by the allowable stresses for these elements which are limited to a smaller percentage of the concrete compressive strength. The purpose of ACI 318 Equation (10-5) is to provide significant residual compressive strength for concentrically loaded spiral columns subjected to very large axial compression. The amount of spiral reinforcement required by Equation (10-5) is intended to provide additional load-carrying strength for concentrically loaded columns equal to or slightly greater than the strength lost when the shell spalls off. The concrete cover required for uncased deep foundation elements is much greater than that for columns. As a result, application of Equation (10-5) to deep foundation elements results in amounts of transverse reinforcement that are unwarranted and often unplaceable. The primary concern in providing transverse confinement reinforcement for deep foundations is flexural ductility. Because proper design for deep foundation elements differs from that for columns in several respects, this provision relaxes an overly restrictive code requirement.

1810.3.2.2 Prestressing steel. Prestressing steel shall conform to ASTM A 416.

❖ Rods, strands or wires conforming to the requirements of ASTM A 416 are used as prestressing steel in the manufacture of precast, prestressed concrete elements of both the pretensioned and the post- tensioned type. The most commonly used prestressing steel is the seven wire, uncoated, stress-relieved strand.

1810.3.2.3 Structural steel. Structural steel piles, steel pipe and fully welded steel piles fabricated from plates shall conform to ASTM A 36, ASTM A 252, ASTM A 283, ASTM A 572, ASTM A 588, ASTM A 690, ASTM A 913 or ASTM A 992.

❖ The materials used in the manufacture of steel piles must comply with requirements of one of the listed ASTM standards.

While H-piles are ordinarily made with steel materials conforming to ASTM A 36 or ASTM A 572 requirements, the employment of other standards is not excluded, provided it can be shown that such steels meet the applicable chemical and mechanical properties of one of the listed specifications to establish the suitability of the material for pile use.

Welded or seamless pipe piles must conform to ASTM A 252 requirements.

1810.3.2.4 Timber. Timber deep foundation elements shall be designed as piles or poles in accordance with AF&PA NDS. Round timber elements shall conform to ASTM D 25. Sawn timber elements shall conform to DOC PS-20.

❖ Round timber elements are best suited as friction-type piles. They are not recommended to be driven through dense gravel, boulders or till, or for end-bearing piles on rock. While timber elements do not have load capacities of structural steel or concrete elements, they are probably the most commonly used type of deep foundation element throughout the United States, mainly because of their availability, ease of handling and, sometimes, for economic reasons.

Timber elements are shaped from tree trunks and their lengths are dependent on the heights that the various species of trees used for making piles will grow. Timber piles are made as tapered sections because of the natural taper of tree trunks.

Round timber piles are usually made from Southern pine in lengths up to about 80 feet (24 384 mm) and from Pacific Coast Douglas fir in lengths up to about 125 feet (38 100 mm). Other species commonly used for piles are red oak and red pine. Timber piles that are 40 feet (12 192 mm) to 60 feet (18 288 mm) in length are common, but longer lengths cannot be obtained economically in all areas of the country. Untreated timber piles that are embedded below the permanent ground-water level (fresh water only) may last indefinitely. Under conditions where piles are required to extend above ground-water level but still remain totally embedded in the ground, the untreated wood material may be subject to decay. In cases where untreated timber piles extend above the ground surface into the air, they are exposed to decay and insect attack. The durability of timber piles is best served by applying treatment with an approved preservative.

This section requires timber deep foundation elements to be designed in accordance with the AF&PA NDS. Section 6 of AF&PA NDS provides the appropriate design values, different species of timber piles and the applicable adjustment factors. While timber piles have to be designed in accordance with the AF&PA NDS, they are usually only considered for design loads of 10 (89 kN) to 50 tons (445 kN).

One of the significant problems associated with timber pile installations is the possibility of damage caused by overdriving. Overdriving of wood piles may cause failure by bending, brooming at the tip, crushing or brooming at the butt end or splitting or breaking along the pile section (see Figure 1810.4.1.5). Another problem is the difficulties encountered when splicing timber piles to achieve greater lengths.

The ASTM D 25 standard referenced in this section covers the physical characteristics of treated and untreated round timber piles. Essentially, the standard divides timber piles into two classifications: friction and end-bearing piles. The dimensional requirements for each classification are tabulated, giving the minimum circumference requirements for pile heads and the corresponding smaller tip circumference based on pile taper and lengths between 20 (6096 mm) and 120 feet (36 576 mm), measured in increments of 10 feet (3048 mm). The ASTM D 25 standard also includes requirements for the quality of wood, tolerances on pile straightness, twist of grain, knots, holes, scars, wood checks, shakes, splits and other necessary information. However, the requirements stated in the standard do not relate to the several species of wood used for making timber piles, nor to their relative strengths. A timber pile with typical dimensions from ASTM D 25 is shown in Figure 1810.3.2.4.

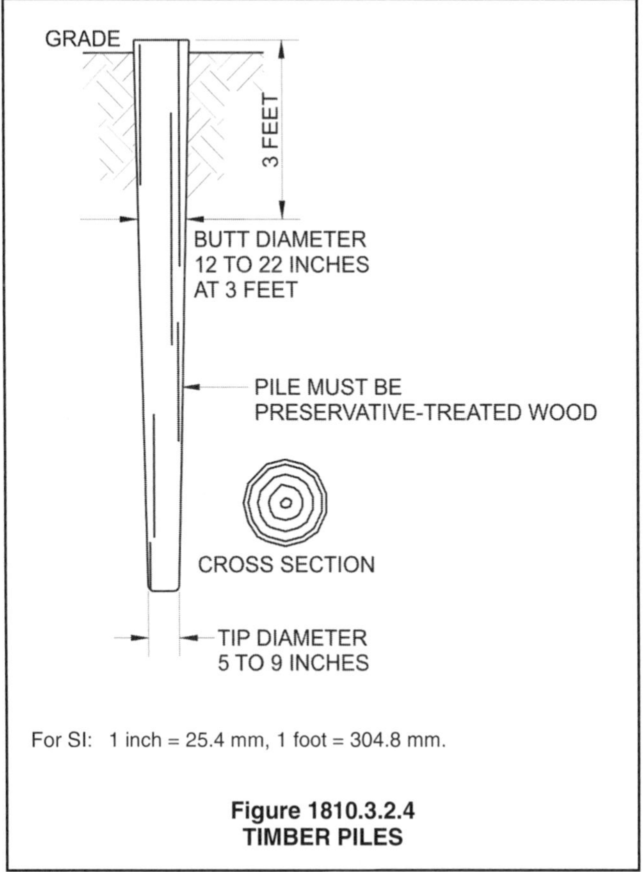

Figure 1810.3.2.4
TIMBER PILES

1810.3.2.4.1 Preservative treatment. Timber deep foundation elements used to support permanent structures shall be treated in accordance with this section unless it is established that the tops of the untreated timber elements will be below the lowest ground-water level assumed to exist during the life of the structure. Preservative and minimum final retention shall be in accordance with AWPA U1 (Commodity Specification E, Use Category 4C) for round timber elements and AWPA U1 (Commodity Specification A, Use Category 4B) for sawn timber elements. Preservative-treated timber elements shall be subject to a quality control program administered by an *approved agency*. Element cutoffs shall be treated in accordance with AWPA M4.

❖ For a general discussion on the need for treating wood with preservatives, see the commentary to Sections 1809.12,1810.3.2.4 and 1810.3.2.5. This section of the code specifically requires that the preservative treatment of round timber piles and sawn timber piles by pressure processes conforms to the applicable requirements of AWPA U1. Round timber pile requirements are covered in Commodity Specification E. Use Category 4C refers to conditions of very severe ground contact. Sawn timber piles are covered under Commodity Specification A and Use Category 4B, which refer to ground contact in severe environments. Note that round timber piles conform to ASTM D 25, while sawn timber piles comply with the grading rules of the species for strength.

This section also requires that, for timber piles subjected to saltwater exposure, the treatment of piles with water-borne preservatives and creosote complies with quality control procedures administered by an approved agency. Guides used in establishing these procedures are those contained in the American Wood Preservers' Bureau (AWPB) publications such as:

- MP1, *Quality Control and Inspection Procedures for Dual Treatment of Marine Piling Pressure Treated with Waterborne Preservatives and Creosote for Use in Marine Waters*;
- MP2, *Quality Control and Inspection Procedures for Marine Piling Pressure Treated with Creosote for Use in Marine Waters*; or
- MP4, *Quality Control and Inspection Procedures for Marine Piling Pressure Treated with Waterborne Preservatives for Use in Marine Waters.*

The field treatment of cuts and injuries to timber piles, including the treatment at pile cutoffs, with preservatives made for applications of creosote and creosote mixtures or with water-borne preservatives is covered by AWPA M4.

1810.3.2.5 Protection of materials. Where boring records or site conditions indicate possible deleterious action on the materials used in deep foundation elements because of soil constituents, changing water levels or other factors, the elements shall be adequately protected by materials, methods or processes *approved* by the *building official*. Protective materials shall be applied to the elements so as not to be rendered ineffective by installation. The effectiveness of such protective measures for the particular purpose shall have been thoroughly established by satisfactory service records or other evidence.

❖ Deep foundation elements are often exposed to the deteriorating effects of biological, chemical and physical actions caused by a hostile underground environment that may exist at the time of their installation or that may later develop at the site. Under such conditions, deep foundation element materials must be properly protected to ensure their expected durability.

The problems associated with element durability relate directly to the type of pile materials used. For example, concrete elements that are entirely embedded in undisturbed soil are deemed to be permanent installations. The level of the ground-water table is generally not a factor affecting the durability of concrete elements. Ground water that contains deleterious substances and readily flows through disturbed or granular soils, such as sand and gravel (regardless of the level of the water table), can have a deteriorating effect on concrete piles. Concrete elements embedded in fine-grained, impervious soils, such as clay, generally are not adversely affected by ground water containing harmful substances. Concrete can also be affected by exposure to soils having a high sulfate content, unless Type II or V portland cement is used in making the concrete mixture.

Concrete piles installed in saltwater, such as for buildings or other structures in waterfront construction, are subject to chemical action from polluted waters, frost action on porous concrete, spalling and rusting of steel reinforcement. Spalling action may become particularly critical under tidal conditions where alternate wetting and drying of the concrete occurs in conjunction with cycles of freezing and thawing. Generally, concrete can be protected from damage by such adverse conditions with the use of special cements, dense concrete mixtures rich in cement content, adequate concrete cover over the reinforcement, air entrainment, suitable concrete admixtures or special surface coatings.

Elements made of steel materials and embedded entirely in undisturbed soil (regardless of its type) are not significantly affected by corrosion due to oxidation, mainly because undisturbed soil is so deficient in oxygen that progressive corrosion is repressed. However, steel may be subject to serious corrosion and structural deterioration where ground water contains deleterious substances from coal piles, alkali soils, active cinder fills, chemical waste from manufacturing operations, etc. Under such conditions, steel piles may be protected by encasement in concrete or by applying protective coatings, such as coal-tar or other suitable materials. Steel piles installed in saltwater or exposed to a saltwater environment can corrode severely and should be protected by encasement in concrete or by the application of approved coatings. Elements that extend above ground level and are exposed to air should be painted in the same way as any type of structural steel construction to prevent rusting. Corrosion of load-bearing steel can also occur because of electrolytic action, and in such cases, cathodic protection may be required.

Timber piles totally embedded in the earth below the low point of ground-water level or entirely submerged in fresh water will last indefinitely without preservative treatment. However, timber piles that extend above the ground-water level or are exposed to air or saltwater are subject to decay as well as attacks by insects and marine borers. The piles may also be subjected to damage by the percolation of ground water heavily charged with alkali and acids. Under such conditions, timber piles must be pressure treated with preservatives in accordance with AWPA U1 listed in Chapter 35.

Generally, at any site where piles are to be installed and where the soil is suspect or there is sufficient evidence of an adverse underground environment, a soils investigation should be conducted to determine the need and method to protect the piles against possible deterioration.

1810.3.2.6 Allowable stresses. The allowable stresses for materials used in deep foundation elements shall not exceed those specified in Table 1810.3.2.6.

❖ This section refers the code user to the table of allowable stresses in order to identify the correct values that apply to various types of deep foundations. Note that Section 1810.1.4 allows "special types of piles" using the allowable stresses for materials that are specified herein.

TABLE 1810.3.2.6. See below.

❖ This table provides a complete list of the relevant allowable stresses for deep foundation element materials including concrete, reinforcing steel and structural steel.

1810.3.2.7 Increased allowable compressive stress for cased cast-in-place elements. The allowable compressive stress in the concrete shall be permitted to be increased as specified in Table 1810.3.2.6 for those portions of permanently cased cast-in-place elements that satisfy all of the following conditions:

1. The design shall not use the casing to resist any portion of the axial load imposed.
2. The casing shall have a sealed tip and be mandrel driven.
3. The thickness of the casing shall not be less than manufacturer's standard gage No.14 (0.068 inch) (1.75 mm).
4. The casing shall be seamless or provided with seams of strength equal to the basic material and be of a configuration that will provide confinement to the cast-in-place concrete.
5. The ratio of steel yield strength (F_y) to specified compressive strength (f'_c) shall not be less than six.
6. The nominal diameter of the element shall not be greater than 16 inches (406 mm).

❖ For cased cast-in-place concrete elements formed by driving permanent steel casings, the allowable design compressive stress in Table 1810.3.2.6 is generally not to exceed $0.33f'_c$. When the permanent casing complies with the requirements of this section, the allowable concrete compressive stress may, be increased to $0.40f'_c$. The basis for this increase in allowable concrete stress is the added strength given to the concrete by the confining action of the steel

TABLE 1810.3.2.6
ALLOWABLE STRESSES FOR MATERIALS USED IN DEEP FOUNDATION ELEMENTS

MATERIAL TYPE AND CONDITION	MAXIMUM ALLOWABLE STRESS[a]
1. Concrete or grout in compression[b]	
Cast-in-place with a permanent casing in accordance with Section 1810.3.2.7	$0.4f'_c$
Cast-in-place in a pipe, tube, other permanent casing or rock	$0.33f'_c$
Cast-in-place without a permanent casing	$0.3f'_c$
Precast nonprestressed	$0.33f'_c$
Precast prestressed	$0.33f'_c - 0.27f_{pc}$
2. Nonprestressed reinforcement in compression	$0.4f_y \le 30{,}000$ psi
3. Structural steel in compression	
Cores within concrete-filled pipes or tubes	$0.5\,F_y \le 32{,}000$ psi
Pipes, tubes or H-piles, where justified in accordance with Section 1810.3.2.8	$0.5\,F_y \le 32{,}000$ psi
Pipes or tubes for micropiles	$0.4\,F_y \le 32{,}000$ psi
Other pipes, tubes or H-piles	$0.35\,F_y \le 16{,}000$ psi
Helical piles	$0.6\,F_y \le 0.5\,F_u$
4. Nonprestressed reinforcement in tension	
Within micropiles	$0.6f_y$
Other conditions	$0.5f_y \le 24{,}000$ psi
5. Structural steel in tension	
Pipes, tubes or H-piles, where justified in accordance with Section 1810.3.2.8	$0.5\,F_y \le 32{,}000$ psi
Other pipes, tubes or H-piles	$0.35\,F_y \le 16{,}000$ psi
Helical piles	$0.6\,F_y \le 0.5\,F_u$
6. Timber	In accordance with the AF&PA NDS

a. f'_c is the specified compressive strength of the concrete or grout; f_{pc} is the compressive stress on the gross concrete section due to effective prestress forces only; f_y is the specified yield strength of reinforcement; F_y is the specified minimum yield stress of structural steel; F_u is the specified minimum tensile stress of structural steel.

b. The stresses specified apply to the gross cross-sectional area within the concrete surface. Where a temporary or permanent casing is used, the inside face of the casing shall be considered the concrete surface.

casing. The general formula for increased allowable stress caused by confinement is:

$$f_c 0.33 f'_c \left(\frac{1 + 7.5 t f_y}{D f'_c} \right)$$

where:

f_c = Allowable concrete stress.

f'_c = Specified concrete strength.

t = Thickness of steel shell.

f_y = Yield strength of steel.

D = Diameter of steel shell.

This formula is from the Portland Cement Association's (PCA) *Report on Allowable Stresses in Concrete Piles.*

When values for the various terms required by Items 3 through 6 are inserted in the given formula, the resulting allowable stress is $0.40f'_c$. Higher allowable stresses would result if, for example, the shell thickness was increased or the shell diameter was decreased. This increased allowable stress caused by confinement applies only to nonaxial load-bearing steel where the stress in the steel is taken in hoop tension instead of axial compression.

Steel pile shells are to be No. 14 gage (U.S. standard) or thicker, but are not to be considered in the design of the pile to carry a portion of the pile load. The equivalent thickness for No. 14 gage material is approximately 0.068 inches (1.7 mm).

The shell for this type of pile must be seamless or have spirally welded seams and be of the strength and configuration required to provide structural confinement of the concrete fill. "Confinement" is the technical qualification that permits the use of increased allowable compressive stresses. Simply stated, the pile casing restrains the concrete in directions perpendicular to the applied stresses.

Item 5 requires that the ratio of the yield strength (f_y) of the steel used in pile casings to the design compressive strength of concrete (f'_c) is not to be less than six. The yield strength of the steel used for pile casings of the type specified in this section is normally 30,000 psi (207 MPa) or greater. For example, in selecting a casing with a yield strength (f_y) of 30,000 psi (207 MPa) and a concrete compressive strength of 3,000 psi (21 MPa), the resulting ratio (f_y/f'_c) would be 10, which is greater than the minimum required ratio of six; therefore, the strengths of the pile materials are acceptable. For comparison, use the same steel casing and a specified concrete compressive strength (f'_c) of 5,000 psi (34 MPa). The resulting ratio would be exactly six. It can readily be seen that for concrete strengths greater than 5,000 psi (34 MPa), the type of steel used for the casing material would need to yield strengths greater than 30,000 psi (207 MPa). For example, in order to meet the minimum ratio of six as required by this section, if the concrete compressive strength (f'_c) was specified at 6,000 psi (41 MPa), the material to be used for the casing would require a steel yield strength of at least 36,000 psi (248 MPa) conforming to ASTM A 36

Item 6 limits the nominal diameter of the element to 16 inches (406 mm) in order to qualify for an increase in the allowable design compressive stress.

Item 1 is consistent with current practice as well as the requirements of ACI 543R. Item 2 requires a sealed tip that produces a displacement pile with increased capacity.

1810.3.2.8 Justification of higher allowable stresses. Use of allowable stresses greater than those specified in Section 1810.3.2.6 shall be permitted where supporting data justifying such higher stresses is filed with the *building official*. Such substantiating data shall include:

1. A geotechnical investigation in accordance with Section 1803; and
2. Load tests in accordance with Section 1810.3.3.1.2, regardless of the load supported by the element.

The design and installation of the deep foundation elements shall be under the direct supervision of a *registered design professional* knowledgeable in the field of soil mechanics and deep foundations who shall submit a report to the *building official* stating that the elements as installed satisfy the design criteria.

❖ In other parts of this chapter, limitations are specified for the stress values used for design purposes. These allowable stresses are stated as a percentage of some limiting strength property of the element's material. For example, allowable design stresses for deep foundation elements made of steel are stated as a percentage of the yield strengths of the several grades of steel typically used for pile construction. For concrete, the allowable design stress is prescribed as a percentage of the specified compression strength. The allowable design stresses permitted for timber elements are based on the natural strengths of the different species of wood used for deep foundations. The values have been developed and tabulated by AF&PA and include reductions in element strengths because of preservative treatment.

The allowable design stresses stipulated in the code for the different types of elements provide an adequate factor of safety against the dynamic forces of pile driving that may cause damage to the elements and prevent overstresses because of loading and subsoil conditions.

This section allows the use of higher allowable stresses when evidence supporting the values is submitted and approved by the building official. The data submitted to the building official should include analytical evaluations and findings from a foundation investigation as specified in Section 1803.5.5, and the results of load tests performed in accordance with the requirements of Section 1810.3.3.1.2. The technical data and the recommendation for the use of higher stress values must come from a registered engineer who is knowledgeable in soil mechanics and experienced in the design of pile foundations. This engineer is to supervise the deep foundation design work and witness the installation of the deep foundation so as to certify to the building official that the construction satisfies the design criteria. In any case, the use of greater design stresses is not to result in permitting design loads that are larger than one-half of the test loads (see Section 1810.3.3.1.2).

1810.3.3 Determination of allowable loads. The allowable axial and lateral loads on deep foundation elements shall be determined by an *approved* formula, load tests or method of analysis.

❖ There are two general considerations for determining capacity as required for the design and installation of deep foundations. The first consideration involves the determination of the underlying soil or rock characteristics. The second is the application of approved driving formulas, load tests or accepted methods of analysis to determine the element capacities required to resist the axial and lateral loads they will be subjected to, as well as to provide the basis for the proper selection of driving equipment.

1810.3.3.1 Allowable axial load. The allowable axial load on a deep foundation element shall be determined in accordance with Sections 1810.3.3.1.1 through 1810.3.3.1.9.

❖ This section states the criteria for determining the capacity of deep foundation elements.

1810.3.3.1.1 Driving criteria. The allowable compressive load on any driven deep foundation element where determined by the application of an *approved* driving formula shall not exceed 40 tons (356 kN). For allowable loads above 40 tons (356 kN), the wave equation method of analysis shall be used to estimate driveability for both driving stresses and net displacement per blow at the ultimate load. Allowable loads shall be verified by load tests in accordance with Section 1810.3.3.1.2. The formula or wave equation load shall be determined for gravity-drop or power-actuated hammers and the hammer energy used shall be the maximum consistent with the size, strength and weight of the driven elements. The use of a follower is permitted only with the approval of the *building official*. The introduction of fresh hammer cushion or pile cushion material just prior to final penetration is not permitted.

❖ It has been accepted practice for many decades to predict the load capacity of an element by its resistance to driving as determined by a pile-driving formula. The simple premise upon which a pile-driving formula is founded is that as the resistance of a pile to driving increases, the pile's capacity to support loads also increases. While several pile formulas have been developed over the years, none have been completely dependable.

The Engineering News formula is the simplest and probably the most widely used in the United States. This calculation method, as well as other formulas in common use today, have generally shown poor correlations with load test results. However, the comparative differences between pile capacities as determined by driving formulas and the results of load tests are much smaller for soils consisting of free-draining, coarse-grained materials, such as sand and gravel, than for soils consisting of silt, clay or fine, dense sand.

The use of pile-driving formulas to determine pile capacities should generally be avoided, except on small jobs where the piles are to be driven in well-drained granular soils, and the cost of load testing cannot be justified.

1810.3.3.1.2 Load tests. Where design compressive loads are greater than those determined using the allowable stresses specified in Section 1810.3.2.6, where the design load for any deep foundation element is in doubt, or where cast-in-place deep foundation elements have an enlarged base formed either by compacting concrete or by driving a precast base, control test elements shall be tested in accordance with ASTM D 1143 or ASTM D 4945. At least one element shall be load tested in each area of uniform subsoil conditions. Where required by the *building official*, additional elements shall be load tested where necessary to establish the safe design capacity. The resulting allowable loads shall not be more than one-half of the ultimate axial load capacity of the test element as assessed by one of the published methods listed in Section 1810.3.3.1.3 with consideration for the test type, duration and subsoil. The ultimate axial load capacity shall be determined by a *registered design professional* with consideration given to tolerable total and differential settlements at design load in accordance with Section 1810.2.3. In subsequent installation of the balance of deep foundation elements, all elements shall be deemed to have a supporting capacity equal to that of the control element where such elements are of the same type, size and relative length as the test element; are installed using the same or comparable methods and equipment as the test element; are installed in similar subsoil conditions as the test element; and, for driven elements, where the rate of penetration (e.g., net displacement per blow) of such elements is equal to or less than that of the test element driven with the same hammer through a comparable driving distance.

❖ The most reliable method for determining pile capacity is by a load test. A load test should be conducted wherever feasible and used where the pile capacity is intended to exceed 40 tons (356 kN) per pile (see Section 1810.3.3.1.1). Test piles are to be of the same type and size intended for use in the permanent foundation and installed with the same equipment, by the same procedure and in the same soils intended or specified for the work. Load tests are to be conducted in accordance with the requirements of ASTM D 1143 or D 4945, which covers procedures for testing vertical or batter foundation piles, individually or in groups, to determine the ultimate pile load (pile capacity) and whether the pile or pile group is capable of supporting the load(s) without excessive or continuous settlement. Recognition, however, must be given to the fact that load-settlement characteristics and pile capacity determinations are based on data derived at the time and under conditions of the test. The long-term performance of a pile or group of piles supporting actual loads may produce behaviors that are different than those indicated by load test results. Judgement based on experience must be used to predict pile capacity and expected behavior.

The load-bearing capacity of all piles, except those seated on rock, does not reach the ultimate load until after a period of rest. The results of load tests cannot be deemed accurate or reliable unless there is an allowance for a period of adjustment. For piles driven in permeable soils, such as coarse-grained sand and gravel, the waiting period may be as little as two or three days. For test piles driven in silt, clay or fine sand, the waiting period may be 30 days or longer. The waiting period may be determined by testing (i.e., by redriving piles) or from previous experience.

This section requires that at least one pile be tested in each area of uniform subsoil conditions. The statement should not be misconstrued to mean that the tested area is to have only one uniform stratum of subsurface material, but rather that the soil profile, which may consist of several layers (strata) of different materials, must represent a substantially unchanging cross section in each area to be tested.

The allowable pile load to be used for design purposes is not to be more than one-half of the test pile's ultimate axial load capacity, as determined in Section 1810.3.3.1.3. In establishing the pile capacity, the registered design professional must consider the tolerable settlement that can be structure dependent (see Section 1810.2.3).

The rate of penetration of production piles must be equal to or less than that of the test pile(s). All production piles should be of the same type, size and approximate length as the prototype test pile, as well as installed with comparable equipment and methods. Driven pile capacities are only valid when the same hammer is used, because different hammers of the same model, though comparable, can actually have different efficiencies. Production piles should also be installed in soils similar to those for the test pile.

1810.3.3.1.3 Load test evaluation methods. It shall be permitted to evaluate load tests of deep foundation elements using any of the following methods:

1. Davisson Offset Limit.
2. Brinch-Hansen 90% Criterion.
3. Butler-Hoy Criterion.
4. Other methods *approved* by the *building official*.

❖ This section lists generally accepted methods that can be used to determine the ultimate axial load capacity of test elements. Since no single method applies to all situations that are encountered, this list provides the necessary latitude to select a method of load test evaluation that is appropriate for the type of element being tested, the test procedure and the subsurface conditions. The Davisson Offset Limit is perhaps the most widely used method of test load evaluation. It has proven to provide capacities that are conservative, yet reasonable. The Brinch-Hansen 90% Criterion and Butler-Hoy Criterion are considered a little less conservative than Davisson, but are viable methods of evaluating test elements.

1810.3.3.1.4 Allowable frictional resistance. The assumed frictional resistance developed by any uncased cast-in-place deep foundation element shall not exceed one-sixth of the bearing value of the soil material at minimum depth as set forth in Table 1806.2, up to a maximum of 500 psf (24 kPa), unless a greater value is allowed by the *building official* on the basis of a geotechnical investigation as specified in Section 1803 or a greater value is substantiated by a load test in accordance with Section 1810.3.3.1.2. Frictional resistance and bearing resistance shall not be assumed to act simultaneously unless determined by a geotechnical investigation in accordance with Section 1803.

❖ Under certain circumstances, such as when a deep foundation element extends through cohesive soils, like clays, to a bearing stratum of compact sand and gravels, both skin friction and end bearing act together to support the pile. However, the nature of load sharing between the two and whether in fact both act simultaneously can be determined only by a soils investigation. Thus, the code requires that in order to allow the design to be based on both skin friction and end bearing acting simultaneously, the assumption must be justified by a geotechnical investigation.

1810.3.3.1.5 Uplift capacity of a single deep foundation element. Where required by the design, the uplift capacity of a single deep foundation element shall be determined by an *approved* method of analysis based on a minimum factor of safety of three or by load tests conducted in accordance with ASTM D 3689. The maximum allowable uplift load shall not exceed the ultimate load capacity as determined in Section 1810.3.3.1.2, using the results of load tests conducted in accordance with ASTM D 3689, divided by a factor of safety of two.

Exception: Where uplift is due to wind or seismic loading, the minimum factor of safety shall be two where capacity is determined by an analysis and one and one-half where capacity is determined by load tests.

❖ Deep foundation elements subjected to uplift forces act in tension and are actually friction piles. The amount of tension that can be sustained by an element depends on the strength of the element material and the frictional or cohesive properties of the soil. Tensile resistance is not necessarily correlated with the bearing capacity of a deep foundation element under compressive load. For example, the tensile resistance of a friction pile in clay will usually be about the same value as its bearing capacity because the skin friction developed in cohesive soils is very large. In comparison, a friction pile installed in granular materials (noncohesive), such as sand, will develop a tensile resistance that is considerably less than its bearing capacity.

Analytical methods can be used to determine the ultimate uplift resistance of a deep foundation element, provided that the properties of the soil are well known. When the ultimate uplift resistance is established by analysis, a safety factor of three must be applied to determine the allowable uplift load of the element.

The response of a vertical or battered element to an axially applied uplift force is best determined by an extraction test performed in accordance with the requirements of ASTM D 3689 and in accordance with the provisions of this section.

Deep foundation elements must be well anchored into the pile cap by adequate connection devices in order to be effective in resisting uplift forces. In turn, the pile cap must also be designed to resist uplift stresses. Sometimes, it is necessary to give special consideration in the design of the element itself to take the tensile stresses imposed by uplift conditions. For example, a cast-in-place or precast concrete element must be designed so that the tensile reinforcement will extend the full length of the element. Special consideration should also be given to the design of splices that must resist tension.

The exception reduces the required factor of safety for uplift due to wind or seismic loading. This is analogous to the long-standing practice of allowing stress increases for earthquake and wind loads.

1810.3.3.1.6 Uplift capacity of grouped deep foundation elements. For grouped deep foundation elements subjected to uplift, the allowable working uplift load for the group shall be calculated by an *approved* method of analysis where the deep foundation elements in the group are placed at a center-to-center spacing of at least 2.5 times the least horizontal dimension of the largest single element, the allowable working uplift load for the group is permitted to be calculated as the lesser of:

1. The proposed individual uplift working load times the number of elements in the group.
2. Two-thirds of the effective weight of the group and the soil contained within a block defined by the perimeter of the group and the length of the element, plus two-thirds of the ultimate shear resistance along the soil block.

❖ There are two limitations on the capacity of grouped deep foundation elements and the lesser of the two is taken as the allowable uplift. The first limitation is the single-element capacity as determined in the previous section multiplied by the number of elements in the group. The second of these is limited by the weight of the group plus the weight of the soil within the perimeter of the group, in addition to the soil's shear resistance that will be developed during an uplift loading event.

1810.3.3.1.7 Load-bearing capacity. Deep foundation elements shall develop ultimate load capacities of at least twice the design working loads in the designated load-bearing layers. Analysis shall show that no soil layer underlying the designated load-bearing layers causes the load-bearing capacity safety factor to be less than two.

❖ The bearing capacity of a deep foundation element, whether it is a single acting element or part of a group, is determined as a deep foundation soil system. In this respect, element bearing capacity is a function of either the strength properties of the deep foundation element or the supporting strength of the soil. Obviously, the bearing capacity is controlled by the smaller value obtained in the two considerations.

In most cases, the supporting strength of the soil governs the bearing capacity of a deep foundation element. This section requires that the bearing capacity of an individual element or group of elements must not be more than one-half of the ultimate load capacities of the elements as a function of the bearing capacity of the soil.

Sometimes, soils investigations show that weaker layers of soil underlie the intended bearing strata. To avoid damaging settlements, the weaker soils must have a safety factor of 2 or more as determined by analytical methods. Where the safety factor is less than 2, elements must either be driven to deeper bearing soils to obtain adequate and safe support or the design capacity of the elements must be reduced, thus increasing the total number of elements in the foundation system.

1810.3.3.1.8 Bent deep foundation elements. The load-bearing capacity of deep foundation elements discovered to have a sharp or sweeping bend shall be determined by an *approved* method of analysis or by load testing a representative element.

❖ This section requires that deep foundation elements that are discovered to have sharp or sweeping bends, usually occurring because of obstructions encountered during driving operations, must be analyzed by an approved method or load tested by a representative element to determine their load-carrying capacity. Where acceptable, such deep foundation elements may be used at a reduced capacity; otherwise, they should be abandoned and replaced.

1810.3.3.1.9 Helical piles. The allowable axial design load, P_a, of helical piles shall be determined as follows:

$$P_a = 0.5\ P_u \qquad \textbf{(Equation 18-4)}$$

where P_u is the least value of:

1. Sum of the areas of the helical bearing plates times the ultimate bearing capacity of the soil or rock comprising the bearing stratum.
2. Ultimate capacity determined from well-documented correlations with installation torque.
3. Ultimate capacity determined from load tests.
4. Ultimate axial capacity of pile shaft.
5. Ultimate axial capacity of pile shaft couplings.

6. Sum of the ultimate axial capacity of helical bearing plates affixed to pile.

❖ The allowable load on a helical pile is limited to one-half of the ultimate load. The ultimate load is taken as the least of the six criteria that are listed. Helical pile systems having ICC Evaluation Service reports can be viewed at www.icc-es.org (also see ICC Evaluation Service Acceptance Criteria 358).

1810.3.3.2 Allowable lateral load. Where required by the design, the lateral load capacity of a single deep foundation element or a group thereof shall be determined by an *approved* method of analysis or by lateral load tests to at least twice the proposed design working load. The resulting allowable load shall not be more than one-half of the load that produces a gross lateral movement of 1 inch (25 mm) at the lower of the top of foundation element and the ground surface, unless it can be shown that the predicted lateral movement shall cause neither harmful distortion of, nor instability in, the structure, nor cause any element to be loaded beyond its capacity.

❖ Because of wind loads, unbalanced building loads, earth pressures and the like, it is inevitable that piers, individual elements or groups of elements supporting buildings or other structures will be subjected to lateral forces. The distribution of these lateral forces to deep foundation elements largely depends on how the loads are carried down through the structural framing system and transferred through the supporting foundation to the deep foundation elements. The amount of lateral load that can be taken by the deep foundation is a function of the type of element used; the soil characteristics, particularly in the upper 10 feet (3048 mm) of the elements; the embedment of the element head (fixity); the magnitude of the axial compressive load on the deep foundation element; the nature of the lateral forces and the amount of horizontal element movement deemed acceptable.

The degree of fixity of the deep foundation element head is an important design consideration under very high lateral loading unless some other method, such as the use of batter piles, is employed to resist lateral loads. The fixing of the deep foundation element head against rotation reduces the lateral deflection. In general, pile butts are embedded 3 inches (76 mm) to 4 inches (102 mm) into the pile cap (see Section 1810.3.11) with no ties to the cap. These pile heads are neither fixed nor free, but somewhere in the middle. Such construction is satisfactory for most loading conditions.

The magnitude of friction developed between the surfaces of two structural elements in contact with each other is a function of the weight of loads applied. The larger the weight, the greater the frictional resistance developed. In the design of pile foundations, frictional resistance between the soil and the bottom of the pile caps (footings) should not be relied on to provide lateral restraint, since the vertical loads are transmitted through the elements to the supporting soil below and not to the ground immediately under the pile caps. Only the weights of the pile caps can supply some frictional resistance because such footings are constructed by placing fresh concrete on the soil, thus providing a positive contact. The weights of the pile caps in comparison to the magnitude of loads and lateral forces transmitted to the piles is nominal, however, and not too significant from a structural design standpoint. Also, in rare occurrences, soil has been known to settle under pile caps, leaving open spaces and thus eliminating the development of any frictional restraint.

Generally, about $^1/_4$-inch (6.4 mm) horizontal movement of a deep foundation element is considered acceptable without tests. Deep foundation elements with their upper sections embedded in deep strata of very soft or soft clays and silts should not be relied on to resist lateral forces of more than 1,000 pounds (4.45 kN) per element.

Where vertical elements are subjected to lateral forces exceeding acceptable limitations, the use of batter piles may be required. Lateral forces on many structures are also resisted by the embedded foundation walls and the sides of the pile caps.

The allowable lateral load capacity of a pier, single element or group of elements is to be determined either by approved analytical methods or load tests. Load tests are to be conducted to produce lateral forces that are twice the proposed design load; however, in no case is the allowable deep foundation element load to exceed one-half of the test load, which produces a gross lateral deep foundation element movement of 1 inch (25 mm) as measured at the ground surface.

1810.3.4 Subsiding soils. Where deep foundation elements are installed through subsiding fills or other subsiding strata and derive support from underlying firmer materials, consideration shall be given to the downward frictional forces that may be imposed on the elements by the subsiding upper strata.

Where the influence of subsiding fills is considered as imposing loads on the element, the allowable stresses specified in this chapter shall be permitted to be increased where satisfactory substantiating data are submitted.

❖ Where compacted fill or other surcharge loads are placed over compressible soils, the underlying material will consolidate because of the added load. The depth to suitable bearing material will, over a period of time, be shifted downward by the forces of the subsiding soil. Such forces caused by the weight of the fill are transmitted to the elements by skin friction and, in effect, serve as added loads on the elements. The magnitude of such loads must be determined by accepted analytical methods and be taken into account in the design of foundations.

1810.3.5 Dimensions of deep foundation elements. The dimensions of deep foundation elements shall be in accordance with Sections 1810.3.5.1 through 1810.3.5.3, as applicable.

❖ This section groups together any dimensional limitations that apply to various types of elements.

1810.3.5.1 Precast. The minimum lateral dimension of precast concrete deep foundation elements shall be 8 inches (203 mm). Corners of square elements shall be chamfered.

❖ This section prescribes the minimum dimension for precast concrete elements based on the size required to withstand the driving operation. Chamfered corners consist of rounding off or smoothing the corners of a square pile. The triangular portions of the corners are typically weak spots and chamfering them reduces the risk of concrete spalling, cracking or breakage during driving.

1810.3.5.2 Cast-in-place or grouted-in-place. Cast-in-place and grouted-in-place deep foundation elements shall satisfy the requirements of this section.

❖ Dimensional limitations applicable to deep foundation elements that are cast-in-place are dependent on whether the element is cased or uncased.

1810.3.5.2.1 Cased. Cast-in-place deep foundation elements with a permanent casing shall have a nominal outside diameter of not less than 8 inches (203 mm).

❖ Steel-cased piles are the most widely used type of cast-in-place concrete deep foundation element. Essentially, they consist of mandrel-driven, light-gage steel shells or thin-walled pipes that are left permanently in place, reinforced as required by the design and filled with concrete. The shell is either a constant section for the full length of the element or a step-tapered shape. The presence of the casing permits a higher allowable stress in the concrete than for an uncased pile (see Table 1810.3.2.6).

1810.3.5.2.2 Uncased. Cast-in-place deep foundation elements without a permanent casing shall have a diameter of not less than 12 inches (305 mm). The element length shall not exceed 30 times the average diameter.

> **Exception:** The length of the element is permitted to exceed 30 times the diameter, provided the design and installation of the deep foundations are under the direct supervision of a *registered design professional* knowledgeable in the field of soil mechanics and deep foundations. The *registered design professional* shall submit a report to the *building official* stating that the elements were installed in compliance with the *approved construction documents*.

❖ The dimensional relationship between the diameter and length of a deep foundation element has been utilized for many decades and is based on the premise that an element under axial load behaves somewhat as a column that could be susceptible to buckling. On the other hand, it has also been established through technological advancements, research and experience that deep foundation elements embedded in soils, even in soft materials, do not behave as free-standing columns and the risk of buckling is extremely low. Notwithstanding this kind of evidence, the code limitations placed on dimensional requirements based on diameter-to-length ratios have become accepted practices.

1810.3.5.2.3 Micropiles. Micropiles shall have an outside diameter of 12 inches (305 mm) or less. The minimum diameter set forth elsewhere in Section 1810.3.5 shall not apply to micropiles.

❖ The micropile does not have a minimum required size—only the maximum diameter is specified.

1810.3.5.3 Steel. Steel deep foundation elements shall satisfy the requirements of this section.

❖ These dimensional limits are established for steel H-piles, pipes and tubes.

1810.3.5.3.1 H-piles. Sections of H-piles shall comply with the following:

1. The flange projections shall not exceed 14 times the minimum thickness of metal in either the flange or the web and the flange widths shall not be less than 80 percent of the depth of the section.
2. The nominal depth in the direction of the web shall not be less than 8 inches (203 mm).
3. Flanges and web shall have a minimum nominal thickness of $^{3}/_{8}$ inch (9.5 mm).

❖ Dimensional requirements for manufacturing structural steel H-piles are provided in this section. H-piles are proportioned to withstand the large impact stresses imposed on piles during hard driving. The thicknesses of the flanges and the web of a rolled-steel H-pile section are made equal in order to avoid damage that could occur during hard driving if the piles were proportioned with a mixture of thick and thinner parts. Flange widths are proportioned in relation to the depth of the pile section to provide rigidity in the weak axis.

It would seem unnecessary to repeat the dimensioning requirements for H-piles in the code, since they were originally created by the steel industry and have for decades been the industry standard used in the manufacture of hot-rolled steel shapes. The main purpose is to provide the dimensional basis for the fabrication of similar pile products made principally of welded steel plates and other necessary steel parts.

While it is the general preference and practice to use rolled-steel piles, occasionally it becomes necessary to fabricate special pile sections because of time problems imposed by mill scheduling and delivery; a special need for heavier H-pile sections than are customarily available; an immediate need for replacement piles; or for any other reason. The dimensional requirements contained in this section regulate the fabrication of such special piles and provide the building official with a basis for approval. Fabricated pile materials are to comply with the requirements of Section 1810.3.2.3.

This section requires that the flange projection not exceed 14 times the minimum thickness of metal in either the flange or the web. The measurement of the flange projection is shown in Figure 1810.3.5.3.1 and can be calculated by the indicated formula:

$$P = \frac{W - t_w}{2} \text{ or } \frac{W - t_f}{2}$$

where W is the flange width, tw is the web thickness, tf is the flange thickness and P is the flange projection.

For example, the dimensions of an HP 14 x 73 pile section are as follows:

The width of flange, W = 14.585 inches.
The thickness of the flange, t_f = 0.505 inch.
The thickness of the web, t_w = .505 inch.
Therefore, the actual flange projection is:

$$P = \frac{14.585 - 0.505}{2} = 7.04 \text{ inches (179mm)}$$

The maximum allowable flange projection is 14 × t_f or 14 × t_w (whichever is smaller).

Allowable flange projection: 14 × 0.505 = 7.07 inches (179.5 mm), which is greater than the actual flange projection of 7.04 inches (178.8 mm); therefore, an HP 14 × 73 pile is acceptable.

This section also specifies that flange widths are not to be less than 80 percent of the depth of the pile section. In actuality, practically all H-piles are manufactured so that their flanges are slightly greater in width (by fractions of an inch) than the depths of the sections. In fact, piles are squared so that their flange widths are about equal to their depth.

Flange and web thickness is not to be less than $^3/_8$ inch (9.5 mm). In H-piles, the flange thicknesses are equal to the web thickness.

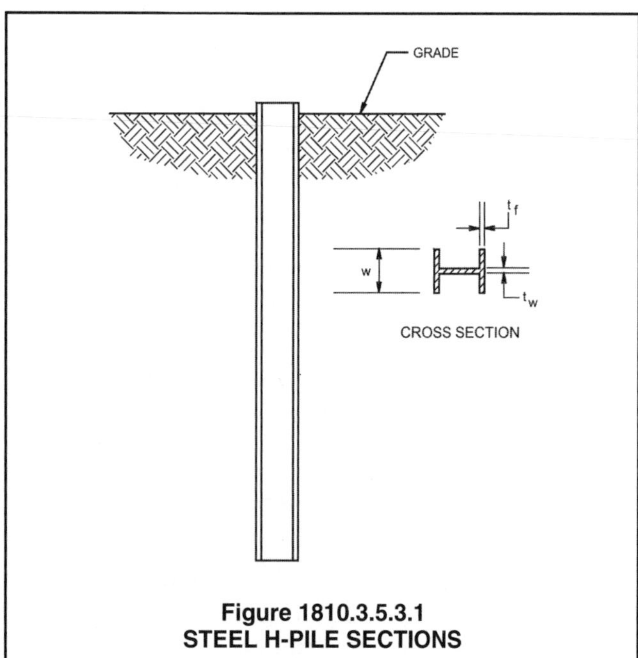

Figure 1810.3.5.3.1
STEEL H-PILE SECTIONS

1810.3.5.3.2 Steel pipes and tubes. Steel pipes and tubes used as deep foundation elements shall have a nominal outside diameter of not less than 8 inches (203 mm). Where steel pipes or tubes are driven open ended, they shall have a minimum of 0.34 square inch (219 mm^2) of steel in cross section to resist each 1,000 foot-pounds (1356 Nm) of pile hammer energy, or shall have the equivalent strength for steels having a yield strength greater than 35,000 psi (241 MPa) or the wave equation analysis shall be permitted to be used to assess compression stresses induced by driving to evaluate if the pile section is appropriate for the selected hammer. Where a pipe or tube with wall thickness less than 0.179 inch (4.6 mm) is driven open ended, a suitable cutting shoe shall be provided. Concrete-filled steel pipes or tubes in structures assigned to *Seismic Design Category* C, D, E or F shall have a wall thickness of not less than $^3/_{16}$ inch (5 mm). The pipe or tube casing for socketed drilled shafts shall have a nominal outside diameter of not less than 18 inches (457 mm), a wall thickness of not less than $^3/_8$ inch (9.5 mm) and a suitable steel driving shoe welded to the bottom; the diameter of the rock socket shall be approximately equal to the inside diameter of the casing.

Exceptions:

1. There is no minimum diameter for steel pipes or tubes used in micropiles.
2. For mandrel-driven pipes or tubes, the minimum wall thickness shall be $^1/_{10}$ inch (2.5 mm).

❖ Because of their uniform cross section from butt to tip, steel pipe elements provide unvarying resistance to bending and lateral forces applied in any direction.

This section requires that seamless or welded-steel pipe and tube elements have a nominal outside diameter of no less than 8 inches (203 mm). The 8-inch (203 mm) minimum diameter is necessary in order to inspect the inside of the pile to observe any damage that may have occurred during the driving process. Smaller diameter pipes could allow soil to bridge the sides and cause plugging at their tips. The minimum diameter also provides stiffness for driving purposes.

A pipe to be driven open ended must be a minimum cross-sectional area which can be determined by one of two methods. The first is a long-standing rule of thumb that relates pile area to the hammer energy. The second approach recognizes the more up-to-date wave equation. The latter approach is commonly used to evaluate the induced driving stresses in piles.

Exception 1 clarifies that the minimum diameter is not applicable to micropiles. Exception 2 permits a thinner pipe thickness when it is mandrel-driven.

1810.3.5.3.3 Helical piles. Dimensions of the central shaft and the number, size and thickness of helical bearing plates shall be sufficient to support the design loads.

❖ The dimensions of the helical pile component are not explicitly limited. Rather the code merely requires that the dimensions are established by the required capacity. Helical pile systems having ICC Evaluation Service reports can be viewed at www.icc-es.org

(also see ICC Evaluation Service Acceptance Criteria 358).

1810.3.6 Splices. Splices shall be constructed so as to provide and maintain true alignment and position of the component parts of the deep foundation element during installation and subsequent thereto and shall be designed to resist the axial and shear forces and moments occurring at the location of the splice during driving and for design load combinations. Where deep foundation elements of the same type are being spliced, splices shall develop not less than 50 percent of the bending strength of the weaker section. Where deep foundation elements of different materials or different types are being spliced, splices shall develop the full compressive strength and not less than 50 percent of the tension and bending strength of the weaker section. Where structural steel cores are to be spliced, the ends shall be milled or ground to provide full contact and shall be full-depth welded.

Splices occurring in the upper 10 feet (3048 mm) of the embedded portion of an element shall be designed to resist at allowable stresses the moment and shear that would result from an assumed eccentricity of the axial load of 3 inches (76 mm), or the element shall be braced in accordance with Section 1810.2.2 to other deep foundation elements that do not have splices in the upper 10 feet (3048 mm) of embedment.

❖ While it is physically and economically better to drive deep foundation elements in one piece, site conditions sometimes necessitate that elements be driven in spliced sections. For example, when the soil or rock- bearing stratum is located so deep below the ground that the leads on the driving equipment will not receive full-length elements, it becomes necessary to install the elements sectionally or, where possible, to take up the extra length by setting the tip in a preexcavated hole (see commentary, Section 1810.4.4). When elements are installed in areas such as existing buildings with restricted headroom, they are also required to be placed in spliced sections. There are a number of other reasons for field-splicing elements, such as restrictions on shipping lengths, the use of composite elements, etc.

This provision requires that splices be constructed so as to provide and maintain true alignment and position of the element sections during installation. Furthermore, splices must be of sufficient strength to transmit safely the vertical and lateral loads on the elements, as well as to resist the bending stresses that may occur at splice locations during the driving operations and under long-term service loads. Where the sections being spliced are the same type of element, splices are to develop at least 50 percent of the bending strength of the weaker section.

There are different methods employed in splicing elements based on the different materials used in pile construction. For example, timber piles are spliced by one of two commonly used methods. The first method uses a pipe sleeve with a length of about four to five times the diameter of the pile. The butting ends of the pile are sawn square for full contact of the two pile sections, and the spliced portions of the timber pile are trimmed smoothly around their periphery so as to fit tightly into the pipe sleeve. The other splicing method involves the use of steel straps and bolts. The butting ends of the pile sections are sawn square for full contact and proper alignment and the four sides are planed flat to receive the splicing straps. This type of splicing can resist some uplift forces.

Splicing of precast concrete piles usually occurs at the head portions of the piles where, after the piles are driven to their required depth, pile heads are cut off or spliced to the desired elevation for proper embedment in the concrete pile caps. Any portion of the pile that is cracked or shattered caused by the driving operations or cutting off of pile heads should be removed and spliced with fresh concrete. To cut off a precast concrete pile section, a deep groove is chiseled around the pile exposing the reinforcing bars, which are then cut off (by torch) to desired heights or extensions. The pile section above the groove is snapped off (by crane) and a new pile section is freshly cast to tie in with the precast pile.

Steel H-piles are spliced in the same manner as steel columns, normally by welding the sections together. Welded splices may be welded-plate or bar splices, butt-welded splices, special welded splice fittings or a combination of these. Spliced materials should be kept on the inner faces of the H-pile sections to avoid forcing a hole in the ground larger than the pile, causing at least a temporary loss in frictional value and lateral support that might result in excessive bending stresses.

Steel pipe piles may be spliced by butt welding, sometimes using straps to guide the sections and provide more strength to the welded joint. Another method is to use inside sleeves having a driving fit, with a flange extending between the pipe sections. By applying bituminous cement or compound on the outside of the ring before driving, a water-tight joint is obtained.

The strength of a composite pile is governed not only by the weaker section, but also by the strength and details of the splice that joins and holds the pile sections together.

Guidance is also provided for splicing sections of different element types—i.e., a composite element. There are several problems associated with the splicing of sections comprising a composite pile. For example, if the length of a composite pile is of such dimension that it will fit in the leaders of the driving equipment, then the full length of the spliced pile can be driven as a continuous operation. However, if the pile is too long to fit in the leaders as a single unit, the pile must be installed in sections, and the driving would have to be interrupted in order to make the splice in the leaders.

It is most important that splicing devices be made in such a way that their installation will be simple and quick. When pile driving must be stopped to make a splice in the leaders, the time expended should not be of a duration that would allow the soil to set up (soil freeze) so that continued driving would produce excessive stresses in the pile.

Sometimes when the predetermined length of a composite pile is too long to fit in the leaders, but not overly long, the difference in length between the pile and the leaders may be made up by inserting the lower end of the spliced unit in a preexcavated hole, thus allowing for continuous driving of the pile (see Section 1810.4.4).

Another problem related to splicing occurs in trying to accurately align the pile sections. When the lower section of a composite pile is driven nearly full length into the ground, it is very difficult to determine its true direction, particularly if the pile has drifted off line because of soil pressures, subsurface obstructions, improper driving or for any other reason. When a pile section is connected to a section already driven in the ground, the upper section should be installed in the same direction as the longitudinal axis of the lower section, even though the direction may not be vertical. It is better to maintain a misdirection rather than to try to correct a situation and create a bend at the pile joint.

The selection of the splicing method to be used for assembling a composite pile should be based on the driving and service load stresses that must be resisted by the pile. Splices should be designed to prevent separation of the pile sections during construction and thereafter. Special consideration must be given to the design of splices to resist uplift forces, whether the piles are subject to heaving or specifically designed as tension piles. Designing the splice strong enough to develop about one and one-half to two times the design uplift force is proper engineering practice. Pile splices must also be designed to resist compressive, bending and shear stresses imposed by construction and service loads.

Additionally, splices that occur in the upper 10 feet (3048 mm) of element embedment are to be designed to resist the bending moments and shears at the allowable stress levels of the element material, based on an assumed element load eccentricity of 3 inches (76 mm), unless the element is properly braced. Proper bracing of a spliced element is deemed to exist if stability of the element group is present in accordance with Section 1810.2.2, provided that the other elements in the group do not have splices in the upper 10 feet (3048 mm) of their embedded length.

1810.3.6.1 Seismic Design Categories C through F. For structures assigned to *Seismic Design Category* C, D, E or F splices of deep foundation elements shall develop the lesser of the following:

1. The nominal strength of the deep foundation element; and
2. The axial and shear forces and moments from the seismic load effects including overstrength factor in accordance with Section 12.4.3 or 12.14.3.2 of ASCE 7.

❖ More stringent minimum strength requirements are stated in this section that apply to structures with moderate to high seismic risk.

1810.3.7 Top of element detailing at cutoffs. Where a minimum length for reinforcement or the extent of closely spaced confinement reinforcement is specified at the top of a deep foundation element, provisions shall be made so that those specified lengths or extents are maintained after cutoff.

❖ This section accounts for the condition where an element encounters refusal at a shallower depth than intended and an unanticipated portion of the element is cut off. It is imperative that the required reinforcement be provided at the top of the element even when excess length is cut off.

1810.3.8 Precast concrete piles. Precast concrete piles shall be designed and detailed in accordance with Sections 1810.3.8.1 through 1810.3.8.3.

❖ Precast concrete piles are manufactured as nonprestressed (conventionally reinforced) or prestressed. Both types can be formed by bedcasting, spinning (centrifugal casting), vertical casting, slip forming or extrusion methods. They are usually made in square, octagonal or round shapes (see Figure 1810.3.8). Precast piles are manufactured as solid units or may be made with a hollow core. They can also be made with internal jet pipes or inspection ducts.

Precast piles are generally ordered in predetermined lengths based on the findings and analysis of soil exploratory work conducted at the project site.

This type of element is a displacement pile that is normally installed with pile-driving equipment.

Precast concrete piles are to be designed in accordance with ACI 318. Nonprestressed concrete piles are usually considered for lengths of 40 to 50 feet (12 192 to 15 240 mm), while prestressed concrete piles are usually considered for lengths of 60 to 100 feet (18 288 to 30 480 mm). The general loading range for precast concrete piles is 40 to 400 tons (355 to 3558 kN).

Advantages of using precast concrete piles include their high load capacities and corrosion resistance. Disadvantages include their vulnerability to damage associated with handling, high breakage rates, particularly when spliced, and the considerable displacement of soil.

1810.3.8.1 Reinforcement. Longitudinal steel shall be arranged in a symmetrical pattern and be laterally tied with steel ties or wire spiral spaced center to center as follows:

1. At not more than 1 inch (25 mm) for the first five ties or spirals at each end; then
2. At not more than 4 inches (102 mm), for the remainder of the first 2 feet (610 mm) from each end; and then

3. At not more than 6 inches (152 mm) elsewhere.

The size of ties and spirals shall be as follows:

1. For piles having a least horizontal dimension of 16 inches (406 mm) or less, wire shall not be smaller than 0.22 inch (5.6 mm) (No. 5 gage).
2. For piles having a least horizontal dimension of more than 16 inches (406 mm) and less than 20 inches (508 mm), wire shall not be smaller than 0.238 inch (6 mm) (No. 4 gage).
3. For piles having a least horizontal dimension of 20 inches (508 mm) and larger, wire shall not be smaller than $^1/_4$ inch (6.4 mm) round or 0.259 inch (6.6 mm) (No. 3 gage).

❖ For precast piles, the longitudinal steel must be set in a symmetrical arrangement in the pile. To resist high-impact stresses, lateral steel ties used for confining the longitudinal reinforcement are to be provided at each end of the pile for a distance of 2 feet (610 mm) or more and be closely spaced at 4 inches (102 mm) o.c., except that the first five ties from each end are to be spaced at 1-inch (25 mm) centers. Between these two closely tied ends, the longitudinal steel must be similarly tied at spacings not to exceed 6 inches (152 mm) o.c. so as to provide a reinforcing cage that will keep the pile from buckling or cracking during handling.

1810.3.8.2 Precast nonprestressed piles. Precast nonprestressed concrete piles shall comply with the requirements of Sections 1810.3.8.2.1 through 1810.3.8.2.3.

❖ Precast nonprestressed (conventionally reinforced) concrete piles are manufactured from concrete and have reinforcement consisting of a steel reinforcing cage made up of several longitudinal bars or tie steel in the form of individual hoops or spirals.

1810.3.8.2.1 Minimum reinforcement. Longitudinal reinforcement shall consist of at least four bars with a minimum longitudinal reinforcement ratio of 0.008.

❖ This section specifies the minimum amount of longitudinal reinforcement where seismic effects are minimal or low. See Sections 1810.3.8.2.2 and 1810.3.8.2.3 for reinforcement requirements for moderate and high seismic regions.

1810.3.8.2.2 Seismic reinforcement in Seismic Design Categories C through F. For structures assigned to *Seismic Design Category* C, D, E or F, precast nonprestressed piles shall be reinforced as specified in this section. The minimum longitudinal reinforcement ratio shall be 0.01 throughout the length. Transverse reinforcement shall consist of closed ties or spirals with a minimum $^3/_8$ inch (9.5 mm) diameter. Spacing of transverse reinforcement shall not exceed the smaller of eight times the diameter of the smallest longitudinal bar or 6 inches (152 mm) within a distance of three times the least pile dimension from the bottom of the pile cap. Spacing of transverse reinforcement shall not exceed 6 inches (152 mm) throughout the remainder of the pile.

❖ This section includes moderately ductile detailing requirements for precast nonprestressed piles in buildings assigned to Seismic Design Categories C and higher. The minimum longitudinal and transverse reinforcement requirements are so that piles will be able to accommodate seismically induced ground deformations. The 1-percent minimum longitudinal reinforcement is a standard requirement for rein-

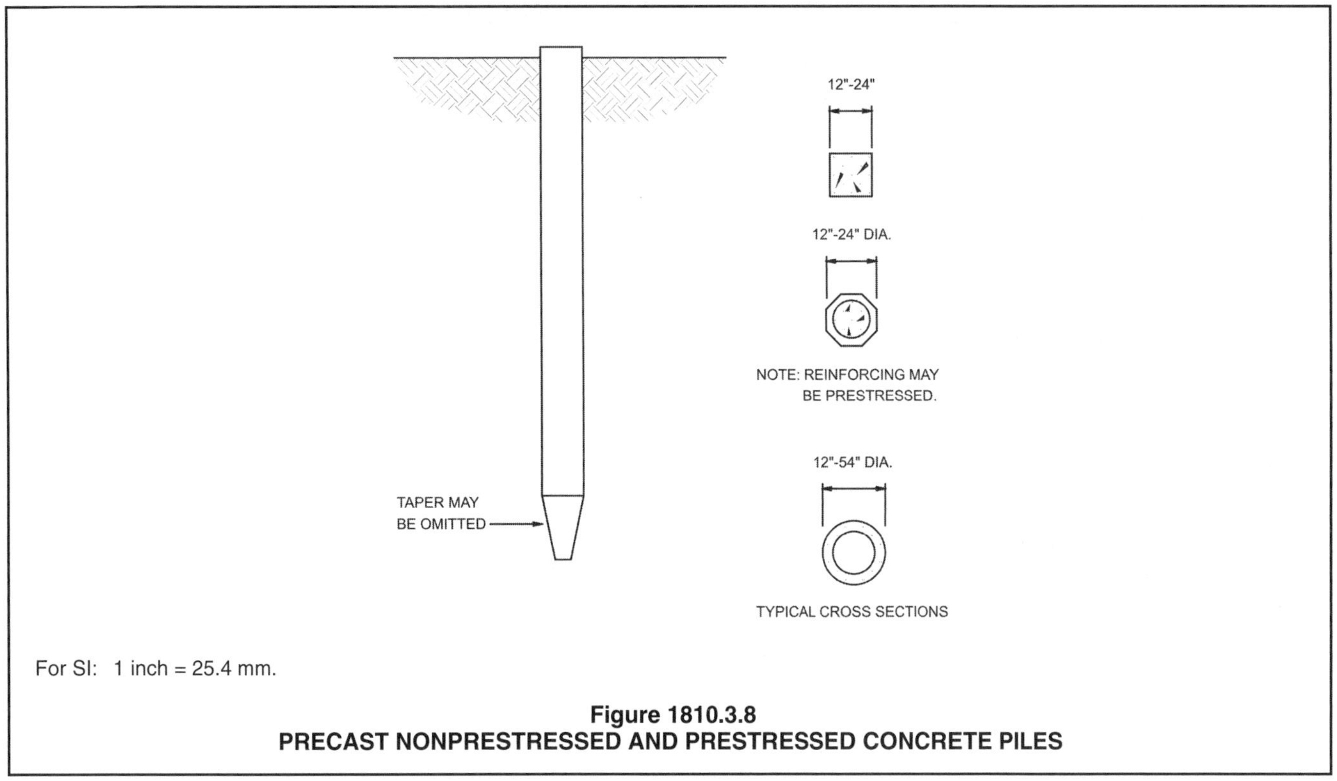

Figure 1810.3.8
PRECAST NONPRESTRESSED AND PRESTRESSED CONCRETE PILES

forced concrete columns. A 6-bar-diameter or 6-inch (152 mm) spacing of transverse reinforcement is a fairly common requirement to prevent buckling of longitudinal compression reinforcement. The transverse reinforcement spacing requirement of this section for the confinement region of the element adjacent to the pile cap is somewhat less stringent than that, allowing an 8-bar-diameter or 6-inch (152 mm) spacing. Outside of this region, only the 6-inch spacing applies as already required by Section 1810.3.8.1.

1810.3.8.2.3 Additional seismic reinforcement in Seismic Design Categories D through F. For structures assigned to *Seismic Design Category* D, E or F, transverse reinforcement shall be in accordance with Section 1810.3.9.4.2.

❖ The confinement region of the element requires the more stringent and demanding detailing provided in the referenced sections of ACI 318, which are the requirements for special moment frames. These increased transverse reinforcement requirements are intended to provide a high degree of ductility in the upper portion (confinement region) of the element. Experience has shown that concrete elements tend to hinge or sustain damage immediately below the pile cap; therefore, tie spacing is reduced in this area to better confine the concrete.

1810.3.8.3 Precast prestressed piles. Precast prestressed concrete piles shall comply with the requirements of Sections 1810.3.8.3.1 through 1810.3.8.3.3.

❖ Precast prestressed concrete piles are either of the pretensioned or post-tensioned type. Pretensioned type of piles are normally cast in a plant in their full lengths, as predetermined by soils investigations and engineering analyses. The reinforcement in the concrete pile consists of tendons (stressing steel) that are tensioned before the concrete is placed. After casting and when the concrete has attained sufficient strength, the stretched tendons are released from their anchorage, thus placing the pile in continuous compression. In comparison, post-tensioned type of piles are made in a plant or on the job site, and the tendons are released after the concrete has hardened. In addition to the tendons, this type of pile also contains mild reinforcing steel to resist the handling stresses before the stressing steel is tensioned.

One of the primary advantages of prestressed over nonprestressed concrete piles is durability. Since the concrete is under continuous compression, hairline cracks are kept tightly closed and, thus, are more durable. Another advantage is that the tensile stresses that can develop in the concrete under certain driving conditions are less critical. Prestressed concrete piles are best suited for friction piles in sand, gravel and clays.

The purpose for prestressing concrete piles is to place the concrete under continuous compression so that any hairline cracks that may develop will be kept tightly closed and to prevent possible injury to the pile from tensile stresses that may occur during installation operations. The handling and driving of prestressed piles do not require the same degree of care as needed to install conventionally reinforced concrete piles. As a general rule, prestressed piles are more durable than precast reinforced piles.

1810.3.8.3.1 Effective prestress. The effective prestress in the pile shall not be less than 400 psi (2.76 MPa) for piles up to 30 feet (9144 mm) in length, 550 psi (3.79 MPa) for piles up to 50 feet (15 240 mm) in length and 700 psi (4.83 MPa) for piles greater than 50 feet (15 240 mm) in length.

Effective prestress shall be based on an assumed loss of 30,000 psi (207 MPa) in the prestressing steel. The tensile stress in the prestressing steel shall not exceed the values specified in ACI 318.

❖ For prestressed concrete piles, the effective prestress (the stress remaining in the pile after all losses have occurred—excluding the effect of superimposed loads and the weight of the pile) is not to be less than the specified values for the various lengths. Experience has shown that an effective prestress less than the minimum value prescribed herein is sometimes inadequate in preventing or controlling cracking of the concrete during the handling and installation operations. This section also requires the effective prestress to be based on an assumed loss of 30,000 psi (207 MPa) in the prestressing steel and the tensile stresses in the steel not to exceed the values set forth in ACI 318.

1810.3.8.3.2 Seismic reinforcement in Seismic Design Category C. For structures assigned to *Seismic Design Category* C, precast prestressed piles shall have transverse reinforcement in accordance with this section. The volumetric ratio of spiral reinforcement shall not be less than the amount required by the following formula for the upper 20 feet (6096 mm) of the pile.

$$\rho_s = 0.12 f'_c / f_{yh} \qquad \textbf{(Equation 18-5)}$$

where:

f'_c = Specified compressive strength of concrete, psi (MPa).

f_{yh} = Yield strength of spiral reinforcement $\leq$ 85,000 psi (586 MPa).

ρ_s = Spiral reinforcement index (vol. spiral/vol. core).

At least one-half the volumetric ratio required by Equation 18-5 shall be provided below the upper 20 feet (6096 mm) of the pile.

❖ This section gives requirements for precast prestressed piles supporting structures assigned to Seismic Design Category C. The minimum spiral reinforcement requirement that results in ductile prestressed concrete piles is based on Precast/Prestressed Concrete Institute's (PCI) *Recommended Practice for Design, Manufacture and Installation of Prestressed Concrete Piling*. This document incorporates work done on piles in the United States and New Zealand.

These piles exhibit larger curvatures in the top 20 feet (6096 mm). This section requires (confinement) spiral reinforcement as determined by Equation 18-5. Based on ACI 318, this formula is deemed to provide for moderate ductility. In the lower portion of the element, the required reinforcing is reduced by one-half.

1810.3.8.3.3 Seismic reinforcement in Seismic Design Categories D through F. For structures assigned to *Seismic Design Category* D, E or F, precast prestressed piles shall have transverse reinforcement in accordance with the following:

1. Requirements in ACI 318, Chapter 21, need not apply, unless specifically referenced.
2. Where the total pile length in the soil is 35 feet (10 668 mm) or less, the lateral transverse reinforcement in the ductile region shall occur through the length of the pile. Where the pile length exceeds 35 feet (10 668 mm), the ductile pile region shall be taken as the greater of 35 feet (10 668 mm) or the distance from the underside of the pile cap to the point of zero curvature plus three times the least pile dimension.
3. In the ductile region, the center-to-center spacing of the spirals or hoop reinforcement shall not exceed one-fifth of the least pile dimension, six times the diameter of the longitudinal strand or 8 inches (203 mm), whichever is smallest.
4. Circular spiral reinforcement shall be spliced by lapping one full turn and bending the end of each spiral to a 90-degree hook or by use of a mechanical or welded splice complying with Section 12.14.3 of ACI 318.
5. Where the transverse reinforcement consists of circular spirals, the volumetric ratio of spiral transverse reinforcement in the ductile region shall comply with the following:

$$\rho_s = 0.25(f'_c/f_{yh})(A_g/A_{ch} - 1.0)[0.5 + 1.4P/(f'_c A_g)]$$

(Equation 18-6)

but not less than

$$\rho_s = 0.12(f'_c/f_{yh})[0.5 + 1.4P/(f'_c A_g)]\ ^3\ 0.12f'_c/f_{yh}$$

(Equation 18-7)

and need not exceed:

$$\rho_s = 0.021$$

(Equation 18-8)

where:

A_g = Pile cross-sectional area, square inches (mm^2).

A_{ch} = Core area defined by spiral outside diameter, square inches (mm^2).

f'_c = Specified compressive strength of concrete, psi (MPa).

f_{yh} = Yield strength of spiral reinforcement ≤ 85,000 psi (586 MPa).

P = Axial load on pile, pounds (kN), as determined from Equations 16-5 and 16-7.

ρ_s = Volumetric ratio (vol. spiral/vol. core).

This required amount of spiral reinforcement is permitted to be obtained by providing an inner and outer spiral.

6. Where transverse reinforcement consists of rectangular hoops and cross ties, the total cross-sectional area of lateral transverse reinforcement in the ductile region with spacing, s, and perpendicular dimension, h_c, shall conform to:

$$A_{sh} = 0.3s\,h_c(f'_c/f_{yh})(A_g/A_{ch} - 1.0)[0.5 + 1.4P/(f'_c A_g)]$$

Equation 18-9)

but not less than:

$$A_{sh} = 0.12s\,h_c(f'_c/f_{yh})[0.5 + 1.4P/(f'_c A_g)]$$

(Equation 18-10)

where:

f_{yh} = yield strength of transverse reinforcement ≤ 70,000 psi (483 MPa).

h_c = Cross-sectional dimension of pile core measured center to center of hoop reinforcement, inch (mm).

s = Spacing of transverse reinforcement measured along length of pile, inch (mm).

A_{sh} = Cross-sectional area of tranverse reinforce-ment, square inches (mm^2).

f'_c = Specified compressive strength of concrete, psi (MPa).

The hoops and cross ties shall be equivalent to deformed bars not less than No. 3 in size. Rectangular hoop ends shall terminate at a corner with seismic hooks.

Outside of the length of the pile requiring transverse confinement reinforcing, the spiral or hoop reinforcing with a volumetric ratio not less than one-half of that required for transverse confinement reinforcing shall be provided.

❖ This section clarifies that Chapter 21 of ACI 318 does not generally apply to precast prestressed concrete elements unless it is explicitly referenced. That ACI 318 chapter was never intended for deep foundation elements.

The provisions found in this section are based on PCI's *Recommended Practice for Design, Manufacture and Installation of Prestressed Concrete Piling*. This document incorporates work done on piles in the United States and New Zealand.

The maximum volumetric ratio set forth in Equation 18-8 is based on testing that has shown the 0.021 maximum is sufficient for the smaller square precast prestressed concrete elements to perform in a ductile manner.

1810.3.9 Cast-in-place deep foundations. Cast-in-place deep foundation elements shall be designed and detailed in accordance with Sections 1810.3.9.1 through 1810.3.9.6.

❖ This section sets forth the design and detailing requirements for cast-in-place concrete deep foundation elements.

1810.3.9.1 Design cracking moment. The design cracking moment (ϕM_n) for a cast-in-place deep foundation element not enclosed by a structural steel pipe or tube shall be determined using the following equation:

$$\varphi M_n = 3\sqrt{f'_c} S_m \quad \textbf{(Equation 18-11)}$$

For SI: $\varphi M_n = 0.25\sqrt{f'_c} S_m$

where:

f'_c = Specified compressive strength of concrete or grout, psi (MPa).

S_m = Elastic section modulus, neglecting reinforcement and casing, cubic inches (mm^3).

❖ For both uncased and cased cast-in-place deep foundation elements (but not concrete-filled pipes and tubes), reinforcement must be provided where moments exceed a reasonable lower boundary for the capacity of the plain concrete section. This criterion is consistent with ACI 318.

1810.3.9.2 Required reinforcement. Where subject to uplift or where the required moment strength determined using the load combinations of Section 1605.2 exceeds the design cracking moment determined in accordance with Section 1810.3.9.1, cast-in-place deep foundations not enclosed by a structural steel pipe or tube shall be reinforced.

❖ This section lists two conditions under which reinforcing must be provided in cast-in-place elements. Note the cracking moment is checked using the strength level load effects.

1810.3.9.3 Placement of reinforcement. Reinforcement where required shall be assembled and tied together and shall be placed in the deep foundation element as a unit before the reinforced portion of the element is filled with concrete.

Exceptions:

1. Steel dowels embedded 5 feet (1524 mm) or less shall be permitted to be placed after concreting, while the concrete is still in a semifluid state.
2. For deep foundation elements installed with a hollow-stem auger, tied reinforcement shall be placed after elements are concreted, while the concrete is still in a semifluid state. Longitudinal reinforcement without lateral ties shall be placed either through the hollow stem of the auger prior to concreting or after concreting, while the concrete is still in a semifluid state.
3. For Group R-3 and U occupancies not exceeding two stories of light-frame construction, reinforcement is permitted to be placed after concreting, while the concrete is still in a semifluid state, and the concrete cover requirement is permitted to be reduced to 2 inches (51 mm), provided the construction method can be demonstrated to the satisfaction of the *building official.*

❖ Main reinforcement consisting of a cage of longitudinal deformed reinforcing bars tied together with individual steel hoops or a spiral must be placed and securely held in position in the pile opening (cased or uncased) prior to the placement of concrete. Reinforcement required for resisting tensile stresses imposed by uplift forces may consist of a single bar or other structural steel unit or a cluster of bars placed at the center for the full length of the element.

Exception 1 recognizes the typical procedure for constructing cast-in-place concrete elements is to tie the head of the pile with the pile cap, inserting dowels in the freshly placed concrete pile to obtain an embedment of about 5 feet (1524 mm).

Exception 2 recognizes this restriction is not applicable to auger-injected concrete piles. Unlike drilled cast-in-place concrete piles, caged reinforcement for auger-placed piles cannot be installed prior to filling the pile hole with concrete because the hollow-stem auger must be positioned in the hole at all times during drilling and concreting operations. To facilitate this, the cage must be pushed through the concrete in fluid form after the auger has been withdrawn. The problem associated with the placement of caged reinforcement in this way, particularly where it involves long cages, is that it cannot be determined if the assembly has been positioned appropriately within the filled hole such that the reinforcement will have the required minimum cover at all places.

Exception 3 exempts up to two stories of light-frame construction in Group R-3 and U occupancies.

1810.3.9.4 Seismic reinforcement. Where a structure is assigned to *Seismic Design Category* C, reinforcement shall be provided in accordance with Section 1810.3.9.4.1. Where a structure is assigned to *Seismic Design Category* D, E or F, reinforcement shall be provided in accordance with Section 1810.3.9.4.2.

Exceptions:

1. Isolated deep foundation elements supporting posts of Group R-3 and U occupancies not exceeding two stories of light-frame construction shall be permitted to be reinforced as required by rational analysis but with not less than one No. 4 bar, without ties or spirals, where detailed so the element is not subject to lateral loads and the soil provides adequate lateral support in accordance with Section 1810.2.1.
2. Isolated deep foundation elements supporting posts and bracing from decks and patios appurtenant to Group R-3 and U occupancies not exceeding two stories of light-frame construction shall be permitted to be reinforced as required by rational analysis but with not less than one No. 4 bar, without ties or spirals, where the lateral load, E, to the top of the element does not exceed 200 pounds (890 N) and the soil provides adequate lateral support in accordance with Section 1810.2.1.
3. Deep foundation elements supporting the concrete foundation wall of Group R-3 and U occupancies not exceeding two stories of light-frame construction shall be permitted to be reinforced as required

by rational analysis but with not less than two No. 4 bars, without ties or spirals, where the design cracking moment determined in accordance with Section 1810.3.9.1 exceeds the required moment strength determined using the load combinations with overstrength factor in Section 12.4.3.2 or 12.14.3.2 of ASCE 7 and the soil provides adequate lateral support in accordance with Section 1810.2.1.

4. Closed ties or spirals where required by Section 1810.3.9.4.2 shall be permitted to be limited to the top 3 feet (914 mm) of deep foundation elements 10 feet (3048 mm) or less in depth supporting Group R-3 and U occupancies of *Seismic Design Category* D, not exceeding two stories of light-frame construction.

❖ This section prescribes the necessary seismic reinforcement for cast-in-place deep foundation elements. Exception 1 exempts isolated deep foundation elements from the transverse reinforcement requirements of this section. Exceptions 2 through 4 allow lesser reinforcement than what is required by this section for residential structures, given certain conditions.

1810.3.9.4.1 Seismic reinforcement in Seismic Design Category C. For structures assigned to *Seismic Design Category* C, cast-in-place deep foundation elements shall be reinforced as specified in this section. Reinforcement shall be provided where required by analysis.

A minimum of four longitudinal bars, with a minimum longitudinal reinforcement ratio of 0.0025, shall be provided throughout the minimum reinforced length of the element as defined below starting at the top of the element. The minimum reinforced length of the element shall be taken as the greatest of the following:

1. One-third of the element length;
2. A distance of 10 feet (3048 mm);
3. Three times the least element dimension; and
4. The distance from the top of the element to the point where the design cracking moment determined in accordance with Section 1810.3.9.1 exceeds the required moment strength determined using the load combinations of Section 1605.2.

Transverse reinforcement shall consist of closed ties or spirals with a minimum $^{3}/_{8}$ inch (9.5 mm) diameter. Spacing of transverse reinforcement shall not exceed the smaller of 6 inches (152 mm) or 8-longitudinal-bar diameters, within a distance of three times the least element dimension from the bottom of the pile cap. Spacing of transverse reinforcement shall not exceed 16 longitudinal bar diameters throughout the remainder of the reinforced length.

Exceptions:

1. The requirements of this section shall not apply to concrete cast in structural steel pipes or tubes.
2. A spiral-welded metal casing of a thickness not less than manufacturer's standard gage No.14 gage (0.068 inch) is permitted to provide concrete confinement in lieu of the closed ties or spirals. Where used as such, the metal casing shall be protected against possible deleterious action due to soil constituents, changing water levels or other factors indicated by boring records of site conditions.

❖ Longitudinal and transverse reinforcement requirements prescribed by this section result in moderate ductility in buildings assigned to Seismic Design Category C to withstand seismically induced ground deformations that they can encounter. Separate transverse reinforcement is specified for the portions of the element within and outside of the potential plastic hinge zone.

The purpose of this section is to include element bending, which is a result of ground horizontal movement during an earthquake, in the structural design. The reinforcement in the element, required to resist the tension caused by element bending, increases the ductility of the foundation such that bending or shear failure is precluded.

1810.3.9.4.2 Seismic reinforcement in Seismic Design Categories D through F. For structures assigned to *Seismic Design Category* D, E or F, cast-in-place deep foundation elements shall be reinforced as specified in this section. Reinforcement shall be provided where required by analysis.

A minimum of four longitudinal bars, with a minimum longitudinal reinforcement ratio of 0.005, shall be provided throughout the minimum reinforced length of the element as defined below starting at the top of the element. The minimum reinforced length of the element shall be taken as the greatest of the following:

1. One-half of the element length;
2. A distance of 10 feet (3048 mm);
3. Three times the least element dimension; and
4. The distance from the top of the element to the point where the design cracking moment determined in accordance with Section 1810.3.9.1 exceeds the required moment strength determined using the load combinations of Section 1605.2.

Transverse reinforcement shall consist of closed ties or spirals no smaller than No. 3 bars for elements with a least dimension up to 20 inches (508 mm), and No. 4 bars for larger elements. Throughout the remainder of the reinforced length outside the regions with transverse confinement reinforcement, as specified in Section 1810.3.9.4.2.1 or 1810.3.9.4.2.2, the spacing of transverse reinforcement shall not exceed the least of the following:

1. 12 longitudinal bar diameters;
2. One-half the least dimension of the element; and
3. 12 inches (305 mm).

Exceptions:

1. The requirements of this section shall not apply to concrete cast in structural steel pipes or tubes.
2. A spiral-welded metal casing of a thickness not less than manufacturer's standard gage No. 14 gage

(0.068 inch) is permitted to provide concrete confinement in lieu of the closed ties or spirals. Where used as such, the metal casing shall be protected against possible deleterious action due to soil constituents, changing water levels or other factors indicated by boring records of site conditions.

❖ This section is similar in intent to Section 1810.3.9.4.1, with increased minimum reinforcement requirements for ductility during earthquake ground motion that may be experienced by buildings assigned to Seismic Design Category D or higher. Separate transverse reinforcement requirements are given for portions of an element within and beyond the potential plastic hinge zone.

The purpose of this section is to include pile bending, which is a result of ground horizontal movement during an earthquake, in the structural design. The reinforcement in the pile, required to resist tension caused by pile bending, increases the ductility of the foundation such that bending or shear failure is precluded. The shear strength and confining ability of spiral-welded metal casing eliminates the need for special pile ties, which is the basis for Exception 2.

1810.3.9.4.2.1 Site Classes A through D. For *Site Class* A, B, C or D sites, transverse confinement reinforcement shall be provided in the element in accordance with Sections 21.6.4.2, 21.6.4.3 and 21.6.4.4 of ACI 318 within three times the least element dimension of the bottom of the pile cap. A transverse spiral reinforcement ratio of not less than one-half of that required in Section 21.6.4.4(a) of ACI 318 shall be permitted.

❖ The specified confinement reinforcing similar to ACI 318 requirements for special moment frames is reduced in competent soils in recognition of the confinement attributed to those soils.

1810.3.9.4.2.2 Site Classes E and F. For *Site Class* E or F sites, transverse confinement reinforcement shall be provided in the element in accordance with Sections 21.6.4.2, 21.6.4.3 and 21.6.4.4 of ACI 318 within seven times the least element dimension of the pile cap and within seven times the least element dimension of the interfaces of strata that are hard or stiff and strata that are liquefiable or are composed of soft- to medium-stiff clay.

❖ The specified confinement reinforcing is similar to ACI 318 requirements for special moment frames.

1810.3.9.5 Belled drilled shafts. Where drilled shafts are belled at the bottom, the edge thickness of the bell shall not be less than that required for the edge of footings. Where the sides of the bell slope at an angle less than 60 degrees (1 rad) from the horizontal, the effects of vertical shear shall be considered.

❖ Deep foundation elements are sometimes belled out at their bottoms to increase bearing areas. Belled bottoms can be made by either hand excavation or a mechanical method involving underreaming with a belling bucket, as is commonly used in the construction of drilled elements.

The soil should be sufficiently cohesive so the roof of the bell will not collapse during excavation, cleanout operations and the placement of fresh concrete. Where the character of the soil is such that attempts at belling out the shaft at the bottom might not produce acceptable results, it would be preferable to continue the shaft into soil strata with better load-bearing values, allowing the loads to be carried by the smaller belled bottoms or to such depths that would permit the soils to carry the loads by side friction. The building official should ascertain the suitability of the soil for bell construction through geotechnical investigations and reports and by the recommendations of the design professional.

Bells must have vertical edges at their bottoms that are equal to the thickness requirements for concrete footings. The purpose of this specification is to prevent shear breaks in angled edges caused by soil pressures. Bell slopes (sides) are not to have sides that are less than 60 degrees (1 rad) from the horizontal, unless the effects of vertical shear are considered in the design. Figure 1810.3.9.5 illustrates these requirements.

1810.3.9.6 Socketed drilled shafts. Socketed drilled shafts shall have a permanent pipe or tube casing that extends down to bedrock and an uncased socket drilled into the bedrock, both filled with concrete. Socketed drilled shafts shall have reinforcement or a structural steel core for the length as indicated by an *approved* method of analysis.

The depth of the rock socket shall be sufficient to develop the full load-bearing capacity of the element with a minimum safety factor of two, but the depth shall not be less than the outside diameter of the pipe or tube casing. The design of the rock socket is permitted to be predicated on the sum of the allowable load-bearing pressure on the bottom of the socket plus bond along the sides of the socket.

Where a structural steel core is used, the gross cross-sectional area of the core shall not exceed 25 percent of the gross area of the drilled shaft.

❖ The socketed drilled shaft is a high-load capacity, end-bearing type of deep foundation element. Essentially, it is constructed as a cased, cast-in-place concrete element formed by driving a thick-walled, open-ended steel pipe down to suitable rock material, cleaning out the soil materials within the pipe, drilling a socket into the rock, inserting a structural steel core or reinforcement into the pipe and then filling the pipe and drilled socket with concrete. The core material used in the pipe casing is a structural steel shape such as a wide flange section. Steel rails are also used. Core material is installed to extend from the bottom of the rock socket up to the head of the pile or part of the way up the casing, as determined by analysis. Figure 1810.3.9.6 depicts a socketed drilled shaft with a full-length structural core.

The principal structural feature of the socketed drilled shaft is the rock socket, which is filled with concrete and designed to take the full load of the element by end bearing and the frictional resistance offered by the rough walls of the socket.

The relationship between the depth of the socket and the bearing capacity of the rock cannot be determined with any great accuracy because of the natural joints, bedding planes and fissures generally found in rock formations. Experience has shown, however, that the frictional resistance in hard rock is often sufficient to carry the load. In the case of soft rock, the shear strength of the socket walls should be determined by testing rock samples. The purpose of the steel core is to bond with the concrete fill to act as a composite section and provide additional load-bearing capacity.

1810.3.10 Micropiles. Micropiles shall be designed and detailed in accordance with Sections 1810.3.10.1 through 1810.3.10.4.

❖ This section provides minimum requirements for the design and installation of micropiles. This type of pile has been a popular alternative to more conventional pile types where headroom or equipment access is otherwise limited.

1810.3.10.1 Construction. Micropiles shall develop their load-carrying capacity by means of a bond zone in soil, bedrock or a combination of soil and bedrock. Micropiles shall be grouted and have either a steel pipe or tube or steel reinforcement at every section along the length. It shall be permitted to transition from deformed reinforcing bars to steel pipe or tube reinforcement by extending the bars into the pipe or tube section by at least their development length in tension in accordance with ACI 318.

❖ Section 1802.1 defines "Micropiles" as bored or grouted in-place elements. This section provides a description of micropile construction. It requires incorporating steel pipe casing or steel reinforcement. Either of which is required to extend the full length of the pile. Note the further requirement in Section 1810.3.10.4 for a permanent steel casing in structures that are classified as Seismic Design Category C. Steel pipe casings are typically manufactured in segments with threaded ends.

This section clarifies the intent to permit the pipe or tube casing to terminate above the bond zone, with deformed bar reinforcement continuing below. It also specifies a splice condition for that transition.

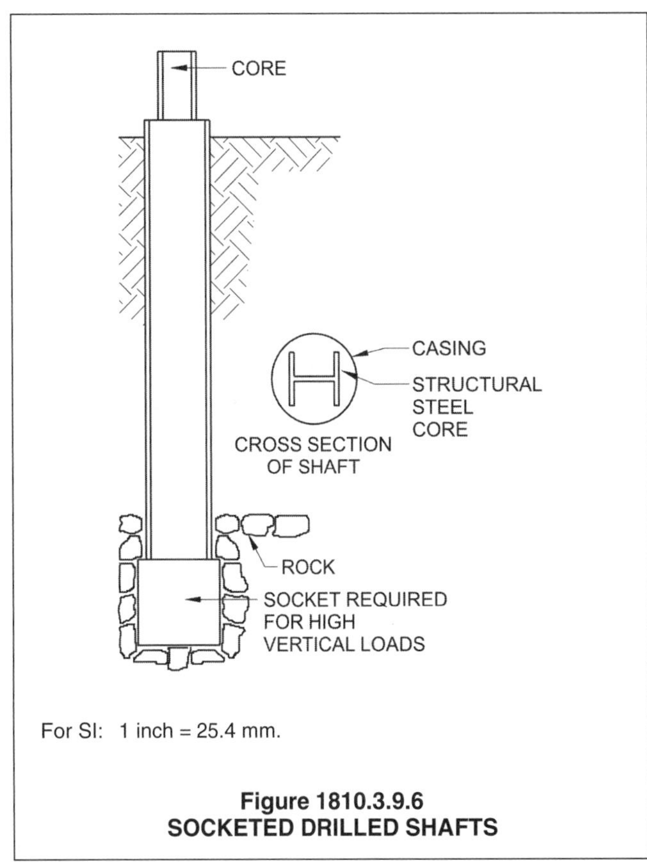

For SI: 1 inch = 25.4 mm.

Figure 1810.3.9.6
SOCKETED DRILLED SHAFTS

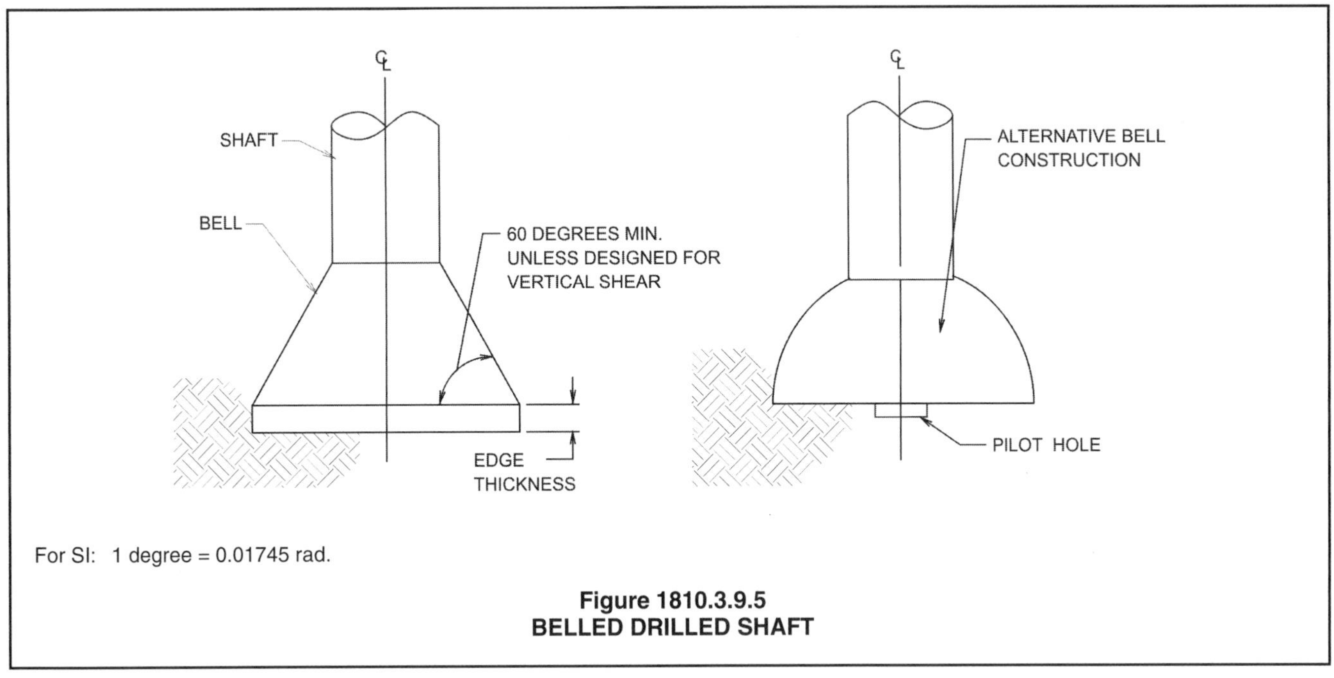

For SI: 1 degree = 0.01745 rad.

Figure 1810.3.9.5
BELLED DRILLED SHAFT

1810.3.10.2 Materials. Reinforcement shall consist of deformed reinforcing bars in accordance with ASTM A 615 Grade 60 or 75 or ASTM A 722 Grade 150.

The steel pipe or tube shall have a minimum wall thickness of $^3/_{16}$ inch (4.8 mm). Splices shall comply with Section 1810.3.6. The steel pipe or tube shall have a minimum yield strength of 45,000 psi (310 MPa) and a minimum elongation of 15 percent as shown by mill certifications or two coupon test samples per 40,000 pounds (18 160 kg) of pipe or tube.

❖ This section provides the minimum specifications for the component materials of the micropile, such as reinforcing steel and steel pipe casing. It also establishes a minimum wall thickness for the steel pipe that is virtually identical to the limits for other steel pipe piles.

1810.3.10.3 Reinforcement. For micropiles or portions thereof grouted inside a temporary or permanent casing or inside a hole drilled into bedrock or a hole drilled with grout, the steel pipe or tube or steel reinforcement shall be designed to carry at least 40 percent of the design compression load. Micropiles or portions thereof grouted in an open hole in soil without temporary or permanent casing and without suitable means of verifying the hole diameter during grouting shall be designed to carry the entire compression load in the reinforcing steel. Where a steel pipe or tube is used for reinforcement, the portion of the grout enclosed within the pipe is permitted to be included in the determination of the allowable stress in the grout.

❖ Micropiles with steel pile casing in place at the time grout is placed are designed with 40 percent of the compression load carried by the steel casing or steel reinforcing. Micropiles with grout that is placed without a casing must be designed so that 100 percent of the compression load is carried by the reinforcing steel.

1810.3.10.4 Seismic reinforcement. For structures assigned to *Seismic Design Category* C, a permanent steel casing shall be provided from the top of the micropile down to the point of zero curvature. For structures assigned to *Seismic Design Category* D, E or F, the micropile shall be considered as an alternative system in accordance with Section 104.11. The alternative system design, supporting documentation and test data shall be submitted to the *building official* for review and approval.

❖ In structures that are classified as Seismic Design Category A or B, there are no additional micropile requirements. In structures that are classified as Seismic Design Category C, permanent steel pipe casing must be provided for the length of the micropile noted. In structures that are classified as Seismic Design Category D, or higher, these micropile provisions are not applicable; approval of micropiles can only be as an alternative method of design and construction.

1810.3.11 Pile caps. Pile caps shall be of reinforced concrete, and shall include all elements to which vertical deep foundation elements are connected, including grade beams and mats. The soil immediately below the pile cap shall not be considered as carrying any vertical load. The tops of vertical deep foundation elements shall be embedded not less than 3 inches (76 mm) into pile caps and the caps shall extend at least 4 inches (102 mm) beyond the edges of the elements. The tops of elements shall be cut or chipped back to sound material before capping.

❖ Pile caps include all elements to which the piles are connected and are to be of reinforced concrete and designed in accordance with the requirements of ACI 318. For footings (pile caps) on piles, computations for moments and shears may be based on the assumption that the load reaction from any pile is concentrated at the pile center (see ACI 318 for loads and reactions of footings on piles).

The soil immediately under the pile cap is not considered to provide any support for vertical loads. The heads of all piles are to be embedded no less than 3 inches (76 mm) into pile caps and the edges of the pile caps are to extend at least 4 inches (102 mm) beyond the closest sides of all piles. The degree of fixity between a pile head and the concrete cap depends on the method of connection required to satisfy design considerations.

1810.3.11.1 Seismic Design Categories C through F. For structures assigned to *Seismic Design Category* C, D, E or F, concrete deep foundation elements shall be connected to the pile cap by embedding the element reinforcement or field-placed dowels anchored in the element into the pile cap for a distance equal to their development length in accordance with ACI 318. It shall be permitted to connect precast prestressed piles to the pile cap by developing the element prestressing strands into the pile cap provided the connection is ductile. For deformed bars, the development length is the full development length for compression, or tension in the case of uplift, without reduction for excess reinforcement in accordance with Section 12.2.5 of ACI 318. Alternative measures for laterally confining concrete and maintaining toughness and ductile-like behavior at the top of the element shall be permitted provided the design is such that any hinging occurs in the confined region.

The minimum transverse steel ratio for confinement shall not be less than one-half of that required for columns.

For resistance to uplift forces, anchorage of steel pipes, tubes or H-piles to the pile cap shall be made by means other than concrete bond to the bare steel section. Concrete-filled steel pipes or tubes shall have reinforcement of not less than 0.01 times the cross-sectional area of the concrete fill developed into the cap and extending into the fill a length equal to two times the required cap embedment, but not less than the development length in tension of the reinforcement.

❖ This section prescribes special detailing requirements between the deep foundation element and pile cap, including required development of the longitudinal pile reinforcement in the cap and associated transverse reinforcement. This reinforcement is required to extend into the pile cap to tie the elements together and to assist in load transfer at the top of the element to the pile cap. The connection must consist of embedment of the element reinforcement in the pile cap for a distance equal to the development length, as specified in ACI 318. Field-placed dowels anchored in the plastic concrete elements are acceptable. The development length to be provided is that for compression or, where uplift is indicated by analysis, tension without reduction in length for excess area. Where seismic confinement reinforcement at the top of the pile is required, alternative measures for laterally confining concrete and maintaining toughness and ductile-like behavior at the top of the pile are permitted, provided the design would force the hinge to occur in the confined region.

1810.3.11.2 Seismic Design Categories D through F. For structures assigned to *Seismic Design Category* D, E or F, deep foundation element resistance to uplift forces or rotational restraint shall be provided by anchorage into the pile cap, designed considering the combined effect of axial forces due to uplift and bending moments due to fixity to the pile cap. Anchorage shall develop a minimum of 25 percent of the strength of the element in tension. Anchorage into the pile cap shall comply with the following:

1. In the case of uplift, the anchorage shall be capable of developing the least of the following:
   1.1. The nominal tensile strength of the longitudinal reinforcement in a concrete element;
   1.2. The nominal tensile strength of a steel element; and
   1.3. The frictional force developed between the element and the soil multiplied by 1.3.

   Exception: The anchorage is permitted to be designed to resist the axial tension force resulting from the seismic load effects including overstrength factor in accordance with Section 12.4.3 or 12.14.3.2 of ASCE 7.

2. In the case of rotational restraint, the anchorage shall be designed to resist the axial and shear forces, and moments resulting from the seismic load effects including overstrength factor in accordance with Section 12.4.3 or 12.14.3.2 of ASCE 7; or shall be capable of developing the full axial, bending and shear nominal strength of the element.

Where the vertical lateral force-resisting elements are columns, the pile cap flexural strengths shall exceed the column flexural strength. The connection between batter piles and pile caps shall be designed to resist the nominal strength of the pile acting as a short column. Batter piles and their connection shall be designed to resist forces and moments that result from the application of seismic load effects including overstrength factor in accordance with Section 12.4.3 or 12.14.3.2 of ASCE 7.

❖ The requirements of this section address the need for conservatism in the design of connections between deep foundation elements and pile caps. They are intended to allow energy dissipating mechanisms, such as rocking, to occur in the soil without failure of the element.

This section requires that the pile cap flexural strength exceed that of the supported column flexural strength if the column is a part of the lateral-force-resisting system.

Additional requirements are specified in this section for batter piles in order to limit earthquake damage to these systems. By their nature, batter pile systems have limited ductility and have performed poorly under strong ground motions. They are required to be designed using the overstrength load combinations, which consider the maximum force expected to be developed in these elements during a seismic event.

1810.3.12 Grade beams. For structures assigned to *Seismic Design Category* D, E or F, grade beams shall comply with the provisions in Section 21.12.3 of ACI 318 for grade beams, except where they are designed to resist the seismic load effects including overstrength factor in accordance with Section 12.4.3 or 12.14.3.2 of ASCE 7.

❖ This section references Chapter 21 of ACI 318 for designing the grade beams. It also allows grade beams to be designed to have the strength required by the overstrength load combinations instead of designing the grade beams as beams in accordance with Chapter 21 of ACI 318. The overstrength load combinations estimate the maximum forces that can realistically develop in the grade beam in an earthquake situation.

1810.3.13 Seismic ties. For structures assigned to *Seismic Design Category* C, D, E or F, individual deep foundations shall be interconnected by ties. Unless it can be demonstrated that equivalent restraint is provided by reinforced concrete beams within slabs on grade or reinforced concrete slabs on grade or confinement by competent rock, hard cohesive soils or very dense granular soils, ties shall be capable of carrying, in tension or compression, a force equal to the lesser of the product of the larger pile cap or column design gravity load times the seismic coefficient, S_{DS}, divided by 10, and 25 percent of the smaller pile or column design gravity load.

Exception: In Group R-3 and U occupancies of light-frame construction, deep foundation elements supporting foundation walls, isolated interior posts detailed so the element is not subject to lateral loads or exterior decks and patios are not subject to interconnection where the soils are of adequate stiffness, subject to the approval of the *building official*.

❖ The purpose of this section is to preclude excessive movement of one group of deep foundation elements with respect to another. The section is similar to Section 1809.13. One of the prerequisites of adequate structural performance during an earthquake is that the foundation of the structure must act as a unit. This is typically accomplished by tying together the pile caps of deep foundation elements with ties capable of carrying, both in tension and compression, a force equal to the lesser of 10 percent of the larger pile cap or column load multiplied by S_{DS}, or 25 percent of the smaller pile cap or column load. This can be accomplished through the use of equivalent types of restraint, such as a concrete floor slab, where restraint can be substantiated. Reliance upon passive soil pressure is limited to competent rock, dense granular soil or hard cohesive soil.

If soils are shown to be of adequate stiffness, the building official may allow the use of the exception for the following elements in Group R-3 and U occupancies not exceeding two stories of light-frame construction:

1. Elements supporting foundation walls.
2. Elements supporting isolated interior posts detailed so the pier is not subject to lateral loads.
3. Elements supporting lightly loaded exterior decks and patios.

1810.4 Installation. Deep foundations shall be installed in accordance with Section 1810.4. Where a single deep foundation element comprises two or more sections of different materials or different types spliced together, each section shall satisfy the applicable conditions of installation.

❖ These provisions establish minimum requirements for the installation of deep foundation elements.

1810.4.1 Structural integrity. Deep foundation elements shall be installed in such a manner and sequence as to prevent distortion or damage that may adversely affect the structural integrity of adjacent structures or of foundation elements being installed or already in place and as to avoid compacting the surrounding soil to the extent that other foundation elements cannot be installed properly.

❖ The placement of deep foundation elements is often by driving, vibrating, jacking, jetting, direct weight or a combination of these methods. Because of the generally harsh nature of deep foundation installation operations, elements can experience some degree of damage during placement; however, damage can be prevented or minimized by selecting the proper type of deep foundation placement methods and techniques, as well as the right equipment to accomplish the work, all based on adequate knowledge of the soil conditions obtained from a foundation investigation (see Section 1803.5.5).

Piles must be placed in a manner that maintains their structural integrity and installed to such depths as determined by foundation investigation and engineering analysis to safely resist the design loads that are to be imposed upon them. Care must be exercised during deep foundation placement operations to provide for the safety of adjacent piles or other structures—leaving their strength and load capacity unimpaired. Any pile damaged during installation to the extent that its structural integrity is affected must be satisfactorily repaired or rejected.

1810.4.1.1 Compressive strength of precast concrete piles. A precast concrete pile shall not be driven before the concrete has attained a compressive strength of at least 75 percent of the specified compressive strength (f'_c), but not less than the strength sufficient to withstand handling and driving forces.

❖ This section requires that precast reinforced concrete piles must not be driven until the concrete has acquired at least three-quarters of its specified compressive strength. For example, if a pile is manufactured at the minimum compressive strength of 4,000 psi (27.6 MPa), it must not be driven until it has obtained a strength of at least 3,000 psi (20.68 MPa) ($^3/_4$ x 4,000). In all cases, however, concrete strength must have developed to the point that it is sufficient to sustain the stresses imposed on the pile by handling and driving operations.

1810.4.1.2 Casing. Where cast-in-place deep foundation elements are formed through unstable soils and concrete is placed in an open-drilled hole, a casing shall be inserted in the hole prior to placing the concrete. Where the casing is withdrawn during concreting, the level of concrete shall be maintained above the bottom of the casing at a sufficient height to offset any hydrostatic or lateral soil pressure. Driven casings shall be mandrel driven their full length in contact with the surrounding soil.

❖ This section requires a casing where a a cast-in-place deep foundation element is formed through unstable soil.

1810.4.1.3 Driving near uncased concrete. Deep foundation elements shall not be driven within six element diameters center to center in granular soils or within one-half the element length in cohesive soils of an uncased element filled with concrete less than 48 hours old unless *approved* by the *building official*. If the concrete surface in any completed element rises or drops, the element shall be replaced. Driven uncased deep foundation elements shall not be installed in soils that could cause heave.

❖ Casings driven in granular soils are to be spaced at least six times the diameter of the element. Casings driven in cohesive soils are not to be spaced less than one-half of the element length when the concrete (in adjacent elements) is less than 48 hours old. This latter requirement reflects the fact that deep foundation elements driven in cohesive materials can cause significant soil displacement that can induce high lateral pressures on adjacent elements. Unless the concrete in these adjacent elements has had sufficient time to set (48 hours or more) and acquire some of its ultimate strength, considerable damage can be caused by the earth pressures. Because of the voids between the particles in granular soils, the effects of soil displacement are not as serious as for cohesive soils, such as clay.

The 48-hour concrete time-set requirement as stated in this section should not be confused with the 12-hour requirement referred to in Section 1810.4.8 for drilled or augered elements. While both types are cast-in-place concrete elements, unlike the driven uncased elements, drilled or augered elements are not of the displacement type, which essentially compacts the soil and may cause great lateral pressures and earth movement. The soil in the drilled method of installation is removed rather than displaced, and the subsurface influence on adjacent deep foundation elements is far less than on the driven element, particularly in cohesive soils.

Driven uncased elements are exposed to various types of potential damage. Besides the possibility of concrete contamination by the surrounding soil and surface water at the top of the element, its cross section may be subjected to squeezing or necking from lateral soil pressures and intrusions of displaced soils or other obstructions. Also, the concrete may be damaged by loss of support caused by removal of adjacent element casings from the soil surrounding the element. Driven uncased deep foundation elements are not recommended under conditions where significant ground heave could occur or where highly unstable soils exist.

Subsurface investigations and load testing may be required to establish capacity of deep foundation element, since there can be no correlation between the driving resistance of the element casing and its capacity. This is because the casing is eventually removed.

1810.4.1.4 Driving near cased concrete. Deep foundation elements shall not be driven within four and one-half average diameters of a cased element filled with concrete less than 24 hours old unless *approved* by the *building official*. Concrete shall not be placed in casings within heave range of driving.

❖ One advantage to this type of pile is that after the driving and removal of the mandrel before the concrete is placed, the steel casing can be inspected internally along its full length; therefore, any damage that has resulted from the pile-driving operation can be readily discovered and corrected.

Steel casings are not to be driven closer than four and one-half pile diameters of any adjacent piles filled with concrete until the concrete in the shells has cured for at least 24 hours and achieved an early strength. The basic reason for requiring four and one-half pile diameters between driving and concreting is a general reluctance to expose fresh concrete to vibrations as it sets. Several independent tests have shown, however, that there is no detrimental effect caused by vibrations on setting concrete. In some cases, there was a strength gain. The code requirement of four and one-half diameters is considered reasonable. It is often impractical to place concrete in element shells right next to deep foundation elements being driven. Casings can be driven earlier than the minimum 24-hour period when approved by the building official.

Driven light-gage steel shells that are left open can also be damaged by earth pressures when driving other piles in the close vicinity. Under such circumstances, the shells are sometimes protected by inserting dummy mandrels.

Shells that are in place, adjacent to and within heave range of a deep foundation element being driven are normally left open if the soil condition is such that it can cause heave. This condition is particularly critical in cohesive soils, such as clay. Heaved elements must be redriven before filling with concrete. If a concreted element has heaved; however, it can be safely redriven if proper techniques and a suitably designed cushion are used.

For a general description of steel-cased cast-in-place concrete piles, refer to the commentary to Section 1810.3.1.6.

1810.4.1.5 Defective timber piles. Any substantial sudden increase in rate of penetration of a timber pile shall be investigated for possible damage. If the sudden increase in rate of penetration cannot be correlated to soil strata, the pile shall be removed for inspection or rejected.

❖ A small drop in blow count is not usually a cause for concern and is not uncommon in layered soils. It is only a substantial change in the rate of pile penetration that is not related to soil properties, which indicates a potential problem. If the penetration resistance suddenly decreases, serious damage to the pile could have occurred (including breaking), and the condition must be investigated. The pile can be withdrawn and inspected, and if required, abandoned and replaced.

Another significant problem associated with the use of timber piles is the possibility of structural damage caused by overdriving. Overdriving usually occurs upon reaching rock or hard soil stratas and may result in pile bending, brooming at the tip, crushing or brooming at the head or splitting or breaking along the pile section (see Figure 1810.4.1.5). To avoid such damage, if the penetration resistance suddenly increases, pile driving should be stopped immediately.

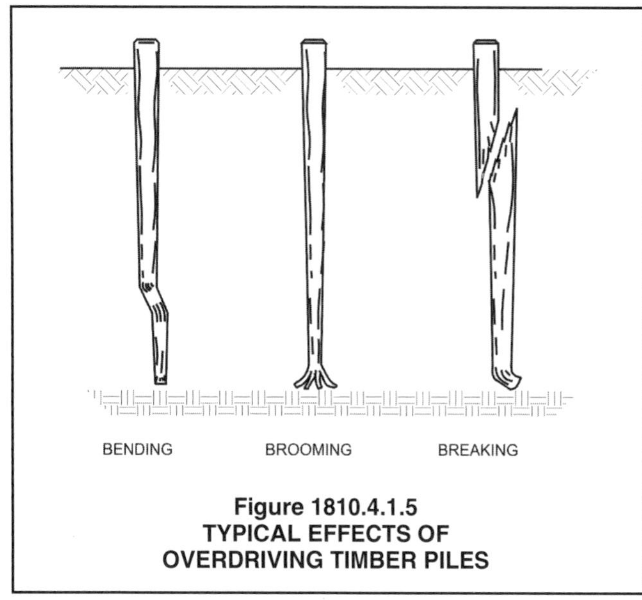

Figure 1810.4.1.5
TYPICAL EFFECTS OF OVERDRIVING TIMBER PILES

1810.4.2 Identification. Deep foundation materials shall be identified for conformity to the specified grade with this identity maintained continuously from the point of manufacture to the point of installation or shall be tested by an *approved agency* to determine conformity to the specified grade. The *approved agency* shall furnish an affidavit of compliance to the *building official*.

❖ All deep foundation element materials must be identified for conformity to construction specifications, providing information such as strength (species and grade for timber piles) and dimensions, and other pertinent information as may be required. Such identification must be provided for all piers and piles, whether they are taken from the manufacturer's stock or made for a particular project. Identification is to be maintained from the place of manufacture to the shipment, on-site handling, storage and installation of the deep foundation elements. Manufacturers, upon request, usually furnish certificates of compliance with construction specifications. In the absence of adequate data, piers and piles are to be tested to prove conformity to the specified grade.

In addition to mill certificates (steel piles), identification is made through plant manufacturing or inspection reports (precast concrete deep foundation elements and timber piles) and delivery tickets (concrete). Timber piles are stamped (branded) with information, such as producer, species, treatment and length.

Identification is essential when high-yield strength steel is specified. Frequently, pile cutoff lengths are reused and pile material may come from a jobber, a contractor's yard or a material supplier. In such cases, mill certificates are not available and the steel should be tested to see if it complies with the specifications.

1810.4.3 Location plan. A plan showing the location and designation of deep foundation elements by an identification system shall be filed with the *building official* prior to installation of such elements. Detailed records for elements shall bear an identification corresponding to that shown on the plan.

❖ This section requires that a deep foundation element location plan clearly showing the designation of all piers or piles in the foundation system be submitted to the building official prior to installation of piles. Preferably, such plans should be submitted before delivery of the piers or piles to the construction site and prior to staking the deep foundation element locations.

The deep foundation element location plan is not only an important tool for the installation of piers or piles, but also serves to communicate information between the owner, engineer, contractor, manufacturer, special inspector, building official and other interested persons. The special inspector (see commentary, Section 1810.4.12) must keep piling logs, records and reports based on this identification system. The use of the deep foundation element plan is particularly important at sites where the variations in soil profiles are so extensive that it becomes necessary to manufacture piers or piles of different lengths to reach proper bearing levels. The building official should receive revised copies of the deep foundation element location plan whenever field changes are made that add, delete or relocate piles.

1810.4.4 Preexcavation. The use of jetting, augering or other methods of preexcavation shall be subject to the approval of the *building official*. Where permitted, preexcavation shall be carried out in the same manner as used for deep foundation elements subject to load tests and in such a manner that will not impair the carrying capacity of the elements already in place or damage adjacent structures. Element tips shall be driven below the preexcavated depth until the required resistance or penetration is obtained.

❖ The use of preexcavation methods to facilitate the installation of deep foundation elements is often necessary for a number of reasons, including:

1. To drive elements through upper layers of hard soil;
2. To penetrate through subsurface obstructions;
3. To eliminate or reduce the possibility of ground heave that could result in lifting adjacent elements already driven;
4. To reduce ground pressures that could result in the lateral movement of adjacent piles or structures;
5. To reduce the amount of driving necessary to seat the elements in the required bearing stratum;
6. To reduce the possibility of damaging vibrations;
7. To reduce the amount of noise associated with pile-driving operations; and
8. To accommodate the placement of elements that are longer than the leads of the driving equipment.

Methods commonly used for preexcavation are jetting and predrilling. The jetting method has been found to be more effective in granular soils, such as sand and fine gravel, than in cohesive materials, such as clay. However, jetting should not be done in granular soils containing very coarse gravel, cobbles and small boulders because such material cannot be removed by the jet stream and will result in a collection of stones at the bottom of the hole, making it very difficult, if not impossible, to drive a pile through the mass of stone material. The jetting operations must be controlled to avoid excessive losses of soil that could affect the stability of adjacent structures or the required bearing capacity of previously installed piles.

Jetting operations must be carefully controlled to avoid excessive loss of soil, which could affect the load-bearing capacity of deep foundation elements already installed or the stability of adjacent structures.

Deep foundation elements should be driven below the depth of the jetted hole until the required resistance or penetration is obtained. Before this preexcavation method is used, consideration should be given to the possibility that jetting, unless strictly controlled, can adversely affect load transfer, particularly as it involves the placement of nontapered piles.

In comparison to the jetting method described, a more controllable form of preexcavation is by predrilling or coring. This method greatly reduces the possibility of detrimental effects on adjacent piles or structures and can be performed as a dry operation or a wet rotary process. Dry drilling can be done by the use of a continuous-flight auger or a short-flight auger attached to the end of a drill stem or kelly bar. Wet drilling requires a hollow-stem, continuous-flight auger or a hollow drill stem employing the use of spade bits. When the wet rotary process of predrilling is used, bentonite slurry or plain water is circulated to keep the hole open. As in the case of jetting, deep foundation elements should be driven with tips below the predrilled hole. This is necessary to prevent any voids or very loose or soft soils from occurring below the element tip.

There are other methods used for preexcavation purposes, such as the dry tube method and spudding, but such procedures are seldom used. The methods to be employed for preexcavation are subject to the approval of the building official.

1810.4.5 Vibratory driving. Vibratory drivers shall only be used to install deep foundation elements where the element load capacity is verified by load tests in accordance with Section 1810.3.3.1.2. The installation of production elements shall be controlled according to power consumption, rate of penetration or other *approved* means that ensure element capacities equal or exceed those of the test elements.

❖ The use of vibratory drivers for the installation of piles is not applicable to all types of soil conditions. They are most effective in granular soils with the use of nondisplacement piles, such as steel H-piles and pipe piles driven open ended. Vibratory drivers are also used for extracting piles or temporary casings employed in the construction of cast-in-place concrete piles.

Vibratory drivers, either low or high frequency, cause the pile to penetrate the soil by longitudinal vibrations. While this type of pile driver can produce remarkable results in the installation of nondisplacement piles under favorable soil conditions, the greatest difficulty is the lack of a reliable method of estimating the load- bearing capacity. After the pile has been installed with a vibratory driver, pile capacity can best be determined by using an impact-type hammer to set the pile in its final position.

An acceptable means of controlling pile capacity is by determining the power consumption in relation to the rate of penetration. Nonetheless, the use of a vibratory driver is only permitted where the pile load capacity is established by load tests in accordance with the requirements of Section 1810.3.3.1.2.

1810.4.6 Heaved elements. Deep foundation elements that have heaved during the driving of adjacent elements shall be redriven as necessary to develop the required capacity and penetration, or the capacity of the element shall be verified by load tests in accordance with Section 1810.3.3.1.2.

❖ When deep foundation elements are driven into cohesive soils, particularly saturated plastic clay materials, they often displace a volume of soil equal to that of the elements themselves. Such soil displacement usually occurs as ground heaves and may cause adjacent elements already driven to lift up and become unseated, causing a loss of load capacity. When this happens, heaved piles must be redriven to firm bearing in order to regain their required capacity. If heaved elements are not redriven, their capacities must be verified by means of load tests.

It should be noted that not all types of elements can be redriven. For example, heaved uncased cast-in-place concrete elements, or sectional elements whose splices cannot take tension, should be abandoned and replaced.

In redriving heaved elements, the same or comparable driving equipment should be employed as used in the original installation; however, there are exceptions to this rule. For example, during the installation of concrete-filled pipe elements, only the empty pipes were first driven. In redriving this type of element, the pipes may now be filled with concrete resulting in much stiffer and heavier sections than were driven initially. In such cases, the driving technique must be adjusted to accommodate a considerably lesser driving resistance.

One method commonly used to prevent or reduce objectionable displacements caused by deep foundation installations in soils subject to heaving is to preexcavate element holes in accordance with the requirements of Section 1810.4.4.

1810.4.7 Enlarged base cast-in-place elements. Enlarged bases for cast-in-place deep foundation elements formed by compacting concrete or by driving a precast base shall be formed in or driven into granular soils. Such elements shall be constructed in the same manner as successful prototype test elements driven for the project. Shafts extending through peat or other organic soil shall be encased in a permanent steel casing. Where a cased shaft is used, the shaft shall be adequately reinforced to resist column action or the annular space around the shaft shall be filled sufficiently to reestablish lateral support by the soil. Where heave occurs, the element shall be replaced unless it is demonstrated that the element is undamaged and capable of carrying twice its design load.

❖ An enlarged base cast-in-place deep foundation element is often installed by first driving a steel casing into the ground. The casing can either be temporary or permanent. After the required depth has been reached, the enlarged base is formed in the granular bearing soil by progressively adding and driving out small batches of zero-slump concrete using a drop weight.

Another method of installing an enlarged base element is by use of an enlarged precast concrete base at the end of a mandrel-driven steel casing. A major problem associated with this type of installation is that, in driving the permanent casing into the ground, the precast base creates a hole that is larger than the diameter of the shaft, leaving an open space around the element and thus losing the lateral support usually provided by the surrounding soil. In such cases, the annular space around the shaft must be filled to provide the necessary lateral support. The customary practice is to fill the annular space by pumping grout or washing in granular material. An illustration of these enlarged base elements is provided in Figure 1810.4.7.

The spacings of enlarged base piles are normally greater than most other types of conventional piles in order to avoid base interferences and the close overlapping of soil-bearing areas that would serve to reduce pile capacity.

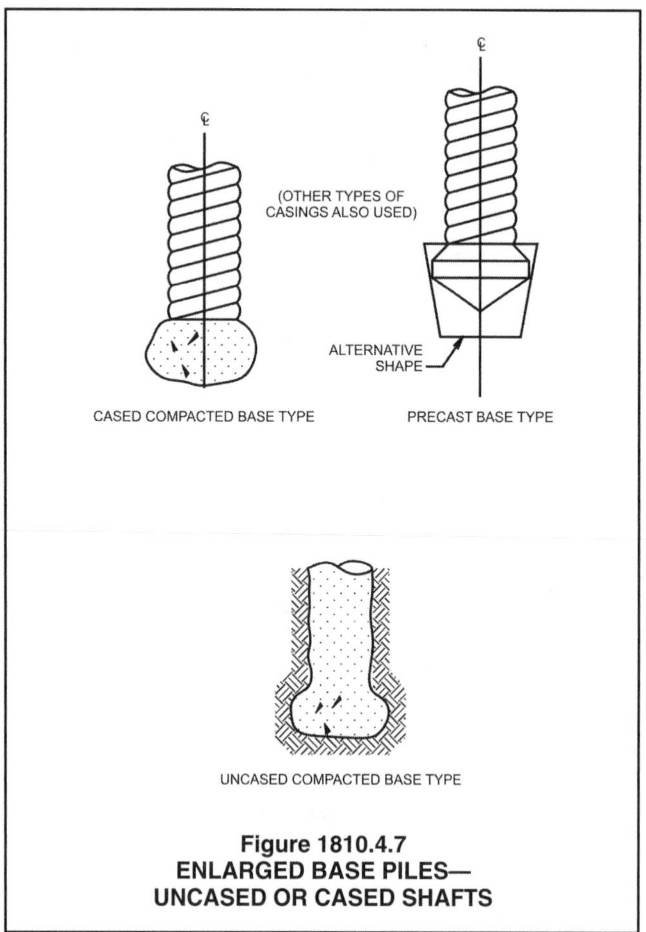

Figure 1810.4.7
ENLARGED BASE PILES—UNCASED OR CASED SHAFTS

1810.4.8 Hollow-stem augered, cast-in-place elements. Where concrete or grout is placed by pumping through a hollow-stem auger, the auger shall be permitted to rotate in a clockwise direction during withdrawal. As the auger is withdrawn at a steady rate or in increments not to exceed 1 foot (305 mm), concreting or grouting pumping pressures shall be measured and maintained high enough at all times to offset hydrostatic and lateral earth pressures. Concrete or grout volumes shall be measured to ensure that the volume of concrete or grout placed in each element is equal to or greater than the theoretical volume of the hole created by the auger. Where the installation process of any element is interrupted or a loss of concreting or grouting pressure occurs, the element shall be redrilled to 5 feet (1524 mm) below the elevation of the tip of the auger when the installation was interrupted or concrete or grout pressure was lost and reformed. Augered cast-in-place elements shall not be installed within six diameters center to center of an element filled with concrete or grout less than 12 hours old, unless *approved* by the *building official*. If the concrete or grout level in any completed element drops due to installation of an adjacent element, the element shall be replaced.

❖ There are many potential problems associated with drilled-hole piles. Most of these problems are related to soil conditions, including soil or rock debris accumulating at the base of the pile or occurring in the pile shaft; reductions in the shaft cross section caused by the necking of soil walls because of soft materials or earth pressures; discontinuities in the pile shaft; hollows on the surface of the shaft and other problems linked to drilling operations. Such problems are usually addressed by the use of proper installation techniques.

Whenever unstable soils are encountered in drilling operations, such as loose granular soils, organic soils, very soft silts or clays and water-bearing subsurface materials, a temporary steel liner is to be placed in the hole to prevent the collapse of the earth walls or sloughing off of the soil during concrete placement.

In placing concrete in temporarily lined holes, the top of the concrete should be kept well above the bottom edge of the steel liner as it is withdrawn in order to offset any hydrostatic or lateral soil pressures. Stiff (low slump) concrete should not be used, so as to avoid the potential problem of concrete arching in the liner tube and causing discontinuities or voids to occur in the pile shaft as the liner is withdrawn. For a discussion on concrete consistency (slump values), see the commentary to Section 1808.8.1.

While many of the problems associated with auger-placed concrete piles are similar to those described above for drilled-hole piles, other problems may also be introduced because of the particular installation technique. The pile installation procedure using a hollow stem continuous-flight auger, both for drilling and concrete placement purposes, does not permit the use of temporary steel liners. Damage to this type of pile, because of improper installation procedures, can result in the incomplete filling of the pile hole, concrete discontinuities along the pile shaft, reductions in the cross-sectional area of the shaft (necking) and soil inclusions. Also, the loss of side support of the drilled hole or vertical displacement caused by ground pressures or soil movement may cause serious damage to the pile shaft.

In drilling the pile hole, the hollow-stem auger should be rotated and advanced continuously until the required tip elevation is reached. At that point in the drilling operation, rotation of the auger should be stopped to avoid removing excess soils that could result in damaging adjacent piles and to keep the auger flights full of soil as a means of retaining the hole walls. Concrete is then pumped through the hollow stem, filling the hole from the bottom up as the auger is withdrawn.

The volume of placed concrete must be monitored and it should equal or exceed the theoretical volume of the augered hole. An excess concrete volume of 10 percent is not uncommon, but even larger excesses can be expected where the soils are more porous. If the amount of the concrete pumped into the hole is considerably more than the anticipated volume, then the cause should be investigated. It could mean, for instance, that concrete (grout) is being pumped into soft soil strata, solution cavities, underground pipelines or other underground structures.

Concrete should be pumped under continuous pressure and the rate of withdrawal of the auger should be carefully controlled for a continuous and full-sized shaft. Discontinuities in the concrete shaft can develop if the auger is improperly withdrawn. A smooth, continuous auger withdrawal should help to prevent the surrounding soil from squeezing into the hole (necking) and possibly contaminating the concrete.

Concrete pumping pressures at the auger outlet should be greater than any hydrostatic or lateral pressures occurring in the hole. Excessively high pumping pressures, however, should be avoided while placing concrete surrounded by soft soils because it could cause upward or lateral movement of adjacent piles.

If, for any reason, the pile-concreting operation is interrupted, a pressure drop occurs while pumping the concrete, the auger is improperly handled and is raised too fast or for any other cause that could damage the pile or cause a reduction in the required size of the pile section, then the hollow-stem auger must be redrilled to the required tip elevation and the pile reformed from the bottom up.

Unless approved by the building official, no new pile holes are to be drilled or injected with concrete if they are located within a center-to-center distance of six pile diameters from other adjacent piles containing fresh concrete that have not been allowed to set for a period of 12 hours or more. This is to reduce the possibility of damage to adjacent piles. If the concrete surface of a completed pile is observed to drop below its cast elevation as a result of the drilling of adjacent piles, then the completed pile should be rejected and replaced. Such an occurrence could indicate some deformation along the pile shaft that could affect its structural integrity.

1810.4.9 Socketed drilled shafts. The rock socket and pipe or tube casing of socketed drilled shafts shall be thoroughly cleaned of foreign materials before filling with concrete. Steel cores shall be bedded in cement grout at the base of the rock socket.

❖ A brief description of the installation sequence of caisson piles is given in the commentary to Section 1810.3.9.6. One of the problems in the construction of caisson piles is that some water will seep into the bottom of the pile opening through the rock joints or fissures in the socket. Sometimes water may seep into the opening at the end of the pipe casing because of incomplete seating of the pipe into the rock material. If the water in the hole cannot be controlled by ejection or other methods rendering the placement of concrete under reasonably dry conditions, the concrete may be placed by a tremie or other methods when approved by the building official.

1810.4.10 Micropiles. Micropile deep foundation elements shall be permitted to be formed in holes advanced by rotary or percussive drilling methods, with or without casing. The elements shall be grouted with a fluid cement grout. The grout shall be pumped through a tremie pipe extending to the bottom of the element until grout of suitable quality returns at the top of the element. The following requirements apply to specific installation methods:

1. For micropiles grouted inside a temporary casing, the reinforcing bars shall be inserted prior to withdrawal of the casing.The casing shall be withdrawn in a controlled manner with the grout level maintained at the top of the element to ensure that the grout completely fills the drill hole. During withdrawal of the casing, the grout level inside the casing shall be monitored to verify that the flow of grout inside the casing is not obstructed.

2. For a micropile or portion thereof grouted in an open drill hole in soil without temporary casing, the minimum design diameter of the drill hole shall be verified by a suitable device during grouting.

3. For micropiles designed for end bearing, a suitable means shall be employed to verify that the bearing surface is properly cleaned prior to grouting.

4. Subsequent micropiles shall not be drilled near elements that have been grouted until the grout has had sufficient time to harden.

5. Micropiles shall be grouted as soon as possible after drilling is completed.

6. For micropiles designed with a full-length casing, the casing shall be pulled back to the top of the bond zone and reinserted or some other suitable means employed to assure grout coverage outside the casing.

❖ This section provides the basic micropile installation requirements. Micropile boreholes are typically advanced by either of the listed methods, rotary drilling or rotary percussive drilling. Installation requirements differ based on whether a steel casing is permanent or temporary or not provided (Item 6, 1 or 2, respectively).

1810.4.11 Helical piles. Helical piles shall be installed to specified embedment depth and torsional resistance criteria as determined by a *registered design professional*. The torque applied during installation shall not exceed the maximum allowable installation torque of the helical pile.

❖ Helical pile shafts are screwed into the ground by application of torsion (also see ICC Evaluation Service AC 358).

1810.4.12 Special inspection. *Special inspections* in accordance with Sections 1705.7 and 1705.8 shall be provided for driven and cast-in-place deep foundation elements, respectively. *Special inspections* in accordance with Section 1705.9 shall be provided for helical piles.

❖ This section requires special inspections (see definition, Section 1702) in accordance with Sections 1704.8 and 1704.9 to verify proper and safe installations. To be approved by the building official, an inspector should be experienced in deep foundation element installation work. An inspector must be able to read and understand foundation plans and specifications, maintain accurate records, have a thorough understanding of the scope of the pile work to be performed, give concise and timely reports to the owner or engineer and submit in writing to the building official whatever field information is required.

The duties of the deep foundation element inspector include the inspection and approval of pile-driving equipment; the inspection and acceptance of piles furnished for the work; the verification of all pile location stakes and spacings; the observation and recording of required information relating to the installation of the deep foundation element foundation, as well as keeping a log for each pile showing either the number of hammer blows for each foot of penetration and the final penetration in blows per inch or the driving record for only the last few feet of penetration. The driving log should also include information on the duration and cause of any delays, such as splicing time, equipment breakdown, changing cushions, etc. The record should also show the final tip and butt (cutoff) elevations, as well as the depths of preexcavation holes (see Section 1810.4.4). Finally, there should also be a record of any pile damage and repair work; pile extractions and replacements; observations on pile heaving and depth of redriving and information on pile alignment.

Bibliography

The following resource materials are referenced in this chapter or are relevant to the subject matter addressed in this chapter.

ACI 229R-99, *Controlled Low-strength Materials*. Farmington Hills, MI: American Concrete Institute, 1999.

ACI 306R-88, *Cold Weather Concreting*. Farmington Hills, MI: American Concrete Institute, 1988.

ACI 306.1-90, *Standard Specification for Cold Weather Concrete*. Farmington Hills, MI: American Concrete Institute, 1990.

ACI 318-11, *Building Code Requirements for Structural Concrete*. Farmington Hills, MI: American Concrete Institute, 2011.

ACI 543R-74 (1980), *Recommendations for Design, Manufacture and Installation of Concrete Piles*. Farmington Hills, MI: American Concrete Institute, 1980.

AF&PA NDS-12, *National Design Specification for Wood Construction—with 2012 Supplement*. Washington, DC: American Forest and Paper Association, 2012.

AF&PA PWF-07, *Permanent Wood Foundation Design Specification*. Washington, DC: American Forest and Paper Association, 2007.

ASCE 7-10, *Minimum Design Loads for Buildings and Other Structures*. New York: American Society of Civil Engineers, 2010.

ASCE 24-05, *Flood Resistant Design and Construction Standard*. Reston, VA: American Society of Civil Engineers, 2005.

ASCE 32-01, *Design of Frost Protected Shallow Foundations*. Reston, VA: American Society of Civil Engineers, 2001.

ASTM A 36/A 36M-08, *Specification for Carbon Structural Steel*. West Conshohocken, PA: ASTM International, 2008.

ASTM A 252-98(2007), *Specification for Welded and Seamless Steel Pipe Piles*. West Conshohocken, PA: ASTM International, 2007.

ASTM A 416/416M-06, *Specification for Steel Strand, Uncoated Seven-wire for Prestressed Concrete*. West Conshohocken, PA: ASTM International, 2006.

ASTM A 572/A 572M-07, *Specification for High-strength Low-alloy Columbian-vanadium Structural Steel*. West Conshohocken, PA: ASTM International, 2007.

ASTM C 270-08a, *Specification for Mortar for Unit Masonry*. West Conshohocken, PA: ASTM International, 2008.

ASTM C 887-05, *Specification for Packaged, Dry, Combined Materials for Surface Bonding Mortar*. West Conshohocken, PA: ASTM International, 2005.

ASTM D 25-99(2005), *Specification for Round Timber Piles*. West Conshohocken, PA: ASTM International, 2005.

ASTM D 422-63(2007), *Standard Test Method for Particle Size Analysis of Soils*. West Conshohocken, PA: ASTM International, 2007.

ASTM D 1143/D 1143M-07e1, *Standard Test Method for Piles Under Static Axial Compressive Load.* West Conshohocken, PA: ASTM International, 2007.

ASTM D 1668-97a, *Standard Specification for Glass Fabrics (Woven and Treated) for Roofing and Waterproofing*. West Conshohocken, PA: ASTM International, 1997.

ASTM D 2487-06e1, *Standard Practice for Classification of Soils for Engineering Purposes*. West Conshohocken, PA: ASTM International, 2006.

ASTM D 3689-07, *Specification of Individual Piles Under Static Axial Tensile Load.* West Conshohocken, PA: ASTM International, 2007.

ASTM D 4318-05, *Standard Test Method for Liquid Limit, Plastic Limit, and Plasticity Index of Soils*. West Conshohocken, PA: ASTM International, 2005.

ASTM D 4829-08a, *Test Method for Expansion Index of Soils*. West Conshohocken, PA: ASTM International, 2008.

ASTM D 4832-02, *Standard Test Method for Preparation and Testing of Controlled Low-strength Material (CLSM) Test Cylinders*. West Conshohocken, PA: ASTM International, 2002.

ASTM D 4945-08, *Test Method for High-strain Dynamic Testing of Piles*. West Conshohocken, PA: ASTM International, 2008.

ASTM D 5971-01, *Standard Practice for Sampling Freshly Mixed Controlled Low-strength Material*. West Conshohocken, PA: ASTM International, 2001.

ASTM D 6023-02, *Standard Test Method for Unit Weight, Yield, Cement Content, and Air Content (Gravimetric) of Controlled Low-strength Material (CLSM)*. West Conshohocken, PA: ASTM International, 2002.

ASTM D 6024-02, *Standard Test Method for Ball Drop on Controlled Low-strength Material (CLSM) to Determine Suitability for Load Application*. West Conshohocken, PA: ASTM International, 2002.

ASTM D 6103-97, *Standard Test Method for Flow Consistency of Controlled Low-strength Material (CLSM).* West Conshohocken, PA: ASTM International, 1997.

AWPA M4-08, *Standard for the Care of Preservative-treated Wood Products*. Grandbury, TX: American Wood Preservers' Association, 2008.

AWPA U1-11, *Use Category System: User Specification for Treated Wood, except Section 6, Commodity Specification H.* Grandbury, TX: American Wood Preservers' Association, 2011.

AWPB MP1, *Quality Control and Inspection Procedures for Dual Treatment of Marine Piling Pressure Treated with Waterborne Preservatives and Creosote for Use in Marine Waters*. American Wood Preservers Bureau.

AWPB MP2, *Quality Control and Inspection Procedures for Dual Treatment of Marine Piling Pressure Treated with Waterborne Preservatives and Creosote for Use in Marine Waters*. American Wood Preservers' Bureau.

AWPB MP4, *Quality Control and Inspection Procedures for Dual Treatment of Marine Piling Pressure Treated with Waterborne Preservatives and Creosote for Use in Marine Waters*. American Wood Preservers Bureau.

Bowles, J. E. *Foundation Analysis and Design, 5th edition*. New York: McGraw Hill, 1996.

FEMA 450-1, *NEHRP Recommend Provisions for Seismic Regulations for New Buildings and Other Structures, Part* 1. Washington, DC: Prepared by the Building Seismic Safety Council for the Federal Emergency Management Agency, 2004.

FEMA TB11-01, *Crawl Space Construction for Buildings Located in Special Flood Hazard Areas*. Washington, DC: Federal Emergency Management Agency, 2001.

Hanson, W.E., R.B. Peck, and T.H. Thornburn. *Foundation Engineering, 2nd edition*. New York: John Wiley & Sons, 1974.

ICC-ES AC 358, *Acceptance Criteria for Helical Foundation Systems and Devices*. Whittier, CA: International Code Council Evaluation Service, Inc. 2007.

IPC-12, *International Plumbing Code*. Washington, DC: International Code Council, 2011.

IRC-12, *International Residential Code*. Washington, DC: International Code Council, 2011.

NCMA Report, *Research Evaluation of the Flexural Strength of Concrete Masonry, Project No. 93-172, Dec. 7, 1993*. Herndon VA: National Concrete Masonry Association, 1993.

PCI, *Recommended Practice for Design, Manufacture and Installation of Prestressed Concrete Piling*. PCI Journal. 38, No. 2, March-April 1993, pp. 14-14.

PTI, *Standard Requirements for Analysis of Shallow Concrete Foundations on Expansive Soils, 1st edition*. Phoenix: Post-tensioning Institute, 2007.

PTI, *Standard Requirements for Design of Shallow Post-tensioned Concrete Foundations on Expansive Soils, 1st edition*. Phoenix: Post-tensioning Institute, 2007.

"Recommended Practice for Design, Manufacture and Installation of Prestressed Concrete Piling." *PCI Journal*, V. 38, No. 2, March-April 1993, pp. 14-41.

Report on Allowable Stresses in Concrete Piles. Portland Cement Association, Skokie, IL., 19671.

Terzaghi, Karl and R. B. Peck. *Soil Mechanics in Engineering Practice, 2nd edition*. New York: John Wiley & Sons, 1967.

TMS 402/ACI 530/ASCE 5-11, *Building Code Requirements for Masonry Structures*. Farmington Hills, MI: American Concrete Institute; Reston, VA: American Society of Civil Engineers; Boulder, CO: The Masonry Society, 2011.

WRI/CRSI-96, *Design of Slab-on-ground Foundations*. Leesburg, VA: Wire Reinforcing Institute, 1996.

Chapter 19: Concrete

General Comments

The words "cement" and "concrete" are used in several sections in this chapter. At the outset of this commentary, the distinction between cement and concrete should be made. Often, one may hear of references to cement floors, cement sidewalks and driveways, or of cement walls in describing walls made of cast-in-place or precast concrete or of concrete masonry, and the like. Cement and concrete are not interchangeable words. Cement is only one of four basic ingredients (cement, water, sand and stone) that make up the concrete material used in construction. In the end, there are only concrete structures and concrete building components. Structures of cement alone simply do not exist.

While concrete is the most versatile of the structural materials, it is also one of the most variable and complex materials used in building construction. Concrete is used in a number of different ways. It is used for building substructures, spread footings, mats, piles and foundation walls. Above-ground concrete is used for building walls and enclosures, for floors and roof construction, and for structural framework. Concrete also has a number of ancillary uses in building facilities: the construction of sidewalks, patios and driveways; for sewers and basins; for equipment bases; for fire protection purposes; as an insulating material; and so on. Concrete can be cast into any desired shape that can be structurally formed in the field or in a precasting plant with a variety of architectural finishes and surface configurations to satisfy aesthetic requirements. Mixtures can be varied to achieve strengths [2,500 to 20,000 pounds per square inch (psi) (17 238 to 137 908 kPa)] that meet the structural needs of the design. Concrete mixes can also be made to achieve good workability of the fresh material and to provide durability of the hardened product exposed to adverse environmental conditions, as well as to the conditions of wear imposed by the use of the facilities.

Because of the variable properties of the material and the numerous design and construction options available in the uses of concrete, due care and control throughout the entire construction process is necessary. This process includes: selection of materials and design of the mixture; mixing and transport of the concrete; proper construction and use of formwork; setting of the reinforcement and, in prestressed construction, the tensioning of the tendons and finally, the placement, finishing, curing and testing of the concrete material.

Purpose

The provisions of this chapter are intended to set the minimum accepted practices that apply to the design and construction of buildings or structural components using concrete.

Italics are used for text within Sections 1903 through 1905 of this code to indicate provisions that differ from ACI 318.

SECTION 1901 GENERAL

1901.1 Scope. The provisions of this chapter shall govern the materials, quality control, design and construction of concrete used in structures.

❖ Section 1901 provides information on the application and organization of the concrete provisions of Chapter 19. For additional background information on the provisions of this chapter, particularly as they relate to the inspection of concrete construction, consult PCA-EB115, *Concrete Inspection Handbook*, listed in the bibliography. Section 1901.1 provides a definitive statement on the important aspects of concrete construction covered by Chapter 19. If proper attention is given to these aspects, it will improve the chances of a quality concrete structure that is code compliant.

1901.2 Plain and reinforced concrete. Structural concrete shall be designed and constructed in accordance with the requirements of this chapter and ACI 318 as amended in Section 1905 of this code. Except for the provisions of Sections 1904 and 1907, the design and construction of slabs on grade shall not be governed by this chapter unless they transmit vertical loads or lateral forces from other parts of the structure to the soil.

❖ The structural concrete provisions of the code comply strictly with the provisions of ACI 318/318R, except for the modifications contained in Section 1905. Slabs on grade are another important exception. Slabs on grade typically only convey the gravity dead load of the slab and the gravity live load superimposed on the slab directly to the soil; thus, slabs are not considered to be the "structural concrete" that must comply with the design provisions of ACI 318. Slabs on grade, however, must comply with the durability provisions of Section 1904 and the minimum thickness and vapor retarder provisions of Section 1907.

The last sentence of Section 1901.2 clarifies that slabs on grade used to transmit "vertical loads," "lateral forces" or both from other parts of the structure to the soil must comply with all the applicable portions of this chapter and ACI 318. In many cases, slabs on grade are used to transmit lateral forces due to wind or earthquakes into the ground. A common example is a typical preengineered metal building in which the shear at the bottom of the columns is resisted by placing "hairpins" around the column anchor bolts in the footing and extending the hairpins into the slab on grade.

1901.3 Construction documents. The *construction documents* for structural concrete construction shall include:

1. The specified compressive strength of concrete at the stated ages or stages of construction for which each concrete element is designed.
2. The specified strength or grade of reinforcement.
3. The size and location of structural elements, reinforcement and anchors.
4. Provision for dimensional changes resulting from creep, shrinkage and temperature.
5. The magnitude and location of prestressing forces.
6. Anchorage length of reinforcement and location and length of lap splices.
7. Type and location of mechanical and welded splices of reinforcement.
8. Details and location of contraction or isolation joints specified for plain concrete.
9. Minimum concrete compressive strength at time of posttensioning.
10. Stressing sequence for post-tensioning tendons.
11. For structures assigned to *Seismic Design Category* D, E or F, a statement if slab on grade is designed as a structural diaphragm.

❖ This section provides a list of important information that must be included in construction documents. These items are specifically related to concrete construction and are in addition to the structural design information required by Section 1603 and additional requirements of Section 107. The list is not necessarily all-inclusive and may be expanded at the discretion of the building official.

1901.4 Special inspection. The *special inspection* of concrete elements of buildings and structures and concreting operations shall be as required by Chapter 17.

❖ This section provides a cross reference to the code's special inspection provisions for concrete found in Section 1705, which specifies that most structural concrete, with the exception of some foundations, needs to have a special inspection. ACI 318 does not contain provisions for the special inspection of concrete, except that Section 1.3.5 requires continuous inspection of the placement of reinforcement and concrete in special-moment frames that resist seismic forces in regions of high seismic risk, which correspond to Seismic Design Category D, E or F of the code.

SECTION 1902 DEFINITIONS

1902.1 General. The words and terms defined in ACI 318 shall, for the purposes of this chapter and as used elsewhere in this code for concrete construction, have the meanings shown in ACI 318 as modified by Section 1905.1.1.

❖ Section 1902 refers to the definitions of terms in ACI 318 that are associated with concrete construction. See Section 1905.1.1 for several ACI 318 definitions that are modified under the code.

SECTION 1903 SPECIFICATIONS FOR TESTS AND MATERIALS

1903.1 General. Materials used to produce concrete, concrete itself and testing thereof shall comply with the applicable standards listed in ACI 318. *Where required, special inspections and tests shall be in accordance with Chapter 17.*

❖ Section 1903 regulates the materials commonly used for both reinforced and plain concrete construction and requires the selection and acceptance of such materials to conform to the provisions of ACI 318. Chapter 3 of ACI 318 establishes the basis for the selection, specification and acceptance of materials used in concrete construction, including cement, aggregate, water, steel reinforcement and admixtures.

Cement: The words "cement" and "concrete" are used in several sections of this chapter. It is important to understand the distinction between cement and concrete. Often, one may hear of references to cement floors; cement sidewalks and driveways, cement walls in describing walls made of cast-in-place or precast concrete or of concrete masonry. Cement and concrete are not interchangeable words. Cement is only one of three basic ingredients (cement, water and aggregate) that make up the concrete material used in construction. In the end use, there are only concrete structures and building components. Structures of cement alone simply do not exist. The code permits the use of cements that meet ASTM C 150, ASTM C 595, ASTM C 845 and ASTM C 1157. General information on each of these cements is provided below.

Portland Cement—ASTM C 150:

ASTM C 150 specifies the chemical composition,

physical properties and test methods for producing eight types of portland cement as follows:

1. Type I portland cement is for general use where the special properties of other types of cement are not required. While Type I is the most widely used type of portland cement produced, it should not be used in concrete structures that will be subject to sulfate attack, such as from soil or ground water, or in large concrete placements where the temperature rises from the heat of hydration produced by the chemical reaction of cement and water and can adversely affect the construction process.
2. Type IA portland cement is an air-entraining cement for general use, as is Type I portland cement, but it is used specifically where the benefits of air entrainment are desired. The purpose of air entrainment is discussed in other parts of this commentary.
3. Type II portland cement is for general use when moderate sulfate resistance or moderate heat of hydration is desired.
4. Type IIA portland cement is for general use, as is Type II portland cement, but is used where the benefits of air entrainment are also desired.
5. Type III portland cement is a high early strength cement especially used to accelerate the hardening of wet concrete to facilitate construction scheduling.
6. Type IIIA portland cement has the same properties and is used for the same purposes as Type III portland cement, but is also used where the benefits of air entrainment are desired.
7. Type IV portland cement is used especially where a low heat of hydration is required. It is used principally in mass concrete construction such as dams, bridge abutments, heavy foundations, etc. It is seldom used for building construction.
8. Type V portland cement is for special use where high sulfate resistance is desired, and is principally used in areas where the soils or ground waters have high concentrations of sulfate.

Blended Hydraulic Cement—ASTM C 595:

ASTM C 595 specifies five types of blended cements as follows:

1. Portland blast-furnace slag cement, Type IS, is for use in general concrete construction. Portland cement (ASTM C 150) may be modified by blending fine-granulated blast-furnace slag to provide moderate sulfate resistance; Type IS (MS); air entrainment, Type IS (A); moderate heat of hydration, Type IS (MH) or any combination by adding the indicated suffixes.
2. Portland-pozzolan cement, Type IP, is for general use and will perform similar to Type I portland cement, except that it can have slightly lower 28-day concrete compressive strength and can be less resistant to deicer scaling. The cement can be modified to provide moderate sulfate resistance, Type IP (MS); air entrainment, Type IP (A); moderate heat of hydration, Type IP (MH) or any combination by adding the indicated suffixes.
3. Slag cement, Type S, is used in combination with portland cement in making concrete or in combination with hydrated lime in making masonry mortar. It can be modified to provide air entrainment, Type S (A). Section 3.2.1 of ACI 318 prohibits the use of slag cement in structural concrete.
4. Pozzolan-modified portland cement, Type I (PM), is used in general concrete construction and can be modified to provide moderate sulfate resistance, Type I (PM)(MS); air entrainment, Type I (PM)(A); or moderate heat of hydration, Type I (PM)(MH), respectively.
5. Slag-modified portland cement, Type I (SM), is used for general concrete construction and may be modified to provide moderate sulfate resistance, air entrainment or moderate heat of hydration by adding the suffixes (MS), (A) or (MH), respectively.

Expansive Hydraulic Cements—ASTM C 845:

Expansive hydraulic cements expand during the early hardening period after the concrete has set. The cements are essentially composed of hydraulic calcium aluminates and calcium sulfates. These cements are used on projects where shrinkage-compensating concrete is specified. When restrained, the expansion of shrinkage-compensating concrete induces compressive stresses that can offset tensile stresses due to drying shrinkage in the concrete member. Typically, this type of concrete is used for floor slabs and pavements to minimize or eliminate drying-shrinkage cracking. The cement is classified as Type E-1 and will include the letter-designating suffix (K), (M) or (S), depending on the chemical composition of the cement.

Performance Blended Hydraulic Cement—ASTM C 1157:

ASTM C 1157 differs from ASTM C 150 and ASTM C 595 in that it does not establish the chemical composition of the different types of cements; however, individual constituents used to manufacture ASTM C 1157 cements must comply with the requirements specified in the standard. The standard also provides for several optional requirements, including one for cement with low reactivity to alkali-reactive aggregates.

Aggregates: Section 3.3 of ACI 318 requires that concrete aggregates conform to ASTM C 33 or ASTM

C 330. All fine and coarse aggregates used in producing normal-weight concrete [135 to 160 pounds per square foot (psf) (6.5 to 7.6 kNm^2)] must comply with the provisions of ASTM C 33. All aggregates used in making lightweight structural concrete [85 to 115 psf (4 to 5.5 kNm^2)] must comply with the requirements of ASTM C 330. Both ASTM C 33 and ASTM C 330 specify physical property requirements and provide for sampling and testing of aggregate materials.

Fine aggregates used in normal-weight concrete consist of natural sand, manufactured sand or a combination of both materials. Coarse aggregates consist of either gravel, crushed gravel, crushed stone, air-cooled blast-furnace slag, crushed hydraulic-cement concrete or combinations of any of these materials. Aggregates used for making lightweight structural concrete are expanded shale, clay, slate and slag. Lightweight aggregates, such as pumice, perlite and vermiculite, are generally used in making insulating concrete, but their use is prohibited in structural concrete.

The performance characteristics of concrete are greatly affected by the selection and proportioning of aggregates in the concrete mixture. Aggregates not only affect the workability of fresh concrete, but in the hardened state they are a factor in providing adequate resistance to abrasion; freezing and thawing, drying shrinkage and fire exposure. They also affect the strength of concrete.

Harmful substances may be present in aggregates in such quantities that the performance of the concrete could be seriously affected. ASTM C 33 and ASTM C 330 set content limitations for such deleterious materials.

Sometimes the practicalities of construction necessitate deviations from normal practices. Aggregates that conform to ASTM International (ASTM) material standards are not always economically or readily available, and local, nonconforming aggregates may be successfully used in concrete. In such cases, ACI 318 permits evidence of a long history of satisfactory performance or special tests conducted to show adequate strength and durability of the concrete to serve as the basis for acceptance of nonconforming aggregate materials.

Water: The water used in concrete mixtures is required to be clear and clean. It must not contain injurious amounts of oils, acids, alkalis, salts, organic materials or other substances that are harmful to concrete or concrete reinforcement.

Potable water obtained from municipal water supply systems is normally suitable for use as mixing water for concrete; however, this does not mean that nonpotable water cannot be used in making concrete. ACI 318 permits the use of nonpotable water under specific conditions of acceptance. Water that appears discolored or has an unusual or repulsive taste or odor must undergo an analysis using applicable ASTM test methods. Evidence of its acceptability can be provided by checking service records of concrete made with such water or from other information that demonstrates the water will not be detrimental to the concrete. In the absence of such evidence, water suspected of being of questionable quality can be subject to the acceptance criteria for concrete strength and time set contained in ASTM C 94.

Environmental concerns have caused many ready-mixed producers to reuse washout water (i.e., water used to reclaim aggregate or clean out returning trucks) as mix water. This is permitted by ASTM C 94, provided the water meets the criteria outlined in the standard. In addition, the standard permits the concrete purchaser to require optional testing to determine if chloride ion, sulfate, alkali and total solids are in the water. This is particularly important when the chloride ion content of the concrete must be limited for corrosion protection when aggregate susceptible to alkali reactivity will be used, or both (see ACI 318, Section 4.3.1).

Steel reinforcement: Steel reinforcement conforming to the requirements of ACI 318 must be of the deformed type (with lugs or protrusions), except that plain steel reinforcement can be used for spirals and prestressing tendons. Under certain conditions, reinforcement can also consist of structural steel, steel pipe or steel tubing.

Section 3.5 of ACI 318, requires that reinforcement for concrete must comply with one of the following ASTM material standards:

- Deformed reinforcement:
    1. Deformed and plain billet-steel bars, ASTM A 615;
    2. Deformed and plain low-alloy steel bars, ASTM A 706 (this special reinforcement is used primarily in structures assigned to Seismic Design Category D, E or F and is not readily available in all parts of the country.);
    3. Stainless steel bars, ASTM A 955; or
    4. Rail-steel and axle-steel deformed bars, ASTM A 996.

Bar mats for concrete reinforcement must conform to ASTM A 184 and be fabricated from reinforcing bars that conform to either ASTM A 615 or ASTM A 706.

Deformed wire must conform to ASTM A 1064 but the wire must not be smaller than Size D-4 and welded deformed wire fabric for concrete reinforcement must conform to ASTM A 1064. Welded intersections must not be spaced farther apart than 16 inches (406 mm) in the direction of the principal reinforcement, except when used as stirrups.

Welded plain wire fabric for concrete reinforcement must conform to ASTM A 1064. The code considers welded plain fabric to be deformed reinforcement.

Reinforcing bars may be zinc coated (galvanized) or epoxy coated to provide corrosion resistance.

Galvanized bars must conform to ASTM A 767. Epoxy coated bars must conform to ASTM A 775 or to ASTM A 934. Zinc and epoxy dual-coated reinforcing bars shall conform to ASTM A 1055.

Steel wire and welded wire fabric may be epoxy coated in accordance with ASTM A 884 to provide corrosion resistance.

ASTM A 996 is a replacement for ASTM A 616 and ASTM A 617. Unlike its predecessors, ASTM A 996 does not provide for plain bars. Although manufacturers will be producing bars in accordance with the new standard, there are no additional changes that will affect those involved in the design, construction and inspection of reinforced concrete structures.

- Plain reinforcement:

  Plain bars for spiral reinforcement must conform to either ASTM A 615 or ASTM A 706. Smooth wire for spiral reinforcement must conform to ASTM A 1064.

- Prestressing steel:

  Steel used for prestressed concrete must conform to one of the following ASTM standards:

  1. Wire, including low-relaxation wire, must be uncoated, stress-relieved steel wire, ASTM A 421;
  2. Strand, including low-relaxation wire, must be uncoated, seven-wire, stress-relieved steel, ASTM A 416;
  3. High-strength steel bars, ASTM A 722; or
  4. Wire, strands and bars that are not specifically listed in ASTM A 421, ASTM A 416 or ASTM A 722 can be used, provided they conform to the minimum requirements of these specifications and do not have properties that make the materials less satisfactory than intended by these standard specifications.

- Structural steel:

  Structural steel used with reinforcing bars in composite compression members must conform to one of the following ASTM standards:

  1. Carbon steel shapes of structural quality, ASTM A 36;
  2. High-strength, low-alloy structural steel shapes, ASTM A 242;
  3. High-strength, low-alloy columbium-vanadium steels of structural quality, ASTM A 572; or
  4. High-strength, low-alloy structural steel with 50,000 pounds per square inch (psi) (344 750 kPa) minimum yield point to 4 inches (102 mm) thick, ASTM A 588.

- Steel pipe or tubing:

  Steel pipe or tubing for composite compression members consisting of a steel-encased concrete core must comply with one of the following ASTM standards:

  1. Grade B seamless and welded black and hot-dipped galvanized steel pipe, ASTM A 53;
  2. Cold-formed welded and seamless carbon steel structural tubing of round, square, rectangular or special shape, ASTM A 500; or
  3. Hot-formed welded and seamless carbon steel structural tubing of round, square, rectangular or special shape, ASTM A 501.

See Section 1912 for concrete-filled pipe columns.

The most important considerations before welding reinforcing bars are the weldability of the steel and the compatibility of the welding process. The American Welding Society (AWS) specifies the proper welding practices in the AWS D1.4 standard.

The ASTM standards for concrete reinforcement permitted in the code each contain the material properties of the specific types of metal. The steel chemistry set forth in any of these standards must be restricted to a given range to provide compatibility with the welding requirements of AWS D1.4. Therefore, to provide the weldability of the steel and conformance to AWS requirements, this section stipulates that the ASTM reinforcing bar specifications must be supplemented with a report addressing the material properties of the particular type of steel to be used in the welding process. However, the steel produced using ASTM A 706 is exempt from the requirement for a supplemental report because it was purposely developed for welding and already has a restricted chemistry.

Where there is a need to weld existing reinforcing bars in making alterations to existing buildings, and mill reports of the steel are not available, AWS D1.4 requires that a chemical analysis be made to determine the composition of the steel. As a practical alternative, an assumption can be made that the carbon equivalent is above 0.75 percent for the reinforcement and the steel can be welded after a preheat temperature of 500°F (260°C) is applied. While this procedure solves the welding problem, consideration must be given to the possible effects of heat damage to the concrete and the stress levels in the reinforcing steel.

Admixtures: By definition, an admixture is a material other than water, aggregates or hydraulic cement, used as an ingredient in concrete (or mortar) and added to the concrete immediately before or during mixing. There are four reasons for using admixtures:

1. To achieve certain desired properties in concrete more readily and effectively than can be obtained by other means;
2. To ensure the quality of concrete during the mixing, transporting, placing and curing operations under adverse weather conditions;
3. To overcome certain adverse conditions during concreting operations; and
4. To reduce the cost of concrete construction.

While each of the several different types of admixtures used in concrete is intended to have a beneficial effect on certain properties of concrete, at the same time, the admixture can have an adverse effect on other properties. Therefore, to make sure the overall quality of concrete intended by the design engineer and the code is met, the use of admixtures must have prior approval of the engineer.

The effectiveness of an admixture in concrete is dependent on such factors as the type and the amount of cement; the shape, gradation and proportions of the aggregate used in the concrete mixture; the water content and the mixing time, slump and temperatures of the surrounding air and concrete. To verify proper proportioning of the concrete in accordance with the requirements of ACI 318, trial mixtures should be made with the admixture and the job materials at temperatures and humidities anticipated at the job site. Trial mixtures will establish the compatibility of the admixture with the other admixtures that may be used in the concrete, as well as determine the effects of the admixture on the properties of fresh and hardened concrete. The amount of admixture used in concrete should not exceed the amount recommended by the manufacturer or the optimum amount determined by laboratory tests.

Because concentrations of chloride ions can cause severe corrosive effects on metallic materials, admixtures containing chloride as part of the formulation, including the use of calcium chloride as an antifreeze admixture during cold weather, are not allowed in prestressed concrete, concrete with aluminum embedment or concrete that is in contact with permanent installations of galvanized formwork.

Admixtures used in concrete must comply with the applicable ASTM material specifications referenced in ACI 318. There are several classes of admixtures covered by the ASTM specifications. Each class of admixture, as well as each type of admixture within a class, serve different purposes and affect the properties of concrete in different ways.

Air-entraining Admixtures—ASTM C 260:

Air-entraining admixtures incorporate microscopic air bubbles in the concrete mixture. Air entrainment serves three main purposes:

1. It improves the resistance of moist concrete to the effects of freezing and thawing cycles;
2. It improves the resistance of concrete to the actions of deicing chemicals; and
3. It improves the workability of fresh concrete.

Entrained air can be introduced in concrete by means of air-entraining cements specified in Section 1903.2, or as a separate ingredient (i.e., an admixture) in the concrete mixture (ASTM C 260) introduced before or during the mixture operations. If desired, both methods can be used together, but difficulties may arise in controlling the amount of air entrainment.

Water-reducing, Retarding and Accelerating Admixtures—ASTM C 494 and ASTM C 1017:

The ASTM C 494 material specification for chemical admixtures gives the physical requirements for each of seven types of admixture—Types A through G listed below—and prescribes the applicable test methods and other information necessary for the acceptance of admixture materials:

Type A—Water-reducing admixtures
Type B—Retarding admixtures
Type C—Accelerating admixtures
Type D—Water-reducing and retarding admixtures
Type E—Water-reducing and accelerating admixtures
Type F—Water-reducing and high-range admixtures
Type G—Water-reducing, high-range and retarding admixtures

Type A water-reducing (and set-controlling) admixtures can be used in several ways in proportioning concrete mixtures to achieve certain desired results:

1. The cement content can be maintained while reducing the amount of water, thus lowering the water-cementitious materials (w/cm) ratio and increasing the strength of concrete, as well as its durability.
2. The w/cm ratio can be maintained by reducing both the water and cement contents of the concrete resulting in a cost savings (less cement used); and
3. The slump of the concrete can be increased without increasing the w/cm ratio, thus maintaining the same strength level while improving the workability of the concrete.

The purpose of Type B chemical admixtures is to retard the rate of the setting of concrete. Such admixtures are normally used in hot weather when an accelerated rate of hardening of concrete occurs, making placing and finishing operations very difficult.

On the negative side, the use of retarding admixtures will cause reductions in concrete strength at very early ages. More important, the effects of retarding admixtures on other properties of concrete, particularly shrinkage, are unpredictable.

Type C accelerating admixtures are used mainly in cold-weather operations to shorten the setting time of concrete and accelerate the strength development at early ages. The problem is that calcium chloride is the prevalent ingredient used in producing these admixtures. Calcium chloride can cause an increase in drying shrinkage, reinforcement corrosion, discoloration of concrete (darkens concrete), loss of strength at later ages and possibly scaling. This type of admixture is prohibited in prestressed concrete construction.

Type D chemical admixtures perform as a combination of the properties of Types A and B. They are used to reduce the quantity of mixing water required to produce concrete of a given consistency (slump) and to retard the setting time of the concrete.

Type E chemical admixtures perform as a combination of the properties of Type A and C admixtures. They serve to reduce the quantity of mixing water required to produce concrete of a given consistency (slump) and accelerate the setting time and early strength development of the concrete.

Type F chemical admixtures are water-reducing admixtures similar to Type A, except that they can reduce the quantity of mixing water by 12 percent or more to produce concrete of a given consistency. Up to 30-percent water reduction can be achieved with these admixtures.

Type G chemical admixtures are high-range, water-reducing and retarding admixtures similar in function to Types F and B, except they can reduce the quantity of mixing water by 12 percent or more to produce concrete of a given consistency and retard the setting time of the concrete.

ASTM C 1017 regulates admixtures that are similar to a high-range water reducer, but are used in a different manner. These admixtures are typically referred to as "superplasticizers." Instead of being used to reduce the amount of mix water and maintain the same consistency (i.e., as a water reducer), the superplasticizer is used with the same amount of mix water that would be necessary to produce relatively low-slump concrete [e.g., $3^1/_2$ inches (89 mm)]. The superplasticizer causes an increase in the slump, which makes the concrete flowable. Two types of admixtures are provided for under ASTM C 1017: Type I, plasticizing; and Type II, plasticizing and retarding. The requirements of ASTM C 1017 are similar to those of ASTM C 494 (Types F and G); however, in order to comply with ASTM C 1017, the admixture must increase the slump at least $3^1/_2$ inches (89 mm) compared to the control specimen.

Pozzolans—ASTM C 618:

Admixtures made of raw or natural pozzolans and fly ash are covered by ASTM C 618 material specifications. While these mineral admixtures will benefit as well as adversely affect the properties of concrete, they are widely used as partial replacements for portland cement and often result in more economical concrete construction.

By definition, a pozzolan is a siliceous or aluminosiliceous material that in itself possesses little or no cementitious value, but will, in finely divided form and in the presence of water, chemically react with the calcium hydroxide released by the hydration of portland cement to form compounds possessing cementitious properties. Pozzolanic materials used for making admixtures come from a number of natural sources, such as diatomaceous earth, opaline cherts, clays, shales, volcanic tuffs and pumicites. The ASTM C 618 standard designates these materials as Class N pozzolans.

Fly ash is the most widely used of all the mineral admixtures in concrete. Essentially, it is a finely divided residue (a powder) that results from the combustion of pulverized coal. Large quantities of fly ash are produced in coal-burning electric power-generating plants. Fly ash is primarily silicate glass containing silica, aluminum, iron and calcium. It also contains small amounts of magnesium, sulfur, sodium, potassium and carbon. ASTM C 618 designates these materials as Class F and C fly ashes, depending on the percentages of calcium and carbon contents. Each class of fly ash affects the properties of concrete in different ways. It is a complex subject, and for more information on fly ash and other mineral admixtures, refer to the Portland Cement Association publications cited in the bibliography.

Aside from economic reasons, pozzolan and fly ash admixtures are used in concrete for any one or more of the following purposes: increase sulfate resistance, reduce expansions in concrete due to alkali-silica reactions, reduce the permeability of concrete, decrease heat generation or improve the workability and finishing qualities of the concrete.The properties of pozzolans and fly ash may vary widely. Fly ash, in particular, can have varying amounts of carbon, silica, sulfur, alkalis and other ingredients that can adversely affect the strength, air content and durability of concrete. Because it is a residue and not a controlled product, the properties of fly ash can vary widely, not only between productions coming from different electric power-generating plants, but also between batches coming from the same plant.

Slag—ASTM C 989:

Ground-granulated blast-furnace slag complying with the requirements of ASTM C 989 provides characteristics similar to those exhibited by pozzolan admixtures. Ground-granulated blast-furnace slag and fly ash are often used in combination as admixtures in portland cement (ASTM C 150) concrete. The practice has been growing in recent years due to energy conservation considerations, but mainly because of savings in the cost of concrete when fly ash or slag or both fly ash and slag are used to partially replace cement.

Silica Fume—ASTM C 1240:

Silica fume is a byproduct of the manufacture of elemental silicon or ferro-silicon alloys in electric arc furnaces. When used in conjunction with a high-range water-reducing admixture, it is possible to produce concrete with compressive strengths of 20,000 psi (138 MPa) or higher. It is also used to achieve a very dense cement paste matrix to reduce the permeability of concrete, thus providing better corrosion protection to reinforcing steel in concrete subjected to deicing chemicals (e.g., parking garages). Tests have shown that silica fume greatly improves the sulfate resistance of concrete. For this reason, the code requires that concrete to be subjected to very severe sulfate exposure be made with Type V ASTM C 150 cement plus a pozzolan, such as silica fume, that has shown through tests or service records to improve sulfate resistance (see Table 4.3.1, ACI 318).

Storage of materials: Section 3.7 of ACI 318 regulates the storage of concrete ingredients and prohibits use of such materials if they have suffered contamination or deterioration. Cement stored in contact with damp air or moisture sets more slowly and has less strength than cement that is maintained in a dry condition. Bulk cement can be stored in weather-tight containers, such as steel bins or concrete silos.

Bags of cement are typically stored in warehouses or sheds where the relative humidity is kept as low as possible. Bags of cement must not be stored on damp floors or on earth (job-site conditions; dirt), but should rest on pallets and be stacked close together to reduce air circulation. Bags should be covered with a waterproof covering to prevent deterioration when exposed to weather.

Aggregates are normally stockpiled on the ground in areas that have been stripped of all vegetation and debris and have been leveled to facilitate handling operations. When two or more types or sizes of aggregates are involved, the stockpiles must be so arranged as to avoid crowding or overlapping with other materials or other sizes of like aggregates. In removing material from a stockpile, a layer of aggregate of sufficient thickness must be left on the ground to keep the handling equipment, such as a front-end loader or clamshell, from picking up earth along with the aggregate. Stockpiles typically are built up in centric uniform layers to prevent coarse aggregates from rolling down the sides and segregating the material.

Ordinarily, cement does not remain in storage long, but it can be stored for long periods without deterioration. At the time of use, however, cement is required to be free flowing and relatively free of lumps. If the lumps do not readily break up or there is other reason to doubt the quality of the cement, the material can be tested for strength or loss on ignition in accordance with the ASTM standard under which it was manufactured before it is used in concrete.

If there is reason to doubt that any aggregate taken from a stockpile is not clean or appears to be contaminated, any of the applicable standard ASTM tests that would detect impurities should be made before the material is used for making concrete.

1903.2 Glass fiber reinforced concrete. *Glass fiber reinforced concrete (GFRC) and the materials used in such concrete shall be in accordance with the PCI MNL 128 standard.*

❖ This section references a document that addresses the quality of materials used in glass fiber reinforced concrete panels. Glass fiber reinforced concrete is utilized in the prescriptive fire-resistance ratings for wall assemblies [see Table 721.1(2), Item 15].

1903.3 Flat wall insulating concrete form (ICF) systems. *Insulating concrete form material used for forming flat concrete walls shall conform to ASTM E 2634.*

❖ This section references ASTM E 2634 for the material used to make forms for insulating concrete form (ICF) systems using molded expanded polystyrene (EPS) insulation panels. ICFs are rigid plastic foam forms used in the construction of cast-in-place concrete structural members. The ICFs remain in place as permanent building insulation for energy-efficient, cast-in-place, reinforced concrete walls, floors, roofs, beams and columns. The forms are interlocking modular units that are dry-stacked (without mortar) and filled with concrete. ICF are generally manufactured from polystyrene or polyurethane. Reinforcing steel is placed in the forms and then concrete is pumped into the cavity to form the structural portion of the walls. The forms provide thermal and acoustic insulation, space to run electrical conduit and plumbing and backing for interior and exterior finishes.

ASTM E 2634 applies to ICF systems that consist of molded EPS insulation panels that are connected by cross ties and act as permanent formwork for cast-in-place reinforced concrete structural members such as reinforced concrete beams; lintels; exterior and interior bearing and nonbearing walls; foundations and retaining walls. ASTM E 2634 lists test methods appropriate for establishing ICF system performance as a permanent concrete forming system.

SECTION 1904 DURABILITY REQUIREMENTS

1904.1 Exposure categories and classes. Concrete shall be assigned to exposure classes in accordance with the durability requirements of ACI 318 based on:

1. Exposure to freezing and thawing in a moist condition or deicer chemicals;
2. Exposure to sulfates in water or soil;
3. Exposure to water where the concrete is intended to have low permeability; and

4. Exposure to chlorides from deicing chemicals, salt, saltwater, brackish water, seawater or spray from these sources, where the concrete has steel reinforcement.

❖ This section states the criteria that are the basis for assigning concrete to the appropriate exposure category, either F, S, P or C, in Table 4.2.1 of ACI 318. The table further assigns a class within each exposure category, based on the severity of the exposure.

1904.2 Concrete properties. Concrete mixtures shall conform to the most restrictive maximum water-cementitious materials ratios, maximum cementitious admixtures, minimum air-entrainment and minimum specified concrete compressive strength requirements of ACI 318 based on the exposure classes assigned in Section 1904.1.

Exception: *For occupancies and appurtenances thereto in Group R occupancies that are in buildings less than four stories above grade plane, normal-weight aggregate concrete is permitted to comply with the requirements of Table 1904.2 based on the weathering classification (freezing and thawing) determined from Figure 1904.2 in lieu of the durability requirements of ACI 318.*

❖ Concrete subject to the exposures listed in Table 4.3.1 of ACI 318 must comply with the limitations on the w/cm ratio, cementitious admixtures and air content limitations for concretes subject to freezing and specified compressive strength. The intent of the upper limit on the w/cm ratio and lower limit on specified compressive strength is to achieve dense, impermeable or water-tight concrete. Note that the minimum specified compressive strength applies to both normal-weight and lightweight aggregate concrete, while the maximum w/cm ratio applies only to normal-weight concrete. The variability of moisture absorption in lightweight aggregate makes calculation of the w/cm ratio uncertain. The intent of an upper limit on cementitious admixtures and lower limit of air content is to control cracking in concrete exposed to freeze-thaw cycles.

In the exception, in climatic areas where concrete construction is exposed to freeze-thaw cycles or deicer chemicals, concrete must be of the quality necessary to resist the harmful effects of such weather conditions. In all buildings that are subject to special exposure conditions, except those of Group R occupancies that are less than four stories in height, the w/cm ratio and specified compressive strength of the concrete is required to be designed to conform to the requirements of Table 4.3.1 of ACI 318. The specified compressive strengths of concrete in low-rise Group R occupancies (less than four stories) need only comply with Table 1904.2, which mandates a minimum specified compressive strength as a function of the concrete element and exposure.

Table 4.3.1 of ACI 318 provides w/cm ratios for normal-weight aggregate concrete in conjunction with various exposure conditions. The values for w/cm ratios are the maximum permissible without regard for the relative strength of concrete. A control on the w/cm ratio is provided in addition to the minimum specified compressive strength values (f'_c) for normal-weight concrete, because the w/cm ratio is a more reliable indication of durability and water tightness of concrete. The table only gives a minimum specified compressive strength value for proportioning concrete mixtures using lightweight aggregates because the amount of water absorbed by lightweight aggregates can be difficult to determine, making the calculation of w/cm ratios unreliable.

Table 4.3.1 of ACI 318 references Table 4.4.1 of ACI 318 for air content requirements for concrete exposed to freeze-thawing. Limits of air content in Table 4.4.1 are +/- 1.5 percent. The air content is a percentage of the total concrete mixture for normal-weight or lightweight concrete.

Type F3 concrete has additional cementitious material limitations placed on the concrete mix. Cementitious materials that may be added to Type F3 concrete–concrete subject to frequent freeze-thaw cycles with moisture present and exposure to deicing chemicals is anticipated—include fly ash, other pozzolans, silica fume and slag. Table 4.3.1 of ACI 318 references Table 4.4.2 of ACI 318 for maximum cementitious material quantities for Type F3 concretes. Other exposure categories do not limit the maximum amount of cementitious material; rather the type of material is limited by reference to ASTM C 150, C 595, or C 1157.

TABLE 1904.2. See page 19-10.

❖ This table gives the minimum specified compressive strength values (f'_c) required for concrete used in components, such as walls, slabs and foundations, in Group R buildings that are less than four stories, where they are subject to either freezing and thawing while wet or exposed to deicer chemicals. The main purpose of this table is to provide adequate performance of the concrete by requiring a minimum compressive strength (f'_c) based on the classifications of weathering determined from code Figure 1904.2. Note b requires air-entrained concrete for garage floor slabs located in moderate and severe-weather regions. The purpose of this air-entrainment is to protect the slab from deterioration by deicer chemicals that may be carried into the garage on vehicle tires and the underside of the vehicle. Most building owners prefer a garage slab with a smooth trowel finish to facilitate cleaning, but experience has shown that it is difficult to get a smooth, durable finish by using a steel trowel on concrete with total air content in the range of 5 to 7 percent. Note d provides an option to air-entrained concrete that has been successfully applied to garage floors in moderate and severe-weather regions. Field experience has shown that durable concrete may be obtained with a lesser amount of air-entrainment if the specified compressive strength of the concrete is increased to 4,000 psi (27.6 MPa). The higher concrete strength is accompanied by a denser cement-paste matrix, which results in the concrete being less permeable.

FIGURE 1904.2. See page 19-11.

❖ Figure 1904.2 shows the geographic regions within the U.S. mainland, which are classified as having negligible, moderate and severe weather exposures. The sole purpose of the map is to locate those weathering areas in the country where the requirements for minimum specified compressive strength of concrete of Table 1904.2 apply. Note that the boundary lines defining the weather regions are only approximate. Therefore, building officials in communities located near a boundary line must establish applicable exposure classifications (severe, moderate or negligible) based on their knowledge of weather conditions in the area and the relative performance of concrete.

SECTION 1905 MODIFICATIONS TO ACI 318

1905.1 General. The text of ACI 318 shall be modified as indicated in Sections 1905.1.1 through 1905.1.10.

❖ Section 1905 contains modifications to ACI 318, the standard adopted by reference in Section 1901.2. Many of the modifications are related to the seismic design of concrete structures. Text that differs from or is not found in ACI 318 is printed in italics to identify those differences.

1905.1.1 ACI 318, Section 2.2. Modify existing definitions and add the following definitions to ACI 318, Section 2.2.

DESIGN DISPLACEMENT. Total lateral displacement expected for the design-basis earthquake, *as specified by Section 12.8.6 of ASCE 7.*

❖ This International Building Code® (IBC®) modification to the definition given in ACI 318 provides a more specific code reference. Design displacement is defined to be the same as δ_X (or interstory drift based on δ_X) as given in Section 12.8.6 of ASCE 7.

DETAILED PLAIN CONCRETE STRUCTURAL WALL. *A wall complying with the requirements of Chapter 22, including 22.6.7.*

❖ The classification of shear wall systems is indicative of the amount of reinforcement and the type of detailing to be provided. Shear wall system classifications are also used to identify the appropriate earthquake design coefficients and system limitations in ASCE 7. This definition describes a type of shear wall system permitted in the earthquake load requirements of ASCE 7, but not defined under ACI 318. The definition provides a specific reference to requirements added to ACI 318 by Section 1905.1.7. This supplements ACI 318 provisions for plain concrete by providing additional requirements for "detailed plain concrete structural walls." Note that ACI 318 refers to a "shear wall" as a "structural wall."

ORDINARY PRECAST STRUCTURAL WALL. *A precast wall complying with the requirements of Chapters 1 through 18.*

❖ This definition adds an explanation of shear wall system that is not explicitly defined in ACI 318. The classification of shear wall systems reflects the amount of reinforcement and the type of detailing to be provided. Shear wall system classifications are also used to identify the appropriate earthquake design coefficients and limitations in ASCE 7. The ACI 318 definition of "Ordinary reinforced concrete structural wall" includes both cast-in-place and precast walls. Because ordinary reinforced concrete structural walls constructed of precast elements are not expected to perform as well as a cast-in-place wall when subjected to the same earthquake ground motion, a precast ordinary reinforced concrete structural wall does not warrant the same response modification coefficient, *R*, as a comparable cast-in-place wall. In order to assign distinct design coefficients to precast shear wall systems, they must be defined separately. Note that ACI 318 refers to a "shear wall" as a "structural wall."

TABLE 1904.2
MINIMUM SPECIFIED COMPRESSIVE STRENGTH (f'_c)

TYPE OR LOCATION OF CONCRETE CONSTRUCTION	MINIMUM SPECIFIED COMPRESSIVE STRENGTH (f'_c at 28 days, psi)		
	Negligible exposure	Moderate exposure	Severe exposure
Basement walls[c] and foundations not exposed to the weather	2,500	2,500	2,500[a]
Basement slabs and interior slabs on grade, except garage floor slabs	2,500	2,500	2,500[a]
Basement walls[c], foundation walls, exterior walls and other vertical concrete surfaces exposed to the weather	2,500	3,000[b]	3,000[b]
Driveways, curbs, walks, patios, porches, carport slabs, steps and other flatwork exposed to the weather, and garage floor slabs	2,500	3,000[b, d]	3,500[b, d]

For SI: 1 pound per square inch = 0.00689 MPa.

a. Concrete in these locations that can be subjected to freezing and thawing during construction shall be of air-entrained concrete in accordance with Section 1904.2.

b. Concrete shall be air entrained in accordance with ACI 318.

c. Structural plain concrete basement walls are exempt from the requirements for exposure conditions of Section 1904.2.

d. For garage floor slabs where a steel trowel finish is used, the total air content required by ACI 318 is permitted to be reduced to not less than 3 percent, provided the minimum specified compressive strength of the concrete is increased to 4,000 psi.

ORDINARY REINFORCED CONCRETE STRUCTURAL WALL. A *cast-in-place* wall complying with the requirements of Chapters 1 through 18.

❖ The classification of shear wall systems is indicative of the amount of reinforcement and the type of detailing to be provided. Shear wall system classifications are also used to identify the appropriate earthquake design coefficients and limitations in ASCE 7. The definition of "Ordinary reinforced concrete structural wall" in ACI 318 includes both cast-in-place and precast walls. This modification clarifies that under the code the term "ordinary reinforced concrete structural wall" applies only to cast-in-place shear wall systems. Note that ACI 318 refers to a "shear wall" as a "structural wall."

ORDINARY STRUCTURAL PLAIN CONCRETE WALL. A wall complying with the requirements of Chapter 22, *excluding 22.6.7.*

❖ The classification of shear wall systems reflects the amount of reinforcement and the type of detailing to be provided. Shear wall system classifications are also used to identify the appropriate earthquake design coefficients and limitations in ASCE 7. This modification clarifies that the additional code requirements intended for detailed plain concrete shear walls do not apply to ordinary plain concrete shear walls. Note that ACI 318 refers to a "shear wall" as a "structural wall."

SPECIAL STRUCTURAL WALL. A cast-in-place or precast wall complying with the requirements of 21.1.3 through 21.1.7, 21.9 and 21.10, as applicable, in addition to the requirements for ordinary reinforced concrete structural walls *or ordinary precast structural walls, as applicable. Where ASCE 7 refers to a "special reinforced concrete structural wall," it shall be deemed to mean a "special structural wall."*

❖ This modified definition coordinates terminology used in ACI 318 with that used in ASCE 7. Under ACI 318-05, a "special reinforced concrete structural wall" referred to a cast-in-place wall, while a separate definition used for a "Special precast structural wall." In

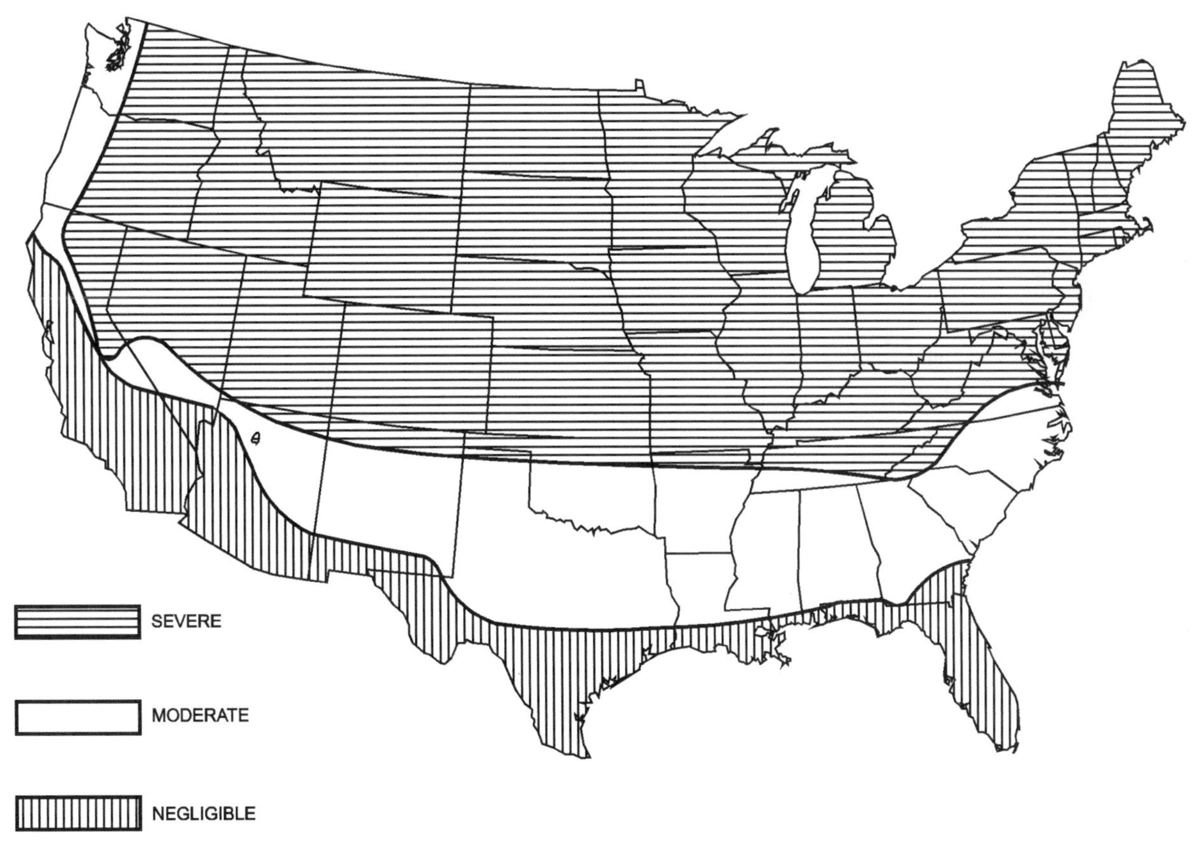

FIGURE 1904.2
WEATHERING PROBABILITY MAP FOR CONCRETE[a, b, c]

a. Lines defining areas are approximate only. Local areas can be more or less severe than indicated by the region classification.

b. A "severe" classification is where weather conditions encourage or require the use of deicing chemicals or where there is potential for a continuous presence of moisture during frequent cycles of freezing and thawing. A "moderate" classification is where weather conditions occasionally expose concrete in the presence of moisture to freezing and thawing, but where deicing chemicals are not generally used. A "negligible" classification is where weather conditions rarely expose concrete in the presence of moisture to freezing and thawing.

c. Alaska and Hawaii are classified as severe and negligible, respectively.

ACI 318-08, the separate definition for "Special precast structural wall" was removed and the former definition of "Special reinforced concrete structural wall" was revised to "Special structural wall." The current definition includes both cast-in-place and precast walls. This modification to the ACI 318 definition of "Special structural wall" was necessary since Table 12.2-1 of ASCE 7 uses the former term "Special reinforced concrete structural wall." In addition, the ASCE 7 table does not refer to "special structural wall."

WALL PIER. *A wall segment with a horizontal length-to-thickness ratio of at least 2.5, but not exceeding 6, whose clear height is at least two times its horizontal length.*

❖ See Figure 1905.1.1.

1905.1.2 ACI 318, Section 21.1.1. Modify ACI 318 Sections 21.1.1.3 and 21.1.1.7 to read as follows:

> *21.1.1.3 - Structures assigned to Seismic Design Category A* shall satisfy requirements of Chapters 1 to 19 and 22*; Chapter 21 does not apply*. Structures assigned to *Seismic Design Category* B, C, D, E or F also shall satisfy 21.1.1.4 through 21.1.1.8, as applicable. *Except for structural elements of plain concrete complying with Section 1905.1.8 of the International Building Code, structural elements of plain concrete are prohibited in structures assigned to Seismic Design Category C, D, E or F.*
>
> 21.1.1.7 - Structural systems designated as part of the seismic force-resisting system shall be restricted to those *permitted by ASCE 7*. Except for *Seismic Design Category* A, for which Chapter 21 does not apply, the following provisions shall be satisfied for each structural system designated as part of the seismic force-resisting system, regardless of the *Seismic Design Category*:
>
> (a) Ordinary moment frames shall satisfy 21.2.
>
> (b) Ordinary reinforced concrete structural walls *and ordinary precast structural walls* need not satisfy any provisions in Chapter 21.
>
> (c) Intermediate moment frames shall satisfy 21.3.
>
> (d) Intermediate precast *structural* walls shall satisfy 21.4.
>
> (e) Special moment frames shall satisfy 21.5 through 21.8.
>
> (f) Special structural walls shall satisfy 21.9.
>
> (g) Special structural walls constructed using precast concrete shall satisfy 21.10.
>
> All special moment frames and special structural walls shall also satisfy 21.1.3 through 21.1.7.

❖ This modification provides clarifications that are necessary for correct application of the earthquake design provisions.

1905.1.3 ACI 318, Section 21.4. Modify ACI 318, Section 21.4, by renumbering Section 21.4.3 to become 21.4.4 and adding new Sections 21.4.3, 21.4.5, 21.4.6 and 21.4.7 to read as follows:

> *21.4.3 - Connections that are designed to yield shall be capable of maintaining 80 percent of their design strength at the deformation induced by the design displacement or shall use Type 2 mechanical splices.*
>
> 21.4.4 - Elements of the connection that are not designed to yield shall develop at least 1.5 S_y.
>
> *21.4.5 - Wall piers in Seismic Design Category D, E or F shall comply with Section 1905.1.4 of the International Building Code.*
>
> *21.4.6 - Wall piers not designed as part of a moment frame in buildings assigned to Seismic Design Category C shall have transverse reinforcement designed to resist the shear forces determined from 21.3.3. Spacing of transverse reinforcement shall not exceed 8 inches (203 mm). Transverse reinforcement shall be extended beyond the pier clear height for at least 12 inches (305 mm).*
>
> ***Exceptions:***
>
> 1. *Wall piers that satisfy 21.13.*
> 2. *Wall piers along a wall line within a story where other shear wall segments provide lateral support to the wall piers and such segments have a total stiffness of at least six times the sum of the stiffnesses of all the wall piers.*
>
> *21.4.7 - Wall segments with a horizontal length-to-thickness ratio less than 2.5 shall be designed as columns.*

❖ This modification to ACI 318 affects connections between precast concrete wall panels in intermediate precast structural walls. Section 21.4.2 of ACI 318 restricts yielding to steel elements or reinforcement in connections between wall panels, or between wall

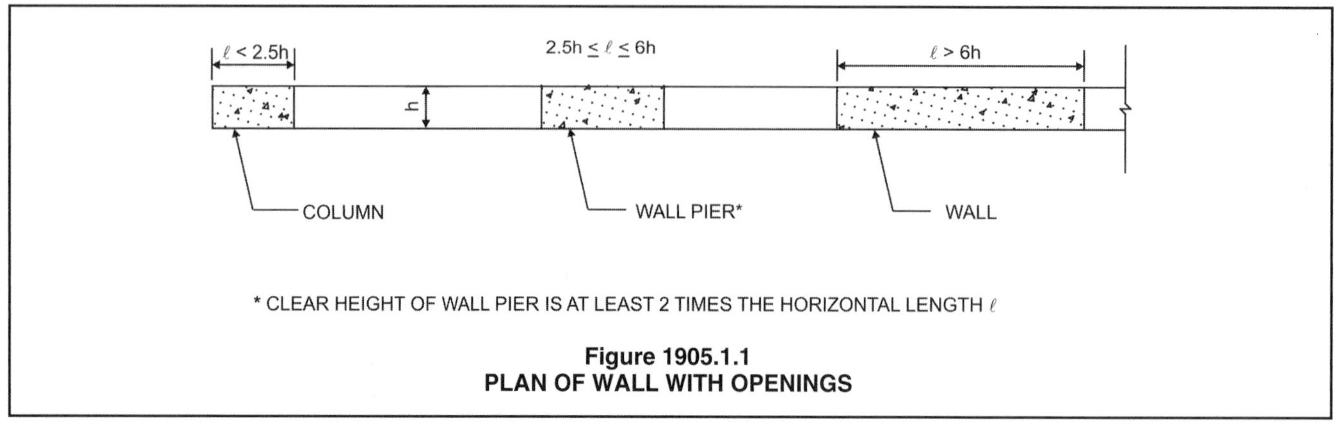

Figure 1905.1.1
PLAN OF WALL WITH OPENINGS

panels and the foundation of intermediate precast structural walls. This restriction is also made applicable by Section 21.10.2 to special structural walls of precast concrete (see Section 1905.1.5).

The modification to Section 21.4.3 of ACI 318 is consistent with a clarification of ductile elements that is made in the 2003 National Earthquake Hazards Reduction Program (NEHRP) provisions. Connection elements that are designed to yield must be capable of maintaining 80 percent of their design strength up to the stage where a structure is subjected to the design displacement as defined in ACI 318 (see Section 1905.1.1).

Section 21.4.5 clarifies that wall piers in buildings assigned to Seismic Design Category D, E or F are required to comply with Section 1905.1.4.

Section 21.4.6 adds provisions for wall piers for intermediate precast structural walls that parallel those for special structural walls in Section 1905.1.4 and allows transverse reinforcement spacing of 8 inches (203 mm).

1905.1.4 ACI 318, Section 21.9. Modify ACI 318, Section 21.9, by deleting Section 21.9.8 and replacing with the following:

21.9.8 - Wall piers and wall segments.

21.9.8.1 - Wall piers not designed as a part of a special moment frame shall have transverse reinforcement designed to satisfy the requirements in 21.9.8.2.

Exceptions:

1. *Wall piers that satisfy 21.13.*
2. *Wall piers along a wall line within a story where other shear wall segments provide lateral support to the wall piers and such segments have a total stiffness of at least six times the sum of the stiffnesses of all the wall piers.*

21.9.8.2 - Transverse reinforcement with seismic hooks at both ends shall be designed to resist the shear forces determined from 21.6.5.1. Spacing of transverse reinforcement shall not exceed 6 inches (152 mm). Transverse reinforcement shall be extended beyond the pier clear height for at least 12 inches (305 mm).

21.9.8.3 - Wall segments with a horizontal length-to-thickness ratio less than 2.5 shall be designed as columns.

❖ This ACI 318 modification is based on the 2003 NEHRP provisions. It replaces Section 21.9.8 of the ACI 318 provisions for special structural walls. "Wall piers" are defined in Section 1905.1.1 and illustrated in Figure 1905.1.1. Wall segments that meet the definition of "Wall pier" must be designed as required by this section.

Wall pier detailing requirements are stated in Section 21.9.8.2, but three alternatives are provided in Section 21.9.8.1. The first alternative is that the wall pier may be designed as part of a special moment frame. Two other alternatives are provided by the exceptions. Exception 1 applies to wall piers that are not part of the lateral-force-resisting system and that satisfy the requirement for deformation compatibility. Exception 2 applies to wall piers laterally supported by much stiffer shear walls along the same line within a story.

Wall pier requirements were first introduced in the 1988 *Uniform Building Code*® (UBC™) out of concern that thin column-like elements between openings in shear walls were being designed without proper transverse reinforcement. In order to prevent premature shear failure, a wall pier must be designed for the shear force computed in accordance with ACI 318 requirements for special moment frame members subjected to axial load and bending. Section 21.9.8.2 requires transverse reinforcement to be spaced at no more than 6 inches (152 mm), and to be extended beyond the clear pier height by at least 12 inches (305 mm).

Section 21.9.8.3 clarifies that a segment of wall with a horizontal length-to-thickness ratio of less than $2^1/_2$ is considered a column and must be designed accordingly.

1905.1.5 ACI 318, Section 21.10. Modify ACI 318, Section 21.10.2, to read as follows:

21.10.2 - Special structural walls constructed using precast concrete shall satisfy all the requirements of 21.9 for cast-in-place special structural walls in addition to Sections 21.4.2 *through* 21.4.4.

❖ The modification made to ACI 318 in this section coordinates with the sections that are referenced—and are themselves modified in Section 1905.1.3.

1905.1.6 ACI 318, Section 21.12.1.1. Modify ACI 318, Section 21.12.1.1, to read as follows:

21.12.1.1 - Foundations resisting earthquake-induced forces or transferring earthquake-induced forces between a structure and ground shall comply with the requirements of Section 21.12 and other applicable provisions of ACI 318 *unless modified by Chapter 18 of the International Building Code.*

❖ This ACI 318 modification points out that there are foundation requirements in the code that, in accordance with Section 102.4 of the IBC, would take precedence over those of ACI 318.

1905.1.7 ACI 318, Section 22.6. Modify ACI 318, Section 22.6, by adding new Section 22.6.7 to read as follows:

22.6.7 - Detailed plain concrete structural walls.

22.6.7.1 - Detailed plain concrete structural walls are walls conforming to the requirements of ordinary structural plain concrete walls and 22.6.7.2.

22.6.7.2 - Reinforcement shall be provided as follows:

(a) *Vertical reinforcement of at least 0.20 square inch (129 mm^2) in cross-sectional area shall be provided continuously from support to support at each corner, at each side of each opening and at the ends of walls. The continuous vertical bar* *required beside an opening is permitted to substitute for one of the two No. 5 bars required by 22.6.6.5.*

(b) Horizontal reinforcement at least 0.20 square inch (129 mm^2) in cross-sectional area shall be provided:

1. Continuously at structurally connected roof and floor levels and at the top of walls;

2. At the bottom of load-bearing walls or in the top of foundations where doweled to the wall; and

3. At a maximum spacing of 120 inches (3048 mm).

Reinforcement at the top and bottom of openings, where used in determining the maximum spacing specified in Item 3 above, shall be continuous in the wall.

❖ This ACI 318 modification provides requirements for detailed plain concrete structural walls that are based on the 2003 NEHRP provisions (see definition in Section 1905.1.1). In addition to conforming to the requirements for ordinary plain concrete shear walls, detailed plain concrete shear walls must have some reinforcement to provide limited ductility and to keep the concrete from disintegrating should cracking occur before or during a seismic event. Detailed plain concrete shear walls may only be used in structures assigned to Seismic Design Category A or B. The area of vertical reinforcement of 0.20 square inch (129 mm^2) corresponds to the area of a No. 4 reinforcing bar. The provision requiring vertical reinforcement at the "ends of walls" is intended to also require a bar on each side of a contraction (i.e., control) joint.

Item 1 requires that a bar be located at the top of a wall. Where the provisions of Item 1 require a closer spacing of horizontal reinforcement than Item 3 [i.e., 120 inches or 10 feet (3048 mm)], then the closer spacing applies.

1905.1.8 ACI 318, Section 22.10. Delete ACI 318, Section 22.10, and replace with the following:

22.10 - Plain concrete in structures assigned to Seismic Design Category C, D, E or F.

22.10.1 - Structures assigned to Seismic Design Category C, D, E or F shall not have elements of structural plain concrete, except as follows:

(a) Structural plain concrete basement, foundation or other walls below the base are permitted in detached one- and two-family dwellings three stories or less in height constructed with stud-bearing walls. In dwellings assigned to Seismic Design Category D or E, the height of the wall shall not exceed 8 feet (2438 mm), the thickness shall not be less than 7$^1/_2$ inches (190 mm), and the wall shall retain no more than 4 feet (1219 mm) of unbalanced fill. Walls shall have reinforcement in accordance with 22.6.6.5.

(b) Isolated footings of plain concrete supporting pedestals or columns are permitted, provided the projection of the footing beyond the face of the supported member does not exceed the footing thickness.

***Exception:** In detached one- and two-family dwellings three stories or less in height, the projection of the footing beyond the face of the supported member is permitted to exceed the footing thickness.*

(c) Plain concrete footings supporting walls are permitted, provided the footings have at least two continuous longitudinal reinforcing bars. Bars shall not be smaller than No. 4 and shall have a total area of not less than 0.002 times the gross cross-sectional area of the footing. For footings that exceed 8 inches (203 mm) in thickness, a minimum of one bar shall be provided at the top and bottom of the footing. Continuity of reinforcement shall be provided at corners and intersections.

Exceptions:

1. In Seismic Design Categories A, B and C, detached one- and two-family dwellings three stories or less in height constructed with stud-bearing walls, are permitted to have plain concrete footings without longitudinal reinforcement.

2. For foundation systems consisting of a plain concrete footing and a plain concrete stemwall, a minimum of one bar shall be provided at the top of the stemwall and at the bottom of the footing.

3. Where a slab on ground is cast monolithically with the footing, one No. 5 bar is permitted to be located at either the top of the slab or bottom of the footing.

❖ Structural plain concrete walls are prohibited in structures assigned to Seismic Design Category C due to their inability to respond inelastically, which is considered to be essential if they are to perform satisfactory during the design earthquake. Item (a) permits this in one- and two-family dwellings where the stated conditions are met.

Plain concrete footings supporting columns or pedestals are permitted if the projection of the footing beyond the face of the column or pedestal does not exceed the footing thickness. The limitation on the projection was judged to be prudent because of the empirical manner in which the vertical component of earthquake ground motion is considered. With the limitation on the projection, a flexural or shear failure in the footing is virtually impossible.

Wall footings of structural plain concrete are permitted in all structures assigned to Seismic Design Category C. Some longitudinal reinforcement is specified in order to keep the footing together should it crack before or during an earthquake. The reinforcement will also allow the footing to act as a "reinforced" concrete member should the earthquake-induced ground motion cause the supported structure to be subjected to differential vertical movement. For footings 8 inches (203 mm) or less in thickness, the bars will typically be located with 3 inches (76 mm) of concrete cover to the bottom of the footing as required by Section 7.7 of ACI 318. For footings thicker than 8 inches (203 mm), one bar will be located with respect to the bottom as noted previously, and the other with $1^1/_2$ or 2 inches (38 or 51 mm) of cover from the top of the footing as required under Section 7.7 of ACI 318. Typically, two No. 4 [total of 0.40 square inches (258 mm^2)] or two No. 5 [total of 0.62 square inches (400 mm^2)] bars will suffice, with the former being adequate for footings up to 200 square inches (0.129 m^2) in cross-sectional area [0.40 in. 2/0.002 = 200 in. 2 (258 mm^2/.002 × 106 = 0.129 m^2)], and the latter being suitable for footings up to 310 square inches (0.200 m^2) in cross-sectional area [0.62 in. 2/0.002 = 310 in. 2 (400 mm^2/.002 × 106 = 0.200 m^2)]. Where one bar will be placed in the bottom of the footing and the other will be placed in the top of the stem wall, the concrete surface of the top of the footing should be intentionally roughened so that a good bond develops between the two concrete placements.

Exception 1 permits structural plain concrete footings without longitudinal reinforcement in buildings assigned to Seismic Design Categories A, B and C for certain detached one- and two-family dwellings. Where a slab on ground is cast monolithically with the footing, typically referred to as a "turned-down slab," Exception 3 allows one No. 5 bar (versus two No. 4 bars) to be located in either the top or bottom of the footing. In this case, the top of the footing is the top of the slab.

1905.1.9 ACI 318, Section D.3.3. Delete ACI 318 Sections D.3.3.4 through D.3.3.7 and replace with the following:

D.3.3.4 - The anchor design strength associated with concrete failure modes shall be taken as $0.75\phi N_n$ and $0.75\phi V_n$, where ϕ is given in D4.3 or D4.4 and N_n and V_n are determined in accordance with D5.2, D5.3, D5.4, D6.2 and D6.3, assuming the concrete is cracked unless it can be demonstrated that the concrete remains uncracked.

D.3.3.5 - Anchors shall be designed to be governed by the steel strength of a ductile steel element as determined in accordance with D.5.1 and D.6.1, unless either D.3.3.6 or D.3.3.7 is satisfied.

Exceptions:

1. *Anchors designed to resist wall out-of-plane forces with design strengths equal to or greater than the force determined in accordance with ASCE 7 Equation 12.11-1 or 12.14-10 need not satisfy* Section D.3.3.5.
2. *D.3.3.5 need not apply and the design shear strength in accordance with D.6.2.1(c) need not be computed for anchor bolts attaching wood sill plates of bearing or nonbearing walls of light-frame wood structures to foundations or foundation stem walls provided all of the following are satisfied:*
   - *2.1. The allowable in-plane shear strength of the anchor is determined in accordance with AF&PA NDS Table 11E for lateral design values parallel to grain.*
   - *2.2. The maximum anchor nominal diameter is $^5/_8$ inches (16 mm).*
   - *2.3. Anchor bolts are embedded into concrete a minimum of 7 inches (178 mm).*
   - *2.4. Anchor bolts are located a minimum of $1^3/_4$ inches (45 mm) from the edge of the concrete parallel to the length of the wood sill plate.*
   - *2.5. Anchor bolts are located a minimum of 15 anchor diameters from the edge of the concrete perpendicular to the length of the wood sill plate.*
   - *2.6. The sill plate is of 2-inch or 3-inch nominal thickness.*
3. *Section D.3.3.5 need not apply and the design shear strength in accordance with Section D.6.2.1(c) need not be computed for anchor bolts attaching cold-formed steel track of bearing or nonbearing walls of light-frame construction to foundations or foundation stem walls provided all of the following are satisfied:*
   - *3.1. The maximum anchor nominal diameter is $^5/_8$ inches (16 mm).*
   - *3.2. Anchors are embedded into concrete a minimum of 7 inches (178 mm).*
   - *3.3. Anchors are located a minimum of $1^3/_4$ inches (45 mm) from the edge of the concrete parallel to the length of the track.*
   - *3.4. Anchors are located a minimum of 15 anchor diameters from the edge of the concrete perpendicular to the length of the track.*
   - *3.5. The track is 33 to 68 mil designation thickness.*

     Allowable in-plane shear strength of exempt anchors, parallel to the edge of concrete shall be permitted to be determined in accordance with AISI S100 Section E3.3.1.
4. *In light-frame construction, design of anchors in concrete shall be permitted to satisfy D.3.3.8.*

D.3.3.6 - Instead of D.3.3.5, the attachment that the anchor is connecting to the structure shall be designed so that the attachment will undergo ductile yielding at a force level cor-

responding to anchor forces no greater than the design strength of anchors specified in D.3.3.4.

Exceptions:

1. *Anchors in concrete designed to support nonstructural components in accordance with ASCE 7 Section 13.4.2 need not satisfy Section D.3.3.6.*
2. *Anchors designed to resist wall out-of-plane forces with design strengths equal to or greater than the force determined in accordance with ASCE 7 Equation 12.11-1 or 12.14-10 need not satisfy Section D.3.3.6.*

D.3.3.7 - As an alternative to D.3.3.5 and D.3.3.6, it shall be permitted to take the design strength of the anchors as 0.4 times the design strength determined in accordance with D.3.3.4.

D.3.3.8 – In light-frame construction, bearing or non-bearing walls, shear strength of concrete anchors less than or equal to 1 inch (25 mm) in diameter of sill plate or track to foundation or foundation stem wall need not satisfy D.3.3.7 when the design strength of the anchors is determined in accordance with D.6.2.1(c).

❖ This modification to Appendix D of ACI 318 is based on the 2008 version of the provision. It clarifies the applicability of the requirements for designing anchors for earthquake forces by adding exceptions. The basic premise of Section D.3.3 is that anchorage design is controlled by the strength of a "ductile steel element," which is defined in the ACI 318 appendix commentary.

Section D.3.3.5 has four exceptions. Exception 1 applies to concrete wall anchorage that is designed for maximum expected seismic forces. Under ASCE 7, higher wall anchorage force levels are required to protect against brittle failure. This exception clarifies that these special wall anchorage design forces need not be compounded with the anchorage ductility requirement in Appendix D of ACI 318.

Exception 2 applies to the anchorage of wood sill plates to concrete foundations in light-frame construction. Based on light-frame shear wall testing, the wood sill plate controls the ductile behavior of the anchorage assembly. If the anchor meets the requirements of the exception, then the anchor need not meet the requirements of Section D.3.3.5. Allowable in-plane shear capacity of the anchor bolt is determined in accordance with National Design Specification (NDS) Table 11E rather than Appendix D of ACI 318.

Exception 3 applies to the anchorage of cold-formed steel track to concrete foundations in light-frame construction. Based on light-frame shear wall testing, the cold-formed steel track controls the ductile behavior of the anchorage assembly. If the anchor meets the requirements of the exception, then the anchor need not meet the requirements of Section D.3.3.5. Allowable in-plane shear capacity of the anchor bolt is determined from Section E3.3.1 of AISI S100 rather than Appendix D of ACI 318.

There are five criteria to meet in order to use Exception 2 or 3 for design of anchor bolts connecting wood sill plates of shear walls of light-frame structures. The maximum nominal diameter of the anchor is $^5/_8$ inches (15.9 mm); anchors must be embedded a minimum of 7 inches into the footing and located a minimum of $1^3/_4$ inches (44 mm) from the edge of the concrete parallel to the wood sill plate or steel track and a minimum of 15 diameters from the edge of the concrete perpendicular to the wood sill plate or steel track; the wood sill plate must be a 2X or 3X and the track must be within a thickness range of 33 mil to 68 mil.

For light-frame construction, Exception 4 allows use of the provision in Section D.3.3.8 rather than Section D.3.3.5. Section 202 defines "light-frame construction" as a system where vertical and horizontal structural elements are primarily formed by repetitive wood or cold-formed steel framing members.

Section D.3.3.6 allows design of the anchor attachment to control the anchorage assembly design. Yielding of the attachment must begin before the anchor begins to yield. Appendix D of ACI 318 defines "Attachment" as the structural assembly external to the surface of the concrete that transmits loads to or from the anchor. As steel today has a higher expected yield strength than its specified yield strength, the attachment must have an expected yield strength no larger than the design strength of the anchor.

There are two exceptions to this provision. Exception 1 deals with nonstructural components designed for earthquake loading in accordance with Section 13.4.2 of ASCE 7. That provision imposes additional nonductile anchor force increases on anchors in structures assigned to Seismic Design Category C and higher. The exception clarifies that it is not intended for nonductile anchor force increase to be compounded with this ACI 318 anchorage requirement.

Exception 2 applies to concrete wall anchorage that is designed for maximum expected seismic forces. Under ASCE 7, higher wall anchorage force levels are required to protect against brittle failure. This exception clarifies that these special wall anchorage design forces need not be compounded with the anchorage ductility requirement in Appendix D of ACI 318.

Section D.3.3.7 allows the design strength of an anchor to be set as 0.4 multiplied by the steel strength of the element as determined in accordance with Section D.5.1 for steel tensile strength or D.6.1 for steel shear strength in ACI 318. This provision was initially added to ACI 318-08 to provide a capacity for nonductile steel anchors. In ACI 318-11, anchors are assumed to be ductile steel elements.

Section D.3.3.8 refers to Section D6.2.1(c) of ACI 318 where shear loads are examined parallel to the length of the sill plate or track. This provision allows use of ductile or nonductile anchors in wood sill plates and tracks without the decrease in design strength required in Section D.3.3.7. This is an acceptable practice since wood plate or steel track failure occurs before the anchor reaches design strength.

1905.1.10 ACI 318, Section D.4.2.2. Delete ACI 318, Section D.4.2.2, and replace with the following:

D.4.2.2 - The concrete breakout strength requirements for anchors in tension shall be considered satisfied by the design procedure of D.5.2 provided Equation D-7 is not used for anchor embedments exceeding 25 inches. The concrete breakout strength requirements for anchors in shear with diameters not exceeding 2 inches shall be considered satisfied by the design procedure of D.6.2. For anchors in shear with diameters exceeding 2 inches, shear anchor reinforcement shall be provided in accordance with the procedures of D.6.2.9.

❖ This modification to Appendix D of ACI 318 is based on the 2008 version of the provision. Section D.4.2.2 of ACI 318-08 allowed determination of the concrete breakout strength in both tension (Section D5.2) and shear (Section D6.2) to be determined according to the referenced sections that are based on the Concrete Capacity Design Method. Based on the range of original test data, it restricted the use of this approach for anchorage exceeding 2 inches (51 mm) in diameter or with a depth of embedment greater than 25 inches (635 mm).

Because there was evidence that the predicted concrete breakout strength for anchors in tension as given in Section D5.2 by Equation D-7 could be safely extended to anchors with embedment depths exceeding 25 inches (635 mm), the modified text allowed this approach, provided the less conservative prediction of concrete breakout strength in tension given by Equation D-8 was not used.

There was also evidence that the breakout strength procedure in Section D.6.2 was not suitable for very large diameter anchors loaded in shear towards a free edge. Where larger anchor diameters were used and where concrete edge breakout would otherwise control the anchor strength, the modified text recognizes the use of anchor reinforcement in accordance with Section D.6.2.9.

SECTION 1906 STRUCTURAL PLAIN CONCRETE

1906.1 Scope. The design and construction of structural plain concrete, both cast-in-place and precast, shall comply with the minimum requirements of ACI 318, as modified in Section 1905.

Exception: For Group R-3 occupancies and buildings of other occupancies less than two stories above grade plane of light-frame construction, the required footing thickness of ACI 318 is permitted to be reduced to 6 inches (152 mm), provided that the footing does not extend more than 4 inches (102 mm) on either side of the supported wall.

❖ Section 1906 references the provisions in Chapter 22 of ACI 318 for the minimum requirements for plain concrete used for structural purposes. By definition, plain concrete contains no reinforcement, or has less reinforcement than necessary to meet the requirements for reinforced concrete in ACI 318.

The exception applies to Section 22.7.4 of ACI 318, which requires the thickness of a structural plain concrete footing to be a minimum of 8 inches (203 mm). This exception permits the thickness of the footing to be reduced to 6 inches (152 mm). The reduced footing thickness applies to any Group R-3 occupancy and to buildings of any other occupancy that are one story in height and of light-frame construction. An additional limitation is that the footing cannot extend more than 4 inches (102 mm) beyond the face of the supported wall. The 4-inch (102 mm) limitation is to reduce the likelihood that the footing will experience a flexural failure at the junction of the footing and the wall supported above.

SECTION 1907 MINIMUM SLAB PROVISIONS

1907.1 General. The thickness of concrete floor slabs supported directly on the ground shall not be less than $3^1/_2$ inches (89 mm). A 6-mil (0.006 inch; 0.15 mm) polyethylene vapor retarder with joints lapped not less than 6 inches (152 mm) shall be placed between the base course or subgrade and the concrete floor slab, or other *approved* equivalent methods or materials shall be used to retard vapor transmission through the floor slab.

Exception: A vapor retarder is not required:

1. For detached structures accessory to occupancies in Group R-3, such as garages, utility buildings or other unheated facilities.
2. For unheated storage rooms having an area of less than 70 square feet (6.5 m^2) and carports attached to occupancies in Group R-3.
3. For buildings of other occupancies where migration of moisture through the slab from below will not be detrimental to the intended occupancy of the building.
4. For driveways, walks, patios and other flatwork which will not be enclosed at a later date.
5. Where *approved* based on local site conditions.

❖ Section 1907 requires that slabs on grade be a minimum thickness so superimposed loads are transmitted to the subgrade without causing structural distress of the slab. Additionally, slabs must be protected on the underside with material that will retard the transmission of water vapor through the floor slab into the enclosed space so moisture levels are not excessive.

The coarse aggregate most commonly used in concrete slabs supported directly on the ground is allowed up to 5 percent (by weight) of the stone larger than 1 inch (25 mm) in size, with all coarse aggregates in this grading classification passing a $1^1/_2$-inch (38 mm) sieve. Section 3.3.2 of ACI 318 requires that the nominal maximum size of coarse aggregate be no larger than one-third the depth of the slab, thus making it necessary that slabs be a minimum thickness of $3^1/_2$ inches (89 mm) to accommodate coarse aggregate stones between 1 and $1^1/_2$ inches (25 and 38 mm) in size. Also, as a matter of experience, a $3^1/_2$-inch (89 mm) minimum slab thickness is needed to support typical concentrated loads without damage to the concrete, particularly at locations where small voids or pockets in the subgrade may develop because of improper compaction during construction or from other causes.

An approved 6-mil vapor retarder (see Section 202 for definition) placed between the concrete slab and the supporting ground is required to prevent the migration of water vapor through the slab into the space above. The five exceptions acknowledge conditions where the migration of limited moisture through the slab will not adversely affect the occupancy of the structure. Exception 3 is intended to give relief to occupancies in which additional moisture will not be detrimental, such as indoor swimming pools, industrial processes generating large amounts of moisture and certain storage facilities. Exception 5 is intended to give relief based on environmental or geological conditions, such as locations either with arid climates where the relative humidity is typically low or with well-drained soils and low water tables. Vapor retarders, while preventing moisture from migrating through the slab from below, can adversely affect the construction and performance of concrete slabs on grade. Finishing operations may be delayed since it will normally take longer for the bleed water to evaporate from the surface. Additionally, the tendency for concrete to develop both plastic and drying shrinkage cracking may be exacerbated. Another problem that occurs is the tendency for the slab to curl or warp upward at the edges. Curling occurs because the bottom of the slab remains moist longer than the top portion due to the vapor retarder; therefore, the top shortens more than the bottom, causing curling.

SECTION 1908 ANCHORAGE TO CONCRETE— ALLOWABLE STRESS DESIGN

1908.1 Scope. The provisions of this section shall govern the *allowable stress design* of headed bolts and headed stud anchors cast in normal-weight concrete for purposes of transmitting structural loads from one connected element to the other. These provisions do not apply to anchors installed in hardened concrete or where load combinations include earthquake loads or effects. The bearing area of headed anchors shall be not less than one and one-half times the shank area. Where strength design is used, or where load combinations include earthquake loads or effects, the design strength of anchors shall be determined in accordance with Section 1909. Bolts shall conform to ASTM A 307 or an *approved* equivalent.

❖ Section 1908 contains requirements for design of anchorage to concrete by the allowable stress design (ASD) Method, formerly called "working stress design." Provisions for strength design of anchorage to concrete are given in Section 1909. The ASD procedure applies to anchors that are cast-in-place in normal-weight concrete. This section cannot be used to design anchors that are post-installed in hardened concrete or where load combinations include seismic forces or their effects. In those cases, the provisions of Section 1909 are to be used.

The code provides two methods of design for anchors cast in concrete. Section 1908 provides for limited use of the ASD method and Section 1909 provides an expanded approach for the strength design method. The ASD provisions of this section are adapted from the 1997 edition of the UBC. They are based on limited test data of headed bolts cast in normal-weight concrete and subjected to static loading. Therefore, its application may not be used for hooked (J- or L-) bolts, lightweight concrete, post-installed anchors (i.e., anchors installed in hardened concrete) or when load combinations include earthquake load effects.

1908.2 Allowable service load. The allowable service load for headed anchors in shear or tension shall be as indicated in Table 1908.2. Where anchors are subject to combined shear and tension, the following relationship shall be satisfied:

$$(P_s / P_t)^{5/3} + (V_s / V_t)^{5/3} \le 1 \qquad \textbf{(Equation 19-1)}$$

where:

P_s = Applied tension service load, pounds (N).

P_t = Allowable tension service load from Table 1908.2, pounds (N).

V_s = Applied shear service load, pounds (N).

V_t = Allowable shear service load from Table 1908.2, pounds (N).

❖ Table 1908.2 gives allowable service loads for tension or shear, but not at the same time. In real-world applications, both forces are frequently acting at the same time. This section provides a formula for determining the maximum values that may be assigned to bolts that are subjected to combinations of shear and tension loading. The term "allowable service load" signifies that unfactored shear and tension loads are to be used. They may include live, dead, wind and other loading conditions, except for earthquake loads.

TABLE 1908.2. See below.

❖ Table 1908.2 provides shear and tension loads that may be assigned to headed bolts or stud anchors that are cast in normal-weight concrete. Note that the table assigns values for either shear or tension. Both values cannot be assigned to the same bolt at the same time. The table values are based on a particular combination of bolt size [from $^1/_4$ to $1^1/_4$ inches (6.4 to 32 mm)] and concrete strength [i.e., 2,500, 3,000 and 4,000 psi (17, 21 and 28 MPa)]. When using the table values, minimum bolt embedments, edge distances and spacing between bolts must be observed. Edge distances are measured from the face of the bolt to the sides and ends of concrete members. Decreases up to 50 percent in edge distance or spacing, or both, are provided for in Section 1908.3.

1908.3 Required edge distance and spacing. The allowable service loads in tension and shear specified in Table 1908.2 are for the edge distance and spacing specified. The edge distance and spacing are permitted to be reduced to 50 percent of the values specified with an equal reduction in allowable service load. Where edge distance and spacing are reduced less than 50 percent, the allowable service load shall be determined by linear interpolation.

❖ This section expands the applicability of Table 1908.2 when bolt locations cannot meet the required minimum edge distances or spacing, or both, specified by the table. Tabulated minimum edge distances and spacings may be reduced up to 50 percent for a corresponding 50-percent reduction of service loads. For reductions in edge distance or spacing, or both, less than 50 percent, corresponding service loads may be determined by interpolation. For example, if $^5/_8$-inch (15.9 mm) bolts are embedded $4^1/_2$ inches (114 mm) in 2,500 psi (17 MPa) concrete, the table assigns shear and tension values of 2,750 and 1,500 pounds (12 100 and 6600 N), respectively, if at least $3^3/_4$-inch (95 mm) edge distance and $7^1/_2$-inch (191 mm) spacing between bolts are provided. If only a 25-percent reduction in edge distance or spacing, or both, is desired, the corresponding shear and tension values (determined by linear interpolation) are 2,063 and 1,125 pounds (9077 and 4950 N), respectively.

1908.4 Increase in allowable load. Increase of the values in Table 1908.2 by one-third is permitted where the provisions of Section 1605.3.2 permit an increase in allowable stress for wind loading.

❖ This section follows a time-honored principle of permitting a one-third increase in allowable stress for wind. The rationale for this increase is that design wind loads occur infrequently and their maximum effects last only a few seconds. Therefore, an overstress of one-third for a very short time at infrequent intervals is considered acceptable.

1908.5 Increase for special inspection. Where *special inspection* is provided for the installation of anchors, a 100-percent increase in the allowable tension values of Table 1908.2 is permitted. No increase in shear value is permitted.

❖ The tension and shear values in Table 1908.2 are based on load test data that has been reduced by a safety factor. The code acknowledges that special inspection improves construction practices sufficiently to allow a 100-percent increase in the tension values provided in the table, thus allowing a lower safety factor when special inspection is provided.

The shear values of the table are not permitted to be increased when special inspection is provided, because the table values reflect a safety factor that is not overly conservative. The provisions of this section were taken from the footnotes of Table 19D (Allowable Service Load on Embedded Bolts) of the 1997 UBC. The footnotes first appeared in the 1976 UBC

TABLE 1908.2
ALLOWABLE SERVICE LOAD ON EMBEDDED BOLTS (pounds)

BOLT DIAMETER (inches)	MINIMUM EMBEDMENT (inches)	EDGE DISTANCE (inches)	SPACING (inches)	MINIMUM CONCRETE STRENGTH (psi)					
				f'_c = 2,500		f'_c = 3,000		f'_c = 4,000	
				Tension	Shear	Tension	Shear	Tension	Shear
$^1/_4$	$2^1/_2$	$1^1/_2$	3	200	500	200	500	200	500
$^3/_8$	3	$2^1/_4$	$4^1/_2$	500	1,100	500	1,100	500	1,100
$^1/_2$	4 4	3 5	6 6	950 1,450	1,250 1,600	950 1,500	1,250 1,650	950 1,550	1,250 1,750
$^5/_8$	$4^1/_2$ $4^1/_2$	$3^3/_4$ $6^1/_4$	$7^1/_2$ $7^1/_2$	1,500 2,125	2,750 2,950	1,500 2,200	2,750 3,000	1,500 2,400	2,750 3,050
$^3/_4$	5 5	$4^1/_2$ $7^1/_2$	9 9	2,250 2,825	3,250 4,275	2,250 2,950	3,560 4,300	2,250 3,200	3,560 4,400
$^7/_8$	6	$5^1/_4$	$10^1/_2$	2,550	3,700	2,550	4,050	2,550	4,050
1	7	6	12	3,050	4,125	3,250	4,500	3,650	5,300
$1^1/_8$	8	$6^3/_4$	$13^1/_2$	3,400	4,750	3,400	4,750	3,400	4,750
$1^1/_4$	9	$7^1/_2$	15	4,000	5,800	4,000	5,800	4,000	5,800

For SI: 1 inch = 25.4 mm, 1 pound per square inch = 0.00689 MPa, 1 pound = 4.45 N.

and were accompanied by a 50-percent increase in the table's shear values, but no increase in the table's tensile values. The footnotes also enabled the tensile values to be increased 50 percent, but only when special inspection is provided.

SECTION 1909 ANCHORAGE TO CONCRETE— STRENGTH DESIGN

1909.1 Scope. The provisions of this section shall govern the strength design of anchors installed in concrete for purposes of transmitting structural loads from one connected element to the other. Headed bolts, headed studs and hooked (J- or L-) bolts cast in concrete and expansion anchors and undercut anchors installed in hardened concrete shall be designed in accordance with Appendix D of ACI 318 as modified by Sections 1905.1.9 and 1905.1.10, provided they are within the scope of Appendix D.

The strength design of anchors that are not within the scope of Appendix D of ACI 318, and as amended in Sections 1905.1.9 and 1905.1.10, shall be in accordance with an *approved* procedure.

❖ Section 1909 is restricted in scope to anchors that transmit structural loads from attachments into concrete members and vice versa under typical in-service conditions. Other standards, such as those applicable to construction and material handling, for instance, can require more stringent safety levels and must be followed where they apply.

The first edition of the code included provisions for the strength design of anchorage to concrete. At that time, the provisions that were being developed for inclusion in ACI 318 were not yet available for reference by the code. With the publication of the 2002 edition of ACI 318, requirements for the strength design of concrete anchorage were included in Appendix D. The code provisions were then replaced with a reference to Appendix D of ACI 318, since it was the most up-to-date version of the same requirements. Based on the content of the newer requirements in Appendix D, the scope of this section also includes post-installed anchors, such as expansion anchors and undercut anchors, provided they are evaluated and categorized in accordance with ACI 355.2. Design of anchors that are not within the scope of this section must be in accordance with an approved procedure. Examples of anchors that fall into the latter category are specialty inserts, through-bolts, adhesive anchors, grouted anchors and powder-actuated fasteners.

Section D.2.4 of the referenced ACI 318 provisions excludes load applications producing high-cycle fatigue or extremely short duration impact (such as blast or shock wave). This statement does not exclude seismic loads and, in fact, Section 1908 for design of anchorage to concrete by the ASD method requires the use of strength design for load combinations that include seismic forces or their effects.

SECTION 1910 SHOTCRETE

1910.1 General. Shotcrete is mortar or concrete that is pneumatically projected at high velocity onto a surface. Except as specified in this section, shotcrete shall conform to the requirements of this chapter for plain or reinforced concrete.

❖ Section 1910 regulates the materials and test procedures for shotcrete construction. Shotcrete is the commonly accepted and generic name for pneumatically projected mortar or concrete. Terms such as gunite, sprayed concrete, spraycrete, pneumatically applied mortar and concrete are often used to refer to shotcrete.

ACI 506R, the guide for shotcreting, defines "Shotcrete" as mortar or concrete conveyed through a hose and pneumatically projected at a high velocity onto a surface. The force of the jet impacting on the shotcrete surface compacts the material and, when applied properly, the shotcrete surface will support itself without sagging or sloughing, even for vertical and overhead applications. Shotcrete must conform to the requirements for structural concrete specified in this chapter and ACI 318 unless modified by Section 1910.

1910.2 Proportions and materials. Shotcrete proportions shall be selected that allow suitable placement procedures using the delivery equipment selected and shall result in finished in-place hardened shotcrete meeting the strength requirements of this code.

❖ Although the compressive strength of shotcrete must be verified prior to beginning work, its resulting strength and acceptability relies more on the suitability of placement procedures and the delivery equipment. Prior to construction, strength is established by the submittal of a mix design that is supported by test results. Mix designs from previous jobs can be used. Product data sheets for any admixtures and gradations for the sand and aggregate need to be included in the submittal. Verification of in-place hardened shotcrete strength is determined by testing in accordance with Section 1910.10.

1910.3 Aggregate. Coarse aggregate, if used, shall not exceed $^{3}/_{4}$ inch (19.1 mm).

❖ For construction sections that are several inches thick and where adequate gunning equipment is available, the use of shotcrete mixtures containing coarse aggregates may be advantageous. Coarse aggregate must not exceed $^{3}/_{4}$ inch (19.1 mm) in size and the gradation limitations for a combination of fine (sand) and coarse aggregates should be in accordance with the requirements of ACI 506.2. Normal-weight coarse aggregate for concrete must comply with the requirements of ASTM C 33. Lightweight coarse aggregate for concrete must comply with the requirements of ASTM C 330 (see commentary, Section 1903).

1910.4 Reinforcement. Reinforcement used in shotcrete construction shall comply with the provisions of Sections 1910.4.1 through 1910.4.4.

❖ The best results will be obtained when the reinforcing steel is designed and placed to cause the least interference with the high-velocity placement of shotcrete. When proper techniques are used, shotcrete can be successfully placed through two or three layers of reinforcing steel. The size, spacing and splicing of reinforcement is specified in Sections 1910.4.1, 1910.4.2 and 1910.4.3, respectively.

1910.4.1 Size. The maximum size of reinforcement shall be No. 5 bars unless it is demonstrated by preconstruction tests that adequate encasement of larger bars will be achieved.

❖ Depending on the thickness and nature of the work, reinforcement will consist of either deformed bars or welded wire fabric. Bar sizes are limited to a No. 5 (16) bar. Larger bar sizes can be used if it can be shown that adequate bonding with and encasement in the concrete can be accomplished. Preconstruction tests may be required by the building official to provide verification that proper bonding can be achieved (see Section 1910.5).

1910.4.2 Clearance. When No. 5 or smaller bars are used, there shall be a minimum clearance between parallel reinforcement bars of $2^1/_2$ inches (64 mm). When bars larger than No. 5 are permitted, there shall be a minimum clearance between parallel bars equal to six diameters of the bars used. When two curtains of steel are provided, the curtain nearer the nozzle shall have a minimum spacing equal to 12 bar diameters and the remaining curtain shall have a minimum spacing of six bar diameters.

Exception: Subject to the approval of the *building official*, required clearances shall be reduced where it is demonstrated by preconstruction tests that adequate encasement of the bars used in the design will be achieved.

❖ Experience has shown that a $2^1/_2$-inch (64 mm) space between No. 5(16) bars, and a six-bar diameter space between larger bars, is sufficient clearance for placement of shotcrete where there is one layer of reinforcement. Experience has also shown that shotcrete application can be effective when the spacing between bars in the closer of two layers (nearer to nozzle) of steel is 12 bar diameters, and in the remaining layer the bars have a spacing of six bar diameters. The exception clarifies that lesser clearance dimensions are permitted if they can be verified by preconstruction tests.

For welded wire fabric, ACI 506R recommends a minimum spacing of 2 inches (51 mm) between wires in both directions. The minimum clearance between the reinforcement and the form or other back-up surface may vary depending on whether a mortar mix (fine aggregate only) or concrete (fine and coarse aggregate) is used for shotcrete and whether conventional reinforcing steel or welded wire fabric is used. This spacing should be specified or shown on the construction documents. ACI 506R recommends that no less than a 2-inch (51 mm) clearance should be used for No. 5 (16) reinforcing bars. The minimum cover for reinforcement must comply with the requirements of ACI 318.

1910.4.3 Splices. Lap splices of reinforcing bars shall utilize the noncontact lap splice method with a minimum clearance of 2 inches (51 mm) between bars. The use of contact lap splices necessary for support of the reinforcing is permitted when *approved* by the *building official*, based on satisfactory preconstruction tests that show that adequate encasement of the bars will be achieved, and provided that the splice is oriented so that a plane through the center of the spliced bars is perpendicular to the surface of the shotcrete.

❖ Bars and welded wire fabric must be lapped in such a manner so as not to create weak sections in the shotcrete. Lapped reinforcing bars should not be tied together, but rather be separated by at least 2 inches (51 mm) wherever possible. The building official should only permit contact lap splices when it can be shown by preconstruction tests that the reinforcement will be completely encased. Welded wire fabric must be lapped $1^1/_2$ squares in all directions. Continuous chair support should not be placed parallel and directly underneath bars because chairs may prevent full shotcrete encasement or cause sand pockets. Individual chairs should also to be offset to prevent a line of weakness in the shotcrete.

1910.4.4 Spirally tied columns. Shotcrete shall not be applied to spirally tied columns.

❖ Typically, plain reinforcement (i.e., no surface deformations) is used for spiral reinforcement in columns. Shotcrete is not permitted to be applied to spirally tied columns because of the lack of these deformations. Additionally, ACI 318 requires that the clear spacing between spirals must not exceed 3 inches (76 mm). These conditions make it difficult to achieve adequate bond between the concrete and the reinforcement.

1910.5 Preconstruction tests. When required by the *building official*, a test panel shall be shot, cured, cored or sawn, examined and tested prior to commencement of the project. The sample panel shall be representative of the project and simulate job conditions as closely as possible. The panel thickness and reinforcing shall reproduce the thickest and most congested area specified in the structural design. It shall be shot at the same angle, using the same nozzleman and with the same concrete mix design that will be used on the project. The equipment used in preconstruction testing shall be the same equipment used in the work requiring such testing, unless substitute equipment is *approved* by the *building official*.

❖ When required by the building official, preconstruction test specimens must be made for each type of construction application (flat, vertical or overhead) using the materials, mix proportions, equipment and the nozzle operator proposed for the project. Test panels should be at least 30 inches (762 mm) square and must have the same reinforcement and thickness [but not less than 3 inches (76 mm)] as specified and shown on the construction documents for permanent construction. Drilled cores should be taken from the panels and tested in accordance with ASTM C 42 procedures for compliance with strength requirements. Preconstruction testing may not be necessary if it can be shown to the satisfaction of the building official that mix, materials, equipment and construction personnel have produced satisfactory results on similar work.

1910.6 Rebound. Any rebound or accumulated loose aggregate shall be removed from the surfaces to be covered prior to placing the initial or any succeeding layers of shotcrete. Rebound shall not be used as aggregate.

❖ Rebound is loose aggregate and cement paste that bounces off application surfaces after colliding with formwork, reinforcement or the shotcrete surface itself. Factors that affect the amount of rebound include the position of the work (flat, vertical or overhead applications); air pressure; cement content; water content; maximum size and grading of aggregates; amount of reinforcement and thickness of the layer. Rebound must not be reused as aggregate or in any way worked back into the construction. If rebound does not fall clear of the work, it must be removed.

1910.7 Joints. Except where permitted herein, unfinished work shall not be allowed to stand for more than 30 minutes unless edges are sloped to a thin edge. For structural elements that will be under compression and for construction joints shown on the *approved construction documents*, square joints are permitted. Before placing additional material adjacent to previously applied work, sloping and square edges shall be cleaned and wetted.

❖ Construction joints are generally tapered to a thin edge over a width of approximately 12 inches (305 mm). Square construction joints (nontapered) should be avoided wherever possible in shotcrete construction because they form a trap for rebound. Where the joint will be subjected to compressive stresses, however, square joints are frequently used and, in such cases, care must be taken to avoid or remove trapped rebound at the joint. All joints, tapered or square, must be thoroughly cleaned of loose material and wetted prior to the application of additional shotcrete.

1910.8 Damage. In-place shotcrete that exhibits sags, sloughs, segregation, honeycombing, sand pockets or other obvious defects shall be removed and replaced. Shotcrete above sags and sloughs shall be removed and replaced while still plastic.

❖ After placement, any shotcrete that lacks uniformity or exhibits segregation, honeycombing or delamination; that contains dry patches, slugs, voids or sand pockets (porous areas low in cement content) or that sags or sloughs should be removed and replaced. As a practical matter, small holes created by cores taken for testing purposes cannot be filled solidly by shooting mortar at a high velocity. The patching mixture to be applied by hand should be made of the same materials and approximately of the same proportions as used for the shotcrete, except that coarse aggregate, if used, should be omitted. ACI 506.4R provides additional recommendations for repairing shotcrete, including filling of holes created where cores were obtained for testing.

1910.9 Curing. During the curing periods specified herein, shotcrete shall be maintained above 40°F (4°C) and in moist condition.

❖ Like cast-in-place concrete, shotcrete must be cured in a moist condition to achieve its potential strength and durability. Good curing practice for shotcrete construction generally requires that surfaces be kept continuously moist for a period of at least seven days. During curing, the air that is in contact with shotcrete surfaces must be maintained at temperatures above freezing, with the temperature of the shotcrete maintained above 40°F (4°C). For recommendations on winter protection, see ACI 318 or refer to ACI 306R.

1910.9.1 Initial curing. Shotcrete shall be kept continuously moist for 24 hours after shotcreting is complete or shall be sealed with an *approved* curing compound.

❖ Immediately after finishing operations are complete, shotcrete must be kept continuously moist for a period of 24 hours by one of the following methods:

1. Ponding or continuous sprinkling;
2. Use of absorptive mat or fabric, sand or other protective covering kept continuously moist;
3. Continuous steam not exceeding 150°F (66°C) or vapor mist bath; or
4. Membrane-forming curing compounds and application conforming to ASTM C 309.

1910.9.2 Final curing. Final curing shall continue for seven days after shotcreting, or for three days if high- early-strength cement is used, or until the specified strength is obtained. Final curing shall consist of the initial curing process or the shotcrete shall be covered with an *approved* moisture-retaining cover.

❖ Immediately following the initial curing specified in Section 1910.9.1, additional curing must be provided using one of the following materials or methods:

1. Continue the method used in the initial curing process (see commentary, Section 1911.9.1);
2. Use sheet materials conforming to ASTM C 171; or
3. Use other moisture-retaining covers acceptable to the registered design professional and approved by the building official.

Curing must be continued for at least the first seven days after shotcreting, or for the first three days if high early strength cement is used (Type III or IIIA portland cement), or until the specified strength is obtained. During the curing period, shotcrete must be maintained at a temperature above 40°F (4°C) and in a moist condition.

1910.9.3 Natural curing. Natural curing shall not be used in lieu of that specified in this section unless the relative humidity remains at or above 85 percent, and is authorized by the *registered design professional* and *approved* by the *building official*.

❖ Natural curing is exposure to the atmosphere without any cover or water spray. If the atmospheric conditions surrounding the shotcrete construction are satisfactory, such as when the relative humidity is at or above 85 percent, the registered design professional, with the approval of the building official, may authorize natural curing. Where natural curing is used, the shotcrete must be maintained above 40°F (4°C) for the period specified in Section 1910.9.2.

1910.10 Strength tests. Strength tests for shotcrete shall be made by an *approved agency* on specimens that are representative of the work and which have been water soaked for at least 24 hours prior to testing. When the maximum-size aggregate is larger than $^3/_8$ inch (9.5 mm), specimens shall consist of not less than three 3-inch-diameter (76 mm) cores or 3-inch (76 mm) cubes. When the maximum-size aggregate is $^3/_8$ inch (9.5 mm) or smaller, specimens shall consist of not less than 2-inch-diameter (51 mm) cores or 2-inch (51 mm) cubes.

❖ This section provides basic strength test criteria for shotcrete. Experience has shown that aggregate size has a direct bearing on the accuracy of compressive tests. Therefore, core and cube specimens containing aggregates that are larger than $^3/_8$ inch (9.5 mm) must be a minimum of 3 inches (76 mm) in diameter (or cross section). Test specimens are required to be soaked in water for at least 24 hours prior to testing to maintain a common environmental condition for all specimens and to ensure uniformity between test results. Cores should be obtained and tested in accordance with ASTM C 42.

1910.10.1 Sampling. Specimens shall be taken from the in-place work or from test panels, and shall be taken at least once each shift, but not less than one for each 50 cubic yards (38.2 m^3) of shotcrete.

❖ Frequency of testing is a significant factor in the effectiveness of quality control of concrete. To ensure that test results will be representative of the concrete in the structure, specimens must be taken at the minimum frequency required by this section.

1910.10.2 Panel criteria. When the maximum-size aggregate is larger than $^3/_8$ inch (9.5 mm), the test panels shall have minimum dimensions of 18 inches by 18 inches (457 mm by 457 mm). When the maximum size aggregate is $^3/_8$ inch (9.5 mm) or smaller, the test panels shall have minimum dimensions of 12 inches by 12 inches (305 mm by 305 mm). Panels shall be shot in the same position as the work, during the course of the work and by the nozzlemen doing the work. The conditions under which the panels are cured shall be the same as the work.

❖ When test panels are used, they are required to meet the criteria in this section. Experience has shown that aggregate size has a direct bearing on the ability to produce test panels that accurately portray the condition of in-place shotcrete. Evidence suggests that test panels containing aggregates larger than $^3/_8$ inch (9.5 mm) must be a minimum of 18 inches (457 mm) square, and that test panels with aggregate $^3/_8$ inch (9.5 mm) or smaller must be a minimum 12 inches (305 mm) square. The ability of the nozzle operator is a significant factor in achieving quality shotcrete. Therefore, it is critical that the nozzle operator doing the work also prepares the test panels. Since the position of the work (flat, vertical or overhead) and the method of curing have an impact on the quality of work, they must also be accounted for when making test panels.

1910.10.3 Acceptance criteria. The average compressive strength of three cores from the in-place work or a single test panel shall equal or exceed 0.85 f'_c with no single core less than 0.75 f'_c. The average compressive strength of three cubes taken from the in-place work or a single test panel shall equal or exceed f'_c with no individual cube less than 0.88 f'_c. To check accuracy, locations represented by erratic core or cube strengths shall be retested.

❖ In evaluating core test results, the fact that core strengths may not equal the specified compressive strength of concrete, f'_c, should not be a cause for concern, provided they are within the limits specified. It is realistic to expect the compressive strength of cores to be less than f'_c because the size of the core differs from that of a standard concrete test cylinder, and procedures for obtaining and curing the cores are different. Therefore, the shotcrete can be considered acceptable from the standpoint of strength if test results from the cores are at least 85 percent of f'_c for the average of three cores and 75 percent of f'_c for a single core. Cores should be obtained and tested in accordance with ASTM C 42. Cores for compressive strength testing must have a length-to-diameter ratio between 1.0 and 2.10. For ratios less than 1.94, the compressive strength is multiplied by a correction factor (less than 1) given in ASTM C 42 to correlate the strength of the core to that of a standard concrete test cylinder that has a length-to-diameter ratio of 2. Experience has shown that test results from cubes can be held to a higher standard, thus requiring the average of three cubes to equal or exceed f'_c with no individual cube less than 0.88 f'_c. Since the height-to-least-lateral dimension of a cube is 1, ACI 506R indicates that the compressive strength of a cube can be correlated to that of a standard test cylinder by multiplying the test results by 0.85. Either correlated or uncorrelated results for cubes can be used to determine if the specified acceptance criteria are met. Because of the less than delicate nature of the coring and sawing process, this section also permits erratic core and cube strengths to be discarded from the sample under consideration.

SECTION 1911 REINFORCED GYPSUM CONCRETE

1911.1 General. Reinforced gypsum concrete shall comply with the requirements of ASTM C 317 and ASTM C 956.

❖ The provisions of Section 1911 apply to poured-in-place reinforced gypsum concrete that is used to construct structural roof decks (when placed over permanent formboards). Gypsum concrete is a manufactured product consisting of calcined gypsum to which the producer may add aggregates complying with ASTM C 35, wood chips or wood shavings. Only water is added at the job site.

Similar proprietary products manufactured under various trade names are intended to be used in nonstructural applications such as floor toppings, where increased sound resistance is desired. These products can also be used as floor levers, and this application is prevalent in rehabilitation of existing structures. This section is not intended to govern the use of these materials where used in nonstructural applications.

Gypsum concrete (the product) used in reinforced gypsum concrete construction must conform to the requirements of ASTM C 317. This standard provides for two classes, A and B, of gypsum concrete based on minimum compressive strength. Class A must achieve a minimum compressive strength of 500 psi (3.5 MPa) and have a density not exceeding 60 pounds per cubic foot (960 kg/m^3). Class B must achieve a compressive strength of 1,000 psi (6.9 MPa) with no restriction on density. Before the development of newer insulating materials suitable for installation on roof decks, reinforced gypsum concrete roof decks were used where energy efficiency was a concern. To increase the energy efficiency of the material, an aggregate complying with ASTM C 35, such as perlite or vermiculite, is used to reduce the density of the gypsum concrete, thus increasing its *R*-value.

The design and installation of reinforced gypsum concrete roof decks must comply with the requirements of ASTM C 956. Gypsum concrete roof decks must be reinforced with zinc-coated (galvanized) welded or woven wire mesh or fabric and must be designed to support anticipated loads without exceeding the allowable stresses specified in Appendix Section X2 of ASTM C 956. ASTM C 956 indicates that "methods of design [of reinforced gypsum concrete] shall follow established principles of mechanics and principles of design for reinforced concrete in accordance with ACI 318." Requirements of Section 7.12 of ACI 318 for shrinkage and temperature reinforcement in reinforced concrete do not apply to reinforced gypsum concrete.

1911.2 Minimum thickness. The minimum thickness of reinforced gypsum concrete shall be 2 inches (51 mm) except the minimum required thickness shall be reduced to $1^1/_2$ inches (38 mm), provided the following conditions are satisfied:

1. The overall thickness, including the formboard, is not less than 2 inches (51 mm).
2. The clear span of the gypsum concrete between supports does not exceed 33 inches (838 mm).
3. Diaphragm action is not required.
4. The design live load does not exceed 40 pounds per square foot (psf) (1915 Pa).

❖ This section provides an exception to the ASTM C 956 minimum 2-inch (51 mm) thickness requirement for reinforced gypsum concrete roof decks. The design live load mentioned in Item 4 should be interpreted to mean the larger of the design roof live load and the design snow load.

SECTION 1912 CONCRETE-FILLED PIPE COLUMNS

1912.1 General. Concrete-filled pipe columns shall be manufactured from standard, extra-strong or double-extra-strong steel pipe or tubing that is filled with concrete so placed and manipulated as to secure maximum density and to ensure complete filling of the pipe without voids.

❖ Section 1912 regulates the design and construction of concrete-filled steel pipe or tube columns that rely on the concrete and steel acting compositely to resist loads.

During placement, the concrete must be thoroughly consolidated with a mechanical vibrator to develop a cohesive mass that fills all voids in the steel pipe or tube. This provides a complete bonding of the concrete to the steel shell, and any reinforcing steel and other embedded items that may be present in the pipe or tube.

Although the code requires concrete-filled pipe columns to be standard, extra-strong or double extra-strong steel pipe, this language is out of date since it only refers to pipe manufactured in accordance with ASTM A 53. Both ACI 318 and AISC 360 require that steel tube or pipe used in composite concrete-filled pipe columns comply with ASTM A 53 Grade B, ASTM A 500 or ASTM A 501. In addition, both ACI 318 and AISC 360 provide limitations on minimum steel wall thickness and either of these referenced standards should be consulted for up-to-date design information.

1912.2 Design. The safe supporting capacity of concrete-filled pipe columns shall be computed in accordance with the *approved* rules or as determined by a test.

❖ The load-carrying capacity of concrete-filled pipe columns should be calculated in accordance with the provisions contained in ACI 318 or AISC 360 for composite compression members. In lieu of using analytical procedures, the load-carrying capacity may be determined by preconstruction load tests in accordance with Section 1710.

1912.3 Connections. Caps, base plates and connections shall be of *approved* types and shall be positively attached to the shell and anchored to the concrete core. Welding of brackets without mechanical anchorage shall be prohibited. Where the pipe is slotted to accommodate webs of brackets or other connections, the integrity of the shell shall be restored by welding to ensure hooping action of the composite section.

❖ In general terms, this section prescribes conditions for making structural connections to concrete-filled pipe columns. For the column to act compositely, all loads must be transferred to the concrete core, not just to the steel shell, as would be the case with a bracket welded to the exterior surface of the shell. Since increased structural capacity is achieved with concrete fill, the integrity of the shell must be preserved so it can function to laterally confine the concrete core. If a connection requires welding to the steel shell, it must be done prior to filling the core with concrete unless it can be shown that the concrete will not be damaged from the heat of welding.

1912.4 Reinforcement. To increase the safe load-supporting capacity of concrete-filled pipe columns, the steel reinforcement shall be in the form of rods, structural shapes or pipe embedded in the concrete core with sufficient clearance to ensure the composite action of the section, but not nearer than 1 inch (25 mm) to the exterior steel shell. Structural shapes used as reinforcement shall be milled to ensure bearing on cap and base plates.

❖ This section does not require steel reinforcement in the concrete core; however, if it is added, it must comply with the requirements of this section and the applicable provisions of ACI 318 or AISC 360. Since the concrete fill and the steel must serve as a composite section to support the loads on the column, construction must be such that reinforcement placed in the core of the pipe has sufficient clearance to allow the concrete to flow around it for proper encasement and bonding.

1912.5 Fire-resistance-rating protection. Pipe columns shall be of such size or so protected as to develop the required fire-resistance ratings specified in Table 601. Where an outer steel shell is used to enclose the fire protective covering, the shell shall not be included in the calculations for strength of the column section. The minimum diameter of pipe columns shall be 4 inches (102 mm) except that in structures of Type V construction not exceeding three *stories above grade plane* or 40 feet (12 192 mm) in *building height*, pipe columns used in basements and as secondary steel members shall have a minimum diameter of 3 inches (76 mm).

❖ Composite pipe columns required to be fire-resistant rated may be designed to achieve the required rating with the steel left unprotected on the exterior. This is accomplished by allowing the concrete to act as a heat sink, and Kodur and MacKinnon have developed a detailed methodology for providing the required fire resistance in this manner.

Some prefabricated steel pipe columns consist of an inner pipe filled with concrete and an outer steel shell encasing conventional concrete or a proprietary insulating concrete that enables the column assembly to achieve a fire-resistance rating. Where such a column is required to have a fire-resistance rating, the structural calculations for the load-carrying capacity of the column must not include the outer steel shell unless it is protected with materials that provide a fire-resistance rating as required in Table 601.

Since the concrete in these columns is completely encased, if it is heated sufficiently by a fire, free moisture in the concrete will be turned into steam. If provisions are not made to relieve the resulting pressure, the pipe or tube may burst causing premature failure of the column. To prevent this, Kodur and MacKinnon recommend that a 1-inch-diameter (25 mm) vent hole be provided near the top and bottom of the column. If the column is continuous through more than one story of the structure, pairs of holes as described previously should be provided in each story. Listed columns may have other venting arrangements. Care must be taken to ensure that the holes are not inadvertently or unintentionally plugged during the construction process.

1912.6 Approvals. Details of column connections and splices shall be shop fabricated by *approved* methods and shall be *approved* only after tests in accordance with the *approved* rules. Shop-fabricated concrete-filled pipe columns shall be inspected by the *building official* or by an *approved* representative of the manufacturer at the plant.

❖ Concrete-filled pipe columns, including their connection details and splices, which are shop fabricated as preengineered sections, are to be inspected at the plant by the building official or an approved manufacturer's representative. Approvals of such manufactured products are to be based on accepted fabrication practices and tests.

Bibliography

The following resource materials are referenced in this chapter or are relevant to the subject matter addressed in this chapter.

ACI 318-11/318R-11, *Building Code Requirements for Structural Concrete and Commentary.* Farmington Hills, MI: American Concrete Institute, 2011.

ACI 355.2-04 *Evaulating the Performance of Post-installed Mechanical Anchors in Concrete.* Farmington Hills, MI: American Concrete Institute, 2009.

ACI 355.2R-04, *Commentary on Evaluating the Performance of Post-installed Mechanical Anchors in Concrete.* Farmington Hills, MI: American Concrete Institute, 2004.

ACI 506R-05, *Guide to Shotcrete*. Farmington Hills, MI: American Concrete Institute, 2005.

ACI 506.2-95, *Specification for Shotcrete*. Farmington Hills, MI: American Concrete Institute, 1995.

AISC 360-10, *Specification for Structural Steel Buildings*. Chicago: American Institute of Steel Construction, 2010.

ANSI A58.1-1982, *Minimum Design Loads for Buildings and Other Structures* (Also, ANSI A58.1-1972). New York: American National Standards Institute, 1972, 1982.

ASCE 7-10, *Minimum Design Loads for Buildings and Other Structures* (also ASCE 7-88, ASCE 7-93, ASCE 7-95). New York: American Society of Civil Engineers, 1990, 1993, 1995; American Soceity of Civil Engineers, Reston, VA: 2010.

ASME B1.1-89, *Unified Inch Screw Threads (UN and UNR Thread Form)*. Fairfield, NJ: American Society of Mechanical Engineers, 1989.

ASME B18.2.1-96, *Square and Hex Bolts and Screws, Inch Series*. Fairfield, NJ: American Society of Mechanical Engineers, 1996.

ASME B18.2.6-96, *Fasteners for Use in Structural Applications*. Fairfield, NJ: American Society of Mechanical Engineers, 1996.

ASTM A 36/A 36M-08, *Standard Specification for Carbon Structural Steel*. West Conshohocken, PA: ASTM International, 2008.

ASTM A 53/A 53M-07, *Standard Specification for Pipe, Steel, Black and Hot-dipped, Zinc-coated Welded and Seamless*. West Conshohocken, PA: ASTM International, 2007.

ASTM A 82-07, *Standard Specification for Steel Wire, Plain, for Concrete Reinforcement*. West Conshohocken, PA: ASTM International, 2007.

ASTM A 184/A 184M-06, *Standard Specification for Fabricated Deformed Steel Bar Mats for Concrete Reinforcement*. West Conshohocken, PA: ASTM International, 2006.

ASTM A 185-07, *Standard Specification for Steel Welded Wire Reinforcement, Plain, for Concrete*. West Conshohocken, PA: ASTM International, 2007.

ASTM A 242/A 242M-04-1, *Standard Specification for High-Strength Low-Alloy Structural Steel*. West Conshohocken, PA: ASTM International, 2004.

ASTM A 416/A 416-06, *Standard Specification for Steel Strand, Uncoated Seven-Wire for Prestressed Concrete*. West Conshohocken, PA: ASTM International, 2006.

ASTM A 421/A 421 M-05, *Standard Specification for Uncoated Stress-Relieved Steel Wire for Prestressed Concrete*. West Conshohocken, PA: ASTM International, 2005.

ASTM A 496/A 496-07, *Standard Specification for Steel Wire, Deformed, for Concrete Reinforcement*. West Conshohocken, PA: ASTM International, 2007.

ASTM A 497-07, *Standard Specification for Steel Welded Wire Fabric, Deformed, for Concrete Reinforcement*. West Conshohocken, PA: ASTM International, 2007.

ASTM A 500-07, *Standard Specification for Cold-Formed Welded and Seamless Carbon Steel Structural Tubing in Rounds and Shapes*. West Conshohocken, PA: ASTM International, 2007.

ASTM A 501-07, *Standard Specification for Hot-formed Welded and Seamless Carbon Steel Structural Tubing*. West Conshohocken, PA: ASTM International, 2007.

ASTM A 572/A 572M-07, *Standard Specification for High-strength Low-alloy Columbium-vanadium Steels of Structural Quality*. West Conshohocken, PA: ASTM International, 2007.

ASTM A 588/A 588M-05, *Standard Specification for High-strength Low-alloy Structural Steel with 50 ksi [345 MPa] Minimum Yield Point With Atmospheric Corrosion Resistance*. West Conshohocken, PA: ASTM International, 2005.

ASTM A 615/A 615M-09, *Standard Specification for Deformed and Plain Billet-steel Bars for Concrete Reinforcement*. West Conshohocken, PA: ASTM International, 2009.

ASTM A 706/A 706M-09, *Standard Specification for Low-alloy Steel Deformed and Plain Bars for Concrete Reinforcement*. West Conshohocken, PA: ASTM International, 2009.

ASTM A 722/A 722M-07, *Standard Specification for Uncoated High-strength Steel Bar for Prestressing Concrete*. West Conshohocken, PA: ASTM International, 2007.

ASTM A 767/A 767M-05, *Standard Specification for Zinc-Coated (Galvanized) Steel Bars for Concrete Reinforcement*. West Conshohocken, PA: ASTM International, 2005.

ASTM A 775/A 775M-07a, *Standard Specification for Epoxy-coated Reinforcing Steel Bars*. West Conshohocken, PA: ASTM International, 2007.

ASTM A 884/A 884 M-06, *Standard Specification for Epoxy-coated Steel Wire and Welded Wire Fabric for Reinforcement*. West Conshohocken, PA: ASTM International, 2006.

ASTM A 955/A 955M-07a, *Specification for Deformed and Plain Stainless Steel Bars for Concrete Reinforcement*. West Conshohocken, PA: ASTM International, 2007.

ASTM A 996/A 996M-09, *Specification for Rail-steel and Axle-steel Deformed Bars for Concrete Reinforcement.* West Conshohocken, PA: ASTM International, 2009.

ASTM C 31/C 31 M-08b, *Standard Practice for Making and Curing Concrete Test Specimens in the Field.* West Conshohocken, PA: ASTM International, 2008.

ASTM C 33-08, *Standard Specification for Concrete Aggregates.* West Conshohocken, PA: ASTM International, 2008.

ASTM C 35-01 (2005), *Standard Specification for Inorganic Aggregates for Use in Gypsum Plaster.* West Conshohocken, PA: ASTM International, 2005.

ASTM C 39-05-1, *Standard Test Method for Compressive Strength of Cylindrical Concrete Specimens.* West Conshohocken, PA: ASTM International, 2005.

ASTM C 42/C 42M-04, *Standard Test Method for Obtaining and Testing Drilled Cores and Sawed Beams of Concrete.* West Conshohocken, PA: ASTM International, 2004.

ASTM C 94/C 94M-09, *Standard Specification for Ready-mixed Concrete.* West Conshohocken, PA: ASTM International, 2009.

ASTM C 142-78(1990), *Standard Test Method for Clay Lumps and Friable Particles in Aggregates.* West Conshohocken, PA: ASTM International, 1990.

ASTM C 150-05, *Standard Specification for Portland Cement.* West Conshohocken, PA: ASTM International, 2005.

ASTM C 171-92, *Standard Specification for Sheet Materials for Curing Concrete.* West Conshohocken, PA: ASTM International, 1992.

ASTM C 172-08, *Standard Practice for Sampling Freshly Mixed Concrete.* West Conshohocken, PA: ASTM International, 2008.

ASTM C 231-04, *Standard Test Method for Air Content of Freshly Mixed Concrete by the Pressure Method.* West Conshohocken, PA: ASTM International, 2004.

ASTM C 260-06, *Standard Specification for Air-entraining Admixtures for Concrete.* West Conshohocken, PA: ASTM International, 2006.

ASTM C 309-93, *Standard Specification for Liquid Membrane-forming Compounds for Curing Concrete.* West Conshohocken, PA: ASTM International, 1993.

ASTM C 317/C 317M-00 (2005), *Standard Specification for Gypsum Concrete.* West Conshohocken, PA: ASTM International, 2005.

ASTM C 330-05, *Standard Specification for Lightweight Aggregates for Structural Concrete.* West Conshohocken, PA: ASTM International, 2005.

ASTM C 494/C 494M-05a, *Standard Specification for Chemical Admixtures for Concrete.* West Conshohocken, PA: ASTM International, 2005.

ASTM C 595-08a, *Standard Specification for Blended Hydraulic Cements.* West Conshohocken, PA: ASTM International, 2008.

ASTM C 618-05, *Standard Specification for Coal Fly Ash and Raw or Calcined Natural Pozzolan for Use in Concrete.* West Conshohocken, PA: ASTM International, 2005.

ASTM C 685/C 685M-07, *Standard Specification for Concrete Made by Volumetric Batching and Continuous Mixing.* West Conshohocken, PA: ASTM International, 2007.

ASTM C 803-90, *Standard Test Method for Penetration Resistance of Hardened Concrete.* West Conshohocken, PA: ASTM International, 1990.

ASTM C 805-94, *Standard Test Method for Rebound Number of Hardened Concrete.* West Conshohocken, PA: ASTM International, 1994.

ASTM C 845-04, *Standard Specification for Expansive Hydraulic Cement.* West Conshohocken, PA: ASTM International, 2004.

ASTM C 873-94, *Standard Test Method for Compressive Strength of Concrete Cylinders Cast in Place in Cylindrical Molds.* West Conshohocken, PA: ASTM International, 1994.

ASTM C 900-87(1993), *Standard Test Method of Pullout Strength of Hardened Concrete.* West Conshohocken, PA: ASTM International, 1993.

ASTM C 956-04, *Standard Specification for Installation of Cast-in-Place Reinforced Gypsum Concrete.* West Conshohocken, PA: ASTM International, 2004.

ASTM C 989-06, *Standard Specification for Ground Granulated Blast-furnace Slag for Use in Concrete and Mortars.* West Conshohocken, PA: ASTM International, 2006.

ASTM C 1017/C 1017M-03, *Standard Specification for Chemical Admixtures for Use in Producing Flowing Concrete.* West Conshohocken, PA: ASTM International, 2003.

ASTM C 1074-93, *Standard Practice of Estimating Concrete Strength by the Maturity Method.* West Conshohocken, PA: ASTM International, 1993.

ASTM C 1077-96, *Standard Practice for Laboratories Testing Concrete and Concrete Aggregates for Use in Construction and Criteria for Laboratory Evaluation.* West Conshohocken, PA: ASTM International, 1996.

ASTM C 1157-08, *Standard Performance Specification for Blended Hydraulic Cement.* West Conshohocken, PA: ASTM International, 2008.

ASTM C 1218/C 1218M-99, *Standard Test Method for Water-soluble Chloride in Mortar and Concrete.* West Conshohocken, PA: ASTM International, 1999.

ASTM C 1240-05, *Standard Specification for Silica Fume Used in Cementitious Materials*. West Conshohocken, PA: ASTM International, 2005.

ASTM D 3665-93, *Standard Practice for Random Sampling of Construction Materials*. West Conshohocken, PA: ASTM International, 1993.

ASTM E 119-08a, *Standard Test Methods for Fire Tests of Building Construction Materials*. West Conshohocken, PA: ASTM International, 2008.

ATC 3-06, *Tentative Provisions for the Development of Seismic Regulations for Buildings*. Applied Technology Council. NBS Special Publication 510, NSF Publication 78-8. Washington, DC: U.S. Government Printing Office, 1978.

AWS D1.1-04, *Structural Welding Code—Steel.* Miami: American Welding Society, 2004.

AWS D1.4-98, *Structural Welding Code—Reinforcing Steel.* Miami: American Welding Society, 1998.

BOCA National Building Code/1999. Country Club Hills, IL: Building Officials and Code Administrators International, Inc., 1999.

Cook, R. A. and R. E. Klingner. "Behavior of Ductile Multiple-anchor Steel-to-Concrete Connections with Surface-mounted Baseplates." *Anchors in Concrete: Design and Behavior.* American Concrete Institute Special Publication SP-130, February 1992, pp. 61-122.

Cook, R. A. and R. E. Klingner. "Ductile Multiple-anchor Steel-to-Concrete Connections" *Journal of Structural Engineering.* ASCE, Vol. 118, No. 6, June 1992, pp. 1645-1665.

Design of Fasteners in Concrete. Comite Euro-International du Beton (CEB). London: Thomas Telford Services Ltd., Jan. 1997.

DeVries, Richard A. *Anchorage of Headed Reinforcement in Concrete.* Ph.D. Dissertation. The University of Texas at Austin, December 1996.

Eligehausen, R. and T. Balogh. "Behavior of Fasteners Loaded in Tension in Cracked Reinforced Concrete." *AC/ Structural Journal,* Vol. 92, No. 3, May-June 1995, pp. 365-379.

Eligehausen, R., W. Fuchs, and B. Mayer. "Load Bearing Behavior of Anchor Fastenings in Tension." Betonwerk + Fertigteiltechnik, December 1987, pp. 826-832, and January 1988, pp. 29-35.

Eligehausen, R. and W. Fuchs. "Load Bearing Behavior of Anchor Fastenings under Shear, Combined Tension and Shear or Flexural Loadings" Betonwerk + Fertigteiltechnik. pp. 48-56, February 1988.

Farrow, C. B. and R. E. Klingner. "Tensile Capacity of Anchors with Partial or Overlapping Failure Surfaces: Evaluation of Existing Formulas on an LRFD Basis." *AC/ Structural Journal,* Vol. 92, No. 6, November-December 1995, pp. 698-710.

Fastenings to Concrete and Masonry Structures, State of the Art Report. Comite Euro-International du Beton, (CEB), Bulletin No. 216. London: Thomas Telford Services Ltd., 1994.

FEMA 450-1, NEHRP *Recommended Provisions for Seismic Regulations for New Buildings and Other Structures, Part 1.* Washington, DC: Prepared by the Building Seismic Safety Council for the Federal Emergency Management Agency, 2004.

FEMA 450-2, NEHRP *Recommended Provisions for Seismic Regulations for New Buildings and Other Structures, Part 2 Commentary.* Washington, DC: Prepared by the Building Seismic Safety Council for the Federal Emergency Management Agency, 2004.

Fuchs, W., R. Eligehausen, and J. Breen. "Concrete Capacity Design (CCD) Approach for Fastening to Concrete." *AC/ Structural Journal,* Vol. 92, No. 1, January-February 1995, pp. 73-93. Discussion—ACI Structural Journal, Vol. 92, No. 6, November-December 1995, pp. 787-802.

Furche, J. and R. Eligehausen. "Lateral Blow-out Failure of Headed Studs Near a Free Edge." *Anchors in Concrete: Design and Behavior.* American Concrete Institute Special Publication SP-130, February 1992, pp. 235-252.

Ghosh, S.K. "Design of Reinforced Concrete Buildings Under the 1997 UBC." *Building Standards.* International Conference of Building Officials, May-June 1998, pp. 20-24.

Ghosh, S.K. "Needed Adjustments in 1997 UBC." *Proceedings,* 1998 Convention, Structural Engineers Association of California, October 7-10, 1998, Reno-Sparks, NV, pp. T9.1-T9.15.

Ghosh, S.K., S.D. Nakaki, and K. Krishnan. "Precast Structures in Regions of High Seismicity: 1997 UBC Design Provisions." *PCI Journal,* Vol. 42, No. 6, November-December 1997, pp. 76-93.

Ishizuka, T. and N.M. Hawkins. "Effect of Bond Deterioration on the Seismic Response of Reinforced and Partially Prestressed Concrete and Ductile Moment Resistant Frames." Report SM87-2. Seattle: Department of Civil Engineering, University of Washington, 1987.

Klingner, R., J. Mendonca, and J. Malik. "Effect of Reinforcing Details on the Shear Resistance of Anchor Bolts under Reversed Cyclic Loading." *Journal of the American Concrete Institute,* Vol. 79, No. 1, 1982, pp. 3-12.

Kodur, V. and D. MacKinnon. "Design of Concrete-Filled Hollow Structural Steel Columns for Fire Endurance." *AISC Engineering Journal.*

Kuhn, D. and F. Shaikh. "Slip-pullout Strength of Hooked Anchors." Research Report. University of Wisconsin—Milwaukee, 1996.

Lotze, D. and R.E. Klingner. "Behavior of Multiple-anchor Attachments to Concrete from the Perspective of Plastic Theory." Report PMFSEL 96-4. Ferguson Structural Engineering Laboratory, the University of Texas at Austin, March 1997.

Lutz, L. "Discussion to Concrete Capacity Design (CCD) Approach for Fastening to Concrete." *ACI Structural Journal*, November/December 1995, pp. 791-792 and author's closure, pp. 798-799.

Manual of Standard Practice, 26th edition. Schaumburg, IL: Concrete Reinforcing Steel Institute, 1997.

Matlock, A. H., J. Yamazaki, and B. T. Kattula. "Comparative Study of Prestressed Concrete Beams, with and without Bond." *ACI Journal*, *Proceedings*, Vol. 68, No. 2, February 1971, pp. 116-125.

Muspratt, M. A. "Behavior of a Prestressed Concrete Waffle Slab with Unbonded Tendons." *ACI Journal, Proceedings*, Vol. 66, No. 12, December 1969, pp. 1001-1004.

Odello, R. J. and B. M. Mehta. "Behavior of a Continuous Prestressed Concrete Slab with Drop Panels." Report. Division of Structural Engineering and Structural Mechanics, University of California, Berkeley, 1967.

PCA-EB001, *Design and Control of Concrete Mixtures*, 13th edition. Skokie, IL: Portland Cement Association, 1988.

PCA-EB070, Notes on ACI 318-99 Building Code Requirements for Structural Concrete—with Design Applications. Skokie, IL: Portland Cement Association, 1999.

PCA-EB080, *Strength Design of Anchorage to Concrete*. Skokie, IL: Portland Cement Association, 2002.

PCA-EB115, *Concrete Inspection Handbook*. Skokie, IL: Portland Cement Association, 2000.

PCA-IS521, *Strength Design Load Combinations for Concrete Elements*. Skokie, IL: Portland Cement Association, 1998.

PCA-SPO40, *Fly Ash in Cement and Concrete*. Skokie, IL: Portland Cement Association, 1987.

PCI Design Handbook, 4th edition. Chicago: Precast/ Prestressed Concrete Institute, 1992.

Placing Reinforcing Bars, 7th edition. Schaumburg, IL: Concrete Reinforcing Steel Institute, 1997.

Primavera, E.J ., J. P. Pinelli, and E. H. Kalajian. "Tensile Behavior of Cast-in-Place and Undercut Anchors in High-strength Concrete." *ACI Structural Journal*, Vol. 94, No. 5, September-October 1997, pp. 583-594.

SDI NC-1.0-10, *Standard for Noncomposite Steel Floor Deck*. Fox River Grove, IL: Steel Deck Institute, 2010.

Shaikh, A. F. and W. Yi. "In-Place Strength of Welded Studs." *PCI Journal*, Vol. 30 (2), March-April 1985.

Standard Building Code, 1999 ed. Birmingham, AL: Southern Building Code Congress, International, 1999.

Stanton, J., W. C. Stone, and G. S. Cheok. "A Hybrid Reinforced Precast Frame for Seismic Regions." *PCI Journal*, Vol. 42, No. 2, March-April 1997, pp. 20-32.

Uniform Building Code, 1997 ed. Whittier, CA: International Conference of Building Officials, 1997.

Wong, T. L. "Stud Groups Loaded in Shear." M.S. Thesis, Oklahoma State University, 1988.

Zhang, Y. "Dynamic Behavior of Multiple Anchor Connections in Cracked Concrete." Ph.D. Dissertation, the University of Texas at Austin, August 1997.

Chapter 20: Aluminum

General Comments

Chapter 20 contains standards for the use of aluminum in building construction. Only provisions for structural applications of these materials are included.

This chapter does not seek to establish standards for aluminum specialty products, such as storefront framing, architectural hardware, etc. The use of aluminum in heating, ventilating and air-conditioning (HVAC) and plumbing systems is addressed in the *International Mechanical Code*® (IMC®) and the *International Plumbing Code*® (IPC®), respectively. This chapter applies to the structural requirements.

Purpose

Aluminum has certain physical properties, structural characteristics and nonstructural characteristics that set it apart as a building material. By utilizing the standards set forth in this chapter, a proper application of this material can be obtained.

SECTION 2001 GENERAL

2001.1 Scope. This chapter shall govern the quality, design, fabrication and erection of aluminum.

❖ The scope for Chapter 20 is provided in this section.

SECTION 2002 MATERIALS

2002.1 General. Aluminum used for structural purposes in buildings and structures shall comply with AA ASM 35 and AA ADM 1. The *nominal loads* shall be the minimum design loads required by Chapter 16.

❖ The referenced standards to be applied in the utilization of aluminum in buildings and other structures are established. Parts 1-A and 1-B of the Aluminum Association's (AA) *Aluminum Design Manual* apply to the design of aluminum building-type structural load-carrying members and elements according to allowable stress design (ASD) and load and resistance factor design (LRFD) criteria, respectively. ASM 35 sets criteria for the use and installation of flashings, sheet roofing and similar applications.

Bibliography

The following resource materials are referenced in this chapter or are relevant to the subject matter addressed in this chapter.

AA ADM1-05, *Aluminum Design Manual*; "Part 1-A: Specifications for Aluminum Structures, Allowable Stress Design" and "Part 1-B: Specifications for Aluminum Structures, Load and Resistance Factor Design of Buildings and Similar Type Structures" Washington, DC: The Aluminum Association, 2005.

AA ASM 35-00, *Specification for Aluminum Sheet Metal Work in Building Construction*. Washington, DC: The Aluminum Association, 2000.

IMC-12, *International Mechanical Code*. Washington, DC: International Code Council, 2011.

IPC-12, *International Plumbing Code*. Washington, DC: International Code Council, 2011.

Chapter 21: Masonry

General Comments

Masonry construction has been used for at least 10,000 years in a variety of structures—homes, private and public buildings and historical monuments. The masonry of ancient times involved two major materials: brick manufactured from sun-dried mud or burned clay and shale; and natural stone.

The first masonry structures were unreinforced and intended to support mainly gravity loads. The weight of these structures stabilized them against lateral loads from wind and earthquakes.

Masonry construction has progressed through several stages of development. Fired clay brick became the principal building material in the United States during the middle 1800s. Concrete masonry was introduced to construction during the early 1900s and, along with clay masonry, expanded in use to all types of structures.

Historically, "rules of thumb" (now termed "empirical design") were the only available methods of masonry design. Only in recent times have masonry structures been engineered using structural calculations. In the last 45 years, the introduction of engineered reinforced masonry has resulted in structures that are stronger and more stable against lateral loads, such as wind pressure and seismic ground motion.

Masonry consists of a variety of materials. Raw materials are made into masonry units of different sizes and shapes, each having specific physical and mechanical properties. Both the raw materials and the method of manufacture affect masonry unit properties.

The word "masonry" is a general term that applies to construction using hand-placed units of clay, concrete, structural clay tile, glass block, natural stones and the like. One or more types of masonry units are bonded together with mortar, metal ties, reinforcement and accessories to form walls and other structural elements.

Proper masonry construction depends on correct design, materials, handling, installation and workmanship.

With a fundamental understanding of the functions and properties of the materials that comprise masonry construction and with proper design and construction, quality masonry structures are not difficult to obtain.

During the pioneer era of U.S. history, the fireplace was the central focus of residential cooking and heating.

Today, the fireplace is essentially a decorative feature of residential construction. For energy conservation, existing fireplaces are sometimes converted and new fireplaces are designed to provide supplemental heat.

Of the many types of fireplaces, the most common are single face. Multifaced fireplaces, such as a corner fireplace with two adjacent open sides, fireplaces with two opposite faces open (common exposure to two rooms) or fireplaces with three or all faces open also occur, but are less common.

While the provisions of this chapter are for single-faced fireplaces, almost all types of masonry fireplaces include the same basic construction features: the base assembly, which consists of a foundation and hearth support; the firebox assembly, which consists of a fireplace opening, a hearth, a firebox or combustion chamber; and the throat and the smoke chamber, which supports the chimney liner.

Masonry fireplaces are made primarily of clay brick or natural stones, but also of concrete masonry or cast-in-place concrete. Chimneys for medium- and high-heat appliances require special attention for fire safety.

Purpose

Chapter 21 provides comprehensive and practical requirements for masonry construction, based on the latest state of technical knowledge. The provisions of Chapter 21 require minimum accepted practices and the use of standards for the design and construction of masonry structures and elements of structures. The provisions address: material specifications and test methods; types of wall construction; criteria for engineered design (by the working stress and strength design methods, prestressed masonry and direct design method); criteria for empirical design; required details of construction and other aspects of masonry, including execution of construction. The provisions are intended to result in safe and durable masonry. The provisions of Chapter 21 are also intended to prescribe minimum accepted practices for the design and construction of glass unit masonry, masonry fireplaces, masonry heaters and masonry chimneys.

SECTION 2101 GENERAL

2101.1 Scope. This chapter shall govern the materials, design, construction and quality of masonry.

❖ Section 2101 prescribes general requirements for masonry designed in accordance with Chapter 21 of the code. It identifies masonry design methods and the conditions required for the use of each method. The methods are intended as a practical means for safety under a variety of potential service conditions.

Minimum requirements for construction documents and fireplace drawings are also included in Section 2101.

Chapter 21 contains the minimum code requirements for acceptance of masonry design and construction by the building official. Compliance with these requirements is intended to result in masonry construction with the minimum required structural adequacy and durability. Requirements more stringent than these are appropriate where mandated by sound engineering and judgment. Less restrictive requirements, however, are not permitted.

2101.2 Design methods. Masonry shall comply with the provisions of one of the following design methods in this chapter as well as the requirements of Sections 2101 through 2104. Masonry designed by the *allowable stress design* provisions of Section 2101.2.1, the strength design provisions of Section 2101.2.2, the prestressed masonry provisions of Section 2101.2.3, or the direct design requirements of Section 2101.2.7 shall comply with Section 2105.

❖ This section requires masonry to comply with one of seven design methods and the requirements contained in Sections 2101 through 2104 for construction documents, materials and construction.

Other provisions of the code also apply to masonry. For example, fire-resistant construction using masonry is required to comply with Chapter 7. Design loads and related requirements, including seismic forces and detailing, are required to comply with Chapter 16 and ASCE 7. Masonry foundations are required to comply with the provisions of Chapter 18. Special inspections of masonry construction are required in Chapter 17, which references the quality assurance program requirements in TMS 402/ACI 530/ASCE 5 and TMS 602/ACI 530.1/ASCE 6. Masonry veneer is addressed in Chapter 14.

2101.2.1 Allowable stress design. Masonry designed by the *allowable stress design* method shall comply with the provisions of Sections 2106 and 2107.

❖ This section requires that masonry designed by the allowable stress design (ASD) method meets both the ASD requirements in Section 2107 and the seismic design requirements in Section 2106. Section 2106 references Section 1.18 of the Masonry Standards Joint Committee Code (MSJC) for seismic requirements based on the seismic design category of the structure. Section 2107 requires ASD to comply with Chapters 1 and 2 of TMS 402/ACI 530/ASCE 5 with several modifications. Additional information on these procedures is given in the commentaries to Section 2107 and TMS 402/ACI 530/ASCE 5.

TMS 402/ACI 530/ASCE 5 and TMS 602/ACI 530.1/ ASCE 6 are referenced throughout Chapter 21. A description of the standard and specification is warranted here. Both are joint publications of the MSJC, which includes The Masonry Society (TMS), the American Concrete Institute (ACI) and the Structural Engineering Institute of the American Society of Civil Engineers (ASCE). These standards are typically referred to as the MSJC Code and MSJC Specification, respectively, to reflect their joint authorship and sponsorship of the committee that oversees their development. The standards are developed through an ANSI-accredited consensus process and reflect the current state of technical knowledge on masonry design and construction.

The MSJC Code (TMS 402/ACI 530/ASCE 5) contains minimum design requirements for masonry elements of structures. Topics include: construction documents; quality assurance; materials; analysis and design; strength and serviceability; flexural and axial stresses; shear; reinforcement; walls; columns; pilasters; beams and lintels and empirical design.

The ASD method in the MSJC Code assumes linearly elastic material behavior and properties, and uses nominal loads (see Chapter 16). The standard also includes provisions for strength design, prestressed masonry design, empirical design, veneer, glass unit masonry and design of autoclaved aerated concrete (AAC) masonry.

The MSJC Specification (TMS 602/ACI 530.1/ ASCE 6) establishes minimum acceptable levels of construction. It includes minimum requirements for composition; preparation and placement of materials; quality assurance for materials and masonry; execution of masonry construction and inspection and verification of quality. The MSJC Specification contains both mandatory requirements, as well as optional requirements that may be project specific. The mandatory requirements are enforceable code requirements; the optional requirements may be invoked by the design professional. The specification is meant to be modified for use with the particular project under design.

2101.2.2 Strength design. Masonry designed by the strength design method shall comply with the provisions of Sections 2106 and 2108, except that autoclaved aerated concrete (AAC) masonry shall comply with the provisions of Section 2106 and Chapters 1 and 8 of TMS 402/ACI 530/ASCE 5.

❖ Masonry is required to meet the strength design provisions referenced in Section 2108 and the seismic design requirements referenced in Section 2106. Section 2106 references Section 1.18 of the MSJC Code for seismic requirements based on the seismic design category of the structure. Additional information on the strength design requirements is given in the commentaries to Section 2108 and to Chapter 3 of the MSJC Code. Chapter 8 of MSJC Code is referenced for strength design provisions for AAC masonry.

2101.2.3 Prestressed masonry. Prestressed masonry shall be designed in accordance with Chapters 1 and 4 of TMS 402/ ACI 530/ASCE 5 and Section 2106. *Special inspection* during construction shall be provided as set forth in Section 1705.4.

❖ Prestressed masonry must comply with Chapters 1 and 4 of the MSJC Code, as well as the seismic detailing requirements in Section 2106. Section 2106 references Section 1.18 of the MSJC Code for seismic requirements based on the seismic design category of the structure. Additional information on prestressed masonry design is given in the MSJC Code commentary for Chapter 4. The section refers to Section 1705.4 for special inspection requirements for prestressed masonry, which in turn references the quality assurance program contained in TMS 402/ACI 530/ASCE 5 and TMS 602/ACI 530.1/ASCE 6.

2101.2.4 Empirical design. Masonry designed by the empirical design method shall comply with the provisions of Sections 2106 and 2109 or Chapter 5 of TMS 402/ACI 530/ ASCE 5.

❖ This section permits the empirical design of masonry by either the provisions of Section 2109 or Chapter 5 of the MSJC Code. The provisions of Section 2109 reference the MSJC Code, with some additional provisions added to that section that address surface-bonded masonry construction and adobe masonry, which are not included in the MSJC Code. Additional information on these provisions is given in the commentaries to Section 2109 and the MSJC Code.

2101.2.5 Glass unit masonry. Glass unit masonry shall comply with the provisions of Section 2110 or Chapter 7 of TMS 402/ACI 530/ASCE 5.

❖ The provisions in Section 2110 reference the requirements in Chapter 7 of the MSJC Code. Additional information on these provisions is given in the commentaries to Section 2110 and the MSJC Code.

2101.2.6 Masonry veneer. Masonry veneer shall comply with the provisions of Chapter 14 or Chapter 6 of TMS 402/ ACI 530/ASCE 5.

❖ This section requires masonry veneer to comply with the provisions of Chapter 14; specifically, Sections 1405.5 for anchored masonry veneer and 1405.9 for adhered masonry veneer. These sections reference the provisions in Chapter 6 of the MSJC Code. Additional information on these provisions is given in the commentaries to Chapter 14 and the MSJC Code.

2101.2.7 Direct design. Masonry designed by the direct design method shall comply with the provisions of TMS 403.

This section references TMS 403, *Direct Design Handbook for Masonry Structures*. TMS 403 was developed by TMS's Design Practices Committee to provide a direct procedure for the structural design of single-story, concrete masonry structures using both reinforced and unreinforced masonry. The handbook was written so that architects, engineers, contractors, building officials, researchers, educators, suppliers, manufacturers and others can readily use the standard for both residential and commercial structures. The procedure is based on the strength design provisions of the 2008 MSJC Code and ASCE 7-05 and can be used as an alternative design method for masonry structures that meet the same scope and limitations of TMS 403. TMS 403 can be used to design typical masonry structures in the U.S. with mapped ground snow loads up to 60 lb/ft^2, nominal design wind speeds (V_{asd}) up to 150 mph, mapped seismic 0.2 second spectral response accelerations up to 3.0g and mapped seismic 1.0 second spectral response accelerations up to 1.25g. Topics covered include referenced standards; definitions and notations; site limitations; architectural limitations; loading limitations; material and construction requirements; the direct design procedure; optional modifications to the direct design procedure; specifications and details. The commentary presents background analysis, details and committee considerations used to develop the handbook. An appendix follows the commentary that provides an example of how to use the direct design procedure for a typical masonry building.

2101.3 Construction documents. The *construction documents* shall show all of the items required by this code including the following:

1. Specified size, grade, type and location of reinforcement, anchors and wall ties.
2. Reinforcing bars to be welded and welding procedure.
3. Size and location of structural elements.
4. Provisions for dimensional changes resulting from elastic deformation, creep, shrinkage, temperature and moisture.
5. Loads used in the design of masonry.
6. Specified compressive strength of masonry at stated ages or stages of construction for which masonry is designed, except where specifically exempted by this code.
7. Details of anchorage of masonry to structural members, frames and other construction, including the type, size and location of connectors.
8. Size and permitted location of conduits, pipes and sleeves.
9. The minimum level of testing and inspection as defined in Chapter 17, or an itemized testing and inspection program that meets or exceeds the requirements of Chapter 17.

❖ Construction documents are a part of the submittal documents required by Section 107. In addition to the construction document requirements prescribed in Sections 107.2 and 1603, the items listed in this section are specific to masonry construction. Construction requirements must be clearly identified in the contract documents so that the structure is properly constructed using appropriate materials and methods. This section requires that, at a minimum, critical items required by the code and the particular design are shown in the construction documents. The list is a minimum and should not necessarily be considered all-inclusive by the design professional. Both the design professional and the building official are permitted to require additional items as needed for a particular structure.

2101.3.1 Fireplace drawings. The *construction documents* shall describe in sufficient detail the location, size and construction of masonry fireplaces. The thickness and characteristics of materials and the clearances from walls, partitions and ceilings shall be indicated.

❖ This section requires the submission of construction documents for all fireplaces so that compliance with appropriate code sections can be properly determined during plan review. The type of information and its format for plan review are established in this section. Construction documents are required showing relationships of components, as well as details related to the specific characteristics of the materials and techniques to be used when erecting the fireplace and chimney system. Such details are to include the type of brick or stone; refractory brick; concrete masonry; mortar requirements; wall thicknesses; clearances; dimensions of openings; and dimensions of the firebox and hearth extension.

SECTION 2102 DEFINITIONS AND NOTATIONS

2102.1 General. The following terms are defined in Chapter 2:

❖ Definitions facilitate the understanding of code provisions and minimize potential confusion. To that end, this section lists definitions of terms associated with roof assemblies, roof coverings and rooftop structures. Note that these definitions are found in Chapter 2. The use and application of defined terms, as well as undefined terms, are set forth in Section 201.

AAC MASONRY.

ADOBE CONSTRUCTION.

Adobe, stabilized.

Adobe, unstabilized.

ANCHOR.

ARCHITECTURAL TERRA COTTA.

AREA.

Gross cross-sectional.

Net cross-sectional.

AUTOCLAVED AERATED CONCRETE (AAC).

BED JOINT.

BOND BEAM.

BRICK.

Calcium silicate (sand lime brick).

Clay or shale.

Concrete.

CAST STONE.

CELL.

CHIMNEY.

CHIMNEY TYPES.

High-heat appliance type.

Low-heat appliance type.

Masonry type.

Medium-heat appliance type.

CLEANOUT.

COLLAR JOINT.

COMPRESSIVE STRENGTH OF MASONRY.

DIMENSIONS.

Nominal.

Specified.

FIREPLACE.

FIREPLACE THROAT.

FOUNDATION PIER.

HEAD JOINT.

MASONRY.

Ashlar masonry.

Coursed ashlar.

Glass unit masonry.

Plain masonry.

Random ashlar.

Reinforced masonry.

Solid masonry.

Unreinforced (plain) masonry.

MASONRY UNIT.

Hollow.

Solid.

MORTAR.

MORTAR, SURFACE-BONDING.

PRESTRESSED MASONRY.

PRISM.

RUBBLE MASONRY.

Coursed rubble.

Random rubble.

Rough or ordinary rubble.

RUNNING BOND.

SHEAR WALL.

Detailed plain masonry shear wall.

Intermediate prestressed masonry shear wall.

Intermediate reinforced masonry shear wall.

Ordinary plain masonry shear wall.

Ordinary plain prestressed masonry shear wall.

Ordinary reinforced masonry shear wall.

Special prestressed masonry shear wall.

Special reinforced masonry shear wall.

SPECIFIED.

SPECIFIED COMPRESSIVE STRENGTH OF MASONRY, f'_m.

STACK BOND.

STONE MASONRY.

Ashlar stone masonry.

Rubble stone masonry.

STRENGTH.

Design strength.

Nominal strength.

Required strength.

THIN-BED MORTAR.

TIE, WALL.

TILE, STRUCTURAL CLAY.

WALL.

Cavity wall.

Composite wall.

Dry-stacked, surface-bonded wall.

Masonry-bonded hollow wall.

Parapet wall.

WYTHE.

NOTATIONS.

d_b = Diameter of reinforcement, inches (mm).

F_s = Allowable tensile or compressive stress in reinforcement, psi (MPa).

f_r = Modulus of rupture, psi (MPa).

f'_{AAC} = Specified compressive strength of AAC masonry, the minimum compressive strength for a class of AAC masonry as specified in ASTM C 1386, psi (MPa).

f'_m = Specified compressive strength of masonry at age of 28 days, psi (MPa).

f'_{mi} = Specified compressive strength of masonry at the time of prestress transfer, psi (MPa).

K = The lesser of the masonry cover, clear spacing between adjacent reinforcement, or five times d_b, inches (mm).

L_s = Distance between supports, inches (mm).

l_d = Required development length or lap length of reinforcement, inches (mm).

P = The applied load at failure, pounds (N).

S_t = Thickness of the test specimen measured parallel to the direction of load, inches (mm).

S_w = Width of the test specimen measured parallel to the loading cylinder, inches (mm).

❖ Explanations of the notations used in Chapter 21 that are listed above clarify the meaning of these terms and the appropriate units, if any, that apply to them.

SECTION 2103
MASONRY CONSTRUCTION MATERIALS

2103.1 Concrete masonry units. Concrete masonry units shall conform to the following standards: ASTM C 55 for concrete brick; ASTM C 73 for calcium silicate face brick; ASTM C 90 for load-bearing concrete masonry units or ASTM C 744 for prefaced concrete and calcium silicate masonry units.

❖ Proper selection of materials is essential to produce masonry with adequate strength and durability. This section sets forth prescriptive and performance-based requirements (referenced standards) for masonry materials. Test procedures and criteria for establishing and verifying quality are included. Concrete masonry refers to solid and hollow concrete units, including concrete brick, concrete block, split-face block, slump block and other special units. This section requires conformance to ASTM International (ASTM) standards for each specific type of concrete masonry unit. The standards include requirements for materials; manufacture; physical properties; strength; absorption; minimum dimensions and permissible variations; inspection; testing and rejection.

Concrete masonry units are selected based on the desired use and appearance. Units are typically specified by density class and compressive strength.

Hollow load-bearing concrete masonry units are manufactured in accordance with the requirements of ASTM C 90 using portland cement, water and mineral aggregates. Other suitable materials, such as approved admixtures, are permitted in accordance with ASTM C 90.

Hollow load-bearing concrete masonry units have three weight classifications: normal-weight units of 125 pounds per cubic foot (pcf) (2000 kg/m^3) or more; medium-weight units of between 105 pcf and 125 pcf (1680 and 2000 kg/m^3) and lightweight units of less than 105 pcf (1680 kg/m^3). Lightweight aggregates generally are expanded blast-furnace slag or similar suitable materials. Normal-weight units are made from sand, gravel, crushed stone or air-cooled, blast-furnace-slag aggregates.

Concrete brick is required to comply with ASTM C 55. Concrete building brick and other solid concrete veneer and facing units are typically smaller than concrete masonry units conforming to ASTM C 90.

They are made from portland cement, water and mineral aggregates, with or without inclusion of other approved materials.

Concrete brick is manufactured as lightweight, medium-weight and normal-weight units as described above for hollow concrete masonry units. Concrete brick is manufactured in two grades: Grade N and Grade S. Grade N units are used as architectural veneer and facing units in exterior walls, where high strength and resistance to moisture penetration, and freeze-thaw cycling are required. Grade S units are used when moderate strength and resistance to freeze-thaw action and moisture penetration are required.

Calcium silicate face brick is a solid masonry unit complying with ASTM C 73. These units are manufactured principally from silica sand, hydrated lime and water.

Calcium silicate face brick is manufactured in two grades: Grade SW and Grade MW. Grade SW is required where exposure to moisture in the presence of freezing temperatures is anticipated. Grade MW is permitted where the anticipated exposure has freezing temperatures, but without water saturation.

Grade SW units have a minimum permitted compressive strength on the gross area of 4,500 psi (31.0 MPa) (average of three units), but not less than 3,500 psi (24.1 MPa) for any individual unit. Grade MW units have a minimum permitted compressive strength on the gross area of 2,500 psi (17.2 MPa) (average of three units), but not less than 2,000 psi (13.8 MPa) for any individual unit.

ASTM C 744 is referenced for the manufacture of prefaced concrete and calcium silicate masonry units, commonly referred to as "glazed" concrete masonry units. The specified exposed surfaces of these units are covered during their manufacture with resin, resin and inert filler, or cement and inert filler to produce a smooth resinous tile-like facing.

Facing requirements of that standard address resistance to chemicals; failure of adhesion of the facing material; abrasion surface-burning characteristics, color and color change; soiling and cleansability. The standard also covers dimensional tolerances, including face dimensions and distortions.

2103.2 Clay or shale masonry units. Clay or shale masonry units shall conform to the following standards: ASTM C 34 for structural clay *load-bearing wall* tile; ASTM C 56 for structural clay nonload-bearing wall tile; ASTM C 62 for building brick (solid masonry units made from clay or shale); ASTM C 1088 for solid units of thin veneer brick; ASTM C 126 for ceramic-glazed structural clay facing tile, facing brick and solid masonry units; ASTM C 212 for structural clay facing tile; ASTM C 216 for facing brick (solid masonry units made from clay or shale); ASTM C 652 for hollow brick (hollow masonry units made from clay or shale) or ASTM C 1405 for glazed brick (single-fired solid brick units).

> **Exception:** Structural clay tile for nonstructural use in fireproofing of structural members and in wall furring shall not be required to meet the compressive strength specifications. The fire-resistance rating shall be determined in accordance with ASTM E 119 or UL 263 and shall comply with the requirements of Table 602.

❖ Section 2103.2 requires conformance with ASTM standards for masonry units manufactured from clay or shale. The various standards also include requirements for materials; manufacture; physical properties; minimum dimensions and permissible variations; inspections and testing.

Clay or shale masonry units are manufactured from clay and shale, which are compounds of silica or alumina. Shale is simply a hardened clay. The raw materials are formed into the desired shape by extrusion and cutting, molding or pressing while in the plastic state. The units are then fired in a kiln. The raw materials and the manufacturing process influence the physical properties of the manufactured unit.

Clay or shale masonry units are selected for their intended use from a variety of shapes, sizes and strengths. Units are specified based on grades and type.

Solid face brick units are required to conform to ASTM C 216. This standard covers clay brick intended to be used in masonry and supplying structural or facing components, or both, to the structure. These units are available in a variety of sizes, textures, colors and shapes.

ASTM C 216 contains requirements for two grades of durability: Grade SW and Grade MW. Grade SW brick is intended for use where high and uniform resistance to damage caused by freeze-thaw cycling is desired and where the brick may be subjected to such cycling while saturated with water. Grade MW brick is intended for use where moderate resistance to cyclic freeze-thaw damage is permissible or where the brick may be damp, but not saturated with water, when such cycling occurs. ASTM C 216 further classifies face brick into three types of appearance: Types FBS, FBX and FBA. Type FBS (face brick standard) units are permitted for general use in masonry. Type FBX (face brick select) units are also for general use, but have a higher degree of precision and lower permissible variation in size than Type FBS units. Type FBA (face brick architectural) units are also for general use, but are intentionally manufactured to produce characteristic architectural effects resulting from nonuniformity in size and texture of the units.

Solid units of building brick are required to conform to ASTM C 62. This specification covers brick intended for use in both structural and nonstructural masonry where external appearance is not critical. This brick was formerly called "common brick," and the standard does not contain appearance requirements.

ASTM C 62 contains requirements for three grades of durability: Grade SW, Grade MW and Grade NW. Grade SW brick is intended for use where high and uniform resistance to freeze-thaw damage is desired and where the brick may be subjected to such cycling while saturated with water. Grade MW brick is intended for use where moderate resistance to cyclic freeze-thaw damage is permissible or where the brick may be damp but not saturated with water when such cycling occurs. Grade NW is used where little resistance to cyclic freeze-thaw damage is required and is acceptable for applications protected from water absorption and freezing.

Hollow brick is required to conform to ASTM C 652, which regulates hollow building and hollow facing brick. Hollow brick differs from structural clay tile in that it has more stringent physical property requirements, such as thicker shell and web dimensions, and higher minimum compressive strengths.

Hollow brick has two grades of durability: Grade SW and Grade MW. Grade SW is used where a high and uniform degree of resistance to freeze-thaw degradation and disintegration by weathering is desired and when the brick may be saturated while frozen. Grade MW is intended for use where a moderate and somewhat nonuniform degree of resistance to freeze-thaw degradation is permissible.

ASTM C 652 also classifies hollow brick into four types of appearance: Types HBS, HBX, HBA and HBB. Type HBS (hollow brick standard) units are permitted for general use. Type HBX (hollow brick select) units are also for general use, but have a higher degree of dimensional precision than Type HBS units. Type HBA (hollow brick architectural) units are also for general use, but are intentionally manufactured to produce characteristic architectural effects resulting from nonuniformity in size and texture of the units. Type HBB (hollow building brick) units are for general use in masonry where a particular color, texture, finish, uniformity or limits on cracks, warpage or other imperfections detracting from their appearance are not a consideration.

ASTM C 1088 regulates thin brick used in adhered veneer having a maximum actual thickness of $1^{3}/_{4}$ inches (44 mm). Thin veneer brick units are specified in three types of appearance and two grades of durability.

Exterior-grade units are for exposure to weather; interior-grade units are not. Type TBS (standard) units are permitted for general use. Type TBX (select) units are also for general use, but have a higher degree of precision and lower permissible variation in size than Type TBS units. Type TBA (architectural) units are also for general use, but are intentionally manufactured to produce characteristic effects in variations of size, color and texture.

ASTM C 34 is referenced for the manufacture of structural clay load-bearing wall tile. This type of tile has two grades: Grade LBX and LB. Grade LBX is suitable for general use in masonry construction and can be used in masonry exposed to weathering, provided that the units meet durability requirements for Grade SW of ASTM C 216 for solid units of face brick. This tile grade is also suitable for the direct application of stucco.

Grade LB is suitable for general use in masonry not exposed to freeze-thaw action, or for exposed masonry where protected with a minimum 3-inch (76 mm) facing of stone, brick or other masonry materials.

Fire-resistant tile intended for use in load-bearing masonry is required to conform to ASTM C 34. Tile intended for use in fireproofing structural members is required to be of such sizes and shapes as to cover completely the exposed surfaces of the members.

ASTM C 212, the referenced standard regulating structural clay facing tile, covers two types of structural clay load-bearing facing tile: Types FTX and FTS.

Type FTX clay facing tile is a smooth-face tile suitable for use in exposed exterior and interior masonry walls, and partitions where low absorption, easy cleaning and resistance to staining are required. The physical characteristics of this tile require a high degree of mechanical perfection, narrow color range and minimum variation in face dimensions.

Type FTS clay facing tile is a smooth- or rough-textured face tile suitable for general use in exposed exterior and interior masonry walls and partitions where moderate absorption, variation in face dimensions, minor defects in surface finish and moderate color variations are permissible.

There are two classes of tile for the types stated above: standard (for general use) and special duty (having superior resistance to impact and moisture transmission and supporting greater lateral and vertical loads).

ASTM C 56 is the standard regulating the manufacture of nonload-bearing tile, which is made from the same materials as other types of tile previously mentioned and is used for partitions, fireproofing and furring.

ASTM C 126 prescribes requirements for ceramic-glazed structural clay load-bearing facing tile and brick and other solid masonry units made from clay, shale, fire clay or combinations thereof. The standard specifies two grades and two types of ceramic-glazed masonry: Grade S is used with comparatively narrow mortar joints, while Grade SS is used where variations in face dimensions are very small; Type I units are used where only one finished face is to be exposed, while Type II units are used where both finished faces are to be exposed.

The finish of glazed units is important. Requirements and test methods are prescribed for imperviousness, opacity, resistance to fading and crazing, hardness, abrasion resistance and flame spread smoke density. Compressive strength requirements, dimensional tolerances and permissible distortions are also prescribed by the standard.

The exception to this section allows structural clay tile that does not meet the compressive strength specifications to be used as nonstructural fireproofing for structural members. The fire-resistance rating must be determined in accordance with ASTM E 119.

2103.3 AAC masonry. AAC masonry units shall conform to ASTM C 1386 for the strength class specified.

❖ AAC masonry units are a low-density cementitious product first introduced into the 2005 edition of the MSJC Code and MSJC Specification, although its use has been prevalent in other countries for several decades. AAC masonry units must comply with ASTM C 1386 for the strength class specified.

2103.4 Stone masonry units. Stone masonry units shall conform to the following standards: ASTM C 503 for marble building stone (exterior); ASTM C 568 for limestone building stone; ASTM C 615 for granite building stone; ASTM C 616 for sandstone building stone; or ASTM C 629 for slate building stone.

❖ Many types of natural stone, including marble, granite, slate, limestone and sandstone, are used in building construction. This section covers natural stones that are sawed, cut, split or otherwise shaped for masonry purposes. Natural stones for building purposes are specified in various grades, textures and finishes, and are required to have physical properties appropriate to their intended use.

Where applicable to specific kinds of natural stone, the finished units are required to be sound and free from spalls, cracks, open seams, pits and other defects that would impair structural strength and durability and, when applicable, fire resistance.

This section requires conformance to the following ASTM standards for natural stone masonry: ASTM C 503 for marble building stone (exterior); ASTM C 568 for limestone building stone; ASTM C 615 for granite building stone; ASTM C 616 for sandstone building stone or ASTM C 629 for slate building stone. The standards contain requirements for absorption, density (except for slate), compressive strength (except for slate), modulus of rupture, abrasion resistance (except for granite) and acid resistance (slate only).

2103.5 Architectural cast stone. Architectural cast stone shall conform to ASTM C 1364.

❖ Architectural cast stone is a masonry product that provides ornamental or functional features to buildings and other structures. Cast stone is made from fine and coarse aggregates, portland cement, mineral oxide color pigments, chemical admixtures and water. The products are available in virtually any color and can give the appearance of a variety of natural building stones, such as limestone, granite, slate, travertine or marble. Applications range from simple window sills to the most complicated architectural elements and features. When properly manufactured and installed, cast stone has similar strength and physical properties as most dimensional building stone. ASTM 1364 establishes the physical requirements, sampling, testing and visual inspection of architectural cast stone that may be made from facing and backup mixtures or from a homogeneous mixture, produced either by wet or dry casting. The standard requires the raw materials, such as portland cement, aggregates, coloring pigment, reinforcement, chemical admixtures, air-entraining admixtures, water- reducing and accelerating admixtures, ground slag, fly ash or natural pozzolan, and other constituents to be precisely observed. The product must also conform to specified physical and performance requirements such as compressive strength, cold water absorption, boiling water absorption, air content, field testing, resistance to freezing and thawing and linear drying shrinkage.

2103.6 Ceramic tile. Ceramic tile shall be as defined in, and shall conform to the requirements of, ANSI A137.1.

❖ Ceramic tile is made from clay, possibly mixed with other ceramic materials. Metallic oxides may be included for glaze coloring. Ceramic tile products are available in a broad range of sizes, appearances, characteristics and functions.

ANSI A137.1 is the recognized industry standard for the manufacture, testing and labeling of ceramic tile. According to this standard, tile should be shipped in sealed cartons with the grade of contents indicated by grade seals with a distinctive coloring: blue for standard grade (units as perfect and free from defects as is possible in the manufacturing process) and yellow for "seconds" (units having slight defects, but free from structural defects and cracks).

ANSI A137.1 groups ceramic tile into four major types: glazed wall tile, mosaic tile, quarry tile and paving tile.

2103.7 Glass unit masonry. Hollow glass units shall be partially evacuated and have a minimum average glass face thickness of $^{3}/_{16}$ inch (4.8 mm). Solid glass-block units shall be provided when required. The surfaces of units intended to be in contact with mortar shall be treated with a polyvinyl butyral coating or latex-based paint. Reclaimed units shall not be used.

❖ Consensus national standards have not been written to establish minimum material properties or test methods for glass blocks. Reliance must be placed on the manufacturer's specifications and glass block is required to meet the minimum specified dimensions of those specifications. Glass block has generally performed well in service.

2103.8 Second-hand units. Second-hand masonry units shall not be reused unless they conform to the requirements of new units. The units shall be of whole, sound materials and free from cracks and other defects that will interfere with proper laying or use. Old mortar shall be cleaned from the unit before reuse.

❖ This section allows for the use of salvaged brick and other second-hand masonry units, provided that their quality and condition meet the requirements for new masonry units. Second-hand units must be whole, of sound material, clean and free from defects that would interfere with proper laying or use.

Most second-hand masonry units come from the demolition of old buildings. Masonry units manufactured in the past do not generally compare with the quality of masonry made by modern manufacturing methods under controlled conditions. Therefore, designers should expect salvaged masonry units to have lower strength and durability than new units.

Generally, the difference between walls laid up with new masonry units and second-hand units of the same type is the adhesion of the mortar to the masonry surfaces. When new masonry units are laid in fresh mortar, water and fine cementitious particles are absorbed into the masonry, thereby improving bond strength. In contrast, pores in the bed faces of second-hand masonry units, regardless of cleaning, are filled with particles of cement, lime and deleterious substances that impede adequate absorption, thereby adversely affecting bond between the mortar and the masonry.

2103.9 Mortar. Mortar for use in masonry construction shall conform to ASTM C 270 and Articles 2.1 and 2.6 A of TMS 602/ACI 530.1/ASCE 6, except for mortars listed in Sections 2103.10, 2103.11 and 2103.12. Type S or N mortar conforming to ASTM C 270 shall be used for glass unit masonry.

❖ Masonry mortar bonds masonry units to form an integral structure. Mortar provides a tight and weather-resistant seal between units; bonds with steel joint reinforcement, ties and other metal accessories; and compensates for dimensional variations in masonry units. It can also serve aesthetic purposes through contrasts of color, texture and shadow lines created by different types of tooled joints.

This section of the code establishes ASTM C 270 as the standard regulating mortar to be used in masonry construction. The standard covers mortars for use in the construction of nonreinforced and reinforced unit masonry structures. It specifies types of mortar and two alternative specifications—the proportion specification and the property specification.

Type M mortar has the highest compressive and tensile bond strength. Its use should be limited only to those conditions where it is needed, such as high compressive loads or extreme environments.

Type S mortar is a general-purpose mortar with high compressive and tensile bond strengths. It is often used in reinforced masonry and in unreinforced masonry that requires high strength to resist out-of-plane lateral loads.

Type N mortar is a general-purpose mortar with medium compressive and tensile bond strengths. It is used where high vertical and lateral loads are not expected.

Type O is a low-strength mortar suitable for use in nonload-bearing construction and where the masonry will not be subject to severe weathering or freeze-thaw cycling.

The materials used in mortar are required to conform to the following standard specifications referenced in ASTM C 5, ASTM C 91, ASTM C 150, ASTM C 207, ASTM C 270 and ASTM C 595.

In addition to type, mortar for masonry is further classified according to its primary cementitious materials. In portland cement-lime mortar, the cementitious materials are portland cement and hydrated mason's lime. In masonry-cement mortar and mortar-cement mortar, the principal cementitious material is portland cement, which is contained in the masonry cement and the mortar cement.

Regardless of a masonry mortar's principal cementitious constituents, each ingredient—cement, lime (when used), sand and water—contributes to the mortar's overall performance. Portland cement, mortar cement, masonry cement or blended hydraulic cement make the hardened mortar strong and durable. Lime contributes to the workability and water retention of fresh mortar and to the flexibility, compressive strength and tensile bond strength of hardened mortar. Sand serves as a filler material, increases strength and reduces shrinkage. Water determines the consistency of fresh mortar and also hydrates the cementitious materials, resulting in hardening of the mortar.

Tables 1 and 2 of ASTM C 270 provide requirements for the production of mortar by the proportion method (by volume) and by the property method (by compressive strength and other properties), respectively. Whether the proportion or property specification governs depends on the contract documents. When neither proportion nor property specification is prescribed, the proportion specification governs, unless data are presented to and accepted by the specifier to show that mortar meets the requirements of the property specification.

ASTM C 270 also covers materials and testing requirements for water retention, compressive strength and air content.

The types of mortar recommended for use in various types of masonry construction are shown in the appendix to ASTM C 270. The performance of masonry is influenced by mortar workability, water retentivity, bond strength, compressive strength and long-term deformability (creep). Each mortar type has corresponding use limitations.

Glass-block units are required to be laid in Type S or N mortar. Admixtures containing set accelerators or antifreeze compounds are not permitted in mortars for glass-block masonry.

2103.10 Surface-bonding mortar. Surface-bonding mortar shall comply with ASTM C 887. Surface bonding of concrete masonry units shall comply with ASTM C 946.

❖ Specifications for materials to be used in premixed surface-bonding mortar and for the properties of the mortar are contained in ASTM C 887. Requirements for masonry units and other materials used in constructing dry-stacked, surface-bonded masonry for walls and for the construction itself are contained in ASTM C 946.

In addition to its primary function of bonding masonry units, surface-bonding mortar can provide resistance to penetration by wind-driven rain.

2103.11 Mortars for ceramic wall and floor tile. Portland cement mortars for installing ceramic wall and floor tile shall comply with ANSI A108.1A and ANSI A108.1B and be of the compositions indicated in Table 2103.11.

❖ This section pertains to cement mortars and organic adhesives used for setting ceramic wall and floor tiles. Each mortar type has certain qualities that make it suitable for installing tile over different kinds of backing materials or under a certain set of conditions.

Cement mortars are "thick-bed" mortars applied in thicknesses of $^3/_4$ to $1^1/_4$ inches (19.1 to 32 mm) on floors and $^3/_4$ to 1 inch (19.1 to 25 mm) on walls to achieve the specified slopes and flatness in the finished tile work. Portland cement mortars are suitable for setting ceramic tile in most installations. They can be applied over properly prepared backings of clay or concrete masonry; concrete; wood frame; rough wood floors and plywood floors; foam insulation board; gypsum wallboard; and portland cement or gypsum plaster. Cement mortars can be reinforced with metal lath or wire mesh; in such cases, however, additional mortar thickness may be required. Cement mortars have good structural strength and are not affected by prolonged contact with water.

Complete material and installation specifications are contained in ANSI A108.1A and A108.1B, which include required specifications for: installation of wire lath and scratch coats; mortar mixes, bond coat mixes and mortar application; installation methods on floors, walls and countertops; grouting of tile and general requirements for tile installations.

TABLE 2103.11
CERAMIC TILE MORTAR COMPOSITIONS

LOCATION	MORTAR	COMPOSITION
Walls	Scratchcoat	1 cement; $^1/_5$ hydrated lime; 4 dry or 5 damp sand
	Setting bed and leveling coat	1 cement; $^1/_2$ hydrated lime; 5 damp sand to 1 cement 1 hydrated lime, 7 damp sand
Floors	Setting bed	1 cement; $^1/_{10}$ hydrated lime; 5 dry or 6 damp sand; or 1 cement; 5 dry or 6 damp sand
Ceilings	Scratchcoat and sand bed	1 cement; $^1/_2$ hydrated lime; $2^1/_2$ dry sand or 3 damp sand

❖ Portland cement-lime mortars are a mixture of portland cement (ASTM C 150), hydrated lime (ASTM C 207) and damp mason's sand. Proportions of these ingredients are given in Table 2103.11, dealing with the portland cement-lime mortars most commonly used for setting ceramic tile on wall, floor and ceiling construction and measured in parts by volume. Complete material and installation specifications are contained in ANSI A108.1A and A108.1B.

2103.11.1 Dry-set Portland cement mortars. Premixed prepared Portland cement mortars, which require only the addition of water and are used in the installation of ceramic tile, shall comply with ANSI A118.1. The shear bond strength for tile set in such mortar shall be as required in accordance with ANSI A118.1. Tile set in dry-set Portland cement mortar shall be installed in accordance with ANSI A108.5.

❖ Dry-set portland cement mortar for ceramic tile is required to comply with ANSI A118.1, which describes test methods and minimum requirements for dry-set mortar, including sampling, free water content, setting characteristics, shrinkage, shear strength and staining.

Dry-set mortar is a mixture of portland cement, sand and perhaps water-retention admixtures. Dry-set mortars are suitable for use over properly prepared backings of clay or concrete masonry; concrete; cut-cell expanded polystyrene or rigid closed-cell urethane insulation board; gypsum wallboard; lean portland cement mortar; and hardened wall and floor setting beds.

Installation must comply with ANSI A108.5. The "thin-bed" mortars are applied in a single layer as thin as $^3/_{32}$ inch (2.4 mm), but usually in thicknesses of $^1/_8$ to $^1/_4$ inch (3.2 to 6.4 mm). The use of dry-set mortar for leveling work is limited to a maximum thickness of $^1/_4$ inch (6.4 mm).

This material has excellent water and impact resistance. It is also water-cleanable, nonflammable, good for exterior use and requires no presoaking of the tile. Dry-set mortar is intended for use with glazed wall tile, ceramic mosaics, pavers and quarry tile.

Shear bond strength (of tile set in mortar) is required to be tested in accordance with standards applicable to the mortar used. Tile set with dry-set mortar must be installed in accordance with ANSI A108.5.

2103.11.2 Latex-modified Portland cement mortar. Latex-modified Portland cement thin-set mortars in which latex is added to dry-set mortar as a replacement for all or part of the gauging water that are used for the installation of ceramic tile shall comply with ANSI A118.4. Tile set in latex-modified Portland cement shall be installed in accordance with ANSI A108.5.

❖ Latex-modified portland cement mortar is a mixture of portland cement, sand and special latex admixtures. It is applied as a thin-bed material, like dry-set portland cement mortar (see commentary, Section 2103.11.1). Since the latex used in these mortars varies among manufacturers, instructions for mixing and use must be followed carefully. Applicable material and installation standards are ANSI A108.5 and A118.4.

2103.11.3 Epoxy mortar. Ceramic tile set and grouted with chemical-resistant epoxy shall comply with ANSI A118.3. Tile set and grouted with epoxy shall be installed in accordance with ANSI A108.6.

❖ Epoxy mortar is a two-part system consisting of an epoxy resin and hardener. It is applied as a layer of $^1/_{16}$ to $^1/_8$ inch (1.6 to 3.2 mm) in thickness and is suitable for use on properly prepared floors of concrete, wood, plywood, steel plate or ceramic tile. It is particularly suitable where chemical resistance or high bond strength is required. Deflection control is critical for the successful use of this material, and floor systems should not deflect more than $^{1}/_{360}$ of their span. Epoxy mortar is recommended for setting ceramic mosaics, quarry tile and paver tile. Applicable material and installation standards are ANSI A108.6 and A118.3.

2103.11.4 Furan mortar and grout. Chemical-resistant furan mortar and grout that are used to install ceramic tile shall comply with ANSI A118.5. Tile set and grouted with furan shall be installed in accordance with ANSI A108.8.

❖ Furan mortar is a two-part system consisting of a furan resin and hardener. It is suitable where chemical resistance is critical. It is used primarily on floors in laboratories and industrial plants. Acceptable subfloors include concrete, steel plate and ceramic tile. Furan grout is intended for quarry tile and pavers, mainly in industrial areas requiring maximum chemical resistance.

ANSI A118.5 is the standard regulating both furan mortar and grout. Installation must comply with ANSI A108.8.

2103.11.5 Modified epoxy-emulsion mortar and grout. Modified epoxy-emulsion mortar and grout that are used to install ceramic tile shall comply with ANSI A118.8. Tile set and grouted with modified epoxy-emulsion mortar and grout shall be installed in accordance with ANSI A108.9.

❖ Modified epoxy-emulsion mortar and grout used to install ceramic tile are required to comply with ANSI A118.8 and to be installed in accordance with ANSI A108.9.

ANSI A118.8 describes the test methods and the minimum requirements for modified epoxy-emulsion mortar and grout. The chemical and solvent resistance of these mortars and grouts tends to exceed those of organic adhesives and equal those of latex-modified portland cement mortars. They are not, however, designed to meet the requirements of ANSI A108.6 or A118.3.

These types of mortars and grouts are three-part systems that include emulsified epoxy resins and hardeners, preblended portland cement and silica sand. They are used as a bond-coat setting mortar or grout, and can be cleaned from wall and floor surfaces using a wet sponge prior to initial set.

ANSI A118.8 regulates water absorption, flexural strength, thermal expansion, linear shrinkage, tensile strength and compressive strength.

2103.11.6 Organic adhesives. Water-resistant organic adhesives used for the installation of ceramic tile shall comply with ANSI A136.1. The shear bond strength after water immersion shall not be less than 40 psi (275 kPa) for Type I adhesive and not less than 20 psi (138 kPa) for Type II adhesive when tested in accordance with ANSI A136.1. Tile set in organic adhesives shall be installed in accordance with ANSI A108.4.

❖ Organic adhesives are prepared materials that cure or set by evaporation and that are ready to use without adding liquid or powder. Adhesives are suitable for installing tiles on prepared wall and floor surfaces, including: brick and concrete masonry; concrete; gypsum wallboard; portland cement or gypsum plaster; and wood-flooring systems.

Organic adhesives (mastics) are applied as a thin layer approximately $^{1}/_{16}$ inch (1.6 mm) thick. An underlayment is used to level and true surfaces. Organic adhesive does not permit the soaking of tiles and is not suitable for exterior use. Bond strength varies greatly among the numerous brands of organic adhesives available for use in construction. Adhesives must meet minimum bond-strength requirements of this section of ANSI A136.1 for Type I and II adhesives. The installation of tile with organic adhesives is required to conform to ANSI A108.4.

2103.11.7 Portland cement grouts. Portland cement grouts used for the installation of ceramic tile shall comply with ANSI A118.6. Portland cement grouts for tile work shall be installed in accordance with ANSI A108.10.

❖ Portland cement grouts are the most commonly used grouts for tile walls. The mixture of portland cement and other ingredients is water resistant and uniform in color. The water in the grout is essential to promote a good bond and to develop full grout strength. This type of grout is required to comply with ANSI A118.6 and to be installed in accordance with ANSI A108.10.

2103.12 Mortar for AAC masonry. Thin-bed mortar for AAC masonry shall comply with Article 2.1 C.1 of TMS 602/ACI 530.1/ASCE 6. Mortar used for the leveling courses of AAC masonry shall comply with Article 2.1 C.2 of TMS 602/ACI 530.1/ASCE 6.

❖ Except for the initial leveling courses of AAC masonry, which uses Type M or S masonry mortar, a special thin-bed mortar is used to bind AAC masonry units.

2103.13 Grout. Grout shall comply with Article 2.2 of TMS 602/ACI 530.1/ASCE 6.

❖ Grout intended for use in the construction of engineered and empirically designed masonry structures is required to comply with ASTM C 476, which regulates materials, measurement, mixing and storage of materials.

Two types of grout are used in masonry: fine and coarse. Fine grout is made of cement, sand and water, with optional small quantities of lime. Coarse grout includes the same ingredients, plus pea gravel, which is limited to $^{3}/_{8}$ inch (9.5 mm) in size.

Whether fine or coarse grout is used depends on the size of the grout space to be filled and the height of the grout pour to be used in construction. Table 1.19.1 of the MSJC Code contains the limiting cavity and cell sizes for a given grout pour height for both fine and coarse grouts.

The materials used in masonry grout are required to be listed in ASTM C 476 and must conform to the following standard specifications: ASTM C 5; ASTM C 150; ASTM C 207 and ASTM C 595.

2103.14 Metal reinforcement and accessories. Metal reinforcement and accessories shall conform to Article 2.4 of TMS 602/ACI 530.1/ASCE 6. Where unidentified reinforcement is *approved* for use, not less than three tension and three bending tests shall be made on representative specimens of the reinforcement from each shipment and grade of reinforcing steel proposed for use in the work.

❖ This section contains standards and material requirements for anchors, joint reinforcement, wire accessories, ties and wire fabric, and for required corrosion protection of these items.

SECTION 2104 CONSTRUCTION

2104.1 Masonry construction. Masonry construction shall comply with the requirements of Sections 2104.1.1 through 2104.4 and with TMS 602/ACI 530.1/ASCE 6.

❖ This section establishes the requirements, based on accepted practice and referenced standards, regulating materials and construction methods used in engineered and empirically designed masonry construction. Engineered masonry construction is further regulated by Sections 2101.2.1, 2101.2.2, 2101.2.3 and 2101.2.7.

2104.1.1 Tolerances. Masonry, except masonry veneer, shall be constructed within the tolerances specified in TMS 602/ACI 530.1/ASCE 6.

❖ The MSJC Specification (TMS 602/ACI 530.1/ASCE 6) prescribes tolerances for placement of reinforcement and masonry units. Specific tolerances listed include cross sections of elements; thicknesses of mortar joints; widths of grout spaces; variation from level; variation from plumb; trueness to a line; alignment of columns and walls; location of elements; and placement of reinforcement and accessories. These required tolerances are intended to inhibit corrosion; provide placement locations reflecting those assumed in design; provide clearance for mortar and grout; and maintain compatible lateral deflections of parallel wythes. Tolerances are relatively strict since masonry is a structural material and is also exposed to weather. Aesthetics are not a factor in these requirements.

2104.1.2 Placing mortar and units. Placement of mortar, grout, and clay, concrete, glass, and AAC masonry units shall comply with TMS 602/ACI 530.1/ASCE 6.

❖ Workmanship is of primary importance in providing strong and durable masonry construction. This is especially true for the placement of mortar and masonry units. Water penetration can often be traced to bond breaks at the mortar-unit interface. Additionally, masonry strength relies on the mortar bedded areas required in this section and the MSJC Specification.

2104.1.3 Installation of wall ties. Wall ties shall be installed in accordance with TMS 602/ACI 530.1/ASCE 6.

❖ Installation of wall ties requires adequate embedment in mortar, and adequate strength and durability of ties. Wall ties are not permitted to be bent after being embedded in grout or mortar, since movement of the ties will reduce the effectiveness of the embedment and subsequent bonding.

2104.1.4 Chases and recesses. Chases and recesses shall be constructed as masonry units are laid. Masonry directly above chases or recesses wider than 12 inches (305 mm) shall be supported on lintels.

❖ Chases and recesses are designed so that they do not reduce the required strength of the walls or affect the required fire-resistance rating of the wall. Any required chases must be formed during construction of the masonry wall and not by cutting out a section of the finished wall, which could create a weak point and jeopardize the wall strength.

This empirical limitation requires lintels for chases and recesses wider than 12 inches (305 mm). Although arching action reduces lintel moments, the potential for weaknesses created by chases and recesses makes them subject to engineered design. The supporting lintel is permitted to be of masonry, concrete or steel and must be engineered as required in Section 2104.1.5.

2104.1.5 Lintels. The design for lintels shall be in accordance with the masonry design provisions of either Section 2107 or 2108.

❖ Masonry lintels are required to be engineered as load-bearing beam elements by either the allowable stress design method of Section 2107 or the strength design method of Section 2108. This requirement is consistent with the empirical provisions, which do not include horizontally spanning elements, such as beams.

2104.1.6 Support on wood. Masonry shall not be supported on wood girders or other forms of wood construction except as permitted in Section 2304.12.

❖ Masonry is only permitted to be supported on wood construction in accordance with Section 2304.12 because of the serviceability considerations noted below.

Masonry is brittle and relatively weak in tension. It can crack when subjected to deformations. Wood is flexible and usually exhibits elastic deformation and creep (additional deflection from long-term loads). Because of this, masonry supported on wood tends to crack. Therefore, masonry that is permanently supported by wood members must consider long-term deflection effects in accordance with Section 2304.12 which references the American Forest and Paper Association (AF&PA) *National Design Specification* (NDS). The exception permits concrete nonstructural floor or roof surfacing not more than 4 inches (102 mm) thick to be used without being checked for long-term deflection.

2104.2 Corbeled masonry. Corbeled masonry shall comply with the requirements of Section 1.12 of TMS 402/ACI 530/ ASCE 5.

❖ This section covers the limitations and the construction features of corbeled masonry.

A corbel is formed by projecting courses of masonry, with the first or lowest course projecting out from the face of the wall and each successive course projecting out from the supporting course below. While corbels may be constructed using various load-bearing masonry units, they are most commonly applied to clay masonry and are used for aesthetic purposes and to support offset or thickened walls above them.

The permitted extent of corbeling is illustrated in Figure 2104.2. The total horizontal projection of a corbel from the face of the wall is limited to no more than one-half the wall thickness of a solid wall or more than one-half the thickness of the wythe in a cavity wall. Furthermore, the projection of a single course is not to exceed one-half of the unit height, nor one-third of the unit bed depth, whichever is less.

These limitations establish a maximum angle of corbelled brick masonry, measured from the plane of the wall, of about 26.5 degrees (0.46 rad). A smaller angle can be achieved by decreasing the brick projections to less than the maximum allowed by this section.

Corbels produce and often transfer eccentric loads, which must be considered in the design of masonry walls.

Corbels constructed in walls of hollow units or in hollow masonry walls are required to be of solid masonry. This requirement can be satisfied by using solid units, grouting the cores of hollow units or grouting open spaces in hollow walls.

2104.2.1 Molded cornices. Unless structural support and anchorage are provided to resist the overturning moment, the center of gravity of projecting masonry or molded cornices shall lie within the middle one-third of the supporting wall. Terra cotta and metal cornices shall be provided with a structural frame of *approved* noncombustible material anchored in an *approved* manner.

❖ Figure 2104.2.1 illustrates this requirement. Unless cornices or other projections are specifically designed to be supported and anchored to the wall or other adequate structural supports, such as columns, beams, spandrels or floor slabs, the center of gravity of the projecting element is required to fall within the middle third (kern) of the supporting wall.

Cornices are usually ornamental and may or may

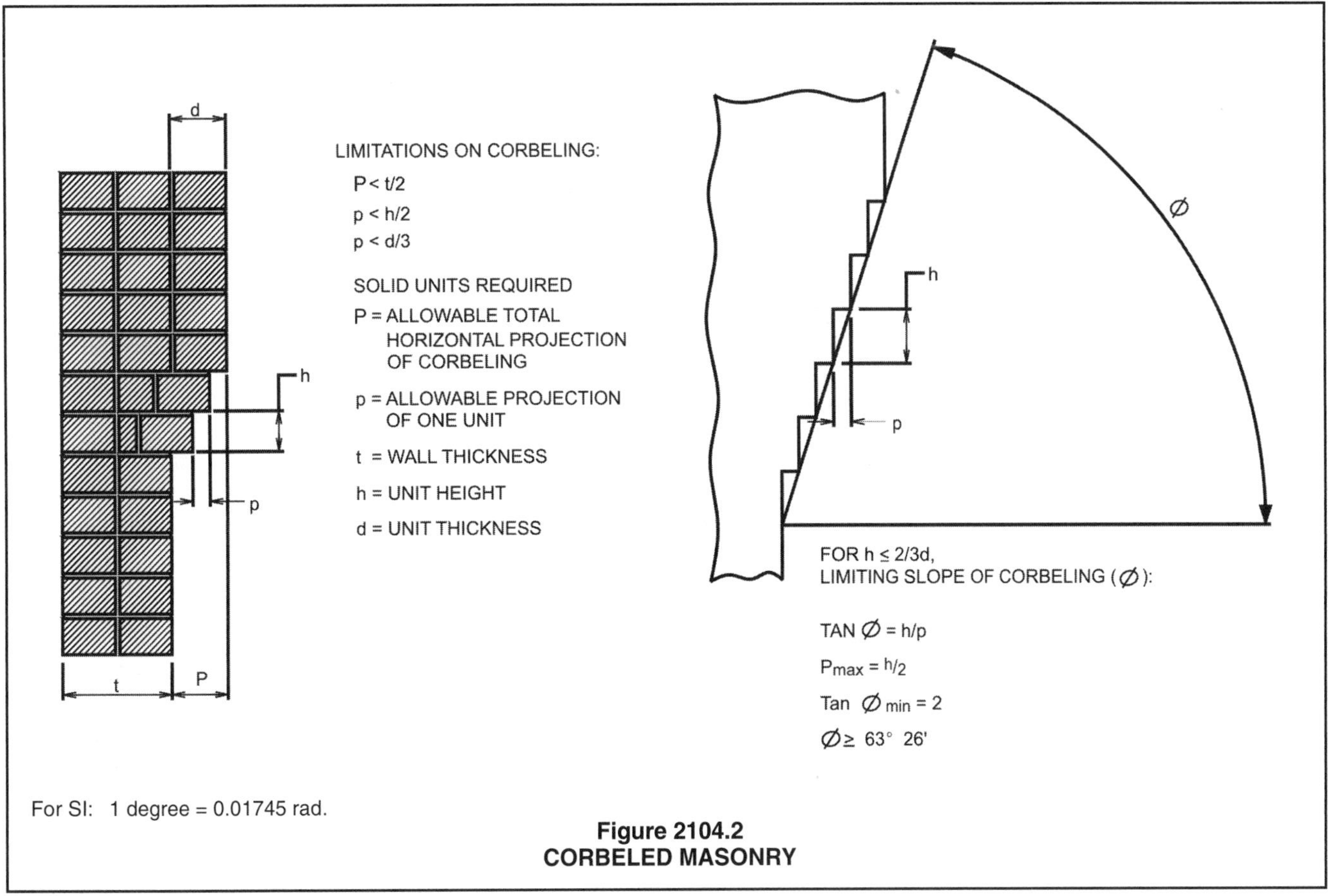

For SI: 1 degree = 0.01745 rad.

Figure 2104.2
CORBELED MASONRY

not support other loads. They can be made of cast-in-place or precast concrete; cast or natural stone; terra cotta; brick or even metal.

Metal and terra-cotta cornices are commonly furnished with a structural frame of noncombustible material that can be anchored to the masonry wall or to other building supports as approved by the building official.

2104.3 Cold weather construction. The cold weather construction provisions of TMS 602/ACI 530.1/ASCE 6, Article 1.8 C, shall be implemented when the ambient temperature falls below 40°F (4°C).

❖ Provisions in this Section are based on Article 1.8 C of the MSJC Specification. Readers are referred to the commentary on the MSJC Specification for additional information on these provisions.

When masonry construction is conducted in temperatures below 40°F (4°C), both the masonry components and the structure are required to be protected in accordance with this section. One of the main goals of these provisions is to prevent fresh mortar from freezing.

During cold-weather construction, particularly in temperatures below freezing, the curing and subsequent performance of masonry are influenced by the temperature and properties of the mortar and the masonry units, as well as the severity of the exposure (temperature and wind).

2104.4 Hot weather construction. The hot weather construction provisions of TMS 602/ACI 530.1/ASCE 6, Article 1.8 D, shall be implemented when the ambient air temperature exceeds 100°F (37.8°C), or 90°F (32.2°C) with a wind velocity greater than 8 mph (12.9 km/hr).

❖ Provisions in this Section are based on Article 1.8 C of the MSJC Specification. Readers are referred to the commentary on the MSJC Specification for additional information on these provisions.

Temperature, solar radiation, humidity and wind influence the rate of absorption of masonry units and the rate of mortar set. During hot-weather conditions, special precautions must be taken for adequate strength gain of the masonry.

SECTION 2105 QUALITY ASSURANCE

2105.1 General. A quality assurance program shall be used to ensure that the constructed masonry is in compliance with the *construction documents*.

The quality assurance program shall comply with the inspection and testing requirements of Chapter 17.

❖ This section requires that a quality assurance program be used so that the masonry is constructed according to the construction documents. In addition, inspection and testing complying with Chapter 17 also requires verification of the masonry compressive strength.

The quality assurance program shall comply with the inspection and testing requirements of Chapter 17, which in turn references the quality assurance provisions in the MSJC Code and Specification.

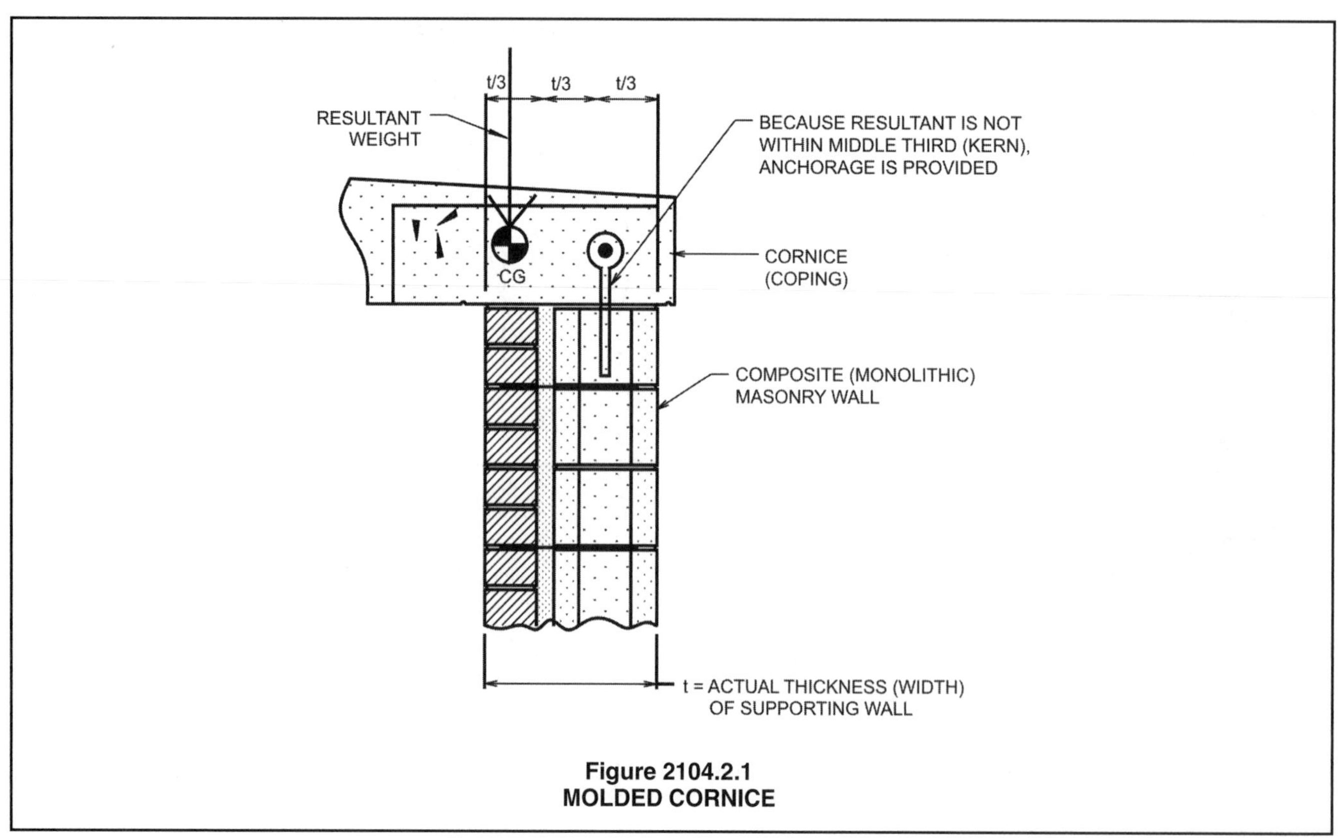

Figure 2104.2.1
MOLDED CORNICE

2105.2 Acceptance relative to strength requirements. Where required by Chapter 17, verification of the strength of masonry shall be in accordance with Sections 2105.2.1 and 2105.2.2.

❖ The quality assurance provisions in this section emphasize verification of masonry compressive strengths. This is accomplished by comparing conservatively estimated strengths (based on unit strength and mortar type) or tested prism strengths to the specified compressive strength of the masonry, f'_m, and, when required, by mortar, grout or both to see that they are in compliance.

These quality assurance methods are for general consistency of the constructed masonry. Masonry is relatively strong in compression and thus rarely fails in that manner, but rather in flexural tension. Compression tests are required, however, because they are simple ways of assessing quality.

As implied in the above paragraph, two methods are prescribed in Section 2105.2.2 to judge acceptance relative to the compressive strength of the masonry assemblage: the unit strength method in Section 2105.2.2.1 and the prism test method in Section 2105.2.2.2. These are described in Section 2105.2 and in this commentary. When the strength of constructed masonry is questioned, testing of prisms that have been saw cut from the masonry is permitted in accordance with Section 2105.3.

2105.2.1 Compliance with f'_m and f'_{AAC}. Compressive strength of masonry shall be considered satisfactory if the compressive strength of each masonry wythe and grouted collar joint equals or exceeds the value of f'_m for clay and concrete masonry and f'_{AAC} for AAC masonry. For partially grouted clay and concrete masonry, the compressive strength of both the grouted and ungrouted masonry shall equal or exceed the applicable f'_m. At the time of prestress, the compressive strength of the masonry shall equal or exceed f'_{mi}, which shall be less than or equal to f'_m.

❖ Design of structural masonry is based on the specified compressive strength of the masonry, f'_m, and the specified strength of AAC masonry, f'_{AAC}. This strength is required to be shown on the contract documents, since structural design is based on it. Strength of the constructed masonry determined by the unit or the prism strength method is required to equal or exceed the specified compressive strength of the masonry.

In a multiwythe wall designed as a composite wall, the compressive strength of masonry for each wythe or grouted collar joint must equal or exceed the specified compressive strength.

2105.2.2 Determination of compressive strength. The compressive strength for each wythe shall be determined by the unit strength method or by the prism test method as specified herein.

❖ There are two means of determining the compressive strength of masonry: the unit strength method and the prism test method. The first eliminates the expense of prism tests, but is more conservative.

2105.2.2.1 Unit strength method. The determination of compressive strength by the unit strength method shall be in accordance with Section 2105.2.2.1.1 for clay masonry, Section 2105.2.2.1.2 for concrete masonry and Section 2105.2.2.1.3 for AAC masonry.

❖ This section describes a prescriptive procedure to estimate the expected compressive strength of the masonry based on the compressive strength of masonry units and the mortar type. This so-called unit strength method was generated using prism test data. Mortar joint thickness is limited because it influences the compressive strength of masonry.

2105.2.2.1.1 Clay masonry. The compressive strength of masonry shall be determined based on the strength of the units and the type of mortar specified using Table 2105.2.2.1.1, provided:

1. Units are sampled and tested to verify compliance with ASTM C 62, ASTM C 216 or ASTM C 652.
2. Thickness of bed joints does not exceed $^5/_8$ inch (15.9 mm).
3. For grouted masonry, the grout meets one of the following requirements:
   3.1. Grout conforms to Article 2.2 of TMS 602/ACI 530.1/ASCE 6.
   3.2. Minimum grout compressive strength equals or exceeds f'_m but not less than 2,000 psi (13.79 MPa). The compressive strength of grout shall be determined in accordance with ASTM C 1019.

❖ This section prescribes the conditions under which the unit strength method can be used for clay masonry. These conditions include requirements for clay masonry units, maximum bed joint thickness and grout.

TABLE 2105.2.2.1.1
COMPRESSIVE STRENGTH OF CLAY MASONRY

NET AREA COMPRESSIVE STRENGTH OF CLAY MASONRY UNITS (psi)		NET AREA COMPRESSIVE STRENGTH OF MASONRY (psi)
Type M or S mortar	Type N mortar	
1,700	2,100	1,000
3,350	4,150	1,500
4,950	6,200	2,000
6,600	8,250	2,500
8,250	10,300	3,000
9,900	—	3,500
11,500	—	4,000

For SI: 1 pound per square inch = 0.00689 MPa.

❖ Table 2105.2.2.1.1 lists the compressive strength of masonry in terms of the strength of the clay masonry unit and the mortar type. This table is based on the research results cited in the commentary to the MSJC

Specification. A similar table has been used successfully in both the MSJC Specification and the *Uniform Building Code*® (UBC™) since 1988.

The designer can use this table to estimate a specified compressive strength of masonry to use in design, based on the expected strength of the clay masonry units and the specified mortar type. The contractor can use the table to find what clay unit masonry strength and mortar type are needed to comply with the specified strength of the masonry, f'_m, given in the contract documents. The column entitled "Net Area Compressive Strength of Masonry" must equal or exceed the specified strength of the masonry, f'_m.

2105.2.2.1.2 Concrete masonry. The compressive strength of masonry shall be determined based on the strength of the unit and type of mortar specified using Table 2105.2.2.1.2, provided:

1. Units are sampled and tested to verify compliance with ASTM C 55 or ASTM C 90.
2. Thickness of bed joints does not exceed $^5/_8$ inch (15.9 mm).
3. For grouted masonry, the grout meets one of the following requirements:
   3.1. Grout conforms to Article 2.2 of TMS 602/ACI 530.1/ASCE 6.
   3.2. Minimum grout compressive strength equals or exceeds f'_m but not less than 2,000 psi (13.79 MPa). The compressive strength of grout shall be determined in accordance with ASTM C 1019.

❖ This section prescribes the conditions under which the unit strength method can be used for concrete masonry. These conditions include requirements for the concrete masonry units, maximum bed joint thickness and grout. Concrete masonry units must be tested in accordance with ASTM C 140.

TABLE 2105.2.2.1.2
COMPRESSIVE STRENGTH OF CONCRETE MASONRY

NET AREA COMPRESSIVE STRENGTH OF CONCRETE MASONRY UNITS (psi)		NET AREA COMPRESSIVE STRENGTH OF MASONRY (psi)[a]
Type M or S mortar	Type N mortar	
1,250	1,300	1,000
1,900	2,150	1,500
2,800	3,050	2,000
3,750	4,050	2,500
4,800	5,250	3,000

For SI: 1 inch = 25.4 mm, 1 pound per square inch = 0.00689 MPa.
a. For units less than 4 inches in height, 85 percent of the values listed.

❖ Table 2105.2.2.1.2 lists the compressive strength of masonry in terms of the strengths of the concrete masonry units and the mortar type. This table is based on the research cited in the commentary to the MSJC Specification. A similar table has been used successfully in both the MSJC Specification and the UBC since 1988.

The designer can use this table to estimate a specified compressive strength of masonry for use in design, based on the expected strength of the concrete masonry units and the specified mortar type. The contractor can use the table to find what concrete masonry strength and mortar type are needed to comply with the specified strength of the masonry, f'_m, given in the contract documents. The column entitled "Net Area Compressive Strength of Masonry" must equal or exceed the specified strength of the masonry, f'_m.

2105.2.2.1.3 AAC masonry. The compressive strength of AAC masonry shall be based on the strength of the AAC masonry unit only and the following shall be met:

1. Units conform to ASTM C 1386.
2. Thickness of bed joints does not exceed $^1/_8$ inch (3.2 mm).
3. For grouted masonry, the grout meets one of the following requirements:
   3.1. Grout conforms to Article 2.2 of TMS 602/ACI 530.1/ASCE 6.
   3.2. Minimum grout compressive strength equals or exceeds f'_{AAC} but not less than 2,000 psi (13.79 MPa). The compressive strength of grout shall be determined in accordance with ASTM C 1019.

❖ Due to the relative difference between the physical properties of AAC masonry units and the thin-bed mortar used to construct AAC masonry, the mortar does not have a significant contribution to the compressive strength of AAC masonry assemblies. Therefore, the compressive strength of AAC masonry is considered equal to the compressive strength of the AAC masonry units alone.

2105.2.2.2 Prism test method. The determination of compressive strength by the prism test method shall be in accordance with Sections 2105.2.2.2.1 and 2105.2.2.2.2.

❖ The prism test method is used when required in the project specifications or when the restrictions of Section 2105.2.2.1 do not apply. Prisms are required to be constructed in accordance with ASTM C 1314 using the same materials and workmanship as in the structure.

2105.2.2.2.1 General. The compressive strength of clay and concrete masonry shall be determined by the prism test method:

1. Where specified in the *construction documents*.

2. Where masonry does not meet the requirements for application of the unit strength method in Section 2105.2.2.1.

❖ Prism tests are required whenever specified and the masonry does not meet the restrictions for the unit strength method.

2105.2.2.2.2 Number of prisms per test. A prism test shall consist of three prisms constructed and tested in accordance with ASTM C 1314.

❖ Whenever prism testing is specified or used, three prism specimens must be constructed and tested in accordance with ASTM C 1314.

2105.3 Testing prisms from constructed masonry. When *approved* by the *building official*, acceptance of masonry that does not meet the requirements of Section 2105.2.2.1 or 2105.2.2.2 shall be permitted to be based on tests of prisms cut from the masonry construction in accordance with Sections 2105.3.1, 2105.3.2 and 2105.3.3.

❖ While uncommon, there are times when the strength of masonry determined by the unit strength method or prism test method may be questioned or may be lower than the specified strength. Because low strengths could result from inappropriate testing procedures or unintentional damage to the test specimens, prisms may be saw cut from the completed masonry wall and tested. This section prescribes procedures for such tests.

Such testing is difficult, requires at least 28 days and requires replacement of the affected wall area. Therefore, every effort should be taken so that strengths determined by the unit strength method or the prism test method are adequate.

2105.3.1 Prism sampling and removal. A set of three masonry prisms that are at least 28 days old shall be saw cut from the masonry for each 5,000 square feet (465 m^2) of the wall area that is in question but not less than one set of three masonry prisms for the project. The length, width and height dimensions of the prisms shall comply with the requirements of ASTM C 1314. Transporting, preparation and testing of prisms shall be in accordance with ASTM C 1314.

❖ Removal of prisms from a constructed wall requires care so that the prism is not damaged and that damage to the wall is minimal. Prisms must be representative of the wall, yet not contain reinforcing steel, which would bias the results. As with a prism test of newly constructed masonry, a prism test from existing masonry requires three prism specimens.

2105.3.2 Compressive strength calculations. The compressive strength of prisms shall be the value calculated in accordance ASTM C 1314, except that the net cross-sectional area of the prism shall be based on the net mortar bedded area.

❖ Compressive strength calculations from saw-cut specimens must be based on the net mortar bedded area, which must be determined before the prism is tested. The testing agency must determine this area accurately.

2105.3.3 Compliance. Compliance with the requirement for the specified compressive strength of masonry, f'_m, shall be considered satisfied provided the modified compressive strength equals or exceeds the specified f'_m. Additional testing of specimens cut from locations in question shall be permitted.

❖ Strengths determined from saw-cut prisms must equal or exceed the specified strength of masonry, f'_m.

SECTION 2106
SEISMIC DESIGN

2106.1 Seismic design requirements for masonry. Masonry structures and components shall comply with the requirements in Section 1.18 of TMS 402/ACI 530/ASCE 5 depending on the structure's *seismic design category*.

❖ Section 2106 references Section 1.18 of the MSJC Code for minimum seismic design requirements for masonry structures based upon their seismic design category. This section requires the use of the MSJC Code for specific seismic design criteria. Requirements established for various seismic risk categories are cumulative from lower to higher categories. These prescriptive and design-oriented provisions have been established to improve the performance of masonry structures during seismic events by providing additional structural strength, ductility and stability against the dynamic effects of earthquake ground motion. As the seismic demand increases, the provisions require more seismic detailing, increased ductility and greater material reliability.

To comply with the provisions in Section 2106, the seismic design category must be determined for the building or structure under consideration. Section 1613 contains criteria for determining a structure's seismic design category and references ASCE 7 for calculating seismic load effects. The requirements for seismic resistance (for example, design forces and detailing) remain applicable and are discussed in the commentary for Chapter 16. Compliance with Chapter 21 is not a substitute for compliance with the seismic provisions required by Section 1613. More information on seismic design is contained in the commentaries to ASCE 7-10 and the 2009 National Earthquake Hazards Reduction Program (NEHRP) *Recommended Provisions for Seismic Regulations for New Buildings and Other Structures* (FEMA P-750). The reader is also referred to the commentary to the MSJC Code for additional background on seismic design requirements for masonry structures.

SECTION 2107
ALLOWABLE STRESS DESIGN

2107.1 General. The design of masonry structures using *allowable stress design* shall comply with Section 2106 and the requirements of Chapters 1 and 2 of TMS 402/ACI 530/

ASCE 5 except as modified by Sections 2107.2 through 2107.4.

❖ Section 2107 adopts the ASD method of Chapters 1 and 2 of the MSJC Code with modifications deemed necessary by the International Building Code Structural Committee. This method of engineered masonry design has been used successfully for years and remains a popular method for designing masonry structures.

Refer to the commentary to Chapters 1 and 2 of the MSJC Code for additional information on the allowable stress design method for masonry. This section also requires conformance to the seismic design provisions of Section 2106.

2107.2 TMS 402/ACI 530/ASCE 5, Section 2.1.8.7.1.1, lap splices. In lieu of Section 2.1.8.7.1.1, it shall be permitted to design lap splices in accordance with Section 2107.2.1.

❖ The section provides an alternative to the MSJC lap splice requirement.

2107.2.1 Lap splices. The minimum length of lap splices for reinforcing bars in tension or compression, l_d, shall be

$l_d = 0.002 d_b f_s$ **(Equation 21-1)**

For SI: $l_d = 0.29 d_b f_s$

but not less than 12 inches (305 mm). In no case shall the length of the lapped splice be less than 40 bar diameters.

where:

d_b = Diameter of reinforcement, inches (mm).

f_s = Computed stress in reinforcement due to design loads, psi (MPa).

In regions of moment where the design tensile stresses in the reinforcement are greater than 80 percent of the allowable steel tension stress, F_s, the lap length of splices shall be increased not less than 50 percent of the minimum required length. Other equivalent means of stress transfer to accomplish the same 50 percent increase shall be permitted. Where epoxy coated bars are used, lap length shall be increased by 50 percent.

❖ This alternative provision is based on allowable stress detailing requirements that were used in the UBC. Lap splice and development length requirements for allowable stress and strength design of masonry differ, due in part to the slightly more conservative nature of the ASD requirements. Note that Section 2107.3 prohibits the lap splicing of reinforcement larger than No. 9 (32 mm) bars.

2107.3 TMS 402/ACI 530/ASCE 5, Section 2.1.8.7, splices of reinforcement. Modify Section 2.1.8.7 as follows:

2.1.8.7 Splices of reinforcement. Lap splices, welded splices or mechanical splices are permitted in accordance with the provisions of this section. All welding shall conform to AWS D1.4. Welded splices shall be of ASTM A 706 steel reinforcement. Reinforcement larger than No. 9 (M #29) shall be spliced using mechanical connections in accordance with Section 2.1.8.7.3.

❖ Research has shown that effectively lap splicing large reinforcing bars in masonry is difficult and impractical because of the excessive lap lengths required. This section adds a provision not in the MSJC Code requiring mechanical splices for all bars larger than No. 9 (32 mm) in diameter. If splices are welded, the use of ASTM A 706 reinforcing steel is mandatory, because this steel specification has controlled chemistry and can, therefore, be more reliably welded than steel produced without such controls (see commentary, Section 2108.3 for additional information).

2107.4 TMS 402/ACI 530/ASCE 5, Section 2.3.7, maximum bar size. Add the following to Chapter 2:

2.3.7 Maximum bar size. The bar diameter shall not exceed one-eighth of the nominal wall thickness and shall not exceed one-quarter of the least dimension of the cell, course or collar joint in which it is placed.

❖ Allowing large bars in masonry walls can result in an over-reinforced section that increases the risk of a brittle (nonductile) failure. This section specifies limits on the diameter of reinforcement that are not in the MSJC Code in order to prevent potential problems with over reinforcement and congestion of reinforcement. These requirements are based on practical guidelines and empirical judgment rather than research.

SECTION 2108 STRENGTH DESIGN OF MASONRY

2108.1 General. The design of masonry structures using strength design shall comply with Section 2106 and the requirements of Chapters 1 and 3 of TMS 402/ACI 530/ASCE 5, except as modified by Sections 2108.2 through 2108.3.

Exception: AAC masonry shall comply with the requirements of Chapters 1 and 8 of TMS 402/ACI 530/ASCE 5.

❖ The 2000 edition of the code contained extensive strength design requirements for masonry structures because at that time a consensus standard for strength design of masonry did not yet exist. Starting with the 2003 edition of the code, this section has required that strength design be in accordance with Chapters 1 and 3 of the MSJC Code with some modifications to specific sections. Like ASD, this section also invokes the minimum seismic requirements of Section 2106 for all masonry designed by the strength method. Additionally, masonry designed by this method (as with any engineered masonry construction) must be inspected during construction in accordance with the special inspection provisions of Section 1705.4, which in turn references the quality assurance provisions of the MSJC Code and Specification.

2108.2 TMS 402/ACI 530/ASCE 5, Section 3.3.3.3 development. Modify the second paragraph of Section 3.3.3.3 as follows:

The required development length of reinforcement shall be determined by Equation (3-16), but shall not be less than 12 inches (305 mm) and need not be greater than 72 d_b.

❖ This section modifies the MSJC Code by placing a limit on the calculated development length, which in turn limits the maximum required splice length. The limit is based upon historical limits used in the UBC. It is intended to address concerns that the MSJC Code, at times, requires excessive lap splice lengths.

2108.3 TMS 402/ACI 530/ASCE 5, Section 3.3.3.4, splices. Modify items (c) and (d) of Section 3.3.3.4 as follows:

3.3.3.4 (c). A welded splice shall have the bars butted and welded to develop at least 125 percent of the yield strength, f_y, of the bar in tension or compression, as required. Welded splices shall be of ASTM A 706 steel reinforcement. Welded splices shall not be permitted in plastic hinge zones of intermediate or special reinforced walls or special moment frames of masonry.

3.3.3.4 (d). Mechanical splices shall be classified as Type 1 or 2 according to Section 21.2.6.1 of ACI 318. Type 1 mechanical splices shall not be used within a plastic hinge zone or within a beam-column joint of intermediate or special reinforced masonry shear walls or special moment frames. Type 2 mechanical splices are permitted in any location within a member.

❖ This section modifies the strength design splice requirements for consistency with the NEHRP Recommended Provisions. Splices for reinforcement can be achieved by lapping the reinforcement, welding the reinforcement or mechanical splicing.

Two modifications are made to welded splice requirements [Item (b)]. Splices in reinforcing steel used in the lateral-force-resisting system subjected to high seismic strains must be able to develop the strength of the steel in order to achieve the required performance. To be successfully welded, the chemistry of the steel must be controlled to limit carbon content, as well as other elements, such as sulfur and phosphorus. Since the chemistry of reinforcing steel conforming to ASTM A 615, for example, is not controlled and is likely to be unknown, a weld that develops the strength of the steel is not guaranteed. ASTM A 706 steel, on the other hand, has controlled chemistry and can be more reliably welded; therefore, if splices are to be accomplished by welding, the use of ASTM A 706 reinforcing steel becomes mandatory.

Welded splices are required to be able to develop at least 125 percent of the yield strength of the spliced rebars; however, reinforcing steel conforming to ASTM A 706 or, for that matter, ASTM A 615 or A 996 can have an actual yield strength greater than 125 percent of the minimum required yield strength. This means a code-conforming welded splice may conceivably fail before the spliced bars have yielded, thereby limiting the inelastic deformability of that structural member. The use of welded splices is, therefore, prohibited at locations of where there is the potential for plastic hinging of members in structural systems that are expected to undergo significant inelastic response in resisting forces due to earthquake ground motion.

The modification to mechanical splices in Item (c) requires that the splice be classified in accordance with ACI 318 as Class 1 or Class 2. This is due to the fact that reinforcing steel is predominantly produced from remelted steel scrap, making it difficult to control the strength. The resulting products tend to have a strength considerably higher than the specified yield strength. This is similar to the situation that has occurred in structural steel where the actual yield strength can be much greater than the specified yield strength. Since there is no upper limit on the yield strength (except for ASTM A 706) and only a minimum required yield strength, most reinforcing steel will have a higher yield point than that specified.

Testing by the California Department of Transportation (CALTRANS) indicates the overstrength can be as much as 60 percent over the specified strength. Cyclic tests of splices meeting only the 125-percent criterion (i.e., Class 1) show that, in many cases, they cannot survive several excursions in the post yield range as imposed by cyclic testing. Splices in reinforcing steel used in the seismic-force-resisting system in plastic-hinge zones and beam-column joints are subjected to high seismic strains so they must be able to develop the strength of the steel in order to achieve the required performance; hence, the requirement for a Type-2 splice that develops the specified tensile strength of the bar.

SECTION 2109
EMPIRICAL DESIGN OF MASONRY

2109.1 General. Empirically designed masonry shall conform to the requirements of Chapter 5 of TMS 402/ACI 530/ASCE 5, except where otherwise noted in this section.

❖ This section permits empirical design of masonry utilizing Chapter 5 of the MSJC Code. The only exceptions are surface-bonded walls in accordance with Section 2109.2 and adobe construction in accordance with Section 2109.3. Additional information can be found in the commentary to the MSJC Code.

Empirical provisions are design rules developed by experience rather than engineering analysis. The empirical rules in these provisions are based on records dating back as far as 1889 in *A Treatise on Masonry Construction* by Ira Baker. The most recent publication providing the basis for these empirical provisions is ANSI A41.1.

This empirical design method is based on several premises: gravity loads are reasonably centered on bearing walls; effects of reinforcement are neglected; buildings have limited height, seismic risk and wind loading. The requirements of and limitations regarding the use of empirical design reflect these assumptions.

2109.1.1 Limitations. The use of empirical design of masonry shall be limited as noted in Section 5.1.2 of TMS 402/ACI 530/ASCE 5. The use of dry-stacked, surface-bonded masonry shall be prohibited in *Risk Category* IV structures. In buildings that exceed one or more of the limitations of Section 5.1.2 of TMS 402/ACI 530/ASCE 5, masonry shall be designed in accordance with the engineered design provisions of Section 2101.2.1, 2101.2.2 or 2101.2.3 or the foundation wall provisions of Section 1807.1.5.

Section 5.1.2.2 of TMS 402/ACI 530/ASCE 5 shall be modified as follows:

5.1.2.2 *Wind* – Empirical requirements shall not apply to the design or construction of masonry for buildings, parts of buildings, or other structures to be located in areas where V_{asd} as determined in accordance with Section 1609.3.1 of the *International Building Code* exceeds 110 mph.

❖ Empirical design is permitted for structures having limited seismic risk, wind loading and height. These limitations are justified, since buildings that were representative of the historically based empirical provisions are uncommon today. For example, buildings of the past were smaller, had more interior masonry walls and typically had different floor construction than modem buildings.

Where any one of the limitations in Chapter 5 of the MSJC Code is exceeded, the masonry structure is not permitted to be empirically designed. Engineered design in accordance with Section 2107, 2108 or TMS 403 is required in such instances. Masonry foundation walls complying with Section 1807.1.5 are also acceptable.

The empirical provisions are adequate for the level of risk associated with the seismic-force-resisting systems of buildings located in Seismic Design Category A. Where masonry is used for purposes other than the seismic-force-resisting system, the empirical design method may be used in buildings assigned to Seismic Design Category A, B or C. Engineered design is required for buildings in higher seismic design categories.

In addition, empirical design of the lateral-load-resisting system is limited to locations where the nominal design wind speed, V_{asd}, determined in accordance with Section 1609.3.1 does not exceed 110 mph (145 km/hr). Other limits on various building elements are based on the type of element, its location and the building height, as well as the basic wind speed.

The use of empirical design for AAC masonry is not permitted. Additional discussion of the empirical design method can be found in the commentary for the MSJC Code.

2109.2 Surface-bonded walls. Dry-stacked, surface-bonded concrete masonry walls shall comply with the requirements of Chapter 5 of TMS 402/ACI 530/ASCE 5, except where otherwise noted in this section.

❖ Dry-stacked, surface-bonded masonry walls consist of courses of concrete masonry units without mortar joints assembled to form unreinforced walls. Both sides of the walls are coated with a $^1/_{16}$- to $^1/_8$- inch-thick layer (1.6 to 3.2 mm) of cementitious mortar reinforced with glass fibers capable of increasing the tensile strength of the masonry and unifying the construction.

2109.2.1 Strength. Dry-stacked, surface-bonded concrete masonry walls shall be of adequate strength and proportions to support all superimposed loads without exceeding the allowable stresses listed in Table 2109.2.1. Allowable stresses not specified in Table 2109.2.1 shall comply with the requirements of TMS 402/ACI 530/ASCE 5.

❖ In surface-bonded masonry construction, strength is lower than conventional masonry construction because of the lack of solid contact between the units.

Where stresses are not specified in Table 2109.2.1, engineered design in accordance with Section 2107 or 2108 is required.

TABLE 2109.2.1
ALLOWABLE STRESS GROSS CROSS-SECTIONAL AREA FOR DRY-STACKED, SURFACE-BONDED CONCRETE MASONRY WALLS

DESCRIPTION	MAXIMUM ALLOWABLE STRESS (psi)
Compression standard block	45
Flexural tension	
Horizontal span	30
Vertical span	18
Shear	10

For SI: 1 pound per square inch = 0.006895 MPa.

❖ The values for allowable stresses based on the gross cross-sectional area of masonry units are given in this table. The gross cross-sectional area is the actual area of a section perpendicular to the direction of the load, without subtraction of the core areas of hollow masonry units.

The flexural strength of surface-bonded walls is about the same as conventional masonry with mortar joints. In the vertical direction, where walls are supported top and bottom, Table 2109.2.1 allows a maximum flexural tensile stress of 18 psi (0.12 MPa) based on the gross area. In the horizontal direction, where walls span laterally between supports, a maximum flexural stress of 30 psi (0.21 MPa) is permitted based on the gross area when units are dry stacked in running bond. The shear strength of surface-bonded walls is less than that of conventional, mortar-jointed walls. This table allows a shear strength of 10 psi (0.069 MPa) based on the gross area.

2109.2.2 Construction. Construction of dry-stacked, surface-bonded masonry walls, including stacking and leveling of units, mixing and application of mortar and curing and protection shall comply with ASTM C 946.

❖ The construction of dry-stacked, surface-bonded walls must conform to the requirements of ASTM C 946.

It is not practical to construct the horizontal surface of a wall footing or foundation level enough to receive the base (first) course of masonry without additional leveling. Therefore, it is customary to lay the base course of masonry on a mortar bed so that the remainder of the dry-stacked units will be erected level. As the units are erected, their ends should be butted together as tightly as possible.

If the bearing surfaces of the concrete units are not ground smooth and flat, shims may be required between the units to erect the wall plumb and level. Such shims should be of metal, mortar or plastic.

Because dry-stacked walls have no mortar joints, it is not possible to use horizontal steel joint reinforcement to reduce the size of cracks associated with temperature and moisture movements. Reinforced bond beams and sufficient control joints in surface-bonded masonry can be used to reduce the widths of such cracks.

The joints of dry-stacked units are tight, with no space for connectors to be embedded in the wall. The face shells or cross webs of such concrete masonry units, therefore, should be notched or depressed to accommodate ties and anchors that must be embedded in grout.

Packaged dry-bonding mortar should be mixed with water at the job site in accordance with ASTM C 946 or the manufacturer's recommendations, including curing and protection procedures after application of the material. Surface-bonding mortars are usually applied by hand troweling to thicknesses between $^{1}/_{16}$ and $^{1}/_{8}$ inch (1.6 to 3.2 mm). While they may also be sprayed on, this is usually followed by hand or mechanical troweling to obtain the desired finish.

2109.3 Adobe construction. Adobe construction shall comply with this section and shall be subject to the requirements of this code for Type V construction, Chapter 5 of TMS 402/ ACI 530/ASCE 5, and this section.

❖ Adobe masonry was popular in the southwest United States due to the availability of soil for units, the limited rainfall and low humidity to dry the units, the thermal mass provided by the completed adobe structure and the low cost of this form of construction. Requirements for adobe construction are a combination of empirical provisions and rudimentary engineering. Since there are no ASTM standards for adobe materials, test methods have been included in the code. Design is based on gross cross-sectional dimensions.

Requirements for unstabilized adobe are contained in Section 2109.3.1. Requirements for stabilized adobe are contained in Section 2109.3.2. Requirements in Sections 2109.3.3 and 2109.3.4 apply to both unstabilized and stabilized adobe. This is one of the few sources for such design information.

2109.3.1 Unstabilized adobe. Unstabilized adobe shall comply with Sections 2109.3.1.1 through 2109.3.1.4.

❖ Unstabilized adobe does not contain stabilizers and is generally not as durable or dimensionally stable as stabilized adobe.

2109.3.1.1 Compressive strength. Adobe units shall have an average compressive strength of 300 psi (2068 kPa) when tested in accordance with ASTM C 67. Five samples shall be tested and no individual unit is permitted to have a compressive strength of less than 250 psi (1724 kPa).

❖ Average compressive strength, based on five specimens tested in accordance with ASTM C 67, must be at least 300 psi (2068 kPa).

2109.3.1.2 Modulus of rupture. Adobe units shall have an average modulus of rupture of 50 psi (345 kPa) when tested in accordance with the following procedure. Five samples shall be tested and no individual unit shall have a modulus of rupture of less than 35 psi (241 kPa).

❖ Average modulus of rupture, based on five specimens tested in accordance with Sections 2109.3.1.2.1 through 2109.3.2.1.4, must be at least 50 psi (345 kPa).

2109.3.1.2.1 Support conditions. A cured unit shall be simply supported by 2-inch-diameter (51 mm) cylindrical supports located 2 inches (51 mm) in from each end and extending the full width of the unit.

❖ These required support conditions are typical for modulus of rupture tests.

2109.3.1.2.2 Loading conditions. A 2-inch-diameter (51 mm) cylinder shall be placed at midspan parallel to the supports.

❖ Loading through a hydraulic cylinder at midspan is common for these tests.

2109.3.1.2.3 Testing procedure. A vertical load shall be applied to the cylinder at the rate of 500 pounds per minute (37 N/s) until failure occurs.

❖ The required application of vertical load is easily controlled in testing laboratories.

2109.3.1.2.4 Modulus of rupture determination. The modulus of rupture shall be determined by the equation:

$$f_r = 3\,PL_s / 2\,S_w (S_t^2) \qquad \textbf{(Equation 21-2)}$$

where, for the purposes of this section only:

S_w = Width of the test specimen measured parallel to the loading cylinder, inches (mm).

f_r = Modulus of rupture, psi (MPa).

L_s = Distance between supports, inches (mm).

S_t = Thickness of the test specimen measured parallel to the direction of load, inches (mm).

P = The applied load at failure, pounds (N).

❖ Equation 21-2 is based on basic engineering mechanics and is valid for all rectangular specimens tested in this fashion.

2109.3.1.3 Moisture content requirements. Adobe units shall have a moisture content not exceeding 4 percent by weight.

❖ This section limits the moisture content of unstabilized adobe units to acceptable levels.

2109.3.1.4 Shrinkage cracks. Adobe units shall not contain more than three shrinkage cracks and any single shrinkage crack shall not exceed 3 inches (76 mm) in length or $^1/_8$ inch (3.2 mm) in width.

❖ As adobe units dry, they shrink and can crack. This section places limits on those potential cracks to keep the masonry structurally sound and reasonably water resistant.

2109.3.2 Stabilized adobe. Stabilized adobe shall comply with Section 2109.3.1 for unstabilized adobe in addition to Sections 2109.3.2.1 and 2109.3.2.2.

❖ This type of adobe is manufactured with stabilizers to increase its durability and decrease its water absorption. Stabilized adobe must also meet the requirements for unstabilized adobe prescribed in Section 2109.3.1.

2109.3.2.1 Soil requirements. Soil used for stabilized adobe units shall be chemically compatible with the stabilizing material.

❖ The soil and stabilizing materials must be chemically compatible, so that the stabilized units will be durable.

2109.3.2.2 Absorption requirements. A 4-inch (102 mm) cube, cut from a stabilized adobe unit dried to a constant weight in a ventilated oven at 212°F to 239°F (100°C to 115°C), shall not absorb more than $2^1/_2$ percent moisture by weight when placed upon a constantly water-saturated, porous surface for seven days. A minimum of five specimens shall be tested and each specimen shall be cut from a separate unit.

❖ This section prescribes a test method to verify that stabilized adobe units meet absorption limits.

2109.3.3 Allowable stress. The allowable compressive stress based on gross cross-sectional area of adobe shall not exceed 30 psi (207 kPa).

❖ This section prescribes the allowable compressive stress of adobe based on its gross cross-sectional area.

2109.3.3.1 Bolts. Bolt values shall not exceed those set forth in Table 2109.3.3.1.

❖ This section requires the capacity of bolts to be based on Table 2109.3.3.1. Specific types of bolts are not identified, but headed, bent-bar and plate anchors should all be acceptable.

**TABLE 2109.3.3.1
ALLOWABLE SHEAR ON BOLTS IN ADOBE MASONRY**

DIAMETER OF BOLTS (inches)	MINIMUM EMBEDMENT (inches)	SHEAR (pounds)
$^1/_2$	—	—
$^5/_8$	12	200
$^3/_4$	15	300
$^7/_8$	18	400
1	21	500
$1^1/_8$	24	600

For SI: 1 inch = 25.4 mm, 1 pound = 4.448 N.

❖ The allowable shear values in this table are based on the capacity of the adobe masonry. The capacity of the anchor-bolt steel is much higher, so the lower strength of the adobe controls.

2109.3.4 Detailed requirements. Adobe construction shall comply with Sections 2109.3.4.1 through 2109.3.4.9.

❖ This section contains detailed construction requirements including height restrictions, mortar restrictions, mortar joint construction and water-resistance requirements for parapet walls and also specific requirements for wall thickness; foundations; isolated piers and columns; tie beams; exterior finish and lintels.

2109.3.4.1 Number of stories. Adobe construction shall be limited to buildings not exceeding one *story*, except that two-*story* construction is allowed when designed by a *registered design professional*.

❖ Because of the low strength of adobe masonry, it is limited to use in single-story buildings, unless designed by a registered design professional, in which case two-story buildings are permitted.

2109.3.4.2 Mortar. Mortar for adobe construction shall comply with Sections 2109.3.4.2.1 and 2109.3.4.2.2.

❖ A variety of mortars are acceptable for stabilized adobe as noted in Section 2109.3.4.2.1.

2109.3.4.2.1 General. Mortar for stabilized adobe units shall comply with Chapter 21 or adobe soil. Adobe soil used as mortar shall comply with material requirements for stabilized adobe. Mortar for unstabilized adobe shall be Portland cement mortar.

❖ A variety of mortars are acceptable for stabilized adobe as noted; however, portland cement-lime mortars are required for unstabilized adobe. Selection of a relatively weak mortar that is compatible with the units is appropriate.

2109.3.4.2.2 Mortar joints. Adobe units shall be laid with full head and bed joints and in full running bond.

❖ Full mortar joints, the same as for other solid units, are required for adobe construction. Units are required to be laid in running bond.

2109.3.4.3 Parapet walls. Parapet walls constructed of adobe units shall be waterproofed.

❖ Waterproofing parapets reduces moisture infiltration into the adobe.

2109.3.4.4 Wall thickness. The minimum thickness of *exterior walls* in one-story buildings shall be 10 inches (254 mm). The walls shall be laterally supported at intervals not exceeding 24 feet (7315 mm). The minimum thickness of interior *load-bearing walls* shall be 8 inches (203 mm). In no case shall the unsupported height of any wall constructed of adobe units exceed 10 times the thickness of such wall.

❖ Because of the low strength of adobe masonry walls, thicker walls with more closely spaced supports are required.

2109.3.4.5 Foundations. Foundations for adobe construction shall be in accordance with Sections 2109.3.4.5.1 and 2109.3.4.5.2.

❖ This section prescribes foundation requirements for adobe masonry.

2109.3.4.5.1 Foundation support. Walls and partitions constructed of adobe units shall be supported by foundations or footings that extend not less than 6 inches (152 mm) above adjacent ground surfaces and are constructed of solid masonry (excluding adobe) or concrete. Footings and foundations shall comply with Chapter 18.

❖ So that adobe masonry is properly supported, solid masonry or concrete foundations are required.

2109.3.4.5.2 Lower course requirements. Stabilized adobe units shall be used in adobe walls for the first 4 inches (102 mm) above the finished first-floor elevation.

❖ Because of their greater durability, stabilized adobe units are required at the base of adobe walls. Conventional masonry units can also be used to satisfy this requirement.

2109.3.4.6 Isolated piers or columns. Adobe units shall not be used for isolated piers or columns in a load-bearing capacity. Walls less than 24 inches (610 mm) in length shall be considered isolated piers or columns.

❖ Adobe units are not strong enough to carry significant loads and are, therefore, not permitted to be used as isolated piers or columns.

2109.3.4.7 Tie beams. *Exterior walls* and interior *load-bearing walls* constructed of adobe units shall have a continuous tie beam at the level of the floor or roof bearing and meeting the following requirements.

❖ To distribute loads more evenly into the adobe, tie beams are required at the floor or roof levels. Tie beams can be constructed of concrete or wood, as described in Sections 2109.3.4.7.1 and 2109.3.4.7.2, respectively.

2109.3.4.7.1 Concrete tie beams. Concrete tie beams shall be a minimum depth of 6 inches (152 mm) and a minimum width of 10 inches (254 mm). Concrete tie beams shall be continuously reinforced with a minimum of two No. 4 reinforcing bars. The specified compressive strength of concrete shall be at least 2,500 psi (17.2 MPa).

❖ This section provides requirements for concrete tie beams to be cast above adobe masonry walls to distribute loads from floors and roofs.

2109.3.4.7.2 Wood tie beams. Wood tie beams shall be solid or built up of lumber having a minimum nominal thickness of 1 inch (25 mm), and shall have a minimum depth of 6 inches (152 mm) and a minimum width of 10 inches (254 mm). Joints in wood tie beams shall be spliced a minimum of 6 inches (152 mm). No splices shall be allowed within 12 inches (305 mm) of an opening. Wood used in tie beams shall be *approved* naturally decay-resistant or preservative-treated wood.

❖ This section provides requirements for wood tie beams to be constructed above adobe masonry walls to distribute loads from floors and roofs.

2109.3.4.8 Exterior finish. *Exterior walls* constructed of unstabilized adobe units shall have their exterior surface covered with a minimum of two coats of Portland cement plaster having a minimum thickness of $^3/_4$ inch (19.1 mm) and conforming to ASTM C 926. Lathing shall comply with ASTM C 1063. Fasteners shall be spaced at 16 inches (406 mm) o.c. maximum. Exposed wood surfaces shall be treated with an *approved* wood preservative or other protective coating prior to lath application.

❖ Unstabilized adobe must be coated with plaster to increase its durability.

2109.3.4.9 Lintels. Lintels shall be considered structural members and shall be designed in accordance with the applicable provisions of Chapter 16.

❖ Lintels over door and window openings are required to be structurally designed to carry imposed loads and to distribute those loads into the supporting adobe.

SECTION 2110
GLASS UNIT MASONRY

2110.1 General. Glass unit masonry construction shall comply with Chapter 7 of TMS 402/ACI 530/ASCE 5 and this section.

❖ Section 2110 contains provisions for glass unit masonry walls, which are nearly identical to the glass unit masonry provisions in Chapter 7 of the MSJC Code. Because those provisions are essentially the same, Section 2101.2.5 permits glass unit masonry to comply with the provisions of Chapter 7 of the MSJC Code or of Section 2110.

Glass unit masonry panels are permitted to be used in interior or exterior walls, provided that they are nonload bearing and comply with the requirements of Section 2110, which are partly empirical and partly based on tests.

2110.1.1 Limitations. Solid or hollow *approved* glass block shall not be used in fire walls, party walls, fire barriers, fire partitions or smoke barriers, or for load-bearing construction. Such blocks shall be erected with mortar and reinforcement in metal channel-type frames, structural frames, masonry or concrete recesses, embedded panel anchors as provided for both exterior and interior walls or other *approved* joint materials. Wood strip framing shall not be used in walls required to have a fire-resistance rating by other provisions of this code.

Exceptions:

1. Glass-block assemblies having a fire protection rating of not less than $^3/_4$ hour shall be permitted as opening protectives in accordance with Section 716 in fire barriers, fire partitions and smoke barriers that have a required fire-resistance rating of 1 hour or less and do not enclose exit stairways, exit ramps or exit passageways.
2. Glass-block assemblies as permitted in Section 404.6, Exception 2.

❖ Structural glass blocks are not permitted in fire walls, party walls, fire barrier walls or fire partitions, with two exceptions. Exception 1 permits glass blocks that have been tested and classified for a $^3/_4$-hour fire protection rating in openings to be used in smoke barriers, fire barrier walls or fire partitions with a required fire-resistance rating of 1 hour or less. Since 1-hour fire barrier walls can be utilized to enclose interior exit stairways and exit ramps (see Section 1022.1), as well as exit passageways (see Section 1023.3), the exception does not apply to those locations. This is consistent with Sections 1022.4 and 1023.5, which limit openings in these exit components to those that are necessary for egress purposes. Exception 2 permits glass blocks to be installed in accordance with the requirements in Section 404.6, Exception 2, as the fire barrier enclosure of atriums. The two exceptions differ in that Exception 1 addresses the use of glass block for opening protectives (fire windows) while Exception 2 permits the acceptance of complying glass block assemblies as 1-hour fire barriers.

SECTION 2111 MASONRY FIREPLACES

2111.1 Definition. A masonry fireplace is a fireplace constructed of concrete or masonry. Masonry fireplaces shall be constructed in accordance with this section.

❖ The provisions of this section apply to the design and construction of concrete and masonry fireplaces, which for simplicity are referred to as "masonry fireplaces." Figure 2111.1 shows a cross section of a typical masonry fireplace. In accordance with the definition in Section 2102.1, a fireplace consists of a hearth and fire chamber and it is built in conjunction with a chimney.

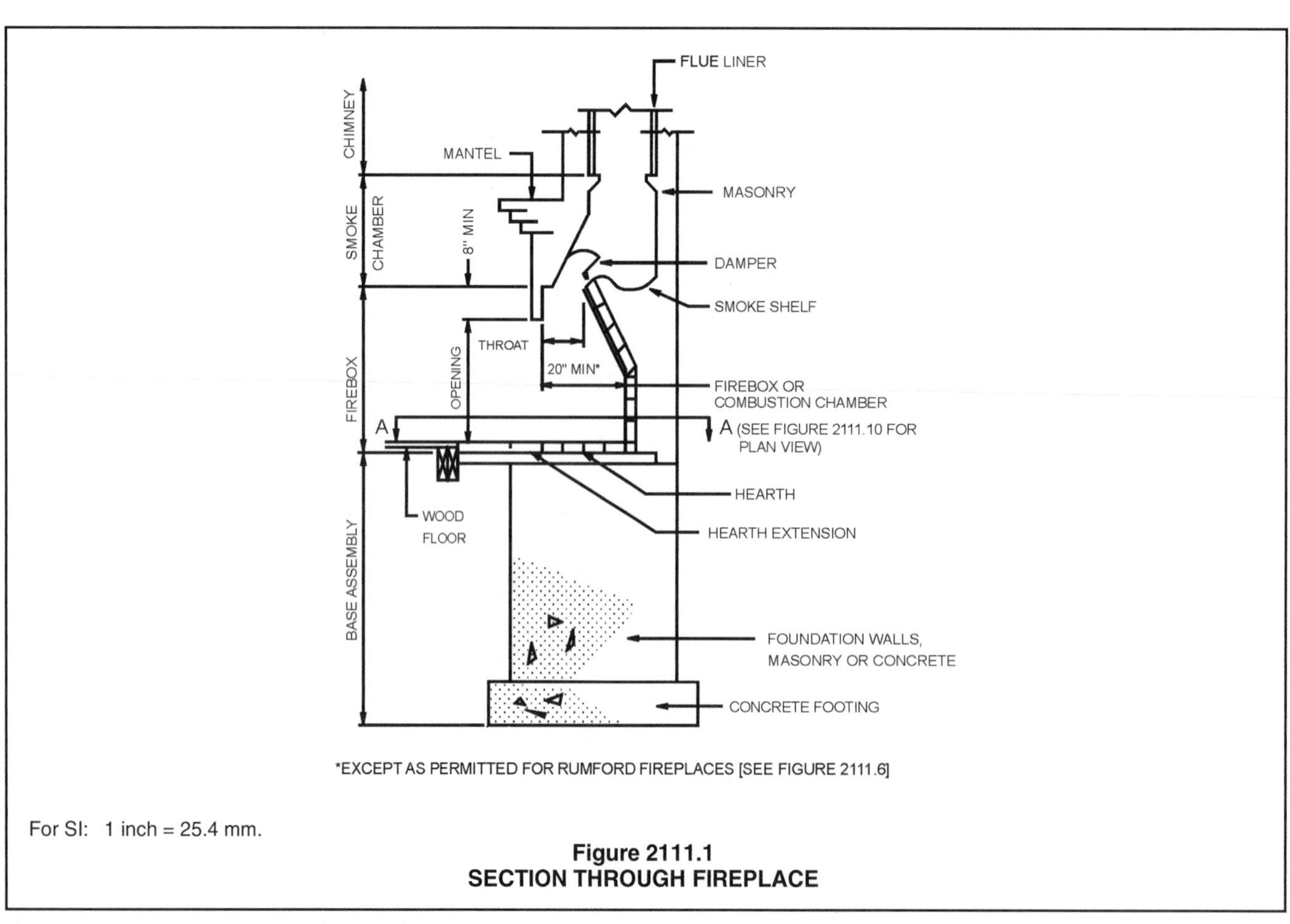

For SI: 1 inch = 25.4 mm.

Figure 2111.1
SECTION THROUGH FIREPLACE

2111.2 Footings and foundations. Footings for masonry fireplaces and their chimneys shall be constructed of concrete or solid masonry at least 12 inches (305 mm) thick and shall extend at least 6 inches (153 mm) beyond the face of the fireplace or foundation wall on all sides. Footings shall be founded on natural undisturbed earth or engineered fill below frost depth. In areas not subjected to freezing, footings shall be at least 12 inches (305 mm) below finished grade.

❖ Masonry fireplaces and chimneys must be supported on adequate footings due to their weight and the forces imposed on them by wind, earthquakes and other effects. This section prescribes minimum footing requirements that are typically adequate for standard fireplaces and chimneys. Extremely large, tall or heavy fireplaces and chimneys, however, may need more substantial foundations (see Figure 2111.1).

2111.2.1 Ash dump cleanout. Cleanout openings, located within foundation walls below fireboxes, when provided, shall be equipped with ferrous metal or masonry doors and frames constructed to remain tightly closed, except when in use. Cleanouts shall be accessible and located so that ash removal will not create a hazard to combustible materials.

❖ Noncombustible, tightly sealed cleanout doors are required to reduce the danger of fire spread through the cleanout openings. Cleanout openings are required to be easily accessible to allow ash to be readily removed.

2111.3 Seismic reinforcing. In structures assigned to *Seismic Design Category* A or B, reinforcement and seismic anchorage are not required. Masonry or concrete fireplaces shall be constructed, anchored, supported and reinforced as required in this chapter. In structures assigned to *Seismic Design Category* C or D, masonry and concrete fireplaces shall be reinforced and anchored as detailed in Sections 2111.3.1, 2111.3.2, 2111.4 and 2111.4.1 for chimneys serving fireplaces. In structures assigned to *Seismic Design Category* E or F, masonry and concrete chimneys shall be reinforced in accordance with the requirements of Sections 2101 through 2108.

❖ Unreinforced masonry fireplaces and chimneys subjected to moderate ground motion have sustained damage in past earthquakes. The requirements in this section provide minimum reinforcement in an effort to increase structural integrity in structures classified as Seismic Design Category C or D. More substantial reinforcement may be indicated in structures having a higher seismic risk or for atypical fireplaces and chimneys. The section requires masonry fireplaces to be engineered in structures assigned to Seismic Design Category E or F.

2111.3.1 Vertical reinforcing. For fireplaces with chimneys up to 40 inches (1016 mm) wide, four No. 4 continuous vertical bars, anchored in the foundation, shall be placed in the concrete between wythes of solid masonry or within the cells of hollow unit masonry and grouted in accordance with Section 2103.12. For fireplaces with chimneys greater than 40 inches (1016 mm) wide, two additional No. 4 vertical bars shall be provided for each additional 40 inches (1016 mm) in width or fraction thereof.

❖ These requirements are traditional minimum prescriptive provisions to help maintain the integrity of fireplaces and chimneys during earthquakes. To resist strong earthquakes, however, more substantial reinforcement may be required.

2111.3.2 Horizontal reinforcing. Vertical reinforcement shall be placed enclosed within $^1/_4$-inch (6.4 mm) ties or other reinforcing of equivalent net cross-sectional area, spaced not to exceed 18 inches (457 mm) on center in concrete; or placed in the bed joints of unit masonry at a minimum of every 18 inches (457 mm) of vertical height. Two such ties shall be provided at each bend in the vertical bars.

❖ These requirements are traditional minimum prescriptive provisions to help maintain the integrity of fireplaces and chimneys during earthquakes. The vertical reinforcement required by Section 2111.3.1 must be enclosed within the horizontal reinforcement required by this section. To resist strong earthquakes, however, more substantial reinforcement may be required.

2111.4 Seismic anchorage. Masonry and concrete chimneys in structures assigned to *Seismic Design Category* C or D shall be anchored at each floor, ceiling or roof line more than 6 feet (1829 mm) above grade, except where constructed completely within the *exterior walls*. Anchorage shall conform to the following requirements.

❖ Fireplace chimneys can fail by overturning during earthquakes. This section duplicates chimney anchorage requirements that are in Section 2113.4. Seismic anchorage to floors and roof diaphragms is required for exterior chimneys in order to prevent detachment from the structure. While it is not required for a chimney that is enclosed within the exterior walls, anchorage could be utilized to minimize the risk of damage to floor or roof framing due to pounding.

2111.4.1 Anchorage. Two $^3/_{16}$-inch by 1-inch (4.8 mm by 25.4 mm) straps shall be embedded a minimum of 12 inches (305 mm) into the chimney. Straps shall be hooked around the outer bars and extend 6 inches (152 mm) beyond the bend. Each strap shall be fastened to a minimum of four floor joists with two $^1/_2$-inch (12.7 mm) bolts.

❖ The prescriptive requirements in this section are traditional for typical chimney construction. More substantial anchorage may be required in structures having an elevated seismic risk or for chimneys where the distance between floor and roof diaphragms is large.

2111.5 Firebox walls. Masonry fireboxes shall be constructed of solid masonry units, hollow masonry units grouted solid, stone or concrete. When a lining of firebrick at least 2 inches (51 mm) in thickness or other *approved* lining is provided, the minimum thickness of back and sidewalls shall each be 8 inches (203 mm) of solid masonry, including the lining. The width of joints between firebricks shall not be greater than $^1/_4$ inch (6.4 mm). When no lining is provided, the total minimum thickness of back and sidewalls shall be 10 inches (254 mm) of solid masonry. Firebrick shall conform to ASTM C 27 or ASTM C 1261 and shall be laid with medium-duty refractory mortar conforming to ASTM C 199.

❖ This section specifies the minimum thicknesses of refractory brick or solid masonry necessary to contain the generated heat.

Solid masonry walls forming the firebox are required to have a minimum total thickness of 8 inches (204 mm), including the refractory lining.

The refractory lining is to consist of a low-duty, fire-clay refractory brick with a minimum thickness of 2 inches (52 mm), laid with medium-duty refractory mortar. Mortar joints are generally $^1/_{16}$ to $^3/_{16}$ inch (1.6 to 4.8 mm) thick, but not thicker than 1/4 inch (6.4 mm), to reduce thermal movements and prevent joint deterioration.

Where a firebrick lining is not used in firebox construction, the wall thickness is not to be less than 10 inches (254 mm) of solid masonry. Firebrick is required to conform to ASTM C 27 or ASTM C 1261. Firebrick must be laid with medium-duty refractory mortar conforming to ASTM C 199.

2111.5.1 Steel fireplace units. Steel fireplace units are permitted to be installed with solid masonry to form a masonry fireplace provided they are installed according to either the requirements of their listing or the requirements of this section. Steel fireplace units incorporating a steel firebox lining shall be constructed with steel not less than $^1/_4$ inch (6.4 mm) in thickness, and an air-circulating chamber which is ducted to the interior of the building. The firebox lining shall be encased with solid masonry to provide a total thickness at the back and sides of not less than 8 inches (203 mm), of which not less than 4 inches (102 mm) shall be of solid masonry or concrete. Circulating air ducts employed with steel fireplace units shall be constructed of metal or masonry.

❖ This section provides minimum requirements for steel linings used in masonry fireplaces to improve heat flow.

2111.6 Firebox dimensions. The firebox of a concrete or masonry fireplace shall have a minimum depth of 20 inches (508 mm). The throat shall not be less than 8 inches (203 mm) above the fireplace opening. The throat opening shall not be less than 4 inches (102 mm) in depth. The cross-sectional area of the passageway above the firebox, including the throat, damper and smoke chamber, shall not be less than the cross-sectional area of the flue.

Exception: Rumford fireplaces shall be permitted provided that the depth of the fireplace is at least 12 inches (305 mm) and at least one-third of the width of the fireplace opening, and the throat is at least 12 inches (305 mm) above the lintel, and at least $^1/_{20}$ the cross-sectional area of the fireplace opening.

❖ The proper functioning of the fireplace depends on the size of the face opening and the chimney dimensions, which in turn are related to the room size (see Figure 2111.1). This section specifies a minimum depth of 20 inches (508 mm) for the combustion chamber because that depth influences the draft requirement. The dimensions of the firebox (depth, opening size and shape) are usually based on two considerations: aesthetics and the need to prevent the room from over-heating. Suggested dimensions for single-opening fireboxes are given in technical publications of the Brick Institute of America (BIA) and the National Concrete Masonry Association (NCMA).

This section also provides additional criteria for the throat's location and minimum cross-sectional area. Those criteria are based on many years of construction of successfully functioning fireplaces.

The exception permits the use of Rumford-style fireplaces and the dimensions associated with this design style. Rumford fireplaces are tall and shallow to reflect more heat, and they have streamlined throats to eliminate turbulence and to carry away the smoke with little loss of heated room air.

The code reference to the depth of the fireplace is interpreted as the depth of the firebox (see Figure 2111.6). The throat is required to be made at least 12 inches (305 mm) above the lintel and at least one-twen-tieth of the cross-sectional area of the fireplace opening. Smoke chambers and flues for Rumford fireplaces should be sized and built like other masonry fireplaces. While those who build Rumford fireplaces do not com-pletely agree about how they work, many books and guides address their construction.

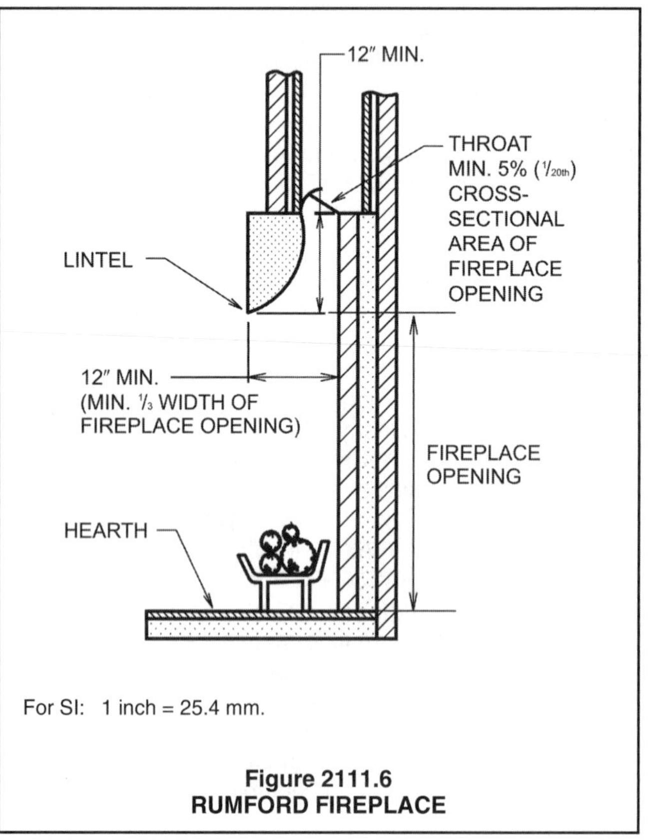

For SI: 1 inch = 25.4 mm.

Figure 2111.6
RUMFORD FIREPLACE

2111.7 Lintel and throat. Masonry over a fireplace opening shall be supported by a lintel of noncombustible material. The minimum required bearing length on each end of the fireplace opening shall be 4 inches (102 mm). The fireplace throat or damper shall be located a minimum of 8 inches (203 mm) above the top of the fireplace opening.

❖ Permanent support for the masonry above the fireplace opening is provided by noncombustible lintels of steel, masonry or concrete. Combustible lintels (for example, those made from wood) are not appropriate due to the risk of fire damage and the probable collapse of the masonry above the opening.

The minimum bearing requirement of 4 inches (102 mm) is empirical, based on typical masonry fireplace openings. Lintels that support more than the typical weight of masonry above the fireplace opening may require a larger bearing area.

The throat of a fireplace is the slot-like opening above the firebox, through which flames, smoke and hot combustion gases pass into the smoke chamber. The throat is as wide as the combustion chamber and is required to be located at least 8 inches (204 mm) above the lintel for conventional fireplaces. The back wall of the combustion chamber extends up to the throat, which is provided with a metal damper.

2111.7.1 Damper. Masonry fireplaces shall be equipped with a ferrous metal damper located at least 8 inches (203 mm) above the top of the fireplace opening. Dampers shall be installed in the fireplace or at the top of the flue venting the fireplace, and shall be operable from the room containing the fireplace. Damper controls shall be permitted to be located in the fireplace.

❖ A damper is used to close the chimney flue when the fireplace is not in use. This section provides guidance on its location and construction.

2111.8 Smoke chamber walls. Smoke chamber walls shall be constructed of solid masonry units, hollow masonry units grouted solid, stone or concrete. The total minimum thickness of front, back and sidewalls shall be 8 inches (203 mm) of solid masonry. The inside surface shall be parged smooth with refractory mortar conforming to ASTM C 199. When a lining of firebrick at least 2 inches (51 mm) thick, or a lining of vitrified clay at least $^{5}/_{8}$ inch (15.9 mm) thick, is provided, the total minimum thickness of front, back and sidewalls shall be 6 inches (152 mm) of solid masonry, including the lining. Firebrick shall conform to ASTM C 1261 and shall be laid with refractory mortar conforming to ASTM C 199. Vitrified clay linings shall conform to ASTM C 315.

❖ The minimum wall thickness specified for the throat and smoke chamber is required to provide support for the flue construction, as well as adequate thermal insulation for the adjacent combustible construction.

In conventional fireplace construction, the smoke chamber is a tapering section whose vertical dimension is measured from the damper or throat to the bottom of the chimney flue. The actual height is a function of the fireplace opening.

2111.8.1 Smoke chamber dimensions. The inside height of the smoke chamber from the fireplace throat to the beginning of the flue shall not be greater than the inside width of the fireplace opening. The inside surface of the smoke chamber shall not be inclined more than 45 degrees (0.76 rad) from vertical when prefabricated smoke chamber linings are used or when the smoke chamber walls are rolled or sloped rather than corbeled. When the inside surface of the smoke chamber is formed by corbeled masonry, the walls shall not be corbeled more than 30 degrees (0.52 rad) from vertical.

❖ This section specifies the smoke chamber configuration that is needed for the proper function of the masonry chimney (see Figure 2111.1).

2111.9 Hearth and hearth extension. Masonry fireplace hearths and hearth extensions shall be constructed of concrete or masonry, supported by noncombustible materials, and reinforced to carry their own weight and all imposed loads. No combustible material shall remain against the underside of hearths or hearth extensions after construction.

❖ The fireplace hearth consists of two parts. The hearth, commonly called the "inner hearth," is the floor of the combustion chamber and is obviously constructed of noncombustible material. The outer hearth, commonly known as the "hearth extension," projects beyond the face of the fireplace into the room and also consists of noncombustible materials, such as brick or concrete masonry, concrete, floor tile or stone (see Figure 2111.1). The hearth extension may continue out from the inner hearth at the same level or be stepped down to a lower level. Minimum dimensions for the hearth extension are prescribed in Section 2111.10 and are discussed in the commentary to that section.

Combustible forms and centers could ignite from exposure to heat from the adjacent fireplace and from burning embers that escape the firebox; these and other similar concealed, combustible components must be removed.

2111.9.1 Hearth thickness. The minimum thickness of fireplace hearths shall be 4 inches (102 mm).

❖ The required minimum thickness of 4 inches (102 mm) is an empirical requirement that has historically been successful.

2111.9.2 Hearth extension thickness. The minimum thickness of hearth extensions shall be 2 inches (51 mm).

Exception: When the bottom of the firebox opening is raised at least 8 inches (203 mm) above the top of the hearth extension, a hearth extension of not less than $^{3}/_{8}$-inch-thick (9.5 mm) brick, concrete, stone, tile or other *approved* noncombustible material is permitted.

❖ These requirements are empirical and have historically been successful.

2111.10 Hearth extension dimensions. Hearth extensions shall extend at least 16 inches (406 mm) in front of, and at least 8 inches (203 mm) beyond, each side of the fireplace opening. Where the fireplace opening is 6 square feet (0.557 m^2) or larger, the hearth extension shall extend at least 20 inches (508 mm) in front of, and at least 12 inches (305 mm) beyond, each side of the fireplace opening.

❖ The hearth extension is required to extend the full width of the fireplace opening and at least 8 inches (203 mm) beyond each side of the opening. It is also required to extend at least 16 inches (406 mm) out from the face of the fireplace. For fireplace openings larger than 6 square feet (0.557 m^2) in area, the hearth extension is required to extend at least 20 inches (508 mm) beyond the face of the fireplace and at least 12 inches (305 mm) beyond each side of the fireplace opening (see Figure 2111.10).

The hearth extension is intended to serve as a fire-protective separation between the firebox and adjacent combustible flooring or furnishings and to prevent accidental spills of hot embers or logs from the fire.

2111.11 Fireplace clearance. Any portion of a masonry fireplace located in the interior of a building or within the *exterior wall* of a building shall have a clearance to combustibles of not less than 2 inches (51 mm) from the front faces and sides of masonry fireplaces and not less than 4 inches (102 mm) from the back faces of masonry fireplaces. The airspace shall not be filled, except to provide fireblocking in accordance with Section 2111.12.

Exceptions:

1. Masonry fireplaces *listed* and labeled for use in contact with combustibles in accordance with UL 127 and installed in accordance with the manufacturer's installation instructions are permitted to have combustible material in contact with their exterior surfaces.
2. When masonry fireplaces are constructed as part of masonry or concrete walls, combustible materials shall not be in contact with the masonry or concrete walls less than 12 inches (306 mm) from the inside surface of the nearest firebox lining.
3. Exposed combustible *trim* and the edges of sheathing materials, such as wood siding, flooring and drywall, are permitted to abut the masonry fireplace sidewalls and hearth extension, in accordance with Figure 2111.11, provided such combustible *trim* or sheathing is a minimum of 12 inches (306 mm) from the inside surface of the nearest firebox lining.
4. Exposed combustible mantels or *trim* is permitted to be placed directly on the masonry fireplace front surrounding the fireplace opening, provided such combustible materials shall not be placed within 6 inches (153 mm) of a fireplace opening. Combustible material directly above and within 12 inches (305 mm) of the fireplace opening shall not project more than $^1/_8$ inch (3.2 mm) for each 1-inch (25 mm) distance from such opening. Combustible materials located along the sides of the fireplace opening that project more than $1^1/_2$ inches (38 mm) from the face of the fireplace shall have an additional clearance equal to the projection.

❖ Combustible materials, such as framing studs and joists, must not be installed closer than 2 inches (51 mm) from the front or side surfaces of fireplace walls because of the fire hazard to materials in this location. Heat transmitted through fireplace walls can ignite combustible structural materials in contact with the walls. For this reason, a minimum required clearance has been established from the fireplace to combustibles.

The exceptions cover specific instances where the clearance requirements vary. For a discussion and

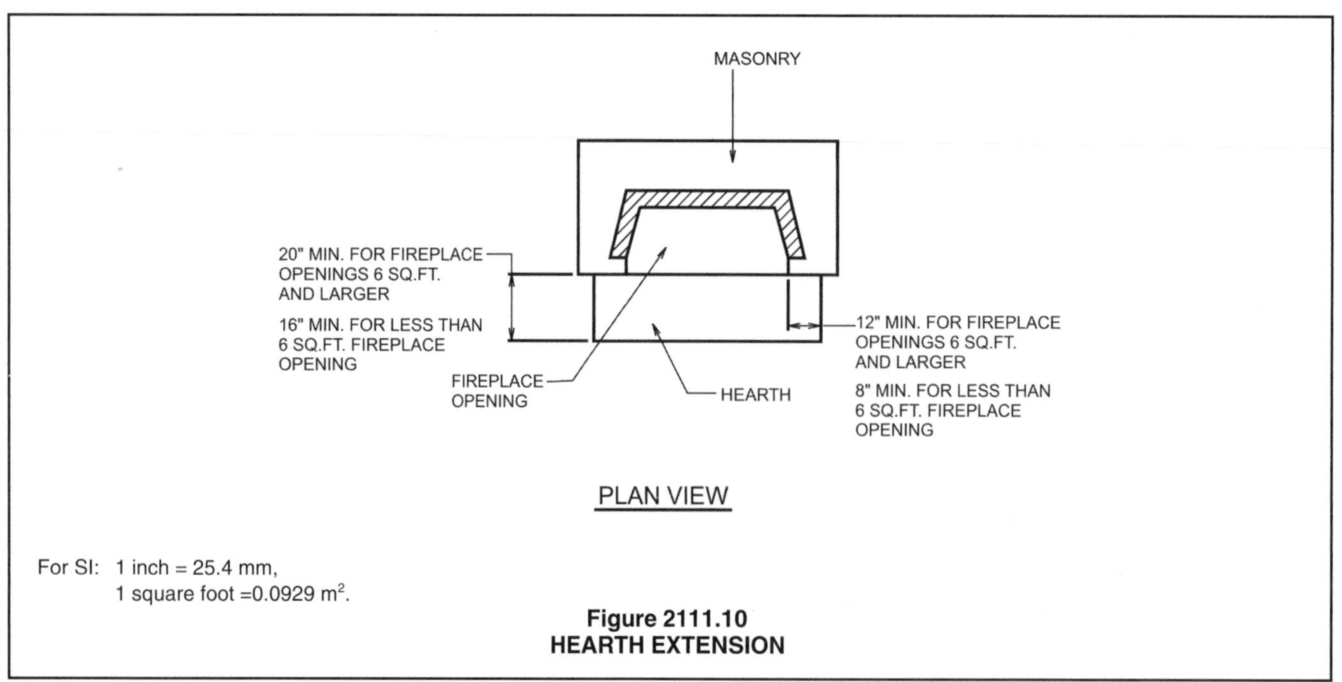

Figure 2111.10
HEARTH EXTENSION

illustration of Exception 3, see Code Figure 2111.11 and commentary.

Exception 4 regulates the location of mantels and trim adjacent to fireplace openings. Figure 2111.11(1) illustrates the limits above the opening. A less restrictive projection limit is applicable to combustible materials along the side of the fireplace opening, because radiant heat and convection heat are a greater problem for combustible material above the fireplace opening than for combustible material at the side of the fireplace opening.

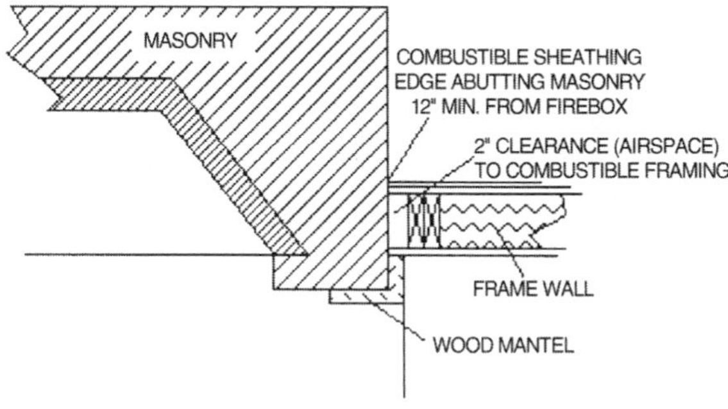

For SI: 1 inch = 25.4 mm

FIGURE 2111.11
ILLUSTRATION OF EXCEPTION TO FIREPLACE CLEARANCE PROVISION

❖ This figure depicts the provisions of Exception 3 to Section 2111.11 for combustible trim and sheathing adjacent to masonry fireplaces and hearth extensions.

2111.12 Fireplace fireblocking. All spaces between fireplaces and floors and ceilings through which fireplaces pass shall be fireblocked with noncombustible material securely fastened in place. The fireblocking of spaces between wood joists, beams or headers shall be to a depth of 1 inch (25 mm) and shall only be placed on strips of metal or metal lath laid across the spaces between combustible material and the chimney.

❖ Fireblocking is required to prevent the travel of flames, smoke or hot gases to other areas of the building through gaps between the chimney and the floor or ceiling assemblies. The 1-inch (25 mm) depth requirement is intended to be both a minimum and maximum.

2111.13 Exterior air. Factory-built or masonry fireplaces covered in this section shall be equipped with an exterior air supply to ensure proper fuel combustion unless the room is mechanically ventilated and controlled so that the indoor pressure is neutral or positive.

❖ Adequate airflow is needed to provide exterior oxygen for the fire and to maintain draft through the chimney so that toxic combustion gases can be exhausted. Adequate airflow is especially important in modern, tightly sealed buildings. Note that this requirement applies to factory-built fireplaces, as well as masonry fireplaces.

2111.13.1 Factory-built fireplaces. Exterior combustion air ducts for factory-built fireplaces shall be *listed* components of the fireplace, and installed according to the fireplace manufacturer's instructions.

❖ Factory-built fireplaces are required to include exterior combustion air ducts (see Figure 2111.13.1). To function properly, these units need to be installed according to the manufacturer's recommendations.

2111.13.2 Masonry fireplaces. *Listed* combustion air ducts for masonry fireplaces shall be installed according to the terms of their listing and manufacturer's instructions.

❖ For proper functioning and airflow, air ducts for masonry fireplaces must be installed according to the manufacturer's recommendations (also see commentary, Section 2111.13).

2111.13.3 Exterior air intake. The exterior air intake shall be capable of providing all combustion air from the exterior of the *dwelling*. The exterior air intake shall not be located within a garage, *attic*, basement or crawl space of the *dwell-*

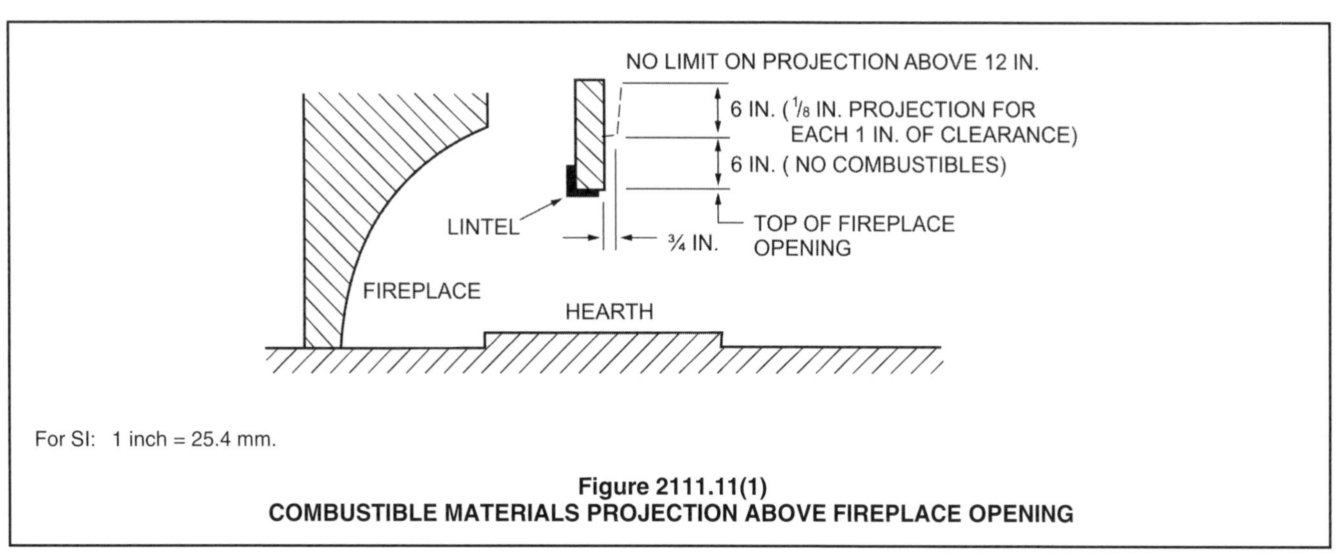

For SI: 1 inch = 25.4 mm.

Figure 2111.11(1)
COMBUSTIBLE MATERIALS PROJECTION ABOVE FIREPLACE OPENING

ing nor shall the air intake be located at an elevation higher than the firebox. The exterior air intake shall be covered with a corrosion-resistant screen of $^1/_4$-inch (6.4 mm) mesh.

❖ Air intakes are required to provide air from outside of the building. The air intakes are not permitted to be placed in garages, basements, crawl spaces or other areas where gases from the exhaust of automobiles, furnaces and other sources could be brought into the building.

Air intakes should be covered with screens to prevent pests from entering the building when the fireplace is not functioning.

2111.13.4 Clearance. Unlisted combustion air ducts shall be installed with a minimum 1-inch (25 mm) clearance to combustibles for all parts of the duct within 5 feet (1524 mm) of the duct outlet.

❖ The 1-inch (25 mm) clearance is required to reduce the risk of fire through the air intakes.

2111.13.5 Passageway. The combustion air passageway shall be a minimum of 6 square inches (3870 mm^2) and not more than 55 square inches (0.035 m^2), except that combustion air systems for *listed* fireplaces or for fireplaces tested for emissions shall be constructed according to the fireplace manufacturer's instructions.

❖ These minimum requirements are intended to provide adequate airflow through the fireplace.

2111.13.6 Outlet. The exterior air outlet is permitted to be located in the back or sides of the firebox chamber or within 24 inches (610 mm) of the firebox opening on or near the floor. The outlet shall be closable and designed to prevent burning material from dropping into concealed combustible spaces.

❖ The requirements on the location of the outlet in the firebox chamber are needed to ensure that adequate airflow to the fire is provided, while avoiding the direct flow of air into the adjacent room. Such openings must not become clogged with ash, which would restrict airflow. Even more important, burning material must not drop into concealed combustible spaces due to the risk of fire damage to the building. Outlets must therefore be both closable and adequately designed to prevent this.

SECTION 2112 MASONRY HEATERS

2112.1 Definition. A masonry heater is a heating appliance constructed of concrete or solid masonry, hereinafter referred to as "masonry," which is designed to absorb and store heat from a solid fuel fire built in the firebox by routing the exhaust gases through internal heat exchange channels in which the flow path downstream of the firebox may include flow in a horizontal or downward direction before entering

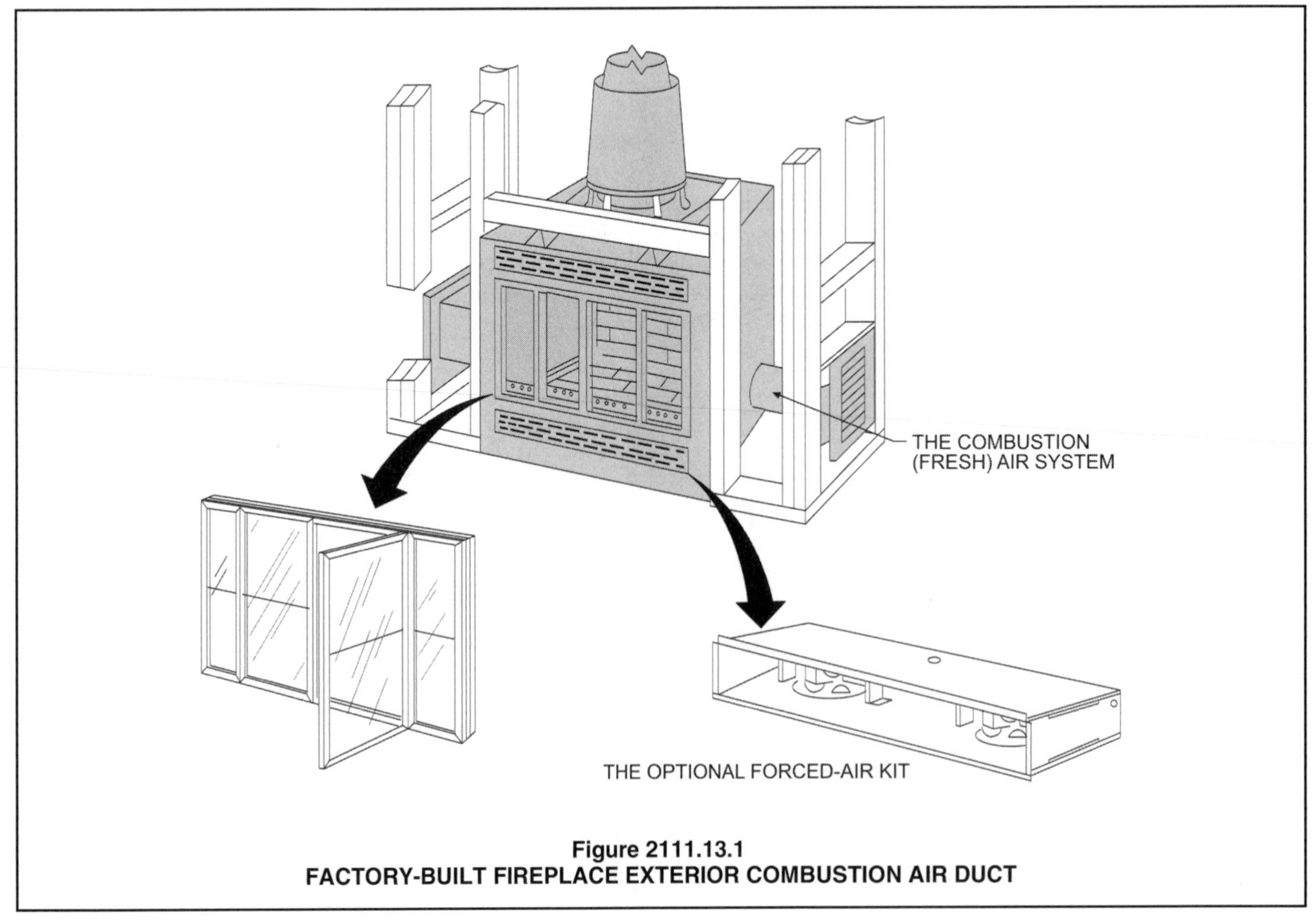

Figure 2111.13.1
FACTORY-BUILT FIREPLACE EXTERIOR COMBUSTION AIR DUCT

the chimney and which delivers heat by radiation from the masonry surface of the heater.

❖ Masonry heaters are appliances designed to absorb and store heat from a relatively small fire and to radiate that heat into the building interior. They are thermally more efficient than traditional fireplaces because of their design. Interior passageways through the heater allow hot exhaust gases from the fire to transfer heat into the masonry, which then radiates into the building.

2112.2 Installation. Masonry heaters shall be installed in accordance with this section and comply with one of the following:

1. Masonry heaters shall comply with the requirements of ASTM E 1602; or
2. Masonry heaters shall be *listed* and labeled in accordance with UL 1482 and installed in accordance with the manufacturer's installation instructions.

❖ The listed referenced standards provide guidelines for installing masonry heaters.

2112.3 Footings and foundation. The firebox floor of a masonry heater shall be a minimum thickness of 4 inches (102 mm) of noncombustible material and be supported on a noncombustible footing and foundation in accordance with Section 2113.2.

❖ This section refers the user to the foundation requirements for masonry chimneys.

2112.4 Seismic reinforcing. In structures assigned to *Seismic Design Category* D, E or F, masonry heaters shall be anchored to the masonry foundation in accordance with Section 2113.3. Seismic reinforcing shall not be required within the body of a masonry heater with a height that is equal to or less than 3.5 times its body width and where the masonry chimney serving the heater is not supported by the body of the heater. Where the masonry chimney shares a common wall with the facing of the masonry heater, the chimney portion of the structure shall be reinforced in accordance with Section 2113.

❖ Because of the large bulk and squat geometry of these heaters, seismic reinforcement is not typically required. Flexural tensile stresses, which typically cause damage to unreinforced masonry, rarely occur. Where chimneys extend above these heaters, however, seismic reinforcement is required by Section 2113. See the commentary to that section for additional information on this requirement.

2112.5 Masonry heater clearance. Combustible materials shall not be placed within 36 inches (765 mm) of the outside surface of a masonry heater in accordance with NFPA 211, Section 8-7 (clearances for solid fuel-burning appliances), and the required space between the heater and combustible material shall be fully vented to permit the free flow of air around all heater surfaces.

Exceptions:

1. When the masonry heater wall thickness is at least 8 inches (203 mm) thick of solid masonry and the wall thickness of the heat exchange channels is at least 5 inches (127 mm) thick of solid masonry, combustible materials shall not be placed within 4 inches (102 mm) of the outside surface of a masonry heater. A clearance of at least 8 inches (203 mm) shall be provided between the gas-tight capping slab of the heater and a combustible ceiling.
2. Masonry heaters *listed* and labeled in accordance with UL 1482 and installed in accordance with the manufacturer's instructions.

❖ Heat conducted through masonry heater walls can ignite combustible structural materials in contact with these walls. For this reason, a minimum required clearance to combustibles from masonry heaters has been established. Because masonry heaters typically generate more heat for a longer period of time than traditional fireplaces, greater clearances to combustible materials are needed to reduce the risk of fire.

SECTION 2113 MASONRY CHIMNEYS

2113.1 Definition. A masonry chimney is a chimney constructed of solid masonry units, hollow masonry units grouted solid, stone or concrete, hereinafter referred to as "masonry." Masonry chimneys shall be constructed, anchored, supported and reinforced as required in this chapter.

❖ A masonry chimney is a field-constructed assembly that can consist of masonry units, grout, reinforced concrete, rubble stone, fire-clay liners and mortars. The *International Fuel Gas Code*® (IFGC®) and the *International Mechanical Code*® (IMC®) address the installation of chimneys and venting systems that are required to convey products of combustion from fuel-burning appliances to the outdoors so that the operation of the appliance does not adversely affect building occupants. A masonry chimney is permitted to serve residential (low-heat), medium- and high-heat appliances. A masonry chimney is required for masonry fireplaces constructed in accordance with Section 2111. This section provides the construction requirements for all masonry chimneys.

2113.2 Footings and foundations. Footings for masonry chimneys shall be constructed of concrete or solid masonry at least 12 inches (305 mm) thick and shall extend at least 6 inches (152 mm) beyond the face of the foundation or support wall on all sides. Footings shall be founded on natural undisturbed earth or engineered fill below frost depth. In areas not subjected to freezing, footings shall be at least 12 inches (305 mm) below finished grade.

❖ Masonry chimneys must be supported on adequate foundations due to their weight and the forces imposed on them by wind, earthquakes and other effects. This section prescribes minimum foundation requirements that are typically adequate for standard chimneys.

Extremely large, tall or heavy chimneys, however, may need more substantial foundations. Also, a chimney foundation probably will support a larger load than the adjacent building foundations. For this reason, chimney footings and foundations are often separated from the building foundation. A chimney foundation monolithic with the building foundation is permitted, provided the soil pressure and anticipated settlement are approximately uniform.

2113.3 Seismic reinforcing. Masonry or concrete chimneys shall be constructed, anchored, supported and reinforced as required in this chapter. In structures assigned to *Seismic Design Category* C or D, masonry and concrete chimneys shall be reinforced and anchored as detailed in Sections 2113.3.1, 2113.3.2 and 2113.4. In structures assigned to *Seismic Design Category* A or B, reinforcement and seismic anchorage is not required. In structures assigned to *Seismic Design Category* E or F, masonry and concrete chimneys shall be reinforced in accordance with the requirements of Sections 2101 through 2108.

❖ Unreinforced masonry fireplaces and chimneys subjected to moderate ground motion have been severely damaged in past earthquakes. The requirements in this section provide minimum reinforcement in an effort to keep these structures intact during such events. More substantial reinforcement, however, may be required in areas of high seismicity or for atypical chimneys. The section requires masonry chimneys to be engineered in structures assigned to Seismic Design Category E or F.

2113.3.1 Vertical reinforcing. For chimneys up to 40 inches (1016 mm) wide, four No. 4 continuous vertical bars anchored in the foundation shall be placed in the concrete between wythes of solid masonry or within the cells of hollow unit masonry and grouted in accordance with Section 2103.12. Grout shall be prevented from bonding with the flue liner so that the flue liner is free to move with thermal expansion. For chimneys greater than 40 inches (1016 mm) wide, two additional No. 4 vertical bars shall be provided for each additional 40 inches (1016 mm) in width or fraction thereof.

❖ These requirements are traditional minimum prescriptive provisions to help maintain the structural integrity of fireplaces and chimneys during earthquakes. More reinforcement may be required in areas of high seismicity or for atypical chimneys.

2113.3.2 Horizontal reinforcing. Vertical reinforcement shall be placed enclosed within $^1/_4$-inch (6.4 mm) ties, or other reinforcing of equivalent net cross-sectional area, spaced not to exceed 18 inches (457 mm) o.c. in concrete, or placed in the bed joints of unit masonry, at a minimum of every 18 inches (457 mm) of vertical height. Two such ties shall be provided at each bend in the vertical bars.

❖ These requirements are traditional minimum prescriptive provisions intended to help maintain the integrity of fireplaces and chimneys during earthquakes. The vertical reinforcement required by Section 2113.3.1 must be enclosed within the horizontal reinforcement required by this section. More reinforcement may be required in areas of high seismicity or for atypical chimneys.

2113.4 Seismic anchorage. Masonry and concrete chimneys and foundations in structures assigned to *Seismic Design Category* C or D shall be anchored at each floor, ceiling or roof line more than 6 feet (1829 mm) above grade, except where constructed completely within the *exterior walls*. Anchorage shall conform to the following requirements.

❖ Chimneys must be connected to floor and roof diaphragms to prevent overturning during earthquakes. Chimneys must be anchored at the ceiling line of roof or ceiling assemblies and at floor levels below the roof. Such anchorage is of lesser importance where the floor assembly is 6 feet (1829 mm) or less above grade.

2113.4.1 Anchorage. Two $^3/_{16}$-inch by 1-inch (4.8 mm by 25 mm) straps shall be embedded a minimum of 12 inches (305 mm) into the chimney. Straps shall be hooked around the outer bars and extend 6 inches (152 mm) beyond the bend. Each strap shall be fastened to a minimum of four floor joists with two $^1/_2$-inch (12.7 mm) bolts.

❖ The prescriptive requirements in this section are traditional for typical fireplaces and chimneys. More substantial anchorage may be required in areas of high seismicity, for large fireplaces or where the distance between floor and roof diaphragms is large.

2113.5 Corbeling. Masonry chimneys shall not be corbeled more than half of the chimney's wall thickness from a wall or foundation, nor shall a chimney be corbeled from a wall or foundation that is less than 12 inches (305 mm) in thickness unless it projects equally on each side of the wall, except that on the second *story* of a two-story *dwelling*, corbeling of chimneys on the exterior of the enclosing walls is permitted to equal the wall thickness. The projection of a single course shall not exceed one-half the unit height or one-third of the unit bed depth, whichever is less.

❖ A corbel is formed by projecting courses of masonry with the first or lowest course projecting out from the face of the wall and each successive course projecting out from the supporting course below. The angle of a corbel is limited so that the structural capacities of the masonry are not exceeded.

Figure 2113.5 illustrates the limitations of corbeling in an eccentrically loaded masonry chimney. The total corbeled projection cannot exceed 6 inches (152 mm). Where corbeling projects from the wall on one side only, the minimum allowable thickness of the wall is 12 inches (305 mm). The maximum allowable horizontal projection of an individual course of brick cannot exceed one-half the height and one-third the thickness of the masonry unit.

A chimney that projects equally on each side of the wall loads the wall concentrically. The requirement for a minimum wall thickness of 12 inches (305 mm) does not apply to this chimney configuration.

This section contains an exception to the limitations on corbel projection. In single-family homes, traditional chimneys originated in the second story and commonly served heating appliances in bedrooms, studies and sewing rooms. The exception allows corbeling to project from the exterior masonry wall a distance equivalent to the masonry wall thickness. This exception, however, does not affect the limitation on minimum wall thickness stated in the second sentence of this section. For example, a chimney in the second story of a two-story building cannot be corbeled at all if the wall is less than 12 inches (305 mm) thick and the corbeling is to project from one side only. If the corbeling is to project equally from both sides of the wall, the minimum wall thickness does not apply.

2113.6 Changes in dimension. The chimney wall or chimney flue lining shall not change in size or shape within 6 inches (152 mm) above or below where the chimney passes through floor components, ceiling components or roof components.

❖ At changes in shape or direction, masonry chimneys can have thinner walls, making them susceptible to leakage of water from the outside, and to hot spots and leakage of combustion gases from the inside. This is a fire hazard. The intent of this provision is to prohibit changes in shape and size near combustible construction. This section prohibits any change in dimension or direction of a masonry chimney from 6 inches (152 mm) below the combustible construction to 6 inches (152 mm) above it (see Figure 2113.6).

2113.7 Offsets. Where a masonry chimney is constructed with a fireclay flue liner surrounded by one wythe of masonry, the maximum offset shall be such that the centerline of the flue above the offset does not extend beyond the center of the chimney wall below the offset. Where the chimney offset is supported by masonry below the offset in an *approved* manner, the maximum offset limitations shall not apply. Each individual corbeled masonry course of the offset shall not exceed the projection limitations specified in Section 2113.5.

❖ An offset requires two changes in direction and causes two vertical sections of a chimney to be misaligned (offset) from each other. The intent of this section is to provide an upper portion of an offset that is structurally stable with respect to the lower portion.

The offset limitation does not apply when the inclined portion of the chimney and the portion above the offset are supported in an approved manner by underlying masonry construction.

2113.8 Additional load. Chimneys shall not support loads other than their own weight unless they are designed and constructed to support the additional load. Masonry chimneys are permitted to be constructed as part of the masonry walls or concrete walls of the building.

❖ Because the requirements for chimneys in Section 2113 are based on past performance, chimneys should not be used to support other loads unless they are specifically designed to do so.

2113.9 Termination. Chimneys shall extend at least 2 feet (610 mm) higher than any portion of the building within 10 feet (3048 mm), but shall not be less than 3 feet (914 mm) above the highest point where the chimney passes through the roof.

❖ Chimneys must be terminated well above adjacent portions of the building so that flue gases are exhausted away from combustible materials and for proper airflow through the chimney. The termination requirements are illustrated in Figure 2113.9. These requirements are applicable to chimneys serving masonry fireplaces or residential-type appliances.

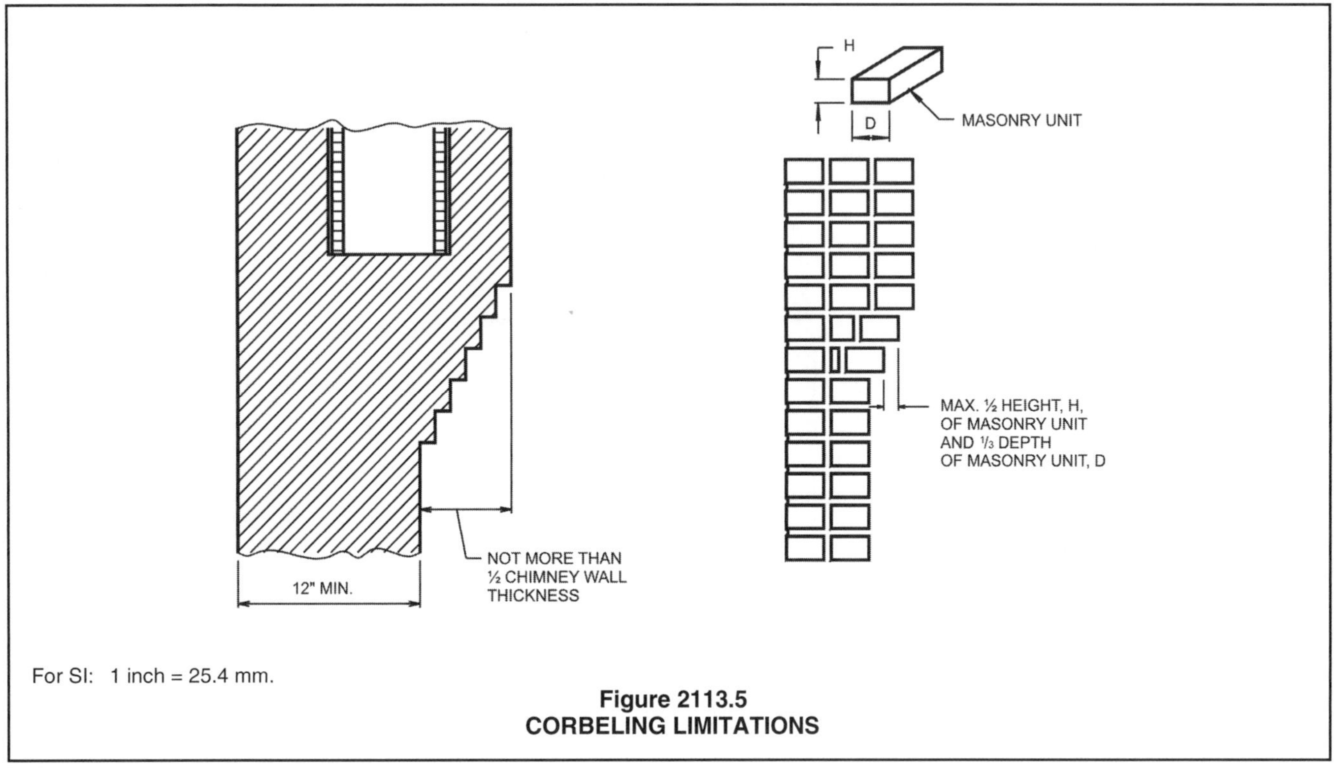

For SI: 1 inch = 25.4 mm.

Figure 2113.5
CORBELING LIMITATIONS

More restrictive termination requirements are specified for medium- and high-heat appliances.

2113.9.1 Chimney caps. Masonry chimneys shall have a concrete, metal or stone cap, sloped to shed water, a drip edge and a caulked bond break around any flue liners in accordance with ASTM C 1283.

❖ The section requires masonry chimneys to have caps and references ASTM C 1283, which covers the minimum requirements for installing a clay flue lining for residential concrete or masonry chimneys. A chimney cap protects the top of the masonry surrounding the flue. Although masonry chimneys typically have some type of weatherproof cap, this section specifically requires a chimney cap and prescribes the minimum criteria for its construction. The cap must be sloped to the outside, overhang the face of the masonry chimney and provide a drip edge. The prescribed caulking of the joint between the masonry and the flue serves as both a sealant and a bond break for any differential movement. The requirements for chimney caps are consistent with the referenced standard ASTM C 1283, *Standard Practice for Installing Clay Flue Lining,* and consistent with ASTM C 315, *Standard Specification for Clay Flue Liners and Chimney Pots* (see Figure 2113.9.1). See also Section 2113.9.3 for requirements for rain caps.

2113.9.2 Spark arrestors. Where a spark arrestor is installed on a masonry chimney, the spark arrestor shall meet all of the following requirements:

1. The net free area of the arrestor shall not be less than four times the net free area of the outlet of the chimney flue it serves.
2. The arrestor screen shall have heat and corrosion resistance equivalent to 19-gage galvanized steel or 24-gage stainless steel.
3. Openings shall not permit the passage of spheres having a diameter greater than $^1/_2$ inch (12.7 mm) nor block the passage of spheres having a diameter less than $^3/_8$ inch (9.5 mm).
4. The spark arrestor shall be accessible for cleaning and the screen or chimney cap shall be removable to allow for cleaning of the chimney flue.

❖ This section provides specifications for spark arrestors, if they are provided. Their use is not mandated by the model code, but they may be required by local regulations and owners or builders often install them.

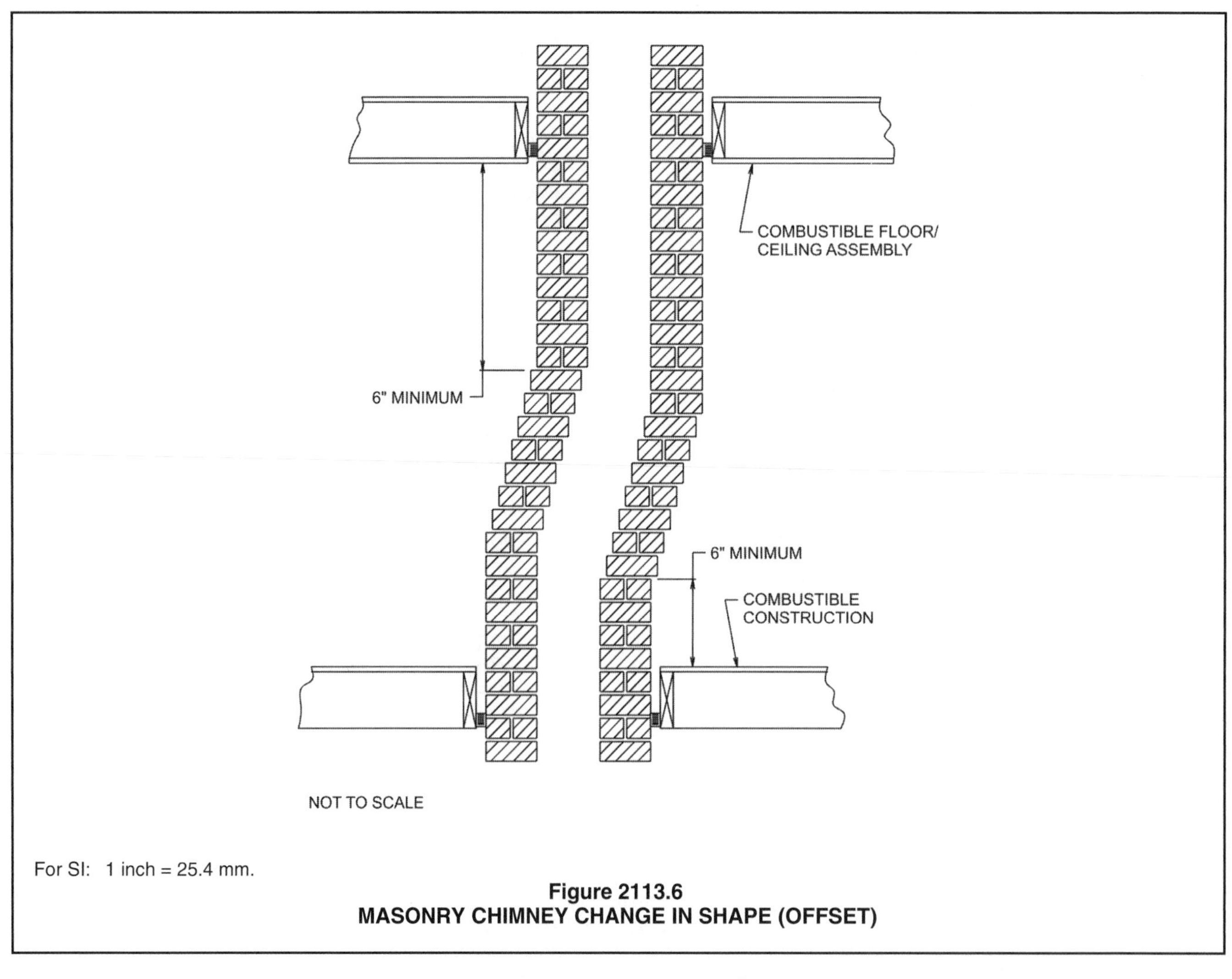

Figure 2113.6
MASONRY CHIMNEY CHANGE IN SHAPE (OFFSET)

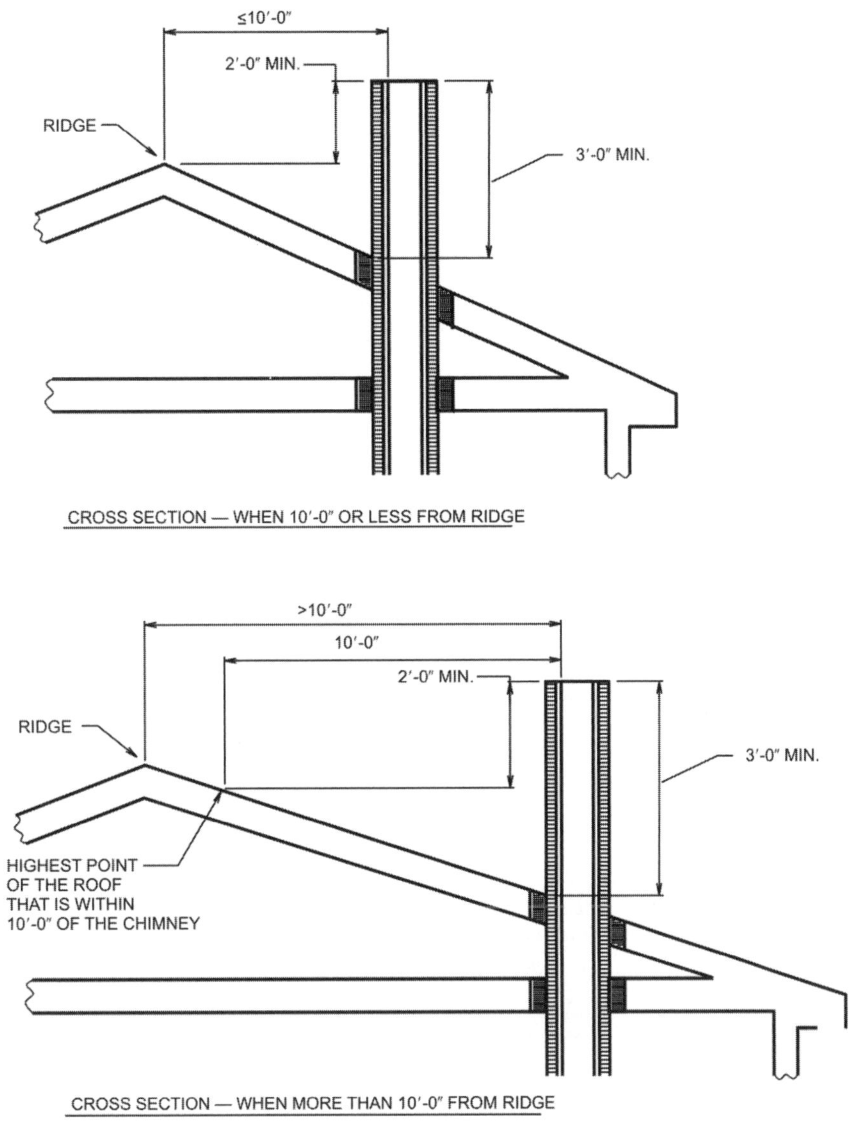

≤10′-0″
2′-0″ MIN.
RIDGE
3′-0″ MIN.
CROSS SECTION — WHEN 10′-0″ OR LESS FROM RIDGE
>10′-0″
10′-0″
2′-0″ MIN.
RIDGE
3′-0″ MIN.
HIGHEST POINT OF THE ROOF THAT IS WITHIN 10′-0″ OF THE CHIMNEY
CROSS SECTION — WHEN MORE THAN 10′-0″ FROM RIDGE

For SI: 1 foot = 304.8 mm.

Figure 2113.9
MINIMUM TERMINATION OF CHIMNEYS

FLUE
SEALANT
SLOPE ALL SIDES
FACE OF CHIMNEY
REQUIRED CHIMNEY CAP
RAIN CAP (OPTIONAL)
FLUE
SEALANT
CHIMNEY CAP
DRIP LEDGE

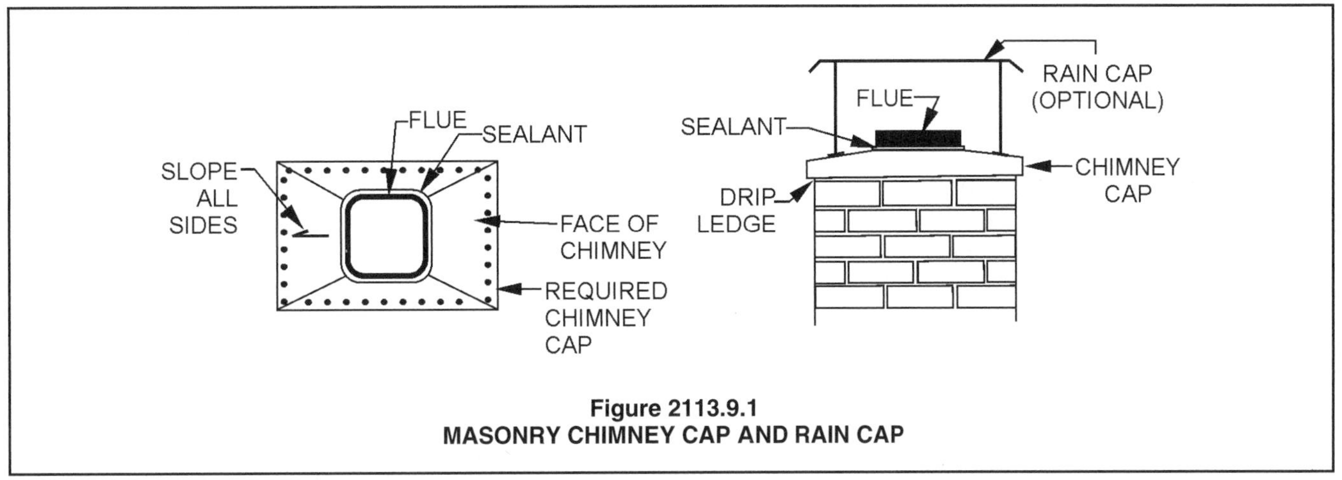

Figure 2113.9.1
MASONRY CHIMNEY CAP AND RAIN CAP

2113.9.3 Rain caps. Where a masonry or metal rain cap is installed on a masonry chimney, the net free area under the cap shall not be less than four times the net free area of the outlet of the chimney flue it serves.

❖ Rain caps are installed a prescribed distance above a chimney flue termination to limit the amount of rain entering the flue. Because an improperly designed chimney rain cap can reduce the draft of a chimney, the section requires the net free area under the rain cap to be at least four times the area of the outlet of the flue opening. Rain caps are also used to keep animals such as rodents and birds from nesting in the chimney or fireplace. As the language indicates, rain caps are not required by the code, but when they are installed they must have the minimum clearance above the flue termination to provide the minimum net free area required by the section (see Figure 2113.9.1).

2113.10 Wall thickness. Masonry chimney walls shall be constructed of concrete, solid masonry units or hollow masonry units grouted solid with not less than 4 inches (102 mm) nominal thickness.

❖ The minimum chimney wall thickness is necessary to achieve a thermal mass that will predictably control heat transmission through the walls of the chimney. This thickness is applicable to chimneys serving masonry fireplaces or residential-type appliances. More restrictive thicknesses are specified for medium- and high-heat appliances.

2113.10.1 Masonry veneer chimneys. Where masonry is used as veneer for a framed chimney, through flashing and weep holes shall be provided as required by Chapter 14.

❖ While the definition of "Masonry chimney" would not necessarily include a framed chimney with masonry veneer, this cross reference to Chapter 14 is provided as a precaution.

2113.11 Flue lining (material). Masonry chimneys shall be lined. The lining material shall be appropriate for the type of appliance connected, according to the terms of the appliance listing and the manufacturer's instructions.

❖ The liner forms the flue passageway and is the actual conduit for all products of combustion. The temperature and composition of flue gases will vary based on the type of appliance being vented. The flue lining is required to withstand exposure to high temperatures and corrosive chemicals. It protects the masonry construction of the chimney walls and allows the chimney to be gas tight. A mismatch between an appliance and the type of chimney can result in a hazardous operating condition.

2113.11.1 Residential-type appliances (general). Flue lining systems shall comply with one of the following:

1. Clay flue lining complying with the requirements of ASTM C 315.
2. *Listed* chimney lining systems complying with UL 1777.
3. Factory-built chimneys or chimney units *listed* for installation within masonry chimneys.
4. Other *approved* materials that will resist corrosion, erosion, softening or cracking from flue gases and condensate at temperatures up to 1,800°F (982°C).

❖ This section requires that the lining for residential-type appliances comply with ASTM C 315 or other approved materials that are capable of resisting degradation from flue gas. Chimney liner systems that are tested and labeled by an approved agency in accordance with UL 1777 are also permitted.

2113.11.1.1 Flue linings for specific appliances. Flue linings other than those covered in Section 2113.11.1 intended for use with specific appliances shall comply with Sections 2113.11.1.2 through 2113.11.1.4 and Sections 2113.11.2 and 2113.11.3.

❖ This section identifies flue lining materials to be used for specific appliances.

2113.11.1.2 Gas appliances. Flue lining systems for gas appliances shall be in accordance with the *International Fuel Gas Code*.

❖ The IFGC must be used to determine appropriate flue-lining systems for gas appliances.

2113.11.1.3 Pellet fuel-burning appliances. Flue lining and vent systems for use in masonry chimneys with pellet fuel-burning appliances shall be limited to flue lining systems complying with Section 2113.11.1 and pellet vents *listed* for installation within masonry chimneys (see Section 2113.11.1.5 for marking).

❖ Flue lining and vent systems in masonry chimneys of pellet fuel-burning appliances can either conform to Section 2113.11.1 or be an approved listed system.

2113.11.1.4 Oil-fired appliances approved for use with L-vent. Flue lining and vent systems for use in masonry chimneys with oil-fired appliances *approved* for use with Type L vent shall be limited to flue lining systems complying with Section 2113.11.1 and *listed* chimney liners complying with UL 641 (see Section 2113.11.1.5 for marking).

❖ Flue lining and vent systems in masonry chimneys of oil-fired appliances with Type L vents can either conform to Section 2113.11.1 or be an approved listed system (UL 641).

2113.11.1.5 Notice of usage. When a flue is relined with a material not complying with Section 2113.11.1, the chimney shall be plainly and permanently identified by a *label* attached to a wall, ceiling or other conspicuous location adjacent to where the connector enters the chimney. The *label* shall include the following message or equivalent language: "This chimney is for use only with (type or category of appliance) that burns (type of fuel). Do not connect other types of appliances."

❖ Clearly displayed information on the appropriate use and fuel for appliances is required to protect against the use of improper types of fuel that could cause fire.

2113.11.2 Concrete and masonry chimneys for medium-heat appliances.

❖ This section establishes requirements for masonry chimneys serving medium-heat appliances, including the chimney materials, lining, termination height and proper clearances to combustibles.

2113.11.2.1 General. Concrete and masonry chimneys for medium-heat appliances shall comply with Sections 2113.1 through 2113.5.

❖ Chimneys serving a medium-heat appliance must comply with the general chimney requirements in Sections 2113.1 through 2113.5. These include minimum requirements for footings and foundations; seismic reinforcement; anchorage and corbeling.

2113.11.2.2 Construction. Chimneys for medium-heat appliances shall be constructed of solid masonry units or of concrete with walls a minimum of 8 inches (203 mm) thick, or with stone masonry a minimum of 12 inches (305 mm) thick.

❖ Masonry chimneys for medium-heat appliances must be constructed of concrete in accordance with Chapter 19, or of solid masonry units in accordance with this chapter. Chimneys constructed of solid masonry units or reinforced concrete are required to have a minimum wall thickness of 8 inches (203 mm). The minimum thickness requirement is necessary to achieve a thermal mass that will predictably control heat transmission through the walls of the chimney. Chimneys constructed of rubble stone masonry (irregular or roughly shaped stones) are required to have a wall thickness of no less than 12 inches (305 mm). The rate of heat transmission through stone walls is not predictable because of the varying thickness of the individual stones and the nonuniform mortar joints between the stones. For these reasons, stone wall chimneys are required to be thicker to provide a reasonable margin of safety.

2113.11.2.3 Lining. Concrete and masonry chimneys shall be lined with an *approved* medium-duty refractory brick a minimum of $4^1/_2$ inches (114 mm) thick laid on the $4^1/_2$-inch bed (114 mm) in an *approved* medium-duty refractory mortar. The lining shall start 2 feet (610 mm) or more below the lowest chimney connector entrance. Chimneys terminating 25 feet (7620 mm) or less above a chimney connector entrance shall be lined to the top.

❖ A medium-heat appliance produces flue-gas temperatures up to 2,000°F (1093°C). The chimney lining reduces heat transmission to the walls of the chimney and contains flue gases within a continuous duct until they are away from the building. This section requires at least a medium-duty refractory brick. A $4^1/_2$-inch (114 mm) medium-duty refractory brick lining tested and classified in accordance with ASTM C 27 meets this criterion. Each course of the refractory brick liner is required to be installed on a full bed of refractory mortar to form a structurally stable, continuous, gas-tight liner.

Figure 2113.11.2.3 depicts a masonry chimney serving a medium-heat appliance. The liner extends from 2 feet (610 mm) below the lowest inlet to the top of the chimney, which is located less than 25 feet (7620 mm) above the highest inlet.

A chimney liner is required in all portions of the chimney that are exposed to flue gases. The liner is required to extend below the lowest inlet to provide protection to the masonry from flue gases, which can deteriorate the masonry and the mortar joints.

2113.11.2.4 Multiple passageway. Concrete and masonry chimneys containing more than one passageway shall have the liners separated by a minimum 4-inch-thick (102 mm) concrete or solid masonry wall.

❖ When a chimney requires multiple passageways, a 4-inch (102 mm) partition of solid masonry is required between them to act as a barrier and to enhance structural integrity.

2113.11.2.5 Termination height. Concrete and masonry chimneys for medium-heat appliances shall extend a minimum of 10 feet (3048 mm) higher than any portion of any building within 25 feet (7620 mm).

❖ Chimneys for medium-heat appliances are required to extend at least 10 feet (3048 mm) above the highest portion of the building within 25 feet (7620 mm) horizontally. This is intended to provide an acceptable height to carry away the flue gases safely and to provide adequate clearance to the roof and surrounding structures to allow any burning embers to extinguish before landing.

2113.11.2.6 Clearance. A minimum clearance of 4 inches (102 mm) shall be provided between the exterior surfaces of a concrete or masonry chimney for medium-heat appliances and combustible material.

❖ A 4-inch (102 mm) minimum airspace clearance to combustibles is required for a medium-heat masonry chimney. This large clearance is needed because of the higher temperatures produced by medium-heat appliances.

2113.11.3 Concrete and masonry chimneys for high-heat appliances.

❖ This section establishes the requirements for concrete and masonry chimneys serving high-heat appliances, including chimney materials, lining, termination height and proper clearances to combustibles.

2113.11.3.1 General. Concrete and masonry chimneys for high-heat appliances shall comply with Sections 2113.1 through 2113.5.

❖ Chimneys serving high-heat appliances are required to comply with the general chimney requirements in Sections 2113.1 through 2113.5, including minimum provisions for footings and foundations; seismic reinforcement; anchorage and corbeling.

2113.11.3.2 Construction. Chimneys for high-heat appliances shall be constructed with double walls of solid masonry units or of concrete, each wall to be a minimum of 8 inches

(203 mm) thick with a minimum airspace of 2 inches (51 mm) between the walls.

❖ The flue gases produced by a high-heat appliance can have temperatures above 2,000°F (1,093°C). To prevent fire, such chimneys must be enclosed by two wythes of masonry with an airspace between them. Thus, the chimney liner is required to be protected by two solid masonry unit walls or reinforced concrete walls, each a minimum of 8 inches (204 mm) thick with an airspace between them of at least 2 inches (51 mm). The airspace is intended to provide thermal insulation and to allow for thermal expansion of the chimney components.

2113.11.3.3 Lining. The inside of the interior wall shall be lined with an *approved* high-duty refractory brick, a minimum of $4^1/_2$ inches (114 mm) thick laid on the $4^1/_2$-inch bed (114 mm) in an *approved* high-duty refractory mortar. The lining shall start at the base of the chimney and extend continuously to the top.

❖ High-heat appliances can produce flue gases with temperatures exceeding 2,000°F (1,093°C). The chimney lining reduces heat transmission to the walls of the chimney and contains flue gases within a continuous passageway until they are outside the building. This section permits only a high-duty refractory brick having a minimum thickness of $4^1/_2$ inches (114 mm) to be used in a masonry chimney serving a high-heat appliance. An approved high-duty refractory lining usually consists of brick tested and is classified in accordance with ASTM C 27. To provide a structurally stable and continuous gas-tight liner, each course of the refractory brick liner must be laid on a full bed of refractory mortar approved for high-heat appliances.

2113.11.3.4 Termination height. Concrete and masonry chimneys for high-heat appliances shall extend a minimum of 20 feet (6096 mm) higher than any portion of any building within 50 feet (15 240 mm).

❖ Chimneys for high-heat appliances must terminate at least 20 feet (6096 mm) above the highest portion of the building within 50 feet (15 240 mm) horizontally. This allows flue gases to be safely carried away and provides enough clearance to allow any burning embers to extinguish before landing on combustibles.

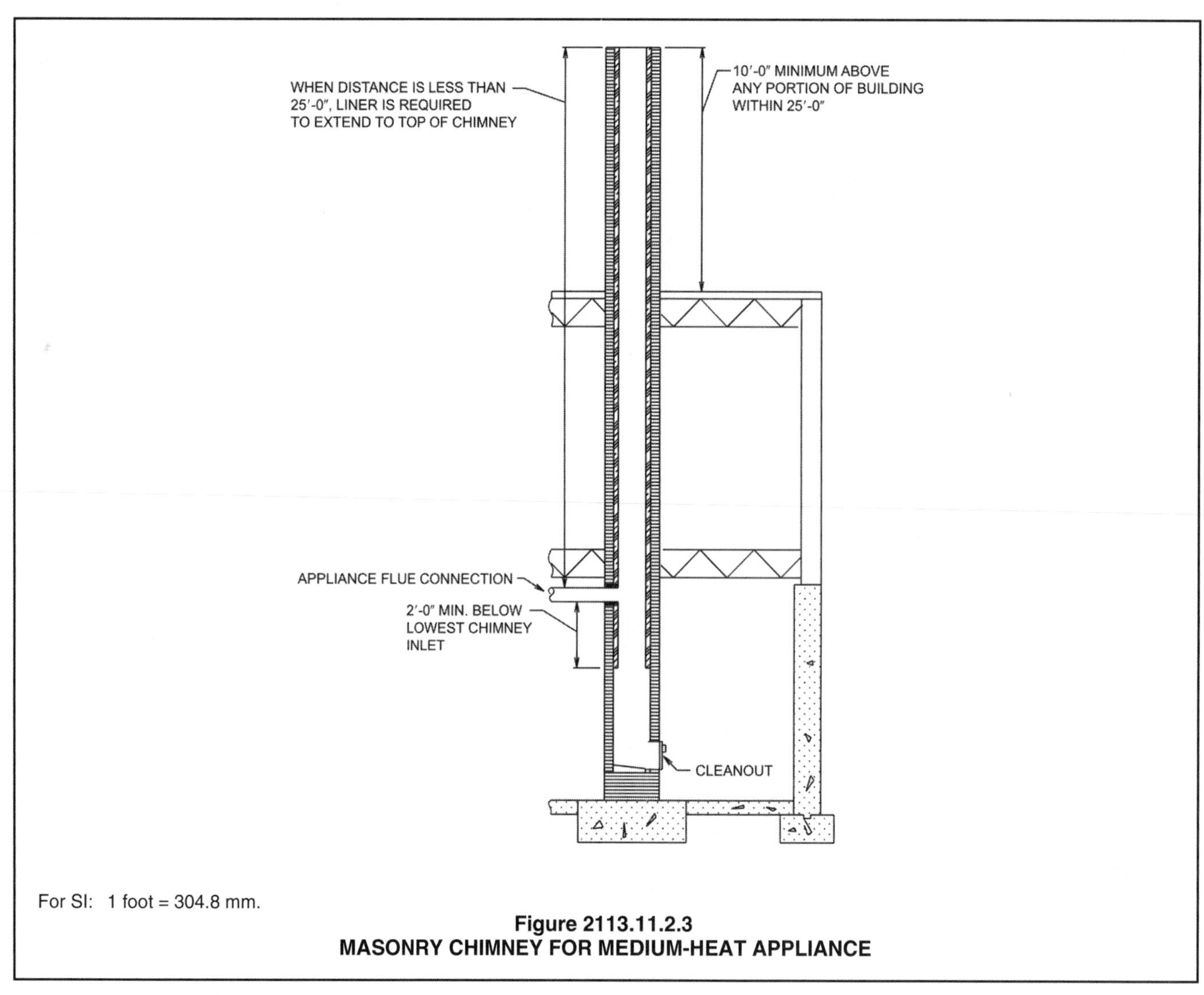

Figure 2113.11.2.3
MASONRY CHIMNEY FOR MEDIUM-HEAT APPLIANCE

2113.11.3.5 Clearance. Concrete and masonry chimneys for high-heat appliances shall have *approved* clearance from buildings and structures to prevent overheating combustible materials, permit inspection and maintenance operations on the chimney and prevent danger of burns to persons.

❖ The extremely high-temperature flue gas produced by a high-heat appliance is a major fire safety concern, and requiring sufficient clearance from buildings protects nearby combustible material, permits inspection and maintenance of the chimney and minimizes danger to persons.

2113.12 Clay flue lining (installation). Clay flue liners shall be installed in accordance with ASTM C 1283 and extend from a point not less than 8 inches (203 mm) below the lowest inlet or, in the case of fireplaces, from the top of the smoke chamber to a point above the enclosing walls. The lining shall be carried up vertically, with a maximum slope no greater than 30 degrees (0.52 rad) from the vertical.

Clay flue liners shall be laid in medium-duty nonwater-soluble refractory mortar conforming to ASTM C 199 with tight mortar joints left smooth on the inside and installed to maintain an air space or insulation not to exceed the thickness of the flue liner separating the flue liners from the interior face of the chimney masonry walls. Flue lining shall be supported on all sides. Only enough mortar shall be placed to make the joint and hold the liners in position.

❖ Section 2113.11.1 lists clay flue liners as an acceptable flue-lining system for masonry chimneys serving residential-type appliances, as well as masonry fireplaces. This section provides the installation requirements for clay flue liners. The liner forms the flue passageway and is the actual conduit of all products of combustion. It must withstand exposure to high temperatures and corrosive chemicals from the flue gases. The chimney lining protects the masonry construction of the chimney walls and allows the chimney to be constructed gas tight.

Installation must comply with ASTM C 1283. The liner is required to extend below the lowest inlet to provide protection to the masonry from flue gases, which can deteriorate the masonry and mortar joints.

Refractory mortar must comply with ASTM C 199. Since they are exposed to the weather, nonwater-soluble mortar must be used in the joints of chimneys so that it does not wash out of the joints.

2113.13 Additional requirements.

❖ This section contains requirements for listed materials to be used as flue linings and for spaces surrounding chimney lining systems.

2113.13.1 Listed materials. *Listed* materials used as flue linings shall be installed in accordance with the terms of their listings and the manufacturer's instructions.

❖ Listed materials for flue linings must be installed in accordance with the manufacturer's instructions. Such instructions are not listed in the code since they vary with each system.

2113.13.2 Space around lining. The space surrounding a chimney lining system or vent installed within a masonry chimney shall not be used to vent any other appliance.

Exception: This shall not prevent the installation of a separate flue lining in accordance with the manufacturer's instructions.

❖ Where a listed flue lining system, such as a flexible metallic pipe, is installed in a chimney flue, an annular space can be created. This annular space will have an irregular shape that is not conducive to the flow of flue gas. In addition, the annular space surrounding a chimney lining system provides a thermal buffer between the flue and the surrounding masonry. Using it for another purpose is prohibited.

2113.14 Multiple flues. When two or more flues are located in the same chimney, masonry wythes shall be built between adjacent flue linings. The masonry wythes shall be at least 4 inches (102 mm) thick and bonded into the walls of the chimney.

Exception: When venting only one appliance, two flues are permitted to adjoin each other in the same chimney with only the flue lining separation between them. The joints of the adjacent flue linings shall be staggered at least 4 inches (102 mm).

❖ A single masonry chimney can contain any number of flue-gas passageways. When more than two flues are contained in the same chimney, the flue liners are required to be subdivided by masonry wythes into groups of not more than two (see Figure 2113.14). The purpose of the masonry wythes is to unify the chimney structurally and to isolate pairs of flues serving dissimilar appliances.

2113.15 Flue area (appliance). Chimney flues shall not be smaller in area than the area of the connector from the appliance. Chimney flues connected to more than one appliance shall not be less than the area of the largest connector plus 50 percent of the areas of additional chimney connectors.

Exceptions:

1. Chimney flues serving oil-fired appliances sized in accordance with NFPA 31.
2. Chimney flues serving gas-fired appliances sized in accordance with the *International Fuel Gas Code*.

❖ Flues are sized to match the opening at the top of the appliance so that appliances function properly. Smaller flues are not permitted since they may constrict the passage of gases out from the appliance.

2113.16 Flue area (masonry fireplace). Flue sizing for chimneys serving fireplaces shall be in accordance with Section 2113.16.1 or 2113.16.2.

❖ This section provides requirements for the net cross-sectional area of the flue and throat between the firebox and the smoke chamber. Airflow through the fireplace is affected by the dimensions of the firebox opening, the shape and cross-sectional area of the flue and the height of the chimney (see Code Figure 2113.16). For proper fireplace operation, the required cross-sectional area is about one-tenth that of the fireplace opening. This ratio may vary somewhat with the height of the chimney and the configuration (round or rectangular) of the chimney flue.

This section states that either of the two options may be used to determine the flue area. Section 2113.16.1 prescribes the minimum flue area based on the fireplace opening alone, while Section 2113.16.2 prescribes it based on the height of the fireplace, the fireplace opening area and flue type.

FIGURE 2113.16. See page 21-41.

❖ This figure prescribes minimum flue sizes as a function of chimney height and the fireplace opening area. For example, for a 20-foot-high (6096 mm) chimney and a fireplace opening area of 2,000 square inches (1.3 m^2), the minimum cross-sectional area is 164 square inches (1.05 m^2) for a round flue or 190 square inches (0.12 m^2) for a square or rectangular flue. Using Table 2113.16(1), that minimum required cross-sectional area can be met by a circular flue with a diameter of 15 inches (381 mm). Using Table 2113.16(2), the minimum cross-sectional area would require a rectangular flue with a nominal outside dimension of 16 by 20 inches (405 by 508 mm).

2113.16.1 Minimum area. Round chimney flues shall have a minimum net cross-sectional area of at least $^1/_{12}$ of the fireplace opening. Square chimney flues shall have a minimum net cross-sectional area of at least $^1/_{10}$ of the fireplace opening. Rectangular chimney flues with an aspect ratio less than 2 to 1 shall have a minimum net cross-sectional area of at least $^1/_{10}$ of the fireplace opening. Rectangular chimney flues with an aspect ratio of 2 to 1 or more shall have a minimum net cross-sectional area of at least $^1/_8$ of the fireplace opening.

❖ This section prescribes the minimum flue area based on the fireplace opening and the shape of the flue. For a given size fireplace opening, circular chimney flues are permitted to have a smaller area than rectangular flues. Wide, narrow flues require a larger area for adequate airflow.

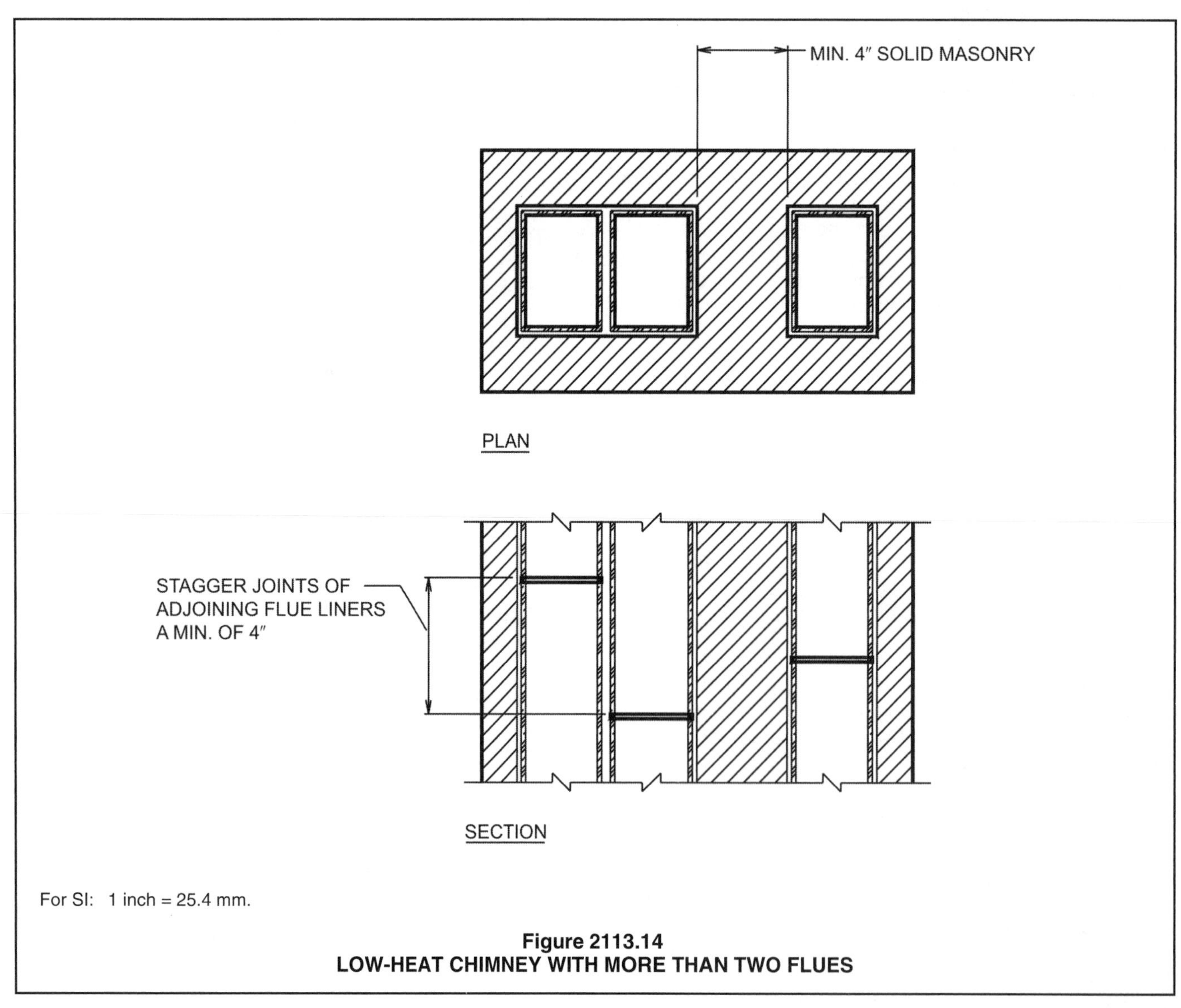

Figure 2113.14
LOW-HEAT CHIMNEY WITH MORE THAN TWO FLUES

2113.16.2 Determination of minimum area. The minimum net cross-sectional area of the flue shall be determined in accordance with Figure 2113.16. A flue size providing at least the equivalent net cross-sectional area shall be used. Cross-sectional areas of clay flue linings are as provided in Tables 2113.16(1) and 2113.16(2) or as provided by the manufacturer or as measured in the field. The height of the chimney shall be measured from the firebox floor to the top of the chimney flue.

❖ This section provides an alternative method to determine flue size based on the height of the chimney, the fireplace opening area and the flue type [see commentary, Tables 2113.16(1) and 2113.16(2), and Figure 2113.16].

TABLE 2113.16(1). See next column.

❖ This table gives the areas of circular flues of standard sizes so that users of the code can easily determine a flue size complying with Figure 2113.16. The flue areas shown in this table are determined by the equation $A = \pi \times (d/2)^2$, where A is the cross-sectional area, d is the inside diameter of the flue and π is approximated as 3.14. Alternatively, a minimum diameter can be calculated as $d = [(A/\pi)^{1/2}]/4$.

TABLE 2113.16(1)
NET CROSS-SECTIONAL AREA OF ROUND FLUE SIZES[a]

FLUE SIZE, INSIDE DIAMETER (inches)	CROSS-SECTIONAL AREA (square inches)
6	28
7	38
8	50
10	78
$10^3/_4$	90
12	113
15	176
18	254

For SI: 1 inch = 25.4 mm, 1 square inch = 645.16 mm^2.
a. Flue sizes are based on ASTM C 315.

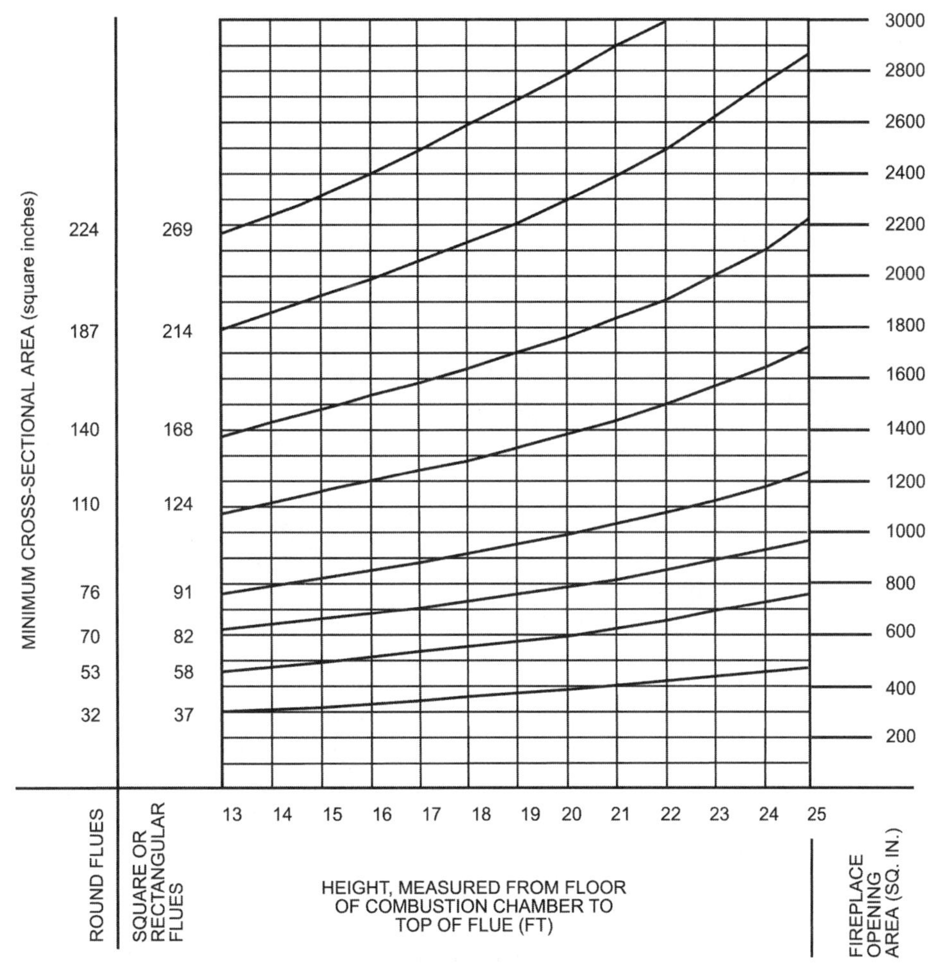

For SI: 1 inch = 25.4 mm, 1 square inch = 645 mm^2

FIGURE 2113.16
FLUE SIZES FOR MASONRY CHIMNEYS

**TABLE 2113.16(2)
NET CROSS-SECTIONAL AREA OF SQUARE AND RECTANGULAR FLUE SIZES**

FLUE SIZE, OUTSIDE NOMINAL DIMENSIONS (inches)	CROSS-SECTIONAL AREA (square inches)
4.5 × 8.5	23
4.5 × 13	34
8 × 8	42
8.5 × 8.5	49
8 × 12	67
8.5 × 13	76
12 × 12	102
8.5 × 18	101
13 ×13	127
12 × 16	131
13 × 18	173
16 × 16	181
16 × 20	222
18 × 18	233
20 × 20	298
20 × 24	335
24 × 24	431

For SI: 1 inch = 25.4 mm, 1 square inch = 645.16 mm^2.

❖ Clay flue liners are designated according to their nominal outside dimensions. This table gives the net cross-sectional area of standard size square and rectangular flues so that code users can readily determine a flue size that provides the minimum required area.

2113.17 Inlet. Inlets to masonry chimneys shall enter from the side. Inlets shall have a thimble of fireclay, rigid refractory material or metal that will prevent the connector from pulling out of the inlet or from extending beyond the wall of the liner.

❖ The inlet must be noncombustible and strong enough not to be pulled out.

2113.18 Masonry chimney cleanout openings. Cleanout openings shall be provided within 6 inches (152 mm) of the base of each flue within every masonry chimney. The upper edge of the cleanout shall be located at least 6 inches (152 mm) below the lowest chimney inlet opening. The height of the opening shall be at least 6 inches (152 mm). The cleanout shall be provided with a noncombustible cover.

Exception: Chimney flues serving masonry fireplaces, where cleaning is possible through the fireplace opening.

❖ This section requires a cleanout to be installed in a chimney to facilitate cleaning and inspection. A fireplace inherently provides access to its chimney through the firebox, throat and smoke chamber. The cleanout cover and opening frame are to be of an approved noncombustible material, such as cast iron or precast concrete, and must be arranged to remain tightly closed. The requirement for placing the cleanout at least 6 inches (152 mm) below the lowest connection to the chimney is intended to minimize the possibility of combustion products exiting the chimney through the cleanout.

2113.19 Chimney clearances. Any portion of a masonry chimney located in the interior of the building or within the *exterior wall* of the building shall have a minimum airspace clearance to combustibles of 2 inches (51 mm). Chimneys located entirely outside the *exterior walls* of the building, including chimneys that pass through the soffit or cornice, shall have a minimum airspace clearance of 1 inch (25 mm). The airspace shall not be filled, except to provide fireblocking in accordance with Section 2113.20.

Exceptions:

1. Masonry chimneys equipped with a chimney lining system *listed* and labeled for use in chimneys in contact with combustibles in accordance with UL 1777, and installed in accordance with the manufacturer's instructions, are permitted to have combustible material in contact with their exterior surfaces.
2. Where masonry chimneys are constructed as part of masonry or concrete walls, combustible materials shall not be in contact with the masonry or concrete wall less than 12 inches (305 mm) from the inside surface of the nearest flue lining.
3. Exposed combustible *trim* and the edges of sheathing materials, such as wood siding, are permitted to abut the masonry chimney sidewalls, in accordance with Figure 2113.19, provided such combustible *trim* or sheathing is a minimum of 12 inches (305 mm) from the inside surface of the nearest flue lining. Combustible material and *trim* shall not overlap the corners of the chimney by more than 1 inch (25 mm).

❖ Clearance between the external surfaces of the masonry chimney and all combustible materials must be maintained. The intent of this section is to require a 2-inch (51 mm) minimum airspace clearance between combustibles and surfaces of all chimneys located in the interior of a building or within an exterior wall assembly. A 1-inch (25 mm) airspace clearance is allowed only where the chimney is located entirely outside of the building.

If any portion of a chimney is located in an exterior wall, that chimney must be considered as an interior chimney and must have a 2-inch (51 mm) minimum airspace clearance. The 1-inch (25 mm) clearance is allowed because the exterior surface of an outdoor chimney can dissipate heat. An outdoor chimney is exposed to outdoor ambient temperatures, allowing it to operate with cooler surface temperatures. Like all airspace clearances, the 1-inch (25 mm) airspace clearance is not permitted to be filled with any material except the noncombustible material necessary for fireblocking. The required clearance for exterior chimneys applies to all combustible materials, including sheathing, siding, insulation, framing and trim.

The exception recognizes a variety of chimney liners that are tested and labeled in accordance with UL 1777.

UL 1777 covers metallic and nonmetallic chimney liners intended for field installation into new and existing masonry chimneys used for natural draft venting of gas, oil and solid fuel-burning appliances having maximum continuous flue-gas temperatures not exceeding 1,000°F (538°C). Some lining systems are labeled for reduced clearance applications and could allow for the construction or rehabilitation of chimneys in contact with combustibles, without compromising the safety normally provided by the code-prescribed airspace clearances.

Chimney liner systems are typically metal or poured-in-place concrete and incorporate insulation to retard the transfer of heat to the surrounding masonry walls of the chimney.

This figure clarifies Exception 3 to the clearance requirements for masonry chimneys. The edge abutment of combustible sheathing or trim where there is an adequate thickness of masonry is a long-standing practice that is considered safe, provided the minimum clearance to the flue lining is maintained.

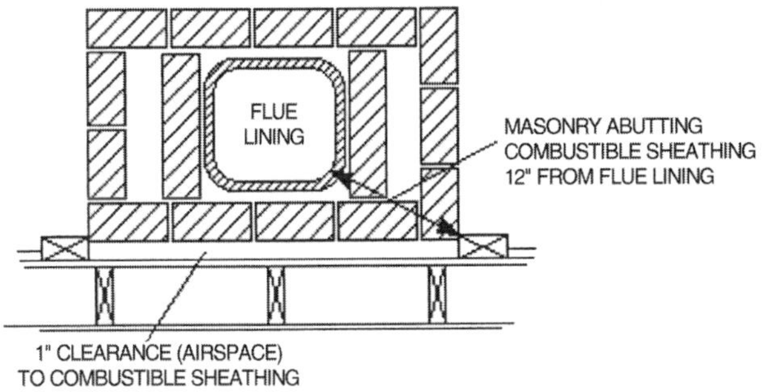

FIGURE 2113.19
ILLUSTRATION OF EXCEPTION THREE CHIMNEY CLEARANCE PROVISION

2113.20 Chimney fireblocking. All spaces between chimneys and floors and ceilings through which chimneys pass shall be fireblocked with noncombustible material securely fastened in place. The fireblocking of spaces between wood joists, beams or headers shall be self-supporting or be placed on strips of metal or metal lath laid across the spaces between combustible material and the chimney.

❖ Fireblocking is required to prevent the travel of flames, smoke and hot gases to other areas of the building through the gaps between the chimney and the floor or ceiling assemblies [see Figures 2113.20(1) and (2)].

Bibliography

The following resource materials are referenced in this chapter or are relevant to the subject matter addressed in this chapter.

ACI 318-11, *Building Code Requirements for Structural Concrete*. Farmington Hills, MI: American Concrete Institute, 2011.

ANSI A41.1-53, *American Standard Building Code Requirements for Masonry*. New York: American National Standards Institute, 1953.

ANSI A108.1A-99, *Glazed Wall Tile, Ceramic Mosaic Tile, Quarry Tile and Paver Tile Installed with Portland Cement Mortar*. New York: American National Standards Institute, 1999.

ANSI A108.1B-99, *Glazed Wall Tile, Ceramic Mosaic Tile, Quarry Tile and Paver Tile Installed with Portland Cement Mortar*. New York: American National Standards Institute, 1999.

ANSI A108.4-99, *Ceramic Tile Installed with Organic Adhesives or Water Cleanable Tile Setting Epoxy Adhesive*. New York: American National Standards Institute, 1999.

ANSI A108.5-99, *Ceramic Tile Installed with Dry-set Portland Cement Mortar or Latex Portland Cement Mortar*. New York: American National Standards Institute, 1999.

ANSI A108.6-99, *Ceramic Tile Installed with Chemical-resistant, Water-cleanable Tile Setting and Grouting Epoxy*. New York: American National Standards Institute, 1999.

ANSI A108.8-99, *Ceramic Tile Installed with Chemical-resistant Furan Mortar and Grout*. New York: American National Standards Institute, 1999.

ANSI A108.9-99, *Ceramic Tile Installed with Modified Epoxy-emulsion Mortar/Grout*. New York: American National Standards Institute, 1999.

ANSI A108.10-99, *Installation of Grout in Tilework*. New York: American National Standards Institute, 1999.

ANSI A118.1-99, *Dry-set Portland Cement Mortar*. New York: American National Standards Institute, 1999.

ANSI A118.3-99, *Chemical-resistant, Water-cleanable Tile Setting and Grouting Epoxy, and Water-cleanable Tile Setting Epoxy Adhesive*. New York: American National Standards Institute, 1999.

ANSI A118.4-99, *Latex Portland Cement Mortar*. New York: American National Standards Institute, 1999.

ANSI A118.5-99, *Chemical-resistant Furan*. New York: American National Standards Institute, 1999.

ANSI A118.6-99, *Ceramic Tile Grouts*. New York: American National Standards Institute, 1999.

ANSI A118.8-99, *Modified Epoxy-emulsion Mortar/ Grout*. New York: American National Standards Institute, 1999.

ANSI A136.1-99, *Organic Adhesives for Installation of Ceramic Tile*. New York: American National Standards Institute, 1999.

ANSI A137.1-88, *Ceramic Tile*. New York: American National Standards Institute, 1988.

ASCE 7-10, *Minimum Design Loads for Buildings and Other Structures*. Reston, VA: American Society of Civil Engineers, 2010.

ASTM A 36/A 36-08, *Specification for Structural Steel*. West Conshohocken, PA: ASTM International, 2005.

ASTM A 153-05, *Specification for Zinc Coating (Hot Dip) on Iron and Steel Hardware*. West Conshohocken, PA: ASTM International, 2005.

ASTM A 615-09a, *Specification for Deformed and Plain Billet-steel Bars for Concrete Reinforcement*. West Conshohocken, PA: ASTM International, 2004.

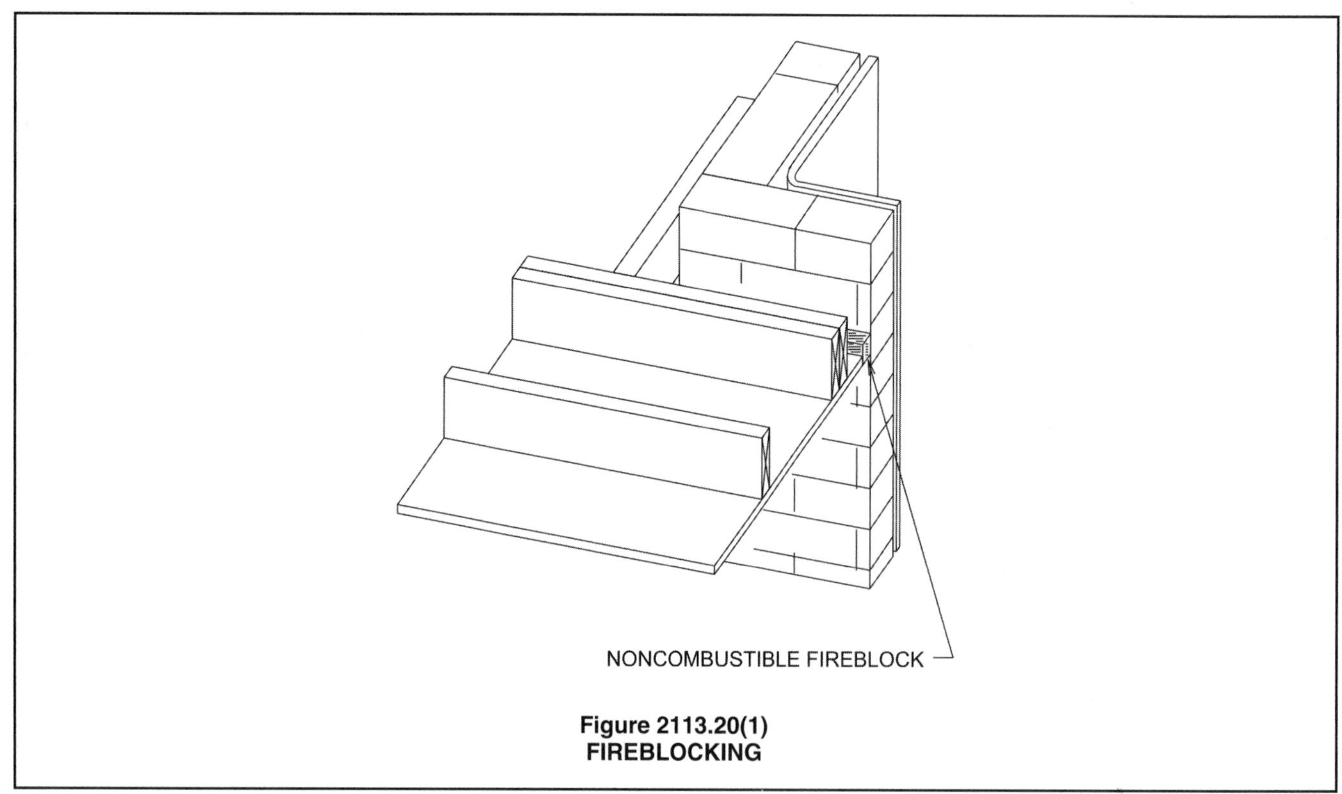

Figure 2113.20(1)
FIREBLOCKING

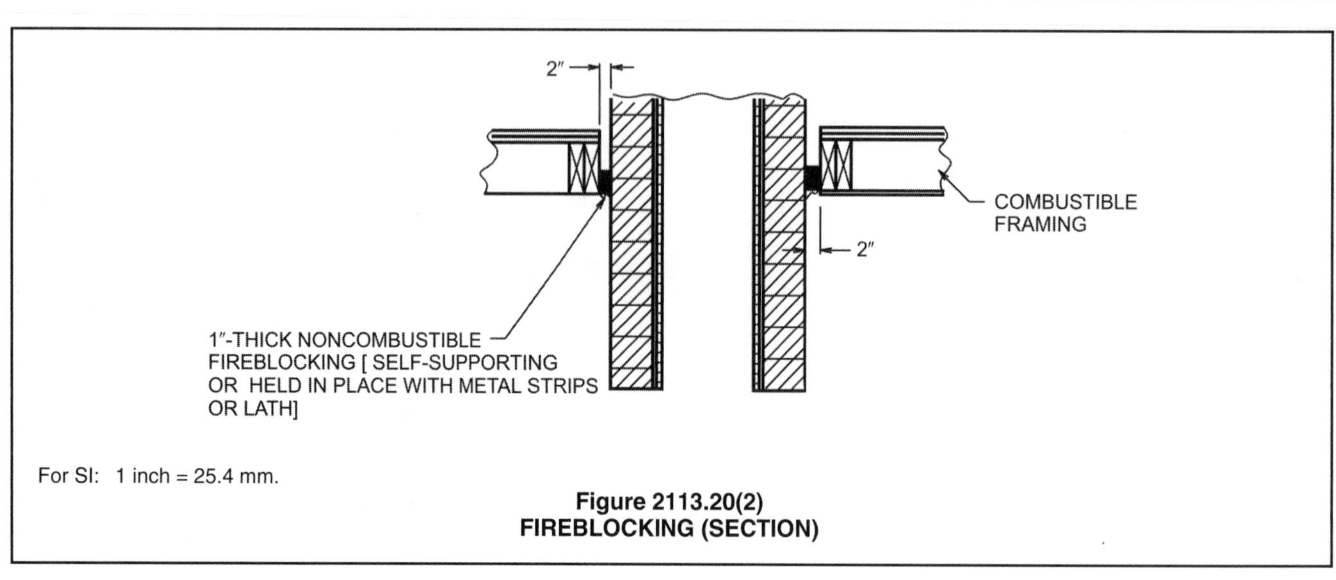

Figure 2113.20(2)
FIREBLOCKING (SECTION)

ASTM A 706/A 706M-09a, *Specification for Low-alloy Steel Deformed Bars for Concrete Reinforcement*. West Conshohocken, PA: ASTM International, 2004.

ASTM A 996-09, *Specification for Rail Steel and Axel-steel Deformed Bars for Concrete Reinforcement*. West Conshohocken, PA: ASTM International, 2006.

ASTM C 5-03, *Specification for Quicklime for Structural Purposes*. West Conshohocken, PA: ASTM International, 2003.

ASTM C 27-98 (2008), *Standard Classification of Fire-clay and High-alumina Refractory Brick*. West Conshohocken, PA: ASTM International, 2008.

ASTM C 34-03, *Specification for Structural Clay Load-bearing Wall Tile*. West Conshohocken, PA: ASTM International, 2003.

ASTM C 55-06e01, *Specification for Concrete Building Brick*. West Conshohocken, PA: ASTM International, 2006.

ASTM C 56-05, *Specification for Structural Clay Non-load-bearing Tile*. West Conshohocken, PA: ASTM International, 2005.

ASTM C 62-08, *Specification for Building Brick (Solid Masonry Units Made from Clay or Shale)*. West Conshohocken, PA: ASTM International, 2008.

ASTM C 67-08, *Test Methods of Sampling and Testing Brick and Structural Clay Tile*. West Conshohocken, PA: ASTM International, 2008.

ASTM C 73-99a (2005), *Specification for Calcium Silicate Face Brick (Sand Lime Brick)*. West Conshohocken, PA: ASTM International, 2005.

ASTM C 90-08, *Specification for Load-bearing Concrete Masonry Units*. West Conshohocken, PA: ASTM International, 2008.

ASTM C 91-05, *Standard Specification for Masonry Cement*. West Conshohocken, PA: ASTM International, 2005.

ASTM C 126-99 (2005), *Specification for Ceramic Glazed Structural Clay Facing Tile, Facing Brick and Solid Masonry Units*. West Conshohocken, PA: ASTM International, 2005.

ASTM C 140-07, *Methods of Sampling and Testing Concrete Masonry Units*. West Conshohocken, PA: ASTM International, 2007.

ASTM C 150-07, *Specification for Portland Cement*. West Conshohocken, PA: ASTM International, 2007.

ASTM C 199-84 (2005), *Standard Test Method for Pier Test for Refractory Mortars*. West Conshohocken, PA: ASTM International, 2005.

ASTM C 207-06, *Specification for Hydrated Lime*. West Conshokhocken, PA: ASTM International, 2006.

ASTM C 212-00 (2006), *Specification for Structural Clay Facing Tile*. West Conshohocken, PA: ASTM International, 2006.

ASTM C 216-07a, *Specification for Facing Brick (Solid Masonry Units Made from Clay or Shale)*. West Conshohocken, PA: ASTM International, 2007.

ASTM C 270-08a, *Specification for Mortar for Unit Masonry*. West Conshohocken, PA: ASTM International, 2008.

ASTM C 315-07, *Specification for Clay Flue Linings*. West Conshohocken, PA: ASTM International, 2007.

ASTM C 331-05, *Standard Specification for Lightweight Aggregates for Concrete Masonry Units*. West Conshohocken, PA: ASTM International, 2005.

ASTM C 476-08, *Specification for Grout for Masonry*. West Conshohocken, PA: ASTM International, 2008.

ASTM C 503-08a, *Specification for Marble Dimension Stone (Exterior)*. West Conshohocken, PA: ASTM International, 2008.

ASTM C 568-08a, *Specification for Limestone Dimension Stone*. West Conshohocken, PA: ASTM International, 2008.

ASTM C 595-08a, *Specification for Blended Hydraulic Cements*. West Conshohocken, PA: ASTM International, 2008.

ASTM C 615-03, *Specification for Granite Dimension Stone*. West Conshohocken, PA: ASTM International, 2003.

ASTM C 616-08a, *Specification for Quartz-based Dimension Stone*. West Conshohocken, PA: ASTM International, 2008.

ASTM C 629-08, *Specification for Slate Dimension Stone*. West Conshohocken, PA: ASTM International, 2008.

ASTM C 652-09, *Specification for Hollow Brick (Hollow Masonry Units Made from Clay or Shale)*. West Conshohocken, PA: ASTM International, 2009.

ASTM C 744-08, *Specification for Prefaced Concrete and Calcium Silicate Masonry Units*. West Conshohocken, PA: ASTM International, 2008.

ASTM C 887-05, *Specification for Packaged, Dry, Combined Materials for Surface-bonding Mortar*. West Conshohocken, PA: ASTM International, 2005.

ASTM C 946-91 (2001), *Practice for Construction of Dry-stacked, Surface-bonded Walls*. West Conshohocken, PA: ASTM International, 2001.

ASTM C 1019-09, *Test Method for Sampling and Testing Grout*. West Conshohocken, PA: ASTM International, 2009.

ASTM C 1088-09, *Specification for Thin Veneer Brick Units Made from Clay or Shale*. West Conshohocken, PA: ASTM International, 2009.

ASTM C 1261-07, *Firebox Brick for Residential Fireplaces*. West Conshohocken, PA: ASTM International, 2007.

ASTM C 1283-07a, *Standard Practice for Installing Clay Flue Liners*. West Conshohocken, PA: ASTM International, 2007.

ASTM C 1314-07, *Standard Method for Constructing and Testing Masonry Prisms Used to Determine Compliance with Specified Compressive Strength of Masonry*. West Conshohocken, PA: ASTM International, 2007.

ASTM C 1327-97, *Specification for Mortar Cement*. West Conshohocken, PA: ASTM International, 1997.

ASTM C 1386-98, *Specification for Precast Autoclad Aerated Concrete (PAAC) Wall Construction Units*. West Conshohocken, PA: ASTM International, 1998.

ASTM E 72-05, *Standard Test Methods of Conducting Strength Tests of Panels for Building Construction*. West Conshohocken, PA: ASTM International, 2005.

ASTM E 84-09, *Test Method for Surface-burning Characteristics of Building Materials*. West Conshohocken, PA: ASTM International, 2009.

ASTM E 119-08a, *Test Methods for Fire Tests of Building Construction and Materials*. West Conshohocken, PA: ASTM International, 2008.

ASTM E 518-03, *Standard Test Methods for Flexural Bond Strength of Masonry*. West Conshohocken, PA: ASTM International, 2003.

ASTM E 519-07, *Standard Test Method for Diagonal Tension (Shear) in Masonry Assemblages*. West Conshohocken, PA: ASTM International, 2007.

ASTM E 1602-03, *Standard Guide for Construction of Solid Fuel-burning Masonry Heaters*. West Conshohocken, PA: ASTM International, 2003.

AWS D 1.4-98, *Structural Welding Code—Reinforced Steel*. Miami, FL: American Welding Society, 1998.

Baker, Ira. *A Treatise on Masonry Construction*. Chicago: University of Illinois, 1889.

Commentary on Building Code Requirements for Masonry Structures (TMS 402-11/ACI 530-11/ASCE 5-11). Farmington Hills, MI: American Concrete Institute; Reston, VA: Structural Engineering Institute of the American Society of Civil Engineers; Boulder, CO: The Masonry Society, 2011.

Commentary on Specifications for Masonry Structures (TMS 602-11/ACI 530.1-11/ASCE 6-11). Farmington Hills, MI: American Concrete Institute; Reston, VA: Structural Engineering Institute of the American Society of Civil Engineers; Boulder, CO: The Masonry Society, 2011.

Compressive Strength Testing of Masonry Mortar, A TMS Monograph. Boulder, CO: The Masonry Society, 1996.

FEMA P750, *NEHRP Recommended Provisions for Seismic Regulations for New Buildings and Other Structures Part 1*. Washington, DC: Federal Emergency Management Agency, 2009.

FEMA P750, *NEHRP Recommended Provisions for Seismic Regulations for New Buildings and Other Structures Part 2: Commentary*. Washington, DC: Federal Emergency Management Agency, 2009.

IFC-12, *International Fire Code*. Washington, DC: International Code Council, 2011.

IFGC-12, *International Fuel Gas Code*. Washington, DC: International Code Council, 2011.

IMC-12, *International Mechanical Code*. Washington, DC: International Code Council, 2011.

IRC-12, *International Residential Code*. Washington, DC: International Code Council, 2011.

TMS 402-11/ACI 530-11/ASCE 5-11, *Building Code Requirements for Masonry Structures*. Farmington Hills, MI: American Concrete Institute; Reston, VA: Structural Engineering Institute of the American Society of Civil Engineers; Boulder, CO: The Masonry Society, 2011.

TMS 602-11/ACI 530.1-11/ASCE 6-11, *Specifications for Masonry Structures*. Farmington Hills, MI: American Concrete Institute; Reston, VA: Structural Engineering Institute of the American Society of Civil Engineers; Boulder, CO: The Masonry Society, 2011.

UBC-97, *Uniform Building Code*. Whittier, CA: International Conference of Building Officials, 1997.

UL 641-95, *Type L Low-temperature Venting Systems*. Northbrook, IL: Underwriters Laboratories Inc., 1995.

UL 1777-07, *Chimney Liners*. Northbrook, IL: Underwriters Laboratories Inc., 2007.

Chapter 22: Steel

General Comments

Chapter 22 contains provisions governing the materials, design, construction and quality of steel structural members.

Section 2201 contains the general requirements.

Section 2202 lists the definitions that are related to steel construction.

Section 2203 includes provisions for the identification and protection of steel for structures.

Section 2204 addresses requirements for welded and bolted connections.

Section 2205 references the design standards for structural steel construction.

Section 2206 covers requirements for composite structural steel and concrete structures in a manner similar to Section 2205 for structural steel.

Section 2207 addresses steel joists.

Section 2208 specifies the design standard for steel cable structures.

Section 2209 references the specification for the design and installation of steel storage racks.

Section 2210 specifies the appropriate referenced standard for cold-formed steel.

Section 2211 references standards for light-frame construction utilizing cold-formed steel.

Steel is a noncombustible material commonly associated with Type I and II construction; however, it is permitted to be used and commonly found in all types of construction. There are two main families of steel members: the first is hot-rolled structural shapes and members that are made up of a combination of rolled shapes and plates, and the second is composed of sections that are cold formed from steel sheets, strips, plates or flat bars in roll-forming machines, or by bending in brake-press operations.

The code requires that materials used in the design of structural steel members conform to designated national standards. Chapter 22 is involved largely with the design and use of steel materials using the specifications and standards of the American Institute of Steel Construction (AISC), the American Iron and Steel Institute (AISI) and the Steel Joist Institute (SJI).

Purpose

The purpose of Chapter 22 is to provide the requirements necessary for the design and construction of structural steel, cold-formed steel, steel joists, steel cable structures, steel storage racks and composite construction. This chapter specifies the appropriate design and construction standards for these types of structures. It also provides a road map of the applicable technical requirements for steel structures.

SECTION 2201 GENERAL

2201.1 Scope. The provisions of this chapter govern the quality, design, fabrication and erection of steel used structurally in buildings or structures.

❖ The provisions of this chapter govern the quality, design, fabrication and erection of steel used structurally in buildings or structures.

Chapter 22 is formatted in a very similar fashion to the legacy model codes, since the code is essentially a road map for the use of design standards appropriate for various types of steel construction.

While this section indicates that Chapter 22 governs "quality, design, fabrication and erection," this is a general statement and the code user should take note of the scope of reference used to adopt specific standards since it may be different. Each section is intended to reflect the scope of the adopted standard and it is this specific scope of reference that establishes how each standard must be applied. The following text will discuss the background of each section.

SECTION 2202 DEFINITIONS

2202.1 Definitions. The following terms are defined in Chapter 2:

STEEL CONSTRUCTION, COLD-FORMED.

STEEL JOIST.

STEEL MEMBER, STRUCTURAL.

❖ Definitions facilitate the understanding of code provisions and minimize potential confusion. To that end, this section lists definitions of terms associated with steel construction. Note that these definitions are found in Chapter 2. The use and application of defined terms, as well as undefined terms, are set forth in Section 201.

SECTION 2203 IDENTIFICATION AND PROTECTION OF STEEL FOR STRUCTURAL PURPOSES

2203.1 Identification. Identification of structural steel members shall comply with the requirements contained in AISC 360. Identification of cold-formed steel members shall comply with the requirements contained in AISI S100. Identification of cold-formed steel light-frame construction shall also comply with the requirements contained in AISI S200. Other steel furnished for structural load-carrying purposes shall be properly identified for conformity to the ordered grade in accordance with the specified ASTM standard or other specification and the provisions of this chapter. Steel that is not readily identifiable as to grade from marking and test records shall be tested to determine conformity to such standards.

❖ A wide variety of steel types and grades are available and important properties, such as tensile strength and weldability, vary accordingly. For that reason, it is imperative that the properties of the steel that is furnished be verifiable. This section clarifies which standards govern the identification of steel construction. The identification of structural steel members should be in accordance with the requirements of AISC 360, *Specification for Structural Steel Buildings*, specifically Section M5.5. The identification of cold-formed steel members should be in accordance with the requirements of AISI S100, specifically Section A. It also clarifies that cold-formed steel light-frame construction should be identified in accordance with the additional requirements of AISI S200, specifically Section A5. Steel must be identified so that both the engineer and inspector can verify compliance with the approved plans, as well as the code. All steel that is not marked for these identification purposes shall be tested to verify compliance.

2203.2 Protection. Painting of structural steel members shall comply with the requirements contained in AISC 360. Painting of open-web steel joists and joist girders shall comply with the requirements of SJI CJ-1.0, SJI JG-1.1, SJI K-1.1 and SJI LH/DLH-1.1. Individual structural members and assembled panels of cold-formed steel construction shall be protected against corrosion in accordance with the requirements contained in AISI S100. Protection of cold-formed steel light-frame construction shall also comply with the requirements contained in AISI S200.

❖ The intent of this provision is to clarify which standards govern the protection of the various types of steel construction. This section clarifies that it is structural steel members that are to be protected in accordance with AISC 360. This section further clarifies that the protection of cold-formed steel members should be in accordance with the general requirements of AISI S100, specifically Section A. The code also clarifies that cold-formed steel light-frame construction should be protected in accordance with the additional requirements of AISI S200, specifically Section A4. This section also addresses the protection of joists and joist girders, referring to the SJI specifications, which cover this subject for each type of joist and joist girder.

This section cautions the user that protection of steel members is a consideration. The need to provide protection, however, is dependent upon the potential for corrosion in a given installation. Typically, unprotected carbon steel can perform adequately in a dry, interior exposure, while weathering steel can tolerate atmospheric exposure without the need for additional protection. Therefore, the AISC provisions require consideration of corrosion where the strength or serviceability of structural steel could be impaired. A design that can tolerate the anticipated corrosion is permitted as an alternative to providing protection against corrosion. Many types of coating systems are available, yet painting remains the most common type of corrosion protection. Where painting of structural steel is called for, it must be in accordance with the AISC specification.

SECTION 2204 CONNECTIONS

2204.1 Welding. The details of design, workmanship and technique for welding, inspection of welding and qualification of welding operators shall conform to the requirements of the specifications listed in Sections 2205, 2206, 2207, 2208, 2210 and 2211. *Special inspection* of welding shall be provided where required by Section 1705.

❖ This section does not directly adopt specifications for the design and details of welding; rather, it requires that welding be accomplished in accordance with the requirements of the appropriate design specification. This is done because, in the case of structural steel and cold-formed steel, the referenced standards adopt different American Welding Society (AWS) specifications, while in the case of steel joists, welding requirements are contained within the SJI specifications. The SJI welding criteria are, for the most part, consistent with AWS specifications, but there are some differences that are specific to the manufacture of steel joists.

The processes and techniques for welding heavy structural steel are different from those necessary for welding thinner cold-formed steel. Therefore, the AISC specification for structural steel construction requires welding to be in accordance with AWS D1.1, *Structural Welding Code—Steel*, which has a thickness limitation of not less than 0.125 inch (3.2 mm). The AISI standards on cold-formed steel refer to AWS D1.3, *Structural Welding Code—Sheet Steel*, which is limited to thicknesses of not more than 0.188 inch (4.8 mm). The overlapping thickness limitations provide the flexibility to allow the sole use of AWS D1.1 or D1.3 on projects where all the welded steel meets the corresponding thickness limitations.

Reference is made to the special inspection requirements in Section 1705. These include special inspections of structural steel and cold-formed steel that are triggered by the requirements for seismic resistance.

2204.2 Bolting. The design, installation and inspection of bolts shall be in accordance with the requirements of the specifications listed in Sections 2205, 2206, 2207, 2210 and 2211. *Special inspection* of the installation of high-strength bolts shall be provided where required by Section 1705.

❖ This section is very similar to the section on welding in that it does not directly reference bolt installation standards. The requirements for the design, installation and inspection for bolting are found in standards that are referenced in the sections that are noted.

2204.2.1 Anchor rods. Anchor rods shall be set in accordance with the *construction documents*. The protrusion of the threaded ends through the connected material shall fully engage the threads of the nuts, but shall not be greater than the length of the threads on the bolts.

❖ This section provides the requirements for the setting and projection of anchor rods. Guidance is provided on the length of threads so that the anchor is effective.

SECTION 2205
STRUCTURAL STEEL

2205.1 General. The design, fabrication and erection of structural steel for buildings and structures shall be in accordance with AISC 360. Where required, the seismic design of structural steel structures shall be in accordance with the additional provisions of Section 2205.2.

❖ This section requires that structural steel be designed in accordance with AISC 360. It is the successor to the separate AISC specifications previously published and combines design provisions for both allowable stress design (ASD) and load and resistance factor design (LRFD) in what is referred to as a "unified format." AISC 360 also incorporates requirements that were previously published in a separate standard for the design of round, square and rectangular steel tubular structural sections.

2205.2 Seismic requirements for structural steel structures. The design of structural steel structures to resist seismic forces shall be in accordance with the provisions of Section 2205.2.1 or 2205.2.2, as applicable.

❖ The code references state-of-the-art seismic detailing requirements based on AISC 341, *Seismic Provisions for Structural Steel Buildings*. It incorporates significant research results from the Structural Engineers Association of California (SEAC) Research Program, which began as a result of the damage to steel moment frames that was caused by the Northridge earthquake.

2205.2.1 Seismic Design Category B or C. Structural steel structures assigned to *Seismic Design Category* B or C shall be of any construction permitted in Section 2205. Where a response modification coefficient, *R*, in accordance with ASCE 7, Table 12.2-1 is used for the design of structural steel structures assigned to *Seismic Design Category* B or C, the structures shall be designed and detailed in accordance with the requirements of AISC 341.

Exception: The response modification coefficient, *R*, designated for "Steel systems not specifically detailed for seismic resistance, excluding cantilever column systems" in ASCE 7, Table 12.2-1 shall be permitted for systems designed and detailed in accordance with AISC 360, and need not be designed and detailed in accordance with AISC 341.

❖ The earthquake provisions in Section 1613 require all structures to be assigned to a seismic design category, which is a function of the building's risk category, as well as the design earthquake ground motions. Seismic design category is an indicator of the seismic risk and it is used to establish, among other things, the complexity of the seismic analysis and the types of structural systems that may be utilized to resist the effects of earthquakes. Structures classified as Seismic Design Category B or C are considered to have a low-to-moderate seismic risk and this section of the code clarifies the options that are available in designing a seismic-force-resisting system for these structures utilizing structural steel.

The base requirement makes the AISC 341 seismic detailing provisions mandatory. If a steel special concentrically braced frame (SCBF) is utilized, for example, Table 12.2-1 of ASCE 7 specifies an *R* coefficient of 6 for this system and the detailing requirements for SCBF members and connections in Section 13 of AISC 341 must be satisfied. The only exception to this is for systems that fall under Line H of ASCE 7, Table 12.2-1, which is for "steel systems not specifically detailed for seismic resistance, excluding cantilever column systems." Such systems are assigned a response modification coefficient, *R*, of 3 in Table 12.2-1 of ASCE 7. The system is then designed for the resulting earthquake load effects using only AISC 360. Under AISC 360, the assumption is elastic behavior as opposed to AISC 341, which requires a level of detailing that assumes inelastic response.

While this section gives minimum requirements for structural steel systems in buildings that are assigned to Seismic Design Category B or C, it is worth noting that for structures classified as Seismic Design Category A (i.e., low earthquake risk), compliance with this section is not necessary. The ASCE 7 standard requires design for a minimum lateral load for structural integrity in any structure and that the minimum requirement provides sufficient lateral resistance for buildings that are classified as Seismic Design Category A.

2205.2.2 Seismic Design Category D, E or F. Structural steel structures assigned to *Seismic Design Category* D, E or F shall be designed and detailed in accordance with AISC 341, except as permitted in ASCE 7, Table 15.4-1.

❖ In buildings classified as Seismic Design Categories D, E and F, the provisions of AISC 341 are mandatory for seismic-force-resisting systems of structural steel. Where the seismic risk is considered high, the earthquake load requirements that are referenced in Section 1613 limit the permitted seismic-force-resisting systems to those considered to be capable of inelastic response. The design earthquake forces are a function of the response modification coefficient, *R*, for the seismic-force-resisting system that is utilized. The design of the steel members and connections, in accordance with AISC 341 requirements for the selected seismic-force-resisting system, provides the appropriate level of energy dissipation by inelastic response.

The only exception to the above provisions recognizes ASCE 7, Table 15.4-1 requirements for nonbuilding structures. It permits certain steel seismic systems that utilize a lower response modification coefficient, *R*, to be designed and detailed using AISC 360 alone (see commentary, Section 2205.2.1).

SECTION 2206 COMPOSITE STRUCTURAL STEEL AND CONCRETE STRUCTURES

2206.1 General. Systems of structural steel acting compositely with reinforced concrete shall be designed in accordance with AISC 360 and ACI 318, excluding ACI 318 Chapter 22. Where required, the seismic design of composite steel and concrete systems shall be in accordance with the additional provisions of Section 2206.2.

❖ This section provides requirements for composite structural steel and concrete structures similar to the requirements in Section 2205 for structural steel. Composite construction consists of steel and concrete members that act in unison to resist the effects of applied loads. The use of composite action in gravity load applications has been well established and the referenced standards for steel and concrete are sufficient.

2206.2 Seismic requirements for composite structural steel and concrete construction. Where a response modification coefficient, *R*, in accordance with ASCE 7, Table 12.2-1 is used for the design of systems of structural steel acting compositely with reinforced concrete, the structures shall be designed and detailed in accordance with the requirements of AISC 341.

❖ Unlike structural steel, no distinction is made between seismic design categories: no matter what category a composite structure is assigned to, it must be designed and detailed in accordance with AISC 341. Only composite structures classified as Seismic Design Category A are permitted to be designed in accordance with ACI 318 and AISC 360 (see commentary, Section 2205.2.1).

In structures classified as Seismic Design Category B or above a composite seismic-force-resisting system is required to be one of the types listed in Table 12.2-1 of ASCE 7. The corresponding response modification coefficient, *R*, is used to determine the seismic base shear and the system is designed in accordance with AISC 341. AISC 341 provides detailed requirements for testing where appropriate, specifically in Section G3.6b (Composite Special Moment Frames), Section G4.6b (Composite Partially Restrained Moment Frames) and Section H3 (Composite Eccentrically Braced Frames).

SECTION 2207 STEEL JOISTS

2207.1 General. The design, manufacture and use of open web steel joists and joist girders shall be in accordance with one of the following Steel Joist Institute (SJI) specifications:

1. SJI CJ-1.0
2. SJI K-1.1
3. SJI LH/DLH-1.1
4. SJI JG-1.1

Where required, the seismic design of buildings shall be in accordance with the additional provisions of Section 2205.2 or 2211.6.

❖ Steel joists are required to be designed and manufactured in accordance with the specifications published by the SJI. Seismic loading must always be evaluated and the code user is referred to the sections governing the design of steel seismic-force-resisting systems. Open-web steel joists are primarily used as gravity-load-carrying members, and where a steel joist transfers earthquake loads or carries other unique loads, this information needs to be provided to the joist manufacturer.

2207.2 Design. The *registered design professional* shall indicate on the *construction documents* the steel joist and/or steel joist girder designations from the specifications listed in Section 2207.1 and shall indicate the requirements for joist and joist girder design, layout, end supports, anchorage, non-SJI standard bridging, bridging termination connections and bearing connection design to resist uplift and lateral loads. These documents shall indicate special requirements as follows:

1. Special loads including:
   - 1.1. Concentrated loads;
   - 1.2. Nonuniform loads;
   - 1.3. Net uplift loads;
   - 1.4. Axial loads;
   - 1.5. End moments; and
   - 1.6. Connection forces.
2. Special considerations including:
   - 2.1. Profiles for nonstandard joist and joist girder configurations (standard joist and joist girder

configurations are as indicated in the SJI catalog);

2.2. Oversized or other nonstandard web openings; and

2.3. Extended ends.

3. Deflection criteria for live and total loads for non-SJI standard joists.

❖ Sections 2207.2 through 2207.5 are meant to delineate the responsibilities of the steel joist manufacturer and the registered design professional who performs the overall structural design. Like precast concrete components and wood I-joists, the design of steel joists is typically relegated to the component manufacturer by contractual arrangements on any given project. Since the steel joist manufacturer designs the steel joists, this section lists the necessary steel joist design criteria that the registered design professional must furnish in the construction documents.

2207.3 Calculations. The steel joist and joist girder manufacturer shall design the steel joists and/or steel joist girders in accordance with the current SJI specifications and load tables to support the load requirements of Section 2207.2. The *registered design professional* may require submission of the steel joist and joist girder calculations as prepared by a *registered design professional* responsible for the product design. If requested by the *registered design professional*, the steel joist manufacturer shall submit design calculations with a cover letter bearing the seal and signature of the joist manufacturer's *registered design professional.* In addition to standard calculations under this seal and signature, submittal of the following shall be included:

1. Non-SJI standard bridging details (e.g.for cantilevered conditions, net uplift, etc.).
2. Connection details for:

   2.1. Non-SJI standard connections (e.g.flush-framed or framed connections);

   2.2. Field splices; and

   2.3. Joist headers.

❖ This section requires the steel joist manufacturer to design the steel joists in accordance with the criteria provided by the registered design professional and also the applicable SJI specification(s). Where a more complete accounting of the steel joist design is desired, the registered design professional is permitted to request sealed design calculations from the steel joist manufacturer. The list of nonstandard items that are required for submittal is intended to minimize the chance that key details could be overlooked.

2207.4 Steel joist drawings. Steel joist placement plans shall be provided to show the steel joist products as specified on the *construction documents* and are to be utilized for field installation in accordance with specific project requirements as stated in Section 2207.2. Steel placement plans shall include, at a minimum, the following:

1. Listing of all applicable loads as stated in Section 2207.2 and used in the design of the steel joists and joist girders as specified in the *construction documents.*
2. Profiles for nonstandard joist and joist girder configurations (standard joist and joist girder configurations are as indicated in the SJI catalog).
3. Connection requirements for:

   3.1. Joist supports;

   3.2. Joist girder supports;

   3.3. Field splices; and

   3.4. Bridging attachments.

4. Deflection criteria for live and total loads for non-SJI standard joists.
5. Size, location and connections for all bridging.
6. Joist headers.

Steel joist placement plans do not require the seal and signature of the joist manufacturer's *registered design professional.*

❖ The steel joist placement plans are provided by the steel joist manufacturer and illustrate how the steel joists are to be installed in more detail than the project's construction documents that are prepared by the registered design professional. The list of information that must be included is intended to minimize the possibility of overlooking critical details. The steel joist placement plan is considered to be a type of shop drawing that does not require the seal of a registered design professional.

2207.5 Certification. At completion of manufacture, the steel joist manufacturer shall submit a *certificate of compliance* in accordance with Section 1704.2.5.2 stating that work was performed in accordance with *approved construction documents* and with SJI standard specifications.

❖ The certificate of compliance provides a written declaration from the steel joist manufacturer that attests to its compliance with the SJI specifications, as well as the construction documents.

SECTION 2208 STEEL CABLE STRUCTURES

2208.1 General. The design, fabrication and erection including related connections, and protective coatings of steel cables for buildings shall be in accordance with ASCE 19.

❖ The code references ASCE 19 for the design of steel cable structures. This standard addresses the use of carbon steel or stainless steel cables formed from coated or uncoated helically twisted wire strand, wire rope or a parallel wire strand.

2208.2 Seismic requirements for steel cable. The design strength of steel cables shall be determined by the provisions of ASCE 19 except as modified by these provisions.

1. A load factor of 1.1 shall be applied to the prestress force included in T_3 and T_4 as defined in Section 3.12.
2. In Section 3.2.1, Item (c) shall be replaced with "1.5 T_3" and Item (d) shall be replaced with "1.5 T_4."

❖ This section provides the user with appropriate modifications for structural cables that are intended for use in seismic applications. These modifications were first developed by the Building Seismic Safety Council for incorporation into the first edition of the National Earthquake Hazard Reduction Program (NEHRP) *Recommended Provisions for Seismic Regulations for New Buildings and Other Structures.*

SECTION 2209 STEEL STORAGE RACKS

2209.1 Storage racks. The design, testing and utilization of industrial steel storage racks made of cold-formed or hot-rolled steel structural members, shall be in accordance with RMI/ANSI MH 16.1. Where required by ASCE 7, the seismic design of storage racks shall be in accordance with the provisions of Section 15.5.3 of ASCE 7, except that the mapped acceleration parameters, S_s and S_1, shall be determined in accordance with Section 1613.3.1.

❖ The code references ANSI/MH16 for the design of steel storage racks and the additional requirements in Section 15.5.3 of ASCE 7 for seismic design of steel storage. Because ANSI/MH16.1 is a self-contained document that includes seismic maps that are based on the prior editions of the code and ASCE 7, this provision further requires that the storage racks be designed using the latest seismic maps that have been adopted in the code (see Section 1613.3.1).

The design of steel storage racks in accordance with the Rack Manufacturers Institute (RMI) standard is limited to the types of racks that are actually included within the scope of that standard. Rack types that are beyond the scope of the standard include drive-in or drive-through racks, cantilever racks, portable racks or rack buildings. These rack structures should be designed in accordance with other applicable referenced specifications, such as AISC 360 or AISI S100.

SECTION 2210 COLD-FORMED STEEL

2210.1 General. The design of cold-formed carbon and low-alloy steel structural members shall be in accordance with AISI S100. The design of cold-formed stainless-steel structural members shall be in accordance with ASCE 8. Cold-formed steel light-frame construction shall also comply with Section 2211. Where required, the seismic design of cold-formed steel structures shall be in accordance with the additional provisions of Section 2210.2.

❖ This section references AISI S100, which has been developed by AISI in cooperation with the Canadian Standards Association (CSA) and Camara Nacional de la Industria del Hierro y del Acero (CANACERO). It supercedes the standards previously published individually by AISI and CSA, and is intended for use throughout North America. The latter is accomplished by including three appendix chapters with the country-specific provisions applicable in either the United States, Canada or Mexico. It incorporates LRFD and ASD into a single document.

Cold-formed stainless steel is required to comply with ASCE 8. This standard applies to structural members that are cold formed from annealed and cold-rolled stainless steels. ASCE 8 includes both the LRFD and ASD methods.

2210.1.1 Steel decks. The design and construction of cold-formed steel decks shall be in accordance with this section.

❖ This section references the appropriate standards for steel decks based on their application.

2210.1.1.1 Noncomposite steel floor decks. Noncomposite steel floor decks shall be permitted to be designed and constructed in accordance with ANSI/SDI-NC1.0.

❖ This section references the appropriate Steel Deck Institute (SDI) standard that governs the materials and design of cold-formed steel deck used in noncomposite floor construction. The deck can serve as a form for the structural concrete floor slab.

2210.1.1.2 Steel roof deck. Steel roof decks shall be permitted to be designed and constructed in accordance with ANSI/SDI-RD1.0.

❖ This section references the appropriate SDI standard that governs the materials and design of cold-formed steel decks used in roof construction.

2210.2 Seismic requirements for cold-formed steel structures. Where a response modification coefficient, R, in accordance with ASCE 7, Table 12.2-1 is used for the design of cold-formed steel structures, the structures shall be designed and detailed in accordance with the requirements of AISI S100, ASCE 8, and, for cold-formed steel special-bolted moment frames, AISI S110.

❖ This section references AISI S110, *Standard For Seismic Design Of Cold-Formed Steel Structural Systems – Special Bolted Moment Frames.* The "Cold-formed steel – special bolted moment frame," or CFS-SBMF, is expected to experience substantial inelastic deformation during significant seismic events. It is intended that most of the inelastic deformation will take place at the bolted connections, due to slip and bearing. In order to develop the designated mechanism, requirements based on the capacity design principles are provided for the design of the beams, columns and associated connections. The system design coefficients and limits are provided in ASCE 7. The height is limited to 35 feet (10.67 m) for all seismic design categories, based primarily on practical considerations rather than any limits on the system strength. As an alternative to the conventional computation of the seismic load effect with overstrength, E_{mh}, ASCE 7 permits the expected strength to be in accordance with AISI S110, which allows the designer to explicitly calculate E_{mh} at the design story drift level.

SECTION 2211 COLD-FORMED STEEL LIGHT-FRAME CONSTRUCTION

2211.1 General. The design and installation of structural members and nonstructural members utilized in cold-formed steel light-frame construction where the specified minimum base steel thickness is between 0.0179 inches (0.455 mm) and 0.1180 inches (2.997 mm) shall be in accordance with AISI S200 and Sections 2211.2 through 2211.7, as applicable.

❖ Section 2210 applies to light-frame construction utilizing cold-formed steel. Light-frame construction is characterized by vertical (i.e., walls) and horizontal (i.e., floors and roofs) structural elements formed by a system of repetitive members (see Chapter 2 definition). This section references AISI S200, which covers general requirements for the design and installation of cold-formed steel framing members in walls, floors and roof assemblies, and it is intended to be a supplement to AISI S100. For specific design information, AISI S200 refers the reader to AISI S100.

2211.2 Header design. Headers, including box and back- to-back headers, and double and single L-headers shall be designed in accordance with AISI S212 or AISI S100.

❖ The code references AISI S212, which has design and installation requirements for the commonly used header cross sections depicted in Figure 2211.2. Box and back-to-back headers consist of two cold-formed steel C sections. Single and double L-headers consist of cold-formed steel angle sections connected to the top track.

2211.3 Truss design. Cold-formed steel trusses shall be designed in accordance with AISI S214, Sections 2211.3.1 through 2211.3.4 and accepted engineering practice.

❖ The code references AISI S214, which covers truss design, quality criteria, installation and bracing (note that the version of the standard referenced in Chapter 35 includes Supplement Number 2). Cold-formed steel trusses are engineered components and the complete design of cold-formed steel trusses can be a complex process. The designer must keep the entire system in mind, as cold-formed steel trusses are not typically installed as single members in light-frame construction, but rather work together to form a complete structural system. Cold-formed steel truss designs should address connections, including the truss-to-wall plate.

2211.3.1 Truss design drawings. The truss design drawings shall conform to the requirements of Section B2.3 of AISI S214 and shall be provided with the shipment of trusses delivered to the job site. The truss design drawings shall include the details of permanent individual truss member restraint/bracing in accordance with Section B6(a) or B6(c) of AISI S214 where these methods are utilized to provide restraint/bracing.

❖ The code references the section of AISI S214 that specifies the information required on the truss design drawings, and clarifies that the bracing details must be included. This section also makes it clear that the design drawings must be delivered to the project location.

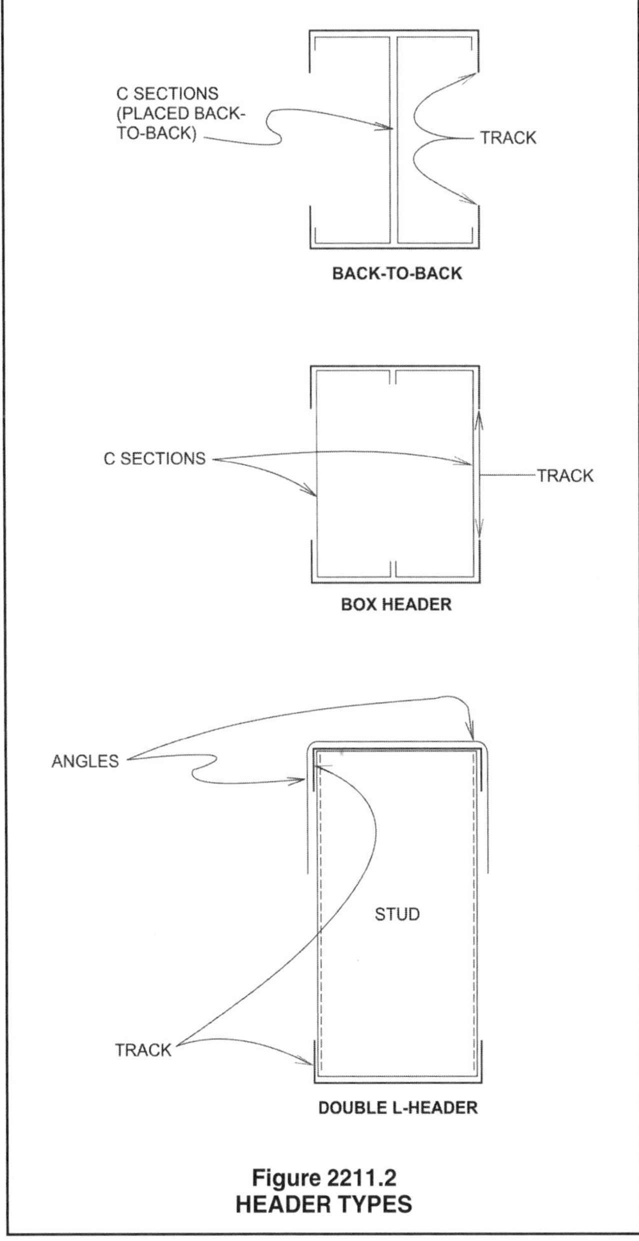

Figure 2211.2 HEADER TYPES

2211.3.2 Deferred submittals. AISI S214 Section B4.2 shall be deleted.

❖ This section excludes a provision in the referenced standard that deals with deferred submittals.

2211.3.3 Trussses spanning 60 feet or greater. The owner shall contract with a *registered design professional* for the design of the temporary installation restraint/bracing and the permanent individual truss member restraint/bracing for trusses with clear spans 60 feet (18 288 mm) or greater. *Special inspection* of trusses over 60 feet (18 288 mm) in length shall conform to Section 1705.

❖ This provision addresses the need for temporary bracing during the installation of trusses having longer spans. AISI S214 only covers the permanent truss bracing. Due to their size and weight, these longer span trusses pose greater stability concerns, making the proper design of the bracing, both temporary and permanent, a key safety measure.

2211.3.4 Truss quality assurance. Trusses not part of a manufacturing process that provides requirements for quality control done under the supervision of a third-party quality control agency, shall be manufactured in compliance with Sections 1704.2.5 and 1705.2, as applicable.

❖ Chapter E of AISI S214 addresses the quality assurance of trusses produced under a manufacturing process. This section applies to any truss that is not covered under the standard.

2211.4 Wall stud design. Wall studs shall be designed in accordance with either AISI S211 or AISI S100.

❖ AISI S211 provides design and installation requirements for cold-formed steel studs. The more formal design approach of AISI S100 is also an option.

2211.5 Floor and roof system design. Framing for floor and roof systems in buildings shall be designed in accordance with either AISI S210 or AISI S100.

❖ The section refers to AISI S210 for the design of cold-formed steel framing in floors and roofs, but also allows a more formal design approach in accordance with AISI S100 as an alternative. AISI S210 contains criteria specific to designing floor joists, ceiling joists and roof rafters.

2211.6 Lateral design. Light-frame shear walls, diagonal strap bracing that is part of a structural wall and diaphragms used to resist wind, seismic and other in-plane lateral loads shall be designed in accordance with AISI S213.

❖ Where shear walls, diagonal strap bracing or diaphragms of light-frame cold-formed steel provide resistance to the effects of earthquakes and wind loads, AISI S213 provides the design requirements for the lateral-force-resisting system. Shear wall design provisions include both segmented shear walls (Type I) and perforated shear walls (Type II).

2211.7 Prescriptive framing. Detached one- and two-family *dwellings* and *townhouses*, less than or equal to three *stories above grade plane*, shall be permitted to be constructed in accordance with AISI S230 subject to the limitations therein.

❖ AISI S230 contains prescriptive requirements for light-frame cold-formed steel construction that can be applied in one- and two-family dwellings as well as townhouses not more than three stories.

Bibliography

The following resource materials are referenced in this chapter or are relevant to the subject matter addressed in this chapter.

ACI 318-11, *Building Code Requirements for Structural Concrete*. Farmington Hills, MI: American Concrete Institute, 2011.

AISC 341-10, *Seismic Provisions for Structural Steel Buildings*. Chicago, IL: American Institute of Steel Construction, 2010.

AISC 360-10, *Specification for Structural Steel Buildings*. Chicago, IL: American Institute of Steel Construction, 2010.

AISI S100-07, *North American Specification for the Design of Cold-formed Steel Structural Members, including Supplement 1, dated 2010*. Washington, DC: American Iron and Steel Institute, 2007.

AISI S110-07, *Standard for Seismic Design Of Cold-Formed Steel Structural Systems—Special Bolted Moment Frames, with Supplement 1, dated 2009*. Washington, DC: American Iron and Steel Institute, 2007.

AISI S200-07, *North American Standard for Cold-formed Steel Framing—General Provisions*. Washington, DC: American Iron and Steel Institute, 2007.

AISI S210-07, *North American Standard for Cold-formed Steel Framing—Floor and Roof Framing System Design*. Washington, DC: American Iron and Steel Institute, 2007.

AISI S211-07, *North American Standard for Cold-formed Steel Framing—Wall Stud Design*. Washington, DC: American Iron and Steel Institute, 2007.

AISI S212-07, *North American Standard for Cold-formed Steel Framing—Header Design*. Washington, DC: American Iron and Steel Institute, 2007.

AISI S213-07, *North American Standard for Cold-formed Steel Framing—Lateral Design, including Supplement 1, dated 2010*. Washington, DC: American Iron and Steel Institute, 2007.

AISI S214-07, *North American Standard for Cold-formed Steel Framing—Truss Design*. Washington, DC: American Iron and Steel Institute, 2007.

AISI S230-07, *Standard for Cold-formed Steel Framing—Prescriptive Method for One- and Two-family*

Dwellings. Washington, DC: American Iron and Steel Institute, 2007.

ASCE 7-10, *Minimum Design Loads for Buildings and Other Structures*. New York: American Society of Civil Engineers, 2010.

ASCE 8-02, *Standard Specification for the Design of Cold-formed Stainless Steel Structural Members*. Reston, VA: American Society of Civil Engineers, 2002.

ASCE 19-09, *Structural Applications of Steel Cables for Buildings*. Reston, VA: American Society of Civil Engineers, 2009.

AWS D1.1-04, *Structural Welding Code—Steel*. Miami, FL: American Welding Society, 2004.

AWS D1.3-98, *Structural Welding Code—Sheet Steel*. Miami, FL: American Welding Society, 1998.

FEMA 450, *NEHRP Recommended Provisions for Seismic Regulations for New Buildings and Other Structures*. Washington, DC: Federal Emergency Management Agency, 2004.

Naeim, Farzad. *The Seismic Design Handbook*. Norwell, MA: Kluwer Academic Publishers, 2001.

RMI ANSI/MH16.1-08, *Specification for the Design, Testing and Utilization of Industrial Steel Storage Racks*. Charlotte, NC: Rack Manufacturers Institute, 2008.

SJI CJ-10 *Standard Specification for Composite Steel Joists, CJ-series*. Myrtle Beach, SC: Steel Joist Institute, 2010.

SJI JG-10, *Standard Specification for Joist Girders*. Myrtle Beach, SC: Steel Joist Institute, 2010.

SJI K-10, *Standard Specification for Open-web Steel Joists, K Series*. Myrtle Beach, SC: Steel Joist Institute, 2010.

SJI LH/DLH-10, *Standard Specification for Longspan Steel Joists, LH Series, and Deep Longspan Steel Joists, DLH Series*. Myrtle Beach, SC: Steel Joist Institute, 2010.

Williams, Alan. *Seismic and Wind Forces Structural Design Examples*. Falls Church, VA: International Code Council, 2005.

Williams, Alan. *Structural Steel Design LRFD*. Falls Church, VA: International Code Council, 2005.

Chapter 23: Wood

General Comments

This chapter contains information required to design and construct buildings or structures that include wood or wood-based structural elements, and is organized around the application of three design methodologies: allowable stress design (ASD), load and resistance factor design (LRFD) and conventional construction. Included are references to design and manufacturing standards for various wood and wood-based products; general construction requirements; design criteria for lateral-force-resisting systems and specific requirements for the application of the three design methods (ASD, LRFD and conventional construction). Chapter 23 includes elements of all three previous regional model codes—the *BOCA® National Building Code* (BNBC), the *Standard Building Code* (SBC) and the *Uniform Building Code®* (UBC™). It most closely follows the format of the UBC.

Acceptable standards for the manufacture of wood or wood-based products include provisions for sizes, grades (labels), quality control and certification programs, or similar methods of identification. Specific requirements and tables have been developed using referenced design methods to provide a minimum level of safety. This chapter also contains requirements both for the use of products in conjunction with wood and wood-based structural elements, and for prevention of decay.

In general, only Type III, IV or V buildings may be constructed of wood. Accordingly, Chapter 23 is referenced when the combination of the occupancy (determined in Chapter 3) and the height and area of the building or structure (determined in Chapter 5) indicate that the construction (specified in Chapter 6) can be Type III, IV or V. Another basis for referencing Chapter 23 is when wood elements are used in Type I or II structures as permitted in Section 603. This chapter gives information on the application of fire-retardant-treated wood, interior wood elements and trim in these structures. All structural criteria for application of referenced standards and procedures included in Chapter 23 are based on the loading requirements of Chapter 16 or on historical performance.

Chapter 23 is not a textbook on construction. It is assumed that the reader has both the training and experience needed to understand the principles and practices of wood design and construction. Without such understanding, some sections may be misunderstood and misapplied. This commentary should help to promote better understanding of the structure and application of the methods specified in Chapter 23.

Section 2301 identifies three methods of design, and compliance with one or more is required.

Section 2302 contains a list of the most commonly used terms to describe wood and wood-based products. Some engineering terms are also included. Refer to Chapter 2 for the definitions of the terms listed.

Section 2303 provides reference to manufacturing standards, necessary specification criteria and use and application provisions.

Section 2304 contains general provisions for the proper design and construction of all wood structures and the use of all wood products. Note that the general provisions in Section 2304 apply to all design methods. This section also includes the typical fastening schedule which is the minimum requirement for fastening various wood members.

Section 2305 references the American Forest & Paper Association's (AF&PA) *Special Design Provisions for Wind and Seismic* for design of lateral-force-resisting systems. It also contains provisions not found in the standard, such as design values for staples. Whether the structure is engineered using ASD or LRFD, the provisions of this section apply to the design of the lateral-force-resisting system.

Section 2306 contains provisions for the design of structures using ASD and references applicable standards. The two primary design standards are the AF&PA *National Design Specification for Wood Construction* (NDS) and SDPWS. Historically, all of the industry publications have been developed for ASD. More recently, LRFD has been introduced, making it necessary to distinguish clearly which provisions are appropriate for ASD or LRFD. As the section title implies, the provisions of this section only apply to ASD and are not appropriate for LRFD.

Because the AF&PA NDS and SDPWS are dual format standards permitting both ASD and LRFD design procedures, Section 2307 also references these consensus standards for the design of structures using the LRFD methodology.

Although fairly limited in application, Section 2308 contains the prescriptive provisions for conventional construction that may be used to construct certain wood-frame structures that conform to the restrictions and limitations. Limitations on the use of the conventional construction provisions in this section are provided in Section 2308.2 for a quick determination. Note that structures of otherwise conventional construction are allowed to contain portions or elements that are designed by the engineering provisions of Chapter 23 (see commentary, Sections 2308.1.1 and 2308.4).

Purpose

This chapter provides minimum guidance for the design of buildings and structures that use wood and wood-based products in their framing and fabrication. Alternative methods and materials can be used where justified by engineering analysis and testing. In all cases, the provisions of Section 2304 apply to all elements of wood-frame construction.

SECTION 2301 GENERAL

2301.1 Scope. The provisions of this chapter shall govern the materials, design, construction and quality of wood members and their fasteners.

❖ Section 2301 includes specifications for use of and standards for production of wood and wood-based products such as boards, dimensional lumber and engineered wood products, such as I-joists, glued-laminated timber, structural panels, trusses, particleboard, fiberboard and hardboard. Also included are criteria and specifications for the use of other materials such as connectors used in conjunction with wood or wood-based products. Other chapters of the code also affect the use of wood materials in buildings and should be referenced prior to making final decisions on the use of any product.

The scope of this chapter is established in Section 2301.1, and broadly encompasses wood products and the limitations placed on them and their various applications within the code.

2301.2 General design requirements. The design of structural elements or systems, constructed partially or wholly of wood or wood-based products, shall be in accordance with one of the following methods:

1. *Allowable stress design* in accordance with Sections 2304, 2305 and 2306.
2. *Load and resistance factor design* in accordance with Sections 2304, 2305 and 2307.
3. *Conventional light-frame construction* in accordance with Sections 2304 and 2308.

   Exception: Buildings designed in accordance with the provisions of the AF&PA WFCM shall be deemed to meet the requirements of the provisions of Section 2308.

4. The design and construction of log structures shall be in accordance with the provisions of ICC 400.

❖ This chapter includes three methods of designing with wood or wood-based products. This section limits designs to one of these three methods unless an alternative method has been proven to be acceptable as permitted in Section 104.11. It is not uncommon for only one element of a structure to require engineered design. This section recognizes "partial" design.

2301.3 Nominal sizes. For the purposes of this chapter, where dimensions of lumber are specified, they shall be deemed to be nominal dimensions unless specifically designated as actual dimensions (see Section 2304.2).

❖ The use of nominal sizes for lumber is part of the traditional nomenclature for grading and identification of manufactured pieces. "Nominal" simply refers to the short-hand term such as "2 × 4" when the actual piece of lumber has a real dimension of $1^1/_2$ inches by $3^1/_2$ inches (38 mm by 89 mm)—not 2 inches by 4 inches (51 mm by 102 mm). Section 2304.2, however, prescribes that in determining the required size for design purposes, computations must be based on the actual size, rather than the nominal size, of the lumber.

SECTION 2302 DEFINITIONS

2302.1 Definitions. The following terms are defined in Chapter 2:

ACCREDITATION BODY.

BRACED WALL LINE.

BRACED WALL PANEL.

COLLECTOR.

CONVENTIONAL LIGHT-FRAME CONSTRUCTION.

CRIPPLE WALL.

DIAPHRAGM, UNBLOCKED.

DRAG STRUT.

FIBERBOARD.

GLUED BUILT-UP MEMBER.

GRADE (LUMBER).

HARDBOARD.

NAILING, BOUNDARY.

NAILING, EDGE.

NAILING, FIELD.

NOMINAL SIZE (LUMBER).

PARTICLEBOARD.

PERFORMANCE CATEGORY.

PREFABRICATED WOOD I-JOIST.

SHEAR WALL.

Shear wall, perforated.

Shear wall segment, perforated.

STRUCTURAL COMPOSITE LUMBER.

Laminated strand lumber (LSL).

Laminated veneer lumber (LVL).

Oriented strand lumber (OSL).

Parallel strand lumber (PSL).

STRUCTURAL GLUED-LAMINATED TIMBER.

SUBDIAPHRAGM.

TIE-DOWN (HOLD-DOWN).

TREATED WOOD.

Fire-retardant-treated wood.

Preservative-treated wood.

WOOD SHEAR PANEL.

WOOD STRUCTURAL PANEL.

Composite panels.

Oriented strand board (OSB).

Plywood.

❖ Definitions facilitate the understanding of code provisions and minimize potential confusion. To that end, this section lists definitions of terms associated with wood construction. Note that these definitions are found in Chapter 2. The use and application of defined terms, as well as undefined terms, are set forth in Section 201. In addition to the terms listed in this chapter, other notations and definitions related to wood construction can be found in the various referenced standards, such as the AF&PA NDS and SDPWS.

SECTION 2303 MINIMUM STANDARDS AND QUALITY

2303.1 General. Structural sawn lumber; end-jointed lumber; prefabricated wood I-joists; structural glued-laminated timber; wood structural panels, fiberboard sheathing (when used structurally); hardboard siding (when used structurally); particleboard; *preservative-treated wood*; structural log members; structural composite lumber; round timber poles and piles; *fire-retardant-treated wood*; hardwood plywood; wood trusses; joist hangers; nails; and staples shall conform to the applicable provisions of this section.

❖ When the components of a wood structure comply with the various standards listed in Section 2303, a building or structure is deemed to comply with the minimum standards of quality prescribed by the code. When combined with the construction requirements of Section 2304 and the design standards referenced in Sections 2305 and 2306, these standards contain most of the information needed to adequately design a structure. For engineered structures, it is necessary for the designer to have a working knowledge of engineering principles and experience with construction to properly interpret the recommendations and meet the provisions of other applicable sections of the code. For conventional wood-frame structures, the standards in Section 2303 combined with the construction requirements of Section 2304 and the prescriptive construction provisions in Section 2308 can be used to construct code-complying wood-frame buildings.

Section 2303.1 lists the various materials that have production and quality control standards. The section covers minimum standards of quality for sawn lumber; end-jointed lumber; wood I-joists; glued-laminated timber; wood structural panels; fiberboard sheathing; hardboard siding; particleboard; preservative-treated wood; log members; composite lumber; round timber poles and piles; fire-retardant-treated wood; hardwood plywood; wood trusses; joist hangers and nails and staples. The use of these standards is fundamental for manufacturers in producing products and maintaining quality control procedures. Designers, owners and building officials must understand these standards and the methods prescribed in them to be able to properly identify products that have been produced in accordance with their criteria. Without the knowledge that the various products and components meet the applicable standards, there is little assurance that a safe and efficient building or structure will be constructed.

2303.1.1 Sawn lumber. Sawn lumber used for load-supporting purposes, including end-jointed or edge-glued lumber, machine stress-rated or machine-evaluated lumber, shall be identified by the grade *mark* of a lumber grading or inspection agency that has been approved by an accreditation body that complies with DOC PS 20 or equivalent. Grading practices and identification shall comply with rules published by an agency approved in accordance with the procedures of DOC PS 20 or equivalent procedures.

❖ All lumber used to support loads in a building or structure is required to be properly identified. Every species and grade of lumber has a unique inherent strength value. These values are further modified in sawn timber by the presence of growth characteristics that vary from piece to piece, such as knots, slope of grain, checks, etc. Without adequate identification, it would be impossible to verify that the proper material is being used in the field. The required grade mark must identify the species or species grouping; grade and moisture content at the time of surfacing; the grading agency and the mill name or grader's number. Figure 2303.1.1 illustrates typical grade mark labels.

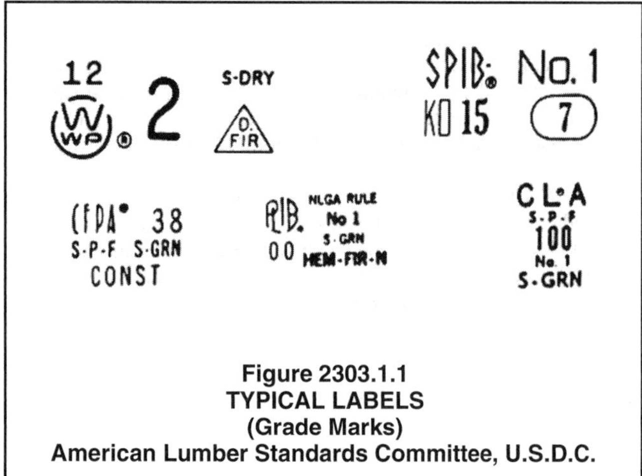

Figure 2303.1.1
TYPICAL LABELS
(Grade Marks)
American Lumber Standards Committee, U.S.D.C.

2303.1.1.1 Certificate of inspection. In lieu of a grade *mark* on the material, a certificate of inspection as to species and grade issued by a lumber grading or inspection agency meeting the requirements of this section is permitted to be accepted for precut, remanufactured or rough-sawn lumber and for sizes larger than 3 inches (76 mm) nominal thickness.

❖ Certification is an acceptable alternative to a grade mark from both United States and Canadian grading agencies. Grading agencies are certified by the American Lumber Standards Committee (ALSC).

Design values are published by lumber grade rules-writing agencies for both individual and grouped species. A grouped species is lumber that is cut and marketed in lots containing two or more species, such as Spruce-Pine-Fir. These species grow together in large areas. It is more economical to market the lumber as a species group than attempt segregation. The assigned strength values include those applicable to the weaker species in the group.

The code also allows certain types of structural lumber to have a certificate of inspection instead of a grade mark. A certificate of inspection is acceptable for precut, remanufactured or rough-sawn lumber and for sizes larger than 3 inches (76 mm) nominal in thickness. It is industry practice to place only one label (grade mark) on a piece of lumber, which may be removed on precut and remanufactured lumber. Each piece of lumber is graded after it has been cut to a standard size. The grade of the piece is determined based on its size, number and location of strength-reducing characteristics; therefore, one log may produce lumber of two or more different grades.

It is also industry practice not to label lumber having a nominal thickness larger than 3 inches (76 mm), or rough-sawn material where the label may be illegible. A certificate of inspection from an approved agency is acceptable instead of the label for these types of lumber. The certificate should be filed with the permanent records of the building or structure.

If defects exceeding those permitted for the allegedly installed grade are visible, then a grader would be able to determine that the wood is definitely not of a suitable grade. To determine if the wood in question is definitely of a suitable grade, the grader must inspect all four faces of the piece. This cannot happen once the lumber is installed in the building, as other components of the building will be covering up some of the faces of the pieces.

2303.1.1.2 End-jointed lumber. *Approved* end-jointed lumber is permitted to be used interchangeably with solid-sawn members of the same species and grade. End-jointed lumber used in an assembly required to have a fire-resistance rating shall have the designation "Heat Resistant Adhesive" or "HRA" included in its grade mark.

❖ End-joined or edge-glued lumber is acceptable when identified by an appropriate grade mark. Section 4.1.6 of the AF&PA NDS permits the use of such lumber for light framing, studs, joists, planks and decking. Where finger-jointed lumber is marked "Stud Use Only," then it is limited to applications where bending or tension stresses are subjected to short-term loading only. Where end-jointed lumber is used in fire-resitant assemblies, it must be joined with heat-resistant adhesive and must be indicated on the grade stamp.

2303.1.2 Prefabricated wood I-joists. Structural capacities and design provisions for prefabricated wood I-joists shall be established and monitored in accordance with ASTM D 5055.

❖ This section specifies that the shear, moment and stiffness capacities of prefabricated wood I-joists be established and monitored by ASTM D 5055. This standard also specifies that application details, such as bearing length and web openings, are to be considered in determining structural capacity. Wood I-joists are structural members typically used in floor and roof construction manufactured out of sawn or structural composite lumber flanges and structural panel webs, bonded together with exterior adhesives forming an "I" cross section (see definition of "Prefabricated wood I-joist" in Section 2302.1). The standard requires I-joist manufacturers to employ an independent inspection agency to monitor the procedures for quality assurance. Finally, the standard specifies that proper installation instructions accompany the product to the job site. The instructions are required to include weather protection, handling requirements and, where required, web reinforcement, connection details, lateral support, bearing details, web hole-cutting limitations and any special situation.

2303.1.3 Structural glued-laminated timber. Glued-laminated timbers shall be manufactured and identified as required in ANSI/AITC A 190.1 and ASTM D 3737.

❖ Glued-laminated timbers are required by this section to be manufactured following ANSI/AITC 190.1 and ASTM D 3737 for procedures to establish allowable structural properties. Knowing the standards these products must meet makes it easier to determine that the product found in the field will meet the design requirements.

2303.1.4 Wood structural panels. Wood structural panels, when used structurally (including those used for siding, roof and wall sheathing, subflooring, diaphragms and built-up members), shall conform to the requirements for their type in DOC PS 1, DOC PS 2 or ANSI/APA PRP 210. Each panel or member shall be identified for grade, bond classification, and Performance Category by the trademarks of an *approved* testing and grading agency. The Performance Category value shall be used as the "nominal panel thickness" or "panel thickness" whenever referenced in this code. Wood structural panel components shall be designed and fabricated in accordance with the applicable standards listed in Section 2306.1 and identified by the trademarks of an *approved* testing and inspection agency indicating conformance to the applicable standard. In addition, wood structural panels when permanently exposed in outdoor applications shall be of Exterior type, except that wood structural panel roof sheathing exposed to the outdoors on the underside is permitted to be Exposure 1 type.

❖ "Wood structural panels" is a collective term referring to plywood, oriented strand board (OSB) and other composite panels of wood-based materials (see definition of "Wood structural panel" in Chapter 2). As noted in this section, wood structural panels must conform to the specific requirements of Department of Commerce (DOC) PS 1, PS 2 or ANSI/APA PRP 210. A new American National Standards Institute (ANSI) standard for engineered wood panel siding, ANSI/APA PRP-210, *Standard for Performance-Rated Engineered Wood Siding*, is included in the 2012 *International Building Code*® (IBC®) and *International Residential Code*® (IRC®). The standard was developed by American Plywood Association (APA) under the ANSI consensus process based on APA's PRP-108, *Performance Standards and Policies for Structural-Use Panels.* ANSI/APA PRP-210 provides requirements and test methods for qualification and quality assurance for performance-rated engineered wood siding intended for use in construction applications as exterior siding.

Identification of grade for plywood includes N, A, B, C Plugged, C and D (from no knots or patches to large knots and knotholes).

Plywood is manufactured with an odd number of layers that have their grain direction placed perpendicular to each other and are bonded together with a strong adhesive. Each individual layer may consist of an assembly of several different plies laminated together with their grains running in the same direction. Alternating the grain direction in successive layers gives a plywood panel dimensional stability across its width. Stamped on top of most panels is a span rating that indicates the maximum roof and floor joist spacings that can be accommodated. Additionally, an exposure rating is assigned as Exterior (designed for applications subject to permanent exposure to the weather or moisture), Exposure 1 (designed for applications where long construction delays may be expected prior to providing protection against moisture or weather extremes) or Exposure 2 (intended solely for protected construction applications where only moderate delays in providing protection from moisture are expected).

APA-rated Stud-I-floor panels are commonly used for thicker sheathing applications where added stiffness is desired. This type of panel has a single span rating, indicating that it is specifically engineered for floors, as the name would suggest. Adjoining panel edges may be blocked or may be ordered as tongue-and-groove, which provides added stability in an unblocked diaphragm. If a diaphragm has the necessary load capacity without the addition of blocking at all panel edges, such a specification can lower installation costs and decrease framing mistakes.

2303.1.5 Fiberboard. Fiberboard for its various uses shall conform to ASTM C 208. Fiberboard sheathing, when used structurally, shall be identified by an *approved* agency as conforming to ASTM C 208.

❖ All fiberboard must meet the requirements of ASTM C 208, as well as being verminproof, resistant to rot-producing fungi and water repellent. This standard gives physical requirements for construction grades of fiberboard, including sheathing grade and roof-insulating grade. The sheathing grade of fiberboard is further broken down into regular and intermediate densities.

2303.1.5.1 Jointing. To ensure tight-fitting assemblies, edges shall be manufactured with square, shiplapped, beveled, tongue-and-groove or U-shaped joints.

❖ Tight-fitting joints in the fiberboard are required for all applications, including insulation, siding and wall sheathing.

2303.1.5.2 Roof insulation. Where used as roof insulation in all types of construction, fiberboard shall be protected with an *approved* roof covering.

❖ Fiberboard is not intended for prolonged exposure to sunlight, wind, rain or snow. Where fiberboard is used as roof insulation, it must be protected with an approved roof covering to prevent water saturation and subsequent delamination and to avoid decay and destruction of the glue bond by moisture.

2303.1.5.3 Wall insulation. Where installed and fireblocked to comply with Chapter 7, fiberboards are permitted as wall insulation in all types of construction. In fire walls and fire barriers, unless treated to comply with Section 803.1 for Class A materials, the boards shall be cemented directly to the concrete, masonry or other noncombustible base and shall be protected with an *approved* noncombustible veneer anchored to the base without intervening airspaces.

❖ Fiberboard is permitted without any fire-resistance treatment in the walls of all types of construction (see Section 603.1). When used in fire walls and fire barrier walls, fiberboard must be either treated to comply with Class A flame spread or adhered directly to a noncombustible base and protected by a tight-fitting, noncombustible veneer that is fastened through the fiberboard to the base. This is intended to prevent the fiberboard from contributing to the spread of fire.

2303.1.5.3.1 Protection. Fiberboard wall insulation applied on the exterior of foundation walls shall be protected below ground level with a bituminous coating.

❖ Fiberboard insulation applied to the exterior side of foundation walls is required to be protected from the weather to improve its service life and maintain its performance characteristics. Of particular concern is foundation insulation that is in close proximity to grade and has the risk of being damaged by a lawn mower; rocks or soil that is kicked up against it; water from a garden hose; rainwater splash back, etc. Protection is required for all fiberboard insulation on the exterior face of foundation walls.

2303.1.6 Hardboard. Hardboard siding used structurally shall be identified by an *approved agency* conforming to CPA/ANSI A135.6. Hardboard underlayment shall meet the strength requirements of $^{7}/_{32}$-inch (5.6 mm) or $^{1}/_{4}$-inch (6.4 mm) service class hardboard planed or sanded on one side to a uniform thickness of not less than 0.200 inch (5.1 mm). Prefinished hardboard paneling shall meet the requirements of CPA/ANSI A135.5. Other basic hardboard products shall meet the requirements of CPA/ANSI A135.4. Hardboard products shall be installed in accordance with manufacturer's recommendations.

❖ Hardboard siding that is to be used structurally must be manufactured in accordance with CPA/ANSI A135.6 and marked to indicate conformance with the standard, whether primed or unprimed, and to identify the producer and the type, either lap or panel. Hardboard products are produced primarily from interfelted lignocellulosic fibers. There are five classes based on strength values. Underlayments are limited to $^{7}/_{32}$-inch (5.6 mm) or $^{1}/_{4}$-inch (6.4 mm) service class.

Prefinished hardboard is required to be manufactured to the CPA/ANSI A135.5 standard, and must be marked to indicate the standard and to identify the producer, flame spread index, finish class, type of gloss and type of substrate, or must be accompanied by written certification of the same information.

2303.1.7 Particleboard. Particleboard shall conform to ANSI A208.1. Particleboard shall be identified by the grade *mark* or certificate of inspection issued by an *approved agency*. Particleboard shall not be utilized for applications other than indicated in this section unless the particleboard complies with the provisions of Section 2306.3.

❖ Sponsored by the Composite Panel Association (CPA), ANSI A208.1 is the basic specification for the manufacture of particleboard, which establishes a system of marks for the boards' grade, density, and strength.

Particleboard used in construction is medium density and is first designated by an "M." The second digit or letter in the designation is related to grade. The designations range from 1 to 3, with higher designations being the strongest. Grade M-S refers to medium density, "special grade" particleboard. This grade was added to ANSI A208.1 after the M-1, M-2 and M-3 grades had been established. Grade M-S falls between M-1 and M-2 in physical properties.

An optional third part of the grade designation indicates that the particleboard has a special characteristic. The grades of particleboard specified in Table 2306.5, M-S and M-2 "Exterior Glue," are manufactured with exterior glue to increase their durability characteristics.

While ANSI A208.1 has provisions for Grade M-3 particleboard, panels that meet the requirements of this higher grade are more commonly evaluated and used as wood structural panels, commonly referred to as oriented strand board, in accordance with DOC PS-2. Therefore, Grade M-3 material is not addressed in Table 2306.5.

2303.1.7.1 Floor underlayment. Particleboard floor underlayment shall conform to Type PBU of ANSI A208.1. Type PBU underlayment shall not be less than $^{1}/_{4}$-inch (6.4 mm) thick and shall be installed in accordance with the instructions of the Composite Panel Association.

❖ Although similar to medium density, Grade 1 particleboard—particleboard intended for use as floor underlayment—is designated "PBU" and has stricter limits on levels of formaldehyde emission permitted than those placed on Grade "M" particleboard. Particleboard intended for use as floor underlayment is not commonly manufactured with exterior glue, which could emit higher levels of formaldehyde than that permitted by ANSI A208.1 for Grade "PBU" floor underlayment.

Particleboard underlayment is often applied over a structural subfloor to provide a smooth surface for resilient-finish or textile floor coverings. The minimum $^{1}/_{4}$-inch (6.4 mm) thickness is applicable over panel-type subflooring. Particleboard underlayment installed over board or deck subflooring that has multiple joints should have a thickness of $^{3}/_{8}$ inch (9.5 mm). Joints in the underlayment should not be over joints in the subflooring.

All particleboard underlayment with thicknesses of $^{1}/_{4}$ through $^{3}/_{4}$ inch (6.4 through 19.1 mm) should be attached with a minimum of 6d annular threaded nails spaced 6 inches (152 mm) o.c. on the edges and 10 inches (254 mm) o.c. for intermediate supports.

2303.1.8 Preservative-treated wood. Lumber, timber, plywood, piles and poles supporting permanent structures required by Section 2304.11 to be preservative treated shall conform to the requirements of the applicable AWPA Standard U1 and M4 for the species, product, preservative and end use. Preservatives shall be listed in Section 4 of AWPA U1. Lumber and plywood used in wood foundation systems shall conform to Chapter 18.

❖ Wood is able to absorb chemicals because of its cellular characteristics. Preservative treatment procedures that will repel termites and destroy decay-causing fungus utilize this capability. The process includes placing wood in a large cylinder and applying a vacuum to remove as much air as possible from the wood. The chemical solution is then introduced and pressure is applied to force the solution into the wood. The pressure is maintained until the desired absorption is obtained. A final vacuum is often applied to remove as much excess preservative as possible.

There are several variations of this basic process, all of which must provide the required retention of preservatives. AWPA U1, Section 4, prescribes the requirements for the preservatives that are used, while AWPA U1 specifies the minimum results of the treatment process based on a commodity specification and use category. These requirements include the depth of penetration of the chemical into the wood, and the amount of chemical retained by the wood once the process is completed. Additional requirements for wood used in foundation systems are given in Section 1807.1.4.

2303.1.8.1 Identification. Wood required by Section 2304.11 to be preservative treated shall bear the quality *mark* of an inspection agency that maintains continuing supervision, testing and inspection over the quality of the *preservative-treated wood.* Inspection agencies for *preservative-treated wood* shall be *listed* by an accreditation body that complies with the requirements of the American Lumber Standards Treated Wood Program, or equivalent. The quality *mark* shall be on a stamp or *label* affixed to the *preservative-treated wood*, and shall include the following information:

1. Identification of treating manufacturer.
2. Type of preservative used.
3. Minimum preservative retention (pcf).
4. End use for which the product is treated.
5. AWPA standard to which the product was treated.
6. Identity of the accredited inspection agency.

❖ Quality marks are necessary to determine that preservative-treated wood conforms to applicable standards. The identifying mark must be by an approved inspection agency that has continuous follow-up services. Additionally, the inspection agency must be listed and certified as being competent by an approved organization. The American Lumber Standards Committee (ALSC) provides certification of treating agencies. Facsimiles of its quality marks are available by contacting the ALSC. The required quality mark is not a substitute for a grade mark. When wood or wood-based materials are being used structurally, both the quality mark and grade mark must be displayed on the piece.

2303.1.8.2 Moisture content. Where *preservative-treated wood* is used in enclosed locations where drying in service cannot readily occur, such wood shall be at a moisture content of 19 percent or less before being covered with insulation, interior wall finish, floor covering or other materials.

❖ Water-borne preservatives are subject to leaching unless properly dried and protected. The requirement for reducing the moisture content to 19 percent or less is intended to aid in preventing such leaching. Also, all structural members are presumed to have a moisture content of 19 percent or less.

Preservative treatment does not require any adjustment of design values. Some species have a high surface tension, which means that it is difficult to penetrate the surface of the lumber. These species are often incised to break the surface tension and allow the chemicals to penetrate the member. Incising requires a reduction in the design values for the member.

2303.1.9 Structural composite lumber. Structural capacities for structural composite lumber shall be established and monitored in accordance with ASTM D 5456.

❖ The purpose of this section is to specify the appropriate standard for establishing structural capacities of structural composite lumber. Included within the standard are criteria for laminated veneer lumber and parallel strand lumber. The ASTM International (ASTM) standard includes requirements for testing, criteria for determining allowable stresses, requirements for independent inspection and quality assurance procedures (also see definition of "Structural composite lumber" in Chapter 2).

2303.1.10 Structural log members. Stress grading of structural log members of nonrectangular shape, as typically used in log buildings, shall be in accordance with ASTM D 3957. Such structural log members shall be identified by the grade *mark* of an *approved* lumber grading or inspection agency. In lieu of a grade *mark* on the material, a certificate of inspection as to species and grade issued by a lumber grading or inspection agency meeting the requirements of this section shall be permitted.

❖ This section addresses grading requirements for logs used as structural members by referencing ASTM D 3957 methods for establishing structural capacities of logs. It also specifies the requirement for a grading stamp or an alternative certification on structural logs.

2303.1.11 Round timber poles and piles. Round timber poles and piles shall comply with ASTM D 3200 and ASTM D 25, respectively.

❖ This section provides references to the appropriate material standards for round timber poles and piles.

2303.2 Fire-retardant-treated wood. *Fire-retardant-treated wood* is any wood product which, when impregnated with chemicals by a pressure process or other means during manufacture, shall have, when tested in accordance with ASTM E 84 or UL 723, a *listed* flame spread index of 25 or less and show no evidence of significant progressive combustion when the test is continued for an additional 20-minute period. Additionally, the flame front shall not progress more than $10^1/_2$ feet (3200 mm) beyond the centerline of the burners at any time during the test.

❖ Fire-retardant-treated wood is plywood and lumber that has been pressure impregnated with chemicals to improve its flame spread characteristics beyond that of untreated wood. Since fire-retardant-treated wood is allowed in some applications where noncombustible materials are otherwise required, it is important that these products meet rigorous requirements. The effectiveness of the pressure-impregnated fire-retardant treatment is determined by subjecting the material to tests conducted in accordance with ASTM E 84, with the modification that the test is extended to 30 minutes rather than 15 minutes. Using this procedure, a flame spread index is established during the standard 10-minute test period. The test is continued for an additional 20 minutes. During this added time period, there must not be any significant flame spread. At no time must the flame spread more than $10^1/_2$ feet (3200 mm) past the centerline of the burners.

The result of impregnating wood with fire-retardant chemicals is a chemical reaction at certain temperature ranges. This reaction reduces the release of certain intermediate products that contribute to the flaming of wood, and also results in the formation of a greater percentage of charcoal and water. Some chemicals are also effective in reducing the oxidation rate for charcoal residue. Fire-retardant chemicals also reduce the heat release rate of fire-retardant-treated wood when burning over a wide range of temperatures. This section gives provisions for the treatment and use of fire-retardant-treated wood.

2303.2.1 Pressure process. For wood products impregnated with chemicals by a pressure process, the process shall be performed in closed vessels under pressures not less than 50 pounds per square inch gauge (psig) (345 kPa).

❖ This section elaborates on the requirements of treatment using a pressure process and specifies a minimum pressure of 50 pound per square inch gauge (psig) (345 kPa).

2303.2.2 Other means during manufacture. For wood products produced by other means during manufacture, the treatment shall be an integral part of the manufacturing process of the wood product. The treatment shall provide permanent protection to all surfaces of the wood product.

❖ This section elaborates on the requirements of treatment using other means during manufacture and requires treatment to be an integral part of the manufacturing process.

2303.2.3 Testing. For wood products produced by other means during manufacture, other than a pressure process, all sides of the wood product shall be tested in accordance with and produce the results required in Section 2303.2. Wood structural panels shall be permitted to test only the front and back faces.

❖ This section provides for added testing of treatments not impregnated by a pressure process. Requiring equivalent performance from all sides of the wood product eliminates any concern over the orientation when it is installed. Only the front and back faces of wood structural panels need to be tested.

2303.2.4 Labeling. Fire-retardant-treated lumber and wood structural panels shall be labeled. The *label* shall contain the following items:

1. The identification *mark* of an *approved agency* in accordance with Section 1703.5.
2. Identification of the treating manufacturer.
3. The name of the fire-retardant treatment.
4. The species of wood treated.
5. Flame spread and smoke-developed index.
6. Method of drying after treatment.
7. Conformance with appropriate standards in accordance with Sections 2303.2.2 through 2303.2.5.
8. For *fire-retardant-treated wood* exposed to weather, damp or wet locations, include the words "No increase in the *listed* classification when subjected to the Standard Rain Test" (ASTM D 2898).

❖ For continued quality, each piece of fire-retardant-treated wood must be identified by an approved agency having a reinspection service. The identification must show the performance rating of the material, including the 30-minute ASTM E 84 test results determined in Section 2303.2, and design adjustment values determined in Section 2303.2.5. The third-party agency that provides the fire-retardant-treated wood label is also required to state on the label that the fire-retardant-treated wood complies with the requirements of Section 2303.2, and that design adjustment values have been determined for the fire-retardant-treated wood in compliance with the provisions of Section 2303.2.5. The fire-retardant-treated wood label must be distinct from the grading label to avoid confusing the two. The grading label provides information about the properties of wood before it is pressure treated with fire-retardant chemicals. The label provides properties of the wood after fire-retardant- treated wood treatment. It is imperative that the fire-retardant-treated wood label be presented in such a manner that it complements the grading label, and does not create confusion over which label takes precedence.

2303.2.5 Strength adjustments. Design values for untreated lumber and wood structural panels, as specified in Section 2303.1, shall be adjusted for *fire-retardant-treated wood*. Adjustments to design values shall be based on an *approved* method of investigation that takes into consideration the effects of the anticipated temperature and humidity to which the *fire-retardant-treated wood* will be subjected, the type of treatment and redrying procedures.

❖ Experience has shown that certain factors can affect the physical properties of fire-retardant-treated wood. Among these factors are the pressure treatment and redrying process and the extremes of temperature and humidity that the fire-retardant-treated wood will be subjected to once installed. The design values for all fire-retardant-treated wood must be adjusted for the effects of the treatment and environmental conditions, such as high temperature and humidity in attic installations. This section requires the determination of these design adjustment values, based on an investigation procedure that includes subjecting the fire-retardant-treated wood to similar temperatures and humidities and approval by the building official. The tested fire-retardant-treated wood must be identical to that which is produced. Items to be considered by the building official reviewing the test procedure include species and grade of the untreated wood and conditioning of wood, such as drying before the fire-retardant treatment process. A fire-retardant wood treater may choose to have its treatment process evaluated by model code evaluation services.

The fire-retardant-treated wood is required by Section 2303.2.1 to be labeled with the design adjustment values. These can take the form of factors that are multiplied by the original design values of the untreated wood to determine its allowable stresses or new allowable stresses can be used that have already been factored down in consideration of the fire-retardant treatment.

2303.2.5.1 Wood structural panels. The effect of treatment and the method of redrying after treatment, and exposure to high temperatures and high humidities on the flexure properties of fire-retardant-treated softwood plywood shall be determined in accordance with ASTM D 5516. The test data developed by ASTM D 5516 shall be used to develop adjustment factors, maximum loads and spans, or both, for untreated plywood design values in accordance with ASTM D 6305. Each manufacturer shall publish the allowable maximum loads and spans for service as floor and roof sheathing for its treatment.

❖ This section references the test standard developed to evaluate the flexural properties of fire-retardant-treated plywood that is exposed to high temperatures. Note that while the section title refers to wood structural panels, the referenced standard is limited to softwood plywood. Therefore, judgment is required in determining the effects of elevated temperature and humidity on other types of wood structural panels. Each product manufacturer is required to publish the allowable maximum loads and spans for floor and roof sheathing for the particular type of treatment.

2303.2.5.2 Lumber. For each species of wood that is treated, the effects of the treatment, the method of redrying after treatment and exposure to high temperatures and high humidities on the allowable design properties of fire-retardant-treated lumber shall be determined in accordance with ASTM D 5664. The test data developed by ASTM D 5664 shall be used to develop modification factors for use at or near room temperature and at elevated temperatures and humidity in accordance with ASTM D 6841. Each manufacturer shall publish the modification factors for service at temperatures of not less than 80°F (27°C) and for roof framing. The roof framing modification factors shall take into consideration the climatological location.

❖ This section references the test standard developed to determine the necessary adjustments to design values for lumber that has been fire-retardant treated and includes the effects of elevated temperatures. Each lumber manufacturer is required to publish the modification factors to design values.

2303.2.6 Exposure to weather, damp or wet locations. Where *fire-retardant-treated wood* is exposed to weather, or damp or wet locations, it shall be identified as "Exterior" to indicate there is no increase in the *listed* flame spread index as defined in Section 2303.2 when subjected to ASTM D 2898.

❖ Some fire-retardant treatments are soluble when exposed to the weather or used under high-humidity conditions. Note that the humidity threshold established for interior applications in Section 2303.2.7 is 92 percent. Therefore, fire-retardant-treated wood used in an interior location that will exceed this threshold should comply with this section. When a fire-retardant-treated wood product is to be exposed to any of the conditions noted in this section, it must be further tested in accordance with ASTM D 2898. Testing requires the material to meet the performance criteria listed in Section 2303.2. The material is then subjected to the ASTM weathering test and retested after drying. There must not be any significant differences in the performance recorded before and after the weathering test.

2303.2.7 Interior applications. Interior *fire-retardant-treated wood* shall have moisture content of not over 28 percent when tested in accordance with ASTM D 3201 procedures at 92-percent relative humidity. Interior *fire-retardant-treated wood* shall be tested in accordance with Section 2303.2.5.1 or 2303.2.5.2. Interior *fire-retardant-treated wood* designated as Type A shall be tested in accordance with the provisions of this section.

❖ The environment in which the fire-retardant-treated wood is used can affect its performance. In order to make sure that the performance will be adequate in a humid interior condition, testing in accordance with ASTM D 3201 as well as that specified in Section 2303.2.5.1 or 2303.2.5.2 is required for all interior wood. Requiring all interior wood to be tested for the effects of high temperature and humidity reduces the chance of premature failure due to improper use of fire-retardant-treated wood.

2303.2.8 Moisture content. *Fire-retardant-treated wood* shall be dried to a moisture content of 19 percent or less for lumber and 15 percent or less for wood structural panels before use. For wood kiln dried after treatment (KDAT), the kiln temperatures shall not exceed those used in kiln drying the lumber and plywood submitted for the tests described in Section 2303.2.5.1 for plywood and 2303.2.5.2 for lumber.

❖ These moisture content thresholds are necessary to prevent leaching of the fire retardant from the wood.

Section 2303.2.5 requires that the strength adjustments consider the redrying procedure, and this section clarifies that the drying temperatures for wood kiln dried after treatment must be consistent with those on which the adjustment factors were based.

2303.2.9 Type I and II construction applications. See Section 603.1 for limitations on the use of *fire-retardant-treated wood* in buildings of Type I or II construction.

❖ Although Type I and II construction typically would only allow noncombustible materials, Section 603.1 allows various parts of a building to be combustible. Specifically, there are 22 items listed that are exceptions to the strict limits on Type I and II construction. There are three conditions listed in that section that will allow fire-retardant-treated wood to be used: in nonbearing partitions where their fire resistance is 2 hours or less; in nonbearing exteriors that do not require a fire-resistance rating; and in roof construction of buildings not over two stories high.

2303.3 Hardwood and plywood. Hardwood and decorative plywood shall be manufactured and identified as required in HPVA HP-1.

❖ Hardwood and decorative plywood, like any other product that undergoes a change in the process of being configured for use in construction, has standard criteria for the manufacturing process and the quality of the materials that are to be used. HPVA HP-1 is the standard for hardwood and decorative plywood, and is one of the family of standards for this type of product.

2303.4 Trusses. Wood trusses shall comply with Sections 2303.4.1 through 2303.4.7.

❖ Trusses are engineered components. They can also be incorporated into a building of otherwise conventional light-frame construction in accordance with Section 2308.4. All types of trusses are included within these provisions, including those with parallel, pitched or curved chords; bowstring, sawtooth or camelback profiles; heavy-bolted timber or light-gage metal-plate-connected elements. Other trusses that are somewhat common in the industry include those with wood chords and metal tubular webs connected with bolts or pins.

2303.4.1 Design. Wood trusses shall be designed in accordance with the provisions of this code and accepted engineering practice. Members are permitted to be joined by nails, glue, bolts, timber connectors, metal connector plates or other *approved* framing devices.

❖ Trusses generally consist of top chord, web members and bottom chord. Under normal loading conditions the top chord is in compression and bending, the bottom chord is in tension and bending and the web members are in tension or compression. The complete design of trusses can be a complex process involving a variety of load combinations. The truss designer must keep the entire system in mind, as trusses do not typically exist as single members, but rather components that must work together to form a complete structural system. Truss designs should address required connections such as truss-to-truss and truss-to-wall plate since they are not covered in the fastening schedule provided by Table 2304.9.1. A variety of references are available to assist in the design of wood trusses, including the American Institute of TImber Construction (AITC) *Timber Construction Manual and Structural Design in Wood.* Helpful design resources are also available from industry associations such as the Structural Building Components Association/Wood Truss Council of America (WTCA) and truss manufacturers, such as Mitek Industries and Alpine Engineered Products.

2303.4.1.1 Truss design drawings. The written, graphic and pictorial depiction of each individual truss shall be provided to the *building official* for approval prior to installation. Truss design drawings shall also be provided with the shipment of trusses delivered to the job site. Truss design drawings shall include, at a minimum, the information specified below:

1. Slope or depth, span and spacing;
2. Location of all joints and support locations;
3. Number of plies if greater than one;
4. Required bearing widths;
5. Design loads as applicable, including;
   - 5.1. Top chord live load;
   - 5.2. Top chord dead load;
   - 5.3. Bottom chord live load;
   - 5.4. Bottom chord dead load;
   - 5.5. Additional loads and locations; and
   - 5.6. Environmental design criteria and loads (wind, rain, snow, seismic, etc.).
6. Other lateral loads, including drag strut loads;
7. Adjustments to wood member and metal connector plate design value for conditions of use;
8. Maximum reaction force and direction, including maximum uplift reaction forces where applicable;
9. Metal-connector-plate type, size and thickness or gage, and the dimensioned location of each metal connector plate except where symmetrically located relative to the joint interface;
10. Size, species and grade for each wood member;
11. Truss-to-truss connections and truss field assembly requirements;
12. Calculated span-to-deflection ratio and maximum vertical and horizontal deflection for live and total load as applicable;
13. Maximum axial tension and compression forces in the truss members; and

14. Required permanent individual truss member restraint location and the method and details of restraint/bracing to be used in accordance with Section 2303.4.1.2.

❖ Because engineered trusses are essential structural components of the building, the code requires truss drawings to be submitted to the building official for review and approval prior to installation. To be consistent regarding the specific information that is required to be included on truss drawings, the code lists 14 items that are to be included in order to effectively review and approve truss submittals. Engineers and contractors frequently depend on the fabricator to design the truss. Sometimes special supports or connections for trusses to the structure below may be required, such as at girder trusses or those necessary to resist high uplift forces, and the truss designer must coordinate these with the building designer. The truss engineer knows and understands how connectors can be applied to resist imposed forces and that knowledge is critical to providing an adequate design to transfer the forces to the supporting structure below. The building engineer can then use that information to properly complete the connection and provide a continuous load path to the foundation.

The majority of the items listed in this section are readily available from any truss designer. Since the documentation is required to be included with the submitted plans and the trusses delivered to the jobsite, the inspector should be able to readily verify that the field installation meets the intent of the truss engineer.

2303.4.1.2 Permanent individual truss member restraint. Where permanent restraint of truss members is required on the truss design drawings, it shall be accomplished by one of the following methods:

1. Permanent individual truss member restraint/bracing shall be installed using standard industry lateral restraint/bracing details in accordance with generally accepted engineering practice. Locations for lateral restraint shall be identified on the truss design drawing.
2. The trusses shall be designed so that the buckling of any individual truss member is resisted internally by the individual truss through suitable means (i.e., buckling reinforcement by T-reinforcement or L-reinforcement, proprietary reinforcement, etc.). The buckling reinforcement of individual members of the trusses shall be installed as shown on the truss design drawing or on supplemental truss member buckling reinforcement details provided by the truss designer.
3. A project-specific permanent individual truss member restraint/bracing design shall be permitted to be specified by any *registered design professional.*

❖ This section emphasizes the need for permanent truss bracing to be installed and lists the options for permanent truss member bracing. There are two different types of lateral bracing for trusses and their systems: temporary and permanent bracing. Temporary lateral bracing is typically required for safe installation before the sheathing is applied and nailed to the top and bottom chords. All forms of lateral bracing are critical and are well described within common truss publications, such as BCSI 1-08, *Guide to Good Practice for Handling, Installing, Restraining & Bracing of Metal Plate Connected Wood Trusses.* Permanent lateral bracing will remain in place for the life of the structure and is intended to stabilize entire trusses so that they can function as the truss designer intends. Item 14 of Section 2303.4.1.1 requires that permanent individual truss member bracing be provided on the truss design drawings.

The truss designer provides engineering for the entire truss system, which includes the design of individual trusses, girder trusses, truss-to-truss connections and both temporary and permanent lateral bracing. In some cases, permanent lateral bracing applied to the webs or the bottom chord may require a connection to the building, such as at a balloon-framed wall, which would then require the involvement of the building designer. In these special cases, the truss designer and building engineers must work together to arrive at acceptable details that will fulfill the truss bracing requirements and also be compatible with both the truss system and the building system.

A type of permanent bracing commonly recognized is the plywood sheathing applied to the top chords or the gypsum wallboard that may be applied to the bottom chords (provided the stiffness of the gypsum wallboard is adequate enough to meet the needs of the truss designer). Since the truss designer is thoroughly familiar with the lateral bracing forces and capabilities of the system that he or she has designed, that engineer is typically the one who must specify whether the sheathing can be used for permanent lateral bracing.

BCSI 1-08, *Guide to Good Practice for Handling, Installing and Bracing Metal Plate Connected Wood Trusses,* jointly published by the WTCA and the Truss Plate Institute (TPI), describes different methods that may be used to provide permanent lateral bracing for web members, including continuous lateral bracing, T-reinforcement, L-reinforcement, scab reinforcement, stacked web reinforcement and proprietary metal reinforcement products. BCSI 1-08 is available from TPI.

2303.4.1.3 Trusses spanning 60 feet or greater. The owner shall contract with any qualified *registered design professional* for the design of the temporary installation restraint/bracing and the permanent individual truss member restraint/bracing for all trusses with clear spans 60 feet (18 288 mm) or greater.

❖ Due to their size, weight and potentially large support reactions, these longer span trusses pose greater stability concerns, making the proper design of the bracing, both temporary and permanent, a key safety measure.

2303.4.1.4 Truss designer. The individual or organization responsible for the design of trusses.

❖ This section defines a term unique to this section, "truss designer." Truss design is a specialty and the designer should be thoroughly familiar with member forces, connection capabilities and behavior as well as what is necessary to cause the entire truss system to function in a structurally sound fashion. Engineers and contractors often depend on the fabricator to design the truss. Sometimes special connections for trusses to the supporting structure may be required, such as those necessary to resist high uplift forces, and the truss designer must coordinate these with the building designer. The truss designer is familiar with the manner in which those connectors can be applied to the truss (i.e., to engage the whole section, attach only to the chord or attach only to the web). In some cases, the truss designer must work under the responsible charge of a registered design professional and the truss drawings must be stamped by such person (see Section 2304.1.4.1).

2303.4.1.4.1 Truss design drawings. Where required by the *registered design professional*, the *building official* or the statutes of the jurisdiction in which the project is to be constructed, each individual truss design drawing shall bear the seal and signature of the truss designer.

Exceptions:

1. Where a cover sheet and truss index sheet are combined into a single sheet and attached to the set of truss design drawings, the single cover/truss index sheet is the only document required to be signed and sealed by the truss designer.
2. When a cover sheet and a truss index sheet are separately provided and attached to the set of truss design drawings, the cover sheet and the truss index sheet are the only documents required to be signed and sealed by the truss designer.

❖ In some cases the truss drawings must be stamped by the registered design professional. The requirement for signing and sealing each truss design drawing, versus signing and sealing only a cover sheet or index sheet, is the recognized approach since it is often used in current construction practice.

2303.4.2 Truss placement diagram. The truss manufacturer shall provide a truss placement diagram that identifies the proposed location for each individually designated truss and references the corresponding truss design drawing. The truss placement diagram shall be provided as part of the truss submittal package, and with the shipment of trusses delivered to the job site. Truss placement diagrams that serve only as a guide for installation and do not deviate from the *permit* submittal drawings shall not be required to bear the seal or signature of the truss designer.

❖ This section describes and defines the term "truss placement diagram." It is intended to minimize the confusion that exists in the construction industry between a variety of terms that are often used interchangeably, such as "installation documents," "construction documents," "shop drawings," etc. The term "truss placement diagram" is preferred by the truss industry and is very specific. The section requires a truss placement diagram to identify the location of each truss and references the corresponding truss design drawing to facilitate inspection and proper installation.

2303.4.3 Truss submittal package. The truss submittal package provided by the truss manufacturer shall consist of each individual truss design drawing, the truss placement diagram, the permanent individual truss member restraint/bracing method and details and any other structural details germane to the trusses; and, as applicable, the cover/truss index sheet.

❖ This section lists the minimum information that must be submitted to the building official by the truss manufacturer.

2303.4.4 Anchorage. The design for the transfer of loads and anchorage of each truss to the supporting structure is the responsibility of the *registered design professional.*

❖ This section clarifies that the transfer of all design loads from the trusses through the building and the connections of the trusses to the supporting structure to resist those loads are the responsibility of the registered design professional who designed the building.

2303.4.5 Alterations to trusses. Truss members and components shall not be cut, notched, drilled, spliced or otherwise altered in any way without written concurrence and approval of a *registered design professional*. Alterations resulting in the addition of loads to any member (e.g., HVAC equipment, piping, additional roofing or insulation, etc.) shall not be permitted without verification that the truss is capable of supporting such additional loading.

❖ In any structural member, the existence of holes or notches can affect its ability to resist design loads. The effects of any alterations on the capacity of truss members must be assessed by a registered design professional. This requirement applies not only to the truss chords and webs, but also to permanent lateral bracing components. Additionally, trusses are designed to carry a specific load and any loads that are added must be assessed by a registered design professional. The roof live load must also be considered in any review of existing trusses, as that live load also covers reroofing operations that may take place in the future.

2303.4.6 TPI 1 specifications. In addition to Sections 2303.4.1 through 2303.4.5, the design, manufacture and quality assurance of metal-plate-connected wood trusses shall be in accordance with TPI 1. Job-site inspections shall be in compliance with Section 110.4, as applicable.

❖ Metal-plate-connected wood trusses have been a popular product for the construction of wood-frame build-ings for many years. There are significant benefits to building with these engineered systems of wood and metal connectors; however, there are also criteria as-sociated with the design, manufacturing, transportation and placement of trusses that are critical to their proper use in a structure. TPI 1, *National Design Standard for Metal-plate-connected Wood Truss Construction*, is the industry developed standard for managing this process. The standard establishes minimum requirements for the design and construction of metal-plate-connected wood trusses and describes the lumber and steel materials used and design procedures for truss members and joints. Responsibilities, methods for evaluating metal connector plates and truss manufacturing quality assurance are also contained in the standard.

The section also requires a truss manufacturer to have an agency review its operations to determine compliance with this standard. This type of evaluation by a qualified person is necessary to verify the quality of the product and its ability to perform as designed.

2303.4.7 Truss quality assurance. Trusses not part of a manufacturing process in accordance with either Section 2303.4.6 or a referenced standard, which provides requirements for quality control done under the supervision of a third-party quality control agency, shall be manufactured in compliance with Sections 1704.2.5 and 1705.5, as applicable.

❖ Trusses fabricated under the TPI 1 standard are subject to third-party supervision for quality assurance purposes. Likewise, trusses fabricated under other standards that are referenced in the code are generally subject to quality assurance by third-party supervision. Trusses that are not manufactured under such quality assurance processes require special inspection as required by Chapter 17. In this case, the section references Section 1704.2.5 for the inspection of fabricators and Section 1705.5 for special inspection of wood fabrication processes.

2303.5 Test standard for joist hangers. For the required test standards for joist hangers see Section 1711.1.

❖ Metal joist hangers are very common in modern wood-frame construction. Tests for joist hangers and connections are included in the criteria for special inspections for materials and test standards in Section 1711. This section references ASTM D 1761 and establishes the criteria for use of lumber with a specific gravity of 0.49 or larger, but not more than 0.55 in accordance with the AF&PA NDS. Additionally, Section 1711.2 adds limits on the vertical load capacity for joist hangers by requiring that there be consistency of the tested capacity within 20 percent of the average ultimate load. If this cannot be achieved, additional testing is required (see Section 1711).

2303.6 Nails and staples. Nails and staples shall conform to requirements of ASTM F 1667. Nails used for framing and sheathing connections shall have minimum average bending yield strengths as follows: 80 kips per square inch (ksi) (551 MPa) for shank diameters larger than 0.177 inch (4.50 mm) but not larger than 0.254 inch (6.45 mm), 90 ksi (620 MPa) for shank diameters larger than 0.142 inch (3.61 mm) but not larger than 0.177 inch (4.50 mm) and 100 ksi (689 MPa) for shank diameters of at least 0.099 inch (2.51 mm) but not larger than 0.142 inch (3.61 mm).

❖ The code specifies the material standard for driven nails and staples by reference to ASTM F 1667, *Specification for Driven Fasteners: Nails, Spikes and Staples*. ASTM F 1667 specifies dimensions, etc., for nails, spikes and staples. Minimum nail strengths are specified based upon the stiffness of the nail shank. Minimum fastening is provided in the nailing schedule in Table 2304.9.1.

Just as the capacity of nailed connections is dependent upon the stiffness of the nail shank, the capacity of stapled connections is dependent upon the stiffness of the staple legs. Research was performed at Virginia Tech on stapled connections paralleling the research that was performed there on nailed connections. The research on nailed connections was incorporated into the AF&PA NDS. The research on stapled connections has been incorporated into the ICC Evaluation Services (ICC ES) *Acceptance Criteria 201* for steel wire staples used to connect wood or metal elements into wood and wood-based materials. The determination of staple-bending moments is in accordance with ASTM F 1575 as well as the modifications of that standard that were made in Sections 3.5.1 through 3.5.4 of ICC ES *Acceptance Criteria 201*.

2303.7 Shrinkage. Consideration shall be given in design to the possible effect of cross-grain dimensional changes considered vertically which may occur in lumber fabricated in a green condition.

❖ Lumber in a "green condition" has a moisture content higher than that used to define a "dry condition" under the applicable grading rules, which is reflected on the grade mark (see Figure 2303.1.1). Because it will shrink more than dry lumber, use of lumber that is fabricated in a green condition requires consideration of the effects of cross-grain shrinkage. The extent of cross-grain shrinkage is a function of the lumber's moisture at the time of construction, the amount of drying that occurs after construction and the in-service moisture content.

SECTION 2304
GENERAL CONSTRUCTION REQUIREMENTS

2304.1 General. The provisions of this section apply to design methods specified in Section 2301.2.

❖ This section is intended to provide some of the more important details of wood-framed construction and, as a result, provide consistency in building inspections. The methods of design are noted in Section 2301.2 and in the industry documents specified therein. Regardless of the design method chosen, the construction requirements of this section must be followed and enforced. In general, Section 2304 is related more to construction requirements than it is to design requirements.

Wood-framed construction has not changed much over the years, which is beneficial for a variety of reasons. First of all, many framers prefer doing things the way they've always done them in the past, most of which are detailed in carpentry manuals or architectural detailing books. To the extent such practices do not violate the need to support or transfer loads, it is preferable to allow them. Another reason that it is important to reach some type of agreement with these framers is to help maintain a consistent product that a building inspector would be familiar with from building to building. This section helps to achieve this goal by having a consistent set of rules that apply to all wood-frame design methods.

2304.2 Size of structural members. Computations to determine the required sizes of members shall be based on the net dimensions (actual sizes) and not nominal sizes.

❖ ASD and LRFD methods require the use of the actual lumber dimensions that occur after the wood has been surfaced and dried. Typically, dimension lumber will shrink $^1/_2$ inch (12.7 mm) in thickness and $^1/_2$ inch to $^3/_4$ inch (12.7 mm to 19 mm) in face width as noted in the American Softwood Lumber (ASL) standard DOC PS 20. This provision is an attempt to make sure that if design principles are applied to a specific element, the designer uses design values consistent with the actual net section properties and load-carrying capacity of the member, as reflected in the referenced design standards. Nominal sizes, standard dressed dimensions and section properties of (S4S) lumber and glued-laminated timber are given in the AF&PA NDS Supplement.

2304.3 Wall framing. The framing of exterior and interior walls shall be in accordance with the provisions specified in Section 2308 unless a specific design is furnished.

❖ The conventional construction criteria in Section 2308 provide great detail and specifications for the construction of typical stud-framed walls. Under the majority of circumstances, the methods and procedures for framing stud walls in accordance with Section 2308.9 will provide the necessary resistance to conventional-level loads for a typical stud wall. Although special loading conditions would need to be considered in the design of nonconventional walls, as indicated by Section 2308.4, the completed product will likely be similar in shape and form to an otherwise conventionally framed wall.

2304.3.1 Bottom plates. Studs shall have full bearing on a 2-inch-thick (actual $1^1/_2$-inch, 38 mm) or larger plate or sill having a width at least equal to the width of the studs.

❖ It is necessary to provide a continuous wood plate at the bottom of the wall with a minimum nominal thickness of 2 inches (51 mm), so that the loads supported by the studs in a wood-framed wall are adequately transferred to the supporting structural members or foundation below. The plate is required to be at least the same width as the studs to achieve full bearing to ensure that the capacity of each stud is realized. This requirement is commonly known in the wood construction industry and will typically be met unless there is some type of problem in the field or such a condition occurs in existing construction. If the bottom plate of the wall is found to be narrower than the width of the studs for some reason, the structural issue that needs to be checked is the perpendicular-to-grain bearing stress in that plate and axial load capacity of the stud. The AF&PA NDS provides guidelines for calculating and checking these stresses.

2304.3.2 Framing over openings. Headers, double joists, trusses or other *approved* assemblies that are of adequate size to transfer loads to the vertical members shall be provided over window and door openings in load-bearing walls and partitions.

❖ Stress in stud-framed walls need to be continuous in order to carry the imposed loads from the top plate to the bottom plate and eventually to the foundation. When there is an opening in the framed wall, this continuity is interrupted. It is necessary to transfer the loads from above the opening to adjacent structural members, which could be a single stud, multiple studs or a solid post. The loads above the opening are typically carried by a solid wood beam or header, a combination of cripple studs and plates or some other combination of members that provides the needed capacity to adequately span between the supporting studs on each side of the opening.

An important structural consideration is the connection between the member spanning the opening and the supporting member. Whether that connection is a simple direct bearing support, a hanger or some other type, it is necessary to provide a mechanism that will transfer not only the vertical loads but also forces that occur perpendicular to the wall plane where such a consideration is warranted.

2304.3.3 Shrinkage. Wood walls and bearing partitions shall not support more than two floors and a roof unless an analysis satisfactory to the *building official* shows that shrinkage of the wood framing will not have adverse effects on the structure or any plumbing, electrical or mechanical systems, or other equipment installed therein due to excessive shrinkage or differential movements caused by shrinkage. The analysis shall also show that the roof drainage system and the foregoing systems or equipment will not be adversely affected or, as an alternate, such systems shall be designed to accommodate the differential shrinkage or movements.

❖ The adverse effects of wood shrinkage are greatest in multistory buildings having nonstructural components and supported equipment that may not be able to tolerate significant differential movements caused by shrinkage. This section states conditions under which shrinkage of the wood framing must be considered in the construction of wood-framed wall systems. An analysis of this phenomenon is required in order to determine the effect of shrinkage on the building system that is supported by the wall below. A typical wood-framed building will be constructed of mostly dry lumber, which has a moisture content of 19 percent or less. When the supply of dry lumber is limited, the contractor may need to use green lumber, which has a moisture content greater than 19 percent. The overall effect of lumber shrinkage on a multistory wood-framed building is described in AF&PA, *Allowable Stress Design Manual for Wood Construction Design of Wood Structures–ASD.*

Another example of the adverse effects of shrinkage is the impact it can have on shear wall performance. Shrinkage has been recognized as a factor in loose hold-down anchor bolt nuts in earthquake-damaged buildings. In this situation, the hold-down cannot resist tension until the slack is taken up, resulting in excessive movement of the shear wall under lateral loading. Since in most cases the shear wall sheathing conceals the hold-downs, this problem often goes undetected. Additional slippage or looseness due to wood shrinkage can also contribute to the overall wall lateral displacement, which results in more story drift than expected. This is especially the case for narrow shear walls (e.g., shear walls with a height-to-width ratio more than 1:1) since the variable d_a in Equation 23-2 plays a larger role in the lateral displacement of such walls based upon wall geometry.

2304.4 Floor and roof framing. The framing of wood-joisted floors and wood framed roofs shall be in accordance with the provisions specified in Section 2308 unless a specific design is furnished.

❖ Similar to the criteria for stud wall construction, floors may be constructed using the conventional construction methods prescribed in Section 2308, subject to the limitations described therein. One such limitation is a restriction on the magnitude of the applied dead, live and snow loads that may be applied to the members described in the rafter span tables. Span Tables 2308.8(1) and (2); 2308.9.5; 2308.9.6; 2308.10.2(1) and (2) and 2308.10.3(1), (2), (3), (4), (5) and (6) each describe a different loading condition for joists, rafters, headers and girders at the ceiling, floor and roof. The maximum floor load covered by Table 2308.8 is 40 pounds per square foot (psf) (1.9 kN/m^2) live load. Table 2308.10.2 covers only dead loads of 5 psf (0.24 kN/m^2) for ceiling rafters and Table 2308.10.3 only provides for up to a 50 psf (2.4 kN/m^2) ground snow load. The range of required loading is much larger, as demonstrated in Table 1607.1. The only structures for which a 40 psf (1.9 kN/m^2) live load on the floor is allowed are catwalks; fire escapes on single-family dwellings; private rooms and wards in a hospital; cell blocks; school classrooms and all areas of residential occupancies other than public areas. All other uses require a higher design floor load, in which case these tables do not apply and an engineered solution in accordance with Section 2308.4 is required.

2304.5 Framing around flues and chimneys. Combustible framing shall be a minimum of 2 inches (51 mm), but shall not be less than the distance specified in Sections 2111 and 2113 and the *International Mechanical Code*, from flues, chimneys and fireplaces, and 6 inches (152 mm) away from flue openings.

❖ This provision is duplicated in the *International Mechanical Code*® (IMC®), but must be included in both places in order to alert the framer, installer and inspector to the requirement. When exposed to heat for a prolonged period of time, the characteristics of wood may change. In particular, the temperature at which wood will ignite can be lowered. Repeated exposure to temperature extremes has a cumulative effect on the mechanical properties of wood, whereby sustained heating at temperatures above 212°F (100°C) may cause negative nonreversible effects on these properties. In order to help mitigate this phenomenon, a separation is provided so that the wood will not be directly exposed or noncombustible firestopping material is placed between the wood and the heat source. Some type of separation has been required in legacy model codes and is detailed in commonly used construction manuals.

2304.6 Wall sheathing. Except as provided for in Section 1405 for weatherboarding or where stucco construction that complies with Section 2510 is installed, enclosed buildings shall be sheathed with one of the materials of the nominal thickness specified in Table 2304.6 or any other *approved* material of equivalent strength or durability.

❖ Wall sheathing material can be any of those listed in Table 2304.6. There are two main reasons for providing sheathing to at least one face of wood-framed walls: (1) to stabilize the studs against buckling when the wall carries vertical loads and (2) to provide a nominal ability to transfer lateral forces from the roof or floor deck to the sill plate and anchor bolts. Studs nailed to the top and bottom plate are stable to a degree, but will collapse with significant force placed on them unless there is some type of stabilizing membrane or brace in the plane of the wall. Sheathing, whether it is a heavy structural membrane found on exterior bearing walls or a light one that would be used for nonbearing interior partitions, provides necessary in-plane stability for the walls. Some sheathing may also serve as insulation for the building, in addition to providing lateral bracing. Other sheathing may provide a finish for the wall. Sheathing is not the only means of bracing a wall.

This section establishes that only the sheathing types listed in Table 2304.6, which include board and various panel products (structural panels, boards, gypsum and reinforced cement mortar) are to be used for walls. There is an allowance that other materials may be used if it is shown they are capable of providing equivalent strength and durability. Let-in braces are common for temporary bracing of the wall against in-plane racking forces, but they provide only minimal resistance to lateral forces. Since many let-in braces failed in tension or at the bottom plate connection during the 1971 San Fernando earthquake, their use is limited to structures with low seismic risk for conventional braced wall panels in accordance with Section 2308.9.3.

2304.6.1 Wood structural panel sheathing. Where wood structural panel sheathing is used as the exposed finish on the outside of exterior walls, it shall have an exterior exposure durability classification. Where wood structural panel sheathing is used elsewhere, but not as the exposed finish, it shall be of a type manufactured with exterior glue (Exposure 1 or Exterior). Wood structural panel wall sheathing or siding used as structural sheathing shall be capable of resisting wind pressures in accordance with Section 1609. Maximum wind speeds for wood structural panel sheathing used to resist wind pressures shall be in accordance with Table 2304.6.1 for enclosed buildings with a mean roof height not greater than 30 feet (9144 mm) and a topographic factor (K_{zt}) of 1.0.

❖ Wood structural panels are defined in Chapter 2 as being plywood, OSB and composite panels. Each of these types of panel products has identical structural characteristics when manufactured to meet the DOC PS1 or PS 2 standards in accordance with Section 2303.1.4. This section requires equivalent exposure durability performance. It requires the classification to be Exposure 1 or Exterior when the material is likely to be exposed to the weather. According to the APA, "Exterior"-type panels are suitable for applications permanently ex-posed to weather, "Exposure 1"-type panels may be used for extended temporary exposure to weather (such as during construction when long delays are expected prior to providing permanent protection), "Exposure 2"-type panels apply to protected applications not exposed to high humidity and may be briefly subjected to the weather while "Interior"-type panels are to be used only where permanent protection against the weather is provided.

Assessment of high-wind events has shown that failure of wall sheathing, in winds as low as 60 mph (27 m/s), has led to damage due to breaching of the wall envelope. This section provides guidance for using wood structural panels to resist wind pressures. Table 2304.6.1 provides the required fastening, span rating, panel thickness and stud spacing for wood structural panel cladding based on wind speed and exposure category.

TABLE 2304.6. See below.

❖ Table 2304.6 reinforces Section 2304.6, which requires all wall sheathing to meet a minimum set of requirements, including minimum material thickness and maximum wall stud spacing, depending on the type of sheathing material chosen for a particular project. Because of loading conditions or the design of the assembly to which the wall sheathing is attached, the sheathing may have to be thicker or the wall studs may need to be placed closer together.

TABLE 2304.6.1. See page 23-17.

❖ This table provides guidelines for using wood structural panel wall sheathing to resist component and cladding wind pressures normal to the wall. The table clearly specifies the maximum wind speed and exposure category permitted for wood structural panel wall sheathing. It was developed using the component and cladding wind pressures (wind speed) given in Figure 30.5-1 of ASCE 7 with the wood structural panel capacities based on the DOC PS 2 standard, engineering calculations as well as the APA panel design specification that is referenced in Section 2306.1. Nail head pull through and withdrawal were also considered in addition to panel stiffness and bending strength. The panel-fastener capacity was based on the area tributary to a single critical nail.

2304.6.2 Interior paneling. Softwood wood structural panels used for interior paneling shall conform to the provisions of Chapter 8 and shall be installed in accordance with Table 2304.9.1. Panels shall comply with DOC PS 1, DOC PS 2 or ANSI/APA PRP 210. Prefinished hardboard paneling shall meet the requirements of CPA/ANSI A135.5. Hardwood plywood shall conform to HPVA HP-1.

❖ Interior paneling, similar to exterior paneling, must meet minimum standards of acceptance. DOC PS 1 and PS 2 establish the general standards for all softwood veneer panel products. CPA/ANSI A135.5 and HPVA HP-1 address the criteria for hardboard paneling and hardwood veneered plywood, respectively.

TABLE 2304.6
MINIMUM THICKNESS OF WALL SHEATHING

SHEATHING TYPE	MINIMUM THICKNESS	MAXIMUM WALL STUD SPACING
Wood boards	$^5/_8$ inch	24 inches on center
Fiberboard	$^1/_2$ inch	16 inches on center
Wood structural panel	In accordance with Tables 2308.9.3(2) and 2308.9.3(3)	—
M-S "Exterior Glue" and M-2 "Exterior Glue" Particleboard	In accordance with Section 2306.3 and Table 2308.9.3(4)	—
Gypsum sheathing	$^1/_2$ inch	16 inches on center
Gypsum wallboard	$^1/_2$ inch	24 inches on center
Reinforced cement mortar	1 inch	24 inches on center

For SI: 1 inch = 25.4 mm.

2304.7 Floor and roof sheathing. Structural floor sheathing and structural roof sheathing shall comply with Sections 2304.7.1 and 2304.7.2, respectively.

❖ Floor and roof sheathing perform several functions. They are the primary base for both the walking surface for the floor and the main support of the roofing materials. Additionally, the sheathing transfers horizontally applied lateral forces to the vertical lateral-force-resisting elements. Floor sheathing is the structural element in a wood floor that transfers live loads to the floor joists.

Roof sheathing is typically covered by the roofing material. The sheathing typically provides the only means by which loads from wind and snow are transferred to the structural framing members so the fastening that completes this load path is critical. Panels are typically "span rated," which provides the designer guidance as to the maximum spacing that the framing members can be placed to adequately support imposed loads. The tables are for both lumber and wood structural panel floor and roof sheathing.

2304.7.1 Structural floor sheathing. Structural floor sheathing shall be designed in accordance with the general provisions of this code and the special provisions in this section.

Floor sheathing conforming to the provisions of Table 2304.7(1), 2304.7(2), 2304.7(3) or 2304.7(4) shall be deemed to meet the requirements of this section.

❖ The tables referenced are used for both floor and roof applications (see commentary, Section 2304.7.2). In a two-layer floor sheathing system, the structural panels are placed at the bottom ("subfloor") and fastened directly to the framing members, and the top layer ("underlayment") is typically a different grade of material, occasionally specified to resist indentations depending on what floor finish material is to be installed.

2304.7.2 Structural roof sheathing. Structural roof sheathing shall be designed in accordance with the general provisions of this code and the special provisions in this section.

Roof sheathing conforming to the provisions of Table 2304.7(1), 2304.7(2), 2304.7(3) or 2304.7(5) shall be deemed to meet the requirements of this section. Wood structural panel roof sheathing shall be bonded by exterior glue.

❖ There are four tables referenced in this section that establish either allowable load or allowable span of structural sheathing typically produced by mills in North America. Using the tables is straightforward: simply locate with the span between structural framing members, and depending on the direction of the framing, determine the minimum thickness required by the code. It may be necessary to install thicker material if the loads require additional support.

Lumber and wood structural panel sheathing are required to be marked with grade stamps indicating the lumber grade, panel span rating or panel grade required by the appropriate table based on the conditions of use.

TABLE 2304.7(1). See page 23-18.

❖ Table 2304.7(1) applies to "surfaced dry" or "surfaced unseasoned" lumber used for floor or roof sheathing. These terms refer to the condition of the lumber at the time it was surfaced, or finished. Surfaced dry indicates a maximum moisture content of 19 percent at the time of surfacing. This is critical when determining the actual dimension of the lumber. A piece that is

TABLE 2304.6.1
MAXIMUM NOMINAL DESIGN WIND SPEED, V_{asd} PERMITTED FOR WOOD STRUCTURAL PANEL WALL SHEATHING USED TO RESIST WIND PRESSURES[a, b, c]

MINIMUM NAIL		MINIMUM WOOD STRUCTURAL PANEL SPAN RATING	MINIMUM NOMINAL PANEL THICKNESS (inches)	MAXIMUM WALL STUD SPACING (inches)	PANEL NAIL SPACING		MAXIMUM NOMINAL DESIGN WIND SPEED, V_{asd}[d] (MPH)		
Size	Penetration (inches)				Edges (inches o.c.)	Field (inches o.c.)	Wind exposure category		
							B	C	D
6d common (2.0" × 0.113")	1.5	24/0	$^3/_8$	16	6	12	110	90	85
		24/16	$^7/_{16}$	16	6	12	110	100	90
						6	150	125	110
8d common (2.5" × 0.131")	1.75	24/16	$^7/_{16}$	16	6	12	130	110	105
						6	150	125	110
				24	6	12	110	90	85
						6	110	90	85

For SI: 1 inch = 25.4 mm, 1 mile per hour = 0.447 m/s.

a. Panel strength axis shall be parallel or perpendicular to supports. Three-ply plywood sheathing with studs spaced more than 16 inches on center shall be applied with panel strength axis perpendicular to supports.

b. The table is based on wind pressures acting toward and away from building surfaces in accordance with Section 30.7 of ASCE 7. Lateral requirements shall be in accordance with Section 2305 or 2308.

c. Wood structural panels with span ratings of wall-16 or wall-24 shall be permitted as an alternative to panels with a 24/0 span rating. Plywood siding rated 16 o.c. or 24 o.c. shall be permitted as an alternative to panels with a 24/16 span rating. Wall-16 and plywood siding 16 o.c. shall be used with studs spaced a maximum of 16 inches o.c.

d. V_{asd} shall be determined in accordance with Section 1609.3.1.

surfaced dry is narrower than a piece that is surfaced unseasoned (often referred to as surfaced green). The grade stamp will indicate whether the lumber was surfaced dry or surfaced green.

TABLE 2304.7(2). See below.

❖ The different agencies that publish grading rules are listed and are an important part of the criteria used in determining the adequacy of structural lumber sheathing. All of these agencies are regulated by the U.S. DOC through the ALSC in order to maintain some type of standard by which stress-grading principles are used to create grading rules. These requirements are outlined in ALSC standard DOC PS 20. The rules-writing agencies included in this table are the National Lumber Grades Authority (NLGA), West Coast Lumber Inspection Bureau (WCLIB), Western Wood Products Association (WWPA), Northern Softwood Lumber Bureau (NSLB), Northeastern Lumber Manufacturers Association (NELMA), Southern Pine Inspection Bureau (SPIB) and Redwood Inspection Service (RIS).

TABLE 2304.7(3). See page 23-19.

❖ Wood structural panels are also graded, but the type of grade stamp used identifies the thickness and the span capabilities of the piece. The panel span rating (first column of the table), also known as a "panel index," will appear on the grade stamp. The first number on the panel span rating indicates the allowable spacing of support members (rafters or trusses) in a roof application. The second number indicates the allowable spacing of floor joists or trusses in a floor application. For instance, if the span rating is $^{24}/_{16}$, the panel may be used on roof joists spaced 24 inches (610 mm) on center (o.c.) and on floor joists spaced 16 inches (406 mm) o.c. Span ratings were developed for easy identification and to assist in verifying the correct use of panels in the field. According to *Design of Wood Structures—ASD/LRFD*, typical wood structural panel applications are not controlled so much by the uniform load requirements, but depend more on the basis of panel deflection under concentrated loads and judged by the panel's performance under passing traffic.

For the panel span rating to be applicable, other conditions of the table for loading and use must be met. For instance, for sheathing with a panel index of 24/16 to be used on roof rafters spaced 16 inches (406 mm) o.c., the snow load for the roof cannot exceed 40 psf (1.92 MPa) (listed in the sixth column of the table). The allowable spacing of support members is also dependent on the type of edge support provided for the panels (see Note f). The table is applicable only when the strength axis of the panel is perpendicular to the supports, in other words, when the long edge of the panel is perpendicular to the joists or rafters.

TABLE 2304.7(4). See page 23-20.

❖ The subfloor is the structural element that supports the floor loads and is the substrate (supporting layer) for the underlayment. Underlayment is the smooth surface used as the substrate for the floor covering. Combination subfloor-underlayment is a single layer of sheathing that meets the structural requirements for subfloor and has a smooth surface suitable as underlayment for the floor covering. Wood species are grouped on the basis of their mechanical properties in DOC PS 1. Because bending stiffness and bending strength are the most important properties

TABLE 2304.7(1)
ALLOWABLE SPANS FOR LUMBER FLOOR AND ROOF SHEATHING[a, b]

SPAN (inches)	MINIMUM NET THICKNESS (inches) OF LUMBER PLACED			
	Perpendicular to supports		Diagonally to supports	
	Surfaced dry[c]	Surfaced unseasoned	Surfaced dry[c]	Surfaced unseasoned
Floors				
24	$^{3}/_{4}$	$^{25}/_{32}$	$^{3}/_{4}$	$^{25}/_{32}$
16	$^{5}/_{8}$	$^{11}/_{16}$	$^{5}/_{8}$	$^{11}/_{16}$
Roofs				
24	$^{5}/_{8}$	$^{11}/_{16}$	$^{3}/_{4}$	$^{25}/_{32}$

For SI: 1 inch = 25.4 mm.

a. Installation details shall conform to Sections 2304.7.1 and 2304.7.2 for floor and roof sheathing, respectively.

b. Floor or roof sheathing conforming with this table shall be deemed to meet the design criteria of Section 2304.7.

c. Maximum 19-percent moisture content.

TABLE 2304.7(2)
SHEATHING LUMBER, MINIMUM GRADE REQUIREMENTS: BOARD GRADE

SOLID FLOOR OR ROOF SHEATHING	SPACED ROOF SHEATHING	GRADING RULES
Utility	Standard	NLGA, WCLIB, WWPA
4 common or utility	3 common or standard	NLGA, WCLIB, WWPA, NSLB or NELMA
No. 3	No. 2	SPIB
Merchantable	Construction common	RIS

for many plywood uses, the species groups are based on bending stiffness and bending strength. Single layer combined subfloor and underlayment is currently the most common method used for wood floor construction.

TABLE 2304.7(5). See page 23-20.

❖ The table provides criteria for panels that may be used on roofs with significant live loads, the maximum live load being listed at 90 psf (4.31 MPa). That being the case, it is clear that the live loads in excess of 20 psf (0.958 MPa) (which is generally the maximum roof live loading determined in Section 1607.12.2) is primarily referring to the roof snow load rather than a live load, strictly speaking. For further discussion on required snow loads, see the commentary to Section 1608. Required panel grade and thickness is based on the spacing of rafters or trusses and the maximum allowable load at that span. Note that the table is based on 5 ply plywood. In some cases, the allowable loads for four ply plywood and composite panels must be reduced by 15 percent.

2304.8 Lumber decking. Lumber decking shall be designed and installed in accordance with the general provisions of this code and Sections 2304.8.1 through 2304.8.5.3.

❖ Section 2304.8 has provisions for lumber decking that were previously contained in the AITC 112 standard, which has been discontinued.

2304.8.1 General. Each piece of lumber decking shall be square-end trimmed. When random lengths are furnished, each piece shall be square end trimmed across the face so that at least 90 percent of the pieces are within 0.5 degrees (0.00873 rad) of square. The ends of the pieces shall be permitted to be beveled up to 2 degrees (0.0349 rad) from the vertical with the exposed face of the piece slightly longer than the opposite face of the piece. Tongue-and-groove decking shall be installed with the tongues up on sloped or pitched roofs with pattern faces down.

❖ Along with design requirements in Section 2306.1.4, this section provides design and installation requirements for lumber decking, including tongue-and-groove as well as mechanically laminated decking.

TABLE 2304.7(3)
ALLOWABLE SPANS AND LOADS FOR WOOD STRUCTURAL PANEL SHEATHING AND SINGLE-FLOOR GRADES CONTINUOUS OVER TWO OR MORE SPANS WITH STRENGTH AXIS PERPENDICULAR TO SUPPORTS[a, b]

SHEATHING GRADES		ROOF[c]				FLOOR[d]
Panel span rating roof/ floor span	Panel thickness (inches)	Maximum span (inches)		Load[e] (psf)		Maximum span (inches)
		With edge support[f]	Without edge support	Total load	Live load	
16/0	$^3/_8$	16	16	40	30	0
20/0	$^3/_8$	20	20	40	30	0
24/0	$^3/_8$, $^7/_{16}$, $^1/_2$	24	20[g]	40	30	0
24/16	$^7/_{16}$, $^1/_2$	24	24	50	40	16
32/16	$^{15}/_{32}$, $^1/_2$, $^5/_8$	32	28	40	30	16[h]
40/20	$^{19}/_{32}$, $^5/_8$, $^3/_4$, $^7/_8$	40	32	40	30	20[h,i]
48/24	$^{23}/_{32}$, $^3/_4$, $^7/_8$	48	36	45	35	24
54/32	$^7/_8$, 1	54	40	45	35	32
60/32	$^7/_8$, $1^1/_8$	60	48	45	35	32
SINGLE FLOOR GRADES		ROOF[c]				FLOOR[d]
Panel span rating	Panel thickness (inches)	Maximum span (inches)		Load[e] (psf)		Maximum span (inches)
		With edge support[f]	Without edge support	Total load	Live load	
16 o.c.	$^1/_2$, $^{19}/_{32}$, $^5/_8$	24	24	50	40	16[h]
20 o.c.	$^{19}/_{32}$, $^5/_8$, $^3/_4$	32	32	40	30	20[h,i]
24 o.c.	$^{23}/_{32}$, $^3/_4$	48	36	35	25	24
32 o.c.	$^7/_8$, 1	48	40	50	40	32
48 o.c.	$1^3/_{32}$, $1^1/_8$	60	48	50	40	48

For SI: 1 inch = 25.4 mm, 1 pound per square foot = 0.0479 kN/m^2.

a. Applies to panels 24 inches or wider.

b. Floor and roof sheathing conforming with this table shall be deemed to meet the design criteria of Section 2304.7.

c. Uniform load deflection limitations $^1/_{180}$ of span under live load plus dead load, $^1/_{240}$ under live load only.

d. Panel edges shall have approved tongue-and-groove joints or shall be supported with blocking unless $^1/_4$-inch minimum thickness underlayment or $1^1/_2$ inches of approved cellular or lightweight concrete is placed over the subfloor, or finish floor is $^3/_4$-inch wood strip. Allowable uniform load based on deflection of $^1/_{360}$ of span is 100 pounds per square foot except the span rating of 48 inches on center is based on a total load of 65 pounds per square foot.

e. Allowable load at maximum span.

f. Tongue-and-groove edges, panel edge clips (one midway between each support, except two equally spaced between supports 48 inches on center), lumber blocking or other. Only lumber blocking shall satisfy blocked diaphragm requirements.

g. For $^1/_2$-inch panel, maximum span shall be 24 inches.

h. Span is permitted to be 24 inches on center where $^3/_4$-inch wood strip flooring is installed at right angles to joist.

i. Span is permitted to be 24 inches on center for floors where $1^1/_2$ inches of cellular or lightweight concrete is applied over the panels.

TABLE 2304.7(4)
ALLOWABLE SPAN FOR WOOD STRUCTURAL PANEL COMBINATION SUBFLOOR-UNDERLAYMENT (SINGLE FLOOR)[a, b]
(Panels Continuous Over Two or More Spans and Strength Axis Perpendicular to Supports)

IDENTIFICATION	MAXIMUM SPACING OF JOISTS (inches)				
	16	20	24	32	48
Species group[c]	Thickness (inches)				
1	$^1/_2$	$^5/_8$	$^3/_4$	—	—
2, 3	$^5/_8$	$^3/_4$	$^7/_8$	—	—
4	$^3/_4$	$^7/_8$	1	—	—
Single floor span rating[d]	16 o.c.	20 o.c.	24 o.c.	32 o.c.	48 o.c.

For SI: 1 inch = 25.4 mm, 1 pound per square foot = 0.0479 kN/m^2.

a. Spans limited to value shown because of possible effects of concentrated loads. Allowable uniform loads based on deflection of $^1/_{360}$ of span is 100 pounds per square foot except allowable total uniform load for 1$^1/_8$-inch wood structural panels over joists spaced 48 inches on center is 65 pounds per square foot. Panel edges shall have approved tongue-and-groove joints or shall be supported with blocking, unless $^1/_4$-inch minimum thickness underlayment or 1$^1/_2$ inches of approved cellular or lightweight concrete is placed over the subfloor, or finish floor is $^3/_4$-inch wood strip.

b. Floor panels conforming with this table shall be deemed to meet the design criteria of Section 2304.7.

c. Applicable to all grades of sanded exterior-type plywood. See DOC PS 1 for plywood species groups.

d. Applicable to Underlayment grade, C-C (Plugged) plywood, and Single Floor grade wood structural panels.

TABLE 2304.7(5)
ALLOWABLE LOAD (PSF) FOR WOOD STRUCTURAL PANEL ROOF SHEATHING CONTINUOUS OVER TWO OR MORE SPANS AND STRENGTH AXIS PARALLEL TO SUPPORTS
(Plywood Structural Panels Are Five-Ply, Five-Layer Unless Otherwise Noted)[a, b]

PANEL GRADE	THICKNESS (inch)	MAXIMUM SPAN (inches)	LOAD AT MAXIMUM SPAN (psf)	
			Live	Total
Structural I sheathing	$^7/_{16}$	24	20	30
	$^{15}/_{32}$	24	35[c]	45[c]
	$^1/_2$	24	40[c]	50[c]
	$^{19}/_{32}$, $^5/_8$	24	70	80
	$^{23}/_{32}$, $^3/_4$	24	90	100
Sheathing, other grades covered in DOC PS 1 or DOC PS 2	$^7/_{16}$	16	40	50
	$^{15}/_{32}$	24	20	25
	$^1/_2$	24	25	30
	$^{19}/_{32}$	24	40[c]	50[c]
	$^5/_8$	24	45[c]	55[c]
	$^{23}/_{32}$, $^3/_4$	24	60[c]	65[c]

For SI: 1 inch = 25.4 mm, 1 pound per square foot = 0.0479 kN/m^2.

a. Roof sheathing conforming with this table shall be deemed to meet the design criteria of Section 2304.7.

b. Uniform load deflection limitations $^1/_{180}$ of span under live load plus dead load, $^1/_{240}$ under live load only. Edges shall be blocked with lumber or other approved type of edge supports.

c. For composite and four-ply plywood structural panel, load shall be reduced by 15 pounds per square foot.

2304.8.2 Layup patterns. Lumber decking is permitted to be laid up following one of five standard patterns as defined in Sections 2304.8.2.1 through 2304.8.2.5. Other patterns are permitted to be used provided they are substantiated through engineering analysis.

❖ This section describes typically used layup patterns for lumber decking. It also allows other patterns to be used where the engineer provides substantiation. These layup patterns are determined, in part, by the location of end joints and the pattern is the basis for determining the allowable deck load in Section 2306.1.4.

2304.8.2.1 Simple span pattern. All pieces shall be supported on their ends (i.e., by two supports).

❖ See the commentary to Section 2304.8.2 and Figure 2304.8.2.1.

2304.8.2.2 Two-span continuous pattern. All pieces shall be supported by three supports, and all end joints shall occur in line on alternating supports. Supporting members shall be designed to accommodate the load redistribution caused by this pattern.

❖ See the commentary to Section 2304.8.2 and Figure 2304.8.2.2.

2304.8.2.3 Combination simple and two-span continuous pattern. Courses in end spans shall be alternating simple-span pattern and two-span continuous pattern. End joints shall be staggered in adjacent courses and shall bear on supports.

❖ See the commentary to Section 2304.8.2 and Figure 2304.8.2.3.

2304.8.2.4 Cantilevered pieces intermixed pattern. The decking shall extend across a minimum of three spans. Pieces in each starter course and every third course shall be simple span pattern. Pieces in other courses shall be cantilevered over the supports with end joints at alternating quarter or third points of the spans. Each piece shall bear on at least one support.

❖ See commentary to Section 2304.8.2 and Figure 2304.8.2.4.

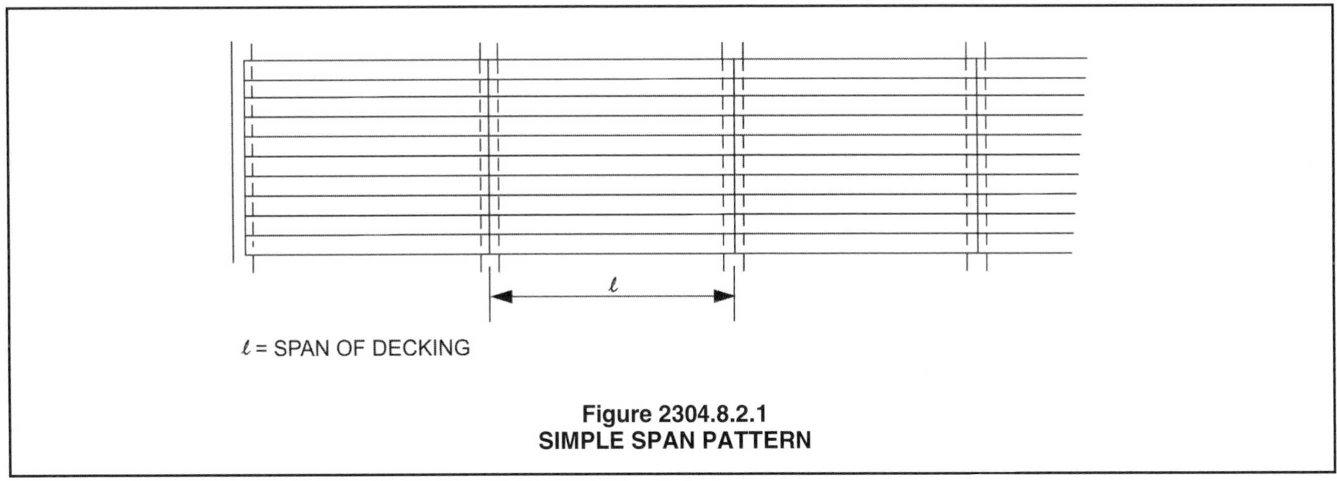

Figure 2304.8.2.1
SIMPLE SPAN PATTERN

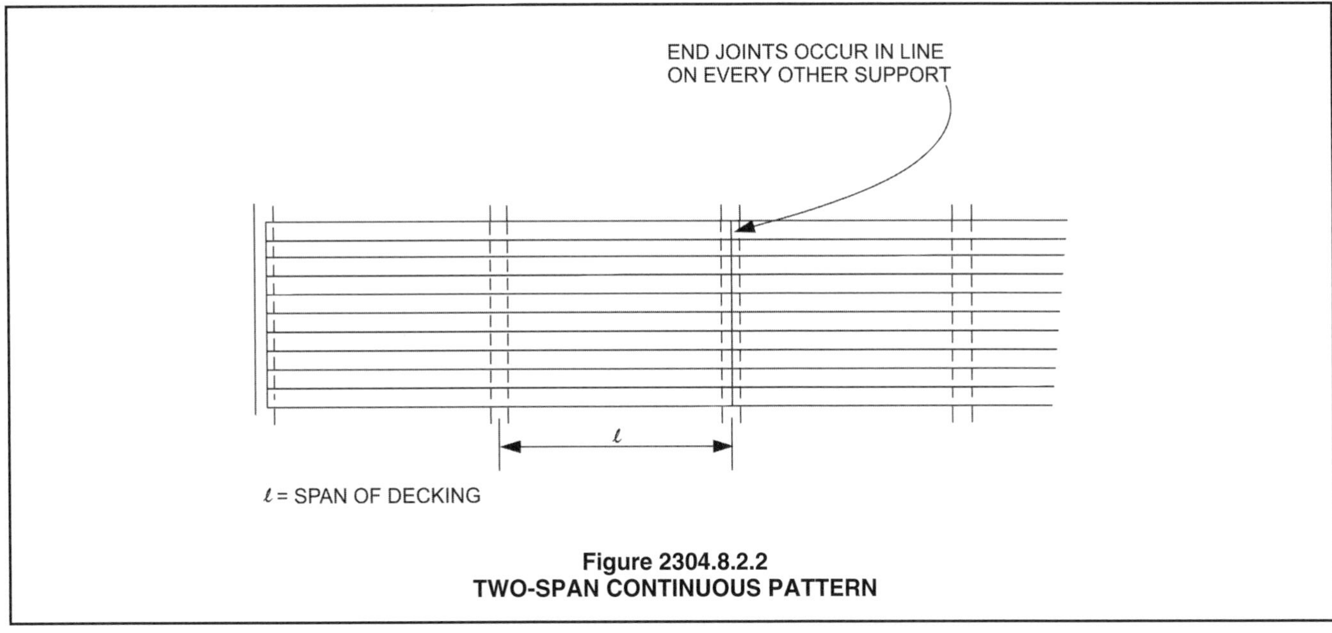

Figure 2304.8.2.2
TWO-SPAN CONTINUOUS PATTERN

2304.8.2.5 Controlled random pattern. The decking shall extend across a minimum of three spans. End joints of pieces within 6 inches (152 mm) of the end joints of the adjacent pieces in either direction shall be separated by at least two intervening courses. In the end bays, each piece shall bear on at least one support. Where an end joint occurs in an end bay, the next piece in the same course shall continue over the first inner support for at least 24 inches (610 mm). The details of the controlled random pattern shall be as specified for each decking material in Section 2304.8.3.3, 2304.8.4.3 or 2304.8.5.3.

Decking that cantilevers beyond a support for a horizontal distance greater than 18 inches (457 mm), 24 inches (610 mm) or 36 inches (914 mm) for 2-inch (51 mm), 3-inch (76 mm) and 4-inch (102 mm) nominal thickness decking, respectively, shall comply with the following:

1. The maximum cantilevered length shall be 30 percent of the length of the first adjacent interior span.
2. A structural fascia shall be fastened to each decking piece to maintain a continuous, straight line.
3. There shall be no end joints in the decking between the cantilevered end of the decking and the centerline of the first adjacent interior span.

❖ General requirements for controlled random layup are stated in this section. They apply along with the additional specific layup requirements subsequently given under mechanically laminated and tongue-and-

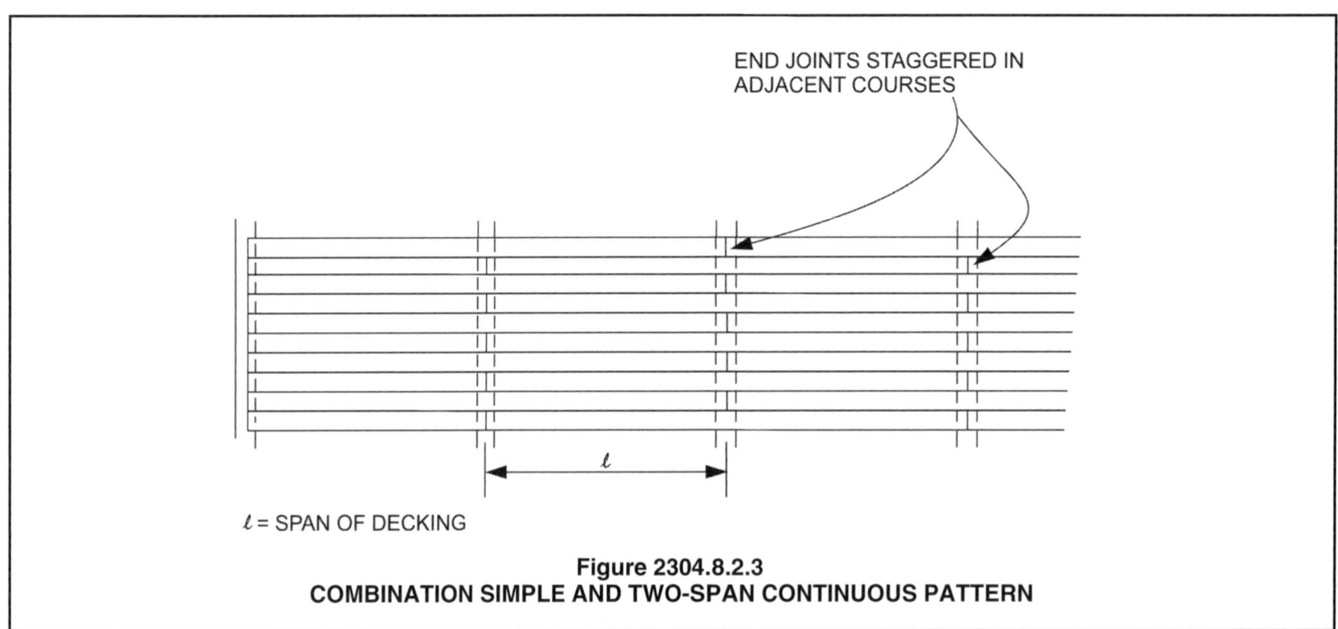

Figure 2304.8.2.3
COMBINATION SIMPLE AND TWO-SPAN CONTINUOUS PATTERN

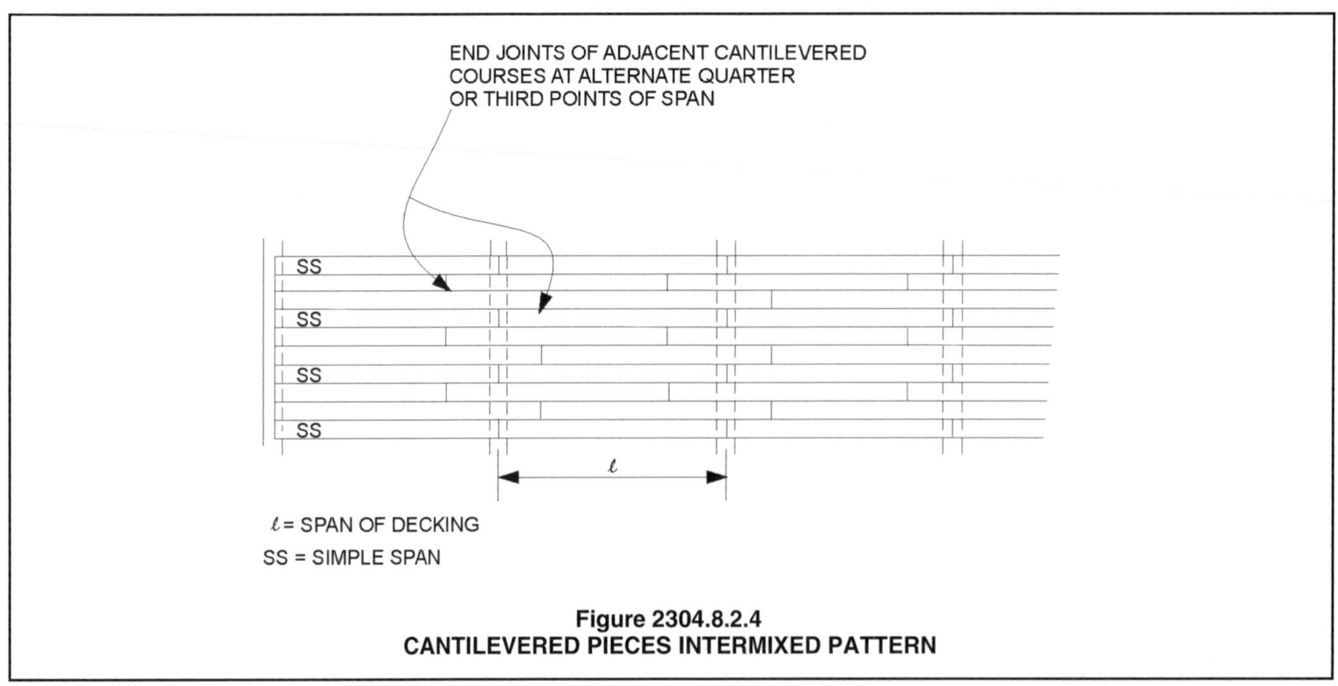

Figure 2304.8.2.4
CANTILEVERED PIECES INTERMIXED PATTERN

groove decking. The second paragraph states requirements where there is a cantilevered span such as at a roof overhang (also see commentary, Section 2304.8.2).

2304.8.3 Mechanically laminated decking. Mechanically laminated decking shall comply with Sections 2304.8.3.1 through 2304.8.3.3.

❖ This section provides construction requirements for square-edged lumber that is set on edge and mechanically fastened together with nails to produce a solid flooring.

2304.8.3.1 General. Mechanically laminated decking consists of square-edged dimension lumber laminations set on edge and nailed to the adjacent pieces and to the supports.

❖ One method of constructing floor decking is the use of individual wood members set on edge, connected with nails and mechanically laminated together to produce a solid monolithic surface.

This section gives the fastening requirements for floors that are constructed of individual wood members set on edge and mechanically laminated to produce a continuous surface. For the floor to act as a unit, fastening is critical, and therefore it is prescribed in this section. If the floor is constructed in accordance with these fastening requirements, then for purposes of design it can be considered a solid wood floor acting as one element.

2304.8.3.2 Nailing. The length of nails connecting laminations shall not be less than two and one-half times the net thickness of each lamination. Where decking supports are 48 inches (1219 mm) on center (o.c.) or less, side nails shall be installed not more than 30 inches (762 mm) o.c. alternating between top and bottom edges, and staggered one-third of the spacing in adjacent laminations. Where supports are spaced more than 48 inches (1219 mm) o.c., side nails shall be installed not more than 18 inches (457 mm) o.c. alternating between top and bottom edges and staggered one-third of the spacing in adjacent laminations. Two side nails shall be installed at each end of butt-jointed pieces.

Laminations shall be toenailed to supports with 20d or larger common nails. Where the supports are 48 inches (1219 mm) o.c. or less, alternate laminations shall be toenailed to alternate supports; where supports are spaced more than 48 inches (1219 mm) o.c., alternate laminations shall be toenailed to every support.

❖ This section provides requirements for nailing the deck laminations together (see commentary, Section 2304.8.3.1).

2304.8.3.3 Controlled random pattern. There shall be a minimum distance of 24 inches (610 mm) between end joints in adjacent courses. The pieces in the first and second courses shall bear on at least two supports with end joints in these two courses occurring on alternate supports. A maximum of seven intervening courses shall be permitted before this pattern is repeated.

❖ These requirements are specific to the layup of tongue-and-groove decking and are in addition to the general limitations stated in Section 2304.8.2.5 (see Figure 2304.8.4.3).

2304.8.4 Two-inch sawn tongue-and-groove decking. Two-inch (51 mm) sawn tongue-and-groove decking shall comply with Sections 2304.8.4.1 through 2304.8.4.3.

❖ The section covers 2-inch-thick (51 mm) tongue-and-groove boards used for decking.

2304.8.4.1 General. Two-inch (51 mm) decking shall have a maximum moisture content of 15 percent. Decking shall be machined with a single tongue-and-groove pattern. Each decking piece shall be nailed to each support.

❖ Another method of floor construction is the use of sawn lumber tongue-and-groove decking. The projecting tongue on one side fits into the matching groove of the adjacent board, causing these two elements to deflect as a unit when subjected to loads. As Figure 2304.8.4.1 illustrates, 2-inch-thick (51 mm) tongue-and-groove boards have one tongue and one groove per board.

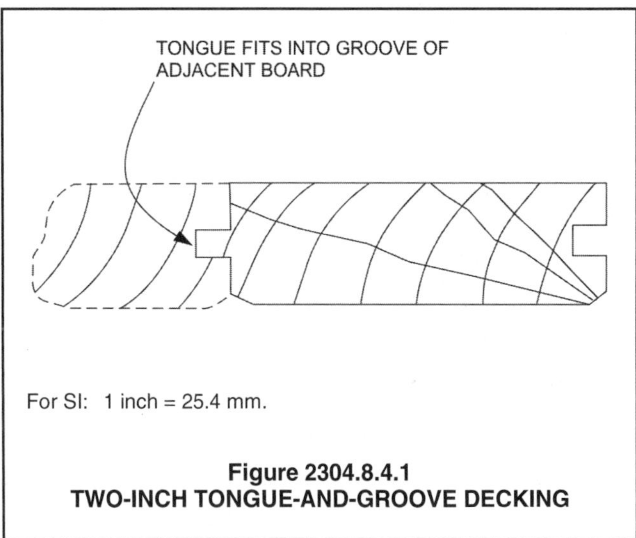

For SI: 1 inch = 25.4 mm.

Figure 2304.8.4.1
TWO-INCH TONGUE-AND-GROOVE DECKING

2304.8.4.2 Nailing. Each piece of decking shall be toenailed at each support with one 16d common nail through the tongue and face-nailed with one 16d common nail.

❖ For the floor to distribute loads and act as a unit, fastening is critical; therefore, it is prescribed in this section.

2304.8.4.3 Controlled random pattern. There shall be a minimum distance of 24 inches (610 mm) between end joints in adjacent courses. The pieces in the first and second courses shall bear on at least two supports with end joints in these two courses occurring on alternate supports. A maximum of seven intervening courses shall be permitted before this pattern is repeated.

❖ These requirements are specific to the layup of tongue-and-groove decking and are in addition to the general limitations stated in Section 2304.8.2.5.

2304.8.5 Three- and four-inch sawn tongue-and-groove decking. Three- and four-inch (76 mm and 102 mm) sawn tongue-and-groove decking shall comply with Sections 2304.8.5.1 through 2304.8.5.3.

❖ The section covers 3- and 4-inch-thick (76 and 102 mm) tongue-and-groove boards used for decking (see Figure 2304.8.5.1).

2304.8.5.1 General. Three-inch (76 mm) and four-inch (102 mm) decking shall have a maximum moisture content of 19 percent. Decking shall be machined with a double tongue-and-groove pattern. Decking pieces shall be interconnected and nailed to the supports.

❖ As Figure 2304.8.5.1 illustrates, 3- and 4-inch-thick (76 and 102 mm) tongue-and-groove boards have a pair of tongues and a pair of grooves in each board.

2304.8.5.2 Nailing. Each piece shall be toenailed at each support with one 40d common nail and face-nailed with one 60d common nail. Courses shall be spiked to each other with 8-inch (203 mm) spikes at maximum intervals of 30 inches (762 mm) through predrilled edge holes penetrating to a depth of approximately 4 inches (102 mm). One spike shall be installed at a distance not exceeding 10 inches (254 mm) from the end of each piece.

❖ For the floor to distribute loads and act as a unit, fastening is critical; therefore, it is prescribed in this section.

2304.8.5.3 Controlled random pattern. There shall be a minimum distance of 48 inches (1219 mm) between end joints in adjacent courses. Pieces not bearing on a support are permitted to be located in interior bays provided the adjacent pieces in the same course continue over the support for at least 24 inches (610 mm). This condition shall not occur more than once in every six courses in each interior bay.

❖ These requirements are specific to the layup of tongue-and-groove decking and are in addition to the general limitations stated in Section 2304.8.2.5.

2304.9 Connectors and fasteners. Connectors and fasteners shall comply with the applicable provisions of Sections 2304.9.1 through 2304.9.7.

❖ The section contains requirements for connectors and fasteners used to connect wood members, including nails, staples, joist hangers, clips and framing anchors.

Section 2301.2 gives alternative design methodologies for wood construction: (1) the ASD method; (2) the LRFD method, and (3) conventional light-frame methodologies that are embodied in Section 2308 of the code. Both ASD and LRFD are covered in the AF&PA NDS and SDPWS. If conventional construction provisions are used and there is no design provided by means of the ASD or LRFD methodologies, then the prescriptive fastening of Table 2304.9.1 is

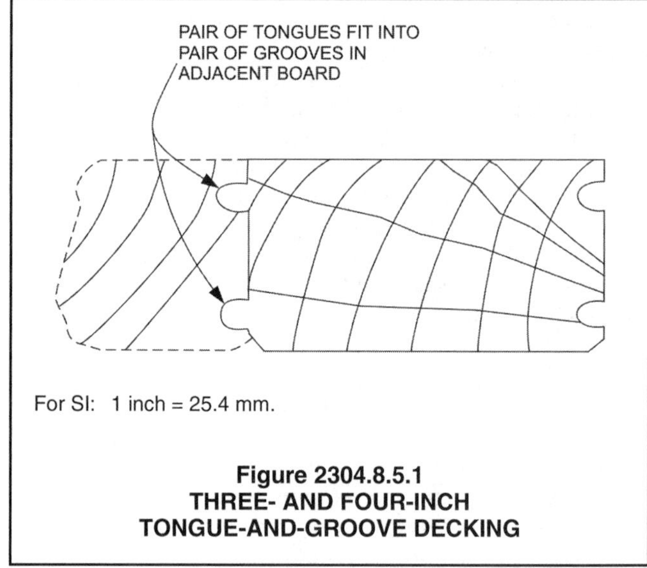

For SI: 1 inch = 25.4 mm.

Figure 2304.8.5.1
THREE- AND FOUR-INCH TONGUE-AND-GROOVE DECKING

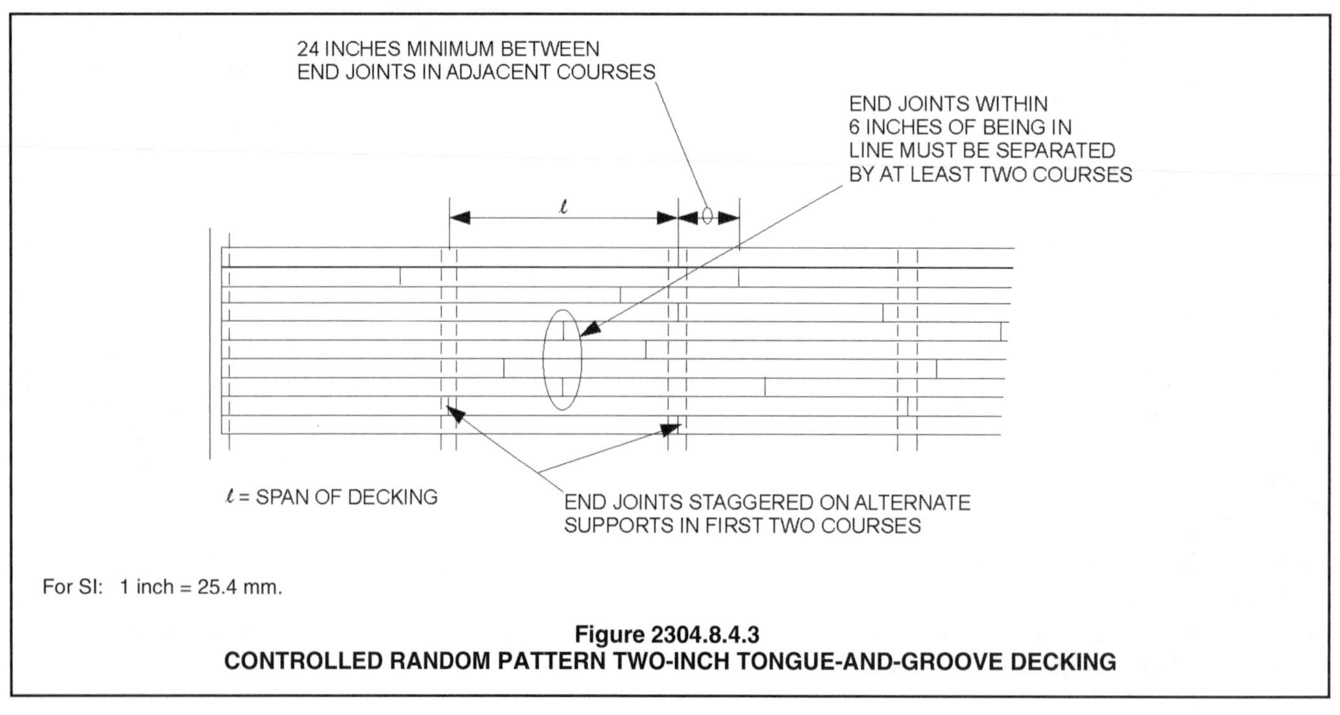

For SI: 1 inch = 25.4 mm.

Figure 2304.8.4.3
CONTROLLED RANDOM PATTERN TWO-INCH TONGUE-AND-GROOVE DECKING

typically all that is required; however, when the structure is designed by the ASD or LRFD methods, fastening must comply with the applicable design standard as well as the minimum prescriptive fastenings specified in Table 2304.9.1.

2304.9.1 Fastener requirements. Connections for wood members shall be designed in accordance with the appropriate methodology in Section 2301.2. The number and size of fasteners connecting wood members shall not be less than that set forth in Table 2304.9.1.

❖ Section 2301.2 gives alternative design methodologies for wood construction: (1) the ASD method; (2) the LRFD method and (3) conventional light-frame methodologies that are embodied in Section 2308 of the code. If conventional provisions are used and there is no design being conducted by means of the ASD or LRFD methodologies, then the prescriptive fastening of Table 2304.9.1 is all that is required; however, when the ASD or LRFD methods are used, fastening must comply with the associated design standard as well as the prescriptive fastenings specified in Table 2304.9.1.

TABLE 2304.9.1. See page 23-26.

❖ This table provides the minimum fastening requirements for all light wood-framed construction, regardless of the design method used. Familiarity with the terms and configurations of conventional frame construction is necessary to apply this table. All members identified within this table are commonly described in any available wood design or construction manual. See Figure 2304.9.1 for an illustration of certain nail types.

2304.9.2 Sheathing fasteners. Sheathing nails or other *approved* sheathing connectors shall be driven so that their head or crown is flush with the surface of the sheathing.

❖ This requirement is a matter of workmanship (see Figure 2304.9.2 for an illustration of a nail driven to fasten sheathing properly). Protruding nails that are not fully driven do not provide the intended connecting capacity and could be hazardous. Likewise, nails that are overdriven into structural sheathing may not perform as expected. Framing installation in the field is often less than perfect and fasteners can be overdriven to a point where the top layer of sheathing is fractured and crushed beneath the nail head. An occasional overdriven nail is common and may not be significant; however, as the percentage of overdriven fasteners increases, the issue raised is at what point does this adversely affect the stiffness and shear capacity of a diaphragm or shear wall elements. The APA has recognized that this is a common occurrence and has made a guideline available at no cost on its website (www.apawood.org). Another condition to be aware of in sheathing nailing is where the depth of the supporting member is less than the length of a commonly used fastener, such as in a case where sheathing will be applied over the top of flat decking. Shorter nails are available for these situations, but the holding and shear capacities are typically reduced.

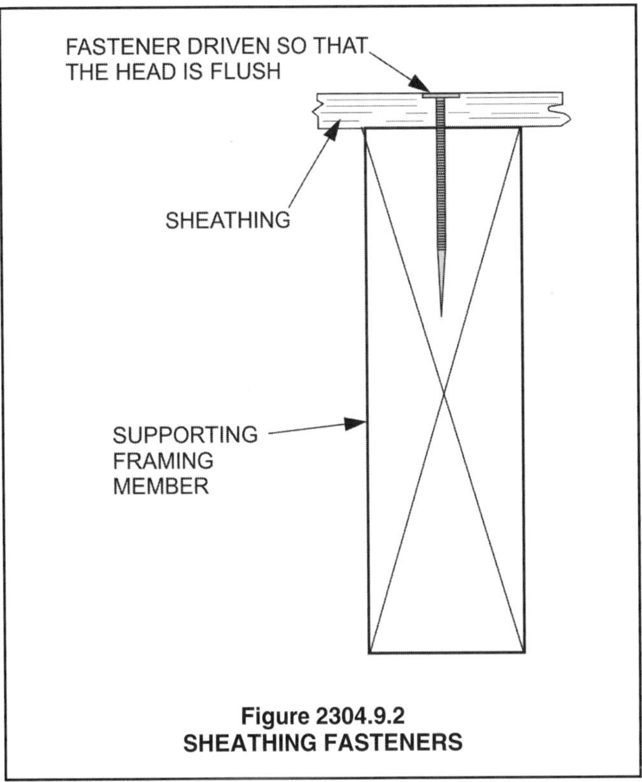

Figure 2304.9.2
SHEATHING FASTENERS

2304.9.3 Joist hangers and framing anchors. Connections depending on joist hangers or framing anchors, ties and other mechanical fastenings not otherwise covered are permitted where *approved*. The vertical load-bearing capacity, torsional moment capacity and deflection characteristics of joist hangers shall be determined in accordance with Section 1716.1.

❖ This section recognizes the use of premanufactured joist and framing anchors as an alternative to conventional fastening. These items are engineered components and must be used in accordance with the manufacturer's instructions and design specifications. Hangers and framing anchors specified by the manufacturer must have load capacities and torsional requirements developed in accordance with the test methods outlined in Section 1711.1. All fasteners required to make the framing anchor or hanger adequately support the design load are specified by the manufacturer, who also must describe any limitations that the designer or contractor needs to consider when using the product. In order to facilitate approval of these products, many hardware manufacturers have ICC Evaluation Service Reports (ESRs) that provide design values and other detailed information for properly installing their framing anchors.

TABLE 2304.9.1
FASTENING SCHEDULE

CONNECTION	FASTENING[a, m]	LOCATION
1. Joist to sill or girder	3 - 8d common ($2^1/_2$" × 0.131") 3 - 3" × 0.131" nails 3 - 3" 14 gage staples	toenail
2. Bridging to joist	2 - 8d common ($2^1/_2$" × 0.131") 2 - 3" × 0.131" nails 2 - 3" 14 gage staples	toenail each end
3. 1" × 6" subfloor or less to each joist	2 - 8d common ($2^1/_2$" × 0.131")	face nail
4. Wider than 1" × 6" subfloor to each joist	3 - 8d common ($2^1/_2$" × 0.131")	face nail
5. 2" subfloor to joist or girder	2 - 16d common ($3^1/_2$" × 0.162")	blind and face nail
6. Sole plate to joist or blocking	16d ($3^1/_2$" × 0.135") at 16" o.c. 3" × 0.131" nails at 8" o.c. 3" 14 gage staples at 12" o.c.	typical face nail
Sole plate to joist or blocking at braced wall panel	3 - 16d ($3^1/_2$" × 0.135") at 16" o.c. 4 - 3" × 0.131" nails at 16" o.c. 4 - 3" 14 gage staples at 16" o.c.	braced wall panels
7. Top plate to stud	2 - 16d common ($3^1/_2$" × 0.162") 3 - 3" × 0.131" nails 3 - 3" 14 gage staples	end nail
8. Stud to sole plate	4 - 8d common ($2^1/_2$" × 0.131") 4 - 3" × 0.131" nails 3 - 3" 14 gage staples	toenail
	2 - 16d common ($3^1/_2$" × 0.162") 3 - 3" × 0.131" nails 3 - 3" 14 gage staples	end nail
9. Double studs	16d ($3^1/_2$" × 0.135") at 24" o.c. 3" × 0.131" nail at 8" o.c. 3" 14 gage staple at 8" o.c.	face nail
10. Double top plates	16d ($3^1/_2$" × 0.135") at 16" o.c. 3" × 0.131" nail at 12" o.c. 3" 14 gage staple at 12" o.c.	typical face nail
Double top plates	8 - 16d common ($3^1/_2$" × 0.162") 12 - 3" × 0.131" nails 12 - 3" 14 gage staples	lap splice
11. Blocking between joists or rafters to top plate	3 - 8d common ($2^1/_2$" × 0.131") 3 - 3" × 0.131" nails 3 - 3" 14 gage staples	toenail
12. Rim joist to top plate	8d ($2^1/_2$" × 0.131") at 6" o.c. 3" × 0.131" nail at 6" o.c. 3" 14 gage staple at 6" o.c.	toenail
13. Top plates, laps and intersections	2 - 16d common ($3^1/_2$" × 0.162") 3 - 3" × 0.131" nails 3 - 3" 14 gage staples	face nail
14. Continuous header, two pieces	16d common ($3^1/_2$" × 0.162")	16" o.c. along edge
15. Ceiling joists to plate	3 - 8d common ($2^1/_2$" × 0.131") 5 - 3" × 0.131" nails 5 - 3" 14 gage staples	toenail
16. Continuous header to stud	4 - 8d common ($2^1/_2$" × 0.131")	toenail

(continued)

TABLE 2304.9.1—continued
FASTENING SCHEDULE

CONNECTION	FASTENING[a, m]	LOCATION
17. Ceiling joists, laps over partitions (see Section 2308.10.4.1, Table 2308.10.4.1)	3 - 16d common (3$^1/_2$" × 0.162") minimum, Table 2308.10.4.1 4 - 3" × 0.131" nails 4 - 3" 14 gage staples	face nail
18. Ceiling joists to parallel rafters (see Section 2308.10.4.1, Table 2308.10.4.1)	3 - 16d common (3$^1/_2$" × 0.162") minimum, Table 2308.10.4.1 4 - 3" × 0.131" nails 4 - 3" 14 gage staples	face nail
19. Rafter to plate (see Section 2308.10.1, Table 2308.10.1)	3 - 8d common (2$^1/_2$" × 0.131") 3 - 3" × 0.131" nails 3 - 3" 14 gage staples	toenail
20. 1" diagonal brace to each stud and plate	2 - 8d common (2$^1/_2$" × 0.131") 2 - 3" × 0.131" nails 3 - 3" 14 gage staples	face nail
21. 1" × 8" sheathing to each bearing	3 - 8d common (2$^1/_2$" × 0.131")	face nail
22. Wider than 1" × 8" sheathing to each bearing	3 - 8d common (2$^1/_2$" × 0.131")	face nail
23. Built-up corner studs	16d common (3$^1/_2$" × 0.162") 3" × 0.131" nails 3" 14 gage staples	24" o.c. 16" o.c. 16" o.c.
24. Built-up girder and beams	20d common (4" × 0.192") 32" o.c. 3" × 0.131" nail at 24" o.c. 3" 14 gage staple at 24" o.c.	face nail at top and bottom staggered on opposite sides
	2 - 20d common (4" × 0.192") 3 - 3" × 0.131" nails 3 - 3" 14 gage staples	face nail at ends and at each splice
25. 2" planks	16d common (3$^1/_2$" × 0.162")	at each bearing
26. Collar tie to rafter	3 - 10d common (3" × 0.148") 4 - 3" × 0.131" nails 4 - 3" 14 gage staples	face nail
27. Jack rafter to hip	3 - 10d common (3" × 0.148") 4 - 3" × 0.131" nails 4 - 3" 14 gage staples	toenail
	2 - 16d common (3$^1/_2$" × 0.162") 3 - 3" × 0.131" nails 3 - 3" 14 gage staples	face nail
28. Roof rafter to 2-by ridge beam	2 - 16d common (3$^1/_2$" × 0.162") 3 - 3" × 0.131" nails 3 - 3" 14 gage staples	toenail
	2 -16d common (3$^1/_2$" × 0.162") 3 - 3" × 0.131" nails 3 - 3" 14 gage staples	face nail
29. Joist to band joist	3 - 16d common (3$^1/_2$" × 0.162") 4 - 3" × 0.131" nails 4 - 3" 14 gage staples	face nail

(continued)

TABLE 2304.9.1—continued
FASTENING SCHEDULE

CONNECTION	FASTENING[a, m]		LOCATION
30. Ledger strip	3 - 16d common ($3^{1}/_{2}$" × 0.162") 4 - 3" × 0.131" nails 4 - 3" 14 gage staples		face nail at each joist
31. Wood structural panels and particleboard[b] Subfloor, roof and wall sheathing (to framing)	$^{1}/_{2}$" and less $^{19}/_{32}$" to $^{3}/_{4}$" $^{7}/_{8}$" to 1" $1^{1}/_{8}$" to $1^{1}/_{4}$"	6d[c, l] $2^{3}/_{8}$" × 0.113" nail[n] $1^{3}/_{4}$" 16 gage[o] 8d[d] or 6d[e] $2^{3}/_{8}$" × 0.113" nail[p] 2" 16 gage[p] 8d[c] 10d[d] or 8d[e]	
Single floor (combination subfloor-underlayment to framing)	$^{3}/_{4}$" and less $^{7}/_{8}$" to 1" $1^{1}/_{8}$" to $1^{1}/_{4}$"	6d[e] 8d[e] 10d[d] or 8d[e]	
32. Panel siding (to framing)	$^{1}/_{2}$" or less $^{5}/_{8}$"	6d[f] 8d[f]	
33. Fiberboard sheathing[g]	$^{1}/_{2}$" $^{25}/_{32}$"	No. 11 gage roofing nail[h] 6d common nail (2" × 0.113") No. 16 gage staple[i] No. 11 gage roofing nail[h] 8d common nail ($2^{1}/_{2}$" × 0.131") No. 16 gage staple[i]	
34. Interior paneling	$^{1}/_{4}$" $^{3}/_{8}$"	4d[j] 6d[k]	

For SI: 1 inch = 25.4 mm.

a. Common or box nails are permitted to be used except where otherwise stated.

b. Nails spaced at 6 inches on center at edges, 12 inches at intermediate supports except 6 inches at supports where spans are 48 inches or more. For nailing of wood structural panel and particleboard diaphragms and shear walls, refer to Section 2305. Nails for wall sheathing are permitted to be common, box or casing.

c. Common or deformed shank (6d - 2" × 0.113"; 8d - $2^{1}/_{2}$" × 0.131"; 10d - 3" × 0.148").

d. Common (6d - 2" × 0.113"; 8d - $2^{1}/_{2}$" × 0.131"; 10d - 3" × 0.148").

e. Deformed shank (6d - 2" × 0.113"; 8d - $2^{1}/_{2}$" × 0.131"; 10d - 3" × 0.148").

f. Corrosion-resistant siding (6d - $1^{7}/_{8}$" × 0.106"; 8d - $2^{3}/_{8}$" × 0.128") or casing (6d - 2" × 0.099"; 8d - $2^{1}/_{2}$" × 0.113") nail.

g. Fasteners spaced 3 inches on center at exterior edges and 6 inches on center at intermediate supports, when used as structural sheathing. Spacing shall be 6 inches on center on the edges and 12 inches on center at intermediate supports for nonstructural applications.

h. Corrosion-resistant roofing nails with $^{7}/_{16}$-inch-diameter head and $1^{1}/_{2}$-inch length for $^{1}/_{2}$-inch sheathing and $1^{3}/_{4}$-inch length for $^{25}/_{32}$-inch sheathing.

i. Corrosion-resistant staples with nominal $^{7}/_{16}$-inch crown or 1-inch crown and $1^{1}/_{4}$-inch length for $^{1}/_{2}$-inch sheathing and $1^{1}/_{2}$-inch length for $^{25}/_{32}$-inch sheathing. Panel supports at 16 inches (20 inches if strength axis in the long direction of the panel, unless otherwise marked).

j. Casing ($1^{1}/_{2}$" × 0.080") or finish ($1^{1}/_{2}$" × 0.072") nails spaced 6 inches on panel edges, 12 inches at intermediate supports.

k. Panel supports at 24 inches. Casing or finish nails spaced 6 inches on panel edges, 12 inches at intermediate supports.

l. For roof sheathing applications, 8d nails ($2^{1}/_{2}$" × 0.113") are the minimum required for wood structural panels.

m. Staples shall have a minimum crown width of $^{7}/_{16}$ inch.

n. For roof sheathing applications, fasteners spaced 4 inches on center at edges, 8 inches at intermediate supports.

o. Fasteners spaced 4 inches on center at edges, 8 inches at intermediate supports for subfloor and wall sheathing and 3 inches on center at edges, 6 inches at intermediate supports for roof sheathing.

p. Fasteners spaced 4 inches on center at edges, 8 inches at intermediate supports.

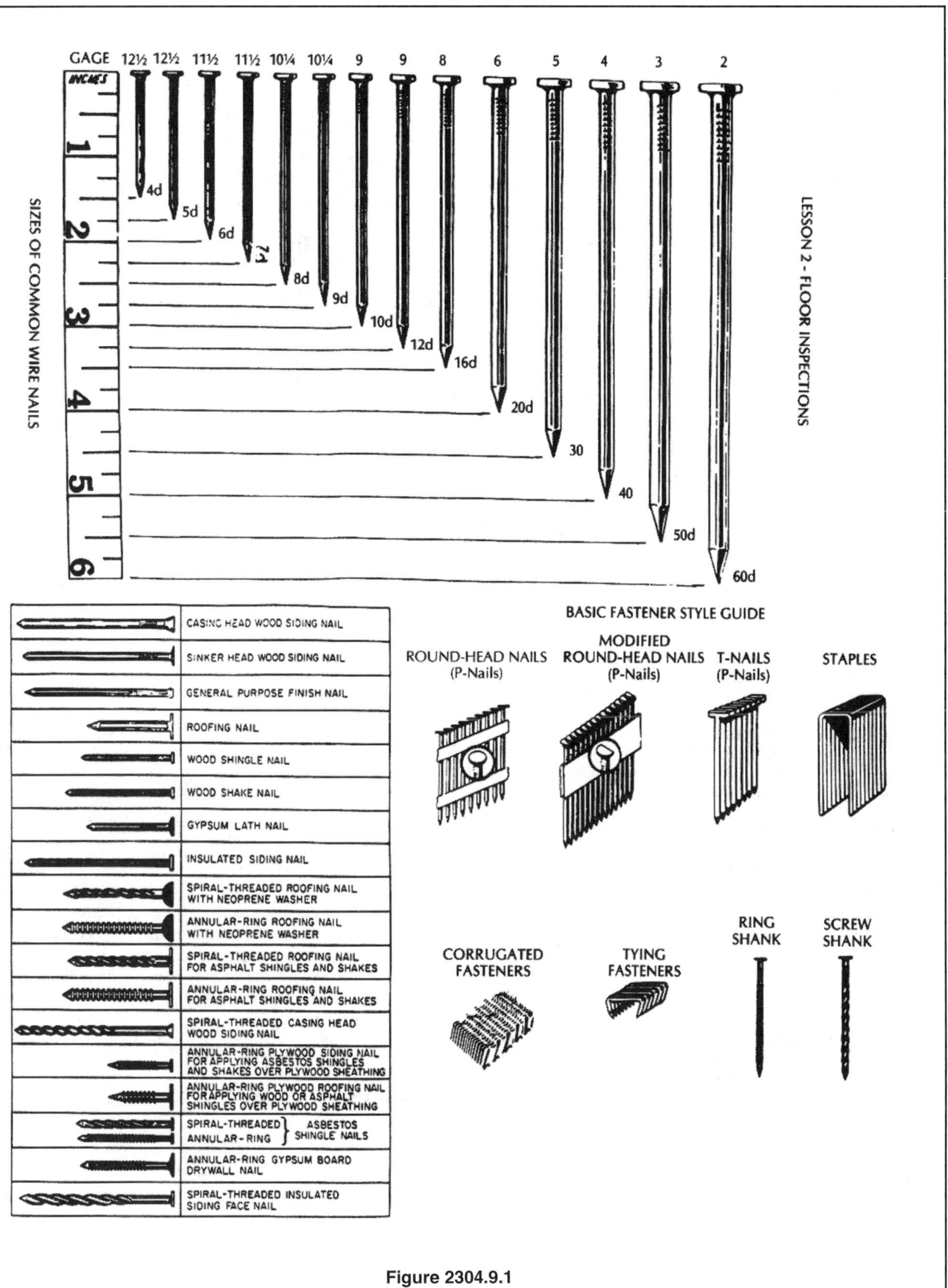

Figure 2304.9.1
STANDARD NAILS AND STAPLES
Independent Nail Packing Co., Bridgewater, MA

2304.9.4 Other fasteners. Clips, staples, glues and other *approved* methods of fastening are permitted where *approved.*

❖ Clips, staples, glues and other methods of fastening are permitted where approved. In addition to joist hangers and typical fastening in accordance with Table 2304.9.1, there are other fastening methods that can be approved by the building official, provided there is sufficient evidence to demonstrate their adequacy. This section is included in the code to address the fact that new fasteners and connection methods are continually being developed, which necessitates that the building official give consideration to viable alternatives. In general, other fasteners or fastening systems that are not specifically covered by the code can be approved under the alternative materials, design and methods of construction specified in Section 104.11. In order to facilitate approval of alternative products that are not specified in the code, many product manufacturers work with ICC Evaluation Service to develop acceptance criteria and, ultimately, ESRs. These include detailed information to assist the building official in determining code compliance and help the inspector verify proper installation in the field.

2304.9.5 Fasteners and connectors in contact with preservative-treated and fire-retardant-treated wood. Fasteners, including nuts and washers, and connectors in contact with *preservative-treated* and *fire-retardant-treated wood* shall be in accordance with Sections 2304.9.5.1 through 2304.9.5.4. The coating weights for zinc-coated fasteners shall be in accordance with ASTM A 153.

❖ Corrosion-resistant fasteners are typically required for use in treated wood. This applies to fire-retardant-treated wood as well as preservative-treated wood. Some chemicals used to preservative treat or fire-retardant treat wood can have a corrosive effect on fasteners. This is a concern since significant corrosion will adversely affect the fastener's strength.

2304.9.5.1 Fasteners and connectors for preservative-treated wood. Fasteners, including nuts and washers, in contact with *preservative-treated wood* shall be of hot-dipped zinc-coated galvanized steel, stainless steel, silicon bronze or copper. Fasteners other than nails, timber rivets, wood screws and lag screws shall be permitted to be of mechanically deposited zinc-coated steel with coating weights in accordance with ASTM B 695, Class 55 minimum. Connectors that are used in exterior applications and in contact with *preservative-treated wood* shall have coating types and weights in accordance with the treated wood or connector manufacturer's recommendations. In the absence of manufacturer's recommendations, a minimum of ASTM A 653, type G185 zinc-coated galvanized steel, or equivalent, shall be used.

Exception: Plain carbon steel fasteners, including nuts and washers, in SBX/DOT and zinc borate *preservative-treated wood* in an interior, dry environment shall be permitted.

❖ Where there are concerns with decay or potential termite damage, the code requires either a naturally durable species of wood or wood that is treated with a preservative. Since the discontinuation of chromated copper arsenate for preservative-treated wood in residential or general consumer products, alternative preservative treatments have been necessary. Studies show that some of these are more corrosive to steel anchors and connectors. Corrosion-resistant fasteners are typically required for use in preservative-treated wood. This includes materials such as stainless steel or copper. Carbon steel typically must be galvanized. The exception is based on reports that show sodium borate disodium octaborate tetrahydrate (SBX/DOT) and zinc borate to be somewhat less corrosive than other alternative preservative treatments. In an interior location where moisture is not a concern, wood treated with these preservatives allows the use of plain carbon steel fasteners.

The base requirements limit galvanized steel to the hot-dipped process. The hot-dip galvanizing process has the relative advantage of being very insensitive to process variables. Though mechanically galvanized fasteners can perform equally to hot-dipped galvanized fasteners, the mechanical galvanizing process, if not performed properly, may produce fasteners whose corrosion resistance does not equal that of hot-dipped galvanized fasteners. Mechanical galvanizing is very common for screws, and in fact is possibly the only way to deposit a thick zinc coating on them. Wood screws and lag screws are frequently installed in predrilled holes, so abrasion of the finish is not the same problem as for nails. Because mechanical galvanizing is a preferred method for certain fastener types, the code recognizes this process where it is appropriate.

Guidance is given for connectors that are in contact with preservative-treated wood and used in exterior applications. The primary requirement is to follow the manufacturer's recommendations. Where none are given, the minimum coating must be G185 in accordance with ASTM A 653, which is the minimum coating currently recommended by many manufacturers.

2304.9.5.2 Fastenings for wood foundations. Fastenings, including nuts and washers, for wood foundations shall be as required in AF&PA PWF.

❖ AF&PA *Permanent Wood Foundation (PWF) Design Specification* requires specific types of corrosion-resistant nails for locations below grade and above grade in wood foundations, based on the increased exposure to moisture in below-grade applications as well as the corrosive nature of preservative treatments.

2304.9.5.3 Fasteners for fire-retardant-treated wood used in exterior applications or wet or damp locations. Fasteners, including nuts and washers, for *fire-retardant-treated wood* used in exterior applications or wet or damp locations shall be of hot-dipped zinc-coated galvanized steel, stainless steel, silicon bronze or copper. Fasteners other than nails, timber rivets, wood screws and lag screws shall be permitted to be of mechanically deposited zinc-coated steel with coating weights in accordance with ASTM B 695, Class 55 minimum.

❖ Corrosion-resistant fasteners are required for fire-retardant-treated wood in exterior applications or other locations subject to moisture. The base requirements limit galvanized steel to the hot-dipped process. Because mechanical galvanizing is a preferred method for certain fastener types, the code recognizes this process where it is deemed appropriate.

2304.9.5.4 Fasteners for fire-retardant-treated wood used in interior applications. Fasteners, including nuts and washers, for *fire-retardant-treated wood* used in interior locations shall be in accordance with the manufacturer's recommendations. In the absence of manufacturer's recommendations, Section 2304.9.5.3 shall apply.

❖ The primary requirement for fasteners in fire-retardant-treated wood used on the interior is to follow the manufacturer's recommendations. Where none are given, the requirements for exterior applications apply.

2304.9.6 Load path. Where wall framing members are not continuous from foundation sill to roof, the members shall be secured to ensure a continuous load path. Where required, sheet metal clamps, ties or clips shall be formed of galvanized steel or other *approved* corrosion-resistant material not less than 0.040 inch (1.01 mm) nominal thickness.

❖ In conventional construction, the minimum fastening required by Table 2304.9.1 is intended to provide a continuous connection between framing elements as required by this section. If the loads on the building exceed the parameters established in Section 2308.2, the building, in total or in part, must be designed in accordance with the ASD or LRFD methodologies and the connections must be designed in accordance with those standards, providing continuity of load support throughout the structure. The designed connections may incorporate engineered tie or strap components with capacities specified by the manufacturer. This section specifies the minimum thickness of sheet metal for fabricated straps or ties to 0.04-inch (1.01 mm) nominal, which is common among manufacturers. Designers and contractors must keep in mind that a wood-framed building is simply a huge assembly of multiple independent parts all connected together to form a complete structural system. When a building is subject to external forces, each independent part has a tendency to move into a different position and this tendency must be controlled by having a complete path to deliver the imposed forces into the ground. This load path may be comprised of hangers from a beam to a column, metal straps from an upper-level stud to a lower-level stud or from hold-down anchors into the concrete, etc. There are countless applications and load path possibilities, but the overall concept is that every individual piece must be adequately fastened to some adjacent piece to form the complete structure.

2304.9.7 Framing requirements. Wood columns and posts shall be framed to provide full end bearing. Alternatively, column-and-post end connections shall be designed to resist the full compressive loads, neglecting end-bearing capacity. Column-and-post end connections shall be fastened to resist lateral and net induced uplift forces.

❖ This section requires that the framing workmanship be such that the ends of wood columns and posts, including studs, bear completely atop the supporting member, whether it be a wood sill plate or a metal base connector. If it does not bear fully on the entire cross section, then design calculations are necessary in order to show that the bearing method provided will be capable of supporting the loads without using end bearing capacity. It also requires a positive connection (not merely a friction connection) to resist incidental lateral forces and any anticipated uplift forces. For instance, a wood column supporting beams in the basement of a structure could not simply bear on the concrete pad beneath it without some means of positive anchorage to the concrete in order to prevent it from accidently being displaced on impact by a person or an object. If the combination of loads on the structure could result in uplift forces, then an adequate connection designed to resist uplift would also be required. This section is a further extension of the philosophy outlined in Section 2304.9.6.

2304.10 Heavy timber construction. Where a structure or portion thereof is required to be of Type IV construction by other provisions of this code, the building elements therein shall comply with the applicable provisions of Sections 2304.10.1 through 2304.10.5.

❖ Heavy timber construction is classified as Type IV construction in Chapter 6. It is described in Section 602.4 as that type of construction where the exterior walls are of noncombustible materials and the interior building elements are of solid or laminated wood without concealed spaces. It originated in New England to serve the needs of the growing textile industry. As the industry modernized, the need for larger open space, unobstructed by columns, gradually reduced the demand for this type of construction.

Today, due primarily to its architectural aesthetic value, heavy timber construction is used in many other occupancies. It is commonly used for assembly and mercantile buildings, such as schools, churches, auditoriums, gymnasiums and supermarkets.

The provisions of this section must be used in conjunction with the minimum dimensional requirements specified in Section 602.4. A structural analysis must be performed to verify that the minimum dimensions are adequate. These final dimensions will always be based on a shrunken section since timbers are ordinarily supplied in a green condition, unless they are special products manufactured to specific tolerances.

2304.10.1 Columns. Columns shall be continuous or superimposed throughout all stories by means of reinforced concrete or metal caps with brackets, or shall be connected by properly designed steel or iron caps, with pintles and base plates, or by timber splice plates affixed to the columns by metal connectors housed within the contact faces, or by other *approved* methods.

❖ The AF&PA publication, WCD 5, *Heavy Timber Construction*, provides details for the proper connection of heavy timber columns and beams. Traditionally, heavy timber structures were designed and constructed using a prescriptive approach, much like today's conventional construction provisions for light-frame buildings. The number and size of bolts, lag screws or connectors should be determined through analysis of the loads to be supported.

Pintles provide a method to fasten the butt ends of beams or columns. A pintle acts like a short column, connecting the ends of the structural elements.

2304.10.1.1 Column connections. Girders and beams shall be closely fitted around columns and adjoining ends shall be cross tied to each other, or intertied by caps or ties, to transfer horizontal loads across joints. Wood bolsters shall not be placed on tops of columns unless the columns support roof loads only.

❖ Where columns and girders or beams intersect, care must be taken for load transfer. Proper detailing at the intersection of columns and girders will limit the effects of shrinkage. Elements that are not continuous must be properly fastened together to transfer design forces. Today's engineered wood products are available in multiple depths and widths, and in lengths up to 60 feet (18 288 mm), so continuous span structural elements are readily available. Bolster blocks are used to reduce the compression perpendicular to grain stresses where roof beams bear on columns. The shrinkage of the bolster block must be considered in the roof design. Roof beams must have a positive connection to the column to resist uplift.

2304.10.2 Floor framing. *Approved* wall plate boxes or hangers shall be provided where wood beams, girders or trusses rest on masonry or concrete walls. Where intermediate beams are used to support a floor, they shall rest on top of girders, or shall be supported by ledgers or blocks securely fastened to the sides of the girders, or they shall be supported by an *approved* metal hanger into which the ends of the beams shall be closely fitted.

❖ Wood members must not bear directly on masonry or concrete, otherwise they will absorb moisture from these elements and lose strength in bearing due to decay. Typically, a steel-bearing plate or custom steel hanger is used to support heavy timber elements. Bearing plates provide an even distribution of load on the wall. Wall plate boxes should be designed and detailed to provide a positive connection to the masonry. Where direct bearing on top of the girders is not possible or desirable, a metal hanger designed to transfer all induced loads is permitted. An additional advantage to using a hanger is less shrinkage. Beams consisting of green lumber should be installed so their top edge is slightly above the top edge of the girder to allow for shrinkage.

2304.10.3 Roof framing. Every roof girder and at least every alternate roof beam shall be anchored to its supporting member; and every monitor and every sawtooth construction shall be anchored to the main roof construction. Such anchors shall consist of steel or iron bolts of sufficient strength to resist vertical uplift of the roof.

❖ Anchorage of roof beams and girders should be in ac-cordance with the engineered design. This section provides minimum acceptable anchorage practices that are especially important in unusual or discontinuous types of roof framing, such as the "sawtooth" arrangement. Sawtooth roof construction is used to provide large amounts of daylight into the floor area of the building. The roof line looks much like the cutting edge of a saw, with regularly spaced peaks and valleys. The vertical or near vertical line connecting the peaks and valleys is filled with glazing. When this unique framing pattern is used, the framing members must be adequately anchored to the main roof supports.

2304.10.4 Floor decks. Floor decks and covering shall not extend closer than $^1/_2$ inch (12.7 mm) to walls. Such $^1/_2$-inch (12.7 mm) spaces shall be covered by a molding fastened to the wall either above or below the floor and arranged such that the molding will not obstruct the expansion or contraction movements of the floor. Corbeling of masonry walls under floors is permitted in place of such molding.

❖ Flooring will expand and contract as the relative humidity in the building changes. To accommodate expansion, a $^1/_2$-inch (12.7 mm) gap is required at the wall. The gap will also prevent moisture in the walls from migrating into the floor boards. The fire integrity of the floor assembly must be maintained, so either molding or corbeling of the wall below the floor is required to prevent the passage of smoke and hot gases.

2304.10.5 Roof decks. Where supported by a wall, roof decks shall be anchored to walls to resist uplift forces determined in accordance with Chapter 16. Such anchors shall consist of steel or iron bolts of sufficient strength to resist vertical uplift of the roof.

❖ The roof deck must be anchored frequently enough to resist uplift forces as prescribed in Chapter 16. This type of anchorage is not covered by the conventional construction provisions of the code; therefore, engineering is required in accordance with the approved methods, ASD or LRFD.

2304.11 Protection against decay and termites. Wood shall be protected from decay and termites in accordance with the applicable provisions of Sections 2304.11.1 through 2304.11.9.

❖ This section requires protection against decay and termites where the stated conditions apply.

2304.11.1 General. Where required by this section, protection from decay and termites shall be provided by the use of naturally durable or *preservative-treated wood*.

❖ Conditions favorable for decay and fungus attack are discussed in Sections 2304.11.2 and 2304.11.4, along with preservative treatments. This section sets forth specific locations of concern and minimum construction practices to prevent such an attack. Termite protection is also defined. According to *The Complete Book of Home Inspection*, termite infestation actually should not be an alarming discovery since termites do their damage very slowly. It will take a colony of 60,000 termites an entire year to destroy only 4 feet (1219 mm) of boards that measure 2 inches by 4 inches (51 mm by 102 mm). Chemically treated lumber, as described in this section, will generally control infestation possibilities.

2304.11.2 Wood used above ground. Wood used above ground in the locations specified in Sections 2304.11.2.1 through 2304.11.2.7, 2304.11.3 and 2304.11.5 shall be naturally durable wood or *preservative-treated wood* using waterborne preservatives, in accordance with AWPA U1 (Commodity Specifications A or F) for above-ground use.

❖ The AWPA is the principal standards developing organization for the wood-preserving industry in the United States. AWPA U1 covers aspects of treated wood from the treating process to appropriate treatment retention levels for specific applications. Wood products used in locations identified in this section are particularly susceptible to deterioration, which can result from continuous or intermittent contact with moisture or termites.

Decay is caused by fungi, which are low forms of plant life that attack the cell walls of wood in order to feed on the contents of the cells. For fungi to attack wood in service, the following conditions must be present: (1) temperature in the range of 35°F to 100°F (1°C to 38°C); (2) adequate supply of oxygen and (3) wood-moisture content in excess of 20 percent. A properly constructed and maintained building will typically remain at a moisture content of less than 15 percent throughout its life, but some special conditions might cause fluctuations leading to problems (see Section 2304.11.2.2).

2304.11.2.1 Joists, girders and subfloor. Where wood joists or the bottom of a wood structural floor without joists are closer than 18 inches (457 mm), or wood girders are closer than 12 inches (305 mm) to the exposed ground in crawl spaces or unexcavated areas located within the perimeter of the building foundation, the floor construction (including posts, girders, joists and subfloor) shall be of naturally durable or *preservative-treated wood*.

❖ The clearances specified in this section are recommended for all buildings and are considered necessary to: (1) maintain wood elements in permanent structures at a safe moisture content by promoting air circulation and ventilation of the crawl space; (2) provide a termite barrier and (3) facilitate periodic inspections (see Figure 2304.11.2.1). When it is not possible or practical to comply with the clearance specified, the use of naturally durable or pressure-preservative-treated wood is necessary. However, when termites are known to exist, space must be provided for periodic inspections. Often termite infestations are the result of poor construction or improper clearance between wood framing and the surrounding soil. Unfortunately, termites are very determined insects and will often build small, moisture-controlled tubes or tunnels from the ground surface to the

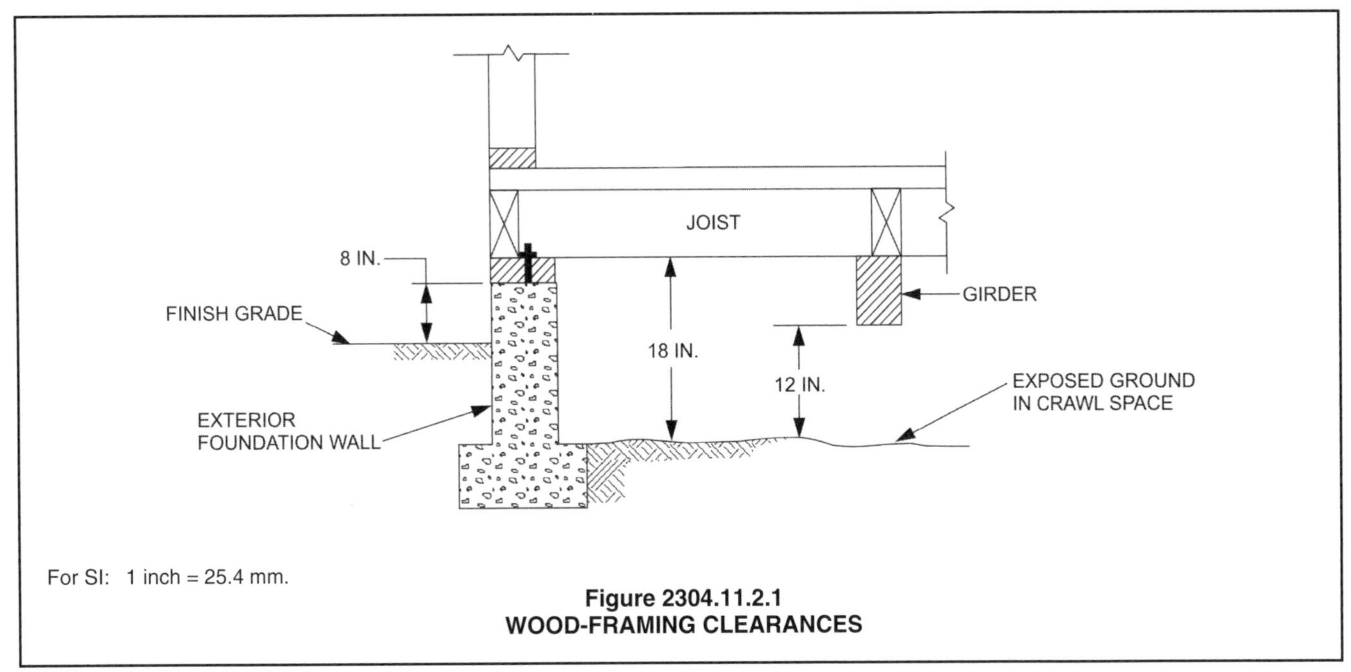

Figure 2304.11.2.1
WOOD-FRAMING CLEARANCES

source of food, bypassing inedible barriers, which is one of the reasons it is important to provide adequate space below floor framing for periodic visual inspections of wood framing.

2304.11.2.2 Wood supported by exterior foundation walls. Wood framing members, including wood sheathing, that rest on exterior foundation walls and are less than 8 inches (203 mm) from exposed earth shall be of naturally durable or *preservative-treated wood.*

❖ Experience has shown that foundation walls can absorb moisture from the ground and convey moisture to wood framing in contact with the foundation. This section requires a clearance (see Figure 2304.11.2.1) that minimizes the chances of moisture being drawn to the framing. The clearance also provides an exposed foundation surface that allows for a visual inspection for termite tubes.

2304.11.2.3 Exterior walls below grade. Wood framing members and furring strips attached directly to the interior of exterior masonry or concrete walls below grade shall be of naturally durable or *preservative-treated wood.*

❖ Wood attached directly to masonry or concrete will draw moisture from these materials, and may reach a moisture content greater than 20 percent, depending on the relative humidity of the basement.

2304.11.2.4 Sleepers and sills. Sleepers and sills on a concrete or masonry slab that is in direct contact with earth shall be of naturally durable or *preservative-treated wood.*

❖ Wood in contact with either the earth or material that retains moisture must be naturally durable or preservative treated. The moisture content of wood will change based on the relative humidity of the air and the moisture of materials with which it is in contact. Interior slabs that are protected from the earth by an impervious moisture barrier should not be considered in direct contact with the earth for purposes of this requirement. Sometimes this moisture barrier becomes damaged in the course of construction, but will remain intact for the most part. Moisture-sensitive flooring materials typically require special measures or care to be taken during placement of the concrete slab and preparation of the subbase, but interior wall sill plates will typically maintain an acceptable moisture content even if special drying measures are not taken for the slab.

2304.11.2.5 Girder ends. The ends of wood girders entering exterior masonry or concrete walls shall be provided with a $^1/_2$-inch (12.7 mm) air space on top, sides and end, unless naturally durable or *preservative-treated wood* is used.

❖ Clearance around the ends of girders prevents moisture from migrating from the concrete into the wood. The minimum clearance that is shown in Figure 2304.11.2.5 allows for air circulation and periodic inspection for termites. This requirement has been in the model codes for many years, proving it to be an effective step in controlling wood deterioration.

2304.11.2.6 Wood siding. Clearance between wood siding and earth on the exterior of a building shall not be less than 6 inches (152 mm) or less than 2 inches (51 mm) vertical from concrete steps, porch slabs, patio slabs and similar horizontal surfaces exposed to the weather except where siding, sheathing and wall framing are of naturally durable or *preservative-treated wood.*

❖ Exterior siding is required to have a minimum 6-inch (152 mm) clearance above exposed earth and a 2-inch (51 mm) clearance above concrete such as slabs and steps. Exterior wood siding should extend at least 1 inch (25 mm) below the top of the foundation to provide a drip line protecting the sill from rainwater. The siding is typi-cally held away from the foundation with a starter strip attached to the sheath-

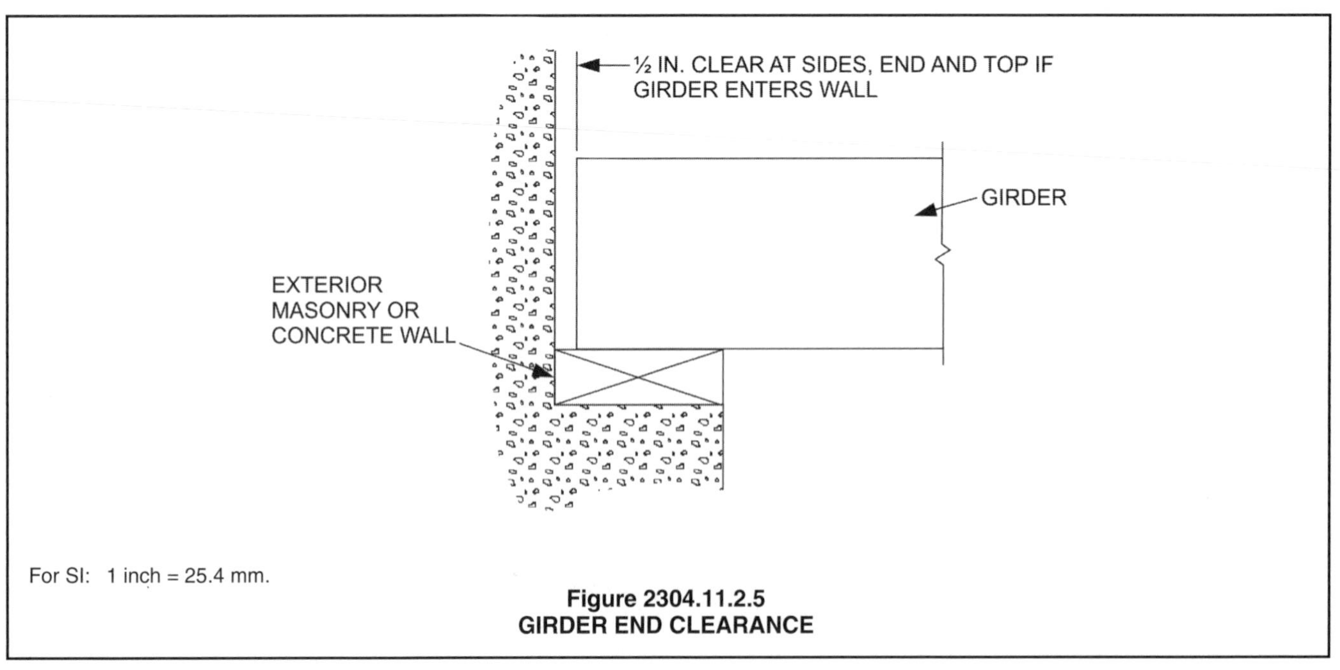

Figure 2304.11.2.5
GIRDER END CLEARANCE

ing that ends at the bottom of the plate. The 1-inch (25 mm) overhang causes the rainwater to drip off the bottom edge of the siding, rather than being absorbed by the underlying sheathing. A minimum clearance illustrated in Figure 2304.11.2.6(1) must be provided for wood siding that is neither naturally durable nor preservative treated. A reduced clearance of 2 inches (51 mm) is permitted at horizontal surfaces such as concrete patios or steps [see Figure 2304.11.2.6(2)].

2304.11.2.7 Posts or columns. Posts or columns supporting permanent structures and supported by a concrete or masonry slab or footing that is in direct contact with the earth shall be of naturally durable or *preservative-treated wood.*

Exceptions:

1. Posts or columns that are either exposed to the weather or located in basements or cellars, supported by concrete piers or metal pedestals projected

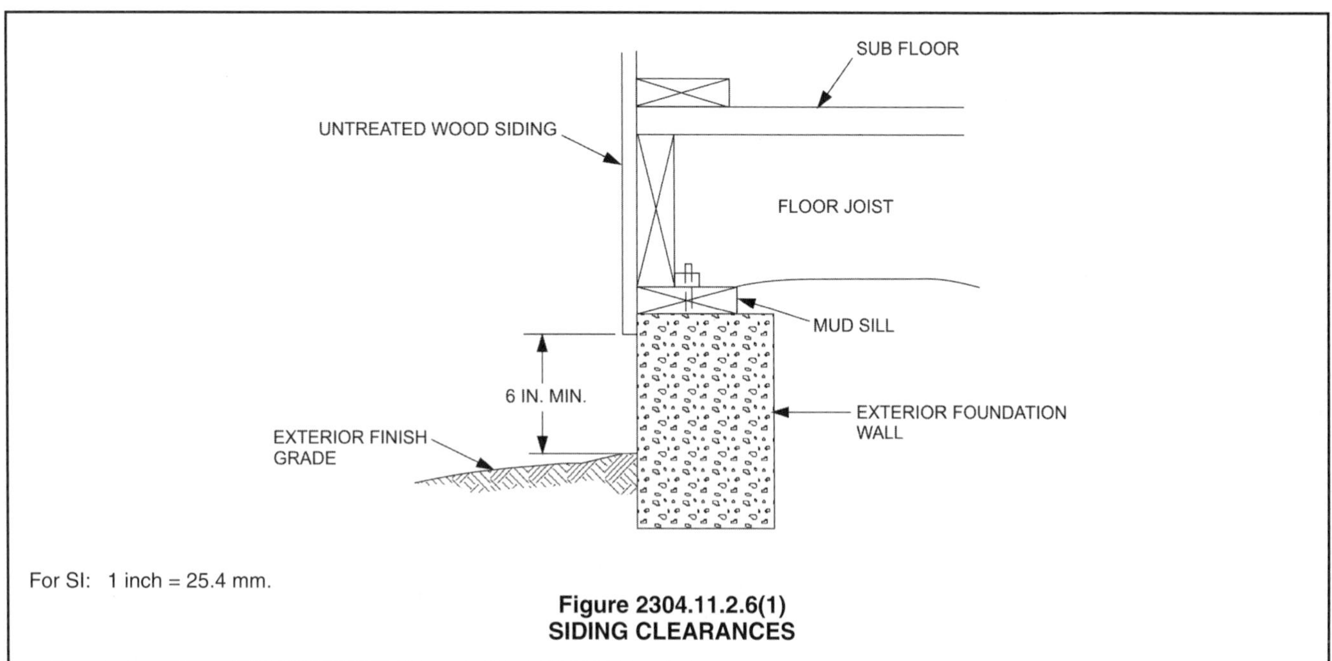

Figure 2304.11.2.6(1)
SIDING CLEARANCES

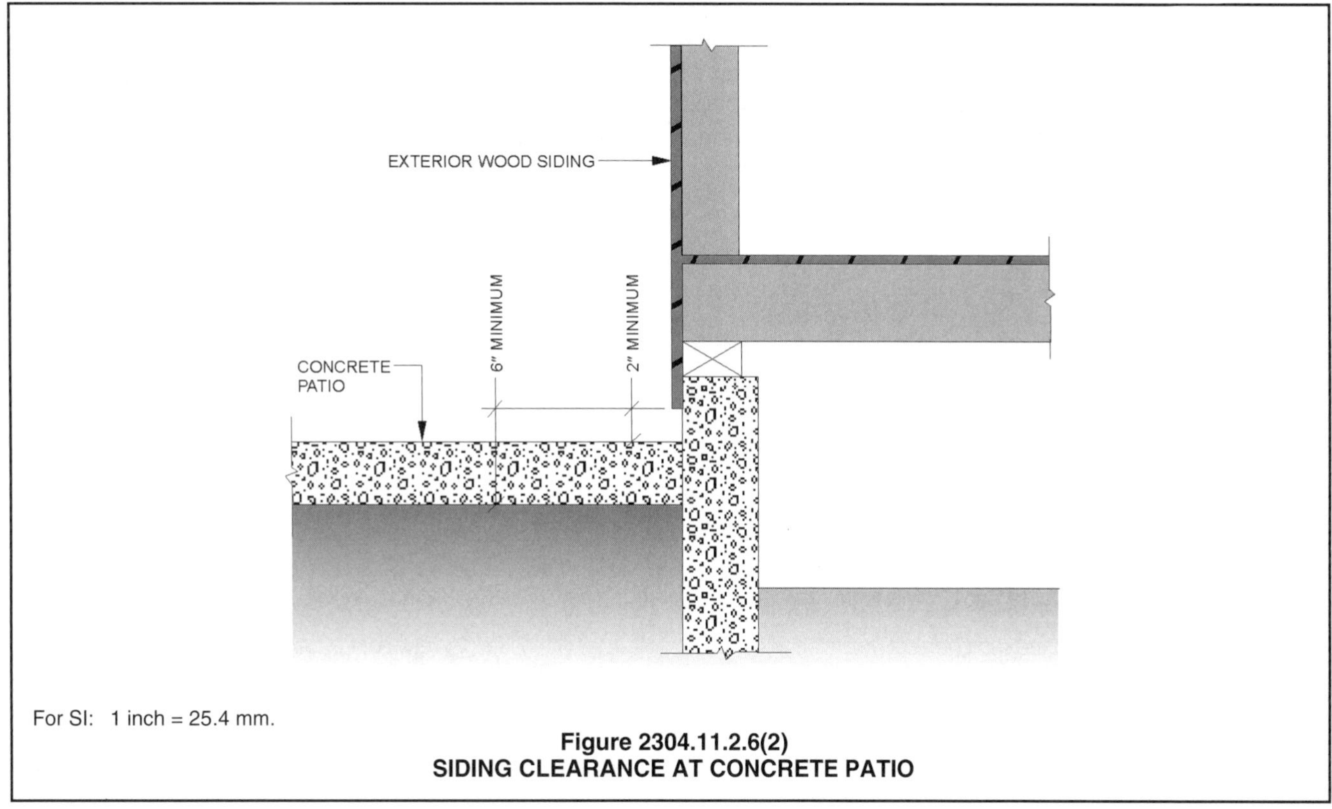

Figure 2304.11.2.6(2)
SIDING CLEARANCE AT CONCRETE PATIO

at least 1 inch (25 mm) above the slab or deck and 6 inches (152 mm) above exposed earth, and are separated therefrom by an impervious moisture barrier.

2. Posts or columns in enclosed crawl spaces or unexcavated areas located within the periphery of the building, supported by a concrete pier or metal pedestal at a height greater than 8 inches (203 mm) from exposed ground, and are separated therefrom by an impervious moisture barrier.

❖ When the post or column is directly in concrete that is in direct contact with earth, it must be naturally durable or preservative treated. The exceptions illustrated in Figure 2304.11.2.7 permit the use of common framing lumber when the post or column is isolated from the earth or concrete. Generally, structural elements in basements or cellars are easier to inspect, which explains the lower clearances provided by Exception 1. Posts or columns in crawl spaces are required to have a greater clearance to safeguard against termites between inspections. In both exceptions, an impervious moisture barrier restricts the absorption of water from the earth or concrete pier.

2304.11.3 Laminated timbers. The portions of glued-laminated timbers that form the structural supports of a building or other structure and are exposed to weather and not fully protected from moisture by a roof, eave or similar covering shall be pressure treated with preservative or be manufactured from naturally durable or *preservative-treated wood.*

❖ It is common practice to design large glued-laminated arches that "spring" or are connected to foundations near ground level. Exterior walls are built inside the span of these arches, leaving the initial few feet of the wood arches exposed to the weather. Experience has shown that covering the tops of these arches with metal or other water seals is not sufficient to prevent decay; therefore, arches and other exposed wood members not protected by roofs or similar covers must be laminated of naturally durable or pressure-treated wood. Creosote solutions are typically used for the most severe outdoor exposures, but they are quite dark, oily and pungent and are, therefore, typically restricted to uses where direct human contact is avoided. Oil-borne and water-borne treatments are slightly friendlier, but they have different effects depending on the timing of the application, either before or after laminating the timbers together.

2304.11.4 Wood in contact with the ground or fresh water. Wood used in contact with the ground (exposed earth) in the locations specified in Sections 2304.11.4.1 and 2304.11.4.2 shall be naturally durable (species for both decay and termite resistance) or preservative treated using water-borne preservatives in accordance with AWPA U1 (Commodity Specifications A or F) for soil or fresh water use.

Exception: Untreated wood is permitted where such wood is continuously and entirely below the ground-water level or submerged in fresh water.

❖ Wood that has been completely submerged below the ground-water table or in fresh water has been known to remain in good condition for many years. This method of preservation is the only exception to requiring preservative treatment of all wood in contact with or embedded in the ground.

2304.11.4.1 Posts or columns. Posts and columns supporting permanent structures that are embedded in concrete that is in direct contact with the earth, embedded in concrete that is exposed to the weather or in direct contact with the earth shall be of *preservative-treated wood.*

❖ This requirement is similar to Section 2304.11.2.7, but emphasizes that posts and columns encased or embedded in concrete must be of preservative-treated wood. The condition is generally more severe than wood placed directly on concrete or embedded in the earth. The embedded portion of the column will most likely be at a moisture content in excess of 20 percent. At some point very near where the concrete ends, the conditions will be ideal for decay—having both a high moisture content and the presence of air. Unless treated wood is used, decay will occur very near where the encasing begins.

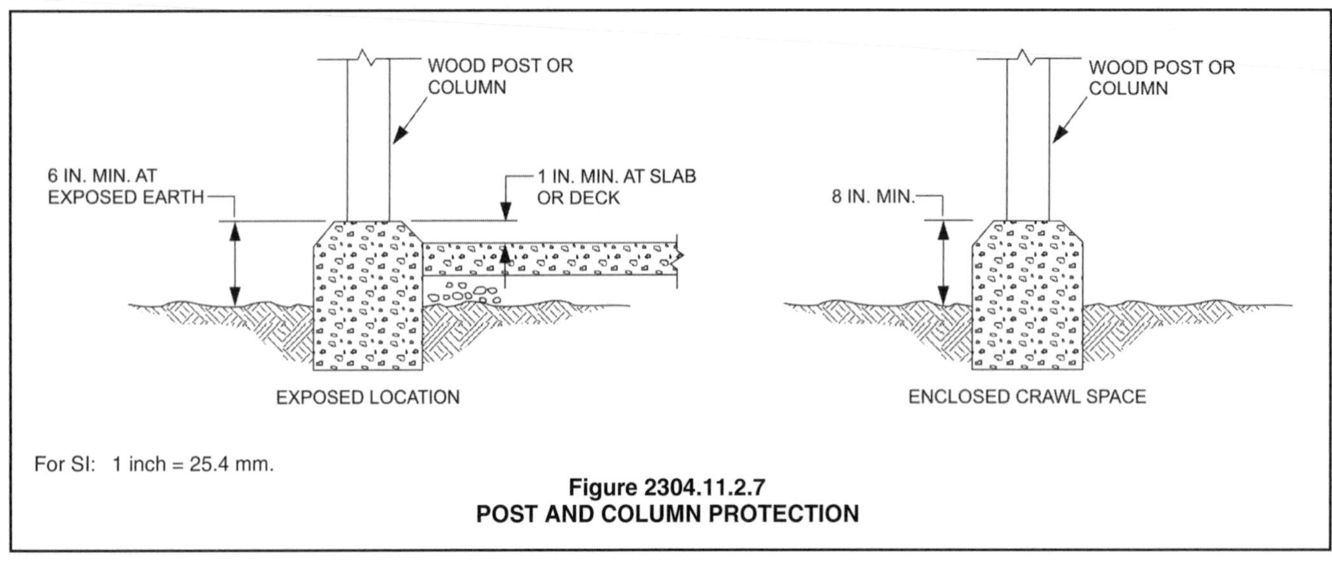

Figure 2304.11.2.7
POST AND COLUMN PROTECTION

2304.11.4.2 Wood structural members. Wood structural members that support moisture-permeable floors or roofs that are exposed to the weather, such as concrete or masonry slabs, shall be of naturally durable or *preservative-treated wood* unless separated from such floors or roofs by an impervious moisture barrier.

❖ The framing condition described is uncommon, but the requirement for naturally durable or preservative-treated wood is appropriate. Concrete or masonry can contain sufficient moisture to raise the moisture content in supporting wood structural elements above 20 percent, therefore; an impervious moisture barrier or naturally durable or preservative-treated wood is required.

2304.11.5 Supporting member for permanent appurtenances. Naturally durable or *preservative-treated wood* shall be utilized for those portions of wood members that form the structural supports of buildings, balconies, porches or similar permanent building appurtenances where such members are exposed to the weather without adequate protection from a roof, eave, overhang or other covering to prevent moisture or water accumulation on the surface or at joints between members.

> **Exception:** When a building is located in a geographical region where experience has demonstrated that climatic conditions preclude the need to use durable materials where the structure is exposed to the weather.

❖ Decks and porches exposed to rain and snow are subject to deterioration. In desert areas where rainfall is slight, there may not be enough moisture to cause any deterioration; therefore, the exception is provided. It is important to confirm climatic conditions with the weather bureau or the building official to determine the proper course of action in protecting these appurtenant structures.

2304.11.6 Termite protection. In geographical areas where hazard of termite damage is known to be very heavy, wood floor framing in the locations specified in Section 2304.11.2.1 and exposed framing of exterior decks or balconies shall be of naturally durable species (termite resistant) or preservative treated in accordance with AWPA U1 for the species, product preservative and end use or provided with *approved* methods of termite protection.

❖ This provision applies to floor framing such as posts, girders, joists, girders and subfloor and exposed framing of exterior decks or balconies where termite damage has been determined to be very heavy.

The termite infestation probability map used in the IRC is included as Figure 2304.11.6. This map provides guidance, but the local jurisdiction should be consulted as it will be more familiar with the threat posed by termites. Most termites found in the United States are classified as subterranean termites, which indicates their need to live in the ground. For many years, metal termite shields were considered the best method of preventing infestation. When properly installed, such shields are still a good defense. Soil poisoning is considered to be a more efficient method of controlling termites; however, there are obvious

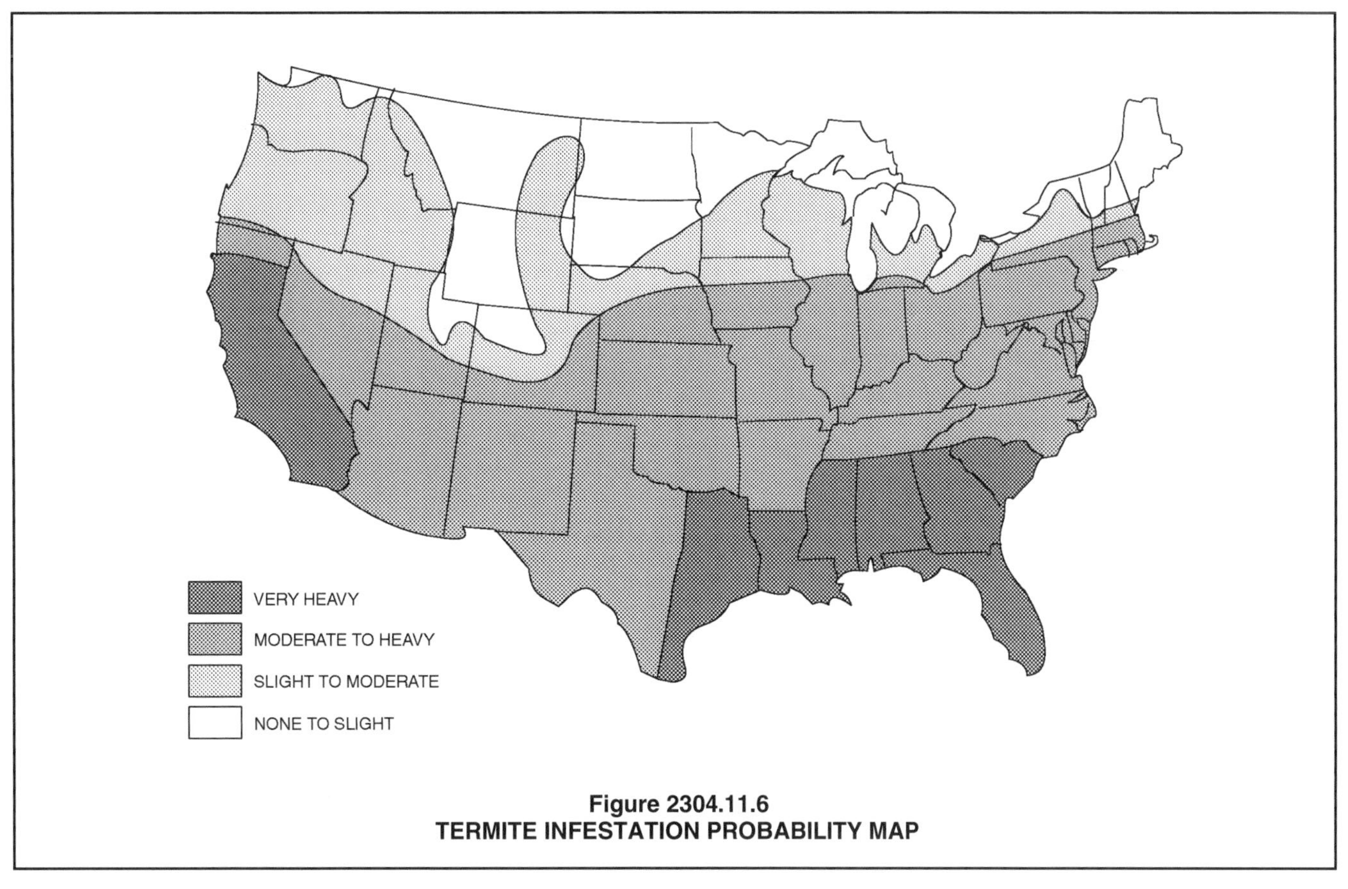

Figure 2304.11.6
TERMITE INFESTATION PROBABILITY MAP

negative environmental implications and some regions may forbid this method altogether.

Additionally, all possible locations of infestation must be poisoned; therefore, the work should be carefully monitored and performed by professionals who are competent and understand applicable standards, laws and procedures. Under no circumstances should soil poisoning be attempted by those unfamiliar with toxic products and their application.

2304.11.7 Wood used in retaining walls and cribs. Wood installed in retaining or crib walls shall be preservative treated in accordance with AWPA U1 (Commodity Specifications A or F) for soil and fresh water use.

❖ Where the structure is regulated by the code, wood used to retain soil is to be preservative treated. Presumably, where the failure of the retaining wall or crib may threaten life safety or result in property damage, the durability provided by preservative-treated wood is necessary.

2304.11.8 Attic ventilation. For *attic* ventilation, see Section 1203.2.

❖ Adequate ventilation allows moisture-laden air to readily escape from the attic space in order to avoid exposure of the framing to excessive moisture (see Section 1203.2 for further discussion).

2304.11.9 Under-floor ventilation (crawl space). For under-floor ventilation (crawl space), see Section 1203.3.

❖ Enclosed spaces under floors of buildings are referred to as crawl spaces when not designed and constructed as basements. The earth is usually exposed or covered with a vapor retarder. Regardless, there is always the potential for a large amount of moisture vapor that comes from the ground to be present. This moisture must be removed to prevent wetting and drying cycles that can cause decay and fungus attack in wood members. Experience has shown that crawl spaces constructed in conformance to this section and ventilated to the exterior as required by Section 1203.3 do not develop problems of decay.

Where a high ground-water table exists or moisture is abnormally high, the crawl space must be adequately drained in accordance with Section 1805.1.2. It should be recognized that the moisture problem may be exaggerated in colder months; therefore, operable vents should be used with caution. Once insulation in such spaces becomes wet, it seldom dries out, thus negating its performance.

2304.12 Long-term loading. Wood members supporting concrete, masonry or similar materials shall be checked for the effects of long-term loading using the provisions of the AF&PA NDS. The total deflection, including the effects of long-term loading, shall be limited in accordance with Section 1604.3.1 for these supported materials.

Exception: Horizontal wood members supporting masonry or concrete nonstructural floor or roof surfacing not more than 4 inches (102 mm) thick need not be checked for long-term loading.

❖ It is common for wood structural elements to support masonry and concrete construction. When properly designed, taking into consideration long-term deflection, wood members can adequately support permanent dead loads of concrete and masonry. The deflection limitations on wood supporting masonry and concrete are recommended by that industry (see Section 3.5.2 and Appendix F, AF&PA NDS). Masonry and concrete are brittle materials, which do not tolerate movement. Wood is hygroscopic, meaning it changes dimensions as a result of absorbing or releasing (drying) water. Dimensional changes in wood due to improper detailing can result in damage to the supported masonry and concrete.

The exception provides some relief. Flooring and roof coverings meeting specific thickness criteria may be supported by wood. The wood must be designed to carry the weight of the supported material, but it is not necessary to consider the effect of long-term loading.

Tabulated modulus of elasticity design values in the AF&PA NDS are intended for the calculation of immediate deformation under load. Under sustained loading, wood members exhibit additional time-dependent formation (creep), which develops at a slow but steady rate over time. Where dead loads or sustained live loads represent a relatively high percentage of total design load, creep is a design consideration and is handled by these provisions in the AF&PA NDS. In such situations, the total deflection under long-term loading is estimated by increasing the initial deflection associated with the long-term load component by the time-dependent creep factor of 1.5 for seasoned or 2.0 for unseasoned or wet lumber or glued-laminated timber (see Section 3.5.2, AF&PA NDS).

SECTION 2305
GENERAL DESIGN REQUIREMENTS FOR LATERAL FORCE-RESISTING SYSTEMS

2305.1 General. Structures using wood-frame shear walls or wood-frame diaphragms to resist wind, seismic or other lateral loads shall be designed and constructed in accordance with AF&PA SDPWS and the applicable provisions of Sections 2305, 2306 and 2307.

❖ General design requirements for wood-framed lateral-force-resisting systems applicable to engineered structures are described in this section. Elements discussed in this section are designed using either ASD or LRFD methodologies. Section 2305 references the AF&PA SDPWS for lateral design of wood structures. Since the referenced standard does not explicitly recognize the use of staples in diaphragms and shear walls, Sections 2305.2 and 2305.3 provide guidance on calculating the deflection-blocked wood structural panel diaphragms and shear walls fastened with staples.

When portions of or elements within structures constructed according to Section 2308 need to be engineered, provisions of Section 2305 can be applied without engineering the entire structure. The extent of engineering design to be provided must be determined by the registered design professional (RDP) and accepted by the building official. However, the minimum acceptable extent is often taken to be force transfer into the element, design of the element and force transfer out of the element. When more than one braced wall line or diaphragm in any area of conventional structures requires engineering analysis, the assumptions used to develop Section 2308 may be affected, and a complete lateral analysis and design may be appropriate for the entire lateral-force-resisting system. For example, the absence of a ceiling may create a nonconventional configuration because conventional construction assumes there is a ceiling diaphragm that contributes some degree of lateral strength and stiffness.

2305.1.1 Openings in shear panels. Openings in shear panels that materially affect their strength shall be detailed on the plans, and shall have their edges adequately reinforced to transfer all shearing stresses.

❖ Openings often occur in shear walls and diaphragms to accomodate stairs, windows, doors, shafts and other purposes. The transfer of shear forces around openings must be effectively addressed in the design to ensure that adequate capacity exists around the perimeter of the opening to effectively transfer shear forces.

If one considers the various approaches to shear wall design, this provision is significant since the framing and connections around the openings are specifically designed and detailed for force transfer. In this approach, design force transfer around the openings involves developing a system of piers and coupling beams within the shear wall. Load paths for the shear and flexure developed in the piers and coupling beams generally require blocking and strapping extending from each corner of the opening to some distance beyond. This design approach is often utilized in highly loaded walls when limited amounts of sheathing are present around openings, often resulting in shear wall detailing that involves many metal tension straps around the openings. The benefit of this additional design and detailing is that the permitted height-to-width ratio of wall piers potentially allows a narrower width pier than would otherwise be required. These are referred to as "force transfer shear walls" in the AF&PA SDPWS.

2305.2 Diaphragm deflection. The deflection of wood-frame diaphragms shall be determined in accordance with AF&PA SDPWS. The deflection (Δ) of a blocked wood structural panel diaphragm uniformly fastened throughout with staples is permitted to be calculated in accordance with Equation 23-1. If not uniformly fastened, the constant 0.188 (For SI: 1/1627) in the third term shall be modified by an approved method.

$$\Delta = \frac{5vL^3}{8EAb} + \frac{vL}{4Gt} + 0.122Le_n + \frac{\Sigma(\Delta_c X)}{2b} \qquad \textbf{(Equation 23-1)}$$

$$\text{For SI: } \Delta = \frac{0.052vL^3}{EAb} + \frac{vL}{4Gt} + \frac{Le_n}{1627} + \frac{\Sigma(\Delta_c X)}{2b}$$

where:

A = Area of chord cross section, in square inches (mm^2).

B = Diaphragm width, in feet (mm).

E = Elastic modulus of chords, in pounds per square inch (N/mm^2).

e_n = Staple deformation, in inches (mm) [see Table 2305.2(1)].

Gt = Panel rigidity through the thickness, in pounds per inch (N/mm) of panel width or depth [see Table 2305.2(2)].

L = Diaphragm length, in feet (mm).

v = Maximum shear due to design loads in the direction under consideration, in pounds per linear foot (plf) (N/mm).

Δ = The calculated deflection, in inches (mm).

$\Sigma(\Delta_c X)$ = Sum of individual chord-splice slip values on both sides of the diaphragm, each multiplied by its distance to the nearest support.

❖ Equation 23-1 provides the estimate of the deflection of a blocked wood structural panel diaphragm that is uniformly fastened using staples. Total deflection represents the sum of four components contributing to the overall deflection: bending deflection, shear deflection, fastener slip and chord splice slip. Guidance on the selection of various equation inputs, such as fastener deformation, effective panel thickness and sum of individual chord-splice values, is provided in Tables 2305.2(1) and 2305.2(2). These tables provide the required parameters needed in the four-part equation. For ease of calculation, the "modulus of rigidity" term (*G*) is combined with the "effective thickness" term (*t*) to form a "panel rigidity capacity" term (*Gt*). Capacity terms, such as *Gt*, are used by the panel industry in lieu of section properties and allowable stresses as it makes it easier to tabulate and compare different panel types.

As noted above, the deflection computation is applicable only to blocked diaphragms. For unblocked diaphragms, *SEAOC Blue Book*, 1999, Section 805.3.2, provides the following guidance:

> Limited testing of diaphragms (APA, 1952, 1954, 1955, 1967) suggests that the deflection of an unblocked diaphragm at its tabulated allowable shear capacity will be about 2.5 times the calculated deflection of a blocked diaphragm of similar

construction and dimensions, at the same shear capacity. If diaphragm framing is spaced more than 24 inches (610 mm) o.c., testing indicates a further increase in deflection of about 20 percent for unblocked diaphragms (e.g., to three times the deflection on a comparable blocked diaphragm). This relationship can be used to develop an estimate of the deflection of unblocked diaphragms.

Note that the unblocked diaphragm testing conducted by the APA mentioned above was conducted on diaphragms with aspect ratios in the 2:1 to 3:1 range. As the aspect ratio decreases, it is reasonable to expect the adjustment factor to similarly decrease.

As noted above, the equation for diaphragm deflection is intended for uniform fastening. The 0.188 constant in the "fastener-slip deflection-contribution" term was derived based on the panel-edge fastener being the same for the entire length of the diaphragm. When fastener spacing becomes less dense near the center of the diaphragm span, ATC 7 recommends that the 0.188 constant should be increased in proportion to the average load on each fastener with a nonuniform fastener spacing, compared to the average load that would be present if a uniform fastener spacing had been maintained. The adjusted constant can be expressed as follows:

$0.188 \times (V_n'/V_n)$

where

V_n' = The average nonuniform load per fastener.

V_n = The average uniform load per fastener.

TABLE 2305.2(1)
e_n VALUES (inches) FOR USE IN CALCULATING DIAPHRAGM AND SHEAR WALL DEFLECTION DUE TO FASTENER SLIP (Structural I)[a, c]

LOAD PER FASTENER[b] (pounds)	FASTENER DESIGNATIONS
	14-Ga staple x 2 inches long
60	0.011
80	0.018
100	0.028
120	0.04
140	0.053
160	0.068

For SI: 1 inch = 25.4 mm, 1 foot = 304.8 mm, 1 pound = 4.448 N.

a. Increase e_n values 20 percent for plywood grades other than Structural I.

b. Load per fastener = maximum shear per foot divided by the number of fasteners per foot at interior panel edges.

c. Decrease e_n values 50 percent for seasoned lumber (moisture content < 19 percent).

❖ Tables 2305.2(1) and 2305.2(2) contain values for the terms e_n, G, and t for use in Equations 23-1 and 23-2. The source of these tabulated values is the 1997 UBC, Volume 3, which is the source for Equations 23-1 and 23-2 as well.

G and *t* are combined into a single table that gives the panel rigidity-through-the-thickness capacity, *Gt*, which allows the use of similar values for wood structural panels other than plywood. These values are recognized by the industry for all products meeting the criteria of either DOC PS 1 or PS 2, both of which are referenced standards in the code.

TABLE 2305.2(2) See page 23-41.

❖ See the commentary for Table 2305.2(1).

2305.3 Shear wall deflection. The deflection of wood-frame shear walls shall be determined in accordance with AF&PA SDPWS. The deflection (Δ) of a blocked wood structural panel shear wall uniformly fastened throughout with staples is permitted to be calculated in accordance with Equation 23-2.

$$\Delta = \frac{8vh^3}{\mathrm{EAb}} + \frac{vh}{Gt} + 0.75he_n + d_a\frac{h}{b} \qquad \textbf{(Equation 23-2)}$$

$$\text{For SI: } \Delta = \frac{vh^3}{3EAb} + \frac{vh}{Gt} + \frac{he_n}{407.6} + d_a\frac{h}{b}$$

where:

A = Area of boundary element cross section in square inches (mm²) (vertical member at shear wall boundary).

b = Wall width, in feet (mm).

d_a = Vertical elongation of overturning anchorage (including fastener slip, device elongation, anchor rod elongation, etc.) at the design shear load (v).

E = Elastic modulus of boundary element (vertical member at shear wall boundary), in pounds per square inch (N/mm²).

e_n = Staple deformation, in inches (mm) [see Table 2305.2(1)].

Gt = Panel rigidity through the thickness, in pounds per inch (N/mm) of panel width or depth [see Table 2305.2(2)].

h = Wall height, in feet (mm).

v = Maximum shear due to design loads at the top of the wall, in pounds per linear foot (N/mm).

Δ = The calculated deflection, in inches (mm).

❖ The equation in this section provides an estimate of the deflection of a blocked wood structural panel shear wall. Total deflection represents the sum of four components contributing to overall deflection: bending deflection, shear deflection, fastener slip and anchorage deformation. Guidance is provided in Tables 2305.2(1) and 2305.2(2), which provide the required parameters needed for the four-part equation. For ease of calculation, the "modulus of rigidity" term (*G*) was combined with the "effective thickness" term (*t*) to form a "panel rigidity capacity" term (*Gt*). Capacity terms such as *Gt* are used by the panel industry in lieu of section properties and allowable stresses as it makes it easier to tabulate and compare different panel types.

SECTION 2306 ALLOWABLE STRESS DESIGN

2306.1 Allowable stress design. The design and construction of wood elements in structures using *allowable stress design* shall be in accordance with the following applicable standards:

❖ This section contains a number of specifications intended to provide guidance to nonprofessional as well as professional users of the code. These specifications, based on accepted engineering practices and experiences, are considered to be the minimum acceptable methods for design and construction of wood elements in structures. When designed and built in accordance with the standards referenced in this section, a building or structure is deemed to comply with the code. These standards contain most of the information needed to adequately design a structure in accordance with the ASD method. It is necessary for the designer to have a working knowledge of wood engineering principles and experience to properly interpret these recommendations and meet the provisions of other applicable sections of the code.

The most common and applicable practices are summarized in the standards listed in Section 2306.1. These standards contain references to additional technical publications that address special problems and design criteria for wood structures. The users of these standards should have training and expertise in engineering and construction in order to properly interpret the standards' requirements. These standards employ ASD—the traditional format for presenting safety checking equations in structural design standards. The basis of ASD is that actual working stresses resulting from specified service loads are compared to allowable stresses. Members and connections are selected so that actual stresses do not

TABLE 2305.2(2)
VALUES OF *Gt* FOR USE IN CALCULATING DEFLECTION OF WOOD STRUCTURAL PANEL SHEAR WALLS AND DIAPHRAGMS

PANEL TYPE	SPAN RATING	VALUES OF Gt (lb/in. panel depth or width)							
		OTHER				STRUCTURAL I			
		3-ply Plywood	4-ply Plywood	5-ply Plywood[a]	OSB	3-ply Plywood	4-ply Plywood	5-ply Plywood[a]	OSB
Sheathing	24/0	25,000	32,500	37,500	77,500	32,500	42,500	41,500	77,500
	24/16	27,000	35,000	40,500	83,500	35,000	45,500	44,500	83,500
	32/16	27,000	35,000	40,500	83,500	35,000	45,500	44,500	83,500
	40/20	28,500	37,000	43,000	88,500	37,000	48,000	47,500	88,500
	48/24	31,000	40,500	46,500	96,000	40,500	52,500	51,000	96,000
Single Floor	16 o.c.	27,000	35,000	40,500	83,500	35,000	45,500	44,500	83,500
	20 o.c.	28,000	36,500	42,000	87,000	36,500	47,500	46,000	87,000
	24 o.c.	30,000	39,000	45,000	93,000	39,000	50,500	49,500	93,000
	32 o.c.	36,000	47,000	54,000	110,000	47,000	61,000	59,500	110,000
	48 o.c.	50,500	65,500	76,000	155,000	65,500	85,000	83,500	155,000

		OTHER			STRUCTURAL I		
	Thickness (in.)	A-A, A-C	Marine	All Other Grades	A-A, A-C	Marine	All Other Grades
Sanded Plywood	$^1/_4$	24,000	31,000	24,000	31,000	31,000	31,000
	$^{11}/_{32}$	25,500	33,000	25,500	33,000	33,000	33,000
	$^3/_8$	26,000	34,000	26,000	34,000	34,000	34,000
	$^{15}/_{32}$	38,000	49,500	38,000	49,500	49,500	49,500
	$^1/_2$	38,500	50,000	38,500	50,000	50,000	50,000
	$^{19}/_{32}$	49,000	63,500	49,000	63,500	63,500	63,500
	$^5/_8$	49,500	64,500	49,500	64,500	64,500	64,500
	$^{23}/_{32}$	50,500	65,500	50,500	65,500	65,500	65,500
	$^3/_4$	51,000	66,500	51,000	66,500	66,500	66,500
	$^7/_8$	52,500	68,500	52,500	68,500	68,500	68,500
	1	73,500	95,500	73,500	95,500	95,500	95,500
	$1^1/_8$	75,000	97,500	75,000	97,500	97,500	97,500

For SI: 1 inch = 25.4 mm, 1 pound/inch = 0.1751 N/mm.

a. Applies to plywood with five or more layers; for five-ply/three-layer plywood, use values for four ply.

exceed allowable stresses.

While providing life safety by preventing structural failure or collapse remains the primary purpose of structural design, the designer must also consider how the design will perform from the perspective of serviceability, durability and fire safety. Serviceability limit states are those that restrict the normal use and occupancy of the structure, such as excessive deflection and vibration. The code establishes deflection limits as a ratio of the member span; for example, *L*/360 computed under live load or *L*/240 under total load (dead plus live) for floor members (see Section 1604.3).

American Forest & Paper Association.

NDS	National Design Specification for Wood Construction
SDPWS	Special Design Provisions for Wind and Seismic

❖ The AF&PA NDS is promulgated and distributed by the AF&PA. The AF&PA NDS was first adopted in 1944 and has been updated periodically to reflect new knowledge under the auspices of AF&PA and its predecessor organizations, the National Lumber Manufacturers Association (NLMA) and the National Forest Products Association. The AF&PA NDS is now published in a dual format, meaning it allows both ASD and LRFD methods.

AF&PA also publishes *Wood Construction Data (WCD) No. 5, Heavy Timber Construction Details*. It is written in code language for easy reference. WCD No. 5 contains detailed specifications for sizes of members required and other provisions that collectively provide a definition of "Heavy timber construction." Recommendations are also included for the fabrication and erection of this unique construction. It should be used in conjunction with the NDS and the American Institute of Timber Construction (AITC) specifications when the members are glued laminated.

The *Supplement to the National Design Specification* NDS supplement is a collation of the reference design values for wood construction. The reference design values for sawn lumber are as published by seven grade rules-writing agencies: National Lumber Grades Authority (Canada), Northern Softwood Lumber Bureau, Northeastern Lumber Manufacturers Association, Redwood Inspection Bureau, Southern Pine Inspection Bureau, West Coast Lumber Inspection Bureau and Western Wood Products Association.

These reference design values have been approved by the Board of Review of the ALSC with advice from the U.S. Forest Products Laboratory of the U.S. Department of Agriculture. They have been certified for conformance with the U.S. DOC PS 20. Regional grading agencies formulate and publish grading rules, and ALSC approves these rules as conforming with DOC PS 20.

DOC PS 20 requires that design values for visually graded lumber be developed in accordance with appropriate ASTM standards or other technically sound criteria. The design values reflected in the NDS Supplement are derived from ASTM D 1990, which is utilized to derive strength values from full-sized samples of lumber.

Design values for visually graded timbers, decking and some species and grades of dimension lumber are based on the provisions of ASTM D 245. The methods in ASTM D 245 involve adjusting the strength properties of small clear specimens of wood for the effects of knots, slope of grain, splits, checks, size, duration of load, moisture content and other influencing factors to obtain design values applicable to normal conditions of service. The NDS Supplement also contains reference design values for structural glued-laminated timber from AITC and APA – The Engineered Wood Association in accordance with principles established by the U.S. Forest Products Laboratory standards. These involve adjustment of strength properties to account for specific factors such as knots, grain slope, density, member size and the number of laminations.

The design values given in the supplement are for normal loading conditions and under dry service conditions in a covered building. If other conditions of loadings or wet end-use conditions exist, these values must be adjusted in accordance with the recommendations of the NDS.

American Institute of Timber Construction.

AITC 104	Typical Construction Details
AITC 110	Standard Appearance Grades for Structural Glued Laminated Timber
AITC 113	Standard for Dimensions of Structural Glued Laminated Timber
AITC 117	Standard Specifications for Structural Glued Laminated Timber of Softwood Species
AITC 119	Standard Specifications for Structural Glued Laminated Timber of Hardwood Species
ANSI/AITC A190.1	Structural Glued Laminated Timber
AITC 200	Inspection Manual

❖ The AITC standards listed are primarily intended for glued-laminated structural members, but also include recom-mendations for sawn timber members. These stan-dards supplement the provisions of the AF&PA NDS and WCD No. 5 with no conflicting provisions. All the standards should be used collectively.

ANSI/AITC A190.1 is intended to provide nationally recognized requirements for the production, inspection, testing and certification of structural glued-laminated timber. Section 6 of the standard provides detailed requirements for in-plant quality control and the qualification of third-party inspection and testing agencies. These specifications include provisions for required testing during and after fabrication. Section 7 outlines the requirements for marking, which must be distinctive and must identify the edition of ANSI/AITC A190.1 used, the inspection and testing agency and the laminating plant.

The marking on other than custom members meeting specific job specifications must also include information on: species or species group of lumber; design values used (if not standard); intended use of member (simple beam, tension or compression member, etc.); type of adhesive (dry-wet-use); appearance grade (industrial-architectural-premium); treatment (if any) and proofloading (if applicable). Many specifiers require a Certificate of Conformance for glued-laminated beams showing compliance with APA/EWS or AITC/ANSI A190.1. Figure 2306.1 shows an example of an AITC certificate of conformance for a glue laminated timber.

American Society of Agricultural and Biological Engineers.

ASABE EP 484.2	Diaphragm Design of Metal-clad, Post-Frame Rectangular Buildings
ASABE EP 486.1	Shallow Post Foundation Design
ASABE 559	Design Requirements and Bending Properties for Mechanically Laminated Columns

❖ The American Society of Agricultural Engineers (ASAE) develops standards for design and construction of agricultural and utility buildings.

APA—The Engineered Wood Association.

Panel Design Specification

Plywood Design Specification Supplement 1—Design & Fabrication of Plywood Curved Panel

Plywood Design Specification Supplement 2—Design & Fabrication of Glued Plywood-lumber Beams

Plywood Design Specification Supplement 3—Design & Fabrication of Plywood Stressed-skin Panels

Plywood Design Specification Supplement 4—Design & Fabrication of Plywood Sandwich Panels

Plywood Design Specification Supplement 5—Design & Fabrication of All-plywood Beams

EWS T300	Glulam Connection Details
EWS S560	Field Notching and Drilling of Glued Laminated Timber Beams
EWS S475	Glued Laminated Beam Design Tables
EWS X450	Glulam in Residential Construction
EWS X440	Product and Application Guide: Glulam
EWS R540	Builders Tips: Proper Storage and Handling of Glulam Beams

❖ The APA Engineered Wood Association recommendations and design standards are considered to provide the best criteria for the design and construction of structures incorporating plywood and other wood structural panel products.

Truss Plate Institute, Inc.

TPI 1	National Design Standard for Metal Plate Connected Wood Truss Construction

❖ TPI 1 is the standard for the design, fabrication, quality control and erection of light-frame metal-plate-connected wood trusses. The standard was prepared through a joint effort by the American National Standards Institute (ANSI) and TPI. This standard should be used in conjunction with other standards for wood construction.

2306.1.1 Joists and rafters. The design of rafter spans is permitted to be in accordance with the *AF&PA Span Tables for Joists and Rafters.*

❖ Spans for joists and rafters may be designed in accordance with the AF&PA *Span Tables for Joists and Rafters* or approved engineering standards. When sizes of joists or rafters are other than standard sizes conforming to DOC PS 20, they must be designed in accordance with the AF&PA NDS. When approved, other span tables may be used, such as those from the IRC, Southern Forest Products Association, Western Wood Products Association and Canadian Wood Council. The spans given in these tables are the same as those published in AF&PA *Span Tables for Joists and Rafters.* The AF&PA span tables were developed to provide uniformity in use of wood joists and rafters for light wood-framed structures where the live and dead loads do not exceed those for which the tables were designed. The tables for joists and rafters were designed with the live load, dead load and deflection criteria shown in Chapter 16.

2306.1.2 Plank and beam flooring. The design of plank and beam flooring is permitted to be in accordance with the *AF&PA Wood Construction Data No. 4.*

❖ The plank-and-beam method for framing floors and roofs has been used in buildings for many years. The adaptation of this system to residential construction has raised technical issues concerning details of application. AF&PA's *Plank and Beam Framing for Residential Buildings, Wood Construction Data No. 4* contains information pertaining to design principles, advantages and limitations, construction details and structural requirements for the plank-and-beam method of framing.

CERTIFICATE OF CONFORMANCE

THE UNDERSIGNED MANUFACTURER HEREBY CERTIFIES that the products identified below and on attached sheets are marked with the Collective Mark of the AMERICAN INSTITUTE OF TIMBER CONSTRUCTION (AITC) and were manufactured in conformance with applicable provisions of the latest revision of American National Standard for wood products – Structural Glued Laminated Timber, ANSI/AITC A190.1, and that such manufacture occurred at our plant in , which plant has a quality control system approved by the INSPECTION BUREAU OF THE AMERICAN INSTITUTE OF TIMBER CONSTRUCTION, and is audited periodically by such Bureau.

Job Name: ____________________

Job Location: ____________________

Customer's Order No. ________ Date: ________ Mfgr's Order No. ________

Order Description:

Signature: ____________________ Company: ____________________

Title: ________ Address: ____________________ Date: ________

AITC HEREBY CERTIFIES that the said company at its said plant is licensed by the AMERICAN INSTITUTE OF TIMBER CONSTRUCTION to use the AITC Collective Mark in respect of products which comply with applicable provisions of said Standard, that the adequacy of the quality control system in effect at said plant is periodically audited and verified by the AITC INSPECTION BUREAU, and that in the judgment of such Bureau, said company is capable of complying with applicable manufacturing and testing provisions of said Standard in respect of products manufactured in said plant. Conformance with the Standard in respect of any specific or particular product is the sole responsibility of the manufacturer; AITC's guarantee hereunder being only that the said company is qualified to produce a product meeting the said Standard and that its plant is periodically audited and verified by the AITC INSPECTION BUREAU.

AMERICAN INSTITUTION OF TIMBER CONSTRUCTION CORPORATE SEAL 1952 DELAWARE

AITC Certificate No. Void without Label

AMERICAN INSTITUTE OF TIMBER CONSTRUCTION

Figure 2306.1
CERTIFICATE OF CONFORMANCE

2306.1.3 Treated wood stress adjustments. The allowable unit stresses for *preservative-treated wood* need no adjustment for treatment, but are subject to other adjustments.

The allowable unit stresses for *fire-retardant-treated wood*, including fastener values, shall be developed from an *approved* method of investigation that considers the effects of anticipated temperature and humidity to which the *fire-retardant-treated wood* will be subjected, the type of treatment and the redrying process. Other adjustments are applicable except that the impact load duration shall not apply.

❖ Allowable unit stresses for preservative-treated wood do not need to be adjusted for treatment, but they are subject to other adjustments as required by the AF&PA NDS. The allowable stresses for fire-retardant-treated wood products must be determined by testing that takes into account anticipated temperatures and humidity levels that will be encountered during service. The allowable stresses for untreated products are not applicable to fire-retardant-treated products because the treatment chemicals can reduce the strength of the materials. This is particularly true for products, such as roof sheathing, that can be subjected to conditions of high temperature and humidity.

Before approving the use of fire-retardant-treated wood products, the building official should require the manufacturer of each subject product to submit its recommended allowable stresses and provide documentation on how those values were determined. This documentation should be required for each chemical that is to be used because performance will vary between formulations. Some manufacturers maintain ICC Evaluation Service Reports for their products that can be used by the engineer and building official to design and approve these products.

Additionally, the building official should enforce any restriction on the use and location of any such treated product recommended by the manufacturer or justified based on the nature of the testing performed. For example, when these products are proposed for use in attics, the building official should make sure that they have been tested under conditions of high temperature and humidity that are comparable to the actual conditions that can occur in attics.

The building official should also be alerted to the fact that the ventilation requirement in the code takes on added importance when such products are used in attics. Ventilation requirements should be strictly enforced, and the building official should consult the manufacturer to determine if any additional ventilation above the code requirements might be necessary. Ventilation is needed to control temperature and humidity levels to which the products will be subjected. A lack of adequate ventilation can change the service conditions to which the products are subjected and, as a result, reduce the strength of the products below the values published by the manufacturer.

2306.1.4 Lumber decking. The capacity of lumber decking arranged according to the patterns described in Section 2304.8.2 shall be the lesser of the capacities determined for flexure and deflection according to the formulas in Table 2306.1.4.

❖ Along with the installation requirements specified in Section 2304.8, this section incorporates allowable load provisions for design of lumber decking.

TABLE 2306.1.4
ALLOWABLE LOADS FOR LUMBER DECKING

PATTERN	ALLOWABLE AREA LOAD[a, b]	
	Flexure	Deflection
Simple span	$\sigma_b = \frac{8F'_b d^2}{l^2 6}$	$\sigma_\Delta = \frac{384\Delta E' d^3}{5l^4\ 12}$
Two-span continuous	$\sigma_b = \frac{8F'_b d^2}{l^2 6}$	$\sigma_\Delta = \frac{185\Delta E' d^3}{l^4\ 12}$
Combination simple- and two-span continuous	$\sigma_b = \frac{8F'_b d^2}{l^2 6}$	$\sigma_\Delta = \frac{131\Delta E' d^3}{l^4\ 12}$
Cantilevered pieces intermixed	$\sigma_b = \frac{20F'_b d^2}{3l^2 6}$	$\sigma_\Delta = \frac{105\Delta E' d^3}{l^4\ 12}$
Controlled random layup		
Mechanically laminated decking	$\sigma_b = \frac{20F'_b d^2}{3l^2 6}$	$\sigma_\Delta = \frac{100\Delta E' d^3}{l^4\ 12}$
2-inch decking	$\sigma_b = \frac{20F'_b d^2}{3l^2 6}$	$\sigma_\Delta = \frac{100\Delta E' d^3}{l^4\ 12}$
3-inch and 4-inch decking	$\sigma_b = \frac{20F'_b d^2}{3l^2 6}$	$\sigma_\Delta = \frac{116\Delta E' d^3}{l^4\ 12}$

For SI: 1 inch = 25.4 mm.

a. σ_b = Allowable total uniform load limited by bending.
σ_Δ = Allowable total uniform load limited by deflection.

b. d = Acutal decking thickness.
l = Span of decking.
F'_b = Allowable bending stress adjusted by applicable factors.
E' = Modulus of elasticity adjusted by applicable factors.

❖ Based on the decking layup patterns in Section 2304.8.2, the applicable equations for computing allowable uniform loads limited by flexural bending and deflection are selected from this table. Note that the allowable loading will be the most restrictive value based on bending stresses and deflection limits. The units are not specified in the equation but do need to be consistent when carrying out the load computation.

2306.2 Wood-frame diaphragms. Wood-frame diaphragms shall be designed and constructed in accordance with AF&PA SDPWS. Where panels are fastened to framing members with staples, requirements and limitations of AF&PA SDPWS shall be met and the allowable shear values set forth in Table 2306.2(1) or 2306.2(2) shall be permitted. The allowable shear values in Tables 2306.2(1) and 2306.2(2) are permitted to be increased 40 percent for wind design.

❖ Diaphragms constructed of wood structural panels or diagonal lumber sheathing must be designed in accordance with the AF&PA SDPWS. Wood structural panel diaphragms fastened with staples may be designed using the values given in Tables 2306.2(1) and 2306.2(2). A diaphragm is a relatively thin structural element (e.g., floor or roof assembly), usually rectangular in plan and capable of resisting shear in its plane and parallel to its edges. The three major elements of a diaphragm are the sheathing (wood structural panels), web members (joists/rafters), the chords (boundary members) and the fasteners (such as nails).

Most floors, roofs and walls can function as diaphragms in distributing lateral (wind or seismic) forces to the vertical resisting elements, such as shear walls. Full diaphragm action can be obtained by detailing the sheathing fasteners and framing connections. Diaphragm construction is used extensively to resist horizontal forces of wind and earthquakes. Diaphragms may be either vertical (often referred to as "shear walls") or horizontal. In special designs, they may be oriented in any direction. A diaphragm acts in a similar manner as an I-beam. The wood structural panel skin acts as the web and the boundary members act as the flanges, which are called diaphragm chords. However, due to the relatively greater depth of a diaphragm, the reactions to stress are somewhat different. The diaphragm design assumes that stresses are uniformly distributed across the panel and the web does not contribute to the resistance of tension and compression stresses as it does in a shallow beam.

A series of diaphragms may be tied together to develop a very rigid structure. Their ability to resist large shear stresses has led to their use in innovative construction methods, such as folded plate roof elements, geodesic domes and space frames.

Diaphragms must be securely connected to supporting members and to foundations that are capable of resisting all design forces. Anchorage for floors and roofs should be detailed, but generally presents no special problems. It is required that the stresses be transferred to the chords, which should be adequately supported. Shear walls, however, often require special attention to their anchorage at the foundation. The foundation and the connections thereto must be designed to resist the uplift, compression and shear forces applied.

Openings in diaphragms, such as skylights, doors and windows, disrupt the uniform distribution of shear stresses across the diaphragm. Provisions must be made for continuity and compensation for this disruption. For instance, reinforcement may be necessary at windows.

As noted, wood structural panel diaphragms should be in accordance with the AF&PA SDPWS or may be designed using the information in Table 2306.2(1) or 2306.2(2). Note that the panel thickness and joist spacings must also comply with the requirements for sheathing of floors and roofs. The AF&PA SDPWS allows wood structural panel diaphragms to be designed using accepted engineering principles. AF&PA NDS provides information on shear resistance of mechanical fasteners and chord design. Working stresses for wood structural panels and other materials are contained in other references in the code.

The design of horizontal diaphragms to resist shear stresses is also dependent on the orientation of continuous panel joints relative to the direction of load and not on the direction of the long dimension of the panel or framing. Six cases of panel orientation are illustrated in Table 2306.2(1). The fastening schedule and the resultant shear resistance will be affected by the panel layout (relative to load) selected. A diaphragm usually must resist loads in both transverse and longitudinal directions. Each direction must be considered separately with a different case of panel orientation; thus, the allowable load will be applicable to each direction.

Blocked diaphragms have all panel edges, including the continuous panel joint, supported by and fastened to framing members. Blocking is used at the edges of panels that are not over joists to provide for connecting units and to assist in the transfer of shear forces to the adjacent panel. Unblocked diaphragms may be used, provided the applicable fastening schedule in Table 2304.9.1 is used and the design shear does not exceed that allowed for unblocked diaphragms in Table 2306.2(1). Unblocked diaphragms are often controlled by the buckling of the panel (skin). With the same fastener spacing, allowable design loads for blocked diaphragms are one and one-half to two times the design loads for unblocked diaphragms. In addition, the maximum loads for which blocked diaphragms may be designed are many times greater. The lateral forces and design shear that must be resisted by the diaphragm usually serve as the basis for choosing whether to use blocked or unblocked diaphragm construction.

The values given in the tables are for seismic and may be increased 40 percent for wind loads. The original diaphragm and shear wall tables were developed in the 1960s when code-prescribed wind and seismic loads were quite different than they are today. In recent reevaluation of the tabulated values, it was determined that they were too low for wind design purposes because code-prescribed wind loads have been refined over the years and the confidence level in their accuracy is high. Because the tables use a 2.8 factor of safety, it was determined that a 2 factor of safety is adequate for wind design, which resulted in a 40-percent increase in the tabulated values. In the diaphragm tables in the AF&PA SDPWS, the nominal unit shear capacities for wind loads are shown in a different column than seismic loads and the wind capacities are 40 percent higher than the seismic capacities.

TABLE 2306.2(1). See page 23-48.

❖ Design values for nailed diaphragms are given in the AF&PA SDPWS. Table 2306.2(1) gives allowable shear values for wood structural panel diaphragms fastened with staples. The table's fifth column specifies the minimum nominal width of framing members at adjoining panel edges and boundaries. Note that 3-inch (76 mm) framing has higher shear capacities than the 2-inch (51 mm) framing. The intent is to provide thicker framing where the fastener size, spacing or number of rows at adjoining panel edges, or boundaries, is dictated. The purpose of using larger framing at these locations is to allow the specified connection details to transfer the higher shear forces from one panel to the next without splitting the framing members. No matter what the edge or boundary fastening requirement is for a specific panel, the minimum fastening requirement at intermediate framing (framing that does not support a panel edge) is always a single row of fasteners at 6 or 12 inches (152 or 305 mm) o.c., as this "field nailing" connection is not part of the shear transfer system between panels. As such, from a shear-transfer perspective, there is no reason to use the "at adjoining panel edges" framing requirements at intermediate framing locations. The normal (usually 2x) framing specified for the vertical load or other engineering considerations is sufficient at the intermediate framing locations.

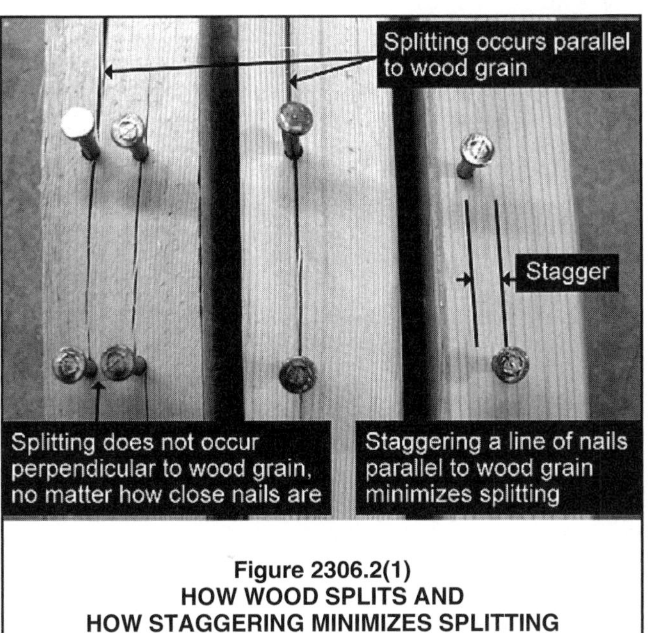

Figure 2306.2(1)
HOW WOOD SPLITS AND HOW STAGGERING MINIMIZES SPLITTING

Because the table values are based on Douglas Fir-Larch or Southern Pine, Note a specifies the adjustment factors to be applied where other species of framing lumber is used.

Note f clarifies the calculation method required to convert table values for applications other than seismic or wind loading. These adjustments reflect the fact that the values in the tables have a built-in load durtion factor (LDF) of 1.6. As such, to convert to a normal load duration (LDF = 1.0), the tabular value must be multiplied by 0.63 (1.0/1.6 = 0.63), and to convert to a permanent load duration (LDF = 0.90), a factor of 0.56 must be used (0.9/1.6 = 0.56).

Note d requires staples to be installed with the crowns of the staple parallel to the long dimension of the framing members. This makes the installation requirement for staples consistent with past practice for staples used in wood-framed diaphragms as required by the National Evaluation Report (NER-272) issued by the National Evaluation Service, Inc.

As Figure 2306.2(1) illustrates, splitting occurs parallel to the grain, so staggering the fasteners helps alleviate this problem. Notes c and d address the concern of splitting of framing members where tighter fastener spacing is specified. Figure 2306.2(2) illustrates staggered nailing that meets the intent of these footnotes (see commentary, Section 2306.2).

TABLE 2306.2(2). See page 23-50.

❖ This table contains values for stapled wood structural panel diaphragm assemblies that have higher capacities compared to those of Table 2306.2.1(1). These high-load diaphragms can be useful for larger flat-roofed structures subjected to high wind or seismic loads. The figures that accompany the table illustrate typical nailing patterns at panel joints. Note g requires high-load diaphragms to have special inspection in accordance with Section 1705. The capacities reflected in this table are based upon calculations using the European Yield Method and were also confirmed by testing [also see commentary, Table 2306.2(1)].

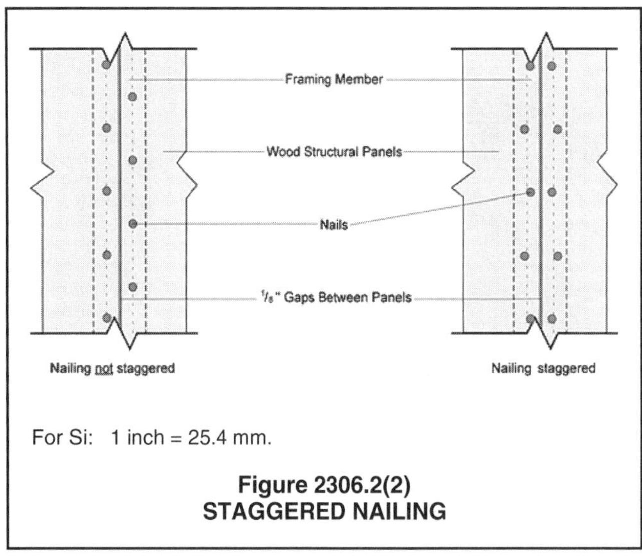

For Si: 1 inch = 25.4 mm.

Figure 2306.2(2)
STAGGERED NAILING

TABLE 2306.2(1)
ALLOWABLE SHEAR VALUES (POUNDS PER FOOT) FOR WOOD STRUCTURAL PANEL DIAPHRAGMS UTILIZING STAPLES WITH FRAMING OF DOUGLAS FIR-LARCH, OR SOUTHERN PINE[a] FOR WIND OR SEISMIC LOADING[f]

PANEL GRADE	STAPLE LENGTH AND GAGE[d]	MINIMUM FASTENER PENETRATION IN FRAMING (inches)	MINIMUM NOMINAL PANEL THICKNESS (inch)	MINIMUM NOMINAL WIDTH OF FRAMING MEMBERS AT ADJOINING PANEL EDGES AND BOUNDARIES[e] (inches)	BLOCKED DIAPHRAGMS				UNBLOCKED DIAPHRAGMS	
					Fastener spacing (inches) at diaphragm boundaries (all cases) at continuous panel edges parallel to load (Cases 3, 4), and at all panel edges (Cases 5, 6)[b]				Fasteners spaced 6 max. at supported edges[b]	
					6	4	$2^1/_2$[c]	2[c]	Case 1 (No unblocked edges or continuous joints parallel to load)	All other configurations (Cases 2, 3, 4, 5 and 6)
					Fastener spacing (inches) at other panel edges (Cases 1, 2, 3 and 4)[b]					
					6	6	4	3		
Structural I grades	$1^1/_2$ 16 gage	1	$^3/_8$	2	175	235	350	400	155	115
				3	200	265	395	450	175	130
			$^{15}/_{32}$	2	175	235	350	400	155	120
				3	200	265	395	450	175	130
Sheathing, single floor and other grades covered in DOC PS 1 and PS 2	$1^1/_2$ 16 gage	1	$^3/_8$	2	160	210	315	360	140	105
				3	180	235	355	400	160	120
			$^7/_{16}$	2	165	225	335	380	150	110
				3	190	250	375	425	165	125
			$^{15}/_{32}$	2	160	210	315	360	140	105
				3	180	235	355	405	160	120
			$^{19}/_{32}$	2	175	235	350	400	155	115
				3	200	265	395	450	175	130

(continued)

TABLE 2306.2(1)—continued
ALLOWABLE SHEAR VALUES (POUNDS PER FOOT) FOR WOOD STRUCTURAL PANEL DIAPHRAGMS UTILIZING STAPLES WITH FRAMING OF DOUGLAS FIR-LARCH, OR SOUTHERN PINE[a] FOR WIND OR SEISMIC LOADING[f]

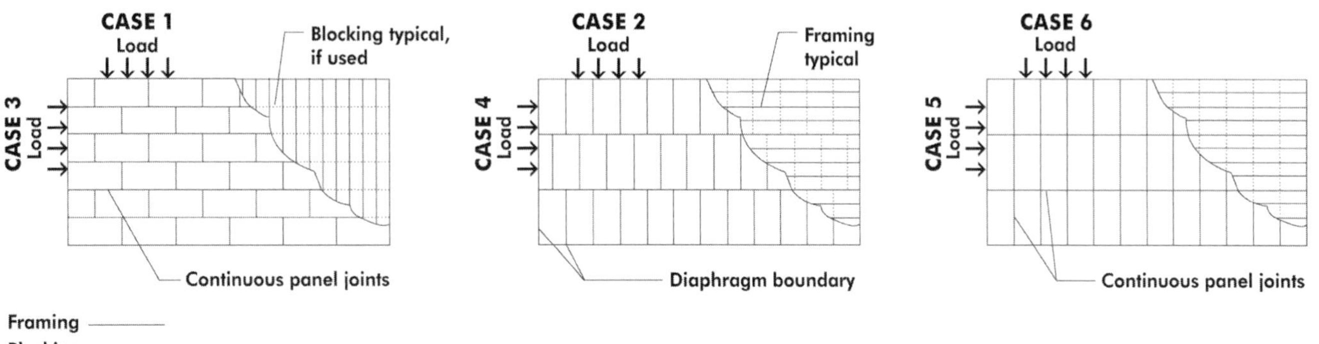

For SI: 1 inch = 25.4 mm, 1 pound per foot = 14.5939 N/m.

a. For framing of other species: (1) Find specific gravity for species of lumber in AF&PA NDS. (2) For staples find shear value from table above for Structural I panels (regardless of actual grade) and multiply value by 0.82 for species with specific gravity of 0.42 or greater, or 0.65 for all other species.

b. Space fasteners maximum 12 inches o.c. along intermediate framing members (6 inches o.c. where supports are spaced 48 inches o.c.).

c. Framing at adjoining panel edges shall be 3 inches nominal or wider.

d. Staples shall have a minimum crown width of $^{7}/_{16}$ inch and shall be installed with their crowns parallel to the long dimension of the framing members.

e. The minimum nominal width of framing members not located at boundaries or adjoining panel edges shall be 2 inches.

f. For shear loads of normal or permanent load duration as defined by the AF&PA NDS, the values in the table above shall be multiplied by 0.63 or 0.56, respectively.

TABLE 2306.2(2)
ALLOWABLE SHEAR VALUES (POUNDS PER FOOT) FOR WOOD STRUCTURAL PANEL BLOCKED DIAPHRAGMS UTILIZING MULTIPLE ROWS OF STAPLES (HIGH-LOAD DIAPHRAGMS) WITH FRAMING OF DOUGLAS FIR-LARCH OR SOUTHERN PINE[a] FOR WIND OR SEISMIC LOADING[b, g, h]

PANEL GRADE[c]	STAPLE GAGE[f]	MINIMUM FASTENER PENETRATION IN FRAMING (inches)	MINIMUM NOMINAL PANEL THICKNESS (inch)	MINIMUM NOMINAL WIDTH OF FRAMING MEMBER AT ADJOINING PANEL EDGES AND BOUNDARIES[e]	LINES OF FASTENERS	BLOCKED DIAPHRAGMS					
						Cases 1 and 2[d]					
						Fastener Spacing Per Line at Boundaries (inches)					
						4		$2^1/_2$		2	
						Fastener Spacing Per Line at Other Panel Edges (inches)					
						6	4	4	3	3	2
Structural I grades	14 gage staples	2	$^{15}/_{32}$	3	2	600	600	860	960	1,060	1,200
				4	3	860	900	1,160	1,295	1,295	1,400
			$^{19}/_{32}$	3	2	600	600	875	960	1,075	1,200
				4	3	875	900	1,175	1,440	1,475	1,795
Sheathing single floor and other grades covered in DOC PS 1 and PS 2	14 gage staples	2	$^{15}/_{32}$	3	2	540	540	735	865	915	1,080
				4	3	735	810	1,005	1,105	1,105	1,195
			$^{19}/_{32}$	3	2	600	600	865	960	1,065	1,200
				4	3	865	900	1,130	1,430	1,370	1,485
			$^{23}/_{32}$	4	3	865	900	1,130	1,490	1,430	1,545

For SI: 1 inch = 25.4 mm, 1 pound per foot = 14.5939 N/m.

a. For framing of other species: (1) Find specific gravity for species of framing lumber in AF&PA NDS. (2) For staples, find shear value from table above for Structural I panels (regardless of actual grade) and multiply value by 0.82 for species with specific gravity of 0.42 or greater, or 0.65 for all other species.

b. Fastening along intermediate framing members: Space fasteners a maximum of 12 inches on center, except 6 inches on center for spans greater than 32 inches.

c. Panels conforming to PS 1 or PS 2.

d. This table gives shear values for Cases 1 and 2 as shown in Table 2306.2(1). The values shown are applicable to Cases 3, 4, 5 and 6 as shown in Table 2306.2(1), providing fasteners at all continuous panel edges are spaced in accordance with the boundary fastener spacing.

e. The minimum nominal depth of framing members shall be 3 inches nominal. The minimum nominal width of framing members not located at boundaries or adjoining panel edges shall be 2 inches.

f. Staples shall have a minimum crown width of $^7/_{16}$ inch, and shall be installed with their crowns parallel to the long dimension of the framing members.

g. High-load diaphragms shall be subject to special inspection in accordance with Section 1705.5.1.

h. For shear loads of normal or permanent load duration as defined by the AF&PA NDS, the values in the table above shall be multiplied by 0.63 or 0.56, respectively.

(continued)

TABLE 2306.2(2)—continued
ALLOWABLE SHEAR VALUES (POUNDS PER FOOT) FOR WOOD STRUCTURAL PANEL BLOCKED DIAPHRAGMS UTILIZING MULTIPLE ROWS OF STAPLES (HIGH-LOAD DIAPHRAGMS) WITH FRAMING OF DOUGLAS FIR-LARCH OR SOUTHERN PINE FOR WIND OR SEISMIC LOADING

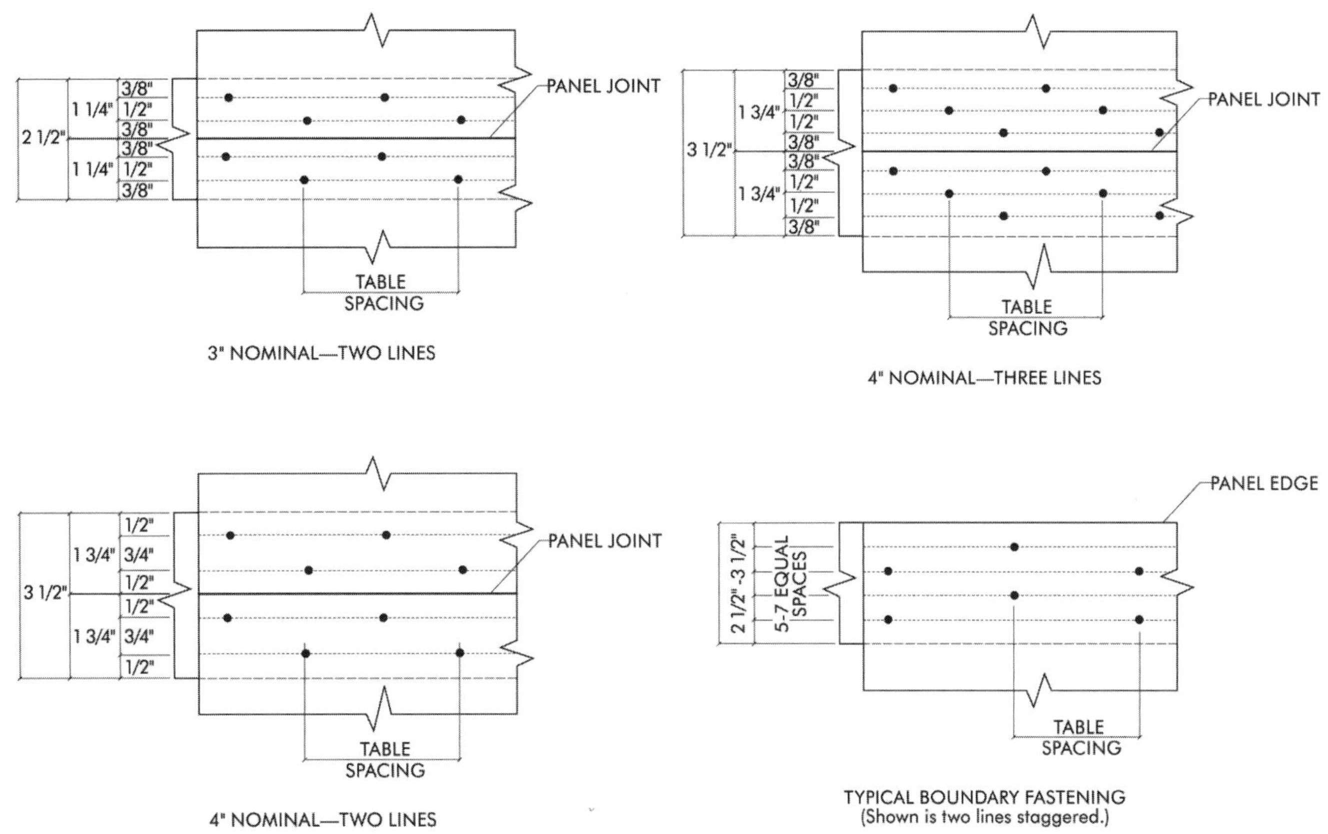

NOTE: SPACE PANEL END AND EDGE JOINT 1/8-INCH. REDUCE SPACING BETWEEN LINES OF NAILS AS NECESSARY TO MAINTAIN MINIMUM 3/8-INCH FASTENER EDGE MARGINS, MINIMUM SPACING BETWEEN LINES IS 3/8-INCH

2306.2.1 Gypsum board diaphragm ceilings. Gypsum board diaphragm ceilings shall be in accordance with Section 2508.5.

❖ See the commentary to Section 2508.5.

2306.3 Wood-frame shear walls. Wood-frame shear walls shall be designed and constructed in accordance with AF&PA SDPWS. Where panels are fastened to framing members with staples, requirements and limitations of AF&PA SDPWS shall be met and the allowable shear values set forth in Table 2306.3(1), 2306.3(2) or 2306.3(3) shall be permitted. The allowable shear values in Tables 2306.3(1) and 2306.3(2) are permitted to be increased 40 percent for wind design. Panels complying with ANSI/APA PRP-210 shall be permitted to use design values for Plywood Siding in the AF&PA SDPWS.

❖ Shear walls constructed of wood framing must be designed in accordance with the AF&PA SDPWS. The AF&PA SDPWS contains provisions for design of shear walls using wood structural panels, particleboard, structural fiberboard, gypsum wall board, portland cement plaster (stucco) and lumber sheathing. Al-though wood structural panel sheathed shear walls are most prevalently used today, diagonally sheathed lumber shear walls are still used for new construction in some regions and are encountered in the alteration of existing structures. Wood structural panel, structural fiberboard, gypsum board and portland cement plaster shear walls fastened with staples may be designed using the values given in Tables 2306.3(1), 2306.3(2) and 2306.3(3). As with diaphragms, the allowable shear values for wood structural panel and fiberboard sheathed shear walls may be increased by 40 percent for wind loads (see commentary, Section 2306.2).

TABLE 2306.3(1). See below.

❖ The allowable shear capacities for wood structural panel shear walls in Table 2306.3(1) were developed based on engineering principles and monotonic testing.

The AF&PA SDPWS allows shear walls to be designed without limitations by a rational method using either the values for nail strength in the AF&PA NDS or wood structural panel design properties in APA's *Panel Design Specification.* For shear walls utilizing wood framing subjected to seismic loading, energy dissipation is almost entirely due to nail bending.

TABLE 2306.3(1)
ALLOWABLE SHEAR VALUES (POUNDS PER FOOT) FOR WOOD STRUCTURAL PANEL SHEAR WALLS UTILIZING STAPLES WITH FRAMING OF DOUGLAS FIR-LARCH OR SOUTHERN PINE[a] FOR WIND OR SEISMIC LOADING[b, f, g, i]

PANEL GRADE	MINIMUM NOMINAL PANEL THICKNESS (inch)	MINIMUM FASTENER PENETRATION IN FRAMING (inches)	PANELS APPLIED DIRECT TO FRAMING					PANELS APPLIED OVER $^{1}/_{2}$" OR $^{5}/_{8}$" GYPSUM SHEATHING				
			Staple size[h]	Fastener spacing at panel edges (inches)				Staple size[h]	Fastener spacing at panel edges (inches)			
				6	4	3	2[d]		6	4	3	2[d]
Structural I sheathing	$^{3}/_{8}$	1	$1^{1}/_{2}$ 16 Gage	155	235	315	400	2 16 Gage	155	235	310	400
	$^{7}/_{16}$			170	260	345	440		155	235	310	400
	$^{15}/_{32}$			185	280	375	475		155	235	300	400
Sheathing, plywood siding[e] except Group 5 Species, ANSI/APA PRP 210 siding	$^{5}/_{16}$[c] or $^{1}/_{4}$[c]	1	$1^{1}/_{2}$ 16 Gage	145	220	295	375	2 16 Gage	110	165	220	285
	$^{3}/_{8}$			140	210	280	360		140	210	280	360
	$^{7}/_{16}$			155	230	310	395		140	210	280	360
	$^{15}/_{32}$			170	255	335	430		140	210	280	360
	$^{19}/_{32}$		$1^{3}/_{4}$ 16 Gage	185	280	375	475	—	—	—	—	—

For SI: 1 inch = 25.4 mm, 1 pound per foot = 14.5939 N/m.

a. For framing of other species: (1) Find specific gravity for species of lumber in AF&PA NDS. (2) For staples find shear value from table above for Structural I panels (regardless of actual grade) and multiply value by 0.82 for species with specific gravity of 0.42 or greater, or 0.65 for all other species.

b. Panel edges backed with 2-inch nominal or wider framing. Install panels either horizontally or vertically. Space fasteners maximum 6 inches on center along intermediate framing members for $^{3}/_{8}$-inch and $^{7}/_{16}$-inch panels installed on studs spaced 24 inches on center. For other conditions and panel thickness, space fasteners maximum 12 inches on center on intermediate supports.

c. $^{3}/_{8}$-inch panel thickness or siding with a span rating of 16 inches on center is the minimum recommended where applied directly to framing as exterior siding. For grooved panel siding, the nominal panel thickness is the thickness of the panel measured at the point of fastening.

d. Framing at adjoining panel edges shall be 3 inches nominal or wider.

e. Values apply to all-veneer plywood. Thickness at point of fastening on panel edges governs shear values.

f. Where panels are applied on both faces of a wall and fastener spacing is less than 6 inches o.c. on either side, panel joints shall be offset to fall on different framing members, or framing shall be 3 inches nominal or thicker at adjoining panel edges.

g. In Seismic Design Category D, E or F, where shear design values exceed 350 pounds per linear foot, all framing members receiving edge fastening from abutting panels shall not be less than a single 3-inch nominal member, or two 2-inch nominal members fastened together in accordance with Section 2306.1 to transfer the design shear value between framing members. Wood structural panel joint and sill plate nailing shall be staggered at all panel edges. See AF&PA SDPWS for sill plate size and anchorage requirements.

h. Staples shall have a minimum crown width of $^{7}/_{16}$ inch and shall be installed with their crowns parallel to the long dimension of the framing members.

i. For shear loads of normal or permanent load duration as defined by the AF&PA NDS, the values in the table above shall be multiplied by 0.63 or 0.56, respectively.

Where shear design values exceed 350 pounds per linear foot, Note g requires that all framing members receiving edge fastening from abutting panels must be a single 3-inch (76 mm) nominal member in Seismic Design Category D, E or F. A pair of nominal 2-inch (51 mm) members can be used in lieu of a single 3-inch (76 mm) nominal member, provided the two 2-inch (51 mm) members are fastened together to transfer the design shear. This provision is based on testing by APA, which showed that two 2-inch (51 mm) members fastened to transfer the shear performed as well as shear walls with a single 3-inch (76 mm) nominal member. See the commentary to Table 2306.2(1) for additional discussion of the footnotes.

TABLE 2306.3(2). See below.

❖ Fiberboard sheathed shear walls fastened with staples may be designed using Table 2306.3(2). The table is restricted to Type V buildings as defined in Chapter 6. Note e limits the use of fiberboard shear walls to buildings in Seismic Design Category A, B and C and prohibits their use in Seismic Design Category D, E or F.

TABLE 2306.3(3). See page 23-54.

❖ This table provides the shear capacities of gypsum board sheathing, expanded metal or woven wire lath and portland cement plaster in shear walls fastened with staples. Because of the brittleness of these materials, they are not permitted to resist seismic forces imposed by masonry or concrete walls and their use is limited to structures classified as Seismic Design Category D. They are prohibited in structures classified as Seismic Design Category E or F (see Table 12.2-1 ASCE 7). In wind design, the shear capacities of these materials can be added to the shear capacity of the wood structural panel or wood sheathing to determine the total shear capacity of the shear wall to resist wind forces (see AF&PA SDPWS).

Note f requires staples to be installed with the crowns of the staple parallel to the long dimension of the framing members. This makes the installation requirement for staples consistent with what has been required for staples used in wood-framed diaphragms and shear walls.

SECTION 2307 LOAD AND RESISTANCE FACTOR DESIGN

2307.1 Load and resistance factor design. The design and construction of wood elements and structures using *load and resistance factor design* shall be in accordance with AF&PA NDS and AF&PA SDPWS.

❖ The LRFD methodology that was originally developed for ASCE 16-95 was subsequently incorporated into the AF&PA NDS and SDPWS. The current edition of the AF&PA NDS is a dual format specification that contains provisions for both ASD and LRFD.

Theoretical reliability-based analysis has been used for many years in the electronics and aerospace industries with great success. LRFD has evolved to become the preferred format for converting structural design standards to a so-called limit states approach.

SECTION 2308 CONVENTIONAL LIGHT-FRAME CONSTRUCTION

2308.1 General. The requirements of this section are intended for *conventional light-frame construction.* Other methods are permitted to be used, provided a satisfactory design is submitted showing compliance with other provisions of this code. Interior nonload-bearing partitions, ceilings and curtain walls of *conventional light-frame construction* are not subject to the limitations of this section. Alternatively, compliance with AF&PA WFCM shall be permitted subject to the limitations therein and the limitations of this code. Detached one- and two-family dwellings and multiple single-family dwellings (townhouses) not more than three *stories above grade plane* in height with a separate *means of egress* and their accessory structures shall comply with the *International Residential Code.*

❖ The provisions of Section 2308 recognize that light-frame wood construction for a limited range of build-

**TABLE 2306.3(2)
ALLOWABLE SHEAR VALUES (plf) FOR WIND OR SEISMIC LOADING ON SHEAR WALLS OF FIBERBOARD SHEATHING BOARD CONSTRUCTION UTILIZING STAPLES FOR TYPE V CONSTRUCTION ONLY[a, b, c, d, e]**

THICKNESS AND GRADE	FASTENER SIZE	ALLOWABLE SHEAR VALUE (pounds per linear foot) STAPLE SPACING AT PANEL EDGES (inches)[a]		
		4	3	2
$^1/_2$″ or $^{25}/_{32}$″ Structural	No. 11 gage galvanized staple, $^7/_{16}$″ crown[f]	150	200	225
	No. 11 gage galvanized staple, 1″ crown[f]	220	290	325

For SI: 1 inch = 25.4 mm, 1 pound per foot = 14.5939 N/m.

a. Fiberboard sheathing shall not be used to brace concrete or masonry walls.

b. Panel edges shall be backed with 2-inch or wider framing of Douglas fir-larch or Southern pine. For framing of other species: (1) Find specific gravity for species of framing lumber in AF&PA NDS. (2) For staples, multiply the shear value from the table above by 0.82 for species with specific gravity of 0.42 or greater, or 0.65 for all other species.

c. Values shown are for fiberboard sheathing on one side only with long panel dimension either parallel or perpendicular to studs.

d. Fastener shall be spaced 6 inches on center along intermediate framing members.

e. Values are not permitted in Seismic Design Category D, E or F.

f. Staple length shall not be less than $1^1/_2$ inches for $^{25}/_{32}$-inch sheathing or $1^1/_4$ inches for $^1/_2$-inch sheathing.

ings will perform to the extent required by the code based on many years of experience and good historical performance. The stated intent of the code is to provide minimum requirements necessary to safeguard the health, safety and general welfare of the public (see the preface and Section 101.3), not to ensure a complete lack of structural damage after a major wind or seismic event. A structure should not collapse, but may require extensive repairs following a major disaster such as an earthquake or high-wind event. The prescriptive details and specifications for conventional construction are deemed to meet these standards. The development of conventional means of construction is not new. The construction of wood-framed structures supported by concrete and masonry foundations came long before engineering design and analysis, which has resulted in a large residual body of building methods and techniques that have withstood the test of time, satisfying standards of structural stability without additional substantiating data. These traditional modes of construction are called "conventional" and are acceptable, within the prescribed limits, without further substantiation. The prescriptive conventional construction provisions in early editions of U.S. codes described the use of repetitive light-frame wood construction, which is typically referred to as "light-frame construction." The conventional construction provisions have always been entirely prescriptive and were intended to apply to repetitive light-wood framing, such as the stud walls, joists and rafters, along with the foundations that support the structure.

Although prescriptive methods have been followed ever since wood has been used as a building material, in 1990 the Conventional Construction Task Force of the Structural Engineers Association of California (SEAOC) was formed for the purpose of reviewing existing code provisions and construction practices in order to develop better and more uniformly enforceable provisions. This work was integrated into the 1994 UBC and much of that work still remains in the code today.

TABLE 2306.3(3)
ALLOWABLE SHEAR VALUES FOR WIND OR SEISMIC FORCES FOR SHEAR WALLS OF LATH AND PLASTER OR GYPSUM BOARD WOOD FRAMED WALL ASSEMBLIES UTILIZING STAPLES

TYPE OF MATERIAL	THICKNESS OF MATERIAL	WALL CONSTRUCTION	STAPLE SPACING[b] MAXIMUM (inches)	SHEAR VALUE[a, c] (plf)	MINIMUM STAPLE SIZE[f, g]
1. Expanded metal or woven wire lath and Portland cement plaster	7/8″	Unblocked	6	180	No. 16 gage galv. staple, 7/8″ legs
2. Gypsum lath, plain or perforated	3/8″ lath and 1/2″ plaster	Unblocked	5	100	No. 16 gage galv. staple, 1 1/8″ long
3. Gypsum sheating	1/2″ × 2′ × 8′	Unblocked	4	75	No. 16 gage galv. staple, 1 3/4″ long
	1/2″ × 4′	Blocked[d]	4	175	
		Unblocked	7	100	
4. Gypsum board, gypsum veneer base or water-resistant gypsum backing board	1/2″	Unblocked[d]	7	75	No. 16 gage galv. staple, 1 1/2″ long
		Unblocked[d]	4	110	
		Unblocked	7	100	
		Unblocked	4	125	
		Blocked[e]	7	125	
		Blocked[e]	4	150	
	5/8″	Unblocked[d]	7	115	No. 16 gage galv. staple, 1 1/2″ legs, 1 5/8″ long
			4	145	
		Blocked[e]	7	145	
			4	175	
		Blocked[e] Two-ply	Base ply: 9 Face ply: 7	250	No. 16 gage galv. staple 1 5/8″ long No. 15 gage galv. staple, 2 1/4″ long

For SI: 1 inch = 25.4 mm, 1 foot = 304.8 mm, 1 pound per foot = 14.5939 N/m.

a. These shear walls shall not be used to resist loads imposed by masonry or concrete walls (see AF & PA SDPWS). Values shown are for short-term loading due to wind or seismic loading. Walls resisting seismic loads shall be subject to the limitations in Section 12.2.1 of ASCE 7. Values shown shall be reduced 25 percent for normal loading.

b. Applies to fastening at studs, top and bottom plates and blocking.

c. Except as noted, shear values are based on a maximum framing spacing of 16 inches on center.

d. Maximum framing spacing of 24 inches on center.

e. All edges are blocked, and edge fastening is provided at all supports and all panel edges.

f. Staples shall have a minimum crown width of 7/16 inch, measured outside the legs, and shall be installed with their crowns parallel to the long dimension of the framing members.

g. Staples for the attachment of gypsum lath and woven-wire lath shall have a minimum crown width of 3/4 inch, measured outside the legs.

These provisions are generally only applicable to light wood-framed building construction having closely spaced repetitive framing, using studs up to 2 inches by 6 inches (51 mm by 152 mm) in size and joists and rafters up to 2 inches by 12 inches (51 mm by 305 mm) in size. With a few exceptions, modern conventional structures are framed in the "platform" style. Repetitive, closely spaced framing is not specifically defined by the code; however, it is usually assumed to include framing that does not exceed a spacing of 24 inches (610 mm) o.c. This is the greatest spacing commonly found in these types of structures, the greatest spacing covered by AF&PA *Span Tables for Joists and Rafters* and the maximum spacing for which referenced standards allow increases in bending stresses for repetitive member use.

The provisions are prescriptive and as such do not typically require a structural design to comply with the code. For structures, portions of structures or elements that are not within the scope of conventional construction, however, other methods of construction are permitted, provided that they are designed to comply with either ASD or LRFD methodology, as discussed in the commentary to Section 2301.2. Methods other than what are prescribed by the section are allowed where a design is submitted and approved by the building official. Structures must either comply with the restrictions, limitations and requirements prescribed in the conventional construction provisions or documentation is to be provided that demonstrates compliance with other provisions of the code and accepted engineering practice. Generally, accepted engineering practice means an engineered analysis has been done based on well-established principles of mechanics and the analysis conforms to accepted principles, tests or nationally recognized standards. In the case of unusual configurations, weight or loading conditions that exceed the prescriptive limitations, the code requires a structural design.

Some interior nonbearing applications are permitted to use the provisions of Section 2308, even though the remainder of the building does not fall within the limitations of Section 2308.2. These buildings would be of light wood-frame construction as well as other types of construction in which wood framing is permitted under other portions of the code.

2308.1.1 Portions exceeding limitations of conventional construction. When portions of a building of otherwise conventional construction exceed the limits of Section 2308.2, these portions and the supporting load path shall be designed in accordance with accepted engineering practice and the provisions of this code. For the purposes of this section, the term "portions" shall mean parts of buildings containing volume and area such as a room or a series of rooms.

❖ In recognition of the complexity of many wood-framed structures, this section of the code allows for a combined use of prescriptive provisions and engineered systems where the necessary limitations of conventional construction are exceeded in portions of a structure. It is the intention that engineered portions within an otherwise conventionally framed building will be integrated in such a way that a complete load path is provided for the engineered system and the forces carried will be properly delivered to resisting elements, possibly even across an area of the building that is not engineered. For example, an engineered shear wall line requires collectors that might extend over the top of conventional beams, studs and columns, which is perfectly acceptable, provided the lateral forces that this engineered line of shear wall is to resist are properly transferred into the shear wall through an engineered load path and ultimately carried to the foundation and ground. The applicable code requirements in terms of design loads, shear walls, collector elements, diaphragm limitations etc., form the basis for design of engineered portions (see also a discussion of engineered elements under Section 2308.4).

2308.2 Limitations. Buildings are permitted to be constructed in accordance with the provisions of *conventional light-frame construction*, subject to the following limitations, and to further limitations of Sections 2308.11 and 2308.12.

1. Buildings shall be limited to a maximum of three *stories above grade plane*. For the purposes of this section, for buildings assigned to *Seismic Design Category* D or E, cripple stud walls shall be considered to be a *story*.

   Exception: Solid blocked cripple walls not exceeding 14 inches (356 mm) in height need not be considered a *story*.

2. Maximum floor-to-floor height shall not exceed 11 feet, 7 inches (3531 mm). Bearing wall height shall not exceed a stud height of 10 feet (3048 mm).

3. Loads as determined in Chapter 16 shall not exceed the following:

   3.1. Average dead loads shall not exceed 15 psf (718 N/m^2) for combined roof and ceiling, exterior walls, floors and partitions.

   Exceptions:

   1. Subject to the limitations of Sections 2308.11.2 and 2308.12.2, stone or masonry veneer up to the lesser of 5 inches (127 mm) thick or 50 psf (2395 N/m^2) and installed in accordance with Chapter 14 is permitted to a height of 30 feet (9144 mm) above a noncombustible foundation, with an additional 8 feet (2438 mm) permitted for gable ends.

   2. Concrete or masonry fireplaces, heaters and chimneys shall be permitted in accordance with the provisions of this code.

   3.2. Live loads shall not exceed 40 psf (1916 N/m^2) for floors.

   3.3. Ground snow loads shall not exceed 50 psf (2395 N/m^2).

4. V_{asd} as determined in accordance with Section 1609.3.1 shall not exceed 100 miles per hour (mph) (44 m/s) (3-second gust).

 Exception: V_{asd} as determined in accordance with Section 1609.3.1 shall not exceed 110 mph (48.4 m/s) (3-second gust) for buildings in Exposure Category B that are not located in a *hurricane-prone region.*

5. Roof trusses and rafters shall not span more than 40 feet (12 192 mm) between points of vertical support.

6. The use of the provisions for *conventional light-frame construction* in this section shall not be permitted for *Risk Category* IV buildings assigned to *Seismic Design Category* B, C, D, E or F.

7. *Conventional light-frame construction* is limited in irregular structures assigned to *Seismic Design Category* D or E, as specified in Section 2308.12.6.

❖ It is acknowledged that conventional construction provisions concerning framing members and sheathing that carry gravity loads are adequate due to historical performance and the fact that the tables for selecting members have been developed from engineering calculations. For resistance to lateral loads (wind and seismic), however, experience has shown that additional requirements—or limits on the application of conventional construction provisions—are needed as follows:

1. The requirements of conventional construction are based on anticipated loads, both gravity and lateral. Buildings greater than three stories should load the lower stories to a level higher than what is safely addressed by these provisions.

2. Tall studs in bearing walls can have a tendency to bow (buckle) laterally out of the plane of the wall. Although this tendency to buckle can be restrained or otherwise accounted for by design of the stud as a column, the prescriptive provisions of Section 2308 do not attempt to address the problem for studs that are taller than 10 feet (3048 mm). This height is assumed to be the maximum stable height using the typical size, grades and species (type) of wood studs in general use in conventional construction (refer to Table 2308.9.1). The code specifies a maximum floor-to-floor height that accommodates typical floor trusses or I-joists.

3. Because the requirements of Section 2308 are based on anticipated loads, this item limits both the dead and live loads in the structure. The prescriptive provisions of conventional construction are based on relatively light dead loads and do not address support of masonry or concrete other than veneer not more than 5 inches (127 mm) thick and with a total weight of not greater than 50 psf (2395 N/m^2). Additionally, the span tables for joists are primarily based on live loads for residential uses, and other elements that support floor loads assume residential loading of 40 psf (1916 N/m^2). The maximum ground snow load of 50 psf (2395 N/m^2) corresponds to the maximum load in the rafter span tables.

4. While some elements of conventional construction are acceptable for use in buildings in high-wind areas (as will be seen in the documents referenced in Section 2308.2.1), the general framing, sheathing and connection requirements are considered to be insufficient. Buildings in Exposure Category B, as defined in Section 1609.4, are somewhat sheltered by other buildings as well as by trees and other obstructions. For that reason, the exception recognizes that conventional construction will provide acceptable lateral resistance for higher basic wind speeds in buildings in those areas when outside of the hurricane prone region (see Section 202 for definition of "Areas vulnerable to hurricanes").

5. In buildings with roof framing spans in excess of 40 feet (12 192 mm), the axial load on studs and horizontal thrust of that framing on the top plate on which it rests is greater than can be resisted by the conventional ceiling joist and rafter connections specified in Section 2308.10.4.1. Note that the limitation is on the span of the truss or rafter and not on the width of the building. The building width could exceed 40 feet (12 192 mm) as long as the actual span of the roof framing is no more than 40 feet (12 192 mm).

6. In order to use conventional construction provisions for a building in Risk Category IV, the building must also be assigned to Seismic Design Category A. Risk Category IV structures that are not classified as Seismic Design Category A would therefore require an engineered design. Because of this restriction on seismic design category, this exclusion applies to essential facilities that are classified as Seismic Design Category C, D or F in accordance with Section 1613.3.5. These buildings are considered essential facilities and the design seismic forces in such structures are likely to be greater than what is expected to be resisted by conventional construction. Because of their role in emergency preparedness, essential facility building plans and calculations are often reviewed more extensively by state and local agencies before a building permit is issued and other required design requirements may apply.

7. Highly irregular portions of buildings (see commentary, Section 2308.12.6) are not permitted to use conventional construction in the higher seismic design categories. One of the reasons for this is that the response characteristics of

irregular building types are more difficult to predict. Irregularities often exacerbate earthquake load effects. Where the irregularities are limited so that the building's performance would be somewhat predictable, such minor irregularities are permissible as described in Section 2308.12.6.

2308.2.1 Nominal design wind speed greater than 100 mph (3-second gust). Where V_{asd} as determined in accordance with Section 1609.3.1 exceeds 100 mph (3-second gust), the provisions of either AF&PA WFCM or ICC 600 are permitted to be used. Wind speeds in Figures 1609A, 1609B, and 1609C shall be converted in accordance with Section 1609.3.1 for use with AF&PA WFCM or ICC 600.

❖ Where the nominal design wind speed, V_{asd}, exceeds the limitation in Section 2308.2, Item 4, the provisions of Section 2308 cannot be used to construct the building. Engineering design, however, is not always required. The engineered yet prescriptive provisions of AF&PA's *Wood Frame Construction Manual for One- and Two-Family Dwellings* (WFCM) and ICC 600 may be used to design buildings that are within the scope of those documents.

2308.2.2 Buildings in Seismic Design Category B, C, D or E. Buildings of *conventional light-frame construction* and assigned to *Seismic Design Category* B or C shall comply with the additional requirements in Section 2308.11.

Buildings of *conventional light-frame construction* and assigned to *Seismic Design Category* D or E shall comply with the additional requirements in Section 2308.12.

❖ In recognition of additional lateral forces anticipated in Seismic Design Category B or C and, more importantly, those in D or E, additional limits are placed on the application of conventional construction (see commentary, Sections 2308.11 and 2308.12). One of the reasons that additional care is given to structures having an elevated seismic risk is that the actual earthquake load effect is cyclic in nature rather than a statically applied force. The seismic-force-resisting system and connections must resist a different character of force that needs to be considered in the provisions, which typically results in more stringent requirements and added restrictions.

2308.3 Braced wall lines. Buildings shall be provided with exterior and interior braced wall lines as described in Section 2308.9.3 and installed in accordance with Sections 2308.3.1 through 2308.3.4.

❖ Buildings that do not require a seismic analysis in accordance with Section 1613.1 and are located in regions where hurricane winds are not anticipated (see Section 2308.2) are required to be braced in accordance with this section. In addition, if the spacing between the braced exterior walls, exceeds 35 feet (10 668 mm) (see Section 2308.3.1), then interior braced wall lines are also required. Braced wall lines and braced wall panels serve the same purpose as shear wall lines and shear walls, but are prescriptive in nature. Braced wall lines consist of a series of braced wall panels.

The requirements for wall bracing vary depending upon the level of seismic activity for the region in which the building under consideration is located and whether the story in question is supporting other stories above. The wall bracing requirements increase for higher seismic design categories and for walls that support the weight of additional stories above. Various types of structural sheathing materials are permitted, although not every material is permitted in every instance. For example, the let-in brace is not typically permitted in high-seismic risk buildings, but may still be used to prevent temporary racking of the wall during construction.

Nailing requirements are critical to the performance of wall bracing. In order for the bracing material to be effective in resisting racking of the wall, it must be installed in the manner prescribed by the code.

2308.3.1 Spacing. Spacing of braced wall lines shall not exceed 35 feet (10 668 mm) o.c. in both the longitudinal and transverse directions in each *story*.

❖ This section establishes the maximum spacing of 35 feet (10 668 mm) between braced wall lines and thereby limits the lateral force acting on individual vertical elements within that line. For buildings in Seismic Design Category D and E, the maximum spacing between braced wall lines is 25 feet (7620 mm) (see Section 2308.12.3 and refer to Figure 2308.9.3 for further understanding of this principle). The philosophy of conventional construction regarding the resistance of lateral forces is to: (1) limit the magnitude of forces that individual elements must resist; (2) streamline the effects of force on the entire structure by restricting the type and extent of irregularities and (3) strengthen resisting elements beyond what might otherwise be installed without such requirements. All of these concepts are addressed in Section 2308.

2308.3.2 Braced wall line connections. Wind and seismic lateral forces shall be transferred from the roof and floor diaphragms to braced wall lines and from the braced wall lines in upper stories to the braced wall lines in the *story* below in accordance with Sections 2308.3.2.1 and 2308.3.2.2.

❖ The section provides prescriptive connection requirements intended to transfer lateral forces from the roof and floor diaphragm to the braced wall lines, as well as from upper stories to stories below and ultimately to the foundation. The requirements provide some continuity in the load path for the braced wall lines and braced wall panels that serve as the lateral-force-resisting system in conventionally constructed wood-framed buildings.

2308.3.2.1 Bottom plate connection. Braced wall line bottom plates shall be connected to joists or full-depth blocking below in accordance with Table 2304.9.1, Item 6, or to foundations in accordance with Section 2308.3.3.

❖ The bottom plate at braced wall lines must be fastened to either joists or blocking between joists or directly to the foundation, such as a sill plate supported by a slab-on-grade footing.

2308.3.2.2 Top plate connection. Where joists and/or rafters are used, braced wall line top plates shall be fastened over the full length of the braced wall line to joists, rafters, rimboards or blocking above in accordance with Table 2304.9.1, Items 11, 12, 15 or 19, as applicable, based on the orientation of the joists or rafters to the braced wall line. Blocking at joists with walls above shall be equal to the depth of the joist at the braced wall line. Blocking at rafters need not be full depth but shall extend to within 2 inches (51 mm) from the roof sheathing above. Blocking shall be a minimum of 2 inches (51 mm) nominal thickness and shall be fastened to the braced wall line top plate as specified in Table 2304.9.1, Item 11. Notching or drilling of holes in blocking in accordance with the requirements of Section 2308.8.2 or Section 2308.10.4.2 shall be permitted.

At exterior gable end walls braced wall panel sheathing in the top *story* shall be extended and fastened to roof framing where the spacing between parallel exterior braced wall lines is greater than 50 feet (15 240 mm).

Where roof trusses are used and are installed perpendicular to an exterior braced wall line, lateral forces shall be transferred from the roof diaphragm to the braced wall over the full length of the braced wall line by blocking of the ends of the trusses or by other *approved* methods providing equivalent lateral force transfer. Blocking shall be minimum 2 inches (51 mm) nominal thickness and shall extend to within 2 inches (51 mm) from the roof sheathing above and shall be fastened to the braced wall line top plate as specified in Table 2304.9.1, Item 11. Notching or drilling of holes in blocking in accordance with the requirements of Section 2308.8.2 or Section 2308.10.4.2 shall be permitted.

❖ Braced wall line top plates are required to be fastened to rafters, joists or blocking, depending on the orientation of the braced wall line to the framing. Where joists support braced wall lines above, full-depth blocking must be used. Where blocking is used at rafters, the blocking need not be the full depth but is required to extend to within 2 inches (51 mm) of the roof sheathing. Where blocking is used at the top plate connection, it must be a minimum 2 inch (51 mm) nominal thickness.

Where the spacing between exterior braced walls is more than 50 feet (15 240 mm), the braced wall panel sheathing at the gable ends must be extended up to and fastened to the roof framing. Note that because the maximum spacing between braced wall lines is 35 feet (10 668 mm) [25 feet (7620 mm) in Seismic Design Category D and E], this requirement applies to the spacing between the exterior braced wall lines.

Where roof trusses are used, the roof diaphragm must be connected the full length of the braced wall line along the eave by 2-inch (51 mm) nominal blocking or other approved means. As noted for rafters, the eave blocking at trusses need not be the full depth of the truss heel but is required to extend to within 2 inches (51 mm) of the roof sheathing.

Where blocking at the top plate connection is notched or drilled, the section refers to the requirements for notching and boring holes in joists and rafters. The provisions essentially limits notches to one-fourth of the depth of the block, the diameter of drilled holes are limited to one-third of the depth of the block and drilled holes cannot be within 2 inches (51 mm) of the top or bottom of the block.

2308.3.3 Sill anchorage. Where foundations are required by Section 2308.3.4, braced wall line sills shall be anchored to concrete or masonry foundations. Such anchorage shall conform to the requirements of Section 2308.6 except that such anchors shall be spaced at not more than 4 feet (1219 mm) o.c. for structures over two *stories above grade plane*. The anchors shall be distributed along the length of the braced wall line. Other anchorage devices having equivalent capacity are permitted.

❖ The typical sill plate anchorage of $^1/_2$ inch (12.7 mm) diameter bolts spaced not more than 6 feet (1826 mm) o.c. is specified in Section 2308.6. This section requires a more restrictive anchor spacing of not more than 4 feet (1219 mm) o.c. for braced wall line sills in buildings greater than two stories above grade plane. This addresses the greater earthquake load effects at the foundation of three-story buildings. Interior braced wall lines in a building with plan dimensions less than or equal to 50 feet (15 240 mm) do not require support on a continuous foundation. Such interior braced wall lines would be fastened to supporting framing in accordance with Table 2304.9.1. This section clarifies that something more substantial is needed where foundations are required below these braced wall lines and allows other types of anchorage devices having "equivalent capacity" to be substituted for this anchorage.

2308.3.3.1 Anchorage to all-wood foundations. Where all-wood foundations are used, the force transfer from the braced wall lines shall be determined based on calculation and shall have a capacity greater than or equal to the connections required by Section 2308.3.3.

❖ There are no prescriptive provisions currently in the code for anchorage of braced wall lines to wood foundations; therefore, transfer of the forces from the wall line to the foundation equivalent to the required sill anchorage must be provided. Connections must be designed to provide capacity equivalent to that of the bolts described in Section 2308.6: $^1/_2$-inch-diameter (12.7 mm) bolts embedded into concrete, spaced at 6 feet (1829 mm) o.c. maximum (two stories or less) or 4 feet (1219 mm) o.c. maximum (over two stories in height).

2308.3.4 Braced wall line support. Braced wall lines shall be supported by continuous foundations.

Exception: For structures with a maximum plan dimension not over 50 feet (15 240 mm), continuous foundations are required at exterior walls only.

❖ In general, braced wall lines should be supported on continuous foundations except as noted in the exception. Buildings with a maximum plan dimension of 50 feet (15 240 mm) have limited lateral forces so that continuous foundations are only required at exterior braced wall lines. The exception allows horizontal dimensions of up to 50 feet (15 240 mm) between continuous foundations. This is intended to permit residential cripple walls to be braced at exterior walls only, within the limitations described, but applies to other conditions as well. At one time in the development of the conventional construction provisions, this requirement was only necessary in the highest seismic regions, but has now become a general requirement.

2308.4 Design of elements. Combining of engineered elements or systems and conventionally specified elements or systems is permitted subject to the following limits:

❖ The conventional construction provisions are intended not only to control the amount of irregularity within a building, but also to make things easier for plans examiners, builders, inspectors and designers. Buildings can be very complex; simple, "box-like" buildings are no longer the norm. Most modern conventional wood-framed buildings have some elements that are not strictly conventional, such as engineered trusses or Laminated Veneer Lumber (LVL) beams. This section makes it possible to design a "hybrid"-type of building that is a mixture of conventional and engineered elements, provided appropriate attention is given to each.

2308.4.1 Elements exceeding limitations of conventional construction. When a building of otherwise conventional construction contains structural elements exceeding the limits of Section 2308.2, these elements and the supporting load path shall be designed in accordance with accepted engineering practice and the provisions of this code.

❖ The intent of this section is to permit individual framing members that are used in applications that exceed the limitations of Section 2308 (load, span, etc.) to be engineered and placed within an otherwise conventional system without requiring the entire structural system to be engineered. Methods that must be used for all elements or systems that require an engineered design are to comply with other requirements of the code. The code also requires that the load path supporting these elements must be engineered. It should be noted that it is not the intent of this section to allow situations that result in load transfer beyond the underlying assumptions of the tables for other conventional members.

2308.4.2 Structural elements or systems not described herein. When a building of otherwise conventional construction contains structural elements or systems not described in Section 2308, these elements or systems shall be designed in accordance with accepted engineering practice and the provisions of this code. The extent of such design need only demonstrate compliance of the nonconventional elements with other applicable provisions of this code and shall be compatible with the performance of the conventionally framed system.

❖ This section permits individual framing members that are not specifically covered in Section 2308 (new or proprietary products) to be engineered and placed within an otherwise conventional construction system without requiring the entire structural system to be engineered. The intent of this section is to recognize that there is a large volume of preengineered products available on the market that might fit into a conventionally framed building, but not necessarily follow this section for their individual designs. Typical examples of elements used in conventional wood-framed buildings that are not conventional are engineered trusses, I-joists, LVL headers, glued laminated beams, etc. Most elements that have an approval report have published design values based on laboratory testing and are approved as alternative materials or methods of construction, provided it is demonstrated that they are compatible with and adequately fit into the building system without violating other requirements of Section 2308. The extent of the design is limited to the element itself, but it also must be compatible with the conventional framing system. An example would be an LVL header in a conventionally framed stud wall. The LVL header would be designed, but obviously the trimmers must be sufficient to support the loads from the LVL header. Another example would be an engineered truss system where the trusses would be designed, but the truss reactions must be adequately supported by the conventionally framed wall.

2308.5 Connectors and fasteners. Connectors and fasteners used in conventional construction shall comply with the requirements of Section 2304.9.

❖ As discussed in the commentary to Section 2304.9, the section on fasteners and connectors applies to all types of wood construction addressed in Chapter 23 and the number of fasteners stipulated in Table 2304.9.1 is the minimum to be used. This section emphasizes that the minimum connection and fastener requirements specified in Section 2304.9 also apply to conventional construction.

2308.6 Foundation plates or sills. Foundations and footings shall be as specified in Chapter 18. Foundation plates or sills resting on concrete or masonry foundations shall comply with Section 2304.3.1. Foundation plates or sills shall be bolted or anchored to the foundation with not less than $^1/_2$-inch-diameter (12.7 mm) steel bolts or *approved* anchors spaced to provide equivalent anchorage as the steel bolts. Bolts shall be embedded at least 7 inches (178 mm) into concrete or masonry, and spaced not more than 6 feet (1829 mm) apart. There shall be a minimum of two bolts or anchor straps per piece with one bolt or anchor strap located not more than 12 inches (305 mm) or less than 4 inches (102 mm) from each end of each piece. A properly sized nut and washer shall be tightened on each bolt to the plate.

❖ This section prescribes the minimum size and spacing of foundation bolts or anchors in structures that comply with conventional construction, but the minimum anchorage requirements indicated would also apply to engineered wood structures unless a design is provided (see Section 2304.3). Note that requirements pertaining to the foundation or footing are found in Chapter 18. A very important requirement where in-plane shear is involved is that one bolt is to be located not more than 12 inches (305 mm) and not less than 4 inches (102 mm) from each end of each piece of plate. The 4 inches (102 mm) corresponds to the minimum seven diameter end distance required by the AF&PA NDS for parallel-to-grain loading. If the bolt is closer than seven diameters to the end, the bolt will not develop its full capacity when loaded parallel to grain.

Anchor types other than bolts, such as anchor straps, are also permitted when approved. Anchor straps come in different shapes and sizes and the allowable load values that can be used with them are specified by the respective manufacturer. The furnished capacity of other types of anchors must be no less than the resistance provided by the anchor bolts specified in this section.

2308.7 Girders. Girders for single-story construction or girders supporting loads from a single floor shall not be less than 4 inches by 6 inches (102 mm by 152 mm) for spans 6 feet (1829 mm) or less, provided that girders are spaced not more than 8 feet (2438 mm) o.c. Spans for built-up 2-inch (51 mm) girders shall be in accordance with Table 2308.9.5 or 2308.9.6. Other girders shall be designed to support the loads specified in this code. Girder end joints shall occur over supports.

Where a girder is spliced over a support, an adequate tie shall be provided. The ends of beams or girders supported on masonry or concrete shall not have less than 3 inches (76 mm) of bearing.

❖ Girders constructed of a single piece of wood must meet the size requirement of this section. Those that are built up—constructed of two or more pieces of 2-inch (51 mm) nominal wide lumber—must comply with the specified tables. All of the span lengths listed in Tables 2308.9.5 and 2308.9.6 are short enough to allow individual pieces of lumber to be continuous between supports (i.e., no splices). Splicing of individual pieces together would require an engineering analysis to design the connectors for shear flow. Other girders are permitted, if properly designed.

The minimum bearing length for girders supported by concrete or masonry walls is 3 inches (76 mm). This is based on the anticipated loads for the construction typical in this chapter, the allowable compressive stresses perpendicular to the grain for beam sizes and grades typical of the requirements in this chapter and the consideration of shear failure of the masonry.

2308.8 Floor joists. Spans for floor joists shall be in accordance with Table 2308.8(1) or 2308.8(2). For other grades and or species, refer to the *AF&PA Span Tables for Joists and Rafters*.

❖ See the commentary to Tables 2308.8(1) and (2).

TABLES 2308.8(1) and (2). See pages 23-63 and 23-65.

❖ The spans shown in Tables 2308.8(1) and (2) were derived from the AF&PA *Span Tables for Joists and Rafters*. For simplicity, spans of only the most common species and grades of wood are provided in the code. When other species or grades of wood are used, the AF&PA tables should be used to determine the proper span. Additionally, span tables published by the Southern Forest Products Association, the Western Wood Products Association and the Canadian Wood Council can be consulted. The spans published in those tables were derived in the same manner as those published in the AF&PA tables.

2308.8.1 Bearing. Except where supported on a 1-inch by 4-inch (25.4 mm by 102 mm) ribbon strip and nailed to the adjoining stud, the ends of each joist shall not have less than $1^1/_2$ inches (38 mm) of bearing on wood or metal, or less than 3 inches (76 mm) on masonry.

❖ This section is needed so that floor joists have adequate bearing at each end. A minimum bearing distance is stipulated for bearing on wood or steel. The 3 inches (76 mm) on masonry, which should also apply to bearing on concrete, is to minimize the potential for shear failure of the masonry or concrete. As an exception to the minimum $1^1/_2$-inch (38 mm) bearing surface on wood, joists are permitted to bear on a ribbon strip nailed to the narrow face of the studs, as long as the ends of the joists are also nailed to the adjoining studs.

2308.8.2 Framing details. Joists shall be supported laterally at the ends and at each support by solid blocking except where the ends of the joists are nailed to a header, band or rim joist or to an adjoining stud or by other means. Solid blocking shall not be less than 2 inches (51mm) in thickness and the full depth of the joist. Notches on the ends of joists shall not exceed one-fourth the joist depth. Holes bored in joists shall not be within 2 inches (51 mm) of the top or bottom of the joist, and the diameter of any such hole shall not exceed one-third the depth of the joist. Notches in the top or bottom of joists shall not exceed one-sixth the depth and shall not be located in the middle third of the span.

Joist framing from opposite sides of a beam, girder or partition shall be lapped at least 3 inches (76 mm) or the opposing joists shall be tied together in an approved manner.

Joists framing into the side of a wood girder shall be supported by framing anchors or on ledger strips not less than 2 inches by 2 inches (51 mm by 51 mm).

❖ The general requirement of this section is that joists be laterally supported by solid blocking at each end and at any intermediate support. The blocking is intended to prevent the joists from rolling over at the supports. The blocking may be omitted where the ends of the joists are restrained either by nailing or by the use of some other means, such as a joist hanger, to accomplish the same action. Most blocking is 2 inches (51 mm) nominal in thickness.

Wood is comprised of relatively small elongated cells oriented parallel to each other and bound together physically and chemically by lignin. The cells are continuous from end to end of a piece of lumber. When a beam or joist is notched or cut, the cell strands are interrupted. In notched beams, the effective depth is reduced equal to the depth of the notch (see Figure 2308.8.2).

Some designs and installation practices require that limited notching and cutting occur. Notching should be avoided when possible. Holes bored in beams and joists create the same problems as notches. When holes are necessary, the holes should be located in areas with the least stress concentration, generally along the neutral axis of the joist and, ideally, at a minimum distance away from the supports.

Joists must be adequately tied together in order to provide continuity and adequately supported by framing anchors or ledgers when not bearing on the top of girders.

2308.8.2.1 Engineered wood products. Cuts, notches and holes bored in trusses, structural composite lumber, structural glue-laminated members or I-joists are not permitted except where permitted by the manufacturer's recommendations or where the effects of such alterations are specifically considered in the design of the member by a *registered design professional.*

❖ The notching and boring allowances that are permitted for solid sawn joist framing cannot be applied to engineered wood products. Because these types of products are commonly incorporated into conventional construction, this section points out that they must be treated differently. However, it does permit notching and boring of engineered wood products when it is considered in the design of the member. Each manufacturer typically publishes guidelines to establish safe notching and boring limitations.

2308.8.3 Framing around openings. Trimmer and header joists shall be doubled, or of lumber of equivalent cross section, where the span of the header exceeds 4 feet (1219 mm). The ends of header joists more than 6 feet (1829 mm) long shall be supported by framing anchors or joist hangers unless bearing on a beam, partition or wall. Tail joists over 12 feet (3658 mm) long shall be supported at the header by framing anchors or on ledger strips not less than 2 inches by 2 inches (51 mm by 51 mm).

❖ Figure 2308.8.3 illustrates the criteria for framing around openings.

2308.8.4 Supporting bearing partitions. Bearing partitions parallel to joists shall be supported on beams, girders, doubled joists, walls or other bearing partitions. Bearing partitions perpendicular to joists shall not be offset from supporting girders, walls or partitions more than the joist depth unless such joists are of sufficient size to carry the additional load.

❖ The floor joist spans provided in Tables 2308.8(1) and (2) assume uniform loading conditions and will support only limited concentrated loads. Unless the floor joists are designed using an engineering analysis which considers these loads, bearing partitions must be supported by other bearing partitions, beams or girders.

A major problem that frequently occurs is the orientation of heavy loads, such as special types of bathtubs, parallel with the floor joists. Additional joists should be installed to support these loads (see Figure 2308.8.4), which brings the design of this portion into an engineering analysis.

Bearing partitions oriented perpendicular to joists cannot be offset from supporting girders, walls or partitions more than the depth of the joists, unless the joists are designed to be of sufficient size to carry the additional load (see Figure 2308.8.4).

Although not specifically required by the code, it is typically advantageous to double up floor joists supporting nonload-bearing partitions oriented parallel to the joists or to provide solid blocking between the joists to transfer the wall load to the supporting joists (see Figure 2308.8.4), unless the floor sheathing is of sufficient strength to carry the load. APA literature addresses common framing issues such as these that can be integrated into a conventionally framed floor without an engineering analysis.

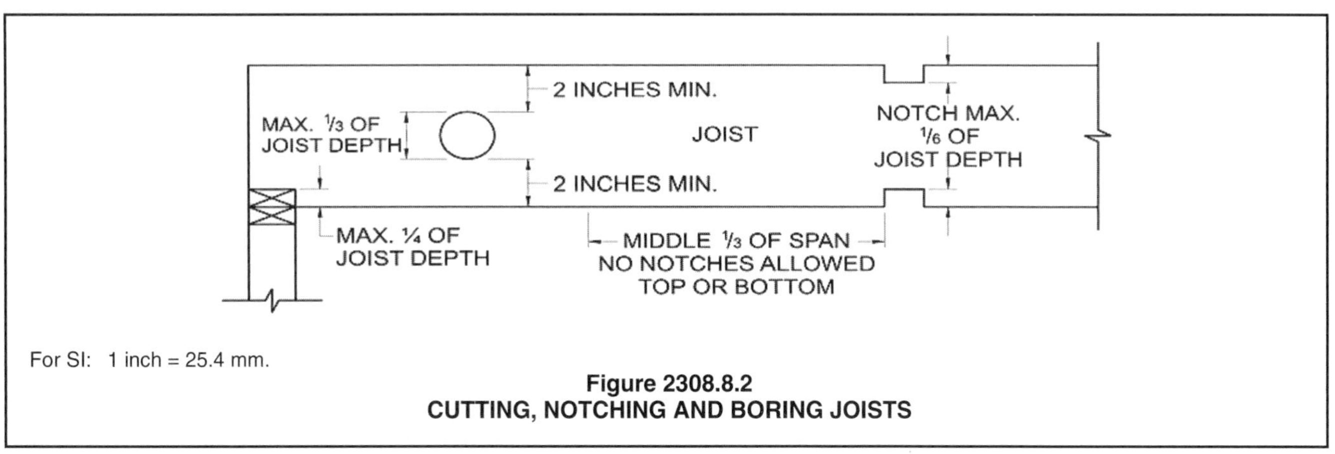

Figure 2308.8.2
CUTTING, NOTCHING AND BORING JOISTS

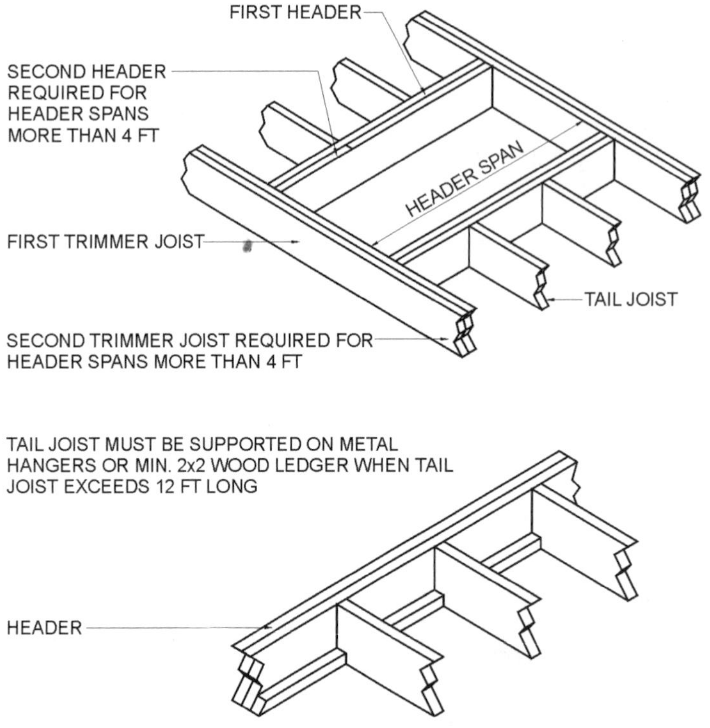

For SI: 1 foot = 304.8 mm.

Figure 2308.8.3
FRAMING AROUND OPENING

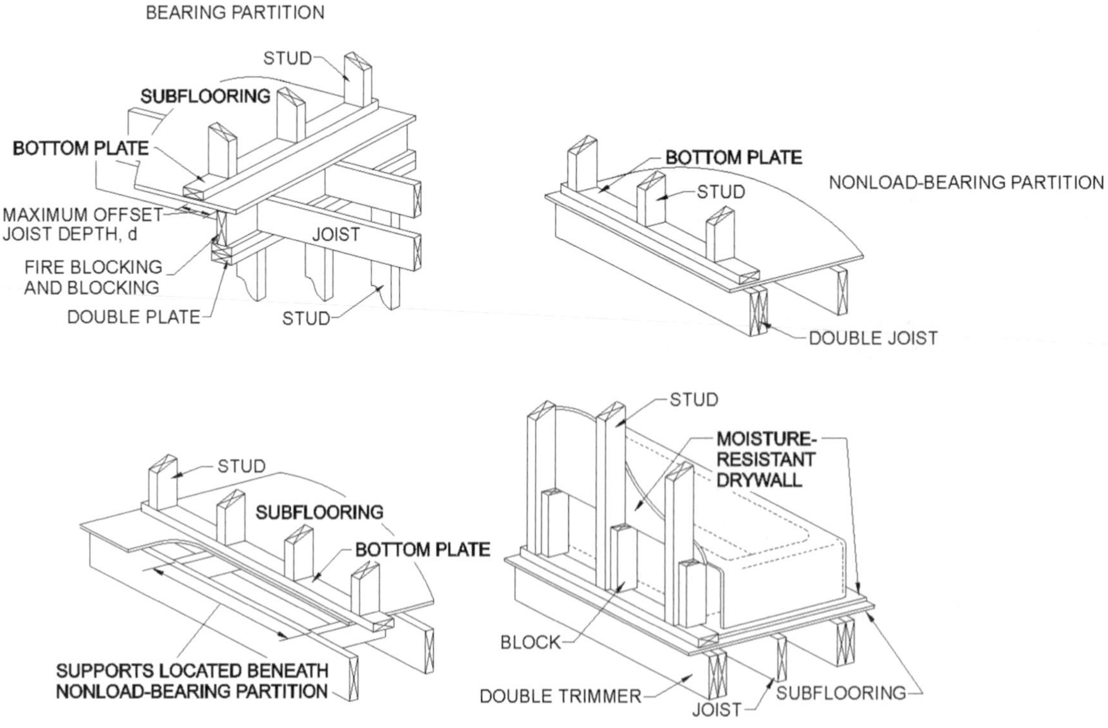

Figure 2308.8.4
JOISTS SUPPORTING PARTITIONS AND HEAVY LOADS

TABLE 2308.8(1)
FLOOR JOIST SPANS FOR COMMON LUMBER SPECIES
(Residential Sleeping Areas, Live Load = 30 psf, L/Δ = 360)

JOIST SPACING (inches)	SPECIES AND GRADE		DEAD LOAD = 10 psf				DEAD LOAD = 20 psf			
			2x6	2x8	2x10	2x12	2x6	2x8	2x10	2x12
			Maximum floor joist spans							
			(ft. - in.)	(ft. - in.)	(ft. - in.)	(ft. - in.)	(ft. - in.)	(ft. - in.)	(ft. - in.)	(ft. - in.)
12	Douglas Fir-Larch	SS	12-6	16-6	21-0	25-7	12-6	16-6	21-0	25-7
	Douglas Fir-Larch	#1	12-0	15-10	20-3	24-8	12-0	15-7	19-0	22-0
	Douglas Fir-Larch	#2	11-10	15-7	19-10	23-0	11-6	14-7	17-9	20-7
	Douglas Fir-Larch	#3	9-8	12-4	15-0	17-5	8-8	11-0	13-5	15-7
	Hem-Fir	SS	11-10	15-7	19-10	24-2	11-10	15-7	19-10	24-2
	Hem-Fir	#1	11-7	15-3	19-5	23-7	11-7	15-2	18-6	21-6
	Hem-Fir	#2	11-0	14-6	18-6	22-6	11-0	14-4	17-6	20-4
	Hem-Fir	#3	9-8	12-4	15-0	17-5	8-8	11-0	13-5	15-7
	Southern Pine	SS	12-3	16-2	20-8	25-1	12-3	16-2	20-8	25-1
	Southern Pine	#1	12-0	15-10	20-3	24-8	12-0	15-10	20-3	24-8
	Southern Pine	#2	11-10	15-7	19-10	24-2	11-10	15-7	18-7	21-9
	Southern Pine	#3	10-5	13-3	15-8	18-8	9-4	11-11	14-0	16-8
	Spruce-Pine-Fir	SS	11-7	15-3	19-5	23-7	11-7	15-3	19-5	23-7
	Spruce-Pine-Fir	#1	11-3	14-11	19-0	23-0	11-3	14-7	17-9	20-7
	Spruce-Pine-Fir	#2	11-3	14-11	19-0	23-0	11-3	14-7	17-9	20-7
	Spruce-Pine-Fir	#3	9-8	12-4	15-0	17-5	8-8	11-0	13-5	15-7
16	Douglas Fir-Larch	SS	11-4	15-0	19-1	23-3	11-4	15-0	19-1	23-0
	Douglas Fir-Larch	#1	10-11	14-5	18-5	21-4	10-8	13-6	16-5	19-1
	Douglas Fir-Larch	#2	10-9	14-1	17-2	19-11	9-11	12-7	15-5	17-10
	Douglas Fir-Larch	#3	8-5	10-8	13-0	15-1	7-6	9-6	11-8	13-6
	Hem-Fir	SS	10-9	14-2	18-0	21-11	10-9	14-2	18-0	21-11
	Hem-Fir	#1	10-6	13-10	17-8	20-9	10-4	13-1	16-0	18-7
	Hem-Fir	#2	10-0	13-2	16-10	19-8	9-10	12-5	15-2	17-7
	Hem-Fir	#3	8-5	10-8	13-0	15-1	7-6	9-6	11-8	13-6
	Southern Pine	SS	11-2	14-8	18-9	22-10	11-2	14-8	18-9	22-10
	Southern Pine	#1	10-11	14-5	18-5	22-5	10-11	14-5	17-11	21-4
	Southern Pine	#2	10-9	14-2	18-0	21-1	10-5	13-6	16-1	18-10
	Southern Pine	#3	9-0	11-6	13-7	16-2	8-1	10-3	12-2	14-6
	Spruce-Pine-Fir	SS	10-6	13-10	17-8	21-6	10-6	13-10	17-8	21-4
	Spruce-Pine-Fir	#1	10-3	13-6	17-2	19-11	9-11	12-7	15-5	17-10
	Spruce-Pine-Fir	#2	10-3	13-6	17-2	19-11	9-11	12-7	15-5	17-10
	Spruce-Pine-Fir	#3	8-5	10-8	13-0	15-1	7-6	9-6	11-8	13-6
19.2	Douglas Fir-Larch	SS	10-8	14-1	18-0	21-10	10-8	14-1	18-0	21-0
	Douglas Fir-Larch	#1	10-4	13-7	16-9	19-6	9-8	12-4	15-0	17-5
	Douglas Fir-Larch	#2	10-1	12-10	15-8	18-3	9-1	11-6	14-1	16-3
	Douglas Fir-Larch	#3	7-8	9-9	11-10	13-9	6-10	8-8	10-7	12-4
	Hem-Fir	SS	10-1	13-4	17-0	20-8	10-1	13-4	17-0	20-7
	Hem-Fir	#1	9-10	13-0	16-4	19-0	9-6	12-0	14-8	17-0
	Hem-Fir	#2	9-5	12-5	15-6	17-1	8-11	11-4	13-10	16-1
	Hem-Fir	#3	7-8	9-9	11-10	13-9	6-10	8-8	10-7	12-4

(continued)

TABLE 2308.8(1)—continued
FLOOR JOIST SPANS FOR COMMON LUMBER SPECIES
(Residential Sleeping Areas, Live Load = 30 psf, L/Δ = 360)

JOIST SPACING (inches)	SPECIES AND GRADE		DEAD LOAD = 10 psf				DEAD LOAD = 20 psf			
			2x6	2x8	2x10	2x12	2x6	2x8	2x10	2x12
			Maximum floor joist spans							
			(ft. - in.)	(ft. - in.)	(ft. - in.)	(ft. - in.)	(ft. - in.)	(ft. - in.)	(ft. - in.)	(ft. - in.)
19.2	Southern Pine	SS	10-6	13-10	17-8	21-6	10-6	13-10	17-8	21-6
	Southern Pine	#1	10-4	13-7	17-4	21-1	10-4	13-7	16-4	19-6
	Southern Pine	#2	10-1	13-4	16-5	19-3	9-6	12-4	14-8	17-2
	Southern Pine	#3	8-3	10-6	12-5	14-9	7-4	9-5	11-1	13-2
	Spruce-Pine-Fir	SS	9-10	13-0	16-7	20-2	9-10	13-0	16-7	19-6
	Spruce-Pine-Fir	#1	9-8	12-9	15-8	18-3	9-1	11-6	14-1	16-3
	Spruce-Pine-Fir	#2	9-8	12-9	15-8	18-3	9-1	11-6	14-1	16-3
	Spruce-Pine-Fir	#3	7-8	9-9	11-10	13-9	6-10	8-8	10-7	12-4
24	Douglas Fir-Larch	SS	9-11	13-1	16-8	20-3	9-11	13-1	16-2	18-9
	Douglas Fir-Larch	#1	9-7	12-4	15-0	17-5	8-8	11-0	13-5	15-7
	Douglas Fir-Larch	#2	9-1	11-6	14-1	16-3	8-1	10-3	12-7	14-7
	Douglas Fir-Larch	#3	6-10	8-8	10-7	12-4	6-2	7-9	9-6	11-0
	Hem-Fir	SS	9-4	12-4	15-9	19-2	9-4	12-4	15-9	18-5
	Hem-Fir	#1	9-2	12-0	14-8	17-0	8-6	10-9	13-1	15-2
	Hem-Fir	#2	8-9	11-4	13-10	16-1	8-0	10-2	12-5	14-4
	Hem-Fir	#3	6-10	8-8	10-7	12-4	6-2	7-9	9-6	11-0
	Southern Pine	SS	9-9	12-10	16-5	19-11	9-9	12-10	16-5	19-11
	Southern Pine	#1	9-7	12-7	16-1	19-6	9-7	12-4	14-7	17-5
	Southern Pine	#2	9-4	12-4	14-8	17-2	8-6	11-0	13-1	15-5
	Southern Pine	#3	7-4	9-5	11-1	13-2	6-7	8-5	9-11	11-10
	Spruce-Pine-Fir	SS	9-2	12-1	15-5	18-9	9-2	12-1	15-0	17-5
	Spruce-Pine-Fir	#1	8-11	11-6	14-1	16-3	8-1	10-3	12-7	14-7
	Spruce-Pine-Fir	#2	8-11	11-6	14-1	16-3	8-1	10-3	12-7	14-7
	Spruce-Pine-Fir	#3	6-10	8-8	10-7	12-4	6-2	7-9	9-6	11-0

For SI: 1 inch = 25.4 mm, 1 foot = 304.8 mm, 1 pound per square foot = 47.8 N/m^2.

TABLE 2308.8(2)
FLOOR JOIST SPANS FOR COMMON LUMBER SPECIES
(Residential Living Areas, Live Load = 40 psf, L/Δ = 360)

JOIST SPACING (inches)	SPECIES AND GRADE		DEAD LOAD = 10 psf				DEAD LOAD = 20 psf			
			2x6	2x8	2x10	2x12	2x6	2x8	2x10	2x12
			Maximum floor joist spans							
			(ft. - in.)	(ft. - in.)	(ft. - in.)	(ft. - in.)	(ft. - in.)	(ft. - in.)	(ft. - in.)	(ft. - in.)
12	Douglas Fir-Larch	SS	11-4	15-0	19-1	23-3	11-4	15-0	19-1	23-3
	Douglas Fir-Larch	#1	10-11	14-5	18-5	22-0	10-11	14-2	17-4	20-1
	Douglas Fir-Larch	#2	10-9	14-2	17-9	20-7	10-6	13-3	16-3	18-10
	Douglas Fir-Larch	#3	8-8	11-0	13-5	15-7	7-11	10-0	12-3	14-3
	Hem-Fir	SS	10-9	14-2	18-0	21-11	10-9	14-2	18-0	21-11
	Hem-Fir	#1	10-6	13-10	17-8	21-6	10-6	13-10	16-11	19-7
	Hem-Fir	#2	10-0	13-2	16-10	20-4	10-0	13-1	16-0	18-6
	Hem-Fir	#3	8-8	11-0	13-5	15-7	7-11	10-0	12-3	14-3
	Southern Pine	SS	11-2	14-8	18-9	22-10	11-2	14-8	18-9	22-10
	Southern Pine	#1	10-11	14-5	18-5	22-5	10-11	14-5	18-5	22-5
	Southern Pine	#2	10-9	14-2	18-0	21-9	10-9	14-2	16-11	19-10
	Southern Pine	#3	9-4	11-11	14-0	16-8	8-6	10-10	12-10	15-3
	Spruce-Pine-Fir	SS	10-6	13-10	17-8	21-6	10-6	13-10	17-8	21-6
	Spruce-Pine-Fir	#1	10-3	13-6	17-3	20-7	10-3	13-3	16-3	18-10
	Spruce-Pine-Fir	#2	10-3	13-6	17-3	20-7	10-3	13-3	16-3	18-10
	Spruce-Pine-Fir	#3	8-8	11-0	13-5	15-7	7-11	10-0	12-3	14-3
16	Douglas Fir-Larch	SS	10-4	13-7	17-4	21-1	10-4	13-7	17-4	21-0
	Douglas Fir-Larch	#1	9-11	13-1	16-5	19-1	9-8	12-4	15-0	17-5
	Douglas Fir-Larch	#2	9-9	12-7	15-5	17-10	9-1	11-6	14-1	16-3
	Douglas Fir-Larch	#3	7-6	9-6	11-8	13-6	6-10	8-8	10-7	12-4
	Hem-Fir	SS	9-9	12-10	16-5	19-11	9-9	12-10	16-5	19-11
	Hem-Fir	#1	9-6	12-7	16-0	18-7	9-6	12-0	14-8	17-0
	Hem-Fir	#2	9-1	12-0	15-2	17-7	8-11	11-4	13-10	16-1
	Hem-Fir	#3	7-6	9-6	11-8	13-6	6-10	8-8	10-7	12-4
	Southern Pine	SS	10-2	13-4	17-0	20-9	10-2	13-4	17-0	20-9
	Southern Pine	#1	9-11	13-1	16-9	20-4	9-11	13-1	16-4	19-6
	Southern Pine	#2	9-9	12-10	16-1	18-10	9-6	12-4	14-8	17-2
	Southern Pine	#3	8-1	10-3	12-2	14-6	7-4	9-5	11-1	13-2
	Spruce-Pine-Fir	SS	9-6	12-7	16-0	19-6	9-6	12-7	16-0	19-6
	Spruce-Pine-Fir	#1	9-4	12-3	15-5	17-10	9-1	11-6	14-1	16-3
	Spruce-Pine-Fir	#2	9-4	12-3	15-5	17-10	9-1	11-6	14-1	16-3
	Spruce-Pine-Fir	#3	7-6	9-6	11-8	13-6	6-10	8-8	10-7	12-4
19.2	Douglas Fir-Larch	SS	9-8	12-10	16-4	19-10	9-8	12-10	16-4	19-2
	Douglas Fir-Larch	#1	9-4	12-4	15-0	17-5	8-10	11-3	13-8	15-11
	Douglas Fir-Larch	#2	9-1	11-6	14-1	16-3	8-3	10-6	12-10	14-10
	Douglas Fir-Larch	#3	6-10	8-8	10-7	12-4	6-3	7-11	9-8	11-3
	Hem-Fir	SS	9-2	12-1	15-5	18-9	9-2	12-1	15-5	18-9
	Hem-Fir	#1	9-0	11-10	14-8	17-0	8-8	10-11	13-4	15-6
	Hem-Fir	#2	8-7	11-3	13-10	16-1	8-2	10-4	12-8	14-8
	Hem-Fir	#3	6-10	8-8	10-7	12-4	6-3	7-11	9-8	11-3
	Southern Pine	SS	9-6	12-7	16-0	19-6	9-6	12-7	16-0	19-6
	Southern Pine	#1	9-4	12-4	15-9	19-2	9-4	12-4	14-11	17-9
	Southern Pine	#2	9-2	12 -1	14-8	17-2	8-8	11-3	13-5	15-8
	Southern Pine	#3	7-4	9-5	11-1	13-2	6-9	8-7	10-1	12-1
	Spruce-Pine-Fir	SS	9-0	11-10	15-1	18-4	9-0	11-10	15-1	17-9
	Spruce-Pine-Fir	#1	8-9	11-6	14-1	16-3	8-3	10-6	12-10	14-10
	Spruce-Pine-Fir	#2	8-9	11-6	14-1	16-3	8-3	10-6	12-10	14-10
	Spruce-Pine-Fir	#3	6-10	8-8	10-7	12-4	6-3	7-11	9-8	11-3

(continued)

TABLE 2308.8(2)—continued
FLOOR JOIST SPANS FOR COMMON LUMBER SPECIES
(Residential Living Areas, Live Load = 40 psf, L/Δ = 360)

JOIST SPACING (inches)	SPECIES AND GRADE		DEAD LOAD = 10 psf				DEAD LOAD = 20 psf			
			2x6	2x8	2x10	2x12	2x6	2x8	2x10	2x12
			Maximum floor joist spans							
			(ft. - in.)	(ft. - in.)	(ft. - in.)	(ft. - in.)	(ft. - in.)	(ft. - in.)	(ft. - in.)	(ft. - in.)
24	Douglas Fir-Larch	SS	9-0	11-11	15-2	18-5	9-0	11-11	14-9	17-1
	Douglas Fir-Larch	#1	8-8	11-0	13-5	15-7	7-11	10-0	12-3	14-3
	Douglas Fir-Larch	#2	8-1	10-3	12-7	14-7	7-5	9-5	11-6	13-4
	Douglas Fir-Larch	#3	6-2	7-9	9-6	11-0	5-7	7-1	8-8	10-1
	Hem-Fir	SS	8-6	11-3	14-4	17-5	8-6	11-3	14-4	16-10[a]
	Hem-Fir	#1	8-4	10-9	13-1	15-2	7-9	9-9	11-11	13-10
	Hem-Fir	#2	7-11	10-2	12-5	14-4	7-4	9-3	11-4	13-1
	Hem-Fir	#3	6-2	7-9	9-6	11-0	5-7	7-1	8-8	10-1
	Southern Pine	SS	8-10	11-8	14-11	18-1	8-10	11-8	14-11	18-1
	Southern Pine	#1	8-8	11-5	14-7	17-5	8-8	11-3	13-4	15-11
	Southern Pine	#2	8-6	11-0	13-1	15-5	7-9	10-0	12-0	14-0
	Southern Pine	#3	6-7	8-5	9-11	11-10	6-0	7-8	9-1	10-9
	Spruce-Pine-Fir	SS	8-4	11-0	14-0	17-0	8-4	11-0	13-8	15-11
	Spruce-Pine-Fir	#1	8-1	10-3	12-7	14-7	7-5	9-5	11-6	13-4
	Spruce-Pine-Fir	#2	8-1	10-3	12-7	14-7	7-5	9-5	11-6	13-4
	Spruce-Pine-Fir	#3	6-2	7-9	9-6	11-0	5-7	7-1	8-8	10-1

For SI: 1 inch = 25.4 mm, 1 foot = 304.8 mm, 1 pound per square foot = 47.8 N/m^2.

a. End bearing length shall be increased to 2 inches.

2308.8.5 Lateral support. Floor, *attic* and roof framing with a nominal depth-to-thickness ratio greater than or equal to 5:1 shall have one edge held in line for the entire span. Where the nominal depth-to-thickness ratio of the framing member exceeds 6:1, there shall be one line of bridging for each 8 feet (2438 mm) of span, unless both edges of the member are held in line. The bridging shall consist of not less than 1-inch by 3-inch (25 mm by 76 mm) lumber, double nailed at each end, of equivalent metal bracing of equal rigidity, full-depth solid blocking or other *approved* means. A line of bridging shall also be required at supports where equivalent lateral support is not otherwise provided.

❖ When the depth-to-thickness ratio (based on nominal dimensions) of joists and rafters exceeds 5:1, as would be the case in members larger than 2 inches by 10 inches (51 mm by 254 mm), the minimum lateral support required by Section 2308.8.2 is not sufficient to prevent lateral buckling between supports. Additional restraint is required. Sheathing, subflooring, decking and similar materials attached to each joist or rafter are considered to provide edge restraint.

These requirements are cumulative. The lateral support required by Section 2308.8.2 applies to all joists. Additionally, members greater than 2 inches by 10 inches (51 mm by 254 mm) must have one edge held in line, and members greater than 2 inches by 12 inches (51 mm by 305 mm) must have one edge held in line as well as a line of bridging at each 8 feet (2438 mm) of span (which may be omitted if both edges are held in line). The section also prescribes acceptable bridging methods.

2308.8.6 Structural floor sheathing. Structural floor sheathing shall comply with the provisions of Section 2304.7.1.

❖ Subflooring is considered to be the structural element that supports floor loads and distributes the loads to the joists. Underlayment is related to serviceability in that it provides a suitable surface for the floor-covering material. Both must comply with the appropriate tables for the maximum span referenced in Section 2304.7.1 (see commentary, Sections 2304.6 and 2304.7).

2308.8.7 Under-floor ventilation. For under-floor ventilation, see Section 1203.3.

❖ Enclosed spaces under floors of buildings with raised floor construction are referred to as crawl spaces when not designed and constructed as basements. The earth is usually exposed or covered with a vapor retarder. Regardless, there is always the potential for a large amount of moisture vapor from the ground. This moisture must be removed to prevent wetting and drying cycles that can cause decay and fungus attack in wood members. Experience has shown that crawl spaces constructed in compliance with Section 2308 and ventilated to the exterior as required by Section 1203.3 do not develop problems. Access to crawl spaces is required as specified in Section 1209.1.

It should be recognized that moisture problems

may be exaggerated in colder months; therefore, operable vents should be used with caution.

2308.9 Wall framing. Walls and partitions shall be constructed in accordance with the applicable provisions of Sections 2308.9.1 through 2308.9.4.2.

❖ Section 2308.9 contains extensive provisions for conventional wall framing including: size, height and spacing of studs; notching and boring of studs; plates and sills; wall bracing; cripple walls; headers and framing at openings.

2308.9.1 Size, height and spacing. The size, height and spacing of studs shall be in accordance with Table 2308.9.1 except that utility-grade studs shall not be spaced more than 16 inches (406 mm) o.c., or support more than a roof and ceiling, or exceed 8 feet (2438 mm) in height for exterior walls and load-bearing walls or 10 feet (3048 mm) for interior nonload-bearing walls. Studs shall be continuous from a support at the sole plate to a support at the top plate to resist loads perpendicular to the wall. The support shall be a foundation or floor, ceiling or roof diaphragm or shall be designed in accordance with accepted engineering practice.

Exception: Jack studs, trimmer studs and cripple studs at openings in walls that comply with Table 2308.9.5.

❖ The specifications in these sections are empirical interpretations of accepted engineering design practices and traditional designs for light wood-framed buildings. When bearing wall stud heights exceed 10 feet (3048 mm) or the structure is outside the scope of conventional construction requirements, studs must be designed in accordance with accepted engineering practice using the forces prescribed in the code. The section requires studs to be continuous from a support at the sole plate to a support at the top plate to resist wind loads perpendicular to the wall. The support for the studs must be a foundation or floor, a ceiling or roof diaphragm or must be designed. The requirement is intended to address conditions where a top plate may not be laterally supported. For example, where scissor trusses are used to create a vaulted ceiling, the top plate is not laterally supported because the ceiling follows the profile of the bottom chord of the truss. This section clarifies that studs are required to be laterally supported at the sole plate and top plate to resist out-of-plane loads perpendicular to the wall. Where scissor trusses are used at the gable end wall, the wall studs should be framed up to the bottom chord of the truss or the gable end wall should be balloon framed up to the roof sheathing. In either of these cases, the studs are supported at the top and bottom as required by this section. If the studs are not supported at the top by a ceiling or roof deck in accordance with the prescriptive provisions, then an engineered design should be provided for the gable end-wall studs to provide support and for combined axial compression and bending. The exception excludes jack studs, trimmer studs and cripple studs at openings in walls that comply with Table 2308.9.5.

TABLE 2308.9.1. See below.

❖ Wood studs in conventional construction are limited to the maximum height and spacing as indicated in this table to prevent buckling. It should be noted that the laterally unsupported stud height refers to the distance between points of lateral support perpendicular to the plane of the wall (see Note a). This is to prevent bucking out of the plane of the wall. In general, the sheathing is considered to provide support in the plane of the wall. The height and spacing limits are based on the dead and live loads associated with conventional construction specified in Section 2308.2, along with a nominal lateral load, to account for wind or earthquakes acting perpendicular to the wall.

2308.9.2 Framing details. Studs shall be placed with their wide dimension perpendicular to the wall. Not less than three studs shall be installed at each corner of an *exterior wall.*

Exception: At corners, two studs are permitted, provided wood spacers or backup cleats of $^3/_8$-inch-thick (9.5 mm) wood structural panel, $^3/_8$-inch (9.5 mm) Type M "Exterior Glue" particleboard, 1-inch-thick (25 mm) lumber or other *approved* devices that will serve as an adequate backing for the attachment of facing materials are used. Where fire-resistance ratings or shear values are involved, wood

**TABLE 2308.9.1
SIZE, HEIGHT AND SPACING OF WOOD STUDS**

STUD SIZE (inches)	BEARING WALLS				NONBEARING WALLS	
	Laterally unsupported stud height[a] (feet)	Supporting roof and ceiling only	Supporting one floor, roof and ceiling	Supporting two floors, roof and ceiling	Laterally unsupported stud height[a] (feet)	Spacing (inches)
		Spacing (inches)				
2 × 3[b]	—	—	—	—	10	16
2 × 4	10	24	16	—	14	24
3 × 4	10	24	24	16	14	24
2 × 5	10	24	24	—	16	24
2 × 6	10	24	24	16	20	24

For SI: 1 inch = 25.4 mm, 1 foot = 304.8 mm.

a. Listed heights are distances between points of lateral support placed perpendicular to the plane of the wall. Increases in unsupported height are permitted where justified by an analysis.

b. Shall not be used in exterior walls.

spacers, backup cleats or other devices shall not be used unless specifically *approved* for such use.

❖ Wood-framed exterior walls support all vertical loads and also must act as braced walls to resist lateral loads in the plane of the wall, such as from wind and earthquakes. Adequately secured and sufficient corner posts are required to assist in this function, in addition to providing a degree of continuity in the exterior walls. Traditionally, three studs were installed in corners. Experience has shown that two studs will adequately perform this function, provided other means are employed to support any interior finish that may be installed. Note that these other means may not be acceptable if they adversely affect the fire resistance or shear-resisting capacity of the walls. Most books on architectural standards contain common corner framing details. One issue with corner framing that becomes sensitive for engineered construction is when the designer chooses to attach one hold-down in such a way that it will provide uplift resistance for both of the perpendicular walls at that corner. A continuous load path is necessary to deliver uplift forces to the foundation, and there is an eccentricity for one of the walls that should be considered. In this case the engineer should detail the corner hold down installation so the builder and inspector understand it.

2308.9.2.1 Top plates. Bearing and *exterior wall* studs shall be capped with double top plates installed to provide overlapping at corners and at intersections with other partitions. End joints in double top plates shall be offset at least 48 inches (1219 mm), and shall be nailed with not less than eight 16d face nails on each side of the joint. Plates shall be a nominal 2 inches (51 mm) in depth and have a width at least equal to the width of the studs.

> **Exception:** A single top plate is permitted, provided the plate is adequately tied at joints, corners and intersecting walls by at least the equivalent of 3-inch by 6-inch (76 mm by 152 mm) by 0.036-inch-thick (0.914 mm) galvanized steel that is nailed to each wall or segment of wall by six 8d nails or equivalent, provided the rafters, joists or trusses are centered over the studs with a tolerance of no more than 1 inch (25 mm).

❖ Double top plates are necessary at exterior walls and bearing walls. The plates are required to be overlapped at corners and intersecting partitions. The joints in the top plate are to be offset 48 inches (1219 mm) from each other and fastened with 8-16d nails on each side of the joint. Minimum nominal 2-inch-thick (51 mm) plates are required and, to achieve full bearing, the plate should be as wide as the studs. The exception permits a single top plate to be used, provided the joints in the plate are strapped and the joists, rafters or trusses are centered within 1 inch (25 mm) over the studs. The tolerance of 1 inch (25 mm) with which the rafters or joists are to be centered over the studs is required to prevent overstress or excessive deflection in the single top plate. The steel strap at the joints provides a minimum level of continuity in the plate.

2308.9.2.2 Top plates for studs spaced at 24 inches (610 mm). Where bearing studs are spaced at 24-inch (610 mm) intervals and top plates are less than two 2-inch by 6-inch (51 mm by 152 mm) or two 3-inch by 4-inch (76 mm by 102 mm) members and where the floor joists, floor trusses or roof trusses that they support are spaced at more than 16-inch (406 mm) intervals, such joists or trusses shall bear within 5 inches (127 mm) of the studs beneath or a third plate shall be installed.

❖ This section has specific plate requirements where studs are spaced 24 inches (610 mm) o.c. Ideally, joists, rafters or trusses should be located di-rectly over bearing studs to avoid excessive bending or deflection of the plate; this is typically not planned for unless the drawings clearly state that it is required. The empirical provisions of this section allow joists or trusses to bear within 5 inches (127 mm) of the stud below when the joists or trusses are spaced more than 16 inches (406 mm) o.c. and the studs are spaced 24 inches (610 mm) o.c. This provision assumes there are double top (bearing) plates that have all joints located over the studs. Additional support is required if the offset is more than 5 inches (127 mm), such as the addition of a third plate.

2308.9.2.3 Nonbearing walls and partitions. In nonbearing walls and partitions, studs shall be spaced not more than 28 inches (711 mm) o.c. and in interior nonbearing walls and partitions, are permitted to be set with the long dimension parallel to the wall. Interior nonbearing partitions shall be capped with no less than a single top plate installed to provide overlapping at corners and at intersections with other walls and partitions. The plate shall be continuously tied at joints by solid blocking at least 16 inches (406 mm) in length and equal in size to the plate or by $^{1}/_{2}$-inch by $1^{1}/_{2}$-inch (12.7 mm by 38 mm) metal ties with spliced sections fastened with two 16d nails on each side of the joint.

❖ A nonbearing wall is any wall that is not a bearing wall (see definition of "Nonload-bearing wall" in Chapter 2). Nonbearing walls do not support any portion of the building or structure except the weight of the wall itself. The term "partition," although not defined in the code, implies an interior wall. Thus, a nonbearing partition is a nonload-bearing interior wall. Nonbearing partitions can have the studs placed flat and parallel to the wall and have a single top plate, provided the requirements of this section are met. In general, nonbearing partitions are not intended to add to the rigidity or structural performance of the building. They are, however, expected to stay in place and have the ability to support applied finished membranes. The code limits stud spacing to 28 inches (711 mm) o.c., but allows the stud to be oriented with the wide dimension parallel to the wall length. Partitions must be able to support the lateral loads to which they are subjected, but not less than 5 psf (0.240 kN/m^2) lateral load, in accordance with Section 1607.14. These walls are often deflection controlled because of their very low loading (see Table 1604.3, Note b). It should be noted that this section is not intended to apply to interior nonbearing walls that are serving as braced wall lines.

Lateral support of partitions can also be achieved by the use of light-gage metal clips produced by wood connector manufacturers. Allowable loads per clip are published in their respective literature and ICC Evaluation Service Reports. These clips typically require a minimum and maximum gap between the top of the wall and the bottom of the framing to avoid loading the partition.

2308.9.2.4 Plates or sills. Studs shall have full bearing on a plate or sill not less than 2 inches (51 mm) in thickness having a width not less than that of the wall studs.

❖ Studs must have full bearing on minimum 2-inch (51 mm) nominal sole or sill plates and the plate must be at least as wide as the supported studs. The single bottom plate serves to anchor the wall to the floor. Studs are attached to the sill by end nailing or toe nailing specified in Table 2304.9.1.

2308.9.3 Bracing. Braced wall lines shall consist of braced wall panels that meet the requirements for location, type and amount of bracing as shown in Figure 2308.9.3, specified in Table 2308.9.3(1) and are in line or offset from each other by not more than 4 feet (1219 mm). Braced wall panels shall start not more than $12^1/_2$ feet (3810 mm) from each end of a braced wall line. Braced wall panels shall be clearly indicated on the plans. Construction of braced wall panels shall be by one of the following methods:

1. Nominal 1-inch by 4-inch (25 mm by 102 mm) continuous diagonal braces let into top and bottom plates and intervening studs, placed at an angle not more than 60 degrees (1.0 rad) or less than 45 degrees (0.79 rad) from the horizontal and attached to the framing in conformance with Table 2304.9.1.
2. Wood boards of $^5/_8$ inch (15.9 mm) net minimum thickness applied diagonally on studs spaced not over 24 inches (610 mm) o.c.
3. Wood structural panel sheathing with a thickness not less than $^3/_8$ inch (9.5 mm) for 16-inch (406 mm) or 24-inch (610 mm) stud spacing in accordance with Tables 2308.9.3(2) and 2308.9.3(3).
4. Fiberboard sheathing panels not less than $^1/_2$ inch (12.7 mm) thick applied vertically or horizontally on studs spaced not over 16 inches (406 mm) o.c. where installed with fasteners in accordance with Section 2306.6 and Table 2306.6.
5. Gypsum board [sheathing $^1/_2$-inch-thick (12.7 mm) by 4-feet-wide (1219 mm) wallboard or veneer base] on studs spaced not over 24 inches (610 mm) o.c. and nailed at 7 inches (178 mm) o.c. with nails as required by Table 2306.7.
6. Particleboard wall sheathing panels where installed in accordance with Table 2308.9.3(4).
7. Portland cement plaster on studs spaced 16 inches (406 mm) o.c.installed in accordance with Section 2510.
8. Hardboard panel siding where installed in accordance with Section 2303.1.6 and Table 2308.9.3(5).

For cripple wall bracing, see Section 2308.9.4.1. For Methods 2, 3, 4, 6, 7 and 8, each panel must be at least 48 inches (1219 mm) in length, covering three stud spaces where studs are spaced 16 inches (406 mm) apart and covering two stud spaces where studs are spaced 24 inches (610 mm) apart.

For Method 5, each panel must be at least 96 inches (2438 mm) in length where applied to one face of a panel and 48 inches (1219 mm) where applied to both faces. All vertical joints of panel sheathing shall occur over studs and adjacent panel joints shall be nailed to common framing members. Horizontal joints shall occur over blocking or other framing equal in size to the studding except where waived by the installation requirements for the specific sheathing materials. Sole plates shall be nailed to the floor framing and top plates shall be connected to the framing above in accordance with Section 2308.3.2. Where joists are perpendicular to braced wall lines above, blocking shall be provided under and in line with the braced wall panels.

❖ This section is a continuation of the provisions in Section 2308.3 for bracing of walls to resist lateral forces from wind or seismic effects. Braced wall panels are portions of braced wall lines, as required by Table 2308.9.3(1), composed of the types of bracing materials discussed below. Portions of braced wall lines are permitted to be offset from each other and braced wall panels are not required to extend to the end of the braced wall line (see Figure 2308.9.3). The philosophy of this prescriptive lateral-force-resisting system is to divide a building into smaller "boxes," similar to the use of stiffening baffles within a storage tank. The roof system delivers the lateral forces to the braced wall system below. Historical analysis has shown that the more regularly defined these "boxes" are, the better the building will perform even though there may not be a complete and continuous load path connecting every single line together.

FIGURE 2308.9.3. See page 23-70.

❖ See the commentary to Section 2308.9.3.

TABLE 2308.9.3(1). See page 23-71.

❖ See the commentary to Section 2308.9.3.

TABLES 2308.9.3(2) through 2308.9.3(5). See pages 23-71 and 23-72.

❖ Tables 2308.9.3(2) through 2308.9.3(5) are for the materials that are prescriptively permitted to be used to construct the braced wall panels in braced wall lines. The location and minimum acceptable length of braced wall panels are specified in Table 2308.9.3(1). Detailed requirements for various materials—grades, thickness, nailing schedule, stud spacing and similar requirements—are also contained in the tables referenced for each material. See the commentary to Section 2303 concerning the minimum standards and quality of these materials.

SEISMIC DESIGN CATEGORY	MAXIMUM WALL SPACING (feet)	REQUIRED BRACING LENGTH, b
A, B and C	35′-0″	Table 2308.9.3(1) and Section 2308.9.3
D and E	25′-0″	Table 2308.12.4

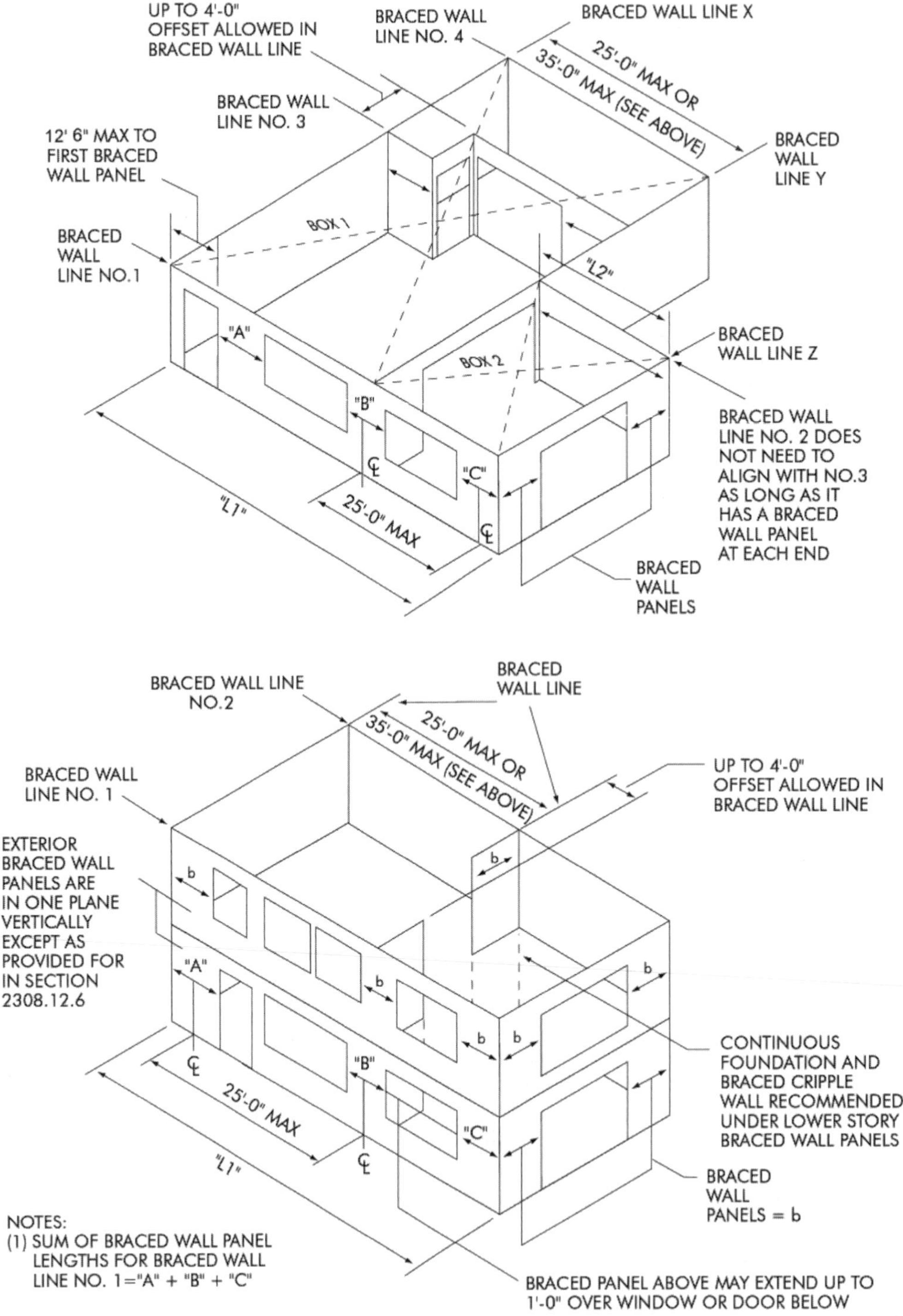

For SI: 1 foot = 304.8 mm.

FIGURE 2308.9.3
BASIC COMPONENTS OF THE LATERAL BRACING SYSTEM

TABLE 2308.9.3(1)
BRACED WALL PANELS[a]

SEISMIC DESIGN CATEGORY	CONDITION	CONSTRUCTION METHODS[b, c]								BRACED PANEL LOCATION AND LENGTH[d]
		1	2	3	4	5	6	7	8	
A and B	One story, top of two or three story	X	X	X	X	X	X	X	X	Located in accordance with Section 2308.9.3 and not more than 25 feet on center.
	First story of two story or second story of three story	X	X	X	X	X	X	X	X	
	First story of three story	—	X	X	X	X[e]	X	X	X	
C	One story or top of two story	—	X	X	X	X	X	X	X	Located in accordance with Section 2308.9.3 and not more than 25 feet on center.
	First story of two story	—	X	X	X	X[e]	X	X	X	Located in accordance with Section 2308.9.3 and not more than 25 feet on center, but total length shall not be less than 25% of building length[f].

For SI: 1 inch = 25.4 mm, 1 foot = 304.8 mm.

a. This table specifies minimum requirements for braced panels that form interior or exterior braced wall lines.

b. See Section 2308.9.3 for full description.

c. See Sections 2308.9.3.1 and 2308.9.3.2 for alternative braced panel requirements.

d. Building length is the dimension parallel to the braced wall length.

e. Gypsum wallboard applied to framing supports that are spaced at 16 inches on center.

f. The required lengths shall be doubled for gypsum board applied to only one face of a braced wall panel.

TABLE 2308.9.3(2)
EXPOSED PLYWOOD PANEL SIDING

MINIMUM THICKNESS[a] (inch)	MINIMUM NUMBER OF PLIES	STUD SPACING (inches) Plywood siding applied directly to studs or over sheathing
$^{3}/_{8}$	3	16[b]
$^{1}/_{2}$	4	24

For SI: 1 inch = 25.4 mm.

a. Thickness of grooved panels is measured at bottom of grooves.

b. Spans are permitted to be 24 inches if plywood siding applied with face grain perpendicular to studs or over one of the following: (1) 1-inch board sheathing, (2) $^{7}/_{16}$-inch wood structural panel sheathing or (3) $^{3}/_{8}$-inch wood structural panel sheathing with strength axis (which is the long direction of the panel unless otherwise marked) of sheathing perpendicular to studs.

TABLE 2308.9.3(3)
WOOD STRUCTURAL PANEL WALL SHEATHING[b]
(Not Exposed to the Weather, Strength Axis Parallel or Perpendicular to Studs Except as Indicated Below)

MINIMUM THICKNESS (inch)	PANEL SPAN RATING	STUD SPACING (inches)		
		Siding nailed to studs	Nailable sheathing	
			Sheathing parallel to studs	Sheathing perpendicular to studs
$^{3}/_{8}$, $^{15}/_{32}$, $^{1}/_{2}$	16/0, 20/0, 24/0, 32/16 Wall—24″ o.c.	24	16	24
$^{7}/_{16}$, $^{15}/_{32}$, $^{1}/_{2}$	24/0, 24/16, 32/16 Wall—24″ o.c.	24	24[a]	24

For SI: 1 inch = 25.4 mm.

a. Plywood shall consist of four or more plies.

b. Blocking of horizontal joints shall not be required except as specified in Sections 2306.3 and 2308.12.4.

TABLE 2308.9.3(4)
ALLOWABLE SPANS FOR PARTICLEBOARD WALL SHEATHING
(Not Exposed to the Weather, Long Dimension of the Panel Parallel or Perpendicular to Studs)

GRADE	THICKNESS (inch)	STUD SPACING (inches)	
		Siding nailed to studs	Sheathing under coverings specified in Section 2308.9.3 parallel or perpendicular to studs
M-S "Exterior Glue" and M-2 "Exterior Glue"	$^{3}/_{8}$	16	—
	$^{1}/_{2}$	16	16

For SI: 1 inch = 25.4 mm.

TABLE 2308.9.3(5)
HARDBOARD SIDING

SIDING	MINIMUM NOMINAL THICKNESS (inch)	2 × 4 FRAMING MAXIMUM SPACING	NAIL SIZE[a, b, d]	NAIL SPACING	
				General	Bracing panels[c]
1. Lap siding					
Direct to studs	$^{3}/_{8}$	16″ o.c.	8d	16″ o.c.	Not applicable
Over sheathing	$^{3}/_{8}$	16″ o.c.	10d	16″ o.c.	Not applicable
2. Square edge panel siding					
Direct to studs	$^{3}/_{8}$	24″ o.c.	6d	6″ o.c. edges; 12″ o.c. at intermediate supports	4″ o.c. edges; 8″ o.c. at intermediate supports
Over sheathing	$^{3}/_{8}$	24″ o.c.	8d	6″ o.c. edges; 12″ o.c. at intermediate supports	4″ o.c. edges; 8″ o.c. at intermediate supports
3. Shiplap edge panel siding					
Direct to studs	$^{3}/_{8}$	16″ o.c.	6d	6″ o.c. edges; 12″ o.c. at intermediate supports	4″ o.c. edges; 8″ o.c. at intermediate supports
Over sheathing	$^{3}/_{8}$	16″ o.c.	8d	6″ o.c. edges; 12″ o.c. at intermediate supports	4″ o.c. edges; 8″ o.c. at intermediate supports

For SI: 1 inch = 25.4 mm.

a. Nails shall be corrosion resistant.

b. Minimum acceptable nail dimensions:

	Panel Siding (inch)	Lap Siding (inch)
Shank diameter	0.092	0.099
Head diameter	0.225	0.240

c. Where used to comply with Section 2308.9.3.

d. Nail length must accommodate the sheathing and penetrate framing $1^{1}/_{2}$ inches.

2308.9.3.1 Alternative bracing. Any bracing required by Section 2308.9.3 is permitted to be replaced by the following:

1. In one-story buildings, each panel shall have a length of not less than 2 feet 8 inches (813 mm) and a height of not more than 10 feet (3048 mm).Each panel shall be sheathed on one face with $^3/_8$-inch-minimum-thickness (9.5 mm) wood structural panel sheathing nailed with 8d common or galvanized box nails in accordance with Table 2304.9.1 and blocked at wood structural panel edges. Two anchor bolts installed in accordance with Section 2308.6 shall be provided in each panel. Anchor bolts shall be placed at each panel outside quarter points. Each panel end stud shall have a tie-down device fastened to the foundation, capable of providing an *approved* uplift capacity of not less than 1,800 pounds (8006 N). The tie-down device shall be installed in accordance with the manufacturer's recommendations. The panels shall be supported directly on a foundation or on floor framing supported directly on a foundation that is continuous across the entire length of the braced wall line. This foundation shall be reinforced with not less than one No. 4 bar top and bottom.

   Where the continuous foundation is required to have a depth greater than 12 inches (305 mm), a minimum 12-inch by 12-inch (305 mm by 305 mm) continuous footing or turned down slab edge is permitted at door openings in the braced wall line. This continuous footing or turned down slab edge shall be reinforced with not less than one No. 4 bar top and bottom. This reinforcement shall be lapped 15 inches (381 mm) with the reinforcement required in the continuous foundation located directly under the braced wall line.

2. In the first *story* of two-story buildings, each wall panel shall be braced in accordance with Section 2308.9.3.1, Item 1, except that the wood structural panel sheathing shall be provided on both faces, three anchor bolts shall be placed at one-quarter points, and tie-down device uplift capacity shall not be less than 3,000 pounds (13 344 N).

❖ As stated in the beginning of the section, any of the braced wall panel methods required by Section 2308.9.3 may be replaced by an alternate braced wall panel that is considered equivalent to 4 feet (1219 mm) of bracing. The alternate braced wall panel provisions allow a minimum 32-inch (813 mm) wall bracing length in lieu of the 48-inch (1219 mm) length required elsewhere. This bracing detail allows more flexibility in constructing braced wall panels adjacent to garage doors and other similar openings. This detail was developed by APA and is supported in APA Research Report No. 157. Item 1 describes the requirements for a one-story building (see Figure 2308.9.3.1).

An alternative panel is allowed on the ground level

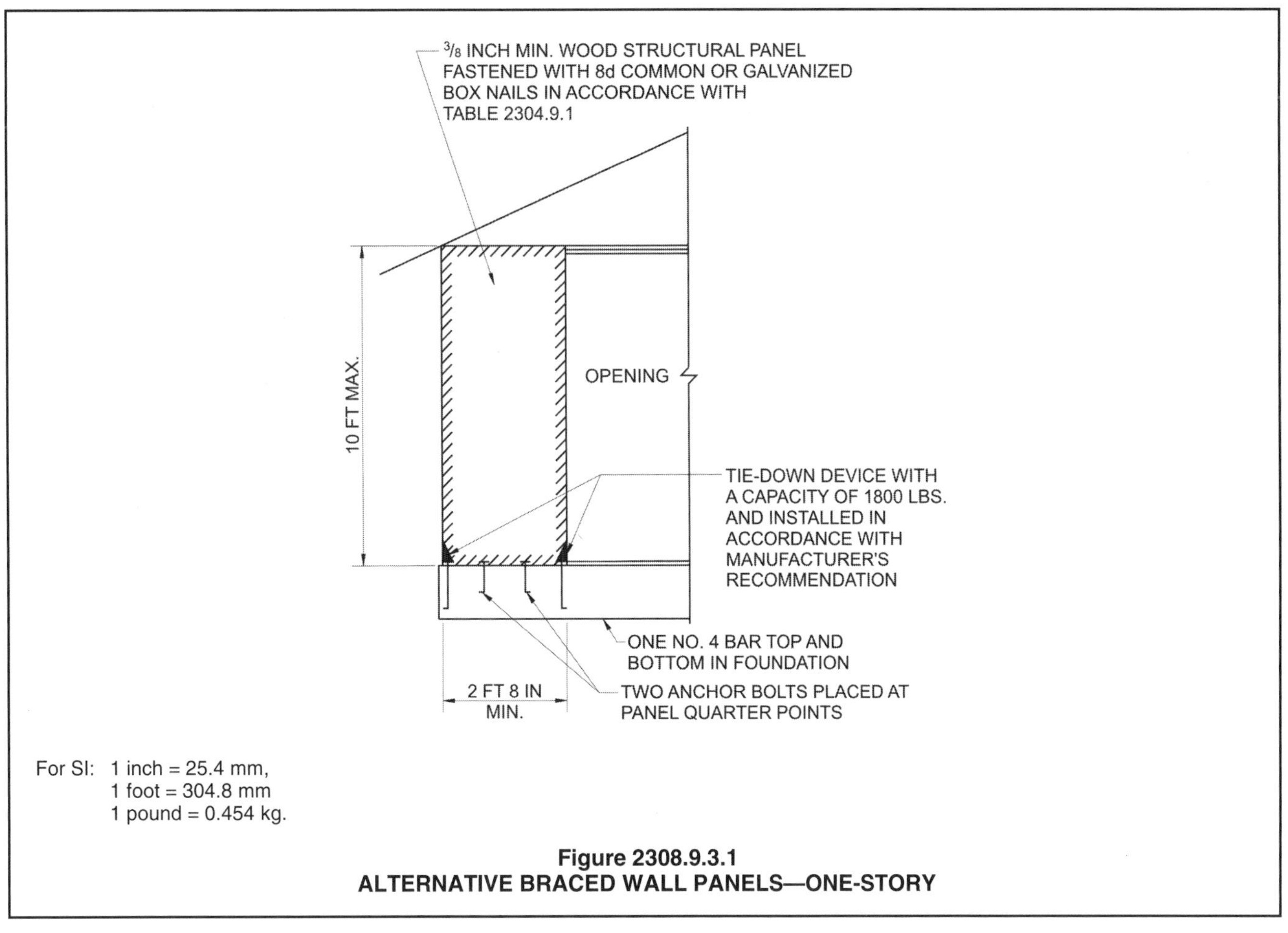

Figure 2308.9.3.1
ALTERNATIVE BRACED WALL PANELS—ONE-STORY

of a two-story structure as well, but the requirements are more stringent—almost double the resistance required for the single-story panel. One important issue is the specific requirement for a continuous foundation for the length of the entire braced wall line that contains the alternate braced wall panels, which would even include those interior lines at a spacing less than 50 feet (15 240 mm) from an adjacent braced wall line as described in Section 2308.3.4. Note that the alternate braced wall panel has a maximum height of 10 feet (3048 mm).

2308.9.3.2 Alternate bracing wall panel adjacent to a door or window opening. Any bracing required by Section 2308.9.3 is permitted to be replaced by the following when used adjacent to a door or window opening with a full-length header:

1. In one-story buildings, each panel shall have a length of not less than 16 inches (406 mm) and a height of not more than 10 feet (3048 mm). Each panel shall be sheathed on one face with a single layer of $^{3}/_{8}$ inch (9.5 mm) minimum thickness wood structural panel sheathing nailed with 8d common or galvanized box nails in accordance with Figure 2308.9.3.2. The wood structural panel sheathing shall extend up over the solid sawn or glued-laminated header and shall be nailed in accordance with Figure 2308.9.3.2. A built-up header consisting of at least two 2 × 12s and fastened in accordance with Item 24 of Table 2304.9.1 shall be permitted to be used. A spacer, if used, shall be placed on the side of the built-up beam opposite the wood structural panel sheathing. The header shall extend between the inside faces of the first full-length outer studs of each panel. The clear span of the header between the inner studs of each panel shall be not less than 6 feet (1829 mm) and not more than 18 feet (5486 mm) in length. A strap with an uplift capacity of not less than 1,000 pounds (4,400 N) shall fasten the header to the inner studs opposite the sheathing. One anchor bolt not less than $^{5}/_{8}$ inch (15.9 mm) diameter and installed in accordance with Section 2308.6 shall be provided in the center of each sill plate. The studs at each end of the panel shall have a tie-down device fastened to the foundation with an uplift capacity of not less than 4,200 pounds (18 480 N).

   Where a panel is located on one side of the opening, the header shall extend between the inside face of the first full-length stud of the panel and the bearing studs at the other end of the opening. A strap with an uplift capacity of not less than 1,000 pounds (4400 N) shall fasten the header to the bearing studs. The bearing studs shall also have a tie-down device fastened to the foundation with an uplift capacity of not less than 1,000 pounds (4400 N).

   The tie-down devices shall be an embedded strap type, installed in accordance with the manufacturer's recommendations. The panels shall be supported directly on a foundation that is continuous across the entire length of the braced wall line. This foundation shall be reinforced with not less than one No. 4 bar top and bottom.

   Where the continuous foundation is required to have a depth greater than 12 inches (305 mm), a minimum 12-inch by 12-inch (305 mm by 305 mm) continuous footing or turned down slab edge is permitted at door openings in the braced wall line. This continuous footing or turned down slab edge shall be reinforced with not less than one No. 4 bar top and bottom. This reinforcement shall be lapped not less than 15 inches (381 mm) with the reinforcement required in the continuous foundation located directly under the braced wall line.

2. In the first *story* of two-story buildings, each wall panel shall be braced in accordance with Item 1 above, except that each panel shall have a length of not less than 24 inches (610 mm).

❖ This provision allows an additional alternative bracing method for use adjacent to a window or door opening. It includes specifications for creating a type of portal frame system with built-up "columns" at each end that are fastened to the header with a nominal amount of rigidity. In this alternative, the header runs the length of the sheathed bracing panel—up to the first full-length stud—and is attached to the full-height sheathing with a grid pattern of nails where the sheathing and header overlap. This provides a partially rigid moment-resisting connection. At the base of the embedded sheathed section, tie-down anchors are nailed to the framing, which provide a more conventional moment-resisting connection. These framing anchors are sized to provide uplift in addition to shear capacity, reducing the plate anchoring requirements. Additional strap anchors, as shown in code Figure 2308.9.3.2, are required. The result of these provisions is a reduced width of the full height segment of the assembly from 32 to 16 inches (813 to 406 mm) for a one-story building, and to 24 inches (610 mm) for the first floor of a two-story building compared to the alternate braced wall panel option permitted in Section 2308.9.3.1. The maximum allowable garage (or other room) size for this application would have to comply with the restrictions for braced wall line spacing and framing member spans noted elsewhere in Section 2308. As stated in the beginning of the section, any of the braced wall panel methods required by Section 2308.9.3 are permitted to be replaced by a portal frame system and be considered equivalent to 4 feet (1219 mm) of bracing. Like the alternate braced wall panel, it has a maximum height of 10 feet (3048 mm).

FIGURE 2308.9.3.2. See page 23-75.

❖ Narrow "portal frame" details such as these have been used successfully in the Pacific Northwest for almost two decades. This alternative was first developed in 1988 and was based on monotonic testing. The popularity of this detail grew until it was used in a number of different jurisdictions in the Pacific Northwest. During a series of cyclic tests conducted by the APA, the single-sided/single-story 32-inch-wide (813 mm) alternative bracing in Section 2308.9.3.1 was tested along with the 16-inch-wide (406 mm) alternative. Similarly, the double-sided/two-story 32-inch-wide (813 mm) assembly was compared with the 24-inch-wide (610 mm) alternative. In both cases, testing demonstrated that alternative bracing in Section 2308.9.3.2 performs significantly better in both strength and stiffness than the 32-inch-wide (813 mm) bracing in Section 2308.9.3.1.

2308.9.4 Cripple walls. Foundation cripple walls shall be framed of studs not less in size than the studding above with a minimum length of 14 inches (356 mm), or shall be framed of solid blocking. Where exceeding 4 feet (1219 mm) in height, such walls shall be framed of studs having the size required for an additional *story*.

❖ Cripple walls (sometimes referred to as "foundation stud walls" or "knee walls") are stud walls usually less than 8 feet (2438 mm) in height that rest on the foundation plate and support the first (lowest) immediate floor above. Cripple walls are often used on sloped lots to frame up to the floor level. Cripple studs should be at least the same size as the supported studs above. The minimum stud length of 14 inches (356 mm) (unless solid blocking is used) is based on the minimum length necessary to properly fasten the studs to the foundation wall plate and the double plate above. Where the cripple studs would be less than 14 inches (356 mm) in length, solid blocking should be used or the studs should be installed with wall plates and with the solid blocking tightly fitted between each stud. This blocking performs two purposes: it provides a level, uniform bearing surface for the support of the floor above and transfers lateral forces from the floor to the foundation.

2308.9.4.1 Bracing. For the purposes of this section, cripple walls having a stud height exceeding 14 inches (356 mm) in structures assigned to *Seismic Design Category* A, B or C shall be considered a story and shall be braced in accordance with Table 2308.9.3(1). See Section 2308.12.4 for cripple walls in structures assigned to *Seismic Design Category* D or E.

❖ When the cripple wall stud height is more than 14 inches (356 mm), the wall is to be considered as a first-story wall and the bracing requirements of Table

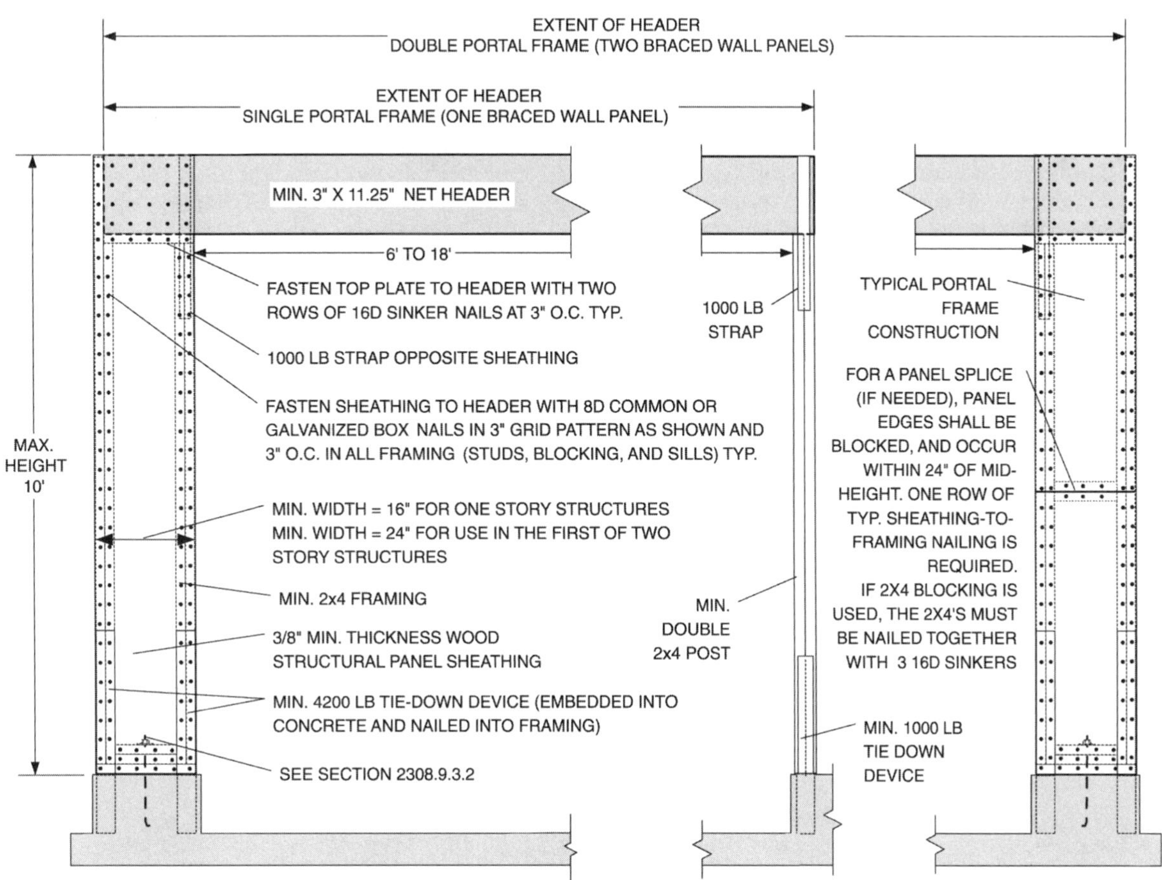

For SI: 1 foot = 304.8 mm; 1 inch = 25.4 mm; 1 pound = 4.448 N.

FIGURE 2308.9.3.2
ALTERNATE BRACED WALL PANEL ADJACENT TO A DOOR OR WINDOW OPENING

2308.9.3(1) will apply for buildings in lower seismic design categories. In Seismic Design Category D or E, the increased bracing requirements of Table 2308.12.4 will apply.

2308.9.4.2 Nailing of bracing. Spacing of edge nailing for required wall bracing shall not exceed 6 inches (152 mm) o.c. along the foundation plate and the top plate of the cripple wall. Nail size, nail spacing for field nailing and more restrictive boundary nailing requirements shall be as required elsewhere in the code for the specific bracing material used.

❖ Cripple wall bracing and 6-inch (152 mm) edge nailing are important in order to avoid the creation of a soft story and to provide a continuous load path to transfer lateral forces from the structure above to the foundation. The minimum edge nailing specified in this section provides a level of shear resistance at the bottom of the structure that is at least equivalent to that provided elsewhere.

2308.9.5 Openings in exterior walls. Openings in exterior walls shall be constructed in accordance with Sections 2308.9.5.1 and 2308.9.5.2.

❖ Prescriptive requirements for framing openings for doors and windows are given in this section, including minimum headers and trimmer (jack) studs.

TABLE 2308.9.5. See page 23-77.

❖ The header support conditions utilized in this table are shown in Figure 2308.9.5 (see commentary, Section 2308.9.5.1).

2308.9.5.1 Headers. Headers shall be provided over each opening in exterior-bearing walls. The spans in Table 2308.9.5 are permitted to be used for one- and two-family *dwellings*. Headers for other buildings shall be designed in accordance with Section 2301.2, Item 1 or 2. Headers shall be of two pieces of nominal 2-inch (51 mm) framing lumber set on edge as permitted by Table 2308.9.5 and nailed together in accordance with Table 2304.9.1 or of solid lumber of equivalent size.

❖ Headers over openings are required to carry loads from the wall, floors and roof above. Header spans for exterior bearing walls of one- and two-family dwellings may be sized and supported in accordance with Table 2308.9.5. Headers for other structures must be designed.

Table 2308.9.5, as well as Table 2308.9.6, provide an empirical procedure for determining header sizes in exterior walls. The table gives the header size, allowable span and number of jack studs required for various ground snow loads and building widths. The 30 psf ground snow load is used for sizing headers with a roof live load of 20 psf (0.96 kN/m^2) (see Note e). The tables were developed using the following design load criteria:

Roof LL	= 20 psf	(0.96 kN/m^2)
Roof DL	= 10 psf	(0.48 kN/m^2)
Ceiling LL	= 5 psf	(0.24 kN/m^2)
Ceiling DL	= 5 psf	(0.24 kN/m^2)
Floor LL	= 40 psf	(1.92 kN/m^2)
Floor DL	= 10 psf	(0.48 kN/m^2)
Wall DL	= 11 psf	(0.53 kN/m^2)

The spans were developed by using the given loads and performing bending moment and deflection calculations to determine the maximum span length for the specific size headers shown in the tables. The span length calculations also included a load combination reduction factor. Current ASD design loads do not address load combination reduction factors. The load combination reduction factors were developed in accordance with LRFD load combinations: the smaller of $1.6L + 0.5S$ or $1.6S + 0.5L$, where S is the snow load and L is the live load. The appropriate factor is then divided by 1.6 to adjust from LRFD to ASD.

The header span in the tables is determined based on the ground snow load in accordance with Figure 1608.2, the building width, the supporting conditions and the size of the header to be used. The header sizes must be comprised of the required lumber sizes shown in the tables or may be of single solid lumber members of equivalent size and capacity.

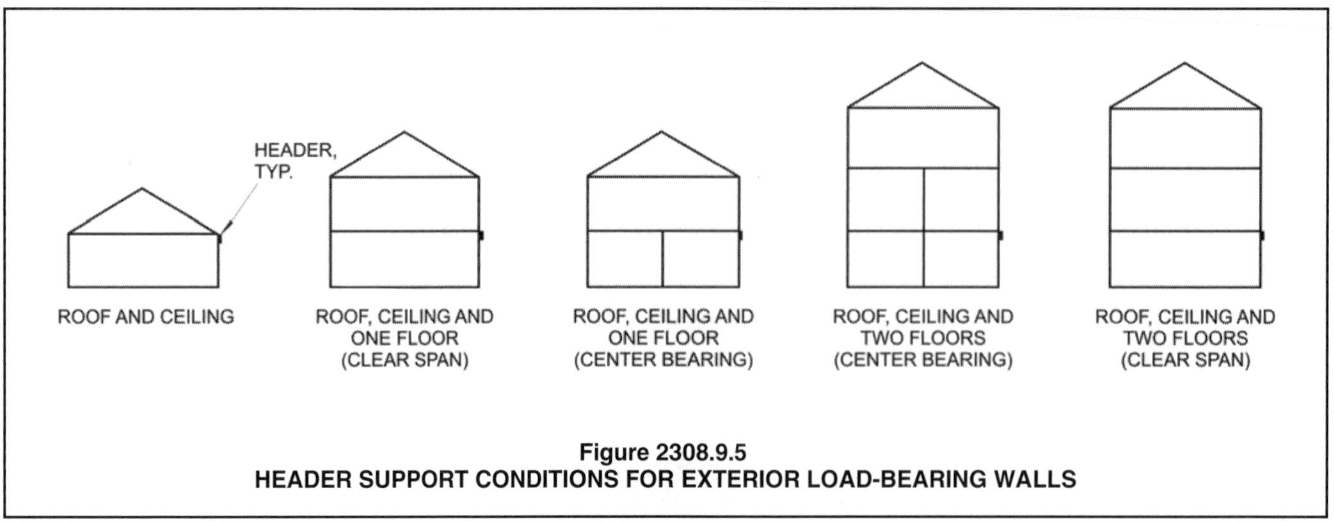

Figure 2308.9.5
HEADER SUPPORT CONDITIONS FOR EXTERIOR LOAD-BEARING WALLS

TABLE 2308.9.5
HEADER AND GIRDER SPANS[a] FOR EXTERIOR BEARING WALLS
(Maximum Spans for Douglas Fir-Larch, Hem-Fir, Southern Pine and Spruce-Pine-Fir[b] and Required Number of Jack Studs)

HEADERS SUPPORTING	SIZE	GROUND SNOW LOAD (psf)[e]											
		30						50					
		Building width[c] (feet)											
		20		28		36		20		28		36	
		Span	NJ[d]	Span	NJ[d]	Span	NJ[d]	Span	NJ[d]	Span	NJ[d]	Span	NJ[d]
Roof & Ceiling	2-2×4	3-6	1	3-2	1	2-10	1	3-2	1	2-9	1	2-6	1
	2-2×6	5-5	1	4-8	1	4-2	1	4-8	1	4-1	1	3-8	2
	2-2×8	6-10	1	5-11	2	5-4	2	5-11	2	5-2	2	4-7	2
	2-2×10	8-5	2	7-3	2	6-6	2	7-3	2	6-3	2	5-7	2
	2-2×12	9-9	2	8-5	2	7-6	2	8-5	2	7-3	2	6-6	2
	3-2×8	8-4	1	7-5	1	6-8	1	7-5	1	6-5	2	5-9	2
	3-2×10	10-6	1	9-1	2	8-2	2	9-1	2	7-10	2	7-0	2
	3-2×12	12-2	2	10-7	2	9-5	2	10-7	2	9-2	2	8-2	2
	4-2×8	9-2	1	8-4	1	7-8	1	8-4	1	7-5	1	6-8	1
	4-2×10	11-8	1	10-6	1	9-5	2	10-6	1	9-1	2	8-2	2
	4-2×12	14-1	1	12-2	2	10-11	2	12-2	2	10-7	2	9-5	2
Roof, Ceiling & 1 Center-Bearing Floor	2-2×4	3-1	1	2-9	1	2-5	1	2-9	1	2-5	1	2-2	1
	2-2×6	4-6	1	4-0	1	3-7	2	4-1	1	3-7	2	3-3	2
	2-2×8	5-9	2	5-0	2	4-6	2	5-2	2	4-6	2	4-1	2
	2-2×10	7-0	2	6-2	2	5-6	2	6-4	2	5-6	2	5-0	2
	2-2×12	8-1	2	7-1	2	6-5	2	7-4	2	6-5	2	5-9	3
	3-2×8	7-2	1	6-3	2	5-8	2	6-5	2	5-8	2	5-1	2
	3-2×10	8-9	2	7-8	2	6-11	2	7-11	2	6-11	2	6-3	2
	3-2×12	10-2	2	8-11	2	8-0	2	9-2	2	8-0	2	7-3	2
	4-2×8	8-1	1	7-3	1	6-7	1	7-5	1	6-6	1	5-11	2
	4-2×10	10-1	1	8-10	2	8-0	2	9-1	2	8-0	2	7-2	2
	4-2×12	11-9	2	10-3	2	9-3	2	10-7	2	9-3	2	8-4	2
Roof, Ceiling & 1 Clear Span Floor	2-2×4	2-8	1	2-4	1	2-1	1	2-7	1	2-3	1	2-0	1
	2-2×6	3-11	1	3-5	2	3-0	2	3-10	2	3-4	2	3-0	2
	2-2×8	5-0	2	4-4	2	3-10	2	4-10	2	4-2	2	3-9	2
	2-2×10	6-1	2	5-3	2	4-8	2	5-11	2	5-1	2	4-7	3
	2-2×12	7-1	2	6-1	3	5-5	3	6-10	2	5-11	3	5-4	3
	3-2×8	6-3	2	5-5	2	4-10	2	6-1	2	5-3	2	4-8	2
	3-2×10	7-7	2	6-7	2	5-11	2	7-5	2	6-5	2	5-9	2
	3-2×12	8-10	2	7-8	2	6-10	2	8-7	2	7-5	2	6-8	2
	4-2×8	7-2	1	6-3	2	5-7	2	7-0	1	6-1	2	5-5	2
	4-2×10	8-9	2	7-7	2	6-10	2	8-7	2	7-5	2	6-7	2
	4-2×12	10-2	2	8-10	2	7-11	2	9-11	2	8-7	2	7-8	2

(continued)

TABLE 2308.9.5—continued
HEADER AND GIRDER SPANS[a] FOR EXTERIOR BEARING WALLS
(Maximum Spans for Douglas Fir-Larch, Hem-Fir, Southern Pine and Spruce-Pine-Fir[b] and Required Number of Jack Studs)

HEADERS SUPPORTING	SIZE	GROUND SNOW LOAD (psf)[e]											
		30						50					
		Building width[c] (feet)											
		20		28		36		20		28		36	
		Span	NJ[d]	Span	NJ[d]	Span	NJ[d]	Span	NJ[d]	Span	NJ[d]	Span	NJ[d]
Roof, Ceiling & 2 Center-Bearing Floors	2-2×4	2-7	1	2-3	1	2-0	1	2-6	1	2-2	1	1-11	1
	2-2×6	3-9	2	3-3	2	2-11	2	3-8	2	3-2	2	2-10	2
	2-2×8	4-9	2	4-2	2	3-9	2	4-7	2	4-0	2	3-8	2
	2-2×10	5-9	2	5-1	2	4-7	3	5-8	2	4-11	2	4-5	3
	2-2×12	6-8	2	5-10	3	5-3	3	6-6	2	5-9	3	5-2	3
	3-2×8	5-11	2	5-2	2	4-8	2	5-9	2	5-1	2	4-7	2
	3-2×10	7-3	2	6-4	2	5-8	2	7-1	2	6-2	2	5-7	2
	3-2×12	8-5	2	7-4	2	6-7	2	8-2	2	7-2	2	6-5	3
	4-2×8	6-10	1	6-0	2	5-5	2	6-8	1	5-10	2	5-3	2
	4-2×10	8-4	2	7-4	2	6-7	2	8-2	2	7-2	2	6-5	2
	4-2×12	9-8	2	8-6	2	7-8	2	9-5	2	8-3	2	7-5	2
Roof, Ceiling & 2 Clear Span Floors	2-2×4	2-1	1	1-8	1	1-6	2	2-0	1	1-8	1	1-5	2
	2-2×6	3-1	2	2-8	2	2-4	2	3-0	2	2-7	2	2-3	2
	2-2×8	3-10	2	3-4	2	3-0	3	3-10	2	3-4	2	2-11	3
	2-2×10	4-9	2	4-1	3	3-8	3	4-8	2	4-0	3	3-7	3
	2-2×12	5-6	3	4-9	3	4-3	3	5-5	3	4-8	3	4-2	3
	3-2×8	4-10	2	4-2	2	3-9	2	4-9	2	4-1	2	3-8	2
	3-2×10	5-11	2	5-1	2	4-7	3	5-10	2	5-0	2	4-6	3
	3-2×12	6-10	2	5-11	3	5-4	3	6-9	2	5-10	3	5-3	3
	4-2×8	5-7	2	4-10	2	4-4	2	5-6	2	4-9	2	4-3	2
	4-2×10	6-10	2	5-11	2	5-3	2	6-9	2	5-10	2	5-2	2
	4-2×12	7-11	2	6-10	2	6-2	3	7-9	2	6-9	2	6-0	3

For SI: 1 inch = 25.4 mm, 1 foot = 304.8 mm, 1 pound per square foot = 47.8 N/m^2.

a. Spans are given in feet and inches (ft-in).

b. Tabulated values are for No. 2 grade lumber.

c. Building width is measured perpendicular to the ridge. For widths between those shown, spans are permitted to be interpolated.

d. NJ - Number of jack studs required to support each end. Where the number of required jack studs equals one, the header is permitted to be supported by an *approved* framing anchor attached to the full-height wall stud and to the header.

e. Use 30 pounds per square foot ground snow load for cases in which ground snow load is less than 30 pounds per square foot and the roof live load is equal to or less than 20 pounds per square foot.

2308.9.5.2 Header support. Wall studs shall support the ends of the header in accordance with Table 2308.9.5. Each end of a lintel or header shall have a length of bearing of not less than $1^1/_2$ inches (38 mm) for the full width of the lintel.

❖ Trimmer or jack studs that support the header at each end should be continuous from the header to the bottom plate (or sill plate). Cutting the trimmers to support a sill is not allowed. Headers should be adequately nailed together and to the wall studs. Minimum required nailing is specified in Table 2304.9.1 [see Figures 2308.9.5.2(1) and (2)].

2308.9.6 Openings in interior bearing partitions. Headers shall be provided over each opening in interior bearing partitions as required in Section 2308.9.5. The spans in Table 2308.9.6 are permitted to be used. Wall studs shall support the ends of the header in accordance with Table 2308.9.5 or 2308.9.6, as appropriate.

❖ It may not be immediately clear that interior bearing walls will require headers to carry loads over openings in the same fashion as headers within exterior bearing walls must; therefore, this section has been included in the code. At interior bearing walls, the tributary width of loading carried from above extends to both sides of the wall instead of just one side, so care is needed to determine whether an interior header meets conventional restrictions or if it requires engineering in accordance with Section 2308.4.

TABLE 2308.9.6. See page 23-80.

❖ The header support conditions that are used in this table are illustrated in Figure 2308.9.6 (see commentary, Section 2308.9.5.1).

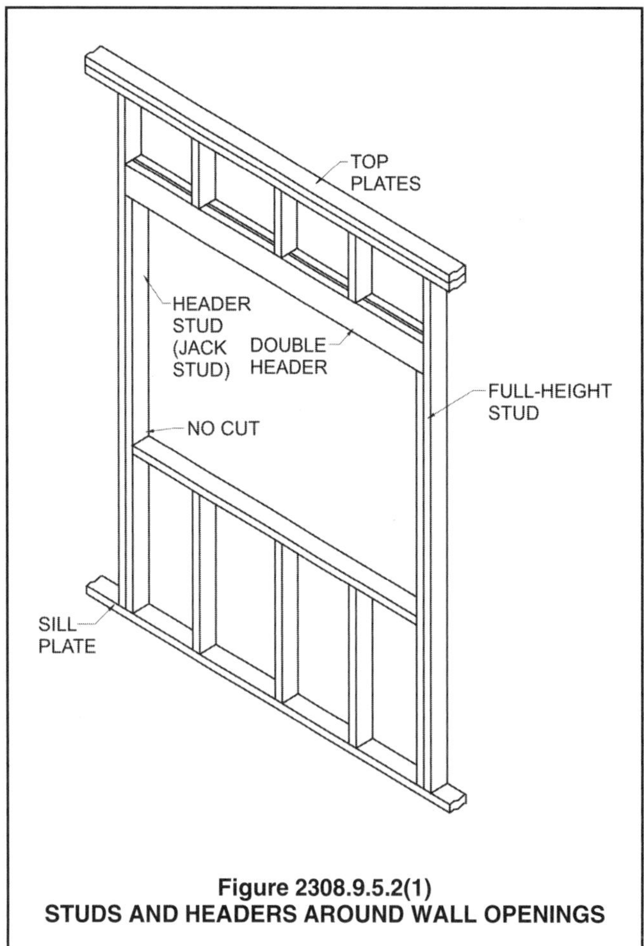

Figure 2308.9.5.2(1)
STUDS AND HEADERS AROUND WALL OPENINGS

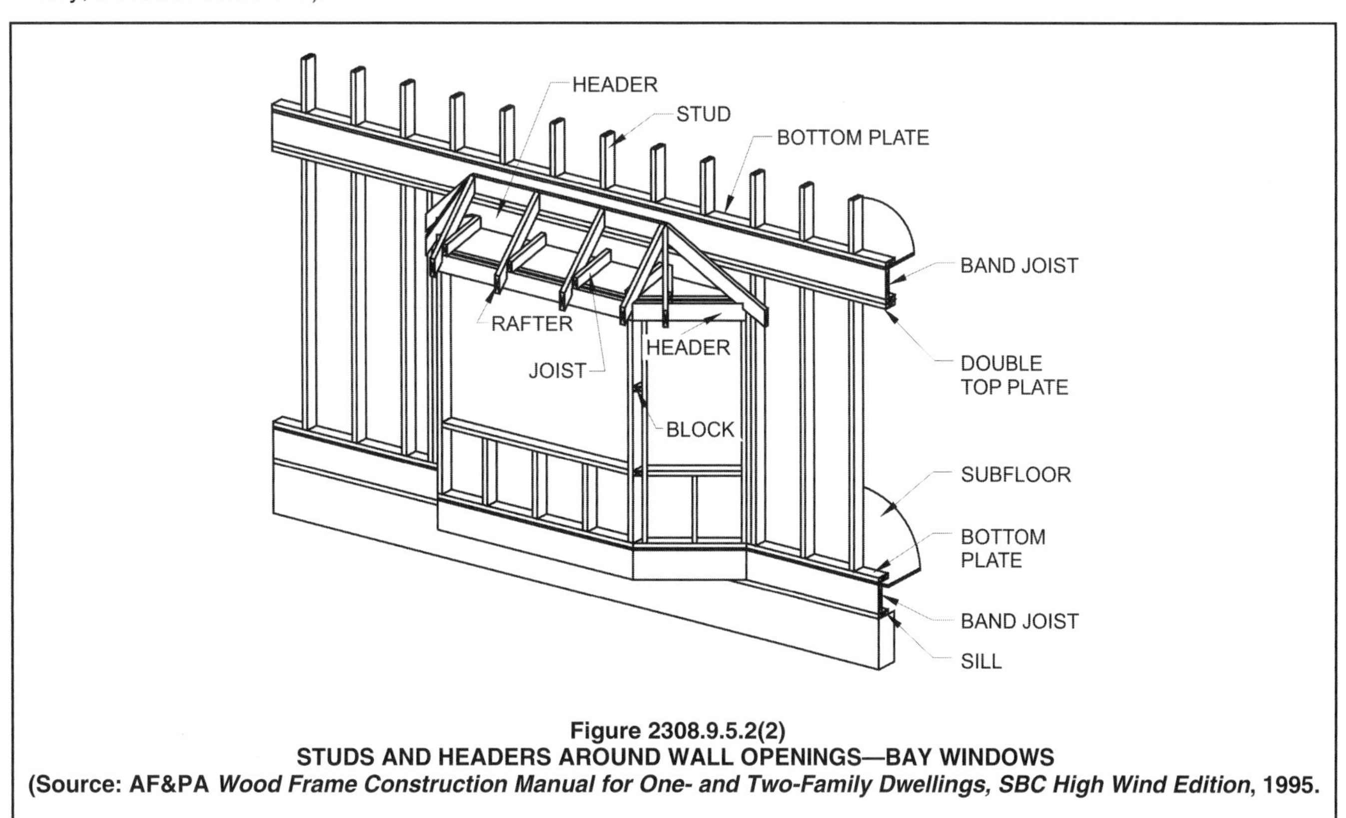

Figure 2308.9.5.2(2)
STUDS AND HEADERS AROUND WALL OPENINGS—BAY WINDOWS
(Source: AF&PA *Wood Frame Construction Manual for One- and Two-Family Dwellings, SBC High Wind Edition*, 1995.

TABLE 2308.9.6
HEADER AND GIRDER SPANS[a] FOR INTERIOR BEARING WALLS
(Maximum Spans for Douglas Fir-Larch, Hem-Fir, Southern Pine and Spruce-Pine-Fir[b] and Required Number of Jack Studs)

HEADERS AND GIRDERS SUPPORTING	SIZE	BUILDING width[c] (feet)					
		20		28		36	
		Span	NJ[d]	Span	NJ[d]	Span	NJ[d]
One Floor Only	2-2×4	3-1	1	2-8	1	2-5	1
	2-2×6	4-6	1	3-11	1	3-6	1
	2-2×8	5-9	1	5-0	2	4-5	2
	2-2×10	7-0	2	6-1	2	5-5	2
	2-2×12	8-1	2	7-0	2	6-3	2
	3-2×8	7-2	1	6-3	1	5-7	2
	3-2×10	8-9	1	7-7	2	6-9	2
	3-2×12	10-2	2	8-10	2	7-10	2
	4-2×8	9-0	1	7-8	1	6-9	1
	4-2×10	10-1	1	8-9	1	7-10	2
	4-2×12	11-9	1	10-2	2	9-1	2
Two Floors	2-2×4	2-2	1	1-10	1	1-7	1
	2-2×6	3-2	2	2-9	2	2-5	2
	2-2×8	4-1	2	3-6	2	3-2	2
	2-2×10	4-11	2	4-3	2	3-10	3
	2-2×12	5-9	2	5-0	3	4-5	3
	3-2×8	5-1	2	4-5	2	3-11	2
	3-2×10	6-2	2	5-4	2	4-10	2
	3-2×12	7-2	2	6-3	2	5-7	3
	4-2×8	6-1	1	5-3	2	4-8	2
	4-2×10	7-2	2	6-2	2	5-6	2
	4-2×12	8-4	2	7-2	2	6-5	2

For SI: 1 inch = 25.4 mm, 1 foot = 304.8 mm.

a. Spans are given in feet and inches (ft-in).

b. Tabulated values are for No. 2 grade lumber.

c. Building width is measured perpendicular to the ridge. For widths between those shown, spans are permitted to be interpolated.

d. NJ - Number of jack studs required to support each end. Where the number of required jack studs equals one, the headers are permitted to be supported by an approved framing anchor attached to the full-height wall stud and to the header.

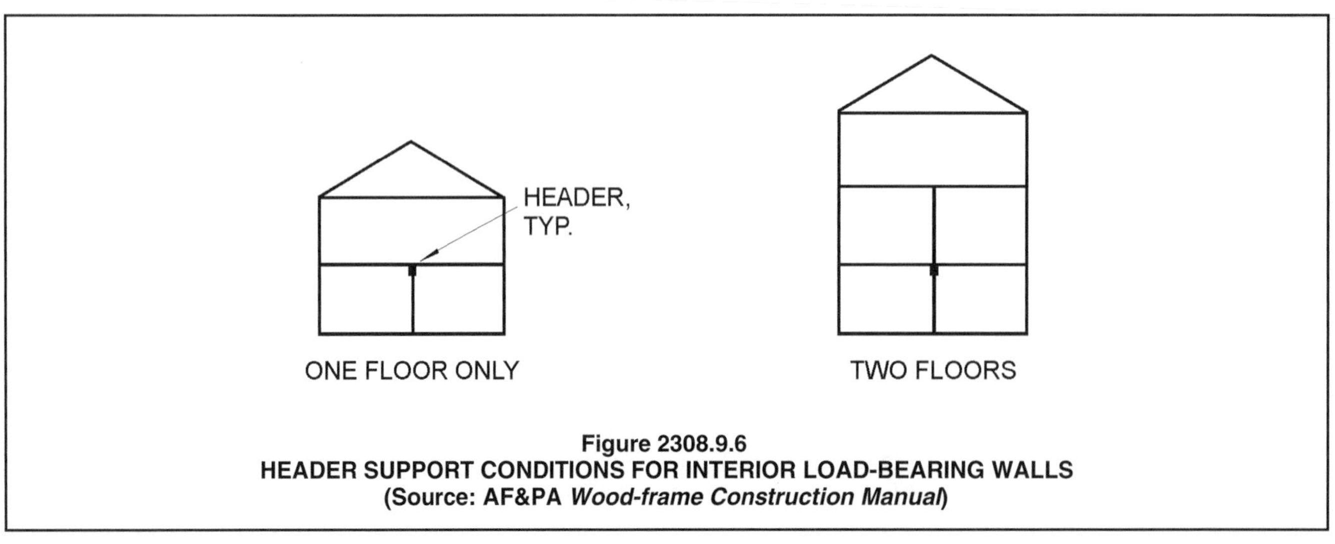

Figure 2308.9.6
HEADER SUPPORT CONDITIONS FOR INTERIOR LOAD-BEARING WALLS
(Source: AF&PA *Wood-frame Construction Manual*)

2308.9.7 Openings in interior nonbearing partitions. Openings in nonbearing partitions are permitted to be framed with single studs and headers. Each end of a lintel or header shall have a length of bearing of not less than $1^1/_2$ inches (38 mm) for the full width of the lintel.

❖ Because interior nonbearing partitions do not carry gravity loads other than their own weight, the prescriptive header requirements over openings are greatly reduced. However, a minimum of $1^1/_2$ inches (38 mm) bearing the width of the header is required at each support.

2308.9.8 Pipes in walls. Stud partitions containing plumbing, heating or other pipes shall be so framed and the joists underneath so spaced as to give proper clearance for the piping. Where a partition containing such piping runs parallel to the floor joists, the joists underneath such partitions shall be doubled and spaced to *permit* the passage of such pipes and shall be bridged. Where plumbing, heating or other pipes are placed in or partly in a partition, necessitating the cutting of the soles or plates, a metal tie not less than 0.058 inch (1.47 mm) (16 galvanized gage) and $1^1/_2$ inches (38 mm) wide shall be fastened to each plate across and to each side of the opening with not less than six 16d nails.

❖ This section requires that the installation of piping in parti-tions and through floors does not compromise the strength or serviceability of the wall framing. This section recognizes that the plumbing trades are not always coordinated with the work of the carpenters so some form of regulation is necessary to guide the installation of plumbing so that it does not adversely affect the performance of a conventionally framed system. If there are piping installations proposed that are different from what is prescribed in this section, an engineering analysis and design will be required.

2308.9.9 Bridging. Unless covered by interior or *exterior wall coverings* or sheathing meeting the minimum requirements of this code, stud partitions or walls with studs having a height-to-least-thickness ratio exceeding 50 shall have bridging not less than 2 inches (51 mm) in thickness and of the same width as the studs fitted snugly and nailed thereto to provide adequate lateral support. Bridging shall be placed in every stud cavity and at a frequency such that no stud so braced shall have a height-to-least-thickness ratio exceeding 50 with the height of the stud measured between horizontal framing and bridging or between bridging, whichever is greater.

❖ In the rare cases where a stud wall will not be covered with a sheathing material specified in the code, solid blocking (bridging) is required at certain intervals. This will enable the studs to act together as a system to maintain stability and prevent buckling. The maximum height-to-least-thickness ratio of 50 is the limit required by the AF&PA NDS for compression members.

The code requires fire-blocking in other sections, which may also be used to meet the bridging requirement stated here.

2308.9.10 Cutting and notching. In exterior walls and bearing partitions, any wood stud is permitted to be cut or notched to a depth not exceeding 25 percent of its width. Cutting or notching of studs to a depth not greater than 40 percent of the width of the stud is permitted in nonbearing partitions supporting no loads other than the weight of the partition.

❖ Studs should not be cut, notched or bored whenever possible to maintain the cross-sectional bearing area. When cuts or notches exceed that specified in Section 2308.9.10, the studs should be doubled or otherwise reinforced to provide the required strength (see Figure 2308.9.10). If overcut, special reinforcing will require an engineered design.

2308.9.11 Bored holes. A hole not greater in diameter than 40 percent of the stud width is permitted to be bored in any wood stud. Bored holes not greater than 60 percent of the width of the stud are permitted in nonbearing partitions or in any wall where each bored stud is doubled, provided not more than two such successive doubled studs are so bored.

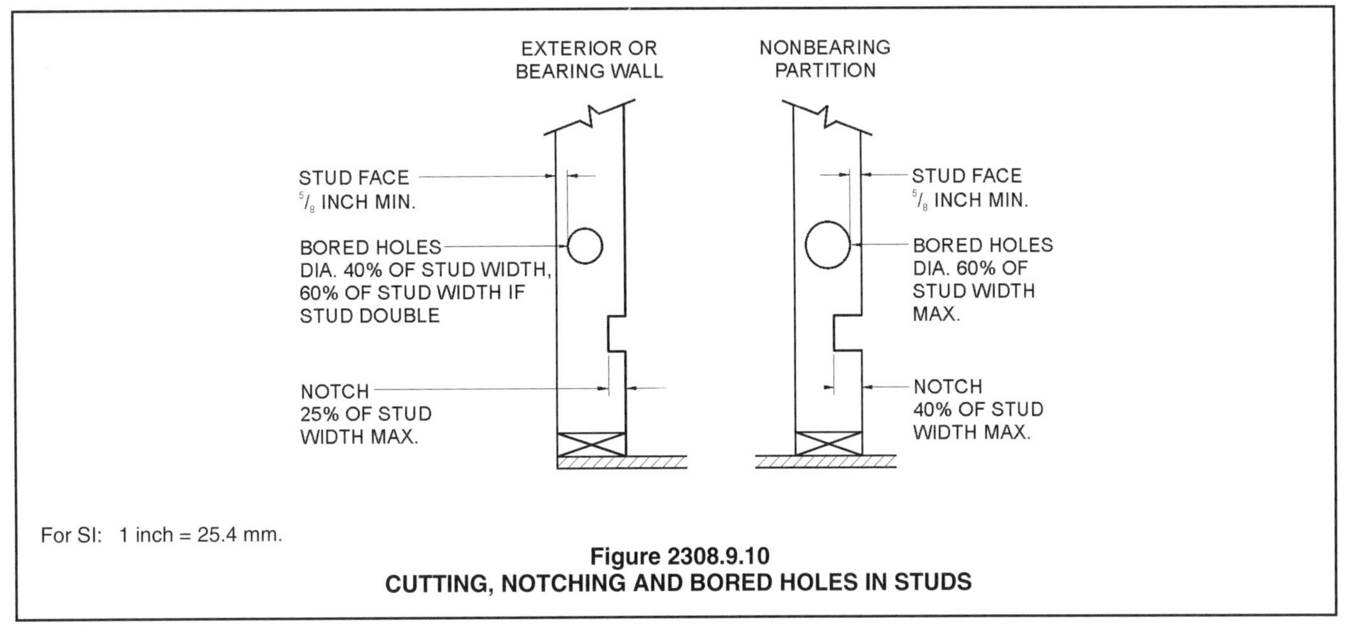

Figure 2308.9.10
CUTTING, NOTCHING AND BORED HOLES IN STUDS

In no case shall the edge of the bored hole be nearer than $^5/_8$ inch (15.9 mm) to the edge of the stud.

Bored holes shall not be located at the same section of stud as a cut or notch.

❖ Due to the redundancy of this type of construction, limited notching and hole boring are allowed (see commentary, Section 2308.9.10). Bored holes must be at least $^5/_8$ inch (15.9 mm) from the edge of the stud and cannot be located at a notch.

2308.10 Roof and ceiling framing. The framing details required in this section apply to roofs having a minimum slope of three units vertical in 12 units horizontal (25-percent slope) or greater. Where the roof slope is less than three units vertical in 12 units horizontal (25-percent slope), members supporting rafters and ceiling joists such as ridge board, hips and valleys shall be designed as beams.

❖ The code intends that the framing details for roofs in Section 2308.10 apply only to roofs having a minimum slope of three units vertical in 12 units horizontal (3:12 or 25-percent slope). For roofs flatter than 3:12, the side thrust at the wall becomes large enough to exceed the capabilities of conventional construction. In that situation, members that support the rafters and ceiling joists, such as the ridge, hips and valleys, must be designed as beams and be supported by exterior or interior bearing walls or posts.

2308.10.1 Wind uplift. The roof construction shall have rafter and truss ties to the wall below. Resultant uplift loads shall be transferred to the foundation using a continuous load path. The rafter or truss to wall connection shall comply with Tables 2304.9.1 and 2308.10.1.

❖ The subject of resisting wind uplift on roofs in locations other than high-wind areas was not addressed in previous legacy model building codes; however, in developing the provisions of Section 2308, it was believed that there was justification for addressing this situation since significant uplift may occur in other areas as well.

This section requires compliance with both Table 2304.9.1, which is the typical fastening schedule, and Table 2308.10.1, which specifies the minimum uplift resistance to be provided between the roof framing and the wall below. Although the toe-nail connections required by Table 2304.9.1 could conceivably provide enough uplift resistance to satisfy Table 2308.10.1 for short roof spans in the lower wind speed areas, in almost all instances, additional uplift resistance is required in the form of approved connectors. When uplift connectors are required by the table, one is required on every rafter or truss to the stud below, assuming the roof framing is spaced 24 inches (610 mm) o.c. in accordance with Note b. Because the table values are based on specific criteria, all of the footnotes should be consulted in order to apply the appropriate adjustments. The intent is to provide a complete and continuous load path from the roof to the foundation.

TABLE 2308.10.1. See page 23-83.

❖ See the commentary to Section 2308.10.1.

2308.10.2 Ceiling joist spans. Allowable spans for ceiling joists shall be in accordance with Table 2308.10.2(1) or 2308.10.2(2). For other grades and species, refer to the *AF&PA Span Tables for Joists and Rafters*.

❖ Allowable spans for ceiling joists are to be in accordance with Table 2308.10.2(1) or 2308.10.2(2) for common species. The tables are for uninhabitable attics without storage (L = 10 psf) and uninhabitable attics with limited storage (L = 20 psf) with a delfection criterion of L/240. For other grades and species, refer to the AF&PA *Span Tables for Joists and Rafters* (see commentary, Section 2308.10.3).

TABLES 2308.10.2(1) and (2). See pages 23-85 and 23-87.

❖ For Tables 2308.10.2(1) and (2), see the commentary to Sections 2308.10.3 and 2304.4.

2308.10.3 Rafter spans. Allowable spans for rafters shall be in accordance with Table 2308.10.3(1), 2308.10.3(2), 2308.10.3(3), 2308.10.3(4), 2308.10.3(5) or 2308.10.3(6). For other grades and species, refer to the *AF&PA Span Tables for Joists and Rafters*.

❖ The spans shown in the referenced tables were derived from the AF&PA *Span Tables for Joists and Rafters*. For simplicity, spans of only the most common species, or types, and grades of wood are shown. When other species or grades of wood are used, the AF&PA tables should be consulted to determine the proper span. Additionally, span tables published by the Southern Forest Products Association (SFPA), the WWPA and the Canadian Wood Council can be consulted. The spans in those tables were derived in the same manner as those published in the AF&PA tables.

Spans are based on species of wood, grade, roof live load or ground snow load, dead load and deflection criteria. With respect to the indicated ground snow loads, the spans shown reflect consideration of unbalanced snow loads, in accordance with established roof snow load criteria. Other considerations such as drifting snow loads are not accounted for in these tables. Where conditions such as a low roof adjacent to a higher roof, projections or roof-mounted equipment are encountered, the need to consider drifting must be assessed in accordance with the snow load criteria in ASCE 7.

TABLES 2308.10.3(1) through (6). See pages 23-89 through 23-99.

❖ For Tables 2308.10.3(1) through (6), see the commentary to Sections 2308.10.3 and 2304.4.

2308.10.4 Ceiling joist and rafter framing. Rafters shall be framed directly opposite each other at the ridge. There shall be a ridge board at least 1-inch (25 mm) nominal thickness at ridges and not less in depth than the cut end of the rafter. At valleys and hips, there shall be a single valley or hip rafter not less than 2-inch (51 mm) nominal thickness and not less in depth than the cut end of the rafter.

❖ Traditional industry practice is to provide a ridge board between opposite rafters to serve as a nailing base and to provide full bearing for the rafter. Rafters must be placed directly opposite each other and the ridge board must have a depth equal to or greater than the cut end of the rafter. The rafter ends must be flush against the ridge board to avoid excessive horizontal shear in the rafters (see Figure 2308.10.4). The ridge board does not act as a beam, but only works in conjunction with thrust-resisting elements to maintain rafter support, which is handled by the ceiling joists and connections thereto (see commentary, Section 2308.10.4.1). Additionally, a prescriptively framed roof system will have a series of purlins and struts that are braced to interior bearing walls that serve to reduce the span of the rafters to the tabulated values.

2308.10.4.1 Ceiling joist and rafter connections. Ceiling joists and rafters shall be nailed to each other and the assembly shall be nailed to the top wall plate in accordance with Tables 2304.9.1 and 2308.10.1. Ceiling joists shall be continuous or securely joined where they meet over interior partitions and fastened to adjacent rafters in accordance with Tables 2308.10.4.1 and 2304.9.1 to provide a continuous rafter tie across the building where such joists are parallel to the rafters. Ceiling joists shall have a bearing surface of not less than $1^1/_2$ inches (38 mm) on the top plate at each end.

Where ceiling joists are not parallel to rafters, an equivalent rafter tie shall be installed in a manner to provide a continuous tie across the building, at a spacing of not more than 4 feet (1219 mm) o.c. The connections shall be in accordance with Tables 2308.10.4.1 and 2304.9.1, or connections of equivalent capacities shall be provided. Where ceiling joists or rafter ties are not provided at the top of the rafter support walls, the ridge formed by these rafters shall also be supported by a girder conforming to Section 2308.4.

Rafter ties shall be spaced not more than 4 feet (1219 mm) o.c. Rafter tie connections shall be based on the equivalent rafter spacing in Table 2308.10.4.1. Where rafter ties are spaced at 32 inches (813 mm) o.c., the number of 16d common nails shall be two times the number specified for rafters spaced 16 inches (406 mm) o.c., with a minimum of four 16d common nails where no snow loads are indicated. Where rafter ties are spaced at 48 inches (1219 mm) o.c., the number of 16d common nails shall be two times the number specified for rafters spaced 24 inches (610 mm) o.c., with a minimum of six 16d common nails where no snow loads are indicated. Rafter/ceiling joist connections and rafter/tie connections

TABLE 2308.10.1
REQUIRED RATING OF APPROVED UPLIFT CONNECTORS (pounds)[a, b, c, e, f, g, h]

NOMINAL DESIGN WIND SPEED, V_{asd}[i]	ROOF SPAN (feet)							OVERHANGS (pounds/feet)[d]
	12	20	24	28	32	36	40	
85	-72	-120	-145	-169	-193	-217	-241	-38.55
90	-91	-151	-181	-212	-242	-272	-302	-43.22
100	-131	-281	-262	-305	-349	-393	-436	-53.36
110	-175	-292	-351	-409	-467	-526	-584	-64.56

	Mean Roof Height (feet)									
EXPOSURE	15	20	25	30	35	40	45	50	55	60
B	1.00	1.00	1.00	1.00	1.05	1.09	1.12	1.16	1.19	1.22
C	1.21	1.29	1.35	1.40	1.45	1.49	1.53	1.56	1.59	1.62
D	1.47	1.55	1.61	1.66	1.70	1.74	1.78	1.81	1.84	1.87

For SI: 1 inch = 25.4 mm, 1 foot = 304.8 mm, 1 mile per hour = 1.61 km/hr, 1 pound = 0.454 Kg, 1 pound/foot = 14.5939 N/m.

a. The uplift connection requirements are based on a 30-foot mean roof height located in Exposure B. For Exposure C or D and for other mean roof heights, multiply the above loads by the adjustment coefficients below.

b. The uplift connection requirements are based on the framing being spaced 24 inches on center. Multiply by 0.67 for framing spaced 16 inches on center and multiply by 0.5 for framing spaced 12 inches on center.

c. The uplift connection requirements include an allowance for 10 pounds of dead load.

d. The uplift connection requirements do not account for the effects of overhangs. The magnitude of the above loads shall be increased by adding the overhang loads found in the table. The overhang loads are also based on framing spaced 24 inches on center. The overhang loads given shall be multiplied by the overhang projection and added to the roof uplift value in the table.

e. The uplift connection requirements are based upon wind loading on end zones as defined in Figure 28.6.3 of ASCE 7. Connection loads for connections located a distance of 20 percent of the least horizontal dimension of the building from the corner of the building are permitted to be reduced by multiplying the table connection value by 0.7 and multiplying the overhang load by 0.8.

f. For wall-to-wall and wall-to-foundation connections, the capacity of the uplift connector is permitted to be reduced by 100 pounds for each full wall above. (For example, if a 500-pound rated connector is used on the roof framing, a 400-pound rated connector is permitted at the next floor level down).

g. Interpolation is permitted for intermediate values of V_{asd} and roof spans.

h. The rated capacity of approved tie-down devices is permitted to include up to a 60-percent increase for wind effects where allowed by material specifications.

i. V_{asd} shall be determined in accordance with Section 1609.3.1.

shall be of sufficient size and number to prevent splitting from nailing.

❖ Upon review of a simple free body diagram, it should be apparent that sloped rafters exert a thrusting force on the walls of buildings in a system that is supported by a ridge board. Therefore, unless the ridge is otherwise supported and designed as a beam, as in post and beam framing, the building must provide horizontal ties to prevent spreading of walls (see Figure 2308.10.4.1). When ceiling joists are parallel to the rafters and adequately connected to the top plate and rafters, the continuous tie requirements of this section are met. If ceiling joists are placed perpendicular to the span of the rafters, the continuity cannot be established as described, so the code simply requires an "equivalent rafter tie" to be installed. Some publications include details or descriptions of acceptable systems that make use of blocking, straps and adequate sheathing to resist these thrusting forces from the rafters. It could be argued that this condition requires an engineered design in order to establish what is an "equivalent" rafter tie.

Ceiling joists are sometimes raised above the wall support of the rafters to provide additional headroom in the space below. Only when such design is accompanied by engineering analysis should this be allowed.

Structural problems are sometimes found in houses having cathedral ceilings. Since no ceiling joists are present to tie the walls together across the building, properly designed beams and posts must be used to support the rafters to satisfy code requirements.

Wind blowing across the roof develops a negative pressure on the rafters. Rafter ties are intended to prevent separation of the rafters at the ridge under such conditions in addition to being used as continuous ties, as explained in this section. Rafter ties are not considered a means of reducing the span of rafters. Purlins and struts braced to bearing walls, as described in Section 2308.10.5, are used to reduce the span of rafters.

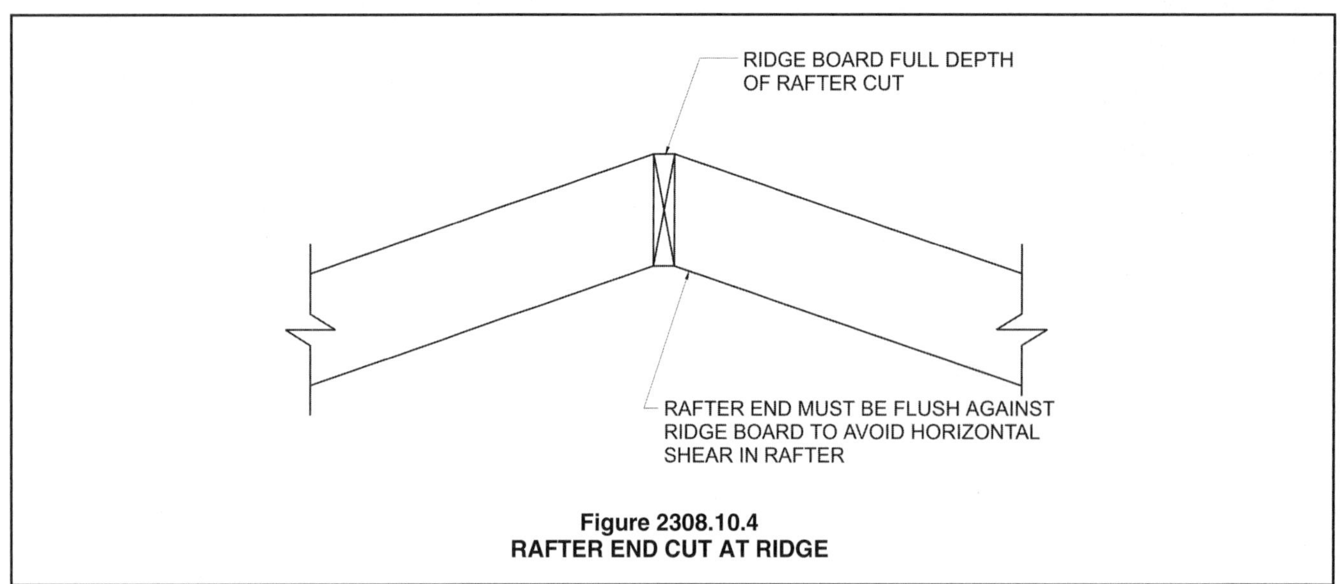

Figure 2308.10.4
RAFTER END CUT AT RIDGE

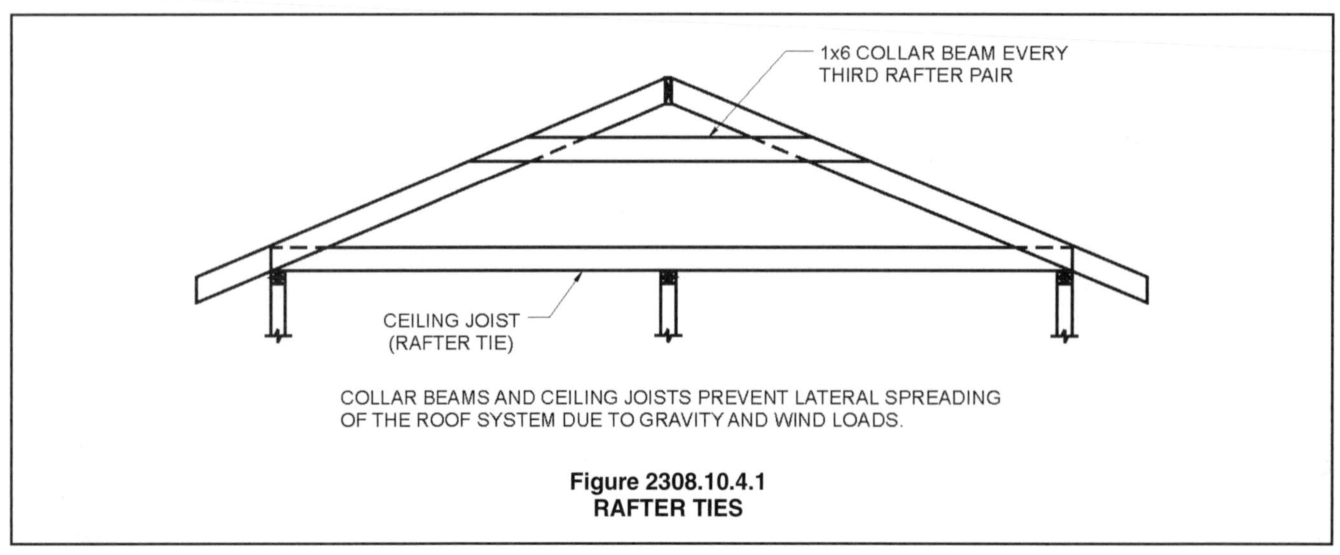

Figure 2308.10.4.1
RAFTER TIES

TABLE 2308.10.2(1)
CEILING JOIST SPANS FOR COMMON LUMBER SPECIES
(Uninhabitable Attics Without Storage, Live Load = 10 pounds psf, L/Δ = 240)

CEILING JOIST SPACING (inches)	SPECIES AND GRADE		DEAD LOAD = 5 pounds per square foot			
			2 × 4	2 × 6	2 × 8	2 × 10
			Maximum ceiling joist spans			
			(ft. - in.)	(ft. - in.)	(ft. - in.)	(ft. - in.)
12	Douglas Fir-Larch	SS	13-2	20-8	26-0	26-0
	Douglas Fir-Larch	#1	12-8	19-11	26-0	26-0
	Douglas Fir-Larch	#2	12-5	19-6	25-8	26-0
	Douglas Fir-Larch	#3	10-10	15-10	20-1	24-6
	Hem-Fir	SS	12-5	19-6	25-8	26-0
	Hem-Fir	#1	12-2	19-1	25-2	26-0
	Hem-Fir	#2	11-7	18-2	24-0	26-0
	Hem-Fir	#3	10-10	15-10	20-1	24-6
	Southern Pine	SS	12-11	20-3	26-0	26-0
	Southern Pine	#1	12-8	19-11	26-0	26-0
	Southern Pine	#2	12-5	19-6	25-8	26-0
	Southern Pine	#3	11-6	17-0	21-8	25-7
	Spruce-Pine-Fir	SS	12-2	19-1	25-2	26-0
	Spruce-Pine-Fir	#1	11-10	18-8	24-7	26-0
	Spruce-Pine-Fir	#2	11-10	18-8	24-7	26-0
	Spruce-Pine-Fir	#3	10-10	15-10	20-1	24-6
16	Douglas Fir-Larch	SS	11-11	18-9	24-8	26-0
	Douglas Fir-Larch	#1	11-6	18-1	23-10	26-0
	Douglas Fir-Larch	#2	11-3	17-8	23-0	26-0
	Douglas Fir-Larch	#3	9-5	13-9	17-5	21-3
	Hem-Fir	SS	11-3	17-8	23-4	26-0
	Hem-Fir	#1	11-0	17-4	22-10	26-0
	Hem-Fir	#2	10-6	16-6	21-9	26-0
	Hem-Fir	#3	9-5	13-9	17-5	21-3
	Southern Pine	SS	11-9	18-5	24-3	26-0
	Southern Pine	#1	11-6	18-1	23-1	26-0
	Southern Pine	#2	11-3	17-8	23-4	26-0
	Southern Pine	#3	10-0	14-9	18-9	22-2
	Spruce-Pine-Fir	SS	11-0	17-4	22-10	26-0
	Spruce-Pine-Fir	#1	10-9	16-11	22-4	26-0
	Spruce-Pine-Fir	#2	10-9	16-11	22-4	26-0
	Spruce-Pine-Fir	#3	9-5	13-9	17-5	21-3

(continued)

TABLE 2308.10.2(1)—continued
CEILING JOIST SPANS FOR COMMON LUMBER SPECIES
(Uninhabitable Attics Without Storage, Live Load = 10 pounds psf, L/Δ = 240)

CEILING JOIST SPACING (inches)	SPECIES AND GRADE		DEAD LOAD = 5 pounds per square foot			
			2 × 4	2 × 6	2 × 8	2 × 10
			Maximum ceiling joist spans			
			(ft. - in.)	(ft. - in.)	(ft. - in.)	(ft. - in.)
19.2	Douglas Fir-Larch	SS	11-3	17-8	23-3	26-0
	Douglas Fir-Larch	#1	10-10	17-0	22-5	26-0
	Douglas Fir-Larch	#2	10-7	16-7	21-0	25-8
	Douglas Fir-Larch	#3	8-7	12-6	15-10	19-5
	Hem-Fir	SS	10-7	16-8	21-11	26-0
	Hem-Fir	#1	10-4	16-4	21-6	26-0
	Hem-Fir	#2	9-11	15-7	20-6	25-3
	Hem-Fir	#3	8-7	12-6	15-10	19-5
	Southern Pine	SS	11-0	17-4	22-10	26-0
	Southern Pine	#1	10-10	17-0	22-5	26-0
	Southern Pine	#2	10-7	16-8	21-11	26-0
	Southern Pine	#3	9-1	13-6	17-2	20-3
	Spruce-Pine-Fir	SS	10-4	16-4	21-6	26-0
	Spruce-Pine-Fir	#1	10-2	15-11	21-0	25-8
	Spruce-Pine-Fir	#2	10-2	15-11	21-0	25-8
	Spruce-Pine-Fir	#3	8-7	12-6	15-10	19-5
24	Douglas Fir-Larch	SS	10-5	16-4	21-7	26-0
	Douglas Fir-Larch	#1	10-0	15-9	20-1	24-6
	Douglas Fir-Larch	#2	9-10	14-10	18-9	22-11
	Douglas Fir-Larch	#3	7-8	11-2	14-2	17-4
	Hem-Fir	SS	9-10	15-6	20-5	26-0
	Hem-Fir	#1	9-8	15-2	19-7	23-11
	Hem-Fir	#2	9-2	14-5	18-6	22-7
	Hem-Fir	#3	7-8	11-2	14-2	17-4
	Southern Pine	SS	10-3	16-1	21-2	26-0
	Southern Pine	#1	10-0	15-9	20-10	26-0
	Southern Pine	#2	9-10	15-6	20-1	23-11
	Southern Pine	#3	8-2	12-0	15-4	18-1
	Spruce-Pine-Fir	SS	9-8	15-2	19-11	25-5
	Spruce-Pine-Fir	#1	9-5	14-9	18-9	22-11
	Spruce-Pine-Fir	#2	9-5	14-9	18-9	22-11
	Spruce-Pine-Fir	#3	7-8	11-2	14-2	17-4

For SI: 1 inch = 25.4 mm, 1 foot = 304.8 mm, 1 pound per square foot = 47.8 N/m^2.

TABLE 2308.10.2(2)
CEILING JOIST SPANS FOR COMMON LUMBER SPECIES
(Uninhabitable Attics With Limited Storage, Live Load = 20 pounds per square foot, L/Δ = 240)

CEILING JOIST SPACING (inches)	SPECIES AND GRADE		DEAD LOAD = 10 pounds per square foot			
			2 × 4	2 × 6	2 × 8	2 × 10
			Maximum ceiling joist spans			
			(ft. - in.)	(ft. - in.)	(ft. - in.)	(ft. - in.)
12	Douglas Fir-Larch	SS	10-5	16-4	21-7	26-0
	Douglas Fir-Larch	#1	10-0	15-9	20-1	24-6
	Douglas Fir-Larch	#2	9-10	14-10	18-9	22-11
	Douglas Fir-Larch	#3	7-8	11-2	14-2	17-4
	Hem-Fir	SS	9-10	15-6	20-5	26-0
	Hem-Fir	#1	9-8	15-2	19-7	23-11
	Hem-Fir	#2	9-2	14-5	18-6	22-7
	Hem-Fir	#3	7-8	11-2	14-2	17-4
	Southern Pine	SS	10-3	16-1	21-2	26-0
	Southern Pine	#1	10-0	15-9	20-10	26-0
	Southern Pine	#2	9-10	15-6	20-1	23-11
	Southern Pine	#3	8-2	12-0	15-4	18-1
	Spruce-Pine-Fir	SS	9-8	15-2	19-11	25-5
	Spruce-Pine-Fir	#1	9-5	14-9	18-9	22-11
	Spruce-Pine-Fir	#2	9-5	14-9	18-9	22-11
	Spruce-Pine-Fir	#3	7-8	11-2	14-2	17-4
16	Douglas Fir-Larch	SS	9-6	14-11	19-7	25-0
	Douglas Fir-Larch	#1	9-1	13-9	17-5	21-3
	Douglas Fir-Larch	#2	8-9	12-10	16-3	19-10
	Douglas Fir-Larch	#3	6-8	9-8	12-4	15-0
	Hem-Fir	SS	8-11	14-1	18-6	23-8
	Hem-Fir	#1	8-9	13-5	16-10	20-8
	Hem-Fir	#2	8-4	12-8	16-0	19-7
	Hem-Fir	#3	6-8	9-8	12-4	15-0
	Southern Pine	SS	9-4	14-7	19-3	24-7
	Southern Pine	#1	9-1	14-4	18-11	23-1
	Southern Pine	#2	8-11	13-6	17-5	20-9
	Southern Pine	#3	7-1	10-5	13-3	15-8
	Spruce-Pine-Fir	SS	8-9	13-9	18-1	23-1
	Spruce-Pine-Fir	#1	8-7	12-10	16-3	19-10
	Spruce-Pine-Fir	#2	8-7	12-10	16-3	19-10
	Spruce-Pine-Fir	#3	6-8	9-8	12-4	15-0

(continued)

TABLE 2308.10.2(2)—continued
CEILING JOIST SPANS FOR COMMON LUMBER SPECIES
(Uninhabitable Attics With Limited Storage, Live Load = 20 pounds per square foot, L/Δ = 240)

CEILING JOIST SPACING (inches)	SPECIES AND GRADE		DEAD LOAD = 10 pounds per square foot			
			2 × 4	2 × 6	2 × 8	2 × 10
			Maximum ceiling joist spans			
			(ft. - in.)	(ft. - in.)	(ft. - in.)	(ft. - in.)
19.2	Douglas Fir-Larch	SS	8-11	14-0	18-5	23-4
	Douglas Fir-Larch	#1	8-7	12-6	15-10	19-5
	Douglas Fir-Larch	#2	8-0	11-9	14-10	18-2
	Douglas Fir-Larch	#3	6-1	8-10	11-3	13-8
	Hem-Fir	SS	8-5	13-3	17-5	22-3
	Hem-Fir	#1	8-3	12-3	15-6	18-11
	Hem-Fir	#2	7-10	11-7	14-8	17-10
	Hem-Fir	#3	6-1	8-10	11-3	13-8
	Southern Pine	SS	8-9	13-9	18-1	23-1
	Southern Pine	#1	8-7	13-6	17-9	21-1
	Southern Pine	#2	8-5	12-3	15-10	18-11
	Southern Pine	#3	6-5	9-6	12-1	14-4
	Spruce-Pine-Fir	SS	8-3	12-11	17-1	21-8
	Spruce-Pine-Fir	#1	8-0	11-9	14-10	18-2
	Spruce-Pine-Fir	#2	8-0	11-9	14-10	18-2
	Spruce-Pine-Fir	#3	6-1	8-10	11-3	13-8
24	Douglas Fir-Larch	SS	8-3	13-0	17-1	20-11
	Douglas Fir-Larch	#1	7-8	11-2	14-2	17-4
	Douglas Fir-Larch	#2	7-2	10-6	13-3	16-3
	Douglas Fir-Larch	#3	5-5	7-11	10-0	12-3
	Hem-Fir	SS	7-10	12-3	16-2	20-6
	Hem-Fir	#1	7-6	10-11	13-10	16-11
	Hem-Fir	#2	7-1	10-4	13-1	16-0
	Hem-Fir	#3	5-5	7-11	10-0	12-3
	Southern Pine	SS	8-1	12-9	16-10	21-6
	Southern Pine	#1	8-0	12-6	15-10	18-10
	Southern Pine	#2	7-8	11-0	14-2	16-11
	Southern Pine	#3	5-9	8-6	10-10	12-10
	Spruce-Pine-Fir	SS	7-8	12-0	15-10	19-5
	Spruce-Pine-Fir	#1	7-2	10-6	13-3	16-3
	Spruce-Pine-Fir	#2	7-2	10-6	13-3	16-3
	Spruce-Pine-Fir	#3	5-5	7-11	10-0	12-3

For SI: 1 inch = 25.4 mm, 1 foot = 304.8 mm, 1 pound per square foot = 47.8 N/m^2.

TABLE 2308.10.3(1)
RAFTER SPANS FOR COMMON LUMBER SPECIES
(Roof Live Load = 20 pounds per square foot, Ceiling Not Attached to Rafters, L/Δ = 180)

RAFTER SPACING (inches)	SPECIES AND GRADE		DEAD LOAD = 10 pounds per square foot					DEAD LOAD = 20 pounds per square foot				
			2 × 4	2 × 6	2 × 8	2 × 10	2 × 12	2 × 4	2 × 6	2 × 8	2 × 10	2 × 12
			Maximum rafter spans									
			(ft. - in.)	(ft. - in.)	(ft. - in.)	(ft. - in.)	(ft. - in.)	(ft. - in.)	(ft. - in.)	(ft. - in.)	(ft. - in.)	(ft. - in.)
12	Douglas Fir-Larch	SS	11-6	18-0	23-9	26-0	26-0	11-6	18-0	23-5	26-0	26-0
	Douglas Fir-Larch	#1	11-1	17-4	22-5	26-0	26-0	10-6	15-4	19-5	23-9	26-0
	Douglas Fir-Larch	#2	10-10	16-7	21-0	25-8	26-0	9-10	14-4	18-2	22-3	25-9
	Douglas Fir-Larch	#3	8-7	12-6	15-10	19-5	22-6	7-5	10-10	13-9	16-9	19-6
	Hem-Fir	SS	10-10	17-0	22-5	26-0	26-0	10-10	17-0	22-5	26-0	26-0
	Hem-Fir	#1	10-7	16-8	21-10	26-0	26-0	10-3	14-11	18-11	23-2	26-0
	Hem-Fir	#2	10-1	15-11	20-8	25-3	26-0	9-8	14-2	17-11	21-11	25-5
	Hem-Fir	#3	8-7	12-6	15-10	19-5	22-6	7-5	10-10	13-9	16-9	19-6
	Southern Pine	SS	11-3	17-8	23-4	26-0	26-0	11-3	17-8	23-4	26-0	26-0
	Southern Pine	#1	11-1	17-4	22-11	26-0	26-0	11-1	17-3	21-9	25-10	26-0
	Southern Pine	#2	10-10	17-0	22-5	26-0	26-0	10-6	15-1	19-5	23-2	26-0
	Southern Pine	#3	9-1	13-6	17-2	20-3	24-1	7-11	11-8	14-10	17-6	20-11
	Spruce-Pine-Fir	SS	10-7	16-8	21-11	26-0	26-0	10-7	16-8	21-9	26-0	26-0
	Spruce-Pine-Fir	#1	10-4	16-3	21-0	25-8	26-0	9-10	14-4	18-2	22-3	25-9
	Spruce-Pine-Fir	#2	10-4	16-3	21-0	25-8	26-0	9-10	14-4	18-2	22-3	25-9
	Spruce-Pine-Fir	#3	8-7	12-6	15-10	19-5	22-6	7-5	10-10	13-9	16-9	19-6
16	Douglas Fir-Larch	SS	10-5	16-4	21-7	26-0	26-0	10-5	16-0	20-3	24-9	26-0
	Douglas Fir-Larch	#1	10-0	15-4	19-5	23-9	26-0	9-1	13-3	16-10	20-7	23-10
	Douglas Fir-Larch	#2	9-10	14-4	18-2	22-3	25-9	8-6	12-5	15-9	19-3	22-4
	Douglas Fir-Larch	#3	7-5	10-10	13-9	16-9	19-6	6-5	9-5	11-11	14-6	16-10
	Hem-Fir	SS	9-10	15-6	20-5	26-0	26-0	9-10	15-6	19-11	24-4	26-0
	Hem-Fir	#1	9-8	14-11	18-11	23-2	26-0	8-10	12-11	16-5	20-0	23-3
	Hem-Fir	#2	9-2	14-2	17-11	21-11	25-5	8-5	12-3	15-6	18-11	22-0
	Hem-Fir	#3	7-5	10-10	13-9	16-9	19-6	6-5	9-5	11-11	14-6	16-10
	Southern Pine	SS	10-3	16-1	21-2	26-0	26-0	10-3	16-1	21-2	26-0	26-0
	Southern Pine	#1	10-0	15-9	20-10	25-10	26-0	10-0	15-0	18-10	22-4	26-0
	Southern Pine	#2	9-10	15-1	19-5	23-2	26-0	9-1	13-0	16-10	20-1	23-7
	Southern Pine	#3	7-11	11-8	14-10	17-6	20-11	6-10	10-1	12-10	15-2	18-1
	Spruce-Pine-Fir	SS	9-8	15-2	19-11	25-5	26-0	9-8	14-10	18-10	23-0	26-0
	Spruce-Pine-Fir	#1	9-5	14-4	18-2	22-3	25-9	8-6	12-5	15-9	19-3	22-4
	Spruce-Pine-Fir	#2	9-5	14-4	18-2	22-3	25-9	8-6	12-5	15-9	19-3	22-4
	Spruce-Pine-Fir	#3	7-5	10-10	13-9	16-9	19-6	6-5	9-5	11-11	14-6	16-10

(continued)

TABLE 2308.10.3(1)—continued
RAFTER SPANS FOR COMMON LUMBER SPECIES
(Roof Live Load = 20 pounds per square foot, Ceiling Not Attached to Rafters, L/Δ = 180)

RAFTER SPACING (inches)	SPECIES AND GRADE		DEAD LOAD = 10 pounds per square foot					DEAD LOAD = 20 pounds per square foot				
			2 × 4	2 × 6	2 × 8	2 × 10	2 × 12	2 × 4	2 × 6	2 × 8	2 × 10	2 × 12
			Maximum rafter spans									
			(ft. - in.)	(ft. - in.)	(ft. - in.)	(ft. - in.)	(ft. - in.)	(ft. - in.)	(ft. - in.)	(ft. - in.)	(ft. - in.)	(ft. - in.)
19.2	Douglas Fir-Larch	SS	9-10	15-5	20-4	25-11	26-0	9-10	14-7	18-6	22-7	26-0
	Douglas Fir-Larch	#1	9-5	14-0	17-9	21-8	25-2	8-4	12-2	15-4	18-9	21-9
	Douglas Fir-Larch	#2	8-11	13-1	16-7	20-3	23-6	7-9	11-4	14-4	17-7	20-4
	Douglas Fir-Larch	#3	6-9	9-11	12-7	15-4	17-9	5-10	8-7	10-10	13-3	15-5
	Hem-Fir	SS	9-3	14-7	19-2	24-6	26-0	9-3	14-4	18-2	22-3	25-9
	Hem-Fir	#1	9-1	13-8	17-4	21-1	24-6	8-1	11-10	15-0	18-4	21-3
	Hem-Fir	#2	8-8	12-11	16-4	20-0	23-2	7-8	11-2	14-2	17-4	20-1
	Hem-Fir	#3	6-9	9-11	12-7	15-4	17-9	5-10	8-7	10-10	13-3	15-5
	Southern Pine	SS	9-8	15-2	19-11	25-5	26-0	9-8	15-2	19-11	25-5	26-0
	Southern Pine	#1	9-5	14-10	19-7	23-7	26-0	9-3	13-8	17-2	20-5	24-4
	Southern Pine	#2	9-3	13-9	17-9	21-2	24-10	8-4	11-11	15-4	18-4	21-6
	Southern Pine	#3	7-3	10-8	13-7	16-0	19-1	6-3	9-3	11-9	13-10	16-6
	Spruce-Pine-Fir	SS	9-1	14-3	18-9	23-11	26-0	9-1	13-7	17-2	21-0	24-4
	Spruce-Pine-Fir	#1	8-10	13-1	16-7	20-3	23-6	7-9	11-4	14-4	17-7	20-4
	Spruce-Pine-Fir	#2	8-10	13-1	16-7	20-3	23-6	7-9	11-4	14-4	17-7	20-4
	Spruce-Pine-Fir	#3	6-9	9-11	12-7	15-4	17-9	5-10	8-7	10-10	13-3	15-5
24	Douglas Fir-Larch	SS	9-1	14-4	18-10	23-4	26-0	8-11	13-1	16-7	20-3	23-5
	Douglas Fir-Larch	#1	8-7	12-6	15-10	19-5	22-6	7-5	10-10	13-9	16-9	19-6
	Douglas Fir-Larch	#2	8-0	11-9	14-10	18-2	21-0	6-11	10-2	12-10	15-8	18-3
	Douglas Fir-Larch	#3	6-1	8-10	11-3	13-8	15-11	5-3	7-8	9-9	11-10	13-9
	Hem-Fir	SS	8-7	13-6	17-10	22-9	26-0	8-7	12-10	16-3	19-10	23-0
	Hem-Fir	#1	8-4	12-3	15-6	18-11	21-11	7-3	10-7	13-5	16-4	19-0
	Hem-Fir	#2	7-11	11-7	14-8	17-10	20-9	6-10	10-0	12-8	15-6	17-11
	Hem-Fir	#3	6-1	8-10	11-3	13-8	15-11	5-3	7-8	9-9	11-10	13-9
	Southern Pine	SS	8-11	14-1	18-6	23-8	26-0	8-11	14-1	18-6	22-11	26-0
	Southern Pine	#1	8-9	13-9	17-9	21-1	25-2	8-3	12-3	15-4	18-3	21-9
	Southern Pine	#2	8-7	12-3	15-10	18-11	22-2	7-5	10-8	13-9	16-5	19-3
	Southern Pine	#3	6-5	9-6	12-1	14-4	17-1	5-7	8-3	10-6	12-5	14-9
	Spruce-Pine-Fir	SS	8-5	13-3	17-5	21-8	25-2	8-4	12-2	15-4	18-9	21-9
	Spruce-Pine-Fir	#1	8-0	11-9	14-10	18-2	21-0	6-11	10-2	12-10	15-8	18-3
	Spruce-Pine-Fir	#2	8-0	11-9	14-10	18-2	21-0	6-11	10-2	12-10	15-8	18-3
	Spruce-Pine-Fir	#3	6-1	8-10	11-3	13-8	15-11	5-3	7-8	9-9	11-10	13-9

For SI: 1 inch = 25.4 mm, 1 foot = 304.8 mm, 1 pound per square foot = 47.9 N/m^2.

TABLE 2308.10.3(2)
RAFTER SPANS FOR COMMON LUMBER SPECIES
(Roof Live Load = 20 pounds per square foot, Ceiling Attached to Rafters, L/Δ = 240)

RAFTER SPACING (inches)	SPECIES AND GRADE		DEAD LOAD = 10 pounds per square foot					DEAD LOAD = 20 pounds per square foot				
			2 × 4	2 × 6	2 × 8	2 × 10	2 × 12	2 × 4	2 × 6	2 × 8	2 × 10	2 × 12
			Maximum rafter spans									
			(ft. - in.)	(ft. - in.)	(ft. - in.)	(ft. - in.)	(ft. - in.)	(ft. - in.)	(ft. - in.)	(ft. - in.)	(ft. - in.)	(ft. - in.)
12	Douglas Fir-Larch	SS	10-5	16-4	21-7	26-0	26-0	10-5	16-4	21-7	26-0	26-0
	Douglas Fir-Larch	#1	10-0	15-9	20-10	26-0	26-0	10-0	15-4	19-5	23-9	26-0
	Douglas Fir-Larch	#2	9-10	15-6	20-5	25-8	26-0	9-10	14-4	18-2	22-3	25-9
	Douglas Fir-Larch	#3	8-7	12-6	15-10	19-5	22-6	7-5	10-10	13-9	16-9	19-6
	Hem-Fir	SS	9-10	15-6	20-5	26-0	26-0	9-10	15-6	20-5	26-0	26-0
	Hem-Fir	#1	9-8	15-2	19-11	25-5	26-0	9-8	14-11	18-11	23-2	26-0
	Hem-Fir	#2	9-2	14-5	19-0	24-3	26-0	9-2	14-2	17-11	21-11	25-5
	Hem-Fir	#3	8-7	12-6	15-10	19-5	22-6	7-5	10-10	13-9	16-9	19-6
	Southern Pine	SS	10-3	16-1	21-2	26-0	26-0	10-3	16-1	21-2	26-0	26-0
	Southern Pine	#1	10-0	15-9	20-10	26-0	26-0	10-0	15-9	20-10	25-10	26-0
	Southern Pine	#2	9-10	15-6	20-5	26-0	26-0	9-10	15-1	19-5	23-2	26-0
	Southern Pine	#3	9-1	13-6	17-2	20-3	24-1	7-11	11-8	14-10	17-6	20-11
	Spruce-Pine-Fir	SS	9-8	15-2	19-11	25-5	26-0	9-8	15-2	19-11	25-5	26-0
	Spruce-Pine-Fir	#1	9-5	14-9	19-6	24-10	26-0	9-5	14-4	18-2	22-3	25-9
	Spruce-Pine-Fir	#2	9-5	14-9	19-6	24-10	26-0	9-5	14-4	18-2	22-3	25-9
	Spruce-Pine-Fir	#3	8-7	12-6	15-10	19-5	22-6	7-5	10-10	13-9	16-9	19-6
16	Douglas Fir-Larch	SS	9-6	14-11	19-7	25-0	26-0	9-6	14-11	19-7	24-9	26-0
	Douglas Fir-Larch	#1	9-1	14-4	18-11	23-9	26-0	9-1	13-3	16-10	20-7	23-10
	Douglas Fir-Larch	#2	8-11	14-1	18-2	22-3	25-9	8-6	12-5	15-9	19-3	22-4
	Douglas Fir-Larch	#3	7-5	10-10	13-9	16-9	19-6	6-5	9-5	11-11	14-6	16-10
	Hem-Fir	SS	8-11	14-1	18-6	23-8	26-0	8-11	14-1	18-6	23-8	26-0
	Hem-Fir	#1	8-9	13-9	18-1	23-1	26-0	8-9	12-11	16-5	20-0	23-3
	Hem-Fir	#2	8-4	13-1	17-3	21-11	25-5	8-4	12-3	15-6	18-11	22-0
	Hem-Fir	#3	7-5	10-10	13-9	16-9	19-6	6-5	9-5	11-11	14-6	16-10
	Southern Pine	SS	9-4	14-7	19-3	24-7	26-0	9-4	14-7	19-3	24-7	26-0
	Southern Pine	#1	9-1	14-4	18-11	24-1	26-0	9-1	14-4	18-10	22-4	26-0
	Southern Pine	#2	8-11	14-1	18-6	23-2	26-0	8-11	13-0	16-10	20-1	23-7
	Southern Pine	#3	7-11	11-8	14-10	17-6	20-11	6-10	10-1	12-10	15-2	18-1
	Spruce-Pine-Fir	SS	8-9	13-9	18-1	23-1	26-0	8-9	13-9	18-1	23-0	26-0
	Spruce-Pine-Fir	#1	8-7	13-5	17-9	22-3	25-9	8-6	12-5	15-9	19-3	22-4
	Spruce-Pine-Fir	#2	8-7	13-5	17-9	22-3	25-9	8-6	12-5	15-9	19-3	22-4
	Spruce-Pine-Fir	#3	7-5	10-10	13-9	16-9	19-6	6-5	9-5	11-11	14-6	16-10

(continued)

TABLE 2308.10.3(2)—continued
RAFTER SPANS FOR COMMON LUMBER SPECIES
(Roof Live Load = 20 pounds per square foot, Ceiling Attached to Rafters, *L*/Δ = 240)

RAFTER SPACING (inches)	SPECIES AND GRADE		DEAD LOAD = 10 pounds per square foot					DEAD LOAD = 20 pounds per square foot				
			2 × 4	2 × 6	2 × 8	2 × 10	2 × 12	2 × 4	2 × 6	2 × 8	2 × 10	2 × 12
			Maximum rafter spans									
			(ft. - in.)	(ft. - in.)	(ft. - in.)	(ft. - in.)	(ft. - in.)	(ft. - in.)	(ft. - in.)	(ft. - in.)	(ft. - in.)	(ft. - in.)
19.2	Douglas Fir-Larch	SS	8-11	14-0	18-5	23-7	26-0	8-11	14-0	18-5	22-7	26-0
	Douglas Fir-Larch	#1	8-7	13-6	17-9	21-8	25-2	8-4	12-2	15-4	18-9	21-9
	Douglas Fir-Larch	#2	8-5	13-1	16-7	20-3	23-6	7-9	11-4	14-4	17-7	20-4
	Douglas Fir-Larch	#3	6-9	9-11	12-7	15-4	17-9	5-10	8-7	10-10	13-3	15-5
	Hem-Fir	SS	8-5	13-3	17-5	22-3	26-0	8-5	13-3	17-5	22-3	25-9
	Hem-Fir	#1	8-3	12-11	17-1	21-1	24-6	8-1	11-10	15-0	18-4	21-3
	Hem-Fir	#2	7-10	12-4	16-3	20-0	23-2	7-8	11-2	14-2	17-4	20-1
	Hem-Fir	#3	6-9	9-11	12-7	15-4	17-9	5-10	8-7	10-10	13-3	15-5
	Southern Pine	SS	8-9	13-9	18-1	23-1	26-0	8-9	13-9	18-1	23-1	26-0
	Southern Pine	#1	8-7	13-6	17-9	22-8	26-0	8-7	13-6	17-2	20-5	24-4
	Southern Pine	#2	8-5	13-3	17-5	21-2	24-10	8-4	11-11	15-4	18-4	21-6
	Southern Pine	#3	7-3	10-8	13-7	16-0	19-1	6-3	9-3	11-9	13-10	16-6
	Spruce-Pine-Fir	SS	8-3	12-11	17-1	21-9	26-0	8-3	12-11	17-1	21-0	24-4
	Spruce-Pine-Fir	#1	8-1	12-8	16-7	20-3	23-6	7-9	11-4	14-4	17-7	20-4
	Spruce-Pine-Fir	#2	8-1	12-8	16-7	20-3	23-6	7-9	11-4	14-4	17-7	20-4
	Spruce-Pine-Fir	#3	6-9	9-11	12-7	15-4	17-9	5-10	8-7	10-10	13-3	15-5
24	Douglas Fir-Larch	SS	8-3	13-0	17-2	21-10	26-0	8-3	13-0	16-7	20-3	23-5
	Douglas Fir-Larch	#1	8-0	12-6	15-10	19-5	22-6	7-5	10-10	13-9	16-9	19-6
	Douglas Fir-Larch	#2	7-10	11-9	14-10	18-2	21-0	6-11	10-2	12-10	15-8	18-3
	Douglas Fir-Larch	#3	6-1	8-10	11-3	13-8	15-11	5-3	7-8	9-9	11-10	13-9
	Hem-Fir	SS	7-10	12-3	16-2	20-8	25-1	7-10	12-3	16-2	19-10	23-0
	Hem-Fir	#1	7-8	12-0	15-6	18-11	21-11	7-3	10-7	13-5	16-4	19-0
	Hem-Fir	#2	7-3	11-5	14-8	17-10	20-9	6-10	10-0	12-8	15-6	17-11
	Hem-Fir	#3	6-1	8-10	11-3	13-8	15-11	5-3	7-8	9-9	11-10	13-9
	Southern Pine	SS	8-1	12-9	16-10	21-6	26-0	8-1	12-9	16-10	21-6	26-0
	Southern Pine	#1	8-0	12-6	16-6	21-1	25-2	8-0	12-3	15-4	18-3	21-9
	Southern Pine	#2	7-10	12-3	15-10	18-11	22-2	7-5	10-8	13-9	16-5	19-3
	Southern Pine	#3	6-5	9-6	12-1	14-4	17-1	5-7	8-3	10-6	12-5	14-9
	Spruce-Pine-Fir	SS	7-8	12-0	15-10	20-2	24-7	7-8	12-0	15-4	18-9	21-9
	Spruce-Pine-Fir	#1	7-6	11-9	14-10	18-2	21-0	6-11	10-2	12-10	15-8	18-3
	Spruce-Pine-Fir	#2	7-6	11-9	14-10	18-2	21-0	6-11	10-2	12-10	15-8	18-3
	Spruce-Pine-Fir	#3	6-1	8-10	11-3	13-8	15-11	5-3	7-8	9-9	11-10	13-9

For SI: 1 inch = 25.4 mm, 1 foot = 304.8 mm, 1 pound per square foot = 47.9 N/m^2.

TABLE 2308.10.3(3)
RAFTER SPANS FOR COMMON LUMBER SPECIES
(Ground Snow Load = 30 pounds per square foot, Ceiling Not Attached to Rafters, L/Δ = 180)

RAFTER SPACING (inches)	SPECIES AND GRADE		DEAD LOAD = 10 pounds per square foot					DEAD LOAD = 20 pounds per square foot				
			2 × 4	2 × 6	2 × 8	2 × 10	2 × 12	2 × 4	2 × 6	2 × 8	2 × 10	2 × 12
			Maximum rafter spans									
			(ft. - in.)	(ft. - in.)	(ft. - in.)	(ft. - in.)	(ft. - in.)	(ft. - in.)	(ft. - in.)	(ft. - in.)	(ft. - in.)	(ft. - in.)
12	Douglas Fir-Larch	SS	10-0	15-9	20-9	26-0	26-0	10-0	15-9	20-1	24-6	26-0
	Douglas Fir-Larch	#1	9-8	14-9	18-8	22-9	26-0	9-0	13-2	16-8	20-4	23-7
	Douglas Fir-Larch	#2	9-5	13-9	17-5	21-4	24-8	8-5	12-4	15-7	19-1	22-1
	Douglas Fir-Larch	#3	7-1	10-5	13-2	16-1	18-8	6-4	9-4	11-9	14-5	16-8
	Hem-Fir	SS	9-6	14-10	19-7	25-0	26-0	9-6	14-10	19-7	24-1	26-0
	Hem-Fir	#1	9-3	14-4	18-2	22-2	25-9	8-9	12-10	16-3	19-10	23-0
	Hem-Fir	#2	8-10	13-7	17-2	21-0	24-4	8-4	12-2	15-4	18-9	21-9
	Hem-Fir	#3	7-1	10-5	13-2	16-1	18-8	6-4	9-4	11-9	14-5	16-8
	Southern Pine	SS	9-10	15-6	20-5	26-0	26-0	9-10	15-6	20-5	26-0	26-0
	Southern Pine	#1	9-8	15-2	20-0	24-9	26-0	9-8	14-10	18-8	22-2	26-0
	Southern Pine	#2	9-6	14-5	18-8	22-3	26-0	9-0	12-11	16-8	19-11	23-4
	Southern Pine	#3	7-7	11-2	14-3	16-10	20-0	6-9	10-0	12-9	15-1	17-11
	Spruce-Pine-Fir	SS	9-3	14-7	19-2	24-6	26-0	9-3	14-7	18-8	22-9	26-0
	Spruce-Pine-Fir	#1	9-1	13-9	17-5	21-4	24-8	8-5	12-4	15-7	19-1	22-1
	Spruce-Pine-Fir	#2	9-1	13-9	17-5	21-4	24-8	8-5	12-4	15-7	19-1	22-1
	Spruce-Pine-Fir	#3	7-1	10-5	13-2	16-1	18-8	6-4	9-4	11-9	14-5	16-8
16	Douglas Fir-Larch	SS	9-1	14-4	18-10	23-9	26-0	9-1	13-9	17-5	21-3	24-8
	Douglas Fir-Larch	#1	8-9	12-9	16-2	19-9	22-10	7-10	11-5	14-5	17-8	20-5
	Douglas Fir-Larch	#2	8-2	11-11	15-1	18-5	21-5	7-3	10-8	13-6	16-6	19-2
	Douglas Fir-Larch	#3	6-2	9-0	11-5	13-11	16-2	5-6	8-1	10-3	12-6	14-6
	Hem-Fir	SS	8-7	13-6	17-10	22-9	26-0	8-7	13-6	17-1	20-10	24-2
	Hem-Fir	#1	8-5	12-5	15-9	19-3	22-3	7-7	11-1	14-1	17-2	19-11
	Hem-Fir	#2	8-0	11-9	14-11	18-2	21-1	7-2	10-6	13-4	16-3	18-10
	Hem-Fir	#3	6-2	9-0	11-5	13-11	16-2	5-6	8-1	10-3	12-6	14-6
	Southern Pine	SS	8-11	14-1	18-6	23-8	26-0	8-11	14-1	18-6	23-8	26-0
	Southern Pine	#1	8-9	13-9	18-1	21-5	25-7	8-8	12-10	16-2	19-2	22-10
	Southern Pine	#2	8-7	12-6	16-2	19-3	22-7	7-10	11-2	14-5	17-3	20-2
	Southern Pine	#3	6-7	9-8	12-4	14-7	17-4	5-10	8-8	11-0	13-0	15-6
	Spruce-Pine-Fir	SS	8-5	13-3	17-5	22-1	25-7	8-5	12-9	16-2	19-9	22-10
	Spruce-Pine-Fir	#1	8-2	11-11	15-1	18-5	21-5	7-3	10-8	13-6	16-6	19-2
	Spruce-Pine-Fir	#2	8-2	11-11	15-1	18-5	21-5	7-3	10-8	13-6	16-6	19-2
	Spruce-Pine-Fir	#3	6-2	9-0	11-5	13-11	16-2	5-6	8-1	10-3	12-6	14-6

(continued)

TABLE 2308.10.3(3)—continued
RAFTER SPANS FOR COMMON LUMBER SPECIES
(Ground Snow Load = 30 pounds per square foot, Ceiling Not Attached to Rafters, L/Δ = 180)

RAFTER SPACING (inches)	SPECIES AND GRADE		DEAD LOAD = 10 pounds per square foot					DEAD LOAD = 20 pounds per square foot				
			2 × 4	2 × 6	2 × 8	2 × 10	2 × 12	2 × 4	2 × 6	2 × 8	2 × 10	2 × 12
			Maximum rafter spans									
			(ft. - in.)	(ft. - in.)	(ft. - in.)	(ft. - in.)	(ft. - in.)	(ft. - in.)	(ft. - in.)	(ft. - in.)	(ft. - in.)	(ft. - in.)
19.2	Douglas Fir-Larch	SS	8-7	13-6	17-9	21-8	25-2	8-7	12-6	15-10	19-5	22-6
	Douglas Fir-Larch	#1	7-11	11-8	14-9	18-0	20-11	7-1	10-5	13-2	16-1	18-8
	Douglas Fir-Larch	#2	7-5	10-11	13-9	16-10	19-6	6-8	9-9	12-4	15-1	17-6
	Douglas Fir-Larch	#3	5-7	8-3	10-5	12-9	14-9	5-0	7-4	9-4	11-5	13-2
	Hem-Fir	SS	8-1	12-9	16-9	21-4	24-8	8-1	12-4	15-7	19-1	22-1
	Hem-Fir	#1	7-9	11-4	14-4	17-7	20-4	6-11	10-2	12-10	15-8	18-2
	Hem-Fir	#2	7-4	10-9	13-7	16-7	19-3	6-7	9-7	12-2	14-10	17-3
	Hem-Fir	#3	5-7	8-3	10-5	12-9	14-9	5-0	7-4	9-4	11-5	13-2
	Southern Pine	SS	8-5	13-3	17-5	22-3	26-0	8-5	13-3	17-5	22-0	25-9
	Southern Pine	#1	8-3	13-0	16-6	19-7	23-4	7-11	11-9	14-9	17-6	20-11
	Southern Pine	#2	7-11	11-5	14-9	17-7	20-7	7-1	10-2	13-2	15-9	18-5
	Southern Pine	#3	6-0	8-10	11-3	13-4	15-10	5-4	7-11	10-1	11-11	14-2
	Spruce-Pine-Fir	SS	7-11	12-5	16-5	20-2	23-4	7-11	11-8	14-9	18-0	20-11
	Spruce-Pine-Fir	#1	7-5	10-11	13-9	16-10	19-6	6-8	9-9	12-4	15-1	17-6
	Spruce-Pine-Fir	#2	7-5	10-11	13-9	16-10	19-6	6-8	9-9	12-4	15-1	17-6
	Spruce-Pine-Fir	#3	5-7	8-3	10-5	12-9	14-9	5-0	7-4	9-4	11-5	13-2
24	Douglas Fir-Larch	SS	7-11	12-6	15-10	19-5	22-6	7-8	11-3	14-2	17-4	20-1
	Douglas Fir-Larch	#1	7-1	10-5	13-2	16-1	18-8	6-4	9-4	11-9	14-5	16-8
	Douglas Fir-Larch	#2	6-8	9-9	12-4	15-1	17-6	5-11	8-8	11-0	13-6	15-7
	Douglas Fir-Larch	#3	5-0	7-4	9-4	11-5	13-2	4-6	6-7	8-4	10-2	11-10
	Hem-Fir	SS	7-6	11-10	15-7	19-1	22-1	7-6	11-0	13-11	17-0	19-9
	Hem-Fir	#1	6-11	10-2	12-10	15-8	18-2	6-2	9-1	11-6	14-0	16-3
	Hem-Fir	#2	6-7	9-7	12-2	14-10	17-3	5-10	8-7	10-10	13-3	15-5
	Hem-Fir	#3	5-0	7-4	9-4	11-5	13-2	4-6	6-7	8-4	10-2	11-10
	Southern Pine	SS	7-10	12-3	16-2	20-8	25-1	7-10	12-3	16-2	19-8	23-0
	Southern Pine	#1	7-8	11-9	14-9	17-6	20-11	7-1	10-6	13-2	15-8	18-8
	Southern Pine	#2	7-1	10-2	13-2	15-9	18-5	6-4	9-2	11-9	14-1	16-6
	Southern Pine	#3	5-4	7-11	10-1	11-11	14-2	4-9	7-1	9-0	10-8	12-8
	Spruce-Pine-Fir	SS	7-4	11-7	14-9	18-0	20-11	7-1	10-5	13-2	16-1	18-8
	Spruce-Pine-Fir	#1	6-8	9-9	12-4	15-1	17-6	5-11	8-8	11-0	13-6	15-7
	Spruce-Pine-Fir	#2	6-8	9-9	12-4	15-1	17-6	5-11	8-8	11-0	13-6	15-7
	Spruce-Pine-Fir	#3	5-0	7-4	9-4	11-5	13-2	4-6	6-7	8-4	10-2	11-10

For SI: 1 inch = 25.4 mm, 1 foot = 304.8 mm, 1 pound per square foot = 47.9 N/m^2.

TABLE 2308.10.3(4)
RAFTER SPANS FOR COMMON LUMBER SPECIES
(Ground Snow Load = 50 pounds per square foot, Ceiling Not Attached to Rafters, L/Δ = 180)

RAFTER SPACING (inches)	SPECIES AND GRADE		DEAD LOAD = 10 pounds per square foot					DEAD LOAD = 20 pounds per square foot				
			2 × 4	2 × 6	2 × 8	2 × 10	2 × 12	2 × 4	2 × 6	2 × 8	2 × 10	2 × 12
			Maximum rafter spans									
			(ft. - in.)	(ft. - in.)	(ft. - in.)	(ft. - in.)	(ft. - in.)	(ft. - in.)	(ft. - in.)	(ft. - in.)	(ft. - in.)	(ft. - in.)
12	Douglas Fir-Larch	SS	8-5	13-3	17-6	22-4	26-0	8-5	13-3	17-0	20-9	24-10
	Douglas Fir-Larch	#1	8-2	12-0	15-3	18-7	21-7	7-7	11-2	14-1	17-3	20-0
	Douglas Fir-Larch	#2	7-8	11-3	14-3	17-5	20-2	7-1	10-5	13-2	16-1	18-8
	Douglas Fir-Larch	#3	5-10	8-6	10-9	13-2	15-3	5-5	7-10	10-0	12-2	14-1
	Hem-Fir	SS	8-0	12-6	16-6	21-1	25-6	8-0	12-6	16-6	20-4	23-7
	Hem-Fir	#1	7-10	11-9	14-10	18-1	21-0	7-5	10-10	13-9	16-9	19-5
	Hem-Fir	#2	7-5	11-1	14-0	17-2	19-11	7-0	10-3	13-0	15-10	18-5
	Hem-Fir	#3	5-10	8-6	10-9	13-2	15-3	5-5	7-10	10-0	12-2	14-1
	Southern Pine	SS	8-4	13-0	17-2	21-11	26-0	8-4	13-0	17-2	21-11	26-0
	Southern Pine	#1	8-2	12-10	16-10	20-3	24-1	8-2	12-6	15-9	18-9	22-4
	Southern Pine	#2	8-0	11-9	15-3	18-2	21-3	7-7	10-11	14-1	16-10	19-9
	Southern Pine	#3	6-2	9-2	11-8	13-9	16-4	5-9	8-5	10-9	12-9	15-2
	Spruce-Pine-Fir	SS	7-10	12-3	16-2	20-8	24-1	7-10	12-3	15-9	19-3	22-4
	Spruce-Pine-Fir	#1	7-8	11-3	14-3	17-5	20-2	7-1	10-5	13-2	16-1	18-8
	Spruce-Pine-Fir	#2	7-8	11-3	14-3	17-5	20-2	7-1	10-5	13-2	16-1	18-8
	Spruce-Pine-Fir	#3	5-10	8-6	10-9	13-2	15-3	5-5	7-10	10-0	12-2	14-1
16	Douglas Fir-Larch	SS	7-8	12-1	15-10	19-5	22-6	7-8	11-7	14-8	17-11	20-10
	Douglas Fir-Larch	#1	7-1	10-5	13-2	16-1	18-8	6-7	9-8	12-2	14-11	17-3
	Douglas Fir-Larch	#2	6-8	9-9	12-4	15-1	17-6	6-2	9-0	11-5	13-11	16-2
	Douglas Fir-Larch	#3	5-0	7-4	9-4	11-5	13-2	4-8	6-10	8-8	10-6	12-3
	Hem-Fir	SS	7-3	11-5	15-0	19-1	22-1	7-3	11-5	14-5	17-8	20-5
	Hem-Fir	#1	6-11	10-2	12-10	15-8	18-2	6-5	9-5	11-11	14-6	16-10
	Hem-Fir	#2	6-7	9-7	12-2	14-10	17-3	6-1	8-11	11-3	13-9	15-11
	Hem-Fir	#3	5-0	7-4	9-4	11-5	13-2	4-8	6-10	8-8	10-6	12-3
	Southern Pine	SS	7-6	11-10	15-7	19-11	24-3	7-6	11-10	15-7	19-11	23-10
	Southern Pine	#1	7-5	11-7	14-9	17-6	20-11	7-4	10-10	13-8	16-2	19-4
	Southern Pine	#2	7-1	10-2	13-2	15-9	18-5	6-7	9-5	12-2	14-7	17-1
	Southern Pine	#3	5-4	7-11	10-1	11-11	14-2	4-11	7-4	9-4	11-0	13-1
	Spruce-Pine-Fir	SS	7-1	11-2	14-8	18-0	20-11	7-1	10-9	13-8	16-8	19-4
	Spruce-Pine-Fir	#1	6-8	9-9	12-4	15-1	17-6	6-2	9-0	11-5	13-11	16-2
	Spruce-Pine-Fir	#2	6-8	9-9	12-4	15-1	17-6	6-2	9-0	11-5	13-11	16-2
	Spruce-Pine-Fir	#3	5-0	7-4	9-4	11-5	13-2	4-8	6-10	8-8	10-6	12-3

(continued)

TABLE 2308.10.3(4)—continued
RAFTER SPANS FOR COMMON LUMBER SPECIES
(Ground Snow Load = 50 pounds per square foot, Ceiling Not Attached to Rafters, L/Δ = 180)

RAFTER SPACING (inches)	SPECIES AND GRADE		DEAD LOAD = 10 pounds per square foot					DEAD LOAD = 20 pounds per square foot				
			2 x 4	2 x 6	2 x 8	2 x 10	2 x 12	2 x 4	2 x 6	2 x 8	2 x 10	2 x 12
			Maximum rafter spans									
			(ft. - in.)	(ft. - in.)	(ft. - in.)	(ft. - in.)	(ft. - in.)	(ft. - in.)	(ft. - in.)	(ft. - in.)	(ft. - in.)	(ft. - in.)
19.2	Douglas Fir-Larch	SS	7-3	11-4	14-6	17-8	20-6	7-3	10-7	13-5	16-5	19-0
	Douglas Fir-Larch	#1	6-6	9-6	12-0	14-8	17-1	6-0	8-10	11-2	13-7	15-9
	Douglas Fir-Larch	#2	6-1	8-11	11-3	13-9	15-11	5-7	8-3	10-5	12-9	14-9
	Douglas Fir-Larch	#3	4-7	6-9	8-6	10-5	12-1	4-3	6-3	7-11	9-7	11-2
	Hem-Fir	SS	6-10	10-9	14-2	17-5	20-2	6-10	10-5	13-2	16-1	18-8
	Hem-Fir	#1	6-4	9-3	11-9	14-4	16-7	5-10	8-7	10-10	13-3	15-5
	Hem-Fir	#2	6-0	8-9	11-1	13-7	15-9	5-7	8-1	10-3	12-7	14-7
	Hem-Fir	#3	4-7	6-9	8-6	10-5	12-1	4-3	6-3	7-11	9-7	11-2
	Southern Pine	SS	7-1	11-2	14-8	18-9	22-10	7-1	11-2	14-8	18-7	21-9
	Southern Pine	#1	7-0	10-8	13-5	16-0	19-1	6-8	9-11	12-5	14-10	17-8
	Southern Pine	#2	6-6	9-4	12-0	14-4	16-10	6-0	8-8	11-2	13-4	15-7
	Southern Pine	#3	4-11	7-3	9-2	10-10	12-11	4-6	6-8	8-6	10-1	12-0
	Spruce-Pine-Fir	SS	6-8	10-6	13-5	16-5	19-1	6-8	9-10	12-5	15-3	17-8
	Spruce-Pine-Fir	#1	6-1	8-11	11-3	13-9	15-11	5-7	8-3	10-5	12-9	14-9
	Spruce-Pine-Fir	#2	6-1	8-11	11-3	13-9	15-11	5-7	8-3	10-5	12-9	14-9
	Spruce-Pine-Fir	#3	4-7	6-9	8-6	10-5	12-1	4-3	6-3	7-11	9-7	11-2
24	Douglas Fir-Larch	SS	6-8	10-3	13-0	15-10	18-4	6-6	9-6	12-0	14-8	17-0
	Douglas Fir-Larch	#1	5-10	8-6	10-9	13-2	15-3	5-5	7-10	10-0	12-2	14-1
	Douglas Fir-Larch	#2	5-5	7-11	10-1	12-4	14-3	5-0	7-4	9-4	11-5	13-2
	Douglas Fir-Larch	#3	4-1	6-0	7-7	9-4	10-9	3-10	5-7	7-1	8-7	10-0
	Hem-Fir	SS	6-4	9-11	12-9	15-7	18-0	6-4	9-4	11-9	14-5	16-8
	Hem-Fir	#1	5-8	8-3	10-6	12-10	14-10	5-3	7-8	9-9	11-10	13-9
	Hem-Fir	#2	5-4	7-10	9-11	12-1	14-1	4-11	7-3	9-2	11-3	13-0
	Hem-Fir	#3	4-1	6-0	7-7	9-4	10-9	3-10	5-7	7-1	8-7	10-0
	Southern Pine	SS	6-7	10-4	13-8	17-5	21-0	6-7	10-4	13-8	16-7	19-5
	Southern Pine	#1	6-5	9-7	12-0	14-4	17-1	6-0	8-10	11-2	13-3	15-9
	Southern Pine	#2	5-10	8-4	10-9	12-10	15-1	5-5	7-9	10-0	11-11	13-11
	Southern Pine	#3	4-4	6-5	8-3	9-9	11-7	4-1	6-0	7-7	9-0	10-8
	Spruce-Pine-Fir	SS	6-2	9-6	12-0	14-8	17-1	6-0	8-10	11-2	13-7	15-9
	Spruce-Pine-Fir	#1	5-5	7-11	10-1	12-4	14-3	5-0	7-4	9-4	11-5	13-2
	Spruce-Pine-Fir	#2	5-5	7-11	10-1	12-4	14-3	5-0	7-4	9-4	11-5	13-2
	Spruce-Pine-Fir	#3	4-1	6-0	7-7	9-4	10-9	3-10	5-7	7-1	8-7	10-0

For SI: 1 inch = 25.4 mm, 1 foot = 304.8 mm, 1 pound per square foot = 47.9 N/m^2.

TABLE 2308.10.3(5)
RAFTER SPANS FOR COMMON LUMBER SPECIES
(Ground Snow Load = 30 pounds per square foot, Ceiling Attached to Rafters, L/Δ = 240)

RAFTER SPACING (inches)	SPECIES AND GRADE		DEAD LOAD = 10 pounds per square foot					DEAD LOAD = 20 pounds per square foot				
			2 × 4	2 × 6	2 × 8	2 × 10	2 × 12	2 × 4	2 × 6	2 × 8	2 × 10	2 × 12
			Maximum rafter spans									
			(ft. - in.)	(ft. - in.)	(ft. - in.)	(ft. - in.)	(ft. - in.)	(ft. - in.)	(ft. - in.)	(ft. - in.)	(ft. - in.)	(ft. - in.)
12	Douglas Fir-Larch	SS	9-1	14-4	18-10	24-1	26-0	9-1	14-4	18-10	24-1	26-0
	Douglas Fir-Larch	#1	8-9	13-9	18-2	22-9	26-0	8-9	13-2	16-8	20-4	23-7
	Douglas Fir-Larch	#2	8-7	13-6	17-5	21-4	24-8	8-5	12-4	15-7	19-1	22-1
	Douglas Fir-Larch	#3	7-1	10-5	13-2	16-1	18-8	6-4	9-4	11-9	14-5	16-8
	Hem-Fir	SS	8-7	13-6	17-10	22-9	26-0	8-7	13-6	17-10	22-9	26-0
	Hem-Fir	#1	8-5	13-3	17-5	22-2	25-9	8-5	12-10	16-3	19-10	23-0
	Hem-Fir	#2	8-0	12-7	16-7	21-0	24-4	8-0	12-2	15-4	18-9	21-9
	Hem-Fir	#3	7-1	10-5	13-2	16-1	18-8	6-4	9-4	11-9	14-5	16-8
	Southern Pine	SS	8-11	14-1	18-6	23-8	26-0	8-11	14-1	18-6	23-8	26-0
	Southern Pine	#1	8-9	13-9	18-2	23-2	26-0	8-9	13-9	18-2	22-2	26-0
	Southern Pine	#2	8-7	13-6	17-10	22-3	26-0	8-7	12-11	16-8	19-11	23-4
	Southern Pine	#3	7-7	11-2	14-3	16-10	20-0	6-9	10-0	12-9	15-1	17-11
	Spruce-Pine-Fir	SS	8-5	13-3	17-5	22-3	26-0	8-5	13-3	17-5	22-3	26-0
	Spruce-Pine-Fir	#1	8-3	12-11	17-0	21-4	24-8	8-3	12-4	15-7	19-1	22-1
	Spruce-Pine-Fir	#2	8-3	12-11	17-0	21-4	24-8	8-3	12-4	15-7	19-1	22-1
	Spruce-Pine-Fir	#3	7-1	10-5	13-2	16-1	18-8	6-4	9-4	11-9	14-5	16-8
16	Douglas Fir-Larch	SS	8-3	13-0	17-2	21-10	26-0	8-3	13-0	17-2	21-3	24-8
	Douglas Fir-Larch	#1	8-0	12-6	16-2	19-9	22-10	7-10	11-5	14-5	17-8	20-5
	Douglas Fir-Larch	#2	7-10	11-11	15-1	18-5	21-5	7-3	10-8	13-6	16-6	19-2
	Douglas Fir-Larch	#3	6-2	9-0	11-5	13-11	16-2	5-6	8-1	10-3	12-6	14-6
	Hem-Fir	SS	7-10	12-3	16-2	20-8	25-1	7-10	12-3	16-2	20-8	24-2
	Hem-Fir	#1	7-8	12-0	15-9	19-3	22-3	7-7	11-1	14-1	17-2	19-11
	Hem-Fir	#2	7-3	11-5	14-11	18-2	21-1	7-2	10-6	13-4	16-3	18-10
	Hem-Fir	#3	6-2	9-0	11-5	13-11	16-2	5-6	8-1	10-3	12-6	14-6
	Southern Pine	SS	8-1	12-9	16-10	21-6	26-0	8-1	12-9	16-10	21-6	26-0
	Southern Pine	#1	8-0	12-6	16-6	21-1	25-7	8-0	12-6	16-2	19-2	22-10
	Southern Pine	#2	7-10	12-3	16-2	19-3	22-7	7-10	11-2	14-5	17-3	20-2
	Southern Pine	#3	6-7	9-8	12-4	14-7	17-4	5-10	8-8	11-0	13-0	15-6
	Spruce-Pine-Fir	SS	7-8	12-0	15-10	20-2	24-7	7-8	12-0	15-10	19-9	22-10
	Spruce-Pine-Fir	#1	7-6	11-9	15-1	18-5	21-5	7-3	10-8	13-6	16-6	19-2
	Spruce-Pine-Fir	#2	7-6	11-9	15-1	18-5	21-5	7-3	10-8	13-6	16-6	19-2
	Spruce-Pine-Fir	#3	6-2	9-0	11-5	13-11	16-2	5-6	8-1	10-3	12-6	14-6

(continued)

TABLE 2308.10.3(5)—continued
RAFTER SPANS FOR COMMON LUMBER SPECIES
(Ground Snow Load = 30 pounds per square foot, Ceiling Attached to Rafters, L/Δ = 240)

RAFTER SPACING (inches)	SPECIES AND GRADE		DEAD LOAD = 10 pounds per square foot					DEAD LOAD = 20 pounds per square foot				
			2 × 4	2 × 6	2 × 8	2 × 10	2 × 12	2 × 4	2 × 6	2 × 8	2 × 10	2 × 12
			Maximum rafter spans									
			(ft. - in.)	(ft. - in.)	(ft. - in.)	(ft. - in.)	(ft. - in.)	(ft. - in.)	(ft. - in.)	(ft. - in.)	(ft. - in.)	(ft. - in.)
19.2	Douglas Fir-Larch	SS	7-9	12-3	16-1	20-7	25-0	7-9	12-3	15-10	19-5	22-6
	Douglas Fir-Larch	#1	7-6	11-8	14-9	18-0	20-11	7-1	10-5	13-2	16-1	18-8
	Douglas Fir-Larch	#2	7-4	10-11	13-9	16-10	19-6	6-8	9-9	12-4	15-1	17-6
	Douglas Fir-Larch	#3	5-7	8-3	10-5	12-9	14-9	5-0	7-4	9-4	11-5	13-2
	Hem-Fir	SS	7-4	11-7	15-3	19-5	23-7	7-4	11-7	15-3	19-1	22-1
	Hem-Fir	#1	7-2	11-4	14-4	17-7	20-4	6-11	10-2	12-10	15-8	18-2
	Hem-Fir	#2	6-10	10-9	13-7	16-7	19-3	6-7	9-7	12-2	14-10	17-3
	Hem-Fir	#3	5-7	8-3	10-5	12-9	14-9	5-0	7-4	9-4	11-5	13-2
	Southern Pine	SS	7-8	12-0	15-10	20-2	24-7	7-8	12-0	15-10	20-2	24-7
	Southern Pine	#1	7-6	11-9	15-6	19-7	23-4	7-6	11-9	14-9	17-6	20-11
	Southern Pine	#2	7-4	11-5	14-9	17-7	20-7	7-1	10-2	13-2	15-9	18-5
	Southern Pine	#3	6-0	8-10	11-3	13-4	15-10	5-4	7-11	10-1	11-11	14-2
	Spruce-Pine-Fir	SS	7-2	11-4	14-11	19-0	23-1	7-2	11-4	14-9	18-0	20-11
	Spruce-Pine-Fir	#1	7-0	10-11	13-9	16-10	19-6	6-8	9-9	12-4	15-1	17-6
	Spruce-Pine-Fir	#2	7-0	10-11	13-9	16-10	19-6	6-8	9-9	12-4	15-1	17-6
	Spruce-Pine-Fir	#3	5-7	8-3	10-5	12-9	14-9	5-0	7-4	9-4	11-5	13-2
24	Douglas Fir-Larch	SS	7-3	11-4	15-0	19-1	22-6	7-3	11-3	14-2	17-4	20-1
	Douglas Fir-Larch	#1	7-0	10-5	13-2	16-1	18-8	6-4	9-4	11-9	14-5	16-8
	Douglas Fir-Larch	#2	6-8	9-9	12-4	15-1	17-6	5-11	8-8	11-0	13-6	15-7
	Douglas Fir-Larch	#3	5-0	7-4	9-4	11-5	13-2	4-6	6-7	8-4	10-2	11-10
	Hem-Fir	SS	6-10	10-9	14-2	18-0	21-11	6-10	10-9	13-11	17-0	19-9
	Hem-Fir	#1	6-8	10-2	12-10	15-8	18-2	6-2	9-1	11-6	14-0	16-3
	Hem-Fir	#2	6-4	9-7	12-2	14-10	17-3	5-10	8-7	10-10	13-3	15-5
	Hem-Fir	#3	5-0	7-4	9-4	11-5	13-2	4-6	6-7	8-4	10-2	11-10
	Southern Pine	SS	7-1	11-2	14-8	18-9	22-10	7-1	11-2	14-8	18-9	22-10
	Southern Pine	#1	7-0	10-11	14-5	17-6	20-11	7-0	10-6	13-2	15-8	18-8
	Southern Pine	#2	6-10	10-2	13-2	15-9	18-5	6-4	9-2	11-9	14-1	16-6
	Southern Pine	#3	5-4	7-11	10-1	11-11	14-2	4-9	7-1	9-0	10-8	12-8
	Spruce-Pine-Fir	SS	6-8	10-6	13-10	17-8	20-11	6-8	10-5	13-2	16-1	18-8
	Spruce-Pine-Fir	#1	6-6	9-9	12-4	15-1	17-6	5-11	8-8	11-0	13-6	15-7
	Spruce-Pine-Fir	#2	6-6	9-9	12-4	15-1	17-6	5-11	8-8	11-0	13-6	15-7
	Spruce-Pine-Fir	#3	5-0	7-4	9-4	11-5	13-2	4-6	6-7	8-4	10-2	11-10

For SI: 1 inch = 25.4 mm, 1 foot = 304.8 mm, 1 pound per square foot = 47.9 N/m^2.

TABLE 2308.10.3(6)
RAFTER SPANS FOR COMMON LUMBER SPECIES
(Ground Snow Load = 50 pounds per square foot, Ceiling Attached to Rafters, L/Δ = 240)

RAFTER SPACING (inches)	SPECIES AND GRADE		DEAD LOAD = 10 pounds per square foot					DEAD LOAD = 20 pounds per square foot				
			2 × 4	2 × 6	2 × 8	2 × 10	2 × 12	2 × 4	2 × 6	2 × 8	2 × 10	2 × 12
			Maximum rafter spans									
			(ft. - in.)	(ft. - in.)	(ft. - in.)	(ft. - in.)	(ft. - in.)	(ft. - in.)	(ft. - in.)	(ft. - in.)	(ft. - in.)	(ft. - in.)
12	Douglas Fir-Larch	SS	7-8	12-1	15-11	20-3	24-8	7-8	12-1	15-11	20-3	24-0
	Douglas Fir-Larch	#1	7-5	11-7	15-3	18-7	21-7	7-5	11-2	14-1	17-3	20-0
	Douglas Fir-Larch	#2	7-3	11-3	14-3	17-5	20-2	7-1	10-5	13-2	16-1	18-8
	Douglas Fir-Larch	#3	5-10	8-6	10-9	13-2	15-3	5-5	7-10	10-0	12-2	14-1
	Hem-Fir	SS	7-3	11-5	15-0	19-2	23-4	7-3	11-5	15-0	19-2	23-4
	Hem-Fir	#1	7-1	11-2	14-8	18-1	21-0	7-1	10-10	13-9	16-9	19-5
	Hem-Fir	#2	6-9	10-8	14-0	17-2	19-11	6-9	10-3	13-0	15-10	18-5
	Hem-Fir	#3	5-10	8-6	10-9	13-2	15-3	5-5	7-10	10-0	12-2	14-1
	Southern Pine	SS	7-6	11-0	15-7	19-11	24-3	7-6	11-10	15-7	19-11	24-3
	Southern Pine	#1	7-5	11-7	15-4	19-7	23-9	7-5	11-7	15-4	18-9	22-4
	Southern Pine	#2	7-3	11-5	15-0	18-2	21-3	7-3	10-11	14-1	16-10	19-9
	Southern Pine	#3	6-2	9-2	11-8	13-9	16-4	5-9	8-5	10-9	12-9	15-2
	Spruce-Pine-Fir	SS	7-1	11-2	14-8	18-9	22-10	7-1	11-2	14-8	18-9	22-4
	Spruce-Pine-Fir	#1	6-11	10-11	14-3	17-5	20-2	6-11	10-5	13-2	16-1	18-8
	Spruce-Pine-Fir	#2	6-11	10-11	14-3	17-5	20-2	6-11	10-5	13-2	16-1	18-8
	Spruce-Pine-Fir	#3	5-10	8-6	10-9	13-2	15-3	5-5	7-10	10-0	12-2	14-1
16	Douglas Fir-Larch	SS	7-0	11-0	14-5	18-5	22-5	7-0	11-0	14-5	17-11	20-10
	Douglas Fir-Larch	#1	6-9	10-5	13-2	16-1	18-8	6-7	9-8	12-2	14-11	17-3
	Douglas Fir-Larch	#2	6-7	9-9	12-4	15-1	17-6	6-2	9-0	11-5	13-11	16-2
	Douglas Fir-Larch	#3	5-0	7-4	9-4	11-5	13-2	4-8	6-10	8-8	10-6	12-3
	Hem-Fir	SS	6-7	10-4	13-8	17-5	21-2	6-7	10-4	13-8	17-5	20-5
	Hem-Fir	#1	6-5	10-2	12-10	15-8	18-2	6-5	9-5	11-11	14-6	16-10
	Hem-Fir	#2	6-2	9-7	12-2	14-10	17-3	6-1	8-11	11-3	13-9	15-11
	Hem-Fir	#3	5-0	7-4	9-4	11-5	13-2	4-8	6-10	8-8	10-6	12-3
	Southern Pine	SS	6-10	10-9	14-2	18-1	22-0	6-10	10-9	14-2	18-1	22-0
	Southern Pine	#1	6-9	10-7	13-11	17-6	20-11	6-9	10-7	13-8	16-2	19-4
	Southern Pine	#2	6-7	10-2	13-2	15-9	18-5	6-7	9-5	12-2	14-7	17-1
	Southern Pine	#3	5-4	7-11	10-1	11-11	14-2	4-11	7-4	9-4	11-0	13-1
	Spruce-Pine-Fir	SS	6-5	10-2	13-4	17-0	20-9	6-5	10-2	13-4	16-8	19-4
	Spruce-Pine-Fir	#1	6-4	9-9	12-4	15-1	17-6	6-2	9-0	11-5	13-11	16-2
	Spruce-Pine-Fir	#2	6-4	9-9	12-4	15-1	17-6	6-2	9-0	11-5	13-11	16-2
	Spruce-Pine-Fir	#3	5-0	7-4	9-4	11-5	13-2	4-8	6-10	8-8	10-6	12-3

(continued)

TABLE 2308.10.3(6)—continued
RAFTER SPANS FOR COMMON LUMBER SPECIES
(Ground Snow Load = 50 pounds per square foot, Ceiling Attached to Rafters, L/Δ = 240)

RAFTER SPACING (inches)	SPECIES AND GRADE		DEAD LOAD = 10 pounds per square foot					DEAD LOAD = 20 pounds per square foot				
			2 × 4	2 × 6	2 × 8	2 × 10	2 × 12	2 × 4	2 × 6	2 × 8	2 × 10	2 × 12
			Maximum rafter spans									
			(ft. - in.)	(ft. - in.)	(ft. - in.)	(ft. - in.)	(ft. - in.)	(ft. - in.)	(ft. - in.)	(ft. - in.)	(ft. - in.)	(ft. - in.)
19.2	Douglas Fir-Larch	SS	6-7	10-4	13-7	17-4	20-6	6-7	10-4	13-5	16-5	19-0
	Douglas Fir-Larch	#1	6-4	9-6	12-0	14-8	17-1	6-0	8-10	11-2	13-7	15-9
	Douglas Fir-Larch	#2	6-1	8-11	11-3	13-9	15-11	5-7	8-3	10-5	12-9	14-9
	Douglas Fir-Larch	#3	4-7	6-9	8-6	10-5	12-1	4-3	6-3	7-11	9-7	11-2
	Hem-Fir	SS	6-2	9-9	12-10	16-5	19-11	6-2	9-9	12-10	16-1	18-8
	Hem-Fir	#1	6-1	9-3	11-9	14-4	16-7	5-10	8-7	10-10	13-3	15-5
	Hem-Fir	#2	5-9	8-9	11-1	13-7	15-9	5-7	8-1	10-3	12-7	14-7
	Hem-Fir	#3	4-7	6-9	8-6	10-5	12-1	4-3	6-3	7-11	9-7	11-2
	Southern Pine	SS	6-5	10-2	13-4	17-0	20-9	6-5	10-2	13-4	17-0	20-9
	Southern Pine	#1	6-4	9-11	13-1	16-0	19-1	6-4	9-11	12-5	14-10	17-8
	Southern Pine	#2	6-2	9-4	12-0	14-4	16-10	6-0	8-8	11-2	13-4	15-7
	Southern Pine	#3	4-11	7-3	9-2	10-10	12-11	4-6	6-8	8-6	10-1	12-0
	Spruce-Pine-Fir	SS	6-1	9-6	12-7	16-0	19-1	6-1	9-6	12-5	15-3	17-8
	Spruce-Pine-Fir	#1	5-11	8-11	11-3	13-9	15-11	5-7	8-3	10-5	12-9	14-9
	Spruce-Pine-Fir	#2	5-11	8-11	11-3	13-9	15-11	5-7	8-3	10-5	12-9	14-9
	Spruce-Pine-Fir	#3	4-7	6-9	8-6	10-5	12-1	4-3	6-3	7-11	9-7	11-2
24	Douglas Fir-Larch	SS	6-1	9-7	12-7	15-10	18-4	6-1	9-6	12-0	14-8	17-0
	Douglas Fir-Larch	#1	5-10	8-6	10-9	13-2	15-3	5-5	7-10	10-0	12-2	14-1
	Douglas Fir-Larch	#2	5-5	7-11	10-1	12-4	14-3	5-0	7-4	9-4	11-5	13-2
	Douglas Fir-Larch	#3	4-1	6-0	7-7	9-4	10-9	3-10	5-7	7-1	8-7	10-0
	Hem-Fir	SS	5-9	9-1	11-11	15-12	18-0	5-9	9-1	11-9	14-5	16-8
	Hem-Fir	#1	5-8	8-3	10-6	12-10	14-10	5-3	7-8	9-9	11-10	13-9
	Hem-Fir	#2	5-4	7-10	9-11	12-1	14-1	4-11	7-3	9-2	11-3	13-0
	Hem-Fir	#3	4-1	6-0	7-7	9-4	10-9	3-10	5-7	7-1	8-7	10-0
	Southern Pine	SS	6-0	9-5	12-5	15-10	19-3	6-0	9-5	12-5	15-10	19-3
	Southern Pine	#1	5-10	9-3	12-0	14-4	17-1	5-10	8-10	11-2	13-3	15-9
	Southern Pine	#2	5-9	8-4	10-9	12-10	15-1	5-5	7-9	10-0	11-11	13-11
	Southern Pine	#3	4-4	6-5	8-3	9-9	11-7	4-1	6-0	7-7	9-0	10-8
	Spruce-Pine-Fir	SS	5-8	8-10	11-8	14-8	17-1	5-8	8-10	11-2	13-7	15-9
	Spruce-Pine-Fir	#1	5-5	7-11	10-1	12-4	14-3	5-0	7-4	9-4	11-5	13-2
	Spruce-Pine-Fir	#2	5-5	7-11	10-1	12-4	14-3	5-0	7-4	9-4	11-5	13-2
	Spruce-Pine-Fir	#3	4-1	6-0	7-7	9-4	10-9	3-10	5-7	7-1	8-7	10-0

For SI: 1 inch = 25.4 mm, 1 foot = 304.8 mm, 1 pound per square foot = 47.9 N/m^2.

TABLE 2308.10.4.1. See page 23-102.

❖ See the commentary to Section 2308.10.4.1.

2308.10.4.2 Notches and holes. Notching at the ends of rafters or ceiling joists shall not exceed one-fourth the depth. Notches in the top or bottom of the rafter or ceiling joist shall not exceed one-sixth the depth and shall not be located in the middle one-third of the span, except that a notch not exceeding one-third of the depth is permitted in the top of the rafter or ceiling joist not further from the face of the support than the depth of the member.

Holes bored in rafters or ceiling joists shall not be within 2 inches (51 mm) of the top and bottom and their diameter shall not exceed one-third the depth of the member.

❖ Wood is comprised of relatively small elongated cells oriented parallel to each other and bound together physically and chemically by lignin. When a rafter is notched or cut, the cell strands are interrupted and the loss in strength needs to be accounted for. This section of the code defines where notches and holes may occur in rafters and ceiling joists without having to perform an engineering analysis on the wood member.

Some designs and installation practices require that limited notching and cutting occur. Although it is permissible to engineer notches into wood framing, notching should be avoided whenever possible. Holes bored in rafters create the same problems as notches and have a more significant effect on the strength of the member if they are located near the top or bottom edge. When necessary, the holes should be located in areas with the least stress concentration, generally along the neutral axis of the rafter. Beams subject to high horizontal shear stress (short span/heavy load) should not be cut. Wood members having a thickness of over 4 inches (102 mm) should not be notched, except at the ends of the members.

2308.10.4.3 Framing around openings. Trimmer and header rafters shall be doubled, or of lumber of equivalent cross section, where the span of the header exceeds 4 feet (1219 mm). The ends of header rafters more than 6 feet (1829 mm) long shall be supported by framing anchors or rafter hangers unless bearing on a beam, partition or wall.

❖ The requirements for framing openings in roofs are similar to the requirements for openings in floors. See the commentary to Figure 2308.8.3 for an illustration of the criteria for framing around openings.

2308.10.5 Purlins. Purlins to support roof loads are permitted to be installed to reduce the span of rafters within allowable limits and shall be supported by struts to bearing walls. The maximum span of 2-inch by 4-inch (51 mm by 102 mm) purlins shall be 4 feet (1219 mm). The maximum span of the 2-inch by 6-inch (51 mm by 152 mm) purlin shall be 6 feet (1829 mm), but in no case shall the purlin be smaller than the supported rafter. Struts shall not be smaller than 2-inch by 4-inch (51 mm by 102 mm) members. The unbraced length of struts shall not exceed 8 feet (2438 mm) and the minimum slope of the struts shall not be less than 45 degrees (0.79 rad) from the horizontal.

❖ This section permits the use of purlins and struts braced to bearing walls to reduce the span of rafters. Where the roof slope is less than three units vertical in 12 units horizontal (3:12 or 25-percent slope), the code requires that members supporting rafters and ceiling joists, such as the ridge, hips and valleys be designed as beams. The section provides prescriptive rules for minimum size and maximum span of purlins and the minimum size and maximum unbraced lengths of struts. The minimum slope of struts is 45 degrees (0.79 rad) from the horizontal (see Figure 2308.10.5).

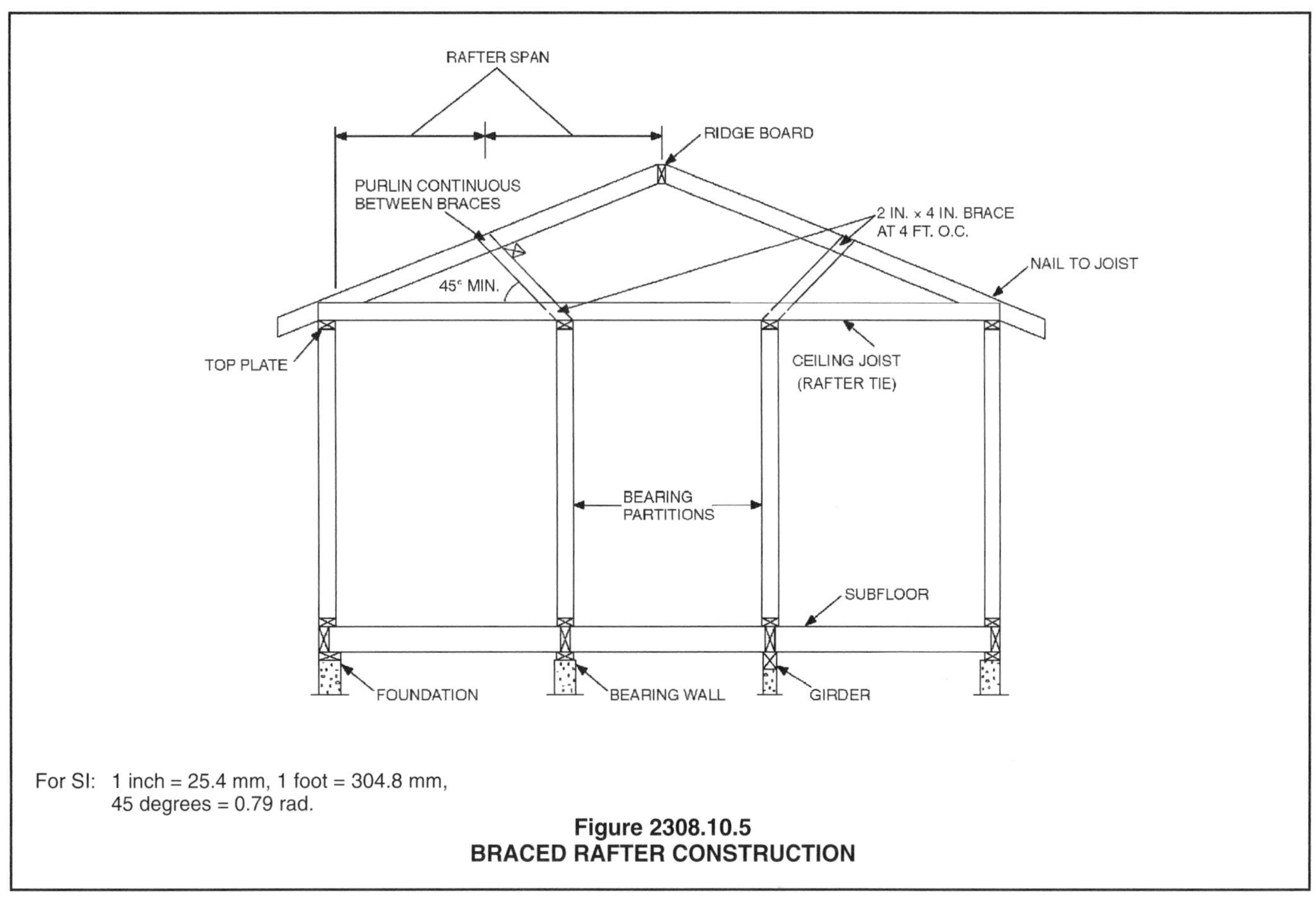

For SI: 1 inch = 25.4 mm, 1 foot = 304.8 mm, 45 degrees = 0.79 rad.

Figure 2308.10.5
BRACED RAFTER CONSTRUCTION

TABLE 2308.10.4.1
RAFTER TIE CONNECTIONS[g]

RAFTER SLOPE	TIE SPACING (inches)	NO SNOW LOAD				GROUND SNOW LOAD (pound per square foot)							
						30 pounds per square foot				50 pounds per square foot			
						Roof span (feet)							
		12	20	28	36	12	20	28	36	12	20	28	36
		Required number of 16d common ($3^{1}/_{2}''$ x 0.162″) nails[a, b] per connection[c, d, e, f]											
3:12	12	4	6	8	10	4	6	8	11	5	8	12	15
	16	5	7	10	13	5	8	11	14	6	11	15	20
	24	7	11	15	19	7	11	16	21	9	16	23	30
	32	10	14	19	25	10	16	22	28	12	27	30	40
	48	14	21	29	37	14	32	36	42	18	32	46	60
4:12	12	3	4	5	6	3	5	6	8	4	6	9	11
	16	3	5	7	8	4	6	8	11	5	8	12	15
	24	4	7	10	12	5	9	12	16	7	12	17	22
	32	6	9	13	16	8	12	16	22	10	16	24	30
	48	8	14	19	24	10	18	24	32	14	24	34	44
5:12	12	3	3	4	5	3	4	5	7	3	5	7	9
	16	3	4	5	7	3	5	7	9	4	7	9	12
	24	4	6	8	10	4	7	10	13	6	10	14	18
	32	5	8	10	13	6	10	14	18	8	14	18	24
	48	7	11	15	20	8	14	20	26	12	20	28	36
7:12	12	3	3	3	4	3	3	4	5	3	4	5	7
	16	3	3	4	5	3	4	5	6	3	5	7	9
	24	3	4	6	7	3	5	7	9	4	7	10	13
	32	4	6	8	10	4	8	10	12	6	10	14	18
	48	5	8	11	14	6	10	14	18	9	14	20	26
9:12	12	3	3	3	3	3	3	3	4	3	3	4	5
	16	3	3	3	4	3	3	4	5	3	4	5	7
	24	3	3	5	6	3	4	6	7	3	6	8	10
	32	3	4	6	8	4	6	8	10	5	8	10	14
	48	4	6	9	11	5	8	12	14	7	12	16	20
12:12	12	3	3	3	3	3	3	3	3	3	3	3	4
	16	3	3	3	3	3	3	3	4	3	3	4	5
	24	3	3	3	4	3	3	4	6	3	4	6	8
	32	3	3	4	5	3	5	6	8	4	6	8	10
	48	3	4	6	7	4	7	8	12	6	8	12	16

For SI: 1 inch = 25.4 mm, 1 foot = 304.8 mm, 1 pound per square foot = 47.8 N/m^2.

a. 40d box (5″ × 0.162″) or 16d sinker ($3^{1}/_{4}''$ × 0.148″) nails are permitted to be substituted for 16d common ($3^{1}/_{2}''$ × 0.16″) nails.

b. Nailing requirements are permitted to be reduced 25 percent if nails are clinched.

c. Rafter tie heel joint connections are not required where the ridge is supported by a load-bearing wall, header or ridge beam.

d. When intermediate support of the rafter is provided by vertical struts or purlins to a load-bearing wall, the tabulated heel joint connection requirements are permitted to be reduced proportionally to the reduction in span.

e. Equivalent nailing patterns are required for ceiling joist to ceiling joist lap splices.

f. Connected members shall be of sufficient size to prevent splitting due to nailing.

g. For snow loads less than 30 pounds per square foot, the required number of nails is permitted to be reduced by multiplying by the ratio of actual snow load plus 10 divided by 40, but not less than the number required for no snow load.

2308.10.6 Blocking. Roof rafters and ceiling joists shall be supported laterally to prevent rotation and lateral displacement in accordance with the provisions of Section 2308.8.5.

❖ Solid blocking must be provided at the ends of rafters to prevent twisting or rolling over, as required by Section 2308.8.5. There are other common ways of providing resistance to this type of action, but solid blocking is always preferred since it also can be used to provide a nominal amount of lateral force transfer from the roof sheathing to the wall or beam below.

2308.10.7 Engineered wood products. Prefabricated wood I-joists, structural glued-laminated timber and structural composite lumber shall not be notched or drilled except where permitted by the manufacturer's recommendations or where the effects of such alterations are specifically considered in the design of the member by a *registered design professional*.

❖ The notching and boring allowances that are permitted for solid-sawn joist framing cannot be applied to engineered wood products. Because these types of products are commonly incorporated into conventional construction, this section (as does Section 2308.8.8.2.1) points out that they must be treated differently. Engineered wood products should not be cut, notched or bored unless those alterations are considered in the design of the member. This section makes it clear that any alterations to an engineered wood product must be considered in the design of the member. The consideration is to be made either by the manufacturer and reflected in use recommendations (which is common in I-joists, permitting some limited openings in webs) or by a registered design professional.

2308.10.8 Roof sheathing. Roof sheathing shall be in accordance with Tables 2304.7(3) and 2304.7(5) for wood structural panels, and Tables 2304.7(1) and 2304.7(2) for lumber and shall comply with Section 2304.7.2.

❖ See the commentary to Section 2304.7.2.

2308.10.8.1 Joints. Joints in lumber sheathing shall occur over supports unless *approved* end-matched lumber is used, in which case each piece shall bear on at least two supports.

❖ When lumber sheathing is used, the ends of the boards must be located over the joists or rafters. Such support is not required when end-matched tongue-and-groove lumber is used or when the ends are otherwise prevented from moving relative to each other.

2308.10.9 Roof planking. Planking shall be designed in accordance with the general provisions of this code.

In lieu of such design, 2-inch (51 mm) tongue-and-groove planking is permitted in accordance with Table 2308.10.9. Joints in such planking are permitted to be randomly spaced, provided the system is applied to not less than three continuous spans, planks are center matched and end matched or splined, each plank bears on at least one support, and joints are separated by at least 24 inches (610 mm) in adjacent pieces.

❖ Lumber planking on roofs must conform to the requirements of Section 2304.7.2 and Table 2304.7(1) for typical lumber roof sheathing. This section provides an alternative using 2-inch (51 mm) tongue-and-groove planking.

TABLE 2308.10.9. See page 23-104.

❖ Table 2308.10.9 gives allowable span (spacing of roof supports) based on the live load and deflection limits (from the code) and the design values for the planking used.

2308.10.10 Wood trusses. Wood trusses shall be designed in accordance with Section 2303.4.

❖ Engineered wood trusses are to be designed in accordance with Section 2303.4 (see the commentary to that section). Most modern trusses are gang-nailed metal plate trusses, but they could also be custom designed with bolted members. They are engineered elements that can be incorporated into a building of otherwise conventional light-frame construction in accordance with Section 2308.4.

2308.10.11 Attic ventilation. For *attic* ventilation, see Section 1203.2.

❖ Large amounts of water vapor migrate by movement of air carrying water or by diffusion through the building envelope materials because of a vapor pressure difference. The sources of water vapor include cooking, laundering, bathing and breathing and perspiration of persons. These can account for an average daily production of 25 pounds (11.3 kg) of water vapor for a family of four in a typical dwelling. This average can be much higher where appliances, such as humidifiers, washers and dryers, are used.

As the vapor moves into the attic, it may reach its dew point, thus condensing on wood roof components. This wetting and drying action will cause rotting and decay. To avoid this, the attic must be ventilated to prevent the accumulation of water on building components. The installation of a vapor retarder acts to prevent the passage of moisture to the attic. An effective vapor retarder allows for a decrease in ventilation requirements. Vapor retarders are ineffective when openings in the barrier are allowed where moisture can be carried by air into the attic. Care should be exercised so that attic vent openings remain unobstructed. Attic ventilation is also required for the removal of excess heat. Access to attic spaces is required by Section 1208.2.

Truss manufacturers may provide special blocking that contains openings along the centerline and a galvanized screen to be shipped with their product that will effectively meet the attic ventilation requirements at the eaves. Structural calculations have shown that the strength of this blocking with openings is n reduced significantly but allows it to effectively transfer forces from the roof diaphragm to the wall below.

2308.11 Additional requirements for conventional construction in Seismic Design Category B or C. Structures of *conventional light-frame construction* and assigned to *Seismic Design Category* B or C shall comply with Sections 2308.11.1 through 2308.11.3, in addition to the provisions of Sections 2308.1 through 2308.10.

❖ This section places additional construction limits on structures where the seismic risk is considered moderate. Additional limitations are imposed on conventional structures in high seismic regions in Section 2308.12. Many of the seismic requirements and limitations on conventional construction are based on FEMA 450, Part 1.

2308.11.1 Number of stories. Structures of *conventional light-frame construction* and assigned to *Seismic Design Category* C shall not exceed two stories above grade plane.

❖ This section limits conventional construction to two stories in height in Seismic Design Category C because of amplified dynamic effects in taller structures.

2308.11.2 Concrete or masonry. Concrete or masonry walls and stone or masonry veneer shall not extend above a basement.

Exceptions:

1. In structures assigned to *Seismic Design Category* B, stone and masonry veneer is permitted to be used in the first two stories above grade plane or the first three stories above grade plane where the lowest story has concrete or masonry walls, provided that structural use panel wall bracing is used and the length of bracing provided is one- and one-half times the required length as determined in Table 2308.9.3(1).
2. In structures assigned to *Seismic Design Category* B or C, stone and masonry veneer is permitted to be used in the first story above grade plane or the first two stories above grade plane where the lowest story has concrete or masonry walls.

TABLE 2308.10.9
ALLOWABLE SPANS FOR 2-INCH TONGUE-AND-GROOVE DECKING

SPAN[a] (feet)	LIVE LOAD (pound per square foot)	DEFLECTION LIMIT	BENDING STRESS (f) (pound per square inch)	MODULUS OF ELASTICITY (E) (pound per square inch)
Roofs				
4	20	1/240 1/360	160	170,000 256,000
	30	1/240 1/360	210	256,000 384,000
	40	1/240 1/360	270	340,000 512,000
4.5	20	1/240 1/360	200	242,000 305,000
	30	1/240 1/360	270	363,000 405,000
	40	1/240 1/360	350	484,000 725,000
5.0	20	1/240 1/360	250	332,000 500,000
	30	1/240 1/360	330	495,000 742,000
	40	1/240 1/360	420	660,000 1,000,000
5.5	20	1/240 1/360	300	442,000 660,000
	30	1/240 1/360	400	662,000 998,000
	40	1/240 1/360	500	884,000 1,330,000
6.0	20	1/240 1/360	360	575,000 862,000
	30	1/240 1/360	480	862,000 1,295,000
	40	1/240 1/360	600	1,150,000 1,730,000

(continued)

3. In structures assigned to *Seismic Design Category* B or C, stone and masonry veneer is permitted to be used in both stories of buildings with two stories above grade plane, provided the following criteria are met:

   3.1. Type of brace per Section 2308.9.3 shall be Method 3 and the allowable shear capacity in accordance with Section 2306.3 shall be a minimum of 350 plf (5108 N/m).

   3.2. Braced wall panels in the second *story* shall be located in accordance with Section 2308.9.3 and not more than 25 feet (7620 mm) on center, and the total length of braced wall panels shall be not less than 25 percent of the braced wall line length. Braced wall panels in the first *story* shall be located in accordance with Section 2308.9.3 and not more than 25 feet (7620 mm) on center, and the total length of braced wall panels shall be not less than 45 percent of the braced wall line length.

   3.3. Hold-down connectors shall be provided at the ends of each braced wall panel for the second *story* to first *story* connection with an allowable capacity of 2,000 pounds (8896 N). Hold-down connectors shall be provided at the ends of each braced wall panel for the first story to foundation connection with an allowable capacity of 3,900 pounds (17 347 N). In all cases, the hold-down connector force shall be transferred to the foundation.

   3.4. Cripple walls shall not be permitted.

❖ Because of their weight, concrete walls, masonry walls and stone or masonry veneer can impose lateral loads from seismic ground motion events that are beyond the typical loading considered for conventional light-frame construction. For this reason, the code requires an engineered design for concrete or masonry walls above the basement with some exceptions. Exception 1 allows masonry veneer either two or three stories above grade plane in structures classified as Seismic Design Category B under certain conditions, such as increasing the amount of wall bracing. It would not be necessary to meet these conditions for veneer on the first story only, since Exception 2 permits this. Exception 2 allows one story of stone or masonry veneer above grade plane and two stories above grade plane if the lower story walls are

TABLE 2308.10.9—continued
ALLOWABLE SPANS FOR 2-INCH TONGUE-AND-GROOVE DECKING

SPAN[a] (feet)	LIVE LOAD (pound per square foot)	DEFLECTION LIMIT	BENDING STRESS (f) (pound per square inch)	MODULUS OF ELASTICITY (E) (pound per square inch)
Roofs				
6.5	20	1/240 1/360	420	595,000 892,000
	30	1/240 1/360	560	892,000 1,340,000
	40	1/240 1/360	700	1,190,000 1,730,000
7.0	20	1/240 1/360	490	910,000 1,360,000
	30	1/240 1/360	650	1,370,000 2,000,000
	40	1/240 1/360	810	1,820,000 2,725,000
7.5	20	1/240 1/360	560	1,125,000 1,685,000
	30	1/240 1/360	750	1,685,000 2,530,000
	40	1/240 1/360	930	2,250,000 3,380,000
8.0	20	1/240 1/360	640	1,360,000 2,040,000
	30	1/240 1/360	850	2,040,000 3,060,000
Floors				
4 4.5 5.0	40	1/360	840 950 1,060	1,000,000 1,300,000 1,600,000

For SI: 1 inch = 25.4 mm, 1 foot = 304.8 mm, 1 pound per square foot = 0.0479 kN/m^2, 1 pound per square inch = 0.00689 N/mm^2.

a. Spans are based on simple beam action with 10 pounds per square foot dead load and provisions for a 300-pound concentrated load on a 12-inch width of decking. Random layup is permitted in accordance with the provisions of Section 2308.10.9. Lumber thickness is $1^1/_2$ inches nominal.

constructed of concrete or masonry, which are engineered for seismic effects. Exception 3 allows up to two stories of stone or masonry veneer above grade plane, provided the added requirements for wall bracing and hold-downs are complied with in order to increase the resistance of the lateral-force-resisting system.

2308.11.3 Framing and connection details. Framing and connection details shall conform to Sections 2308.11.3.1 through 2308.11.3.3.

❖ Sections 2308.11.3.1 through 2308.11.3.3 contain specific connection requirements for conventional wood-framed buildings in Seismic Design Category B or C.

2308.11.3.1 Anchorage. Braced wall lines shall be anchored in accordance with Section 2308.6 at foundations.

❖ The section refers to the typical anchorage requirements in Section 2308.6. Braced wall lines must be anchored to the foundation so that they will not slide due to shear or overturn because of a moment created by lateral loads. Anchorage for overturning will nearly always be more critical for the individual braced walls or shear panels than for the entire wall, though the effectiveness of the entire system cannot be ignored. Conventional construction assumes an entire system that is adequately tied together by all of the requirements noted throughout the code provides a complete load path for imposed forces.

2308.11.3.2 Stepped footings. Where the height of a required braced wall panel extending from foundation to floor above varies more than 4 feet (1219 mm), the following construction shall be used:

1. Where the bottom of the footing is stepped and the lowest floor framing rests directly on a sill bolted to the footings, the sill shall be anchored as required in Section 2308.3.3.
2. Where the lowest floor framing rests directly on a sill bolted to a footing not less than 8 feet (2438 mm) in length along a line of bracing, the line shall be considered to be braced. The double plate of the cripple stud wall beyond the segment of footing extending to the lowest framed floor shall be spliced to the sill plate with metal ties, one on each side of the sill and plate. The metal ties shall not be less than 0.058 inch [1.47 mm (16 galvanized gage)] by $1^1/_2$ inches (38 mm) wide by 48 inches (1219 mm) with eight 16d common nails on each side of the splice location (see Figure 2308.11.3.2). The metal tie shall have a minimum yield of 33,000 pounds per square inch (psi) (227 MPa).
3. Where cripple walls occur between the top of the footing and the lowest floor framing, the bracing requirements for a *story* shall apply.

❖ The primary objectives of footing design are to provide a level surface for construction of the foundation wall; to provide adequate transfer and distribution of the building loads to the underlying soil; to provide adequate strength to prevent differential settlement of the building and to provide adequate anchorage or mass to resist potential uplift and overturning forces from lateral loads imposed on the structure. The most common footing type in residential construction is a continuous concrete spread footing. These concrete footings are recognized in prescriptive footing size tables for most typical conditions. In contrast, special conditions such as stepped footings, which might be required in steeply sloped sites, require special consideration and possibly an engineered design. This section gives prescriptive requirements for constructing stepped foundations where the height of a braced wall panel extending from the foundation to the floor above varies by more than 4 feet (1219 mm).

FIGURE 2308.11.3.2. See page 23-107.

❖ This figure illustrates the code requirements for stepped footing connection details required by Section 2308.11.3.2, Item 2 (see Section 2308.11.3.2 for further discussion). Care should be taken to install the metal strap shown so that the nailing does not interfere with anchor bolts at the sill plate. Additionally, nails used to fasten the strap to pressure-treated plates must be galvanized or otherwise protected as described in Section 2304.9.5.

2308.11.3.3 Openings in horizontal diaphragms. Openings in horizontal diaphragms with a dimension perpendicular to the joist that is greater than 4 feet (1219 mm) shall be constructed in accordance with the following:

1. Blocking shall be provided beyond headers.
2. Metal ties not less than 0.058 inch [1.47 mm (16 galvanized gage)] by $1^1/_2$ inches (38 mm) wide with eight 16d common nails on each side of the header-joist intersection shall be provided (see Figure 2308.11.3.3). The metal ties shall have a minimum yield of 33,000 psi (227 MPa).

❖ Horizontal diaphragms are floor and roof assemblies that are usually clad with wood structural panel sheathing, such as plywood or OSB. Diaphragms support out-of-plane walls and distribute lateral forces to the braced wall lines or shear walls. Though more complicated and difficult to visualize, lateral forces that are applied to a building from wind or seismic ground motion events follow a load path that distributes and transfers shear and overturning forces from the lateral loads. When openings are built into the diaphragm, they disrupt the continuity of load across the diaphragm, requiring the opening to be reinforced to compensate. Another concern is the stiffness of the diaphragm with an opening in it. These provisions are a prescriptive solution to strengthen openings greater than 4 feet (1219 mm) in dimension perpendicular to the joists and provide a general means for a load path in these specific cases in lieu of requiring an engineered design.

FIGURE 2308.11.3.3. See page 23-107.

❖ See the commentary to Section 2308.11.3.3.

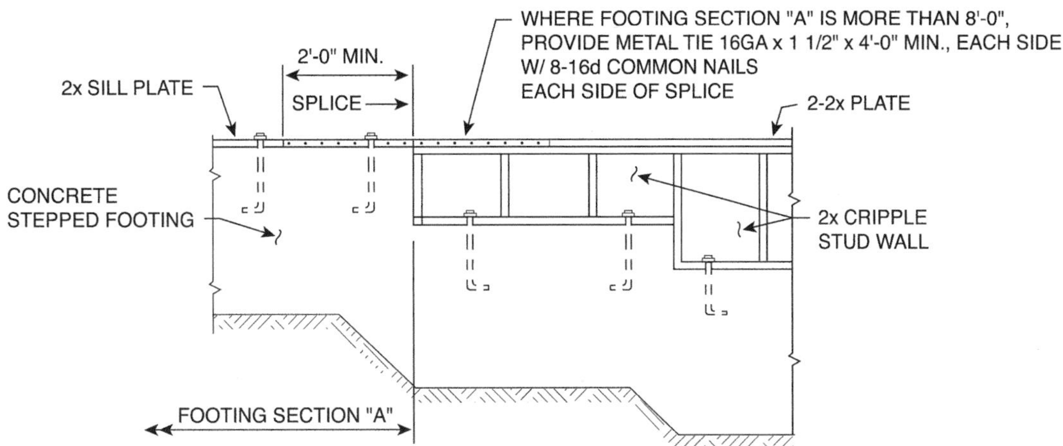

For SI: 1 inch = 25.4 mm, 1 foot = 304.8 mm.

FIGURE 2308.11.3.2
STEPPED FOOTING CONNECTION DETAILS

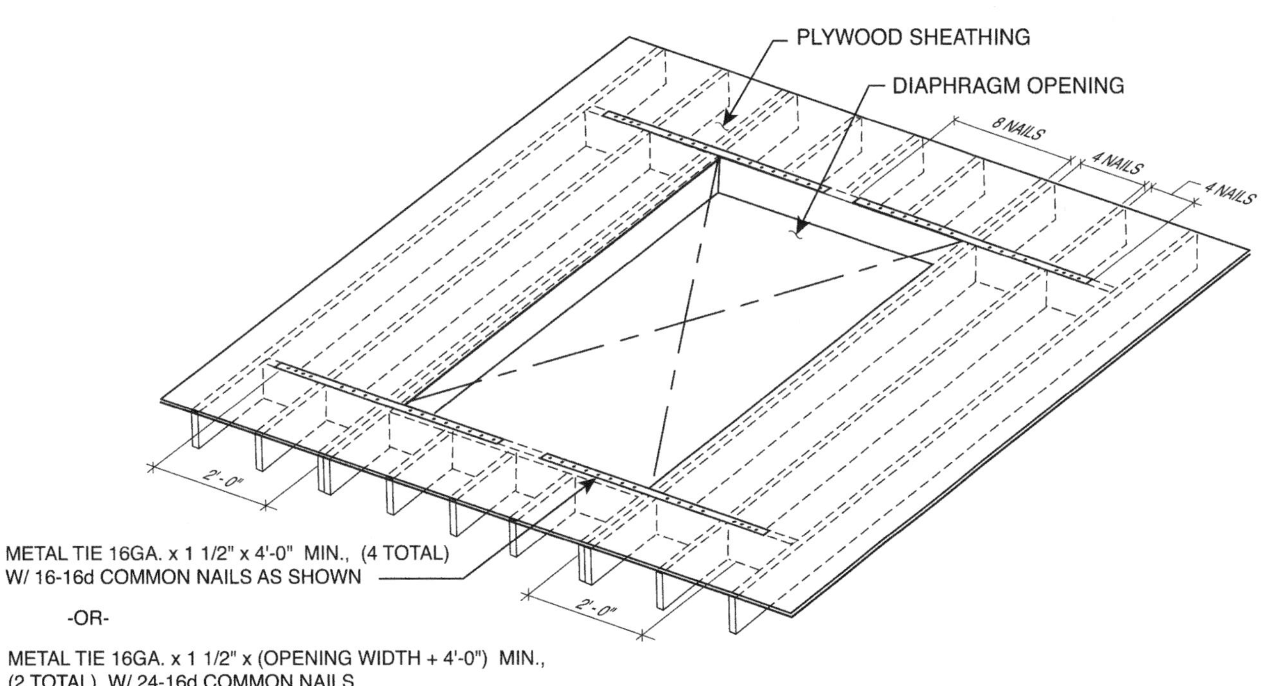

For SI: 1 inch = 25.4 mm, 1 foot = 304.8 mm.

FIGURE 2308.11.3.3
OPENINGS IN HORIZONTAL DIAPHRAGMS

2308.12 Additional requirements for conventional construction in Seismic Design Category D or E. Structures of conventional light-frame construction and assigned to *Seismic Design Category* D or E shall conform to Sections 2308.12.1 through 2308.12.9, in addition to the requirements for structures assigned to *Seismic Design Category* B or C in Section 2308.11.

❖ Conventional construction is a type of construction that has been used successfully for many years and relies on standard practice as governed by prescriptive building code requirements. For conditions where the seismic lateral loading on the building is higher, such as in structures that are assigned to Seismic Design Categories D and E, additional prescriptive requirements or limitations are considered necessary. The basis for the seismic requirements for Seismic Design Category D or E is FEMA 450.

2308.12.1 Number of stories. Structures of *conventional light-frame construction* and assigned to *Seismic Design Category* D or E shall not exceed one story above grade plane.

❖ Similar to buildings classified as Seismic Design Category C, conventional wood-framed structures in areas having a higher seismic risk need to be limited in height due to historic observations of the effects of severe earthquakes. The section limits the height of conventional wood-framed structures to one story in Seismic Design Categories D and E. Structural failures were observed in the aftermath of the 1994 Northridge, California, earthquake of multistory residential buildings that were sheathed with the weakest of materials allowed under the prescriptive methods. As a result of these observations, newer regulations for conventional construction have included more restrictive height limits and a more restrictive placement of braced wall lines and braced wall panels.

2308.12.2 Concrete or masonry. Concrete or masonry walls and stone or masonry veneer shall not extend above a basement.

Exception: In structures assigned to *Seismic Design Category* D, stone and masonry veneer is permitted to be used in the first story above grade plane, provided the following criteria are met:

1. Type of brace in accordance with Section 2308.9.3 shall be Method 3 and the allowable shear capacity in accordance with Section 2306.3 shall be a minimum of 350 plf (5108 N/m).
2. The bracing of the first *story* shall be located at each end and at least every 25 feet (7620 mm) o.c. but not less than 45 percent of the braced wall line.
3. Hold-down connectors shall be provided at the ends of braced walls for the first floor to foundation with an allowable capacity of 2,100 pounds (9341 N).
4. Cripple walls shall not be permitted.

❖ Because of their weight, concrete walls, masonry walls and stone or masonry veneer can impose lateral loads from seismic events in excess of the loading considered acceptable for conventional light-frame construction. For this reason, the code prohibits the use of concrete or masonry walls above the basement. The use of stone or masonry veneer one story above grade plane is permitted for structures classified as Seismic Design Category D in accordance with the exception, provided the requirements for increased wall bracing and hold-downs at braced wall panels are met.

2308.12.3 Braced wall line spacing. Spacing between interior and exterior braced wall lines shall not exceed 25 feet (7620 mm).

❖ This provision limits the spacing between lines of bracing elements to 25 feet (7620 mm) o.c. This controls the lateral forces supported by the elements by limiting the tributary loading to the braced wall lines. A series of braced wall panels comprise a braced wall line. A two-dimensional grid of interior and exterior braced wall lines breaks up the building into a series of boxes, as illustrated in Figure 2308.9.3. The size of the box is either 25 or 35 feet (7620 or 10 668 mm), depending on seismic design category and wind speed. The 25-foot (7620 mm) dimension is related to the maximum framing spans, which are included in other sections of this chapter. For additional information on the development and philosophy of the conventional construction method, see the commentary to Section 2308.1.

2308.12.4 Braced wall line sheathing. Braced wall lines shall be braced by one of the types of sheathing prescribed by Table 2308.12.4 as shown in Figure 2308.9.3. The sum of lengths of braced wall panels at each braced wall line shall conform to the required percentage of wall length required to be braced per braced wall line in Table 2308.12.4. Braced wall panels shall be distributed along the length of the braced wall line and start at not more than 8 feet (2438 mm) from each end of the braced wall line. Panel sheathing joints shall occur over studs or blocking. Sheathing shall be fastened to studs, top and bottom plates and at panel edges occurring over blocking. Wall framing to which sheathing used for bracing is applied shall be nominal 2-inch-wide [actual $1^1/_2$ inch (38 mm)] or larger members.

Cripple walls having a stud height exceeding 14 inches (356 mm) shall be considered a *story* for the purpose of this section and shall be braced as required for braced wall lines in accordance with the required percentage of wall length required to be braced per braced wall line in Table 2308.12.4. Where interior braced wall lines occur without a continuous foundation below, the length of parallel exterior cripple wall bracing shall be one and one-half times the lengths required by Table 2308.12.4. Where the cripple wall sheathing type used is Type S-W and this additional length of bracing cannot be provided, the capacity of Type S-W sheathing shall be increased by reducing the spacing of fasteners along the perimeter of each piece of sheathing to 4 inches (102 mm) o.c.

❖ Braced wall lines in conventional wood-framed buildings in Seismic Design Categories D and E are required to contain braced wall panels sheathed with a minimum percentage of bracing and placed in accordance with one of the methods specified in Table 2308.12.4, as illustrated in Figure 2308.9.3. All vertical joints of the panel are required to occur over studs and horizontal joints are required to occur over blocking. A braced wall panel shall occur within 8 feet (2438 mm) of each end of a braced wall line. If the first braced wall panel must occur beyond 8 feet (2438 mm) from the end of a braced wall line, then that line should be designed as a shear wall line. The minimum length of panel bracing given in the table is based on one face sheathed for Type S-W sheathing, both faces sheathed for Type G-P sheathing and an *h/w* ratio not greater than 2:1 (see Note a for more specifics). See Note c for fastening requirements for Type G-P sheathing methods.

TABLE 2308.12.4. See below.

❖ See the commentary to Section 2308.12.4.

2308.12.4.1 Alternative bracing. An alternate braced wall panel constructed in accordance with Section 2308.9.3.1 or 2308.9.3.2 is permitted to be substituted for a braced wall panel in Section 2308.9.3 Items 2 through 8. For methods 2, 3, 4, 6, 7 and 8, each 48-inch (1219 mm) section or portion thereof required by Table 2308.12.4 is permitted to be replaced by one alternate braced wall panel constructed in accordance with Section 2308.9.3.1 or 2308.9.3.2. For method 5, each 96-inch (2438 mm) section (applied to one face) or 48-inch (1219 mm) section (applied to both faces) or portion thereof required by Table 2308.12.4 is permitted to be replaced by one alternate braced wall panel constructed in accordance with Section 2308.9.3.1 or 2308.9.3.2.

❖ This section clarifies that alternate braced wall panels (Section 2308.9.3.1) and alternate braced wall panels adjacent to an opening (Section 2308.9.3.2) are permitted to be substituted for braced wall panels (other than let-in braces) as described in Section 2308.9.3. In this case, each 4-foot (1219 mm) section required by Table 2308.12.4 is permitted to be replaced by one alternate braced wall panel or one alternate braced wall panel adjacent to an opening. For Method 5, each 8-foot (2438 mm) section (when applied to one face) or 4-foot (1219 mm) section (when applied to both faces) required by Table 2308.12.4 is permitted to be replaced by one alternate braced wall panel or one alternate braced wall panel adjacent to an opening. The alternate panels are to be constructed in accordance with Section 2308.9.3.1 or 2308.9.3.2.

2308.12.5 Attachment of sheathing. Fastening of braced wall panel sheathing shall not be less than that prescribed in Table 2308.12.4 or 2304.9.1. Wall sheathing shall not be attached to framing members by adhesives.

❖ The nailing required by Table 2308.12.4 is specified in Note c. The connections in Table 2304.9.1 are the minimum allowed for conventional construction. Adhesives are not ordinarily used to fasten panels to wall studs and plates and, therefore, have not been cyclically tested for such installations where seismic resistance is concerned. Floor sheathing panels are often glued down to the framing members prior to final nailing in order to increase stiffness and help prevent squeaks, but that need does not currently exist for wall panels.

2308.12.6 Irregular structures. *Conventional light-frame construction* shall not be used in irregular portions of structures assigned to *Seismic Design Category* D or E. Such irregular portions of structures shall be designed to resist the forces specified in Chapter 16 to the extent such irregular features affect the performance of the conventional framing system. A portion of a structure shall be considered to be irregular where one or more of the conditions described in Items 1 through 6 below are present.

1. Where exterior braced wall panels are not in one plane vertically from the foundation to the uppermost *story* in which they are required, the structure shall be considered to be irregular [see Figure 2308.12.6(1)].

   Exception: Floors with cantilevers or setbacks not exceeding four times the nominal depth of the floor

TABLE 2308.12.4
WALL BRACING IN SEISMIC DESIGN CATEGORIES D AND E
(Minimum Percentage of Wall Bracing per each Braced Wall Line[a])

CONDITION	SHEATHING TYPE[b]	$S_{DS} < 0.50$	$0.50 \le S_{DS} < 0.75$	$0.75 \le S_{DS} \le 1.00$	$S_{DS} > 1.00$
One story	G-P[c]	43	59	75	100
	S-W	21	32	37	48

For SI: 1 inch = 25.4 mm, 1 foot = 304.8 mm.

a. Minimum length of panel bracing of one face of the wall for S-W sheathing or both faces of the wall for G-P sheathing; h/w ratio shall not exceed 2:1. For S-W panel bracing of the same material on two faces of the wall, the minimum length is permitted to be one-half the tabulated value but the h/w ratio shall not exceed 2:1 and design for uplift is required. The 2:1 h/w ratio limitation does not apply to alternate braced wall panels constructed in accordance with Section 2308.9.3.1 or 2308.9.3.2.

b. G-P = gypsum board, fiberboard, particleboard, lath and plaster or gypsum sheathing boards; S-W = wood structural panels and diagonal wood sheathing.

c. Nailing as specified below shall occur at all panel edges at studs, at top and bottom plates and, where occurring, at blocking:
For $^1/_2$-inch gypsum board, 5d (0.113 inch diameter) cooler nails at 7 inches on center;
For $^5/_8$-inch gypsum board, No. 11 gage (0.120 inch diameter) at 7 inches on center;
For gypsum sheathing board, $1^3/_4$ inches long by $^7/_{16}$-inch head, diamond point galvanized nails at 4 inches on center;
For gypsum lath, No. 13 gage (0.092 inch) by $1^1/_8$ inches long, $^{19}/_{64}$-inch head, plasterboard at 5 inches on center;
For Portland cement plaster, No. 11 gage (0.120 inch) by $1^1/_2$ inches long, $^7/_{16}$-inch head at 6 inches on center;
For fiberboard and particleboard, No. 11 gage (0.120 inch) by $1^1/_2$ inches long, $^7/_{16}$-inch head, galvanized nails at 3 inches on center.

joists [see Figure 2308.12.6(2)] are permitted to support braced wall panels provided:

1. Floor joists are 2 inches by 10 inches (51 mm by 254 mm) or larger and spaced not more than 16 inches (406 mm) o.c.
2. The ratio of the back span to the cantilever is at least 2:1.
3. Floor joists at ends of braced wall panels are doubled.
4. A continuous rim joist is connected to the ends of cantilevered joists. The rim joist is permitted to be spliced using a metal tie not less than 0.058 inch (1.47 mm) (16 galvanized gage) and $1^1/_2$ inches (38 mm) wide fastened with six 16d common nails on each side. The metal tie shall have a minimum yield of 33,000 psi (227 MPa).
5. Joists at setbacks or the end of cantilevered joists shall not carry gravity loads from more than a single *story* having uniform wall and roof loads, nor carry the reactions from headers having a span of 8 feet (2438 mm) or more.

2. Where a section of floor or roof is not laterally supported by braced wall lines on all edges and connected in accordance with Section 2308.3.2, the structure shall be considered to be irregular [see Figure 2308.12.6(3)].

 Exception: Portions of roofs or floors that do not support braced wall panels above are permitted to extend up to 6 feet (1829 mm) beyond a braced wall line [see Figure 2308.12.6(4)] provided that the framing members are connected to the braced wall line below in accordance with Section 2308.3.2.

3. Where the end of a required braced wall panel extends more than 1 foot (305 mm) over an opening in the wall below, the structure shall be considered to be irregular. This requirement is applicable to braced wall panels offset in plane and to braced wall panels offset out of plane as permitted by the exception to Item 1 above in this section [see Figure 2308.12.6(5)].

 Exception: Braced wall panels are permitted to extend over an opening not more than 8 feet (2438 mm) in width where the header is a 4-inch by 12-inch (102 mm by 305 mm) or larger member.

4. Where portions of a floor level are vertically offset such that the framing members on either side of the offset cannot be lapped or tied together in an *approved* manner, the structure shall be considered to be irregular [see Figure 2308.12.6(6)].

 Exception: Framing supported directly by foundations need not be lapped or tied directly together.

5. Where braced wall lines are not perpendicular to each other, the structure shall be considered to be irregular [see Figure 2308.12.6(7)].

6. Where openings in floor and roof diaphragms having a maximum dimension greater than 50 percent of the distance between lines of bracing or an area greater than 25 percent of the area between orthogonal pairs of braced wall lines are present, the structure shall be considered to be irregular [see Figure 2308.12.6(8)].

❖ Irregular structures are those that generally require engineered design in Seismic Design Category D or E because of their unusual shape or discontinuities in the lateral-force-resisting system. Irregular conditions identified in the code are known to create torsional effects and unbalanced load distribution to the shear walls, which necessitates an engineered design. The registered design professional is left to judge the extent of the portion required to be designed as a result of an irregularity. This often involves design of the nonconforming element, force transfer into the element and a load path from the element to the foundation. A nonconforming portion will sometimes have enough of an impact on the behavior of a structure to warrant that the entire lateral-force-resisting system require an engineered design. Buildings may have slightly irregular features, as identified in this section and in Figures 2308.12.6(1) through 2308.12.6(8), and a designer who wishes to avoid an engineering design must lay out the building in such a way as to meet the restrictions outlined in this section:

1. This limit applies when braced wall panels are offset out of plane from floor to floor. In-plane offsets are discussed in another item. Ideally, braced wall panels should always stack above each other from floor to floor.

 Because cantilevers and setbacks are often incorporated into structures, the exception to this provision defines the restriction on cantilever and set-back length in order to maintain regular building behavior.

2. This limitation applies to open-front structures or portions of structures where a portion of the roof or floor (diaphragm) is not laterally supported by a braced wall line. The conventional construction bracing concept is based on using braced wall lines to divide a structure into a series of "boxes" of limited dimension, with the seismic force to each box being limited by the size. The intent is that each box be supported by braced wall lines on all four sides, limiting the amount of torsion that can occur. The exception, which permits portions of roofs or floors to extend past the braced wall line, is intended to permit common construction such as porch roofs and bay windows.

3. This limitation applies when braced wall panels are offset in plane. Ends of braced wall panels supported on window or door headers can transfer large vertical reactions to the header below that may not be of adequate size to resist these reactions. This section restricts the portion of an upper braced wall panel that may extend over an opening below where a conven-

tional header of limited size may exist. This is allowed on the basis that the vertical reaction is within an approximate 45-degree (0.78 rad) line of the header support and, therefore, will not result in critical shear or flexure. Where a 4-inch by 12-inch nominal (102 mm by 305 mm) header is framed over the opening below a braced wall panel, the panel may extend completely over this header, provided the opening is not greater than 8 feet (2438 mm) in length. All other header conditions require an engineered design.

4. This limitation results from observation of damage that is somewhat unique to split-level wood-framed construction. If floors on either side of an offset move in opposite directions due to an earthquake or wind loading, the short bearing wall in the middle becomes unstable and vertical support for the upper joists can be lost, resulting in a collapse. If the vertical offset is limited to a dimension equal to or less than the joist depth, then a simple strap tie directly connecting joists on different levels can be provided, making the irregularity manageable without engineering.
5. This limitation applies to nonperpendicular braced wall lines. When braced wall lines are not perpendicular to each other, further evaluation is needed to determine force distributions and required bracing.
6. This limitation attempts to place a practical limit on the size of openings in floors and roofs. Because stair openings are essential to residential construction and have long been used without any report of life-safety hazards after a large earthquake, these have been found to be acceptable in conventional construction without engineering.

FIGURES 2308.12.6(1) through (8). See pages 23-112 through 23-114.

❖ The figures illustrate various type of structural irregularities. For Figures 2308.12.6(1) through (8), see the commentary to Section 2308.12.6.

2308.12.7 Anchorage of exterior means of egress components. Exterior egress balconies, exterior exit stairways and similar *means of egress* components shall be positively anchored to the primary structure at not over 8 feet (2438 mm) o.c. or shall be designed for lateral forces. Such attachment shall not be accomplished by use of toenails or nails subject to withdrawal.

❖ Because of the importance of being able to safely get people out of a building after a design seismic event, the exterior means of egress components identified in this section must be positively anchored to the main structure in buildings assigned to Seismic Design Category D or E to provide adequate resistance to lateral forces. The required connection cannot rely on toe-nails or nails in withdrawal because they are not reliable when subjected to cyclic loading demands.

2308.12.8 Sill plate anchorage. Sill plates shall be anchored with anchor bolts with steel plate washers between the foundation sill plate and the nut, or *approved* anchor straps load rated in accordance with Section 1716.1. Such washers shall be a minimum of 0.229 inch by 3 inches by 3 inches (5.82 mm by 76 mm by 76 mm) in size. The hole in the plate washer is permitted to be diagonally slotted with a width of up to $^{3}/_{16}$ inch (4.76 mm) larger than the bolt diameter and a slot length not to exceed $1^{3}/_{4}$ inches (44 mm), provided a standard cut washer is placed between the plate washer and the nut.

❖ In the aftermath of recent California earthquakes, many cases of shear wall sill plate splitting were observed that were attributed to cross-grain bending and cross-grain tension in the sills due to uplift in the sheathing on the tension side of the shear wall. To mitigate the potential for sill splitting, 3-inch by 3-inch (76 mm by 76 mm) steel plate washers are used to restrain the sill plate and prevent splitting before the shear wall experiences inelastic deformation. For engineered shear walls, the AF&PA SDPWS specifies a minimum 3-inch (76 mm) square by 0.229-inch (5.82 mm) thick plate washer for anchor bolts in all seismic design categories. To account for different bottom plate widths and to mitigate the potential for cross-grain bending, the AF&PA SDPWS requires the edge of the square plate washer to extend to within $^{1}/_{2}$ inch (12.7 mm) of the sheathed edge of the sill plate. The $^{1}/_{2}$-inch (12.7 mm) distance from the washer edge to the sheathed edge, in effect, limits the potential for cross-grain bending by limiting the moment arm. The requirement in Section 2308.12.8 for square plate washers in conventional wood-framed structures in Seismic Design Categories D and E has the same intent as the requirement for engineered shear walls.

2308.12.9 Sill plate anchorage in Seismic Design Category E. In structures assigned to *Seismic Design Category* E, steel bolts with a minimum nominal diameter of $^{5}/_{8}$ inch (15.9 mm) or approved anchor straps load rated in accordance with Section 1711.1 and spaced to provide equivalent anchorage shall be used.

❖ Lateral seismic loads on buildings cause the bottom of the walls to slide, causing a tendency for the sill-plate bolts is to rip through the wood. Dowel-type connectors (bolts, nails, screws and pins) rely on the dowel-to-wood bearing to transfer lateral loads (shear) to the foundation. One of the factors that determines dowel bearing strength is the diameter of the bolt. By increasing the minimum diameter of the bolts from $^{1}/_{2}$ inch (12.7 mm) as specified in Section 2308.6, the bearing surface area of the bolt is increased to provide a connection to resist the increased seismic forces anticipated in Seismic Design Category E.

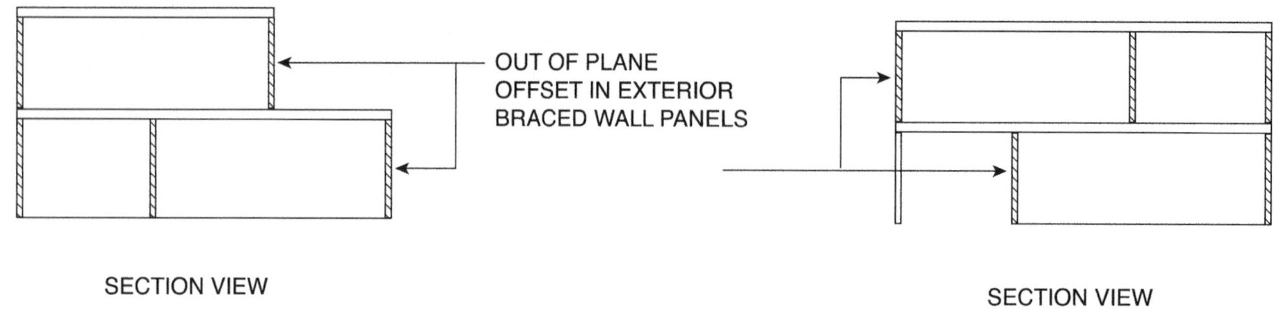

FIGURE 2308.12.6(1)
BRACED WALL PANELS OUT OF PLANE

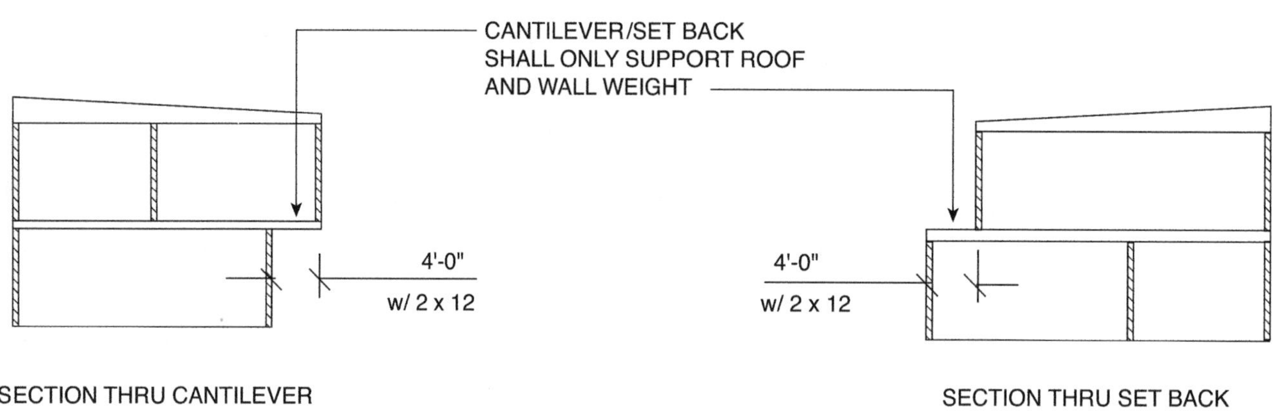

For SI: 1 foot = 304.8 mm.

FIGURE 2308.12.6(2)
BRACED WALL PANELS SUPPORTED BY CANTILEVER OR SET BACK

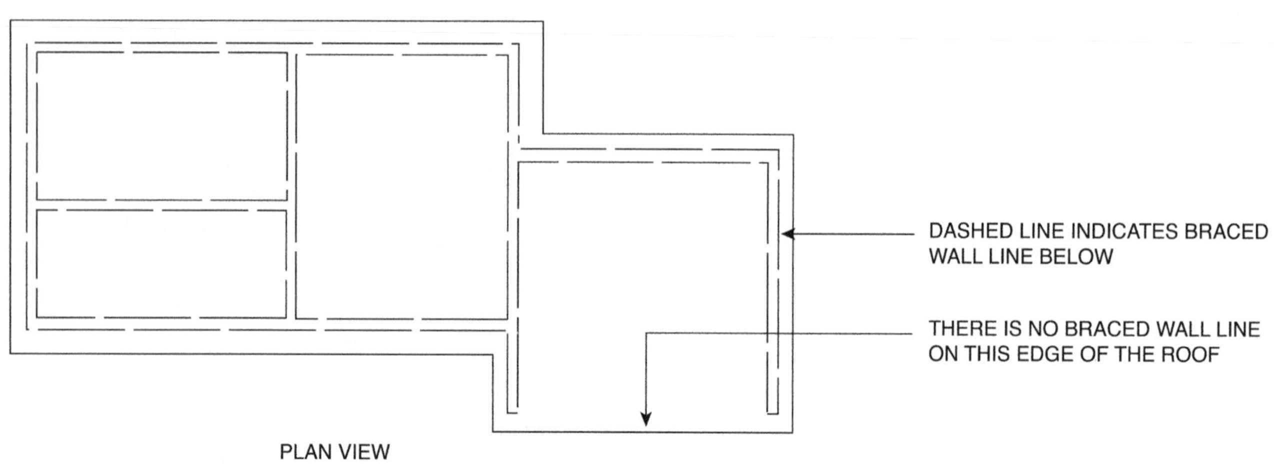

FIGURE 2308.12.6(3)
FLOOR OR ROOF NOT SUPPORTED ON ALL EDGES

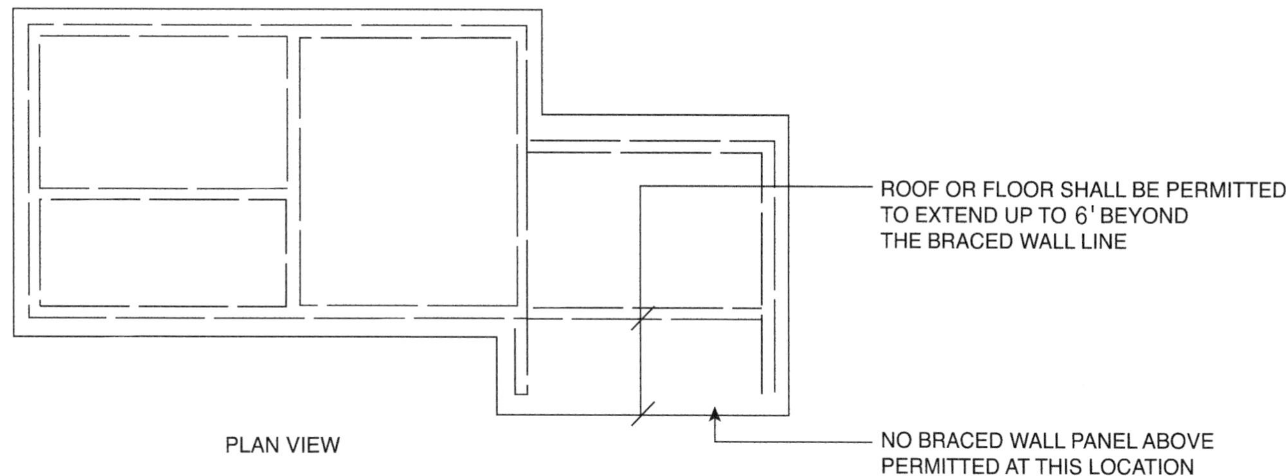

For SI: 1 foot = 304.8 mm.

FIGURE 2308.12.6(4)
ROOF OR FLOOR EXTENSION BEYOND BRACED WALL LINE

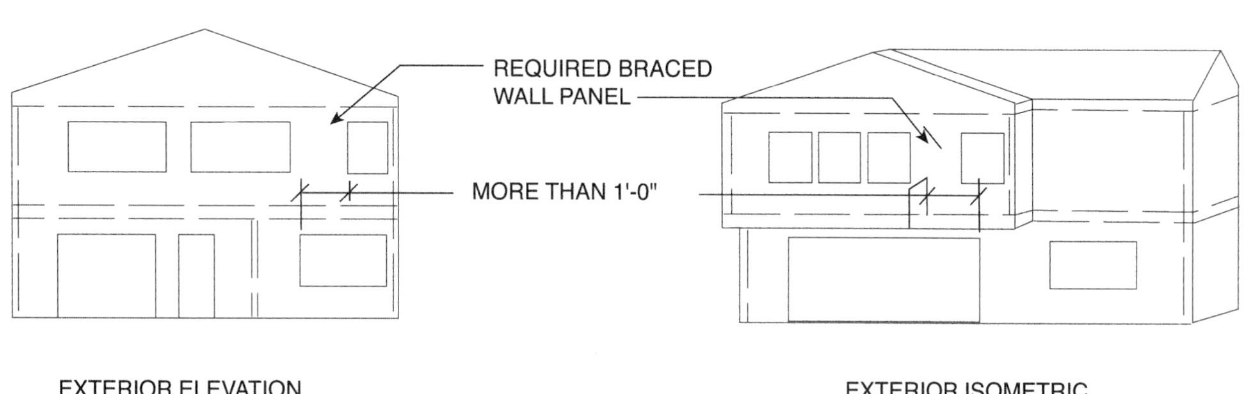

For SI: 1 foot = 304.8 mm.

FIGURE 2308.12.6(5)
BRACED WALL PANEL EXTENSION OVER OPENING

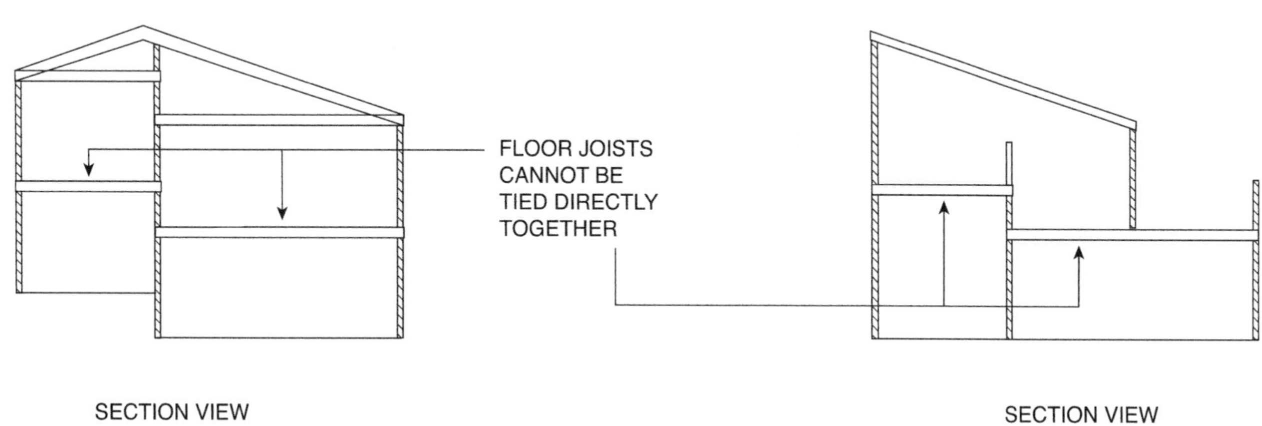

FIGURE 2308.12.6(6)
PORTIONS OF FLOOR LEVEL OFFSET VERTICALLY

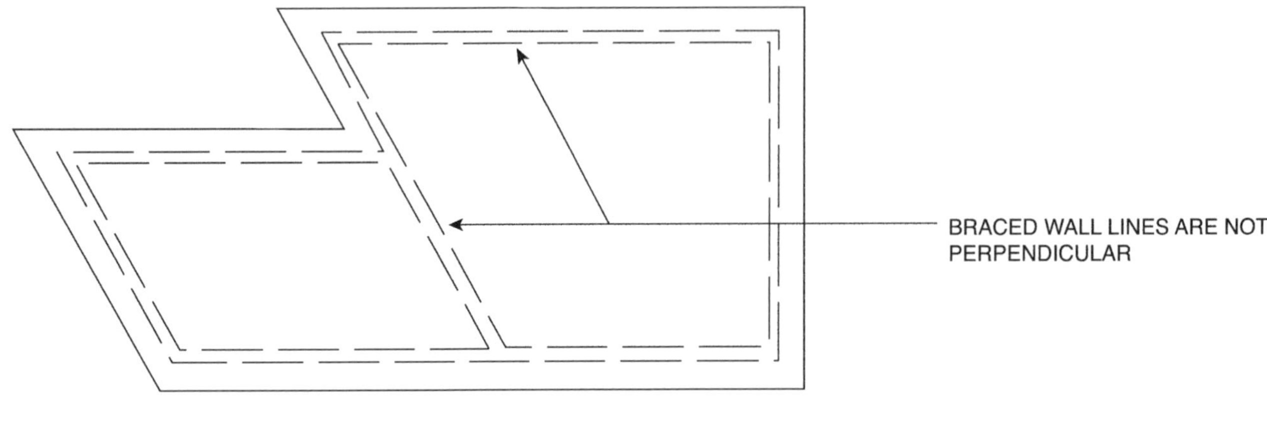

FIGURE 2308.12.6(7)
BRACED WALL LINES NOT PERPENDICULAR

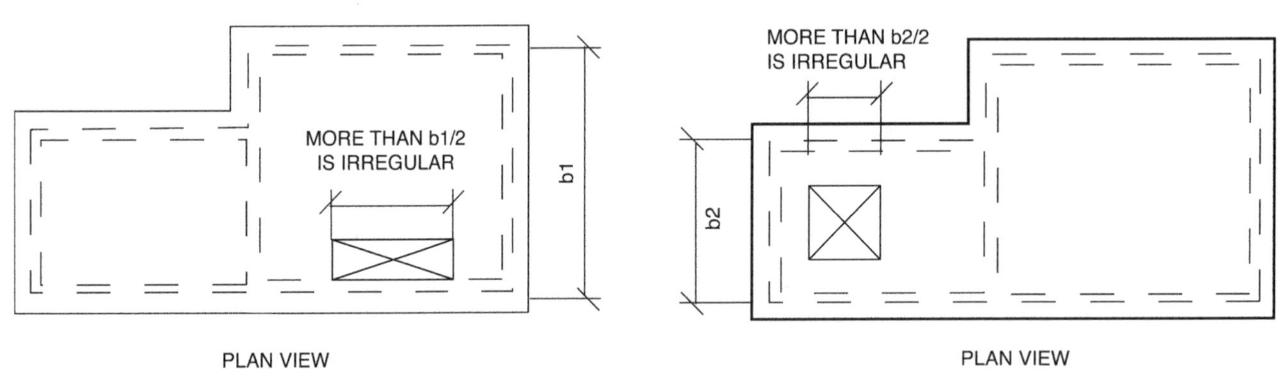

FIGURE 2308.12.6(8)
OPENING LIMITATIONS FOR FLOOR AND ROOF DIAPHRAGMS

Bibliography

The following resource materials are referenced in this chapter or are relevant to the subject matter addressed in this chapter.

1999 *Handbook for Ceramic Tile Installation.* Anderson, SC: TCA, 1999.

AF&PA, ASD/LRFD *Manual for Engineered Wood Construction (ASD/LRFD Manual).* Washington, DC: American Forest & Paper Association, 2005.

AF&PA, *Heavy Timber Construction—Wood Construction Data 5.* Washington, DC: American Forest & Paper Association, 2003.

AF&PA NDS-2012, *National Design Specification for Wood Construction with 2012 Supplement.* Washington, DC: American Forest & Paper Association, 2012.

AF&PA No.4-03, *Plank and Beam Framing for Residential Buildings.* Washington, DC: American Forest & Paper Association, 2003.

AF&PA PWF-07, *Permanent Wood Foundation Design Specification.* Washington, DC: American Forest & Paper Association, 2007.

AF&PA SDPWS, *Special Design Provisions for Wind and Seismic.* Washington, DC: American Forest & Paper Association, 2008.

AF&PA, *Span Tables for Joists and Rafters.* Washington, DC: American Forest & Paper Association, 2012.

AF&PA WCD 5-04, *Heavy Timber Construction Details.* Washington, DC: American Forest & Paper Association, 2004.

AF&PA WFCM-2012, *Wood Frame Construction Manual for One- and Two-family Dwellings (WFCM).* Washington, DC: American Forest and Paper Association, 2012.

AITC 104-03, *Typical Construction Details.* Englewood, CO: American Institute of Timber Construction, 2003.

AITC 110-01, *Standard Appearance Grades for Structural Glued Laminated Timber.* Englewood, CO: American Institute of Timber Construction, 2001.

AITC 112-93, *Standard for Tongue-and-Groove Heavy Timber Roof Decking.* Englewood, CO: American Institute of Timber Construction, 1993.

AITC 113-10, *Standard for Dimensions of Structural Glued Laminated Timber.* Englewood, CO: American Institute of Timber Construction, 2010.

AITC 117-10, *Structural Glued Laminated Timber.* Englewood, CO: American Institute of Timber Construction, 2010.

AITC 119-10, *Standard Specifications for Hardwood Glued-laminated Timber.* Englewood, CO: American Institute of Timber Construction, 2010.

AITC 200-09, *Manufacturing Quality Control Systems Manual for Structural Glued-laminated Timber.* Englewood, CO: American Institute of Timber Construction, 2009.

AITC, *Timber Construction Manual, 5th Edition.* New York: John Wiley and Sons, Inc. 2005.

Andreason, Kenneth R. and John D. Rose. *APA Report T94-5, Northridge, California Earthquake.* American Plywood Association, pp. 2 & 5, March 1994.

ANSI/AITC A 190.1-07, *Structural Glued-laminated Timber.* Englewood, CO: American Institute of Timber Construction, 2007.

ANSI A 208.1-2009, *Particleboard.* New York: American National Standards Institute, 2009.

APA PDS–04, *Panel Design Specification.* Tacoma, WA: APA—Engineered Wood Association, 2004.

APA PDS–90, *Supplement 1—Design and Fabrication of Plywood Curved Panels.* Tacoma, WA: APA—Engineered Wood Association, 1995.

APA PDS–92, *Supplement 2—Design and Fabrication of Plywood-lumber Beams.* Tacoma, WA: APA—Engineered Wood Association, 1998.

APA PDS–90, *Supplement 3—Design and Fabrication of Plywood Stressed-skin Panels.* Tacoma, WA: APA—Engineered Wood Association, 1996.

APA PDS–90, *Supplement 4—Design and Fabrication of Plywood Sandwich Panels.* Tacoma, WA: APA—Engineered Wood Association, 1993.

APA PDS–08, *Supplement 5—Design and Fabrication of All-plywood Beams.* Tacoma, WA: APA—Engineered Wood Association, 2008.

APA Report 157-96, *Wood Structural Panel Shear Walls with Gypsum Wallboard and Window/Door Openings.* Tacoma, WA: APA—The Engineered Wood Association, 1996.

APA Research Report 138, *Plywood Diaphragms.* Tacoma, WA: APA—Engineered Wood Association.

APA Research Report 154, *Wood Structural Panel Shear Wall.* Tacoma, WA: APA—Engineered Wood Association.

ASABE EP 484.2 (1998), *Diaphragm Design of Metal-Clad, Post-Frame Rectangular Buildings.* St. Joseph, MI: American Society of Agricultural and Biological Engineers, 1998, 2008.

ASABE EP 486.1 (1999), *Shallow-Post Foundation Design.* St. Joseph, MI: American Society of Agricultural and Biological Engineers, 2005.

ASABE EP 559 (1997), *Design Requirements and Bending Properties for Mechanically Laminated Columns.* St. Joseph, MI: American Society of Agricultural and Biological Engineers, 1997, 2008.

ASCE 7-10, *Minimum Design Loads for Buildings and Other Structures.* Reston, VA: American Society of Civil Engineers, 2010.

ASTM A 653/A 653M-08, *Specification for Steel Sheet, Zinc-coated Galvanized or Zinc-iron Alloy-coated Galvanealed by the Hot-dip Process.* West Conshohocken, PA: ASTM International, 2008.

ASTM C 208-08a, *Specification for Cellulose Fiber Insulating Board.* West Conshohocken, PA: ASTM International, 2008.

ASTM D 245-06, *Standard Practice for Establishing Structural Grades and Related Allowable Properties for Visually Graded Lumber.* West Conshohocken, PA: ASTM International, 2006.

ASTM D 1761-06, *Test Method for Mechanical Fasteners.* West Conshohocken, PA: ASTM International, 2006.

ASTM D 1990-07, *Standard Practice for Establishing Allowable Properties for Visually Graded Dimension Lumber for In-grade Tests of Full-size Specimens.* West Conshohocken, PA: ASTM International, 2007.

ASTM D 2898-04, *Test Methods for Accelerated Weathering of Fire-retardant-treated Wood for Fire Testing.* West Conshohocken, PA: ASTM International, 2004.

ASTM D 3200-74 (2005), *Standard Specification and Test Method for Establishing Recommended Design Stresses for Rund Timber Construction Poles.* West Conshohocken, PA: ASTM International, 2005.

ASTM D 3201-08a, *Test Method for Hygroscopic Properties of Fire-retardant-treated Wood and Wood-base Products.* West Conshohocken, PA: ASTM International, 2008.

ASTM D 3737-08, *Practice for Establishing Allowable Properties for Structural Glued-laminated Timber (Glulam)*. West Conshohocken, PA: ASTM International, 2008.

ASTM D 3957-06, *Standard Practices for Establishing Stress Grades for Structural Members Used in Log Buildings*. West Conshohocken, PA: ASTM International, 2006.

ASTM D 5055-09, *Specification for Establishing and Monitoring Structural Capacities of Prefabricated Wood I-Joists*. West Conshohocken, PA: ASTM International, 2009.

ASTM D 6841-03, *Standard Practice for Calculating Design Value Treatment Adjustment Factors for Fire-retardant-treated Wood*. West Conshohocken, PA: ASTM International, 2003.

ASTM E 84-09, *Test Method for Surface Burning Characteristics of Building Materials*. West Conshohocken, PA: ASTM International, 2009.

ASTM F 1575-03 (2008), *Standard Test Method for Determining Bending Yield Movement of Nails*. West Conshohocken, PA: ASTM International, 2008.

ASTM F 1667-05, *Specification for Drive Fasteners: Nails, Spikes and Staples*. West Conshohocken, PA: ASTM International, 2005.

ATC 7, *Guidelines for the Design of Horizontal Wood Diaphragms*. Redwood City, CA: Applied Technology Council, 1981.

AWPA U1-11, *Use Category System: User Specification for Treated Wood*. Grandbury, TX: American Wood-preservers' Association, 2011.

BCSI 1-08, *Guide to Good Practice for Handling, Installing and Bracing of Metal Plate Connected Wood Trusses*. Madison, WI: Wood Truss Council of America (WTCA) & Truss Plate Institute (TPI), 2008.

Becker, Norman. *The Complete Book of Home Inspection, 2nd Edition*. McGraw-Hill, New York: 2002.

Bossi, Robert J. and Kelly E. Cobeen. "Conventional Construction for Today's Buildings." *ICBO Building Standards*, Sept.-Oct. 1996, p. 14.

Breyer, Donald E., Kenneth J. Fridley, David G. Pollock, Jr. and Kelly E. Cobeen. *Design of Wood Structures—ASD/LRFD (Sixth Edition)*. New York: McGraw-Hill, 2007.

Cobeen, Kelly, James Russell and J. Daniel Dolan. CUREE W30a & 30b: *Recommendations for Earthquake Resistance in the Design and Construction of Woodframe Buildings*. Consortium of Universities for Research in Earthquake Engineering (CUREE), 2004.

CPA/ANSI AHA A135.4-04, *Basic Hardboard*. Leesburg, VA: Composite Panel Association, 2004.

CPA/ANSI AHA A135.5-04, *Prefinished Hardboard Paneling*. Leesburg, VA: Composite Panel Association, 2004.

CPA/ANSI AHA A135.6-2006, *Hardboard Siding*. Leesburg, VA: Composite Panel Association, 2006.

Details for Conventional Wood Frame Construction. Washington, DC: American Forest and Paper Association, 2001.

DOC PS 1-09, *Construction and Industrial Plywood*. Gaithersburg, MD: U.S. Department of Commerce, 2009.

DOC PS 2-10, *Performance Standard for Wood-based Structural-Use Panels*. Gaithersburg, MD: U.S. Department of Commerce, 2010.

DOC PS 20-10, *American Softwood Lumber Standard*. Gaithersburg, MD: U.S. Department of Commerce, 2010.

EWS R540-96, *Builders Tips: Proper Storage and Handling of Glulam Beams*. Tacoma, WA: APA—Engineered Wood Association, 1996.

EWS S475-01, *Glued Laminated Beam Design Tables*. Tacoma, WA: APA—Engineered Wood Association, 2001.

EWS S560-03, *Field Notching and Drilling of Glued Laminated Timber Beams*. Tacoma, WA: APA—Engineered Wood Association, 2003.

EWS T300-02, *Glulam Connection Details*. Tacoma, WA: APA—Engineered Wood Association, 2002.

EWS X440-00, *Product and Application Guide*: Gluman. Tacoma, WA: APA—Engineered Wood Association, 2000.

EWS X450-01, *Glulam in Residential Construction*. Tacoma, WA: APA—Engineered Wood Association, 2001.

FEMA 450, Part 1, NEHRP *Recommended Provisions for Seismic Regulations for New Buildings and Other Structures*. Washington, DC: Federal Emergency Management Agency, 2004.

FEMA 450, Part 2, NEHRP *Recommended Provisions for Seismic Regulations for New Buildings and Other Structures Part 2: Commentary*. Washington, DC: Federal Emergency Management Agency, 2004.

Henry, John R. "A Better Way to Think: LRFD for Engineered Wood Construction." *Structural Engineer*, February 2003, pp 20-25.

HPVA HP-1-2009, *Standard for Hardwood and Decorative Plywood*. Reston, VA: Hardwood Plywood Veneer Association, 2009.

Hoke, John Ray, Jr., editor. *Ramsay/Sleeper Architectural Graphic Standards, 8th Edition*. New York: John Wiley& Sons, 1988.

ICC-400-12, *Standard on Design and Construction of Log Structures*. Washington, DC: International Code Council, 2012.

ICC-600-08, *Standard for Residential Construction in High Wind Regions*. Washington, DC: International Code Council, 2008.

ICC-ES, *Acceptance Criteria for Staples (AC 201)*. Whittier, CA: ICC Evaluation Service, Inc., 2005, 2008.

IMC-12, *International Mechanical Code*. Washington, DC: International Code Council, 2011.

IRC-12, *International Residential Code*. Washington, DC: International Code Council, 2011.

Recommended Lateral Force Requirements and Commentary (SEAOC Blue Book). Sacramento, CA: Structural Engineers Association of California, 1999.

SEAOC Blue Book: *Seismic Design Recommendations of the SEAOC Seismology Committee*. Sacramento, CA: Structural Engineers Association of California, 2006.

Stalnaker, Judith J. and Ernest Harris. *Structural Design in Wood*. New York: Van Nostrand Reinhold, 1989.

Technical Note B429F, *Nonload-bearing Partitions on APA Structural Panel Floors*. Washington, DC: American Plywood Association, September 1999.

The Wood Truss Handbook. Miami, FL: Gang-Nail Systems, Inc., 1981.

Timber Construction Manual, 5th Edition. New York: John Wiley & Sons, 2005.

TPI 1-07, *National Design Standard for Metal Plate Connected Wood Trusses Construction*. New York: American National Standards Institute, 2007.

Williams, T. Jeff. *Basic Carpentry Techniques*. San Ramon, CA: Orth Books (Chevron Chemical Company), 1981.

Wood Handbook, Wood as an Engineering Material. Washington, DC: USDA Forest Products Laboratory, U.S. Government Printing Office, 2010.

Chapter 24: Glass and Glazing

General Comments

Chapter 24 includes design provisions and quality requirements for glazing consisting of glass and, to a limited extent, light-transmitting plastics.

Section 2401 contains the general contents and scope of this chapter.

Section 2402 lists definitions of words and phrases that are necessary for a better understanding of the requirements of this chapter.

Section 2403 establishes the identification requirements for safety and nonsafety glazing, glass support and framing systems, glass thicknesses and requirements for glass at interior locations. This section also addresses special provisions for louvered and jalousie windows.

Section 2404 provides the design requirements and limitations for vertical glass exposed to wind loads, and sloped glass exposed to wind, snow and dead loads. Design factors for the more common types of glass are included in the design tables.

Section 2405 defines the allowable glazing types and related requirements for sloped glazing and skylights.

Section 2406 provides a detailed listing of the applications where safety glazing is required. Test standards category classifications for safety glazing are also included in this section.

Section 2407 considers glass used as either structural balustrades or in-fill panels in handrail and guardrail systems.

Section 2408 clarifies the special requirements for glass areas exposed to extraordinary impact from people. These are applicable to walls and doors in certain athletic facilities.

Section 2409 covers glass in elevator cars and hoistway enclosures.

As is true with other chapters in the code that deal with building materials and components, this chapter evolved from the considerations of safety, public welfare and the integrity of a building's construction. There is an ever- growing demand from building owners and architects to incorporate larger quantities of glass in buildings. This in turn increases the concerns for such things as safety and energy conservation. The glass industry is being called on to meet these demands. Model building codes are changing almost daily to stay abreast of the ever-changing technology being employed in the glass industry. Recognizing the risks should sloped glass fail, the model code groups and the glass and glazing industry have developed changes that only allow safety glazing materials over occupiable areas. Complying materials include laminated, fully tempered and wired glasses, and acrylic and polycarbonate plastics. Except in residential occupancies, fully tempered glass is required to have a screen below the glass. In addition to the concern for overhead glazing, a need developed to codify and address glazing materials in locations that are exposed to human impact, including but not necessarily limited to: sliding patio doors; side-swinging storm and primary use doors; tub and shower enclosures; and doors and certain fixed and operable glass panels. Building officials and the glass industry joined together to develop and enact safety glazing requirements to fill this need. As a result, safety glazing laws have been adopted in many states. Based on the recommendation of the Consumer Product Safety Commission (CPSC), a federal law was enacted in 1977 that mandated safety glazing in all areas that might reasonably be exposed to impact from people. The requirements of this federal law were similar to the laws adopted by these various states. This chapter also addresses the design and selection of glass to withstand wind loads, snow loads, dead loads and other imposed loads.

The code uses the terms "glass" and "glazing" interchangeably as if there is no difference between the two. This can cause confusion when talking about "safety glazing" or "safety glass."

Glass is a hard, brittle, inorganic material that is manufactured from a mixture of silica, flux and a stabilizer to form a transparent, translucent or opaque material. While in its molten (semiliquid) form, it can be blown, drawn, rolled, pressed or cast into a variety of shapes. Most of the glass referenced in the code is in sheet form. Glass, safety or not, as well as other materials can be used as glazing.

On the other hand, to "glaze" means to furnish or fit with glass, such as to install glass in an opening in a door. Glass installed in an exterior wall would be properly identified as a glazed opening, or more commonly, a window. An enclosure around a shower can be constructed of many materials, but it is considered to be a glazed enclosure if the enclosing material is glass. The code does permit the limited use of some plastics as glazing materials.

To assist in evaluating the requirements for glass in buildings, properties of the various types are listed below:

Annealed glass—has been subjected to the annealing process, which consists of controlled cooling in a cooling or annealing oven. Types of annealed glass include float, plate and sheet.

Float glass—is produced by floating a continuous ribbon of molten glass on molten tin. As the glass flows downstream, the temperature of the glass and tin is reduced. When the glass ribbon solidifies, it is lifted from the tin onto rollers and slowly cooled in a controlled cooling oven or annealing lehr. There is no further treatment of the glass. Most clear, gray, bronze and green-tinted glass for architectural use is presently produced by the float process.

Plate glass—is produced by grinding and polishing rough glass blanks. The glass blanks are formed by drawing semimolten glass through textured rollers, which are then fed through an annealing lehr prior to grinding and polishing. It should be noted that there is limited domestic production of plate glass material.

Sheet glass—is manufactured by continuously drawing molten glass from a bath, horizontally or vertically over rollers, through an annealing lehr. Because of the manufacturing process, sheet glass has an inherent waviness characteristic, and is, therefore, not typically used in applications where good optical quality is desired, such as architectural glazing, reflective glazing or show windows. The most common use of sheet glass is in residential applications.

Tempered or fully tempered glass—is annealed glass that has been heated and quenched in a controlled operation to provide a high level of surface compression. For most types of loading conditions, it has approximately four times the strength of regular annealed glass. Tempered glass breaks into small, relatively harmless fragments, and is used as safety glazing material. After tempering, this glass cannot be cut, drilled or otherwise altered without breaking.

Heat-strengthened glass—is produced in much the same way as tempered glass, except the level of surface compression is less. Heat-strengthened glass has approximately twice the strength of regular annealed glass and breaks into relatively large pieces. It is not a safety glazing material.

Laminated glass—is a sandwich of two or more plies of glass bonded together with resilient plastic interlayers, usually polyvinyl butyral. The glass plies may be regular, tempered or heat-strengthened glass. The plastic interlayer thickness is generally 0.030 inch (0.76 mm), but may range up to 0.090 inch (2.3 mm) for some applications. When this glass breaks, the fragments are held together by the plastic interlayer.

Wired glass—is a monolithic glass with wire mesh embedded in approximately the center of the glass thickness. Coupled with a suitable framing system, some wire glass is fire resistant. The surface of wire glass may be smooth or textured. Wire glass cannot be tempered or heat strengthened.

Insulating glass—is a factory-assembled glazing unit consisting of two or more panes of glass separated by airspaces. The edges of the finished glazing unit, including the airspaces, are hermetically sealed. The entrapped air is removed, creating a vacuum, which is necessary to prevent condensation within the sealed airspaces. The airspaces between the glass can be filled with specialized gas to add to the energy efficiency of the glazing unit. The panes of glass may be any type, except for patterned glasses with deep patterns.

Patterned glass—has a texture or pattern embossed on one or both surfaces during the manufacturing process. Some glass with shallow patterns can be tempered and heat strengthened.

Spandrel glass—is manufactured from either float glass or plate glass that has an opaque coating of fired-on ceramic glaze on the interior surface (or room side). This glass is typically heat strengthened or tempered to resist the thermal stresses (movement) caused by the high absorption of solar radiation by the opaque coating.

Purpose

The primary purpose of this chapter is to provide information to establish the adequacy of glazing from the standpoint of life safety and performance. The sections on glass in handrails and guards, glazing in athletic facilities and safety glazing dictate a performance standard that will minimize risks to a level consistent with other provisions of the code.

SECTION 2401
GENERAL

2401.1 Scope. The provisions of this chapter shall govern the materials, design, construction and quality of glass, light-transmitting ceramic and light-transmitting plastic panels for exterior and interior use in both vertical and sloped applications in buildings and structures.

❖ This chapter establishes regulations for glass and glazing used in buildings and structures that, when installed, are subjected to wind, snow and dead loads. Engineering and design requirements in the form of tables, formulas and design loads are contained in this chapter. Additional structural requirements are found in Chapter 16.

A second concern of this chapter is glass and glazing used in locations where it is likely to be impacted by building occupants. Requirements for identification of safety glass, human impact loads, identification of hazardous locations and the types of glass that are allowed in railings and walls are included in this chapter.

This chapter details certain prescriptive and performance requirements for glazing used in buildings, including thickness limitations, design of glass supports and allowable sizes based on wind, snow and dead loads. The limitations on glass used in sloped glazing and areas exposed to human impact are also included. The information contained in this chapter on light-transmitting plastics is limited. For additional information on plastic materials, referenced standards and other code requirements for plastics, refer to Chapter 26.

2401.2 Glazing replacement. The installation of replacement glass shall be as required for new installations.

❖ Any glazing installed in existing construction is required to meet all the requirements and standards of the code. This includes installing new glass in an existing window, door or other type of opening, even where the glass being replaced did not comply with the standards of the code.

Relocating any type of manufactured unit containing glass, such as a door or window, from one location in an existing building or structure to another location in the same building or structure, or even a different building or structure, is to be considered new glazing, even though the unit containing the glass is existing.

SECTION 2402 DEFINITIONS

2402.1 Definitions. The following terms are defined in Chapter 2:

DALLE GLASS.

DECORATIVE GLASS.

❖ Some terms are used in the code to describe something very specific to the related subject matter while other terms may have multiple meanings that vary based on the subject matter and context. Definitions facilitate the understanding of code provisions and minimize potential confusion. To that end, this section lists definitions of terms associated with glazing. Note that these definitions are found in Chapter 2. The use and application of defined terms, as well as undefined terms, are set forth in Section 201.

SECTION 2403 GENERAL REQUIREMENTS FOR GLASS

2403.1 Identification. Each pane shall bear the manufacturer's *mark* designating the type and thickness of the glass or glazing material. The identification shall not be omitted unless *approved* and an affidavit is furnished by the glazing contractor certifying that each light is glazed in accordance with *approved construction documents* that comply with the provisions of this chapter. Safety glazing shall be identified in accordance with Section 2406.3.

Each pane of tempered glass, except tempered spandrel glass, shall be permanently identified by the manufacturer. The identification *mark* shall be acid etched, sand blasted, ceramic fired, laser etched, embossed or of a type that, once applied, cannot be removed without being destroyed.

Tempered spandrel glass shall be provided with a removable paper marking by the manufacturer.

❖ Basic marking for identifying type and thickness of glass, as well as requirements and precautionary statements regarding the glass support system, are prescribed in this section. Detailed performance requirements are addressed in other sections.

Glass is a product that is manufactured and fabricated as assemblies at locations other than the building site. Identification of glass and glazing with a mark bearing the type and thickness is needed by the building inspector to perform a visual inspection (see Figure 2403.1). The inspection is needed to verify that the glazing is installed in the proper location in the building in compliance with this chapter for hazardous locations and structural loads. Other than verifying compliance with wind load requirements for glazing exposed to such loading, the glass thickness itself is not necessarily an indication of code compliance. For example, where a specific performance is required, such as for safety glazing in Section 2406,

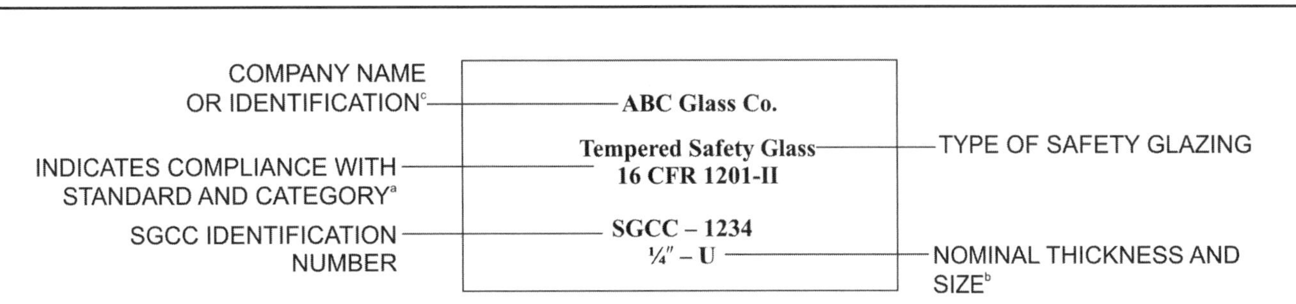

For SI: 1 inch = 25.4 mm, 1 square foot = 0.0929 m^2.

a. "I" after "16 CFR 1201" indicates sizes 9 square feet and less are covered; "II" indicates all sizes are covered.
b. "U" indicates all sizes are covered; "L" indicates only sizes listed by the Safety Glazing Certification Council (SGCC) are covered.
c. Numbers or letters for date coding and plant identification may also be included.

Figure 2403.1
EXAMPLE IDENTIFYING MARK

the manufacturer's mark must identify the standard to which the glass has been tested (see Section 2406.3) which is sufficient to determine compliance with the standard. Note that CPS 16 CFR, Part 1201, does not require the glass thickness to be part of the marking while in accordance with ANSI Z97.1 the glass thickness is an optional marking.

The code requires that each light, which is an individual pane of glass, be marked. Only tempered glass is required by the code to be permanently identified; the identification is to be visible after installation. Tempered glass, both heat strengthened and fully tempered, is produced by reheating and rapidly cooling annealed glass. Heat-strengthened glass and fully tempered glass have increased mechanical strength and resistance to thermal stresses. Fully tempered glass is twice as strong as heat-strengthened glass. Fully tempered glass will pulverize into numerous small, cube-like fragments when broken at any point. Heat-strengthened glass breaks similarly to annealed glass. Heat-strengthened and fully tempered glass cannot be altered after fabrication. Tempered glass cannot be cut, drilled, ground or polished without fracture. The manufacturer of tempered glass must furnish the exact size and shape.

Laminated, annealed, float and spandrel glass, unlike tempered glass, can be cut and fabricated without fracturing. Laminated glass is stockpiled in standard-size sheets by its manufacturers. It consists of two or more sheets of glass held together by an intervening layer of plastic material. The stock sheets are marked by the manufacturer in one corner, which is cut off in the fabrication process. A permanent mark is therefore impractical, so the building official has to rely on an affidavit from the installer of the glazing.

Permanent identification of spandrel glass is not required by the code, since a permanent label may alter the performance of the glass and cause breakage. Spandrel glass is an opaque glass that is heat strengthened through the process of fire fusing an opaque ceramic color to the interior surface of the glass. Spandrel glass is used as exterior windows and curtain walls to conceal internal construction and to provide solar control and energy conservation. An exposed permanent mark on tinted and reflective spandrel glass could change the coefficient of expansion of the glass. The label reacts differently to the heating and cooling effects of the atmosphere than the coated glass surface, which can cause increased incidence of glass breakage. The code, therefore, permits a removable paper mark for identification of spandrel glass.

The glazing contractor is responsible for the final installation of glass and glazing in the building and, with the approval of the building official, may provide an affidavit as an alternative to marking. The affidavit must certify that each light is glazed under the approved plans and specifications. The affidavit relieves the building inspector of the responsibility for inspecting the glass and glazing for type and thickness; however, it does not relieve the plans examiner from the responsibility of reviewing the plans and specifications for compliance with the code. An affidavit may not be submitted for tempered glass that must be permanently identified in compliance with the requirements of this section.

2403.2 Glass supports. Where one or more sides of any pane of glass are not firmly supported, or are subjected to unusual load conditions, detailed *construction documents*, detailed shop drawings and analysis or test data assuring safe performance for the specific installation shall be prepared by a *registered design professional*.

❖ Most glass in buildings is firmly supported on all four edges. With many contemporary designs, glass is installed with support at the top and bottom edges only and, in some cases, with support on three edges. The free edges usually abut adjacent glass panels. When one or more edges of the glass are unsupported, the resistance to wind load is reduced. In this case, or in other cases where glass is subjected to unusual loading conditions, the code requires special engineering analysis and detailed construction documents to demonstrate compliance.

2403.3 Framing. To be considered firmly supported, the framing members for each individual pane of glass shall be designed so the deflection of the edge of the glass perpendicular to the glass pane shall not exceed $^1/_{175}$ of the glass edge length or $^3/_4$ inch (19.1 mm), whichever is less, when subjected to the larger of the positive or negative load where loads are combined as specified in Section 1605.

❖ Framing for glazing systems includes, but is not necessarily limited to: aluminum storefront framing; hollow metal framing; job-fabricated custom wood framing; and factory-manufactured window units constructed of wood, metal, vinyl or a combination of these materials, most of which are industry standards recognized by the code. Regardless of the framing system, it is necessary to comply with the structural requirements of Chapter 16.

It is important to note that "all edges" of each piece of glass must be supported within the parameters described in this section and elsewhere in the code. If not, detailed engineering analysis must be done as indicated in Section 2403.2.

2403.4 Interior glazed areas. Where interior glazing is installed adjacent to a walking surface, the differential deflection of two adjacent unsupported edges shall not be greater than the thickness of the panels when a force of 50 pounds per linear foot (plf) (730 N/m) is applied horizontally to one panel at any point up to 42 inches (1067 mm) above the walking surface.

❖ This provision addresses the installation of glass panels in interior locations adjacent to walking surfaces, such as storefronts within covered mall buildings. Unlike exterior locations, interior installations do not require the placement of mullions between adjacent panels to provide weather resistance. Without intermediate mullions, individual glass panels are free to move independently of each other. When only one panel is subjected to horizontal pressure, the deflection of the panel may cause gaps between adjacent panels. These gaps pose a potential pinching hazard to anyone who may contact the glass panels and cause the differential deflection. This hazard is greater when the glass panels are not mounted in the same plane and the meeting edges form an angle.

The design criteria provided in this section limit the use of glass panels to those that will not deflect more than the thickness of the glass panel itself when the specified force is applied at a height up to 42 inches (1067 mm) above the walking surface. This limitation protects against the occurrence of gaps between adjacent panels that would otherwise create a pinching hazard. The 42-inch (1067 mm) dimension represents the height range in which a person's body will likely contact the glass. The load must be assumed to occur at the point anywhere within the 42-inch (1067 mm) range that will cause the greatest differential deflection between the glass panels.

2403.5 Louvered windows or jalousies. Float, wired and patterned glass in louvered windows and jalousies shall be no thinner than nominal $^3/_{16}$ inch (4.8 mm) and no longer than 48 inches (1219 mm). Exposed glass edges shall be smooth.

Wired glass with wire exposed on longitudinal edges shall not be used in louvered windows or jalousies.

Where other glass types are used, the design shall be submitted to the *building official* for approval.

❖ Louvered and jalousie windows consist of a series of overlapping horizontal glass louvers that pivot simultaneously in the window frame. During opening, the bottom edge of each louver swings toward the exterior and the top edge swings toward the interior. The glass panels are supported on two vertical sides with the horizontal edges of the glass exposed.

For safety reasons, the exposed edges of the louvered panels are required to be smooth; thus, wire glass panels, if used, must have no exposed projecting wire. To provide structural strength, the panels are limited to a minimum thickness of $^3/_{16}$ inch (4.8 mm) and a maximum length of 48 inches (1219 mm).

Louvered and jalousie windows are exempt from safety glazing requirements in all applications, including those where a flat pane of glass is otherwise required to be safety glass. This exemption is based on records that show the injuries associated with this use of glass are primarily from persons impacting the glass edge with no cutting or piercing injuries resulting from glass breakage. Safety glass would not have an effect on this type of injury. There are also practical production reasons associated with fabricating safety glazing for the relatively long, thin slats.

SECTION 2404 WIND, SNOW, SEISMIC AND DEAD LOADS ON GLASS

2404.1 Vertical glass. Glass sloped 15 degrees (0.26 rad) or less from vertical in windows, curtain and window walls, doors and other exterior applications shall be designed to resist the wind loads in Section 1609 for components and cladding. Glass in glazed curtain walls, glazed storefronts and glazed partitions shall meet the seismic requirements of ASCE 7, Section 13.5.9. The load resistance of glass under uniform load shall be determined in accordance with ASTM E 1300.

The design of vertical glazing shall be based on the following equation:

$$F_{gw} \le F_{ga} \quad \textbf{(Equation 24-1)}$$

where:

F_{gw} = Wind load on the glass computed in accordance with Section 1609.

F_{ga} = Short duration load on the glass as determined in accordance with ASTM E 1300.

❖ The procedures for determining compliance of the more common types of glass used in vertical applications are explained in this section. For a given glass type, thickness, plate length and plate width, allowable loads are determined in accordance with ASTM E 1300 using nonfactored loads and established factors that are based on glass type. ASTM E 1300 also considers the influence of width-to-height ratios, effects of weathering and exposure, and statistical influence of glass size. The standard contains more than 40 nonfactored charts covering one-, two-, three- and four-sided support conditions. The application of ASTM E 1300 results in an allowable uniform wind load for a specified glass type, thickness and size. The allowable wind load must equal or exceed the required wind loads determined from Section 1609 for components and cladding. If the glass allowable wind load is less than the required wind load from Section 1609, thicker glass is needed.

In buildings having an elevated seismic risk, glazed partitions, glazed storefronts and glazed curtain walls must be capable of withstanding the relative seismic displacements between the points of attachment to the supporting structure(s). The objective of the referenced ASCE 7 earthquake load provision is to limit the hazard of falling glass during and immediately after an earthquake.

2404.2 Sloped glass. Glass sloped more than 15 degrees (0.26 rad) from vertical in skylights, sunrooms, sloped roofs and other exterior applications shall be designed to resist the most critical of the following combinations of loads.

$$F_g = W_o - D \quad \textbf{(Equation 24-2)}$$

$$F_g = W_i + D + 0.5\,S \quad \textbf{(Equation 24-3)}$$

$F_g = 0.5\ W_i + D + S$ **(Equation 24-4)**

where:

D = Glass dead load psf (kN/m^2).

For glass sloped 30 degrees (0.52 rad) or less from horizontal,

= 13 t_g (For SI: 0.0245 t_g).

For glass sloped more than 30 degrees (0.52 rad) from horizontal,

= 13 t_g cos θ (For SI: 0.0245 t_g cos θ).

F_g = Total load, psf (kN/m^2) on glass.

S = Snow load, psf (kN/m^2) as determined in Section 1608.

t_g = Total glass thickness, inches (mm) of glass panes and plies.

W_i = Inward wind force, psf (kN/m^2) as calculated in Section 1609.

W_o = Outward wind force, psf (kN/m^2) as calculated in Section 1609.

θ = Angle of slope from horizontal.

Exception: Unit skylights shall be designed in accordance with Section 2405.5.

The design of sloped glazing shall be based on the following equation:

$F_g \leq F_{ga}$ **(Equation 24-5)**

where:

F_g = Total load on the glass determined from the load combinations above.

F_{ga} = Short duration load resistance of the glass as determined according to ASTM E 1300 for Equations 24-2 and 24-3; or the long duration load resistance of the glass as determined according to ASTM E 1300 for Equation 24-4.

❖ This section provides the procedure for determining the maximum allowable load that may be supported by a sloped glazing system. The maximum allowable load must be greater than the required design loads. The required design loads are first determined from the three load combinations listed as Equations 24-2, 24-3 and 24-4. The applicable wind loads from the components and cladding requirements of Section 1609 and the snow loads from Section 1608 are used to determine the required design loads from the three load combinations.

After the required design loads are determined, the maximum allowable load resistance of the specified glass is determined in accordance with ASTM E 1300. See the commentary to Section 2404.1 for a discussion of ASTM E 1300.

If the determined maximum allowable load resistance of the glass is not greater than that required from the load combinations, an option would be to choose a greater glass thickness and determine the maximum allowable load resistance of the glass in accordance with ASTM E 1300.

For load combinations given by Equations 24-2 and 24-3, wind is the primary loading, and the short-term load resistance from ASTM E 1300 applies. In Equation 24-4, however, snow is a primary load and wind becomes a "companion load" (factor of 0.5 is applied to wind) requiring use of the long duration load resistance in satisfying Equation 24-5. In the latter case, the nominal snow loads should consider drifts, partial snow load, etc., in accordance with the requirements under Section 1608.

The exception refers to design provisions that are specific to unit skylights (i.e., single-panel, factory-assembled units).

2404.3 Wired, patterned and sandblasted glass.

2404.3.1 Vertical wired glass. Wired glass sloped 15 degrees (0.26 rad) or less from vertical in windows, curtain and window walls, doors and other exterior applications shall be designed to resist the wind loads in Section 1609 for components and cladding according to the following equation:

$F_{gw} < 0.5\ F_{ge}$ **(Equation 24-6)**

where:

F_{gw} = Is the wind load on the glass computed per Section 1609.

F_{ge} = Nonfactored load from ASTM E 1300 using a thickness designation for monolithic glass that is not greater than the thickness of wired glass.

❖ Wired glass is not within the scope of ASTM E 1300 and the requirements for vertical wired glass are therefore specified in the code. As with other vertical glazing, the wind loading on the glass is required to be determined in accordance with Section 1609 of the code. The allowable load on vertical wired glass is based on requirements for monolithic glass, as contained in ASTM E 1300, of a thickness less than or equal to that specified for the wired glass. ASTM E 1300 provides the method for determining values of short-duration and long-duration load resistance for specific glazing types by taking a nonfactored load from the load chart and adjusting that value for the glass type, as well as the load duration. For wired glass, patterned glass or sandblasted glass, the nonfactored load from ASTM E 1300 is utilized. The factors in the corresponding code equations adjust the nonfactored load to establish a corresponding resistance value for these glass types. Note that a factor of 0.5 is used with the nonfactored load determined in accordance with ASTM E 1300. This equates to a factor of 2 on the wind loads determined in Section 1609.

2404.3.2 Sloped wired glass. Wired glass sloped more than 15 degrees (0.26 rad) from vertical in skylights, sunspaces, sloped roofs and other exterior applications shall be designed to resist the most critical of the combinations of loads from Section 2404.2.

For Equations 24-2 and 24-3:

$F_g < 0.5\ F_{ge}$ **(Equation 24-7)**

For Equation 24-4:

$F_g < 0.3\ F_{ge}$ **(Equation 24-8)**

where:

F_g = Total load on the glass.

F_{ge} = Nonfactored load from ASTM E 1300.

❖ Wired glass is not within the scope of ASTM E 1300 and the requirements for sloped wired glass are therefore specified in the code. As with other sloped glazing, the maximum allowable load must be greater than the required loads. The required design loads are first determined from the three load combinations listed as Equations 24-2, 24-3 and 24-4. The applicable wind loads from the components and cladding requirements of Section 1609 and the snow loads from Section 1608 are used to determine the required design loads from the three load combinations.

After the required design loads are determined, the maximum allowable load resistance of the specified glass is determined in accordance with ASTM E 1300. ASTM E 1300 provides the method for determining values of short-duration and long-duration load resistance for specific glazing types by taking a nonfactored load from the load chart and adjusting that value for the glass type, as well as the load duration. For wired glass, patterned glass or sandblasted glass, the nonfactored load from ASTM E 1300 is utilized. The factors in the corresponding code equations adjust the nonfactored load to establish a corresponding resistance value for these glass types. Note that factors of either 0.5 or 0.3 are used with the nonfactored load determined in accordance with ASTM E 1300. This equates to a factor of 2 or 3.33 on the loads determined in the corresponding load combination. The resistance for sloped wired glass, patterned glass or sandblasted glass is to be determined using the nonfactored load from ASTM E 1300. The code fails to specify which type of glazing is appropriate to establish this resistance, but Section 2404.3.1 uses the values for monolithic glass; therefore, it is likely that the intent is to use the nonfactored load for monolithic glass in this section as well.

2404.3.3 Vertical patterned glass. Patterned glass sloped 15 degrees (0.26 rad) or less from vertical in windows, curtain and window walls, doors and other exterior applications shall be designed to resist the wind loads in Section 1609 for components and cladding according to the following equation:

$F_{gw} < 1.0\ F_{ge}$ **(Equation 24-9)**

where:

F_{gw} = Wind load on the glass computed per Section 1609.

F_{ge} = Nonfactored load from ASTM E 1300. The value for patterned glass shall be based on the thinnest part of the glass. Interpolation between nonfactored load charts in ASTM E 1300 shall be permitted.

❖ Patterned glass is not within the scope of ASTM E 1300 and the requirements for vertical patterned glass are, therefore, specified in the code. As with other vertical glazing, the wind loading on the glass is required to be determined in accordance with Section 1609 of the code. The allowable load on vertical patterned glass is to be as contained in ASTM E 1300 for a thickness less than or equal to the thinnest portion specified for the patterned glass. ASTM E 1300 provides the method for determining values of short-duration and long-duration load resistance for specific glazing types by taking a nonfactored load from the load chart and adjusting that value for the glass type, as well as the load duration. For wired glass, patterned glass or sandblasted glass, the nonfactored load from ASTM E 1300 is utilized. The factors in the corresponding code equations adjust the nonfactored load to establish a corresponding resistance value for these glass types. The resistance for sloped wired glass, patterned glass or sandblasted glass is to be determined using the nonfactored load from ASTM E 1300. The code fails to specify which type of glazing is appropriate to establish this resistance, but Section 2404.3.1 uses the values for monolithic glass; therefore, it is likely that the intent is to use the nonfactored load for monolithic glass in this section as well.

2404.3.4 Sloped patterned glass. Patterned glass sloped more than 15 degrees (0.26 rad) from vertical in skylights, sunspaces, sloped roofs and other exterior applications shall be designed to resist the most critical of the combinations of loads from Section 2404.2.

For Equations 24-2 and 24-3:

$F_g < 1.0\ F_{ge}$ **(Equation 24-10)**

For Equation 24-4:

$F_g < 0.6F_{ge}$ **(Equation 24-11)**

where

F_g = Total load on the glass.

F_{ge} = Nonfactored load from ASTM E 1300. The value for patterned glass shall be based on the thinnest part of the glass. Interpolation between the nonfactored load charts in ASTM E 1300 shall be permitted.

❖ Patterned glass is not within the scope of ASTM E 1300 and the requirements for sloped patterned glass are therefore specified in the code. As with other sloped glazing, the maximum allowable load must be greater than the required loads. The required design loads are first determined from the three load combinations listed as Equations 24-2, 24-3 and 24-4. The applicable wind loads from the components and cladding requirements of Section 1609 and the snow loads from Section 1608 are used to determine the required design loads from the three load combinations.

The allowable load on vertical patterned glass is to be as contained in ASTM E 1300 for a thickness less than or equal to the thinnest portion specified for the patterned glass. ASTM E 1300 provides the method for determining values of short-duration and long-duration load resistance for specific glazing types by taking a nonfactored load from the load chart and adjusting that value for the glass type as well as the load duration. For wired glass, patterned glass or sandblasted glass, the nonfactored load from ASTM E 1300 is utilized. The factors in the corresponding code equations adjust the nonfactored load to establish a corresponding resistance value for these glass types. Note that factors of either 1.0 or 0.6 are used with the nonfactored load determined in accordance with ASTM E 1300. This equates to a factor of 1 or 1.66 on the loads determined in the corresponding load combinations. The resistance for sloped wired glass, patterned glass or sandblasted glass is to be determined using the nonfactored load from ASTM E 1300. The code fails to specify which type of glazing is appropriate to establish this resistance, but Section 2404.3.1 uses the values for monolithic glass; therefore, it is likely that the intent is to use the nonfactored load for monolithic glass in this section as well.

2404.3.5 Vertical sandblasted glass. Sandblasted glass sloped 15 degrees (0.26 rad) or less from vertical in windows, curtain and window walls, doors, and other exterior applications shall be designed to resist the wind loads in Section 1609 for components and cladding according to the following equation:

$$F_g < 0.5\, F_{ge} \qquad \textbf{(Equation 24-12)}$$

where:

F_g = Total load on the glass.

F_{ge} = Nonfactored load from ASTM E 1300. The value for sandblasted glass is for moderate levels of sandblasting.

❖ Sandblasted glass is not within the scope of ASTM E 1300 and the requirements for vertical sandblasted glass are, therefore, specified in the code. As with other vertical glazing, the wind loading on the glass is required to be determined in accordance with Section 1609 of the code. The allowable load on vertical patterned glass is to be as contained in ASTM E 1300. ASTM E 1300 provides the method for determining values of short-duration and long-duration load resistance for specific glazing types by taking a nonfactored load from the load chart and adjusting that value for the glass type as well as the load duration. For wired glass, patterned glass or sandblasted glass, the nonfactored load from ASTM E 1300 is utilized. The factors in the corresponding code equations adjust the nonfactored load to establish a corresponding resistance value for these glass types. Note that a factor of 0.5 is used with the nonfactored load determined in accordance with ASTM E 1300. This equates to a factor of 2 on the wind loads determined in Section 1609. The resistance for sloped wired glass, patterned glass or sandblasted glass is to be determined using the nonfactored load from ASTM E 1300. The code fails to specify which type of glazing is appropriate to establish this resistance, but Section 2404.3.1 uses the values for monolithic glass; therefore, it is likely that the intent is to use the nonfactored load for monolithic glass in this section as well.

2404.4 Other designs. For designs outside the scope of this section, an analysis or test data for the specific installation shall be prepared by a *registered design professional*.

❖ This section recognizes that neither the code nor the referenced standard includes all types of glass and all support conditions. Therefore, if a designer proposes an installation that falls outside the scope of the code and referenced standard, this section requires an analysis to be prepared and submitted to the building official for specific approval.

SECTION 2405
SLOPED GLAZING AND SKYLIGHTS

2405.1 Scope. This section applies to the installation of glass and other transparent, translucent or opaque glazing material installed at a slope more than 15 degrees (0.26 rad) from the vertical plane, including glazing materials in skylights, roofs and sloped walls.

❖ This section covers all applications of glazing and light-transmitting plastics installed on a slope more than 15 degrees (0.26 rad) from vertical. The provisions for determining the combination of wind, snow and dead loads that sloped glazing must resist are found in Section 2404. For additional information on the design of sloped glazing and skylights, refer to *Glass Design for Sloped Glazing*, available from the American Architectural Manufacturers Association (AAMA).

"Sloped glazing" is defined as any glazing system or material installed on a slope more than 15 degrees (0.26 rad) from vertical. Where glass is installed within 15 degrees (0.26 rad) of vertical, the component of the weight (dead load) that acts normal to the plane of the glass is relatively small and does not contribute materially to the total load. From an engineering standpoint, glass within 15 degrees (0.26 rad) of vertical will perform essentially the same as glass installed vertically.

Sloped glazing is not intended to specifically exclude any type of glass or glazing system from the requirements of Chapter 24. Opaque glass, such as spandrel glass, as well as transparent and translucent glass, both opaque and light-transmitting plastics, decorative glass and glazing, and any other glazing-type material contained in the chapter are subject to the design and structural requirements of this chapter, Chapter 16 and related chapters elsewhere in the code.

2405.2 Allowable glazing materials and limitations. Sloped glazing shall be any of the following materials, subject to the listed limitations.

1. For monolithic glazing systems, the glazing material of the single light or layer shall be laminated glass with a minimum 30-mil (0.76 mm) polyvinyl butyral (or

equivalent) interlayer, wired glass, light-transmitting plastic materials meeting the requirements of Section 2607, heat-strengthened glass or fully tempered glass.

2. For multiple-layer glazing systems, each light or layer shall consist of any of the glazing materials specified in Item 1 above.

Annealed glass is permitted to be used as specified within Exceptions 2 and 3 of Section 2405.3.

For additional requirements for plastic skylights, see Section 2610. Glass-block construction shall conform to the requirements of Section 2101.2.5.

❖ All configurations of glazing and plastic are covered in this section. This includes single (monolithic) glass and plastic, insulating glass units consisting of any number of glass panes (but usually two), double-walled plastic skylights and any combination of glass or plastic panels separately framed and inserted into a single opening (such as a prime window and a storm window).

Laminated glass is required to have a minimum 30-mil-thick (0.76 mm) plastic interlayer. Approved plastic materials are those that comply with the provisions of the code for strength, durability, sanitation and fire performance required for installation. Acrylics (e.g., Plexiglas) and polycarbonates (e.g., Lexan) are examples of plastics that are commonly found in sloped glazing installations.

2405.3 Screening. Where used in monolithic glazing systems, heat-strengthened glass and fully tempered glass shall have screens installed below the glazing material. The screens and their fastenings shall: (1) be capable of supporting twice the weight of the glazing; (2) be firmly and substantially fastened to the framing members and (3) be installed within 4 inches (102 mm) of the glass. The screens shall be constructed of a noncombustible material not thinner than No. 12 B&S gage (0.0808 inch) with mesh not larger than 1 inch by 1 inch (25 mm by 25 mm). In a corrosive atmosphere, structurally equivalent noncorrosive screen materials shall be used. Heat-strengthened glass, fully tempered glass and wired glass, when used in multiple-layer glazing systems as the bottom glass layer over the walking surface, shall be equipped with screening that conforms to the requirements for monolithic glazing systems.

Exception: In monolithic and multiple-layer sloped glazing systems, the following applies:

1. Fully tempered glass installed without protective screens where glazed between intervening floors at a slope of 30 degrees (0.52 rad) or less from the vertical plane shall have the highest point of the glass 10 feet (3048 mm) or less above the walking surface.

2. Screens are not required below any glazing material, including annealed glass, where the walking surface below the glazing material is permanently protected from the risk of falling glass or the area below the glazing material is not a walking surface.

3. Any glazing material, including annealed glass, is permitted to be installed without screens in the sloped glazing systems of commercial or detached noncombustible greenhouses used exclusively for growing plants and not open to the public, provided that the height of the greenhouse at the ridge does not exceed 30 feet (9144 mm) above grade.

4. Screens shall not be required within individual *dwelling units* in Groups R-2, R-3 and R-4 where fully tempered glass is used as single glazing or as both panes in an insulating glass unit, and the following conditions are met:

   4.1. Each pane of the glass is 16 square feet (1.5 m^2) or less in area.

   4.2. The highest point of the glass is 12 feet (3658 mm) or less above any walking surface or other accessible area.

   4.3. The glass thickness is $^3/_{16}$ inch (4.8 mm) or less.

5. Screens shall not be required for laminated glass with a 15-mil (0.38 mm) polyvinyl butyral (or equivalent) interlayer used within individual *dwelling units* in Groups R-2, R-3 and R-4 within the following limits:

   5.1. Each pane of glass is 16 square feet (1.5 m^2) or less in area.

   5.2. The highest point of the glass is 12 feet (3658 mm) or less above a walking surface or other accessible area.

❖ In sloped applications, certain glass types have a tendency to fall from the opening when broken. As a result, a retaining net or screening is required in specific locations. Regular (annealed) glass may not be used in sloped glazing installations, except as defined in Section 2405.3. Heat-strengthened and fully tempered glazing are two types of glass that may fall from an opening when broken by wind or snow loads. For these glasses, a retaining screen is required on the interior side of the glazing unit. This would apply to single glazing or when the glass is the inboard pane in an insulating glass unit. Other exceptions for fully tempered glass are listed in Section 2405.3.

Wired glass, when used as the inboard pane in an insulating glass unit, has a tendency to crack from thermal stresses. This is due to weakening at the edges where the wire protrudes. Where fire-resistance-rated glazing material is needed, wired glass is commonly used. In this case, because of the increased likelihood of breakage, a retaining screen is also required.

There are both performance and prescriptive requirements for the retaining screen. Any screen that meets the performance requirements is acceptable.

Five exceptions to the requirements for a retaining screen are listed. For the purposes of this section, patterned glass that is annealed is to be considered regular (annealed) glass:

1. Fully tempered glass does not require a protective screen when the glass is sloped 30 degrees (0.52 rad) or less from vertical, and the top edge is less than 10 feet (3048 mm) above a walking surface. The sloped walls in many commercial buildings are within these limitations. The glass has been satisfactory with no known risks. When broken, fully tempered glass will break into small fragmented pieces, which are unlikely to cause harm upon impact [see Figure 2405.3(1)].
2. In cases where the area below the glass is either permanently protected or inaccessible, falling glass does not present a risk. When these conditions are met, screens are not required below the sloped glazing [see Figure 2405.3(2)].
3. Glass in greenhouses does not pose a serious risk to the general public. The failure of a sloped glazing system in a private greenhouse is not considered a high risk, as the majority of floor space is taken up by plants and the building is not occupied on a regular basis by a large number of people. Where a greenhouse is not attached to any other structure, has restricted access and meets the specified height limitation, screens are not required below the sloped glazing.
4. Sloped glazing installations using fully tempered glass are not required to have screens installed below the glass when the glazing units or skylights are installed within individual dwelling units of Group R-2, R-3 and R-4 occupancies, and the glazing complies with the size, thickness and location requirements specified in Items 4.1, 4.2 and 4.3. This exception is applicable only to glazing installations that are located within a dwelling unit; it does not apply to any public or common areas. It is also not applicable to areas in buildings containing two-family and multiple- family dwellings that are common to all occupants (e.g., entry foyers, main corridors, community storage and laundry rooms).

   This exception is based on a survey conducted by the glass and glazing industry. The information produced from the survey indicated a 0.081-percent incidence of breakage among fully tempered sun spaces and skylights. More importantly, the survey found no incidences of injury were reported when the glazing was fully tempered glass installed within the limitations specified in the exception.
5. This exception is permitted because of the protection afforded by the polyvinyl butyral interlayer located within the laminated glass panel.

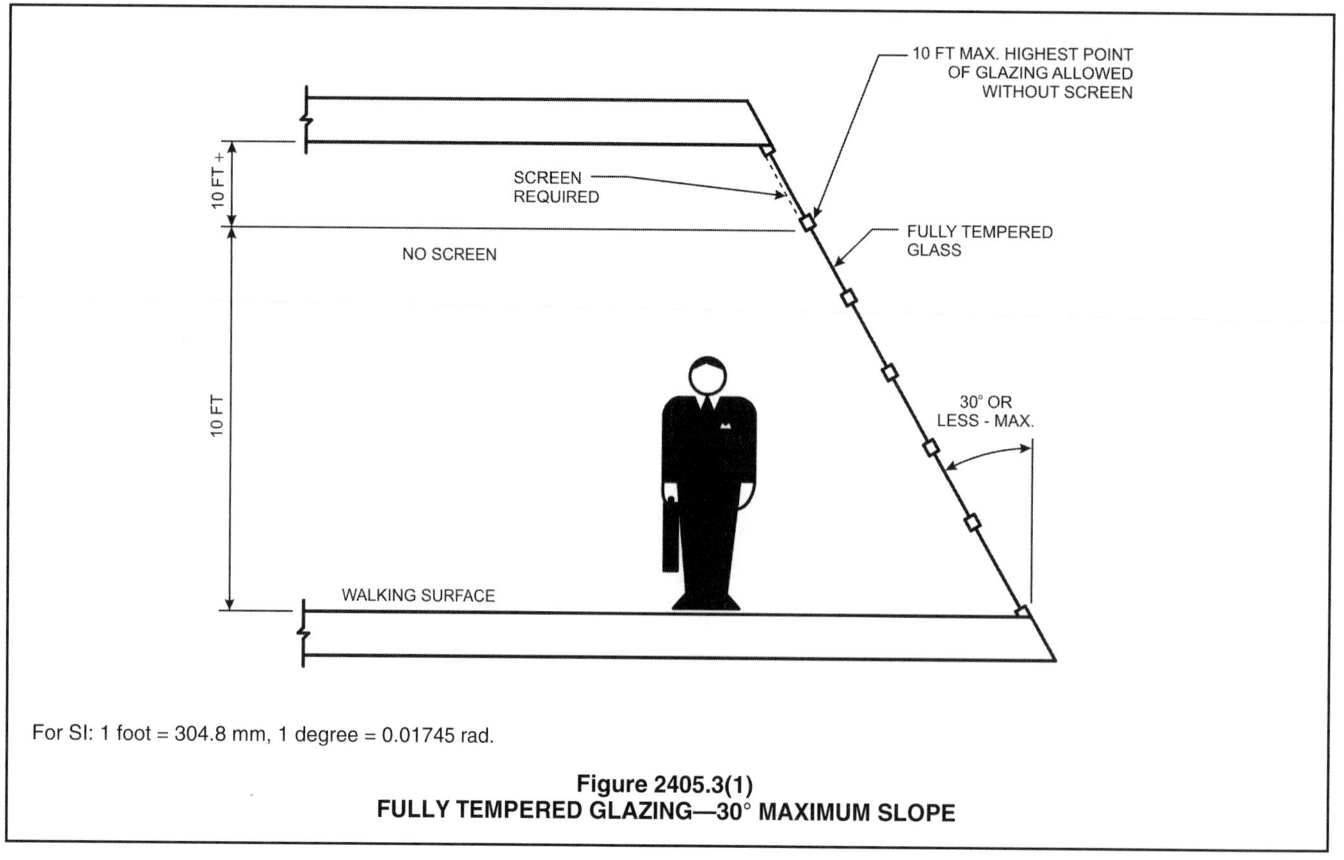

For SI: 1 foot = 304.8 mm, 1 degree = 0.01745 rad.

Figure 2405.3(1)
FULLY TEMPERED GLAZING—30° MAXIMUM SLOPE

The interlayer serves to keep the broken pieces of glass in place so that little or no glass falls when broken by impact. This exception applies only to individual dwelling units within Group R-2, R-3 and R-4 occupancies when sized and located as prescribed in Items 5.1 and 5.2.

2405.4 Framing. In Type I and II construction, sloped glazing and skylight frames shall be constructed of noncombustible materials. In structures where acid fumes deleterious to metal are incidental to the use of the buildings, *approved* pressure-treated wood or other *approved* noncorrosive materials are permitted to be used for sash and frames. Framing supporting sloped glazing and skylights shall be designed to resist the tributary roof loads in Chapter 16. Skylights set at an angle of less than 45 degrees (0.79 rad) from the horizontal plane shall be mounted at least 4 inches (102 mm) above the plane of the roof on a curb constructed as required for the frame. Skylights shall not be installed in the plane of the roof where the roof pitch is less than 45 degrees (0.79 rad) from the horizontal.

Exception: Installation of a skylight without a curb shall be permitted on roofs with a minimum slope of 14 degrees (three units vertical in 12 units horizontal) in Group R-3 occupancies. All unit skylights installed in a roof with a pitch flatter than 14 degrees (0.25 rad) shall be mounted at least 4 inches (102 mm) above the plane of the roof on a curb constructed as required for the frame unless otherwise specified in the manufacturer's installation instructions.

❖ The requirement for noncombustible (metal) framing for glazing systems in Type I and II construction must be consistent with the requirements for the remainder of the structure. In cases where acidic fumes harmful to metal framing may occur, pressure-treated wood or noncorrosive material may be used.

Regardless of the type of construction, when installed on surfaces having a slope of less than 45 degrees (0.79 rad) from the horizontal, all sloped glazing and skylights are required to be mounted on curbs at least 4 inches (102 mm) high. The raised curb will provide additional protection to the skylight from burnt or burning debris in the event of a fire. Burnt or burning pieces of wood are known technically as "brands." When air borne, they are known as "flying brands." For slopes 45 degrees (0.79 rad) and greater, the protection provided by the curbs is not necessary. In these cases, the glass may be installed flush with the surrounding roof.

The exception for Group R-3 occupancies is based on field experience and general practice for one- and two-family dwellings. For the 3:12 slope typical for Group R-3 occupancies neither flying brands nor water runoff has been a problem.

2405.5 Unit skylights. Unit skylights shall be tested and labeled as complying with AAMA/WDMA/CSA 101/I.S./A440. The *label* shall state the name of the manufacturer, the *approved* labeling agency, the product designation and the performance grade rating as specified in AAMA/WDMA/CSA 101/I.S.2/A440. If the product manufacturer has chosen to have the performance grade of the skylight rated separately for positive and negative design pressure, then the *label* shall state both performance grade ratings as specified in AAMA/WDMA/CSA 101/I.S.2/A440 and the skylight shall comply

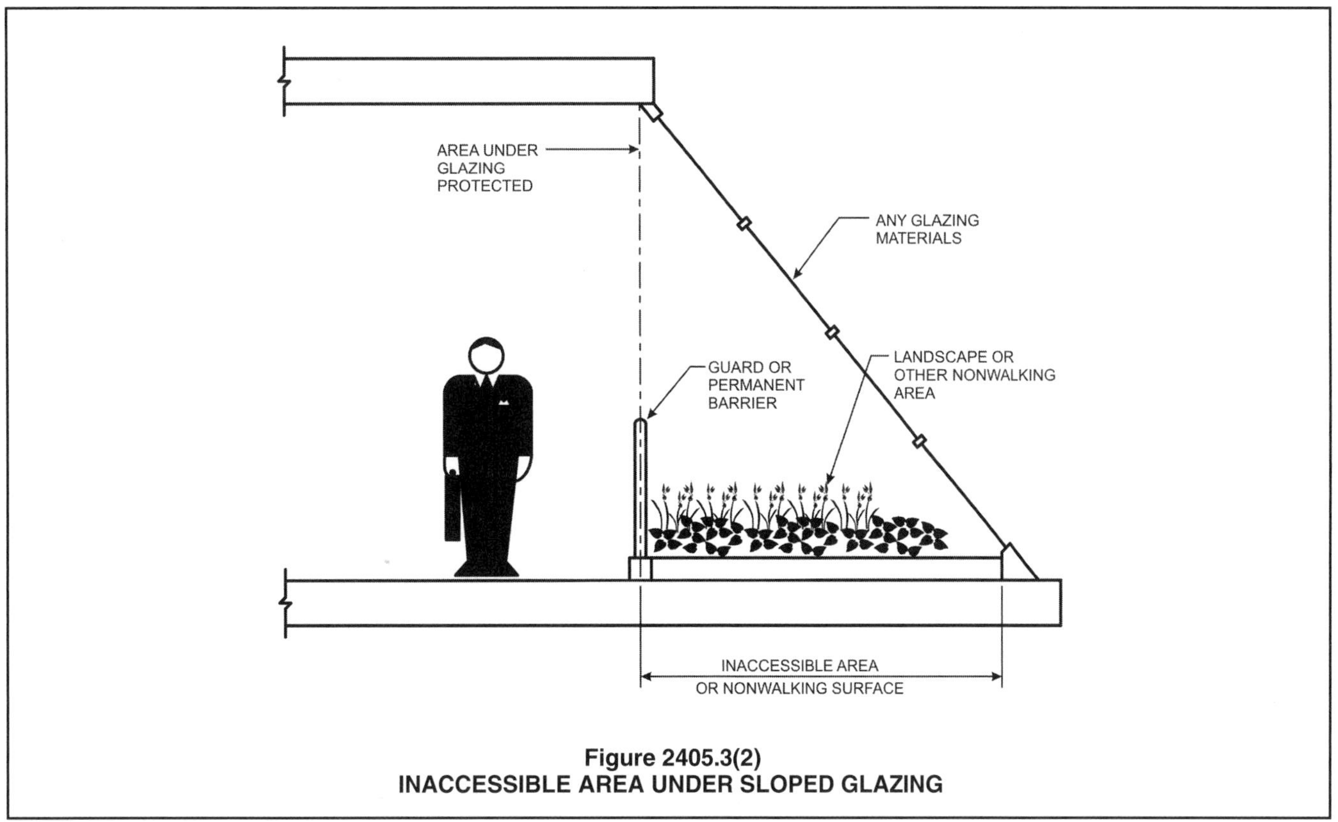

Figure 2405.3(2)
INACCESSIBLE AREA UNDER SLOPED GLAZING

with Section 2405.5.2. If the skylight is not rated separately for positive and negative pressure, then the performance grade rating shown on the *label* shall be the performance grade rating determined in accordance with AAMA/WDMA/CSA 101/I.S.2/A440 for both positive and negative design pressure and the skylight shall conform to Section 2405.5.1.

❖ This section establishes design provisions for unit skylights (single-panel, factory-assembled units) that are consistent with those provided for sloped glazing in Section 2404.2, but permit the unit skylights to be designed separately for maximum positive and negative pressure. The most critical load on a skylight is determined by the climate in which it is installed. In a colder climate with heavier snow loads and moderate design wind speeds, the positive load on a skylight from the combined snow and dead loads will be more critical than the negative load from wind uplift. The opposite will be the case in warmer, coastal climates with higher design wind speeds and little or no snow load.

This method of rating skylights is addressed in the referenced standard AAMA/WMDA/CSA 101/I.S.2/A440, *Specification for Windows, Doors and Unit Skylights*. The separate rating system for positive and negative pressure on skylights allows the manufacturer to design and fabricate products that are best suited for the climate in which they will be used. AAMA/ WDMA/CSA 101/I.S.2/A440 establishes the performance requirements for skylights based on the desired performance grade rating, which includes minimum requirements for resistance to air leakage, water infiltration and design load pressures. The resulting performance grade rating states the design load pressure used to rate the product, but also includes consideration of these additional performance characteristics. For skylights certified for only one performance grade, the rating is based on the minimum requirements met for both positive and negative design pressure. Skylights certified for two performance grades are rated separately for positive and negative design pressure.

2405.5.1 Unit skylights rated for the same performance grade for both positive and negative design pressure. The design of unit skylights shall be based on the following equation:

$F_g \leq PG$ **(Equation 24-13)**

where:

F_g = Maximum load on the skylight determined from Equations 24-2 through 24-4 in Section 2404.2.

PG = Performance grade rating of the skylight.

❖ For skylights that are not rated separately for positive and negative design pressure, this section requires that the minimum performance grade rating of the skylight must not be exceeded by any of the load combinations given in Section 2404.2.

2405.5.2 Unit skylights rated for separate performance grades for positive and negative design pressure. The design of unit skylights rated for performance grade for both positive and negative design pressures shall be based on the following equations:

$F_{gi} \leq PG_{Po}$ **(Equation 24-14)**

$F_{go} \leq PG_{Ne}$ **(Equation 24-15)**

where:

PG_{Pos} = Performance grade rating of the skylight under positive design pressure;

PG_{Neg} = Performance grade rating of the skylight under negative design pressure; and

F_{gi} and F_{go} are determined in accordance with the following:

For $W_o \geq D$,

where:

W_o = Outward wind force, psf (kN/m^2) as calculated in Section 1609.

D = The dead weight of the glazing, psf (kN/m^2) as determined in Section 2404.2 for glass, or by the weight of the plastic, psf (kN/m^2) for plastic glazing.

F_{gi} = Maximum load on the skylight determined from Equations 24-3 and 24-4 in Section 2404.2.

F_{go} = Maximum load on the skylight determined from Equation 24-2.

For $W_o < D$,

where:

W_o = Is the outward wind force, psf (kN/m^2) as calculated in Section 1609.

D = The dead weight of the glazing, psf (kN/m^2) as determined in Section 2404.2 for glass, or by the weight of the plastic for plastic glazing.

F_{gi} = Maximum load on the skylight determined from Equations 24-2 through 24-4 in Section 2404.2.

F_{go} = 0.

❖ This section establishes requirements for skylights that are rated separately for positive and negative design pressure. The performance grade rating for positive design pressure is not to be exceeded by load combinations that include dead loads, snow and wind acting towards the face of the skylight, and the performance grade rating for negative design pressure is not to be exceeded by the load combination that considers wind acting away from the face of the skylight.

SECTION 2406
SAFETY GLAZING

2406.1 Human impact loads. Individual glazed areas, including glass mirrors, in hazardous locations as defined in

Section 2406.4 shall comply with Sections 2406.1.1 through 2406.1.4.

Exception: Mirrors and other glass panels mounted or hung on a surface that provides a continuous backing support.

❖ With the increased use of sliding patio doors, glass tub and shower enclosures and similar applications where large pieces of glass are used, cutting and piercing injuries from contact with broken glass have become a safety concern. In recognition of a lack of code provisions controlling these risks, codes included limited requirements that regulated the use of glass. It was recognized that safety glazing materials must be required for glazed areas that would reasonably be exposed to human impact. Some questions arose as to the proper definition of these areas (Section 2406.4 identifies these "hazardous locations"). In response, Section 2406 provides comprehensive regulations for the use of safety glazing materials and installation parameters.

Safety glazing requirements apply to both replacement glass and new construction. Replacement glass includes windows, doors and other assemblies containing glass that are moved, partially or intact, from one location in a building to another location in the same building, or to another building on the same site or a different site.

While glass mirrors are included within the scope of glazing, mirrors with a continuous backing support are not required to be safety glazed according to the exception. This exception recognizes the potential for minimizing cutting or piercing injuries when the glass or glazing material is completely supported by a solid, continuous backing material, as is often the case for mounted mirrors or other glass panels. When glazing is fully supported with a solid backing material, it still may be susceptible to breakage by human impact; however, a person's limb or other body part would be protected because the backing material would prevent it from being in the path of falling glass.

2406.1.1 Impact test. Except as provided in Sections 2406.1.2 through 2406.1.4, all glazing shall pass the impact test requirements of Section 2406.2.

❖ Sections 2406.1.2 through 2406.1.4 are essentially exceptions to the performance standards for safety glazing that are given in Section 2406.2.

2406.1.2 Plastic glazing. Plastic glazing shall meet the weathering requirements of ANSI Z97.1.

❖ This section has requirements for weather testing plastic safety glazing that have been eliminated from CPSC 16 CFR, Part 1201, necessitating the use of ANSI Z97.1 for plastic glazing. Requirements for plastic glazing are in Chapter 26.

2406.1.3 Glass block. Glass-block walls shall comply with Section 2101.2.5.

❖ The design and installation requirements for glass-block walls are covered in Chapter 21. Glass-block walls and panels are also not required to meet the impact test requirements; however, that does not mean there are no safety requirements placed on the installation of glass block. Structural limitations to prevent failure caused by excessive loading, size limitations of panels, location and extent of panels in fire-resistance-rated assemblies and other safety issues are defined elsewhere in the code.

2406.1.4 Louvered windows and jalousies. Louvered windows and jalousies shall comply with Section 2403.5.

❖ Louvered and jalousie windows are discussed in detail in Section 2403.5. Jalousie and louvered windows are not classified as hazardous locations and are not required to be glazed with safety glass or other approved safety glazing material.

2406.2 Impact test. Where required by other sections of this code, glazing shall be tested in accordance with CPSC 16 CFR Part 1201. Glazing shall comply with the test criteria for Category II, unless otherwise indicated in Table 2406.2(1).

Exception: Glazing not in doors or enclosures for hot tubs, whirlpools, saunas, steam rooms, bathtubs and showers shall be permitted to be tested in accordance with ANSI Z97.1. Glazing shall comply with the test criteria for Class A, unless otherwise indicated in Table 2406.2(2).

❖ In 1977, the CPSC adopted as a mandatory federal safety regulation CPSC 16 CFR, Part 1201, *Safety Standard for Architectural Glazing Materials*. The standard is not only a test method and a procedure for determining the safety performance of architectural glazing, but also a federal standard that mandates where safety glazing must be used in architectural applications. The code specifically requires the installation of safety glazing materials meeting CPSC 16 CFR, Part 1201 in storm doors, combination doors, entrance-exit doors, sliding patio doors, closet doors and shower and tub doors and enclosures (see Section 2406.4.1).

The exception permits glazing complying with ANSI Z97.1 at locations where the code does not mandate compliance solely with CPSC 16 CFR, Part 1201 (see Sections 2406.4.2 through 2406.4.7). The primary difference between the CPSC 16 CFR, Part 1201 standard and the ANSI Z97.1 standard relates to their scope and function. ANSI Z97.1 is a voluntary safety performance specification and test method that does not dictate where safety glazing materials must be used, leaving that determination up to the code.

Not all of the hazardous locations are indicated in Tables 2406.2(1) and 2406.2(2). The code uses the more stringent test criteria as the default classification and the tables relax the requirement by permitting a lower classification for specific applications.

TABLE 2406.2(1). See page 24-14.

❖ The classifications in this table are according to the CPSC 16 CFR, Part 1201, regulation. The tests in the federal regulation vary with the category classification. Category I glass is impacted from a drop height of 18 inches (457 mm) and Category II is impacted from a drop height of 48 inches (1219 mm). The higher drop height for Category II results in a higher impact test load, making Category II the more stringent classification.

TABLE 2406.2(2). See below.

❖ ANSI Z97.1 identifies three separate classes that are based upon impact performance. Class A glazing materials are comparable to the CPSC's Category II glazing materials, passing a 48-inch (1219 mm) drop height test, and Class B glazing materials are comparable to the CPSC's Category I glazing materials, passing the 18-inch (457 mm) drop height test. ANSI Z97.1 also has a product-specific Class C impact test, which uses a 12-inch (303 mm) drop height test and is applicable only for fire-resistant glazing materials; however, Section 2406.2 does not recognize Class C as an acceptable product for use in locations requiring safety glazing.

2406.3 Identification of safety glazing. Except as indicated in Section 2406.3.1, each pane of safety glazing installed in hazardous locations shall be identified by a manufacturer's designation specifying who applied the designation, the manufacturer or installer and the safety glazing standard with which it complies, as well as the information specified in Section 2403.1. The designation shall be acid etched, sand blasted, ceramic fired, laser etched, embossed or of a type that once applied, cannot be removed without being destroyed. A label as defined in Section 202 and meeting the requirements of this section shall be permitted in lieu of the manufacturer's designation.

Exceptions:

1. For other than tempered glass, manufacturer's designations are not required, provided the *building official* approves the use of a certificate, affidavit or other evidence confirming compliance with this code.
2. Tempered spandrel glass is permitted to be identified by the manufacturer with a removable paper designation

❖ This section requires that each individual piece of safety glazing must be permanently marked with the manufacturer's or installer's designation. The designation is applied directly to the glass by the manufacturer or installer and states that the product or material complies with the specified standards or requirements as defined in Section 2406.2.

Many manufacturers of safety glazing materials have their products certified by the Safety Glazing Certification Council (SGCC) to meet CPSC 16 CFR, Part 1201. This standard provides the basis for the marking of safety glazing materials.

2406.3.1 Multi-pane assemblies. Multi-pane glazed assemblies having individual panes not exceeding 1 square foot (0.09 m^2) in exposed areas shall have at least one pane in the assembly marked as indicated in Section 2406.3. Other panes in the assembly shall be marked "CPSC 16 CFR, Part 1201" or "ANSI Z97.1," as appropriate.

❖ In the case of glass material, many times the glass is purchased by the fabricator/installer in large sheets from the manufacturer and cut into many pieces to fit various openings. This is particularly true with laminated or plastic glazing material. In this instance, it is impracticable to apply the designation label in such a manner that each cut piece will have a manufacturer's permanent designation. Where this occurs, the code permits multilight glazed assemblies to have only one light marked with the required information.

2406.4 Hazardous locations. The locations specified in Sections 2406.4.1 through 2406.4.7 shall be considered specific hazardous locations requiring safety glazing materials.

❖ The provisions of this section apply to all occupancies and building types, except as specifically excluded in

TABLE 2406.2(1)
MINIMUM CATEGORY CLASSIFICATION OF GLAZING USING CPSC 16 CFR PART 1201

EXPOSED SURFACE AREA OF ONE SIDE OF ONE LITE	GLAZING IN STORM OR COMBINATION DOORS (Category class)	GLAZING IN DOORS (Category class)	GLAZED PANELS REGULATED BY SECTION 2406.4.3 (Category class)	GLAZED PANELS REGULATED BY SECTION 2406.4.2 (Category class)	DOORS AND ENCLOSURES REGULATED BY SECTION 2406.4.5 (Category class)	SLIDING GLASS DOORS PATIO TYPE (Category class)
9 square feet or less	I	I	No requirement	I	II	II
More than 9 square feet	II	II	II	II	II	II

For SI: 1 square foot = 0.0929 m^2.

TABLE 2406.2(2)
MINIMUM CATEGORY CLASSIFICATION OF GLAZING USING ANSI Z97.1

EXPOSED SURFACE AREA OF ONE SIDE OF ONE LITE	GLAZED PANELS REGULATED BY SECTION 2406.4.3 (Category class)	GLAZED PANELS REGULATED BY SECTION 2406.4.2 (Category class)	DOORS AND ENCLOSURES REGULATED BY SECTION 2406.4.5[a] (Category class)
9 square feet or less	No requirement	B	A
More than 9 square feet	A	A	A

For SI: square foot = 0.0929 m^2.

a. Use is only permitted by the exception to Section 2406.2.

this chapter. This section identifies locations that are considered to be hazardous and require the use of safety glass for glazed openings or partitioning. Several figures [(2406.4.1, 2406.4.2(1), 2406.4.2(3), 2406.4.3(1), 2406.4.3(2), 2406.4.3(3), 2406.4.5(1) and 2406.4.5(2)] are included to help illustrate how and where safety glazing must be used.

2406.4.1 Glazing in doors. Glazing in all fixed and operable panels of swinging, sliding, and bifold doors shall be considered a hazardous location.

Exceptions:

1. Glazed openings of a size through which a 3-inch-diameter (76 mm) sphere is unable to pass.
2. Decorative glazing.
3. Glazing materials used as curved glazed panels in revolving doors.
4. Commercial refrigerated cabinet glazed doors.

❖ This section makes it clear that any door containing glazing must be glazed with safety glass or other safety glazing material recognized by the code for that intended purpose. The requirement for safety glazing in not limited to egress or exit doors (see Figure 2406.4.1). Jalousie assemblies in doors, as described in Section 2403.5, are not required to have safety glazing.

There are four exceptions to this section:

1. Glass view panels 3 inches (76 mm) or less in diameter installed in doors are not likely to create a safety hazard if the glass is broken.
2. Decorative glass assemblies in doors are readily visible and easily identified; therefore, they are less likely to be inadvertently impacted. As a rule, stained glass assemblies are made from small pieces of glass that would cause little or no damage if broken.
3. The reason for not classifying revolving doors, which are generally made from curved glass panels, as a hazardous location is that the two primary safety concerns are either not present or pose minimal or no unsafe condition in the case of human impact.

   Several factors must be considered when classifying revolving doors. The first thing to be considered is that they cannot be used as part of the required means of egress. Another thing to consider is that revolving doors are not fixed glass panels that can be mistaken for a passageway or clear opening. In addition, the individual pieces of glass to make the large door

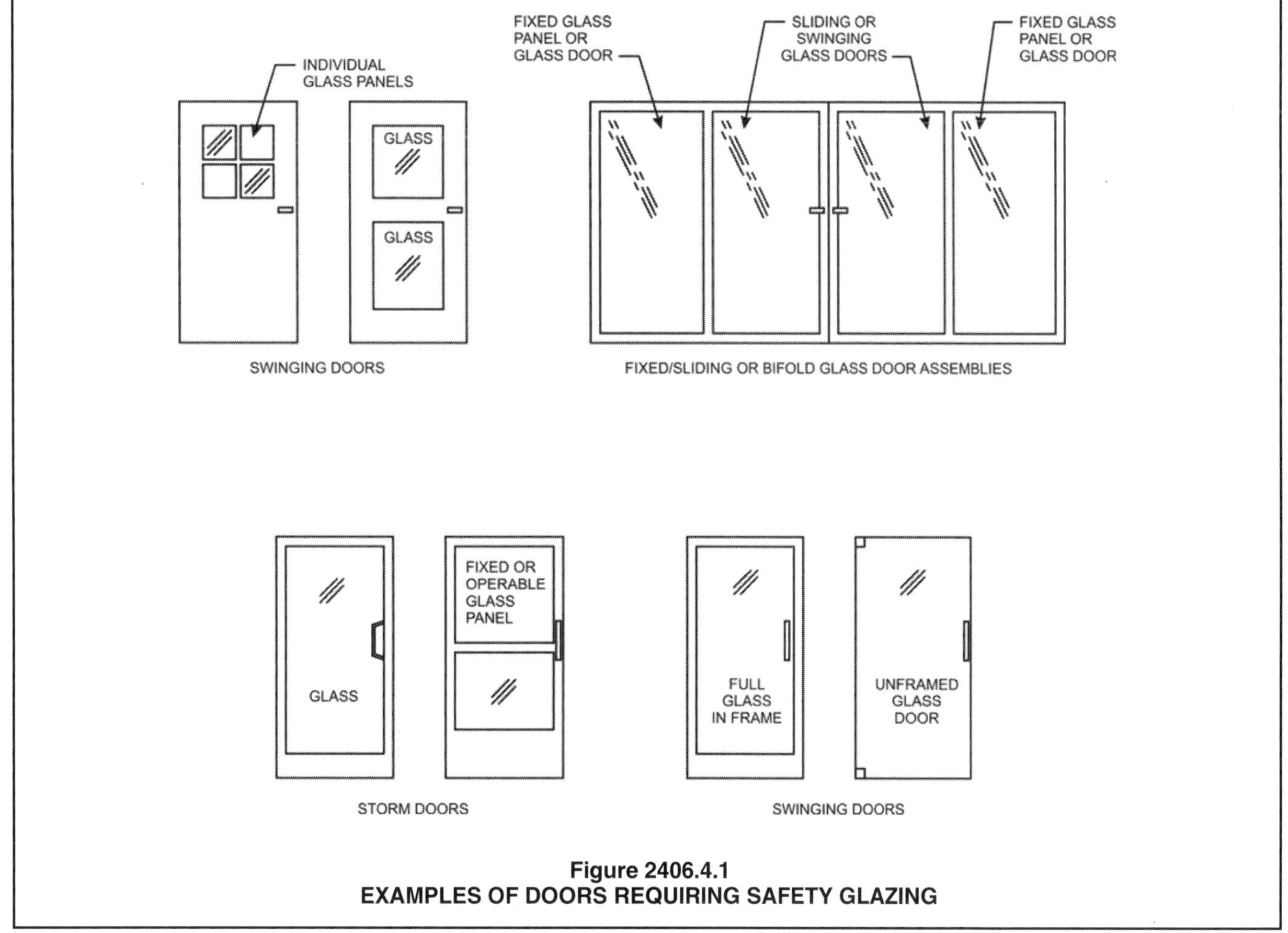

Figure 2406.4.1
EXAMPLES OF DOORS REQUIRING SAFETY GLAZING

panels, by necessity, are required to be very thick. The extra thickness of glass, coupled with the added strength of the curved form and the movable panel, make it highly unlikely that injury could result from glass breakage caused by human impact.

4. This exception is based on CPSC 16 CFR, Part 1201. Technically, glass in doors of refrigerator cabinets should comply with the standard when the door is open. The intent, however, was not to cover doors a person would not ordinarily use for egress. Refrigerated cabinets in food markets would be in this category and are, therefore, exempt. There are also no records of injuries from glass in this application.

2406.4.2 Glazing adjacent to doors. Glazing in an individual fixed or operable panel adjacent to a door where the nearest vertical edge of the glazing is within a 24-inch (610 mm) arc of either vertical edge of the door in a closed position and where the bottom exposed edge of the glazing is less than 60 inches (1524 mm) above the walking surface shall be considered a hazardous location.

Exceptions:

1. Decorative glazing.
2. Where there is an intervening wall or other permanent barrier between the door and glazing.
3. Where access through the door is to a closet or storage area 3 feet (914 mm) or less in depth. Glazing in this application shall comply with Section 2406.4.3.
4. Glazing in walls on the latch side of and perpendicular to the plane of the door in a closed position in one- and two-family dwellings or within dwelling units in Group R-2.

❖ The purpose for identifying the area adjacent to a door as a hazardous location is to provide protection in cases where a person may slip or mistake a glass panel adjacent to a door for a passageway and walk into the glass or where a person may push against the sidelight with one hand for support while opening the door with the other hand. There are reported accidents where a person's hand has slipped from the doorknob, impacting and breaking the glass adjacent to the door, thereby causing injury. This item is applicable to glass adjacent to both exterior and interior doors used for passage for all occupancies and types of buildings.

It is not necessary for an entire piece of glass in a glazed wall or opening to be within the 24-inch (610 mm) arc to require it to be safety glass. If any portion of an individual piece of glass is within the arc, that piece of glass must be safety glass [see Figure 2406.4.2(1)].

There are four exceptions that eliminate the need

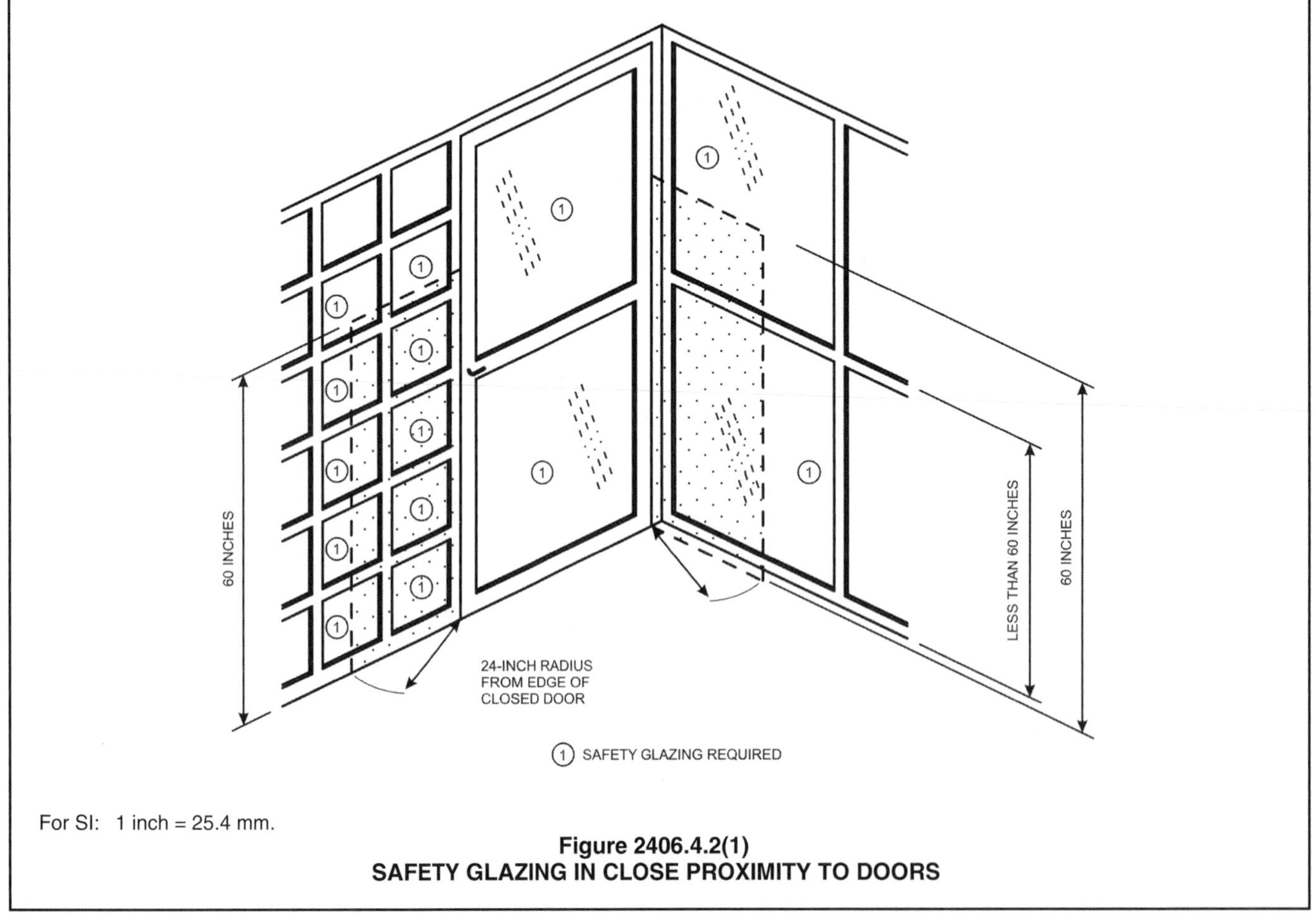

Figure 2406.4.2(1)
SAFETY GLAZING IN CLOSE PROXIMITY TO DOORS

for safety glazing adjacent or in close proximity to glass doors. These exceptions are included because they eliminate or greatly reduce the possibility of human impact with fixed glazing:

1. Decorative glass and decorative glass assemblies are readily visible and easily identified; therefore, they are not likely to be mistaken for an open passageway when they are adjacent to a door. As a rule, stained glass assemblies are made from small pieces of glass that would cause little or no damage if broken.
2. An intervening wall provides a permanent barrier that prevents people from having physical contact with any glass that is beyond the intervening wall, but is still within the 24-inch (610 mm) arc. This barrier eliminates many of the safety issues addressed in this section. This exception may be applied to an interior or exterior wall/glazing condition [see Figure 2406.4.2(2)]. The code does not specify a minimum height requirement for the intervening wall; therefore, where the wall does not extend to the full height of the rooms or spaces where the door/wall is located, the 60-inch (1524 mm) requirement must be applied. If, for example, the top of the wall is 60 inches (1524 mm) above the floor, any individual piece of glass that extends below the top of the wall must be safety glass or the top of the wall must be extended to provide a barrier for any piece of nonsafety glazing within 60 inches (1524 mm) of the floor. The same must be applied to the 24-inch (610 mm) arc in the plan dimension. Any glazing within the arc must be safety glazing, or the wall must be extended to include glazing not protected by the wall [see Figure 2406.4.2(3)].
3. This exception addresses closets and other storage areas that are not considered "walk-ins." The assumption is that at such locations a building occupant would not need to pass through the door opening to access the storage area, thus reducing the risk of impact with any adjacent glazing.
4. Because the occupants of one- and two-family dwellings are presumed to be familiar with their environment, and walls that are perpendicular to the plane of a door are, by definition, parallel to the direction of travel of an occupant using the door, the risk of impact with glazing is considered to be low. This decreased risk is recognized by Exception 4. Since there is a chance that an opening door could push a person into the wall toward which the door opens, the exception does not apply and safety glazing would be required at that location.

   Individual dwelling units within Group R-2 occupancies are allowed the same exemption; however, it does not extend to the use of glazing in close proximity to doors in public-use areas.

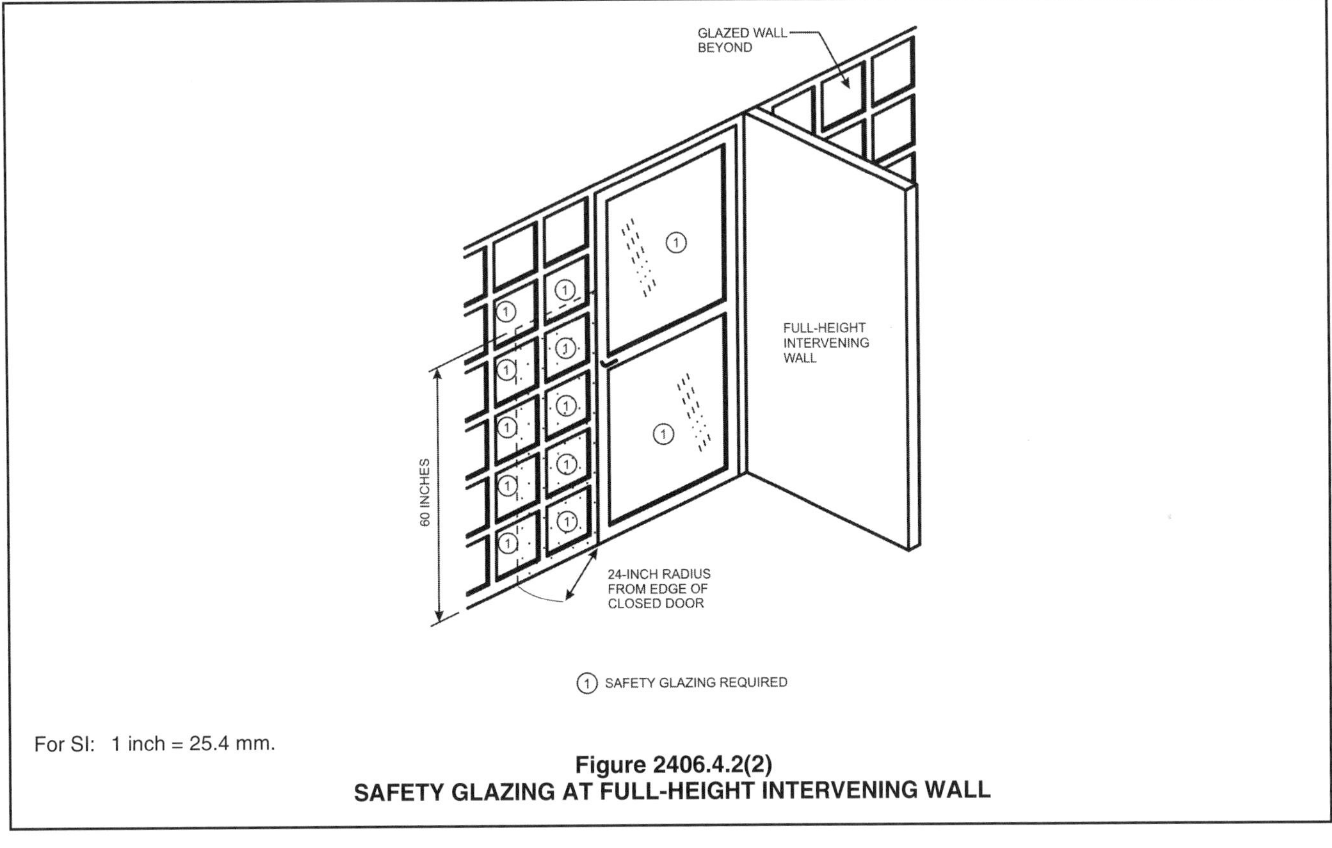

For SI: 1 inch = 25.4 mm.

Figure 2406.4.2(2)
SAFETY GLAZING AT FULL-HEIGHT INTERVENING WALL

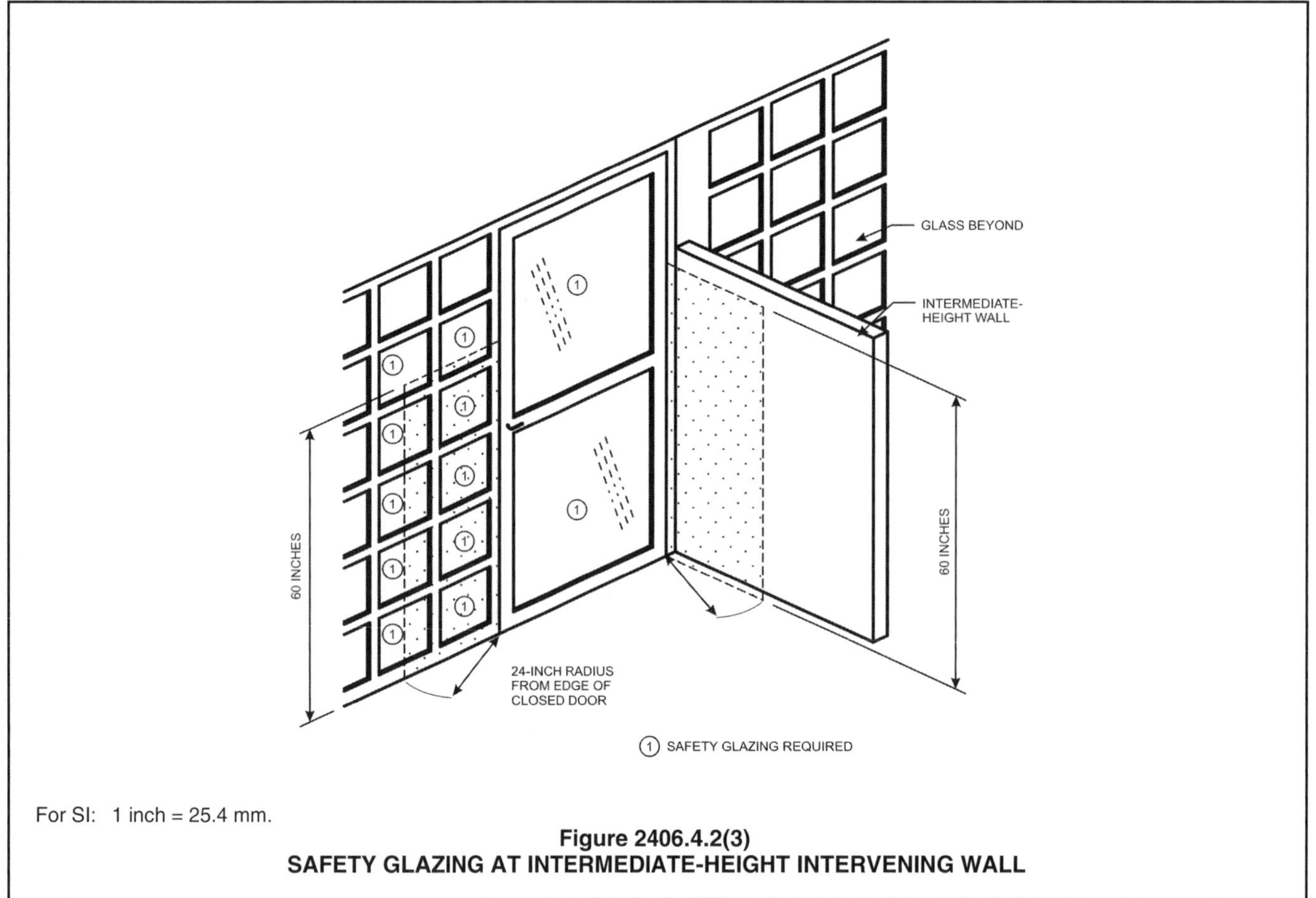

For SI: 1 inch = 25.4 mm.

Figure 2406.4.2(3)
SAFETY GLAZING AT INTERMEDIATE-HEIGHT INTERVENING WALL

2406.4.3 Glazing in windows. Glazing in an individual fixed or operable panel that meets all of the following conditions shall be considered a hazardous location:

1. The exposed area of an individual pane is greater than 9 square feet (0.84 m^2);
2. The bottom edge of the glazing is less than 18 inches (457 mm) above the floor;
3. The top edge of the glazing is greater than 36 inches (914 mm) above the floor; and
4. One or more walking surface(s) are within 36 inches (914 mm), measured horizontally and in a straight line, of the plane of the glazing.

Exceptions:

1. Decorative glazing.
2. Where a horizontal rail is installed on the accessible side(s) of the glazing 34 to 38 inches (864 to 965 mm) above the walking surface. The rail shall be capable of withstanding a horizontal load of 50 pounds per linear foot (730 N/m) without contacting the glass and be a minimum of $1^1/_2$ inches (38 mm) in cross-sectional height.
3. Outboard panes in insulating glass units or multiple glazing where the bottom exposed edge of the glass is 25 feet (7620 mm) or more above any grade, roof, walking surface or other horizontal or sloped (within 45 degrees of horizontal) (0.78 rad) surface adjacent to the glass exterior.

❖ The reason for including this type of glazed opening as a hazardous location is to provide protection where the glazed opening could be mistaken for a passageway or a clear opening that someone might be able to walk through, fall into or otherwise be accidentally forced into. In the case of a child, the opening would not need to be very large to provide the necessary setting to encourage an accident. A glazed opening sized and located within the four listed criteria and not protected with safety glazing presents an unsafe condition [see Figure 2406.4.3(1)].

The requirements in this section are modeled after the criteria established in CPSC 16 CFR, Part 1201, which requires the use of safety glazing where all four of the listed conditions occur.

In Item 1, the 9-square-foot (0.84 m^2) requirement refers to an individual piece or pane of glass. It is not intended to limit the number of panes that can be placed next to each other within a single glazed opening. It just requires that every individual piece exceeding the size specified be safety glass.

Item 2 applies when the bottom edge of a glazed opening is less than 18 inches (457 mm) above a floor or walking surface. This type of glazed opening does not present enough of a visual barrier to provide ample warning to prevent someone from mistaking the glazed opening for a passageway or clear opening. People have a natural tendency to look forward while walking rather than looking down; therefore, a more substantial visual barrier is needed at the floor level.

Where Item 2 addresses the sill height of glazed openings with no regard to how high the head of the opening may be, it assumes the opening is high enough for an adult to walk through. Item 3 addresses the head height of the glazed opening with the assumption that the sill is at the floor level. Item 3 is aimed at the same basic set of safety concerns; however, It is directing the focus at a different group of people: children. Just as adults can mistake a tall glazed opening for a passageway or clear opening, a small child can do the same for a glazed opening with a low head height. Other examples of potential accidents from impacting glass at this type of opening might include: people in wheelchairs, shopping carts, baby strollers (with or without baby inside) and children on tricycles or other small riding toys. The accident potential for this type of glazed opening is high; therefore, safety glass or other approved safety glazing material is demanded accordingly.

The focus of Item 4 for requiring safety glass is no different than it has been throughout this section, which is preventing unsafe glazed openings that people can mistake for passageways or clear openings. In Items 1 through 3, it states that where a walking surface is adjacent to the wall where the glazed opening is located, human contact with the glass in the opening is inevitable. Item 4 states that if there is not a walking surface within 36 inches (914 mm) horizontally, in conjunction with the other requirements, the risks described throughout this section are minimal, so safety glass is not needed. It is only when walking surfaces are within 36 inches (914 mm) that human contact becomes a safety concern that must be addressed.

The three exceptions are allowed because the concerns of this section are resolved or they do not exist. When these exceptions are met, there is no need for the use of safety glass or other approved safety glazing:

1. Decorative glass window assemblies are readily visible and easily identified, so they are much less likely to be mistaken for an open passageway. As a rule, stained glass assemblies are made from small pieces of glass that would cause little or no damage if broken.

2. The horizontal rail described in this exception will provide a visual, as well as a physical barrier, both of which will prevent someone from attempting to walk through a glazed opening [see Figure 2406.4.3(2)].

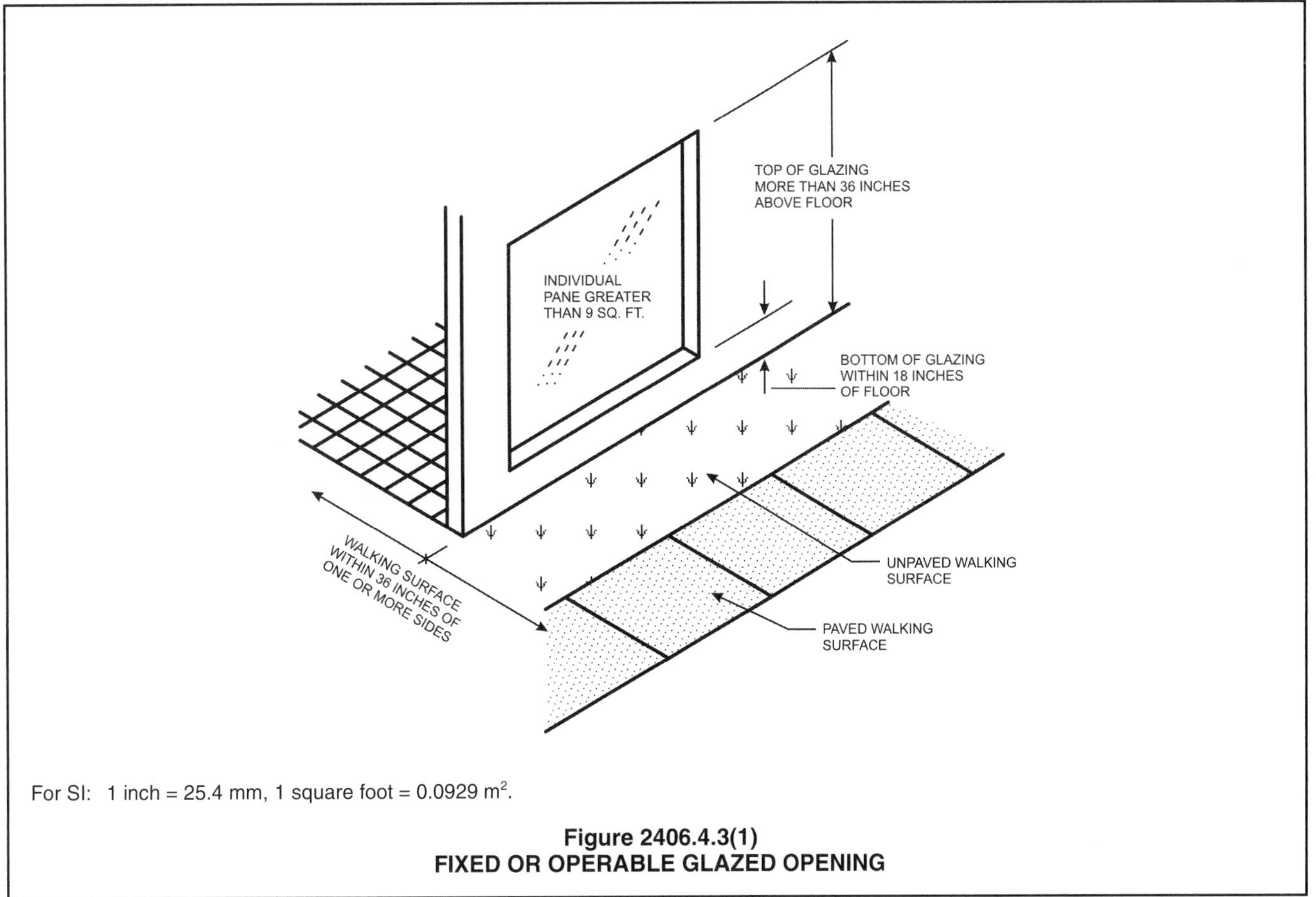

For SI: 1 inch = 25.4 mm, 1 square foot = 0.0929 m^2.

Figure 2406.4.3(1)
FIXED OR OPERABLE GLAZED OPENING

3. The outboard pane of insulating glass as described in this exception is well outside the area that any person could possibly come into contact with and certainly would not be mistaken for a passageway or clear opening. It should be noted that this exception applies only to the outboard pane. The criteria for requiring the inboard pane to be safety glass would be as prescribed in this section as well as elsewhere in this chapter [see Figure 2406.4.3(3)].

2406.4.4 Glazing in guards and railings. Glazing in *guards* and railings, including structural baluster panels and nonstructural in-fill panels, regardless of area or height above a walking surface shall be considered a hazardous location.

❖ This section states that, regardless of any provision for safety glazing or exemption from requiring safety glazing at any location or application contained in this section or elsewhere in the code—where glass or other approved glazing material is used in guards or railings, either wholly or partly and structural or nonstructural, regardless of size, location, type or any other circumstances not specifically mentioned herein—it must be safety glass or other approved safety glazing material. As there are no exceptions listed, any glazing material that is not available as an approved safety glazing material is not permitted in guards or railings.

2406.4.5 Glazing and wet surfaces. Glazing in walls, enclosures or fences containing or facing hot tubs, spas, whirlpools, saunas, steam rooms, bathtubs, showers and indoor or outdoor swimming pools where the bottom exposed edge of the glazing is less than 60 inches (1524 mm) measured vertically above any standing or walking surface shall be considered a hazardous location. This shall apply to single glazing and all panes in multiple glazing.

Exception: Glazing that is more than 60 inches (1524 mm), measured horizontally and in a straight line, from the water's edge of a bathtub, hot tub, spa, whirlpool, or swimming pool.

❖ The reason for requiring safety glazing at walls or fencing used for enclosing swimming pools, hot tubs and spas is the high amount of activity coupled with the wet walking surface conditions that are typically present in these types of facilities or spaces. The conditions presented in this section are focused on glazing in openings as viewed from the pool or spa side of the wall, fence or enclosure. Glazed openings in pool or spa walls viewed from the side of the wall away from the pool or spa are discussed elsewhere in this section. As a point of clarification, a wall, fence or enclosure refers to any wall that encloses a pool or spa area. It can be a wall constructed specifically to enclose an outdoor pool or spa; an existing or new exterior building wall; a residential garage wall; a dressing cabana or a shed for pool equipment. It can also be an interior or exterior wall. It applies to any wall that can support a glazed opening.

Glass wall partitions and glass doors that are not

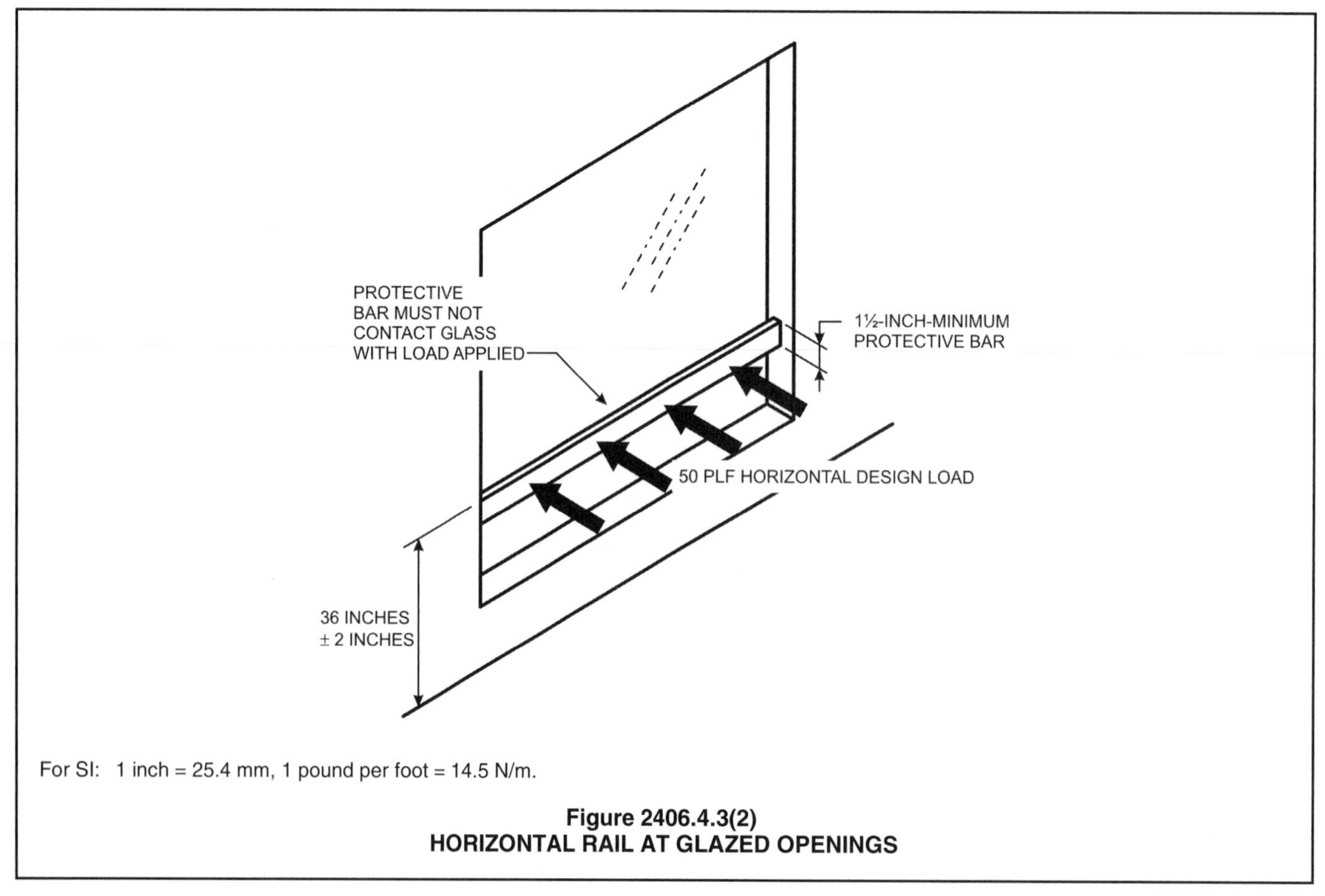

For SI: 1 inch = 25.4 mm, 1 pound per foot = 14.5 N/m.

Figure 2406.4.3(2)
HORIZONTAL RAIL AT GLAZED OPENINGS

part of the primary building wall construction, interior or exterior, used to enclose these areas must be constructed from safety glass or other safety glazing material allowed by the code. Glazed openings in primary building walls, interior or exterior, located within the enclosed compartment areas identified in this section, where the bottom edge of the glass is within 60 inches (1524 mm) of the standing or walking surface, must be glazed with safety glass or other safety glazing material allowed by the code [see Figure 2406.4.5(1)]. Glazed openings (i.e., windows) located outside the identified enclosures are not required to be glazed with safety glass, unless required by other sections in this chapter or elsewhere in the code.

When comparing the 18-inch (457 mm) bottom edge height threshold in Section 2406.4.3 to the 60 inches (1524 mm) required herein, one will find that the 18-inch (457 mm) height is intended to serve as a visual warning in order to avoid contact with glass, whereas the 60-inch (1524 mm) threshold is based

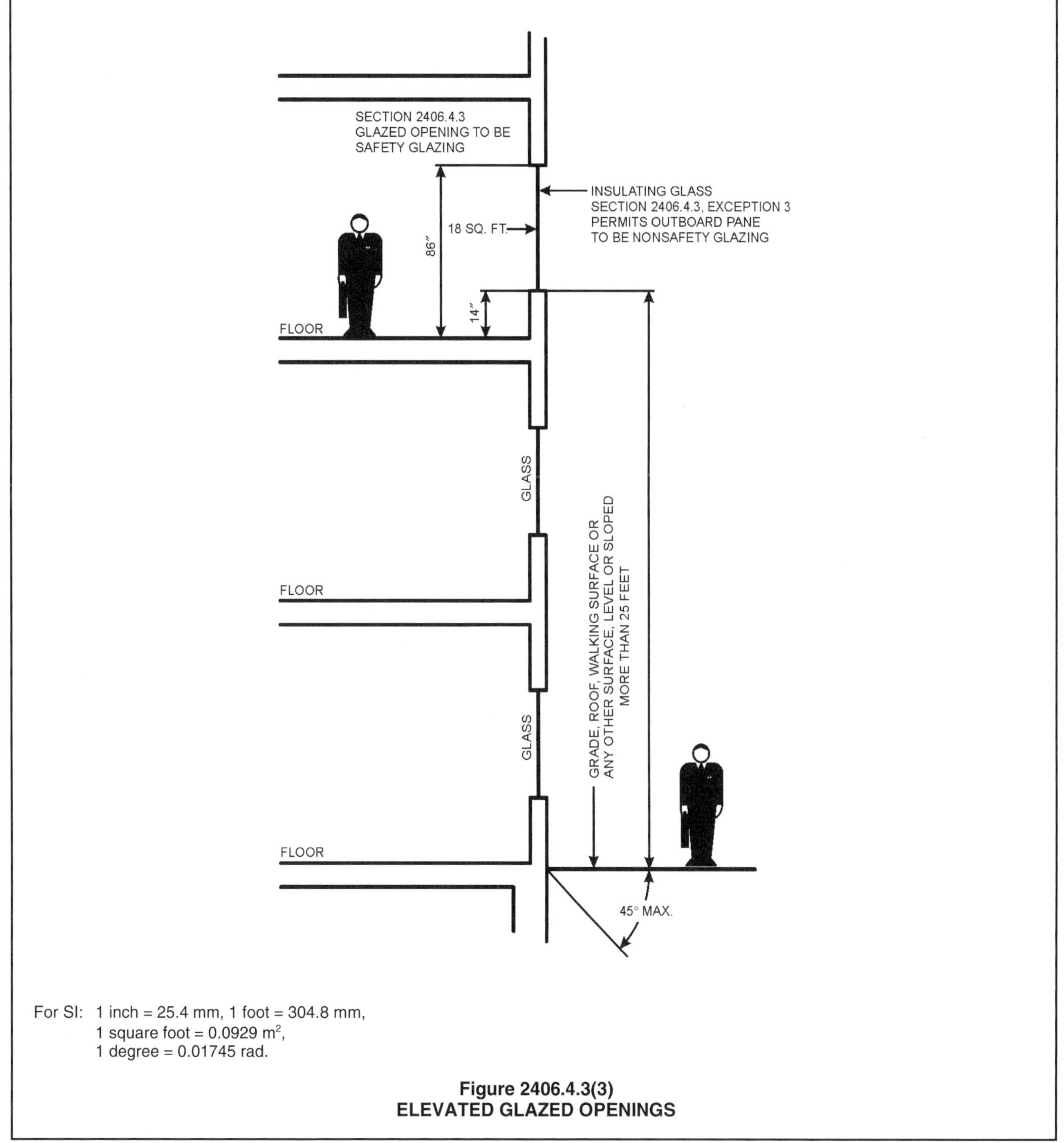

For SI: 1 inch = 25.4 mm, 1 foot = 304.8 mm,
1 square foot = 0.0929 m^2,
1 degree = 0.01745 rad.

Figure 2406.4.3(3)
ELEVATED GLAZED OPENINGS

on an impact consideration [see Figure 2406.4.5(2)].

The exception states that any glazing that is more than 60 inches (1524 mm) from the edge of the pool, tub or spa is exempt from the requirements in this section. Beyond this distance, any glazing should be evaluated against the hazardous location criteria stated in other sections (for example, for windows or doors).

2406.4.6 Glazing adjacent to stairs and ramps. Glazing where the bottom exposed edge of the glazing is less than 60 inches (1524 mm) above the plane of the adjacent walking surface of stairways, landings between flights of stairs, and ramps shall be considered a hazardous location.

Exceptions:

1. The side of a stairway, landing or ramp that has a guard complying with the provisions of Sections 1013 and 1607.8, and the plane of the glass is greater than 18 inches (457 mm) from the railing.
2. Glazing 36 inches (914 mm) or more measured horizontally from the walking surface.

❖ Stairways and ramps present users with a greater risk for injury caused by falling than a flat surface. Not only is the risk of falling greater when using a stairway, but the injuries are potentially more severe. Unlike falling on a flat surface where the floor will, for the most part, break a person's fall, there is nothing to stop someone from continuing to fall until he or she reaches the bottom of the stairway. The increased risks inherent in stairways, as well as attempting to be consistent with other chapters in the code that mandate more restrictive requirements when addressing safety issues involving stairways and ramps, account for the more restrictive requirements for glazing adjacent to stairways or ramps.

Exception 1 relieves the requirement for safety glazing because the risk of a cutting or piercing injury caused by broken glass upon human contact is effectively eliminated. Guards and railings that comply with the means of egress requirements prescribed in Chapter 10 as well as the structural requirements of Chapter 16 will effectively serve to prevent someone from impacting glazed walls or glazed openings in walls located in close vertical proximity to stairways or ramps, provided the glazing is more than 18 inches (457 mm) from the guards or railings.

Exception 2 excludes any glazing that is at least 36 inches (914 mm) horizontally from any walking surface. The walking surface in question would be part of a stairway or ramp itself, including top, bottom and intermediate landings. It does not include adjacent floors or other walking surfaces.

2406.4.7 Glazing adjacent to the bottom stair landing. Glazing adjacent to the landing at the bottom of a stairway where the glazing is less than 36 inches (914 mm) above the landing and within 60 inches (1524 mm) horizontally of the bottom tread shall be considered a hazardous location.

Exception: Glazing that is protected by a guard complying with Sections 1013 and 1607.8 where the plane of the glass is greater than 18 inches (457 mm) from the guard.

❖ This section addresses glass that is located within 60 inches (1524 mm) from the bottom tread in a run of

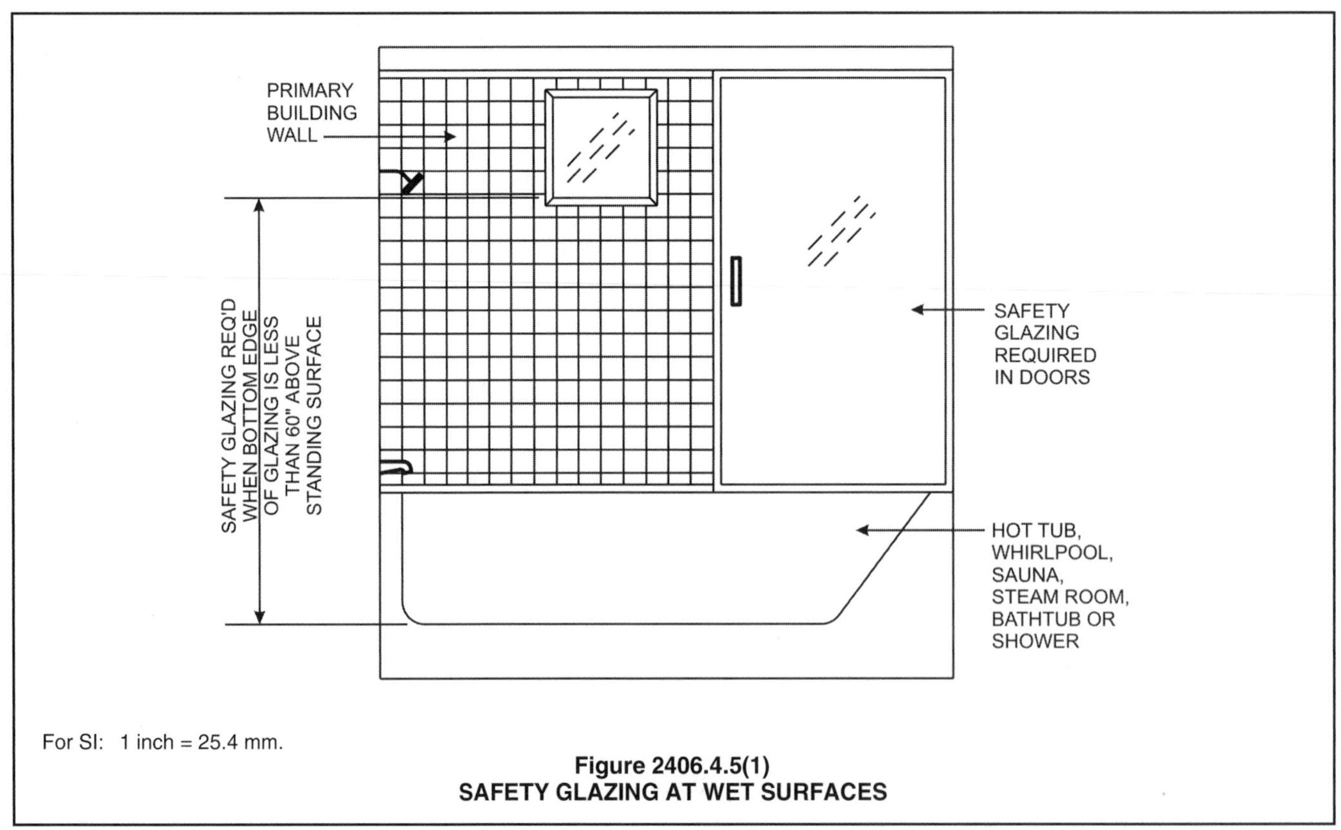

For SI: 1 inch = 25.4 mm.

Figure 2406.4.5(1)
SAFETY GLAZING AT WET SURFACES

stairs and within 36 inches (914 mm) vertically of the landing surface. The code does not distinguish between a bottom tread at a floor level or at an intermediate landing between floors. The last tread in a run of steps is the bottom tread. The 60-inch (1524 mm) dimension is from any point on the bottom tread, horizontally in any direction to the surface of any glazing within that range.

The exception relieves the requirement or safety glazing because it effectively eliminates the risk of a cutting or piercing injury caused by broken glass upon human contact. Guards and railings that comply with the means of egress requirements prescribed in Chapter 10 and the structural requirements of Chapter 16 effectively prevent someone from crashing into glazed walls or glazed openings in walls located in close vertical proximity to stairways or ramps, provided the glazing is more than 18 inches (457 mm) from the guard or railing.

2406.5 Fire department access panels. Fire department glass access panels shall be of tempered glass. For insulating glass units, all panes shall be tempered glass.

❖ Because of the need to provide safety for fire-fighting access, this section requires that all glass be safety glazing. When fire-fighting operations include breaking through openings, all glass will obviously be broken, which will subject fire fighters to the hazards of broken glass from all panes.

SECTION 2407
GLASS IN HANDRAILS AND GUARDS

2407.1 Materials. Glass used as a handrail assembly or a *guard* section shall be constructed of either single fully tempered glass, laminated fully tempered glass or laminated heat-strengthened glass. Glazing in railing in-fill panels shall be of an *approved* safety glazing material that conforms to the provisions of Section 2406.1.1. For all glazing types, the minimum nominal thickness shall be $^1/_4$ inch (6.4 mm). Fully tempered glass and laminated glass shall comply with Category II of CPSC 16 CFR Part 1201 or Class A of ANSI Z97.1.

❖ This section provides requirements for glass applications in handrail and guard assemblies.

Glass in handrail and guard applications is often exposed to impact from people, carried objects and other items. Early code requirements were vague regarding the appropriate types of glass for these installations.

Glass used as structural balustrade panels must resist the design loads applied to the handrail component of the guard system with an adequate factor of safety (see Figure 2407.1). On a practical basis, the only glass types that are structurally adequate are single and laminated tempered glass and laminated heat-strengthened glass. Other glass types and all safety plastics would need to be excessively thick to resist the design loads on the railing.

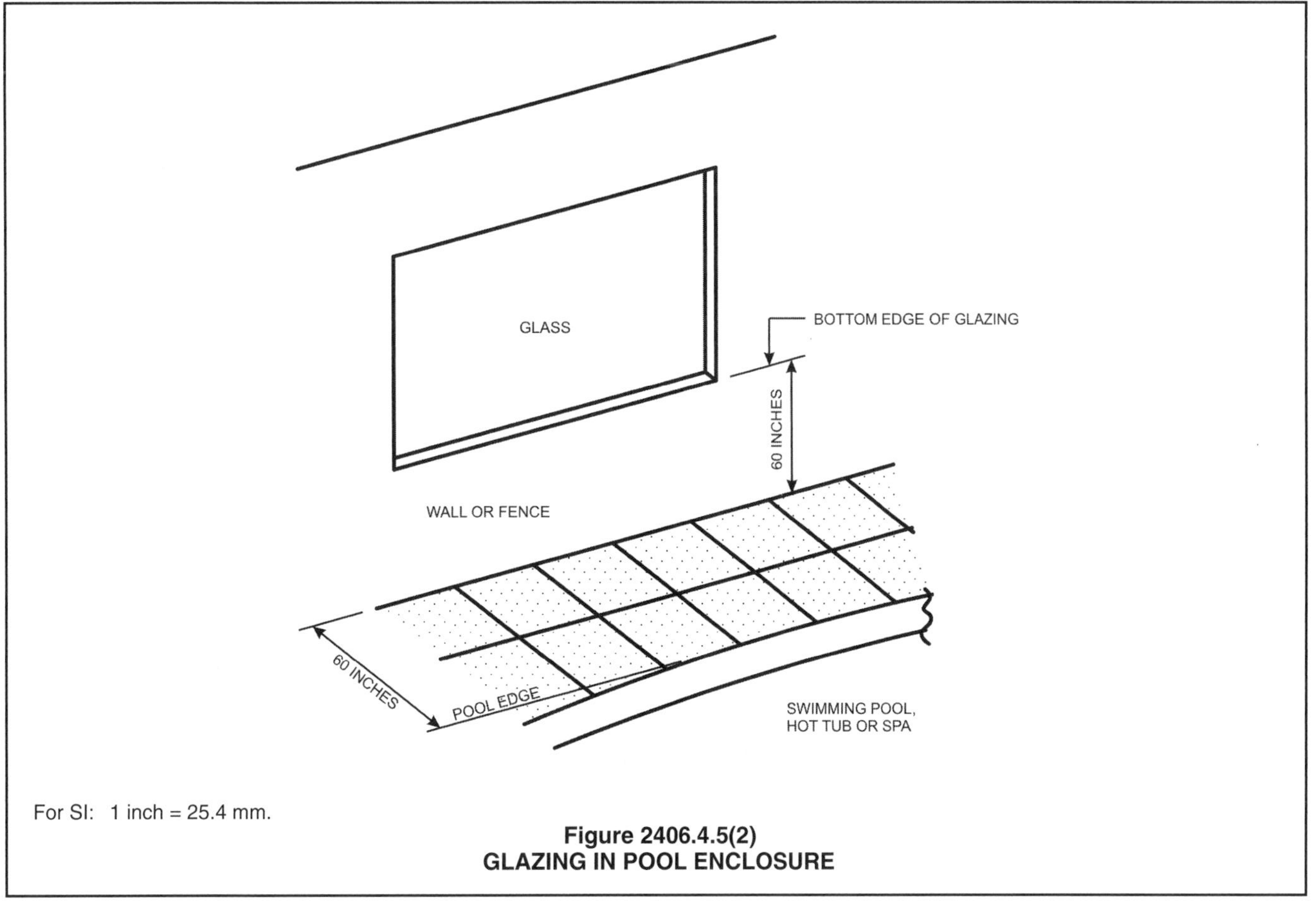

For SI: 1 inch = 25.4 mm.

Figure 2406.4.5(2)
GLAZING IN POOL ENCLOSURE

Glass guard in-fill panels are almost always installed in a hard-setting expanding cement. This is to provide a rigid, secure support for the glass. Large compressive stresses, transferred from the handrail component above, are exerted on the glass at the bottom edge. Only the glass types listed in the previous paragraph will routinely resist these forces without breakage. This is another reason why only single and laminated tempered glass and laminated heat-strengthened glass are allowed.

In-fill panels that do not support the railing have less stringent requirements. All safety glazing materials are allowed. A minimum thickness of $^1/_4$ inch (6.4 mm) is listed to verify reasonable penetration and breakage resistance. All types must meet the requirements of CPSC 16 CFR, Part 1201 or Class A in accordance with ANSI Z97.1.

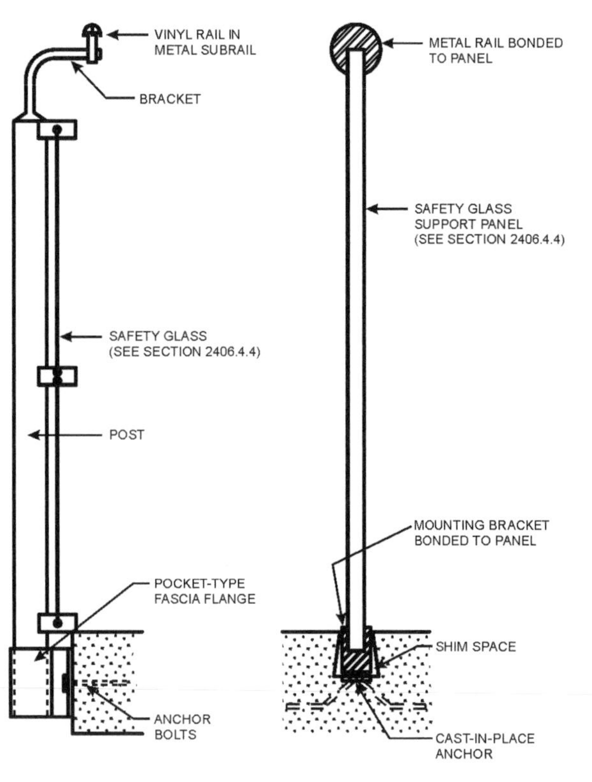

Figure 2407.1
GLASS GUARDS

2407.1.1 Loads. The panels and their support system shall be designed to withstand the loads specified in Section 1607.8. A safety factor of four shall be used.

❖ This section requires that the support system for glass guard or handrail assemblies be designed based on a safety factor of four. Nominally identical panes of glass inherently have a wide variation in strength. The safety factor of four is used in the design to minimize the likelihood that breakage will occur below the design loads. It is not intended that an in-place glass guard or handrail system be tested for or capable of withstanding four times the design load.

2407.1.2 Support. Each handrail or *guard* section shall be supported by a minimum of three glass balusters or shall be otherwise supported to remain in place should one baluster panel fail. Glass balusters shall not be installed without an attached handrail or *guard*.

Exception: A top rail shall not be required where the glass balusters are laminated glass with two or more glass plies of equal thickness and the same glass type when *approved* by the *building official*. The panels shall be designed to withstand the loads specified in Section 1607.8.

❖ Glass balustrade panels have been used for many decades without a significant history of problems. Glass of any type may break, however, if exposed to severe impact or other unusual or unexpected loads. A guard or handrail system should be designed so that the failure of a single glass panel does not cause the collapse of the assembly. Accordingly, the code does not allow glass balusters to be used without a guard or handrail. It also requires that guards or handrails be fastened to a minimum of three glass balustrade panels or supported such that the guard or handrail remains in place if one panel fails.

The exception allows an option where a top rail is an undesirable visual barrier. An example is the guard at the front of the spectator levels of sport arenas and theaters. The balusters must be laminated glass complying with the live load requirements for guards and handrails.

2407.1.3 Parking garages. Glazing materials shall not be installed in handrails or *guards* in parking garages except for pedestrian areas not exposed to impact from vehicles.

❖ This section is included to remove any ambiguity. The large impact loads from vehicles cannot be resisted by glass balustrades. As a result, they are not allowed under any circumstances where the guard or handrail might reasonably be impacted by vehicles. Glass and plastic in-fill panels are also not allowed because of the high likelihood of failure if impacted by vehicles.

2407.1.4 Glazing in wind-borne debris regions. Glazing installed in in-fill panels or balusters in *wind-borne debris regions* shall comply with the following:

❖ This section intends to minimize the glass debris in a windstorm. Reportedly, a dangerous situation occurred during hurricanes in South Florida where top handrails fell off the buildings. It was noted that rails located in portions of the building requiring the small missile test disengaged from the building and became large missile debris in those areas. This provision intends the glass railing system to provide retention and structural integrity and to continue to support the rail in the event the glass is impacted.

2407.1.4.1 Ballusters and in-fill panels. Glass installed in exterior railing in-fill panels or balusters shall be laminated glass complying with Category II of CPSC 16 CFR Part 1201 or Class A of ANSI Z97.1.

❖ Laminated glass installed in in-fill panels or balusters of exterior railings tends to remain integral after breakage occurs.

2407.1.4.2 Glass supporting top rail. When the top rail is supported by glass, the assembly shall be tested according to the impact requirements of Section 1609.1.2. The top rail shall remain in place after impact.

❖ In the case of all-glass-type railings, structural integrity can be maintained by incorporating impact-resistant glass. It is the intent to require only the missile impact test and not the cycling test requirements contained in the test standard referenced in Section 1609.1.2.

SECTION 2408
GLAZING IN ATHLETIC FACILITIES

2408.1 General. Glazing in athletic facilities and similar uses subject to impact loads, which forms whole or partial wall sections or which is used as a door or part of a door, shall comply with this section.

❖ In most applications of safety glazing materials, human impact into glass is an unexpected and infrequent occurrence. Safety glazing is required to minimize injury associated with the unintentional impact and subsequent breakage of fixed glass or glazing. Glass in walls or doors of enclosures for athletic activity can be exposed to frequent and intentional impact. For proper performance, the glass not only must break in a relatively safe manner, but must also resist breakage under normal use. A further requirement is that the glass be sufficiently rigid to provide a resilient playing surface. Therefore, safety glazing must be used in any athletic facility, as well as similar areas where the glass is subject to frequent human impact in the opinion of the building official. This section includes the performance requirements for glass in these specialized applications

2408.2 Racquetball and squash courts.

❖ The primary focus of this section is the walls and doors of racquetball, handball, squash and volleyball courts. In these applications, the glass is intentionally and frequently impacted in the normal course of its use.

2408.2.1 Testing. Test methods and loads for individual glazed areas in racquetball and squash courts subject to impact loads shall conform to those of CPSC 16 CFR Part 1201 or ANSI Z97.1 with impacts being applied at a height of 59 inches (1499 mm) above the playing surface to an actual or simulated glass wall installation with fixtures, fittings and methods of assembly identical to those used in practice.

Glass walls shall comply with the following conditions:

1. A glass wall in a racquetball or squash court, or similar use subject to impact loads, shall remain intact following a test impact.
2. The deflection of such walls shall not be greater than $1^1/_2$ inches (38 mm) at the point of impact for a drop height of 48 inches (1219 mm).

Glass doors shall comply with the following conditions:

1. Glass doors shall remain intact following a test impact at the prescribed height in the center of the door.
2. The relative deflection between the edge of a glass door and the adjacent wall shall not exceed the thickness of the wall plus $^1/_2$ inch (12.7 mm) for a drop height of 48 inches (1219 mm).

❖ The acceptance testing for the glass walls is similar in some respects to that included in CPSC 16 CFR, Part 1201, and ANSI Z97.1. There are some significant variations in the impact testing and end-point criteria. There are two major differences in the testing: the glass is impacted at a height of 59 inches (1499 mm) and the actual (or simulated) construction, complete with fixtures and attachments, is tested. The impact level of 59 inches (1499 mm) is the average shoulder height for an adult male as listed in handbooks for architectural design.

The end-point criteria require that the glass assembly not fail upon impact and the maximum deflection not exceed $1^1/_2$ inches (38 mm). Although the latter is not related to safety, it is included to provide a reasonable playing surface. Both are unique and are not ordinary requirements for safety glazing materials.

Glass doors in the applications cited have been particularly hazardous in some designs. A player may impact and deflect the access door, while the adjacent wall remains in its original position. Occasionally, a player's fingers have become trapped between the door and the wall when the force on the door is released. This section of the code is directed at reducing this hazard.

Though not stated, the intent is that the impact for the glass panels be at the midpoint of the horizontal dimension.

2408.3 Gymnasiums and basketball courts. Glazing in multipurpose gymnasiums, basketball courts and similar athletic facilities subject to human impact loads shall comply with Category II of CPSC 16 CFR Part 1201 or Class A of ANSI Z97.1.

❖ Unlike racquetball and squash courts, the intentional impacting of gymnasium and basketball court enclosures is not anticipated. Nevertheless, due to the nature of their use, these facilities have a higher incidence of unintentional glazing impacts. This section requires any glazing that is subject to human impact loads in gymnasiums, basketball courts and similar athletic facilities to be CPSC 16 CFR, Part 1201, Category II, or Class A in accordance with ANSI Z97.1 [see commentary, Tables 2406.2(1) and 2406.2(2)].

SECTION 2409 GLASS IN ELEVATOR HOISTWAYS AND ELEVATOR CARS

2409.1 Glass in elevator hoistway enclosures. Glass in elevator hoistway enclosures and hoistway doors shall be laminated glass conforming to ANSI Z97.1 or CPSC 16 CFR Part 1201.

❖ The purpose of this section is to provide requirements for glass in elevator cars and elevator hoistways comparable to those in ASME A17.1. This section provides a level of protection to building occupants from the hazards associated with broken glass as it relates to glass in elevator enclosures. This glazing is susceptible to impact from components during elevator operation or servicing.

2409.1.1 Fire-resistance-rated hoistways. Glass installed in hoistways and hoistway doors where the hoistway is required to have a fire-resistance rating shall also comply with Section 716.

❖ This section provides a cross reference to the Chapter 7 requirements for a hoistway that must be fire-resistance rated.

2409.1.2 Glass hoistway doors. The glass in glass hoistway doors shall be not less than 60 percent of the total visible door panel surface area as seen from the landing side.

❖ This section establishes a minimum area threshold for glass hoistway doors.

2409.2 Glass visions panels. Glass in vision panels in elevator hoistway doors shall be permitted to be any transparent glazing material not less than $^{1}/_{4}$ inches (0.64 mm) in thickness conforming to Class A in accordance with ANSI Z97.1 or Category II in accordance with CPSC 16 CFR Part 1201. The area of any single vision panel shall not be less than 24 square inches (15 484 mm^2) and the total area of one or more vision panels in any hoistway door shall be not more than 85 square inches (54 839 mm^2).

❖ This section requires CPSC 16 CFR, Part 1201, Category II safety glazing or Class A in accordance with ANSI Z97.1 in vision panels in hoistway doors in a manner that is consistent with similar hazardous locations.

2409.3 Glass in elevator cars.

❖ This section establishes the permitted glazing type for glass elevator cars and sets a minimum area of glazing in glass doors of elevator cars.

2409.3.1 Glass types. Glass in elevator car enclosures, glass elevator car doors and glass used for lining walls and ceilings of elevator cars shall be laminated glass conforming to Class A in accordance with ANSI Z97.1 or Category II in accordance with CPSC 16 CFR Part 1201.

Exception: Tempered glass shall be permitted to be used for lining walls and ceilings of elevator cars provided:

1. The glass is bonded to a nonpolymeric coating, sheeting or film backing having a physical integrity to hold the fragments when the glass breaks.
2. The glass is not subjected to further treatment such as sandblasting; etching; heat treatment or painting that could alter the original properties of the glass.
3. The glass is tested to the acceptance criteria for laminated glass as specified for Class A in accordance with ANSI Z97.1 or Category II in accordance with CPSC 16 CFR Part 1201.

❖ The requirement for laminated glass is consistent with the requirements of ASME A17.1.

2409.3.2 Surface area. The glass in glass elevator car doors shall be not less than 60 percent of the total visible door panel surface area as seen from the car side of the doors.

❖ This section establishes a minimum area threshold for glass doors in elevator cars.

Bibliography

The following resource materials are referenced in this chapter or are relevant to the subject matter addressed in this chapter.

AAMA/WDMA/CSA 101/I.S.2/A440-11, *Specifications for Windows, Doors and Unit Skylights*. Schaumburg, IL: American Architectural Manufacturers Association, 2011.

ANSI Z97.1-09, *Safety Performance Specifications and Methods of Test for Safety Glazing Materials Used in Buildings*. New York: American National Standards Institute, 2009.

ASCE 7-10, *Minimum Design Loads for Buildings and Other Structures*. New York: American Society of Civil Engineers, 2010.

ASME A17.1/CSA B44-07, *Safety Code for Elevators and Escalators*. American Society of Mechanical Engineers, 2007.

ASTM C 1036-91, *Specification for Flat Glass*. West Conshohocken, PA: ASTM International, 1991.

ASTM C 1048-92, *Heat-treated Flat Glass: Kind HS, Kind FT Coated and Uncoated Glass*. West Conshohocken, PA: ASTM International, 1992.

ASTM E 1300-07e01, *Standard Practice for Determining Load Resistance of Glass in Buildings*. West Conshohocken, PA: ASTM International, 2007.

CPSC 16 CFR, Part 1201 (2002), *Safety Standard for Architectural Glazing*. Washington, DC: Consumer Product Safety Commission, 2002.

Glass Design for Sloped Glazing. Des Plaines, IL: American Architectural Manufacturers Association, 1987.

Chapter 25: Gypsum Board and Plaster

General Comments

Chapter 25 contains the provisions and referenced standards that regulate the design, construction and quality of gypsum board and plaster. These represent the most common interior and exterior finish materials in the building industry, and the provisions dealing with them are contained within this single chapter. This chapter deals primarily with quality-control-related issues with regards to material specifications and installation requirements. This chapter addresses weather ratings, and also provides cross references to the requirements for shear walls constructed of either wood stud or light-gage steel stud framing. In addition, provisions for constructing horizontal gypsum board ceiling diaphragms are included.

Where materials described in this chapter are used or required for fire-resistant construction, the code requires that they also comply with the provisions of Chapter 7.

Section 2501 addresses the scope of this chapter.

Section 2502 provides a list of defined terms that are used throughout this chapter.

Section 2503 explains the responsibility of the contractor to have work inspected by the building official.

Section 2504 defines the construction and application requirements for gypsum board and plaster for use on walls and ceilings.

Section 2505 allows for the use of lightweight shear walls that are sheathed with gypsum board or lath and plaster.

Section 2506 outlines the quality standards for gypsum board and related materials and accessories.

Section 2507 outlines the quality standards for lath, plaster and related materials and accessories.

Section 2508 includes material and construction standards for the use of gypsum board and gypsum plaster materials with reference to fire-resistance-rated assemblies. It also provides requirements for constructing gypsum board ceiling diaphragms using wood joists.

Section 2509 establishes the requirements for the use of gypsum board in wet areas.

Section 2510 describes the material standards and construction requirements for cement plaster and related accessories.

Section 2511 defines the material standards, construction requirements and limitations for plaster and related accessories when applied to an interior surface.

Section 2512 defines the material standards, construction requirements and limitations for plaster and related accessories when applied to exterior and weather-exposed surfaces.

Section 2513 addresses the use of applied aggregate finishes to interior and exterior plaster.

Gypsum board and plaster products represent a line of building products that are, for the most part, manufactured under the control of industry standards. The building official or inspector only needs to verify that the appropriate product has been used and properly installed for the intended use and location. Several key points of installation have been addressed in this chapter where these practices have been known to create problems or when not addressed by other gypsum or plaster product standards.

Purpose

Chapter 25 provides the minimum design requirements through referenced and industry standards for the use of gypsum board, gypsum plaster and cement plaster materials in construction.

SECTION 2501 GENERAL

2501.1 Scope.

❖ Chapter 25 includes the requirements that govern the materials, design, construction, quality and application for both interior and exterior gypsum board and plaster products. Although these products are most commonly used as interior or exterior finished wall and ceiling covering the surface of buildings, proper design and application are necessary to provide weather resistiveness and required fire protection for both structural and nonstructural building components, as well as structural requirements for gypsum board and plaster products and assemblies.

Provisions for the materials, quality, construction and labeling of all gypsum board and plaster materials used in the construction of wall and ceiling coverings are included, along with their method of fastening and, in the case of plaster, the permitted materials for lath, plaster and aggregate in all buildings and structures. In general, all gypsum board and plaster materials are required to conform to the applicable standards as referenced in this chapter.

2501.1.1 General. Provisions of this chapter shall govern the materials, design, construction and quality of gypsum board, lath, gypsum plaster and cement plaster.

❖ Although plaster has many uses in construction, including ornamental and decorative work, its use in the code is regulated purely as a wall and ceiling covering material. The code regulates the installation of wall and ceiling covering materials, as well as the quality standards for the materials themselves. Chapter 25 regulates gypsum board, lath and plaster.

2501.1.2 Performance. Lathing, plastering and gypsum board construction shall be done in the manner and with the materials specified in this chapter, and when required for fire protection, shall also comply with the provisions of Chapter 7.

❖ The extensive use of gypsum board and plaster throughout the construction industry for many years has led to the development of a comprehensive index of tested and proven performance standards. This chapter is directed at identifying the performance standards required to ensure a quality of construction that will provide a safe, habitable environment for the life of the building or structure. The standards contained and referenced in this chapter and elsewhere in the code establish the minimum performance standards and requirements for all methods and materials contained in this chapter.

Gypsum plaster, cement plaster and gypsum board products and materials have inherent fire-resistant qualities that provide outstanding passive fire protection of building components and assemblies. When these materials are used in the construction of fire-resistant assemblies, the details of construction are to comply with the requirements established by the official fire test reports conducted by recognized testing laboratories in accordance with Chapter 7 and the referenced standards listed in Chapter 35.

2501.1.3 Other materials. Other *approved* wall or ceiling coverings shall be permitted to be installed in accordance with the recommendations of the manufacturer and the conditions of approval.

❖ Gypsum board and plaster are currently the primary building materials being used for wall and ceiling covering, but there are many other materials in use today for covering interior and exterior walls, ceilings and soffits. This chapter makes no attempt to address the use of any other building materials, such as wood, concrete, concrete masonry, brick, metal or vinyl, nor is it intended to limit the use of these or any other materials for similar installations. The material standards and construction regulations for other approved building materials are addressed elsewhere in the code.

SECTION 2502 DEFINITIONS

2502.1 Definitions. The following terms are defined in Chapter 2:

CEMENT PLASTER.

EXTERIOR SURFACES.

GYPSUM BOARD.

GYPSUM PLASTER.

GYPSUM VENEER PLASTER.

INTERIOR SURFACES.

WEATHER-EXPOSED SURFACES.

WIRE BACKING.

❖ Definitions facilitate the understanding of code provisions and minimize potential confusion. To that end, this section lists definitions of terms associated with gypsum board and plaster. Note that these definitions are found in Chapter 2. The use and application of defined terms, as well as undefined terms, are set forth in Section 201.

SECTION 2503 INSPECTION

2503.1 Inspection. Lath and gypsum board shall be inspected in accordance with Section 110.3.5.

❖ This section establishes guidelines and requirements for the inspection of all components of gypsum board and plaster products, assemblies and installations as outlined in the code and as required or requested by the building official. The building official is allowed to inspect the installation of all plaster lathing system components prior to being concealed from view by the application of a plaster base coat or finished coats. It is essential not only that the lath be the proper type for the plaster system being used, but also that the components of the system are installed properly.

Plaster wall and ceiling assemblies as well as gypsum board wall and ceiling assemblies are quite often, by design, part of a building's passive fire protection system, as they provide not only the fire-resistant construction necessary for the proper separation of defined areas within a structure, but also fire-protected components of the structure. The building official is to be given the opportunity by the permit holder to verify that the assemblies will serve the purpose for which they are designed. This can only be done by allowing the building official to visually inspect the components of the various systems while they are still accessible.

SECTION 2504 VERTICAL AND HORIZONTAL ASSEMBLIES

2504.1 Scope. The following requirements shall be met where construction involves gypsum board, lath and plaster in vertical and horizontal assemblies.

❖ Section 2504, along with the referenced standards and structural limitations and requirements addressed elsewhere, define the minimum requirements for the structural members supporting the weight of gypsum board and plaster where either or both are used in the construction of walls and ceilings.

Vertical assemblies (walls and partitions) must be designed in accordance with the engineering chapters of the code for proper structural support of gypsum board or plaster wall covering. If the wall or partition construction and assembly are not designed to resist the loads prescribed in the code, the wall covering materials will perform inadequately. The general design provisions of Chapter 16 and the deflection limitations are particularly important in this regard.

The code requirements for the design and framing of horizontal assemblies (ceilings) are intended to achieve the same results as described for vertical assemblies. Gypsum board and plaster ceilings are generally supported by a suspended metal grid framing system of metal channels or T-bars suspended by steel wires tied to the structure above. The ceiling covering may also be fastened directly to the structural framing member or furring, which is fastened to the structural frame. All ceiling support systems must comply with the structural load requirements of Chapter 16.

2504.1.1 Wood framing. Wood supports for lath or gypsum board, as well as wood stripping or furring, shall not be less than 2 inches (51 mm) nominal thickness in the least dimension.

> **Exception:** The minimum nominal dimension of wood furring strips installed over solid backing shall not be less than 1 inch by 2 inches (25 mm by 51 mm).

❖ The minimum dimensions required by this section for wood supports for gypsum board or lath and plaster are intended to provide adequate support and reduce surface distortion due to warping of wood supports.

Wood furring strips attached to a solid backing may be used to support wall or ceiling covering. The furring must be installed with the wide face against the solid backing. This will provide a nominal 2-inch-wide (51 mm) nailing surface equal to the nailing surface of a standard wood stud, joist or rafter.

2504.1.2 Studless partitions. The minimum thickness of vertically erected studless solid plaster partitions of $^3/_8$-inch (9.5 mm) and $^3/_4$-inch (19.1 mm) rib metal lath or $^1/_2$-inch thick (12.7 mm) long-length gypsum lath and gypsum board partitions shall be 2 inches (51 mm).

❖ As the name implies, a studless partition is constructed of gypsum board lath or ribbed metal lath with plaster applied to each face, without the stud frame wall that typically provides support for the wall covering [see Figures 2504.1.2(1) and 2504.1.2(2)].

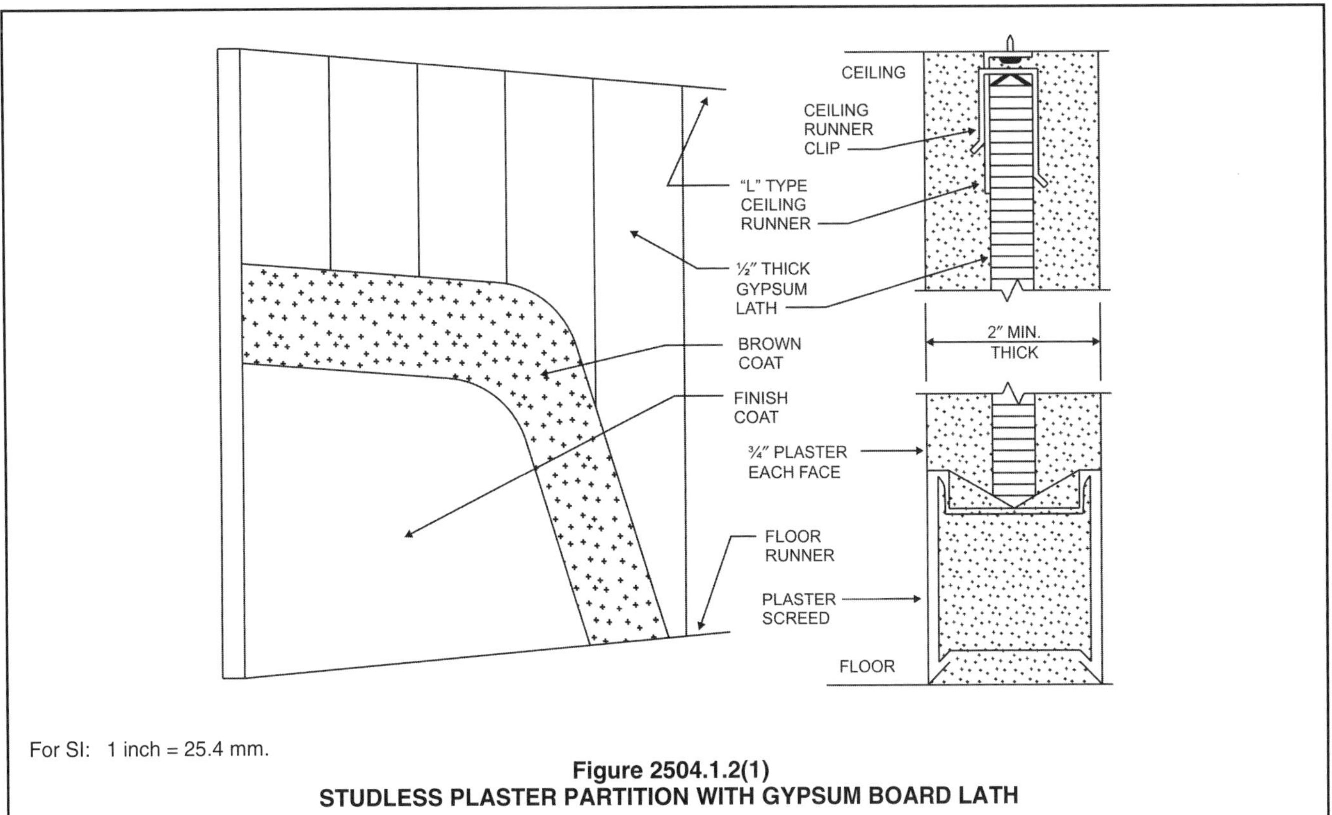

Figure 2504.1.2(1)
STUDLESS PLASTER PARTITION WITH GYPSUM BOARD LATH

The lath and plaster form the structural elements of the partition that must resist the lateral loads applied to the wall. Studless partitions may only be used as nonload-bearing partitions; however, they must comply with the deflection and general design requirements specified in Chapter 16.

SECTION 2505 SHEAR WALL CONSTRUCTION

2505.1 Resistance to shear (wood framing). Wood-framed shear walls sheathed with gypsum board, lath and plaster shall be designed and constructed in accordance with Section 2306.3 and are permitted to resist wind and seismic loads. Walls resisting seismic loads shall be subject to the limitations in Section 12.2.1 of ASCE 7.

❖ Shear walls designed to resist wind and seismic loads and constructed of wood-framing members and sheathed with gypsum wallboard or lath and plaster are permitted by the code (see Chapter 23 for wood-framing requirements).

Section 2306.3 includes seismic application limitations for wood-framed shear walls using gypsum board or lath and plaster as the finished wall surface. The limitations for basic seismic-force-resisting systems are found in Section 12.2.1 of ASCE 7.

Care must be taken when selecting gypsum board or plaster materials if they are being used as part of an exterior wall that will be exposed to the weather. Gypsum board, gypsum board lath, gypsum plaster and gypsum veneer plaster are not water resistant and are intended for interior applications only. Specific gypsum board materials covered with an appropriate exposed finish material or metal lath with cement plaster base coats and finish coat (stucco) are appropriate for exterior applications exposed to the weather.

2505.2 Resistance to shear (steel framing). Cold-formed steel-framed shear walls sheathed with gypsum board and constructed in accordance with the materials and provisions of Section 2211.6 are permitted to resist wind and seismic loads. Walls resisting seismic loads shall be subject to the limitations in Section 12.2.1 of ASCE 7.

❖ Shear walls designed to resist wind and seismic loads and constructed of light-gage steel stud framing members and sheathed with gypsum board are permitted by the code. Structural design guidelines and construction requirements for these shear walls are provided in Section 2211.6. Limitations for basic seismic-force-resisting systems are found in Section 12.2.1 of ASCE 7.

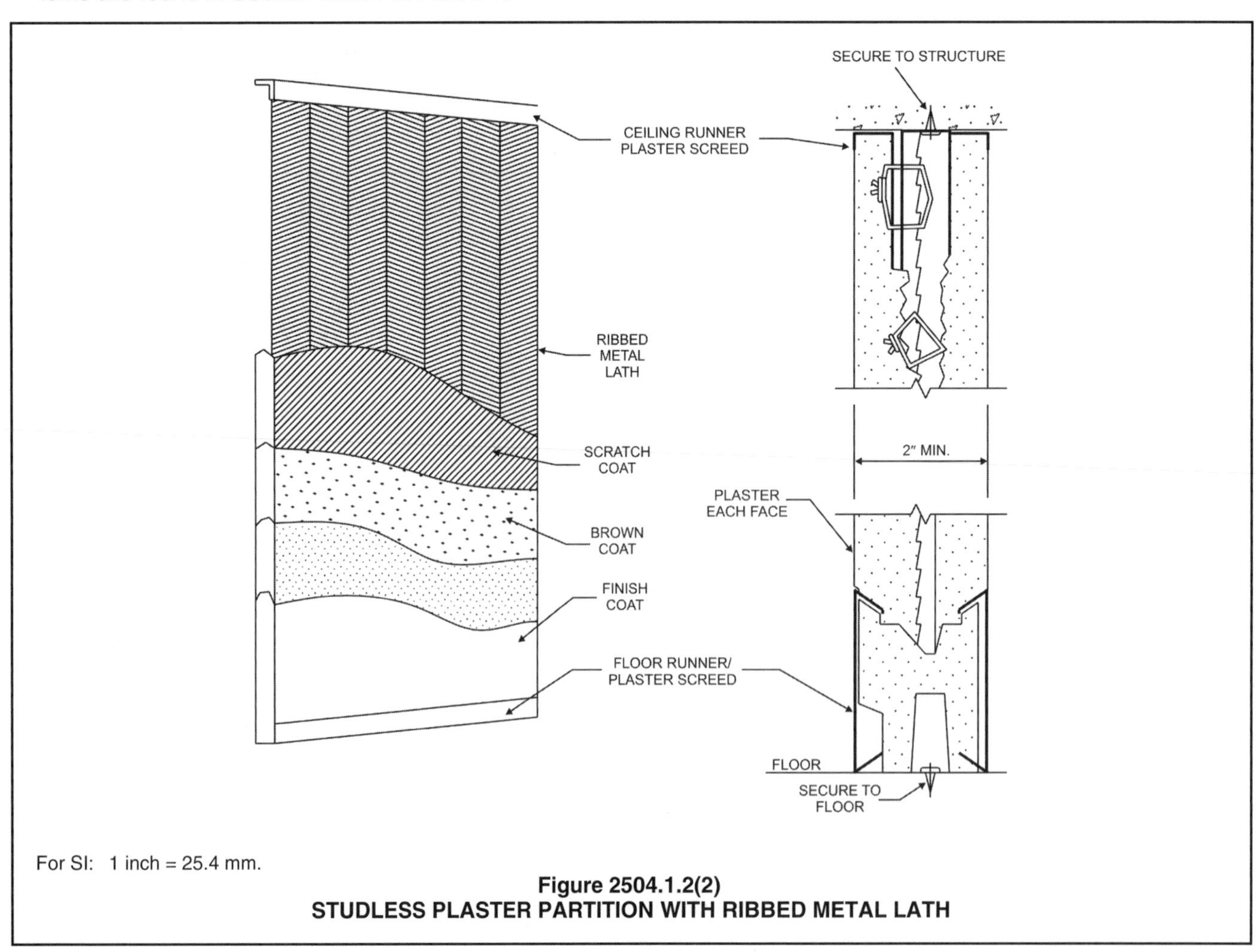

Figure 2504.1.2(2)
STUDLESS PLASTER PARTITION WITH RIBBED METAL LATH

SECTION 2506
GYPSUM BOARD MATERIALS

2506.1 General. Gypsum board materials and accessories shall be identified by the manufacturer's designation to indicate compliance with the appropriate standards referenced in this section and stored to protect such materials from the weather.

❖ This section contains the appropriate referenced standards for the proper use and quality of gypsum board materials and accessories. Gypsum board is the most commonly used interior wall covering. Gypsum board is also used for exterior sheathing, plaster lath and ceiling covering. Because it is installed in sheet form, it is less labor intensive and generally considered more cost effective than other wall and ceiling materials, such as plaster. Gypsum board requires a minimal amount of finishing and will readily accept paint, wallpaper, vinyl fabric, special textured paint and similar surface-finish materials.

All gypsum products, including gypsum board materials, are required to be identified with the manufacturer's designation. The manufacturer is required to certify that its product complies with the applicable standard for the gypsum product. This enables the installer and inspector to verify that the product is being properly installed and utilized.

Generally, all gypsum products and accessories are delivered to the job site in their original packages, containers or bundles. These products will be identified with the manufacturer's name, trademark and any applicable standard designations. For example, for gypsum board products, the thickness and type are shown on the end-bundling tapes or the board panel. The premixed paste products are identified on their container and bag products are appropriately identified on the face of the bag. With regard to weather protection, all gypsum products must be kept dry because of the deleterious effect of moisture. If properly stored above ground and fully protected from the weather and exposure to direct sunlight, gypsum products and accessories may be stored outside; however, the storage of gypsum board is not recommended in exterior locations. Exposure to the elements may result in water stain, light discoloration, mildew, loss of strength or sagging, all of which could hinder the structural and fire performance of the product, as well as the ability to be finished in the intended fashion. Gypsum board should be stored flat—never on edge or end.

2506.2 Standards. Gypsum board materials shall conform to the appropriate standards listed in Table 2506.2 and Chapter 35 and, where required for fire protection, shall conform to the provisions of Chapter 7.

❖ Table 2506.2 lists the appropriate material standards for gypsum board materials and accessories. The most common building material used in Table 2506.2 is gypsum wallboard (also referred to as "gypsum board," "gyp. board" or "sheetrock"). ASTM C 1396 provides the design specifications for gypsum board used for walls and ceilings. Gypsum board consists of a noncombustible core, which is primarily gypsum, and is surfaced with paper firmly bonded to the core. The paper surfacing material is less than 0.125-inch (3.18 mm) thick and has a flame spread index less than 25 when tested in accordance with ASTM E 84. This requirement is in accordance with the fire-resistance requirements of Chapter 7 for composite materials classified as "noncombustible." Gypsum materials that are required to have a fire-resistance rating in accordance with the code must also comply with the applicable provisions of Chapter 7. In addition, Chapter 7 establishes and defines the requirements for the fire-resistance ratings of building assemblies and structural components. Such assemblies are to be constructed in accordance with the test procedures as prescribed in the appropriate referenced standards listed in Chapter 35.

TABLE 2506.2
GYPSUM BOARD MATERIALS AND ACCESSORIES

MATERIAL	STANDARD
Accessories for gypsum board	ASTM C 1047
Adhesives for fastening gypsum wallboard	ASTM C 557
Elastomeric joint sealants	ASTM C 920
Fiber-reinforced gypsum panels	ASTM C 1278
Glass mat gypsum backing panel	ASTM C 1178
Glass mat gypsum panel	ASTM C 1658
Glass mat gypsum substrate	ASTM C 1177
Joint reinforcing tape and compound	ASTM C 474; C 475
Nails FOR gypsum boards	ASTM C 514, F 547, F 1667
Steel screws	ASTM C 954; C 1002
Steel studs, load-bearing	ASTM C 955
Steel studs, nonload-bearing	ASTM C 645
Standard specification for gypsum board	ASTM C 1396
Testing gypsum and gypsum products	ASTM C 22; C 472; C 473

❖ Table 2506.2 provides a list of the materials and accessories recognized by the code for use in gypsum board construction, and makes reference to the appropriate ASTM standard for each. The installation of each of these materials must conform to all the requirements of the referenced standard. It is not the intent of the code, however, by way of this table, to limit or restrict alternative materials or accessories in the use of gypsum board and its associated incidentals as a building material acknowledged and approved by the code. Any and all materials not specifically mentioned, described or referenced, however, are subject to the requirements of the code, as well as the entire family of *International Codes*®.

In recent years, the gypsum industry has replaced several individual gypsum board product standards with the "umbrella" standard, ASTM C 1396, *Specification for Gypsum Board*.

By including ASTM C 920, the code recognizes

elastomeric joint sealants as an acceptable alternative to the traditional joint tape and compound. Some fire-resistance-rated assemblies have been successfully tested utilizing joints sealed with the elastomeric joint sealant material. The use of these materials must also comply with all applicable requirements in other parts of the code.

2506.2.1 Other materials. Metal suspension systems for acoustical and lay-in panel ceilings shall conform with ASTM C 635 listed in Chapter 35 and Section 13.5.6 of ASCE 7 for installation in high seismic areas.

❖ Suspended ceiling systems are widely used in building construction of all types, as they provide for a cost-effective finished ceiling system that is versatile and adaptable to other building systems, such as heating ventilating and air-conditioning systems (HVAC), fire suppression systems, lighting systems, etc. The standard developed for the manufacture and installation of suspended ceiling systems is ASTM C 635. This standard contains minimum specifications for the metal suspension system, including system types (direct hung, indirect hung and furring bar type), structural classification (light-duty, intermediate-duty and heavy-duty), dimensional tolerances, coatings and physical testing for structural capacities.

Additional installation requirements are applicable for high seismic areas. These requirements include installation requirements for Seismic Design Categories C through F and are found in Section 13.5.6 of ASCE 7.

SECTION 2507 LATHING AND PLASTERING

2507.1 General. Lathing and plastering materials and accessories shall be marked by the manufacturer's designation to indicate compliance with the appropriate standards referenced in this section and stored in such a manner to protect them from the weather.

❖ These basic requirements apply to lathing and plastering materials, products and accessories referenced in this chapter. This section requires that all lath and plaster materials be identified with the manufacturer's designation and indicate compliance with the appropriate standards. This enables the installer and the inspector to verify that a product is being installed in a manner that is consistent with its intended use. All accessories must also be marked with a reference to the appropriate standard. The manufacturer is required to certify that its product complies with the applicable standard for lathing and plastering material. Lath and plaster products and accessories are delivered to the job site in their original packages, containers or bundles. These products must be identified with the manufacturer's name, trademark and any applicable standard designations. With regard to weather protection, plaster products must be kept dry because of the deleterious effect of moisture. If properly stored above ground, fully protected from the weather and exposure to direct sunlight and away from condensation and damp surfaces, lath and plaster products and accessories may be stored outside; however, it is not recommended.

2507.2 Standards. Lathing and plastering materials shall conform to the standards listed in Table 2507.2 and Chapter 35 and, where required for fire protection, shall also conform to the provisions of Chapter 7.

❖ The appropriate referenced standards that regulate all lath and plastering materials are contained in this section. It also addresses the proper marking of materials, protection from weather and installation requirements. Additionally, provisions of Chapter 7 are referenced for building assemblies utilizing lath and plastering materials that require a fire-resistance rating; therefore, the assemblies must comply with the provisions of Chapter 7.

TABLE 2507.2 LATH, PLASTERING MATERIALS AND ACCESSORIES

MATERIAL	STANDARD
Accessories for gypsum veneer base	ASTM C 1047
Blended cement	ASTM C 595
Exterior plaster bonding compounds	ASTM C 932
Gypsum casting and molding plaster	ASTM C 59
Gypsum Keene's cement	ASTM C 61
Gypsum plaster	ASTM C 28
Gypsum veneer plaster	ASTM C 587
Interior bonding compounds, gypsum	ASTM C 631
Lime plasters	ASTM C 5; C 206
Masonry cement	ASTM C 91
Metal lath	ASTM C 847
Plaster aggregates	
Sand	ASTM C 35; C 897
Perlite	ASTM C 35
Vermiculite	ASTM C 35
Plastic cement	ASTM C 1328
Portland cement	ASTM C 150
Steel screws	ASTM C 1002; C 954
Steel studs and track	ASTM C 645; C 955
Welded wire lath	ASTM C 933
Woven wire plaster base	ASTM C 1032

❖ Table 2507.2 provides a listing of the materials and accessories that are recognized in the code for use in plaster construction, along with the appropriate ASTM standard for each. The installation of each of these materials must conform to all the requirements of the referenced standard. It is not the intent of the code to prevent alternative materials or accessories in the use of plaster and its associated incidentals as a building material acknowledged and approved by the code. Any materials not specifically mentioned, described or referenced herein are subject to the requirements of the code, as well as the entire family of *International Codes*.

SECTION 2508 GYPSUM CONSTRUCTION

2508.1 General. Gypsum board and gypsum plaster construction shall be of the materials listed in Tables 2506.2 and 2507.2. These materials shall be assembled and installed in compliance with the appropriate standards listed in Tables 2508.1 and 2511.1.1, and Chapter 35.

❖ This section contains the referenced standards and requirements for the proper installation of gypsum construction.

This section references Table 2506.2 for the appropriate materials required in gypsum board and gypsum plaster construction, as well as the corresponding material standard. Table 2508.1 addresses the appropriate installation standards required for use in gypsum construction. Gypsum board in fire-resistance-rated assemblies is to be installed in accordance with the fastening schedule utilized in the tested assembly. The type and spacing of the fasteners are reflected in the fire test reports of various assemblies, including the summary of fire tests contained in various resource documents that compile summaries of fire test results, such as Underwriters Laboratories' *Fire Resistance Directory or the Gypsum Association's Fire Resistance Design Manual.* Refer to Chapter 7 and its commentary for additional information related to fire-resistance-rated assemblies.

Gypsum plaster is available in several forms and is designed for a variety of applications, including those where fire protection of building components, construction of fire-resistance-rated assemblies within the building or control of sound transmission is needed. Gypsum plaster can be applied using a two- or three-coat method, depending upon the backing or lathing system. Gypsum plaster provides a hard, smooth finish surface that will readily receive most decorative finishes, such as paint, vinyl wall fabric, wallpaper and textured paint. Various aggregates can be added to the plaster composition to achieve different surface textures, from a smooth troweled to a heavily textured finish.

TABLE 2508.1
INSTALLATION OF GYPSUM CONSTRUCTION

MATERIAL	STANDARD
Gypsum board	GA-216; ASTM C 840
Gypsum sheathing	ASTM C 1280
Gypsum veneer base	ASTM C 844
Interior lathing and furring	ASTM C 841
Steel framing for gypsum boards	ASTM C 754; C 1007

❖ Table 2508.1 lists the installation requirements for the use of gypsum and gypsum-related construction products, along with the appropriate referenced standard for each. The installation of each of these materials must conform to all the requirements of the referenced standard. There is no intent to limit or restrict alternative methods of construction acknowledged and approved by the code. Any methods not specifically mentioned, described or referenced herein are subject to the requirements of the code, as well as the entire family of *International Codes*.

2508.2 Limitations. Gypsum wallboard or gypsum plaster shall not be used in any exterior surface where such gypsum construction will be exposed directly to the weather. Gypsum wallboard shall not be used where there will be direct exposure to water or continuous high humidity conditions. Gypsum sheathing shall be installed on exterior surfaces in accordance with ASTM C 1280.

❖ Because of the detrimental effect of moisture on gypsum products, gypsum-based materials and construction must be protected from the weather. See Section 202 for the definition of "Weather-exposed surfaces," which includes some limited conditions where gypsum board is permitted to be used on the outside of a building or structure.

2508.2.1 Weather protection. Gypsum wallboard, gypsum lath or gypsum plaster shall not be installed until weather protection for the installation is provided.

❖ This section relates to the necessary environmental conditions for proper plaster application and hydration. Temperature limitations are needed to provide a uniform heating condition for a minimum of one week prior to the start of plastering, and are the same as the requirements stated in ASTM C 842. This minimizes the movements or stresses caused by thermal changes which in turn lowers the chances of plaster cracking.

2508.3 Single-ply application. Edges and ends of gypsum board shall occur on the framing members, except those edges and ends that are perpendicular to the framing members. Edges and ends of gypsum board shall be in moderate contact except in concealed spaces where fire-resistance-rated construction, shear resistance or diaphragm action is not required.

❖ The application of gypsum board is specified in this chapter for nonfire-resistant construction or construction where diaphragm (shear wall) action is not required. Chapter 7 and fire test reports will establish the means of fastening and supporting the ends and edges of gypsum board for nonfire-resistant assemblies. In the case of stud walls required for diaphragm action, Chapters 22 and 23 establish structural requirements such as sizes and spacing fasteners, allowable loads and the conditions of end and edge supports. Gypsum board is the generic name for a family of sheet products consisting of a noncombustible core, primarily composed of gypsum, with a paper surfacing on each face. The installation and finishing of gypsum board materials must comply with the referenced standards in Table 2508.1. GA 216, *Applica-*

tion and Finishing of Gypsum Panel Products, published by the Gypsum Association, is the most commonly recommended specification for the installation and finishing of gypsum board.

- **Framing:** As with any gypsum board or gypsum sheet product, it is critical to inspect the framing on which the gypsum material is to be attached. Finishing of gypsum board will be made extremely difficult when installed over unlevel, unplumb, out-of-square or structurally inadequate framing. All of these concerns must be satisfied before the installation of any gypsum board product can begin. Gypsum board can be attached to a variety of framing members, including but not necessarily limited to, wood or metal studs; joists and rafters; suspended metal-frame ceiling systems and wood or metal furring on concrete or masonry walls. Framing systems that are not installed square, level and plumb will create stress points where the gypsum board is attached to the framing. Stress points can cause the gypsum board to crack, shear or delaminate. This can become a significant issue when fire-resistance-rated assemblies are compromised.
- **Fastening:** The type of fastening devices and methods for attaching gypsum board to framing are dependent upon the type of framing system being used and the type of assembly being constructed. Screws, nails and construction adhesives are approved types of fasteners. Fasteners are to be selected and installed as required by the specifications referenced in GA 216, except as required elsewhere in the code for fire-resistance-rated assemblies, shear wall construction or where more restrictive requirements are provided by the referenced standards listed in Chapter 35.
- **Finishing:** Gypsum must be finished with materials and accessories designed and approved for the application for which they are being used. Joint tape, joint compound and all accessories are to be approved and installed as recommended by the manufacturer and as required by the referenced standards.
- **Multiple-ply application:** This application of gypsum board is generally utilized to increase the fire-resistance rating of an assembly, the resistance of a shear wall assembly or the sound transmission class for an assembly. The construction and technical requirements for these and similar types of specialized assemblies are covered elsewhere in the International Codes. The requirements for gypsum board itself are as defined in this chapter, except where more restrictive requirements are mandated for a specialized assembly.

2508.3.1 Floating angles. Fasteners at the top and bottom plates of vertical assemblies, or the edges and ends of horizontal assemblies perpendicular to supports, and at the wall line are permitted to be omitted except on shear resisting elements or fire-resistance-rated assemblies. Fasteners shall be applied in such a manner as not to fracture the face paper with the fastener head.

❖ Floating angles are the edges of gypsum board that are not directly fastened to a framing member where a change in the plane of the gypsum board occurs. Gypsum board used as part of a shear wall assembly or as a component part of any fire-resistance-rated assembly must have all edges fastened to a framing member. It is given that gypsum board used as part of a vertical assembly (wall) is fastened to the wall studs along the edges and in the field of the gypsum board as defined in GA 216. Adequate support for the weight of the gypsum board itself will be provided, therefore eliminating the need for additional fastening at the top and bottom plates of the wall framing. Fastening is not to be omitted at any inside or outside corner of a wall assembly.

Fasteners may be omitted at the edges of horizontal assemblies (ceiling and soffits) when the edge of the gypsum board is perpendicular to the framing member. As with wall assemblies, it is given that fasteners located along the edges and in the field of the gypsum board in accordance with GA 216 will provide adequate support for the weight of the gypsum board itself. Edges of horizontal gypsum board assemblies that are parallel to framing members must be fastened to the framing member except where the parallel edge is at the intersection of the wall and ceiling. Typically, the gypsum board will be installed on the ceiling before it is installed on the wall. The edge of the gypsum board on the wall will be butted against the underside edge of the ceiling board, providing additional edge support for the ceiling.

This section permits the elimination of unnecessary fastening of gypsum board in specific locations and conditions without compromising the integrity of the gypsum board support or fastening system. Gypsum board must always be fastened to prevent sagging, warping, waviness or any visual defects.

2508.4 Joint treatment. Gypsum board fire-resistance-rated assemblies shall have joints and fasteners treated.

Exception: Joint and fastener treatment need not be provided where any of the following conditions occur:

1. Where the gypsum board is to receive a decorative finish such as wood paneling, battens, acoustical finishes or any similar application that would be equivalent to joint treatment.
2. On single-layer systems where joints occur over wood framing members.
3. Square edge or tongue-and-groove edge gypsum board (V-edge), gypsum backing board or gypsum sheathing.
4. On multilayer systems where the joints of adjacent layers are offset from one to another.

5. Assemblies tested without joint treatment.

❖ This section requires joint and fastener treatment for all gypsum board fire-resistance-rated assemblies. The code also identifies and permits exceptions where the gypsum board is to receive a decorative finish or any similar application that is considered to be equivalent to the joint or fastener treatment.

These exceptions indicate that joint treatment is not required for assemblies tested without it:

1. Where the finish layer of gypsum board is to receive a decorative covering or similar type of applied material that will cover exposed joints, finishing of joints is not required; however, the applied material must provide the same fire-resistance rating of the assembly that it is being applied to.
2. Joint treatment is not required in single-layer applications where joints occur directly over wood-framing members. In effect, the additional joint treatment does not materially increase the fire-resistance rating of the assembly, and many partitions have been tested and passed the fire test without the added protection of joint and fastener treatment as part of the tested design.
3. Gypsum board with factory-designed edges that will prevent the penetration of smoke or fire into a rated assembly does not have to be finished. The factory edges listed will minimize gaps at joints and maintain the integrity of the assembly.
4. Multilayer assemblies with offset joints will eliminate any gaps in the system; therefore, fire smoke and harmful gases cannot readily penetrate the fire-resistance-rated assembly. This makes joint treatment unnecessary.
5. Where approved fire tests indicate that joint and fastener treatment is included as a part of the tested assembly, the code requires that joint and fastener treatment be applied in the same manner and extent as the approved test assembly. Where the approved test assembly does not specifically state whether joint or fastener treatment is part of the test assembly, the code requires that joint and fastener treatment is to be applied to any fire-resistance-rated assembly. Assemblies tested without joint treatment need not be finished.

As indicated elsewhere in this section, with regard to gypsum board application, joint and fastener treatment is primarily applied for aesthetic reasons and to provide a satisfactory finish surface for the application of decorative finishes, such as paint, wallpaper, vinyl wall fabric, textured paint or a number of other finishes used for the appearance of a rated or nonrated assembly.

2508.5 Horizontal gypsum board diaphragm ceilings. Gypsum board shall be permitted to be used on wood joists to create a horizontal diaphragm ceiling in accordance with Table 2508.5.

❖ Generally, a gypsum board ceiling does not serve as a load-carrying structural element for other than its own weight. This section provides installation requirements [refer to Figure 2508.5(1)] that allow the use of a horizontal gypsum board ceiling diaphragm. This permits a wood-framed ceiling to carry limited lateral loads, such as a portion of the component earthquake force on a partition as shown in Figure 2508.5(2).

TABLE 2508.5. See below.

❖ Table 2508.5 provides the shear capacity that is to be used in designing a horizontal gypsum board ceiling diaphragm. Since the material thickness and fastener size do not vary, the capacity is strictly a function of the framing member spacing. Further qualifications on the tabulated shear capacities that are stated in the footnotes include: ceiling diaphragm capacities are not cumulative with other diaphragm capacities; the capacities given are only for short-term loads due to wind and earthquakes; in structures classified as Seismic Design Category D, E or F the capacities must be reduced by 50 percent.

TABLE 2508.5
SHEAR CAPACITY FOR HORIZONTAL WOOD FRAMED GYPSUM BOARD DIAPHRAGM CEILING ASSEMBLIES

MATERIAL	THICKNESS OF MATERIAL (MINIMUM) (inches)	SPACING OF FRAMING MEMBERS (MAXIMUM) (inches)	SHEAR VALUE[a, b] (plf of ceiling)	MIMIMUM FASTENER SIZE
Gypsum board	$^1/_2$	16 o.c.	90	5d cooler or wallboard nail; $1^5/_8$-inch long; 0.086-inch shank; $^{15}/_{64}$-inch head[c]
Gypsum board	$^1/_2$	24 o.c.	70	5d cooler or wallboard nail; $1^5/_8$-inch long; 0.086-inch shank; $^{15}/_{64}$-inch head[c]

For SI: 1 inch = 25.4 mm, 1 pound per foot = 14.59 N/m.

a. Values are not cumulative with other horizontal diaphragm values and are for short-term loading due to wind or seismic loading. Values shall be reduced 25 percent for normal loading.

b. Values shall be reduced 50 percent in Seismic Design Categories D, E and F.

c. $1^1/_4$-inch, No. 6 Type S or W screws are permitted to be substituted for the listed nails.

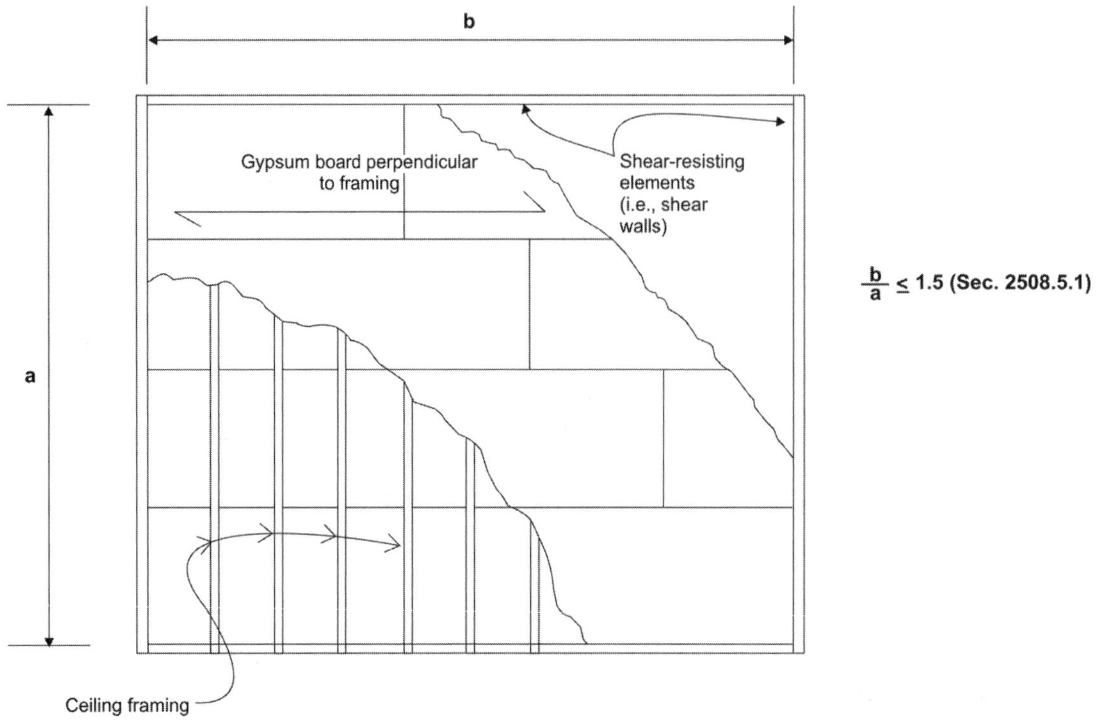

b

Gypsum board perpendicular to framing

Shear-resisting elements (i.e., shear walls)

a

$\frac{b}{a} \leq 1.5$ (Sec. 2508.5.1)

Ceiling framing

Figure 2508.5(1)
HORIZONTAL GYPSUM BOARD CEILING DIAPHRAGM CONSTRUCTION

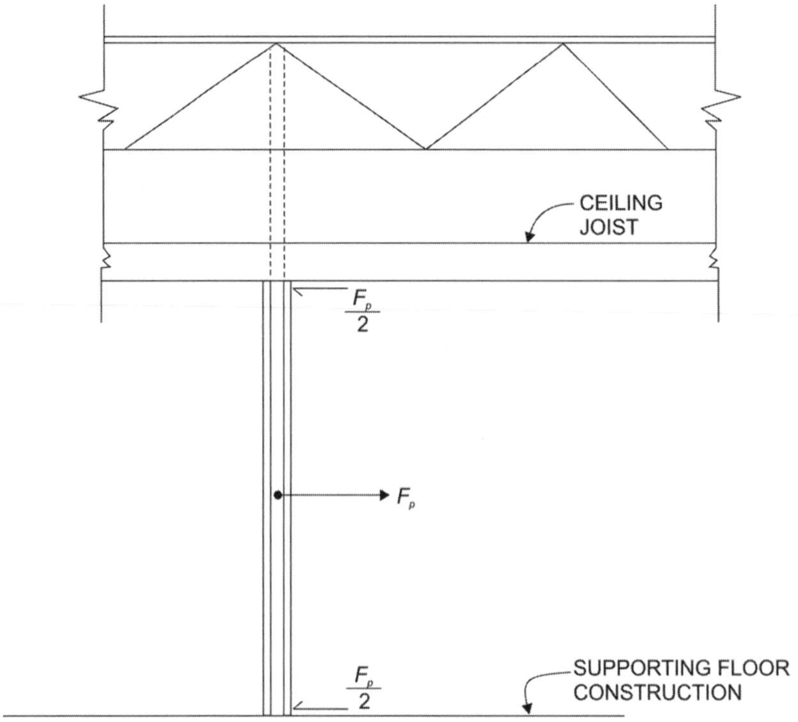

CEILING JOIST

$\frac{F_p}{2}$

F_p

SUPPORTING FLOOR CONSTRUCTION

$\frac{F_p}{2}$

Figure 2508.5(2)
PARTITION SUPPORTED LATERALLY BY GYPSUM BOARD CEILING DIAPHRAGM

2508.5.1 Diaphragm proportions. The maximum allowable diaphragm proportions shall be $1^1/_2$:1 between shear resisting elements. Rotation or cantilever conditions shall not be permitted.

❖ Limiting the diaphragm proportions and prohibiting cantilevers and rotation serves to limit the deflection and distortion of the ceiling diaphragm [see Figure 2508.5(1)].

2508.5.2 Installation. Gypsum board used in a horizontal diaphragm ceiling shall be installed perpendicular to ceiling framing members. End joints of adjacent courses of gypsum board shall not occur on the same joist.

❖ The installation requirements of this section are necessary to obtain the shear capacities indicated in Table 2508.5.

2508.5.3 Blocking of perimeter edges. All perimeter edges shall be blocked using a wood member not less than 2-inch by 6-inch (51 mm by 159 mm) nominal dimension. Blocking material shall be installed flat over the top plate of the wall to provide a nailing surface not less than 2 inches (51 mm) in width for the attachment of the gypsum board.

❖ This section states the minimum required edge blocking that is illustrated in Figure 2508.5.3.

2508.5.4 Fasteners. Fasteners used for the attachment of gypsum board to a horizontal diaphragm ceiling shall be as defined in Table 2508.5. Fasteners shall be spaced not more than 7 inches (178 mm) on center (o.c.) at all supports, including perimeter blocking, and not more than $^3/_8$ inch (9.5 mm) from the edges and ends of the gypsum board.

❖ This provision establishes maximum fastener spacing and minimum edge distances for gypsum board ceiling diaphragms.

2508.5.5 Lateral force restrictions. Gypsum board shall not be used in diaphragm ceilings to resist lateral forces imposed by masonry or concrete construction.

❖ This section prohibits the use of a gypsum board ceiling diaphragm in resisting lateral loads due to masonry or concrete.

SECTION 2509
GYPSUM BOARD IN SHOWERS AND WATER CLOSETS

2509.1 Wet areas. Showers and public toilet walls shall conform to Section 1210.2.

❖ Special caution must be taken when gypsum board is used in applications where it will be exposed to moisture, because gypsum is not naturally resistant to water, humidity or other forms of moisture. Failure of gypsum board can be caused by direct or indirect exposure to all forms of moisture. Moisture-resistant gypsum board, which is faced with moisture-resistant paper, must be used in high moisture areas, and additional protection such as paint must be applied to prevent gypsum board from being exposed to moisture. Moisture-resistant gypsum board is known and referred to as "green board," because of the distinctive green moisture-resistant paper that is applied to it, which also makes it readily identifiable as moisture resistant.

Wet areas at interior locations, such as those identified in Section 1210.2, must be protected from moisture by a nonabsorbent wall or ceiling material. Cement plaster (gypsum plaster is not permitted in wet areas) and tile or other like finish materials will provide moisture protection for the wall surface. An additional waterproofing membrane must be provided where the wall is constructed of wood framing (see related commentary, Section 2511.5).

2509.2 Base for tile. Glass mat water-resistant gypsum backing panels, discrete nonasbestos fiber-cement interior substrate sheets or nonasbestos fiber-mat reinforced cementitious backer units in compliance with ASTM C 1178, C 1288 or C 1325 and installed in accordance with manufacturer recommendations shall be used as a base for wall tile in tub and shower areas and wall and ceiling panels in shower areas. Water-resistant gypsum backing board shall be used as a base for tile in water closet compartment walls when installed in accordance with GA-216 or ASTM C 840 and manufacturer recommendations. Regular gypsum wallboard is permitted

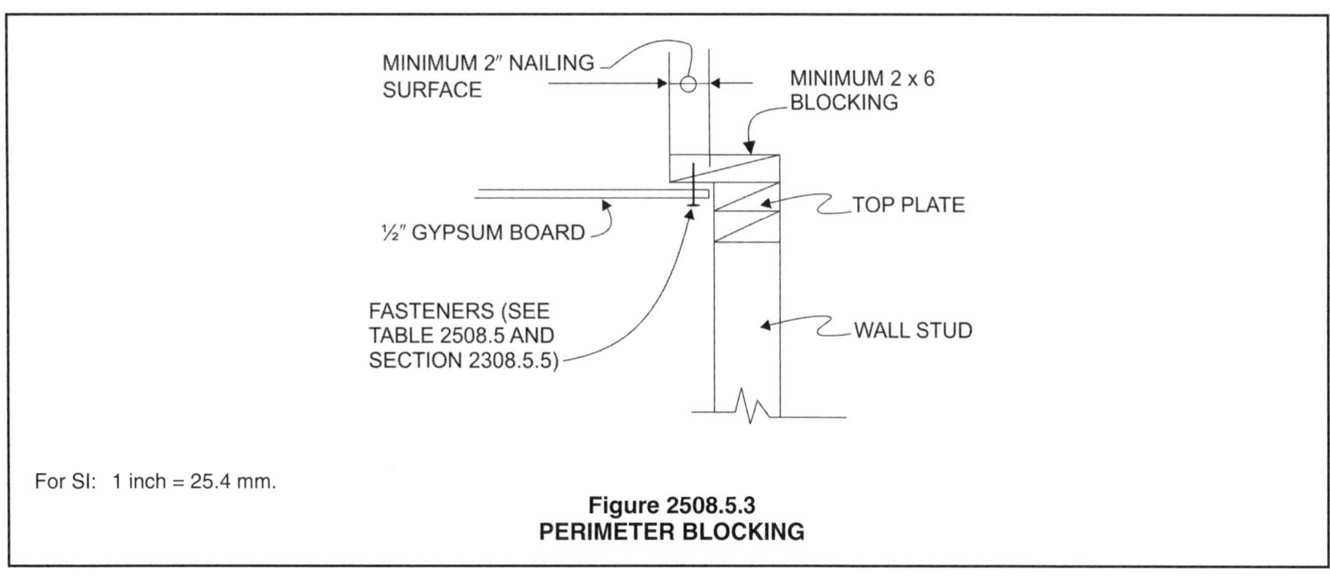

Figure 2508.5.3
PERIMETER BLOCKING

under tile or wall panels in other wall and ceiling areas when installed in accordance with GA-216 or ASTM C 840.

❖ In areas that have routine exposure to heavy concentrations of water, it is necessary to use materials that are resistant to moisture as a backing for tile. This section lists the materials recognized for use as a base for wall tile in tub and shower areas and wall and ceiling panels in shower areas. The code also requires that moisture-resistant gypsum board (green board) is to be used as a base for ceramic and other hard tiles for water-closet compartment walls. It can be used in these locations because of its resistance to incidental moisture exposure. All joints in gypsum board that will be covered by tile or wall panels must be taped and finished even though they will be unexposed to view when the installation of the wall-covering material is complete. This is necessary to prevent moisture from migrating through the unfinished joints and causing damage to the gypsum board and the wall-framing system. Corrosion-resistant fasteners must be used in wet areas.

2509.3 Limitations. Water-resistant gypsum backing board shall not be used in the following locations:

1. Over a vapor retarder in shower or bathtub compartments.
2. Where there will be direct exposure to water or in areas subject to continuous high humidity.
3. On ceilings where frame spacing exceeds 12 inches (305 mm) o.c. for $^1/_2$-inch thick (12.7 mm) water-resistant gypsum backing board and more than 16 inches (406 mm) o.c. for $^5/_8$-inch thick (15.9 mm) water-resistant gypsum backing board.

❖ Although there are many gypsum board sheet products that are manufactured and approved for use in wet areas or areas exposed to moisture or humidity, there are still some extreme conditions where even water-resistant gypsum board will not provide the level of moisture protection necessary:

1. Gypsum board installed on the walls and ceilings at shower and bathtub areas must be finished to prevent moisture from penetrating the wall or ceiling finish and contacting the gypsum board. The finish applied to the exposed face of the gypsum board, in effect, creates a water-resistant barrier that not only stops water from getting to the gypsum board, but also prevents the release of moisture from within the wall or the gypsum board itself. For this reason, gypsum board must not be installed over the outboard side of any vapor barrier or retarder. This will create a waterproof membrane on both faces of the gypsum board, causing moisture to be trapped in the gypsum board that will ultimately cause it to decompose and fail.
2. Water-resistant gypsum board is not to be used in areas that will be subject to direct exposure to water or continuous exposure to high humidity at locations such as saunas, steam rooms, gang showers or indoor pools. Gypsum board products, including the water-resistant type, are not intended for these extreme conditions and will provide unsatisfactory performance. Non-gypsum wall and ceiling materials such as concrete masonry, ceramic tile on cement backer board, cement plaster (stucco) or other materials designed and recommended for high oisture exposure must be used in these locations.
3. The additional weight, due to moisture, that is commonly applied as live loads to ceilings in high moisture or wet areas must be considered when selecting the thickness of the gypsum board and the spacing of the structural support members necessary to prevent sagging of the ceiling. The more restrictive frame spacing and gypsum board thicknesses required in this section are established to allow for the additional weight. Where other structural requirements in the code are less restrictive for the conditions contained in this section, these requirements apply.

SECTION 2510
LATHING AND FURRING FOR CEMENT PLASTER (STUCCO)

2510.1 General. Exterior and interior cement plaster and lathing shall be done with the appropriate materials listed in Table 2507.2 and Chapter 35.

❖ This section contains the appropriate referenced standards for materials, proper application and protection of interior and exterior lath and furring for the installation of cement plaster (often referred to as "stucco"). Cement plaster is a cementitious-based plaster material with excellent water-resistant properties. It is the only type of plaster that the code recognizes as an exterior wall covering. Cement plaster can also be used as an interior finish wall covering.

Table 2507.2 lists the appropriate ASTM standards for portland cement, metal lath and plastering materials. These referenced standards are the minimum material specifications required by the code to meet the requirements of this section, as well as related requirements elsewhere in the code for the proper application and procedures for lath and furring for cement plaster.

2510.2 Weather protection. Materials shall be stored in such a manner as to protect such materials from the weather.

❖ Like gypsum products, cement stucco lathing and plastering materials must be stored to prevent the deleterious effect of the weather.

2510.3 Installation. Installation of these materials shall be in compliance with ASTM C 926 and ASTM C 1063.

❖ Table 2511.1.1 refers to ASTM C 1063 as the standard specification for interior and exterior lathing and furring for cement-based plastering, as specified in

ASTM C 926. Where a fire-resistance rating is required for plastered assemblies and construction, details of the construction must be in accordance with the reports for the fire test assemblies that have met the requirements of the fire-resistance rating imposed. Where a specific degree of sound control is required for plastered assemblies and construction, details of construction must be in accordance with official test reports conducted in recognized testing laboratories in accordance with the applicable requirements of ASTM E 90.

Table 2511.1.1 also refers to ASTM C 1063 as the standard specification for the installation of portland cement-based plaster. For exterior (stucco) and interior work, ASTM C 1063 provides the requirements for proportioning various plaster mixes and thicknesses (refer to Section 2512 for additional requirements for cement plaster).

2510.4 Corrosion resistance. Metal lath and lath attachments shall be of corrosion-resistant material.

❖ Because of the amount of water that is present in plaster when it is mixed and applied to a surface, all metal components and accessories, such as metal lath; screws; screeds; grounds; metal runners at the edges of the plaster; expansion and control joint molding or any other metal component that will be in contact with the plaster, must be protected from corrosion and rusting from prolonged contact with wet plaster. Corrosion of metal components will cause failure and create voids in the plaster that will allow moisture to get into the wall and cause damage to the structural and nonstructural elements of the building. For these reasons, the code requires metal components be made of corrosion-resistant materials. Table 2507.2 provides a listing of the required material standards for lath and related accessories.

2510.5 Backing. Backing or a lath shall provide sufficient rigidity to permit plaster applications.

❖ Proper application of exterior plaster requires that the base coat is thoroughly pushed through the lath to provide the required mechanical bond. For this reason, it is required by the code that the lath and its backing provide sufficient rigidity to permit the proper application of the plaster. A common method of providing a rigid backing for the lath for exterior plaster over open wood or metal stud framing is to fasten 18-gage wire backing (see Figure 2510.5) to the studs at 6 inches (152 mm) on center horizontally. It is critical to stretch the wire tightly across the face of the studs to provide the desired rigidity for the backing. The wire backing must be attached to each stud to prevent slippage or movement of the wire.

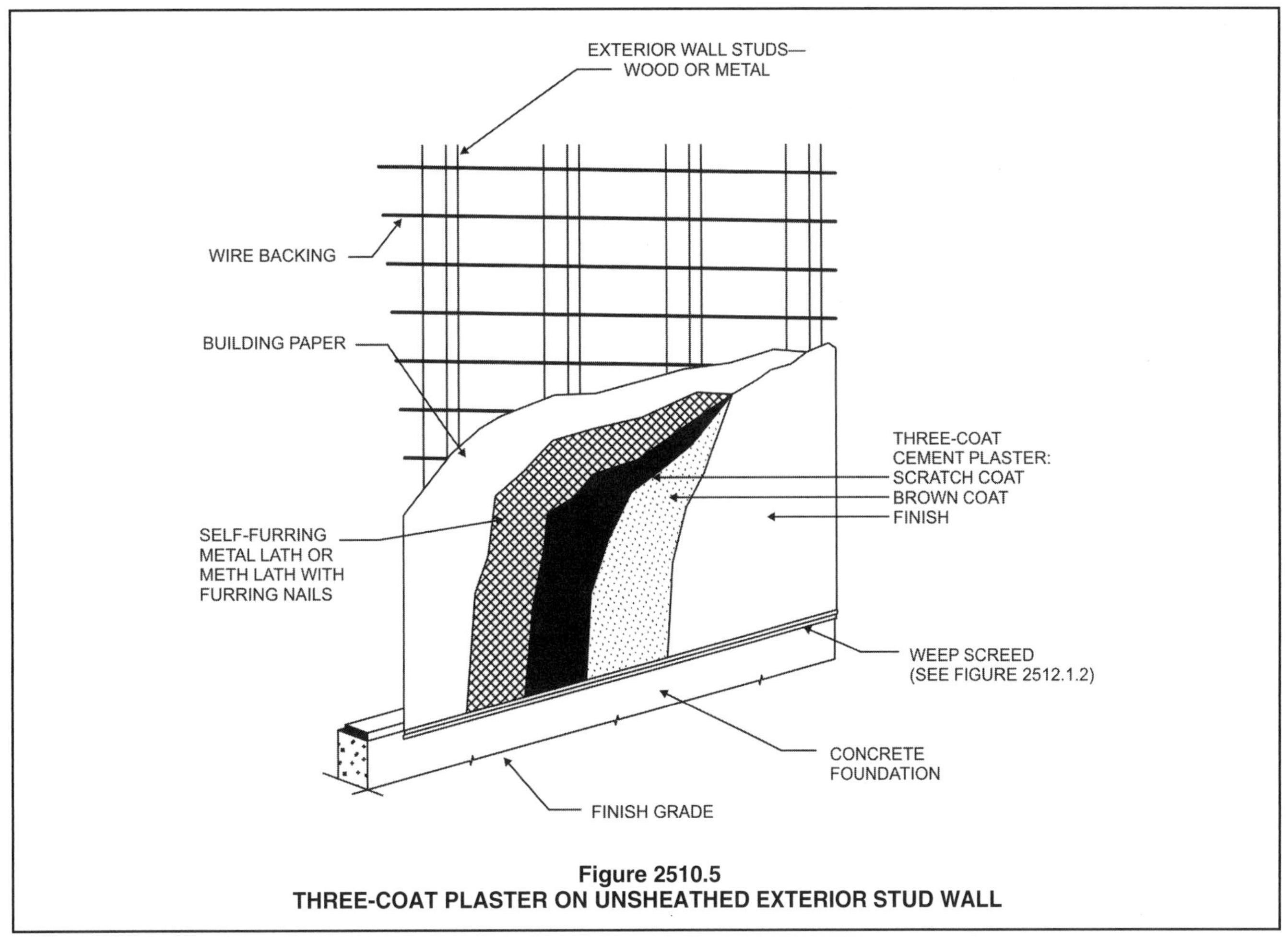

Figure 2510.5
THREE-COAT PLASTER ON UNSHEATHED EXTERIOR STUD WALL

2510.5.1 Support of lath. Where lath on vertical surfaces extends between rafters or other similar projecting members, solid backing shall be installed to provide support for lath and attachments.

❖ It is important for the lath to be supported continuously along all edges. Continuous edge support is necessary to prevent excessive movement of the finished plaster, which can result in unwanted cracking. This is particularly critical on exterior vertical surfaces of cement plaster because cracking provides an entry point for water into the building. At locations where the wall plane is interrupted at regular intervals, such as eaves and soffits where rafter ends project beyond the face of the wall, the code requires that the lath be supported by solid backing at all edges. In addition to supporting the edges of the lath, the code requires that all plaster trim, edging, screeds, grounds and other accessories be rigidly supported with solid backing at all edges.

The damage to finished plaster on an exterior vertical surface due to unsupported edges is much more obvious than damage that would be expected on an interior surface; however, failure of interior wall surfaces is also a concern. For this reason, the code makes no distinction between interior and exterior surfaces. Solid backing is required for lath at all locations, whether interior and exterior.

2510.5.2 Use of gypsum backing board.

❖ The code does not permit the use of gypsum board or gypsum lath as a backing for cement plaster, except as noted in Section 2510.5.2.1.

2510.5.2.1 Use of gypsum board as a backing board. Gypsum lath or gypsum wallboard shall not be used as a backing for cement plaster.

Exception: Gypsum lath or gypsum wallboard is permitted, with a *water-resistive barrier*, as a backing for self-furred metal lath or self-furred wire fabric lath and cement plaster where either of the following conditions occur:

1. On horizontal supports of ceilings or roof soffits.
2. On interior walls.

❖ Because of its excellent water-resistant properties, cement plaster is the only type permitted by the code for use as a finished exterior surface exposed to the weather, or at an interior location subject to excessive moisture. Gypsum has none of the water-resistant properties possessed by cement-based plasters or cementitious backer boards, which are designed specifically for use in high moisture areas. Due to the inability of gypsum-based materials to resist moisture, the code does not permit the use of gypsum board or gypsum lath as an interior or exterior backing in direct contact with cement plaster. Gypsum board lath subjected to moisture will decompose and cause failure to the entire plaster system, including the outer or finished coats. Exceptions to this restriction are permitted when gypsum is protected from contact with cement plaster as allowed in specific locations described below.

Gypsum lath and gypsum wallboard are permitted to be used as a backing for cement plaster when a water-resistive barrier, such as Grade D building paper, is installed at the back of the lath or to the face of the gypsum board to prevent the cement plaster from contacting the gypsum backing. The building paper must be securely attached to the face of the gypsum board before installing the self-furring metal lath or self-furring wire fabric to create a water-resistive barrier. Items 1 and 2 of the exception indicate locations where this may be applied.

2510.5.2.2 Use of gypsum sheathing backing. Gypsum sheathing is permitted as a backing for metal or wire fabric lath and cement plaster on walls. A *water-resistive barrier* shall be provided in accordance with Section 2510.6.

❖ Gypsum board sheathing is permitted to be used as a backer for cement plaster when installed with a water-resistive barrier as described in Section 2510.5.2.1 and as required in Section 2510.6.

2510.5.3 Backing not required. Wire backing is not required under expanded metal lath or paperbacked wire fabric lath.

❖ Because metal lath or paper-backed wire fabric lath has reinforcing ribs to provide necessary rigidity, the code requires no additional backing for these materials.

2510.6 Water-resistive barriers. *Water-resistive barriers* shall be installed as required in Section 1404.2 and, where applied over wood-based sheathing, shall include a water-resistive vapor-permeable barrier with a performance at least equivalent to two layers of Grade D paper. The individual layers shall be installed independently such that each layer provides a separate continuous plane and any flashing (installed in accordance with Section 1405.4) intended to drain to the water-resistive barrier is directed between the layers.

Exception: Where the *water-resistive barrier* that is applied over wood-based sheathing has a water resistance equal to or greater than that of 60-minute Grade D paper and is separated from the stucco by an intervening, substantially nonwater-absorbing layer or drainage space.

❖ The code requires that a water-resistive barrier be installed behind exterior plaster, as it does for any exterior wall veneer, in Section 1404.2. The code also requires that when the barrier is applied over wood-based sheathing, such as plywood, the barrier performance must be equivalent to two layers of Grade D building paper. Note that there is a reference to "Grade D paper" without defining the term or providing a reference to a standard. The reference to Grade D paper can be traced to Federal Specification UU-B790a, *Federal Specifications for Building Paper, Vegetable Fiber–Kraft, Waterproofed, Water Repellent, and Fire Resistant.* Grade D paper is a building paper with a water resistance of 10 minutes (or more) and meets the classification for Type I, Grade D when tested and manufactured in accordance with Federal

Specification UU-B790a. The code references Grade D paper because it exhibits the appropriate water vapor permeability to prevent entrapment of moisture between the paper and the sheathing.

The requirement for equivalency to two layers of Grade D paper is based on the problems that have been observed when only one layer is applied over wood sheathing. The wood sheathing eventually exhibits dry rot due to the penetration of moisture. Cracking is then created in the plaster due to movement of the sheathing caused by alternate expansion and contraction. Field experience has shown that where two layers of building paper are used, the penetration of moisture is considerably decreased, as is the cracking of the plaster due to movement of the sheathing caused by cycles of wetting and drying.

The code also requires each layer of the water-resistive barrier to be independently installed in a manner that provides a continuous drainage plane. The primary function of the inboard layer is to keep water away from the sheathing and also from penetrating into the stud cavity. This layer should be integrated with window and door flashings, the weep screed at the bottom of the wall as well as any through-wall flashings or expansion joints. The primary function of the outboard layer which comes in contact with the stucco is to separate the stucco from the inboard layer of the water-resistive barrier. This layer is at times referred to as a, "sacrificial layer," an "intervening layer" or a "bond break layer."

The exception permits the application of alternative standard practices by recognizing stucco systems in which one of the two layers of the water-resistive barrier is replaced by a layer that, although not a Grade D paper, provides separation from the wet stucco application and provides a barrier to moisture from the stucco to the sheathing.

2510.7 Preparation of masonry and concrete. Surfaces shall be clean, free from efflorescence, sufficiently damp and rough for proper bond. If the surface is insufficiently rough, *approved* bonding agents or a Portland cement dash bond coat mixed in proportions of not more than two parts volume of sand to one part volume of Portland cement or plastic cement shall be applied. The dash bond coat shall be left undisturbed and shall be moist cured not less than 24 hours.

❖ The masonry surface must be properly prepared to receive the cement plaster base coat to ensure proper bond between the masonry base and the plaster. Masonry surfaces must be clean, damp, free of dust and efflorescence and reasonably flat, without excessive protrusions or indentations. These irregularities will affect the evenness and uniform thickness of the finished plaster. Mortar joints should be struck flush to help minimize any surface irregularities.

Masonry bases for plaster have varying degrees of suction (water absorption rate) but generally provide a good adhesive as well as mechanical bond. Most masonry bases have enough suction so that special bonding agents will not be required to achieve the adequate bond between the plaster and the base. Where denser and harder masonry materials that have very low suction characteristics are used, however, preparation of the surface, similar to that of concrete, may be necessary. Such preparation can include mechanical scoring or the use of chemical bonding agents.

Concrete generally has a denser and tighter finished surface than masonry. Concrete does not have the same suction properties as masonry; therefore, the bond between concrete and plaster must be completely mechanical. Like masonry, the concrete surface must be clean, rough, damp and free of efflorescence and form-releasing agents. If the surface of the concrete is not adequately rough to provide the required mechanical bond, it may be prepared as follows:

- If preparation of the surface is to begin before the concrete is completely set, the surface can be roughened with a metal tool or by using a stiff, heavy-duty wire brush.
- If preparation of the concrete surface must be done after the concrete has hardened, the surface may be bush hammered or roughened with chisels or tools designed for that purpose.
- Another common method, which is specified in this section, is to apply a rich mixture of portland cement, sand and water as a dash coat. This application, in which the material is "dashed" onto the concrete surface with a course brush, may only be used where portland cement plaster is going to be applied to the concrete surface.
- A liquid bonding agent is applied.

SECTION 2511 INTERIOR PLASTER

2511.1 General. Plastering gypsum plaster or cement plaster shall not be less than three coats where applied over metal lath or wire fabric lath and not less than two coats where applied over other bases permitted by this chapter.

Exception: Gypsum veneer plaster and cement plaster specifically designed and *approved* for one-coat applications.

❖ This section is intended to define the standards and guidelines for the materials and installation of interior plaster. The requirements of this section cover gypsum plaster, gypsum veneer plaster and cement plaster. This section also addresses plaster applied to nonlath surfaces and the environmental conditions for which interior plaster is stored and installed.

Interior surfaces are all surfaces, exposed and unexposed to view, on the inside of a building or structure that are not subject to damage or failure due to the effects of uncontrollable weather conditions, such as rain, sleet, snow and wind. Interior surfaces are generally exposed to mechanically controlled climatic conditions through air-conditioning, heating and humidity control. Because of these controls, interior surfaces require far less restrictions than exterior surfaces; therefore, the quantities and types of materials and finishes permitted for use on interior surfaces are greatly increased.

It is the consensus of the industry that multicoat work is necessary for control of plaster thickness and density, particularly where plaster application is done by hand because most of the materials used for plaster compact under hand application as a result of pressure applied to the trowel. Experience has proven that this change in density is more controllable and will be more uniform when the plaster is applied in thin, successive layers. For this reason, the code requires three-coat plastering over metal lath or wire fabric lath and two-coat work applied over other plaster bases [see Figures 2511.1(1) and 2511.1(2)]. Reducing the requirements for plaster bases other than metal or wire lath depends on the rigidity of the plaster base itself. More rigid plaster bases are not as susceptible to variations in thickness and flatness of the surface. The first coat in three-coat work on a flexible base, such as wire lath, is used to stiffen that base to provide the rigidity necessary to attain uniform thickness and surface flatness.

Gypsum veneer plaster, as specified in ASTM C 843, is specifically designed to be applied as a one-coat plaster and is recognized by the code for that application. Gypsum veneer plaster must be applied over a solid base such as gypsum board lath, gypsum base for veneer plaster, concrete or concrete unit masonry. Although gypsum veneer plaster is designed to be applied in one coat, it can be applied in two coats, as long as the maximum thickness does not exceed $^{1}/_{4}$ inch (6.4 mm). Gypsum veneer plaster provides a hard, smooth finish surface, much like three-coat gypsum plaster, that will readily receive a variety of decorative finishes.

Any one-coat system must be applied over a solid base in accordance with the code, a referenced standard or the product manufacturer's instructions.

2511.1.1 Installation. Installation of lathing and plaster materials shall conform with Table 2511.1.1 and Section 2507.

❖ This section refers to Section 2507 for the appropriate standards to regulate the proper installation of gypsum lath and related gypsum board materials and accessories. The expected performance of a product or material is dependent on proper installation. Table 2511.1.1 lists the appropriate ASTM standards for plaster construction.

TABLE 2511.1.1
INSTALLATION OF PLASTER CONSTRUCTION

MATERIAL	STANDARD
Cement plaster	ASTM C 926
Gypsum plaster	ASTM C 842
Gypsum veneer plaster	ASTM C 843
Interior lathing and furring (gypsum plaster)	ASTM C 841
Lathing and furring (cement plaster)	ASTM C 1063
Steel framing	ASTM C 754; C 1007

❖ Table 2511.1.1 provides the installation standards required by the code for plaster construction. This

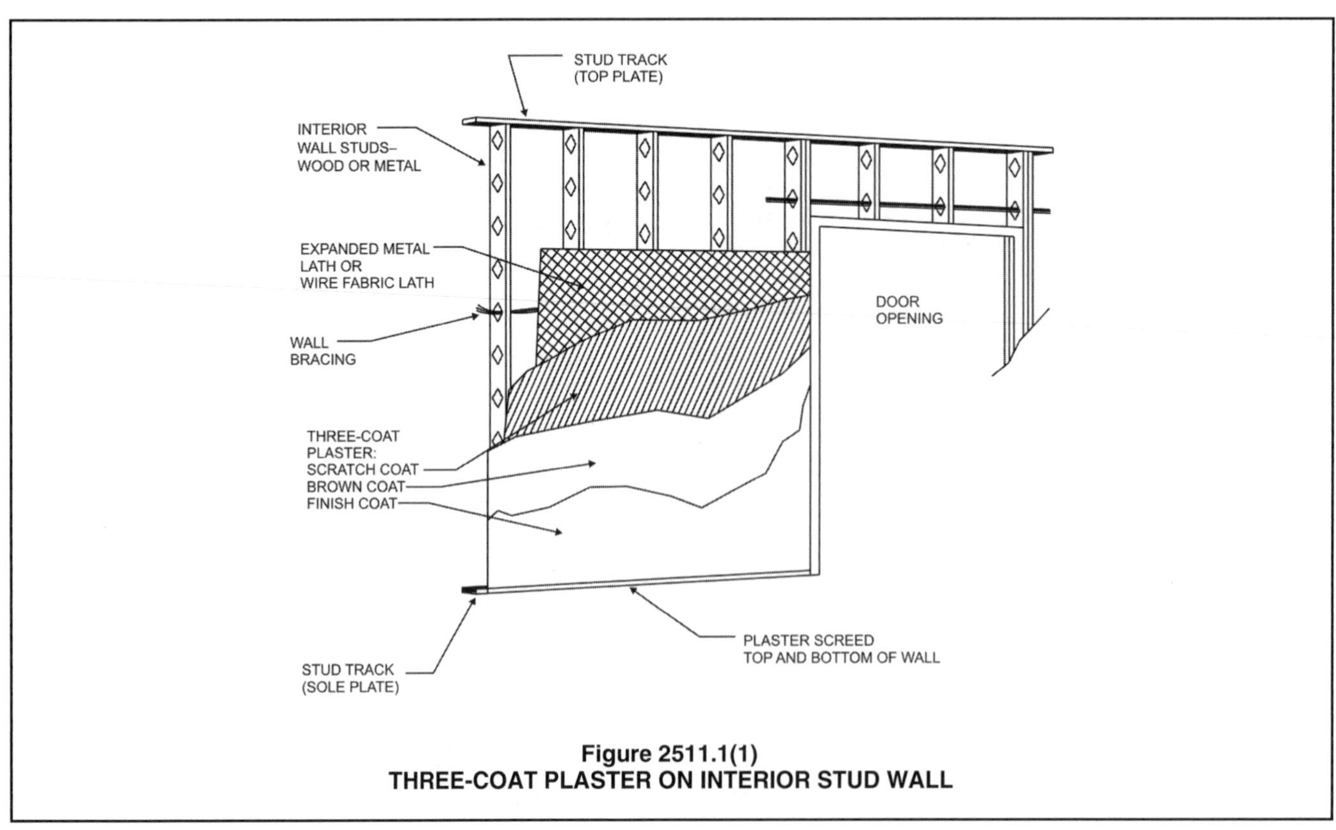

Figure 2511.1(1)
THREE-COAT PLASTER ON INTERIOR STUD WALL

table also includes installation standards for related construction materials.

2511.2 Limitations. Plaster shall not be applied directly to fiber insulation board. Cement plaster shall not be applied directly to gypsum lath or gypsum plaster except as specified in Sections 2510.5.1 and 2510.5.2.

❖ Fiber insulation board does not have the qualities necessary to provide an acceptable base for the application of plaster. It absorbs excessive moisture from the plaster mix that creates problems of workmanship, and it does not have the structural properties to provide the stability and rigidity required for a proper functioning plaster base. In colder and damper climates, fiberboard insulation retains the moisture absorbed from the plaster for a relatively longer period of time than other more dense bases. This lack of density and excessive retention of moisture will cause the premature failure of the plaster. For these reasons, the code prohibits the use of fiber insulation as a plaster base.

Cement plaster will not bond properly to gypsum plaster bases; therefore, the code prohibits the use of cement plaster on this material. The code does, however, permit the use of exterior plaster to be applied on horizontal surfaces, such as ceilings and soffits, over gypsum lath and gypsum board when used as a backing for metal lath.

2511.3 Grounds. Where installed, grounds shall ensure the minimum thickness of plaster as set forth in ASTM C 842 and ASTM C 926. Plaster thickness shall be measured from the face of lath and other bases.

❖ Plaster grounds are utilized to establish and maintain the required thickness of plaster. Grounds are usually wood or metal strips of a known thickness attached to the plaster base. The intent is that plaster grounds are used as a guide for the straightedge in determining the thickness of the plaster. Door or window frames are used as plaster grounds. Dollops of plaster of the desired thickness may also be used as grounds.

2511.4 Interior masonry or concrete. Condition of surfaces shall be as specified in Section 2510.7. *Approved* specially prepared gypsum plaster designed for application to concrete surfaces or *approved* acoustical plaster is permitted. The total thickness of base coat plaster applied to concrete ceilings shall be as set forth in ASTM C 842 or ASTM C 926. Should ceiling surfaces require more than the maximum thickness permitted in ASTM C 842 or ASTM C 926, metal lath or wire fabric lath shall be installed on such surfaces before plastering.

❖ Special gypsum plaster and acoustical plaster that is specifically designed and manufactured for direct application to concrete masonry unit walls, concrete walls and ceilings without a gypsum board or metal lath base are permitted by the code. Special care must be taken to properly prepare the masonry or concrete surfaces that are to receive a plaster finish to ensure both proper bond of the plaster and the

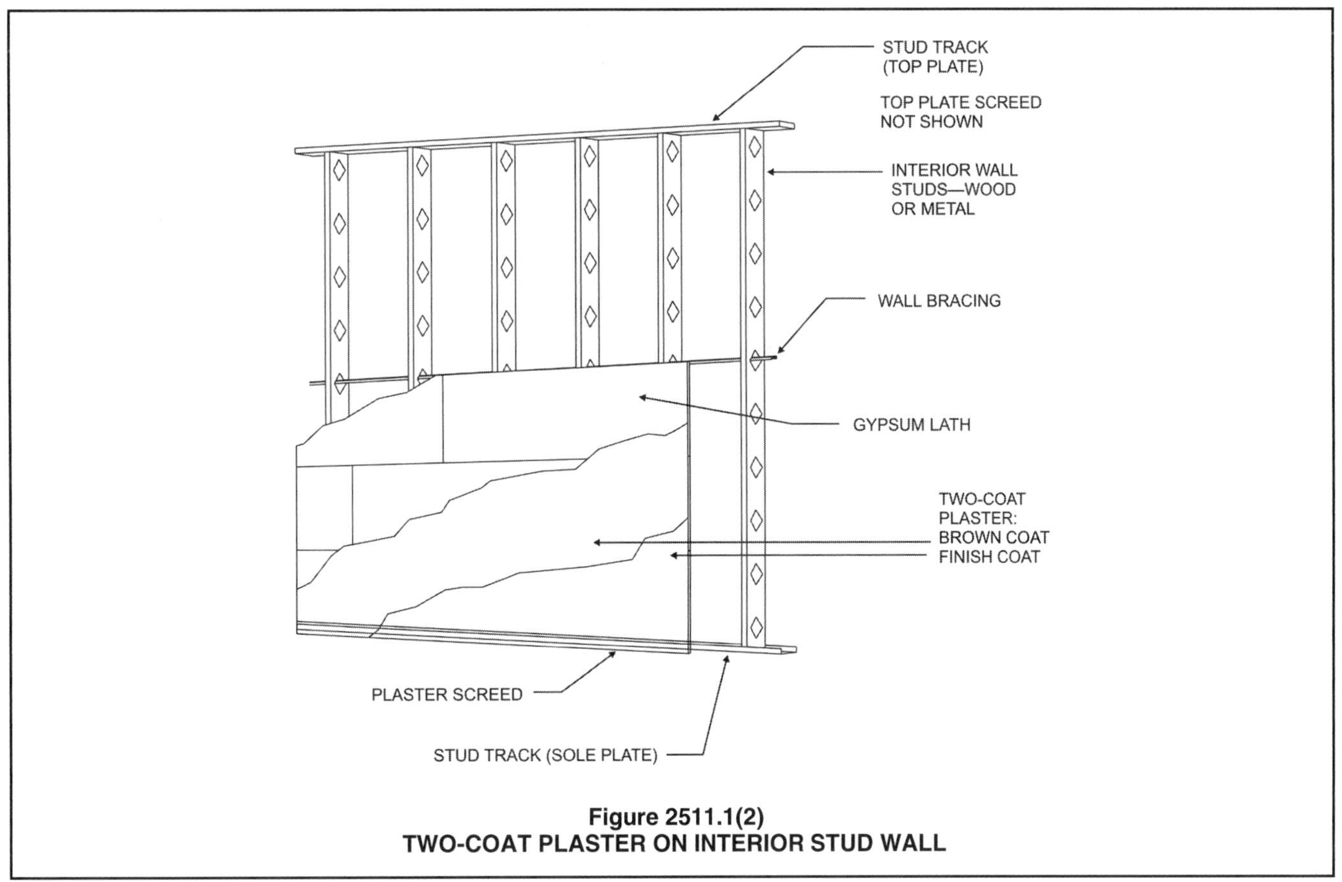

Figure 2511.1(2)
TWO-COAT PLASTER ON INTERIOR STUD WALL

desired or required levelness of the finished surface. Preparation of masonry and concrete surfaces is discussed in Section 2510.7.

Where a concrete ceiling is to receive a plaster finish, the code requires a total thickness of the base coat plaster as required by the referenced standards in this section. For applications of plaster on concrete ceiling surfaces, a thinner coat is required than for other applications. This is due primarily to the difficulty of achieving an adequate bond between the plaster and the concrete surface. Because of the horizontal application, the weight of the plaster base coat must be kept to a minimum. If it becomes necessary to increase the thickness of the plaster beyond the maximum allowed, due for example to the unevenness of the concrete surface or for some similar reason, metal lath or wire fabric lath must be attached to the concrete ceiling surface to provide anchorage for the weight of the additional plaster.

2511.5 Wet areas. Showers and public toilet walls shall conform to Sections 1210.2 and 1210.3. When wood frame walls and partitions are covered on the interior with cement plaster or tile of similar material and are subject to water splash, the framing shall be protected with an *approved* moisture barrier.

❖ Wet areas at interior locations such as those identified in Sections 1210.2 and 1210.3 must be protected from moisture by nonabsorbent wall or ceiling material. Cement plaster (gypsum plaster is not permitted in wet areas) and tile or other like materials will provide moisture protection for the wall surface, provided an additional waterproofing membrane is installed where the wall is constructed of wood framing.

SECTION 2512 EXTERIOR PLASTER

2512.1 General. Plastering with cement plaster shall be not less than three coats when applied over metal lath or wire fabric lath or gypsum board backing as specified in Section 2510.5 and shall be not less than two coats when applied over masonry or concrete. If the plaster surface is to be completely covered by veneer or other facing material, or is completely concealed by another wall, plaster application need only be two coats, provided the total thickness is as set forth in ASTM C 926.

❖ This section is intended to define the standards and guidelines for the type of materials and installation of exterior plaster. The requirements of this section include the limitations of gypsum plaster and gypsum veneer plaster. This section will also address plaster applied to solid backings, alternative application methods, curing and plaster additives. Guidelines and requirements for preventing water infiltration into the building are discussed in this section.

Exterior surfaces are the vertical and horizontal surfaces located on the outside of a building or structure that are subject to weather conditions, such as wind, changes in temperature, humidity, moisture and dampness. The most common way to protect exterior surfaces from damage caused by rain, sleet, wind, ice, snow or weather from other sources is by the use of building materials that are designed, manufactured and installed to resist or withstand these extreme weather conditions. There are isolated surfaces located on the exterior of a building or structure that are not directly exposed to or in contact with the damaging effects of the weather (see definition of "Weather exposed surfaces").

Cement plaster (stucco) is the only type approved by the code for the use of exterior plaster finish. Exterior cement plasters are required by the code to be applied in no less than three coats when applied over expanded metal, gypsum board backing or wire fabric lath for the same reasons discussed for interior plaster (see commentary, Section 2511.1.1 and Figure 2512.1). When cement plaster is applied over solid bases that are rigid, the code permits the application of a two-coat plaster system. The code also allows plaster work that is going to be completely concealed to be applied as a two-coat system, provided the total thickness meets the requirements of ASTM C 926, because the finish coat (second coat) of plaster will be the surface that will receive the exterior finish, such as paint, and it is critical for the finish coat to provide a clean, smooth and visually acceptable surface. Where the plaster surface is to be completely concealed, it is not necessary to provide a finish coat.

2512.1.1 On-grade floor slab. On wood framed or steel stud construction with an on-grade concrete floor slab system, exterior plaster shall be applied in such a manner as to cover, but not to extend below, the lath and paper. The application of lath, paper and flashing or drip screeds shall comply with ASTM C 1063.

❖ The code requires that the exterior plaster be installed to completely cover, but not extend below, the lath and paper (moisture barrier membrane) on wood or metal stud exterior wall construction supported by a concrete slab-on-grade floor. This requirement, combined with the requirement for a continuous weep screed as described in Section 2512.1.2, are intended to prevent the entrapment and subsequent channeling of free moisture to the interior of the building.

2512.1.2 Weep screeds. A minimum 0.019-inch (0.48 mm) (No. 26 galvanized sheet gage), corrosion-resistant weep screed with a minimum vertical attachment flange of $3^1/_2$ inches (89 mm) shall be provided at or below the foundation plate line on exterior stud walls in accordance with ASTM C 926. The weep screed shall be placed a minimum of 4 inches (102 mm) above the earth or 2 inches (51 mm) above paved areas and be of a type that will allow trapped water to drain to the exterior of the building. The *water-resistive barrier* shall lap the attachment flange. The exterior lath shall cover and terminate on the attachment flange of the weep screed.

❖ Water and moisture can penetrate an exterior plaster wall for a variety of reasons and in a number of ways. It is expected that some moisture will penetrate the plaster in an exterior wall; therefore, the design of the wall should include a weep screed, which will provide a way to release the moisture (see Figure 2512.1.2). Once water or moisture penetrates the plaster, it will migrate down the exterior wall face of the water-resistive barrier until it reaches the sill plate or mud sill. At this point, the water will seek a way out of the wall. If the exterior plaster system is not detailed and constructed with provisions to allow the moisture to escape to the exterior, it will find its own way out. This exit will almost certainly be through the interior of the wall and cause leaking, and damage to the interior of the building. For this reason, the code requires a con-

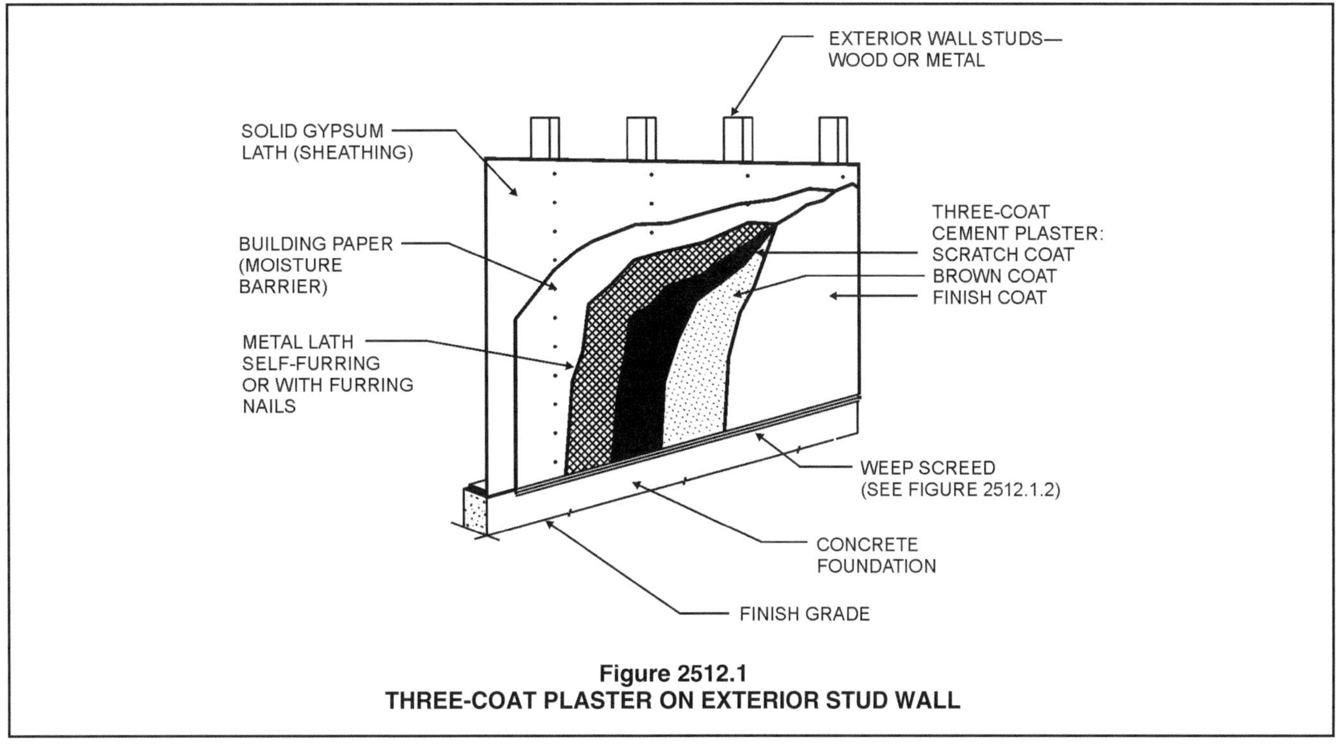

Figure 2512.1
THREE-COAT PLASTER ON EXTERIOR STUD WALL

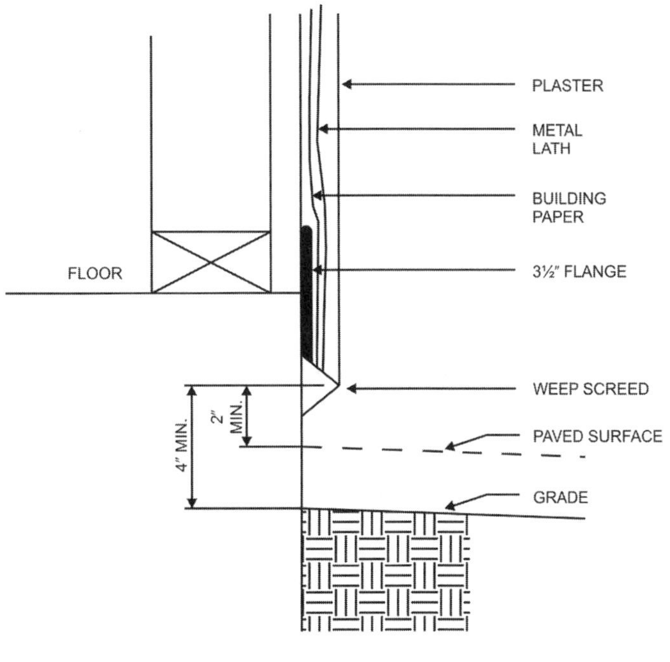

PLASTER
METAL LATH
BUILDING PAPER
FLOOR
3½″ FLANGE
WEEP SCREED
2″ MIN.
4″ MIN.
PAVED SURFACE
GRADE

For SI: 1 inch = 25.4 mm.

Figure 2512.1.2
WEEP SCREED FOR EXTERIOR PLASTER ON STUD WALL AT SLAB ON GRADE

tinuous weep screed at the bottom of exterior walls to permit the moisture to escape to the exterior of the building.

2512.2 Plasticity agents. Only *approved* plasticity agents and *approved* amounts thereof shall be added to Portland cement or blended cements. When plastic cement or masonry cement is used, no additional lime or plasticizers shall be added. Hydrated lime or the equivalent amount of lime putty used as a plasticizer is permitted to be added to cement plaster or cement and lime plaster in an amount not to exceed that set forth in ASTM C 926.

❖ Admixtures, such as plasticizers, should not be added to portland cement or blended cement unless approved by the building official. Some admixtures can create harmful effects that more than offset the desired improvement in plasticity. It is preferable that plasticizers be added during the manufacturing of the cement for product uniformity and proper proportions. When plastic cement is used, the code does not allow any further additions of plasticizers because the amount added during the manufacturing process is adequate and is the maximum amount permitted. Hydrated lime and lime putty are time-tested plasticizers used with cement plaster, and their use is permitted by the code in the amounts specified in ASTM C 926 as referenced herein.

2512.3 Limitations. Gypsum plaster shall not be used on exterior surfaces.

❖ Gypsum veneer plaster, gypsum plaster or any other similar gypsum-based plaster materials are not permitted for use as an exterior finish material. Gypsum-based plaster is subject to deterioration and ultimate failure when it is exposed to exterior weather conditions such as excessive heat, humidity and moisture, which are unavoidable on exterior surfaces of a building.

2512.4 Cement plaster. Plaster coats shall be protected from freezing for a period of not less than 24 hours after set has occurred. Plaster shall be applied when the ambient temperature is higher than 40°F (4°C), unless provisions are made to keep cement plaster work above 40°F (4°C) during application and 48 hours thereafter.

❖ Cement plaster contains a high percentage of water when it is mixed and applied, and before the process of curing is complete. For this reason, it is adversely affected by cold weather and freezing in much the same way that cement mortar and concrete are affected. The application of cement plaster to a frozen base or to a base covered with ice or frost will not only weaken the bond of the plaster to its base, but also cause the plaster itself to freeze.

When cement plaster is applied in freezing weather, to a frozen base or is mixed with frozen ingredients, it loses a high proportion of its strength and, therefore, does not comply with the requirements of the code. Plaster materials must be stored to protect them from adverse weather conditions and should only be installed when weather conditions are within the requirements of the code and will remain so for the duration of the curing process.

2512.5 Second-coat application. The second coat shall be brought out to proper thickness, rodded and floated sufficiently rough to provide adequate bond for the finish coat. The second coat shall have no variation greater than $^{1}/_{4}$ inch (6.4 mm) in any direction under a 5-foot (1524 mm) straight edge.

❖ In plaster work, a base coat is any coat beneath the finish coat. This is true whether the plaster is a two-coat or three-coat application. In a three-coat application, the first coat is usually referred to as the "scratch" coat. It is usually applied over a flexible base, such as expanded metal or wire fabric lath, and is intended to stiffen the base and provide a mechanical bond to the base. As the name implies, the first coat is scratched with a scarifying tool to provide a series of horizontal ridges or scratches that are intended to provide mechanical keys for the application of the second coat (or brown coat). The brown coat usually constitutes the major bulk of the plaster and, consequently, materially affects the membrane strength. As a result, proportioning and workability are critical, and the mix should have high plasticity for proper application. The term "brown coat" is utilized by the trade to differentiate the color of the second coat from the finish coat. The finish coat is usually much lighter in color than the second coat.

2512.6 Curing and interval. First and second coats of cement plaster shall be applied and moist cured as set forth in ASTM C 926 and Table 2512.6.

❖ Sufficient time between coats is to be allowed to permit each coat to cure or develop enough rigidity to resist cracking or other physical damage when the next coat is applied. The proper amount of moisture must be provided to the plaster mix to allow for the proper curing time for each coat. The most effective procedure for curing and the time required between each coat will depend on the climate and job conditions. Moist or fog curing will permit and control the continuous hydration of cementitious materials.

TABLE 2512.6. See page 25-21.

❖ The timing between coats will vary with climate conditions and the type of plaster base being used. Table 2512.6 specifies the minimum curing times between coats; however, temperature and relative humidity will extend or reduce the curing time required between consecutive coats. Cold or wet weather will increase the required curing time, while hot or dry weather will shorten it. Moderate changes in temperature and humidity can be controlled by providing additional heating during cold and damp weather, and by reducing the loss of moisture by prewetting during hot or dry weather.

All the conditions mentioned above must be taken into consideration when selecting the most appropriate method of curing that will provide the best results for the application being considered as well as the job conditions present during the curing process. Any of the three curing methods listed below, or any combination of the three, are permitted by the code:

1. Moist curing is accomplished by applying a fine fog spray of water as often as required to retain moisture and control the rate of curing. Moisture is generally applied two times a day—once in the morning and again in the late afternoon or evening—although job conditions are more of a factor than a predetermined time schedule. Care must be taken to avoid erosion damage to cement plaster surfaces due to excessive amounts of water running down vertical surfaces. Except in extreme or severe drying conditions, the wetting of the finish coat should be avoided.
2. Plastic film, when taped or weighted down around the perimeter of the plastered work area, will provide a vapor barrier to retain the moisture between the membrane and the plaster. Care must be taken when placing the plastic film around the work area. If the film is placed too soon before the plaster is allowed to stiffen, the film can come into contact with the wet plaster and damage the surface texture. If too much time is allowed before placing the film, an excessive amount of moisture will have escaped causing premature curing, which can cause cracking to the finished plaster.
3. Canvas, cloth or sheet material barriers can be erected to deflect sunlight and wind, both of which will affect the rate of evaporation. If the humidity is very low, this method alone may not provide adequate protection.

TABLE 2512.6
CEMENT PLASTERS

COAT	MINIMUM PERIOD MOIST CURING	MINIMUM INTERVAL BETWEEN COATS
First	48 hours[a]	48 hours[b]
Second	48 hours	7 days[c]
Finish	—	Note c

a. The first two coats shall be as required for the first coats of exterior plaster, except that the moist-curing time period between the first and second coats shall not be less than 24 hours. Moist curing shall not be required where job and weather conditions are favorable to the retention of moisture in the cement plaster for the required time period.

b. Twenty-four-hour minimum interval between coats of interior cement plaster. For alternative method of application, see Section 2512.8.

c. Finish coat plaster is permitted to be applied to interior cement plaster base coats after a 48-hour period.

2512.7 Application to solid backings. Where applied over gypsum backing as specified in Section 2510.5 or directly to unit masonry surfaces, the second coat is permitted to be applied as soon as the first coat has attained sufficient hardness.

❖ When plaster is applied directly to a solid backing, such as gypsum board lath, concrete or concrete masonry, the minimum time intervals listed in Table 2512.6 are not required between the first (scratch) coat and the second (brown) coat. The brown coat may be applied as soon as the scratch coat has reached sufficient hardness to support the weight of the additional plaster without causing damage to the scratch coat.

2512.8 Alternate method of application. The second coat is permitted to be applied as soon as the first coat has attained sufficient rigidity to receive the second coat.

❖ While the majority of experts on the application of plaster generally agree that the first two coats of cement plaster should be cured as required by Section 2512.8.2, there is an increasingly large number who feel that the brown coat can be applied without requiring moist curing of the scratch coat, as long as the scratch coat has hardened to a condition of sufficient rigidity to accept the second coat. This alternative method of application is also considered by its proponents to be stronger than when the scratch coat has been cured as required by the code. This is due to a manufacturer's claim that a superior bond is produced between the two coats of plaster when the second coat is applied to a set, but uncured, coat.

2512.8.1 Admixtures. When using this method of application, calcium aluminate cement up to 15 percent of the weight of the Portland cement is permitted to be added to the mix.

❖ In order to develop a quicker set, the code permits the addition of calcium aluminate cement up to 15 percent of the weight of the portland cement. These high-alumina cements are characterized by a faster set time than portland cement. Because high-alumina cements do not possess the same high strength and qualities as portland cement in plaster mixes, the code limits their proportion of the total mix.

2512.8.2 Curing. Curing of the first coat is permitted to be omitted and the second coat shall be cured as set forth in ASTM C 926 and Table 2512.6.

❖ When applied as specified in ASTM C 926, the complete curing of the first (scratch) coat may be omitted. The second (brown) coat may be applied as soon as the scratch coat achieves sufficient rigidity; however, the second coat must be fully cured as indicated in Table 2512.6 prior to the application of the finish coat (for additional commentary on curing, see Section 2512.6).

2512.9 Finish coats. Cement plaster finish coats shall be applied over base coats that have been in place for the time periods set forth in ASTM C 926. The third or finish coat shall be applied with sufficient material and pressure to bond and to cover the brown coat and shall be of sufficient thickness to conceal the brown coat.

❖ The base coats in plaster work provide the strength for the plaster membrane but generally do not provide a proper texture for a finished surface; therefore, a thin, veneer-like coat of plaster or stucco is applied to the base coats as a finish coat. The finish coat can be applied as a textured, ornamental or decorative finish, or it can be applied as a smooth, flat surface appropriate to receive paint, wallpaper, applied textured finishes or a variety of architectural finishes. Finish plaster coats must be allowed to properly cure before applying any adhered or wet finishes.

SECTION 2513 EXPOSED AGGREGATE PLASTER

2513.1 General. Exposed natural or integrally colored aggregate is permitted to be partially embedded in a natural or colored bedding coat of cement plaster or gypsum plaster, subject to the provisions of this section.

❖ This section is provided in the code to establish the requirements for the use of exposed aggregate plaster, both cement plaster and gypsum plaster. Exposed aggregate plaster is approved by the code for both interior and exterior applications, although it is more commonly used with cement plaster as an exterior wall finish surface. The plaster systems, accessories, materials, assemblies, installation procedures and installation applications are essentially the same as other plastering systems with the exception of the bedding coat that is prepared to receive the exposed aggregate material.

Exposed aggregate plaster is a plaster finish, which as the name implies, has small chips of exposed aggregate embedded in its finish surface. Exposed aggregate finish is used most commonly on exterior cement plaster walls, although its use is not limited to exterior applications or cement plaster. Exposed aggregate finishes may also be applied to gypsum plaster, but like all other forms of gypsum plaster, it is limited to interior and nonwet areas. The purpose of using an exposed aggregate plaster is to provide a variety of textured and colored surfaces.

2513.2 Aggregate. The aggregate shall be applied manually or mechanically and shall consist of marble chips, pebbles or similar durable, moderately hard (three or more on the Mohs hardness scale), nonreactive materials.

❖ Two of the more common exposed aggregate finishes are a washed pebble finish and a marble chip finish. There are other aggregates available, but the material requirements and application procedures are the same as those contained in this section. The mention of these two aggregates is not meant to exclude other materials from being acceptable.

The code requires that the materials used for the aggregate be nonreactive and moderately hard (three or harder on the Mohs hardness scale for minerals). The mineral that is used as the identifier for the hardness of three is calcite, which in its common form is represented by limestone and marble, to name only two. By requiring at least this hardness, the intent of the code is to approximate ordinary sand aggregate plaster in durability.

The aggregate is applied by rubbing or dashing it into or against the bedding coat and then using a trowel or other suitable tool to embed the aggregate firmly into the bedding coat. The aggregate often requires a light tamping to ensure the proper bedding.

2513.3 Bedding coat proportions. The bedding coat for interior or exterior surfaces shall be composed of one part Portland cement and one part Type S lime; or one part blended cement and one part Type S lime; or masonry cement; or plastic cement, and a maximum of three parts of graded white or natural sand by volume. The bedding coat for interior surfaces shall be composed of 100 pounds (45.4 kg) of neat gypsum plaster and a maximum of 200 pounds (90.8 kg) of graded white sand. A factory-prepared bedding coat for interior or exterior use is permitted. The bedding coat for exterior surfaces shall have a minimum compressive strength of 1,000 pounds per square inch (psi) (6895 kPa).

❖ The plaster mix proportions for the exposed aggregate bedding coat are to be as defined in this section.

For the same reason the code establishes the hardness requirements for the aggregate used in exposed aggregate plaster, it also establishes durability requirements for the bedding coat. The minimum compressive strength of the exterior bedding coat is to be 1,000 pounds per square inch (psi) (6895 kPa).

Type S lime is a special hydrated lime for masonry purposes, and is required in the exterior bedding coat to increase high early plasticity and water retention.

For the increased plasticity required of the bedding coat on an interior application, the code requires the use of a relatively richer gypsum plaster.

2513.4 Application. The bedding coat is permitted to be applied directly over the first (scratch) coat of plaster, provided the ultimate overall thickness is a minimum of $^7/_8$ inch (22 mm), including lath. Over concrete or masonry surfaces, the overall thickness shall be a minimum of $^1/_2$ inch (12.7 mm).

❖ The requirements for the scratch coat used as part of the exposed aggregate plaster are the same as those for other scratch coats described and specified elsewhere in the code. The bedding coat may be used as the second coat, provided the total plaster thicknesses required in this section are maintained through the entire system.

2513.5 Bases. Exposed aggregate plaster is permitted to be applied over concrete, masonry, cement plaster base coats or gypsum plaster base coats installed in accordance with Section 2511 or 2512.

❖ The requirements for the base coat used as part of the exposed aggregate plaster are the same as those for other base coats described and specified elsewhere in the code.

2513.6 Preparation of masonry and concrete. Masonry and concrete surfaces shall be prepared in accordance with the provisions of Section 2510.7.

❖ See the commentary for Section 2510.7 for the preparation of masonry and concrete.

2513.7 Curing of base coats. Cement plaster base coats shall be cured in accordance with ASTM C 926. Cement plaster bedding coats shall retain sufficient moisture for hydration (hardening) for 24 hours minimum or, where necessary, shall be kept damp for 24 hours by light water spraying.

❖ The curing requirements for cement plaster base coat are required to be the same as that shown in Table 2512.6 and as described in the related commentary, except that moist curing is only necessary for 24 hours in lieu of the 48 hours shown in Table 2512.6.

Bibliography

The following resource materials are referenced in this chapter or are relevant to the subject matter addressed in this chapter.

1995 Certification Listing. Ontario, Canada: Warnock Hersey, 1995.

ASCE 7-10, *Minimum Design Loads for Building's and Other Structures*. New York: American Society of Civil Engineers, 2010.

ASTM C 22/C 22M-00 (2005)e01, *Specification for Gypsum*. West Conshohocken, PA: ASTM International, 2005.

ASTM C 28/C 28M-00 (2005), *Specification for Gypsum Plasters*. West Conshohocken, PA: ASTM International, 2005.

ASTM C 35-01(2005), *Specification for Inorganic Aggregates for Use in Gypsum Plaster*. West Conshohocken, PA: ASTM International, 2005.

ASTM C 59/C 59M-00 (2006), *Specification for Gypsum Casting and Molding Plaster*. West Conshohocken, PA: ASTM International, 2006.

ASTM C 61/C 61 M-00 (2006), *Specification for Gypsum Keene's Cement*. West Conshohocken, PA: ASTM International, 2006.

ASTM C 91-05, *Specification for Masonry Cement*. West Conshohocken, PA: ASTM International, 2005.

ASTM C 150-07, *Specification for Portland Cement*. West Conshohocken, PA: ASTM International, 2007.

ASTM C 206-03, *Specification for Finishing Hydrated Lime*. West Conshohocken, PA: ASTM International, 2003.

ASTM C 472-99 (2004), *Specification for Standard Test Methods for Physical Testing of Gypsum, Gypsum Plasters and Gypsum Concrete*. West Conshohocken, PA: ASTM International, 2004.

ASTM C 473-07, *Standard Test Methods for Physical Testing of Gypsum Panel Products*. West Conshohocken, PA: ASTM International, 2007.

ASTM C 474-05, *Test Methods for Joint Treatment Materials for Gypsum Board Construction*. West Conshohocken, PA: ASTM International, 2005.

ASTM C 475-02(2007), *Specification for Joint Compound and Joint Tape for Finishing Gypsum Board*. West Conshohocken, PA: ASTM International, 2007.

ASTM C 514-04, *Specification for Nails for the Application of Gypsum Board*. West Conshohocken, PA: ASTM International, 2004.

ASTM C 557-03e01, *Specification for Adhesives for Fastening Gypsum Wallboard to Wood Framing*. West Conshohocken, PA: ASTM International, 2003.

ASTM C 587-04, *Specification for Gypsum Veneer Plaster*. West Conshohocken, PA: ASTM International, 2004.

ASTM C 595-08a, *Specification for Blended Hydraulic Cements*. West Conshohocken, PA: ASTM International, 2008.

ASTM C 631-09, *Specification for Bonding Compounds for Interior Gypsum Plastering*. West Conshohocken, PA: ASTM International, 2009.

ASTM C 635-07, *Specification for the Manufacture, Performance, and Testing of Metal Suspension Systems for Acoustical Tile and Lay-in Panel Ceilings*. West Conshohocken, PA: ASTM International, 2007.

ASTM C 645-08a, *Specification for Nonstructural Steel Framing Members*. West Conshohocken, PA: ASTM International, 2008.

ASTM C 754-08, *Specification for Installation of Steel Framing Members to Receive Screw-attached Gypsum Panel Products*. West Conshohocken, PA: ASTM International, 2008.

ASTM C 840-08, *Specification for Application and Finishing of Gypsum Board*. West Conshohocken, PA: ASTM International, 2008.

ASTM C 841-03(2008)e1, *Specification for Installation of Interior Lathing and Furring*. West Conshohocken, PA: ASTM International, 2008.

ASTM C 842-05, *Specification for Application of Interior Gypsum Plaster*. West Conshohocken, PA: ASTM International, 2005.

ASTM C 843-99 (2006), *Specification for Application of Gypsum Veneer Plaster*. West Conshohocken, PA: ASTM International, 2006.

ASTM C 844-04, *Specification for Application of Gypsum Base to Receive Gypsum Veneer Plaster*. West Conshohocken, PA: ASTM International, 2004.

ASTM C 847-09, *Specification for Metal Lath*. West Conshohocken, PA: ASTM International, 2009.

ASTM C 897-05, *Specification for Aggregate for Job-mixed Portland Cement-based Plaster*. West Conshohocken, PA: ASTM International, 2005.

ASTM C 920-08, *Standard for Specification for Elastomeric Joint Sealants.* West Conshohocken, PA: ASTM International, 2008.

ASTM C 926-06, *Specification for Application of Portland Cement-Based Plaster.* West Conshohocken, PA: ASTM International, 2006.

ASTM C 932-06, *Specification for Surface-applied Bonding Agents for Exterior Plastering.* West Conshohocken, PA: ASTM International, 2006.

ASTM C 933-07b, *Specification for Welded Wire Lath.* West Conshohocken, PA: ASTM International, 2007.

ASTM C 954-07, *Specification for Steel Drill Screws for the Application of Gypsum Panel Products or Metal Plaster Bases to Steel Studs From 0.033 in. (0.84 mm) to 0.112 in. (2.84 mm) in Thickness.* West Conshohocken, PA: ASTM International, 2007.

ASTM C 955-09, *Specification for Load-bearing (Transverse and Axial) Steel Studs, Runners (Track), and Bracing or Bridging for Screw Application of Gypsum Panel Products and Metal Plaster Bases.* West Conshohocken, PA: ASTM International, 2009.

ASTM C 1002-07, *Specification for Steel Drill Screws for the Application of Gypsum Panel Products or Metal Plaster Bases.* West Conshohocken, PA: ASTM International, 2007.

ASTM C 1007-08a, *Specification for Installation of Load-Bearing (Transverse and Axial) Steel Studs and Related Accessories.* West Conshohocken, PA: ASTM International, 2008.

ASTM C 1032-06, *Specification for Woven Wire Plaster Base.* West Conshohocken, PA: ASTM International, 2006.

ASTM C 1047-09, *Specification for Accessories for Gypsum Wallboard and Gypsum Veneer Base.* West Conshohocken, PA: ASTM International, 2009.

ASTM C 1063-08, *Specification for Installation of Lathing and Furring to Receive Interior and Exterior Portland Cement-based Plaster.* West Conshohocken, PA: ASTM International, 2008.

ASTM C 1177/C 1177M-08, *Specification for Glass Mat Gypsum Substrate for Use as Sheathing.* West Conshohocken, PA: ASTM International, 2008.

ASTM C 1178/C 1178M-08, *Specification for Glass Mat Water-resistant Gypsum Backing Panel.* West Conshohocken, PA: ASTM International, 2008.

ASTM C 1280-09, *Specification for Application of Gypsum Sheathing.* West Conshohocken, PA: ASTM International, 2009.

ASTM C 1328-05, *Specification for Plastic (Stucco) Cement.* West Conshohocken, PA: ASTM International, 2005.

ASTM C 1396/C 1396M-06a, *Standard Specification for Gypsum Board.* West Conshohocken, PA: ASTM International, 2006.

ASTM E 84-09, *Test Method for Surface Burning Characteristics of Building Materials.* West Conshohocken, PA: ASTM International, 2009.

ASTM E 90-04, *Test Method for Laboratory Measurement of Airborne Sound Transmission Loss of Building Partitions and Elements.* West Conshohocken, PA: ASTM International, 2004.

ASTM F 547-06, *Terminology of Nails for Use with Wood and Wood-base Materials.* West Conshohocken, PA: ASTM International, 2006.

ASTM F 1667-05, *Specification for Driven Fasteners: Nails, Spikes, and Staples.* West Conshohocken, PA: ASTM International, 2005.

Federal Specification UU-B790a, *Federal Specifications for Building Paper, Vegetable Fiber—Kraft, Waterproofed, Water Repelent, and Fire Resistant.* U.S. Department of Defense, Washington, DC, 1992.

Fire Resistance Directory. Northbrook, IL: Underwriters Laboratories Inc., 2008.

GA 216-07, *Application and Finishing of Gypsum Panel Products.* Washington, DC: Gypsum Association, 2007.

GA 600-09, *Fire Resistance Design Manual.* Washington, DC: Gypsum Association, 2009.

Guidelines for Determining the Fire Resistance of Building Elements. Country Club Hills, IL: Building Officials and Code Administrators International, Inc. 1993.

Chapter 26: Plastic

General Comments

Plastics are a very large group of synthetic materials with structures that are based on the chemistry of carbon. Plastics are also called polymers because they are made of extremely long chains of carbon atoms. Since plastics can be readily molded, extruded, cast into various shapes and forms, or drawn into filaments and fibers, they are increasingly being used in the construction of buildings and structures. While progress in polymer technology makes it increasingly difficult to make general statements about these materials, the following properties are characteristic of many plastics:

- Low strength (in comparison to steel and other metals commonly used in construction);
- Low stiffness (modulus of elasticity) as compared to metals (except for reinforced plastics);
- Tendency to creep (increase in length under tensile stress);
- Low hardness (except for formaldehyde plastics);
- Low density (varies, plastics can be expanded or unexpanded);
- Brittleness at low temperatures, and loss of strength and hardness at moderately elevated temperatures;
- Flammability (although many plastics do not burn);
- Outstanding electrical characteristics, such as electrical resistance; and
- Degradation of some plastics, due to environmental agents, such as ultraviolet radiation (although most plastics are highly resistant to chemical attack).

When plastics and foam plastics were first proposed to be included in the code, consideration was given to several factors.

The following is the evolution of plastics being introduced into the model codes, ultimately leading to inclusion in the code.

It was recognized that most buildings classified as noncombustible construction contained combustible elements. These elements were not primary structural members, but included materials such as: asphalt and pitch (in roof coverings); wood (in flooring systems, as wall and ceiling interior finish and trim, and as furring and lath in walls); paper products (as slip sheets and vapor retarders); and cork (as insulation).

There was a sensitivity to the burning characteristics of all plastic materials based on the history of fires involving plastics in warehouses and building applications not covered by any of the model codes. It was agreed, therefore, that plastic materials should be required to meet minimum fire performance properties and each application would have appropriate requirements related to characteristics, such as rate of burning, surface flammability or smoke-development index.

This position is somewhat different from the regulation of traditional materials, such as wood, paper and asphalt, which frequently are not required to be separately fire tested when used within an assembly.

Plastic materials to be left exposed on the interior or exterior of buildings would have to meet the existing requirements for other finish materials, as well as be limited in area and subject to separation requirements between adjoining areas. The primary concern was that plastic materials would be used in applications in which glass had traditionally been used, such as skylights and light-diffusing ceilings. Since little experience was available at the time, a conservative approach was taken.

Foam plastics used as thermal insulation would be separated from the interior of buildings by a thermal barrier equivalent to $^1/_2$-inch-thick (12.7 mm) gypsum wallboard. This position was supported by the industry because most of the applications at that time included plaster or wallboard finishes of $^1/_2$ inch (12.7 mm) or greater thickness, and the fire record for such applications was good.

Specific examples of commonly used acceptable applications would be included in the code, such as the use of foam plastics within wood frame and masonry walls.

Finally, the code would include a statement that "diversified tests" would be necessary for approval of any application or material not meeting the prescriptive requirements of the code. The reason for including diversified tests was because it was clear that it would not be possible or even desirable to include all applications specifically in the code. Diversified tests were intended to mean tests related to the end use and usually were full-scale assemblies exposed to fire conditions that were representative of actual fires that might be expected to occur.

Purpose

This chapter establishes minimum requirements for all light-transmitting and foam plastics in all applications regulated by the code. Some plastics exhibit rapid flame spread and heavy smoke density characteristics when exposed to fire. Additionally, exposure to the heat generated by a fire can cause some plastics to deform and undergo a change in their structural stability. The requirements and limitations of this chapter are necessary to control the use of these combustible products such that they do not compromise the safety of building occupants.

SECTION 2601 GENERAL

2601.1 Scope. These provisions shall govern the materials, design, application, construction and installation of foam plastic, foam plastic insulation, plastic veneer, interior plastic finish and *trim* and light-transmitting plastics. See Chapter 14 for requirements for *exterior wall* finish and *trim.*

❖ Chapter 26 governs foam plastics, plastic veneers and light-transmitting plastic materials regardless of their use. Requirements for foam plastic insulation are found in Section 2603. Requirements for interior finish and trim are found in Section 2604. Requirements for plastic veneer are found in Section 2605. The main requirements for light-transmitting plastics are found in Section 2606. Sections 2607 through 2611 cover a variety of applications of light-transmitting plastics. Requirements for fiber and fiberglass reinforced polymer are found in Section 2612. Lastly, requirements for reflective plastic core insulation are found in Section 2613. Essentially, this chapter goes into detail as to how such combustible materials can be used in building construction. This includes wall panels, glazing, roof panels, skylights and interior signs.

SECTION 2602 DEFINITIONS

2602.1 Definitions. The following terms are defined in Chapter 2:

FIBER-REINFORCED POLYMER.

FOAM PLASTIC INSULATION.

LIGHT-DIFFUSING SYSTEM.

LIGHT-TRANSMITTING PLASTIC ROOF PANELS.

LIGHT-TRANSMITTING PLASTIC WALL PANELS.

PLASTIC, APPROVED.

PLASTIC GLAZING.

THERMOPLASTIC MATERIAL.

THERMOSETTING MATERIAL.

❖ Definitions facilitate the understanding of code provisions and minimize potential confusion. To that end, this section lists definitions of terms associated with roof assemblies, roof coverings and rooftop structures. Note that these definitions are found in Chapter 2. The use and application of defined terms, as well as undefined terms, are set forth in Section 201.

SECTION 2603 FOAM PLASTIC INSULATION

2603.1 General. The provisions of this section shall govern the requirements and uses of foam plastic insulation in buildings and structures.

❖ Section 2603 contains all of the life-safety provisions for foam plastic insulation products, such as rigid insulation board, spray-foam insulation and pour-in-place insulation. "Foam plastic" is a general term given to insulating products that have been manufactured by intentionally expanding plastic by the use of a foaming agent. Extruded polystyrene (XPS), expanded polystyrene (EPS), polyisocyanurate (PIR or ISO), open-cell polyisocyanurate, phenolic, polyurethane and polypropylene foam are among the many types of foam plastic insulation products subject to the requirements found in this section. In addition to technical requirements listed in the following paragraph, all packages and containers of foam plastic insulation must bear a label of an approved agency listing the manufacturer's name, the product listing, product identification and information sufficient to determine that the end use will comply with the code requirements.

Compliance with this section shall be achieved through either a prescriptive method or the performance method. The prescriptive method requires compliance with Sections 2603.4 through 2603.7. The performance method requires compliance with Section 2603.10. Note that both methods require compliance with Section 2603.3, which covers surface-burning characteristics and Section 2603.9, which covers protection against termites.

Figure 2603.1 documents code compliance when foam plastic is used.

2603.2 Labeling and identification. Packages and containers of foam plastic insulation and foam plastic insulation components delivered to the job site shall bear the *label* of an *approved agency* showing the manufacturer's name, product listing, product identification and information sufficient to determine that the end use will comply with the code requirements.

❖ All foam plastics or packages of foam plastics delivered to the construction site must be labeled. Also, labels are required on containers [usually two components in 55-gallon (208 L) drums] of ingredients delivered for the production of foam plastic at the construction site. The label must include identification of the approved agency and detailed product identification or information describing the performance characteristics of the product. It is intended that labeling printed on board stock or on packaging be acceptable. It is also intended that the label information include the name of the manufacturer or distributor, type of foam plastic, performance characteristics required and name of the approved testing agency.

2603.3 Surface-burning characteristics. Unless otherwise indicated in this section, foam plastic insulation and foam plastic cores of manufactured assemblies shall have a flame spread index of not more than 75 and a smoke-developed index of not more than 450 where tested in the maximum thickness intended for use in accordance with ASTM E 84 or UL 723. Loose fill-type foam plastic insulation shall be tested as board stock for the flame spread and smoke-developed indexes.

Exceptions:

1. Smoke-developed index for interior *trim* as provided for in Section 2604.2.
2. In cold storage buildings, ice plants, food plants, food processing rooms and similar areas, foam plas-

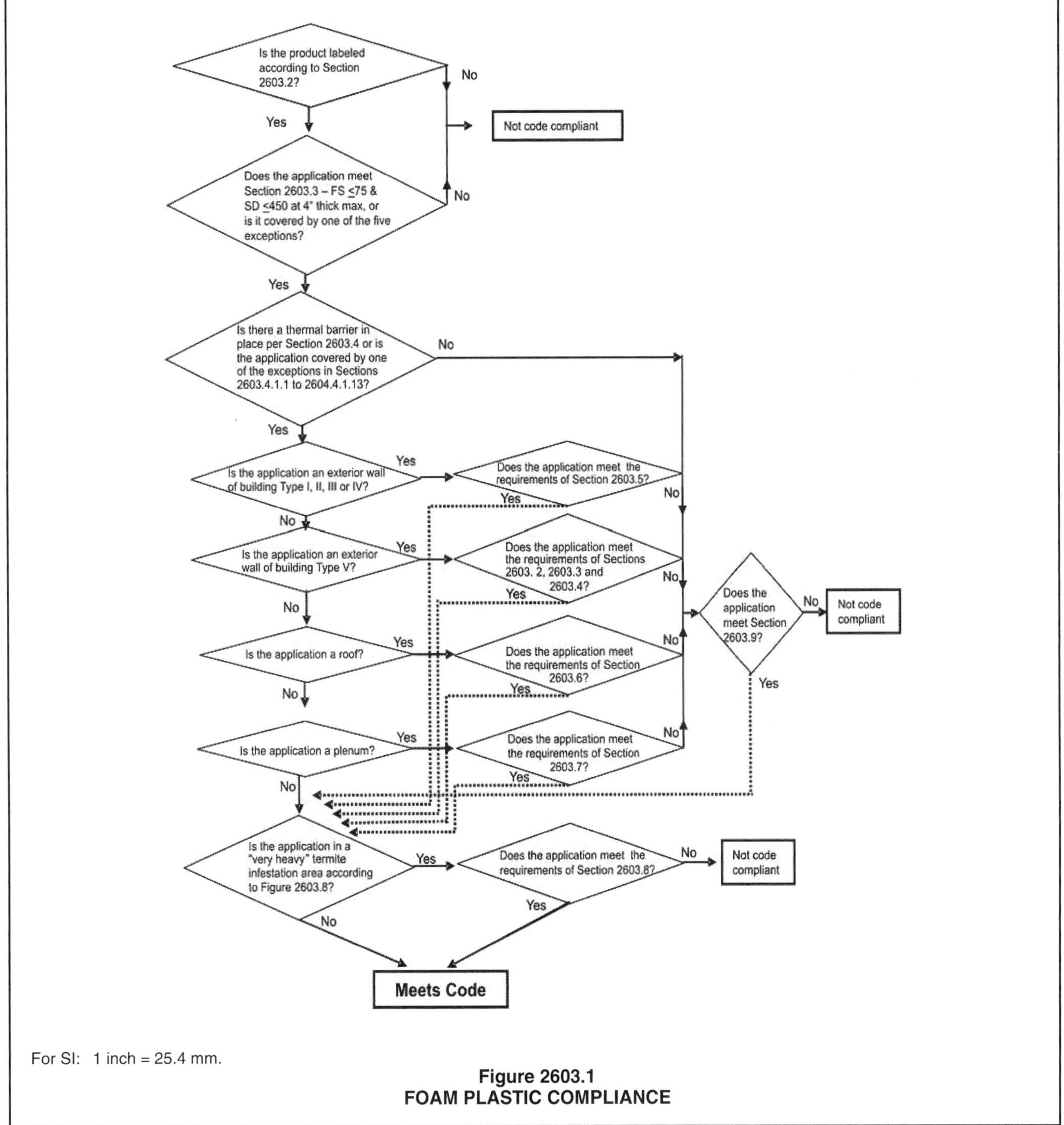

For SI: 1 inch = 25.4 mm.

Figure 2603.1
FOAM PLASTIC COMPLIANCE

tic insulation where tested in a thickness of 4 inches (102 mm) shall be permitted in a thickness up to 10 inches (254 mm) where the building is equipped throughout with an automatic fire sprinkler system in accordance with Section 903.3.1.1. The approved *automatic sprinkler system* shall be provided in both the room and that part of the building in which the room is located.

3. Foam plastic insulation that is a part of a Class A, B or C roof-covering assembly provided the assembly with the foam plastic insulation satisfactorily passes FM 4450 or UL 1256. The smoke-developed index shall not be limited for roof applications.

4. Foam plastic insulation greater than 4 inches (102 mm) in thickness shall have a maximum flame spread index of 75 and a smoke-developed index of 450 where tested at a minimum thickness of 4 inches (102 mm), provided the end use is approved in accordance with Section 2603.10 using the thickness and density intended for use.

5. Flame spread and smoke-developed indexes for foam plastic interior signs in *covered and open mall buildings* provided the signs comply with Section 402.6.4.

❖ Foam plastic insulation or foam plastic cores used as a component in a manufactured assembly are combustible and must be assessed for flame spread index and smoke-developed index. Testing in accordance with ASTM E 84 or UL 263 is required. The foam plastic must be tested in the maximum thickness to be used [up to 4-inch (102 mm) thickness; see Exception 4 for foam thicker than 4 inches (102 mm)] and limited to a flame spread index of 75 or less and a smoke-developed index of 450 or less. The 4-inch (102 mm) thickness for testing is specifically related to the limitation on testing thickness in ASTM E 84 or UL 263. All foam plastic materials are required to be tested in accordance with ASTM E 84 or UL 263, unless specifically exempted or modified by one of the listed exceptions.

Five exceptions relate to whether the limitations—flame spread index of 75 and smoke-developed index of 450—apply or if the thickness of material is allowed to exceed 4 inches (102 mm) as limited by ASTM E 84 or UL 263:

1. Use as interior trim is not required to meet the smoke-developed index limitation of 450, but must still meet the flame spread index requirement of less than or equal to 75.

2. In fully sprinklered buildings, cold storage construction using foam plastic is allowed to exceed the thickness of 4 inches (102 mm) up to 10 inches (254 mm) even when the material has only been tested at 4 inches (102 mm) with ASTM E 84 or UL 263.

3. When used as part of a Class A, B or C roof-covering assembly in which the assembly has successfully passed FM 4450 or UL 1256 testing, foam plastic insulation is not required to meet the flame spread index rating of 75. Also, smoke-developed index ratings are not required or limited for any foam plastic insulation used in roof assemblies. The intent is to recognize that the overall hazards of roofing have already been addressed and additional testing requirements for flame spread would be unnecessary.

4. Foam plastic insulation thicker than 4 inches (102 mm) is allowed if tested in accordance with Section 2603.10. This provision would require the full as-installed thickness of the foam plastic to be tested. The tests referenced in Section 2603.10 are generally full-scale room-corner tests. Such material would also still be required to be tested to ASTM E 84 or UL 263 at a 4-inch (102 mm) thickness and result in a flame spread index no greater than 75 and a smoke-developed index of 450.

5. Use as interior signs in covered mall buildings is required to comply with Section 402.6.4. Section 402.6.4 requires compliance with UL 1975, which measures the rate of heat release of burning materials used in the manufacture of such signs. Section 402.6.4 sets the pass/fail criteria at a maximum heat release rate of 150 kW. Although this test method assesses heat release rate rather than surface-burning characteristics, the testing and criteria specified provide material characteristics that address the goal of limiting the spread of fire.

The maximum flame spread index value of 75 was chosen on the basis that it is lower than untreated wood (which usually is 100 to 165). The maximum smoke-developed index rating of 450 was selected because, at the time, the code permitted interior finish materials that gave off "smoke no more dense than that given off by untreated wood." In selecting the maximum flame spread and smoke-developed index values, it was believed that a conservative approach was being taken by requiring an insulation material to meet the same requirements as interior finish, even though the insulation was intended to be covered with an interior finish material. The requirements for surface-burning characteristics of foam plastic apply to foam plastics used as cores of manufactured assemblies. The intent is that, while the finished assemblies are not required to be tested for surface-burning characteristics, the foam plastic core is not exempt from the general requirement; therefore, foam plastic is regulated in factory-manufactured assemblies the same as it is in field-fabricated applications.

2603.4 Thermal barrier. Except as provided for in Sections 2603.4.1 and 2603.10, foam plastic shall be separated from the interior of a building by an approved thermal barrier of ½-inch (12.7 mm) gypsum wallboard or a material that is tested in accordance with and meets the acceptance criteria of both the Temperature Transmission Fire Test and the Integrity Fire Test of NFPA 275. Combustible concealed spaces shall comply with Section 718.

❖ The use of an approved thermal barrier to separate foam plastics from the interior of a building is a basic requirement for the use of foam plastics as shown in this section. Sections 2603.4.1 and 2603.9 describe circumstances where the thermal barrier is modified or eliminated for specific uses. The job of a thermal barrier is to limit the temperature rise that the foam plastic will be exposed to. An approved thermal barrier is defined as $^1/_2$-inch (12.7 mm) gypsum wallboard or the equivalent. This section sets forth the test method by which alternative thermal barriers are to be qualified.

Before 1975, experience had shown that foam plastics covered with plaster or $^1/_2$-inch (12.7 mm) gypsum wallboard had performed satisfactorily in building fires. For this reason, $^1/_2$-inch (12.7 mm) gypsum wallboard was included in the code as a minimum requirement. It was then recognized that specifying a single material would not be desirable in a performance code; therefore, testing in accordance with NFPA 275 is allowed to qualify other materials. This test method was developed to specifically address the testing of materials to qualify as a thermal barrier. The test method provides specific sample construction, fire exposures and acceptance criteria to qualify a material to be a 15-minute thermal barrier. The test method addresses both the capability of the material to retard heat transfer via a fire-resistance test and to remain in place via a full-scale fire test.

Gypsum wallboard is still the most commonly used thermal barrier, but other materials are permitted when qualified using NFPA 275.

2603.4.1 Thermal barrier not required. The thermal barrier specified in Section 2603.4 is not required under the conditions set forth in Sections 2603.4.1.1 through 2603.4.1.14.

❖ A thermal barrier is not required for applications described in Sections 2603.4.1.1 through 2603.4.1.13. The prescriptive installation methods described in these sections must be strictly followed in order to serve as an alternative to the thermal barrier. Furthermore, installations of foam plastics in accordance with these sections do not have to be tested via the alternative approval testing requirements of Section 2603.9.

2603.4.1.1 Masonry or concrete construction. A thermal barrier is not required for foam plastic installed in a masonry or concrete wall, floor or roof system where the foam plastic insulation is covered on each face by a minimum of 1-inch (25 mm) thickness of masonry or concrete.

❖ No thermal barrier is required when 1 inch (25 mm) or more of masonry or concrete is placed between the foam plastic and interior of the building. The intent is to accept 1 inch (25 mm) of masonry or concrete as equal to (or better than) $^1/_2$-inch (12.7 mm) gypsum wallboard. This condition can arise when foam plastics are installed either within a wall or on the exterior side of a masonry wall. Some common examples are when foam plastics are installed:

- In the cavity of a hollow masonry wall;
- As the core of a concrete-faced panel;
- On the exterior face of a masonry wall and covered with an exterior finish;
- Within the cores of hollow masonry units; or
- Encapsulated within a minimum of 1-inch (25 mm) concrete or masonry wall, floor or roof system, such as in insulated tilt-up or pour-in-place concrete panels.

Note that the exterior surface would be required to comply with Section 2603.5.

2603.4.1.2 Cooler and freezer walls. Foam plastic installed in a maximum thickness of 10 inches (254 mm) in cooler and freezer walls shall:

1. Have a flame spread index of 25 or less and a smoke-developed index of not more than 450, where tested in a minimum 4-inch (102 mm) thickness.
2. Have flash ignition and self-ignition temperatures of not less than 600°F and 800°F (316°C and 427°C), respectively.
3. Have a covering of not less than 0.032-inch (0.8 mm) aluminum or corrosion-resistant steel having a base metal thickness not less than 0.0160 inch (0.4 mm) at any point.
4. Be protected by an *automatic sprinkler system* in accordance with Section 903.3.1.1. Where the cooler or freezer is within a building, both the cooler or freezer and that part of the building in which it is located shall be sprinklered.

❖ A thermal barrier is not required in cooler and freezer walls insulated with 10 inches (254 mm) or less of foam plastic when all four criteria in Section 2603.4.1.2 are met

2603.4.1.3 Walk-in coolers. In nonsprinklered buildings, foam plastic having a thickness that does not exceed 4 inches (102 mm) and a maximum flame spread index of 75 is permitted in walk-in coolers or freezer units where the aggregate floor area does not exceed 400 square feet (37 m^2) and the foam plastic is covered by a metal facing not less than 0.032-inch-thick (0.81 mm) aluminum or corrosion-resistant steel having a minimum base metal thickness of 0.016 inch (0.41 mm). A thickness of up to 10 inches (254 mm) is permitted where protected by a thermal barrier.

❖ A thermal barrier is not required for foam plastics up to 4 inches (102 mm) thick when the minimum thickness metal facings are used in coolers or freezers no greater than 400 square feet (37 m^2) in floor area. The required metal facing is intended to act as a barrier against ignition of the foam plastic. It is not intended to serve as a thermal barrier. This section recognizes the relatively small size of walk-in coolers and the accompanying lower hazard. Foam plastic up to 10 inches thick (254 mm) is permitted when protected by a thermal barrier.

2603.4.1.4 Exterior walls-one-story buildings. For one-*story* buildings, foam plastic having a flame spread index of 25 or less, and a smoke-developed index of not more than 450, shall be permitted without thermal barriers in or on *exterior walls* in a thickness not more than 4 inches (102 mm) where the foam plastic is covered by a thickness of not less than 0.032-inch-thick (0.81 mm) aluminum or corrosion-resistant steel having a base metal thickness of 0.0160 inch (0.41 mm) and the building is equipped throughout with an *automatic sprinkler system* in accordance with Section 903.3.1.1.

❖ Foam plastics may be used without a thermal barrier in one-story buildings provided that:

- The material is not more than 4 inches (102 mm) thick;
- It has a flame spread index not greater than 25;
- It has a smoke-developed index of not more than 450;
- It is covered with specified metal facings; and
- The building is equipped throughout with an automatic sprinkler system.

This provision is intended to permit the use of metal-faced panels, primarily in storage buildings other than cold storage construction. Cold storage construction was also covered in this section (see also Sections 2603.4.1.2 and 2603.4.1.3, and Section 2603.3, Exception 2). The limitations are based on the results of large-scale tests performed by independent laboratories and sponsored by the plastics industry.

2603.4.1.5 Roofing. Foam plastic insulation under a roof assembly or roof covering that is installed in accordance with the code and the manufacturer's instructions shall be separated from the interior of the building by wood structural panel sheathing not less than 0.47 inch (11.9 mm) in thickness bonded with exterior glue, with edges supported by blocking, tongue-and-groove joints or other approved type of edge support, or an equivalent material. A thermal barrier is not required for foam plastic insulation that is a part of a Class A, B or C roof-covering assembly, provided the assembly with the foam plastic insulation satisfactorily passes FM 4450 or UL 1256.

❖ No thermal barrier is required when foam plastic is used in a roof assembly over wood structural panel sheathing given that the wood product meets all of the following:

- At least 0.47 inch (11.9 mm) thick;
- Manufactured utilizing exterior-grade glue; and
- Properly installed to provide adequate edge support.

Equivalent materials to the wood structural panel sheathing described above are specifically allowed by this section. In any situation, the material must be compatible with the type of construction.

Also, no thermal barrier is required when foam plastic is used with a Class A, B or C roof assembly (as prescribed in Section 1505) when the assembly complies with FM 4450 or UL 1256. The intent is to recognize that roof assemblies tested to the criteria of the referenced standards have adequately demonstrated a resistance to fire from the underside of the roof deck; therefore, no additional testing is necessary, nor is an additional thermal barrier needed. It is important to note that this section applies to a specific roof assembly (i.e., specific roof deck, specific foam insulation, fasteners and specific roof membrane). Passing the prescribed tests with one specific roof assembly does not cover all combinations of roof systems where foam plastic insulation is used. Users must confirm that the specific roof assembly has achieved a Class A, B or C rating. This coordinates with Section 2603.3, Exception 4, for surface-burning characteristics

2603.4.1.6 Attics and crawl spaces. Within an attic or crawl space where entry is made only for service of utilities, foam plastic insulation shall be protected against ignition by $1^1/_2$-inch-thick (38 mm) mineral fiber insulation; $^1/_4$-inch-thick (6.4 mm) wood structural panel, particleboard or hardboard; $^3/_8$-inch (9.5 mm) gypsum wallboard, corrosion-resistant steel having a base metal thickness of 0.016 inch (0.4 mm) or other approved material installed in such a manner that the foam plastic insulation is not exposed. The protective covering shall be consistent with the requirements for the type of construction.

❖ In an attic or crawl space where entry is made only for service of utilities, and when foam plastics are used, an ignition barrier may be used in place of a thermal barrier to cover the foam plastic. Multiple materials are listed, which can be used as the ignition barrier:

- $1^1/_2$-inch-thick (38 mm) mineral fiber insulation;
- $^1/_4$-inch-thick (6.4 mm) wood structural panels;
- $^3/_8$-inch (9.5 mm) particleboard;
- $^1/_4$-inch (6.4 mm) hardboard;
- $^3/_8$-inch (9.5 mm) gypsum board; or
- Corrosion-resistant steel having a base metal thickness of 0.016 inch (0.406 mm).

Note that the ignition barrier must be consistent with the type of construction, which means that combustible coverings are permitted only where combustible materials are otherwise allowed.

The foam plastic material covered with the ignition barrier can be on the floor, wall or the ceiling of the attic or crawl space. The phrase "where entry is made only for service of utilities" applies to attics or crawl spaces that only contain mechanical equipment, electrical wiring, fans, plumbing, gas or electric hot water heaters, gas or electric furnaces, etc. Such provisions would not be allowed for such spaces used for storage or general occupancy. The reduced provision (from a thermal barrier to an ignition barrier) provides a protective cover that was the sole purpose to prevent the direct impingement of flame on the foam plastic insulation (see Figure 2603.4.1.6). Note that Section 2603.4 would still require a thermal barrier between the interior space above (crawl space) or below (attic) in a structure and the foam plastic.

If the foam plastic insulation has passed large-scale testing in the thickness and density intended for use, in accordance with Section 2603.10, no thermal barrier or ignition barrier is required over the foam plastic insulation in an attic or crawl space and this section of the code does not apply. However, it is important to note that the actual configuration must be tested. For example, foam insulation applied to the ceiling of the attic or crawl space must be tested. During the test, the foam must be applied to the ceiling in a room corner test or in an assembly that reflects end use. The same restrictions would apply to those insulations applied to the walls, floors or a combination of surfaces.

2603.4.1.7 Doors not required to have a fire protection rating. Where pivoted or side-hinged doors are permitted without a fire protection rating, foam plastic insulation, having a flame spread index of 75 or less and a smoke-developed index of not more than 450, shall be permitted as a core material where the door facing is of metal having a minimum thickness of 0.032-inch (0.8 mm) aluminum or steel having a base metal thickness of not less than 0.016 inch (0.4 mm) at any point.

❖ A thermal barrier is not required when foam plastic is used as a core material for doors, provided that a specified metal facing is used and the door is not required to have a fire-resistance rating. The intent is that doors required to have a fire protection rating must be tested with the foam plastic cores in place in order to qualify. This section recognizes the limited area of doors compared to the walls in which they are located and regulates foam plastics similar to other materials used in doors.

2603.4.1.8 Exterior doors in buildings of Group R-2 or R-3. In occupancies classified as Group R-2 or R-3, foam-filled exterior entrance doors to individual *dwelling units* that do not require a fire-resistance rating shall be faced with wood or other approved materials.

❖ Wood or other approved facings on foam-filled exterior entrance doors of Group R-2 and R-3 buildings are permitted when the doors are not required to have a fire-resistance rating. Such doors do not need to comply with Section 2603.4.1.7.

2603.4.1.9 Garage doors. Where garage doors are permitted without a fire-resistance rating and foam plastic is used as a core material, the door facing shall be metal having a minimum thickness of 0.032-inch (0.8 mm) aluminum or 0.010-inch (0.25 mm) steel or the facing shall be minimum 0.125-inch-thick (3.2 mm) wood. Garage doors having facings other than those described above shall be tested in accordance with, and meet the acceptance criteria of, DASMA 107.

Exception: Garage doors using foam plastic insulation complying with Section 2603.3 in detached and attached garages associated with one- and two-family dwellings need not be provided with a thermal barrier.

❖ The prescriptive requirements of this section regulate garage doors with foam plastic cores used in nonfire-resistance-rated applications. The requirements in Section 2603.4.1.9, which governs overhead sectional, coiling or vertical lift-type garage doors, are slightly different from the requirements of Sections

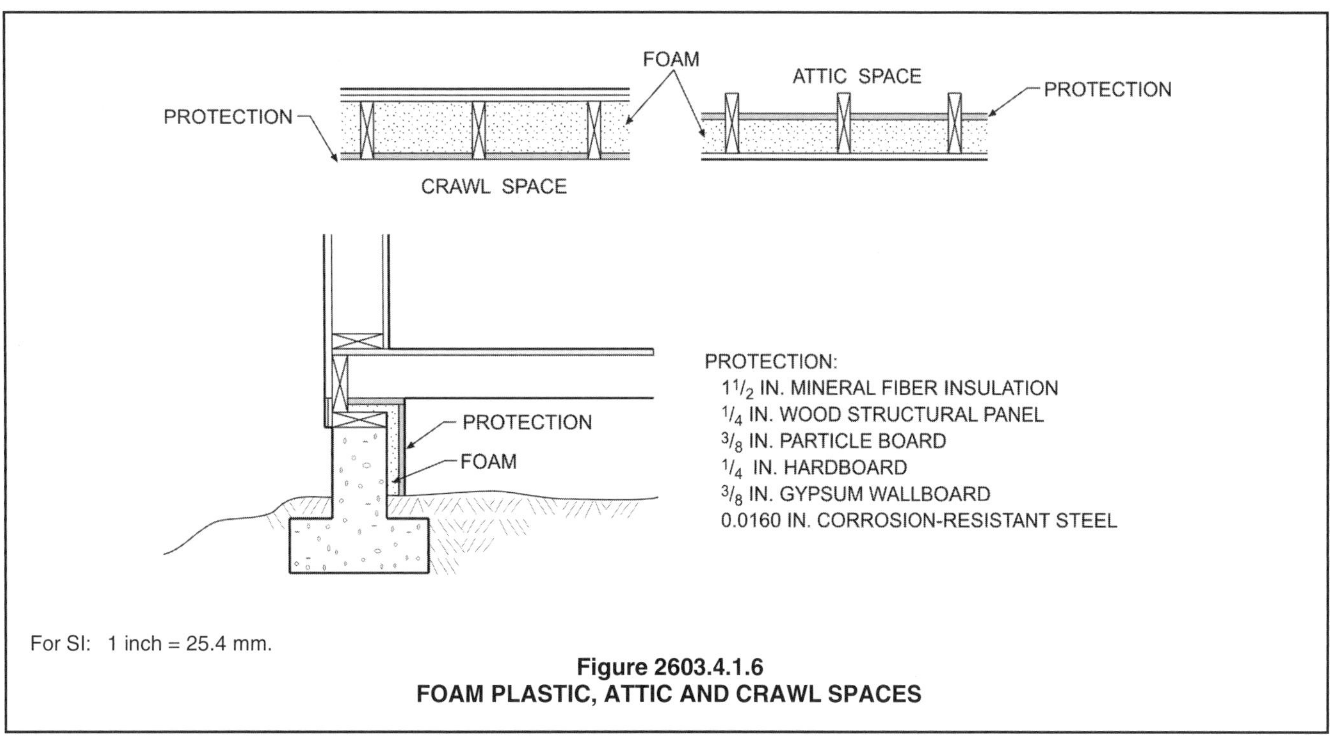

Figure 2603.4.1.6
FOAM PLASTIC, ATTIC AND CRAWL SPACES

2603.4.1.7 and 2603.4.1.8, which govern means of egress doors, such as side-swinging doors. These garage doors are not limited to such doors used in conjunction with a private or public garage occupancy but refer to doors for vehicles in any occupancies.

The minimum door-facing material and thickness requirements for garage doors with foam plastic cores stated in this section were determined based on the results of a fire testing program sponsored by the National Association of Garage Door Manufacturers (NAGDM). This program exposed a variety of commercially manufactured garage doors to a room corner fire test. The results of the testing indicated that garage doors having facings in the minimum thicknesses stated did not spread fire to the edge of the specimen and did not cause flashover in the test room. Other garage door constructions using foam plastics that do not meet minimum code requirements can be brought into compliance by meeting the requirements of testing the complete garage door assembly to ANSI/DASMA 107, *Room Fire Test Standard for Garage Doors Using Foam Plastic Insulation.* This standard includes a description of the test method, performance data to be obtained and the acceptance criteria to use in evaluating the performance data.

This section exempts exterior garage doors used in garages attached to one- and two-family dwellings and detached garages (Group R-3) from the thermal barrier requirements of Section 2603.4. This exemption is similar to that of Section 2603.4.1.8 for exterior egress doors in Group R-2 and R-3 occupancies.

As noted within the section, such doors are not intended for installation in fire-rated exterior walls. There are other products, such as rolling steel fire doors, manufactured for this purpose.

2603.4.1.10 Siding backer board. Foam plastic insulation of not more than 2,000 British thermal units per square feet (Btu/sq. ft.) (22.7 mJ/m^2) as determined by NFPA 259 shall be permitted as a siding backer board with a maximum thickness of $^1/_2$ inch (12.7 mm), provided it is separated from the interior of the building by not less than 2 inches (51 mm) of mineral fiber insulation or equivalent or where applied as insulation with residing over existing wall construction.

❖ Foam plastic is frequently used in residing applications to provide a leveling surface for new siding, while also bringing additional insulation value to the wall assembly. Also available in the marketplace are products that combine exterior siding material and foam plastic. If these siding/residing products are used on the exterior of a wall and the requirements of Sections 2603.3 and 2603.4 are met, this section of the code does not apply. If a thermal barrier is not used on the interior of the building, then there are limitations placed on the products and their use. In addition to the flame spread index limitations of Section 2603.3, other property limitations of the foam plastic or foam plastic portion of the product include a maximum thickness of $^1/_2$ inch (12.7mm) and potential heat of less than 2,000 Btu [1 square foot (22.7 mJ/m^2)] when tested using NFPA 259. Limita-tions in foam and siding/foam combination product use includes separation from the interior of the building by 2 inches (51 mm) of mineral fiber insulation, installation over an existing wall finish as part of residing or when the foam plastic insulation is tested in accordance with Section 2603.10. The removal of the thermal barrier requirement in this section is reasonable considering the separation provided by the existing construction and limitation of the potential heat of the foam plastic imposed by the code. Also, the code requires fire-resistance ratings from the exterior side of walls only when the fire separation distance is 10 feet (1524 mm) or less (see Section 705.5).

2603.4.1.11 Interior trim. Foam plastic used as interior *trim* in accordance with Section 2604 shall be permitted without a thermal barrier.

❖ This section specifies that a thermal barrier is not needed for interior trim that meets the limitations of Section 2604 and related subsections, which place limitations on foam plastic density, thickness, wall and ceiling area coverage and flame spread.

2603.4.1.12 Interior signs. Foam plastic used for interior signs in *covered mall buildings* in accordance with Section 402.6.4 shall be permitted without a thermal barrier. Foam plastic signs that are not affixed to interior building surfaces shall comply with Chapter 8 of the *International Fire Code.*

❖ A thermal barrier is not required for interior signs that meet the limits stated in Section 402.6.4, since the hazard from the interior signs is appropriately addressed by test standard UL1975 (see commentary, Section 402.6.4). If the foam plastic sign is not affixed to an interior surface of the building, then the requirements of Chapter 8 of the *International Fire Code*® (IFC®) must be met. More specifically, Section 808.3 of the IFC requires compliance with UL 1975 when such signs exceed 10 percent of the wall or ceiling area, whichever is less.

2603.4.1.13 Type V construction. Foam plastic spray applied to a sill plate and header of Type V construction is subject to all of the following:

1. The maximum thickness of the foam plastic shall be $3^1/_4$ inches (82.6 mm).
2. The density of the foam plastic shall be in the range of 1.5 to 2.0 pcf (24 to 32 kg/m^3).
3. The foam plastic shall have a flame spread index of 25 or less and an accompanying smoke-developed index of 450 or less when tested in accordance with ASTM E 84 or UL 723.

❖ A thermal barrier is not required when foam plastic is spray applied to the sill plate and joist header in Type V construction when all of the conditions listed in Section 2603.4.1.13 are met. Because foam plastic insulation in this application is left exposed, the four conditions listed limit the hazards of the spray-applied foam plastic used in this application:

- Thickness greater than or equal to $3^1/_4$ inches (82.6 mm);
- Density is between 1.5 to 2.0 pounds per cubic foot (pcf) (24 to 32 kg/m^3);
- Flame spread index is less than or equal to 25; and
- Smoke-developed index is less than or equal to 450.

This particular allowance was based upon testing in a room corner fire test that compared the performance of an all-wood floor system to that of a wood floor system with foam plastic sprayed on sill plates and headers (see Figure 2603.4.1.13).

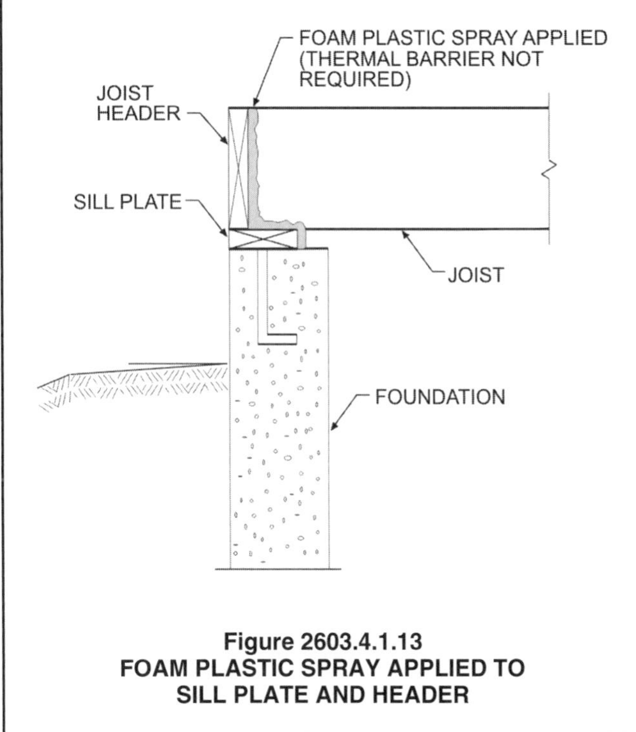

Figure 2603.4.1.13
FOAM PLASTIC SPRAY APPLIED TO SILL PLATE AND HEADER

2603.4.1.14 Floors. The thermal barrier specified in Section 2603.4 is not required to be installed on the walking surface of a structural floor system that contains foam plastic insulation when the foam plastic is covered by a minimum nominal ½-inch-thick (12.7 mm) wood structural panel or approved equivalent. The thermal barrier specified in Section 2603.4 is required on the underside of the structural floor system that contains foam plastic insulation when the underside of the structural floor system is exposed to the interior of the building.

Exception: Foam plastic used as part of an interior floor finish.

❖ Many types of products are being used in construction that incorporate foam plastic insulation for energy reasons. One example is structural insulated panels (SIPS) where the foam plastic is laminated between two structural wood facings. This type of panel can be used as a wall, floor or roof.

Foam plastic is required to be protected by a thermal barrier that typically is $^1/_2$-inch (12.7 mm) gypsum wallboard. In the case of flooring, gypsum wallboard or other common thermal barrier materials cannot be used on the walking surfaces due to their friability to load, etc.

This section allows for the $^1/_2$-inch-thick plywood or equivalent to provide thermal protection to the foam plastic insulation. While $^1/_2$-inch plywood is not by itself a thermal barrier, in the case of a floor, the plywood provides sufficient protection since in the event of an interior fire, the floor is typically the last building element to be significantly exposed by the fire.

If the floor is used in multistory construction, then the underside of the floor system (ceiling of the room below) must be covered by the required thermal barrier.

The exception addresses items such as carpet padding, etc., that have not and do not need to be covered by a thermal barrier.

2603.5 Exterior walls of buildings of any height. *Exterior walls* of buildings of Type I, II, III or IV construction of any height shall comply with Sections 2603.5.1 through 2603.5.7. *Exterior walls* of cold storage buildings required to be constructed of noncombustible materials, where the building is more than one *story* in height, shall also comply with the provisions of Sections 2603.5.1 through 2603.5.7. *Exterior walls* of buildings of Type V construction shall comply with Sections 2603.2, 2603.3 and 2603.4.

❖ All foam plastics used on exterior walls of all types of buildings, except wood frame, are to be installed in accordance with Sections 2603.5.1 through 2603.5.7. Installations on one-story, noncombustible walls of cold storage buildings are required to also comply with the provisions of Sections 2603.5.1 through 2603.5.7. Foam plastics used on exterior walls of Type V buildings are required to comply with Sections 2603.2, 2603.3 and 2603.4. The intent is to regulate the use of an insulating envelope over the exterior of a structure when the envelope provides no structural support other than the transfer of wind loads. It is recognized that some envelopes will be constructed in place by installing a rigid foam plastic and covering it with an exterior finish while others will be installed as prefabricated panels complete with exterior finish.

2603.5.1 Fire-resistance-rated walls. Where the wall is required to have a fire-resistance rating, data based on tests conducted in accordance with ASTM E 119 or UL 263 shall be provided to substantiate that the fire-resistance rating is maintained.

❖ Foam plastics are permitted in walls that are required to have fire-resistance ratings. Such assemblies must be fire tested with the foam plastic (at the intended maximum thickness and density) in place.

2603.5.2 Thermal barrier. Any foam plastic insulation shall be separated from the building interior by a thermal barrier meeting the provisions of Section 2603.4, unless special approval is obtained on the basis of Section 2603.10.

> **Exception:** One-story buildings complying with Section 2603.4.1.4.

❖ A thermal barrier is required unless an alternative approval is obtained in accordance with Section 2603.10. The intent is to make it clear that the thermal barrier requirement of Section 2603.4 is applicable. The focus of this section is upon external exposure. The exception provides for the reduced requirements of Section 2603.4.1.4 for one-story buildings.

2603.5.3 Potential heat. The potential heat of foam plastic insulation in any portion of the wall or panel shall not exceed the potential heat expressed in Btu per square feet (mJ/m^2) of the foam plastic insulation contained in the wall assembly tested in accordance with Section 2603.5.5. The potential heat of the foam plastic insulation shall be determined by tests conducted in accordance with NFPA 259 and the results shall be expressed in Btu per square feet (mJ/m^2).

> **Exception:** One-story buildings complying with Section 2603.4.1.4.

❖ This section limits the combustible content of exterior walls based on the potential heat of the foam plastic insulation. Potential heat is essentially the amount of energy potential a particular material contains in terms of its ability to burn. Generally, the higher the potential heat, the higher the fire hazard. The potential heat must not exceed that as found in the tested wall assembly as required in Section 2603.5.5. It is important to note that NFPA 268 addresses flammability not fire resistance as addressed by Chapter 7.

2603.5.4 Flame spread and smoke-developed indexes. Foam plastic insulation, exterior coatings and facings shall be tested separately in the thickness intended for use, but not to exceed 4 inches (102 mm), and shall each have a flame spread index of 25 or less and a smoke-developed index of 450 or less as determined in accordance with ASTM E 84 or UL 723.

> **Exception:** Prefabricated or factory-manufactured panels having minimum 0.020-inch (0.51 mm) aluminum facings and a total thickness of $^1/_4$ inch (6.4 mm) or less are permitted to be tested as an assembly where the foam plastic core is not exposed in the course of construction.

❖ Each component, including the foam plastic core, must be tested and have a maximum flame spread index of 25 and a maximum smoke-developed index of 450 when tested in accordance with ASTM E 84 or UL 263. The material should be tested with the same thickness for use, but not to exceed a 4-inch (102 mm) thickness for testing. Full-scale testing can address the actual thickness and is required by Section 2603.5.5. Materials should not be restricted by the testing limitations of the referenced standards when they perform well at a higher thickness in the full-scale test. The exception applies to prefabricated panels that can be installed without exposing the foam plastic.

2603.5.5 Vertical and lateral fire propagation. The exterior wall assembly shall be tested in accordance with and comply with the acceptance criteria of NFPA 285.

> **Exception:** One-story buildings complying with Section 2603.4.1.4.

❖ This section is the only provision dealing with the propagation of fire due to exposure from an exterior source. Other combustibles are not allowed as components of noncombustible building exterior walls. The focus is flammability versus fire resistance.

2603.5.6 Label required. The edge or face of each piece, package or container of foam plastic insulation shall bear the *label* of an *approved agency*. The *label* shall contain the manufacturer's or distributor's identification, model number, serial number or definitive information describing the product or materials' performance characteristics and *approved agency*'s identification.

❖ Each piece, package or container of foam plastic is required to be labeled. This requirement is somewhat different than the general marking requirement of Section 2603.2, which only requires a manufacturer's identifying mark on either the product or the packaging. The reason for this is that properties other than fire performance of the foam plastic, such as bond strength to the exterior finish, are of significance in this type of application. Different types of foam plastic insulations are being used in building construction. These foam plastics include manufactured foam boards and spray polyurethane foam. Spray foam products are delivered to the job site in drums, which contain the components of the foam plastic insulation. Typically, materials from two drums are mixed as they are sprayed and the resultant foam plastic insulation is applied to the wall surfaces. The component (drums or packaging) bears a label with the information required in the section.

2603.5.7 Ignition. *Exterior walls* shall not exhibit sustained flaming where tested in accordance with NFPA 268. Where a material is intended to be installed in more than one thickness, tests of the minimum and maximum thickness intended for use shall be performed.

> **Exception:** Assemblies protected on the outside with one of the following:
>
> 1. A thermal barrier complying with Section 2603.4.
> 2. A minimum 1 inch (25 mm) thickness of concrete or masonry.
> 3. Glass-fiber-reinforced concrete panels of a minimum thickness of $^3/_8$ inch (9.5 mm).

4. Metal-faced panels having minimum 0.019-inch-thick (0.48 mm) aluminum or 0.016-inch-thick (0.41 mm) corrosion-resistant steel outer facings.
5. A minimum $^7/_8$-inch (22.2 mm) thickness of stucco complying with Section 2510.

❖ The foam plastic is not to support continued flaming when tested in accordance with NFPA 268, which looks at ignitability when exposed to a radiant heat source. Where the foam plastic is to be used in more than one thickness, the test is to be performed on minimum and maximum thickness. If the assembly is protected on the outside with one of the specified materials, the test is not required.

2603.6 Roofing. Foam plastic insulation meeting the requirements of Sections 2603.2, 2603.3 and 2603.4 shall be permitted as part of a roof-covering assembly, provided the assembly with the foam plastic insulation is a Class A, B or C roofing assembly where tested in accordance with ASTM E 108 or UL 790.

❖ Foam plastic insulation that meets the labeling and identification, surface-burning characteristics and thermal barrier criteria are permitted as part of a Class A, B or C roof-covering assembly that has been tested in accordance with ASTM E 108 or UL 790. Both ASTM E 108 and UL 790 cover exterior spread of flame.

2603.7 Interior finish in plenums. Foam plastic insulation used as interior wall or ceiling finish in plenums shall comply with one or more of the following:

1. The foam plastic insulation shall be separated from the plenum by a thermal barrier complying with Section 2603.4 and shall exhibit a flame spread index of 75 or less and a smoke-developed index of 450 or less when tested in accordance with ASTM E 84 or UL 723 at the thickness and density intended for use.
2. The foam plastic insulation shall exhibit a flame spread index of 25 or less and a smoke-developed index of 50 or less when tested in accordance with ASTM E 84 or UL 723 at the thickness and density intended for use and shall meet the acceptance criteria of Section 803.1.2 when tested in accordance with NFPA 286.
3. The foam plastic insulation shall be covered by corrosion-resistant steel having a base metal thickness of not less than 0.0160 inch (0.4 mm) and shall exhibit a flame spread index of 75 or less and a smoke-developed index of 450 or less when tested in accordance with ASTM E 84 or UL 723 at the thickness and density intended for use.

❖ This section requires exposed foam plastic insulation (i.e.; foam plastic left unprotected), whether used as interior finish or as interior trim (see Section 2603.8), to exhibit a flame spread index of 25 or less, a smoke-developed index of 50 and meet the qualifications of the full-scale room-corner fire test (NFPA 286) with requirements for flame spread, heat release, no flashover and smoke release. Additionally, two alternatives are provided in Items 1 and 2 to protect foam plastics and thus allow them even when they have a higher flame spread index and smoke-developed index. This protection is provided by a thermal barrier or a corrosion-resistant steel barrier.

2603.8 Interior trim in plenums. Foam plastic insulation used as interior trim in plenums shall comply with the requirements of Section 2603.7.

❖ This section requires foam plastic insulation used as interior trim in plenums to comply with the requirements of Section 2603.7. Either full-scale room-corner testing or covering of the foam plastic by a thermal barrier or ignition protection is required (see commentary, Section 2603.7).

2603.9 Protection against termites. In areas where the probability of termite infestation is very heavy in accordance with Figure 2603.9, extruded and expanded polystyrene, polyisocyanurate and other foam plastics shall not be installed on the exterior face or under interior or exterior foundation walls or slab foundations located below grade. The clearance between foam plastics installed above grade and exposed earth shall be at least 6 inches (152 mm).

Exceptions:

1. Buildings where the structural members of walls, floors, ceilings and roofs are entirely of noncombustible materials or preservative-treated wood.
2. An approved method of protecting the foam plastic and structure from subterranean termite damage is provided.
3. On the interior side of basement walls.

❖ This section of the code addresses the use of foam plastics in areas of "very heavy" termite infestation. Termites have been reported to cause damage to foam insulation in very heavy termite infestation areas. These probabilities are referenced in Figure 2603.9 of this code. Foam plastic used in the "very heavy" infestation area is prohibited on the exterior face of below-grade foundation walls or slab foundations; under exterior or interior foundation walls; or slab foundations below grade or where located within 6 inches (152 mm) of exposed earth, unless the requirement of one of the three specific exceptions is met. Foam plastics are permitted on the interior side of basement walls, where the structural members of the building are either noncombustible or pressure preservative-treated wood or where the foam plastic is adequately protected from subterranean termite damage.

FIGURE 2603.9. See page 26-12.

❖ Figure 2603.9 establishes the probabilities of termite infestation for use in applying the requirements of Section 2603.9 (see also commentary, Section 2603.8).

2603.10 Special approval. Foam plastic shall not be required to comply with the requirements of Sections 2603.4 through

2603.8 where specifically approved based on large-scale tests such as, but not limited to, NFPA 286 (with the acceptance criteria of Section 803.2), FM 4880, UL 1040 or UL 1715. Such testing shall be related to the actual end-use configuration and be performed on the finished manufactured foam plastic assembly in the maximum thickness intended for use. Foam plastics that are used as interior finish on the basis of special tests shall also conform to the flame spread and smoke-developed requirements of Chapter 8. Assemblies tested shall include seams, joints and other typical details used in the installation of the assembly and shall be tested in the manner intended for use.

❖ Foam plastic does not have to comply with the installa-tion and use requirements of Sections 2603.4 through 2603.7 when specific approval is obtained in accordance with this section. This section lists examples of specific large-scale tests, such as: FM 4880, NFPA 286, UL 1040 or UL 1715. Also, other large-scale fire tests related to actual end-use configuration can be used. The intent is to require testing based on the proposed end-use configuration of the foam plastic assembly with a fire exposure that is appropriate in size and location for the proposed application. These tests are to be performed on full-scale assemblies. The tested assemblies must include typical seams, joints and other details that will occur in the finished installation. The foam plastic must be tested in the maximum thickness and density intended for use. Thorough testing provides an accurate depiction of the in-place fire performance of assemblies and systems using foam plastics.

There are two ways to show code compliance under Section 2603.10. One method is to provide the actual test report that contains a description of the assembly and test results showing that the foam plastic, in the end-use application, has passed the test. The second method is to obtain, from an evaluation agency such as ICC ES, an evaluation report that covers the end-use application. Materials that are being used as interior finish would still need to additionally pass the flame spread index and smoke-developed index requirements of Chapter 8. It should be noted that NFPA 286 is one way of demonstrating compliance with the interior finish requirements of Chapter 8 and with this section.

2603.10.1 Exterior walls. Testing based on Section 2603.10 shall not be used to eliminate any component of the construction of an exterior wall assembly when that component was included in the construction that has met the requirements of Section 2603.5.5.

❖ This section simply reminds the code user that if both Sections 2603.5.5 and 2603.10 are used to qualify the use of a foam plastic material without a thermal barrier and for use in an exterior wall assembly, then the most critical assembly would be required. For example, typically, testing in accordance with Section 2603.10 would not include gypsum board on the interior of the assembly, whereas gypsum board is typi-

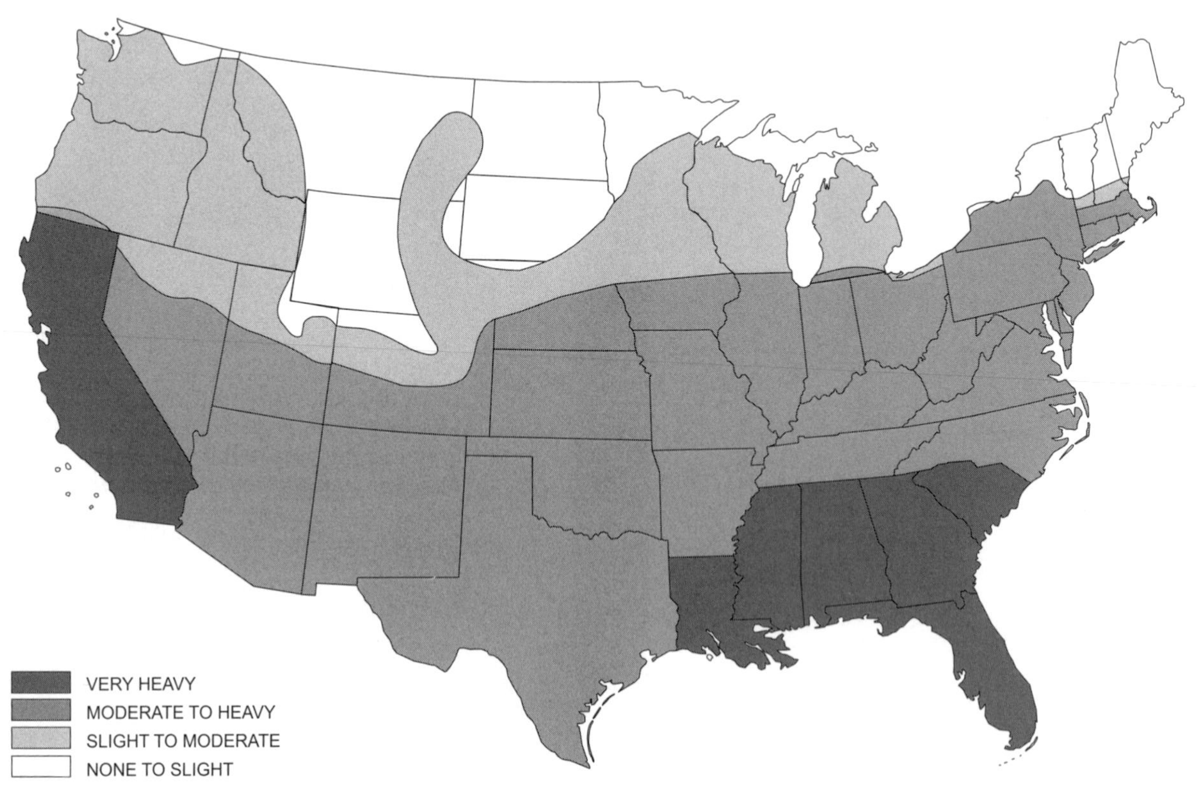

FIGURE 2603.9
TERMITE INFESTATION PROBABILITY MAP

cally used on the interior of an assembly tested in accordance with Section 2603.5.5 (NFPA 285). In this case, the gypsum board would be required on the interior face of the assembly.

SECTION 2604 INTERIOR FINISH AND TRIM

2604.1 General. Plastic materials installed as interior finish or *trim* shall comply with Chapter 8. Foam plastics shall only be installed as interior finish where approved in accordance with the special provisions of Section 2603.10. Foam plastics that are used as interior finish shall also meet the flame-spread index requirements for interior finish in accordance with Chapter 8. Foam plastics installed as interior *trim* shall comply with Section 2604.2.

❖ Plastic materials used as interior finish that are neither light transmitting nor foam are required to comply with Chapter 8. Foam plastic used as interior finish is to be in accordance with the special approval provisions of Section 2603.10 and meet the flame spread index requirements of Chapter 8 for interior finish. Foam plastics installed as interior trim are required to comply with Sectior 2604.2. It should be noted that the IFC addresses foam plastic trim in existing buildings in the same manner.

[F] 2604.2 Interior trim. Foam plastic used as interior *trim* shall comply with Sections 2604.2.1 through 2604.2.4.

❖ This section identifies the applicable subsections that apply to the interior trim requirements.

[F] 2604.2.1 Density. The minimum density of the interior *trim* shall be 20 pcf (320 kg/m^3).

❖ In order to qualify as interior trim, the density of materials must be at least 20 pcf (320 kg/m^3). The intent was to separate those materials used for trim from those intended for use as insulation. The 20-pcf (320 kg/m^3) value was selected because it was approximately midway between the densities commonly used for insulations and for trim when this provision was added to the code. Most foam plastic insulation was, and still is, in the range of 1 to 2$^1/_2$ pcf (16.02 to 40 kg/m^3), with very few materials over 5 pcf (81 kg/m^3). Foam plastic trim was manufactured in a density range of 30 to 45 pcf (481 to 721 kg/m^3) when the density provision was originally incorporated in the code.

[F] 2604.2.2 Thickness. The maximum thickness of the interior *trim* shall be $^1/_2$ inch (12.7 mm) and the maximum width shall be 8 inches (204 mm).

❖ Even though other trim materials are not limited in dimension, the maximum thickness and width of foam plastic trim is limited to $^1/_2$ inch (12.7 mm) and 8 inches (203 mm), respectively. These dimensions were selected because they were typical of the maximums being produced at the time this provision was included in the code.

[F] 2604.2.3 Area limitation. The interior *trim* shall not constitute more than 10 percent of the specific wall or ceiling areas to which it is attached.

❖ Trim cannot constitute more than 10 percent of the of the wall or ceiling area of a room. This limitation is simply a restatement of the general requirement for all combustible trim, which appears in Section 806.5.

[F] 2604.2.4 Flame spread. The flame spread index shall not exceed 75 where tested in accordance with ASTM E 84 or UL 723. The smoke-developed index shall not be limited.

Exception: When the interior *trim* material has been tested as an interior finish in accordance with NFPA 286 and complies with the acceptance criteria in Section 803.1.2.1, it shall not be required to be tested for flame spread index in accordance with ASTM E 84 or UL 723.

❖ The flame spread index must not be greater than 75 when tested in accordance with ASTM E 84 or UL 263. The value of 75 was selected to be consistent with the requirement for foam plastic insulation, even though other materials used as trim are permitted to have flame spread indexes of up to 200 in many locations. The smoke-developed index is not regulated.

The exception allows for foam plastic to be used as an interior trim under the same conditions that other materials are allowed as interior finishes. This is based on the fact that historically the codes have required less stringent flame spread and smoke-development index criteria for interior trim as compared to those for interior finish, based on the area restrictions on interior trim. The conditions for acceptance are based on several acceptance criteria applied after testing of the foam plastic in accordance with NFPA 286 (see commentary, Section 803.1.2.1 for more information).

SECTION 2605 PLASTIC VENEER

2605.1 Interior use. Where used within a building, plastic veneer shall comply with the interior finish requirements of Chapter 8.

❖ Plastic veneer may be used inside a building if it meets the finish requirements of Chapter 8.

2605.2 Exterior use. Exterior plastic veneer, other than plastic siding, shall be permitted to be installed on the *exterior walls* of buildings of any type of construction in accordance with all of the following requirements:

1. Plastic veneer shall comply with Section 2606.4.
2. Plastic veneer shall not be attached to any exterior wall to a height greater than 50 feet (15 240 mm) above grade.
3. Sections of plastic veneer shall not exceed 300 square feet (27.9 m^2) in area and shall be separated by a minimum of 4 feet (1219 mm) vertically.

Exception: The area and separation requirements and the smoke-density limitation are not applicable to plastic veneer applied to buildings constructed of Type VB construction, provided the walls are not required to have a fire-resistance rating.

❖ Exterior plastic veneer, not including plastic siding, must be an approved Class CC1 or CC2 plastic material in accordance with the specifications in Section 2606.4, and meet all three of the requirements in this section for height, area and separation. The classifications within Section 2606.4 are combustibility classifications. Note that plastic siding requirements are contained in Section 2605.3.

2605.3 Plastic siding. Plastic siding shall comply with the requirements of Sections 1404 and 1405.

❖ Plastic siding is used as an exterior veneer finish material. The purpose of this section is to refer the code user back to Chapter 14, specifically Sections 1404 and 1405, for the requirements related to plastic siding used on the exterior side of exterior walls.

SECTION 2606 LIGHT-TRANSMITTING PLASTICS

2606.1 General. The provisions of this section and Sections 2607 through 2611 shall govern the quality and methods of application of light-transmitting plastics for use as light-transmitting materials in buildings and structures. Foam plastics shall comply with Section 2603. Light-transmitting plastic materials that meet the other code requirements for walls and roofs shall be permitted to be used in accordance with the other applicable chapters of the code.

❖ Section 2606 contains the provisions for light-transmitting plastics. It specifies both the requirements for an approved light-transmitting plastic and the fire properties of materials. It also requires adequate structural properties of materials and assemblies, and specific installation methods for various applications.

2606.2 Approval for use. Sufficient technical data shall be submitted to substantiate the proposed use of any light-transmitting material, as approved by the *building official* and subject to the requirements of this section.

❖ Before the material can be used, sufficient data is to be submitted to the building official for approval.

2606.3 Identification. Each unit or package of light-transmitting plastic shall be identified with a *mark* or decal satisfactory to the *building official*, which includes identification as to the material classification.

❖ The material or its packaging must be marked Class CC1 or CC2 plastic.

2606.4 Specifications. Light-transmitting plastics, including thermoplastic, thermosetting or reinforced thermosetting plastic material, shall have a self-ignition temperature of 650°F (343°C) or greater where tested in accordance with ASTM D 1929; a smoke-developed index not greater than 450 where tested in the manner intended for use in accordance with ASTM E 84 or UL 723, or a maximum average smoke density rating not greater than 75 where tested in the thickness intended for use in accordance with ASTM D 2843 and shall conform to one of the following combustibility classifications:

Class CC1: Plastic materials that have a burning extent of 1 inch (25 mm) or less where tested at a nominal thickness of 0.060 inch (1.5 mm), or in the thickness intended for use, in accordance with ASTM D 635.

Class CC2: Plastic materials that have a burning rate of $2^1/_2$ inches per minute (1.06 mm/s) or less where tested at a nominal thickness of 0.060 inch (1.5 mm), or in the thickness intended for use, in accordance with ASTM D 635.

❖ An approved material may be either thermoplastic, thermosetting or reinforced thermosetting. Approved plastics require a minimum self-ignition temperature of 650°F (343°C) when tested in accordance with ASTM D 1929. This temperature excludes the use of easily ignited plastic materials, such as cellulose nitrate, which ignites at a temperature of about 300°F (149°C). For comparative purposes, untreated wood and paper ignite at about 500°F (260°C).

In order for a plastic to be approved, its smoke-development index or smoke density rating is limited and must be evaluated by one of two test methods prescribed (smoke-developed index limited to 450 by ASTM E 84 or UL 263, or a maximum average smoke density rating of 75 by ASTM D 2843). For comparative purposes, these smoke-developed values are the maximum that can be expected from some wood products tested under similar conditions; however, there is no intent to indicate a correlation between the results from the two test methods.

Approved light-transmitting plastics are to be tested in accordance with ASTM D 635 and be of combustibility Class CC1 or CC2. Class CC1 plastic materials burn at a slower rate than Class CC2 plastics. Class CC1 plastics generally consist of polycarbonate materials whereas Class CC2 plastics consist of acrylics.

Thermoplastic materials are those plastics that will melt when heated and are formed into their final shape utilizing this property. Most light-transmitting plastic products are flat or formed sheets.

Thermosetting materials are those plastics that cure with heat and may generate heat during curing but will not melt by heating after curing. They are formed into their final product shape before or during curing, except for possible machining.

2606.5 Structural requirements. Light-transmitting plastic materials in their assembly shall be of adequate strength and durability to withstand the loads indicated in Chapter 16. Technical data shall be submitted to establish stresses, maximum unsupported spans and such other information for the various thicknesses and forms used as deemed necessary by the *building official*.

❖ Light-transmitting plastic applications must meet the necessary structural load requirements. For example, glazing assemblies must be able to withstand the imposed wind loads, and skylight assemblies must be able to support expected snow loads. It is the intent of this section that durability includes weatherability for those applications subjected to outdoor exposure. Technical information to support the structural integrity is to be submitted to the building official as he or she deems necessary.

2606.6 Fastening. Fastening shall be adequate to withstand the loads in Chapter 16. Proper allowance shall be made for expansion and contraction of light-transmitting plastic materials in accordance with accepted data on the coefficient of expansion of the material and other material in conjunction with which it is employed.

❖ The design loads for the assembly, members and fasteners are specified in Chapter 16. Adequate provision must be made for movement caused by the differential expansion expected from the different types of materials connected. This is especially important with plastics since they usually have coefficients of expansion that are greater than the materials to which they are connected.

2606.7 Light-diffusing systems. Unless the building is equipped throughout with an *automatic sprinkler system* in accordance with Section 903.3.1.1, light-diffusing systems shall not be installed in the following occupancies and locations:

1. Group A with an *occupant load* of 1,000 or more.
2. Theaters with a stage and proscenium opening and an *occupant load* of 700 or more.
3. Group I-2.
4. Group I-3.
5. Interior exit stairways and ramps and *exit* passageways

❖ Light-diffusing systems are not permitted in the four occupancies listed in Items 1 through 4 in this section, nor in exits in any group, unless the building is protected by an automatic sprinkler system in accordance with NFPA 13. The intent is to limit the use of combustible materials in occupancies without sprinkler protection. These occupancies are generally recognized as being potentially more hazardous for the building occupants in the event of fire. The specific restriction for nonsprinklered buildings of Groups I-2 and I-3 is based on the physical restraint or incapacitation of the occupants.

2606.7.1 Support. Light-transmitting plastic diffusers shall be supported directly or indirectly from ceiling or roof construction by use of noncombustible hangers. Hangers shall be at least No. 12 steel-wire gage (0.106 inch) galvanized wire or equivalent.

❖ This section specifies that all hangers supporting the plastic diffusers must be of steel wire at least 0.106 inch (2.7 mm) in diameter or equivalent. It is important that noncombustible supports be used because the original test data were generated on these types of assemblies.

2606.7.2 Installation. Light-transmitting plastic diffusers shall comply with Chapter 8 unless the light-transmitting plastic diffusers will fall from the mountings before igniting, at an ambient temperature of at least 200°F (111°C) below the ignition temperature of the panels. The panels shall remain in place at an ambient room temperature of 175°F (79°C) for a period of not less than 15 minutes.

❖ Diffusers must also comply with Chapter 8 unless the plastic panels will fall from their mountings at an ambient temperature of at least 200°F (93°C) below their ignition temperature. The intent is to regulate panels that remain in place in the same manner as interior finish and to eliminate surface flame spread requirements for materials that will not burn while remaining in place.

The code requires that all panels must remain in place for at least 15 minutes at an ambient room temperature of 175°F (79°C). The intent of this requirement is to avoid a panic situation that could arise from the premature dropping of ceiling panels.

2606.7.3 Size limitations. Individual panels or units shall not exceed 10 feet (3048 mm) in length nor 30 square feet (2.79 m^2) in area.

❖ This section states the maximum length [10 feet (3048 mm)] and area [30 square feet (2.8 m^2)] of any single panel. The intent is to limit the size of panels expected to fall from place if the ambient room temperature at the ceiling were to reach 175°F (79°C), in order to avoid possible obstructions and injury caused by the fallen panels.

2606.7.4 Fire suppression system. In buildings that are equipped throughout with an *automatic sprinkler system* in accordance with Section 903.3.1.1, plastic light-diffusing systems shall be protected both above and below unless the sprinkler system has been specifically approved for installation only above the light-diffusing system. Areas of light-diffusing systems that are protected in accordance with this section shall not be limited.

❖ When buildings are required to have approved automatic sprinkler systems throughout, sprinkler protection must be provided both above and below the light-diffusing system. A sprinkler system is permitted to be installed above the light-diffusing system only when the assembly is designed for this use and specifically approved by the building official. The intent is that when specific approval is granted, it is based on appropriate test results related to end use. When sprinkler protection is provided, the area of the light-diffusing system is not subject to the area limitations of Section 2606.7.3 since the likelihood of such panels dropping is decreased.

2606.7.5 Electrical luminaires. Light-transmitting plastic panels and light-diffuser panels that are installed in approved electrical luminaires shall comply with the requirements of Chapter 8 unless the light-transmitting plastic panels conform to the requirements of Section 2606.7.2. The area of approved light-transmitting plastic materials that are used in required *exits* or *corridors* shall not exceed 30 percent of the aggregate area of the ceiling in which such panels are installed, unless the building is equipped throughout with an *automatic sprinkler system* in accordance with Section 903.3.1.1.

❖ Plastic light-transmitting and diffuser panels in electrical luminaires are required to comply with Chapter 8 or the installation requirements of Section 2606.7.2. Additionally, the aggregate ceiling area of such panels is limited to 30 percent in required exits and corridors, unless the building is equipped throughout with an automatic sprinkler system in accordance with NFPA 13. The intent of this section is to regulate luminaires the same as other ceiling finish panels. It also restricts the use of combustible materials in required exits and corridors while giving recognition to the benefit of automatic sprinkler systems.

2606.8 Partitions. Light-transmitting plastics used in or as partitions shall comply with the requirements of Chapters 6 and 8.

❖ When light-transmitting plastics are used in or as partitions, they must comply with the requirements of Chapters 6 and 8. The intent is to regulate light-transmitting plastics with the same requirements as other materials used within a building.

2606.9 Bathroom accessories. Light-transmitting plastics shall be permitted as glazing in shower stalls, shower doors, bathtub enclosures and similar accessory units. Safety glazing shall be provided in accordance with Chapter 24.

❖ Approved plastics are permitted as glazing in a variety of uses in bathrooms. When an approved Class CC1 or CC2 plastic material is used for glazing shower stalls, shower doors, bathtub enclosures or other bathroom accessories, it must meet the human impact requirements of ANSI Z97.1 under Section 2406.1.2.

2606.10 Awnings, patio covers and similar structures. *Awnings* constructed of light-transmitting plastics shall be constructed in accordance with the provisions specified in Section 3105 and Chapter 32 for projections. Patio covers constructed of light-transmitting plastics shall comply with Section 2606. Light-transmitting plastics used in canopies at motor fuel-dispensing facilities shall comply with Section 2606, except as modified by Section 406.7.2.

❖ Light-transmitting plastics used as awnings and similar structures must meet the general performance requirements of other appropriate sections of the code in addition to the requirements found in Chapter 26.

2606.11 Greenhouses. Light-transmitting plastics shall be permitted in lieu of plain glass in greenhouses.

❖ Greenhouses may be constructed of any type of approved plastic material that meets the requirements of Section 2606.

2606.12 Solar collectors. Light-transmitting plastic covers on solar collectors having noncombustible sides and bottoms shall be permitted on buildings not over three *stories above grade plane* or 9,000 square feet (836.1 m^2) in total floor area, provided the light-transmitting plastic cover does not exceed 33.33 percent of the roof area for CC1 materials or 25 percent of the roof area for CC2 materials.

Exception: Light-transmitting plastic covers having a thickness of 0.010 inch (0.3 mm) or less or shall be permitted to be of any plastic material provided the area of the solar collectors does not exceed 33.33 percent of the roof area.

❖ Approved light-transmitting plastics can be used for solar collector covers that meet the requirements of Section 2606.12. Basically, the size and height of the building is limited and depending on the combustibility two different maximums are provided. Any plastic material may be used as solar collector covers provided that the cover thickness does not exceed 0.010 inch (0.254 mm) and the collector does not exceed 33.33 percent of the roof area.

Essentially the fire hazard only exists on the roof, which poses minimal threats initially to occupants.

SECTION 2607
LIGHT-TRANSMITTING PLASTIC WALL PANELS

2607.1 General. Light-transmitting plastics shall not be used as wall panels in *exterior walls* in occupancies in Groups A-1, A-2, H, I-2 and I-3. In other groups, light-transmitting plastics shall be permitted to be used as wall panels in *exterior walls*, provided that the walls are not required to have a fire-resistance rating and the installation conforms to the requirements of this section. Such panels shall be erected and anchored on a foundation, waterproofed or otherwise protected from moisture absorption and sealed with a coat of mastic or other approved waterproof coating. Light-transmitting plastic wall panels shall also comply with Section 2606.

❖ Section 2607 establishes requirements for the use of plastic panels as exterior walls. Plastic wall panels are permitted to be used in most occupancies except those with dense occupant loads or where the occupants are confined or otherwise impaired. The size and quantity of plastic wall panels are also regulated by this section. These panels must also comply with the applicable provisions of Section 2606, which governs all light-transmitting plastics.

Plastic wall panels are permitted to be used in occupancies classified as Groups A-3, A-4, A-5, B, E, F-1, F-2, I-1, M, R-1, R-2, R-3, R-4, S-1, S-2 and U. They may not be used in occupancies where the generated products of combustion would significantly impair the egress of the impaired, confined or crowded occupants.

As with any exterior covering, plastic wall panels must be adequately fastened in a permanent manner. To reduce the chance of water infiltration into the plastic wall panels, they must be erected and anchored on a base that is protected from water absorption. Additionally, the plastic wall panel itself must be sealed with a waterproof coating. These types of combustible wall panels are not permitted where the exterior walls are required to be fire-resistance rated.

2607.2 Installation. *Exterior wall* panels installed as provided for herein shall not alter the type of construction classification of the building.

❖ Plastic is a combustible material. The use of light-transmitting plastic wall panels is permitted, however, without downgrading the type of construction. This exception is also clarified in Section 603.1, Item 14, for Type I and II construction; therefore, the use of light-transmitting plastic wall panels is permitted in all types of construction. These types of combustible wall panels, however, are not permitted where the exterior walls are required to be fire-resistance rated (see Section 2607.1).

2607.3 Height limitation. Light-transmitting plastics shall not be installed more than 75 feet (22 860 mm) above *grade plane*, except as allowed by Section 2607.5.

❖ Plastic wall panels are not permitted to be installed greater than 75 feet (22 860 mm) above grade plane in an unsprinklered building. The 75-foot (22 860 mm) threshold is the highest elevation that can be reached by hose lines on the ground. In structures sprinklered throughout, however, Section 2607.5 permits the maximum height to be unlimited. Be aware that Table 503 also places height limitations on structures. In some cases, the limitations of Table 503 may be more restrictive than those of Sections 2607.3 and 2607.5. At all times, the most restrictive requirement prevails.

2607.4 Area limitation and separation. The maximum area of a single wall panel and minimum vertical and horizontal separation requirements for exterior light-transmitting plastic wall panels shall be as provided for in Table 2607.4. The maximum percentage of wall area of any *story* in light-transmitting plastic wall panels shall not exceed that indicated in Table 2607.4 or the percentage of unprotected openings permitted by Section 705.8, whichever is smaller.

Exceptions:

1. In structures provided with approved flame barriers extending 30 inches (760 mm) beyond the *exterior wall* in the plane of the floor, a vertical separation is not required at the floor except that provided by the vertical thickness of the flame barrier projection.
2. Veneers of approved weather-resistant light-transmitting plastics used as exterior siding in buildings of Type V construction in compliance with Section 1406.
3. The area of light-transmitting plastic wall panels in *exterior walls* of greenhouses shall be exempt from the area limitations of Table 2607.4 but shall be limited as required for unprotected openings in accordance with Section 704.8.

❖ Refer to Table 2607.4 for area limitations of single wall panels and the minimum separation requirements between adjacent panels. The maximum percentage of wall area faced with plastic panels is limited by Table 2607.4, as well as by Section 705.8. At all times, the most restrictive requirement prevails.

Vertical separation of panels (see Table 2607.4) is not required where continuous architectural projections create at least a 30-inch (762 mm) barrier (called a flame barrier) between adjacent panels. Under most conditions, this flame barrier will delay or prevent vertical fire spread from one story to the next. This horizontal barrier is seen as protection equivalent to the vertical separation typically required by Table 2607.4 (see Figure 2607.4).

Veneers of approved exterior light-transmitting plastic siding installed in compliance with Sectior 1406 of Type V construction are exempt from the area limitations and separation requirements.

Greenhouse walls can have an unlimited area of approved light-transmitting wall panels, unless limited by the unprotected opening requirements of Section 705.8.

TABLE 2607.4. See below.

❖ This table establishes size and separation limitations for exterior plastic wall panels. The requirements of this table are not applicable to plastic exterior siding meeting the exception to Section 2607.4. The requirements are determined by the type of plastic material (Class CC1 or CC2) and fire separation distance. Plastic wall panels are only permitted where a minimum fire separation distance of 6 feet (1829 mm) is provided. The vertical separation requirements can be modified by Section 2607.4 (see Table 2607.4,

TABLE 2607.4
AREA LIMITATION AND SEPARATION REQUIREMENTS FOR LIGHT-TRANSMITTING PLASTIC WALL PANELS[a]

FIRE SEPARATION DISTANCE (feet)	CLASS OF PLASTIC	MAXIMUM PERCENTAGE AREA OF EXTERIOR WALL IN PLASTIC WALL PANELS	MAXIMUM SINGLE AREA OF PLASTIC WALL PANELS (square feet)	MINIMUM SEPARATION OF PLASTIC WALL PANELS (feet)	
				Vertical	Horizontal
Less than 6	—	Not Permitted	Not Permitted	—	—
6 or more but less than 11	CC1	10	50	8	4
	CC2	Not Permitted	Not Permitted	—	—
11 or more but less than or equal to 30	CC1	25	90	6	4
	CC2	15	70	8	4
Over 30	CC1	50	Not Limited	3[b]	0
	CC2	50	100	6[b]	3

For SI: 1 foot = 304.8 mm, 1 square foot = 0.0929 m^2.

a. For combinations of plastic glazing and plastic wall panel areas permitted, see Section 2607.6.

b. For reductions in vertical separation allowed, see Section 2607.4.

Note b). The maximum area limitations of a single panel and the maximum percentage area of an exterior wall can be doubled when the structure is completely sprinklered in accordance with NFPA 13 (see Section 2607.5).

2607.5 Automatic sprinkler system. Where the building is equipped throughout with an *automatic sprinkler system* in accordance with Section 903.3.1.1, the maximum percentage area of *exterior wall* in any *story* in light-transmitting plastic wall panels and the maximum square footage of a single area given in Table 2607.4 shall be increased 100 percent, but the area of light-transmitting plastic wall panels shall not exceed 50 percent of the wall area in any story, or the area permitted by Section 705.8 for unprotected openings, whichever is smaller. These installations shall be exempt from height limitations.

❖ The extra protection provided by a complete NFPA 13 automatic fire sprinkler system enables the limitations placed on light-transmitting plastic wall panels to be reduced. In order for these modifications to apply, the structure must be sprinklered throughout. When this protection is achieved, the height to which plastic wall panels can be installed is not limited. Furthermore, the single panel size and exterior wall area limitations can be doubled. At no time, however, can the area of wall panels per story exceed 50 percent of the wall area of that story. The percentage limitation of Section 705.8 for unprotected openings is also applicable.

2607.6 Combinations of glazing and wall panels. Combinations of light-transmitting plastic glazing and light-transmitting plastic wall panels shall be subject to the area, height and percentage limitations and the separation requirements applicable to the class of light-transmitting plastic as prescribed for light-transmitting plastic wall panel installations.

❖ Plastic glazing and plastic wall panels display equivalent fire characteristics; therefore, where plastic glazing (regulated by Section 2608) and plastic wall panels are both used, the aggregate area of plastics must be in compliance with the limitations and requirements of Table 2607.4.

SECTION 2608
LIGHT-TRANSMITTING PLASTIC GLAZING

2608.1 Buildings of Type VB construction. Openings in the *exterior walls* of buildings of Type VB construction, where not required to be protected by Section 705, shall be permitted to be glazed or equipped with light-transmitting plastic.

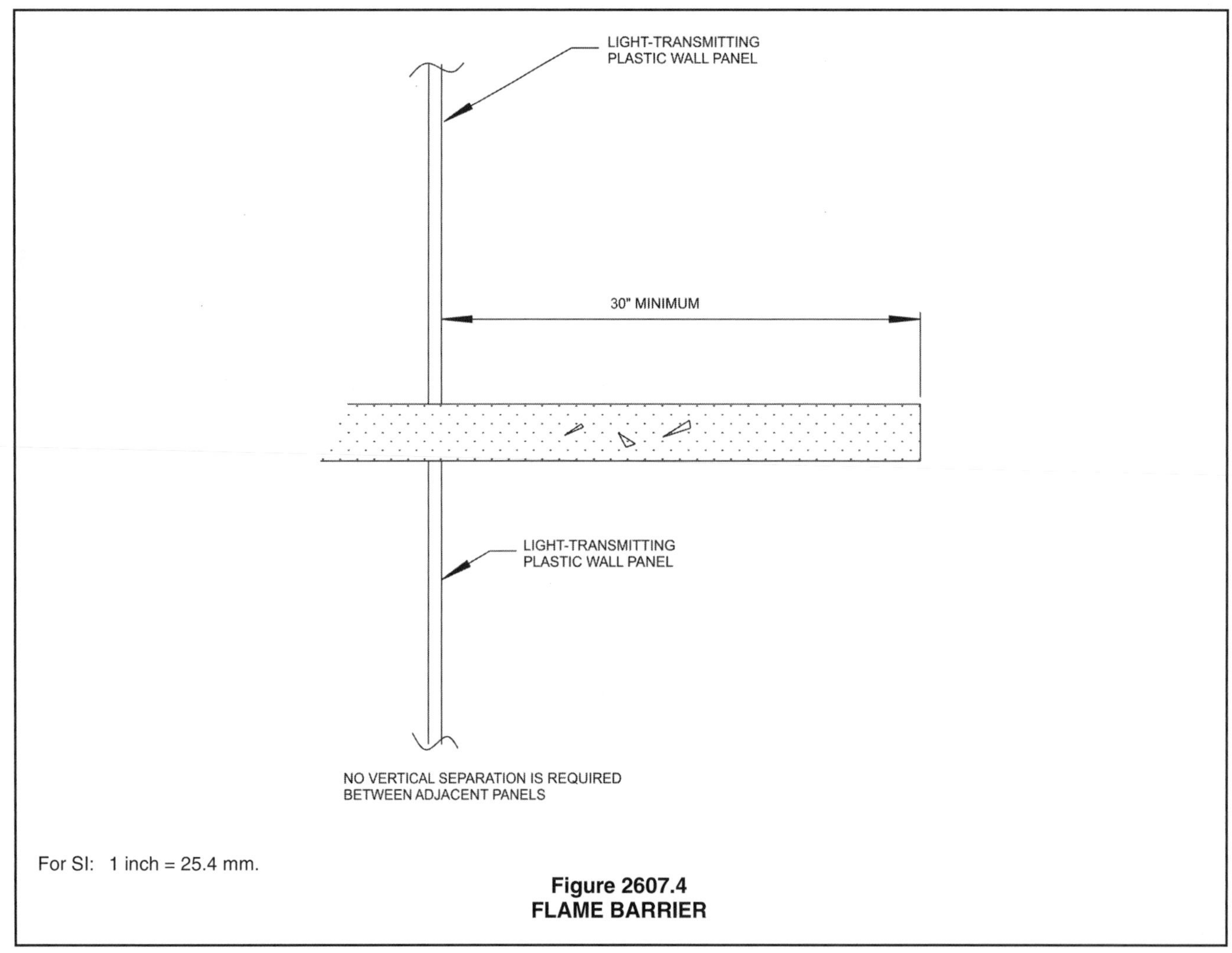

Figure 2607.4
FLAME BARRIER

Light-transmitting plastic glazing shall also comply with Section 2606.

❖ This section restricts the use of plastics to openings that are not required to have a fire protection rating. The requirement for walls, and their openings, to be protected is dependent on their fire separation distance (see Section 705).

2608.2 Buildings of other types of construction. Openings in the *exterior walls* of buildings of types of construction other than Type VB, where not required to be protected by Section 705, shall be permitted to be glazed or equipped with light-transmitting plastic in accordance with Section 2606 and all of the following:

1. The aggregate area of light-transmitting plastic glazing shall not exceed 25 percent of the area of any wall face of the *story* in which it is installed. The area of a single pane of glazing installed above the first *story above grade plane* shall not exceed 16 square feet (1.5 m^2) and the vertical dimension of a single pane shall not exceed 4 feet (1219 mm).

   Exception: Where an *automatic sprinkler system* is provided throughout in accordance with Section 903.3.1.1, the area of allowable glazing shall be increased to a maximum of 50 percent of the wall face of the *story* in which it is installed with no limit on the maximum dimension or area of a single pane of glazing.

2. Approved flame barriers extending 30 inches (762 mm) beyond the *exterior wall* in the plane of the floor, or vertical panels not less than 4 feet (1219 mm) in height, shall be installed between glazed units located in adjacent stories.

   Exception: Buildings equipped throughout with an *automatic sprinkler system* in accordance with Section 903.3.1.1.

3. Light-transmitting plastics shall not be installed more than 75 feet (22 860 mm) above grade level.

   Exception: Buildings equipped throughout with an *automatic sprinkler system* in accordance with Section 903.3.1.1.

❖ In addition to restricting the use of plastic glazing to exterior wall openings that are not required to have a fire protection rating, this section specifies the maximum area within a wall that openings can occupy. An opening that has plastic as the glazing material is essentially treated the same as an unprotected glass opening.

To limit the spread of fire, the allowable area of plastic glazing is limited to the smaller area determined when applying conditions. The first condition restricts the area of plastic glazing to 25 percent of the wall area of the story. The second condition restricts the area of plastics to that of the unprotected openings permitted in Section 705.

The exception doubles the amount of wall area that is allowed to be glazed with plastic when an approved automatic sprinkler system is installed. The allowable areas are increased to 50 percent of the wall area of the story with no limit on the maximum dimension or area of a single pane of glazing.

Consistent with other provisions to limit fire spread, this section places restrictions on the size and vertical separation of plastic glazing. The 4-foot (1219 mm) vertical separation is widely accepted as adequate for spandrels and other intervening construction between floors.

The 75-foot (22 860 mm) height limitation is generally accepted as the maximum height attainable by fire-fighting apparatus. If sprinklers are provided, the need for such height restrictions is lessened.

SECTION 2609 LIGHT-TRANSMITTING PLASTIC ROOF PANELS

2609.1 General. Light-transmitting plastic roof panels shall comply with this section and Section 2606. Light-transmitting plastic roof panels shall not be installed in Groups H, I-2 and I-3. In all other groups, light-transmitting plastic roof panels shall comply with any one of the following conditions:

1. The building is equipped throughout with an *automatic sprinkler system* in accordance with Section 903.3.1.1.
2. The roof construction is not required to have a fire-resistance rating by Table 601.
3. The roof panels meet the requirements for roof coverings in accordance with Chapter 15.

❖ This section establishes general requirements for light-transmitting plastic roof panels. These requirements are similar in nature to the other light-transmitting plastic provisions of this chapter. The requirements are intended to control the use of combustible construction material (i.e., light-transmitting plastics). These provisions specifically detail where roof panels may be used and any accompanying conditions that are required. Occupancies in Group H, I-2 or I-3 may not use light-transmitting plastic roof panels. Buildings or structures containing all other groups may use the roof panels, but at least one of the three conditions listed must be present. The first condition requires the structure under consideration to be equipped with an automatic sprinkler system installed in accordance with NFPA 13. The second condition allows a structure that is not sprinklered to contain plastic roof panels if Table 601 of the code does not require any fire-resistance ratings for the roof construction, including beams, trusses, framing, arches and roof decks (based on the type of construction and the distance between the floor and lowest roof member). The final condition that allows plastic roof panels in certain structures is that panels meet the requirements for roof coverings in Chapter 15.

2609.2 Separation. Individual roof panels shall be separated from each other by a distance of not less than 4 feet (1219 mm) measured in a horizontal plane.

Exceptions:

1. The separation between roof panels is not required in a building equipped throughout with an *automatic*

sprinkler system in accordance with Section 903.3.1.1.

2. The separation between roof panels is not required in low-hazard occupancy buildings complying with the conditions of Section 2609.4, Exception 2 or 3.

❖ The separation requirements for plastic roof panels are established herein. Individual light-transmitting plastic roof panels are limited in area by the provisions of Section 2609.4. This section provides the separation distances necessary between the individual roof panels to isolate the fire hazard associated with this combustible material.

The 4-foot (1219 mm) minimum separation distance required between each individual roof panel must be measured in a horizontal plane (see Figure 2609.2).

There are two exceptions in the code pertaining to the separation requirements. Any structure that is equipped with an automatic sprinkler system installed in accordance with NFPA 13, or low-hazard occupancy buildings complying with Exception 2 or 3 of Section 2609.4, are exempt from the separation requirements.

2609.3 Location. Where *exterior wall* openings are required to be protected by Section 705.8, a roof panel shall not be installed within 6 feet (1829 mm) of such *exterior wall.*

❖ This section provides requirements as to where light-transmitting plastic roof panels may be located in relation to the exterior walls of buildings or structures. Again, limitations are included as to where these panels of combustible construction can be located in relation to adjacent fire hazards. These provisions apply only when they are determined to be applicable from the requirements for fire-resistance-rated exterior walls and their exterior opening protectives found in Section 705.8. When certain openings in the exterior walls are required to be protected, these location requirements must be followed.

When an exterior wall contains required exterior opening protectives, the code text requires plastic roof panels to be installed at least 6 feet (1829 mm) from the wall. There is no exception to this requirement for structures equipped throughout with an approved automatic sprinkler system (see Figure 2609.3).

2609.4 Area limitations. Roof panels shall be limited in area and the aggregate area of panels shall be limited by a percentage of the floor area of the room or space sheltered in accordance with Table 2609.4.

Exceptions:

1. The area limitations of Table 2609.4 shall be permitted to be increased by 100 percent in buildings equipped throughout with an *automatic sprinkler system* in accordance with Section 903.3.1.1.

2. Low-hazard occupancy buildings, such as swimming pool shelters, shall be exempt from the area limitations of Table 2609.4, provided that the buildings do not exceed 5,000 square feet (465 m²) in

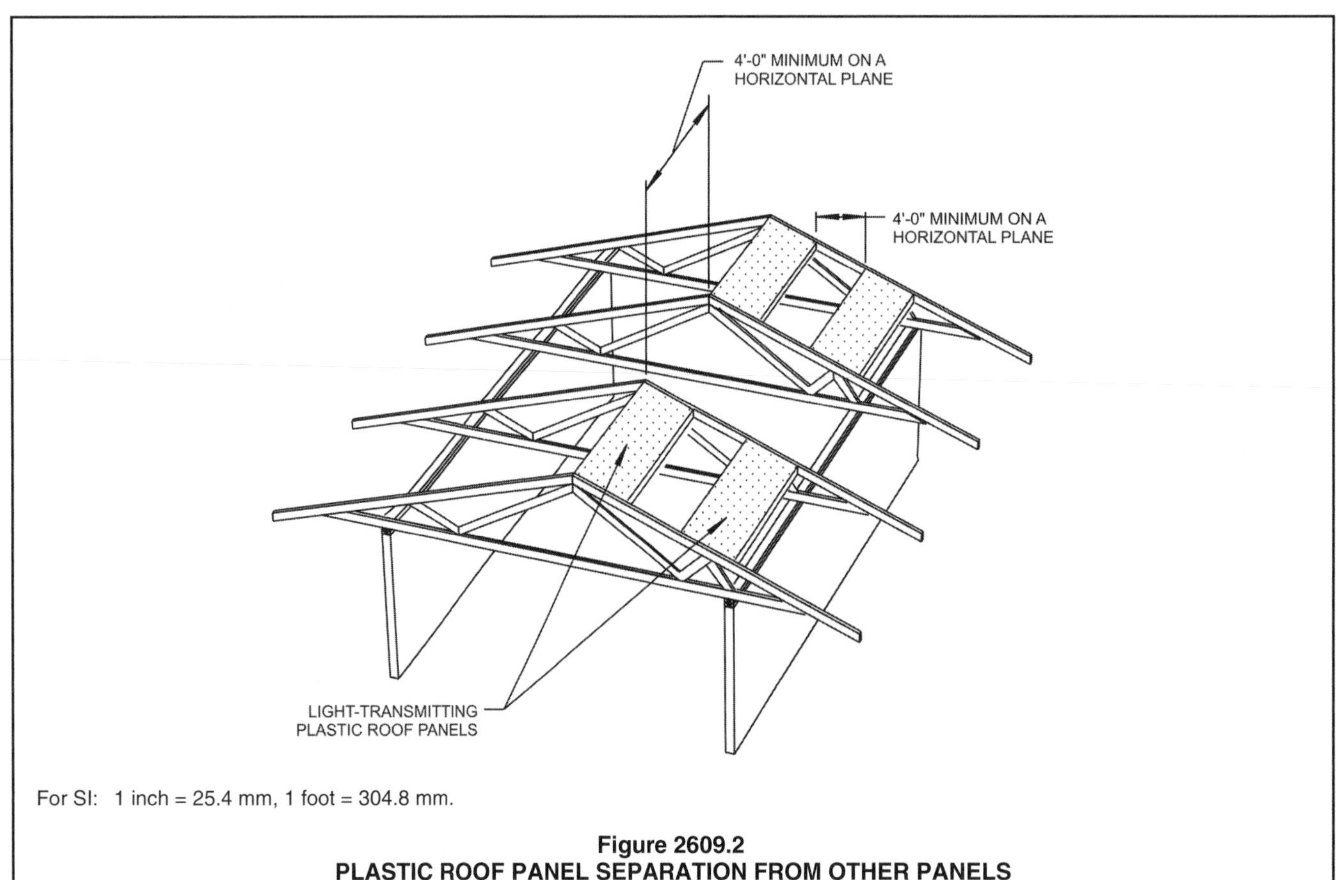

For SI: 1 inch = 25.4 mm, 1 foot = 304.8 mm.

Figure 2609.2
PLASTIC ROOF PANEL SEPARATION FROM OTHER PANELS

area and have a minimum fire separation distance of 10 feet (3048 mm).

3. Greenhouses that are occupied for growing plants on a production or research basis, without public access, shall be exempt from the area limitations of Table 2609.4 provided they have a minimum fire separation distance of 4 feet (1220 mm).

4. Roof coverings over terraces and patios in occupancies in Group R-3 shall be exempt from the area limitations of Table 2609.4 and shall be permitted with light-transmitting plastics.

❖ This section provides certain area limitations for plastic roof panels. Plastic roof panels of combustible plastic construction exhibit high fire-hazard properties and must be limited in terms of the area of the structure. These requirements not only list the area limitations for individual panels, but also limit the aggregate or sum total of all roof panels located in the structure's roof. The aggregate area limitations are based on a floor area percentage of the room or space enclosed by the roof.

The area limitations, which are based on the class of plastic roof panels, are detailed in Table 2609.4.

There are four exceptions to these area limitations. When certain additional protection is provided or the panels are located in structures of limited hazard, the area limitations can be altered. In some cases, these exceptions increase the limitations or, in other cases, exempt certain structures from the limitations. Exception 1 increases the area limitations by 100 percent for those structures equipped throughout with an automatic sprinkler system installed in accordance with NFPA 13. Exception 2 allows certain low-hazard buildings to be exempt from the roof panel limitations. These buildings must not exceed 5,000 square feet (465 m^2) in area and they must have at least 10 feet (3048 mm) of fire separation distance in order for the exception to apply. Exception 3 is for greenhouses that have at least a 4-foot (1219 mm) fire separation distance. Exception 4 allows occupancies in Group R-3 that have roofed-over terraces and patios to be exempt from the area limitations for roof panels. Such roof panels must be constructed with approved plastic meeting the requirements of Section 2606.

TABLE 2609.4
AREA LIMITATIONS FOR LIGHT-TRANSMITTING PLASTIC ROOF PANELS

CLASS OF PLASTIC	MAXIMUM AREA OF INDIVIDUAL ROOF PANELS (square feet)	MAXIMUM AGGREGATE AREA OF ROOF PANELS (percent of floor area)
CC1	300	30
CC2	100	25

For SI: 1 square foot = 0.0929 m^2.

❖ The maximum area limitations for both individual panels and aggregate areas are based on the class of plastic. These limitations are needed to restrict the use of light-transmitting plastic in roofs and roof coverings of buildings and structures. Section 2609.4

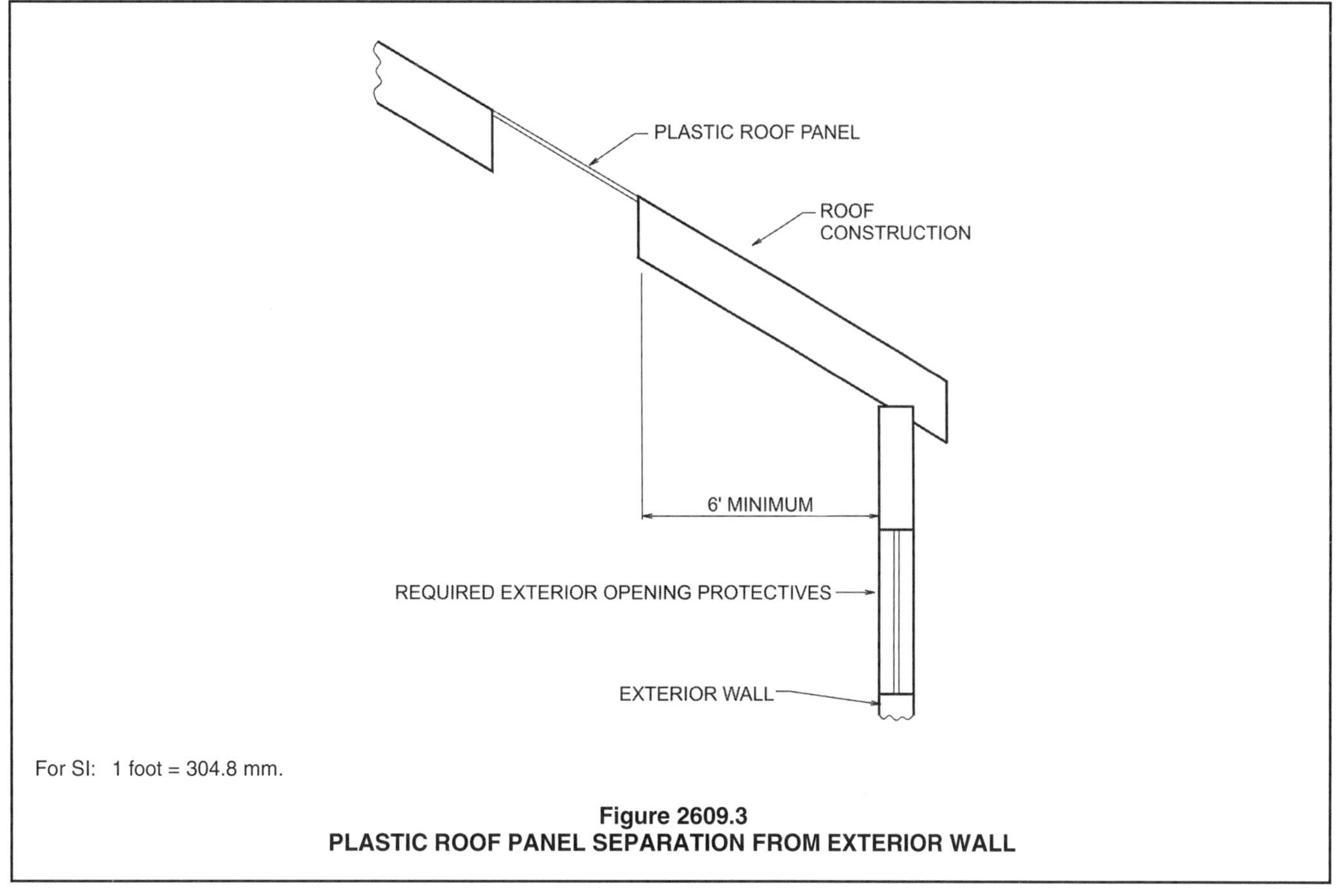

For SI: 1 foot = 304.8 mm.

Figure 2609.3
PLASTIC ROOF PANEL SEPARATION FROM EXTERIOR WALL

provides text that describes when these limitations are to be used. Table 2609.4 contains three different headings and only two lines of entry. The first heading is labeled "Class of plastic" and is to be used in conjunction with Section 2606.4 for determining whether the plastic used in roof panels is either Class CC1 (e.g., polycarbonate) or CC2 (e.g., acrylic). Once the class of plastic of the roof panels in question has been determined, the table is to be used to determine the area limitations for such panels.

The second heading in the table is identified as "Maximum area of individual roof panels (square feet)." These values provide the maximum area limitations for each individual roof panel based on the class of plastic used to manufacture the panel. These limitations are given in units of square feet.

The last heading, labeled "Maximum aggregate area of roof panels (percent of floor area)," provides the aggregate amount of area that can be occupied by plastic panels located in a roof. This limitation is based on a percentage of the floor area of the room or space that is sheltered by the roof in question. The floor area in units of square feet is multiplied by the appropriate percentage for the class of plastic used. This product provides the maximum area for all plastic panels used in the roof.

SECTION 2610 LIGHT-TRANSMITTING PLASTIC SKYLIGHT GLAZING

2610.1 Light-transmitting plastic glazing of skylight assemblies. Skylight assemblies glazed with light-transmitting plastic shall conform to the provisions of this section and Section 2606. Unit skylights glazed with light-transmitting plastic shall also comply with Section 2405.5.

Exception: Skylights in which the light-transmitting plastic conforms to the required roof-covering class in accordance with Section 1505.

❖ This section covers horizontal and sloped applications of light-transmitting plastic skylights. This section contains mounting, size and spacing limitations for these combustible assemblies. Plastic skylights sloped more than 15 degrees (0.26 rad) from vertical are also required to comply with Section 2405.

Acrylic and polycarbonate plastics are commonly used glazing materials in skylight assemblies. The use of combustible skylights is permitted subject to the requirements of Section 2606. The reference to Section 2405.5 is intended to highlight the specific structural considerations for unit skylights, regardless of construction materials.

The exception eliminates any added requirements for plastic skylights that meet the roofing classification for the surrounding roof covering as described in Chapter 15.

2610.2 Mounting. The light-transmitting plastic shall be mounted above the plane of the roof on a curb constructed in accordance with the requirements for the type of construction classification, but at least 4 inches (102 mm) above the plane of the roof. Edges of the light-transmitting plastic skylights or domes shall be protected by metal or other approved noncombustible material, or the light transmitting plastic dome or skylight shall be shown to be able to resist ignition where exposed at the edge to a flame from a Class B brand as described in ASTM E 108 or UL 790. The Class B brand test shall be conducted on a skylight that is elevated to a height as specified in the manufacturer's installation instructions, but not less than 4 inches (102 mm).

Exceptions:

1. Curbs shall not be required for skylights used on roofs having a minimum slope of three units vertical in 12 units horizontal (25-percent slope) in occupancies in Group R-3 and on buildings with a nonclassified roof covering.
2. The metal or noncombustible edge material is not required where nonclassified roof coverings are permitted.

❖ The mounting requirements for plastic skylights are similar to those established in Section 2405.4 for glass skylights. The 4-inch (102 mm) curb requirement protects the skylight from flying brands, provides a wash for water running down the slope and provides a noticeable barrier to caution a person on the roof from stepping on the skylight; however, the skylight must be designed for the structural requirements of Section 2606.5. Since the edge of the plastic is more susceptible to ignition than the surface, noncombustible edge protection is required to reduce the chances of edge ignition. This requirement is only applicable when the plastic is of a type and design that does not resist the flying brand exposure as described in ASTM E 108 or UL 790. This flying brand exposure test is required to be performed on the representative skylight, including mounting at a curb height specified in the manufacturer's installation instructions or 4 inches (102 mm), whichever is greater.

Exception 1 is based on the general field experience with plastic skylights used in one- and two-family dwellings with slopes equal to or greater than three units vertical in 12 units horizontal (3:12). Additionally, skylights in unclassified roofs are not required to be mounted on curbs. Such a skylight need not provide fire performance beyond that required for the surrounding roof covering.

The basis for Exception 2 is similar to that for Exception 1. There is no need to require edge protection for the plastic when the adjacent area is an unclassified roof permitted by Section 1505.5.

2610.3 Slope. Flat or corrugated light-transmitting plastic skylights shall slope at least four units vertical in 12 units horizontal (4:12). Dome-shaped skylights shall rise above the mounting flange a minimum distance equal to 10 percent of the maximum width of the dome but not less than 3 inches (76 mm).

Exception: Skylights that pass the Class B Burning Brand Test specified in ASTM E 108 or UL 790.

❖ As are most of the restrictions for plastic skylights, the minimum slope requirements are based on limiting the risks involved with combustible roof components. Both the slope for flat and corrugated plastic skylights and the rise for domed skylights are selected to shed flying brands. Consistent with other sections, the slope requirements are not applicable when the skylight complies with the specified requirements of ASTM E 108 or UL 790.

2610.4 Maximum area of skylights. Each skylight shall have a maximum area within the curb of 100 square feet (9.3 m^2).

Exception: The area limitation shall not apply where the building is equipped throughout with an *automatic sprinkler system* in accordance with Section 903.3.1.1 or the building is equipped with smoke and heat vents in accordance with Section 910.

❖ The area limitation for unsprinklered buildings minimizes the fire risks for plastic skylights. The 100-square-foot (9 m^2) limitation does not apply to buildings provided throughout with an NFPA 13 automatic sprinkler system or equipped with smoke and heat vents in accordance with Section 910. This exception is largely academic, since contemporary plastic skylights do not normally exceed this area limitation.

2610.5 Aggregate area of skylights. The aggregate area of skylights shall not exceed $33^1/_3$ percent of the floor area of the room or space sheltered by the roof in which such skylights are installed where Class CC1 materials are utilized, and 25 percent where Class CC2 materials are utilized.

Exception: The aggregate area limitations of light-transmitting plastic skylights shall be increased 100 percent beyond the limitations set forth in this section where the building is equipped throughout with an *automatic sprinkler system* in accordance with Section 903.3.1.1 or the building is equipped with smoke and heat vents in accordance with Section 910.

❖ This section includes limitations directed to minimize the risks of fire spread. The $33^1/_3$- and 25-percent limitations based on floor area for Class CC1 (polycarbonate) and CC2 (acrylic) materials, respectively, are judged to inhibit sufficiently the spread of fire. An exception allows the percentage to be increased to $66^2/_3$ percent (Class CC1) and 50 percent (Class CC2) when an approved automatic sprinkler system is installed throughout the building or is equipped with smoke and heat vents in accordance with Section 910. To comply with the intent of this provision, the skylights should be distributed over the entire area of the roof and not concentrated in one area.

2610.6 Separation. Skylights shall be separated from each other by a distance of not less than 4 feet (1219 mm) measured in a horizontal plane.

Exceptions:

1. Buildings equipped throughout with an *automatic sprinkler system* in accordance with Section 903.3.1.1.
2. In Group R-3, multiple skylights located above the same room or space with a combined area not exceeding the limits set forth in Section 2610.4.

❖ As with the other items cited, the minimum 4-foot (1219 mm) spacing minimizes the risk of fire spread for buildings without automatic sprinklers (also see Figure 2609.2). The separation is not necessary when complete fire suppression is provided in accordance with NFPA 13 or in Group R-3, as applicable in Section 2610.4, where the skylights are located above the same room or space with a combined area not to exceed the limits of Section 2610.4.

2610.7 Location. Where *exterior wall* openings are required to be protected in accordance with Section 705, a skylight shall not be installed within 6 feet (1829 mm) of such *exterior wall*.

❖ The minimum 6-foot (1829 mm) spacing to a fire-resistance-rated wall is to minimize the risk of fire spread (also see Section 2609.3).

2610.8 Combinations of roof panels and skylights. Combinations of light-transmitting plastic roof panels and skylights shall be subject to the area and percentage limitations and separation requirements applicable to roof panel installations.

❖ This is a precautionary statement to clarify that skylights and roof panels must be considered together when establishing size and separation limitations. When combinations of plastic skylights and roof panels occur, the area limitations of Section 2609.4 govern. As with wall panels and glazing, roof panels and skylights represent a similar combustible aspect of construction and should be addressed together.

SECTION 2611
LIGHT-TRANSMITTING PLASTIC INTERIOR SIGNS

2611.1 General. Light-transmitting plastic interior wall signs shall be limited as specified in Sections 2611.2 through 2611.4. Light-transmitting plastic interior wall signs in *covered and open mall buildings* shall comply with Section 402.16. Light-transmitting plastic interior signs shall also comply with Section 2606.

❖ This section contains the scoping statement that requires all light-transmitting plastic materials used for interior wall signage applications to comply with the requirements of Sections 2611.2, 2611.3 and 2611.4. These requirements include the maximum allowable aggregate area of signs, maximum individual area of signs and encasement requirements. The only exception to these requirements is light-transmitting plastics that are used for interior wall signage in covered mall buildings, since they are regulated by Section 402.6.4. These provisions help clarify the requirements for plastic signs in all occupancies other than covered mall buildings and exterior applications. One of the most common uses of light-transmitting plastic in interior signage applications is as the facing of internally illuminated or nonilluminated wall signs (sometimes referred to as "box" or "can" signs). Typical interior applications include tenant signage, building directories, promotional or advertising sign boxes and directional signage (other than exit signage).

2611.2 Aggregate area. The sign shall not exceed 20 percent of the wall area.

❖ The provisions of this section limit the maximum amount of wall area that may be covered by all plastic signs located on that wall, whereas Section 2611.3 governs the maximum permitted size of any individual sign. This limitation on the amount of wall area that can be covered by plastic signs is intended to minimize the potential for the spread of fire along a wall surface through involvement of the plastic signage.

2611.3 Maximum area. The sign shall not exceed 24 square feet (2.23 m^2).

❖ When applying the provisions of Sections 2611.2 and 2611.3, the most restrictive of each section must be applied. For example, if a single sign with the maximum area of 24 square feet (2 m^2) exceeds 20 percent of the wall area on which it is located, the area of the sign must be reduced to a size where the 20-percent limitation is not exceeded. Figures 2611.3(1) and 2611.3(2) give two examples of these provisions.

2611.4 Encasement. Edges and backs of the sign shall be fully encased in metal.

❖ The provisions of this section are intended to minimize the amount of surface that could be exposed to direct impingement by fire, and thus reduce the potential for ignition of the plastic.

SECTION 2612
FIBER-REINFORCED POLYMER

2612.1 General. The provisions of this section shall govern the requirements and uses of fiber-reinforced polymer in and on buildings and structures.

❖ Fiber reinforced polymer (FRP) composites are materials consisting of reinforcing fibers (natural or manmade) impregnated with a polymer (thermoset or thermoplastic) and are then molded and hardened into the intended shape. The reinforcement fibers (such as boron, glass, carbon, aramid) impart strength and stiffness to the composite, while the polymer resin matrix binds the fibers, providing bulk stiffness and protecting them from environmental

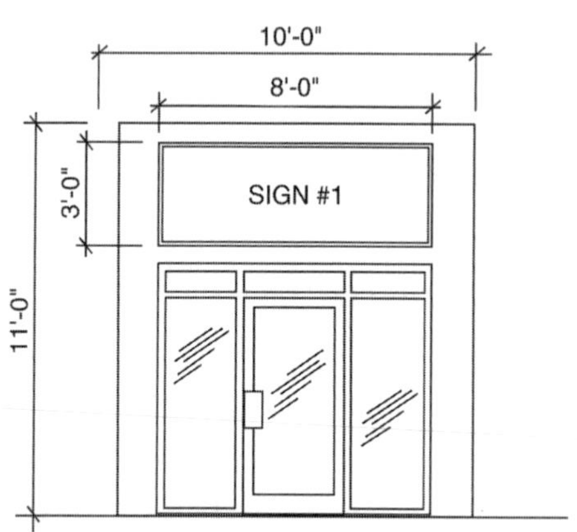

AREA OF SIGN #1 = 24 SQ. FT. (MAX. AREA PER SECTION 2611.3)
AGGREGATE AREA = 24 SQ. FT.

AREA OF WALL = 110 SQ. FT.
20% OF 110 SQ. FT. = 22 SQ. FT. (MAX. AREA PER SECTION 2611.2)
24 SQ. FT. ≤ 22 SQ. FT. ∴ AGGREGATE AREA OF SIGNS EXCEEDS AREA PERMITTED BY SECTION 2611.2

For SI: 1 inch = 25.4 mm, 1 foot = 304.8 mm, 1 square foot = 0.0929 m^2.

Figure 2611.3(1)
MAXIMUM AREA OF LIGHT-TRANSMITTING PLASTIC SIGNS—EXAMPLE 1

exposure. Common terms associated with FRP composites include fiberglass or fiber reinforced plastic, GFRP (glass fiber) or CFRP (carbon fiber).

Since the mid-1950s, FRP has been adapted to building and construction uses as opaque and translucent (light-transmitting) sheet panels; space frame skin structures; structural forms for concrete; sandwich panel structures; and, most recently, a variety of load-bearing and nonload-bearing components. Since the early 1990s, FRP composites have been used to externally strengthen concrete and masonry buildings, as well as to provide seismic strengthening to beams,columns, slabs and walls.

Typical FRP architectural products are manufactured in an open mold. Additives and various fillers incorporated in the composites enable fabricators to provide finished products with special properties, such as resistance to ultraviolet radiation, enhanced fire performance, corrosion resistance and color.

Principal markets served by the composites industry are architectural/construction, such as replication of historic building ornamentation, bathware, marine, automotive/transportation, corrosion-resistant products (tanks and piping) and many others. These products are increasingly being used as building materials and responsible code guidance is imperative.

2612.2 Labeling and identification. Packages and containers of fiber-reinforced polymer and their components delivered to the job site shall bear the *label* of an *approved agency* showing the manufacturer's name, product listing, product identification and information sufficient to determine that the end use will comply with the code requirements.

❖ FRP is required to be labeled and identified in a manner similar to the existing requirements for foam plastic insulation. This requirement will provide assurance to the building official that the product in the field is the same as that tested for compliance.

2612.3 Interior finishes. Fiber-reinforced polymer used as *interior finishes, decorative materials or trim* shall comply with Chapter 8.

❖ This section is intended to point the user to Chapter 8 for additional requirements for use of FRP as an interior finish. Section 2612 is not intended to be the sole location in the code for requirements for FRP. FRP must meet the requirements of the code for specific applications.

2612.3.1 Foam plastic cores. Fiber-reinforced polymer used as interior finish and which contains foam plastic cores shall comply with Chapter 8 and Chapter 26.

❖ Some FRPs may be constructed as sandwich panels. These panels would have FRP on both sides of a

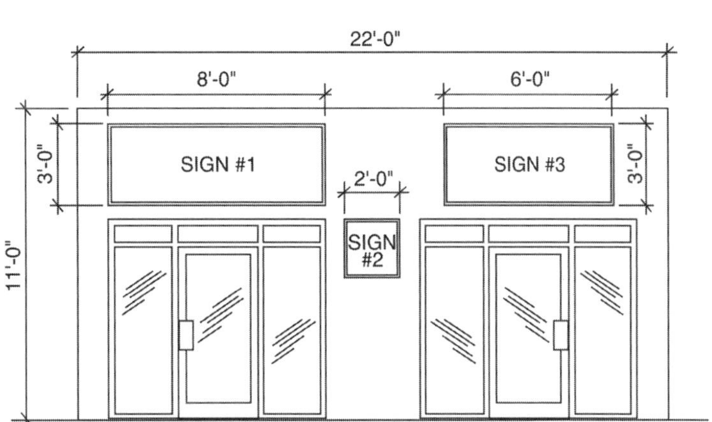

AREA OF SIGN #1 = 24 SQ. FT. (MAX. AREA PER SECTION 2611.3)
AREA OF SIGN #2 = 4 SQ. FT.
AREA OF SIGN #3 = 18 SQ. FT.
AGGREGATE AREA = 46 SQ. FT.

AREA OF WALL = 242 SQ. FT.
20% OF 242 SQ. FT. = 48.4 SQ. FT. (MAX. AREA PER SECTION 2611.2)
46 SQ. FT. $\leq$ 48.4 SQ. FT. $\therefore$ AGGREGATE AREA OF SIGNS COMPLIES WITH SECTION 2611.2

For SI: 1 inch = 25.4 mm, 1 foot = 304.8 mm,
1 square foot = 0.0929 m^2.

Figure 2611.3(2)
MAXIMUM AREA OF LIGHT-TRANSMITTING PLASTIC SIGNS—EXAMPLE 2

core material. The core material provides additional support to the FRP system. Typical core materials can be, but are not limited to, balsa wood, plywood or even foam-plastic materials. The definitions for "Fiber reinforced polymer" addresses these types of materials.

When foam plastic is used as a core material and the resultant FRP is used as interior finish, the resultant FRP must comply with both the requirements of Chapters 8 and 26.

Since foam plastic is regulated by this chapter, it is appropriate that FRP contain foam plastic cores also be regulated by this chapter.

2612.4 Light-transmitting materials. Fiber-reinforced polymer used as light-transmitting materials shall comply with Sections 2606 through 2611 as required for the specific application.

❖ This section is intended to point the user to Section 2606 for additional requirements for use of FRP as a light-transmitting plastic element. Section 2612 is not intended to be the sole location in the code for requirements for FRP. FRP must meet the requirements of the code for specific applications.

2612.5 Exterior use. Fiber-reinforced polymer shall be permitted to be installed on the *exterior walls* of buildings of any type of construction when such polymers meet the requirements of Section 2603.5. Fireblocking shall be installed in accordance with Section 718.

Exceptions:

1. Compliance with Section 2603.5 is not required when all of the following conditions are met:
   - 1.1. The fiber-reinforced polymer shall not exceed an aggregate total of 20 percent of the area of the specific wall to which it is attached, and no single architectural element shall exceed 10 percent of the area of the specific wall to which it is attached, and no contiguous set of architectural elements shall exceed 10 percent of the area of the specific wall to which they are attached.
   - 1.2. The fiber-reinforced polymer shall have a flame spread index of 25 or less. The flame spread index requirement shall not be required for coatings or paints having a thickness of less than 0.036 inch (0.9 mm) that are applied directly to the surface of the fiber-reinforced polymer.
   - 1.3. Fireblocking complying with Section 718.2.6 shall be installed.
   - 1.4. The fiber-reinforced polymer shall be installed directly to a noncombustible substrate or be separated from the exterior wall by one of the following materials: corrosion-resistant steel having a minimum base metal thickness of 0.016 inch (0.41 mm) at any point, aluminum having a minimum thickness of 0.019 inch (0.5 mm) or other approved noncombustible material.
2. Compliance with Section 2603.5 is not required when the fiber-reinforced polymer is installed on buildings that are 40 feet (12 190 mm) or less above grade when all of the following conditions are met:
   - 2.1. The fiber-reinforced polymer shall meet the requirements of Section 1406.2.
   - 2.2. Where the fire separation distance is 5 feet (1524 mm) or less, the area of the fiber-reinforced polymer shall not exceed 10 percent of the wall area. Where the fire separation distance is greater than 5 feet (1524 mm), there shall be no limit on the area of the *exterior wall* coverage using fiber-reinforced polymer.
   - 2.3. The fiber-reinforced polymer shall have a flame spread index of 200 or less. The flame spread index requirements do not apply to coatings or paints having a thickness of less than 0.036 inch (0.9 mm) that are applied directly to the surface of the fiber-reinforced polymer.
   - 2.4. Fireblocking complying with Section 718.2.6 shall be installed.

❖ Section 2612.5 provides requirements that specifically address the use of FRP on the exterior of buildings. This section allows FRP to be used on the exterior of buildings of all types of construction when it meets specific requirements.

The general charging requirements are that FRP meets the requirements of Section 2603.5 and the requirements for fireblocking in accordance with Section 717. Section 2603.5 addresses the use of foam plastics on exterior walls of all types of construction. By using the requirements in this section, FRP must meet tests such as NFPA 285 (multistory fire test), NFPA 268 (radiant heat test) and have Class A flame spread and smoke-developed indexes, as well as meeting other requirements specified in Section 2603.5. As with foam plastics, if a material can meet these requirements, it can be used as an exterior wall covering on buildings of any type of construction.

There are two exceptions to the general requirements. Exception 1 is for when FRP is used as building ornamentation, such as cornices. This set of requirements limits the area of the FRP and its flame spread index based on the percentage of the material on the wall. The area restrictions are based on potential applications and the philosophy that when larger amounts of materials are installed, the fire properties shall be more restrictive. Requirements are also provided whereby FRP shall be installed over noncombustible surfaces where fireblocking is required. This set of requirements provides ensurances that the materials to be used in this application are appropriate for use and do not create any undue hazard.

Exception 2 recognizes that FRP can be used on buildings up to a height of 40 feet (12 190 mm) in a manner consistent with other combustible exterior wall coverings. This is primarily because of the difficulties associated with fire-fighter access to higher portions of a building and to minimize potential injury from flying brands at high elevation. Combustible architectural trim and exterior veneers are permitted below the 40-foot (12 192 mm) height even when the overall building height exceeds 40 feet (12 192 mm). Combustible trim and veneers are also permitted on buildings of Type V construction. It should be noted that although Table 503 limits buildings of Type V construction to 40 feet (12 192 mm) in height, such buildings may be increased in height when protected with an automatic sprinkler system in accordance with Section 504.2.

Additionally, limits with respect to fire-separation distance are also provided in a manner similar to that for metal composite material panels. Further requirements for fireblocking have also been included.

SECTION 2613
REFLECTIVE PLASTIC CORE INSULATION

2613.1 General. The provisions of this section shall govern the requirements and uses of reflective plastic core insulation in buildings and structures. Reflective plastic core insulation shall comply with the requirements of Section 2613.2 and of one of the following: Section 2613.3 or 2613.4.

❖ The definition of "Reflective plastic core insulation" is given in Section 202. As stated, it is: "An insulation material less than 0.5-inch (12.7 mm) thick with at least one exterior low emittance surface and a core material containing voids or cells."

There are a number of reflective insulation materials made using "bubble pack" plastic as the core. This is the same material that is used as a packing material for fragile items.

The current definition of "Foam plastic" does not include bubble pack type materials because the "bubbles" in the materials are not formed with a foaming agent, but are mechanically formed using a stamping process. From a fire performance standpoint, these materials are predominantly plastic containing voids filled with air that support combustion.

2613.2 Identification. Packages and containers of reflective plastic core insulation delivered to the job site shall show the manufacturer's or supplier's name, product identification and information sufficient to determine that the end use will comply with the code requirements.

❖ Unlike foam plastics, the identification required on the reflective plastic core insulation is a manufacturer's self-certification and not a third-party label.

2613.3 Surface-burning characteristics. Reflective plastic core insulation shall have a flame spread index of not more than 25 and a smoke-developed index of not more than 450 when tested in accordance with ASTM E 84 or UL 723. The reflective plastic core insulation shall be tested at the maximum thickness intended for use. Test specimen preparation and mounting shall be in accordance with ASTM E 2599.

❖ As with other plastic materials, reflective plastic core insulation is required to be tested for flame spread and smoke-developed indexes. However, the traditional method of testing reflective plastic core foil insulation in the ASTM E 84 test (using chicken wire or poultry netting and rods) produces misleading results, with the serious potential for leading to severe fires. Testing reflective plastic core foil insulation with the ASTM E 84 test using chicken wire and rods suggests that the reflective plastic core foil insulation is safe and meets a flame spread index of 25 or less. In fact, when reflective plastic core foil insulation is tested in a more realistic way, with fasteners, the flame spread index of some materials can far exceed 25. The referenced test standard, ASTM E 2599, was specifically developed as the test specimen preparation and mounting method for reflective insulation materials when using ASTM E 84 or UL 723.

2613.4 Room corner test heat release. Reflective plastic core insulation shall comply with the acceptance criteria of Section 803.1.2.1 when tested in accordance with NFPA 286 or UL 1715 in the manner intended for use and at the maximum thickness intended for use.

❖ To properly evaluate the flaming, flashover, heat release rate and smoke release characteristics of the reflective plastic core insulation, testing in accordance with NFPA 286 or UL 1715 is required.

Bibliography

The following resource materials are referenced in this chapter or are relevant to the subject matter addressed in this chapter.

ANSI/DASMA 107-97(R 2004), *Room Fire Test Standard for Garage Doors Using Foam Plastic Insulation.* Cleveland, OH: Door and Access Systems Manufactures Association, 2003.

ANSI Z97.1-04, *Safety Glazing Materials Used in Buildings—Safety Performance Specifications and Methods of Test.* New York: American National Standards Institute, 1994.

ASTM D 635-06, *Standard Test Method for Rate of Burning and/or Extent and Time of Burning of Plastics in a Horizontal Position.* West Conshohocken, PA: ASTM International, 2006.

ASTM D 1929-96 (2001)01, *Test Method for Ignition Properties of Plastics.* West Conshohocken, PA: ASTM International, 2001.

ASTM D 2843-99 (2004)e01, *Test for Density of Smoke from the Burning or Decomposition of Plastics.* West Conshohocken, PA: ASTM International, 2004.

ASTM E 84-07, *Test Methods for Surface-burning Characteristics of Building Materials.* West Conshohocken, PA: ASTM International, 2007.

ASTM E 108-07a, *Test Methods for Fire Tests of Roof Coverings*. West Conshohocken, PA: ASTM International, 2007.

ASTM E 119-07, *Test Methods for Fire Tests of Building Construction and Materials*. West Conshohocken, PA: ASTM International, 2007.

ASTM E 2559-09, *Standard Practice for Specimen Preparation and Mounting of Reflective Insulation Materials and Radiant Barrier Materials for Building Applications to Assess Surface Burning Characteristic*. West Conshohocken, PA: ASTM International, 2009.

FM 4450 (1989), *Approval Standard for Class 1 Insulated Steel Deck Roofs—with Supplement (July 1992)*. Norwood, MA: Factory Mutual Standards Laboratories Department, 1989.

FM 4880 (2005), *Approval Standard for Class 1: a) Insulated Wall or Wall and Roof/Ceiling Panels, b) Plastic Interior Finish Materials, c) Plastic Exterior Building Panels, d) Wall/Ceiling Coating Systems and e) Interior or Exterior Finish Systems*. Norwood, MA: Factory Mutual Standards Laboratories Department, 2005.

IFC-12, *International Fire Code*. Washington, DC: International Code Council, 2011.

NFPA 13-07, *Installation of Sprinkler Systems*. Quincy, MA: National Fire Protection Association, 2007.

NFPA 259-03, *Test Method for Potential Heat of Building Materials*. Quincy, MA: National Fire Protection Association, 2003.

NFPA 268-07, *Standard Test Method for Determining Ignitability of Exterior Wall Assemblies Using a Radiant Heat Energy Source*. Quincy, MA: National Fire Protection Association, 2007.

NFPA 285-06, *Standard Method of Test for the Evaluation of Flammability Characteristics of Exterior Nonload-bearing Wall Assemblies Containing Combustible Components*. Quincy, MA: National Fire Protection Association, 2006.

NFPA 286-06, *Standard Method of Fire Test for Evaluating Contribution of Wall and Ceiling Interior Finish to Room Fire Growth*. Quincy, MA: National Fire Protection Association, 2006.

UL 263-03, *Standard for Fire Test of Building Construction and Materials*. Northbrook, IL: Underwriters Laboratories, Inc., 2003.

UL 723-08, *Standard for Test for Surface Burning Characteristics of Building Materials—with Revisions through May 2005*. Northbrook, IL: Underwriters Laboratories, Inc., 2008.

UL 790-04, *Standard Test Methods for Fire Tests of Roof Coverings*. Northbrook, IL: Underwriters Laboratories, Inc., 2004.

UL 1040-96, *Fire Test of Insulated Wall Construction—with Revisions through April 2001*. Northbrook, IL: Underwriters Laboratories, Inc., 2001.

UL 1256-02, *Standard for Safety Fire Test of Roof Deck Constructions*. Northbrook, IL: Underwriters Laboratories, Inc., 2002.

UL 1715-97, *Fire Test of Interior Finish Material*. Northbrook, IL: Underwriters Laboratories, Inc., 1997.

UL 1975-06, *Fire Test of Foamed Plastics Used for Decorative Purposes*. Northbrook, IL: Underwriters Laboratories, Inc., 2006.

Chapter 27: Electrical

General Comments

This chapter references National Fire Protection Association (NFPA) 70 for all requirements pertaining to electrical installations and contains specific provisions for emergency and standby power. NFPA 70 regulates the design, construction, installation, operation, maintenance and use of electrical systems and equipment. Except for the scope section, this chapter is devoted to establishing where emergency and standby power systems are required.

Purpose

Since electrical systems and components are an integral part of almost all structures, it is necessary for the code to address them to protect life, limb, health and property. In addition to general lighting and power needs, structures depend on electricity for operation of many of the life safety systems required by the code, including fire alarm, smoke control, exhaust, fire suppression, communication and fire command systems. Electricity is also depended upon for elements of egress systems, including elevators, exit signage, egress path illumination, power doors, stair pressurization and smokeproof enclosures.

SECTION 2701 GENERAL

2701.1 Scope. This chapter governs the electrical components, equipment and systems used in buildings and structures covered by this code. Electrical components, equipment and systems shall be designed and constructed in accordance with the provisions of NFPA 70.

❖ Chapter 27 regulates the electrical portions of buildings and structures that are included in the scope of the code. Regulation is accomplished by referencing NFPA 70.

SECTION 2702 EMERGENCY AND STANDBY POWER SYSTEMS

[F] 2702.1 Installation. Emergency and standby power systems required by this code or the *International Fire Code* shall be installed in accordance with this code, NFPA 110 and 111.

❖ Emergency power systems are intended to provide electrical power for life safety systems, such as egress illumination, emergency communications, fire pumps, high-rise building elevators and processes involving the handling and use of hazardous materials. In other words, emergency power is required where the loss of normal power would endanger occupants. Such systems are covered in Article 700 of NFPA 70 and one of their key features is the required response time of 10 seconds or less. The time between loss of normal power and the provision of emergency power must be kept very short to prevent putting occupants at risk. This is especially important during an emergency event, such as a building fire, but is important at all times to prevent occupant panic, which could happen if a crowded building is suddenly plunged into darkness.

Standby power systems are covered in Article 701 of NFPA 70 and are intended to provide electrical power for loads not as critical as those requiring emergency power. Standby power loads include smoke control systems; certain elevators; certain hazardous material operations; smokeproof enclosure systems; illumination; heating, ventilating and air-conditioning (HVAC) systems; refrigeration and sewage pumps. Standby power systems must provide power within 60 seconds of failure of primary power.

Sources of power for emergency power systems (NFPA 70, Section 701-11) include storage batteries, generators, uninterruptible power supplies and separate services. Sources of power for standby systems include those allowed for emergency systems plus a source that is taken from a point of connection ahead of the normal service disconnecting means.

NFPA 110 addresses the performance criteria and "nuts and bolts" of emergency and standby power systems and seperates them into types, classes and levels relative to maximum response time, minimum required operation time and life safety importance factor, respectively. NFPA 111 addresses stored emergency power supply systems and is similar in coverage to NFPA 110. Stored energy systems typically rely on batteries that store chemical energy.

[F] 2702.1.1 Stationary generators. Stationary emergency and standby power generators required by this code shall be *listed* in accordance with UL 2200.

❖ UL 2200 was developed at the request of building officials to provide a standard to evaluate the safety and reliability of stationary engine generators. UL 2200 establishes a basis for this evaluation.

[F] 2702.2 Where required. Emergency and standby power systems shall be provided where required by Sections 2702.2.1 through 2702.2.20.

❖ Fires can cause the loss of utility power as a result of either damage to equipment and wiring or fire-fighter action to shut off power to eliminate sources of ignition and danger to personnel.

Sections 2702.2.1 through 2702.2.20 dictate where emergency and standby power systems are required based upon the nature of the electrical loads. Note that emergency and standby power systems have different characteristics, and subsequent sections will require one or the other (see commentary, Section 2702.1).

[F] 2702.2.1 Group A occupancies. Emergency power shall be provided for emergency voice/alarm communication systems in Group A occupancies in accordance with Section 907.5.2.2.4.

❖ Emergency voice/alarm systems are designed to allow the fast, efficient and orderly egress of large numbers of people and are, therefore, considered a life safety system worthy of being provided with emergency power (see commentary, Section 907.2.1.2).

[F] 2702.2.2 Smoke control systems. Standby power shall be provided for smoke control systems in accordance with Section 909.11.

❖ Smoke control systems are intended to maintain a tenable environment in a building to allow the occupants ample time to evacuate or relocate to protected areas. As such, smoke control systems are life safety systems and must be dependable (see commentary, Section 909.11).

[F] 2702.2.3 Exit signs. Emergency power shall be provided for *exit* signs in accordance with Section 1011.6.3.

❖ Emergency power is warranted for exit signage illumination since guiding occupants to the exits is certainly a life safety function (see commentary, Section 1011.5.3).

[F] 2702.2.4 Means of egress illumination. Emergency power shall be provided for *means of egress* illumination in accordance with Section 1006.3.

❖ The path of travel to all exits must be illuminated to guide occupants and allow for safe egress; therefore, emergency power for illumination is warranted (see commentary, Section 1006.3).

[F] 2702.2.5 Accessible means of egress elevators. Standby power shall be provided for elevators that are part of an *accessible means of egress* in accordance with Section 1007.4.

❖ Elevators can be a component of an accessible means of egress in accordance with Section 1007.4 and must, therefore, be dependable at all times. Without backup power, an elevator could be a dead end for someone with physical disabilities who is trying to egress a building (see commentary, Section 1007.4).

[F] 2702.2.6 Accessible means of egress platform lifts. Standby power in accordance with this section or ASME A 18.1 shall be provided for platform lifts that are part of an *accessible means of egress* in accordance with Section 1007.5.

❖ Platform lifts are allowed to be used as part of an accessible means of egress only when they are allowed as part of an accessible route, in accordance with Section 1109.7, Items 1 through 9. When this is the case, the platform lift could be a dead end for someone with physical disabilities if power to the platform lift has been lost; therefore, standby power is required, in accordance with Section 2702 or ASME A18.1.

[F] 2702.2.7 Horizontal sliding doors. Standby power shall be provided for horizontal sliding doors in accordance with Section 1008.1.4.3.

❖ Power-operated doors could be an obstruction to egress if the primary power supply fails; therefore, standby power is required to maintain door operation (see commentary, Section 1008.1.3.3).

[F] 2702.2.8 Semiconductor fabrication facilities. Emergency power shall be provided for semiconductor fabrication facilities in accordance with Section 415.10.10.

❖ Where hazardous materials are utilized, such as in hazardous production materials (HPM) facilities, many systems are depended upon to protect the occupants from exposure to hazardous materials, including exhaust/ventilation systems, gas cabinet exhaust systems, gas detection systems, alarm systems and suppression systems. Loss of power would endanger the occupants; thus, emergency power is essential for these occupancies (see commentary, Sections 415.8.10 and 415.8.10.1).

[F] 2702.2.9 Membrane structures. Standby power shall be provided for auxiliary inflation systems in accordance with Section 3102.8.2. Emergency power shall be provided for *exit* signs in temporary tents and membrane structures in accordance with the *International Fire Code*.

❖ Air-supported and air-inflated structures would collapse on the occupants if the inflation systems fail. Section 3102.8.1.1 requires redundant inflation equipment that would be as worthless as the primary system in the event of power failure; therefore, standby power is required. Emergency power is required for all exit signs in temporary tents and membrane structures (see commentary, Section 3102.8.2).

[F] 2702.2.10 Hazardous materials. Emergency or standby power shall be provided in occupancies with hazardous materials in accordance with Section 414.5.3.

❖ Where hazardous materials and processes are housed, occupant safety could be dependent upon one or more ventilation, treatment, temperature control, alarm and detection system. Thus, emergency or standby power is required, depending upon the load (see commentary, Section 414.5.4).

[F] 2702.2.11 Highly toxic and toxic materials. Emergency power shall be provided for occupancies with highly *toxic* or *toxic* materials in accordance with the *International Fire Code*.

❖ See the commentary to Section 2702.2.9 and Chapter 27 of the *International Fire Code*® (IFC®) for additional information.

[F] 2702.2.12 Organic peroxides. Standby power shall be provided for occupancies with silane gas in accordance with the *International Fire Code*.

❖ See the commentary to Section 2702.2.9 and Chapter 39 of the IFC for additional information.

[F] 2702.2.13 Pyrophoric materials. Emergency power shall be provided for occupancies with silane gas in accordance with the *International Fire Code*.

❖ See the commentary to Section 2702.2.9 and Chapter 41 of the IFC for additional information.

[F] 2702.2.14 Covered and open mall buildings. Standby power shall be provided for voice/alarm communication systems in *covered and open mall buildings* in accordance with Section 402.7.3.

❖ See the commentary to Sections 907.2.20 and 402.13 for additional information.

[F] 2702.2.15 High-rise buildings. Emergency and standby power shall be provided in high-rise buildings in accordance with Sections 403.4.8 and 403.4.9.

❖ Occupants of high-rise buildings are at greater risk due to longer egress travel times, lack of fire-fighter access and the danger of vertical spread of fire and upward smoke migration. In accordance with this chapter and Sections 403.10 and 403.11, some loads in a high-rise building will require standby power and some will require emergency power (see commentary, Sections 403.10 and 403.11).

[F] 2702.2.16 Underground buildings. Emergency and standby power shall be provided in underground buildings in accordance with Sections 405.8 and 405.9.

❖ In the event of power failure, occupants could be underground without light, ventilation and the numerous required life safety systems. These structures are analogous to inverted high-rise buildings. Section 405.9 requires standby power for specified loads while Section 405.10 requires emergency power for specified loads (see commentary, Sections 405.9 and 405.10).

[F] 2702.2.17 Group I-3 occupancies. Emergency power shall be provided for doors in Group I-3 occupancies in accordance with Section 408.4.2.

❖ In an emergency, occupants in detention and correctional facilities are at the mercy of door-locking mechanisms and those who control such locks; thus, emergency power is warranted (see commentary, Section 408.4.2).

[F] 2702.2.18 Airport traffic control towers. Standby power shall be provided in airport traffic control towers in accordance with Section 412.3.4.

❖ This requirement is similar in intent to that for standby power in high-rise buildings, but due to the limited number of occupants and other protective features required by Section 412, only standby power is required (see commentary, Section 412.1.5).

[F] 2702.2.19 Elevators. Standby power for elevators shall be provided as set forth in Sections 3003.1, 3007.9 and 3008.9.

❖ Section 3003.1 provides the "how to" related text for elevator standby power systems, whereas other sections, including Sections 2702.2.5 and 2702.2.15, dictate where such power must be provided (see commentary, Section 3003.1).

[F] 2702.2.20 Smokeproof enclosures. Standby power shall be provided for smokeproof enclosures as required by Section 909.20.6.2.

❖ To protect egress elements from the invasion of smoke, mechanical ventilation/pressurization systems must be dependable and thus provided with standby power (see commentary, Section 909.20.6.2).

[F] 2702.3 Maintenance. Emergency and standby power systems shall be maintained and tested in accordance with the *International Fire Code*.

❖ Emergency and standby power systems must be maintained, serviced and tested to ensure dependability (see commentary, Section 604 of the IFC).

Bibliography

The following resource materials are referenced in this chapter or are relevant to the subject matter addressed in this chapter.

IFC-12, *International Fire Code*. Washington, DC: International Code Council, 2011.

NFPA 70, *National Electrical Code*. Quincy, MA: National Fire Protection Association, 1999.

NFPA 110, *Emergency and Standby Power Systems*. Quincy, MA: National Fire Protection Association, 1999.

NFPA 111, *Stored Electrical Energy Emergency and Standby Power Systems*. Quincy, MA: National Fire Protection Association, 1996.

UL 2200-98, *Stationery Engine Generator Assemblies (Revisions through July 2004)*. Northbrook, IL: Underwriters Laboratories, Inc., 2004.

Chapter 28: Mechanical Systems

General Comments

Chapter 28 provides for the approval, installation, construction, inspection, operation and maintenance of all mechanical appliances, equipment and systems by direct reference to the *International Mechanical Code*® (IMC®) and the *International Fuel Gas Code*® (IFGC®). All mechanical equipment and appliances, other than gas-fired equipment and appliances, must comply with the provisions of the IMC. Fuel-gas piping, gas-fired appliances and gas-fired appliance venting must comply with the IFGC.

Purpose

The purpose of Chapter 28 is to verify that mechanical equipment, appliances and mechanical systems are regulated with respect to design, construction, installation, inspection and maintenance. This chapter references the IMC and IFGC. This eliminates the need to duplicate the text of those codes and utilizes the performance and specification criteria of those codes as the basis for regulation of the construction, inspection and maintenance of all mechanical and fuel gas equipment, appliances and systems.

While this chapter only consists of references to other chapters and codes, it is necessary since almost all newly constructed buildings and structures contain mechanical equipment, appliances and systems. Without this chapter, a necessary link between the code and the IMC and IFGC would not exist.

SECTION 2801 GENERAL

[M] 2801.1 Scope. Mechanical appliances, equipment and systems shall be constructed, installed and maintained in accordance with the *International Mechanical Code* and the *International Fuel Gas Code*. Masonry chimneys, fireplaces and barbecues shall comply with the *International Mechanical Code* and Chapter 21 of this code.

❖ The requirements for all mechanical systems are governed by this chapter. This chapter does not contain any requirements for mechanical equipment, appliances and systems; instead it references the IMC and IFGC. These two codes along with nine others, including this code, make up the family of the 2009 editions of the *International Codes*®.

In the process of producing the code, including actual mechanical equipment, appliances and system provisions was considered unnecessary since such provisions would duplicate and possibly conflict with those of the IMC or IFGC. Since most newly constructed buildings contain mechanical equipment, appliances and systems, this chapter is necessary to reference the IMC and IFGC and tie them together with the code.

The IMC and IFGC work together to regulate all mechanical systems. Therefore, whether designing the mechanical system for a building or enforcing the provisions of the code that deal with mechanical systems, both the IMC and IFGC will be required.

The IMC and IFGC are complete codes in themselves, and set forth necessary administrative and enforcement requirements, as well as the requirements for installation, construction, inspection and maintenance of mechanical equipment, appliances and systems. The IMC regulates all aspects of mechanical systems, including: air distribution systems and ductwork; boilers and hydronic systems; exhaust systems; fuel-oil piping systems; combustion air requirements; chimneys and vents; ventilation air requirements; refrigeration systems; solar-powered systems; and specific appliances and equipment. The IFGC regulates all aspects of fuel-gas piping systems, fuel-gas utilization equipment and related accessories, including: combustion air requirements for gas appliances; chimneys and vents for gas appliances; fuel-gas piping; and specific gas appliances and gas equipment.

A key component of the IMC and IFGC is Chapter 3 of both codes, which requires all mechanical equipment and appliances to be listed and labeled to the standard or standards relevant to that equipment or appliance (see commentary, Figure 2801.1).

Another important element of the IMC and IFGC is a set of unique definitions found in Chapter 2 of both codes. These definitions supersede any others with respect to mechanical and gas systems. Especially noteworthy is the definition of "Noncombustible material" in the IMC, which does not include composite materials as provided for in Section 703.4.2 of this code. In the context of the IMC, such composite materials are considered to be combustible.

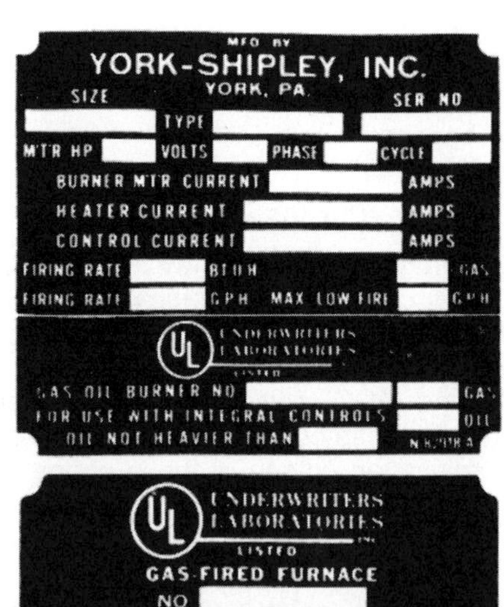

Figure 2801.1
EXAMPLE LABELS

Bibliography

The following resource material is referenced in this chapter or is relevant to the subject matter addressed in this chapter.

IFGC-12, *International Fuel Gas Code*. Washington, DC: International Code Council, 2011.

IMC-12, *International Mechanical Code*. Washington, DC: International Code Council, 2011.

Chapter 29: Plumbing Systems

General Comments

Determining the exact number of plumbing fixtures needed by the maximum number of possible users in any given building continues to be a challenging problem for the plumbing industry. Various methods have been used to evaluate the number and type of plumbing fixtures needed for an occupancy. Studies of office buildings have produced design guidelines based on occupancy times, arrival rates and patterns of fixture use that offer insight into the number of required plumbing fixtures for a desired level of service. The Building Technology Research Laboratory at the Stevens Institute of Technology conducted a study based on the Queueing Theory, which involves probabilities for waiting times and fixture use times based upon a preferred level of service. The study resulted in design guidelines for the quantities of fixtures needed in public toilet facilities. For residential-type buildings and health care facilities, plumbing fixture numbers are based on the minimum need, resulting in at least one water closet and one lavatory for each dwelling unit, guestroom or hospital room.

Studies completed by the U.S. military have been used for dormitories and prisons to determine the number of fixtures required based on a simultaneous need in a regimented society. This assumes that everyone rises at approximately the same time and has a limited amount of time to shower and use the water closet and lavatory.

The National Restaurant Association conducted a study based on the difference in use between a restaurant and a nightclub. It should be noted that the study did not take into account today's fast-food-style restaurant, nor did it allow for restaurants located along heavily traveled routes, such as those at highway rest stops or oases.

Fixture requirements for factory and industrial uses are based on the same requirements as for storage facilities. This method establishes a realistic minimum requirement for factory occupancies. The reasonableness in the number of plumbing fixtures was established through a limited study of factory projects in Henrico County, Virginia.

The fixture needs for the remaining occupancies were determined based on empirical data, experience and tradition. There are no exact studies providing values that are definitively supportable; the values have been modified periodically based on general observations. Various studies will most likely continue to be performed in order to develop a completely rational method for determining the minimum number of plumbing fixtures for any occupancy.

Purpose

The purpose of Chapter 29 is to provide a building with the necessary number of plumbing fixtures of a specific type and quality. The fixtures must be properly installed to be usable by the individuals occupying the building. The quality and design of every fixture must be in accordance with the *International Plumbing Code®* (IPC®).

SECTION 2901 GENERAL

[P] 2901.1 Scope. The provisions of this chapter and the *International Plumbing Code* shall govern the erection, installation, *alteration*, repairs, relocation, replacement, *addition* to, use or maintenance of plumbing equipment and systems. Toilet and bathing rooms shall be constructed in accordance with Section 1210. Plumbing systems and equipment shall be constructed, installed and maintained in accordance with the *International Plumbing Code*. Private sewage disposal systems shall conform to the *International Private Sewage Disposal Code*.

❖ This section contains the scoping requirements for Chapter 29. Compliance with this chapter will result in a building or structure having adequate plumbing fixtures for the sanitary, hygienic, cleaning, washing and food preparation needs of the occupants. This section also refers the reader to Section 1210 of the code, which covers specific requirements about surface finishes within toilet and bathing rooms. The requirements of this chapter are duplicated from the requirements of the IPC, Chapter 4, for the convenience of the architect, who is often responsible for the layout and number of plumbing fixtures when creating the floor plan of a building. All sections of Chapter 29 of the code are under complete control by the IPC Development Committee, thus the "[P]" before each section number.

SECTION 2902 MINIMUM PLUMBING FACILITIES

[P] 2902.1 Minimum number of fixtures. Plumbing fixtures shall be provided for the type of occupancy and in the minimum number shown in Table 2902.1. Types of occupancies not shown in Table 2902.1 shall be considered individually by the *building official*. The number of occupants shall be determined by this code. Occupancy classification shall be determined in accordance with Chapter 3.

❖ This section requires that the type and number of plumbing fixtures be based on the use of an occupancy and the occupant load as determined for the means of egress for a building or space. Occupancy classifications (groups) are determined from Chapter 3 and the occupant load is determined from Section 1004. Table 2902.1 shows ratios indicating the maximum number of occupants that can be served by one fixture of each type. For example, "1 per 75" means one fixture will serve up to 75 occupants. These "ratios" are applied to the occupant load of the building or space in order to determine the total number of each required fixture type.

The "description" of plumbing fixture use determines which set of fixture ratios should be used for calculating plumbing fixture requirements. In most cases, the description of plumbing fixture use will match the occupancy classification. However, there are situations where a description of fixture use different than the occupancy classification might be a more reasonable alternative approach to determining the number of required plumbing fixtures. These alternative methods require approval by the code official in accordance with Section 104.11 of the code. This approach might be useful for certain educational and business facilities as illustrated in the following discussions.

Educational Facilities

Consider an educational facility (Grades 1-12) with a gymnasium with a stage. The code states that assembly areas that are accessory to Group E occupancies are not to be considered as a separate occupancy. Therefore, the number of water closets should be determined from Section 3 of Table 2902.1, which requires one water closet per 50 occupants. Because this gymnasium has a stage, a greater occupant density than for a gymnasium-only space must be chosen, as nonfixed chairs will most likely be set up for viewing on-stage activities. Therefore, for egress purposes, the gymnasium is considered to be an assembly area with an occupant density of 7 square feet (0.65 m^2) per person. The resulting occupant load number applied with the required Group E occupancy plumbing fixture ratio can result in an excessive number of plumbing fixtures for the building. An alternative approach is to consider using the plumbing fixture ratios in the fourth row of Table 2902.1 for the gymnasium space, since these ratios reflect how the space is intended to be used with respect to the chosen occupancy density. In other words, fixture ratios should be chosen to "agree" with the actual use of the space.

A question that is often asked about Group E occupancies is whether the concept of nonsimultaneous occupancy could be considered in order to further reduce the number of required plumbing fixtures. For example, where a school auditorium is occupied by only the students within the school, the students are either in the classroom or the auditorium, but not in both places at the same time. Therefore, why should each area require a number of fixtures as if both areas were occupied simultaneously? The code does not recognize a nonsimultaneous use concept for any building, as simultaneous use could easily occur. For example, while students are in the classrooms, a school auditorium could be used temporarily for blood drives, regional science fairs and town meetings. However, this is not to say that in some circumstances, a nonsimultaneous use could, in some way, be completely guaranteed such that the local code official might entertain a proposed reduction in the required number of fixtures for the building.

Business Facilities

Consider a barber college where the code classifies the entire building as a Group B occupancy. The building has several large assembly rooms where the intent is to have training sessions for large groups of students. Clearly, these areas are used for assembly and, therefore, the use of the fixture ratios in Section 1, Row 4 of Table 2902.1 could be proposed to the code official.

The choice of an occupancy use for the purposes of determining plumbing fixtures does not affect the occupancy group chosen for egress purposes. In other words, using a previous example, the school gymnasium with stage space may be chosen to be "assembly use" for plumbing fixture requirements, but the entire building is still a Group B occupancy for the purposes of egress.

Note that Section 1004.1.1 has an exception, which for the designer to present an "actual" occupancy load number to the code official for approval instead of the calculated load for the occupancy square footage. While this exception would allow a smaller occupancy load to be chosen for the purposes of determining the required number of plumbing fixtures, code officials must carefully consider and appropriately justify the reduction. The code official should consider the potential for the occupancy to be loaded with more persons than stated by the designer, future use by different tenants of the same occupancy classification and the difficulty of enforcement of the maximum occupancy load based upon the limited number of fixtures provided.

[P] TABLE 2902.1. See page 29-3.

❖ The brief wording in the "descriptions" column of Table 2902.1 is not intended to be complete or all-inclusive of all uses for a particular occupancy. However, the descriptions do identify a majority of the types of uses encountered in the design of most buildings. Because the descriptions narrow in on what row of ratios to use, it is logical to enter the table at the description column. The number, classification and occupancy columns are intended only for reference purposes. The requirements for type and number of plumbing fixtures are driven by the actual use of a building or space, not necessarily by the occupancy group classification.

Some use description rows have a different ratio for male and female water closets and, in some cases, lavatories. The larger ratio for female fixtures provides an "equality of fixture availability" in those particular occupancies. The occupancies described by these use descriptions have historically had long lines of females waiting to use toilet facilities while male facilities had no lines. The reasons for this include the following:

1. For a variety of social and physical reasons, women generally take a longer period of time to use the facilities;
2. Women outnumber men in the general population and this becomes especially evident in large groups of people; and

[P] TABLE 2902.1
MINIMUM NUMBER OF REQUIRED PLUMBING FIXTURES[a]
(See Sections 2902.2 and 2902.3)

No.	CLASSIFICATION	OCCUPANCY	DESCRIPTION	WATER CLOSETS (URINALS SEE SECTION 419.2 OF THE *INTERNATIONAL PLUMBING CODE*)		LAVATORIES		BATHTUBS/ SHOWERS	DRINKING FOUNTAINS[e, f] (SEE SECTION 410.1 OF THE *INTERNATIONAL PLUMBING CODE)*	OTHER
				MALE	FEMALE	MALE	FEMALE			
1	Assembly	A-1[d]	Theaters and other buildings for the performing arts and motion pictures	1 per 125	1 per 65	1 per 200		—	1 per 500	1 service sink
		A-2[d]	Nightclubs, bars, taverns, dance halls and buildings for similar purposes	1 per 40	1 per 40	1 per 75		—	1 per 500	1 service sink
			Restaurants, banquet halls and food courts	1 per 75	1 per 75	1 per 200		—	1 per 500	1 service sink
		A-3[d]	Auditoriums without permanent seating, art galleries, exhibition halls, museums, lecture halls, libraries, arcades and gymnasiums	1 per 125	1 per 65	1 per 200		—	1 per 500	1 service sink
			Passenger terminals and transportation facilities	1 per 500	1 per 500	1 per 750		—	1 per 1,000	1 service sink
			Places of worship and other religious services	1 per 150	1 per 75	1 per 200		—	1 per 1,000	1 service sink
		A-4	Coliseums, arenas, skating rinks, pools and tennis courts for indoor sporting events and activities	1 per 75 for the first 1,500 and 1 per 120 for the remainder exceeding 1,500	1 per 40 for the first 1,520 and 1 per 60 for the remainder exceeding 1,520	1 per 200	1 per 150	—	1 per 1,000	1 service sink
		A-5	Stadiums, amusement parks, bleachers and grandstands for outdoor sporting events and activities	1 per 75 for the first 1,500 and 1 per 120 for the remainder exceeding 1,500	1 per 40 for the first 1,520 and 1 per 60 for the remainder exceeding 1,520	1 per 200	1 per 150	—	1 per 1,000	1 service sink

(continued)

3. Women, in general, tend to use the facilities more frequently.

The term "potty parity" was coined in a general sense to mean that a sufficient number of female plumbing fixtures were available such that women did not have to wait any longer that men (in a queue or line) to use an equivalent type of fixture as men. In a specific sense, "potty parity" also indicates the ratio of the number of male to the number of female fixtures of the same type. For example, if 24 male and 48 female water closets are installed in a building, then the potty parity is stated to be "1 to 2."

The code is silent concerning the installation of child-size water closets. In a child care facility having children age 6 years old and under, the provision of child-sized water closets (and sinks mounted at child

[P] TABLE 2902.1—(continued)
MINIMUM NUMBER OF REQUIRED PLUMBING FIXTURES[a]
(See Sections 2902.2 and 2902.3)

No.	CLASSIFICATION	OCCUPANCY	DESCRIPTION	WATER CLOSETS (URINALS SEE SECTION 419.2 OF THE *INTERNATIONAL PLUMBING CODE*)		LAVATORIES		BATHTUBS/ SHOWERS	DRINKING FOUNTAINS[e, f] (SEE SECTION 410.1 OF THE *INTERNATIONAL PLUMBING CODE)*	OTHER
				MALE	FEMALE	MALE	FEMALE			
2	Business	B	Buildings for the transaction of business, professional services, other services involving merchandise, office buildings, banks, light industrial and similar uses	1 per 25 for the first 50 and 1 per 50 for the remainder exceeding 50		1 per 40 for the first 80 and 1 per 80 for the remainder exceeding 80		—	1 per 100	1 service sink[g]
3	Educational	E	Educational facilities	1 per 50		1 per 50		—	1 per 100	1 service sink
4	Factory and industrial	F-1 and F-2	Structures in which occupants are engaged in work fabricating, assembly or processing of products or materials	1 per 100		1 per 100		See Section 411 of the *International Plumbing Code*	1 per 400	1 service sink
5	Institutional	I-1	Residential care	1 per 10		1 per 10		1 per 8	1 per 100	1 service sink
		I-2	Hospitals, ambulatory nursing home care recipient[b]	1 per per room[c]		1 per per room[c]		1 per 15	1 per 100	1 service sink
			Employees, other than residential care[b]	1 per 25		1 per 35		—	1 per 100	—
			Visitors, other than residential care	1 per 75		1 per 100		—	1 per 500	—
		I-3	Prisons[b]	1 per cell		1 per cell		1 per 15	1 per 100	1 service sink
		I-3	Reformatories, detention centers and correctional centers[b]	1 per 15		1 per 15		1 per 15	1 per 100	1 service sink
			Employees[b]	1 per 25		1 per 35		—	1 per 100	—
		I-4	Adult day care and child day care	1 per 15		1 per 15		1	1 per 100	1 service sink
6	Mercantile	M	Retail stores, service stations, shops, salesrooms, markets and shopping centers	1 per 500		1 per 750		—	1 per 1,000	1 service sink[g]

(continued)

height) could be beneficial to a facility's operation. Generally, for children over 6 years old, standard-sized fixtures are suitable because most children have learned to use the same size fixtures in a home setting. Note that the accessibility standard, ICC/ ANSI A117.1 (as referenced by Chapter 11), has specific requirements for accessible child-size water closets.

Some school facilities are designed to have a single-occupant toilet room that can be accessed only from within the classroom. The code is silent on whether the fixtures in this limited-access toilet room can be counted toward the required number of plumbing fixtures for the building.

The intent of table Note d concerning outdoor seating and entertainment areas is to require that outdoor patios, decks, balconies, beer gardens and similar areas, whether those areas have seating or not, be included when calculating the number of plumbing fixtures. While seating does provide a definitive occu-

[P] TABLE 2902.1—continued
MINIMUM NUMBER OF REQUIRED PLUMBING FIXTURES[a]
(See Sections 2902.2 and 2902.3)

No.	CLASSIFICATION	OCCUPANCY	DESCRIPTION	WATER CLOSETS (URINALS SEE SECTION 419.2 OF THE *INTERNATIONAL PLUMBING CODE*)		LAVATORIES		BATHTUBS OR SHOWERS	DRINKING FOUNTAINS[e, f] (SEE SECTION 410.1 OF THE *INTERNATIONAL PLUMBING CODE*)	OTHER
				MALE	FEMALE	MALE	FEMALE			
7	Residential	R-1	Hotels, motels, boarding houses (transient)	1 per sleeping unit		1 per sleeping unit		1 per sleeping unit	—	1 service sink
		R-2	Dormitories, fraternities, sororities and boarding houses (not transient)	1 per 10		1 per 10		1 per 8	1 per 100	1 service sink
		R-2	Apartment house	1 per dwelling unit		1 per dwelling unit		1 per dwelling unit	—	1 kitchen sink per dwelling unit; 1 automatic clothes washer connection per 20 dwelling units
		R-3	One- and two-family dwellings	1 per dwelling unit		1 per 10		1 per dwelling unit	—	1 kitchen sink per dwelling unit; 1 automatic clothes washer connection per dwelling unit
		R-3	Congregate living facilities with 16 or fewer persons	1 per 10		1 per 10		1 per 8	1 per 100	1 service sink
		R-4	Congregate living facilities with 16 or fewer persons	1 per 10		1 per 10		1 per 8	1 per 100	1 service sink
8	Storage	S-1 S-2	Structures for the storage of goods, warehouses, storehouses and freight depots, low and moderate hazard	1 per 100		1 per 100		See Section 411 of the *International Plumbing Code*	1 per 1,000	1 service sink

a. The fixtures shown are based on one fixture being the minimum required for the number of persons indicated or any fraction of the number of persons indicated. The number of occupants shall be determined by this code.

b. Toilet facilities for employees shall be separate from facilities for inmates or care recipients.

c. A single-occupant toilet room with one water closet and one lavatory serving not more than two adjacent patient sleeping units shall be permitted where such room is provided with direct access from each patient sleeping unit and with provisions for privacy.

d. The occupant load for seasonal outdoor seating and entertainment areas shall be included when determining the minimum number of facilities required.

e. The minimum number of required drinking fountains shall comply with Table 2902.1 and Chapter 11.

f. Drinking fountains are not required for an occupant load of 15 or fewer.

g. For business and mercantile occupancies with an occupant load of 15 or fewer, service sinks shall not be required.

pant load for these types of areas, there are many situations where these areas are not provided with seating in order to accommodate as many customers as possible. Therefore, the occupant load must be based upon standing space density. This occupancy load number must be applied to the ratios in the appropriate use description row in Table 2902.1. For example, a restaurant has a fixed seating area with fixed booths and an outdoor entertainment area that has no seating or tables. The fixed seating area load is simply a count of the number of seats according to Section 1004.7. The restaurant fixture ratios (Section 1, Row 3 of Table 2902.1) are then applied to this load to determine the number of plumbing fixtures for the fixed seating area. The outdoor area requires a different approach as it is a standing space area, not fixed seating. The occupant load is calculated from the area using Table 1004.1.1. Even though the building is a restaurant, this outdoor entertainment area might be used, not as a restaurant, but as a nightclub/dance floor. Therefore, the fixture ratios in Row 2 of Table 2902.1 (for nightclubs/dance halls) are applied to the standing occupant load. Using Row 3 fixture ratios for this area would be inappropriate. The use of proper occupant loads and use ratios will ensure that an adequate number of fixtures are provided where the occupant load is substantially increased by the addition of outdoor seating or standing areas.

Choosing the proper use description for a restaurant having a bar or for a nightclub serving food can be challenging. Some restaurant operations evolve into a nightclub/dance hall in the late evening. Bars with dance floors often have kitchens for preparing food to serve to patrons at the bar. The code is silent on how to determine which use description (restaurant or night club) is the best "fit" for the demand on the toilet facilities. The answer to this question lies in determining what use is primary. Is the bar's main purpose to support the restaurant operation? Or is the serving of food in support of the bar operation? In situations where there is no clear answer, perhaps considering the occupancy as a mixed-use arrangement is a practical answer.

The drinking fountain ratio indicated for an occupancy use description does not necessarily imply that every occupancy must have a drinking fountain within that occupancy. Just as toilet facilities are only required to be provided for an occupancy and available when the occupancy is in use, the provision of a drinking fountain follows similar logic. For example, consider a strip center with multiple-tenant spaces, each tenant space having the required number of plumbing fixtures for and within each space. The drinking fountain for that space (and all the adjacent spaces) could be located exterior to all the tenant spaces in a common area. The calculation for the required number of drinking fountains to be used by multiple occupancies is performed in the same manner as for other plumbing fixtures. See the Sample Problems in this commentary section.

The "service sinks" required by Table 2902.1 are intended to be of a type suitable for janitorial and building maintenance purposes. Service sinks include mop sinks/basins, utility tub/sinks, janitor sinks, slop sinks and laundry trays. In all occupancy classifications, except Groups 1-2, R-2 and R-3, only one service sink is required for the entire building, except that hospitals are required to have a service sink on each floor. The one service sink per building must be available from all portions of the building. The intent of the code is that employees and tenants of the building be able to access the required sink. This does not mean that the service sink is required to be in an open area of the building for use by any occupant (e.g., visitors). The service sink can be located in a locked janitor's closet for which all employees have access by key or door lock code to gain entry. However, the required service sink cannot be located in areas that are access-controlled by a single tenant. For example, if the service sink is located on floor one within a tenant's space, it would be possible for the tenant of floor one to lock out (exclude) other tenants from accessing the required service sink. Therefore, that service sink within one tenant-controlled space cannot be the required service sink for the building. For the 2012 edition, Note g was added to the table, which does not require service sinks in business and mercantile occupancies with an occupant load of 15 or fewer persons. Service sinks in these small occupancies are rarely used.

Although the code classifies parking garages as a Group S occupancy, there is no entry in Table 2902.1 for parking structures. It is not the intent of the code to require public toilet facilities in parking garages. A parking garage is similar to a parking lot in that it is not normally the final destination of a person parking a car. The only reason a person is in a parking garage is to park his or her car and then leave. A parking lot, having no enclosing structure, is not required to be provided with toilet facilities for the occupants that are parking their vehicles, so it stands to reason that a parking garage would not be required to be provided with toilet facilities. When people drive into a parking garage, their normal expectation is that the toilet facilities will be located in the building or buildings served by the parking garage. If public toilet facilities were provided in a parking garage, significant maintenance, vandalism and security issues would probably result in the eventual closure of such facilities. Not requiring public toilet facilities in a parking garage does not provide justification for not providing toilet facilities for parking garage employees (such as ticket kiosk at-tendants). These employees must be provided with toilet facilities in accordance with Section 2902.3. However, these facilities are not required to be located in the parking garage but only within the travel limitations of Section 2902.3.2.

Note e clarifies the requirements for drinking fountain accessibility in accordance with Chapter 11. An occupancy that is required to have a minimum of one drinking fountain is required to have either a unit that has two nozzles, one high and one low, or two separate units, one with a high spout and one with a low spout. The accessibility standard ICC/ANSI A117.1 specifies the elevation range for the spouts for both the high and low spout drinking fountains.

Note f allows occupancies with 15 or fewer occupants not to have a drinking fountain. The requirement for drinking fountains in all occupancies put an undue hardship on small establishments where the floor space is usually limited. Adding to this problem is the requirement in Chapter 11 stating that where one drinking fountain is provided, two drinking fountains must be installed, one high (for standing persons) and one low (for wheelchair-seated persons). One required drinking fountain in a small establishment turned into an expensive, space-consuming fixture that, according to the proponents of the change, rarely, if ever, is used.

For the 2012 edition, the term "patient" in the description for the Group I-2 occupancy was replaced by "care recipient" to be in alignment with how other regulations currently designate those who receive care in a hospital or nursing home setting.

[P] 2902.1.1 Fixture calculations. To determine the *occupant load* of each sex, the total *occupant load* shall be divided in half. To determine the required number of fixtures, the fixture ratio or ratios for each fixture type shall be applied to the *occupant load* of each sex in accordance with Table 2902.1. Fractional numbers resulting from applying the fixture ratios of Table 2902.1 shall be rounded up to the next whole number. For calculations involving multiple occupancies, such fractional numbers for each occupancy shall first be summed and then rounded up to the next whole number.

> **Exception:** The total *occupant load* shall not be required to be divided in half where *approved* statistical data indicate a distribution of the sexes of other than 50 percent of each sex.

❖ This section requires that the occupant load be divided by 2 in order to determine the quantities of males and females. This is known as a 50-50 gender distribution, meaning that 50 percent of the occupants are male and 50 percent of occupants are female. While a 50-50 gender distribution is used most of the time, there might be situations where this assumption is inaccurate. The exception to this section allows for a different gender distribution if statistical data is submitted and approved by the code official for an occupancy. Example occupancies that might require a modified distribution would be an all-women's health club, a males-only boarding school, a convent or a monastery.

Once the appropriate use description row in Table 2902.1 is chosen, the ratios for each fixture type and for each gender are used to determine the minimum number of plumbing fixtures for a building or space.

Although not specifically indicated in this code section, prior to performing any plumbing fixture calculations for an occupancy, the exceptions of Section 2902.2 must be reviewed for applicability. For example, a small restaurant could have two employees and 12 seats. If the fixture calculations were performed without consideration of Section 2902.2, this restaurant would be required to have a toilet facility for each gender. However, because the occupancy load is less than 15, a single toilet facility is all that is required.

The following sample problems illustrate calculation methods for various situations.

Sample Problem 1: A mixed-use building has a business-use occupant load of 538, a library-use occupant load of 115 and a storage-use occupant load of 82. The occupancies are not separate tenant spaces. Determine the total number of required plumbing fixtures for the building.

Problem Approach

Check Section 2902.2 for applicability and if not applicable, proceed with calculations.

Apply the gender distribution for each occupancy use, calculate the number of fixtures required by each gender, sum the gender totals for each type of fixture from all occupancies and round up the number of each type of fixture for each gender to a whole number. Note that drinking fountains are not gender specific. Therefore, the total number of occupants for the occupancy is applied to the drinking fountain fixture ratio.

Solution—Part I

For the business-use occupancy, find the required ratios in Table 2902.1, Section 2. Because a gender distribution was not specified, the number of occupants must be divided by 2 to obtain the number of occupants for each gender (refer to Section 2902.1.1). Therefore, in this business-use occupancy, there will be:

538/2 = 269 occupants of each gender

1. Calculate the number of water closets for 269 males:

   For the first 50 males, the ratio of 1 per 25 is applied:

   50 × 1/25 = 2 water closets

   For the remaining number of males, the ratio of 1 per 50 is applied:

   (269 - 50) × 1/50 = 219 × 1/50 = 4.38 water closets

   The required number of water closets for males is:

   2 + 4.38 = 6.38 water closets for males

   Because the gender distribution is equal and the water closet ratio for females is the same as for males, the number of water closets for females is also 6.38.

2. Calculate the number of lavatories for 269 males:

   For the first 80 males, the ratio of 1 per 40 is applied:

   80 × 1/40 = 2 lavatories

   For the remaining males, the ratio of 1 per 80 is applied:

   (269 - 80) × 1/80 = 189 × 1/80 = 2.36 lavatories.

   The required number of lavatories for males is:

   2 + 2.36 = 4.36 lavatories for males

   Because the gender distribution is equal and the lavatory ratio for females is the same as for males, the number of lavatories for females is also 4.36.

3. Calculate the number of drinking fountains for 538 occupants:

   The ratio of 1 per 100 is applied:

   538 × 1/100 = 5.38 drinking fountains for male and female use

Solution—Part II

For the library-use occupancy, find the required ratios in Table 2902.1, Section 1. Because a gender distribution was not specified, the number of occupants must be divided by 2 to obtain the number of occupants for each gender (refer to Section 2902.1.1). Therefore, in this library-use occupancy, there will be:

115/2 = 57.5 occupants of each gender

1. Calculate the number of water closets for males: The ratio of 1 per 125 is applied:

   57.5 × 1/125 = 0.46 water closets for males

2. Calculate the number of water closets for females:

   The ratio of 1 per 65 is applied:

   57.5 × 1/65 = 0.88 water closets for females

3. Calculate the number of lavatories for 57.5 males:

   The ratio of 1 per 200 is applied:

   57.5 × 1/200 = 0.29 lavatories for males

   Because the gender distribution is equal and the lavatory ratio for females is the same as for males, the number of lavatories for females is also 0.29.

4. Calculate the number of drinking fountains for 115 occupants:

   The ratio of 1 per 500 is applied:

   115 × 1/500 = 0.23 drinking fountains of both males and females.

Solution—Part III

For the storage-use occupancy, find the required ratios in Table 2902.1, Section 8. Because a gender distribution was not specified, the number of occupants must be divided by 2 to obtain the number of occupants for each gender (refer to Section 2902.1.1). Therefore, in this storage-use occupancy, there will be:

82/2 = 41 occupants of each gender

1. Calculate the number of water closets for 41 males:

   The ratio of 1 per 100 is applied:

   41 × 1/100 = 0.41 water closets for males

Table 2902.1.1(1)
SOLUTION SUMMARY FOR SAMPLE PROBLEM 1

OCCUPANCY		WATER CLOSETS				LAVATORIES			DF RATIO	DRINKING FOUNTAINS	SERVICE SINK
USE	LOAD	RATIO	MALE	RATIO	FEMALE	RATIO	MALE	FEMALE			
Business	538	1 per 25 for the first 50 and 1 per 50 for the remainder exceeding 50	6.38	1 per 25 for the first 50 and 1 per 50 for the remainder exceeding 50	6.38	1 per 40 for the first 80 and 1 per 80 for the remainder exceeding 80	4.36	4.36	1 per 100	5.38	
Libraries, halls, museums, etc.	115	1/125	0.46	1/65	0.88	1/200	0.29	0.29	1 per 500	0.23	Note 1
Storage	82	1/100	0.41	1/100	0.41	1/100	0.41	0.41	1/1000	0.082	
Subtotals			7.25		7.67		5.06	5.06		5.69	
Required totals			8		8		6	6		6	

Note:
1. The code requires only one service sink per building if all occupancies have access to the service sink at all times.

Because the gender distribution is equal and the water closet ratio for females is the same as for males, the number of water closets for females is also 0.41.

2. Calculate the number of lavatories for 41 males:

   The ratio of 1 per 100 is applied:

   41 × 1/100 = 0.41 lavatories for males

   Because the gender distribution is equal and the lavatory ratio for females is the same as for males, the number of lavatories for females is also 0.41.

3. Calculate the number of drinking fountains for 82 occupants:

   The ratio of 1 per 1000 is applied:

   82 × 1/1000 = 0.082 drinking fountains for males and females.

Solution—Summary

See Commentary Table 2902.1(1) for the solution summary for Sample Problem 1. Note that if the raw fixture numbers were rounded up prior to summation for the building, the calculated minimum fixture requirements would have been increased by one water closet for each gender, one lavatory for each gender and two drinking fountains.

Sample Problem 2: A mixed-use building has a business-use occupant load of 940 and a restaurant-use occupant load of 360. The occupancies are separate tenant spaces. The toilet facilities are to be located in each tenant space. For the business occupancy, bottled water coolers are to be substituted for drinking fountains to the maximum allowable extent. Determine the total number of required plumbing fixtures for the building.

Problem Approach

Check Section 2902.2 for applicability and if not applicable, proceed with calculations.

Note that the plumbing fixtures required by each occupancy will be located in each tenant space. Thus, the required number of plumbing fixtures for each tenant space is calculated as if the space stood alone, in other words, in its own building. Drinking fountain substitution is calculated last.

Solution—Part I

For the business-use occupancy, find the required ratios in Table 2902.1, Section 2. Because a gender distribution was not specified, the number of occupants must be divided by 2 to obtain the number of occupants for each gender (refer to Section 2902.1.1). Therefore, in this business-use occupancy, there will be:

940/2 = 470 occupants of each gender

1. Calculate the number of water closets for 470 males:

   For the first 50 males, the ratio of 1 per 25 is applied:

   50 × 1/25 = 2 water closets

Table 2902.1.1(2)
SOLUTION SUMMARY FOR SAMPLE PROBLEM 2

OCCUPANCY		WATER CLOSETS				LAVATORIES			DF RATIO	DRINKING FOUNTAINS	SERVICE SINK
USE	LOAD	RATIO	MALE	RATIO	FEMALE	RATIO	MALE	FEMALE			
Business	940	1 per 25 for the first 50 and 1 per 50 for the remainder exceeding 50	10.4	1 per 25 for the first 50 and 1 per 50 for the remainder exceeding 50	10.4	1 per 40 for the first 80 and 1 per 80 for the remainder exceeding 80	6.88	6.88	1 per 100	9.4	Note 1
Business required totals			11		11		7	7		10	
Restaurants, banquet halls and food courts	360	1 per 75	2.4	1 per 75	2.4	1 per 200	0.9	0.9	1 per 500	0.72	
Restaurant required totals			3		3		1	1		1	
Building required totals			14		14		8	8		11	

Note:

1. The code only requires one service sink per building if all occupancies have access at all times. If this building has a central area (such as a janitorial closet for the building) that both tenants can access at all times, only one service sink is required for the building. But if the service sink is located in either one of the tenant spaces, one tenant would not have access; therefore, the building would require two service sinks—one for each tenant.

For the remaining number of males, the ratio of 1 per 50 is applied:

(470 - 50) × 1/50 = 420 × 1/50 = 8.4 water closets

The required number of water closets for males is:

2 + 8.4 = 10.4 water closets for males

Because the gender distribution is equal and the water closet ratio for females is the same as for males, the number of water closets for females is also 10.4.

2. Calculate the number of lavatories for 470 males:

For the first 80 males, the ratio of 1 per 40 is applied:

80 × 1/40 = 2 lavatories

For the remaining males, the ratio of 1 per 80 is applied:

(470 - 80) × 1/80 = 390 × 1/80 = 4.88 lavatories.

The required number of lavatories for males is:

2 + 4.88= 6.88 lavatories for males

Because the gender distribution is equal and the lavatory ratio for females is the same as for males, the number of lavatories for females is also 6.88.

3. Calculate the number of drinking fountains for 940 occupants:

The ratio of 1 per 100 is applied:

940 × 1/100 = 9.4 drinking fountains for males and females

Solution—Part II

For the restaurant-use occupancy, find the required ratios in Table 2902.1, Section 1. Because a gender distribution was not specified, the number of occupants must be divided by 2 to obtain the number of occupants for each gender (refer to Section 2902.1.1). Therefore, in this restaurant-use occupancy, there will be:

360/2 = 180 occupants of each gender

1. Calculate the number of water closets for males:

The ratio of 1 per 75 is applied:

180 × 1/75 = 2.4 water closets for males

Therefore, 2.4 male water closets are needed.

Because the gender distribution is equal and the water closet ratio for females is the same as for males, the number of water closets for females is also 2.4.

2. Calculate the number of lavatories for 180 males:

The ratio of 1 per 200 is applied:

180 × 1/200 = 0.9 lavatories for males are required

Therefore, 0.9 male lavatories are needed.

Because the gender distribution is equal and the lavatory ratio for females is the same as for males, the number of lavatories for females is also 0.9.

3. Calculate the number of drinking fountains for 360 occupants:

The ratio of 1 per 500 is applied:

360 × 1/500 = 0.72 drinking fountains of both males and females.

Solution—Part III

The number of required drinking fountains is allowed to be reduced by up to 50 percent by substitution with bottled water coolers or bottled water dispensers (see IPC Section 410.1 and related commentary). In restaurants, drinking fountains are not required if drinking water is served. Chapter 11 requires that any drinking fountains that are provided must be accessible. Where only one drinking fountain is being provided, either two installed at different heights or one combination unit with two bowls (each with a spout) at different heights must be installed. (The spout height requirements are specified in ICC/ANSI A117.1-2003.)

Bottled water cooler (or dispenser) substitution might not be appropriate for applications where located in areas that are completely open to the public. For example, a bottled water cooler to substitute for one of three required drinking fountains that are to be placed outdoors at a large strip center would most likely be subject to vandalism. However, a bottled water cooler substituted for one of three required drinking fountains in a large office building (a more controlled environment) would not be inappropriate. Where this chapter requires two drinking fountains, the accessibility regulations of Chapter 11 require two drinking fountains, one with a high spout and one with a low spout. The following example illustrates how the accessibility requirements for drinking fountains relate to the plumbing code requirements for drinking fountains and bottled water cooler substitution.

Consider an occupancy requiring 10 drinking fountains. Bottled water dispensers can substitute for five of the 10 required drinking fountains. Of the five drinking fountains to be installed, two are mounted at the high spout level, two are mounted at the low spout level and the third can be mounted at either the high or low spout level. A bottled water cooler is not a drinking fountain. While the code provides accessibility requirements for drinking fountains, it does not address bottled water coolers.

Solution—Summary

See Commentary Table 2902.1(2) for the solution summary for Sample Problem 2. Note that if the raw fixture numbers were rounded prior to summation for the building, the calculated minimum fixture requirements would have increased by one water closet for each gender, one lavatory for each gender and two drinking fountains.

Sample Problem 3: A stadium has an occupant load of 5,200. Urinals are to be substituted for male water closets to the maximum extent allowed. Determine the total number of required plumbing fixtures for the stadium.

Problem Approach

Check Section 2902.2 for applicability and if not applicable, proceed with calculations.

Apply the gender distribution, calculate the number of fixtures required by each gender, sum the gender totals for each type of fixture and round up the number of each type of fixture for each gender to a whole number. Apply the urinal substitution (see Section 419.2 of the IPC) to reduce the number of male water closets.

Solution

For the stadium-use occupancy, find the required ratios in Table 2902.1, Row 8. Because a gender distribution was not specified, the number of occupants must be divided by 2 to obtain the number of occupants for each gender (refer to Section 2902.1.1). Therefore, in this stadium-use occupancy, there will be:

5,200/2 = 2,600 occupants of each gender

1. Calculate the number of water closets for 2,600 males:

   For the first 1,500 males, the ratio of 1 per 75 is applied:

   1,500 × 1/75 = 20 water closets

   For the remaining number of males, the ratio of 1 per 120 is applied:

   (2,600 - 1,500) × 1/120 = 1,100 × 1/120 = 9.17 water closets

   The required number of water closets for males is:

   20 + 9.17 = 29.17 → Round to 30 water closets for males

2. Calculate the number of water closets for 2,600 females:

   For the first 1,520 females, the ratio of 1 per 40 is applied:

   1,520 × 1/40 = 38 water closets

   For the remaining number of females, the ratio of 1 per 60 is applied:

   (2,600 - 1,520) × 1/60 = 1,080 × 1/60 = 18 water closets.

   The required number of water closets for females is:

   38 + 18 = 56 water closets for females.

3. Calculate the number of lavatories for 2,600 males:

   The ratio of 1 per 200 is applied:

   2,600 × 1/ 200 = 13 lavatories for males

4. Calculate the number of lavatories for 2,600 females:

   The ratio of 1 per 150 is applied:

   2,600 × 1/150 = 17.3 lavatories → Round to 18 lavatories for females

Table 2902.1.1(3)
SOLUTION SUMMARY FOR SAMPLE PROBLEM 3

OCCUPANCY		WATER CLOSETS				LAVATORIES				DRINKING FOUNTAINS	
USE	LOAD	MALE RATIO	MALE	FEMALE RATIO	FEMALE	MALE RATIO	MALE	FEMALE RATIO	FEMALE	RATIO	ALL GENDERS
Stadiums	5,200	1 per 75 for the first 1,500 and 1 per 120 for the remainder exceeding 1,500	29.17	1 per 40 for the first 1,520 and 1 per 60 for the remainder exceeding 1,520	56	1 per 200	13	1 per 150	17.3	1 per 1,000	5.3
Required totals			30		56		13		18		6

Required totals with urinal substitution	URINALS	MALE W/C	FEMALE W/C	MALE LAV	FEMALE LAV	DRINKING FOUNTAINS	SERVICE SINK
	20	10	56	13	18	6	1

5. Calculate the number of drinking fountains for 5,200 occupants:

   The ratio of 1 per 1,000 is applied:

   5,200 × 1/1,000 = 5.2 → Round to 6 drinking fountains for males and females

6. Calculate the number of urinals to be substituted for male water closets. Because a stadium is in the assembly classification, 67 percent of the water closets can be replaced with urinals.

   The ratio of 67/100 is applied:

   30 × 67/100 = 20.1 urinals → Round down to 20 urinals because rounding up would result in the number of urinals being in excess of 67 percent of the required number of water closets.

Solution—Summary

See Commentary Table 2902.1.1(3) for the solution summary for Sample Problem 3.

[P] 2902.1.2 Family or assisted-use toilet and bath fixtures. Fixtures located within family or assisted-use toilet and bathing rooms required by Section 1109.2.1 are permitted to be included in the number of required fixtures for either the male or female occupants in assembly and mercantile occupancies.

❖ A family or assisted-use toilet room is required by the code in assembly or mercantile occupancies having an aggregate total of six or more required male and female water closets for the assembly or mercantile occupancy. The term "aggregate total" means the combined number of required male and female water closets before any urinal substitutions. A family or assisted-use bathing room is required by the code in recreational facilities having separate sex bathing rooms except where the separate sex bathing rooms only have one shower or fixture. Family or assisted-use toilet (and bathing room) facilities are intended for use by individuals who need the assistance of a family member or care provider to properly use plumbing fixtures in a private and dignified manner. Fixtures located in family or assisted-use toilet or bathing rooms are allowed to reduce the number of required fixtures for the occupancy in either the male or female standard toilet rooms, but not both.

Sample Problem 4: A theater has an occupant load of 520. Urinals are to be substituted for male water closets to the maximum extent allowed. Determine the total number of plumbing fixtures that will be in the building.

Problem Approach

Apply the gender distribution, calculate the number of fixtures required by each gender, sum the gender totals for each type of fixture and round up the number of each type of fixture for each gender to a whole number. Consider the code requirement for a family or assisted-use toilet room. Apply the urinal substitution (IPC Section 419.2) to reduce the number of male water closets.

Solution

Calculation method is similar to Sample Problem 1—Part II.

Solution—Summary

See Commentary Tables 2902.1.1(4) and 2902.1.1(5) for the solution summary for Sample Problem 4.

Because a theater is a use description of Assembly and the aggregate total of required water closets is seven, one family or assisted-use toilet facility is required. The designer has the choice of reducing the number of fixtures in either the male or female toilet rooms by one lavatory and one water closet. In this case, the designer chose to reduce the number of required male fixtures. It may be more logical to reduce the male fixture count because of the greater need for fixtures in the female facilities.

The code is silent as to whether the fixtures in a voluntarily added family or assisted-use toilet facility can be counted toward the required number of fixtures in any occupancy. For example, in Sample Problem 3, perhaps the building owner chooses to have three family or assisted-use toilet facilities in addition to the required number of fixtures in the public toilet facilities. If the fixtures in these "extra" family or assisted-use toilet facilities are allowed to count toward the required number of fixtures, which gender toilet facility should have fixture counts reduced? Note that voluntarily added family or assisted-use toilet facilities should not count toward the required number of fixtures, no matter the occupancy, because the code does not specify a maximum reduction, and any reductions seem to go against the intent of the code to provide ample gender-specific toilet facilities.

Where the calculations require multiple fixtures for a gender, the code is silent as to whether those required fixtures can be provided in multiple single-occupant toilet rooms. For example, if two water closets for each gender are required, could four single-occupant toilet facilities (two labeled male, two labeled female) be provided? While the arrangement of multiple fixtures in one toilet room is cost-efficient and space saving, some business owners might perceive a "marketing advantage" for their business by having all single-occupant toilet facilities. For example, families with young children and elderly customers might be more attracted to a restaurant that has all single-occupant toilet facilities so that a family member can privately assist another family member using the facility. Or simply, some customers might prefer the extra privacy that a single-occupant toilet room offers.

The provision of the required number of plumbing fixtures in single-occupant toilet facilities should be discouraged because the ratios provided in Table

2902.1 never anticipated such arrangements. Users of single-occupant toilet rooms will generally take longer as they have the security of a locking door. Users in multiple-user toilet facility environments are "encouraged" to be quick due to the bustle of activity in the more open environment (people waiting outside stalls or standing in line at urinals). In other words, (four) single-use toilet facilities (marked two male and two female) do not provide the level of use efficiency as does a multiple-user public toilet facility. Even though the code doesn't discuss efficiency with regard to toilet facilities, the desired outcome of Table 2902.1 is to prevent long waiting lines and, thus, use efficiency is a significant factor.

Note that where multiple single-occupancy toilet rooms are provided, Chapter 11 requires that at least 50 percent of the single-occupancy toilet rooms within a cluster of toilet rooms must be of accessible design. If the single-occupant toilet rooms are not clustered but are located individually throughout a building or tenant space, all toilet rooms are required to be of accessible design. For example, four single-occupant toilet rooms, two labeled "men" and two labeled "women" are clustered together. Only one of each gender of those toilet rooms must be of accessible design. However, if the single-occupant toilet rooms are not grouped together in the building, all of those single-occupant toilet rooms must be of accessible design.

[P] 2902.2 Separate facilities. Where plumbing fixtures are required, separate facilities shall be provided for each sex.

Exceptions:

1. Separate facilities shall not be required for *dwelling units* and *sleeping units*.
2. Separate facilities shall not be required in structures or tenant spaces with a total *occupant load*, including both employees and customers, of 15 or less.
3. Separate facilities shall not be required in mercantile occupancies in which the maximum occupant load is 100 or less.

❖ Separate facilities are required for males and females. Exception 1 is redundant as Table 2902.1 already indicates the types of occupancies where one single-occupancy toilet room is sufficient. Exception 2 allows one single-occupant toilet room to be provided in small establishments because the space requirement and installation cost is considered to be a significant hardship for the minimal floor area that results in a 15-person occupant load. Exception 3 allows one single-occupant toilet room to be provided in what are considered to be small floor area mercantile establishments so as to not create a significant hardship in both space and installation cost. There is no specific documentation as to why the 15-person threshold was chosen for Exception 2 other than it was perceived to be reasonable for such small spaces. The 100-person threshold for Exception 3 (increased from 50 in the previous code edition) was justified because small mercantile occupancies [1,500 to 3,000 square feet (139.4 to 278.7 m^2)] were unduly burdened by the requirement for two accessible toilet rooms. For a 1,500-square-foot (139.4 m^2) tenant space, two accessible toilet rooms occupy more than 5 percent of the total square footage. In that it is rare that these small establishments will ever experience the full design occupant load, there is no need to burden the tenant with two toilet rooms.

The separate facilities requirement for males and females addresses two main concerns: privacy and safety. Users of toilet facilities often experience embarrassment if members of the opposite sex are in the same room. This increased embarrassment can lead to difficulty or prevention of waste elimination for many users. While some of these inhibitions are often temporarily "given up" under special conditions such

Table 2902.1.1(4)
SOLUTION SUMMARY FOR SAMPLE PROBLEM 4

OCCUPANCY		WATER CLOSETS				LAVATORIES				DRINKING FOUNTAINS	
USE	LOAD	MALE RATIO	MALE	FEMALE RATIO	FEMALE	MALE RATIO	MALE	FEMALE RATIO	FEMALE	RATIO	ALL GENDERS
Theater	520	1 per 125	2.08	1 per 65	4	1 per 200	1.3	1 per 200	1.3	1 per 500	1.04
Required totals			3		4		2		2		2

Table 2902.1.1(5)
BUILDING PLUMBING FIXTURE TOTALS FOR SAMPLE PROBLEM 4

Theater building totals with urinal substitution	MALE TOILET FACILITY			FEMALE TOILET FACILITY		DRINKING FOUNTAINS	SERVICE SINK	FAMILY OR ASSISTED-USE	
	URINAL	W/C	LAV	W/C	LAV			W/C	LAV
	1	1	1	4	2	2	1	1	1

as in co-ed college dormitories or where one gender's facilities are inadequate for the demand, these inhibitions quickly return for most people as it is innate to the human species to desire privacy during waste elimination. In a public environment, safety for female users of toilet facilities is of paramount concern. While a female restroom placard is no barrier to those intent on harming female occupants, a female user confronted by (or even hearing) a male in the female-only toilet facility will immediately recognize the potential threat and take action. A toilet facility intended to be used by both sexes simultaneously does not offer the same level of immediate situational awareness for females and, therefore, it is not perceived to be as safe as a facility intended for one gender only. Also, some individuals may be inhibited to use a facility that is used by the opposite sex because of notions of cleanliness and perceived appropriateness.

[P] 2902.2.1 Family or assisted-use toilet facilities serving as separate facilities. Where a building or tenant space requires a separate toilet facility for each sex and each toilet facility is required to have only one water closet, two family/assisted-use toilet facilities shall be permitted to serve as the required separate facilities. Family or assisted-use toilet facilities shall not be required to be identified for exclusive use by either sex as required by Section 2902.4.

❖ Family or assisted-use toilets provide inherent potty parity. Two-family or assisted-use toilet rooms increase the overall availability of a toilet room when needed. A single gender-based toilet can be unavailable for periods of up to 15 minutes when, for example, the current occupant is using it for companion care, to change diapers or to change a colostomy bag. Also, there is less impact to availability when one toilet room is being cleaned or serviced. For decades, males and females have used the same toilet facility on airliners.

[P] 2902.3 Employee and public toilet facilities. Customers, patrons and visitors shall be provided with public toilet facilities in structures and tenant spaces intended for public utilization. The number of plumbing fixtures located within the required toilet facilities shall be provided in accordance with Section 2902.1 for all users. Employees shall be provided with toilet facilities in all occupancies. Employee toilet facilities shall either be separate or combined employee and public toilet facilities.

> **Exception:** Public toilet facilities shall not be required in open or enclosed parking garages. Toilet facilities shall not be required in parking garages where there are no parking attendants.

❖ Toilet facilities must be available for all public establishments that are used by persons engaged in activities involved with the purpose of the establishment. Public establishments include but are not limited to restaurants, nightclubs, theaters, offices, retail shops, stadiums, libraries and churches. Persons engaged in the activities of the establishment include, but are not limited to, buyers of merchandise, recipients of services, viewers of displays, receivers of information materials, employees and those persons in attendance with those engaging in the activities. The code is silent about whether the toilet facilities are for use by persons not engaged in the activities involved with the purpose of the establishment, such as passers-by, street maintenance workers or vagrants. However, because the number of plumbing fixtures for any establishment is based only upon either the square footage or number of seats of the establishment's space, the intent of the code is to serve only the people involved with the activities of the establishment. The photo of the front door of the restaurant shown in Figure 2902.3(1) might first appear to be in conflict with the requirements of this section; however, this sign is actually a promotion of the intent of this code section: "If you are engaged with the activities involved with the purpose of this establishment, toilet facilities are available for you."

The quantities of plumbing fixtures required for an establishment are specified in Section 2902 of the code. Employees of the establishment must also have access to the toilet facilities for the establishment whether those facilities are separate from or combined with the facilities for the other users. Except for the specific requirements of Group 1-2 occupancies, Group 1-3 occupancies and Note b of Table 2902.1, employee toilet facilities are not required to be separate from other toilet facilities.

The exception is new for the 2012 edition and clarifies that even though a parking structure is classified as Group S occupancy (storage structure) and Table 403.1 requires storage occupancies to have toilet facilities, parking structures are exempt from this requirement. Note that if a parking structure has attendants, they must be provided with toilet facilities within the travel limitations of the code.

Figure 2902.3(1)
RESTAURANT ENTRY SIGN

[P] 2902.3.1 Access. The route to the public toilet facilities required by Section 2902.3 shall not pass through kitchens, storage rooms or closets. Access to the required facilities shall be from within the building or from the exterior of the building. All routes shall comply with the accessibility requirements of this code. The public shall have access to the required toilet facilities at all times that the building is occupied.

❖ It is inappropriate and dangerous to locate toilet facilities intended for use by persons who are not employees of the establishment such that the path of travel is through kitchens, storage rooms or closets. This provision avoids placing occupants from the general public in an unfamiliar area of the building where they could be confused about the path of egress. Especially during an emergency situation, public users of the toilet facility will instinctively want to try to exit the building along the same path of travel used to access the toilet facilities which could be through an area where the emergency condition exists, such as a kitchen [see Figure 2902.3(2)]. Note that Section 1210.5 prohibits toilet rooms from directly opening into a room used for the preparation of food to be served to the public.

A common practice in many areas required that the toilet facilities were located on the outside of a building, such as is sometimes found in automotive service stations and convenience store buildings. Outlet mall complexes sometimes elect to have a central toilet facility building separate from the tenant spaces. While the code is silent as to whether the required toilet facilities are allowed to be in another building not under the direct control of the owner or tenant of the building from where the toilet user travels to the toilet facilities, the intent of the code is that public toilet facilities will be available (and open for use) when anyone is using a building or tenant space. However, it is unreasonable for toilet facilities in a building under different ownership or in a tenant space under different control to serve as the required toilet facilities for a building or tenant space.

Walk-up and drive-through service windows are unique applications with regard to requirement for toilet facilities. Where these establishments do not allow customers to enter the building, the code is silent on whether toilet facilities for the public are required. However, if facilities were required, how would one calculate the occupant loading to determine the required number of fixtures? Examples are ice cream shops with only a walk-up window, coffee shops with only a drive-up window and 24-hour fast food drive-throughs that close the building entrances to walk-in customers during late night/early morning hours. Where outdoor seating for customers is made available by these types of businesses, public toilet facilities for these walk-up/drive-through establishments should be available (see Figure 2902.3.1). If such outdoor seating areas can be "closed for public use" such as by a perimeter fence and locked entry gates, the seating area is not available for use and no requirement would exist to provide public toilet facilities.

Variations of the above service window applications are dry cleaners, pizza pick-ups and banks that provide service or products to customers from indoor counters or kiosks. Because the public must walk into these buildings for service, public toilet facilities are required. Obviously, these establishments are required to have toilet facilities for the employees; however, those facilities might not be properly located for public access. For example, if the location of the toilet facilities in a dry cleaning establishment

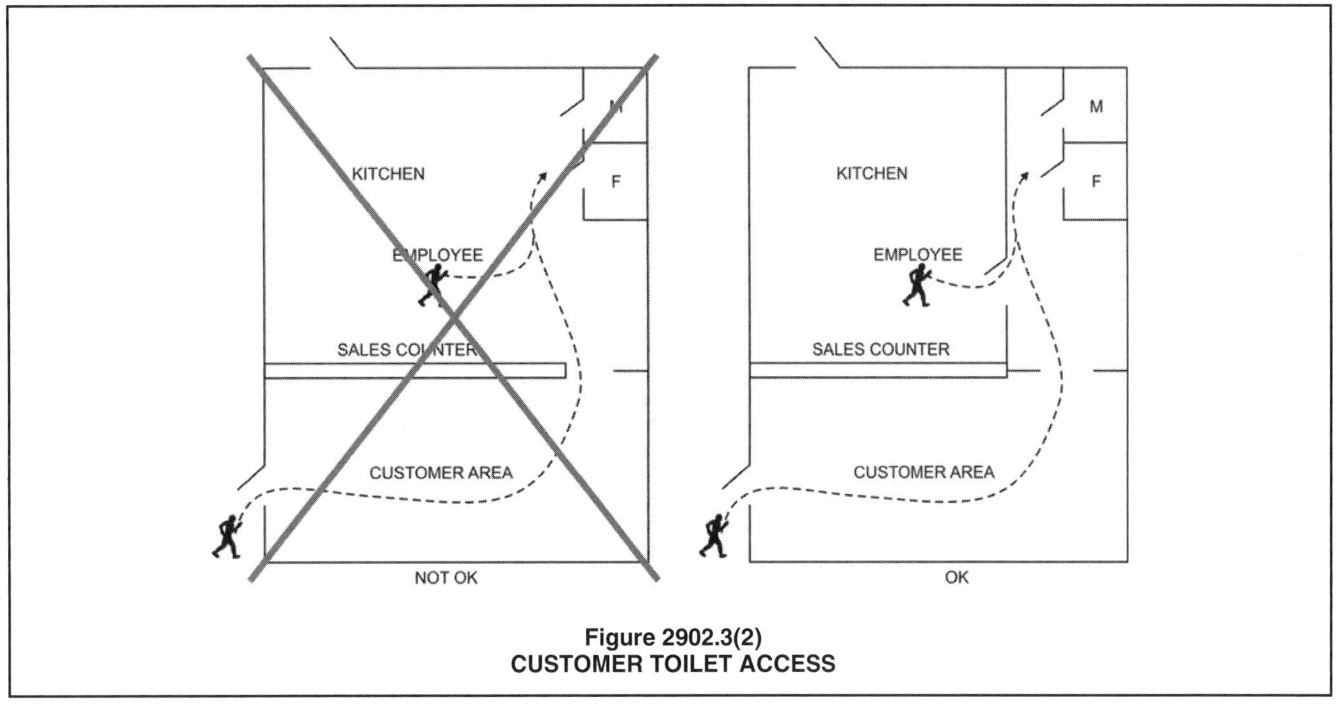

Figure 2902.3(2)
CUSTOMER TOILET ACCESS

requires that the public walk a convoluted path around machinery and hanging clothes racks, the user could become confused about how to safely exit the toilet facilities and return to where he/she started from. Public toilet facilities must be provided along an obvious and safe path of travel. Note that this path of travel must also meet the accessible route requirements for disabled persons. In the pizza pickup situation, because the public is not allowed to pass though a kitchen, the toilet facilities would have to be located on the customer side of the pickup counter. Banks (as well as check cashing establishments) do not have code restrictions for the areas that the public cannot pass through; however, the operators of these establishments might wish to restrict public access to backroom areas for security reasons. These self-imposed restrictions do not justify denying the public access to the toilet facilities located in restricted areas. If the toilet facilities are located in restricted areas, public toilet facilities must be provided outside of those self-imposed restricted areas.

[P] 2902.3.2 Location of toilet facilities in occupancies other than malls. In occupancies other than covered and open mall buildings, the required *public* and employee toilet facilities shall be located not more than one story above or below the space required to be provided with toilet facilities, and the path of travel to such facilities shall not exceed a distance of 500 feet (152 m).

Exception: The location and maximum travel distances to required employee facilities in factory and industrial occupancies are permitted to exceed that required by this section, provided that the location and maximum travel distance are *approved*.

❖ To prevent health problems for building occupants related to difficulty in accessing a needed toilet facility in a timely fashion, the code clearly provides two conditions that must be met: 1) the required toilet facilities must be located within a certain travel distance and 2) the building occupant must not be required to travel beyond the next adjacent story above or below his or her location.

The distribution of plumbing fixtures within a building must satisfy the maximum travel limitations for occupants in the building. Sections 2902.3.2 and 2902.3.3 limit the horizontal and vertical distances that an occupant has to travel to use a toilet facility. A 20-story high-rise office building cannot have all of the required plumbing fixtures located on the 10th floor as the path of travel must neither exceed a travel

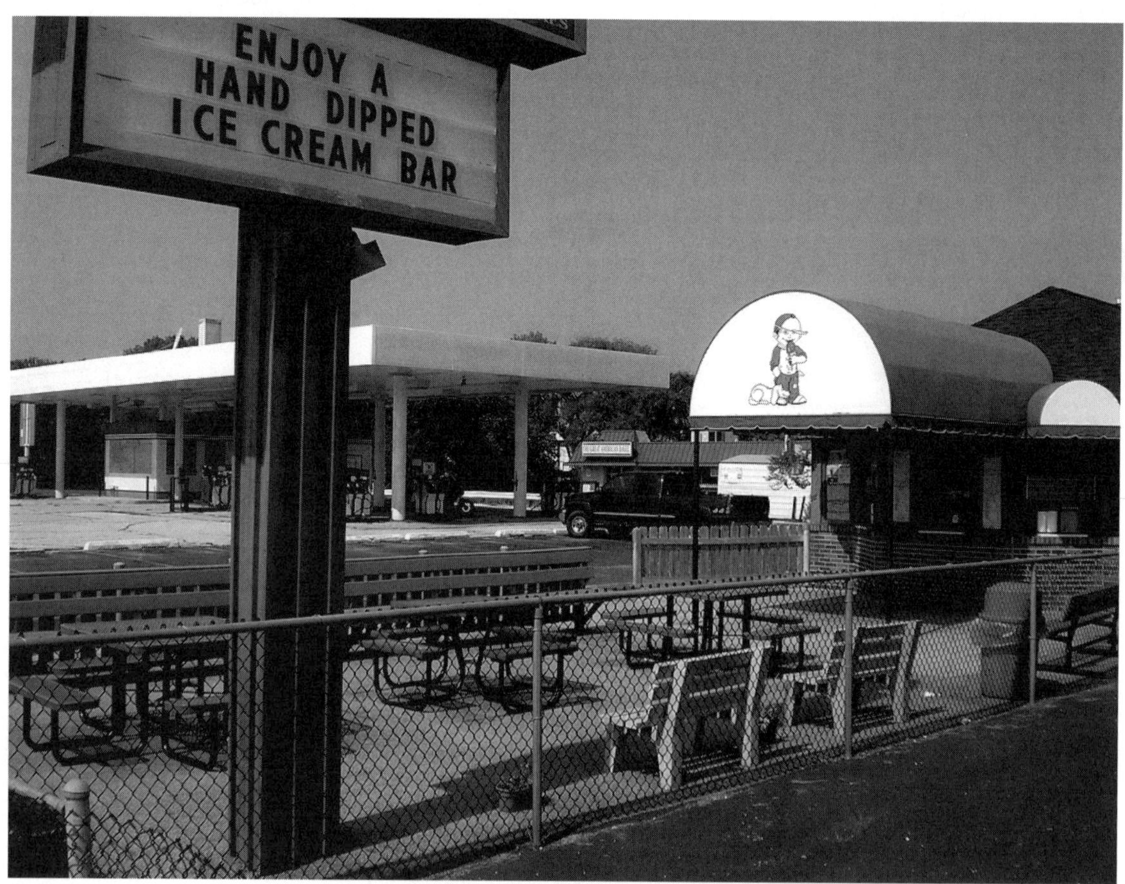

Figure 2902.3.1
WALK-UP ICE CREAM SHOP WITH OUTDOOR SEATING

distance of 500 feet (152 400 mm) nor require travel beyond the adjacent story above or below. While this travel path limitation can result in toilet facilities not being located on every floor of a multistory building, the intent of the code is for a proportional number of fixtures to be available for the intended occupant use. For example, occupants on the 20th, 19th and 18th floors could reasonably be expected to utilize toilet facilities located only on the 19th floor (as long as the 500-foot (152 400 mm) travel distance limitation is not exceeded). However, that toilet facility cannot have just one fixture of each type to serve three floors of occupants. The distribution of fixtures must be proportioned according to the number of occupants that will be using the toilet facility. To illustrate, assume that each of the 20 floors in this high-rise business occupancy building has an occupant load of 100. Applying the business occupancy water closet fixture ratio to the occupant load for the entire building results in 40 male water closets required for the building. Three floors of occupants will require (300 floor occupants/2,000 total occupants) x 40 w/c = 6 male water closets on the 19th floor.

The exception in this section provides for the unique nature of the workplace factory and industrial occupancies where the designer is better equipped to determine the location of the employee toilet facilities.

[P] 2902.3.3 Location of toilet facilities in malls. In covered and open mall buildings, the required *public* and employee toilet facilities shall be located not more than one story above or below the space required to be provided with toilet facilities, and the path of travel to such facilities shall not exceed a distance of 300 feet (91 440 mm). In mall buildings, the required facilities shall be based on total square footage (m^2) within a covered mall building or within the perimeter line of an open mall building, and facilities shall be installed in each individual store or in a central toilet area located in accordance with this section. The maximum travel distance to central toilet facilities in mall buildings shall be measured from the main entrance of any store or tenant space. In mall buildings, where employees' toilet facilities are not provided in the individual store, the maximum travel distance shall be measured from the employees' work area of the store or tenant space.

❖ In covered and open malls, the path of travel to required toilet facilities must not exceed a distance of 300 feet (91 400 mm) and is measured from the main entrance of any store or tenant space. Note, however, that if the tenant space does not have toilets for the employees, the starting point for measuring the travel path is the employee's work area. As for other occupancies, the facilities must be located no more than one story above or below the occupant's location. This shorter travel distance addresses the fact that covered malls are frequently very congested and occupants are unfamiliar with their surroundings. It is also logical to reduce the travel time for employees for security reasons. The minimum number of required plumbing facilities is based on total square footage of the mall, including tenant spaces. This section does not prohibit the installation of separate employee toilet facilities in individual tenant spaces; however, such facilities are not deductible from the total number of common facilities required (see commentary, Section 2902.3).

[P] 2902.3.4 Pay facilities. Where pay facilities are installed, such facilities shall be in excess of the required minimum facilities. Required facilities shall be free of charge.

❖ Pay facilities have been included in some public areas to prevent vagrants from loitering in the public bathrooms. They may also be installed to offer customers facilities with greater amenities where the customer desires such upgrades. This section does not prevent pay facilities from being installed; rather, it requires that such facilities be in excess of those required by Section 2902.

[P] 2902.3.5 Door locking. Where a toilet room is provided for the use of multiple occupants, the egress door for the room shall not be lockable from the inside of the room. This section does not apply to family or assisted-use toilet rooms.

❖ Some multiple-occupant toilet rooms have a locking door to facilitate closure of the toilet facility for maintenance or for security when the toilet room is closed after a building's occupancy hours. However, this section prohibits this type of lock or any other lock from being engaged on the inside of the door so that persons cannot lock themselves in the toilet room for whatever reason. This requirement was put in the 2012 edition to prevent multiple occupant toilet rooms from becoming safe havens for illicit activities. Because family or assisted-use toilet rooms are provided to offer privacy to a user and a family member or caretaker, the lock prohibition does not apply to those types of toilet rooms.

[P] 2902.4 Signage. Required public facilities shall be designated by a legible sign for each sex. Signs shall be readily visible and located near the entrance to each toilet facility. Signs for accessible toilet facilities shall comply with Section 1110.

❖ Public facilities must be designated by a legible sign for each sex and located near the entrance to such facilities. The code is silent on the specifics of the type of designation, making it dependent on the approval of the building official. Where a toilet or bathing room is accessible, the signage must comply with Section 1110 of the code, which requires the International Symbol of Accessibility to be displayed at any single-user accessible toilet or bathing room, including a family or assisted-use toilet or bathing room. In addition, Section 1210 requires directional signage (with the International Symbol of Accessibility) at inaccessible toilet and bathing rooms indicating the direction of accessible toilet or bathing facilities, and at separate sex toilet and bathing facilities indicating the direction of family or assisted-use toilet and bathing facilities, if provided.

[P] 2902.4.1 Directional signage. Directional signage indicating the route to the public facilities shall be posted in accordance with Section 3107. Such signage shall be located in a *corridor* or aisle, at the entrance to the facilities for customers and visitors.

❖ This section strengthens the intent of Section 2902.3, which is to ensure that public toilet facilities are provided for spaces intended for pubic utilization. Where the public facilities are not clearly visible from within the public areas or the facilities are locked, patrons and visitors are sometimes told that public toilet facilities were not available or that they are for employee use only. Mandated directional signs will inform the public that public restrooms do exist and where they are located. It is difficult to tell a customer or patron that there are no public facilities when there is an obvious sign advertising their existence. This code requirement provides to the building official the means for issuing a citation to the building owner if the signs are found to be removed in an attempt by the building owner or tenant to limit public access to toilet facilities. The location of the sign is to be near the entrance to the facilities in either an aisle or corridor where it will be readily seen. The code is silent on specific details of the signage such as wording, size, color or mounting height. Note that Section 1110 covering signage for accessibility requires that directional signage must be provided at the location of separate-sex toilet facilities indicating the location of family or assisted-use toilet facilities, if such facilities are required.

[P] 2902.5 Drinking fountain location. Drinking fountains shall not be required to be located in individual tenant spaces provided that public drinking fountains are located within a travel distance of 500 feet of the most remote location in the tenant space and not more than one story above or below the tenant space. Where the tenant space is in a covered or open mall, such distance shall not exceed 300 feet. Drinking fountains shall be located on an accessible route.

❖ Prior to the 2012 edition, the code was silent concerning the required location for drinking fountains. This new section clarifies their location with respect to the spaces that require access to drinking fountains. Drinking fountains are not required to be near toilet facilities but because of the similar travel distance limitations, they might be in the same area as the toilet facility. Because all installed drinking fountains are required to be of accessible design (either for standing or wheelchair-seated persons), they must be on an accessible route (see Chapter 11).

Previous to the addition of this section to the code, the designer of the building had the choice as to whether drinking fountains were to be installed in individual tenant spaces or in a common public area outside of the tenant space. Because the code did not specify a travel distance limitation, drinking fountains located in the common area for serving the tenant spaces could be located at any distance away from the tenant space. It was, therefore, acceptable to put all of the common-area drinking fountains in a single location instead of spread around in the common area to better serve the demand for drinking water. Now, if common-area drinking fountains are to be installed, they must be apportioned such that they serve the demands of the tenants within the travel distance limitations.

Bibliography

The following resource material is referenced in this chapter or is relevant to the subject matter addressed in this chapter.

ICC/ANSI 117.1-2003, *Accessible and Usable Buildings and Facilities*. Washington, DC: International Code Council, 2004.

IPC-12, *International Plumbing Code*. Washington, DC: International Code Council, 2011.

Chapter 30: Elevators and Conveying Systems

General Comments

Chapter 30 contains the provisions that regulate vertical and horizontal transportation and material-handling systems installed in buildings. The installation must comply with the requirements in this chapter and the standards referenced herein.

Section 3001.1 contains the scope of the chapter.

Section 3001.2 identifies the standards to which elevators and conveying systems must comply.

Section 3001.3 contains requirements for accessible elevators.

Sections 3002.1 through 3002.7 contain the requirements for elevator hoistway enclosures.

Sections 3003.1 through 3003.2 include requirements for emergency operation of elevators.

Sections 3004.1 through 3004.5 identify requirements for hoistway venting.

Sections 3005.1 through 3005.4 address requirements for conveying systems and personnel and material hoists.

Sections 3006.1 through 3006.6 address machine room requirements.

Section 3007.1 through 3007.7 provides the requirements for fire service access elevators.

Section 3008.1 through 3008.15 provides the requirements for occupant evacuation elevators.

Purpose

The purpose of this chapter is to regulate the installation, testing, inspection, maintenance, alteration and repair of vertical and horizontal transportation and material-handling systems installed in buildings. Compliance with the requirements in this chapter provides for life safety and promotes public welfare. The chapter is also intended to be used as a minimum safety standard by architects, engineers, insurance companies, manufacturers and contractors and as standard safety practice for owners and managers of buildings where equipment covered by this chapter is installed and used.

The chapter provides several elements that protect occupants and assist emergency responders during fires which include Section 3003, Emergency Operations, Section 3007, Fire Service Access Elevators and Section 3008, Occupant Evacuation Elevators.

The provisions found in Chapter 30 of the code and associated standards are specifically coordinated to work together.

SECTION 3001
GENERAL

3001.1 Scope. This chapter governs the design, construction, installation, *alteration* and repair of elevators and conveying systems and their components.

❖ This section indicates that the requirements in this chapter are applicable to the design, construction, installation, alteration and repair of any elevator, conveying system or component thereof. In addition to requirements for design, construction and installation, requirements for alteration, repair, testing and inspections of elevators and conveying systems are located in the standards referenced in this chapter, and are, therefore, not included in the code.

3001.2 Referenced standards. Except as otherwise provided for in this code, the design, construction, installation, *alteration*, repair and maintenance of elevators and conveying systems and their components shall conform to ASME A17.1/CSA B44, ASME A90.1, ASME B20.1, ALI ALCTV, and ASCE 24 for construction in *flood hazard areas* established in Section 1612.3.

❖ The enforceability of a standard is established in this section, and applies wherever the provisions of this chapter do not otherwise indicate a requirement. Elevators installed in buildings and structures located in designated flood hazard areas may be subject to additional flood-resistant design and construction requirements. The application of these additional requirements depends on whether the elevator or components of the elevator equipment are to be located below the design flood elevation. Those not familiar with the subtle differences that make a standard applicable to a specific equipment design should review both the standard's scope and definitions. In general, the applications of the referenced standards are as follows:

- ASME A17.1/CSA B44 applies to building transportation equipment that is permanently installed to transport people and freight.

- ASME A90.1 applies specifically to belt-type manlifts.
- ASME B20.1 applies to equipment that is permanently installed to transport freight. People are prohibited from riding this equipment.
- ALI ALCTV applies specifically to automotive lifts, such as those used in automated parking structures.

ASCE 24 applies to the construction of elevators and conveying systems located in flood hazard areas, as established by application of Section 1612.3 of the code. Section 7.5 of ASCE 24-05 and FEMA FIA-TB #4, *Elevator Installation for Buildings Located in Special Flood Hazard Areas* provide extensive guidance on the proper method of installing elevators in buildings and structures located in designated flood hazard areas. In addition to raising elevator service equipment above the design flood elevation, a concern would be determining if the fire fighters' emergency operation requirements of Section 3003 do not require the elevator car to descend automatically to a level below the design flood elevation during conditions of flooding. The safety of occupants may be jeopardized, and the elevator car may be damaged if it descends below the design flood elevation during flooding.

Referenced standards have a long history of providing minimum safety requirements for the equipment covered by their scope. As an example, ASME A17.1 was first published in 1921 and 18 subsequent editions have been published.

Interpretations of the referenced standards can be obtained through the committee responsible for developing and maintaining the standard. The ASME A17.1/CSA B44, ASME A90.1 and ASME B20.1 standards committees have also published interpretations and include copies of all recent interpretations for purchasers of these documents. The publisher of ASME A17.1/CSA B44 also has available a handbook that explains and augments the requirements. While every effort has been made to eliminate conflicts between the code and the referenced standards, such conflicts may occasionally occur. Where differences occur, the requirements of the code take precedence (see Section 102.4).

In terms of the maintenance of elevators, it should be noted that ASME A17.1/CSA B44 contains Appendix N that specifies frequency of maintenance. This appendix must be specifically adopted by a jurisdiction to become effective. Note that Section 606.1 of the *International Property Maintenance Code®* (IPMC®) references Appendix N.

3001.3 Accessibility. Passenger elevators required to be accessible or to serve as part of an *accessible means of egress* shall comply with Sections 1107 and 1109.7.

❖ Section 1109.6 requires elevators that form part of an accessible route to be accessible as required by ICC A117.1. Section 1007 provides that elevators may be required to be part of an accessible means of egress from a building; again; the ICC A117.1 standard addresses the accessibility requirements for elevators. ICC A117.1 contains requirements for elevator location and access; operation and leveling; door operation; door size; door protective and reopening devices; door delay (dwell time) from hall and car calls; car inside dimensions, controls and position indicators and signals; telephone or intercom systems; floor coverings; minimum illumination; hall buttons; hall lanterns; door jamb markings and clearance between sills. This standard provides acceptable and appropriate service for all users and makes multistory buildings accessible and usable by people with physical disabilities. The standard provides requirements for destination-oriented elevators, limited use/limited application elevators (LU/LA), private residential elevators as well as standard passenger elevators. Each type can be used to provide an accessible route. The choice of which type of elevator may be used depends on the limits of their use established in ASME A17.1/CSA B44; for example, private residence elevators may only be used to or in a private dwelling.

3001.4 Change in use. A change in use of an elevator from freight to passenger, passenger to freight, or from one freight class to another freight class shall comply with Section 8.7 of ASME A17.1/CSA B44.

❖ This section intends to require enforcement of Section 8.7 of ASME A17.1/CSA B44 whenever the elevator changes in the use or freight class. The application of these requirements results in an elevator that will operate safely and comply with requirements that are unique to and necessary for the new use or freight class.

SECTION 3002
HOISTWAY ENCLOSURES

3002.1 Hoistway enclosure protection. Elevator, dumbwaiter and other hoistway enclosures shall be *shaft enclosures* complying with Section 713.

❖ Reference is made to Section 713 of the code for the required fire resistance and construction of the hoistway enclosure. Section 713 contains fire-resistance rating requirements, as well as construction requirements for shafts, which are typically the method used to satisfy the requirements for a hoistway enclosure. It should be noted that this section of the code does not require hoistway enclosures for all elevators and dumbwaiters. See Section 2.1.1.3 of ASME A17.1 for the protection requirements for partially enclosed hoistways.

3002.1.1 Opening protectives. Openings in hoistway enclosures shall be protected as required in Chapter 7.

> **Exception:** The elevator car doors and the associated hoistway enclosure doors at the floor level designated for recall in accordance with Section 3003.2 shall be permit-

ted to remain open during Phase I Emergency Recall Operation.

❖ Reference is made to Chapter 7 for protection of openings in hoistway enclosures. Chapter 7 contains minimum hourly rating requirements for openings based on the fire-resistance rating required for the enclosure. For example, if a hoistway enclosure was required to be 2-hour fire-resistance rated, a $1^1/_2$-hour fire door assembly would be required in accordance with Table 716.5. It should be noted that additional requirements may also be included for elevator lobby enclosures in Sections 713.14.1, 3007 and 3008.

When an elevator opens into a corridor that is required to be of fire-resistance-rated construction, the opening between the elevator shaft and the corridor must be protected to meet not only the shaft's fire protection rating but also the additional smoke and draft protection requirements necessary to limit the spread of smoke into the corridor. This additional smoke and draft control requirement is found in Section 716.5.3.1. The elevator hoistway shaft doors opening into such rated corridors will either need to be separated from the corridor by one of the following methods of protection:

1. A lobby (see Section 713.14);
2. An additional door (see Sections 3002.6 and 716.5.3 and its subsections).
3. A shaft door meeting both the smoke and draft protection requirements for corridor doors in Section 716.5.3.1 as well as the appropriate fire-protection rating of Table 716.5 for the shaft.

While many elevator hoistway shaft doors are tested and labeled for the 1-hour or $1^1/_2$-hour fire resistance rating (Section 716.5), very few, if any, of the doors typically sold in the U.S. will also meet the smoke and draft requirements (Section 716.5.3.1) that would allow them to open directly into a fire-resistance-rated corridor. Because of this, Items 1 and 2 above will be the general methods for protecting such openings.

3002.1.2 Hardware. Hardware on opening protectives shall be of an *approved* type installed as tested, except that *approved* interlocks, mechanical locks and electric contacts, door and gate electric contacts and door-operating mechanisms shall be exempt from the fire test requirements.

❖ In order to verify that hardware used with elevator doors will not affect their fire-resistance rating, door hardware not specifically exempt by this section is required to be part of the door assembly during the fire test. The devices that are exempt from the fire test have minimum fire and physical requirements contained in ASME A17.1/CSA B44. Passenger elevator door hardware consists of a header, track hangers, pendant bolts, floor sill with guides, sill support plates, sill brackets, retaining angles and closer assemblies. Each component may bear a separate label, or the assembly may have one label to cover all components. The label should clearly identify the assembly or component to which it applies. Typically, labels are viewed from inside the hoistway.

3002.2 Number of elevator cars in a hoistway. Where four or more elevator cars serve all or the same portion of a building, the elevators shall be located in no fewer than two separate hoistways. Not more than four elevator cars shall be located in any single hoistway enclosure.

❖ The number of elevators provided in a building is not mandated by the code, but is a design decision. The code does limit the maximum number of elevator cars in a single hoistway to four. This limits the potential of a single hoistway fire incident causing the removal of all elevators from service, particularly in larger structures, such as high-rise buildings.

3002.3 Emergency signs. An *approved* pictorial sign of a standardized design shall be posted adjacent to each elevator call station on all floors instructing occupants to use the *exit stairways* and not to use the elevators in case of fire. The sign shall read: IN CASE OF FIRE, ELEVATORS ARE OUT OF SERVICE. USE EXIT STAIRS.

Exceptions:

1. The emergency sign shall not be required for elevators that are part of an *accessible means of egress* complying with Section 1007.4.
2. The emergency sign shall not be required for elevators that are used for occupant self-evacuation in accordance with Section 3008.

❖ Elevators may be unsafe during a fire because:

- Persons may push a corridor button and then wait for an elevator that may never respond. Valuable time in which to leave the building safely is lost.
- Elevators cannot start until the car and hoistway doors are closed. A panic could lead to overcrowding of an elevator and blockage of the doors, thus preventing closing.
- Power failure during a fire can happen at any time and thus lead to entrapment.
- Elevators respond to car and corridor calls. One of these calls may be at the fire floor.

Fatal delivery of the elevator to the fire floor can be caused by:

- An elevator passenger pressing the car button for the fire floor.
- One or both of the corridor call buttons being pushed on the fire floor.
- Heat melting or deforming the corridor push button or its wiring at the fire floor.
- Normal functioning of the elevator, such as high-call or low-call reversal, occurring at the fire floor.

In a fire emergency, an occupant will normally try to exit the building by the route he or she entered (elevators). The pictorial sign will indicate to the occupants not to use the elevators, but rather to use the exit stairways to exit the building.

There are two exceptions to this requirement. Exception 1 recognizes that elevators may be used as part of the accessible means of egress in accordance with Section 1007.4. Exception 2 recognizes that passenger elevators may be used for occupant self-evacuation during fire emergencies in accordance with Section 3008. In either case, the sign required by this section would not be appropriate since it would be communicating a contradictory message. Section 1007.4 dictates specific methods to permit the use of elevators as an element of an accessible means of egress.

Section 1007.4 does not anticipate the unassisted use of an elevator by a person with disabilities but instead anticipates assisted rescue during Phase II operations. At that point, the fire department has control of the elevator and will provide the needed assistance. Section 3008, specifically, permits the use of an elevator by the general public for unassisted evacuation prior to a Phase I recall. Elevators serving floors or even elevators served by different lobbies where smoke detection has not activated the Phase I recall will continue to operate normally and are permitted to be used by the public.

See Figure 3002.3 for an example of an emergency pictorial sign.

3002.4 Elevator car to accommodate ambulance stretcher. Where elevators are provided in buildings four or more *stories* above, or four or more *stories* below, *grade plane*, at least one elevator shall be provided for fire department emergency access to all floors. The elevator car shall be of such a size and arrangement to accommodate an ambulance stretcher 24 inches by 84 inches (610 mm by 2134 mm) with not less than 5-inch (127 mm) radius corners, in the horizontal, open position and shall be identified by the international symbol for emergency medical services (star of life). The symbol shall not be less than 3 inches (76 mm) in height and shall be placed inside on both sides of the hoistway door frame.

❖ A key aspect of the applicability of this section is limited to situations where elevators are already being provided. This section does not require elevators to be installed but simply specifies size requirements when elevators are provided. This particular section is intended to aid rescue efforts in taller buildings (greater than three stories above or below grade plane). One of the provided elevators must be sized to accommodate a typical ambulance stretcher and emergency personnel. The elevator meeting this requirement does not need to be the same elevator throughout the entire building, but each floor must be served by such an elevator. This becomes more of an issue in very tall buildings that generally do not have any elevators that extend the entire height of the

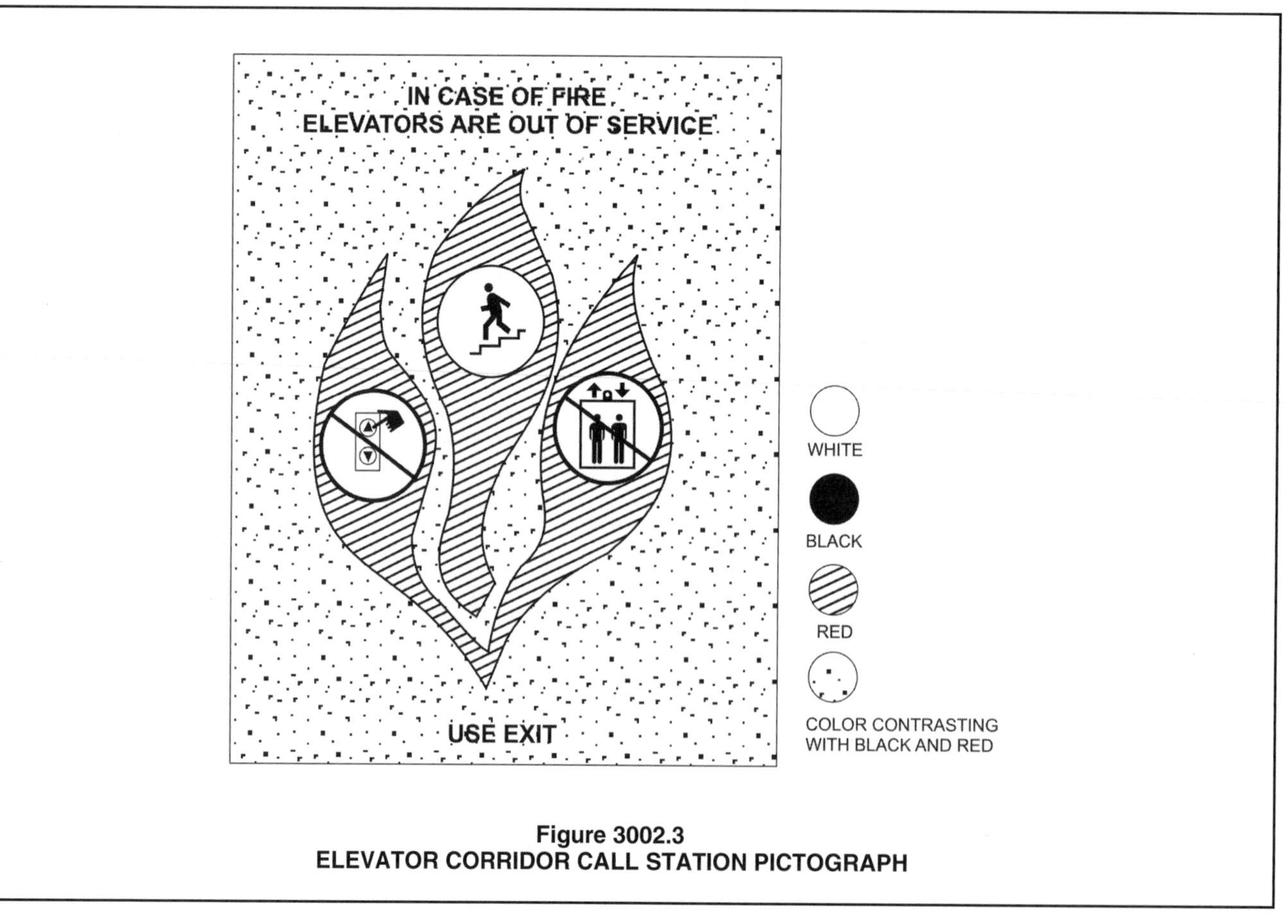

Figure 3002.3
ELEVATOR CORRIDOR CALL STATION PICTOGRAPH

building, or that serve all floors. Often, there are transfer floors to connect different banks of elevators. Additionally, to aid the emergency personnel in identifying which elevator is sized to accommodate a stretcher, the international symbol (star of life) is required to be located on each side of the hoistway door frame.

The size of the ambulance stretcher is based upon the typical size of a stretcher commonly used by fire departments in North America. It is necessary to have the patient in a horizontal position on the elevator to avoid further injury to the patient and to facilitate many of the medical tasks being performed by the paramedics while on the elevator.

This section also recognizes that stretchers do not have right-angle corners but are rounded and thus an allowance for the recognition of 5-inch (127 mm) radius corners is provided. This gives more flexibility in elevator car sizing; for example, 3,500-pound (1585 kg) cars with a 42-inch (1066 mm) side slide door can comply with this requirement. Figures 3002.4(a) and 3002.4(b) demonstrate the application of the radius corners when determining elevator car size.

3002.5 Emergency doors. Where an elevator is installed in a single blind hoistway or on the outside of a building, there shall be installed in the blind portion of the hoistway or blank face of the building, an emergency door in accordance with ASME A17.1/CSA B44.

❖ To allow emergency personnel reasonable access to a single blind elevator hoistway, emergency doors are required to be provided. A single blind hoistway is a hoistway for one elevator without openings. For example, an express elevator that opens at the lobby and serves only floors 10 through 20 is a blind hoistway from floors two through nine. Therefore, emergency doors are required in accordance with ASME A17.1/CSA B44, which contains requirements for opening sizes, opening access, door operation and door location. This requirement does not apply to blind hoistways that have multiple elevators.

3002.6 Prohibited doors. Doors, other than hoistway doors and the elevator car door, shall be prohibited at the point of access to an elevator car unless such doors are readily openable from the car side without a key, tool, special knowledge or effort.

❖ Where a door is installed in front of the hoistway door to meet the hoistway opening protection requirements in Section 713.14.1 (Exception 3) or for security purposes this section requires that occupants be able to open the door. Inappropriate installation of these doors could lead to occupants becoming trapped in the area between the elevator doors and

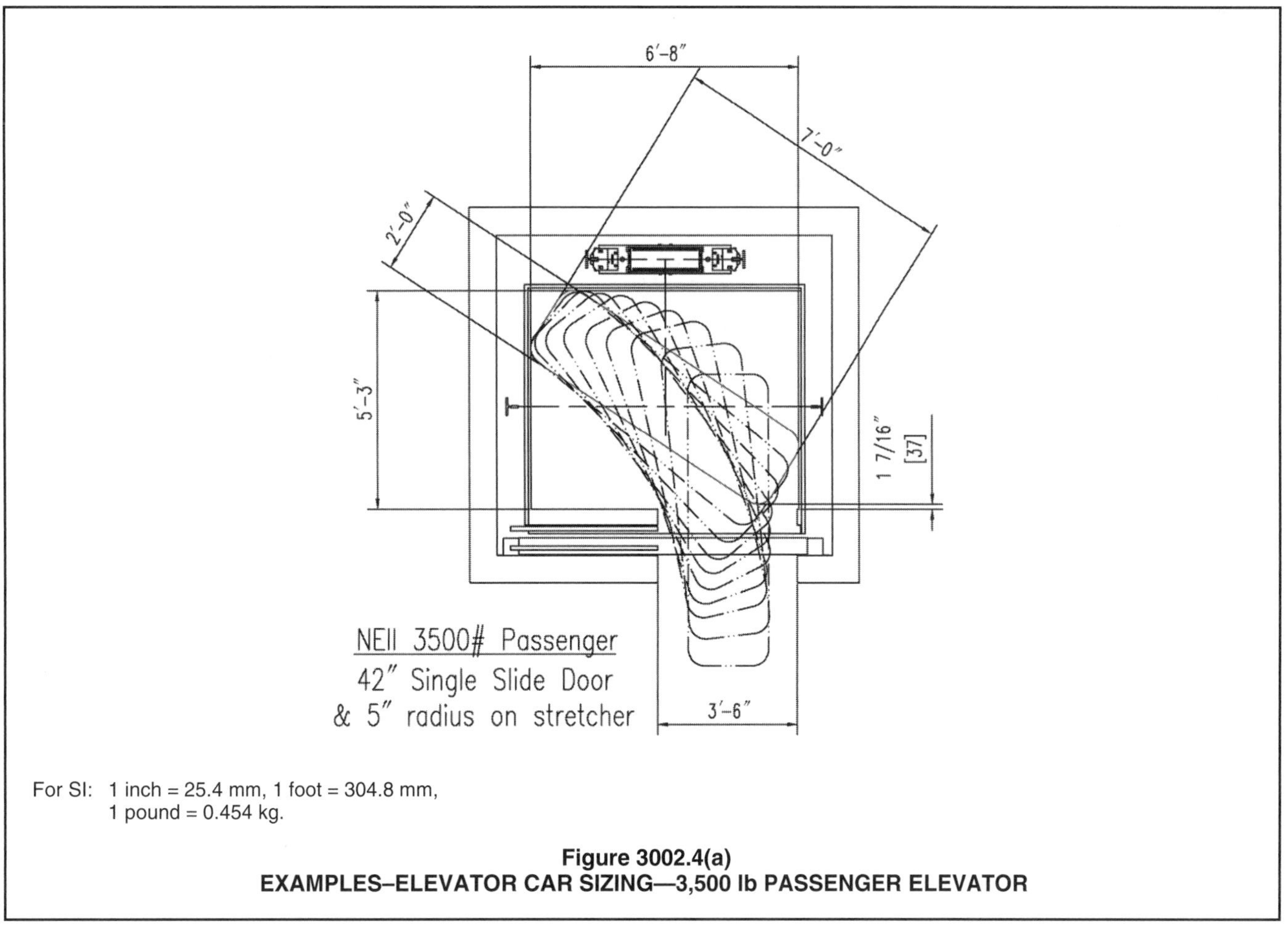

For SI: 1 inch = 25.4 mm, 1 foot = 304.8 mm, 1 pound = 0.454 kg.

Figure 3002.4(a)
EXAMPLES–ELEVATOR CAR SIZING—3,500 lb PASSENGER ELEVATOR

the additional set of doors. See Figure 3002.6 for a picture of an additional door located at the hoistway opening.

3002.7 Common enclosure with stairway. Elevators shall not be in a common *shaft enclosure* with a *stairway*.

Exception: Elevators within *open parking garages need not be separated from stairway enclosures.*

❖ To avoid the potential of a fire spreading from the elevator or elevator hoistway and affecting an exit stairway, separate protective enclosures are required for each. There is an exception to this section that acknowledges that a separate shaft enclosure is not required when in an open parking garage. This correlates with Section 712.1.15 and Exception 5 of Section 1022.1. More specifically, Section 712.1.15 allows elevators in open parking garages without a shaft enclosure. Also, Exception 5 of Section 1022.1 allows stairways in open parking structures to not be enclosed within a shaft enclosure. Therefore, the intent is that elevators need not be separated from stairways in open parking structures as the stairways and elevators would not require a shaft enclosure.

3002.8 Glass in elevator enclosures. Glass in elevator enclosures shall comply with Section 2409.1.

❖ The purpose of this section is simply to remind the code user that Section 2409.1 contains provisions regarding glass in elevator enclosures. Please see

Figure 3002.6
ADDITIONAL DOOR IN FRONT OF HOISTWAY

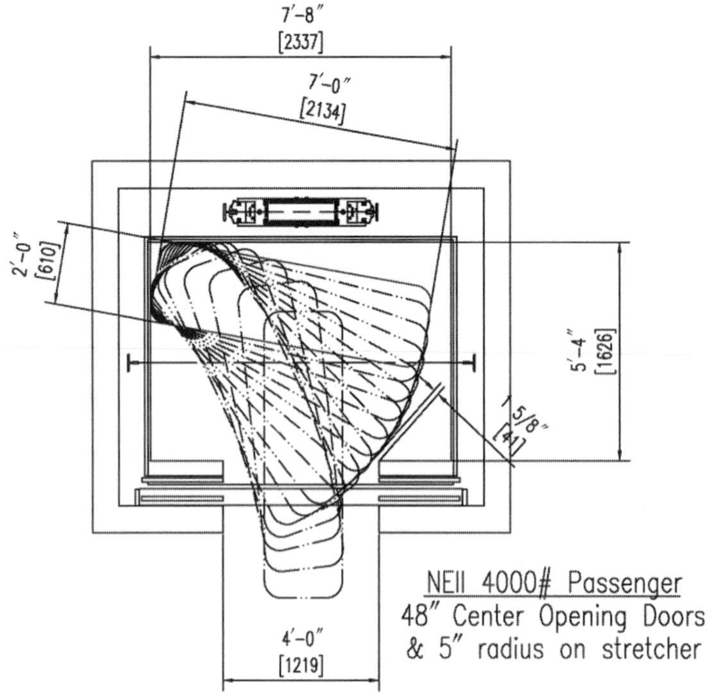

7′-8″
[2337]
7′-0″
[2134]
2′-0″
[610]
5′-4″
[1626]
5/8″
[4]
NEII 4000# Passenger
48″ Center Opening Doors
& 5″ radius on stretcher
4′-0″
[1219]

For SI: 1 inch = 25.4 mm, 1 foot = 304.8 mm,
1 pound = 0.454 kg.

Figure 3002.4(b)
EXAMPLES–ELEVATOR CAR SIZING—4,000 lb PASSENGER ELEVATOR

the commentary to Section 2409.1. It should be noted that other provisions of the code may also be applicable to the glass, such as Section 2406, when human impact loads are possible. Note that Section 2409.2 deals with vision panels in elevator hoistway doors and Section 2409.3 deals with glass in elevator cars.

SECTION 3003
EMERGENCY OPERATIONS

[F] 3003.1 Standby power. In buildings and structures where standby power is required or furnished to operate an elevator, the operation shall be in accordance with Sections 3003.1.1 through 3003.1.4.

❖ The purpose of this section is not to require standby power but to provide the specifics as to how standby power is provided for elevators when required by other sections of the code. The applicable sections that require standby power include:

- High-rise buildings, Section 403.4.8.2.
- Accessible egress via elevators, Section 1007.4.
- Fire service access elevators, Section 3007.9.
- Occupant evacuation elevators, Section 3008.9.

Generally standby power provides both a means to bring all elevators to the designated level and operate elevator(s) during a loss of power.

Once standby power is provided, ASME A17.1/CSA B44 also has specific requirements for standby power operation of elevators, such as absorption of regenerative power (see Section 2.26.10, ASME A17.1/CSA B44) and selection of elevator(s) that are provided standby power (see Section 2.27.2, ASME A17.1/CSA B44).

[F] 3003.1.1 Manual transfer. Standby power shall be manually transferable to all elevators in each bank.

❖ To allow emergency personnel to provide standby power to any elevator(s) at any time, this section requires a manual means of transferring standby power to all elevators in each bank. This does not mean that all elevators need to be running on standby power at the same time but instead that all elevators are capable of being switched to standby power individually. This manual transfer system could assist emergency personnel in evacuating building occupants (see also the requirements in ASME A17.1/CSA B44, Section 2.27.2).

[F] 3003.1.2 One elevator. Where only one elevator is installed, the elevator shall automatically transfer to standby power within 60 seconds after failure of normal power.

❖ In the case where a building is equipped only with one elevator, a standby power system must automatically restore full operation of the elevator within 60 seconds after failure of the primary power supply.

[F] 3003.1.3 Two or more elevators. Where two or more elevators are controlled by a common operating system, all elevators shall automatically transfer to standby power within 60 seconds after failure of normal power where the standby power source is of sufficient capacity to operate all elevators at the same time. Where the standby power source is not of sufficient capacity to operate all elevators at the same time, all elevators shall transfer to standby power in sequence, return to the designated landing and disconnect from the standby power source. After all elevators have been returned to the designated level, at least one elevator shall remain operable from the standby power source.

❖ Where a building is equipped with two or more elevators, this section provides alternate methods of standby power operation. When the standby power system is sized to accommodate all of the elevators, then all are required to automatically switch to standby power 60 seconds after failure of the primary power supply. If a standby power system is sized smaller and is only capable of providing standby power to the elevators in sequence, it must return all elevators to their designated landings by standby power and at least one must remain operable through standby power. This section is not requiring a building to place all elevators on standby power simultaneously. This is only required where the standby power is designed with sufficient capacity to do so. See also requirements the in ASME A17.1/CSA B44, Section 2.27.2 (see also the commentary to Sections 3007.9 and 3008.9).

[F] 3003.1.4 Venting. Where standby power is connected to elevators, the machine room *ventilation* or air conditioning shall be connected to the standby power source.

❖ To allow elevator equipment in the machine room to operate normally, the venting system serving the machine room must be connected to the standby power system that serves the elevators.

[F] 3003.2 Fire-fighters' emergency operation. Elevators shall be provided with Phase I emergency recall operation and Phase II emergency in-car operation in accordance with ASME A17.1/CSA B44.

❖ Two key fire-fighter features are required for essentially all elevators. There are two objectives of these fire-fighter features required by ASME A17.1/CSA B44:

- Phase I emergency recall operation is the operation of an elevator that either automatically recalls it to the designated level due to detection of smoke in the lobby, hoistway or machine room or elevator is manually recalled by the fire fighters and removed from normal service; and
- Phase II emergency in-car operation is the capability of the elevator to be controlled by fire fighters during a fire or other emergency.

Such features are not required by ASME A17.1/CSA B44 in certain cases where the hoistway or a portion thereof is not required to be fire-resistant-rated construction (see Section 2.1.1.1 of the standard), the rise of the elevator does not exceed 6.6 feet (2000 mm),and the hoistway does not penetrate a floor.

Additionally, elevators such as LU/LA elevators would not require Phase I or Phase II emergency operation in accordance with ASME A17.1/CSA B44. More specifically, Section 5.2.1.2.7 of ASME A17.1/CSA B44 exempts LU/LA elevators from such requirements. However, if emergency operation features are installed, they are required to be installed in accordance with ASME A17.1/CSA B44.

Elevators are required to have recall capability to reduce the possibility of taking passengers to the fire floor or trapping them during a fire, and to allow the fire fighters to capture the elevators to use them on Phase II for fire-fighting and rescue operations. Also, Phase II emergency operation allows fire fighters to access the building much faster than depending solely on the stairways. This is especially necessary in high-rise buildings to allow faster access to a fire or other emergencies or to offer aid in rescue or evacuation of persons with disabilities.

[F] 3003.3 Standardized fire service elevator keys. All elevators shall be equipped to operate with a standardized fire service elevator key in accordance with the *International Fire Code*.

❖ When fire departments and other emergency response agencies respond to emergencies, their ability to quickly access the location of the emergency can be the deciding factor of a successful response. Elevators are increasingly being relied upon for emergency operations and their importance has been highlighted by the addition of Section 3007 to the code, which requires the installation of fire service access elevators and provides requirements for the installation of occupant evacuation elevators. One of the difficulties the fire service and other emergency response agencies have when accessing facilities and attempting to use elevators is the number of non-standardized keys that may not be available at the time of response. Even when emergency responders are provided the necessary keys in case of response, the correct key may have to be identified from a large collection of keys for any one building. In larger jurisdictions the sheer number of keys makes the possession of the keys unwieldy for the emergency responders. An elevator key is an important tool to fire fighters. The key allows fire fighters to access the interior of the shaft housing an elevator cabin to initiate rescue in case of a malfunction or the loss of building power. More often, an elevator key is used to capture and control an elevator in emergencies. This section of the 2012 edition of the code establishes new requirements for elevator keys used by the fire service that will only apply to those buildings that have elevators with Phase I or II emergency service or to those buildings equipped with a fire service access elevator.

SECTION 3004 HOISTWAY VENTING

3004.1 Vents required. Hoistways of elevators and dumbwaiters penetrating more than three *stories* shall be provided with a means for venting smoke and hot gases to the outer air in case of fire.

Exception: Venting is not required for the following elevators and hoistways:

1. In occupancies of other than Groups R-1, R-2, I-1, I-2 and similar occupancies with overnight *sleeping units*, where the building is equipped throughout with an *approved automatic sprinkler system* installed in accordance with Section 903.3.1.1 or 903.3.1.2.
2. Sidewalk elevator hoistways.
3. Elevators contained within and serving *open parking garages* only.
4. Elevators within individual residential *dwelling units*.

❖ Hoistway venting is required in buildings where the elevator hoistway extends at least four stories of the building and is intended to prevent the accumulation and spread of hot smoke and gases from a fire to the upper stories of a building. The majority of deaths in fires are a result of smoke. For example, in the fire at the 1980 MGM Grand Hotel in Las Vegas, 70 of the 84 deaths occurred on the upper floors where smoke concentration was the greatest—even though the fire was on the first floor. This could be attributed to stack effect, which is a phenomenon that exists in high-rise buildings. Stack effect describes the movement of air inside and outside a building. Note that the hoistway venting in the MGM Grand Hotel did not comply with the code at the time of the fire. During a fire, the presence of stack effect generally results in the movement of smoke and combustion products from lower levels to upper levels through shafts in the building. Stack effect can be reversed in an air-conditioned building when the outside temperature is relatively high (see Figure 3004.1).

Under the conditions listed in 712.1.2, sprinklers are permitted as an alternative to the hoistway vents in all buildings, except those of the groups noted. The top of sidewalk elevator hoistways is open to the outside; thus, no venting is required as indicated in 712.1.3.

Exception 3 addresses the fact that open parking garages would allow smoke migration throughout and to the exterior by the nature of how such buildings are constructed and an elevator would not contribute significantly to smoke spread within such structures; therefore, venting would not be required. Section 712.1.5 addresses the fact that 712.1.2 would allow floors within residential dwelling units to be open to each other. Therefore, requiring venting would not be necessary as smoke can already freely move throughout the space.

3004.2 Location of vents. Vents shall be located at the top of the hoistway and shall open either directly to the outer air or through noncombustible ducts to the outer air. Noncombustible ducts shall be permitted to pass through the elevator machine room, provided that portions of the ducts located outside the hoistway or machine room are enclosed by construction having not less than the *fire-resistance rating* required for the hoistway. Holes in the machine room floors for the passage of ropes, cables or other moving elevator equipment shall be limited as not to provide greater than 2 inches (51 mm) of clearance on all sides.

❖ The vent location is intended to aid in the exhaust of smoke to the exterior of the building and to limit migration of smoke into the machine room. At one time, hoistways were vented into the machine room, and a grate in the machine room floor vented to the exterior of the building. Allowing smoke and heat into the machine room may cause equipment malfunction. Therefore, vents must be located in the hoistway directly below the ceiling, and if ventilation through the machine room is advantageous by design, it must be provided through the use of noncombustible ductwork. Further, when noncombustible ductwork is utilized for the ventilation system, any portion of the ductwork that is outside of the hoistway or the machine room must be enclosed by fire-resistance-rated construction equivalent to that required for the hoistway. The clearance around elevator ropes and other moving equipment is limited in size in order to limit the passage of smoke into the machine room.

3004.3 Area of vents. Except as provided for in Section 3004.3.1, the area of the vents shall be not less than $3^1/_2$ percent of the area of the hoistway nor less than 3 square feet (0.28 m^2) for each elevator car, and not less than $3^1/_2$ percent nor less than 0.5 square feet (0.047 m^2) for each dumbwaiter car in the hoistway, whichever is greater. Of the total required vent area, not less than one-third shall be permanently open. Closed portions of the required vent area shall consist of openings glazed with annealed glass not greater than $^1/_8$ inch (3.2 mm) in thickness.

Exception: The total required vent area shall not be required to be permanently open where all the vent openings automatically open upon detection of smoke in the elevator lobbies or hoistway, upon power failure and upon activation of a manual override control. The manual override control shall be capable of opening and closing the vents and shall be located in an *approved* location.

❖ Vent sizes are calculated such that there will not be a significant accumulation of smoke at the top of the hoistway. When calculating the required area of the vent openings, the plan view area of the hoistway is used (see Figure 3004.3). In general, one-third of the required vent area needs to be permanently open to ensure that smoke can escape if the openable portions of the vents would not open for some reason. The exception allows that the vents may all be closed, provided that they open automatically upon activation of any of the elevator lobby smoke detectors. This will provide a higher degree of energy conservation while maintaining the ability to exhaust smoke. Such vents must open when the power fails to ensure that the hoistway will be vented even when the power is lost. The fire department will also have the ability to open and close the vents when necessary through manual override. The location of the override must be approved and will depend upon the fire-fighting strategies for the particular building. The override will most likely be located next to the annunciator or in a fire command center when available.

3004.3.1 Reduced vent area. Where mechanical *ventilation* conforming to the *International Mechanical Code* is provided, a reduction in the required vent area is allowed provided that all of the following conditions are met:

1. The occupancy is not in Group R-1, R-2, I-1 or I-2 or of a similar occupancy with overnight *sleeping units.*

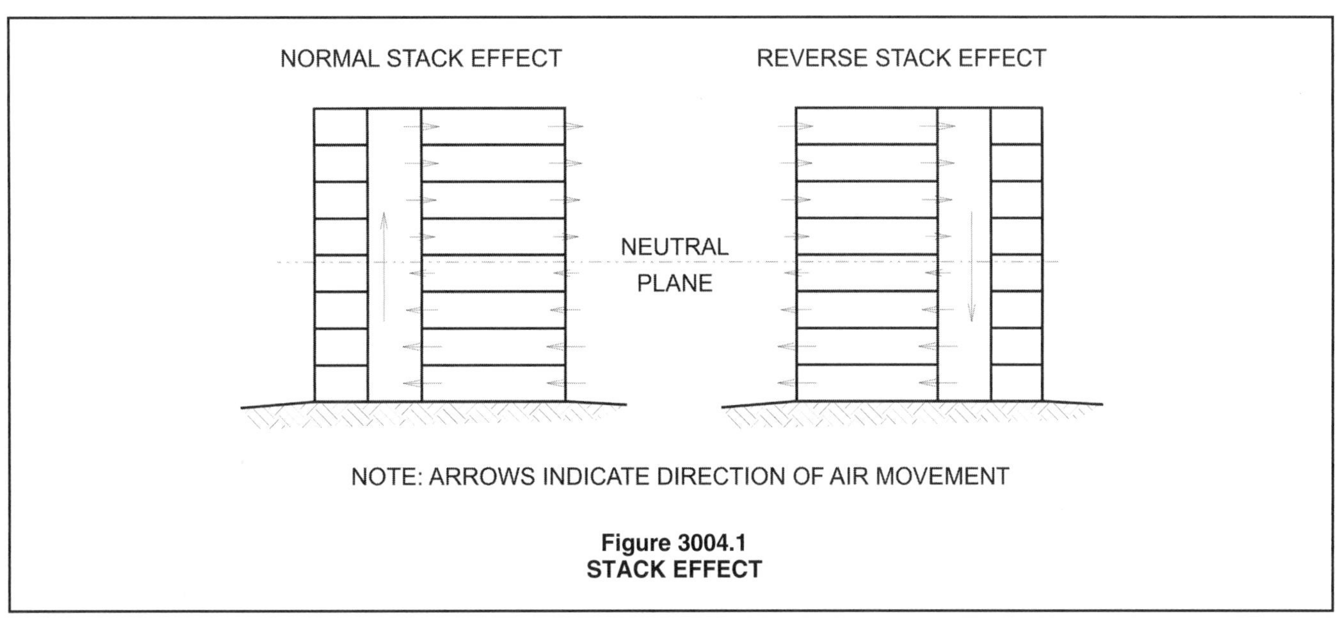

Figure 3004.1
STACK EFFECT

2. The vents required by Section 3004.2 do not have outside exposure.
3. The hoistway does not extend to the top of the building.
4. The hoistway and machine room exhaust fan is automatically reactivated by thermostatic means.
5. Equivalent venting of the hoistway is accomplished.

❖ If a mechanical means of ventilation is provided to the hoistway, then a reduction in the vent area is permitted. This option is not applicable to buildings containing residential or institutional occupancies where people will be sleeping or other occupancies with sleeping units, because mechanical systems are subject to malfunctioning, which could allow smoke to infiltrate areas in which occupants are sleeping. The vent arrangement is to be such that the vents will not be exposed to the elements, and thus not subject to malfunctioning due to weather exposure. Usually, this can be accomplished through the use of louvers or hoods. Engineering calculations should be provided as substantiation of venting that is equivalent to that prescribed in Section 3004.3.

3004.4 Plumbing and mechanical systems. Plumbing and mechanical systems shall not be located in an elevator hoistway enclosure.

Exception: Floor drains, sumps and sump pumps shall be permitted at the base of the hoistway enclosure provided they are indirectly connected to the plumbing system.

❖ If a pipe or duct that conveys gases, vapors or liquids, and passes through a hoistway were to fail, operation of the elevator equipment could be affected in such a manner as to make it unsafe. Water could cause the malfunctioning of hoistway door interlocks, thereby allowing an elevator to leave a floor with the hoistway door open. A wet brake would also cause an extremely hazardous condition. The exception allows for a floor drain, sump or sump pump to be installed at the base of the shaft. However, the discharge is required to indirectly connect to the plumbing system and not be connected directly to any drain. The intent of this requirement is to eliminate the possibility of sewer gases entering into the hoistway, which could cause an explosion hazard.

SECTION 3005
CONVEYING SYSTEMS

3005.1 General. Escalators, moving walks, conveyors, personnel hoists and material hoists shall comply with the provisions of Sections 3005.2 through 3005.4.

❖ Conveying systems include escalators, moving walks, conveyors and personnel and material hoists. Section 3005 contains requirements specific to these types of conveying systems.

3005.2 Escalators and moving walks. Escalators and moving walks shall be constructed of *approved* noncombustible and fire-retardant materials. This requirement shall not apply to electrical equipment, wiring, wheels, handrails and the use of $^1/_{28}$-inch (0.9 mm) wood veneers on balustrades backed up with noncombustible materials.

❖ The intent of this requirement is to limit the fire fuel loading. The presence of large amounts of combustible material would create a large fuel load, and result in an increased migration of fire and smoke from floor to floor, or from space to space.

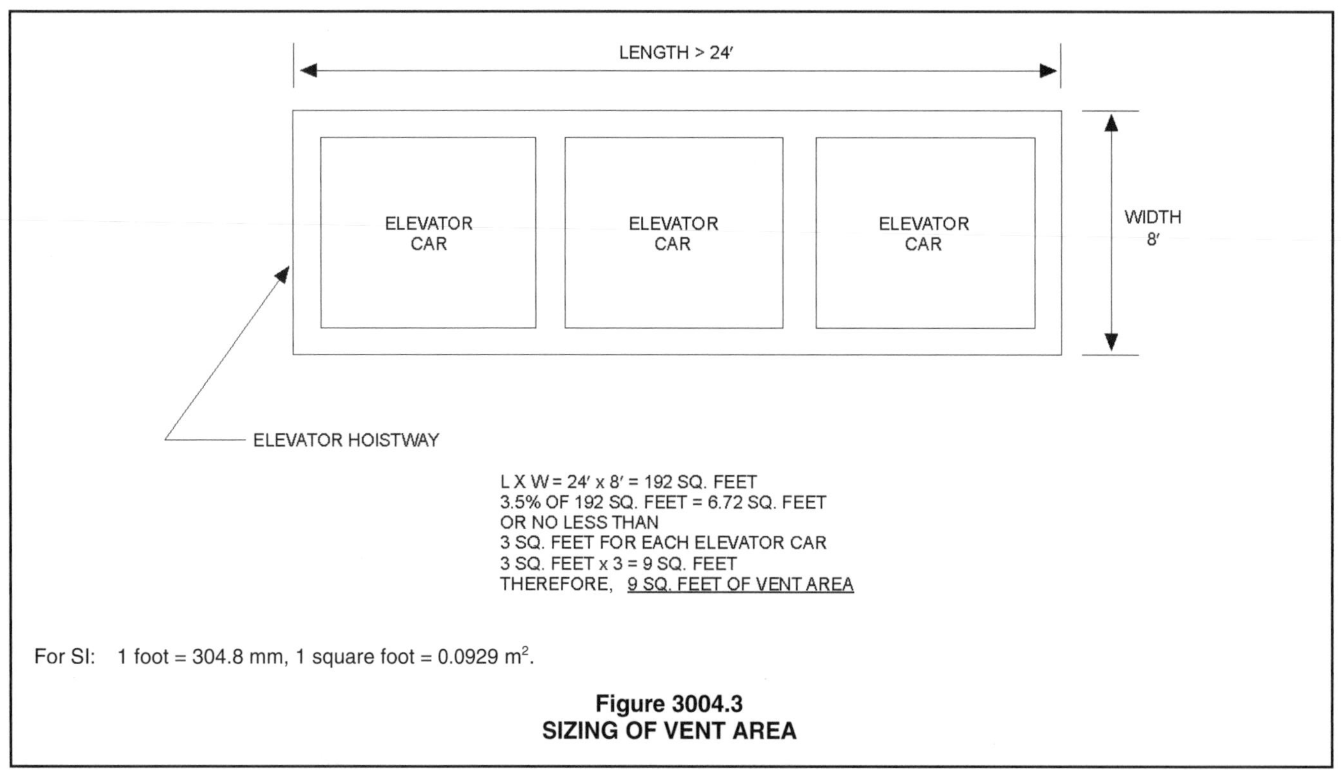

For SI: 1 foot = 304.8 mm, 1 square foot = 0.0929 m^2.

Figure 3004.3
SIZING OF VENT AREA

3005.2.1 Enclosure. Escalator floor openings shall be enclosed with *shaft enclosures* complying with Section 713.

❖ Escalator floor openings are required to be enclosed in shaft enclosures unless they are constructed in accordance with Section 712.1.3. Section 712.1.3 provides two methods of protection that would allow the floor opening without a shaft enclosure.

The first method is to provide draft curtains and sprinklers. The area of the floor opening is also limited to twice the area of the escalator. This is necessary since the protection method is not adequate for every size opening, especially when the floor area within the perimeter of the opening is used for stock storage, displays and similar uses (the atrium provisions of Section 404 are more appropriate for those use conditions). The use of this method is acceptable for connecting an unlimited number of stories in buildings with occupancies of Groups B and M. In buildings other than Groups B and M, the alternative can only be utilized to connect four stories. This provision acknowledges the current uses of this concept but also limits its use in buildings with higher fuel load and more hazardous occupancy characteristics, such as those ocupancies where people sleep (Groups I and R).

The sprinkler draft curtain method is detailed in NFPA 13. The method consists of surrounding the escalator opening, in an otherwise fully sprinklered building, with an 18-inch-deep (457 mm) draft curtain located on the underside of the floor to which the escalator or moving walk ascends. This serves to delay the heat, smoke and combustion gases (developed in the early stages of a fire) on that floor from entering into the escalator or moving walk well. A row of closely spaced automatic sprinklers located outside of the draftstop also surrounds the opening. When activated by heat, the sprinklers provide a water curtain. A typical installation is shown in Figure 3005.2.1. In combination with the sprinkler system in the building, this system delays fire spread to allow time for evacuation.

The second method is to provide power-operated automatic fire shutters at all floor openings. The rolling shutter completely closes off the opening between floors in the event of a fire. When the shutter is actuated, power to the escalator or moving walk must be automatically disconnected. To provide protection against entrapment of an occupant, the shutter is required to have a sensitive leading edge such that it stops closing when it comes in contact with an obstacle. Once the obstacle is removed, the shutter continues to close.

3005.2.2 Escalators. Where provided in below-grade transportation stations, escalators shall have a clear width of not less than 32 inches (815 mm).

> **Exception:** The clear width is not required in existing facilities undergoing *alterations*.

❖ A minimum 32-inch (864 mm) width for escalators servicing below-grade transportation systems, such as subways, is based on minimum accessibility requirements.

3005.3 Conveyors. Conveyors and conveying systems shall comply with ASME B20.1.

❖ ASME B20.1 applies to the design, construction, installation, maintenance, inspection and operation of

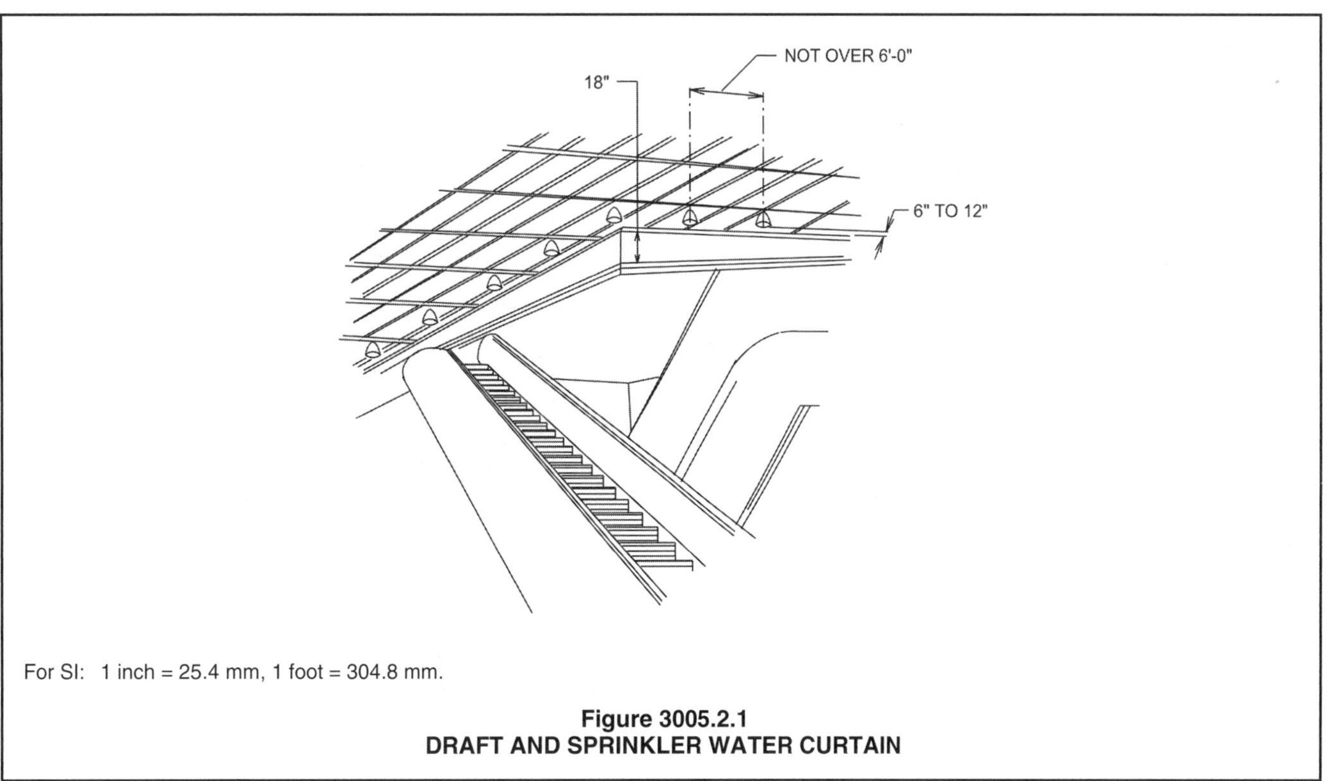

For SI: 1 inch = 25.4 mm, 1 foot = 304.8 mm.

Figure 3005.2.1
DRAFT AND SPRINKLER WATER CURTAIN

conveyors and conveying systems in relation to hazards. The conveyors may be of bulk material, package or unit-handling types where the installation is designed for permanent, temporary or portable operation. This standard must apply, along with the requirements noted below, to all conveyor installations.

This standard specifically excludes any conveyor designed, installed or used primarily for the movement of human beings. It does, however, apply to certain conveying devices that have incorporated platforms or control rooms specifically designed for authorized operating personnel within their supporting structure.

This standard does not apply to equipment, such as industrial trucks, tractors and trailers, tiering machines (except pallet load tiers), cranes, hoists, power shovels, power scoops, bucket drag lines, trenchers, elevators, manlifts, moving walks, escalators, highway or rail vehicles, cableways, tramways or pneumatic conveyors.

3005.3.1 Enclosure. Conveyors and related equipment connecting successive floors or levels shall be enclosed with *shaft enclosures* complying with Section 713.

❖ Protection against the migration of smoke and fire must be maintained whenever a conveyor penetrates a floor/ceiling assembly, and thus connects floor levels. This section references Section 713 for the requirements for shaft enclosures.

3005.3.2 Conveyor safeties. Power-operated conveyors, belts and other material-moving devices shall be equipped with automatic limit switches which will shut off the power in an emergency and automatically stop all operation of the device.

❖ Conveyor controls must be arranged such that in an emergency, the conveyor will stop all operation. All safety devices, including wiring of electrical safety devices, are required to be arranged to operate in a "fail-safe" manner; that is, if power failure or failure of the device itself occurs, a hazardous condition must not result.

3005.4 Personnel and material hoists. Personnel and material hoists shall be designed utilizing an *approved* method that accounts for the conditions imposed during the intended operation of the hoist device. The design shall include, but is not limited to, anticipated loads, structural stability, impact, vibration, stresses and seismic restraint. The design shall account for the construction, installation, operation and inspection of the hoist tower, car, machinery and control equipment, guide members and hoisting mechanism. Additionally, the design of personnel hoists shall include provisions for field testing and maintenance which will demonstrate that the hoist device functions in accordance with the design. Field tests shall be conducted upon the completion of an installation or following a major *alteration* of a personnel hoist.

❖ In lieu of a referenced standard, personnel and material hoists are to be designed and installed using a method that includes structural stability, anticipated loads, impact vibration, stresses and seismic restraint. The building official should examine aspects of the hoists, such as construction, installation and operation of the hoist tower before final approval.

Personnel hoist design must also include field tests upon completion of installation or major alteration and maintenance, which demonstrate that hoist device functions are in accordance with the design.

SECTION 3006
MACHINE ROOMS

3006.1 Access. An *approved* means of access shall be provided to elevator machine rooms and overhead machinery spaces.

❖ A means of access to the elevator machine room is required by this section and must be specifically approved by the building official. Things to consider in providing for this access are the maintenance and replacement of the equipment, and the necessary inspections to be performed. Note that Section 1009.14 requires a stairway to the roof to access elevator equipment and would not allow ladders or alternating tread devices.

3006.2 Venting. Elevator machine rooms that contain solid-state equipment for elevator operation shall be provided with an independent *ventilation* or air-conditioning system to protect against the overheating of the electrical equipment. The system shall be capable of maintaining temperatures within the range established for the elevator equipment.

❖ Elevator machine rooms may contain equipment that is temperature sensitive. If this equipment is not kept within its safe-temperature range, the possibility exists for a malfunction, which could trap occupants within the elevator. Additionally, if this malfunction occurred during an emergency condition, it could impede emergency personnel in their efforts to perform fire-fighting or rescue duties. ASME A17.1/CSA B44, Section 2.7.9.2, requires a tag in the machine room that specifies the temperature and humidity range required to be maintained for operation of the elevator equipment.

Elevator technology has, in some cases, eliminated machine rooms, but equipment still exists that may be temperature sensitive. It is within the intent of this section that such equipment be protected in accordance with the manufacturer's recommendations.

3006.3 Pressurization. The elevator machine room serving a pressurized elevator hoistway shall be pressurized upon activation of a *heat or smoke detector* located in the elevator machine room.

❖ To prevent the migration of smoke and heat into a machine room from a pressurized hoistway enclosure, the machine room is required to be pressurized as well. However, the machine room only needs to be pressurized when heat or smoke is present. The pressurization is required to occur upon activation of a smoke or heat detector located in the machine room. This is only a requirement when the hoistway is pressurized.

3006.4 Machine rooms and machinery spaces. Elevator machine rooms and machinery spaces shall be enclosed with *fire barriers* constructed in accordance with Section 707 or *horizontal assemblies* constructed in accordance with Section 711, or both. The *fire-resistance rating* shall be not less than the required rating of the hoistway enclosure served by the machinery. Openings in the *fire barriers* shall be protected with assemblies having a *fire protection rating* not less than that required for the hoistway enclosure doors.

Exceptions:

1. Where machine rooms and machinery spaces do not abut and have no openings to the hoistway enclosure they serve the *fire barriers* constructed in accordance with Section 707 or *horizontal assemblies* constructed in accordance with Section 711, or both, shall be permitted to be reduced to a 1-hour *fire-resistance rating*.
2. In buildings four *stories* or less above *grade plane* where machine room and machinery spaces do not abut and have no openings to the hoistway enclosure they serve, the machine room and machinery spaces are not required to be fire-resistance rated.

❖ This section acknowledges that there will inherently be openings between the hoistway and the machine room. This protection is required to be both vertical and horizontal; therefore, to maintain the fire-resistance rating required for the hoistway, the machine room must be protected with a fire barrier and horizontal assemblies, both with opening protectives. There are two exceptions to this requirement. Exception 1 allows the reduction of the fire-resistance-rated enclosure to 1 hour where the machine room or space does not abut the hoistway. In other words, the machine room or space is in a location where no connection is made with the hoistway. This reduction only applies in buildings connecting four or more stories because the code only requires a 1-hour enclosure for the hoistway and machine room in buildings with elevator hoistways connecting three stories or less. Exception 2 provides the reduction from 1 hour to no fire-resistance rating for buildings connecting four stories or less. Exception 2 is dependent upon the machine room or machinery space not abutting and having no openings into the hoistway.

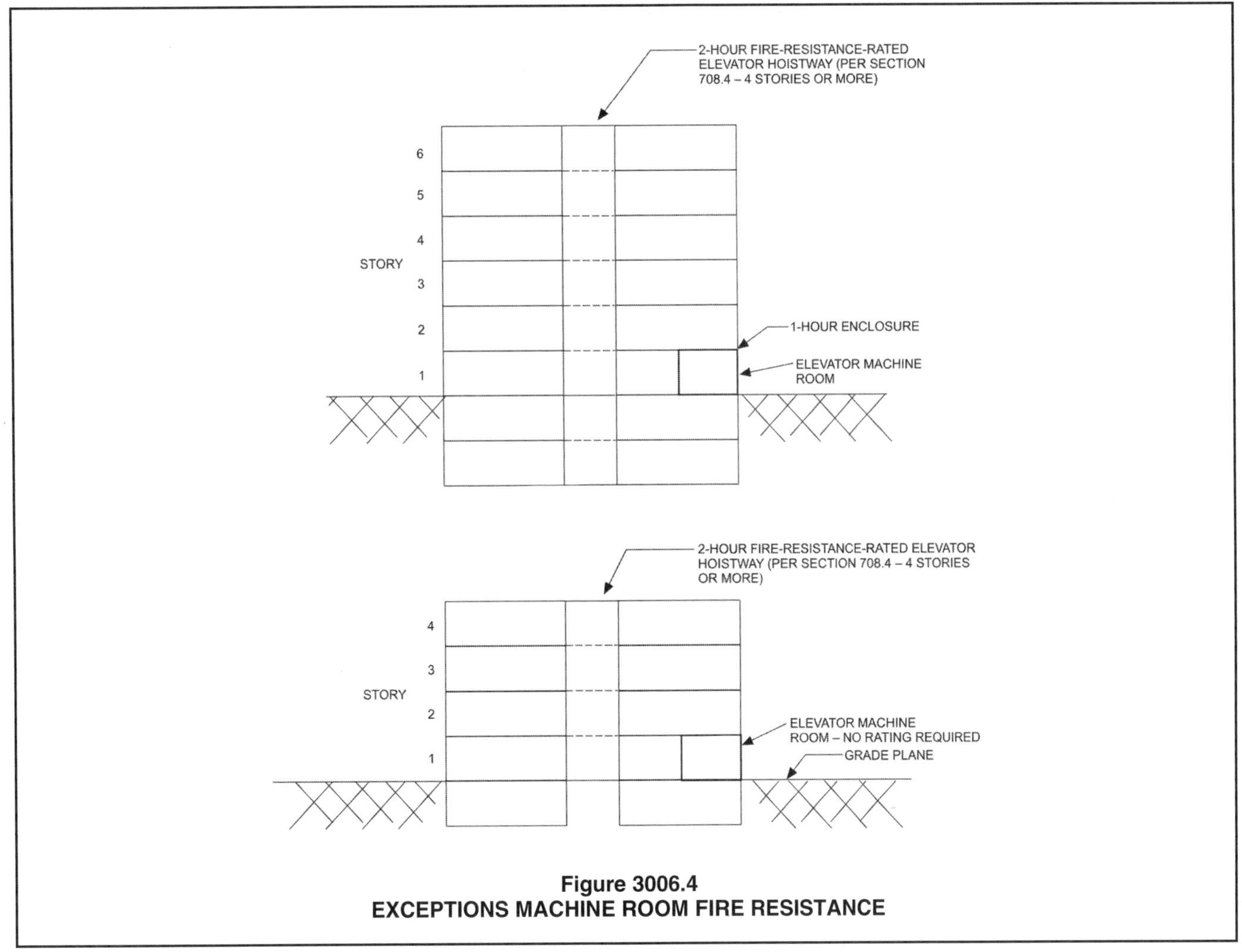

Figure 3006.4
EXCEPTIONS MACHINE ROOM FIRE RESISTANCE

3006.5 Shunt trip. Where elevator hoistways or elevator machine rooms containing elevator control equipment are protected with automatic sprinklers, a means installed in accordance with NFPA 72, Section 6.16.4, Elevator Shutdown, shall be provided to disconnect automatically the main line power supply to the affected elevator prior to the application of water. This means shall not be self-resetting. The activation of sprinklers outside the hoistway or machine room shall not disconnect the main line power supply.

❖ This section acknowledges the hazards of a sprinkler system, mainly electrical malfunction, to an operating elevator and elevator equipment contained in the machine room. Therefore, the main power supply line to the affected elevator(s) is required to automatically disconnect, and not be reset, before the suppression system is activated.

3006.6 Plumbing systems. Plumbing systems shall not be located in elevator equipment rooms.

❖ If a pipe or duct that conveys gases, vapors or liquids and passes through a machine room or machinery space were to fail, the operation of the equipment could be affected in such a manner as to make it unsafe. Water could cause the malfunctioning of a safety circuit, such as hoistway door interlocks, thereby allowing an elevator to leave a floor with the hoistway door open.

SECTION 3007
FIRE SERVICE ACCESS ELEVATOR

3007.1 General. Where required by Section 403.6.1, every floor of the building shall be served by fire service access elevators complying with Sections 3007.1 through 3007.10. Except as modified in this section, fire service access elevators shall be installed in accordance with this chapter and ASME A17.1/CSA B44.

❖ Section 3007 provides a package of requirements for what is termed a fire service access elevator. A minimum of one such elevator is required in buildings with an occupied floor more than 120 feet (36 576 mm) above the lowest level of fire department vehicle access in accordance with Section 403.6.1. These requirements were developed based upon a research initiative by the U.S. General Services Administration (GSA) focused upon the development of performance requirements for the use of elevators for occupant egress and fire service access in buildings. Section 3007 is a by-product of the research by National Institute of Standards and Technology (NIST) as well preliminary information provided by a task group of ASME A17.1/CSA B44 for determining the required system features necessary for safe operation by trained fire fighters during a fire emergency.

The requirements in Section 3007 are intended to provide a reasonable degree of safety for fire fighters operating the fire service access elevator to a location for staging fire fighters and equipment. In a typical situation the staging area will be set up one or two floors below the fire. The staging location already will have access to an exit stairway and standpipe that will allow for fire-fighting operations to be conducted from just above the staging area.

It is important to note that these requirements are in addition to the requirements for Phase I Recall and Phase II Emergency Operation. Therefore these elevators would also have these standard features plus the additional package of requirements.

A fire service access elevator may be an elevator typically used for freight, service elevator, or one of the passenger elevators for general public use. Every floor within a building must be served by at least one fire service access elevator.

The fire service access elevator requirements are intended to work in conjunction with requirements in other codes and standards, especially those that will appear in the elevator code, ASME A17.1/CSA B44. A key part of these requirements is the operation protocol for the fire service access elevator.

3007.2 Phase I Emergency recall operation. Actuation of any building fire alarm-initiating device shall initiate Phase I emergency recall operation on all fire service access elevators in accordance with the requirements in ASME A17.1/CSA B44. All other elevators shall remain in normal service unless Phase I emergency recall operation is manually initiated by a separate, required three-position, key-operated "Fire Recall" switch or automatically initiated by the associated elevator lobby, hoistway or elevator machine room *smoke detectors*. In addition, if the building also contains occupant evacuation elevators in accordance with Section 3008, an independent, three-position, key-operated "Fire Recall" switch conforming to the applicable requirements in ASME A17.1/CSA B44 shall be provided at the designated level for each fire service access elevator.

❖ The intent of this section is to provide further clarification regarding the design and operation of fire service access elevators. This section is included to ensure the subject elevators can be recalled quickly at the designated level by the responding fire fighters.

3007.3 Automatic sprinkler system. The building shall be equipped throughout with an *automatic sprinkler system* in accordance with Section 903.3.1.1, except as otherwise permitted by Section 903.3.1.1.1 and as prohibited by Section 3007.3.1.

❖ This section is intended to provide for specific requirements regarding automatic sprinkler systems for fire service access elevators.

The intent of this section is to provide further clarification in meeting the original intent of Section 3007 regarding prohibiting the installation of automatic sprinklers in the associated elevator machine rooms and elevator machine spaces for fire service access elevators.

3007.3.1 Prohibited locations. Automatic sprinklers shall not be installed in elevator machine rooms, elevator machine spaces, and elevator hoistways of fire service access elevators.

❖ The intent of this section is to provide further clarification in meeting the original intent of Section 3007 regarding prohibiting the installation of automatic sprinklers in the associated elevator machine rooms and elevator machine spaces for fire service access elevators. Specifically, there are a few key locations where sprinklers may be a detriment to the use of elevators for evacuation. If sprinklers are located in a machine room or machinery space and they activate, damage to the elevators may occur and threaten the safety of elevators for use by building occupants. Additionally, such spaces should be protected by smoke detection that would initiate Phase I Recall prior to sprinkler activation (see also commentary, Section 903.3.1.1.1, Exception 6).

3007.3.2 Sprinkler system monitoring. The sprinkler system shall have a sprinkler control valve supervisory switch and waterflow-initiating device provided for each floor that is monitored by the building's *fire alarm system.*

❖ This section allows the origin of the fire to be better determined and will assist in managing a fire emergency. Many of the provisions of this section are in place since using elevators for evacuation is a new concept and many safety factors were deemed necessary. In the future it is anticipated that such monitoring will assist in determining the priority for elevator locations in order to evacuate buildings using elevators more efficiently.

3007.4 Water protection. An *approved* method to prevent water from infiltrating into the hoistway enclosure from the operation of the *automatic sprinkler system* outside the enclosed fire service access elevator lobby shall be provided.

❖ This section provides performance-based language to prevent water from the discharge of automatic sprinklers located outside the elevator lobby from infiltrating into the hoistway opening. It is important to stress that this requirement is only for protection from sprinkler system water and not water from fire-fighter hoses. Also, the language is performance oriented as there are several strategies that could be used to address this problem and a single solution may not fit in all situations. Solutions include concepts such as trench drains, sloped floors, curbs and gasketed openings, which may be used individually or in combination. Other methods may also be available to meet this intent. It should be stressed that the intent of this section is focused upon the actual hoistway opening versus the walls surrounding the hoistway.

3007.5 Shunt trip. Means for elevator shutdown in accordance with Section 3006.5 shall not be installed on elevator systems used for fire service access elevators.

❖ The intent of this section is to provide further clarification in meeting the original intent of Section 3007 regarding prohibiting the installation of shunt trip for fire service access elevators. The subject language is similar to the language in Section 3008.8 for occupant evacuation elevators.

3007.6 Hoistway enclosures. The fire service access elevator hoistway shall be located in a *shaft enclosure* complying with Section 708.

❖ Reference is made to Section 708 of the code for the required fire resistance and construction of the hoistway enclosure. Section 708 contains fire-resistance-rating requirements, as well as construction requirements for shafts, which is typically the method used to satisfy the requirements for a hoistway enclosure. Section 403.2 provides additional standards, and some exceptions, for hoistway construction in high-rise buildings, but the exceptions do not apply to elevator hoistway enclosures.

3007.6.1 Structural integrity of hoistway enclosures. The fire service access elevator hoistway enclosure shall comply with Sections 403.2.3.1 through 403.2.3.4.

❖ This section requires additional structural integrity for elevator hoistway enclosures for fire service access elevators. Please see the commentary for Sections 403.2.3.1 through 403.2.3.4.

Table 3007.4
LOBBY COMPARISON TABLE—FIRE SERVICES ACCESS ELEVATORS AND OCCUPANT EVACUATION ELEVATORS

LOBBY REQUIREMENT COMPARISON		
Element of Lobby	**Fire Service Access Elevator (Section 3007)**	**Occupant Evacuation Elevators (Section 3008)**
Lobby Size	150-sq ft minimum with 8-foot minimum dimension	3 square feet per person for 25% of occupants and 30-inch by 48-inch wheelchair for each 50 occupants
Lobby Enclosure	Not required at street floor	Not required at level(s) of exit discharge
Door	$^3/_4$-hour door tested without bottom seal	Automatic closing $^3/_4$-hour door with a vision panel
Wall	1-hour smoke barrier	1-hour smoke barrier
Access	Direct acccess to exit enclosure	Direct acccess to exit enclosure

For SI: 1 inch = 25.4 mm, 1 foot = 304.8 mm, 1 square foot = 0.0929 m^2.

3007.6.2 Hoistway lighting. When fire-fighters' emergency operation is active, the entire height of the hoistway shall be illuminated at not less than 1 footcandle (11 lux) as measured from the top of the car of each fire service access elevator.

❖ The focus of this section is to provide illumination to assist fire fighters as they advance up into the building. The prescribed procedure without hoistway illumination before leaving the designated level (DL) is to shine a light up into the hoistway to try and detect smoke, flame or water above them. They will repeat this step every five floors until they safely arrive at their staging floor, which is two floors below the lowest reported floor in alarm. If fire fighters become trapped in a stopped elevator and need to self-rescue through the top of the car's emergency exit, they need adequate light to safely escape. Hoistway lighting will make their life-safety maneuver much more effective. The illumination level is the same as that used in Section 1006.

3007.7 Fire service access elevator lobby. The fire service access elevator shall open into a fire service access elevator lobby in accordance with Sections 3007.7.1 through 3007.7.5.

Exception: Where a fire service access elevator has two entrances onto a floor, the second entrance shall be permitted to open into an elevator lobby in accordance with Section 708.14.1.

❖ All fire service access elevators must open into an elevator lobby having specific requirements (e.g., direct access to exit enclosure and minimum fire resistance rating for lobby enclosure and door assemblies) providing a reasonable degree of safety for firefighters operating the fire service access elevator and a location for staging fire fighters and equipment one or two floors below the fire.

The exception clarifies that if the fire service access elevator has two entrances, the second entrance is permitted to open into an enclosed or otherwise protected elevator lobby meeting the requirements in section 708.14.1.

It is possible and the code would not prohibit a fire service access elevator(s) and the occupant evacuation elevator(s) to share the same elevator lobby. See also Table 3007.4 for a comparison of lobby requirements between fire service access elevators and elevators used for occupant self-evacuation.

3007.7.1 Access. The fire service access elevator lobby shall have direct access to an enclosure for an *interior exit stairway*.

❖ As part of the package of requirements, fire service access elevators need to have direct access to an exit enclosure. This relates to the fire fighters' staging requirements during a fire. The stairways are critical to bringing water hose to the fire floor. Keeping the elevators protected by a lobby and then locating an exit enclosure that the fire fighters can directly access will assist in the fire-fighting activities. See also the commentary to Section 3008.7.1 for a discussion regarding using the same exit enclosure for both the fire service access elevator and occupant evacuation elevators.

3007.7.2 Lobby enclosure. The fire service access elevator lobby shall be enclosed with a *smoke barrier* having a *fire-resistance rating* of not less than 1 hour, except that lobby doorways shall comply with Section 3007.7.3.

Exception: Enclosed fire service access elevator lobbies are not required at the *levels of exit discharge*.

❖ The fire service access elevator lobby is required to be enclosed by a smoke barrier having a minimum 1-hour fire-resistance rating which provides both protection from smoke and a level of fire resistance. This smoke barrier is specifically protecting the fire service access elevator lobby and would not be required to be constructed from outside wall to outside wall as required in Section 710.4. Note that the elevator lobby located on the street floor entrance to a building does not require an enclosed lobby. In many cases, such a lobby would be impractical to construct. In the event of a fire on the street level, the elevator would be directed to the alternate level once smoke is detected on that level.

It should be additionally noted that the enclosure created by smoke barriers protects the elevator and the associated lobby from the building. Therefore, protection between the elevator lobby and the elevator is not required.

3007.7.3 Lobby doorways. Other than the door to the hoistway, each doorway to a fire service access elevator lobby shall be provided with a $^{3}/_{4}$-hour *fire door assembly* complying with Section 716.5. The *fire door assembly* shall also comply with the smoke and draft control door assembly requirements of Section 716.5.3.1 with the UL 1784 test conducted without the artificial bottom seal.

❖ The requirement for a $^{3}/_{4}$-hour door is more restrictive than what is required for smoke barriers in Section 716.5. Section 716.5 and Table 716.5 require a 20-minute rating for doors located in smoke barriers. The reference back to Section 716.5 is primarily related to how the door is to be tested and the criteria for acceptance. More specifically, Section 716.5.3.1 requires doors used in smoke and draft control to be tested in accordance with UL 1784 and installed in accordance with NFPA 105. Additionally, a provision is included to address that the test method be performed without an artificial bottom seal installed across the full width of the bottom of the door assembly in order to measure the complete leakage rate for the door assembly on the assumption that leakage could occur under the bottom of the door in this application. This will ensure that the doorways not only have an adequate fire protection rating for protecting the fire service access elevator lobbies, but also protect against smoke intrusion into the lobby where the fire service will stage its operations before attacking a fire on the fire floor. This is a very important component of the overall protection package provided for the fire service access elevators.

3007.7.4 Lobby size. Each enclosed fire service access elevator lobby shall be not less than 150 square feet (14 m^2) in an area with a minimum dimension of 8 feet (2440 mm).

❖ A minimum lobby size of 150 square feet (14 m^2) with a minimum dimension of 8 feet (2438 mm) has been included to ensure that the lobby is of adequate size for conducting necessary fire-fighter operations. A majority of fire departments operate with high-rise fire attack teams comprised of either three or four fire fighters. The lobby should be large enough to accommodate one team preparing to enter the fire area (floor) and one team needing to withdraw from the fire area, while leaving access to the hoistway and exit enclosure doors clear. A minimum lobby size of 150 square feet (14 m^2) was intended to meet the above criteria. The minimum dimension of 8 feet (2438 mm) is included to avoid a long narrow lobby.

3007.7.5 Fire service access elevator symbol. A pictorial symbol of a standardized design designating which elevators are fire service access elevators shall be installed on each side of the hoistway door frame on the portion of the frame at right angles to the fire service access elevator lobby. The fire service access elevator symbol shall be designed as shown in Figure 3007.7.5 and shall comply with the following:

1. The fire service access elevator symbol shall be not less than 3 inches (76 mm) in height.
2. The vertical center line of the fire service access elevator symbol shall be centered on the hoistway door frame. Each symbol shall not be less than 78 inches (1981 mm), and not more than 84 (2134 mm) inches above the finished floor at the threshold.

❖ The "fire hat" symbol is already used inside the elevator cab and is, therefore, instantly recognizable by the fire service. This provision is intended to aid in the quick identification of the fire service access elevators and will assist the fire service in quickly indentifying these elevators and begin using them.

FIGURE 3007.7.5
FIRE SERVICE ACCESS ELEVATOR SYMBOL

3007.8 Elevator system monitoring. The fire service access elevator shall be continuously monitored at the *fire command center* by a standard emergency service interface system meeting the requirements of NFPA 72.

❖ As an added layer of protection, these critical elevators must be specifically monitored in the fire command center by a standard emergency service interface system in accordance with NFPA 72. More specifically, Section 7.11 of NFPA 72 addresses this interface. Basically, it is requiring that the interface such as annunciators, information display systems and controls work with the needs of the fire department in terms of design, arrangement and location. The requirements are fairly general and depend upon the jurisdiction regarding how they will be met. It should be noted that Annex F of NFPA 72 contains NEMA Standards Publication SB 30–2005, *Fire Service Annunciator and Interface*, which provides more detailed information about the design and layout of such systems. This document is written more as a guide and is not directly referenced within NFPA 72.

3007.9 Electrical power. The following features serving each fire service access elevator shall be supplied by both normal power and Type 60/Class 2/Level 1 standby power:

1. Elevator equipment.
2. Elevator hoistway lighting.
3. Elevator machine room *ventilation* and cooling equipment.
4. Elevator controller cooling equipment.

❖ This section provides specific requirements for normal and standby power for fire service access elevators. The standby power would additionally include the power supply for the ventilation equipment for the machine room or machinery space associated with the fire service access elevator and the hoistway lighting as required in Section 3007.6.2.

3007.9.1 Protection of wiring or cables. Wires or cables that are located outside of the elevator hoistway and machine room and that provide normal or standby power, control signals, communication with the car, lighting, heating, air conditioning, *ventilation* and fire-detecting systems to fire service access elevators shall be protected by construction having a *fire-resistance rating* of not less than 2 hours, or shall be circuit integrity cable having a *fire-resistance rating* of not less than 2 hours.

> **Exception:** Wiring and cables to control signals are not required to be protected provided that wiring and cables do not serve Phase II emergency in-car operations.

❖ This section provides additional requirements for the protection of specific wiring and cables for each fire service access elevator. Typical building elevators do not require this level of protection for these wires and cables.

3007.10 Standpipe hose connection. A Class I standpipe hose connection in accordance with Section 905 shall be provided in the *interior exit stairway* and *ramp* having direct access from the fire service access elevator lobby.

❖ Locating the hose connection in the exit enclosure having direct access to the elevator lobby relates directly to the needs of fire fighters when fighting a fire.

Fire fighters typically attack a fire from the floor below, and this allows the shortest path for the fire hose to be laid. If the standpipe were located in the lobby itself, it would simply mean that the exit enclosure door would be propped open on the floor below the fire floor.

Keeping the standpipe within the exit enclosure also provides a protected location for the fire department to gain access to the hose while keeping the lobby, and thus the elevator, more protected. This particular issue has also been reviewed extensively by the ASME task group looking at the use of elevators for fire fighting and similar conclusions in terms of placement within the exit enclosure were made.

3007.10.1 Access. The *exit* enclosure containing the standpipe shall have access to the floor without passing through the fire service access elevator lobby.

❖ Access from the exit enclosure containing the standpipe to the floor is necessary so that the fire department can advance its attack hose onto the fire floor without opening the door between the lobby and the floor, which could permit smoke contamination of the lobby and cause recall of the elevator(s). This access to the floor could be direct or through an access corridor or vestibule between the elevator lobby and the exit enclosure, as long as there is a smoke barrier enclosing the elevator lobby.

SECTION 3008 OCCUPANT EVACUATION ELEVATORS

3008.1 General. Where elevators are to be used for occupant self-evacuation during fires, all passenger elevators for general public use shall comply with Sections 3008.1 through 3008.11. Where other elevators are used for occupant self-evacuation, they shall also comply with these sections.

❖ This section introduces new provisions in the code that are an option for architects to consider, particularly in designing tall buildings. More specifically Section 403.6.2 permits the use of all passenger elevators for general public use to be designated as occupant evacuation elevators in accordance with the provisions in this section. It should be noted that these provisions could be used in buildings other than high-rise buildings as long as all the requirements in Section 3008 are met. The use of these elevators is intended for all building occupants prior to Phase I recall activation of the elevators.

Based on GSA/NIST research and the efforts of the ASME A17/CSA B44, *Task Groups on Elevators and Fire*, the use of these elevators by building occupants will improve overall building evacuation time compared to only utilizing the required exit stairways. Additionally, using elevators will also provide improved evacuation capability for persons with mobility impairments. The requirements provide the necessary safeguards to ensure a reasonable degree of safety for the use of elevators by occupants during building evacuations in normal elevator operating mode prior to Phase I emergency recall.

It should be noted that the provisions are similar to those applied to the fire service access elevator except that the enclosed lobby on each floor is required to be provided with an indicator light system (operated by the building fire alarm system), which displays a green light and associated text indicating that the elevators are available for occupant evacuation. This is to happen whenever there is a fire alarm in the building and the elevators are not in Phase I recall or Phase II emergency operation. A red light and associated text indicate that the elevators are not in service—and that occupants should only use the exit stairways for egress when the elevators go into Phase I emergency recall operation in accordance with the requirements of ASME A17.1/CSA B44. The minimum lobby size is based on what is needed to accommodate 25 percent of the occupant load of the floor at 3 square feet (0.27 m^2) per person, plus one wheelchair space [30 inches by 48 inches (762 mm by 1219 mm)] for each 50 people (or fraction). See Section 3008.7.4 for more information on sizing lobbies.

It should be noted that this section infers that only passenger elevators designed for general public use and not elevators designated as private elevators or freight elevators are required to meet all of the provisions in this section. However, elevators that are not for general public use could be used for occupant self-evacuation but would need to meet the requirements of Section 3008.

There are also some instances where elevators were not intended to be included, such as those that simply travel between the parking garage and the main lobby that are technically passenger elevators. The code does not specifically address this case, but due to the limited use of such elevators, use for evacuation would not provide much benefit in a high-rise building.

Although elevators are specifically being used for evacuation, that would not prohibit building occupants from taking the exit stairways during a fire. Occupant evacuation elevators are an option that will decrease the time needed for evacuation and reduce the likelihood of interfering with the fire-fighting and rescue operations.

3008.1.1 Additional exit stairway. Where an additional *means of egress* is required in accordance with Section 403.5.2, an additional *exit stairway* shall not be required to be installed in buildings provided with occupant evacuation elevators complying with Section 3008.1.

❖ This section simply explains that occupant self-evacuation with elevators is an alternative to the additional exit stairway required for high-rise buildings over 420 feet (128 016 mm) in building height.

3008.1.2 Fire safety and evacuation plan. The building shall have an *approved* fire safety and evacuation plan in accordance with the applicable requirements of Section 404 of the *International Fire Code*. The fire safety and evacuation plan shall incorporate specific procedures for the occupants using evacuation elevators.

❖ This section requires a fire safety and evacuation plan in accordance with Section 404 of the *International Fire Code®* (IFC®). For high-rise buildings, this is already required, but it should be remembered that this section allowing the use of elevators for evacuation is not limited to use with high-rise buildings and could be applied to other building types that may not require such planning. This section also ensures the elevators are specifically accounted for during the planning, as they are not currently listed in Section 404 of the IFC.

Most building occupants are currently unfamiliar with the concept of using elevators for evacuation. Also, in very tall buildings the application of this concept has great benefits but the implementation can be complex. This code and the IFC do not mandate how a building is to be evacuated but instead provide the elements to assist in evacuation, such as proper exit width; limitations on travel distance; life safety features such as sprinklers and many other elements that assist occupants in exiting a building safely. Therefore, the way in which the elevators are used for evacuation must be tailored to the individual building and occupants. In most cases, when a fire occurs in a high-rise building the protocol is not to evacuate the entire building but to phase the evacuation and possibly limit only to the fire floor and the two floors above and below. Therefore, the use of elevators for evacuation needs to be coordinated with that procedure. An example of how such an operation would work is as follows:

A phased evacuation of the fire zone (fire floor and two floors above and below the fire floor) is initiated with appropriate voice announcements and the elevators take occupants of these five floors to the level of exit discharge. While in evacuation operation, the elevators operate in shuttle mode between the five affected floors and the level of exit discharge, with the car buttons disabled and the hall call buttons arranged to tell the controller that there are occupants waiting on the other floors but not controlling the elevator operation.

In most cases, evacuation of the fire zone should be completed at about the time of fire department arrival, and once complete, the occupant evacuation elevators are taken out of service [Fire-Fighter Emergency Operation (FEO) Phase I] until a decision for full evacuation is made by the fire department. Once a full evacuation decision is made, the occupant egress elevators are placed in evacuation mode [Emergency Evacuation Operation (EEO)] and announcements are made on the affected floors. The elevators then begin to move occupants from their floor to the level of exit discharge in shuttle mode, from the highest floors downward, so that occupants with the greatest distance to travel are moved first. Should any occupants left arrive in the lobby on a floor already evacuated and push the hall call button, a single elevator would be dispatched to that floor to pick them up. Alternatively, evacuation assistance can be provided by the fire service using the access elevator under manual (FEO Phase II) operation.

It should be noted that the ASME A17.1/CSA B44, *Elevator Code*, listed in Chapter 35, does not have occupant evacuation elevator requirements that correspond with this new code section. However, technology exists within the elevator industry to program elevators in ways to complement and coordinate with these evacuation requirements, and elevator manufacturers can be expected to provide systems that will meet the evacuation objectives. The elevator industry is currently considering creating a standardized EEO for elevators. The key is understanding how a particular building is to function during a fire.

3008.2 Phase I Emergency recall operation. An independent, three-position, key-operated "Fire Recall" switch complying with ASME A17.1/CSA B44 shall be provided at the designated level for each occupant evacuation elevator.

❖ The use of an occupant self-evacuation elevator is only allowed up and until the elevator is recalled on Phase I. This section ensures that the subject (as specified) elevators can be recalled quickly at the designated level by the responding fire fighters.

3008.2.1 Operation. The occupant evacuation elevators shall be used for occupant self-evacuation only in the normal elevator operating mode prior to Phase I Emergency Recall Operation in accordance with the requirements in ASME A17.1/CSA B44 and the building's fire safety and evacuation plan.

❖ The use of an occupant self-evacuation elevator is only allowed up and until the elevator is recalled on Phase I. Once Phase I has been activated either automatically or manually, the continued use of the elevators for self-evacuation would be considered unsafe and they would no longer be available for such use. After the elevator is recalled, the building occupants served by those elevators would then need to take the exit stairways (and persons with mobility impairments will need evacuation by fire fighters using FEO Phase II). In some buildings, there may be multiple banks of elevators throughout the building. Depending upon the fire safety and evacuation plan and the particular building design, occupants could seek other elevators that have not been recalled. As discussed in the commentary to Section 3008.1.2, the way in which a building is to be evacuated prior to Phase I using the elevators is dependant on the specific building and its associated fire safety and evacuation plan.

3008.2.2 Activation. Occupant evacuation elevator systems shall be activated by any of the following:

1. The operation of an *automatic sprinkler system* complying with Section 3008.3;
2. *Smoke detectors* required by another provision of the code;
3. *Approved* manual controls.

❖ This new section provides a minimum set of initiating devices to activate the automatic operation of occupant evacuation elevator systems. An example of smoke detectors required by another section of this code includes smokeproof enclosures for the mechanical ventilation or stair pressurization alternative. An engineering analysis should be required for occupant evacuation elevators that includes a section on system activation.

3008.3 Automatic sprinkler system. The building shall be protected throughout by an *approved*, electrically supervised *automatic sprinkler system* in accordance with Section 903.3.1.1, except as otherwise permitted by Section 903.3.1.1.1 and as prohibited by Section 3008.3.1.

❖ All high-rise buildings will be required to have a supervised sprinkler system; however, Section 3008 can apply to buildings that may or may not require such systems. This section simply ensures that all buildings using this option contain supervised sprinklered systems.

3008.3.1 Prohibited locations. Automatic sprinklers shall not be installed in elevator machine rooms and elevator machine spaces for occupant evacuation elevators.

❖ There are a few key locations wherein sprinklers may be a detriment to the use of elevators for evacuation. If sprinklers are located in a machine room or machinery space and they activate, damage to the elevators may occur and threaten the safety of elevators for use by building occupants. Additionally, such spaces should be protected by smoke detection which would initiate Phase I recall prior to sprinkler activation. See also the commentary to Section 903.3.1.1.1, Exception 6.

3008.3.2 Sprinkler system monitoring. The sprinkler system shall have a sprinkler control valve supervisory switch and water flow-initiating device provided for each floor that is monitored by the building's *fire alarm system*.

❖ This allows the origin of the fire to be better determined and will assist in managing a fire emergency. Many of the provisions of this section are in place since using elevators for evacuation is a new concept and many safety factors were deemed necessary. In the future it is anticipated that such monitoring will assist in determining the priority for elevator location when evacuating buildings using elevators.

3008.4 Water protection. An *approved* method to prevent water from infiltrating into the hoistway enclosure from the operation of the *automatic sprinkler system* outside the enclosed occupant evacuation elevator lobby shall be provided.

❖ This section provides performance-based language to prevent water from the discharge of automatic sprinklers located outside the elevator lobby from infiltrating into the hoistway opening. It is important to stress that this requirement is only for protection from sprinkler system water and not water from fire-fighter hoses. Also, the language is performance oriented as there are several strategies that could be used to address this problem and a single solution may not fit in all situations. Solutions include concepts such as trench drains, sloped floors, curbs and gasketed openings, which may be used individually or in combination. Other methods may also be available to meet this intent. It should be stressed that the intent of this section is focused upon the actual hoistway opening versus the walls surrounding the hoistway.

3008.5 Shunt trip. Means for elevator shutdown in accordance with Section 3006.5 shall not be installed on elevator systems used for occupant evacuation elevators.

❖ A shunt trip has the possibility of inappropriately shutting down elevators that are removing people from a hazardous situation. Section 3008.3.1 prohibits sprinklers in machine rooms and machine spaces to prevent the shutdown of elevators due to water damage (see commentary, Section 3008.3.1).

3008.6 Hoistway enclosure protection. Occupant evacuation elevator hoistways shall be located in *shaft enclosures* complying with Section 713.

❖ Reference is made to Section 713 of the code for the required fire resistance and construction of the hoistway enclosure. Section 713 contains fire-resistance rating requirements, as well as construction requirements for shafts, which are typically the method used to satisfy the requirements for a hoistway enclosure. Section 403.2 provides additional standards, and some exceptions, for hoistway construction in high-rise buildings but the exceptions do not apply to elevator hoistway enclosures. Also note that Section 403.2.3 would require additional structural integrity for elevator hoistway enclosures in buildings more than 420 feet (128 016 mm) in building height.

3008.6.1 Structural integrity of hoistway enclosures. Occupant evacuation elevator hoistway enclosures shall comply with Sections 403.2.3.1 through 403.2.3.4.

❖ The code requires special structural provisions for exit stairways and elevator shaft enclosures in super high-rise buildings [greater than 420 feet (128 016 mm) in height]. It follows that the structural integrity requirements for super high-rise building exit stairway and elevator hoistway shaft enclosures should also apply to elevator hoistway shaft enclosures provided for occupant evacuation elevators, which are just as critical for life-safety protection. This new technology for evacuation of occupants must be provided with the highest level of fire protection that is reasonably possible in order to ensure that the elevators will be available during a fire emergency to serve their intended purpose of evacuating the occupants. This section requires that the structural integrity of the elevator hoistway shaft enclosures be required to have some reasonable degree of physical protection to ensure that the hoistway shaft enclosures will remain in place when needed during a fire or other emergency.

3008.7 Occupant evacuation elevator lobby. The occupant evacuation elevators shall open into an elevator lobby in accordance with Sections 3008.7.1 through 3008.7.7.

❖ One of the concerns when using elevators for evacuation is the protection of occupants while they wait for elevators. Therefore, a more protected elevator lobby is offered to occupants. As will be seen, these requirements have some similarities to the lobbies required for fire service access elevators. A major difference, however, is that lobbies be provided for all passenger elevators for general public use, not simply for one elevator designated as a fire service access elevator.

Since elevators used for occupant evacuation will be at a minimum all passenger elevators for general public use in buildings, it is very likely that one of these elevators will also be the fire service access elevator. Therefore, they will likely share the same lobby and the most restrictive provisions will apply. See Table 3007.4 for a comparison of lobby requirements.

3008.7.1 Access. The occupant evacuation elevator lobby shall have direct access to an *interior exit stairway* or *ramp*.

❖ As with the fire service access elevators, there must be direct access to an exit enclosure from the lobby. However, the purpose is slightly different. The reason is not to facilitate fire fighting activities but instead provide an exit in close proximity if the elevators are no longer available, thus facilitating a quick evacuation. This also allows those who no longer want to wait to leave via the exit stairways without having to leave the protected environment provided by the lobby enclosure.

Neither Section 3007 nor 3008 prohibits the exit enclosure directly accessed by the lobbies for occupant evacuation from being the same lobby that is for the fire service access elevator. Note that occupant evacuation using elevators will be much faster than using the exit stairways and will also greatly reduce the load on the exit stairways. This decrease in evacuation time will, therefore, reduce the likelihood that the occupants and fire fighters will interfere with one another. This is especially true if the building is only being partially evacuated.

3008.7.2 Lobby enclosure. The occupant evacuation elevator lobby shall be enclosed with a *smoke barrier* having a *fire-resistance rating* of not less than 1 hour, except that lobby doorways shall comply with Section 3008.7.3.

Exception: Enclosed occupant evacuation elevator lobbies are not required at the *levels of exit discharge*.

❖ The enclosure rating is the same as that required for the fire service access elevator lobby, which is a 1-hour smoke barrier. A smoke barrier provides both a fire-resistance rating and an increased level of protection from smoke. This is more specific and restrictive than the elevator lobby requirements found in Section 713.14.1. Again, the use of elevators for evacuation makes lobbies more critical, as they are now part of the life safety systems of a building.

Similar to the fire service access elevators, there is a single exception to the enclosure requirement. In the case of elevators used for egress, this exception is at the level of exit discharge(s).

3008.7.3 Lobby doorways. Other than the door to the hoistway, each doorway to an occupant evacuation elevator lobby shall be provided with a $^{3}/_{4}$-hour *fire door assembly* complying with Section 716.5. The *fire door assembly* shall also comply with the smoke and draft control assembly requirements of Section 716.5.3.1 with the UL 1784 test conducted without the artificial bottom seal.

❖ The lobby doorways are required to be $^{3}/_{4}$-hour protected in accordance with Section 716.5. The requirement for a $^{3}/_{4}$-hour door is more restrictive than what is required for smoke barriers in Section 716.5. Section 716.5 and Table 716.5 require a 20-minute rating for doors located in smoke barriers. The reference back to Section 716.5 in this section is primarily related to how the door is to be tested and the criteria for acceptance. More specifically, Section 716.5.3.1 requires doors used in smoke and draft control to be tested in accordance with UL 1784 and installed in accordance with NFPA 105.

3008.7.3.1 Vision panel. A vision panel shall be installed in each *fire door assembly* protecting the lobby doorway. The vision panel shall consist of fire-protection-rated glazing and shall be located to furnish clear vision of the occupant evacuation elevator lobby.

❖ The vision panel allows occupants located in the elevator lobby to be aware of the conditions outside the lobby and also allows the building occupants outside the lobby to see how crowded the lobby is during a fire. If the lobby is crowded, the occupants can choose to return to the exit stairways for egress. This is not a requirement for fire service access elevators.

3008.7.3.2 Door closing. Each *fire door assembly* protecting the lobby doorway shall be automatic-closing upon receipt of any fire alarm signal from the *emergency voice/alarm communication system* serving the building.

❖ In addition to providing tenability for the lobby, the automatic door closing requirement will provide a longer period of elevator use before Phase I Recall occurs. Essentially, it will delay smoke infiltrating the lobbies and the activation of the smoke detectors in the lobby. Such detectors are one of the mechanisms for the activation of Phase I Recall.

3008.7.4 Lobby size. Each occupant evacuation elevator lobby shall have minimum floor area as follows:

1. The occupant evacuation elevator lobby floor area shall accommodate, at 3 square feet (0.28 m^2) per person, not less than 25 percent of the *occupant load* of the floor area served by the lobby.
2. The occupant evacuation elevator lobby floor area also shall accommodate one *wheelchair space* of 30 inches by 48 inches (760 mm by 1220 mm) for each 50 persons, or portion thereof, of the *occupant load* of the floor area served by the lobby.

Exception: The size of lobbies serving multiple banks of elevators shall have the minimum floor area *approved* on an individual basis and shall be consistent with the building's fire safety and evacuation plan.

❖ This section provides the basic criteria for a minimum amount of lobby floor space. The lobby needs to provide space for all occupants, not just for those with a disability. Therefore, there is wheelchair space provided and also 3 square feet (0.27 m^2) per person for 25 percent of the occupant load. Some percentage of occupants may take the exit stairways, and due to speed of the elevator egress, the need to house the entire occupant load is unnecessary.

The following is an example of how large a lobby in a story containing 200 occupants would need to be:

- 200 occupants/4 (25% of occupants) = 50 occupants
- 50 occupants × 3 square feet per occupant = 150 square feet

In addition, wheelchair space is required as follows:

- 200 occupants/1 wheelchair space for each 50 occupants = 4 wheelchair spaces
- 4 chair spaces × 30 inches × 48 inches = 5,760/144 square inches = 40 square feet
- Total square feet of lobby space = 150 square feet + 40 square feet = 190 square feet

3008.7.5 Signage. An *approved* sign indicating elevators are suitable for occupant self-evacuation shall be posted on all floors adjacent to each elevator call station serving occupant evacuation elevators.

❖ Since this is such a departure from what the general public has been trained to do during a fire emergency, signage is imperative to let occupants know that elevators are available for evacuation during a fire. Occupants have been historically told to take the exit stairways during a fire. Additionally, only some buildings will have this feature, so signage needs to differentiate from buildings that do not anticipate this type of evacuation procedure.

3008.7.6 Lobby status indicator. Each occupant evacuation elevator lobby shall be equipped with a status indicator arranged to display all of the following information:

1. An illuminated green light and the message, "Elevators available for occupant evacuation," when the elevators are operating in normal service and the *fire alarm system* is indicating an alarm in the building.
2. An illuminated red light and the message, "Elevators out of service, use *exit stairs*," when the elevators are in Phase I emergency recall operation in accordance with the requirements in ASME A17.1/CSA B44.
3. No illuminated light or message when the elevators are operating in normal service.

❖ As part of the egress system, using elevator real time information is critical to occupants as the elevators may not be available throughout the entire fire event. As noted in Section 3008.3, elevators will only be available for evacuation prior to Phase I Recall. The signage reflects the different actions the occupants may need to take. The code states that when a fire alarm is activated within a building, the status indicator must be illuminated green.

3008.7.7 Two-way communication system. A two-way communication system shall be provided in each occupant evacuation elevator lobby for the purpose of initiating communication with the *fire command center* or an alternate location *approved* by the fire department.

❖ In addition to informing occupants of whether they can use the elevator in normal service or for evacuation, there needs to be a means of communication between building occupants and emergency responders. This is especially important if all people have not been able to evacuate and Phase I activates. This will assist in prioritizing rescue by the fire department once the elevators are no longer available for those who cannot take the exit stairways.

Please note that the two-way communication system can be the same two-way communication specified in Section 1007.8, as the criteria would meet the intent of this section.

3008.7.7.1 Design and installation. The two-way communication system shall include audible and visible signals and shall be designed and installed in accordance with the requirements in ICC A117.1.

❖ This section provides the basic criteria for a two-way communication system between the lobby and fire command center or alternative approved location. The intent of this section is that the audible and visual signals are consistent with ICC A117.1.

3008.7.7.2 Instructions. Instructions for the use of the two-way communication system along with the location of the station shall be permanently located adjacent to each station. Signage shall comply with the ICC A117.1 requirements for visual characters.

❖ This section simply exists to make sure that instructions on how to use the two-way communication system are posted in a logical location to ensure the use of the system by building occupants.

3008.8 Elevator system monitoring. The occupant evacuation elevators shall be continuously monitored at the *fire command center* or a central control point *approved* by the fire department and arranged to display all of the following information:

1. Floor location of each elevator car.
2. Direction of travel of each elevator car.
3. Status of each elevator car with respect to whether it is occupied.
4. Status of normal power to the elevator equipment, elevator controller cooling equipment, and elevator machine room *ventilation* and cooling equipment.
5. Status of standby or emergency power system that provides backup power to the elevator equipment, elevator controller cooling equipment, and elevator machine room *ventilation* and cooling equipment.
6. Activation of any fire alarm initiating device in any elevator lobby, elevator machine room or machine space, or elevator hoistway.

❖ The monitoring is intended for emergency responders so they can determine if there are any problems occurring where a rescue may be necessary, or when they may determine that the elevators must be recalled and removed from egress service. This monitoring also helps determine where the people are evacuating from and how many through the use of load-weighing capabilities found in elevators currently. Understanding how the people are moving will assist also in prioritizing fire-fighting and rescue operations.

3008.8.1 Elevator recall. The *fire command center* or an alternate location *approved* by the fire department shall be provided with the means to manually initiate a Phase I Emergency Recall of the occupant evacuation elevators in accordance with ASME A17.1/CSA B44.

❖ Buildings with occupant self-evacuation elevators must have elevator recall capability available at the fire command center or similar location in addition to the automatic recall by smoke detection and manual recall ability at the elevator itself. This gives the emergency responders the ability to recall the elevators in the location where they are closely monitoring the elevator movement and associated attributes required to be monitored by Section 3008.8.

3008.9 Electrical power. The following features serving each occupant evacuation elevator shall be supplied by both normal power and Type 60/Class 2/Level 1 standby power:

1. Elevator equipment.
2. Elevator machine room *ventilation* and cooling equipment.
3. Elevator controller cooling equipment.

❖ All passenger elevators used for occupant self-evacuation must be equipped with standby power. This differs from Section 3003.1.3 which does not require all elevators to run simultaneously on standby power.

These requirements are almost identical to those for fire service access elevators. The only difference is that fire service access elevators also require standby power for the hoistway lighting required by Section 3007.6.2.

3008.9.1 Protection of wiring or cables. Wires or cables that are located outside of the elevator hoistway and machine room and that provide normal or standby power, control signals, communication with the car, lighting, heating, air conditioning, *ventilation* and fire-detecting systems to fire service access elevators shall be protected by construction having a *fire-resistance rating* of not less than 2 hours, or shall be circuit integrity cable having a *fire-resistance rating* of not less than 2 hours.

Exception: Wiring and cables to control signals are not required to be protected provided that wiring and cables do not serve Phase II emergency in-car operations.

❖ This section is the same as Section 3007.7.1 except that it is focused upon occupant evacuation elevators (see commentary, Section 3007.7.1). Wiring or cables for all passenger elevators and other elevators that are part of the occupant evacuation system require protection. Typical buildings without occupant evacuation elevators would only provide the protection to the single fire service access elevator.

3008.10 Emergency voice/alarm communication system. The building shall be provided with an *emergency voice/alarm communication system*. The *emergency voice/alarm communication system* shall be accessible to the fire department. The system shall be provided in accordance with Section 907.2.12.2.

❖ Emergency voice/alarm communication systems are required in all high-rise buildings, but the application of Section 3008 is not limited to high-rise buildings. Therefore, this would be an additional requirement in some instances. Using elevators for evacuation makes such systems critical in providing appropriate information to building occupants. The actual requirements are located Chapter 9 of this code.

3008.10.1 Notification appliances. No fewer than one audible and one visible notification appliance shall be installed within each occupant evacuation elevator lobby.

❖ In addition to the normal requirements associated with emergency voice/alarm communication systems, notification appliances must also be placed within elevator lobbies. There is likely to be many people waiting in such lobbies, and it is critical that the occupants receive proper instructions and up-to-date information during an emergency.

3008.11 Hazardous material areas. No building areas shall contain hazardous materials exceeding the maximum allowable quantities per *control area* as addressed in Section 414.2.

❖ This section prohibits any Group H occupancies in buildings that allow the use of elevators for evacuation. This is an added precaution in buildings allowing elevators for evacuation. It simply reduces the risk of fire and other hazardous conditions that could occur in a building.

Bibliography

The following resource materials are referenced in this chapter or are relevant to the subject matter addressed in this chapter.

ALI ALCTV-98, *Standard for Automobile Lifts—Safety Requirements for the Construction, Testing and Validation.* Indialantic, FL: Automotive Lift Institute, 1998.

ASCE 24-05, *Flood Resistant Design and Construction.* Reston, VA: American Society of Civil Engineers, Structural Engineering Institute, 2005.

ASME A17.1/CSA B44-07, *Safety Code for Elevators and Escalators.* New York: American Society of Mechanical Engineers, 2007.

ASME A90.1-04, *Safety Standard for Belt Manlifts.* New York: American Society of Mechanical Engineers, 2003.

ASME B20.1-03, *Safety Standard for Conveyors and Related Equipment.* New York: American Society of Mechanical Engineers, 2003.

FEMA FIA-TB #4, *Elevator Installation for Buildings Located in Special Flood Hazard Areas.* Washington, DC: Federal Emergency Management Agency, 2010.

ICC A117.1-03, *Accessible and Usable Buildings and Facilities.* Washington, DC: International Code Council, 2003.

NFIP Technical Bulletin Series. Washington, DC: National Flood Insurance Program. [Online]. Available: http://www.fema.gov/MITItechbul.htm.

NFPA 13-07, *Installation of Sprinkler Systems.* Quincy, MA: National Fire Protection Association, 2007.

NFPA 72-07, *National Fire Alarm Code.* Quincy, MA: National Fire Protection Association, 2002.

NFPA 105-07, *Standard for the Installation of Smoke Door Assemblies.* Quincy, MA: National Fire Protection Association, 2007.

Tubbs, J. and B. Meacham. *Egress Design Solutions, A Guide to Evacuation and Crowd Management Planning.* Hoboken, NJ: John Wiley and Sons, 2007.

UL 1784-01, *Standard for Air Leakage Tests of Door Assemblies—with Revisions through December 2004.* Northbrook, IL: Underwriters Laboratories, 2001.

Chapter 31: Special Construction

General Comments

Chapter 31 contains provisions that govern the construction and protection of structures with unique characteristics not addressed by other sections of the code and that require special consideration. This chapter includes requirements for signs; membrane structures; temporary structures; pedestrian walkways; telecommunication and broadcast towers; canopies and awnings; marquees, automatic vehicular gates and swimming pool enclosures.

Section 3101 addresses the scope of the chapter.

Section 3102 regulates membrane structures.

Section 3103 regulates structures that are erected for a period of less than 180 days.

Section 3104 regulates pedestrian walkways between buildings.

Section 3105 regulates the construction of awnings and canopies.

Section 3106 regulates the construction of marquees.

Section 3107 addresses the design, construction and maintenance of signs.

Section 3108 regulates the construction of towers and antennas that are used for radio, telephone and television transmission.

Section 3109 regulates swimming pool enclosures.

Section 3110 regulates automatic vehicular gates.

The provisions of Chapter 31 are to be considered as additional requirements to those contained elsewhere in the code. These provisions may alter the general requirements found in other locations, but are only applicable to the specific element regulated therein. For example, the provisions for temporary construction regulate how the structure is to be built, but do not address means of egress. As such, the provisions of Chapter 10 are applicable.

The criteria in this chapter are straightforward and, in turn, lend themselves to straightforward enforcement.

Purpose

The purpose of Chapter 31 is to locate in one place the provisions that regulate construction or protection requirements for structures having unique characteristics, such as structures that are unusually tall and not used for human occupancy, structures occupied for short periods of time or recreational structures (swimming pools).

SECTION 3101 GENERAL

3101.1 Scope. The provisions of this chapter shall govern special building construction including membrane structures, temporary structures, *pedestrian walkways* and tunnels, automatic *vehicular gates*, *awnings* and *canopies*, *marquees*, signs, and towers and antennas.

❖ This section makes it clear that the requirements of this chapter are to be used in conjunction with the general requirements of the code. When a specific provision is contained in Chapter 31, it will take precedence over the general requirement.

SECTION 3102 MEMBRANE STRUCTURES

3102.1 General. The provisions of Sections 3102.1 through 3102.8 shall apply to air-supported, air-inflated, membrane-covered cable and membrane-covered frame structures, collectively known as membrane structures, erected for a period of 180 days or longer. Those erected for a shorter period of time shall comply with the *International Fire Code*. Membrane structures covering water storage facilities, water clarifiers, water treatment plants, sewage treatment plants, greenhouses and similar facilities not used for human occupancy are required to meet only the requirements of Sections 3102.3.1 and 3102.7. Membrane structures erected on a building, balcony, deck or other structure for any period of time shall comply with this section.

❖ This section applies to membrane structures that are erected for a period of 180 days or longer or, in the case of membrane structures erected on buildings, balconies, decks or similar structures, for any amount of time. Some membrane structures are of the inflated type, where the air pressure is within a series of tubes, which then form the structure and keep it upright. In other structures, the interior atmosphere is pressurized and cables may be used to restrain the fabric membrane. Other types include cable-supported membrane structures in which the roof element is supported by cables and air pressure may or may not be provided. These are commonly used in sports stadiums. Membrane structures may also be framework-type structures in which the membrane is stretched over the frame to form the enclosing walls and the roof. Finally, membrane structures may consist of a membrane tensioned to provide weather protection without enclosing walls.

The history of regulating membrane structures dates back to 1944, when efforts were first made to control circus tent flammability after a circus fire in Hartford, Connecticut, claimed 168 lives. Early regulations required that the tent fabric be flame resistant, that adequate exiting from both the grandstands and the tent be provided and that sufficient clear space around the tent be provided to control potential ignition sources and to permit rapid emergency exiting.

Large permanent membrane structures began to see significant use in the United States in the early 1960s. Prior to that time, the only membrane structures in common usage were air-inflated and air-supported covers for participant sports facilities such as tennis courts, and swimming pools. Because of fire prevention code requirements, such structures were often restricted to temporary use, which at that time was defined as less than 90 days. As noted in Section 3102.1, the criteria of 180 days is used to define what is permanent and what is temporary.

The reference to 180 days in Section 3102.1 is intended to establish membrane structures complying with this section as being outside the scope and provisions for temporary structures.

This section includes provisions for construction types, height and area limitations and structural support. Otherwise, the other applicable sections of the code apply. So that a membrane structure will continue to provide its intended function, it must undergo continuous maintenance. Fire safety of such structures and the premises is regulated by the *International Fire Code*® (IFC®).

In general, this section applies to all membrane structures, regardless of the supporting mechanisms or structure, which are erected for a period of 180 days or more. Membrane structures erected for a period of less than 180 days are instead required to comply with the provisions of the IFC. However, there are certain membrane structures that need to be regulated as permanent regardless of how long they are erected. These include membrane structures located on buildings, balconies, decks and similar structures. When a temporary membrane structure is placed in a field or parking lot, several safety features are afforded including fire separation distance from other hazards (buildings, vehicles), separation from other tents, and an unobstructed means of egress path for the uniformly located exits. When a membrane structure is placed upon a building or deck, the temporary membrane structure requirements do not provide regulations for key issues associated with permanent structures, such as the means of egress or the limitation of hazardous materials; therefore, compliance with permanent structure standards is needed. Furthermore, the temporary membrane structure section does not contain requirements on the regulation of the loads that the temporary membrane structures would impose on a structure below.

It is also important to note that the provisions of Section 3102 apply regardless of the use or occupant load of such structures. If the structure is not intended for human occupancy, some relief is provided and compliance with only Sections 3102.3.1 and 3102.7 is required.

3102.2 Definitions. The following terms are defined in Chapter 2:

❖ Definitions of terms that are associated with the content of this section are contained herein. These definitions can help in the understanding and application of the code requirements. It is important to emphasize that these terms are not exclusively related to this section, but are applicable everywhere the term is used in the code. The purpose for including these definitions within this section is to provide more convenient access to them without having to refer back to Chapter 2.

For convenience, these terms are also listed in Chapter 2 with a cross reference to this section. The use and application of all defined terms, including those defined herein, are set forth in Section 201.

AIR-INFLATED STRUCTURE.

❖ This type of membrane structure is characterized by multiple layers arranged such that air-pressurized membrane beams, arches or similar elements are formed. These elements are pressurized with air and form the membrane structure. Note that the occupants of the structure are not subjected to the pressurized areas, because the pressurization is in the structural elements, not within the space used by the occupants.

AIR-SUPPORTED STRUCTURE.

❖ An air-supported structure identifies those membrane structures that are completely pressurized for the purposes of supporting the membrane covering. Most "domed" sports arenas use air pressure within the structure to support the membrane covering. The membrane covering can consist of one layer or multiple layers; thus, air-supported structures are classified as either "single skin" or "double skin."

Double skin.

❖ A double-skin, air-supported structure contains multiple layers of membrane sheathing. The membranes are usually separated by enough distance to allow for pressurized air or other materials to be inserted between the plies. The pressurized air or other materials usually serve to increase the insulating and acoustical properties.

Single skin.

❖ A single-skin, air-supported structure consists of just one membrane covering that is directly supported by the interior pressurized air. No other membranes are provided for insulating or acoustical purposes. If the membrane covering consists of several laminated plies, such an arrangement is still considered a single-skin, air-supported structure.

CABLE-RESTRAINED, AIR-SUPPORTED STRUCTURE.

❖ This definition establishes a variation of the air-supported membrane structure. A single-skin or double-skin membrane is still pressurized by interior air, but does not have the strength or tearing resistance against the internal air pressure. A "fishnet" system of cable wires is placed over the exterior surface of the membrane or is integrally woven into the membrane material. The "fishnet" system is then tied to exterior walls of the building or other structural supports to resist the internal air pressure.

MEMBRANE-COVERED CABLE STRUCTURE.

❖ This definition identifies a structure in which the membrane fabric is draped over a cable framework. These structures do not involve any interior air pressurization. The cable framework system is supported by a mast or other tower system. The cables are usually pretensioned to provide structural support for the rest of the exterior structure.

MEMBRANE-COVERED FRAME STRUCTURE.

❖ This type of membrane structure is the same as the membrane-covered cable structure, except that a trussed framework constructed of structural shapes provides the support instead of pretensioned cables. Again, no air pressurization is required or provided.

NONCOMBUSTIBLE MEMBRANE STRUCTURE.

❖ Any membrane structure that is constructed of all noncombustible materials is classified by this definition. To qualify as noncombustible, the materials, including the membrane itself, must be tested in accordance with ASTM E 136 and must satisfy the criteria in Section 703.4.

3102.3 Type of construction. Noncombustible membrane structures shall be classified as Type IIB construction. Noncombustible frame or cable-supported structures covered by an *approved* membrane in accordance with Section 3102.3.1 shall be classified as Type IIB construction. Heavy timber frame-supported structures covered by an *approved* membrane in accordance with Section 3102.3.1 shall be classified as Type IV construction. Other membrane structures shall be classified as Type V construction.

Exception: Plastic less than 30 feet (9144 mm) above any floor used in greenhouses, where occupancy by the general public is not authorized, and for aquaculture pond covers is not required to meet the fire propagation performance criteria of NFPA 701.

❖ Because of the limited fire-resistance capabilities of the membrane, membrane structures are usually classified as Type IIB, IV or V. These are defined as follows:

Type IIB—Membrane structures in which the membrane is noncombustible as defined by Section 703.4; or membrane-covered structures in which the frame or cable is noncombustible and the membrane is either noncombustible or flame resistant.

Type IV—Membrane-covered structures in which the frame is made from heavy timber and the membrane is noncombustible or flame resistant.

Type V—Membrane structures in which the membrane is flame resistant and combustible; or membrane-covered structures in which the frame or cable is combustible and the membrane is noncombustible or flame resistant.

In accordance with Section 602.2, a membrane-covered structure is to be considered as Type I or II construction, provided that the frame or cable complies with the provisions for Type I or II construction and the membrane structure is noncombustible, used exclusively as a roof and located more than 20 feet (6096 mm) above any floor, balcony or gallery floor level. This is similar to the provisions in Table 601 that do not require fire-resistance-rated roof construction under similar conditions.

The exception indicates that in recent years it has been common practice to build greenhouses with higher ridge lines. Since this increase in height, there has been no record of related property loss.

3102.3.1 Membrane and interior liner material. Membranes and interior liners shall be either noncombustible as set forth in Section 703.5 or meet the fire propagation performance criteria of NFPA 701 and the manufacturer's test protocol.

Exception: Plastic less than 20 mil (0.5 mm) in thickness used in greenhouses, where occupancy by the general public is not authorized, and for aquaculture pond covers is not required to meet the fire propagation performance criteria of NFPA 701.

❖ The membrane is a thin, flexible, impervious material capable of being supported by an air pressure of $1^1/_2$ inches (38 mm) of water column. Membranes may be noncombustible. Except for limited applications of plastic greenhouses and pond covers, combustible membranes are required to be flame resistant.

Noncombustible membranes and liner materials must meet the criteria of Section 703.5, which references ASTM E 136. This test is used in the determination of noncombustibility. See the commentary for Section 703.5.1 for more information on this test.

Materials that are not noncombustible are required to be flame resistant. Flame resistance is determined by testing the membrane in accordance with NFPA 701.

Although typically not applicable to membrane structures, it should be noted that recent tests have indicated a problem with combining two fabrics that individually passed the test procedure, but in combination would not be acceptable; therefore, if multiple layers are used, testing should demonstrate acceptable performance for the combination of materials to be used.

Both greenhouses that are not open to the public and aquaculture pond covers may be of plastic that is less than 20 mil (500 mm) thick and need not be flame resistant or tested to demonstrate compliance with NFPA 701. This provision for greenhouses is based on the limited number of occupants.

3102.4 Allowable floor areas. The area of a membrane structure shall not exceed the limitations set forth in Table 503, except as provided in Section 506.

❖ The allowable area per floor of a membrane structure is determined using Sections 503 and 506, based on the construction types established in Section 3102.3.

3102.5 Maximum height. Membrane structures shall not exceed one *story* nor shall such structures exceed the height limitations in feet set forth in Table 503.

Exception: Noncombustible membrane structures serving as roofs only.

❖ Membrane structures are most commonly used to enclose large, open areas without intermediate structural supports, such as columns. For this reason, as well as the engineering difficulties associated with multistory buildings, membrane structures are limited to one story and the maximum height permitted in Table 503. The specific reference to Table 503 intends that any height increases allowed by Section 504 would not be permitted. The height is not limited, however, if a noncombustible membrane is used as a roof membrane only.

3102.6 Mixed construction. Membrane structures shall be permitted to be utilized as specified in this section as a portion of buildings of other types of construction. Height and area limits shall be as specified for the type of construction and occupancy of the building.

❖ Section 3102.6 applies where the membrane structure is only a portion of a larger structure built of typical "rigid" materials. If a membrane structure is completely created by the membranes, or the membranes supported by a rigid frame or a cable system, then Sections 3102.3 through 3102.5 apply. Where there are additional parts of the building constructed of rigid materials, such as walls, a second story or mezzanine, or something other than a membrane and its supporting structure, then it is a mixed construction membrane structure subject to Section 3102.6. Figure 3102.6.1 illustrates a mixed construction building where a membrane is used as the roof of an otherwise rigid construction building.

For a mixed construction membrane structure, the supporting structure of the membrane must be equal to or better than that of the building to which it is attached. Where air-supported or air-inflated structures are attached as an element of a building, their area limitations are based on the building to which they are attached. For example, an air-supported membrane building for a Group A-1 occupancy of Type VB construction with an allowable area of 22,000 square feet (2045 m^2) would limit the area of the membrane structure to 22,000 square feet (2045 m^2) as well even if a noncombustible membrane structure was used.

3102.6.1 Noncombustible membrane. A noncombustible membrane shall be permitted for use as the roof or as a skylight of any building or atrium of a building of any type of construction provided it is not less than 20 feet (6096 mm) above any floor, balcony or gallery.

❖ See Figure 3102.6.1 for an example of the application of this section. See the commentary for Section 3102.3.1 regarding the determination of a noncombustible membrane versus one complying with NFPA 701.

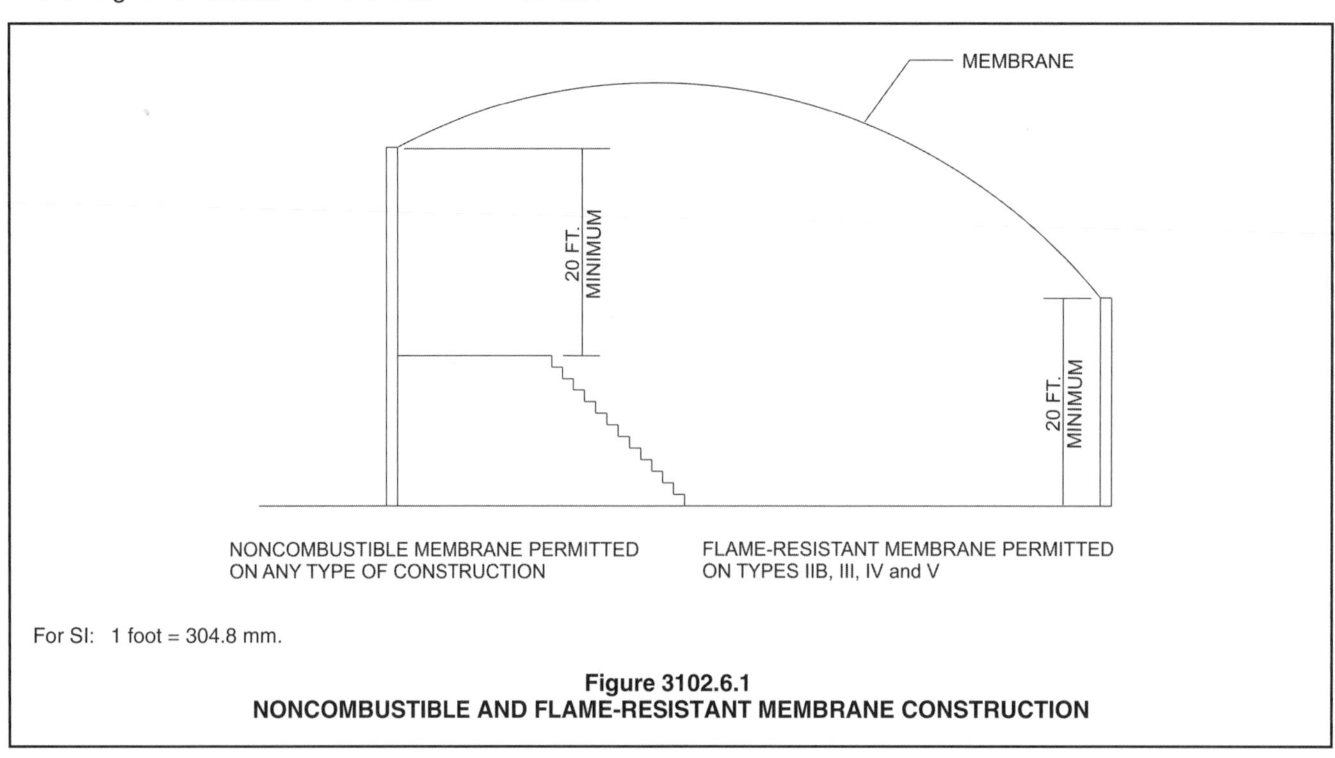

Figure 3102.6.1
NONCOMBUSTIBLE AND FLAME-RESISTANT MEMBRANE CONSTRUCTION

3102.6.1.1 Membrane. A membrane meeting the fire propagation performance criteria of NFPA 701 shall be permitted to be used as the roof or as a skylight on buildings of Types IIB, III, IV and V construction, provided it is not less than 20 feet (6096 mm) above any floor, balcony or gallery.

❖ See Figure 3102.6.1 for an example of the application of this section. See the commentary for Section 3102.3.1 regarding the determination of a noncombustible membrane versus one complying with NFPA 701.

3102.7 Engineering design. The structure shall be designed and constructed to sustain dead loads; loads due to tension or inflation; live loads including wind, snow or flood and seismic loads and in accordance with Chapter 16.

❖ The structure must be designed and constructed to sustain all dead loads, loads caused by tensioning or inflation and live loads, including wind, snow, flood and seismic. Chapter 16 provides criteria for design loads, except those associated with tensioning or inflation.

3102.8 Inflation systems. Air-supported and air-inflated structures shall be provided with primary and auxiliary inflation systems to meet the minimum requirements of Sections 3102.8.1 through 3102.8.3.

❖ While the primary inflation system is essential to the integrity of air-supported and air-inflated buildings, a secondary or auxiliary inflation system is required for all such membrane structures to prevent deflation and to allow the occupants to egress. A back-up structural support system is also required for air-supported and air-inflated structures with an occupant load of more than 50 or that cover a swimming pool because of life safety considerations.

3102.8.1 Equipment requirements. This inflation system shall consist of one or more blowers and shall include provisions for automatic control to maintain the required inflation pressures. The system shall be so designed as to prevent overpressurization of the system.

❖ The structural integrity of air-supported and air-inflated structures depends on a reliable inflation system.

3102.8.1.1 Auxiliary inflation system. In addition to the primary inflation system, in buildings larger than 1,500 square feet (140 m^2) in area, an auxiliary inflation system shall be provided with sufficient capacity to maintain the inflation of the structure in case of primary system failure. The auxiliary inflation system shall operate automatically when there is a loss of internal pressure and when the primary blower system becomes inoperative.

❖ As a safety precaution, such structures that exceed 1,500 square feet (139 m^2) must also be provided with an auxiliary inflation system. The auxiliary system is required to operate automatically upon loss of internal pressure or if the primary system becomes inoperative. In accordance with Section 3102.8.2, when an auxiliary system is required, standby power is also to be provided for at least 4 hours.

3102.8.1.2 Blower equipment. Blower equipment shall meet all of the following requirements:

1. Blowers shall be powered by continuous-rated motors at the maximum power required for any flow condition as required by the structural design.
2. Blowers shall be provided with inlet screens, belt guards and other protective devices as required by the *building official* to provide protection from injury.
3. Blowers shall be housed within a weather-protecting structure.
4. Blowers shall be equipped with backdraft check dampers to minimize air loss when inoperative.
5. Blower inlets shall be located to provide protection from air contamination. The location of inlets shall be *approved*.

❖ Safety features of the blower system for the purposes of reliability, prevention of overpressurization and personal safety include:

- Blowers with motors rated for continuous operation at maximum power;
- Inlet screens, belt guards and other protective devices;
- Weather-protected housing for blowers;
- Backdraft check dampers to minimize air loss; and
- Inlets located to prevent air contamination.

The *International Mechanical Code*® (IMC®) requires makeup air inlets to be at least 10 feet (3048 mm) from any potential source of contamination.

3102.8.2 Standby power. Wherever an auxiliary inflation system is required, an *approved* standby power-generating system shall be provided. The system shall be equipped with a suitable means for automatically starting the generator set upon failure of the normal electrical service and for automatic transfer and operation of all of the required electrical functions at full power within 60 seconds of such service failure. Standby power shall be capable of operating independently for not less than 4 hours.

❖ Air-supported and air-inflated structures exceeding 1,500 square feet (139 m^2) are to be provided with standby power for the auxiliary inflation system. The standby power system must comply with the provisions of Section 2702 and be capable of operating for a minimum of 4 hours.

3102.8.3 Support provisions. A system capable of supporting the membrane in the event of deflation shall be provided for in air-supported and air-inflated structures having an *occupant load* of 50 or more or where covering a swimming pool regardless of *occupant load*. The support system shall be capable of maintaining membrane structures used as a roof for Type I construction not less than 20 feet (6096 mm) above floor or seating areas. The support system shall be capable of maintaining other membranes not less than 7 feet (2134 mm) above the floor, seating area or surface of the water.

❖ To allow safe occupant egress and access for fire fighting, air-supported and air-inflated structures that have an occupant load of more than 50 or cover a swimming pool are required to have a standby support system. Regardless of the occupant load, swimming pools are included because of the additional time necessary to egress when in a pool and because of the hazard that would exist if a membrane collapsed on a pool in which a person was swimming.

The support system is to be capable of maintaining all membrane structures at least 7 feet (2134 mm) above the floor, seating area or surface water, whichever is highest, to permit upright egress travel by the occupants, as well as to aid occupants who may be unable to egress. When the membrane structure is used as a roof on a Type I or II building, the support system must be capable of maintaining the membrane structure at least 20 feet (6096 mm) above the floor or seating area so that the risk of contribution of the roof structure to the fire is minimized.

SECTION 3103
TEMPORARY STRUCTURES

3103.1 General. The provisions of Sections 3103.1 through 3103.4 shall apply to structures erected for a period of less than 180 days. Tents and other membrane structures erected for a period of less than 180 days shall comply with the *International Fire Code*. Those erected for a longer period of time shall comply with applicable sections of this code.

❖ The purpose of Section 3103 is to provide public safety requirements in relation to fire, storm, collapse and crowd behavior in temporary structures. "Temporary use" is defined as a period of less than 180 days. Section 1028 of the code and ICC 300 are applicable to the means of egress and the use of bleachers, grandstands and folding or telescopic seating within the temporary structure. Temporary structures may also include temporary wiring and other fire hazards, which are different from those expected in permanent structures. These items are to be identified in the construction documents.

Any structure that is erected for a period of 180 days or more must comply with other sections of the code applicable to permanent strutures. It should be noted that "temporary" refers to the structure being erected, not the use of the structure; therefore, if a structure is erected for 365 days a year, but is used on a seasonal basis for less than 180 days, this section does not apply. Such structures are considered permanent. It should also be noted that membrane structures, even those erected for less than 180 days, are subject to standards applicable to permanent buildings when they are attached to such buildings (see Section 3102.1).

A temporary structure can be placed on a permanent foundation without being considered as a permanent structure.

Section 107 addresses permitting of such structures. In particular it does give some leeway to the building official to allow extensions of the 180-day criteria.

3103.1.1 Permit required. Temporary structures that cover an area greater than 120 square feet (11.16 m^2), including connecting areas or spaces with a common *means of egress* or entrance which are used or intended to be used for the gathering together of 10 or more persons, shall not be erected, operated or maintained for any purpose without obtaining a *permit* from the *building official*.

❖ Permits are required for temporary structures that are greater than 120 square feet (11 m^2) and are used for the gathering of 10 or more people. This provides notice to the building department for larger temporary structures that pose a higher fire and life safety risk due to their size and presence of occupants as well as relief to the smaller temporary structures that pose a low risk.

3103.2 Construction documents. A *permit* application and *construction documents* shall be submitted for each installation of a temporary structure. The *construction documents* shall include a site plan indicating the location of the temporary structure and information delineating the *means of egress* and the *occupant load*.

❖ Before the installation of any temporary structure, a permit application and construction documents are required to be submitted.

Construction documents, as defined in Chapter 2, include written, graphic and pictorial documents. These documents are to clearly provide pertinent information about the project, including the location, seating capacity, construction and all mechanical and electrical equipment, as a minimum. The code references means of egress and occupant load as items of particular interest for temporary structures becausesuch structures are commonly tents used for special gatherings such as weddings or parties, and therefore the occupant load and clear path of egress are of particular concern. However, it should be noted that these provisions are not intended to obviate the need for submittals that are necessary for all new construction, including structural details, calculations, soils tests and any other construction documents that are pertinent. These documents would need to be signed and sealed by a registered design professional as required for new construction.

3103.3 Location. Temporary structures shall be located in accordance with the requirements of Table 602 based on the *fire-resistance rating* of the *exterior walls* for the proposed type of construction.

❖ With respect to exposure hazards, temporary structures are no different from permanent structures; therefore, the fire-resistance rating of the temporary structure's exterior walls will depend upon the occupancy and the fire separation distance.

3103.4 Means of egress. Temporary structures shall conform to the *means of egress* requirements of Chapter 10 and shall have an *exit access* travel distance of 100 feet (30 480 mm) or less.

❖ The means of egress in temporary structures is to comply with the requirements of Chapter 10, except the maximum exit access travel distance is limited to 100 feet (30 480 mm). The reduced travel distance applies to all temporary uses and is based on concerns associated with the nature of temporary structure construction, which may accelerate fire spread or rapidly decrease the stability of the structure during a fire.

SECTION 3104
PEDESTRIAN WALKWAYS AND TUNNELS

3104.1 General. This section shall apply to connections between buildings such as *pedestrian walkways* or tunnels, located at, above or below grade level, that are used as a means of travel by persons. The *pedestrian walkway* shall not contribute to the *building area* or the number of *stories* or height of connected buildings.

❖ Pedestrian walkways are normally seen as bridges connecting adjacent buildings (see Figure 3104). This section addresses the practical application of pedestrian walkways and tunnels, including walkways between buildings, grade-level connections between buildings or corridors around the perimeter of one building. The area of the walkway itself is not to be considered a part of any building it serves, nor is it considered to add to the number of stories or height of any building it serves, provided that the walkway and its construction comply with this section.

Figure 3104
EXAMPLE—PEDESTRIAN WALKWAY

3104.2 Separate structures. Connected buildings shall be considered to be separate structures.

Exceptions:

1. Buildings on the same *lot* in accordance with Section 503.1.2 shall be considered a single structure.
2. For purposes of calculating the number of Type B units required by Chapter 11, structurally connected buildings and buildings with multiple wings shall be considered one structure.

❖ Exception 1 to this section states that each building to be served by a pedestrian walkway is to be considered as a separate building except where, in accordance with Section 503.1.2, the designer chooses to have separate buildings on the same lot considered as one building. Exception 1 also applies to a connected building that may be of unlimited area as addressed in Section 507. Exception 2 establishes that, for purposes of evaluation of a building complex for the required number of Type B dwelling units, buildings connected by pedestrian walkways must be evaluated as a single building. This provision was installed to coordinate with the *Fair Housing Accessibility Guidelines.*

3104.3 Construction. The *pedestrian walkway* shall be of noncombustible construction.

Exceptions:

1. Combustible construction shall be permitted where connected buildings are of combustible construction.
2. *Fire-retardant-treated wood,* in accordance with Section 603.1, Item 1.3, shall be permitted for the roof construction of the *pedestrian walkway* where connected buildings are a minimum of Type I or II construction.

❖ The walkway must be of noncombustible construction, unless all of the buildings it connects are of combustible construction. It is intended that the construction materials of the walkway be consistent with the type of construction classification of the buildings connected by the walkway. Also note that this section does not require structural members of the walkway to have a fire-resistance rating, even if the connected buildings are of a protected construction type. Exception 2 simply allows fire-retardant-treated wood (FRTW) where it is already allowed for Type I or II construction, because this section does not depend upon Table 601 for determination of the construction requirements for pedestrian walkways. Note that Exception 1 would allow the use of FRTW in all cases, since combustible construction is allowed.

3104.4 Contents. Only materials and decorations *approved* by the *building official* shall be located in the *pedestrian walkway.*

❖ This restriction on the use of the floor area is similar to the restriction on the use of the floor of an atrium, except there is no relaxation of requirements if an automatic sprinkler system is installed.

3104.5 Fire barriers between pedestrian walkways and buildings. Walkways shall be separated from the interior of the building by not less than 2-hour *fire barriers* constructed in accordance with Section 707 or *horizontal assemblies* constructed in accordance with Section 711, or both. This protection shall extend vertically from a point 10 feet (3048 mm) above the walkway roof surface or the connected building roof line, whichever is lower, down to a point 10 feet (3048 mm) below the walkway and horizontally 10 feet (3048 mm) from each side of the *pedestrian walkway*. Openings within the 10-foot (3048 mm) horizontal extension of the protected walls beyond the walkway shall be equipped with devices providing a $^3/_4$-hour *fire protection rating* in accordance with Section 715.

Exception: The walls separating the *pedestrian walkway* from a connected building and the openings within the 10-foot (3048 mm) horizontal extension of the protected walls beyond the walkway are not required to have a *fire-resistance rating* by this section where any of the following conditions exist:

1. The distance between the connected buildings is more than 10 feet (3048 mm). The *pedestrian walkway* and connected buildings, except for *open parking garages*, are equipped throughout with an *automatic sprinkler system* in accordance with Section 903.3.1.1. The wall is capable of resisting the passage of smoke or is constructed of a tempered, wired or laminated glass wall and doors subject to the following:
   - 1.1. The wall or glass separating the interior of the building from the *pedestrian walkway* shall be protected by an *automatic sprinkler system* in accordance with Section 903.3.1.1 and the sprinkler system shall completely wet the entire surface of interior sides of the wall or glass when actuated;
   - 1.2. The glass shall be in a gasketed frame and installed in such a manner that the framing system will deflect without breaking (loading) the glass before the sprinkler operates; and
   - 1.3. Obstructions shall not be installed between the sprinkler heads and the wall or glass.
2. The distance between the connected buildings is more than 10 feet (3048 mm) and both sidewalls of the *pedestrian walkway* are not less than 50 percent open with the open area uniformly distributed to prevent the accumulation of smoke and *toxic* gases.
3. Buildings are on the same *lot* in accordance with Section 503.1.2.
4. Where *exterior walls* of connected buildings are required by Section 705 to have a *fire-resistance rating* greater than 2 hours, the walkway shall be equipped throughout with an *automatic sprinkler system* installed in accordance with Section 903.3.1.1.

The previous exception shall apply to *pedestrian walkways* having a maximum height above grade of three *stories* or 40 feet (12 192 mm), or five *stories* or 55 feet (16 764 mm) where sprinklered.

❖ This requirement for fire-resistance-rated separation allows buildings and walkways to be considered as separate buildings. The extent of this fire-resistance-rated separation is shown in Figure 3104.5. If the wall of the connected building would not normally extend 10 feet (3048 mm) above the walkway roof, there is no requirement in this section to extend it. While Figure 3104.5 shows only the 10-foot (3048 mm) "donut" of rated construction surrounding the intersection of the pedestrian walkway with the building. Regarding the construction supporting fire barriers, Section 707.5 requires that they be supported by construction that has as a minimum the same fire-resistance rating.

The portion of the exterior wall and associated openings that extends 10 feet (3048 mm) horizontally and vertically and separates the walkway from the building is required to have a fire-resistance rating unless one of the exceptions of this section is met. This fire-resistance rating of the wall, including the portion extending 10 feet (3048 mm) horizontally and vertically, separating the walkway from a connected building, is not required when one of the following is met:

1. If the distance between the connected buildings is more than 10 feet (3048 mm) and the walkway and all connected buildings are fully sprin-

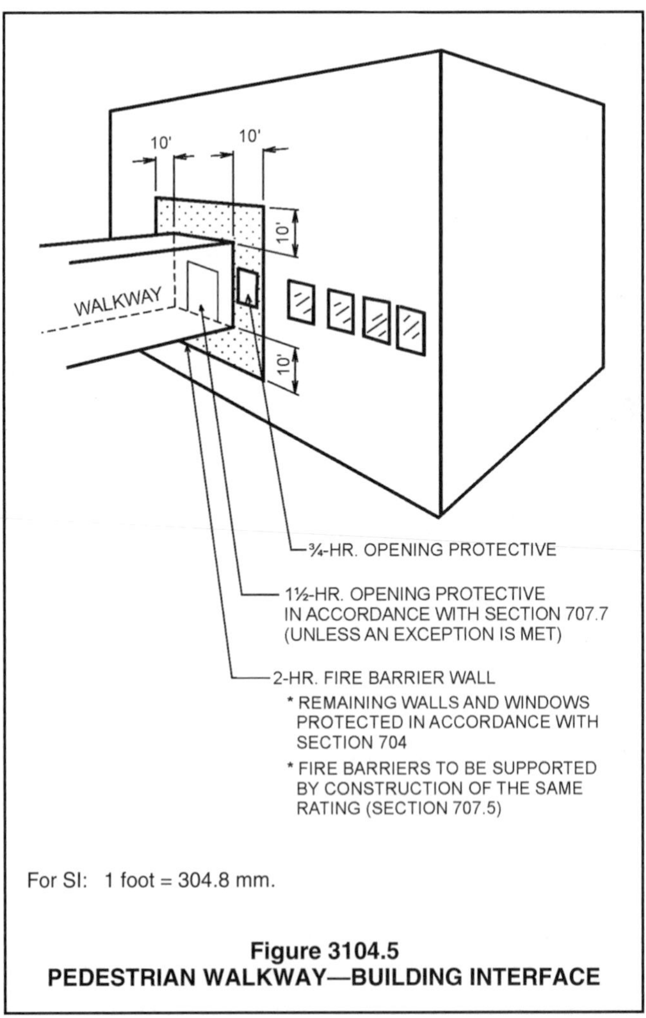

Figure 3104.5
PEDESTRIAN WALKWAY—BUILDING INTERFACE

klered or are open parking garages, then fire-resistant construction is not required as noted in the main section. Any walls must be capable of resisting smoke or be constructed of tempered, wired or laminated glass. Such walls and glass must both be protected as described in the three criteria specified in Items 1.1 through 1.3.

2. The distance between the connected buildings is more than 10 feet (3048 mm) and both sides of the walls of the pedestrian walkway are open for at least 50 percent of each wall area, with the open area uniformly distributed to prevent the accumulation of smoke and toxic gases. This concept is synonymous with that used for open parking structures where free movement of smoke is achieved through the openness of the structure.
3. This exception is based upon the concept that, in Section 503.1.2, two buildings on the same lot are really separate areas of a single building, and therefore there are no fire separation issues that need to be dealt with on the exterior walls of either building facing the other. Keep in mind, however, that the area of the walkway will need to be counted as building area for purposes of area limitations given in Chapter 5.
4. If more than a 2-hour fire-resistance rating is already required by Section 705 for exterior walls and if the walkway is fully sprinklered, no fire-resistance rating would be required at the end wall of the walkway.

These four exceptions apply only to the walkway rather than the buildings that the walkway connects. The height of such allowances was limited based upon the difficulty associated with fire fighting at more extreme heights. These fire-fighting concerns also affect the quality of life safety that can be provided during a fire.

3104.6 Public way. *Pedestrian walkways* over a *public way* shall comply with Chapter 32.

❖ The acceptance of a pedestrian walkway over a public street by the use of a bridge or any other means is required to comply with Chapter 32.

3104.7 Egress. Access shall be provided at all times to a *pedestrian walkway* that serves as a required *exit.*

❖ Pedestrian walkways may be used to provide a portion of the means of egress for one or both of the buildings at either end of the walkway. If the walkway is at grade, it may be that the door from the building is an exterior exit door and the walkway (if just a covering but with open walls) could be treated as exit discharge. Elevated pedestrian walkways could be used as an exit from one building to the other. In this case, one end of the walkway would have to comply with the standards for horizontal exits found in Section 1025. If the two buildings are being treated as one building, the pedestrian walkway might be part of the exit access portion of the means of egress. As such, its length would be included in determining travel distance for areas served by the walkway. Where a walkway is included in the means of egress, it needs to be available at all times the building is occupied. It could not be locked off if one building was closed and the other still open. Finally, any doors included in the pedestrian walkway would need to meet the appropriate requirements of Section 1008 including the direction of swing and the use of appropriate hardware that will allow them to function as required by the code.

3104.8 Width. The unobstructed width of *pedestrian walkways* shall be not less than 36 inches (914 mm). The total width shall be not greater than 30 feet (9144 mm).

❖ A maximum width has been set to prevent entire floors from being extended to adjacent buildings, and thus from being called walkways. This section is intended to encompass bridge-type structures, not full-floor extensions.

The minimum width is also subject to other code requirements. For example, if the walkway provides a means of egress component, it would be sized in accordance with Section 1005.1.

3104.9 Exit access travel. The length of *exit access* travel shall be 200 feet (60 960 mm) or less.

Exceptions:

1. *Exit access* travel distance on a *pedestrian walkway* equipped throughout with an *automatic sprinkler system* in accordance with Section 903.3.1.1 shall be 250 feet (76 200 mm) or less..
2. *Exit access* travel distance on a *pedestrian walkway* constructed with both sides not less than 50 percent open shall be 300 feet (91 440 mm) or less..
3. *Exit access* travel distance on a *pedestrian walkway* constructed with both sides not less than 50 percent open, and equipped throughout with an *automatic sprinkler system* in accordance with Section 903.3.1.1, shall be 400 feet (122 m) or less.

❖ The maximum length of exit access travel distance within the pedestrian walkway has been set as a means of controlling the size of walkways. A 200-foot (60 960 mm) maximum length is consistent with the maximum length of exit access travel distance allowed in unsprinklered buildings for most occupancy classifications (see Table 1016.1). This section gives three means by which the exit access travel distance within the pedestrian walkway can be increased. Exception 1 allows the exit access travel distance to be increased to 250 feet (76 200 mm) if the walkway is equipped with an automatic sprinkler system. This increase in travel distance correlates with that allowed in Table 1016.1. Exception 2 allows the exit access travel distance to be increased to 300 feet (91 440 mm) if both sides of the walkway have openings to the exterior for at least 50 percent of the wall area, with the open area uniformly distributed to prevent the accumulation of smoke and toxic gases. This increase in travel distance correlates to that allowed in Table 1016.1 for an open parking structure (Group S-2). Exception 3 expands on Exception 2 by including the use of an automatic sprinkler system. The exit access travel distance is then increased to 400 feet (122 m^2). This increase in travel distance correlates to the allowances in Table 1016.1.

3104.10 Tunneled walkway. Separation between the tunneled walkway and the building to which it is connected shall be not less than 2-hour fire-resistant construction and openings therein shall be protected in accordance with Table 716.5.

❖ Tunnels present a greater hazard than other walkways. The concern is greater than for enclosed walkways because a tunnel's underground location makes it less accessible to fire-fighting efforts and it relies on a mechanical ventilation system. To prevent the spread of fire, 2-hour fire-resistant construction with $1^1/_2$-hour opening protectives are required at the tunnel and building interface (see Figure 3104.10).

SECTION 3105
AWNINGS AND CANOPIES

3105.1 General. *Awnings* or *canopies* shall comply with the requirements of Sections 3105.2 through 3105.4 and other applicable sections of this code.

❖ This section addresses the variety of special structures that are defined as canopies and awnings. The intent of this section is that if a canopy or awning is used, it should be soundly designed so as not to present a hazard to its users or the public.

3105.2 Definition. The following term is defined in Chapter 2:

❖ Definitions of terms that are associated with the content of this section are contained herein. These definitions can help in the understanding and application of the code requirements. It is important to emphasize that these terms are not exclusively related to this section, but are applicable everywhere the term is used in the code. The purpose for including these definitions within this section is to provide more convenient access to them without having to refer back to Chapter 2.

For convenience, these terms are also listed in

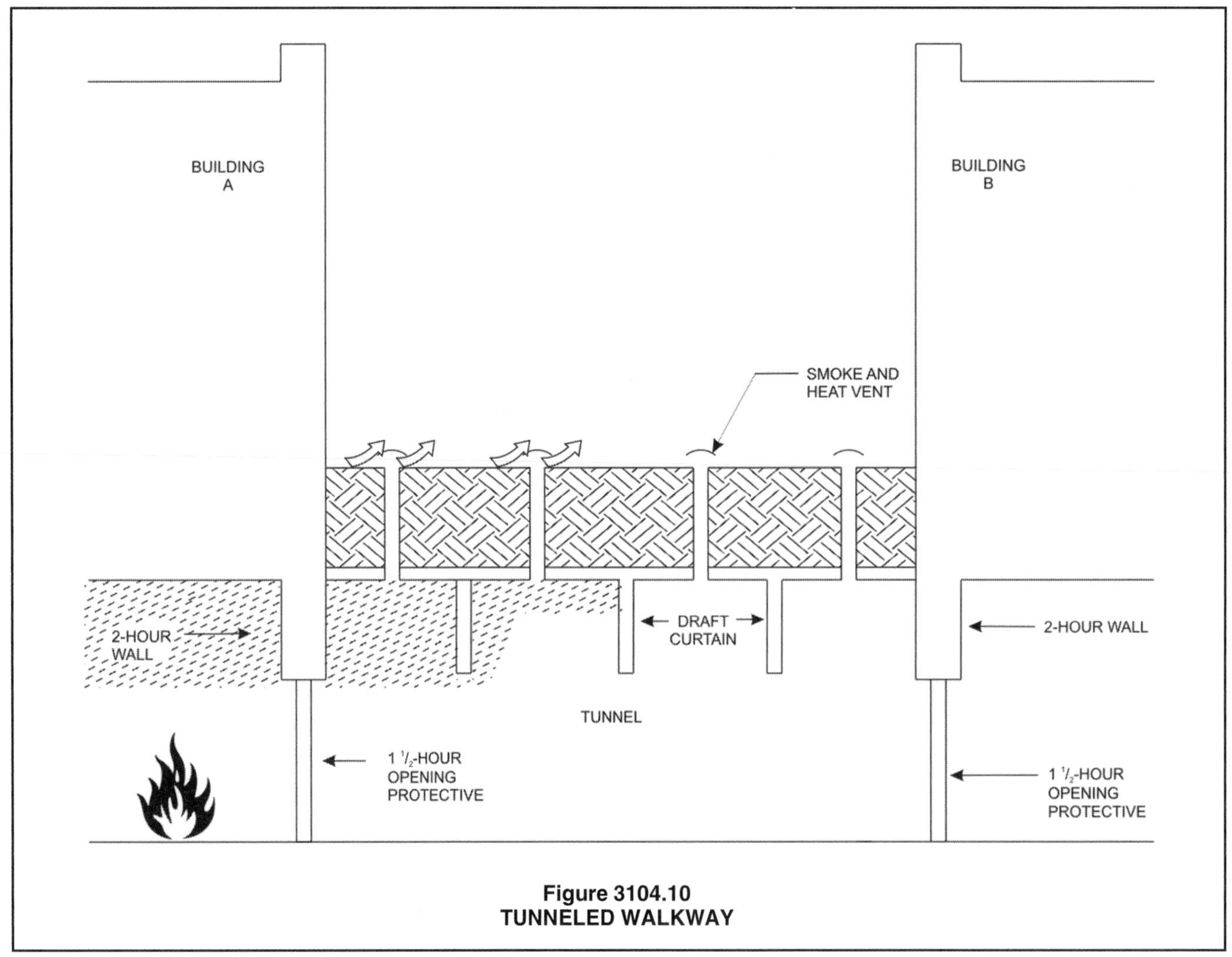

Figure 3104.10
TUNNELED WALKWAY

Chapter 2 with a cross reference to this section. The use and application of all defined terms, including those defined herein, are set forth in Section 201.

RETRACTABLE AWNING.

❖ The key to this definition is that it provides for an element of building construction that is popular in residential construction. The need was to properly identify that this section of the code is applicable to these awnings.

For reference, the terms "canopy" and "awning" are defined in Chapter 2, as follows:

AWNING. An architectural projection that provides weather protection, identity or decoration and is wholly supported by the building to which it is attached. An awning is comprised of a lightweight frame structure over which a covering is attached.

CANOPY. A permanent structure or architectural projection of rigid construction over which a covering is attached that provides weather protection, identity or decoration, and shall be structurally independent or supported by attachment to a building on one end and by not less than one stanchion on the outer end.

Figure 3105(1) shows a typical awning supported by posts to the ground and a fabric cover. Figure 3105(2) shows the underside of a different canopy, highlighting the rigid frame. Figure 3105(3) shows an awning that projects from the facade of the building and has no support directly to the ground.

Figure 3105(1)
EXAMPLE—CANOPY

Figure 3105(2)
EXAMPLE OF CANOPY STRUCTURE
(Photo Courtesy of Edie Industries Inc.)

Figure 3105(3)
EXAMPLE—AWNING
(Photo Courtesy of Edie Industries Inc.)

3105.3 Design and construction. *Awnings* and *canopies* shall be designed and constructed to withstand wind or other lateral loads and live loads as required by Chapter 16 with due allowance for shape, open construction and similar features that relieve the pressures or loads. Structural members shall be protected to prevent deterioration. *Awnings* shall have frames of noncombustible material, *fire-retardant-treated wood*, wood of Type IV size, or 1-hour construction with combustible or noncombustible covers and shall be either fixed, retractable, folding or collapsible.

❖ Any of the structures addressed in Section 3105 are subject to wind and live load considerations under the provisions of Chapter 16. When Chapter 16 is applied, care should be taken to consider the shape of the awning or canopy in question with regard to snow buildup and wind effects.

Structural members of an awning or canopy should be of materials that will resist rust or decay. Chapter 23 should be referenced for wood members that are required to comply with the decay-resistance provision. Steel or aluminum structural members are to be protected to resist rust.

3105.4 Canopy materials. *Canopies* shall be constructed of a rigid framework with an *approved* covering that meets the fire propagation performance criteria of NFPA 701 or has a *flame spread index* not greater than 25 when tested in accordance with ASTM E 84 or UL 723.

❖ This section specifically states the types of materials that are to be used and how a canopy is to be constructed. It limits the frame construction to metal and the coverings to only those that are deemed flame resistant when tested in accordance with NFPA 701, or those having a flame spread of 25 or less when tested in accordance with ASTM E 84 or UL 723. The purpose of these provisions is to limit the possible spread of fire on the exterior of the building due to fire from the interior of the building or from an exterior source on the street or public way.

SECTION 3106 MARQUEES

3106.1 General. Marquees shall comply with Section 3106.2 through 3106.5 and other applicable sections of this code.

❖ The intent of this section is that if a marquee is used, it should be soundly designed so as not to present a hazard to its users or the public. A "Marquee" is defined in Chapter 2 as follows:

MARQUEE. A permanent roofed structure attached to and supported by the building and that projects into the public right-of-way.

Therefore compliance with Chapter 32, "Encroachments into the Public Right-of-way," is also important for marquees. Figures 3106(1) and (2) provide an example of a marquee over a public right-of-way.

3106.2 Thickness. The height or thickness of a marquee measured vertically from its lowest to its highest point shall be not greater than 3 feet (914 mm) where the marquee projects more than two-thirds of the distance from the *lot line* to

Figure 3106(1)
EXAMPLE—MARQUEE

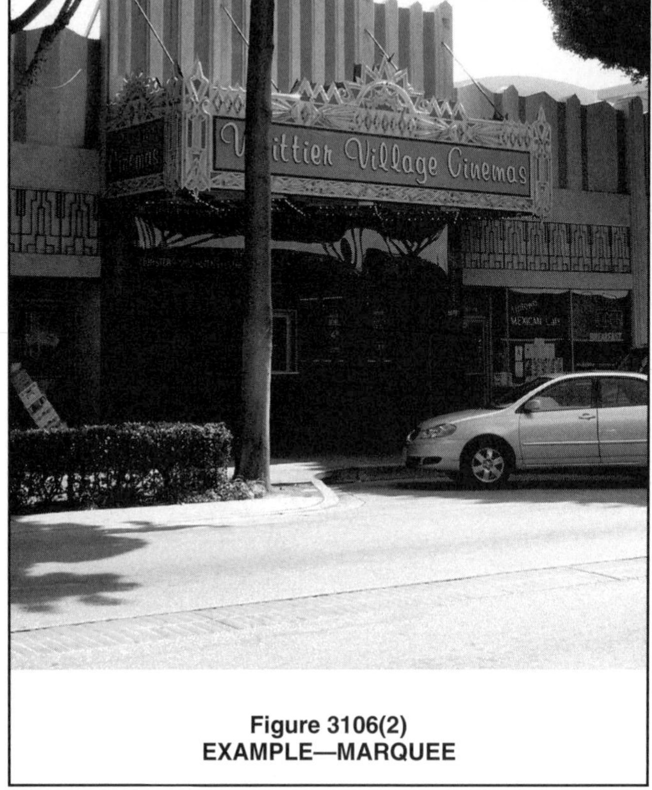

Figure 3106(2)
EXAMPLE—MARQUEE

the curb line, and shall be not greater than 9 feet (2743 mm) where the marquee is less than two-thirds of the distance from the lot line to the curb line.

❖ The restrictions placed on the size, projection and clearances (see Figure 3106.2) for marquees are intended to:

1. Prevent interference with the free movement of pedestrians.
2. Prevent interference with trucks and other tall vehicles using the public street.
3. Prevent interference with fire-fighting operations at a building.
4. Prevent interference with utilities.

3106.3 Roof construction. Where the roof or any part thereof is a skylight, the skylight shall comply with the requirements of Chapter 24. Every roof and skylight of a mar-

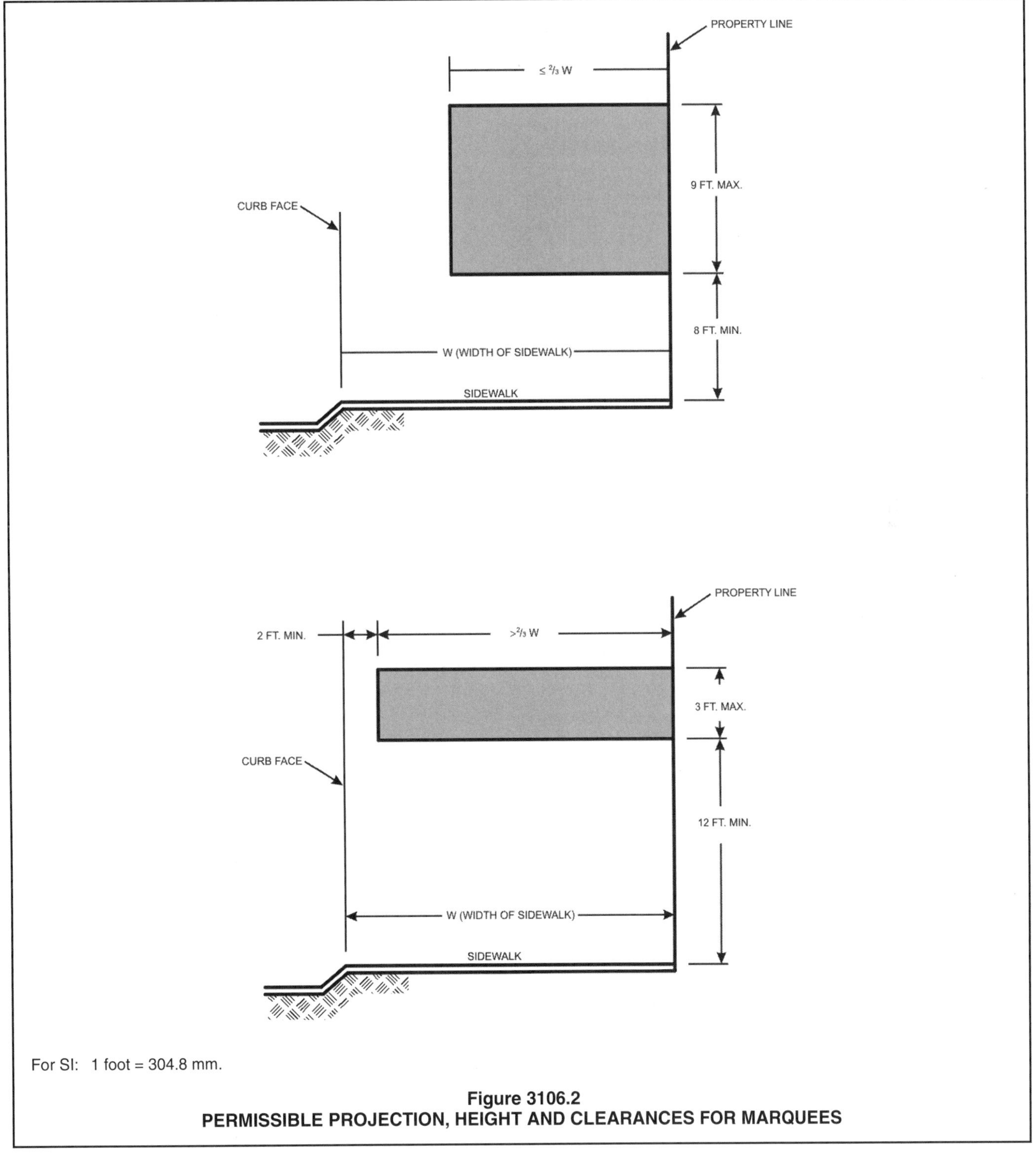

For SI: 1 foot = 304.8 mm.

Figure 3106.2
PERMISSIBLE PROJECTION, HEIGHT AND CLEARANCES FOR MARQUEES

quee shall be sloped to downspouts that shall conduct any drainage from the marquee in such a manner so as not to spill over the sidewalk.

❖ The purpose of this section is to require proper drainage to prevent buildup on the roof due to a marquee or skylight and also to reduce the nuisance of such drainage to those using the sidewalk.

3106.4 Location prohibited. Every marquee shall be so located as not to interfere with the operation of any exterior standpipe, and such that the marquee does not obstruct the clear passage of *stairways* or *exit discharge* from the building or the installation or maintenance of street lighting.

❖ Marquees are an acceptable building feature but they must not interfere or obstruct fire protection features and egress. In addition, such marquees also need to be placed such that of street lights can be maintained.

3106.5 Construction. A marquee shall be supported entirely from the building and constructed of noncombustible materials. Marquees shall be designed as required in Chapter 16. Structural members shall be protected to prevent deterioration.

❖ Damage to both life and property can occur from improperly constructed marquees. Proper design of a marquee involves a structural analysis by qualified persons. Wind and seismic loads must be considered in the design process, as outlined in Chapter 16. Additionally, all material must be noncombustible regardless of the construction type of the building supporting the marquee.

SECTION 3107 SIGNS

3107.1 General. Signs shall be designed, constructed and maintained in accordance with this code.

❖ The design of signs is to be in accordance with the code. If the design is inadequate, no amount of expert construction, repair or maintenance will result in a safe sign. The primary requirements for such signs will be related to structural stability as addressed in Chapter 16. Signs projecting over the public right-of-way will also be subject to the requirements of Chapter 32.

Signs within a covered mall building are specifically addressed by Section 402.16.

Note that Appendix H, when adopted by a jurisdiction, provides more specific requirements related to signs.

SECTION 3108 TELECOMMUNICATION AND BROADCAST TOWERS

3108.1 General. Towers shall be designed and constructed in accordance with the provisions of TIA-222. Towers shall be designed for seismic loads; exceptions related to seismic design listed in Section 2.7.3 of TIA-222 shall not apply. In Section 2.6.6.2 of TIA 222, the horizontal extent of Topographic Category 2, escarpments, shall be 16 times the height of the escarpment.

Exception: Single free-standing poles used to support antennas not greater than 75 feet (22 860 mm), measured from the top of the pole to grade, shall not be required to be noncombustible.

❖ Telecommunication and broadcast towers, which include media such as radio, television and cell phones, are special structures in that they are not normally occupied and are open structures containing little, if any, fuel load. This section references TIA-222, which provides the necessary structural requirements for such telecommunication and broadcast towers. The seismic design exceptions given in TIA-222, Section 2.7.3, do not apply because these exceptions bring TIA 222-G below the minimum standard of care for design established by the code and ASCE 7.

Chapter 15 gives specific requirements for design of communications towers. No exemptions are provided. TIA-222 is not listed as a referenced standard within the *International Codes®* for seismic design, so seismic design requirements must be governed by ASCE 7. TIA 222-G, because of exemptions to seismic design requirements, does not meet the minimum standard of care for design established by the code and ASCE 7.

The exception to this section will allow combustible poles used to support antennas not greater than 75 feet (1905 mm). Such poles are often used for the support of lightweight electrical equipment such as cell phone antennas and have had many years of good performance.

3108.2 Location and access. Towers shall be located such that guy wires and other accessories shall not cross or encroach upon any street or other public space, or over above-ground electric utility lines, or encroach upon any privately owned property without the written consent of the owner of the encroached-upon property, space or above-ground electric utility lines. Towers shall be equipped with climbing and working facilities in compliance with TIA-222. Access to the tower sites shall be limited as required by applicable OSHA, FCC and EPA regulations.

❖ Access must be provided to the tower so that its condition can be surveyed. The features required to access such towers are located in TIA-222. Occupational Safety & Health Administration (OSHA), the Environmental Protection Agency (EPA) and the Federal Communications Commission (FCC) are the regulatory bodies that limit who may access such towers and how.

Guy wires are to be arranged so as not to represent a hazard to the public, and therefore, they shall not cross or encroach a street, public space or electric power line. This is because of both the damage that could occur if a guy wire were suddenly released and the need to prohibit the wire from becoming an obstruction. Guy wires are not to encroach on any other property without previous written consent by the owner of the affected property.

SECTION 3109
SWIMMING POOL ENCLOSURES AND SAFETY DEVICES

3109.1 General. Swimming pools shall comply with the requirements of Sections 3109.2 through 3109.5 and other applicable sections of this code.

❖ The provisions of this section apply to both public and private swimming pools. They are intended to provide for the use of swimming pools as well as to limit or delay unauthorized access to swimming pools by small children, particularly those five years old and younger. All but small pools that are less than 24 inches (610 mm) deep must comply with the provisions of this section. Requirements are included for pool enclosures and barriers.

3109.2 Definition. The following term is defined in Chapter 2:

❖ The definition of a term that is associated with the content of this section is contained herein. This definition can help in the understanding and application of the code requirements. It is important to emphasize that this term is not exclusively related to this section but is applicable everywhere the term is used throughout the code. The purpose for including this definition within this chapter is to provide more convenient access to it without having to refer back to Chapter 2.

For convenience, these terms are also listed in Chapter 2 with a cross reference to this section. The use and application of all defined terms, including those defined herein, are set forth in Section 201.

SWIMMING POOLS.

❖ The small above-ground pools frequently found in residential yards are exempt, provided that they do not exceed 2 feet (610 mm) in depth.

3109.3 Public swimming pools. Public swimming pools shall be completely enclosed by a fence not less than 4 feet (1290 mm) in height or a screen enclosure. Openings in the fence shall not *permit* the passage of a 4-inch-diameter (102 mm) sphere. The fence or screen enclosure shall be equipped with self-closing and self-latching gates.

❖ Whether the pool is privately owned and operated is irrelevant to whether or not it is a public swimming pool. As long as the pool is not associated with an occupancy in Group R-3, it is considered a public swimming pool. Examples of public swimming pools include a pool in an apartment complex located in a common use area, a pool at a health club and a municipal swimming pool.

A public pool requires a barrier no lower than 4 feet (1219 mm). This barrier/wall limits or delays unauthorized access to the pool by small children, particularly those five years old or younger. It is important to note that this barrier is to enclose the swimming pool completely. Any construction or natural element that does not surround the pool will allow access at some point; hence, any gate, door or other access component must be self-closing and self-latching.

3109.4 Residential swimming pools. Residential swimming pools shall comply with Sections 3109.4.1 through 3109.4.3.

Exception: A swimming pool with a power safety cover or a spa with a safety cover complying with ASTM F 1346 need not comply with Section 3109.4.

❖ Although the occupants and guests of a Group R-3 residence have access to the swimming pool regardless of its location, it is important to distinguish what elements separate the pool from the occupants. Pools that are totally within the structure are included in this section. Of critical concern is the easy access afforded to children by an indoor pool. For this reason, Sections 3109.4.1 through 3109.4.3 must be closely followed.

3109.4.1 Barrier height and clearances. The top of the barrier shall be not less than 48 inches (1219 mm) above grade measured on the side of the barrier that faces away from the swimming pool. The vertical clearance between grade and the bottom of the barrier shall be not greater than 2 inches (51 mm) measured on the side of the barrier that faces away from the swimming pool. Where the top of the pool structure is above grade, the barrier is authorized to be at ground level or mounted on top of the pool structure, and the vertical clearance between the top of the pool structure and the bottom of the barrier shall be not greater than 4 inches (102 mm).

❖ The barrier height requirement of 48 inches (1219 mm) above the ground is based on reports that have documented the ability of children under the age of five to climb over barriers that are less than 48 inches (1219 mm) in height. The basis for the 4-inch (102 mm) criterion for an opening between the barrier and the top of pool frame is the same as for guard construction (see Figure 3109.4.1).

3109.4.1.1 Openings. Openings in the barrier shall not allow passage of a 4-inch-diameter (102 mm) sphere.

❖ The basis for the 4-inch (102 mm) criterion is the same as for guard construction (see Figure 3109.4.1.1). It is based on studies of the body measurements of children 13 to 18 months old.

3109.4.1.2 Solid barrier surfaces. Solid barriers which do not have openings shall not contain indentations or protrusions except for normal construction tolerances and tooled masonry joints.

❖ This provision is intended to reduce the potential for gaining a foothold and climbing the barrier.

3109.4.1.3 Closely spaced horizontal members. Where the barrier is composed of horizontal and vertical members and the distance between the tops of the horizontal members is less than 45 inches (1143 mm), the horizontal members shall be located on the swimming pool side of the fence. Spacing between vertical members shall be not greater than $1^{3}/_{4}$ inches (44 mm) in width. Where there are decorative cutouts within vertical members, spacing within the cutouts shall be not greater than $1^{3}/_{4}$ inches (44 mm) in width.

❖ This more stringent $1^{3}/_{4}$-inch (44 mm) provision for spacing between vertical members applies when the spacing between horizontal members is less than 45 inches (1143 mm). It acknowledges the potential for a child to gain both a handhold and a foothold on closely spaced horizontal members, and is intended to reduce that potential by limiting the space between the vertical members on the same barrier. If the horizontal members are spaced less than 45 inches (1143 mm), they must also be located on the swimming pool side of the fence (see Figure 3109.4.1.3) so that they are not available to be used to climb the barriers.

3109.4.1.4 Widely spaced horizontal members. Where the barrier is composed of horizontal and vertical members and the distance between the tops of the horizontal members is 45 inches (1143 mm) or more, spacing between vertical members shall be not greater than 4 inches (102 mm). Where there are decorative cutouts within vertical members, spacing within the cutouts shall be not greater than $1^{3}/_{4}$ inches (44 mm) in width.

❖ This requirement is the counterpart to Section 3109.4.1.3 in that it permits the opening in the barrier to be 4 inches (102 mm), provided the vertical spacing of the horizontal members equals or exceeds 45 inches (1143 mm) (see Figure 3109.4.1.3). It limits openings in the barrier to a 4-inch (102 mm) diameter. The spacing of horizontal members 45 inches (1143 mm) apart precludes them from being used by small children to climb the barrier.

3109.4.1.5 Chain link dimensions. Mesh size for chain link fences shall be not greater than a $2^{1}/_{4}$ inch square (57 mm square) unless the fence is provided with slats fastened at the top or the bottom which reduce the openings to not more than $1^{3}/_{4}$ inches (44 mm).

❖ The $2^{1}/_{4}$-inch (57 mm) dimension is intended to reduce the potential to gain a foothold (see Figure 3109.4.1.5). The opening is permitted to be increased above this $2^{1}/_{4}$-inch (57 mm) dimension if decorative slats are used that reduce the size of the opening to $1^{3}/_{4}$ inches (44 mm), since the potential to obtain a foothold or handhold is reduced when such slats are installed.

3109.4.1.6 Diagonal members. Where the barrier is composed of diagonal members, the opening formed by the diagonal members shall be not greater than $1^{3}/_{4}$ inches (44 mm).

❖ A slightly bigger opening is permitted for barriers composed of diagonal members other than chain-link fences on the basis that such barriers would be more difficult to gain a foothold and handhold on than a chain-link fence. The $1^{3}/_{4}$-inch (44 mm) dimension is consistent with Sections 3109.4.1.3, 3109.4.1.4 and 3109.4.1.5.

3109.4.1.7 Gates. Access doors or gates shall comply with the requirements of Sections 3109.4.1.1 through 3109.4.1.6 and shall be equipped to accommodate a locking device. Pedestrian access doors or gates shall open outward away from the pool and shall be self-closing and have a self-latching device. Doors or gates other than pedestrian access doors or gates shall have a self-latching device. Release mechanisms shall be in accordance with Sections 1008.1.9 and

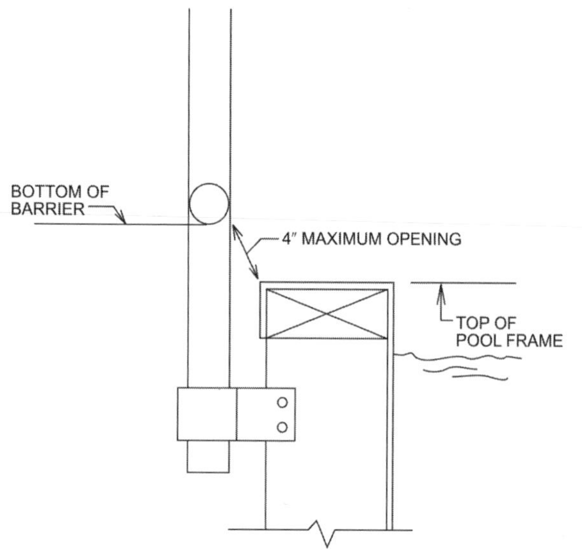

BOTTOM OF BARRIER

4″ MAXIMUM OPENING

TOP OF POOL FRAME

For SI: 1 inch = 25.4 mm.

Figure 3109.4.1
OPENING LIMITATIONS

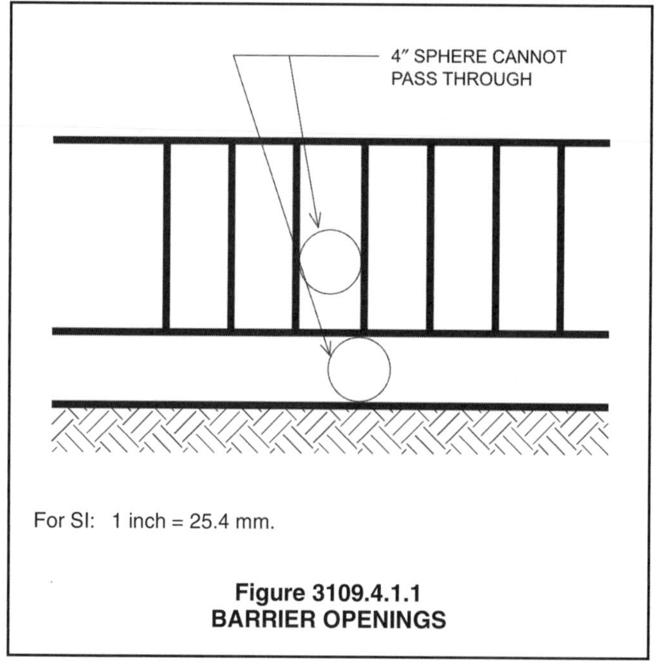

For SI: 1 inch = 25.4 mm.

Figure 3109.4.1.1
BARRIER OPENINGS

1109.13. Where the release mechanism of the self-latching device is located less than 54 inches (1372 mm) from the bottom of the door or gate, the release mechanism shall be located on the pool side of the door or gate 3 inches (76 mm) or more, below the top of the door or gate, and the door or gate and barrier shall be without openings greater than $^1/_2$ inch (12.7 mm) within 18 inches (457 mm) of the release mechanism.

❖ Gates, and some type of doors, represent the same potential hazard relative to climbing as do other portions of the barrier and, therefore, must be constructed in accordance with Sections 3109.4.1.1 through 3109.4.1.6. Additionally, since the door or gate represents a potential breach of the barrier due to the ability to be opened, the code provides prescriptive details for its construction and operation. A self-closing pedestrian door or gate is required to open away from the pool because if the latch fails to operate, a child pushing on the door or gate will not gain immediate access to the pool. Pushing on the door or gate may also engage the latch. Large, non-pedestrian gates are not required to be self-closing due to the prohibitive cost coupled with the fact that these gates are typically operated by persons other than small children. The 54-inch (1372 mm) latch height requirement is intended to limit the potential for small children to reach and activate it. If located less than 54 inches high (1372 mm), the code's prescriptive location requirements are intended to preclude

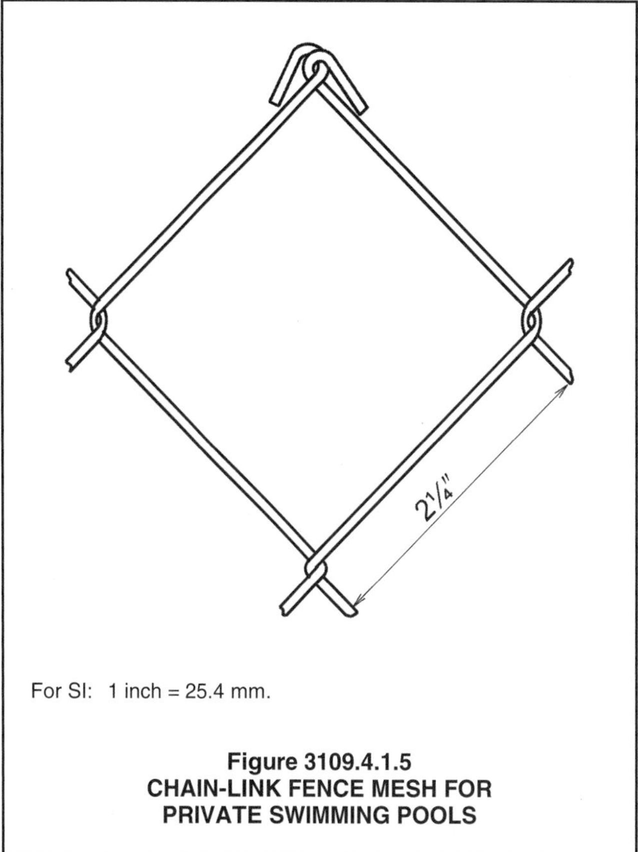

For SI: 1 inch = 25.4 mm.

Figure 3109.4.1.5
CHAIN-LINK FENCE MESH FOR PRIVATE SWIMMING POOLS

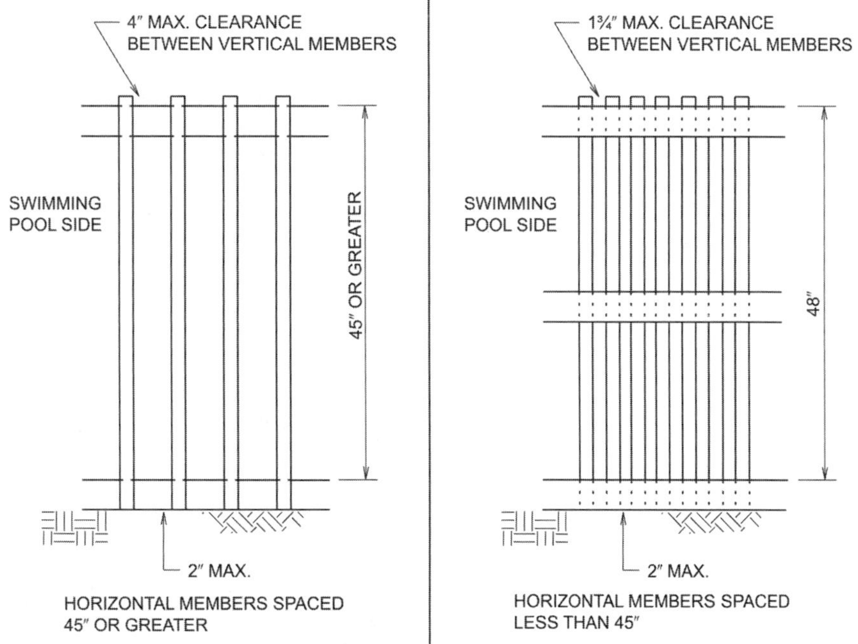

4" MAX. CLEARANCE BETWEEN VERTICAL MEMBERS
SWIMMING POOL SIDE
45" OR GREATER
2" MAX.
HORIZONTAL MEMBERS SPACED 45" OR GREATER

1¾" MAX. CLEARANCE BETWEEN VERTICAL MEMBERS
SWIMMING POOL SIDE
48"
2" MAX.
HORIZONTAL MEMBERS SPACED LESS THAN 45"

For SI: 1 inch = 25.4 mm.

Figure 3109.4.1.3
PRIVATE SWIMMING POOL BARRIER CONSTRUCTION

the latch from being activated by small children who are not on the pool side of the door or gate.

3109.4.1.8 Dwelling wall as a barrier. Where a wall of a *dwelling* serves as part of the barrier, one of the following shall apply:

1. Doors with direct access to the pool through that wall shall be equipped with an alarm that produces an audible warning when the door and/or its screen, if present, are opened. The alarm shall be *listed* and labeled in accordance with UL 2017. In dwellings not required to be *Accessible units*, *Type A units* or *Type B units*, the deactivation switch shall be located 54 inches (1372 mm) or more above the threshold of the door. In dwellings required to be *Accessible units*, *Type A units* or *Type B units*, the deactivation switch shall be located not higher than 54 inches (1372 mm) and not less than 48 inches (1219 mm) above the threshold of the door.
2. The pool shall be equipped with a power safety cover that complies with ASTM F 1346.
3. Other means of protection, such as self-closing doors with self-latching devices, which are *approved*, shall be accepted so long as the degree of protection afforded is not less than the protection afforded by Section 3109.4.1.8, Item 1 or 2.

❖ Many residential settings with backyard pools utilize the dwelling as a portion of the barrier required around the pool (i.e., the fence bounding the property terminates at the dwelling). This limits access to the pool by unsupervised children around the perimeter of the fence but there is still a potential for children to access the pool from within the dwelling. Indeed, almost half the children involved in drowning or near-drowning accidents gained access to the pool from the dwelling.

The provisions of this section are intended to restrict such access by small children and are applicable to all doors in walls that form a portion of the barrier required around private swimming pools, both outdoor and indoor.

Protection of door openings to pool areas can be achieved in any one of the methods described in Items 1 through 3. The alarm is configured to allow adults accessing the house to open the door, enter the house and deactivate the system to prevent a "false alarm." The touchpad permitted to deactivate the system is required to be mounted 54 inches (1372 mm) above the floor, which is presumed to be beyond the reach of small children.

An audible alarm specified in Item 1 on the doors leading to the pool area is intended to provide a warning to a supervising adult that the pool area has been entered.

Item 2 does not require protection of the door itself; but limits access to the pool by means of a power safety cover. The performance criteria specified when this option is selected is to ensure that the power safety cover is an adequate and reliable barrier to the pool.

Item 3 is simply allowing for any innovative technology that provides effective protection for children attempting to gain access to the pool. The device or devices must be approved by the building official.

3109.4.1.9 Pool structure as barrier. Where an aboveground pool structure is used as a barrier or where the barrier is mounted on top of the pool structure, and the means of access is a ladder or steps, then the ladder or steps either shall be capable of being secured, locked or removed to prevent access, or the ladder or steps shall be surrounded by a barrier which meets the requirements of Sections 3109.4.1.1 through 3109.4.1.8. Where the ladder or steps are secured, locked or removed, any opening created shall not allow the passage of a 4-inch-diameter (102 mm) sphere.

❖ The code permits the wall of the pool itself to serve as the barrier to the pool, provided that the wall extends at least 48 inches (1219 mm) above the finished ground level around the perimeter of the pool.

3109.4.2 Indoor swimming pools. Walls surrounding indoor swimming pools shall not be required to comply with Section 3109.4.1.8.

❖ Indoor pools are not required to have the protections required by Section 3109.4.1.8. The pool in an indoor environment is generally under closer scrutiny; therefore, it is less apt to be an area where small children would be allowed to stray into without being noticed.

3109.4.3 Prohibited locations. Barriers shall be located so as to prohibit permanent structures, equipment or similar objects from being used to climb the barriers.

❖ This section is especially important when the wall of the pool itself is used as the barrier (see Section 3109.4.1.9). Pumps and other equipment located adjacent to the pool wall (barrier) present a hazard in that they may provide a means by which small children could climb over the pool wall and access the pool area.

3109.5 Entrapment avoidance. Suction outlets shall be designed and installed in accordance with ANSI/APSP-7.

❖ This section references ANSI/APSP-7, *Standard for Suction Entrapment Avoidance in Swimming Pools, Wading Pools, Spas, Hot Tubs and Catch Basins.* This standard addresses all forms of entrapment, including the underlying causes of entrapment. Body or hair entrapment can cause drowning and evisceration; therefore, it is important to provide protection against possible entrapment at the pool entrances to suction inlets, as well as vacuum relief for the system.

Although rare, entrapment of bathers at suction outlets in pools and spas has gained considerable attention over the last decade, resulting in voluntary standards, building codes and proposed national legislation to prevent these tragic accidents.

A survey of the Epidemiological Reports on Suction Entrapment collected by the U.S. Consumer Product Safety Commission by the Association of Pool and

Spa Professionals (APSP) Technical Committee yielded five distinct modes of entrapment:

ENTRAPMENT TYPE	PERCENTAGE OF INCIDENTS
Hair Entrapment—Hair becomes knotted or snagged in an outlet cover.	33%
Limb Entrapment—A limb sucked or inserted into an opening of a circulation outlet with a broken or missing cover resulting in a mechanical bind or swelling.	28%
Body Entrapment—Suction applied to a large portion of the body or limbs resulting in an entrapment.	33%
Evisceration/Disembowelment—Suction applied directly to the intestines by a circulation outlet with a broken or missing cover.	3%
Mechanical Entrapment—Potential for jewelry, swimsuit, hair decorations, finger, toe or knuckle to be mechanically caught in an opening of a suction outlet or cover.	Included in limb

Early actions to address entrapment were aimed at body entrapment by attempting to control the suction pressure at the drain itself. Unfortunately, these devices do not protect against the major forms of entrapment: hair or evisceration. Additionally, if the pool circulation pump is off—meaning no suction at the outlet—a child can still get a limb trapped if there is a broken or missing cover.

However, suction is only one factor to control in entrapment avoidance.

In order to address avoidance of all forms of entrapment, a comprehensive study of the causes of all types of entrapment was undertaken. It is now known that there are three basic underlying physical phenomena that govern all five modes of entrapment:

- Suction (or delta pressure)
- Water flow rate through the outlet or cover
- Mechanical binding

The Technical Committee of the APSP examined various means to prevent these types of entrapments recognizing the diverse nature of pool construction. Using this knowledge, a new national consensus standard was developed in accordance with the ANSI process. ANSI/APSP-7 provides requirements intended to protect bathers against all modes of entrapment.

The ANSI/APSP-7 standard applies to both commercial and residential pools, for flow rates from a few gallons per minute to thousands of gallons per minute. Although it includes the use of devices or systems that prevent suction, it also expands the lists of options for the pool contractor, while maintaining necessary protective principles.

ANSI/APSP-7 contains design performance criteria including components, devices and related technology installed to protect against entrapment.

ANSI/APSP-7 considers all underlying causes: suction, water flow and mechanical—while recognizing the diverse nature of pool and spa design. It covers all five forms of entrapment.

SECTION 3110
AUTOMATIC VEHICULAR GATES

3110.1 General. *Automatic vehicular gates* shall comply with the requirements of Sections 3110.2 through 3110.4 and other applicable sections of this code.

❖ These provisions are aimed at providing safety from the hazards that could result without the proper requirements for the operation of vehicular gates. Protection is needed from potential entrapment of individuals between an automatically moving gate and a stationary object or surface, in close proximity to such gate. Gates intended for automation require specific design construction and installation to include entrapment protection to minimize or eliminate certain excessive gate gaps, openings and protrusions identified as contributing to the hazard of entrapments that have historically caused numerous serious injuries and deaths. As defined the gates regulated by this section are not those used by pedestrians but instead are intended only for vehicles.

3110.2 Definition. The following term is defined in Chapter 2:

❖ Definitions of terms that are associated with the content of this section are contained herein. These definitions can help in the understanding and application of the code requirements. It is important to emphasize that these terms are not exclusively related to this section, but are applicable everywhere the term is used in the code. The purpose for including these definitions within this section is to provide more convenient access to them without having to refer back to Chapter 2.

For convenience, these terms are also listed in Chapter 2 with a cross reference to this section. The use and application of all defined terms, including those defined herein, are set forth in Section 201.

VEHICULAR GATE.

❖ This definition provides the scope for the particular type of gates intended to be addressed by this section. They are specifically those gates associated with vehicles at a building or facility as a opposed to a parking lot or access road. In addition it clarifies that these gates are not those used by people.

3110.3 Vehicular gates intended for automation. *Vehicular gates* intended for automation shall be designed, constructed and installed to comply with the requirements of ASTM F 2200.

❖ This section simply calls out the design standard to which such gates need to be designed and installed. The standard is fairly specific to this application for automated vehicular gates and has been harmonized with the applicable provisions of UL 325 as referenced in Section 3110.4.

3110.4 Vehicular gate openers. *Vehicular gate* openers, where provided, shall be *listed* in accordance with UL 325.

❖ This section simply references the appropriate standards for the openers of such gates. The standard is UL 325 which is an ANSI recognized safety standard containing provisions governing gate openers. Note that the *International Residential Code*® (IRC®) also references this particular standard.

SECTION 3111
SOLAR PHOTOVOLTAIC PANELS/MODULES

3111.1 General. Solar photovoltaic panels/modules shall comply with the requirements of this code and the *International Fire Code.*

❖ Photovoltaic arrays are increasing in popularity as an alternative energy source. These arrays, which cannot be shut down and retain electrical charges present unique hazards to fire fighters operating on roofs with arrays or nearby circuits. This section references the IFC, which provides general requirements to allow for increased safety of fire fighters working around and near the arrays.

Bibliography

The following resource materials are referenced in this chapter or are relevant to the subject matter addressed in this chapter.

15 USC 8001, *Title XIV—Pool and Spa Safety Act*, 2007.

24 CFR, *Fair Housing Accessibility Guidelines* (FHAG). Washington, DC: Department of Housing and Urban Development, 1991.

ANSI/APSP7-06, *Standard for Suction Entrapment Avoidance in Swimming Pools, Wading Pools, Spas, Hot Tubs and Catch Basins.* Alexandria, VA: The Association of Pool & Spa Professionals, 2006.

ASTM E 84-07, *Test Method for Surface Burning Characteristics of Building Materials.* West Conshohocken, PA: ASTM International, 2004.

ASTM E 136-04, *Test Method for Behavior of Materials in a Vertical Tube Furnace at 750°C.* West Conshohocken, PA: ASTM International, 2004.

ASTM F 2200-05 *Standard Specification for Automated Vehicular Gate Construction.* West Conshohocken, PA: ASTM International, 2004.

ICC 300-07, *ICC Standard on Bleachers, Folding and Telescopic Seating and Grandstands.* Washington, DC: International Code Council, 2007.

IFC-12, *International Fire Code.* Washington, DC: International Code Council, 2011.

IMC-12, *International Mechanical Code.* Washington, DC: International Code Council, 2011.

IRC-12, *International Residential Code.* Washington, DC: International Code Council, 2011.

NFPA 701-04, *Methods of Fire Tests for Flame-resistant Textiles and Films.* Quincy, MA: National Fire Protection Association, 2004.

TIA/EIA 222-G-05, *Structural Standards for Antenna Supporting Structures and Antenna.* Arlington, VA: Electronics Industries Association, 2005.

UL 325-02, *Door, Drapery, Gate, Louver and Window Operators and Systems—with Revisions through February 2006.* Northbrook, IL: Underwriters Laboratory, 2002.

UL 723-03, *Standard for Test for Surface Burning Characteristics of Building Materials—with Revisions through May 2005.* Northbrook, IL: Underwriters Laboratory, 2003.

Chapter 32:
Encroachments Into the Public Right-of-way

General Comments

Chapter 32 contains provisions that regulate a variety of structure encroachments and projections into the public right-of-way, such as a sidewalk, parking lot or street.

Section 3202 addresses encroachments on street lot lines below grade, above grade, below 8 feet (2438 mm) in height and 8 feet (2438 mm) or more above grade.

Encroachments and projections have to be taken into consideration in addition to means of egress, fire resistance, fire protection or even structural and material considerations.

Purpose

The purpose of Chapter 32 is to regulate projections and encroachments of structures into the public right-of-way.

SECTION 3201 GENERAL

3201.1 Scope. The provisions of this chapter shall govern the encroachment of structures into the public right-of-way.

❖ These provisions apply only to encroachments on the public right-of-way, not on interior lot lines. The extent and location of the street lot line (public right-of-way) is determined through a variety of methods based on the regulations of the local jurisdictions and often the state. The location and width of streets is often established when an area is platted and the land is subdivided into a series of blocks, lots and streets. While subdivision and zoning laws may provide for standard street and right-of-way width, many areas of a jurisdiction may predate those standards. Streets and highways are often established by state and local government transportation agencies. Such agencies can create new streets or widen existing streets by either purchasing or condemning land of sufficient width. In other circumstances, public streets are not owned by a city, but they are easements across privately owned lands. Because of the wide variety of ways in which a public right-of-way can be established, it is important for the building official to know how rights-of-way were established within the jurisdiction and to be able to provide this information to permit applicants.

3201.2 Measurement. The projection of any structure or portion thereof shall be the distance measured horizontally from the *lot line* to the outermost point of the projection.

❖ This section defines how the extent of encroachment is determined. The extent of encroachment of any structure or appendage is to be measured from the established street lot line to the outermost (furthest) point of the projection.

3201.3 Other laws. The provisions of this chapter shall not be construed to permit the violation of other laws or ordinances regulating the use and occupancy of public property.

❖ The projections permitted in this chapter are also subject to other limitations identified in applicable laws and ordinances. Other applicable laws, for example, are those established by the applicable governing authority in regulating right-of-way for highways, streets or public areas adjacent to the structure in question.

The street lot line occurs on each side of a structure that faces or fronts on a street. Many property descriptions have the centerline of the public way as the property line. In these cases, the lot line is defined not by the centerline of the public way, but by the line that identifies where the public way exists. This chapter only applies to encroachment across lot lines, however, established between a lot and any adjoining public right-of-way (street, alley, highway) (see commentary, for Section 3201.1). Local zoning codes often establish setbacks or yards on a property, which under the zoning code, must be kept open, or substantially open. Often the yard or setback requirements are minimal in commercial or industrial zones and are more substantial in residential zones. This chapter does not address encroachments into yards or setbacks established by zoning codes unless that setback is actually establishing the public right-of-way. This chapter does not apply to interior lot lines—those separating one lot from another. Encroachments across interior lot lines are not permitted.

3201.4 Drainage. Drainage water collected from a roof, *awning*, canopy or marquee, and condensate from mechanical equipment shall not flow over a public walking surface.

❖ A great number of street projections are canopies or mansard roofs. They have the potential to create a hazardous condition due to improper drainage on the walkways they cover. It is very important that all street projections have proper drainage systems.

SECTION 3202 ENCROACHMENTS

3202.1 Encroachments below grade. Encroachments below grade shall comply with Sections 3202.1.1 through 3202.1.3.

❖ Allowable projections of below-grade structural components are described in Sections 3202.1.1 through 3202.1.3. The term grade as used in this chapter should be considered the finished ground level of the site. Grade should not be confused with grade plane which is used for determining other code standards such as building height and number of stories.

3202.1.1 Structural support. A part of a building erected below grade that is necessary for structural support of the building or structure shall not project beyond the *lot lines*, except that the footings of street walls or their supports which are located not less than 8 feet (2438 mm) below grade shall not project more than 12 inches (305 mm) beyond the street *lot line*.

❖ One of the permissible street lot line encroachments (footings) is defined in this section. Where a structure is built immediately adjacent to the street lot line, the footing is permitted to project across that line (see Figure 3202.1.1). Projection of footings across side (interior) lot lines and onto adjacent property is not permitted.

3202.1.2 Vaults and other enclosed spaces. The construction and utilization of vaults and other enclosed spaces below grade shall be subject to the terms and conditions of the applicable governing authority.

❖ Below-grade vaults and other enclosed spaces that open at the sidewalk level are usually covered with solid structural caps (e.g., concrete slabs) and are permitted to project beyond the building line or street lot line. However, since these structures occur in the public right-of-way, they are subject to local laws governing private use of the public right-of-way as well as the approval of the applicable governing authority.

3202.1.3 Areaways. Areaways shall be protected by grates, *guards* or other *approved* means.

❖ Areaways, which provide natural light or ventilation to below-grade spaces, are required to be protected by grates, guards or other approved means of protection (see Figure 3202.1.3). Local regulations allowing the private use of the public right-of-way may prescribe how areaways are covered and protected. Some jurisdictions allow material lifts to be located in these areaways and open to the public street.

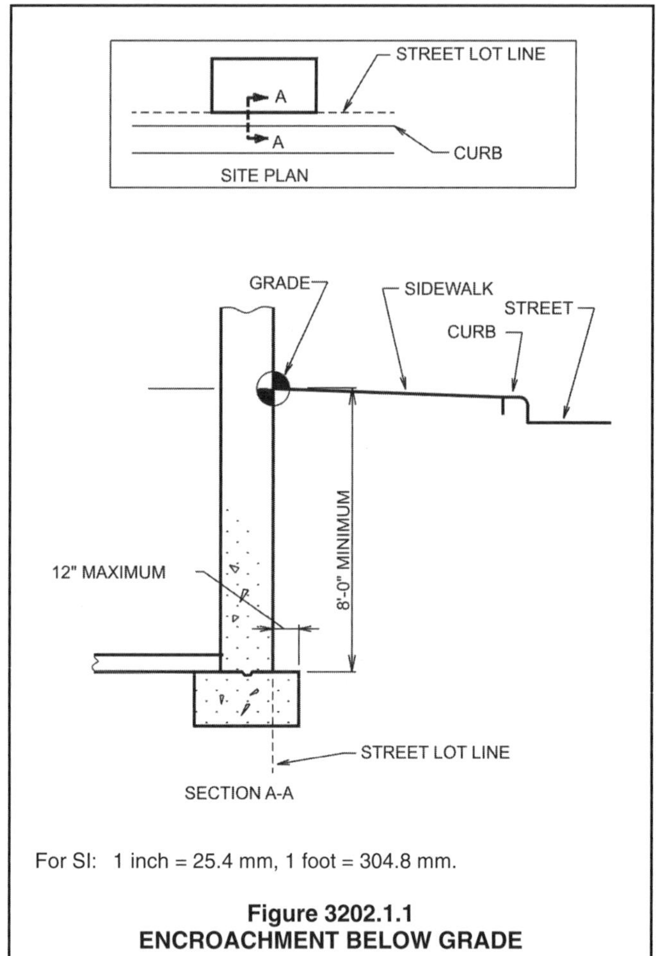

For SI: 1 inch = 25.4 mm, 1 foot = 304.8 mm.

Figure 3202.1.1 ENCROACHMENT BELOW GRADE

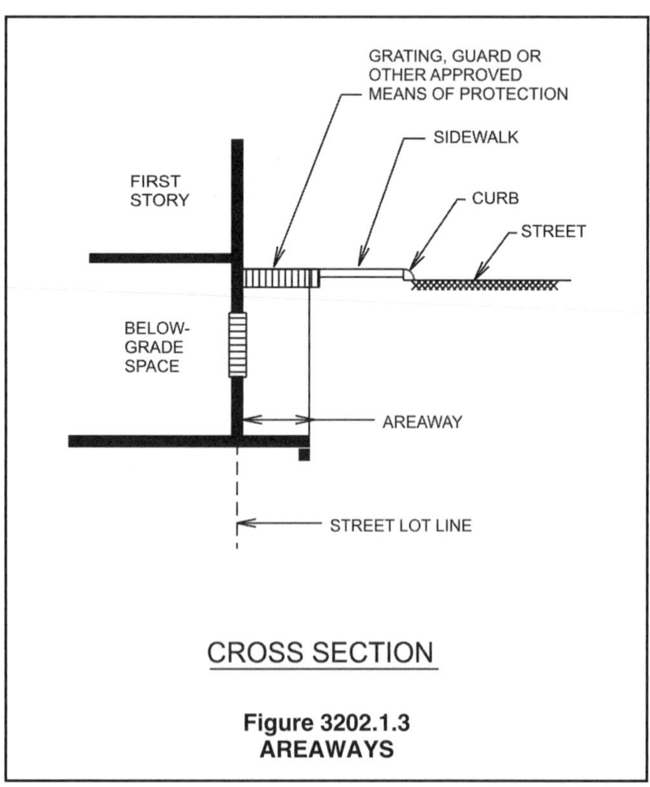

Figure 3202.1.3 AREAWAYS

3202.2 Encroachments above grade and below 8 feet in height. Encroachments into the public right-of-way above grade and below 8 feet (2438 mm) in height shall be prohibited except as provided for in Sections 3202.2.1 through 3202.2.3. Doors and windows shall not open or project into the public right-of-way.

❖ This section prohibits all projections into the public right-of-way that are above grade and below 8 feet (2438 mm), except those specifically allowed by Sections 3202.2.1 through 3202.2.3 and 3202.4 (i.e., steps, architectural features such as columns or pilasters, awnings, temporary vestibules and storm enclosures). The maximum permitted encroachments and minimum clearances, as well as additional protection requirements, for the specified allowed components are detailed in their respective sections. See the commentary to Section 3202.1 regarding the term "grade."

3202.2.1 Steps. Steps shall not project more than 12 inches (305 mm) and shall be guarded by *approved* devices not less than 3 feet (914 mm) in height, or shall be located between columns or pilasters.

❖ Steps are permitted to project a maximum of 12 inches (305 mm) into the public right-of-way by this section. However, since steps may pose a tripping hazard, an approved device (guard), ornamental column or pilaster must be provided at a minimum height of 3 feet (914 mm) to minimize this hazard. Figure 3202.2.1 illustrates compliance with this provision. Where such steps are part of the building's means of egress system, Sections 1009.12 and 1012 may require the placement of handrails and handrail extensions such that the 12-inch (305 mm) encroachment into a right- of-way may not be achievable.

3202.2.2 Architectural features. Columns or pilasters, including bases and moldings shall not project more than 12 inches (305 mm). Belt courses, lintels, sills, architraves, pediments and similar architectural features shall not project more than 4 inches (102 mm).

❖ This section addresses the variety of architectural appurtenances that are part of a structure's exterior wall or roof. Consistent with the projection limitations given for steps in Section 3202.2.1, the maximum encroachment into the public way is 12 inches (305 mm) for columns and pilasters (see Figure 3202.2.1). Other noted architectural features are limited to maximum encroachments of 4 inches (102 mm).

3202.2.3 Awnings. The vertical clearance from the public right-of-way to the lowest part of any *awning*, including valances, shall be not less than 7 feet (2134 mm).

❖ Awnings, including retractable awnings, are permitted to project into the public right-of-way. However, the clearance to any part of an awning, including its valance, is required to be a minimum of 7 feet (2134 mm) above grade (see Figure 3202.2.3).

3202.3 Encroachments 8 feet or more above grade. Encroachments 8 feet (2438 mm) or more above grade shall comply with Sections 3202.3.1 through 3202.3.4.

❖ This section prohibits all projections into the public right-of-way that are 8 feet (2438 mm) or more above grade, except those specifically allowed by Sections 3202.3.1 through 3202.3.4. The maximum permitted encroachments and minimum clearances, as well as

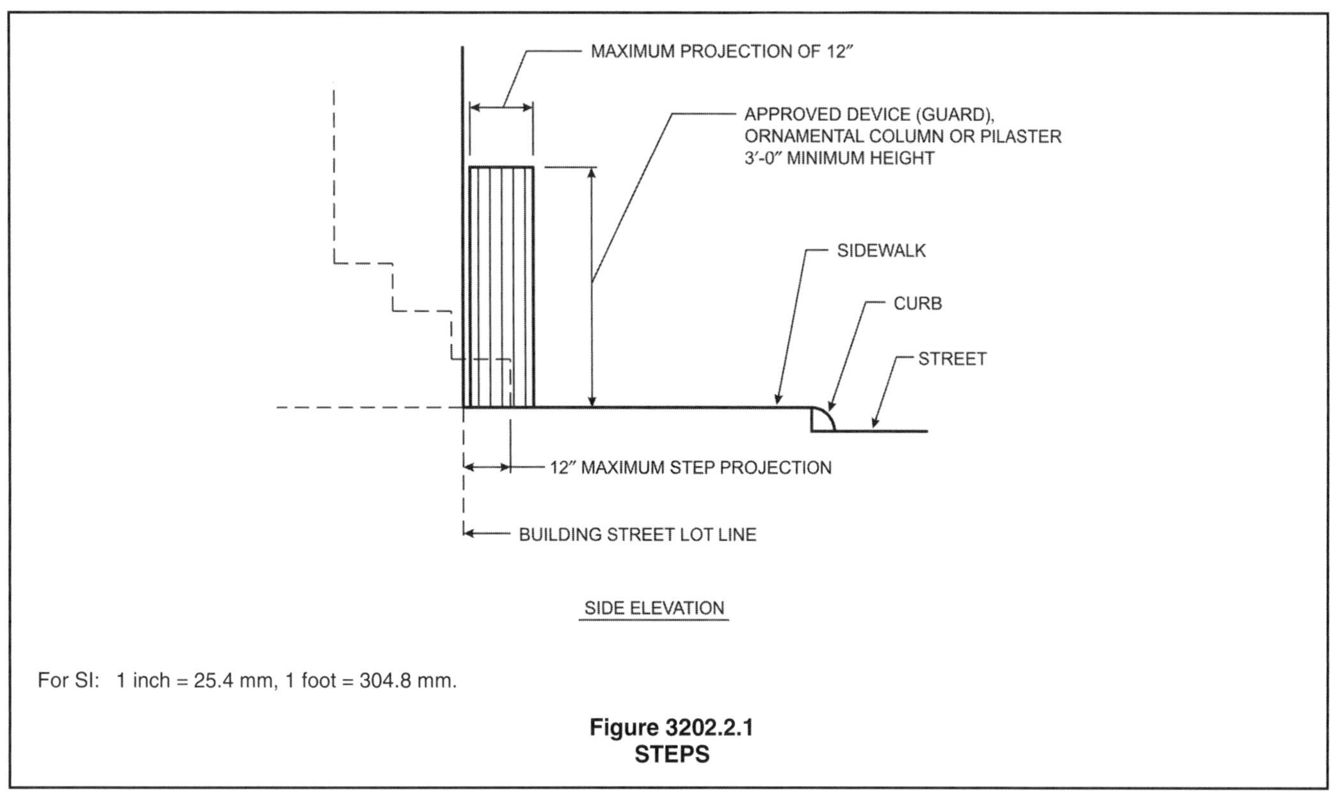

**Figure 3202.2.1
STEPS**

additional approval requirements, for the specified allowed components are detailed in their respective sections.

3202.3.1 Awnings, canopies, marquees and signs. *Awnings*, canopies, marquees and signs shall be constructed so as to support applicable loads as specified in Chapter 16. *Awnings*, canopies, marquees and signs with less than 15 feet (4572 mm) clearance above the sidewalk shall not extend into or occupy more than two-thirds the width of the sidewalk measured from the building. Stanchions or columns that support *awnings*, canopies, marquees and signs shall be located not less than 2 feet (610 mm) in from the curb line.

❖ This section contains three distinct requirements. First, awnings, canopies, marquees and signs must be designed in accordance with the provisions of Chapter 16, with regard to the live loads that may be imposed on such structures.

Second, the allowable projection of awnings, canopies, marquees and signs with less than 15 feet (4572 mm) of clearance above the sidewalk is limited to two-thirds of the width of the sidewalk measured from the building. The standards of these sections apply to retractable awnings (defined in Section 3105.2) as a type of awning.

Finally, the location of vertical structural components used for supporting awnings, canopies, marquees and signs (stanchions or columns) is also regulated. A minimum of 2 feet (610 mm) of clearance between the curb line and the vertical structural support for awnings, canopies, marquees and signs is required (see Figure 3202.3.1).

3202.3.2 Windows, balconies, architectural features and mechanical equipment. Where the vertical clearance above grade to projecting windows, balconies, architectural features or mechanical equipment is more than 8 feet (2438 mm), 1 inch (25 mm) of encroachment is permitted for each additional 1 inch (25 mm) of clearance above 8 feet (2438 mm), but the maximum encroachment shall be 4 feet (1219 mm).

❖ Projections into the public right-of-way for appurtenances such as windows, balconies, architectural features or mechanical equipment more than 8 feet (2438 mm) above grade are allowed by this section. Such projections are limited to 1 inch (25 mm) of encroachment for each additional inch of clearance above the minimum 8-foot (2438 mm) clearance required. However, the maximum allowable projection is limited to 4 feet (1219 mm) (see Figure 3202.3.2).

3202.3.3 Encroachments 15 feet or more above grade. Encroachments 15 feet (4572 mm) or more above grade shall not be limited.

❖ Because encroachments into the public right-of-way that are 15 feet (4572 mm) or more above grade do not interfere with or impede pedestrian or vehicular traffic, they are not regulated by the provisions of this chapter.

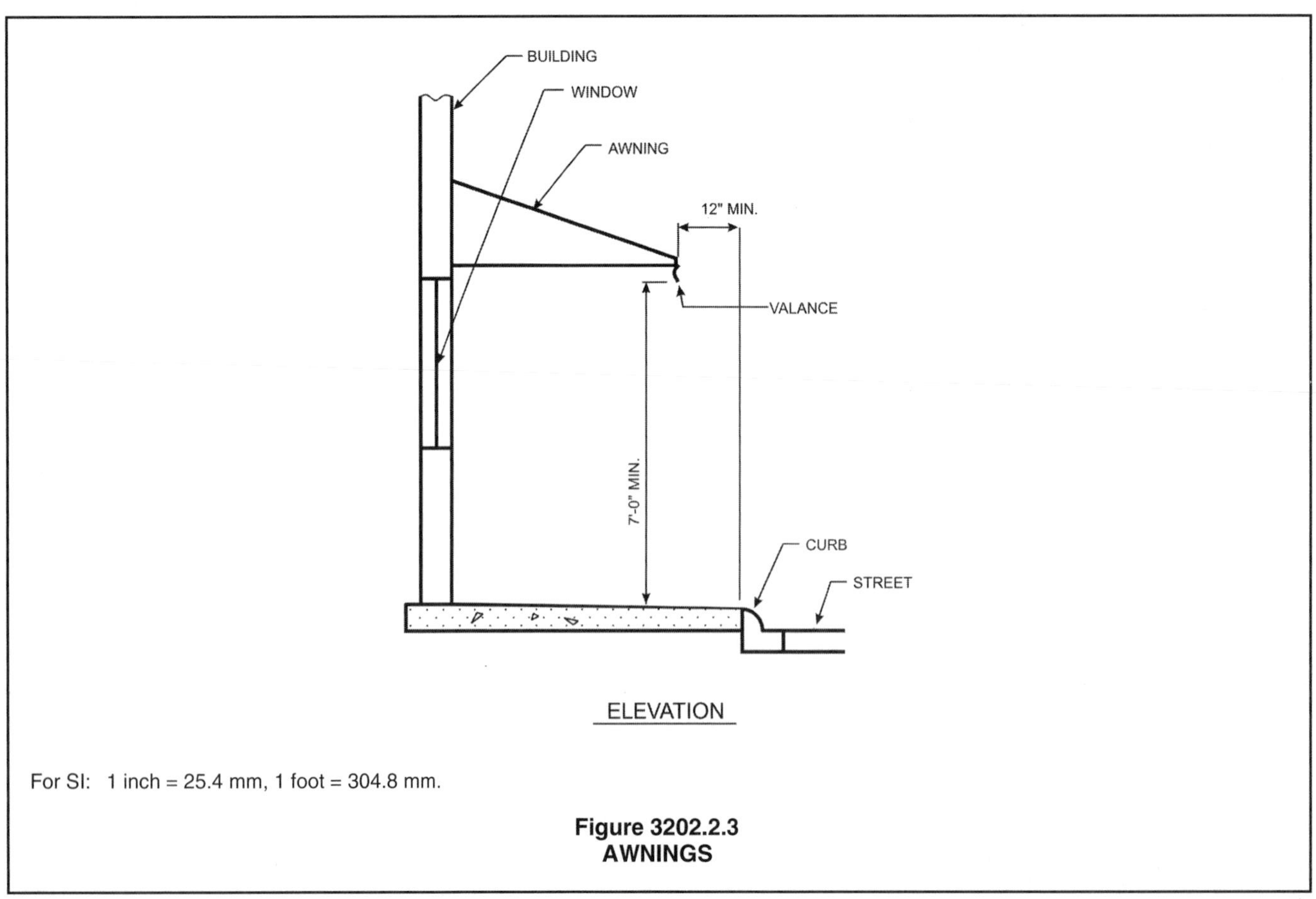

Figure 3202.2.3
AWNINGS

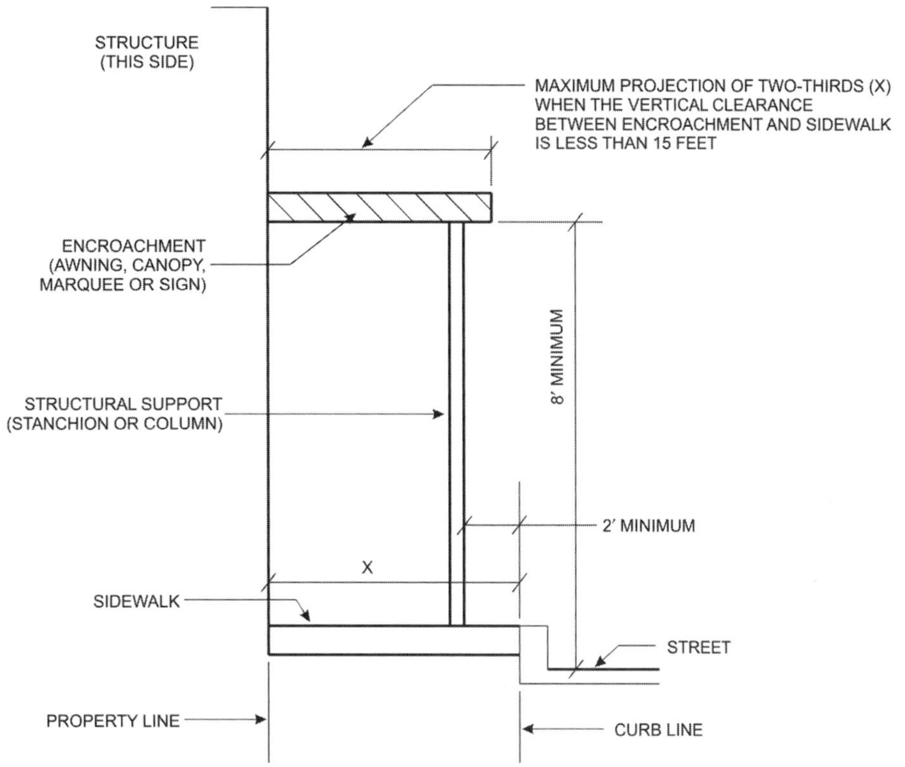

STRUCTURE
(THIS SIDE)
MAXIMUM PROJECTION OF TWO-THIRDS (X)
WHEN THE VERTICAL CLEARANCE
BETWEEN ENCROACHMENT AND SIDEWALK
IS LESS THAN 15 FEET
ENCROACHMENT
(AWNING, CANOPY,
MARQUEE OR SIGN)
8' MINIMUM
STRUCTURAL SUPPORT
(STANCHION OR COLUMN)
2' MINIMUM
X
SIDEWALK
STREET
PROPERTY LINE
CURB LINE

For SI: 1 foot = 304.8 mm.

Figure 3202.3.1
AWNINGS, CANOPIES, MARQUEES AND SIGNS 8 FEET OR MORE ABOVE GRADE

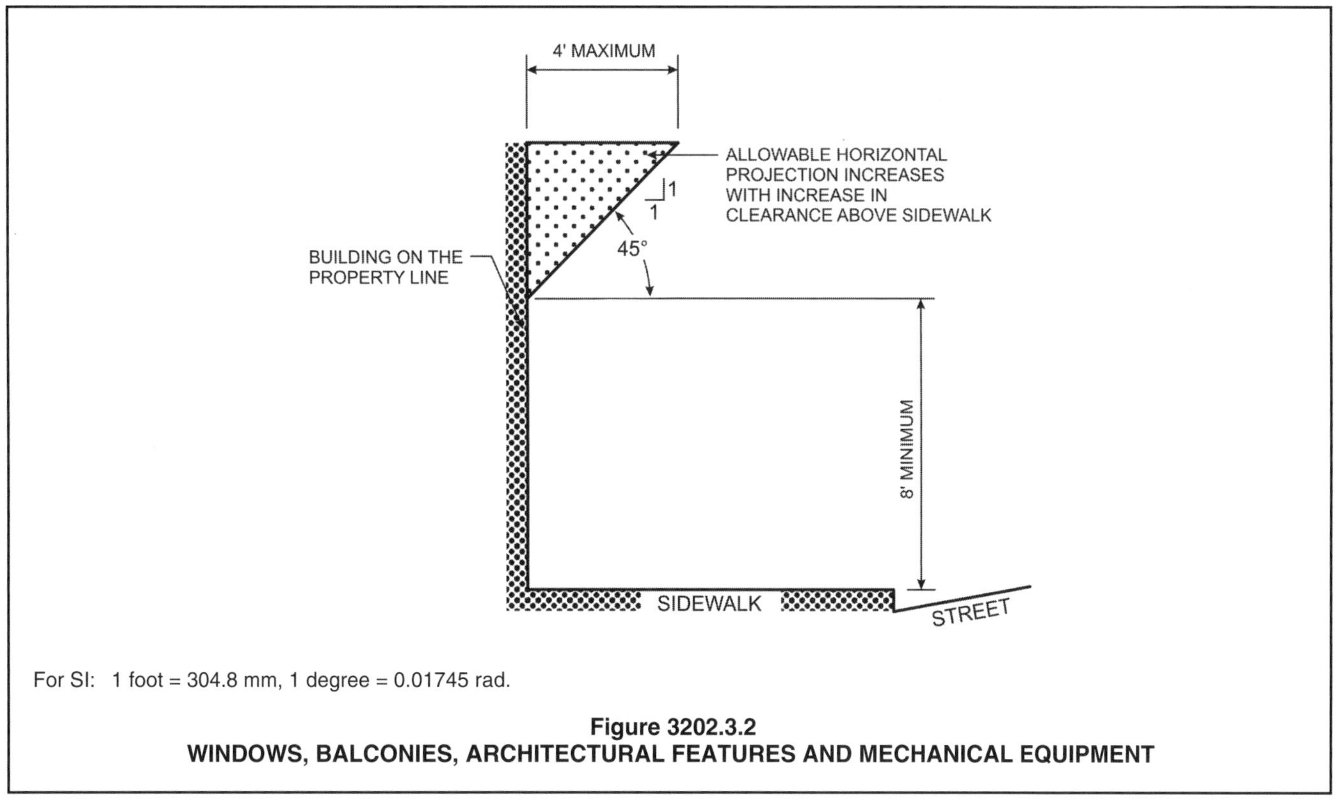

4' MAXIMUM
ALLOWABLE HORIZONTAL
PROJECTION INCREASES
WITH INCREASE IN
CLEARANCE ABOVE SIDEWALK
1
1
45°
BUILDING ON THE
PROPERTY LINE
8' MINIMUM
SIDEWALK
STREET

For SI: 1 foot = 304.8 mm, 1 degree = 0.01745 rad.

Figure 3202.3.2
WINDOWS, BALCONIES, ARCHITECTURAL FEATURES AND MECHANICAL EQUIPMENT

3202.3.4 Pedestrian walkways. The installation of a pedestrian walkway over a public right-of-way shall be subject to the approval of the applicable governing authority. The vertical clearance from the public right-of-way to the lowest part of a *pedestrian walkway* shall be not less than 15 feet (4572 mm).

❖ As in Section 3202.3.3, projections of pedestrian walkways that have a minimum clearance of 15 feet (4572 mm) above grade do not interfere or impede pedestrian or vehicular traffic and thus are not prohibited by this section. The installation of pedestrian walkways over a public right-of-way, however, is required to be approved by the local applicable governing authority. Detailed requirements for pedestrian walkways are provided in Section 3104.

3202.4 Temporary encroachments. Where allowed by the applicable governing authority, vestibules and storm enclosures shall not be erected for a period of time exceeding seven months in any one year and shall not encroach more than 3 feet (914 mm) nor more than one-fourth of the width of the sidewalk beyond the street *lot line*. Temporary entrance *awnings* shall be erected with a clearance of not less than 7 feet (2134 mm) to the lowest portion of the hood or *awning* where supported on removable steel or other *approved* noncombustible support.

❖ This section addresses a special situation in which temporary projections across the street lot line are erected for climatic reasons (i.e., during the winter months some building owners may erect vestibule enclosures or canopies over entrances). The purpose of this section is to allow these energy-saving procedures, yet place limitations on such projections.

Specifically, enclosures temporarily appended to an entranceway are permitted to project not more than 3 feet (914 mm) beyond the street lot line, or up to one-fourth the width of the sidewalk. Whichever dimension results in the lesser projection defines the maximum projection.

Additionally, when temporary awnings are erected, clearance between the sidewalk and the lowest part of the awning must be at least 7 feet (2134 mm), which is consistent with the requirements for permanent awnings in Section 3202.2.3. It is further required that the frame for the hood or awning is to be supported on removable steel or other approved noncombustible material. This last provision of this section is necessitated by unforeseen changes in street lot lines, for example, that would result in the encroachment becoming an obstruction in the public right-of-way.

Chapter 33: Safeguards During Construction

General Comments

The building construction process involves a number of known and unanticipated hazards. Chapter 33 establishes specific regulations in order to minimize the risk to the public and adjacent property. Some construction failures have resulted during the initial stages of grading, excavation and demolition. During these early stages, poorly designed and installed sheeting and shoring have resulted in ditch and embankment cave-ins. Also, inadequate underpinning of adjoining existing structures or careless removal of existing structures has resulted in construction failures.

The most critical period in building construction related to the physical safety of those on the job site is during the ongoing process when all building components have not yet been completed. Compounding this incomplete state is the use of dangerous construction methods, materials and equipment.

The importance of reasonable precautions is evidenced by the federal Occupational Safety and Health Act and related state and local regulations.

Purpose

The purpose of Chapter 33 is to establish safety requirements during construction or demolition of buildings and structures. These requirements are intended to protect the public from injury and adjoining property from damage.

SECTION 3301
GENERAL

3301.1 Scope. The provisions of this chapter shall govern safety during construction and the protection of adjacent public and private properties.

❖ The fundamental rationale behind this chapter is to establish that reasonable safety precautions are to be provided during the construction or demolition process to protect the public from injury and adjoining property from damage. This chapter does not address projects not associated with a building or structure. Section 101.2 of this code establishes the scope as applicable to buildings or structures. For example the widening or deepening of an existing river for flood control purposes or similar project would not be regulated by this chapter.

3301.2 Storage and placement. Construction equipment and materials shall be stored and placed so as not to endanger the public, the workers or adjoining property for the duration of the construction project.

❖ This section requires that construction materials and equipment must be located and protected pursuant to the governing provisions of this chapter so that the public and adjoining property are safeguarded at all times during construction or demolition processes.

SECTION 3302
CONSTRUCTION SAFEGUARDS

3302.1 Alterations, repairs and additions. Required *exits*, existing structural elements, fire protection devices and sanitary safeguards shall be maintained at all times during *alterations*, *repairs* or *additions* to any building or structure.

Exceptions:

1. Where such required elements or devices are being altered or repaired, adequate substitute provisions shall be made.
2. Maintenance of such elements and devices is not required when the existing building is not occupied.

❖ Demolition and construction operations must not create a hazard for the occupants of a building during an alteration or addition. As such, the existing fire protection, means of egress elements and safety systems must remain in place and be functional. Two exceptions, however, are provided in this section that permit the fire protection, means of egress elements and safety systems to be changed, modified or removed. Exception 1 provides an alternative method that must be approved by the building official. Exception 2 is simply to have an unoccupied building, which means that residents of a dwelling unit or employees of a business are not to be present during the construction or demolition process.

3302.2 Manner of removal. Waste materials shall be removed in a manner which prevents injury or damage to persons, adjoining properties and public rights-of-way.

❖ Safe and sanitary procedures for the removal of building construction and demolition waste must be provided. The method of waste removal must be controlled such that debris will not pose a hazard, eyesore or nuisance to the public and neighboring properties. Examples of acceptable practices include: evidence that a professional disposal service will haul away the debris; limiting areas used for storing and handling demolished materials; enclosing storage areas so that only authorized personnel can gain access; establishing routes that waste removal vehicles are permitted to use; covering or tarping debris to prevent flying objects; and providing fully enclosed chutes to control falling objects and dust. Additionally, waste removal should be scheduled when adjoining property and the public will be least exposed to unusual and possibly dangerous situations caused by fumes, noise, dust, as well as unfamiliar events involved during a demolition process. Note that the accepted practice of waste removal is subject to the approval of the applicable governing authority. A growing number of local jurisdictions as part of their sustainability, or green, programs limit the amount of construction and demolition debris which can be sent to landfills.

3302.3 Fire safety during construction. Fire safety during construction shall comply with the applicable requirements of this code and the applicable provisions of Chapter 33 of the *International Fire Code.*

❖ The purpose of this section is to correlate Section 3302 and the *International Fire Code*® (IFC®) requirements on fire safety during construction and demolition. Chapter 14 of the IFC has all the fire-safety requirements provided in one place for the fire-safe operations during construction and demolition.

SECTION 3303 DEMOLITION

3303.1 Construction documents. *Construction documents* and a schedule for demolition shall be submitted where required by the *building official*. Where such information is required, no work shall be done until such *construction documents* or schedule, or both, are *approved*.

❖ Some processes, methods and materials used during demolition can cause damage to adjoining property and increase the risk of injury to the general public. As such, this section indicates that the building official can request: verification that structural stability of existing buildings on the same lot or adjacent properties is maintained and not compromised; to know the materials and equipment necessary to perform the demolition involved; the timetable in which certain demolition activities will be performed; details to establish if the general public and adjoining property are exposed to any unfamiliar or greater hazard than normal; special inspection reports made by qualified firms, agencies or individuals; qualifications of the laborers involved in the operation, handling and removal of demolition equipment and materials; and verification that any federal, state or local statute is followed. Note that the building official must grant approval for any additional information requested before demolition can be started.

3303.2 Pedestrian protection. The work of demolishing any building shall not be commenced until pedestrian protection is in place as required by this chapter.

❖ Demolition must not be started until all of the necessary precautions are taken to protect the general public as indicated throughout this chapter.

3303.3 Means of egress. A *horizontal exit* shall not be destroyed unless and until a substitute *means of egress* has been provided and *approved*.

❖ The exits provided through a party wall must be maintained operational and usable during demolition; however, based upon the extent of work, an alternative means of egress through a party wall is only permitted when approved by the building official.

3303.4 Vacant lot. Where a structure has been demolished or removed, the vacant lot shall be filled and maintained to the existing grade or in accordance with the ordinances of the jurisdiction having authority.

❖ A demolition site must be restored if additional building construction or demolition operations are not scheduled to take place. The site must be filled and graded to the level of the surrounding site or maintained in accordance with local or state statutes, which may set forth other or additional grading requirements, such as provisions for elevations, drainage and flood control.

The time frame imposed or stipulated to abate an imminent hazard created by a vacant lot is at the discretion of the building official unless a time frame is established in other regulations. In addition, footings, foundations, basement walls and floors must be removed when the building official determines that a threat is posed to human life or the public welfare or if any portion of the foundation system prevents proper grading from being done.

3303.5 Water accumulation. Provision shall be made to prevent the accumulation of water or damage to any foundations on the premises or the adjoining property.

❖ A vacant lot must be graded in accordance with Section 3303.4 in such a way that water is prevented from ponding and causing damage to structures on the premises or adjacent properties, particularly foundation systems and building elements in contact with the ground. Footings, foundations, basement walls or floors must be removed if this drainage is prevented.

3303.6 Utility connections. Service utility connections shall be discontinued and capped in accordance with the *approved* rules and the requirements of the applicable governing authority.

❖ The procedures for disconnecting and abandoning utility connections in a safe and satisfactory manner must be in accordance with local, state and federal statutes. Utility providers, such as electrical, telephone, water, gas or sewer, often recommend or require notification and have certain guidelines to follow.

A related aspect is the precaution to investigate for

any on-site underground obstructions, such as utility service lines, and buried oil, gasoline or septic tanks. If the temporary use of some existing service, such as electricity or water, is requested during the demolition project, then temporary permits must be obtained with the appropriate stipulations, including any rerouting or protection requirements from the applicable governing authority.

3303.7 Fire safety during demolition. Fire safety during demolition shall comply with the applicable requirements of this code and the applicable provisions of Chapter 56 of the *International Fire Code.*

❖ The purpose of this section is to correlate Section 3302 and the IFC requirements on fire safety during construction and demolition. Chapter 14 of the IFC has all the fire-safety requirements provided in one place for the fire-safe operations during construction and demolition.

SECTION 3304
SITE WORK

3304.1 Excavation and fill. Excavation and fill for buildings and structures shall be constructed or protected so as not to endanger life or property. Stumps and roots shall be removed from the soil to a depth of not less than 12 inches (305 mm) below the surface of the ground in the area to be occupied by the building. Wood forms which have been used in placing concrete, if within the ground or between foundation sills and the ground, shall be removed before a building is occupied or used for any purpose. Before completion, loose or casual wood shall be removed from direct contact with the ground under the building.

❖ A site that is to be excavated or filled in preparation for new construction must have dead and decaying foliage eliminated where it would be in contact with the new structure. As such, the base of trees, roots and limbs that protrude above the ground as well as 12 inches (305 mm) into the ground, must be taken out or separated from the earth where the building or structure will be erected. Additionally, any wood forms used for the concrete must be removed along with any wood product near the building that may decay or rot.

Appendix J provides a comprehensive set of grading standards which would also apply when the appendix is specifically adopted by the jurisdiction.

3304.1.1 Slope limits. Slopes for permanent fill shall be not steeper than one unit vertical in two units horizontal (50-percent slope). Cut slopes for permanent excavations shall be not steeper than one unit vertical in two units horizontal (50-percent slope). Deviation from the foregoing limitations for cut slopes shall be permitted only upon the presentation of a soil investigation report acceptable to the *building official*.

❖ The final grading around a foundation system must slope away from the building or structure, such that the ratio does not exceed 2:1. A trench or hole dug for a foundation system is required to have a stable embankment. As such, the embankment must not have a slope that exceeds a 2:1 ratio; however, due to the existing soil conditions or particular location, it may be impractical to achieve the required slope. This section, therefore, indicates that an engineering analysis can be submitted with recommendations for alternative methodology and safeguards, subject to the approval of the building official.

3304.1.2 Surcharge. No fill or other surcharge loads shall be placed adjacent to any building or structure unless such building or structure is capable of withstanding the additional loads caused by the fill or surcharge. Existing footings or foundations which can be affected by any excavation shall be underpinned adequately or otherwise protected against settlement and shall be protected against later movement.

❖ Gradient stability must be maintained for existing buildings or structures during the excavation, construction or demolition processes. As such, earth, construction materials, debris and even construction equipment must not impose a load that will cause the ground to give way and collapse or damage the foundation system of adjacent buildings or structures. Therefore, components such as engineered bracing must be provided for protection if adjacent buildings or structures are to be exposed to loads that exceed those of the existing design.

3304.1.3 Footings on adjacent slopes. For footings on adjacent slopes, see Chapter 18.

❖ Any footings located on adjacent slopes must be designed and constructed pursuant to Chapter 18.

3304.1.4 Fill supporting foundations. Fill to be used to support the foundations of any building or structure shall comply with Section 1804.5. *Special inspections* of compacted fill shall be in accordance with Section 1704.7.

❖ This section is simply insuring that the fill that will be installed to support a building or structure be compacted and verified in accordance with Section 1804.5 and the special inspection regulations stated in Section 1704.7.

SECTION 3305
SANITARY

3305.1 Facilities required. Sanitary facilities shall be provided during construction, remodeling or demolition activities in accordance with the *International Plumbing Code.*

❖ Construction employees must have plumbing facilities available during the construction or demolition process of a building. The facilities must conform to the requirements set forth in the *International Plumbing Code®* (IPC®).

SECTION 3306
PROTECTION OF PEDESTRIANS

3306.1 Protection required. Pedestrians shall be protected during construction, remodeling and demolition activities as required by this chapter and Table 3306.1. Signs shall be provided to direct pedestrian traffic.

❖ Safeguards are required to be in place during construction or demolition operations in accordance with Table 3306.1 and this chapter. In addition, since construction operations alter the familiar setting and path of travel, it is necessary to provide some form of visible directional sign to lead the public toward safety and away from potential hazards.

The table establishes the type of protection required based upon the location to overhead hazards and distance in relation to ground hazards. Once the type of protection is determined from the table, the applicable code section, such as Section 3306.4 for construction railings, Section 3306.5 for barriers and Section 3306.7 for covered walkways, must be provided as regulated.

3306.2 Walkways. A walkway shall be provided for pedestrian travel in front of every construction and demolition site unless the applicable governing authority authorizes the sidewalk to be fenced or closed. Walkways shall be of sufficient width to accommodate the pedestrian traffic, but in no case shall they be less than 4 feet (1219 mm) in width. Walkways shall be provided with a durable walking surface. Walkways shall be *accessible* in accordance with Chapter 11 and shall be designed to support all imposed loads and in no case shall the design live load be less than 150 pounds per square foot (psf) (7.2 kN/m^2).

❖ Construction operations must not narrow or impede the normal flow of pedestrian traffic along a walkway by the placement of a fence or other enclosure. Note that the applicable governing authority must approve any construction operations that will cause the narrowing or impedance of a walkway, such as the sidewalk. If a walkway is narrowed or enclosed, the construction of another or wider walkway is required for all pedestrians. The walkway must be able to handle the normal anticipated flow of pedestrian traffic and must not be less than the minimum 4-foot (1219 mm) width. Even though such walkways may be temporary installations, they may still need to be usable by persons with disabilities where such walkway provides an accessible route in accordance with Chapter 11. In addition, the walkway must be a stable surface that is capable of supporting all imposed loads. The minimum design load of the walkway must not be less than 150 pounds per square feet (psf) (7182 Pa).

3306.3 Directional barricades. Pedestrian traffic shall be protected by a directional barricade where the walkway extends into the street. The directional barricade shall be of sufficient size and construction to direct vehicular traffic away from the pedestrian path.

❖ Similar to the local, state and federal guidelines that govern provisions to protect employees while doing road work, this section establishes that a barrier must be erected. The barrier must be capable of redirecting and regulating the flow of vehicular traffic where constructed or projected into a street. An example of how to regulate traffic is to provide the necessary warnings, notice of caution and instructions to the operators of motor vehicles with signs and colors that are common to the local, state and federal guidelines. The intent of this visual barrier is to keep pedestrians out of vehicular traffic and prevent vehicles from encroaching on or using the pedestrian walkway as a roadway.

3306.4 Construction railings. Construction railings shall be not less than 42 inches (1067 mm) in height and shall be sufficient to direct pedestrians around construction areas.

❖ A barrier consisting of a horizontal rail and supports must be constructed with a minimum height of 42 inches (1067 mm). In addition, this barrier must be capable of controlling the flow of pedestrian traffic by routing travel away from the hazards associated with a construction area. Where construction railings are placed such that there is a drop of more than 30 inches (762 mm) adjacent to the walking surface, such railings would also need to comply with Section 1013 for guards. Even where guards are not required, the standards for guards could be useful in the design of construction railings.

3306.5 Barriers. Barriers shall be not less than 8 feet (2438 mm) in height and shall be placed on the side of the walkway nearest the construction. Barriers shall extend the entire length of the construction site. Openings in such barriers shall be protected by doors which are normally kept closed.

❖ When a barrier is required by Table 3306.1, it must be constructed to impede, separate and obstruct passage of pedestrians onto a construction site. The

TABLE 3306.1
PROTECTION OF PEDESTRIANS

HEIGHT OF CONSTRUCTION	DISTANCE FROM CONSTRUCTION TO LOT LINE	TYPE OF PROTECTION REQUIRED
8 feet or less	Less than 5 feet	Construction railings
	5 feet or more	None
More than 8 feet	Less than 5 feet	Barrier and covered walkway
	5 feet or more, but not more than one-fourth the height of construction	Barrier and covered walkway
	5 feet or more, but between one-fourth and one-half the height of construction	Barrier
	5 feet or more, but exceeding one-half the height of construction	None

For SI: 1 foot = 304.8 mm.

structure must be a minimum 8 feet (2438 mm) in height and located continuously along the walkway on the side where construction activities are being done. Openings, such as doors or gates, are permitted as long as the only time they are open is during use by authorized personnel. As such, when the doors or gates are not being used, they must remain closed.

3306.6 Barrier design. Barriers shall be designed to resist loads required in Chapter 16 unless constructed as follows:

1. Barriers shall be provided with 2-inch by 4-inch (51 mm by 102 mm) top and bottom plates.
2. The barrier material shall be boards not less than $^3/_4$-inch (19.1 mm) thick or wood structural panels not less than $^1/_4$-inch (6.4 mm) thick.
3. Wood structural use panels shall be bonded with an adhesive identical to that for exterior wood structural use panels.
4. Wood structural use panels $^1/_4$ inch (6.4 mm) or $^5/_{16}$ inch (23.8 mm) in thickness shall have studs spaced not more than 2 feet (610 mm) on center (o.c.).
5. Wood structural use panels $^3/_8$ inch (9.5 mm) or $^1/_2$ inch (12.7 mm) in thickness shall have studs spaced not more than 4 feet (1219 mm) on center provided a 2-inch by 4-inch (51 mm by 102 mm) stiffener is placed horizontally at midheight where the stud spacing is greater than 2 feet (610 mm) on center.
6. Wood structural use panels $^5/_8$ inch (15.9 mm) or thicker shall not span over 8 feet (2438 mm).

❖ This section establishes two methods that can be used to design barriers. The first method establishes that the barrier can be designed using the same materials and assembly instructions as indicated in Items 1 through 6. The second method provides for a barrier that will resist the loads imposed on it in order to be part of a designed assembly. The loads, which need to be resisted, are the same as those established in Chapter 16.

3306.7 Covered walkways. Covered walkways shall have a clear height of not less than 8 feet (2438 mm) as measured from the floor surface to the canopy overhead. Adequate lighting shall be provided at all times. Covered walkways shall be designed to support all imposed loads. In no case shall the design live load be less than 150 psf (7.2 kN/m^2) for the entire structure.

Exception: Roofs and supporting structures of covered walkways for new, light-frame construction not exceeding two *stories* above *grade plane* are permitted to be designed for a live load of 75 psf (3.6kN/m^2) or the loads imposed on them, whichever is greater. In lieu of such designs, the roof and supporting structure of a covered walkway are permitted to be constructed as follows:

1. Footings shall be continuous 2-inch by 6-inch (51 mm by 152 mm) members.
2. Posts not less than 4 inches by 6 inches (102 mm by 152 mm) shall be provided on both sides of the roof and spaced not more than 12 feet (3658 mm) on center.
3. Stringers not less than 4 inches by 12 inches (102 mm by 305 mm) shall be placed on edge upon the posts.
4. Joists resting on the stringers shall be not less than 2 inches by 8 inches (51 mm by 203 mm) and shall be spaced not more than 2 feet (610 mm) on center.
5. The deck shall be planks not less than 2 inches (51 mm) thick or wood structural panels with an exterior exposure durability classification not less than $^{23}/_{32}$ inch (18.3 mm) thick nailed to the joists.
6. Each post shall be knee braced to joists and stringers by members not less than 2-inch by 4-inch (51 mm by 102 mm); 4 feet (1219 mm) in length.
7. A curb which is not less than 2-inch by 4-inch (51 mm by 102 mm) shall be set on edge along the outside edge of the deck.

❖ Walkways that are provided with a covering that extends out over the pedestrian path of travel, pursuant to Table 3306.1, must be at least 8 feet (2438 mm) in height. The measurement must be taken from the top of the walking surface vertically upward to the underside of the roof covering. The walkway must maintain a satisfactory level of light to the space that will be equal to or better than that provided prior to the installation of the protective covering. The covered walkway must also be structurally capable of resisting the design loads established in Chapter 16, but not less than 150 psf (7182 Pa) for the live load.

Note that this section provides two alternative design options. One option permits the covered walkway to resist the design loads established in Chapter 16, but not less than the minimum 75 psf (3591 Pa) live load if nearby construction does not exceed two stories above grade plane and the construction materials are lightweight, such as light-frame construction. The second option is to construct the covered walkway in accordance with the seven criteria set forth in this section.

3306.8 Repair, maintenance and removal. Pedestrian protection required by this chapter shall be maintained in place and kept in good order for the entire length of time pedestrians are subject to being endangered. The *owner* or the *owner's* agent, upon the completion of the construction activity, shall immediately remove walkways, debris and other obstructions and leave such public property in as good a condition as it was before such work was commenced.

❖ Any safeguards required by this chapter must be kept in good functional condition for the entire duration of the construction or demolition activity so that the public will not be placed in harm's way. Once a building or structure is occupiable and the site is properly graded, the protection must be removed. As such, public property that was affected by the construction activity must be restored to or left in the condition that existed prior to the work.

3306.9 Adjacent to excavations. Every excavation on a site located 5 feet (1524 mm) or less from the street *lot line* shall be enclosed with a barrier not less than 6 feet (1829 mm) in height. Where located more than 5 feet (1524 mm) from the street *lot line*, a barrier shall be erected where required by the *building official*. Barriers shall be of adequate strength to resist wind pressure as specified in Chapter 16.

❖ This section establishes that whenever excavation is to take place less than 5 feet (1524 mm) from the edge of a roadway, a barrier must be erected with a minimum height of 6 feet (1829 mm). If the excavation is located greater than 5 feet (1524 mm) from the edge of the roadway, the building official must evaluate the level of hazard for the public and the necessary precautions to take. As such, the building official has the authority to order the construction of a structural barrier. Any barrier erected must maintain other provisions of the code such that it is capable of handling and resisting design wind loads denoted in Chapter 16.

SECTION 3307 PROTECTION OF ADJOINING PROPERTY

3307.1 Protection required. Adjoining public and private property shall be protected from damage during construction, remodeling and demolition work. Protection shall be provided for footings, foundations, party walls, chimneys, skylights and roofs. Provisions shall be made to control water runoff and erosion during construction or demolition activities. The person making or causing an excavation to be made shall provide written notice to the *owners* of adjoining buildings advising them that the excavation is to be made and that the adjoining buildings should be protected. Said notification shall be delivered not less than 10 days prior to the scheduled starting date of the excavation.

❖ This section emphasizes the need to protect all existing public and private property bordering the proposed construction or demolition operations. The term "property" only alludes to existing buildings. As such, any building element or system must be provided with a safeguard that will limit the damage that could be caused from the processes involved to the equipment and materials used. Additionally, soil erosion and disbursement resulting from the construction or demolition operations must be controlled to prevent spillage and spread of disturbed soil debris. The site must be graded in accordance with Sections 3303.4 and 3303.5 for demolition and must be maintained in a similar manner while there is construction taking place. The owner or owner's agent has the responsibility to provide a written notice 10 days in advance for any demolition or construction activities that may warrant bordering lots to be protected from damage.

SECTION 3308 TEMPORARY USE OF STREETS, ALLEYS AND PUBLIC PROPERTY

3308.1 Storage and handling of materials. The temporary use of streets or public property for the storage or handling of materials or of equipment required for construction or demolition, and the protection provided to the public shall comply with the provisions of the applicable governing authority and this chapter.

❖ In order to protect access to public safety equipment, utilities and public transportation, this section provides standards for the temporary use of public property for construction processes and storage of materials. These are in addition to any local or state regulations.

3308.1.1 Obstructions. Construction materials and equipment shall not be placed or stored so as to obstruct access to fire hydrants, standpipes, fire or police alarm boxes, catch basins or manholes, nor shall such material or equipment be located within 20 feet (6096 mm) of a street intersection, or placed so as to obstruct normal observations of traffic signals or to hinder the use of public transit loading platforms.

❖ This section indicates that precautions for material and equipment storage and placement must be provided so as not to block or obstruct access to fire hydrants; standpipes (including fire department siamese connections for sprinklers and standpipes); fire or police alarm boxes; utility boxes and meters; catch basins or manholes or any other vital facility whose function contributes to the health, safety and welfare of the public. Also, storage must not be placed within 20 feet (6096 mm) of a street intersection if it obstructs the normal observation of traffic signals or hinders the use of any mass-transit loading platforms, such as sidewalk bus stops, taxi waiting areas or light rail platforms.

3308.2 Utility fixtures. Building materials, fences, sheds or any obstruction of any kind shall not be placed so as to obstruct free approach to any fire hydrant, fire department connection, utility pole, manhole, fire alarm box or catch basin, or so as to interfere with the passage of water in the gutter. Protection against damage shall be provided to such utility fixtures during the progress of the work, but sight of them shall not be obstructed.

❖ Utility fixtures, such as those used by electrical, telephone, water, gas or sewer companies, as well as fire protection devices must be hidden from view and not be blocked from access and use. When construction operations are near a utility fixture, precautions must be provided so as not to cause damage. The utility companies and the applicable governing authority, including the fire department, may have specific requirements and guidelines to follow to limit the possibility of damage and obstruction.

SECTION 3309
FIRE EXTINGUISHERS

[F] 3309.1 Where required. All structures under construction, *alteration* or demolition shall be provided with no fewer than one *approved* portable fire extinguisher in accordance with Section 906 and sized for not less than ordinary hazard as follows:

1. At each *stairway* on all floor levels where combustible materials have accumulated.
2. In every storage and construction shed.
3. Additional portable fire extinguishers shall be provided where special hazards exist, such as the storage and use of flammable and combustible liquids.

❖ A means to provide fire protection during the construction and demolition process is required. As such, this section indicates provisions for portable fire extinguishers. In addition to provisions of this section, the regulations of Section 906 and the IFC must be maintained.

[F] 3309.2 Fire hazards. The provisions of this code and the *International Fire Code* shall be strictly observed to safeguard against all fire hazards attendant upon construction operations.

❖ Methods, procedures and construction materials each contribute, in some way, to creating a fire hazard; therefore, in addition to the provisions of the code, the IFC must be used to regulate the proper means of safety that must be provided for a building that is being demolished, altered or constructed.

SECTION 3310
MEANS OF EGRESS

3310.1 Stairways required. Where a building has been constructed to a *building height* of 50 feet (15 240 mm) or four *stories*, or where an existing building exceeding 50 feet (15 240 mm) in *building height* is altered, no fewer than one temporary lighted *stairway* shall be provided unless one or more of the permanent stairways are erected as the construction progresses.

❖ Buildings under construction are determined to require more extensive time for evacuation due to height need to be provided with appropriate stairways for egress. Temporary stairways must be constructed in accordance with Section 1009, including riser, tread, guard and handrail requirements. This section does allow the use of a permanent stairway that has already been erected during construction instead of a temporary stairway.

3310.2 Maintenance of means of egress. Required *means of egress* shall be maintained at all times during construction, demolition, remodeling or *alterations* and *additions* to any building.

Exception: Existing means of egress need not be maintained where *approved* temporary *means of egress* systems and facilities are provided.

❖ While construction or demolition work is being done, the occupants of the existing building still need to have unobstructed access to the minimum number of usable means of egress required for that space, floor and building. Note that depending upon the extent of work, an exit may become obstructed; therefore, an alternative exit must be provided if the blocked exit is a required exit.

SECTION 3311
STANDPIPES

[F] 3311.1 Where required. In buildings required to have standpipes by Section 905.3.1, no fewer than one standpipe shall be provided for use during construction. Such standpipes shall be installed when the progress of construction is not more than 40 feet (12 192 mm) in height above the lowest level of fire department vehicle access. Such standpipe shall be provided with fire department hose connections at accessible locations adjacent to usable stairs. Such standpipes shall be extended as construction progresses to within one floor of the highest point of construction having secured decking or flooring.

❖ The scope of this section is to provide fire safety procedures during the construction operation in accordance with the code and the IFC. Standpipes that are required by Section 905 to be a permanent part of the building must be installed and remain functional as construction or demolition progresses. Functional standpipes are required so that fire-fighting capability is available at all times within a reasonable proximity to all potential fire locations. Standpipes must be in place when a building or structure under construction exceeds 40 feet (12 192 mm) in height above fire department vehicle access and thereafter as it progresses to its completed height. During demolition or construction, the standpipe must be operational no lower than one floor below the highest point of construction. Having a standpipe at least as high as the floor below the highest point of construction reflects typical fire department procedures when fighting a fire within a building, which is to attack the fire from the floor below.

[F] 3311.2 Buildings being demolished. Where a building is being demolished and a standpipe exists within such a building, such standpipe shall be maintained in an operable condition so as to be available for use by the fire department. Such standpipe shall be demolished with the building but shall not be demolished more than one floor below the floor being demolished.

❖ Standpipes in buildings under demolition must remain in service during the demolition process so that fire-fighting capability is maintained. Similar to buildings under construction, the standpipe is to remain in operation up to one floor level below the floor being demolished. The purpose of this section is to establish the requirement for fire safety procedures during the demolition process in accordance with the code and the IFC.

[F] 3311.3 Detailed requirements. Standpipes shall be installed in accordance with the provisions of Chapter 9.

Exception: Standpipes shall be either temporary or permanent in nature, and with or without a water supply, provided that such standpipes conform to the requirements of Section 905 as to capacity, outlets and materials.

❖ During the construction or demolition process, standpipes complying with Section 3311.1 must be provided in a temporary or permanent location so that fire fighters will have sufficient means to supply water to their connection. These standpipes must either be connected to a permanent water supply source or have the capability to be connected to one in accordance with Section 3311.4. All standpipes must comply with Section 905 and must be installed and provided in a building pursuant to the regulations of Chapter 9.

SECTION 3312 AUTOMATIC SPRINKLER SYSTEM

[F] 3312.1 Completion before occupancy. In buildings where an *automatic sprinkler system* is required by this code, it shall be unlawful to occupy any portion of a building or structure until the *automatic sprinkler system* installation has been tested and *approved*, except as provided in Section 111.3.

❖ A certificate of occupancy must not be issued by the building official if a required automatic sprinkler system is not approved; however, a temporary occupancy may be granted at the discretion of the building official.

[F] 3312.2 Operation of valves. Operation of sprinkler control valves shall be permitted only by properly authorized personnel and shall be accompanied by notification of duly designated parties. When the sprinkler protection is being regularly turned off and on to facilitate connection of newly completed segments, the sprinkler control valves shall be checked at the end of each work period to ascertain that protection is in service.

❖ The scope of this section is to provide fire safety procedures during construction operations in accordance with the code and the IFC. A sprinkler system must remain operable unless work is being done on it. In such a case, the water must only be shut off by authorized personnel coupled with the notification of the proper authorities so that a form of check and balance is achieved to make certain the water is turned back on.

SECTION 3313 WATER SUPPLY FOR FIRE PROTECTION

[F] 3313.1 Where required. An *approved* water supply for fire protection, either temporary or permanent, shall be made available as soon as combustible material arrives on the site.

❖ This section is to provide a fire safety provision in accordance with the code and the IFC. As such, when a standpipe system is present during the construction process pursuant to Section 3311.1, a water source must be readily available to provide fire-fighting capabilities at all times.

Bibliography

The following resource materials are referenced in this chapter or are relevant to the subject matter addressed in this chapter.

DOL 29 CFR; Part 1910-74, *Occupational Safety and Health Standards*. Washington, DC: U.S. Department of Labor-Occupational Safety and Health Administration, 1974.

IFC-12, *International Fire Code*. Washington, DC: International Code Council, 2011.

IPC-12, *International Plumbing Code*. Washington, DC: International Code Council, 2011.

Chapter 34: Existing Structures

General Comments

Chapter 34 contains provisions for the alteration, repair, addition and change of occupancy of existing buildings and structures. Chapter 34 includes or refers to the code requirements for existing buildings and structures, exclusive of the administrative provisions presented in Chapter 1 (Section 102.6 regulates unaltered existing buildings). The format of this chapter readily identifies the alternative methods of code compliance for existing buildings and structures.

Section 3401 establishes the provisions for maintenance, repairs and compliance with other technical codes.

Section 3402 provides definitions of terms that are primarily associated with Chapter 34.

Section 3403 states general requirements for additions and also includes provisions and referenced sections that regulate flood hazard areas and existing structural elements.

Section 3404 states general requirements for alterations and also includes provisions and referenced sections that regulate flood hazard areas, existing structural elements, seismic improvements and means of egress capacity.

Section 3405 states general requirements for repairs and also includes provisions and referenced sections that regulate flood hazard areas and the extent of repair for both compliant and noncompliant buildings.

Section 3406 contains the requirements for fire escapes.

Section 3407 contains provisions for the installation of new glass and glass replacement.

Section 3408 describes the conditions that constitute a change of occupancy.

Section 3409 addresses the requirements for historic buildings and structures.

Section 3410 addresses the requirements for structures that are moved into or within a jurisdiction.

Section 3411 contains provisions for accessibility in existing buildings undergoing repair, change of occupancy, additions or alterations. Included in the provisions are requirements for accessible elements within existing buildings and structures, as well as accessible route requirements.

Section 3412 contains provisions for an alternative method of evaluation of additions, alterations or changes of occupancy in existing buildings and structures based on a numerical scoring system involving various safety issues and the degree of code compliance for each issue. This section contains the provisions to be included in the evaluation and the conditions on which that evaluation is based.

There are 19 safety parameters of every existing building that must be evaluated, including building height; building area; compartmentation; tenant and dwelling unit separations; corridor walls; vertical openings; heating, ventilating and air-conditioning (HVAC) systems; automatic fire detection; fire alarm systems; smoke control; means of egress capacity and number; dead-end pockets and corridors; maximum travel distance to an exit; elevator controls; means of egress emergency lighting; mixed occupancies; sprinklers; standpipes and incidental use areas.

The procedure for evaluating the 19 safety parameters in a qualitative and quantitative matrix has evolved from several different sources. A particularly strong proponent for this methodology has been the National Institute of Building Sciences (NIBS). NIBS has published eight rehabilitation guidelines for the U.S. Department of Housing and Urban Development (HUD).

Section 3412 also contains the necessary database information, the form for the evaluation process and the mandatory safety scores by which each occupancy must comply. The procedure for determining whether a building meets the pass/fail criteria and which mandatory score is needed for mixed occupancies is also included in this section.

Purpose

A large number of existing buildings and structures do not comply with the current building code requirements for new construction. Although many of these buildings are potentially salvageable, rehabilitation is often cost prohibitive because they may not be able to comply with all the requirements for new construction. At the same time, it is necessary to regulate construction in existing buildings that undergo additions, alterations, renovations, extensive repairs or change of occupancy. Such activity represents an opportunity to ensure that new construction complies with the current building codes and that existing conditions are maintained, at a minimum, to their current level of compliance or are improved as required. To accomplish this objective, and to make the rehabilitation process easier, this chapter allows for a controlled departure from full compliance with the technical codes, without compromising the minimum standards for fire prevention and life safety features of the rehabilitated building.

SECTION 3401
GENERAL

3401.1 Scope. The provisions of this chapter shall control the *alteration*, *repair*, *addition* and change of occupancy of existing buildings and structures.

Exception: Existing *bleachers*, grandstands and folding and telescopic seating shall comply with ICC 300.

❖ This section states the scope of this chapter and references alternative methods of code compliance for the alteration, repair, addition and change of occupancy of existing buildings and structures. This Section also defines the responsibilities for maintenance, repairs, compliance with other codes and periodic testing. The existing features known as bleachers, grandstands and folding and telescopic seating must comply with ICC 300, *Standard for Bleachers, Folding and Telescopic Seating, and Grandstands*.

3401.2 Maintenance. Buildings and structures, and parts thereof, shall be maintained in a safe and sanitary condition. Devices or safeguards which are required by this code shall be maintained in conformance with the code edition under which installed. The owner or the owner's designated agent shall be responsible for the maintenance of buildings and structures. To determine compliance with this subsection, the *building official* shall have the authority to require a building or structure to be reinspected. The requirements of this chapter shall not provide the basis for removal or abrogation of fire protection and safety systems and devices in existing structures.

❖ This section establishes the owner's responsibility to keep the building maintained in accordance with the code and the other referenced I-Codes.

The building official has the authority to rule on the performance of maintenance work when public health or safety is affected. The building official also has the authority to require a building to be maintained in compliance with the health and safety provisions required by the code.

Fire protection and safety systems which are regulated in Chapter 9 of the code must remain in place and be maintained if they are required by the code. If such existing systems are provided, but are not required by the code, Section 901.6 of the *International Fire Code*® (IFC®) requires such systems to be maintained or removed. For example, if an existing building has a manual fire alarm system, but the code does not require it, the system has to be maintained in operating condition, or it must be disconnected and removed. Leaving the manual pull stations on the wall, but not having them connected to a working system, would give the building occupants the impression that an alarm system was present.

This section requires maintenance of the building in accordance with the standards under which it was built, especially if the system or device is currently required by the code. However, this does not prohibit alteration of the building to meet the current code, and as a result, an existing system not required by the code is removed or altered. For example, an office building was built in the 1990s under a previous code. The building is fully protected by a sprinkler system. The floor layouts include corridors that are built with 1-hour fire-resistance-rated walls and 20-minute-rated doors. Section 1018.1 does not require corridors to be of rated construction in a Group B occupancy where such a building is sprinkler protected. In this situation, an alteration permit could be approved that would replace the rated doors or remove part of the corridor wall. This is permitted because this rated corridor is not required by the code.

Another example is an existing office building in Illinois, two stories and of Type VA construction. The floor area of each story is 17,500 square feet (1626 m^2). This building, also built in the 1990s, was provided with a sprinkler system to allow an increase in floor area per story (the frontage was insufficient to be used for area increase). Neither of the codes used in Illinois at that time would have permitted a building this large without a sprinkler system; however, the code permits a Group B occupied building of Type VA construction to be 18,000 square feet (1672 m^2) in area per story. Again in this case, the sprinkler system would not be required to construct a building of this area under the current code. A permit to decommission and remove the sprinkler system could be approved. Caution needs to be taken before any permit to remove a system is approved. Sprinklers may have been provided in the building for a number of reasons, not just for an increase in area. In the first example, the sprinkler system is allowing a corridor not to be of rated construction. Therefore, any alteration permit to remove an existing protection or system must make sure that all of the reasons such a protection or system might have been provided are no longer applicable and no longer required.

Finally, reviewing alteration permits to allow a change to an existing system or remove a system altogether does not require a complete rereview of an entire building, but does necessitate a review of those building elements, occupancies, systems or other code requirements that might be affected and for which the change or removal of a system would move the building out of compliance with the current code. Only the related systems and building elements need to be reviewed. It would not be necessary to check the classification of the roofing material or the depth of the footings for a permit to alter the rating of a corridor wall.

3401.3 Compliance. *Alterations*, *repairs*, *additions* and changes of occupancy to, or relocation of, existing buildings and structures shall comply with the provisions for *alterations*, *repairs*, *additions* and changes of occupancy or relocation, respectively, in the *International Energy Conservation Code, International Fire Code, International Fuel Gas Code, International Mechanical Code, International Plumbing Code, International Property Maintenance Code, Interna-*

tional Private Sewage Disposal Code, International Residential Code and NFPA 70. Where provisions of the other codes conflict with provisions of this chapter, the provisions of this chapter shall take precedence.

❖ This section clarifies the relationship between this chapter of the code and the IFC, *International Fuel Gas Code*® (IFGC®), *International Mechanical Code*® (IMC®), *International Plumbing Code*® (IPC©), *International Property Maintenance Code*® (IPMC®), *International Private Sewage Disposal Code*® (IPSDC®), *International Residential Code*® (IRC®), *International Energy Conservation Code*® (IECC®) and NFPA 70. When alterations and repairs are made to existing mehanical and plumbing systems, the provisions of the I-Codes and NFPA 70 for alterations and repairs must be followed. Those codes indicate the extent to which existing systems must comply with the stated requirements. Where portions of existing building systems, such as plumbing, mechanical and electrical systems, are not being altered or repaired, those systems may continue to exist without being upgraded as long as they are not hazardous or unsafe to the building occupants.

Another important element of this section is that Chapter 34 will take precedence if a conflict occurs between one of the listed codes and this chapter. This is only as far as it concerns requirements for alteration, repairs, additions and change of occupancy. This, however, would not address a situation where another code, such as the IFC, retroactively required changes to a building regardless if any repairs, alterations, additions or changes of occupancy were occurring.

3401.4 Building materials and systems. Building materials and systems shall comply with the requirements of this section.

❖ This section contains conditions under which existing materials may remain in service and requirements for materials used for the repair or alteration of existing buildings. In addition, this section addresses which baseline seismic-force-resisting system coefficients are required to be used when applying this chapter.

3401.4.1 Existing materials. Materials already in use in a building in compliance with requirements or approvals in effect at the time of their erection or installation shall be permitted to remain in use unless determined by the *building official* to be unsafe per Section 116.

❖ If a material or system had been approved before the code took effect, it can continue to be used as long as it can be shown that the material or system is not unsafe. In other words, the code is not retroactive.

3401.4.2 New and replacement materials. Except as otherwise required or permitted by this code, materials permitted by the applicable code for new construction shall be used. Like materials shall be permitted for repairs and alterations, provided no hazard to life, health or property is created. Hazardous materials shall not be used where the code for new construction would not *permit* their use in buildings of similar occupancy, purpose and location.

❖ There are two options for materials used in repairs to an existing building. Generally, the materials used for repairs should be those that are presently required or permitted for new construction under the I-Codes. It is also acceptable to use materials consistent with those that are already present, except where those materials pose a hazard. This allowance follows the general concept that any repair should not make a building more hazardous than it was prior to the repair. It is generally possible to repair a structure, its components and its systems with materials consistent with those materials that were used previously. However, where materials that are now deemed hazardous are involved in the repair work, they may no longer be used. For example, the code identifies asbestos and lead-based paint as two common hazardous building materials that cannot be used in the repair process. Certain materials previously considered acceptable for building construction are now a threat to the health of the occupants.

3401.4.3 Existing seismic force-resisting systems. Where the existing seismic force-resisting system is a type that can be designated ordinary, values of R, Ω_0, and C_d for the existing seismic force-resisting system shall be those specified by this code for an ordinary system unless it is demonstrated that the existing system will provide performance equivalent to that of a detailed, intermediate or special system.

❖ This section provides guidance to engineers on selecting system-related design coefficients for existing seismic-force systems. The intent is that existing systems should be considered "ordinary" by default. This only applies to systems for which there is a choice of "ordinary," "special," "intermediate" or "detailed" for the permitted seismic-force systems. Some seismic systems, such as light-frame shear walls, are not categorized as "ordinary," "intermediate," "special" or "detailed". Those systems are acceptable, and the system coefficients specified for those systems in ASCE 7 are appropriate. For seismic systems that may be "intermediate," "special" or "detailed," the code requires a demonstration of equivalence.

3401.5 Dangerous conditions. The *building official* shall have the authority to require the elimination of conditions deemed *dangerous*.

❖ This provision provides a global application of authority throughout the application of Chapter 34 to the building official regarding conditions deemed dangerous. In the 2009 edition of the code this was only located in the section dealing with repairs. This authority was felt to be a necessary tool for all of Chapter 34. Conditions that pose a serious safety risk must always be addressed (see definition of "Dangerous" in Section 202). This section provides the building official considerable latitude with respect to mitigating dangerous conditions.

3401.6 Alternative compliance. Work performed in accordance with the *International Existing Building Code* shall be deemed to comply with the provisions of this chapter.

❖ This section allows for alternative methods of compliance with this chapter by allowing repairs, alterations, additions, etc., to meet the requirements of the *International Existing Building Code®* (IEBC®). The IEBC contains several options available to a designer or owner when dealing with construction related to existing buildings: prescriptive compliance method (IEBC Section 301.1.1 and Chapter 4 of the IEBC of the IEBC), work area compliance method (Section 301.1.2 of the IEBC) and performance compliance method (Section 301.1.3 and Chapter 14 of the IEBC). The IEBC also provides procedures for evaluation and design of seismic-force-resisting systems of existing buildings where consideration of seismic forces is required. There is one alternative to using these three compliance methods that allows for compliance with the laws in existence at the time the structure was originally built, unless the building has sustained substantial structural damage or is undergoing more than a limited structural alteration. Repairs and alterations in flood hazard areas have additional requirements to the laws in existence at the time the structure was originally built. Note that Chapters 4 and 14 are duplicated from this chapter.

SECTION 3402 DEFINITIONS

3402.1 Definitions. The following terms are defined in Chapter 2:

❖ This section lists terms that are specifically associated with the subject matter of this chapter. It is important to emphasize that these terms are not exclusively related to this chapter but may or may not also be applicable where the term is used elsewhere in the code.

Definitions of terms can help in the understanding and application of the code requirements. The purpose for including a list within this chapter is to provide more convenient access to terms that may have a specific or limited application within this chapter. For the complete definition and associated commentary, refer back to Chapter 2. Italicizing terms throughout the code indicates that a definition exists for that term. The use and application of all defined terms are set forth in Section 201.

DANGEROUS.

EXISTING STRUCTURE.

PRIMARY FUNCTION.

SUBSTANTIAL STRUCTURAL DAMAGE.

TECHNICALLY INFEASIBLE.

SECTION 3403 ADDITIONS

3403.1 General. *Additions* to any building or structure shall comply with the requirements of this code for new construction. *Alterations* to the existing building or structure shall be made to ensure that the existing building or structure together with the *addition* are no less conforming with the provisions of this code than the existing building or structure was prior to the *addition*. An existing building together with its *additions* shall comply with the height and area provisions of Chapter 5.

❖ The purpose of this section is to establish the guidelines for additions to existing buildings and structures.

An addition is an increase in the area or height of an existing building. When a new building is erected immediately adjacent to an existing building, and they are separated by a fire wall, it is considered a separate building, not an addition to the existing structure. The new building must be designed to comply with the technical provisions of Chapters 1 through 33, not with the provisions of this chapter. The existing building must be evaluated considering the elimination of the adjacent open space now occupied by the new building.

An existing structure that is of a type of construction that does not comply with the height and area limitations of Table 503 may not be added to unless the type of construction is upgraded, or the allowable building height and area are increased through the addition of a sprinkler system throughout both the existing building and the addition.

A building with a proposed addition is to be evaluated based on the type of construction of the existing building or the addition, whichever is the lower type. When reviewing for compliance with Table 503, see Section 602.1.1.

Figure 3403.1 shows an example of this situation. If a sprinkler system is not planned for both the existing building and the addition, the proposed addition does not meet the requirements of Table 503. The area limitation in Table 503 for the existing construction Type VB, Group F-1 is 8,500 square feet (790 m^2) and the total proposed area is 11,000 square feet (1022 m^2). The area evaluation is based on Type VB construction, even though the proposed addition is Type VA construction, because the allowable area in Table 503 for Type VB construction is less than the allowable area for Type VA construction.

One way to satisfy the requirements of Table 503 is to upgrade the type of construction of the existing building to Type VA, which has an allowable area of 14,000 square feet (130 m^2). Another solution is to install an automatic sprinkler system that complies with NFPA 13 throughout the building. The revised allowable area of the building would be:

8,500 square feet + (3 × 8,500 square feet) =
34,000 square feet (3,158.7 m^2)

in accordance with the sprinkler area increase, which is more than the proposed building area of 11,000 square feet (1022 m^2).

For accessibility requirements for additions, see Section 3411.5.

3403.2 Flood hazard areas. For buildings and structures in *flood hazard areas* established in Section 1612.3, any *addition* that constitutes substantial improvement of the existing structure, as defined in Section 1612.2, shall comply with the flood design requirements for new construction, and all aspects of the existing structure shall be brought into compliance with the requirements for new construction for flood design.

For buildings and structures in *flood hazard areas* established in Section 1612.3, any additions that do not constitute substantial improvement of the existing structure, as defined in Section 1612.2, are not required to comply with the flood design requirements for new construction.

❖ Reduction in exposure to flood hazards, including exposure of older buildings, is one of the purposes for regulating flood plain development. Buildings or structures that are located in flood hazard areas are to be brought into compliance with the flood-resistance provisions of Section 1612 when the improvements, including additions, exceed a certain value. See Section 202 for the definition of "Substantial improvement."

Section 105.3 requires the applicant to state the valuation of proposed work, which is to include the total value of work, including materials and labor. If the proposed work will be performed on buildings in flood hazard areas, the building official must determine whether the proposed work constitutes substantial improvement. If applicable, the value of work must include estimates of the value of the property owner's labor and the value of donated labor and materials.

To make a determination about whether a proposed addition constitutes a substantial improvement, the cost of the proposed work is to be compared to the market value of the building or structure before the work is started. In order to determine market value, the building official may require the applicant to provide an appraisal or use other methods acceptable to the Federal Emergency Management Agency (FEMA). For additional guidance, see FEMA P-758, *Substantial Improvement/Substantial Damage Desk Reference.*

3403.3 Existing structural elements carrying gravity load. Any existing gravity load-carrying structural element for which an *addition* and its related alterations cause an increase in design gravity load of more than 5 percent shall be strengthened, supplemented, replaced or otherwise altered as needed to carry the increased gravity load required by this code for new structures. Any existing gravity load-carrying structural element whose gravity load-carrying capacity is decreased shall be considered an altered element subject to the requirements of Section 3404.3. Any existing element that will form part of the lateral load path for any part of the *addition* shall be considered an existing lateral load-carrying structural element subject to the requirements of Section 3403.4.

❖ Wherever an addition to an existing building is made, the affected members of the original structure must be assessed to determine their ability to resist any

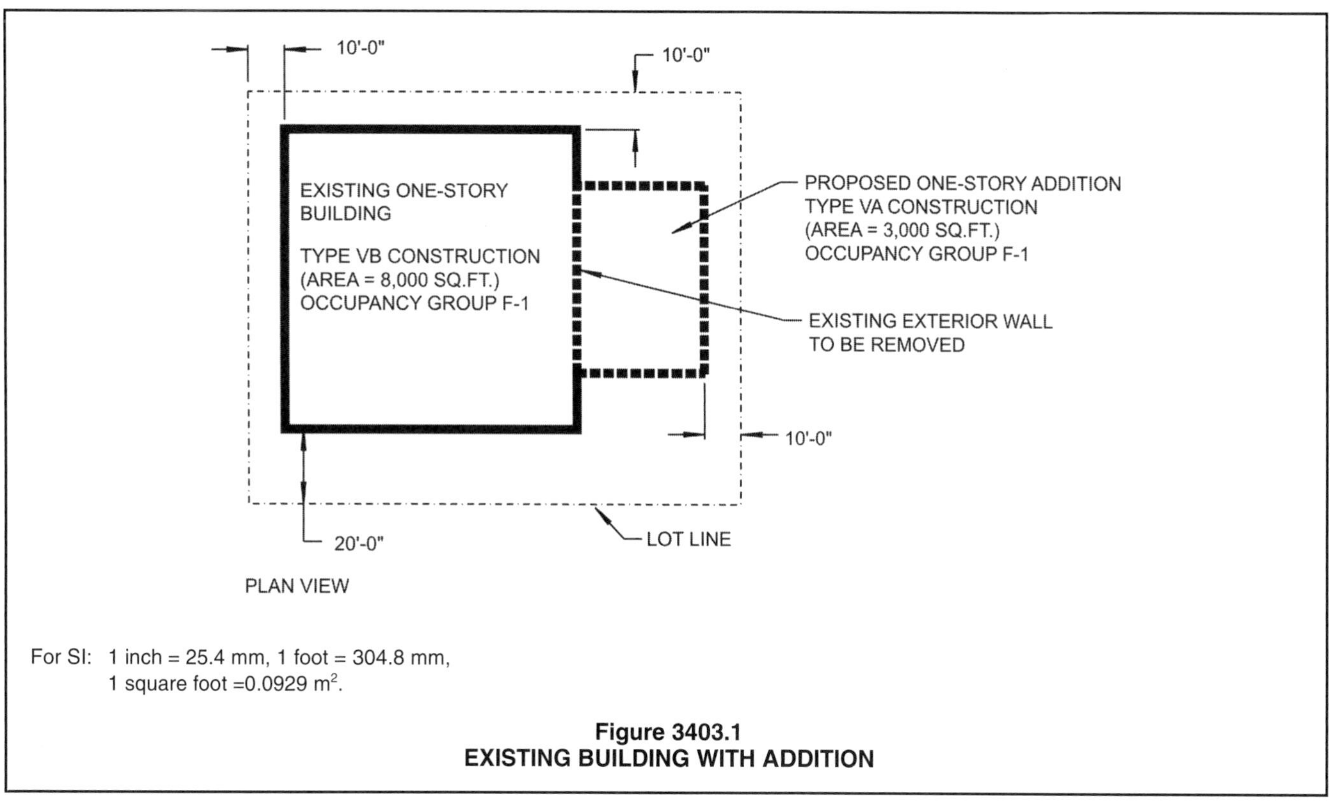

Figure 3403.1
EXISTING BUILDING WITH ADDITION

increased forces. Because an addition typically adds new loads to the existing building, the architect or engineer responsible for the design of the project must analyze the existing gravity load-carrying elements to determine whether there are any structural components having loads increased by more than 5 percent that require reinforcement. Where an existing building was initially designed to support a future expansion, the building official should seek verification that the proposed addition will result in the affected building complying with the current code requirements. Any new construction must comply with current code requirements.

3403.3.1 Design live load. Where the *addition* does not result in increased design live load, existing gravity load-carrying structural elements shall be permitted to be evaluated and designed for live loads *approved* prior to the *addition*. If the *approved* live load is less than that required by Section 1607, the area designed for the nonconforming live load shall be posted with placards of *approved* design indicating the *approved* live load. Where the *addition* does result in increased design live load, the live load required by Section 1607 shall be used.

❖ It is not uncommon for the design live load requirements to change from the time a building is originally designed to when an addition is proposed. The live loads used in the original design may have been adequate for the building's initial use and may have been in compliance with all the code requirements that were in effect at that time. Many years and many code changes can change the status of the structural design criteria and code requirements, which does not mean the existing structural system is inadequate and cannot be used. It just means that when the live loads that were used for the design of the existing building are lower than the those required by current standards, the design live loads used for the original design must be posted.

3403.4 Existing structural elements carrying lateral load. Where the *addition* is structurally independent of the *existing structure*, existing lateral load-carrying structural elements shall be permitted to remain unaltered. Where the *addition* is not structurally independent of the *existing structure*, the *existing structure* and its *addition* acting together as a single structure shall be shown to meet the requirements of Sections 1609 and 1613.

Exception: Any existing lateral load-carrying structural element whose demand-capacity ratio with the *addition* considered is no more than 10 percent greater than its demand-capacity ratio with the *addition* ignored shall be permitted to remain unaltered. For purposes of calculating demand-capacity ratios, the demand shall consider applicable load combinations with design lateral loads or forces in accordance with Sections 1609 and 1613. For purposes of this exception, comparisons of demand-capacity ratios and calculation of design lateral loads, forces and capacities shall account for the cumulative effects of *additions* and *alterations* since original construction.

❖ The requirements of this section do not affect an existing structure where the addition is structurally independent, in which case the addition is required to comply with the seismic requirements for new structures. An existing building or structure that is added on to, where the addition will not be structurally independent, must be carefully evaluated for its ability to withstand earthquake and wind loading. The seismic resistance of an existing building or structure cannot be lessened or pose any undue increase in the fire and life safety hazards of the building or structure. Because the addition and the existing structure resist lateral loads as a single structure, designing the entire structure to resist the total lateral seismic and wind load is necessary.

The exception permits an addition that is not structurally independent without alteration to the seismic-force-resisting system of the existing structure, provided the demand-capacity ratio is increased no more than 10 percent in any structural element. The demand is determined using the load combinations, including wind and seismic effects. By considering the demand-capacity ratio, any decrease in lateral resistance is accounted for as well.

3403.5 Smoke alarms in existing portions of a building. Where an *addition* is made to a building or structure of a Group R or I-1 occupancy, the existing building shall be provided with *smoke alarms* in accordance with Section 1103.8 of the *International Fire Code*.

❖ This section references retroactive smoke alarm requirements in the IFC that apply to Group R and I-1 occupancies regardless of whether alterations, additions or repairs are being made. Section 3403 in general is focused upon additions, but it is also important to realize what may be required for the existing portion of the building even if no changes are being made there. The addition would need to comply with this code for new construction with regard to smoke-alarm requirements.

SECTION 3404 ALTERATIONS

3404.1 General. Except as provided by Section 3401.4 or this section, *alterations* to any building or structure shall comply with the requirements of the code for new construction. *Alterations* shall be such that the existing building or structure is no less complying with the provisions of this code than the existing building or structure was prior to the *alteration*.

Exceptions:

1. An existing *stairway* shall not be required to comply with the requirements of Section 1009 where the

existing space and construction does not allow a reduction in pitch or slope.

2. *Handrails* otherwise required to comply with Section 1009.15 shall not be required to comply with the requirements of Section 1012.6 regarding full extension of the *handrails* where such extensions would be hazardous due to plan configuration.

❖ Alterations include renovations, which implies that something is changed in the structure. For example, the removal, rearrangement or replacement of partition walls in an office building is an alteration because, in part, of possible impact on the means of egress, fire resistance or other life safety features of the building. Conversely, the replacement of damaged trim pieces on a door frame is considered a repair, not an alteration.

Alterations are to conform to the requirements for a new structure. For example, consider the corridor of an office building that is to be extended 18 feet (5486 mm). The existing office building has no sprinkler system (and none is proposed), so the corridor extension walls and doors are to be fire-resistance rated in accordance with Section 1018. This is applicable even if, for some reason, the walls and doors of the existing corridor system are not fire-resistance-rated construction.

This section also indicates that unaltered portions of a structure are not required to comply with code provisions for new construction. However, see also Section 3411, which requires areas outside of the alteration area to be revised in order to provide an accessible route. All requirements applicable to existing, unaltered areas are contained in the code (see Sections 102.6 and 116) or the IPMC and the IFC.

The exceptions address issues that arise with stairways. While not specifically addressed, ramps may have similar concerns. In accordance with Exception 1, if an existing stairway was built with a steeper rise/run ratio than permitted in the current code, the stairway can be replaced with the existing configuration. Enlarging the opening to achieve the current rise/run ratio and headroom would be considered technically infeasible. The principle is that not allowing for this option could result in stairways that were not maintained because they could not be brought up to current codes.

Exception 2 deals with an allowance to reduce or bend handrail extensions when providing those extensions could become a protruding object for circulation or an obstruction for the means of egress. Section 1012.6 requires the handrail extensions to continue in the direction of the stairway run. An example would be where stairways had landings that were part of a corridor.

3404.2 Flood hazard areas. For buildings and structures in *flood hazard areas* established in Section 1612.3, any *alteration* that constitutes *substantial improvement* of the *existing structure*, as defined in Section 1612.2, shall comply with the flood design requirements for new construction, and all aspects of the *existing structure* shall be brought into compliance with the requirements for new construction for flood design.

For buildings and structures in *flood hazard areas* established in Section 1612.3, any *alterations* that do not constitute *substantial improvement* of the *existing structure*, as defined in Section 1612.2, are not required to comply with the flood design requirements for new construction.

❖ Reduction in exposure to flood hazards, including exposure of older buildings, is one of the purposes for regulating flood plain development. Buildings or structures that are located in flood hazard areas are to be brought into compliance with the flood-resistance provisions of Section 1612 when the improvements, including alterations, exceed a certain value. See Section 202 for the definition of "Substantial improvement."

Section 105.3 requires the applicant to state the valuation of proposed work, which is to include the total value of work, including materials and labor. If the proposed work will be performed on buildings in flood hazard areas, the building official must determine whether the proposed work constitutes substantial improvement. If applicable, the value of work must include estimates of the value of the property owner's labor and the value of the donated labor and materials.

To make a determination about whether a proposed alteration constitutes a substantial improvement, the cost of the proposed work is to be compared to the market value of the building or structure before the work is started. In order to determine market value, the building official may require the applicant to provide an appraisal or use other methods acceptable to FEMA. For additional guidance, see FEMA P-758, *Improvement/Substantial Damage Desk Reference*.

3404.3 Existing structural elements carrying gravity load. Any existing gravity load-carrying structural element for which an *alteration* causes an increase in design gravity load of more than 5 percent shall be strengthened, supplemented, replaced or otherwise altered as needed to carry the increased gravity load required by this code for new structures. Any existing gravity load-carrying structural element whose gravity load-carrying capacity is decreased as part of the *alteration* shall be shown to have the capacity to resist the applicable design gravity loads required by this code for new structures.

❖ Wherever an alteration to an existing building is made, the affected members must be assessed to determine their ability to resist the increased forces. Structural element forces may be increased by up to 5 percent without the need to strengthen or replace that element.

3404.3.1 Design live load. Where the *alteration* does not result in increased design live load, existing gravity load-carrying structural elements shall be permitted to be evaluated and designed for live loads *approved* prior to the *alteration*. If the *approved* live load is less than that required by Section 1607, the area designed for the nonconforming live load shall be posted with placards of *approved* design indicating the *approved* live load. Where the *alteration* does result in increased design live load, the live load required by Section 1607 shall be used.

❖ Where an existing building is undergoing alterations requiring changes to the structural loading, the applicable minimum design loads are those required by the code at the time of the original construction. It is not uncommon for the design live load requirements to change from the time a building is originally designed to when it undergoes a renovation. The live loads used in the original design may have been adequate for the building's initial use and may have been in compliance with all the code requirements that were in effect at that time. Many years and many code changes can impact the structural design parameters and code requirements, which does not mean the existing structural system is inadequate and cannot be used when the building is renovated. It just means that when the live loads that were used for the design of the existing building are lower than those required by current standards, the design live loads used for the original design must be posted. If the design live loads of the alteration are greater than those used for the existing building, Section 1607 would apply just as for new construction.

3404.4 Existing structural elements carrying lateral load. Except as permitted by Section 3404.5, where the *alteration* increases design lateral loads in accordance with Section 1609 or 1613, or where the *alteration* results in a structural irregularity as defined in ASCE 7, or where the *alteration* decreases the capacity of any existing lateral load-carrying structural element, the structure of the altered building or structure shall be shown to meet the requirements of Sections 1609 and 1613.

> **Exception:** Any existing lateral load-carrying structural element whose demand-capacity ratio with the *alteration* considered is no more than 10 percent greater than its demand-capacity ratio with the *alteration* ignored shall be permitted to remain unaltered. For purposes of calculating demand-capacity ratios, the demand shall consider applicable load combinations with design lateral loads or forces per Sections 1609 and 1613. For purposes of this exception, comparisons of demand-capacity ratios and calculation of design lateral loads, forces, and capacities shall account for the cumulative effects of *additions* and *alterations* since original construction.

❖ An existing building or structure that is altered must be carefully evaluated for its ability to withstand earthquake and wind loads. The lateral resistance of an existing building or structure cannot be lessened or pose any undue increase in the fire and life safety hazards of the building or structure. This section of the code gives guidance with respect to the impact of an alteration on the lateral resistance of the structure. Alterations that affect existing structural elements to a lesser extent are permitted without requiring the existing structure to comply with the provisions for new structures, as long as the alteration itself complies. The exception allows alterations that increase the demand-capacity ratio on existing lateral load-carrying elements by no more than 10 percent or decrease the lateral resistance of existing structural elements by no more than 10 percent.

3404.5 Voluntary seismic improvements. *Alterations* to existing structural elements or additions of new structural elements that are not otherwise required by this chapter and are initiated for the purpose of improving the performance of the seismic force-resisting system of an *existing structure* or the performance of seismic bracing or anchorage of existing nonstructural elements shall be permitted, provided that an engineering analysis is submitted demonstrating the following:

1. The altered structure and the altered nonstructural elements are no less conforming with the provisions of this code with respect to earthquake design than they were prior to the alteration.
2. New structural elements are detailed as required for new construction.
3. New or relocated nonstructural elements are detailed and connected to existing or new structural elements as required for new construction.
4. The alterations do not create a structural irregularity as defined in ASCE 7 or make an existing structural irregularity more severe.

❖ This provision addresses the issue of upgrading existing structures voluntarily for improved seismic performance. It does not apply to situations where other code sections trigger full compliance with the code. Otherwise, it allows an owner to initiate an improvement to the seismic-force-resisting system to the extent that it is viable to do so and provided the required engineering analysis is furnished. The intent is to encourage building owners to initiate upgrades to seismic systems that are considered prudent without making them cost prohibitive.

3404.6 Smoke alarms. Individual *sleeping units* and individual *dwelling units* in Group R and I-1 occupancies shall be provided with *smoke alarms* in accordance with Section 1103.8 of the *International Fire Code.*

❖ Similar to Section 3403.5, this is simply a reference back to the IFC for retroactive smoke-alarm requirements in sleeping units and dwelling units in Group R and I-1 occupancies. These requirements in the IFC apply regardless of whether an alteration occurs within a building.

SECTION 3405 REPAIRS

3405.1 General. Buildings and structures, and parts thereof, shall be repaired in compliance with Section 3405 and 3401.2. Work on nondamaged components that is necessary for the required *repair* of damaged components shall be considered part of the *repair* and shall not be subject to the requirements for *alterations* in this chapter. Routine maintenance required by Section 3401.2, ordinary repairs exempt from *permit* in accordance with Section 105.2, and abatement of wear due to normal service conditions shall not be subject to the requirements for *repairs* in this section.

❖ In general, the section provides a logical method for evaluating damage and identifying cases where upgrade is warranted.

The provisions require structural improvements, in certain cases to meet "the code for new structures" in the event of damage due to fire, structural overload, settlement, natural hazard or any other cause, no matter how extensive or disproportionate the damage. This section identifies conditions of damage that should warrant improvements to the structural system for purposes of increasing safety.

3405.2 Substantial structural damage to vertical elements of the lateral force-resisting system. A building that has sustained substantial structural damage to the vertical elements of its lateral force-resisting system shall be evaluated and repaired in accordance with the applicable provisions of Sections 3405.2.1 through 3405.2.3.

Exceptions:

1. Buildings assigned to *Seismic Design Category* A, B, or C whose *substantial structural damage* was not caused by earthquake need not be evaluated or rehabilitated for load combinations that include earthquake effects.
2. One- and two-family dwellings need not be evaluated or rehabilitated for load combinations that include earthquake effects.

❖ This section provides requirements that apply where the damage threshold that is based on the extent of damage to the vertical elements of the lateral-force-resisting system in any story is exceeded. Substantial structural damage to the lateral system triggers evaluation of the entire building for wind and seismic loads (see Section 3405.2.1). The emphasis is placed on vertical elements, such as walls and columns, rather than horizontal elements, because it is the vertical elements of the lateral-force-resisting system that primarily determine the structure's response, particularly to earthquakes.

The purpose of the exceptions is to exempt certain combinations of buildings, seismic risk and damage from seismic upgrades. In general, seismic evaluation and possibly upgrade is triggered when damage to the lateral system, independent of cause, reaches the level defined as "Substantial structural damage," in the code, regardless of occupancy, structure type or seismic design category. This broad applicability seems unnecessary to the general purpose of the code and could discourage or delay certain repairs by imposing the additional costs of seismic upgrade. Therefore, the exceptions exempt two classes of buildings from seismic upgrades triggered by substantial structural repairs (Basic repair—that is, restoring the predamaged condition—is still required). The first is for buildings in areas of low or moderate seismicity, where the damage was caused by something other than an earthquake. Where earthquakes are rare, it serves no significant public purpose to require seismic upgrades following damage caused by fire, collision, wind, etc. The second is for one- and two-family dwellings, where the public risk is especially low.

3405.2.1 Evaluation. The building shall be evaluated by a *registered design professional*, and the evaluation findings shall be submitted to the *building official*. The evaluation shall establish whether the damaged building, if repaired to its pre-damage state, would comply with the provisions of this code for wind and earthquake loads.

Wind loads for this evaluation shall be those prescribed in Section 1609. Earthquake loads for this evaluation, if required, shall be permitted to be 75 percent of those prescribed in Section 1613.

❖ The extent of repairs is based on an evaluation prepared by a registered design professional. Generally, the code's approach is that complete structural upgrades should be relatively rare. In this section, a structural upgrade is only triggered upon substantial structural damage to the lateral system, and only when evaluation shows that the predamaged building was substandard.

This provision refers to Sections 1609 and 1613 for wind and earthquake loads, respectively. It permits reduced earthquake loading both for evaluation and any required seismic rehabilitation to recognize that existing buildings can't be expected to perform as well as newer buildings. This concept of using reduced earthquake loading is consistent with FEMA standards, as well as legacy building codes.

3405.2.2 Extent of repair for compliant buildings. If the evaluation establishes compliance of the pre-damage building in accordance with Section 3405.2.1, then repairs shall be permitted that restore the building to its pre-damage state, based on material properties and design strengths applicable at the time of original construction.

❖ Where the evaluation establishes that the predamaged building meets the structural provisions of the code as provided for in Section 3405.2.1, then the repairs may be limited to a restoration of the structural components based upon what was applicable when the structure was originally constructed. This is similar to the requirements in Section 3405.2.3 for wind loads and Section 3405.3 for live loads.

3405.2.3 Extent of repair for noncompliant buildings. If the evaluation does not establish compliance of the pre-damage building in accordance with Section 3404.2.1, then the building shall be rehabilitated to comply with applicable provisions of this code for load combinations that include wind or seismic loads. The wind loads for the repair shall be as required by the building code in effect at the time of original construction, unless the damage was caused by wind, in which case the wind loads shall be as required by this code. Earthquake loads for this rehabilitation design shall be those required for the design of the pre-damage building, but not less than 75 percent of those prescribed in Section 1613. New structural members and connections required by this rehabilitation design shall comply with the detailing provisions of this code for new buildings of similar structure, purpose and location.

❖ If the evaluation of the building in accordance with Section 3405.2.1 shows that in the hypothetically repaired condition the building would not comply with the established requirements, then the building must be rehabilitated as described in this section. The general requirement is to comply with the code's load combinations. These load combinations establish the required strength of structural members.

The effects of wind and seismic loads warrant special consideration. In determining the level of compliance for repairs to buildings that have sustained substantial structural damage, it is important to determine if wind forces have caused that damage. If so, it is considered prudent to require the repairs of wind damage to use wind loading from the current code. Seismic forces for the rehabilitation can be those required by the building code in effect at the time of the building's construction, but this may not be less than the reduced seismic force level, as described in Section 3405.2.1.

3405.3 Substantial structural damage to gravity load-carrying components. Gravity load-carrying components that have sustained *substantial structural damage* shall be rehabilitated to comply with the applicable provisions of this code for dead and live loads. Snow loads shall be considered if the *substantial structural damage* was caused by or related to snow load effects. Existing gravity load-carrying structural elements shall be permitted to be designed for live loads *approved* prior to the damage. Nondamaged gravity load-carrying components that receive dead, live or snow loads from rehabilitated components shall also be rehabilitated or shown to have the capacity to carry the design loads of the rehabilitation design. New structural members and connections required by this rehabilitation design shall comply with the detailing provisions of this code for new buildings of similar structure, purpose and location.

❖ If the evaluation of the building in accordance with Section 3405.2.1 shows that in the hypothetically repaired condition the building would not comply with the established requirements, then the building must be rehabilitated as described in this section. The general requirement is to comply with the current code's load combinations. These load combinations establish the required strength of structural members.

3405.3.1 Lateral force-resisting elements. Regardless of the level of damage to vertical elements of the lateral force-resisting system, if *substantial structural damage* to gravity load-carrying components was caused primarily by wind or earthquake effects, then the building shall be evaluated in accordance with Section 3405.2.1 and, if noncompliant, rehabilitated in accordance with Section 3405.2.3.

Exceptions:

1. One- and two-family dwellings need not be evaluated or rehabilitated for load combinations that include earthquake effects.
2. Buildings assigned to *Seismic Design Category* A, B, or C whose *substantial structural damage* was not caused by earthquake need not be evaluated or rehabilitated for load combinations that include earthquake effects.

❖ Substantial structural damage to gravity load-carrying elements, such as columns or bearing walls, must be repaired so that these members are adequate to resist the dead and live loads in accordance with current code requirements, as must other elements of the load path.

There are two exceptions that acknowledge that applying this requirement in some cases is excessive and may have the effect of discouraging or delaying certain repairs by imposing the additional costs of seismic upgrade. Exception 1 exempts one- and two-family dwellings, where the risk to the public of poor earthquake performance is especially low. Exception 2 exempts buildings in areas of low or moderate seismicity, where the damage was caused by something other than an earthquake. Where earthquakes are rare, it serves no significant public purposes to trigger seismic upgrades following damage caused by fire, collision, wind and other events. Basic repair–that is, restoring to the predamaged condition—is still required.

3405.4 Less than substantial structural damage. For damage less than *substantial structural damage*, *repairs* shall be allowed that restore the building to its pre-damage state, based on material properties and design strengths applicable at the time of original construction. New structural members and connections used for this repair shall comply with the detailing provisions of this code for new buildings of similar structure, purpose and location.

❖ For damage that is not deemed to be substantial structural damage, repairs are allowed that restore the building to its predamaged state using materials and strengths applicable at the time of original construction. New structural members and connections used for this repair are still required to comply with the detailing provisions for new buildings of similar materials, purpose and location.

3405.5 Flood hazard areas. For buildings and structures in *flood hazard areas* established in Section 1612.3, any *repair* that constitutes substantial improvement of the existing structure, as defined in Section 1612.2, shall comply with the flood design requirements for new construction, and all aspects of the existing structure shall be brought into compliance with the requirements for new construction for flood design.

For buildings and structures in *flood hazard areas* established in Section 1612.3, any *repairs* that do not constitute substantial improvement or repair of substantial damage of the existing structure, as defined in Section 1612.2, are not required to comply with the flood design requirements for new construction.

❖ Reduction in exposure to flood hazards, including exposure of older buildings, is one of the purposes for regulating flood plain development. Damaged buildings or structures that are located in flood hazard areas are to be brought into compliance with the flood-resistance provisions of Section 1612 when the cost of repairs necessary to restore damage to its predamaged condition (even if less work is actually proposed) exceeds a certain value. See Section 202 for the definition of "Substantial improvement" and "Substantial damage."

Section 105.3 requires the applicant to state the valuation of proposed work, which is to include the total value of work, including materials and labor. If the proposed work will be performed on buildings in flood hazard areas, the building official must determine whether the proposed work constitutes substantial improvement or repair of substantial damage. If applicable, the value of work must include estimates of the value of the property owner's labor and the value of donated labor and materials.

To make a determination about whether proposed repairs constitute repair of substantial damage, the cost of the proposed work is to be compared to the market value of the building or structure before the work is started. In order to determine market value, the building official may require the applicant to provide an appraisal or use other methods acceptable to FEMA. For additional guidance, see FEMA P-758 *Substantial Improvement/Substantial Damage Desk Reference*. FEMA P-758 suggests that communities with many buildings in flood hazard areas become familiar with software developed to help make substantial damage determinations, especially in the post-disaster period. FEMA P-784, *Substantial Damage Estimator* includes the software, user's manual and workbook.

SECTION 3406
FIRE ESCAPES

3406.1 Where permitted. Fire escapes shall be permitted only as provided for in Sections 3406.1.1 through 3406.1.4.

❖ Sections 3406.1.1 through 3406.1.4 address the current requirements for the use of exterior fire escapes. Their use and features as defined in this section should not be confused with the requirements for exterior exit stairways as defined in Section 1026.

The use of exterior fire escapes as a means of egress was popular in building designs of the past. They were used for economic construction and to "save" usable space in the buildings they served. Being located outside the building, they were also considered safer than the unenclosed interior exit stairways of the time. Over the years, exterior fire escapes lost their appeal for many reasons:

- Fire escapes were never an integral part of the building's design; they were a necessary appendage hung from or attached to the building wall.
- Fire escapes were most commonly constructed of cast-iron or steel, both of which required a high degree of maintenance to protect the corrodible materials from the effects of weather.
- Because fire escapes are "open structures," they are subject to icing during the winter months. This makes them dangerous to use, and in extreme cases they can become completely unusable.
- People with a fear of heights can find exterior fire escapes difficult and sometimes impossible to use.
- Fire escapes were an unpleasant sight and an unwanted element in the architectural design.

The use of exterior fire escapes is all but obsolete except for existing buildings with a clear deficiency in the means of egress that cannot be reasonably rectified in other ways.

This section indicates that fire escapes are permitted only on existing buildings. New fire escapes may be installed on existing buildings only where exterior stairs cannot be used. Section 3406.2 addresses the location of fire escapes, and Section 3406.3 addresses the construction of fire escapes.

See also the requirements for fire escapes in Section 1104.17 of the IFC. Note that these provisions are slightly different than those of Section 3406.

3406.1.1 New buildings. Fire escapes shall not constitute any part of the required *means of egress* in new buildings.

❖ Because of the inherent hazards and lack of dependability, open exterior fire escapes are not permitted as part of the required means of egress in new construction.

3406.1.2 Existing fire escapes. Existing fire escapes shall be continued to be accepted as a component in the *means of egress* in existing buildings only.

❖ Exterior fire escapes in existing buildings are acceptable as a component of the required means of egress because, in most cases, it would be physically impractical and economically prohibitive to retrofit older buildings with interior or exterior exit stairways that comply with all the code requirements for new construction.

3406.1.3 New fire escapes. New fire escapes for existing buildings shall be permitted only where exterior *stairs* cannot be utilized due to lot lines limiting *stair* size or due to the sidewalks, alleys or roads at grade level. New fire escapes shall not incorporate ladders or access by windows.

❖ Continuing corrosion of existing exterior fire escapes, as well as other factors that may affect the safety of the structures, often make replacement of these fire escapes necessary. The building official may determine that the means of egress in an existing building is unsafe, requiring some corrective action, which may be the installation of a new exterior fire escape. This section permits new exterior fire escapes for existing buildings where exterior stairs conforming to the requirements of Section 1026 cannot be used because of either lot lines that limit stair size or encroachments on sidewalks, alleys or roads at grade level.

Because it was the accepted practice of the time, access to exterior fire escapes was quite often through windows located in corridors or individual dwelling units or offices. This practice is not permitted when replacement exterior fire escapes are used. The building must be altered to provide proper exit access to the fire escape.

The code does not permit the use of ladders as a component in the means of egress except as allowed in Section 408.3.5 for control rooms or elevated facility observation rooms for Group I-3 occupancies; therefore, ladders cannot be used either as a component part of an exterior fire escape, or to access an exterior fire escape.

3406.1.4 Limitations. Fire escapes shall comply with this section and shall not constitute more than 50 percent of the required number of exits nor more than 50 percent of the required *exit* capacity.

❖ For reasons of overall life safety, exterior fire escapes may be used as a component of the means of egress only when the total number used does not exceed one-half the required number of exits and more than one-half the total required exit capacity.

3406.2 Location. Where located on the front of the building and where projecting beyond the building line, the lowest landing shall not be less than 7 feet (2134 mm) or more than 12 feet (3658 mm) above grade, and shall be equipped with a counterbalanced stairway to the street. In alleyways and thoroughfares less than 30 feet (9144 mm) wide, the clearance under the lowest landing shall not be less than 12 feet (3658 mm).

❖ In the past, exterior fire escapes were allowed to project beyond the property line and extend over sidewalks, alleyways and roads. Where such conditions are still accepted by the code (see Sections 3406.1.2 and 3406.1.3), the clearance to the lowest landing from grade level must be at least 7 feet (2134 mm), and not more than 12 feet (3658 mm) for exterior fire escapes located in front of the building. In alleyways and public ways less than 30 feet (9144 mm) wide, the clearance under the lowest landing must be at least 12 feet (3658 mm). To facilitate these clearance requirements, exterior fire escapes are usually equipped with a counterbalanced stairway that retracts to a horizontal position when the fire escape is not in use.

3406.3 Construction. The fire escape shall be designed to support a live load of 100 pounds per square foot (4788 Pa) and shall be constructed of steel or other *approved* noncombustible materials. Fire escapes constructed of wood not less than nominal 2 inches (51 mm) thick are permitted on buildings of Type V construction. Walkways and railings located over or supported by combustible roofs in buildings of Type III and IV construction are permitted to be of wood not less than nominal 2 inches (51 mm) thick.

❖ Traditionally and typically, exterior fire escapes are constructed of cast iron or steel, although the code does permit exterior fire escapes to be constructed of other noncombustible materials. The use of nominal 2-inch (51 mm) wood is permitted in Type V construction. Nominal 2-inch (51 mm) wood may also be used in Type III and IV construction when the exterior fire escape is supported by wood construction permitted by Table 601.

3406.4 Dimensions. *Stairs* shall be at least 22 inches (559 mm) wide with risers not more than, and treads not less than, 8 inches (203 mm) and landings at the foot of stairs not less than 40 inches (1016 mm) wide by 36 inches (914 mm) long, located not more than 8 inches (203 mm) below the door.

❖ The normal dimensions for interior and exterior exit stairways (see Section 1009) used in the means of egress do not apply to exterior fire escapes. The dimensional standard for exterior fire escapes has been in place for many years, thus many exterior fire escapes in use were designed to comply with this standard. Many of these fire escapes are installed in places where it would be difficult, and often impossible, to retrofit a new fire escape if a new set of dimensional standards was employed. The minimum tread, riser, stair width and landing size dimensions for fire escape construction remain unchanged from the past.

3406.5 Opening protectives. Doors and windows along the fire escape shall be protected with $^3/_4$-hour opening protectives.

❖ Safe exit access and exits are required for occupants using an exterior fire escape as a means of egress. Door and window openings in exterior walls adjacent to and along the path of travel of the exterior fire escape must be protected from the interior of the building with not less than a $^3/_4$-hour protective. This is similar to the requirements for protection of openings on interior exit stairways and ramps in Section 1022.7 and exterior exit stairways and ramps in Section 1026.6.

Where more restrictive opening protectives are required by other Sections elsewhere in the code, the more restrictive protection must be used. The $^3/_4$-hour requirement for the exterior fire escape is the minimum when no other considerations are present (see Tables 601, 602, 705.8 and 716.5).

SECTION 3407 GLASS REPLACEMENT

3407.1 Conformance. The installation or replacement of glass shall be as required for new installations.

❖ The technical provisions for glass and glazing are detailed in Chapter 24. All technical requirements of the code for glass and glazing that apply to new construction also apply to all repairs, additions or alterations to existing buildings. There are no exceptions to this requirement. When glazing in an existing building is replaced or relocated within the same building, it must comply with current standards.

SECTION 3408 CHANGE OF OCCUPANCY

3408.1 Conformance. No change shall be made in the use or occupancy of any building that would place the building in a different division of the same group of occupancies or in a different group of occupancies, unless such building is made to comply with the requirements of this code for such division or group of occupancies. Subject to the approval of the *building official*, the use or occupancy of existing buildings shall be permitted to be changed and the building is allowed to be occupied for purposes in other groups without conforming to all the requirements of this code for those groups, provided the new or proposed use is less hazardous, based on life and fire risk, than the existing use.

❖ A change in occupancy in an existing structure may change the level of inherent hazards that the code was initially intended to address.

Regardless of whether the change is to an occupancy considered to be more or less hazardous, this section applies the provisions of the code for new construction to an existing structure having a new occupancy. This is done so that the applicable code requirements adequately address the specific hazards of the new occupancy. For example, a change from an existing mercantile occupancy to a business occupancy renders all Group B provisions applicable to all portions of the structure where the occupancy has changed.

This section is one of the most frequently used provisions in the code for application to existing structures, since the occupancy in a building or structure is subject to change during the life of the building. For accessibility requirements, see Section 3411.4

3408.2 Certificate of occupancy. A certificate of occupancy shall be issued where it has been determined that the requirements for the new occupancy classification have been met.

❖ An existing building that has been classified into a new occupancy group must receive a certificate of occupancy before tenancy. The code requirements for one occupancy group are not always the same as those for the new occupancy group. The new occupancy must be inspected to verify that all the applicable code requirements have been meet.

3408.3 Stairways. An existing stairway shall not be required to comply with the requirements of Section 1009 where the existing space and construction does not allow a reduction in pitch or slope.

❖ A stairway in an existing building does not have to be modified to comply with the dimensional provisions for width, treads and riser dimension or landing size for new stairway construction. This provision is for the design of a particular stairway and is aimed at addressing that existing stairway only. This section, which is essentially an exception, is not intended to eliminate any of the other technical provisions that might apply to an occupancy group, such as means of egress, exit capacity, etc.

For example, when a building is altered or there is a change of tenant that results in a change in occupancy, the existing stairway is permitted to remain as it was prior to the alteration or change of occupancy. The means of egress capacity, however, must comply with the requirements for the new occupancy. The capacity may not be lessened to match the existing stairway if it is not sufficient to provide the capacity required by the new occupancy. An additional stairway or another approved means of egress must be provided to comply with the means of egress requirements for the new occupancy as indicated in Chapter 10.

3408.4 Seismic. When a change of occupancy results in a structure being reclassified to a higher risk category, the structure shall conform to the seismic requirements for a new structure of the higher risk category.

Exceptions:

1. Specific seismic detailing requirements of Section 1613 for a new structure shall not be required to be met where the seismic performance is shown to be equivalent to that of a new structure. A demonstration of equivalence shall consider the regularity, overstrength, redundancy and ductility of the structure.
2. When a change of use results in a structure being reclassified from Risk Category I or II to Risk Category III and the structure is located where the seismic coefficient, $S_{DS,}$ is less than 0.33, compliance with the seismic requirements of Section 1613 are not required.

❖ An existing building that undergoes a change of use and occupancy that places the building in a higher risk category must be evaluated for its seismic resistance. For this provision, the code is referring to risk categories listed in Table 1604.5. It is important that an existing building that previously contained ordinary uses and occupancies must be able to meet current code requirements relative to seismic loading if it will be used as an "essential facility."

Without Exception 1, it would be impossible in many instances to make an existing structure comply with the seismic requirements for a new structure.

Exception 1 provides guidance for the building official and designer of areas that need to be investigated when compliance with the seismic requirements set forth in the code cannot be accomplished in a traditional manner.

Exception 2 has its origins in ASCE 7. The purpose is to permit a change from Risk Category I or II to Risk Category III for structures subjected to low earthquake accelerations without meeting the requirements for a new structure.

SECTION 3409 HISTORIC BUILDINGS

3409.1 Historic buildings. The provisions of this code relating to the construction, *repair*, *alteration*, *addition*, restoration and movement of structures, and change of occupancy shall not be mandatory for *historic buildings* where such buildings are judged by the *building official* to not constitute a distinct life safety hazard.

❖ This section provides an exception from code requirements when the building in question has historic value. The most important criterion for application of this section is that the building must be essentially accredited as being of historic significance by a qualified party or agency. Usually, this is done by a state or local authority after careful review of the historical value of the building. Most, if not all, states have such authorities, as do many local jurisdictions. The agencies with such authority can be located at the state or local government level or through the local chapter of the American Institute of Architects (AIA). Other considerations include the structural condition of the building (i.e., is the building structurally sound), its proposed use, its impact on life safety and how the intent of the code, if not the letter, will be achieved. Historically, registered structures have special allowances for accessibility when the historical significance is affected (see Section 3411.9).

3409.2 Flood hazard areas. Within *flood hazard areas* established in accordance with Section 1612.3, where the work proposed constitutes *substantial improvement* as defined in Section 1612.2, the building shall be brought into compliance with Section 1612.

Exception: *Historic buildings* that are:

1. *Listed* or preliminarily determined to be eligible for listing in the National Register of Historic Places;
2. Determined by the Secretary of the U.S. Department of Interior as contributing to the historical significance of a registered historic district or a district preliminarily determined to qualify as an historic district; or
3. Designated as historic under a state or local historic preservation program that is *approved* by the Department of Interior.

❖ With respect to provisions applicable to buildings in flood hazard areas, the discretion given in Section 3409.1 that allows the building official to waive the requirements for historic structures extends only to historic structures that meet the specific limitations in the exception to this section. Location in a historic district does not qualify a structure as a historic one. Care must be taken to ensure that work on such historic structures will not cause them to lose their continued listing or designation. If improvements or repairs cause a structure to lose its listing or designation as a historic building, then the building official must enforce the substantial improvement requirements. Owners of historic structures should consider measures to reduce flood damage to the extent practical. For additional guidance, see FEMA P-467-2, *Floodplain Management Bulletin on Historic Structures*, and FEMA P-758, *Substantial Improvement/ Substantial Damage Desk Reference*.

SECTION 3410 MOVED STRUCTURES

3410.1 Conformance. Structures moved into or within the jurisdiction shall comply with the provisions of this code for new structures.

❖ Moved structures generally are required to comply with the provisions applicable to new construction. The moved structure may comply with the alternative provisions of Section 3412 instead of the code requirements for new structures, which may be particularly useful if the moved structure is older than the effective date of the adoption of building codes within the jurisdiction. The fire separation distance of the moved structure must comply with requirements for new structures, even if the compliance alternative provisions of Section 3412 are used to meet the code requirements.

SECTION 3411 ACCESSIBILITY FOR EXISTING BUILDINGS

3411.1 Scope. The provisions of Sections 3411.1 through 3411.9 apply to maintenance, change of occupancy, *additions* and *alterations* to existing buildings, including those identified as *historic buildings*.

❖ The purpose of Section 3411 is to establish minimum criteria for accessibility when dealing with existing buildings and facilities. The history and efforts involved are similar to that discussed in Chapter 11 (see commentary, Chapter 11, "General Comments"). When a facility or element is altered (including alterations, changes of occupancy or additions), the portion being altered or added must meet new code requirements. For example, if a door and frame are removed and replaced, the door must meet the requirements for width, height, maneuvering clearances and hardware. If just the doorknob is being removed, it must be replaced with lever hardware.

If the area undergoing alteration does not contain a

primary function (see Sections 3411.6 and 3411.7), there are no additional requirements; however, if the area contains a primary function, there are additional criteria that may require work not in the original scope to achieve accessibility. This additional criteria is to provide an accessible route to the altered area, including any toilets and drinking fountains that serve it. Requirements for an accessible route might specify that the door previously discussed be removed and replaced because it did not have adequate width or maneuvering clearances.

The principle behind this approach to upgrading existing buildings is that they will become more accessible over time. A valid time to work towards that goal is when a structure is being altered. Special considerations are offered because of the difficulty involved in dealing with existing facilities that may not have been built with accessibility for physically disabled persons in mind. For example, when a historically registered home is being made into a museum, if changing the front door to allow for wheelchair access would alter the historical significance, alternatives are offered in Section 3411.9. Another example is the limitation offered in Section 3411.6 if it is technically infeasible to provide full accessibility in an existing building. Please note that the term "technically infeasible" refers to either movement of a major structural element or other physical constraints. For example, a ramp to provide entrance or exit from a particular door may not be possible because of property lines or setback constraints.

Appendix E addresses accessibility in existing buildings for items that are not typically enforceable through the normal building code enforcement approach. The items in Appendix E are found in the 2010 ADA Standard for Accessible Design (see the commentary for Appendix E for additional information).

Section 3411.1 addresses accessibility in existing buildings that are being renovated or altered. The exception for Type B dwelling and sleeping units in existing buildings being altered or undergoing a change of occupancy is for consistency with the Fair Housing Amendments Act (FHAA). Additions that contain four or more dwelling and sleeping units would be required to meet Type B dwelling unit requirements unless they meet one of the exceptions in Section 1107.7 (see Section 3411.5). Accessible and Type A dwelling and sleeping units are required in existing institutional and residential buildings undergoing additions, change of occupancy or alterations based on the number of units being altered or added (see Sections 3411.8.7 and 3411.8.8).

3411.2 Maintenance of facilities. A facility that is constructed or altered to be *accessible* shall be maintained *accessible* during occupancy.

❖ Continued compliance with the accessibility requirements of the code is dependent on maintenance of such facilities throughout the life of the building. For example, drinking fountains that are required to be accessible are of little value if they malfunction through deterioration or failure of any of the working parts. In other cases, inoperable elevators, locked accessible doors and obstructed accessible routes are not acceptable. The current level of accessibility must be maintained such that the route and elements are readily usable by individuals with disabilities.

3411.3 Extent of application. An *alteration* of an existing facility shall not impose a requirement for greater accessibility than that which would be required for new construction. Alterations shall not reduce or have the effect of reducing accessibility of a *facility* or portion of a *facility*.

❖ Alterations, including those that are part of additions or changes or occupancy, shall not reduce or have the effect of reducing accessibility of a building, portion of a building or facility. At the same time, existing building should not have to provide a higher level of accessibility than new construction. For example, an addition that included a new entrance might not have to make that entrance accessible if the structure as a whole already met the accessible entrance criteria for new construction in Section 1105.1.

The scoping requirements in Section 3411.8 allow for some additional allowance for dealing with existing structures.

3411.4 Change of occupancy. Existing buildings that undergo a change of group or occupancy shall comply with this section.

Exception: *Type B dwelling units* or *sleeping units* required by Section 1107 of this code are not required to be provided in existing buildings and facilities undergoing a change of occupancy in conjunction with *alterations* where the work area is 50 percent or less of the aggregate area of the building.

❖ When an entire building undergoes a change of occupancy, the building must comply with the provisions in Sections 3411.4.1 and 3411.4.2. If a portion of a building undergoes a change of occupancy, such as when there is a tenant change or a partial renovation where a space changes function, then the level of accessibility is addressed in Section 3411.4.1.

The exception notes when Type B dwelling and sleeping units would not need to be included when there is a change of occupancy to a Group I or R building (see Section 1107). When the change of occupancy does not involve extensive alterations (exceeding 50 percent of the square footage of the building), then Type B accessibility requirements would not need to be provided in the new or altered dwelling or sleeping units. This exception is similar to what is also found in the IEBC for change of occupancy where less than a Level III alteration is occurring. A Level III alteration is greater than 50 percent of the square footage of the building; therefore, this decreases the burden on many smaller projects. The

Type B dwelling units should be provided in larger alterations because where major alterations are being performed there is a prime opportunity to have those buildings move towards being able to serve a wider range of the population. With the population of the United States aging, there will be a steady increase in demand for units that include accessibility features. See the commentary for Sections 3411.4.2 (the exception), 3411.6 (Exception 4), 3411.8.7, 3411.8.8 and 3411.8.9 for additional discussion.

3411.4.1 Partial change in occupancy. Where a portion of the building is changed to a new occupancy classification, any *alterations* shall comply with Sections 3411.6, 3411.7 and 3411.8.

❖ When a building undergoes a partial change in occupancy, such as where there is a tenant change or a change in function of a specific area, then the level of accessibility must be maintained or improved at the same level as if that space was undergoing an alteration. Basically, the intent is that any spaces or elements being altered will meet new accessibility provisions unless technically infeasible (see Section 3411.6). If this area changing occupancy is a primary function of the space, an evaluation of the accessible route, as well as bathrooms and drinking fountains serving this space must be made. If these elements are not accessible, improvements must be made. However, there is a limit to the cost of the additional improvements to a maximum of 20 percent of the cost of the alteration (see Section 3411.7). The reference to Section 3411.8 provides for additional allowances because the designer/owner is still dealing with existing building constraints. For example, the accessible route could be provided by a platform lift (see Section 3411.8.3) where in new construction, this option is limited (see Sections 1007.5 and 1109.8).

3411.4.2 Complete change of occupancy. Where an entire building undergoes a change of occupancy, it shall comply with Section 3411.4.1 and shall have all of the following *accessible* features:

1. At least one *accessible* building entrance.
2. At least one *accessible* route from an *accessible* building entrance to *primary function* areas.
3. Signage complying with Section 1110.
4. *Accessible* parking, where parking is being provided.
5. At least one *accessible* passenger loading zone, when loading zones are provided.
6. At least one *accessible* route connecting *accessible* parking and *accessible* passenger loading zones to an *accessible* entrance.

Where it is *technically infeasible* to comply with the new construction standards for any of these requirements for a change of group or occupancy, the above items shall conform to the requirements to the maximum extent *technically feasible*.

Exception: The *accessible* features listed in Items 1 through 6 are not required for an *accessible* route to *Type B units*.

❖ This section establishes that when an existing building undergoes a complete change of occupancy, full compliance with the six accessible route requirements listed is expected, regardless of cost. That way, a person with mobility impairments would be able to arrive at the building (see Items 4 and 5), get to the accessible entrance (see Items 1, 3 and 6) and have at least one accessible route throughout the building to all the primary function areas (see Item 2). Changes between levels could be via a ramp (see Section 3411.8.5), an elevator (see Section 3411.8.2) or a platform lift (see Section 3411.8.3). Additionally, with the reference back to Section 3411.4.1, bathrooms and drinking fountains serving primary function areas would need to be evaluated. If they are not accessible, they must also be altered, but these elements could use the 20-percent cap on cost offered in Section 3411.7, Exception 1. If full compliance is technically infeasible, the element must be made accessible to the fullest extent that is feasible.

Typically, a building undergoing a complete change of occupancy is being at least partially gutted and undergoing alterations due to changes in function, possibly due to increased occupant load, means of egress requirements, sprinkler requirements and/or mechanical and plumbing changes. The intent is to create a balance between the change of occupancy meeting all new construction requirements for accessibility and the fact that the designer/owner is dealing with some existing building conditions.

This is not based on any specific provisions of the *Americans with Disabilities Act Accessibility Guidelines* (ADAAG), but parallels the intent of the requirements for removal of barriers.

If the change of occupancy results in Type B dwelling units, the exception permits the building to be exempt from providing the additional accessible route requirements listed in this section. This allowance addresses concerns of unpredictabilities in making changes due to existing site conditions. This also reinforces the intent that the inclusion of Type B units is not meant to require elevators when alterations are performed on upper floors in nonelevator buildings (see the exceptions for Section 1107.7). These areas would have been exempted if built new under the Fair Housing Act (FHA) and IBC guidelines, and should continue to be exempted.

3411.5 Additions. Provisions for new construction shall apply to additions. An *addition* that affects the accessibility to, or contains an area of, a *primary function* shall comply with the requirements in Section 3411.7.

❖ Additions must comply with new construction. An addition, however, is also an alteration to an existing building; therefore, accessible route provisions for existing buildings are applicable (see commentary, Section 3411.7). For example, a new dining area is added to a restaurant. All accessible elements within the parameter of the addition must be constructed accessible. If the route to the addition, or the bathrooms that serve the addition are in the existing building, they must be evaluated to see if they need to be altered. The 20-percent maximum cost in Section 3411.7, Exception 1, is applicable.

3411.6 Alterations. A *facility* that is altered shall comply with the applicable provisions in Chapter 11 of this code, unless *technically infeasible*. Where compliance with this section is *technically infeasible*, the *alteration* shall provide access to the maximum extent technically feasible.

Exceptions:

1. The altered element or space is not required to be on an *accessible* route, unless required by Section 3411.7.
2. *Accessible means of egress* required by Chapter 10 are not required to be provided in existing facilities.
3. The *alteration* to Type A individually owned dwelling units within a Group R-2 occupancy shall be permitted to meet the provision for a *Type B dwelling unit*.
4. *Type B dwelling* or *sleeping units* required by Section 1107 of this code are not required to be provided in existing buildings and facilities undergoing a change of occupancy in conjunction with alterations where the work area is 50 percent or less of the aggregate area of the building.

❖ The code approaches application of accessibility provisions to a facility or portion thereof that is altered by broadly requiring full conformance to new construction for altered elements, meaning full accessibility is expected. Exceptions are then provided to indicate the conditions under which less than full accessibility is permitted.

The circumstance under which full compliance with accessibility provisions is not required is when it is deemed to be technically infeasible (see the commentary for the definition of "Technically infeasible" in Section 202). This is considered reasonable since, if not provided for, plans for alterations may be otherwise abandoned by the building owner. The opportunity to upgrade and increase the current level of accessibility in an existing building would then be lost. This concern is also embodied in the requirement that an altered element or space is expected to be made accessible to the extent to which it is technically feasible to do so. In this manner, the code accomplishes the greatest degree of accessibility while recognizing the justifiable difficulties that may be involved in providing full accessibility. Alterations are not required to exceed new construction requirements (see Section 3411.3).

The availability and usability of accessible elements are critically dependent on the presence of an accessible route leading to the accessible elements. The requirement for an accessible route represents one of the potential difficulties in an existing building that does not currently have adequate accessible routes. For example, in a multistory building, the accessible route to upper floors will most often be provided in the form of an elevator. It may be technically infeasible to provide an elevator in an existing building that does not currently contain an elevator. For this reason, Exception 1 indicates that an altered element or space is not required to be on an accessible route, except to the extent required in Section 3411.7.

Exception 2 indicates that accessible means of egress are not required as a result of undertaking alterations to existing buildings. Strict compliance with Section 1007 is often technically infeasible. Note that this is not an exception for accessible entrance requirements. Note that the section for change of occupancy (Section 3411.4.1) references Section 3411.6; therefore, accessible means of egress is not required in buildings undergoing a change of occupancy. However, the need for assisted rescue must be addressed in the fire and safety evacuation plans required by the IFC.

Exception 3 addresses the specific circumstances when an existing Type A dwelling unit is being altered. While Section 3411.2 states that a level of accessibility must be maintained, in the situations where a Type A dwelling unit is "individually owned," such as a condominium, then it only needs to meet the technical requirements for a Type B dwelling unit (see ICC A117.1, Section 1004) when it is altered or remodeled by the owner. Type B units require a lesser level of accessibility than Type A units.

Exception 4 allows some partial building alterations in Group I and R structures to occur without providing Type B dwelling units and sleeping units (see Section 1007). When the alteration does not exceeding 50 percent of the square footage of the building, then Type B accessibility requirements would not need to be provided in the dwelling or sleeping units being altered (see Sections 3411.8.7 and 3411.8.8 for alterations where Accessible units and Type A units are required). This exception is similar to what is also found in the IEBC for alterations where less than a Level-III alteration is occurring. A Level-III alteration is greater than 50 percent of the square footage of the building. This decreases the burden on many smaller projects.

The Type B dwelling units must be provided in larger alterations. Where major alterations are being performed there is a prime opportunity to have those buildings move towards being able to serve a wider range of the population. With the population of the

United States aging, there will be a steady increase in demand for units that include accessibility features. In addition, while the FHA is only applicable to new construction, this law was enacted in 1991. Some buildings built after that time may not be in compliance with the FHA. When a major alteration/renovation is occurring, there is an opportunity to bring those buildings into compliance. See the commentary under Sections 3411.4.2 (the exception), 3411.8.7, 3411.8.8 and 3411.8.9 for additional discussion.

3411.7 Alterations affecting an area containing a primary function. Where an alteration affects the accessibility to, or contains an area of *primary function*, the route to the primary function area shall be *accessible*. The accessible route to the *primary function* area shall include toilet facilities or drinking fountains serving the area of primary function.

Exceptions:

1. The costs of providing the *accessible* route are not required to exceed 20 percent of the costs of the *alterations* affecting the area of *primary function*.
2. This provision does not apply to *alterations* limited solely to windows, hardware, operating controls, electrical outlets and signs.
3. This provision does not apply to *alterations* limited solely to mechanical systems, electrical systems, installation or alteration of fire protection systems and abatement of hazardous materials.
4. This provision does not apply to *alterations* undertaken for the primary purpose of increasing the accessibility of a *facility*.
5. This provision does not apply to altered areas limited to *Type B dwelling* and *sleeping units*.

❖ An area containing a primary function is one in which a major activity for which the building or facility is intended is carried out. For example, the lobby of a hotel in which the registration and check-out desk is located would be a primary function area. Other examples would be the dining area of a restaurant, the meeting rooms or exhibition halls in a conference center, virtually all office and work areas in a business building and retail display areas in a mercantile occupancy. The key concept is that a primary function area is one that contains a major activity of the facility. Areas that contain activities not related to the main purpose of the facility would not be considered a primary function area. For example, a mechanical equipment room, storage closet, toilet facilities, corridors, lounges and locker rooms would not be considered primary function areas. With this background, it is clear that areas containing a primary function are clearly more critical in terms of the purpose for which people enter and use the facility; therefore, this section reflects that when such areas are altered or added, it is important to require that an accessible route to the primary function area be provided. When an accessible route to a primary function area is required by this section, an accessible route to such facilities, including any restrooms and drinking fountains serving the primary function area, must also be made accessible, even though such facilities and areas may not by themselves be considered primary function areas.

There are conditions under which it may not be reasonable to enforce strictly this requirement for an accessible route to an altered or added primary function area. Exception 1 approaches this by utilizing the cost of the alterations or addition as a basis for determining if providing an accessible route is reasonable. The requirement for a complete accessible route does not apply when the cost of providing it exceeds 20 percent of the cost of the alterations or addition to the primary function area. These costs are intended to be based on the actual costs of the planned alterations or addition to the primary function area before consideration of the cost of providing an accessible route. For example, if the planned alterations will cost $100,000, not including the cost of an accessible route to a primary function area, this exception would apply if the additional cost of providing the accessible route would exceed $20,000.

It is not the intent to exempt all requirements for accessibility when the total cost for providing the accessible route exceeds the 20-percent threshold. Improvements to the accessible route are required to the extent that costs do not exceed 20 percent of the cost to the planned alteration or addition. It is not required that the full 20 percent be spent. If the accessible route (including accessible bathrooms and drinking fountains) is already provided, no additional expenditure is required. Note that there is not a priority list given for where money should be spent on improving the accessible route. The logical progression for entering and using a facility is access to the site, accessible exterior routes to accessible entrances, access throughout the facility, access to services within the facility, toilet and bathing rooms and, finally, drinking fountains. However, evaluation on how and where the money available should best be spent must be made on a case-by-case basis. For example, if an accessible route is not available to an upper level, and the cost of an elevator is more than 20 percent of the cost of the renovation, then other alternatives could be investigated, such as a platform lift or limited access elevator, or adding the elevator pit and shaft at this time with elevator equipment added later. If all such items are in excess of the 20-percent limit, perhaps the money available could be spent towards making the toilet rooms accessible. The main idea is that existing buildings would become fully accessible over time.

Exceptions 2 and 3 identify certain alterations that are not intended to trigger the requirement for providing an accessible route to a primary function area. Alterations limited to such elements as windows, hardware, operating controls, electrical outlets, signage, mechanical, electrical and fire protection systems, including alterations for the purpose of abating a hazardous materials circumstance, do not affect the usability of a primary function area in the same manner as alterations that affect the floor plan or the configuration, location or size of rooms or spaces. It is, therefore, considered unreasonable to require the installation of an accessible route when the scope of alterations is limited to that reflected in these exceptions. Note that the costs for these items are not "backed out" of the total cost for the alteration before applying Exception 1. Exceptions 2 and 3 are alterations limited to the specific items referenced.

Exception 4 is intended to avoid penalizing a building owner who is undertaking alterations or additions for the purpose of increasing accessibility. It is appropriate to encourage owners to make such alterations without requiring them to do more work simply because they chose to increase the accessibility of the space. This could otherwise have the opposite effect of discouraging such alterations to avoid the expense of undertaking more work and expense than was originally planned. For example, federal law [Americans with Disabilities Act (ADA)] requires that owners of existing buildings remove certain existing barriers to accessibility. Removal of such barriers may require a permit from the building official. It would be unreasonable to have such activity trigger the mandatory requirement for further alterations to accomplish accessibility beyond the original planned work. In principle, the code takes the view that some extent of greater accessibility is positive progress and should be encouraged, not penalized.

Where the alterations result in Type B dwelling units being provided where none previously existed (see Section 3411.6, Exception 4), Exception 5 would not require any additional money to be spent towards providing an accessible route to those Type B dwelling or sleeping units. This is intended to encourage the creation of such units without penalizing building owners. Similar to Section 3411.4.2, this exception is intended to address the concerns of site impracticality for providing an accessible route to and into existing buildings that contain Type B dwelling units. This also reinforces the intent that the inclusion of Type B units is not meant to require elevators when alterations are performed on upper floors in nonelevator buildings (see the exceptions in Section 1107.7). These areas would have been exempted if built new under FHA and IBC guidelines, and should continue to be exempted.

3411.8 Scoping for alterations. The provisions of Sections 3411.8.1 through 3411.8.14 shall apply to *alterations* to existing buildings and facilities.

❖ The specific provisions of this section are intended to reflect conditions under which less than full accessibility, as would be required in new construction, is permitted in altered areas. As previously discussed, Section 3411.6 requires altered areas to comply with the full range of accessibility-related provisions of the code for new construction. This section reflects a reasonable set of conditions under which a different level of accessibility can be provided. Sections 3411.8.1 through 3411.8.14 are part of the code's coordination effort ICC A117.1 and the recommendations of the ADAAG Review Federal Advisory Committee.

3411.8.1 Entrances. *Accessible* entrances shall be provided in accordance with Section 1105.

> **Exception:** Where an *alteration* includes alterations to an entrance, and the *facility* has an *accessible* entrance, the altered entrance is not required to be *accessible*, unless required by Section 3411.7. Signs complying with Section 1110 shall be provided.

❖ This provision is contained here to point to the accessibility provisions of Chapter 11 for entrances (see commentary, Section 1105) that would require 60 percent of the doors that serve as entrances to the building to be accessible. The exception allows for nonaccessible entrances to remain as such unless the entrance serves as the accessible route to the primary function space. When the building has both accessible and nonaccessible entrances, appropriate signage must be provided so that people using mobility aids know where they can enter.

3411.8.2 Elevators. Altered elements of existing elevators shall comply with ASME A17.1 and ICC A117.1. Such elements shall also be altered in elevators programmed to respond to the same hall call control as the altered elevator.

❖ Requirements for new construction state that all elevators on an accessible route must be fully accessible in accordance with ICC A117.1. When dealing with existing elevators, the nature of the shaft and cab may result in full compliance with ICC A117.1 to be technically infeasible. However, there are elements in the elevator car and lobby that benefit people with vision impairments and who use mobility devices; therefore, the elevators should be made accessible as much as technically feasible (see Section 3411.6). If a passenger elevator is altered, the altered element must be accessible in accordance with the requirements for existing elevators in Section 407 of ICC A117.1, as well as meet the elevator safety code, ASME A17.1. If the altered elevator is part of a bank of elevators, the same element must be made accessible in every elevator that is part of that bank. For example, if new car control panels are installed, Section 407.4.6 of ICC A117.1 includes requirements for the height, configuration and raised and Braille symbols on the control panel. If the control panel is changed in one elevator, then it should be changed in all elevators in the bank. The purpose of this requirement is to have consistency among elevators in a bank so that people with disabilities are not required to wait for a specific elevator when the general population can take the first available elevator. The ICC A117.1 also provides accessibility requirements for limited use/limited access (LULA) elevators (ICC A117.1, Section 408) and private residence elevators (ICC A117.1, Section 409). While these types of elevators can serve as part of an accessible route, where they can be used is limited by ASME A17.1.

3411.8.3 Platform lifts. Platform (wheelchair) lifts complying with ICC A117.1 and installed in accordance with ASME A18.1 shall be permitted as a component of an *accessible* route.

❖ This section provides for the use of platform (wheelchair) lifts in existing buildings. In order to create an accessible route where there are changes in floor levels, the provisions for new construction would most often require the installation of an elevator or ramp. Platform lifts are allowed in new construction for limited conditions (see Section 1109.8). If the space, construction or site constraints in an existing building precludes the installation of an elevator or ramp, a platform lift may be the only practical solution. Given the choice between no accessibility or accessibility by a platform lift, accessibility is preferable. Note that in accordance with Section 1007.5, platform lifts are also permitted for an accessible means of egress; however, accessible means of egress are not required in existing buildings undergoing alterations in accordance with Section 3411.6, Exception 2.

A chair lift is not acceptable as part of an accessible route in either new or existing construction. For examples of the difference between a chair lift and a platform lift, see the commentary to Section 1109.8.

3411.8.4 Stairs and escalators in existing buildings. In *alterations*, change of occupancy or *additions* where an escalator or *stair* is added where none existed previously and major structural modifications are necessary for installation, an *accessible* route shall be provided between the levels served by the escalator or *stairs* in accordance with Sections 1104.4 and 1104.5.

❖ If a stairway or escalator is added as part of an alteration in a location where one did not previously exist, the alteration must also include an accessible route between the same two levels. If an accessible route is already available between the two levels, or if the stairway or escalator is replacing an existing stairway or escalator, this requirement is not applicable. In conjunction with Section 3411.3, if the requirement for the accessible route would be in excess of what is required for new construction, such as an accessible route to an area that was exempted by Section 1103.2, 1104.4, 1107 or 1108, this requirement is not applicable. The intent is that if a route is provided between accessible levels for a nondisabled person to use, it is reasonable to also expect an accessible route.

For comments on when existing stairways are being altered, see Section 3404.1.

3411.8.5 Ramps. Where slopes steeper than allowed by Section 1010.2 are necessitated by space limitations, the slope of ramps in or providing access to existing facilities shall comply with Table 3411.8.5.

❖ This section recognizes the circumstances where because of existing site or configuration constraints, a ramp with a slope of one unit vertical in 12 units horizontal (1:12) may not be feasible. A steeper slope is allowed where the elevation change does not exceed 6 inches (152 mm). The remainder of ramp requirements, such as width, landings, etc., are set forth in Section 1010.

TABLE 3411.8.5
RAMPS

SLOPE	MAXIMUM RISE
Steeper than 1:10 but not steeper than 1:8	3 inches
Steeper than 1:12 but not steeper than 1:10	6 inches

For SI: 1 inch = 25.4 mm.

❖ In existing buildings, ramps that rise 3 inches (76 mm) or less may have a slope as steep as one unit vertical in eight units horizontal (1:8). In existing buildings, ramps that rise 6 inches (152 mm) or less may have a slope as steep as one unit vertical in 10 units horizontal (1:10). If it is possible to provide a lesser slope, it is desirable to do so. These steeper slopes should only be utilized when the one unit vertical in 12 units horizontal (1:12) slope is not possible.

3411.8.6 Performance areas. Where it is *technically infeasible* to alter performance areas to be on an *accessible* route, at least one of each type of performance area shall be made *accessible*.

❖ This section recognizes that, because of the existing arrangement and location of performing areas (e.g., stages, platforms, orchestra pits, etc.), it may be infeasible to alter all performing areas to be on an accessible route. In such cases, it is reasonable to require that a minimum of one of each type of performing area be made accessible.

3411.8.7 Accessible dwelling or sleeping units. Where Group I-1, I-2, I-3, R-1, R-2 or R-4 *dwelling* or *sleeping units* are being altered or added, the requirements of Section 1107 for *Accessible* units apply only to the quantity of spaces being altered or added.

❖ In new construction, the number of Accessible units is required in many types of facilities. The number of Accessible dwelling units and sleeping units provided determines the number of Accessible units, and starts when only one unit is provided (see Table 1107.6.1.1). The number of Accessible units required increased based on the anticipated need (see Section 1107).

This Section sets forth the rate for providing Accessible dwelling or sleeping units in Groups I-1, I-2, I-3, R-1, R-2 and R-4 when such facilities are altered or have units added (i.e., additions, change of use, change of occupancy). Assuming that Accessible units are not already provided, the number of Accessible units to be incorporated into each alteration is based on the number being altered. For example, if a nursing home was being altered a portion at a time, 50 percent of the units being altered each time would be required to be constructed fully wheelchair accessible (i.e., Accessible units). It is not the intent that all units being altered are required to be Accessible units. The total number of Accessible units in the facility is not required to exceed that required for new construction, as indicated in Section 3411.3. It is unreasonable to require a greater level of accessibility in an existing building than is required in new construction.

3411.8.8 Type A dwelling or sleeping units. Where more than 20 Group R-2 *dwelling* or *sleeping units* are being altered or added, the requirements of Section 1107 for *Type A units* apply only to the quantity of the spaces being altered or added.

❖ Type A units are required in new construction when 20 or more apartments (including condominium styles) are constructed on a site. Group R-2 requirements for Type A units would also include convents and monasteries and could include some townhouse-style units (see the commentary for Section 1107.6.2.1.2 for additional information).

Note that this requirement is applicable when 20 or more units are included in the construction project, not when an individual owner performs alterations to his or her condo when there are 20 or more units in the building. This section sets forth the rate for providing Type A dwelling or sleeping units in Group R-2 facilities in an addition; where there is a change to the function of a space (i.e., a change of occupancy, such as creating apartments in an old warehouse, increasing the number of apartments by changing a storage area to apartment units or reconfiguring larger units into smaller units) or an alteration. Assuming that Type A units are not already provided, the number of Type A units required is based on the number being added or altered. For example, if a story was being added onto an apartment building or a story in an apartment building was being altered, the number of Type A units required would be based on the number of units in the new story or the area being altered, not the number of units in the entire building. If Type A units are provided, the total number of Type A units in the facility is not required to exceed that required for new construction, as indicated in Section 3411.3. It is unreasonable to require a greater level of accessibility in an existing building than is required in new construction.

3411.8.9 Type B dwelling or sleeping units. Where four or more Group I-1, I-2, R-1, R-2, R-3 or R-4 *dwelling* or *sleeping units* are being added, the requirements of Section 1107 for *Type B units* apply only to the quantity of the spaces being added. Where Group I-1, I-2, R-1, R-2, R-3 or R-4 *dwelling* or *sleeping units* are being altered and where the work area is greater than 50 percent of the aggregate area of the building, the requirements of Section 1107 for *Type B units* apply only to the quantity of the spaces being altered.

❖ In new construction, Type B units are required when four or more dwelling units or sleeping units are constructed together and those units are "intended to be occupied as a residence" (see commentary, Section 1107). The IRC references this code's Group R-3 accessibility requirements for IRC units (see the definition for "Townhouse" and Section R320.1).

This section sets forth the rate for providing Type B dwelling or sleeping units in Group I-1, I-2, R-2, R-3 or R-4 facilities. Type B units are required when more than four or more units are added in an addition or when major alterations are occurring that affect more the 50 percent of the area of the building. The Type B units in the facility are not required to exceed that required for new construction, as indicated in Sections 1107.7 and 3411.3. Sections 3411.4.2 and 3411.7 also include exceptions for the additional accessible route requirements in structures where Type B units are added as part of an alteration or change of occupancy. It is unreasonable to require a greater level of accessibility in an existing building than is required in new construction.

3411.8.10 Jury boxes and witness stands. In *alterations*, *accessible* wheelchair spaces are not required to be located within the defined area of raised jury boxes or witness stands and shall be permitted to be located outside these spaces where the ramp or lift access restricts or projects into the *means of egress*.

❖ This exception for jury boxes and witness stands is consistent with Section 231 and 808 of the 2010 ADA *Standards for Accessible Design*. The intent is that if ramp access to a jury box or witness stand would have the ramp limiting or blocking the means of egress for the general population in the space, alternative locations for potential jurors or witnesses are viable.

3411.8.11 Toilet rooms. Where it is *technically infeasible* to alter existing toilet and bathing rooms to be *accessible*, an *accessible* family or assisted-use toilet or bathing room constructed in accordance with Section 1109.2.1 is permitted. The family or assisted-use toilet or bathing room shall be located on the same floor and in the same area as the existing toilet or bathing rooms.

❖ This section deals with circumstances in which it is technically infeasible to alter existing toilet facilities to be accessible. In new construction, both the men's and women's facilities would be required to be accessible. An alternative solution when it is technically infeasible to alter the existing toilet rooms would be the creation of a single unisex toilet or bathing room containing accessible facilities. The requirements for family or assisted-use toilet facilities in Sections 1109.2.1 through 1109.2.1.7 provide guidance on what is required within this toilet room. If this alternative is selected, the room must be located on the same floor and in the same area as the existing toilet or bathroom. This is the best alternative to fully complying, separate men's and women's facilities. One might argue that it is technically infeasible to accomplish either of these alternatives, since the alternative to altering existing facilities involves creation of an additional toilet or bathroom that was not otherwise contemplated. This would not be a persuasive argument, since there is likely to be space available somewhere in the facility to commit for use as a toilet or bathroom. In any case, the intent is that some form of accessible toilet room or bathing facility is necessary and must be provided. Signage must be provided at the inaccessible toilet rooms in accordance with Sections 1110.1 and 1110.2 to notify disabled persons when a facility is not accessible and direct them to the nearest accessible facilities. It should be noted that this alternative is not offered as a choice between making the existing separate-sex toilet rooms accessible or providing a family or assisted-use accessible toilet room. The existing separate-sex toilet rooms must be altered when it is technically feasible. Consideration of this alternative is only available when altering the existing toilet rooms is technically infeasible (see the definition for "Technically infeasible" in Section 202).

3411.8.12 Dressing, fitting and locker rooms. Where it is *technically infeasible* to provide *accessible* dressing, fitting or locker rooms at the same location as similar types of rooms, one *accessible* room on the same level shall be provided. Where separate-sex facilities are provided, *accessible* rooms for each sex shall be provided. Separate-sex facilities are not required where only unisex rooms are provided.

❖ This section takes a similar approach for dressing rooms as provided for in Section 3411.8.11 for toilet rooms and bathing facilities. If it is technically infeasible to alter existing dressing rooms to be accessible, then space elsewhere on the level must be committed to providing no less than one accessible dressing room. In this case, if the existing dressing rooms provide separate rooms for each sex, then no less than one accessible dressing room for each sex must be provided. See the commentary in Section 1109.12.1 for additional information on how to make a dressing room accessible.

3411.8.13 Fuel dispensers. Operable parts of replacement fuel dispensers shall be permitted to be 54 inches (1370 mm) maximum measured from the surface of the vehicular way where fuel dispensers are installed on existing curbs.

❖ The requirements for new fuel dispensers (i.e., gas pumps) can be found in Section 1109.14. Basically, the idea is that the controls must be within the reach ranges for someone standing on the parking lot surface. However, many existing facilities have gas pumps located on raised islands as a feature for protection of the pumps from accidental contact. This section would allow the new gas pump with the reach range of 15 inches (380 mm) to 48 inches (1220 mm) to be located on top of a 6-inch (150 mm) curb and still meet the maximum reach of 54 inches (1370 mm).

3411.8.14 Thresholds. The maximum height of thresholds at doorways shall be $^3/_4$ inch (19.1 mm). Such thresholds shall have beveled edges on each side.

❖ Thresholds at doorways may be $^3/_4$-inch (19.1 mm) maximum in existing buildings. In new construction, a typical threshold is $^1/_2$ inch (13 mm) maximum in accordance with Section 1008.1.7. This section recognizes that such things as differences in floor materials may create changes in elevation greater than that allowed in new construction. Edges of thresholds must be beveled to allow passage of a wheelchair.

3411.9 Historic buildings. These provisions shall apply to facilities designated as historic structures that undergo *alterations* or a change of occupancy, unless *technically infeasible*. Where compliance with the requirements for *accessible routes*, entrances or toilet rooms would threaten or destroy the historic significance of the facility, as determined by the applicable governing authority, the alternative requirements of Sections 3411.9.1 through 3411.9.4 for that element shall be permitted.

Exception: *Type B dwelling* or *sleeping units* required by Section 1107 are not required to be provided in historical buildings.

❖ For this section to be applicable, the building must be registered as historic. Historic buildings are treated much the same as provided for in Sections 3411.4 and 3411.6, in that a historic building that is altered or has undergone a change of occupancy is expected to comply with accessibility requirements, unless technical infeasibility can be demonstrated; however, this section also goes on to acknowledge that the historic character of a building may be adversely affected by strict compliance with accessibility provisions. For example, compliance with door width requirements may necessitate the removal of an existing set of doors that is critical to the historic character of the building. This section is intended to exempt such conditions in order to maintain the historic character of the building. Because limited extent of accessibility is desired in all facilities, Sections 3411.9.1 through 3411.9.4 allow for alternatives.

In an effort to coordinate with 2010 ADA *Standards for Accessible Design* requirements, items that cannot be enforced through the typical code enforcement process, but are associated with historical buildings, have been included in Appendix B of the IEBC.

The exception specifically exempts the requirement for Type B dwelling or sleeping units, including when the alterations exceed 50 percent of the building. Historical buildings will not have to prove that the historic significance was in jeopardy in order to exempt Type B units.

3411.9.1 Site arrival points. At least one *accessible* route from a site arrival point to an *accessible* entrance shall be provided.

❖ Full compliance would require an accessible route from all site arrival points. If this requirement would adversely affect the historical significance of the building, the alternative available is to provide an accessible route from one site arrival point to an accessible entrance.

3411.9.2 Multilevel buildings and facilities. An *accessible* route from an *accessible* entrance to public spaces on the level of the *accessible* entrance shall be provided.

❖ Full compliance might require an accessible route to levels above or below, as well as throughout the entrance level. If this requirement would adversely affect the historical significance of the building, the alternative is to provide an accessible route from the accessible entrance to all spaces open to the public on the entrance level. This requirement offers the opportunity to provide equivalent services under the ADA. For example, a historic home that has been made into a museum may offer a video of the other part of the museum, or an office building may offer a means to communicate with other levels or have a meeting room on the accessible level for the offices on nonaccessible levels. If elevators are provided, but are not accessible, signage in accordance with Section 1110.2 is required.

3411.9.3 Entrances. At least one main entrance shall be *accessible*.

Exceptions:

1. If a main entrance cannot be made *accessible*, an *accessible* nonpublic entrance that is unlocked while the building is occupied shall be provided; or
2. If a main entrance cannot be made *accessible*, a locked *accessible* entrance with a notification system or remote monitoring shall be provided.

Signs complying with Section 1110 shall be provided at the primary entrance and the *accessible* entrance.

❖ Full compliance would require 60 percent of the public entrances, as well as any special entrances to be accessible (see Section 1105). If this requirement would adversely affect the historical significance of the building, only one main entrance is required to be made accessible. If a main entrance cannot be made accessible, then an employee or service entrance may serve as the accessible entrance, provided that it remains unlocked when the building is open. Alternatively, a locked entrance, where monitoring and notification system is available, could be provided. Signage must be provided at inaccessible entrances in accordance with Sections 1110.1 and 1110.2.

3411.9.4 Toilet and bathing facilities. Where toilet rooms are provided, at least one *accessible* family or assisted-use toilet room complying with Section 1109.2.1 shall be provided.

❖ Full compliance would require an accessible toilet/bathing facility at each location where toilet/bathing facilities are provided. If altering the existing facilities to be accessible would adversely affect the historical significance of the building, only one accessible toilet/bathing facility is required. The provisions for family or assisted use toilet rooms in Section 1109.2.1 will address what facilities would be required in that toilet room. Signage must be provided at inaccessible toilet rooms in accordance with Section 1110.2.

SECTION 3412
COMPLIANCE ALTERNATIVES

3412.1 Compliance. The provisions of this section are intended to maintain or increase the current degree of public safety, health and general welfare in existing buildings while permitting repair, *alteration*, *addition* and change of occupancy without requiring full compliance with Chapters 2 through 33, or Sections 3401.3, and 3403 through 3409, except where compliance with other provisions of this code is specifically required in this section.

❖ Neither the designer nor the building official can physically inspect and evaluate every aspect of an existing building or structure, since many of its features may be concealed within the construction. Therefore, it is necessary to emphasize those items that can be evaluated. There are 19 critically important elements that can be quantified and evaluated to determine the level of safety for an existing building.

This type of analysis provides the designer and the building official with a rational basis for establishing the safety of an existing building or structure without having physical access to every part of the building, documentation of the original design or the construction history of a building.

3412.2 Applicability. Structures existing prior to [DATE TO BE INSERTED BY THE JURISDICTION. NOTE: IT IS RECOMMENDED THAT THIS DATE COINCIDE WITH THE EFFECTIVE DATE OF BUILDING CODES WITHIN THE JURISDICTION], in which there is work involving *additions*, *alterations* or changes of occupancy shall be made to comply with the requirements of this section or the provisions of Sections 3403 through 3409. The provisions in Sections 3412.2.1 through 3412.2.5 shall apply to existing occupancies that will continue to be, or are proposed to be, in Groups A, B, E, F, M, R, S and U. These provisions shall not apply to buildings with occupancies in Group H or I.

❖ The adopting jurisdiction is to insert the desired date as indicated in this section. Section 3412 applies only to structures existing prior to the established date. The date that construction was first regulated through a comprehensive building code in the jurisdiction is recommended because buildings predating any building regulation are often not equipped with the types of systems and features that modern codes require. Newer buildings are more likely to be in compliance with current codes. These older buildings are assumed to face more difficulty in achieving a minimum level of life safety and are more likely to need the greater flexibility provisions of Section 3412.

The occupancies that qualify for the provisions of Section 3412 are listed in Section 3412.2.

Section 3412 does not apply to Groups H (high hazard) and I (institutional).

3412.2.1 Change in occupancy. Where an existing building is changed to a new occupancy classification and this section is applicable, the provisions of this section for the new occupancy shall be used to determine compliance with this code.

❖ When a building undergoes a change of occupancy classification and Section 3412 is applied, the evaluation method in Section 3412 must be applied to the new occupancy for determining whether the existing building meets the compliance alternative in the code.

3412.2.2 Partial change in occupancy. Where a portion of the building is changed to a new occupancy classification, and that portion is separated from the remainder of the building with *fire barriers* or *horizontal assemblies* having a *fire-resistance rating* as required by Table 508.4 for the separate occupancies, or with *approved* compliance alternatives, the portion changed shall be made to comply with the provisions of this section.

Where a portion of the building is changed to a new occupancy classification, and that portion is not separated from the remainder of the building with *fire barriers* or *horizontal assemblies* having a *fire-resistance rating* as required by Table 508.4 for the separate occupancies, or with *approved* compliance alternatives, the provisions of this section which apply to each occupancy shall apply to the entire building. Where there are conflicting provisions, those requirements which secure the greater public safety shall apply to the entire building or structure.

❖ Where a portion of the building is changed to a new occupancy classification and is designed in accordance with Section 3412, the following options may be employed, dependent upon the fire separation of the portion of the building from the remainder of the existing building.

For a full fire separation, the new occupancy portion must be evaluated with the existing or proposed building design to be in full compliance with the provisions of Section 3412. The remainder of the existing building must also be evaluated in accordance with Section 3412. The mandatory safety scores for the new occupancy portion of the building and the existing occupancy are obtained from those listed in Table 3412.8 and are incorporated in the building's final evaluation score (see Table 3412.9).

For a nonfire-resistance-rated assembly of a new occupancy portion of an existing building, the provisions of Section 3412 for each occupancy must apply to the entire building. The requirements offering the greater public safety are applied to the entire building. Any proposed or existing building attributes and modifications must be reviewed in light of these requirements. See the examples described in Sections 3412.6.16 and 3412.9.1 for mixed occupancies.

3412.2.3 Additions. *Additions* to existing buildings shall comply with the requirements of this code for new construction. The combined height and area of the existing building and the new *addition* shall not exceed the height and area allowed by Chapter 5. Where a *fire wall* that complies with Section 706 is provided between the *addition* and the existing building, the *addition* shall be considered a separate building.

❖ The requirements in this section are applied in the same way as those in Section 3412.1. This section is included in Section 3412 so that provisions for additions are included for both optional methods of code compliance indicated in Section 3401.1. For a discussion of these options, see the commentary to Section 3401.1.

The evaluation method in Section 3412 may be used for an existing building that has been modified by an addition, provided that it meets the applicability requirements of Section 3412.2; however, the addition must comply with all the requirements of the code for new construction.

3412.2.4 Alterations and repairs. An existing building or portion thereof, which does not comply with the requirements of this code for new construction, shall not be altered or repaired in such a manner that results in the building being less safe or sanitary than such building is currently. If, in the *alteration* or repair, the current level of safety or sanitation is to be reduced, the portion altered or repaired shall conform to the requirements of Chapters 2 through 12 and Chapters 14 through 33.

❖ An existing building that is altered or repaired may be designed and evaluated in accordance with Section 3412, provided it meets the applicability provisions of Section 3412.2.

When an existing building is altered or repaired, materials or methods consistent with the original construction must be used. The alteration or repair must not cause the building to be less safe or sanitary than when it was originally constructed. Should the alteration or repair cause a reduction in safety or sanitation, such reductions must be corrected to meet the requirements of Chapters 2 through 12 and Chapters 14 through 33.

3412.2.4.1 Flood hazard areas. For existing buildings located in *flood hazard areas* established in Section 1612.3, if the *alterations* and *repairs* constitute *substantial improvement* of the existing building, the existing building shall be brought into compliance with the requirements for new construction for flood design.

❖ See the commentary to Sections 3404.2 (alterations) and 3505.5 (repairs).

3412.2.5 Accessibility requirements. All portions of the buildings proposed for change of occupancy shall conform to the accessibility provisions of Section 3411.

❖ Any building or part of a building that has a change of occupancy must meet the requirements for accessibility in existing buildings found in Section 3411. This basically requires full accessibility in the portion of the building undergoing a change of occupancy, and in addition, may require some work improving the accessible route to that space and the bathrooms and drinking fountains that serve that space (see the commentary, Section 3411). By the references to new construction and Chapter 11 in Section 3412.2.3 for additions and Section 3412.2.4 for alterations, portions of the building being altered or added would comply with the provisions in Sections 3411.5, 3411.6 and 3411.7 as applicable by the reference in Section 1103.2.2.

3412.3 Acceptance. For *repairs*, *alterations*, *additions* and changes of occupancy to existing buildings that are evaluated in accordance with this section, compliance with this section shall be accepted by the *building official*.

❖ When an owner or designer of an existing building decides to apply Section 3412 and complies with all the provisions of the section, including the applicability requirements in Section 3412.2, the building official must accept for review the proposed work or change of occupancy repairs, alterations and additions. This mandatory acceptance is tempered by one exception, as stated in Section 3412.3.1.

3412.3.1 Hazards. Where the *building official* determines that an unsafe condition exists, as provided for in Section 116, such unsafe condition shall be abated in accordance with Section 116.

❖ When the building official finds an unsafe condition in the building that is not being corrected by the proposed work, he or she must order the abatement or correction of the unsafe condition or hazard just as would be ordered in an existing building that is not being renovated, as stipulated by Section 116. This section sets forth a required comprehensive performance objective of abating any condition that is unsafe, insanitary, illegal or of improper occupancy, egress deficient, creates a fire hazard, poorly maintained or otherwise dangerous to human life or the public welfare. Guidelines for abatement are provided in the code and in Chapter 1 of both the IPMC and the IFC.

3412.3.2 Compliance with other codes. Buildings that are evaluated in accordance with this section shall comply with the *International Fire Code* and the *International Property Maintenance Code*.

❖ This section requires an existing building that is subjected to the evaluation scoring process of Section 3412.6 to also comply with the IFC and IPMC. Those codes provide minimum requirements for health and safety that all existing buildings are expected to meet, regardless of whether there are any changes being made to the building or occupancy. Regardless of an existing building's final safety scores, the requirements of these referenced codes must be followed so occupants are safeguarded from hazards. This provision is similar to the requirements of Section 3401.3.

3412.4 Investigation and evaluation. For proposed work covered by this section, the building owner shall cause the existing building to be investigated and evaluated in accordance with the provisions of this section.

❖ This section and the subsequent sections address what must be done by the owner or designer who elects to employ Section 3412 for a proposed rehabilitation program and the corresponding action by the building official to assess the program objectively for approval or denial. The following actions must be taken:

- Fully investigate the building using both on-site inspections and research of all available building construction documents;
- Evaluate the building for conformance to Sections 3412.5 through 3412.9.1;
- Analyze the structure; and
- Submit to the building official documented results of the investigation and evaluation, plus any proposed compliance alternatives.

The building official must then determine whether the existing building, proposed work or change in occupancy complies with Section 3411.

Thus, the election and implementation of Section 3412 by the designer or owner is part of a code compliance option that may be used for existing buildings that the building official has the responsibility to review and assess. A proper and satisfactorily prepared submission by the owner or designer offering qualitative and quantitative data requires review by the building official for approval. If the submission is rejected, the building official must specifically cite any deficiencies and violations.

The owner is to have the existing building and any proposed work investigated and evaluated for compliance with the 19 parameters specified in Sections 3412.6.1 through 3412.6.19.

3412.4.1 Structural analysis. The owner shall have a structural analysis of the existing building made to determine adequacy of structural systems for the proposed *alteration*, *addition* or change of occupancy. The analysis shall demonstrate that the building with the work completed is capable of resisting the loads specified in Chapter 16.

❖ The owner is required to perform a complete structural analysis of the building to ensure that it can support the required loads. This requires that all interior loads meet the minimum load requirements of Chapter 16. Through this analysis, existing and altered buildings must be shown to support the expected loading. Any existing exterior member not affected by either interior loading or additional exterior loading need not be evaluated, provided that the structural member has, over a period of time, proven its ability to withstand the forces that normally create stress. Loads imposed on existing structural members by alterations, additions or a change of occupancy must be shown to sustain the requirements of Chapter 16, as stated in Sections 3403.3 through 3403.4 and 3404.3 through 3404.4. This structural analysis provides the owner and the building official with reasonable assurance that the building is structurally safe.

3412.4.2 Submittal. The results of the investigation and evaluation as required in Section 3412.4, along with proposed compliance alternatives, shall be submitted to the *building official*.

❖ The results of the investigation, including structural analysis and evaluation, must be submitted to the building official. If alternative methods, materials or equivalency concepts are proposed, these must also be submitted to the building official for review and approval.

3412.4.3 Determination of compliance. The *building official* shall determine whether the existing building, with the proposed *addition*, *alteration* or change of occupancy, complies with the provisions of this section in accordance with the evaluation process in Sections 3412.5 through 3412.9.

❖ When the results of the investigation and evaluation are submitted to the building official, he or she must determine whether the proposed work conforms to the provisions of Section 3412 and whether the evaluation was performed in accordance with Sections 3412.5 through 3412.9.

3412.5 Evaluation. The evaluation shall be comprised of three categories: fire safety, means of egress and general safety, as defined in Sections 3412.5.1 through 3412.5.3.

❖ This section and the subsequent sections address three general areas of safety to be evaluated: fire safety (FS), means of egress (ME) and general safety (GS). Section 3412.6 and the subsequent sections address 19 safety parameters that reflect on those areas. Each of the 19 safety parameters indicated in Sections 3412.6.1 through 3412.6.19 must be carefully reviewed and assigned a numerical value that signifies the degree of safety influence on the three overall general safety categories.

The 19 safety parameters that are given assigned values are:

3412.6.1	Building height
3412.6.2	Building area
3412.6.3	Compartmentation
3412.6.4	Tenant and dwelling unit separations
3412.6.5	Corridor walls
3412.6.6	Vertical openings
3412.6.7	HVAC systems
3412.6.8	Automatic fire detection
3412.6.9	Fire alarm systems
3412.6.10	Smoke control
3412.6.11	Means of egress capacity and number
3412.6.12	Dead ends
3412.6.13	Maximum exit access travel distance
3412.6.14	Elevator control
3412.6.15	Means of egress emergency lighting
3412.6.16	Mixed occupancies
3412.6.17	Automatic sprinklers
3412.6.18	Standpipes
3412.6.19	Incidental uses

These 19 safety parameters are focused on critical factors related to the minimum degree of life safety and property protection needed in an existing building.

When mixed occupancies in an existing building are not separated by fire-resistance-rated assemblies or fire walls meeting the most restrictive fire rating of the different occupancies, the entire evaluation must be based on the occupancy with the most restrictive requirements. The evaluation process considers the score for the various occupancies and applies the lowest score to the entire building.

When the mixed occupancies are separated in compliance with Section 508.4, they are to be evaluated separately and the score for each occupancy will apply to each portion based on its use. Both height and area formulas are computed for each occupancy. See the commentary to Sections 3412.2.2 and 3412.6.16 for a further discussion of application for mixed occupancies.

If there are four different occupancies in the building, the values must be computed for each of the four occupancies. Each separated occupancy is required to meet only its own mandatory building score.

When mixed occupancies are separated by fire walls complying with the requirements of Section 706, separate buildings are created and must be evaluated separately for all 19 safety parameters. The assigning of numerical values to each of the 19 safety parameters establishes a measurable quantity of what each of the parameters contributes to the overall safety of the building. Some evaluated parameters have a negative influence; others have a positive one. In total, the parameters may or may not result in an acceptable building score. The evaluation will determine whether the existing building has enough positive factors to overcome the negative parameters, or will indicate the negative factors that must be upgraded by alternative modifications.

The evaluation is divided into three categories or areas, which are defined in Sections 3412.5.1 through 3412.5.3.

3412.5.1 Fire safety. Included within the fire safety category are the structural fire resistance, automatic fire detection, fire alarm, automatic sprinkler system and fire suppression system features of the facility.

❖ A partial list of the items used to evaluate fire safety in a building is given in this section.

3412.5.2 Means of egress. Included within the means of egress category are the configuration, characteristics and support features for *means of egress* in the facility.

❖ The means of egress features that are evaluated by Section 3412 fall into the general areas of configuration, characteristics and support features. The specific features include travel distance, dead ends, emergency lighting and exit capacity and number.

3412.5.3 General safety. Included within the general safety category are the fire safety parameters and the means of egress parameters.

❖ This category includes every item that is used in either the fire safety or means of egress evaluation.

3412.6 Evaluation process. The evaluation process specified herein shall be followed in its entirety to evaluate existing buildings. Table 3412.7 shall be utilized for tabulating the results of the evaluation. References to other sections of this code indicate that compliance with those sections is required in order to gain credit in the evaluation herein outlined. In applying this section to a building with mixed occupancies, where the separation between the mixed occupancies does not qualify for any category indicated in Section 3412.6.16, the score for each occupancy shall be determined and the lower score determined for each section of the evaluation process shall apply to the entire building.

Where the separation between mixed occupancies qualifies for any category indicated in Section 3412.6.16, the score for each occupancy shall apply to each portion of the building based on the occupancy of the space.

❖ This section is the key to understanding the entire evaluation process. The first sentence of this section clearly states that every one of the 19 safety parameters indicated in Sections 3412.6.1 through 3412.6.19 must be evaluated and nothing may be omitted. After the 19 safety parameters have been evaluated and assigned a numerical value, the values are entered in Table 3412.7. The values must be tabulated to obtain the building score for each of the three general categories.

Some of the 19 safety parameters listed require mandatory compliance with other sections of the code and establish the foundation for determining a proper evaluation of those parameters, regardless of whether the result is positive or negative. This involves a coordination of mandatory basic requirements with the respective existing building conditions to arrive at the numerical evaluations prescribed in Sections 3412.6.1 through 3412.6.19. Section 3412.6.16 also addresses how mixed occupancies are handled in the evaluation. To apply this section, it is necessary to understand Section 508. See the commentary to Sections 508, 3412.2.2, 3412.6.2.2 and 3412.6.16 for a further discussion of the application of the evaluation procedure for mixed occupancies.

3412.6.1 Building height. The value for building height shall be the lesser value determined by the formula in Section 3412.6.1.1. Chapter 5 shall be used to determine the allowable height of the building, including allowable increases due to automatic sprinklers as provided for in Section 504.2. Subtract the actual *building height* in feet from the allowable and divide by $12\,^1/_2$ feet. Enter the height value and its sign (positive or negative) in Table 3412.7 under Safety Parameter 3412.6.1, Building Height, for fire safety, means of egress and general safety. The maximum score for a building shall be 10.

❖ As a starting point for the actual evaluation process, this section and Section 3412.6.1.1 define in detail how to perform the building height evaluation. The exact values are to be computed and compared against the mandatory safety scores. The values are not to be rounded. For calculation purposes, two decimal places would be appropriate. It is not necessary to build in any inaccuracy that occurs through the rounding process.

The maximum score for a building is now 10. Regardless of the building's allowable height as compared to its actual height, the intent of the code is to limit all buildings to a maximum score of 10.

3412.6.1.1 Height formula. The following formulas shall be used in computing the building height value.

$$\text{Height value, feet} = \frac{(AH) - (EBH)}{12.5} \times CF$$

(Equation 34-1)

$$\text{Height value, feet} = (AS - EBS) \times CF$$

(Equation 34-2)

where:

AH = Allowable height in feet from Table 503.

EBH = Existing *building height* in feet.

AS = Allowable height in stories from Table 503.

EBS = Existing building height in stories.

CF = 1 if (*AH*) – (*EBH*) is positive.

CF = Construction-type factor shown in Table 3412.6.6(2) if (*AH*) – (*EBH*) is negative.

Note: Where mixed occupancies are separated and individually evaluated as indicated in Section 3412.6, the values *AH*, *AS*, *EBH* and *EBS* shall be based on the height of the occupancy being evaluated.

❖ Two height formulas for calculating the score to be entered in Table 3412.7 are given. The formulas use the allowable height in both feet and stories from Table 503 and the height of the existing building. The denominator in the formula, 12.5, represents an average story height in feet.

The actual story height and the overall existing building height are to be directly compared to Table 503. Table 503 serves as a datum level that allows for establishing a numerical height value for the existing building by comparing its actual height and its type of construction as represented by a construction factor (*CF*).

If the existing building actual height is less than or equal to the allowable height of Table 503, then the *CF* value is 1 (no negative factor is assigned). If the actual height exceeds the Table 503 allowable height, the building is not in compliance with Table 503 and represents a safety deficiency. A deficiency rating is assigned that is dependent on the type of construction. The construction factors to establish this negative value are given in Table 3412.6.6(2).

When a building is not in compliance with Table 503, it is considered less safe than a building that does comply with Table 503. As a result, deficiency points are assessed. Additional safeguards must be provided to compensate for this condition. This is the primary reason that a different construction factor must be used for buildings not in compliance. The construction factor is the equivalent deficiency that must be overcome by providing additional protection in other areas that are to be evaluated.

Example 1:

A six-story building of Type IB construction is 60 feet (18 288 mm) tall. The building is not sprinklered. It contains a Group B business occupancy.

The allowable height from Table 503 is 11 stories, 160 feet (48 800 mm).

AH = 160 feet
AS = 11 stories
EBH = 60 feet
EBS = 6 stories
CF = 1 (AH - EBH is positive)

Height value in feet $= \frac{160-60}{12.5} \times CF$
$= 8 \times 1$
$= 8$

Height value in stories $= (11-6) \times CF$
$= 5 \times 1$
$= 5$

The building height value will be 5, because it is the lesser of the two height values.

Example 2:

A seven-story building of Type IIIB construction is 80 feet (24 400 mm) tall. The building is sprinklered. It contains a Group M mercantile occupancy.

The allowable height from Table 503 is two stories, 55 feet (16 775 mm).

AH = 55 + 20 (for sprinklers) = 75 feet
AS = 2 + 1 (for sprinklers) = 3 stories
EBS = 7 stories
EBH = 80 feet
CF = 3.5 from Table 3412.6.6(2)

Height value in feet $= \frac{75-80}{12.5} \times CF$
$= -0.4 \times 3.5$
$= -1.4$

Height value in stories $= (3-7) \times CF$
$= -4 \times 3.5$
$= -14$

The building height value will be -14, because it is the lesser of the two height values.

The above values determine the entries for Table 3412.7 under "Safety Parameter 3412.6.1, Building Height."

In Example 1, the height value of 5 is entered into the columns for fire safety (FS), means of egress (ME) and general safety (GS).

In Example 2, the height value of -14 is entered into the columns for fire safety (FS), means of egress (ME) and general safety (GS).

The assessed height value parameter is only one of the 19 safety parameters that need to be evaluated. Example 1 shows a positive contribution; however, this may not be enough to result in a sufficient overall building score. Similarly, the negative value in Example 2 may not in itself result in an overall insufficient building score.

3412.6.2 Building area. The value for building area shall be determined by the formula in Section 3412.6.2.2. Section 503 and the formula in Section 3412.6.2.1 shall be used to determine the allowable area of the building. This shall include any allowable increases due to frontage and automatic sprinklers as provided for in Section 506. Subtract the actual *building area* in square feet from the allowable area and divide by 1,200 square feet. Enter the area value and its sign (positive or negative) in Table 3412.7 under Safety Parameter 3412.6.2, Building Area, for fire safety, means of egress and general safety. In determining the area value, the maximum permitted positive value for area is 50 percent of the fire safety score as *listed* in Table 3412.8, Mandatory Safety Scores.

❖ In this section, the code user is shown how to calculate the building score for the building area.

This section also requires the maximum score of a building in this category. The maximum value is 50 percent of just the fire safety score listed in Table 3412.8 (see commentary, Section 3412.6.2.2). This maximum score is applicable and equal for the three building score categories—fire safety, means of egress and general safety. The positive score is limited to prevent this one parameter from providing enough points to unjustifiably overcome too many deficiencies in other parameters.

3412.6.2.1 Allowable area formula. The following formula shall be used in computing allowable area:

$$A_a = [A_t + (A_t \times I_f) + (A_t \times I_s)] \quad \textbf{(Equation 34-3)}$$

where:

A_a = Allowable *building area* per story (square feet).

A_t = Tabular *building area* per story in accordance with Table 503 (square feet).

I_s = Area increase factor due to sprinkler protection as calculated in accordance with Section 506.3.

I_f = Area increase factor due to for frontage as calculated in accordance with Section 506.2.

❖ The formula used to calculate the allowable area of the building is a direct correlation of the area increases and reductions allowed for new buildings in Section 506. The use of these increases and reductions is addressed in Chapter 5.

3412.6.2.2 Area formula. The following formula shall be used in computing the area value. Determine the area value for each occupancy floor area on a floor-by-floor basis. For each occupancy, choose the minimum area value of the set of values obtained for the particular occupancy

$$\text{Area value } i = \frac{\text{Allowable area}_i}{\text{1, 200squate feet}}\left[1 - \left(\frac{\text{Actual area}_i}{\text{Allowable area}_i} + \ldots + \frac{\text{Actual area}_n}{\text{Allowable area}_n}\right)\right]$$

(Equation 34-4)

where:

i = Value for an individual separated occupancy on a floor.

n = Number of separated occupancies on a floor.

❖ The area formula provides a numerical value for the actual building area to be entered in Table 3412.7 under Safety Parameter 3412.6.2, Building Area. If the area of the existing building is less than the allowable area, it is considered to be safer. Credit for area is given. If the area of the existing building is larger than the allowable area, it represents a condition that is judged to be less safe. The building receives a negative score. This can be overcome by providing additional safeguards.

To use the formula for existing buildings that contain mixed occupancies, an area value must be determined for each story of a building. If a building contains just one occupancy or just one occupancy on a particular floor separated from other floors, the formula simply reduces to the allowable area minus the actual area divided by the constant 1,200. The formula is also applicable to a story containing several separated occupancies. In such a situation, the formula requires each actual area to be divided by its respective allowable area. The resulting fractions are added together, subtracted from the constant of 1 and multiplied by the ratio of the allowable area of the particular use divided by the constant 1,200. This method of determining area values for several occupancies that are separated on a story is directly comparable to the unity formula in Section 508.4.2.

Example 1:

Figure 3412.6.2.2(1) illustrates a Group B building of Type IIA construction. The building is five stories in

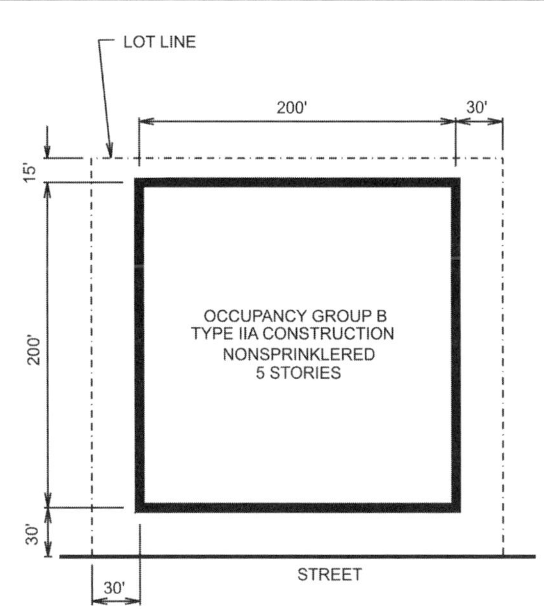

ACTUAL AREA = 200' × 200' = 40,000 SQ.FT.

Sprinkler Increase = 0 (FROM SECTION 506.3)

Frontage Increase = $[\frac{600}{800} - .25]\frac{30}{30}$ = .S = 50% (FROM SECTION 506.2)

ALLOWABLE AREA IN TABLE 503 = 37,500 SQ.FT.

A_a = (1 + .5 + 0) × 37,500 = 56,250 SQ.FT.

For SI: 1 foot = 304.8 mm, 1 square foot = 0.0929 m^2.

Figure 3412.6.2.2(1)
ALLOWABLE AREA—SINGLE OCCUPANCY BUILDING

height, unsprinklered.

The overall building area is 200 feet (60 960 mm) by 200 feet (60 960 mm).

$$\text{Area value} = \frac{56,250}{1,200}\left[1-\left(\frac{40,000}{56,250}\right)\right]$$

$$= 46.88\,[1 - 0.71]$$

$$= 13.60$$

Allowable area = 56,250 square feet

Actual area = 40,000 square feet

In Table 3410.7, enter the following values under Safety Parameter 3410.6.2, Building Area:

Fire safety (FS) = 13.6

Means of egress (ME) = 13.6

General safety (GS) = 13.6

The positive values entered in Table 3412.7 under Safety Parameter 3412.6.2, Building Area (FS = 13.6, ME = 13.6 and GS = 13.6) represent the evaluation for only one of the 19 safety parameters that must be assessed to arriving at the overall building score evaluation of Table 3412.7. As previously described for the building height parameter values under safety parameter 3412.6.1, Building Height of Table 3412.7, a single parameter of the total 19 safety parameters to be assessed does not, in itself, determine the ultimate acceptable building score.

Example 2:

Figure 3410.6.2.2(2) illustrates a separated mixed occupancy building (Groups B, M and S-1) of Type IIIB construction. The building is two stories in height,

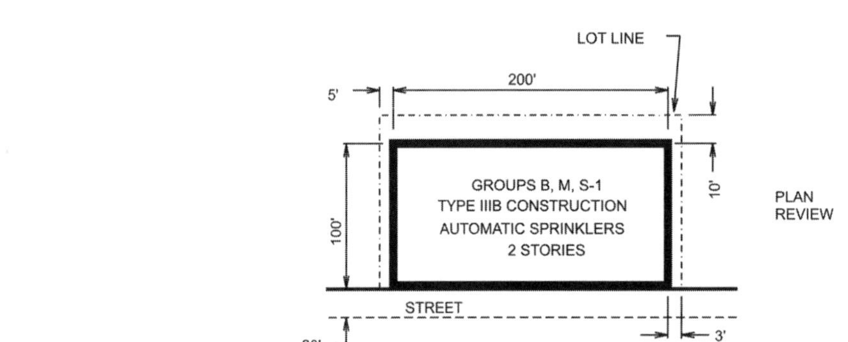

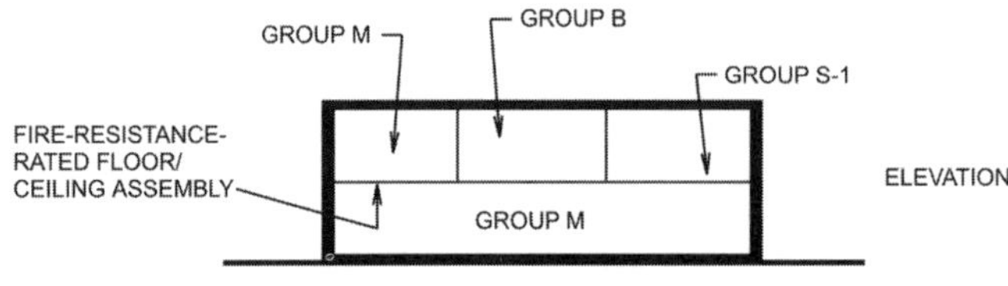

ACTUAL AREA = 100' × 200'
= 20,000 SQ.FT.

Sprinkler Increase = 200% (FROM SECTION 506.3)

Frontage Increased $=[\frac{200}{600} - .25]\frac{20}{30}$ = 5.6% (FROM SECTION 506.2)

ALLOWABLE AREA IN TABLE 503 (GROUP B) = 19,000 SQ.FT.

ALLOWABLE AREA IN TABLE 503 (GROUP M) = 12,500 SQ.FT.

ALLOWABLE AREA IN TABLE 503 (GROUP S-1) = 17,500 SQ.FT.

$$\text{AA GROUP B} = \frac{(200 + 5.6 + 100) \times 19,000}{100} = 58,064 \text{ SQ.FT.}$$

$$\text{AA GROUP M} = \frac{(200 + 5.6 + 100) \times 12,500}{100} = 38,200 \text{ SQ.FT.}$$

$$\text{AA GROUP S-1} = \frac{(200 + 5.6 + 100) \times 17,500}{100} = 53,480 \text{ SQ.FT.}$$

For SI: 1 foot = 304.8 mm, 1 square foot = 0.0929 m^2.

Figure 3412.6.2.2(2)
ALLOWABLE AREA—MIXED OCCUPANCY BUILDING

fully sprinklered.

The overall building area is 100 feet (30 480 mm) by 200 feet (60 960 mm).

Actual area for Group M on first story = 20,000 square feet (1858 m^2).

Actual area for Group M on second story = 4,400 square feet (409 m^2).

Actual area for Group B on second story = 10,000 square feet (929 m^2).

Actual area for Group S-1 on second story = 5,600 square feet (520 m^2).

Sprinkler Increase = 200% (from Section 506.3)

Frontage Increase = 5.6% (from Section 506.2)

Tabular areas from Table 503:

Group B = 19,000 square feet (1765 m^2)

Group M = 12,500 square feet (1161 m^2)

Group S-1 = 17,500 square feet (1626 m^2)

Allowable Areas:

$$AA = \frac{(200 + 5.6 + 100) \times \text{Tabular Area}}{100}$$

AA for Group B = 58,064 square feet

AA for Group M = 38,200 square feet

AA for Group S-1 = 53,480 square feet

Area Values:

For Group M, 1st story

$$\text{Area value} = \frac{38,200}{1,200}\left[1 - \left(\frac{20,000}{38,200}\right)\right]$$

$$= 31.83[1 - 0.52]$$

$$= 15.27$$

For Group B, 2nd story

$$\text{Area value} = \frac{58,064}{1,200}\left[1 - \left(\frac{10,000}{58,064} + \frac{4,400}{38,200} + \frac{5,600}{53,480}\right)\right]$$

$$= 48.39[1 - 0.39]$$

$$\text{Area value} = 29.51$$

For Group M, 2nd story

$$\text{Area value} = \frac{38,200}{1,200}[1 - 0.39]$$

$$\text{Area value} = 19.42$$

For Group S-1, 2nd story

$$\text{Area value} = \frac{53,480}{1,200}[1 - 0.39]$$

$$\text{Area value} = 27.19$$

Since these mixed occupancies are being separated so that one of the categories indicated in Section 3412.6.16 is applicable, a separate score must be computed for each occupancy.

In this example, the area value for the Group B occupancy is calculated to be 29.51, but the maximum area value permitted is 50 percent of the mandatory fire safety (FS) score listed in Table 3412.8. For a Group B occupancy, the mandatory FS score is 30; therefore, the maximum positive value that can be entered into Table 3412.7 is 15.

When an occupancy is located in more than one story, a separate area value must be calculated for each story. For input into Table 3412.7, the area value to be used is the lesser of all the individual area values for that occupancy group, but not greater than 50 percent of the mandatory FS score. In Figure 3412.6.2.2(2), there is an area classified as a Group M occupancy on both the first and second stories. The area value for the Group M occupancy on the first story is 15.27, and the area value for the second story is 19.42. Fifty percent of the mandatory FS score for Group M is 11.5. The lowest value for all the Group M occupant areas in the building is 11.5, which is 50 percent of the mandatory FS score.

The area value for the Group S-1 occupancy portion of the building is 27.19. This value, just as the values for the other occupancies in this example, exceeds the 50-percent maximum permitted by the mandatory FS score. The maximum for Group S-1 occupancy is 9.5, 50 percent of 19.

3412.6.3 Compartmentation. Evaluate the compartments created by *fire barriers* or *horizontal assemblies* which comply with Sections 3412.6.3.1 and 3412.6.3.2 and which are exclusive of the wall elements considered under Sections 3412.6.4 and 3412.6.5. Conforming compartments shall be figured as the net area and do not include shafts, chases, stairways, walls or columns. Using Table 3412.6.3, determine the appropriate compartmentation value (*CV*) and enter that value into Table 3412.7 under Safety Parameter 3412.6.3, Compartmentation, for fire safety, means of egress and general safety.

❖ This section establishes and evaluates the compartments contained within an existing building by the effectiveness of the enclosing fire separation assemblies of both walls and floor/ceiling assemblies. Larger compartments are considered to be a greater safety risk than smaller compartments because the larger areas can become involved in a single fire incident affecting a greater portion of the building at one time.

Fire barriers and horizontal assemblies must comply with Sections 3412.6.3.1 and 3412.6.3.2. The fire barriers and horizontal assemblies considered here are not to include the other separations or enclosures considered under Sections 3412.6.4 and 3412.6.5.

The evaluation of the compartments contained within an existing building is a linear function allowing interpolation between the various categories (see Table 3412.6.3). This approach allows the compartmentation value to increase or decrease consistent with the actual changes in compartment sizes. Such an adjustment removes the previously built-in bias against smaller-sized buildings. Higher compartmentation values are assigned to buildings with smaller compartments.

TABLE 3412.6.3. See below.

❖ See the commentary to Section 3412.6.3.

3412.6.3.1 Wall construction. A wall used to create separate compartments shall be a *fire barrier* conforming to Section 707 with a *fire-resistance rating* of not less than 2 hours. Where the building is not divided into more than one compartment, the compartment size shall be taken as the total floor area on all floors. Where there is more than one compartment within a *story*, each compartmented area on such *story* shall be provided with a *horizontal exit* conforming to Section 1025. The *fire door* serving as the *horizontal exit* between compartments shall be so installed, fitted and gasketed that such *fire door* will provide a substantial barrier to the passage of smoke.

❖ This section states that the walls determining the boundary of the compartment need to have fire barrier ratings of no less than 2 hours. These assemblies must be constructed in accordance with Section 707. For an existing building, this may need to be evaluated by both analyzing available plans and on-site investigations with a professional engineering determination of the required 2-hour fire-resistance rating. If the fire-resistance rating is less than 2 hours or cannot be reasonably determined, such walls should not be considered as compartmenting a fire area. The entire floor of an existing building must then be considered the compartment. If 2-hour-rated floor/ceiling assemblies are not present, the compartment size becomes the total area on all floors.

Openings and continuity requirements of Section 707 must be followed to maintain the integrity of the fire-resistance-rated wall assemblies and, thus, the compartments. Horizontal exits and their fire doors must comply with Section 1025.

The evaluation of an existing door condition requires an investigation of available engineering data and on-site inspections. Any uncertainty as to the performance of such doors may result in the openings being considered unprotected, expanding the compartment area or requiring an upgraded modification of the door to meet the current requirements of Section 1025.

3412.6.3.2 Floor/ceiling construction. A floor/ceiling assembly used to create compartments shall conform to Section 711 and shall have a *fire-resistance rating* of not less than 2 hours.

❖ The building features that provide the horizontal boundaries of the compartment need to provide effective fire-resistant integrity between floors. The existing floor/ceiling assemblies need to be rated for 2 hours and need to be tight against exterior walls. Penetrations in the floor/ceiling assemblies must be protected in accordance with Section 714 to maintain their fire-resistant integrity. The floor/ceiling assemblies must conform to all of the requirements of Section 711 to create the level of compartmentation required for this evaluation parameter.

3412.6.4 Tenant and dwelling unit separations. Evaluate the *fire-resistance rating* of floors and walls separating tenants, including *dwelling units*, and not evaluated under Sections 3412.6.3 and 3412.6.5. Under the categories and occupancies in Table 3412.6.4, determine the appropriate value and enter that value in Table 3412.7 under Safety Parameter 3412.6.4, Tenant and Dwelling Unit Separations, for fire safety, means of egress and general safety.

❖ This parameter is used to evaluate partitions in an existing building other than those used for the creation of compartments in Section 3412.6.3 or the enclosure of corridors in Section 3412.6.5. This section examines the level of separation between tenant spaces and dwelling units. The listed categories specifically reference Sections 707, 708 and 711 not only for fire-resistance ratings but also for continuity and opening purposes. Further credit is provided for existing buildings that have a 2-hour-rated separation between adjacent tenant spaces or dwelling units, which exceeds what is required for new construction.

**TABLE 3412.6.3
COMPARTMENTATION VALUES**

OCCUPANCY	CATEGORIES[a]				
	a Compartment size equal to or greater than 15,000 square feet	b Compartment size of 10,000 square feet	c Compartment size of 7,500 square feet	d Compartment size of 5,000 square feet	e Compartment size of 2,500 square feet or less
A-1, A-3	0	6	10	14	18
A-2	0	4	10	14	18
A-4, B, E, S-2	0	5	10	15	20
F, M, R, S-1	0	4	10	16	22

For SI: 1 square foot = 0.093 m^2.

a. For areas between categories, the compartmentation value shall be obtained by linear interpolation.

**TABLE 3412.6.4
SEPARATION VALUES**

OCCUPANCY	CATEGORIES				
	a	b	c	d	e
A-1	0	0	0	0	1
A-2	-5	-3	0	1	3
A-3, A-4, B, E, F, M, S-1	-4	-3	0	2	4
R	-4	-2	0	2	4
S-2	-5	-2	0	2	4

❖ Table 3412.6.4 provides values for tenant space and dwelling unit separations that resist the spread of flames and smoke. The rationale for considering a nonfire-resistance-rated or incomplete separation as a safety liability is its unassumed role of impeding flame and smoke spread. Separations that compartmentalize the areas with significantly rated fire separation walls and floor/ceiling assemblies are considered the most effective. Buildings containing Group R occupancies have separation values based on the fact that dwelling unit separations are more critical than tenant separations in other occupancies.

This chapter places greater emphasis on compartmentation than on open floor areas when compared to the rest of the code (see commentary, Sections 3412 and 3412.6.4).

3412.6.4.1 Categories. The categories for tenant and *dwelling unit* separations are:

1. Category a—No *fire partitions*; incomplete *fire partitions*; no doors; doors not self-closing or automatic-closing.
2. Category b—*Fire partitions* or floor assemblies with less than a 1-hour *fire-resistance rating* or not constructed in accordance with Sections 708 or 711.
3. Category c—*Fire partitions* with a 1-hour or greater *fire-resistance rating* constructed in accordance with Section 708 and floor assemblies with a 1-hour but less than 2-hour *fire-resistance rating* constructed in accordance with Section 711, or with only one tenant within the floor area.
4. Category d—*Fire barriers* with a 1-hour but less than 2-hour *fire-resistance rating* constructed in accordance with Section 707 and floor assemblies with a 2-hour or greater *fire-resistance rating* constructed in accordance with Section 711.
5. Category e—*Fire barriers* and floor assemblies with a 2-hour or greater *fire-resistance rating* and constructed in accordance with Sections 707 and 711, respectively.

❖ Tenant space and dwelling unit separations are categorized by the partitions being evaluated. The values of each category are listed in Table 3412.6.4 by occupancy classifications. Typical illustrations of the types

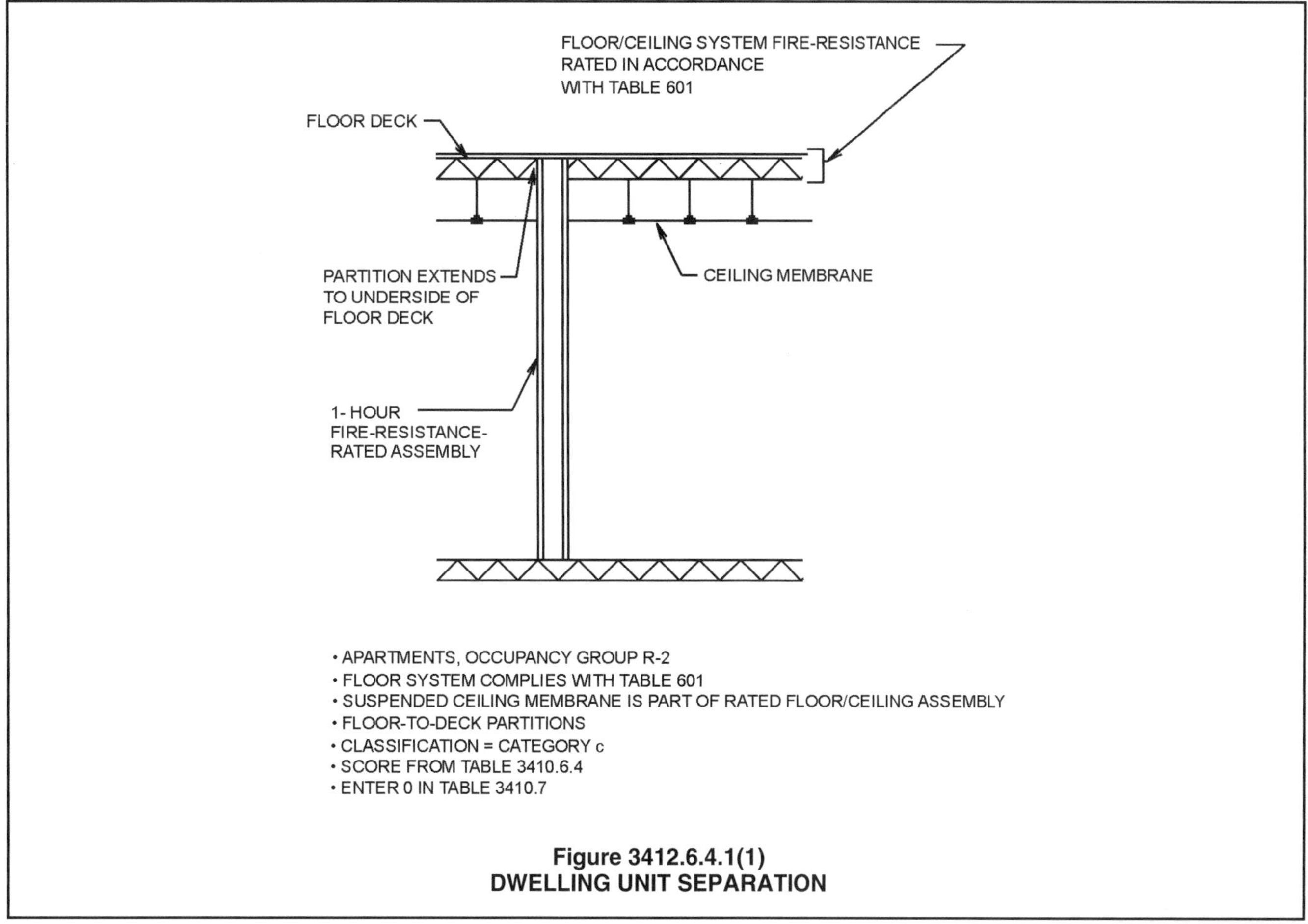

**Figure 3412.6.4.1(1)
DWELLING UNIT SEPARATION**

of partitions are shown in Figures 3412.6.4.1(1) and 3412.6.4.1(2). The listed categories provide a graduated level of separation when compared to new construction requirements.

Category a addresses the situation where there is no separation or there are gaps in the separation provided between tenant spaces or dwelling units.

Category b accounts for those tenant spaces or dwelling units that are separated from one another with less than a 1-hour rating.

Category c represents what a newly constructed building would have for tenant space and dwelling unit separation. An existing building meeting this level of compliance gains no benefit or penalty and, therefore, the separation value is zero.

Category d is given additional credit because the existing building space or dwelling unit has tenant separation that is rated higher than what is required by the code for new construction. Additionally, the walls are required to meet the fire barrier requirements of Section 707.

Category e is an added column that provides increased credit for existing buildings that have walls and floor/ceiling assemblies with fire-resistance ratings exceeding what is required for new buildings. This increased level of compartmentation in an existing building is credited accordingly with high separation values.

3412.6.5 Corridor walls. Evaluate the *fire-resistance rating* and degree of completeness of walls which create *corridors* serving the floor, and constructed in accordance with Section 1018. This evaluation shall not include the wall elements considered under Sections 3412.6.3 and 3412.6.4. Under the categories and groups in Table 3412.6.5, determine the appropriate value and enter that value into Table 3412.7 under Safety Parameter 3412.6.5, Corridor Walls, for fire safety, means of egress and general safety.

❖ Corridor walls are evaluated as fire partitions possessing an adequate fire-resistance rating and completeness to restrict fire and smoke migration into the corridor. Prescribed requirements are stated in Sections 708.4 and 1018. Existing corridor walls require investigation and analysis to determine equivalency to code requirements. Figures 3412.6.5(1) and 3412.6.5(2) illustrate various corridor wall values. Corridor walls contrast with the compartmentation and tenant/dwelling unit separations of Sections 3412.6.3 and 3412.6.4 by requiring an appropriate fire-resistance rating and continuity (see commentary, Sections 708.4, 3412.6.3 and 3412.6.4).

The corridor wall evaluations do not include parti-

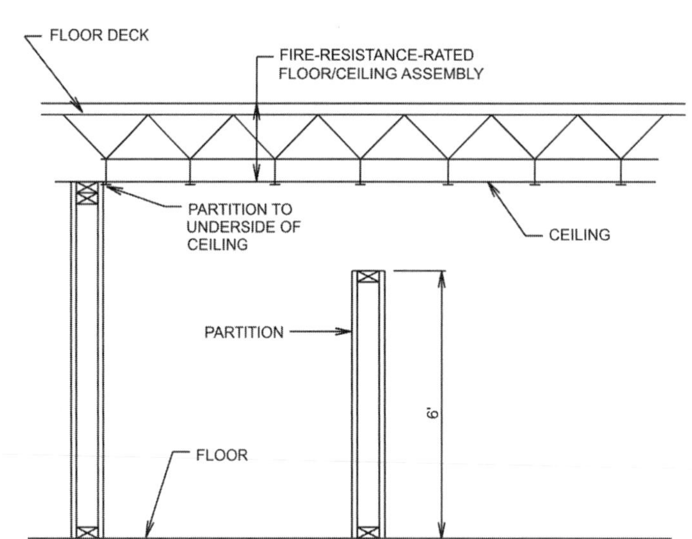

• FOOD COURT RESTAURANTS, OCCUPANCY GROUP A-3
• SOME PARTITIONS EXTEND TO CEILING OF FIRE-RESISTANCE-RATED FLOOR/CEILING ASSEMBLY
• NO CLOSERS ON DOORS BETWEEN ADJACENT TENANT SPACES
• SOME 6' PRIVACY PARTITIONS (PARTIAL PARTITIONS)

• CLASSIFICATION = CATEGORY a
• SCORE FROM TABLE 3412.6.4
• ENTER -4 IN TABLE 3412.7

For SI: 1 foot = 304.8 mm.

Figure 3412.6.4.1(2)
TENANT SEPARATION

tions required to establish a compartment (see Section 3412.6.3) or tenant and dwelling unit separations (see Section 3412.6.4).

TABLE 3412.6.5
CORRIDOR WALL VALUES

OCCUPANCY	CATEGORIES			
	a	b	c[a]	d[a]
A-1	-10	-4	0	2
A-2	-30	-12	0	2
A-3, F, M, R, S-1	-7	-3	0	2
A-4, B, E, S-2	-5	-2	0	5

a. Corridors not providing at least one-half the travel distance for all occupants on a floor shall be category b.

❖ Table 3412.6.5 assigns values to the different uses based on the relative degree of fire resistance and smoke resistance of corridor walls. Since corridors are enclosed (confined) spaces subject to the rapid buildup of smoke and heat, a degree of protection is necessary to minimize this hazard to the occupants egressing the building. The table reflects this emphasis through substantial negative scores in buildings without properly enclosed corridors. Both Section 3412.6.5 and this table place an emphasis on corridor walls, which differs from that expressed elsewhere in the code. For example, Chapter 10 does not require corridors to be provided except in Group I-2 occupancies. The emphasis here is that when corridors are already present, they must provide basic protection. The table is an assessment of the relative risk represented by the corridor.

Note a to the table further controls the application of Categories c and d to existing buildings with certain means of egress arrangements. Although an existing building may have corridors with significant fire-resistance ratings and protected openings, little credit can be granted when the building occupants are protected for only a short period of time. Very short corridors in large floor plans or corridors located in just one tenant space of a floor plan provide a benefit to only a portion of the occupant load for a small portion of the overall exit access travel and, thus, do not accrue any positive points. Unless rated corridors are available for all occupants of that particular floor or they provide a protected path of travel for at least one-half of the occupants' overall travel length, negative corridor wall values are assigned. In existing buildings with very short or limited-use corridors, the code user is directed to use Category b, regardless of the corridors' fire-resistance ratings and opening protectives.

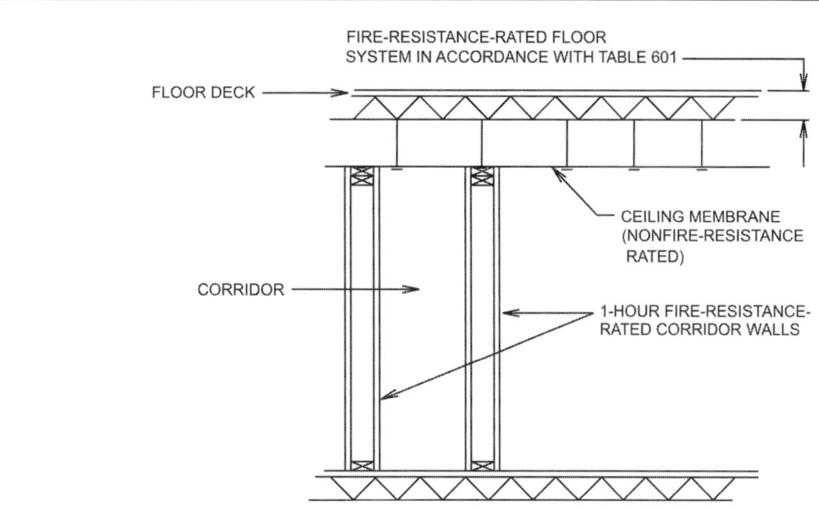

- OCCUPANCY GROUP B
- NO CLOSERS ON DOORS
- FLOOR SYSTEM COMPLIES WITH TABLE 601
- SUSPENDED CEILING MEMBRANE IS NONFIRE-RESISTANCE RATED
- CORRIDOR WALLS ARE 1-HR. FIRE-RESISTANCE RATED

NOTE: THESE CONDITIONS PRODUCE A CATEGORY a BECAUSE OF BOTH INCOMPLETE FIRE PARTITIONS AND NO SELF-CLOSING DOORS

- CLASSIFICATION = CATEGORY a
- SCORE FROM TABLE 3412.6.5 = -5
- ENTER -5 IN TABLE 3412.7

Figure 3412.6.5(1)
CORRIDOR WALL VALUES—CATEGORY a

3412.6.5.1 Categories. The categories for Corridor Walls are:

1. Category a—No *fire partitions*; incomplete *fire partitions*; no doors; or doors not self-closing.
2. Category b—Less than 1-hour *fire-resistance rating* or not constructed in accordance with Section 708.4.
3. Category c—1-hour to less than 2-hour *fire-resistance rating*, with doors conforming to Section 716 or without *corridors* as permitted by Section 1018.
4. Category d—2-hour or greater *fire-resistance rating*, with doors conforming to Section 716.

❖ Corridor walls are categorized by the partitions being evaluated. The values of each category are listed in Table 3412.6.5 by occupancy classification.

3412.6.6 Vertical openings. Evaluate the *fire-resistance rating* of *exit* enclosures, hoistways, escalator openings and other shaft enclosures within the building, and openings between two or more floors. Table 3412.6.6(1) contains the appropriate protection values. Multiply that value by the construction type factor found in Table 3412.6.6(2). Enter the vertical opening value and its sign (positive or negative) in Table 3412.7 under Safety Parameter 3412.6.6, Vertical Openings, for fire safety, means of egress, and general safety. If the structure is a one-story building or if all the unenclosed vertical openings within the building conform to the requirements of Section 708, enter a value of 2. The maximum positive value for this requirement shall be 2.

❖ Vertical openings are used to evaluate the fire-resistance ratings of openings between floors of a building and between shaft enclosures, such as stairways, elevator hoistways and escalator openings. This section also gives the formula for determining the score to be entered in Table 3412.7. This section recognizes the inherent safety of one-story buildings and of buildings where all vertical openings are protected in accordance with one of the methods in Section 713 by allowing the maximum value of 2 for these instances.

TABLE 3412.6.6(1)
VERTICAL OPENING PROTECTION VALUE

PROTECTION	VALUE
None (unprotected opening)	-2 times number floors connected
Less than 1 hour	-1 times number floors connected
1 to less than 2 hours	1
2 hours or more	2

❖ Table 3412.6.6(1) assigns relative protection values based on the fire protection ratings of the vertical

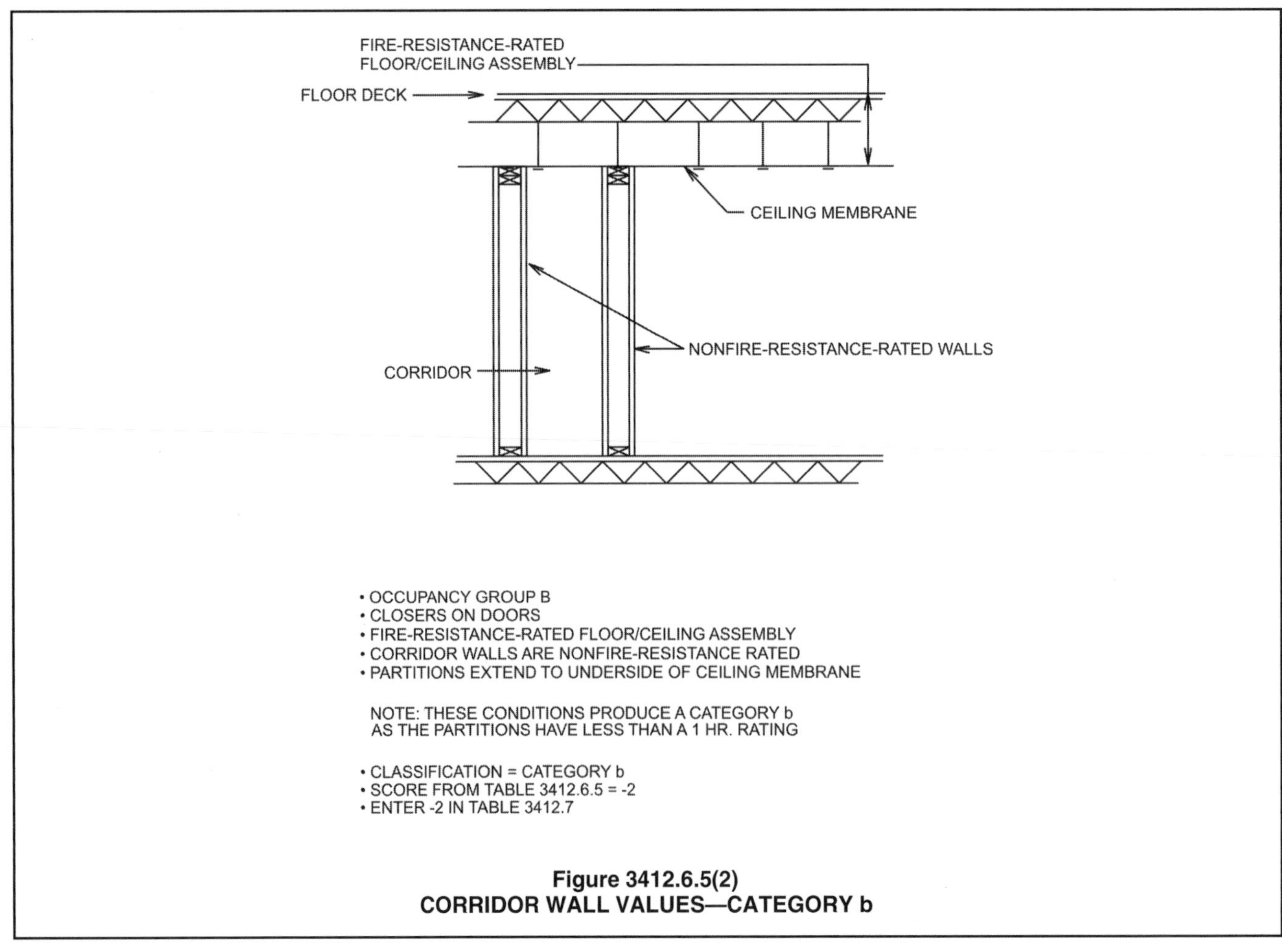

FIRE-RESISTANCE-RATED
FLOOR/CEILING ASSEMBLY
FLOOR DECK
CEILING MEMBRANE
NONFIRE-RESISTANCE-RATED WALLS
CORRIDOR
• OCCUPANCY GROUP B
• CLOSERS ON DOORS
• FIRE-RESISTANCE-RATED FLOOR/CEILING ASSEMBLY
• CORRIDOR WALLS ARE NONFIRE-RESISTANCE RATED
• PARTITIONS EXTEND TO UNDERSIDE OF CEILING MEMBRANE
NOTE: THESE CONDITIONS PRODUCE A CATEGORY b
AS THE PARTITIONS HAVE LESS THAN A 1 HR. RATING
• CLASSIFICATION = CATEGORY b
• SCORE FROM TABLE 3412.6.5 = -2
• ENTER -2 IN TABLE 3412.7

Figure 3412.6.5(2)
CORRIDOR WALL VALUES—CATEGORY b

openings in a building. The lower the fire protection, the greater the hazard to the rest of the building. The table reflects the varying levels of impact to an existing building based on the number of stories that are interconnected by unprotected openings. The greater the number of floors that are interconnected by unprotected openings, the greater the number of negative points assessed in the existing building's evaluation scores. The closer the building comes to meeting code compliance relative to rated shafts, the larger the positive protection value becomes. This table encourages and provides incentives to the existing building renovator to bring noncomplying situations into compliance with new construction requirements.

TABLE 3412.6.6(2)
CONSTRUCTION-TYPE FACTOR

FACTOR	TYPE OF CONSTRUCTION								
	IA	IB	IIA	IIB	IIIA	IIIB	IV	VA	VB
	1.2	1.5	2.2	3.5	2.5	3.5	2.3	3.3	7

❖ Relative values for each type of construction are assigned in Table 3412.6.6(2). These represent the relative degree of fire hazard of each type of construction when compared to other types of construction. Similar factors were considered in the original development of Table 503 for height and area limitations. When one building has two different opening circumstances that individually result in different values, the lower value must be used.

Example 1:

Assume a building of Type IIB construction, three stories in height. The building has a 2-hour fire-resistance-rated exhaust shaft with $1^1/_2$-hour fire dampers (complying with UL 555).

VO = 2 × 3.5 = 7.0

Enter 7.0 in Table 3410.7 under Safety Parameter 3410.6.6, Vertical Openings.

Example 2:

Assume a building of Type IA construction, 12 stories in height. The building has a $^1/_2$-hour return-air shaft with rated dampers.

VO = (-1 × 12) × 1.2 = -14.4

Enter -14.4 in Table 3410.7 under Safety Parameter 3410.6.6, Vertical Openings.

3412.6.6.1 Vertical opening formula. The following formula shall be used in computing vertical opening value.

$$VO = PV \times CF \qquad \textbf{(Equation 34-5)}$$

where:

VO = Vertical opening value.

PV = Protection value [Table 3412.6.6(1)].

CF = Construction type factor [Table 3412.6.6(2)].

❖ See the commentary to Section 3412.6.6

3412.6.7 HVAC systems. Evaluate the ability of the HVAC system to resist the movement of smoke and fire beyond the point of origin. Under the categories in Section 3412.6.7.1, determine the appropriate value and enter that value into Table 3412.7 under Safety Parameter 3412.6.7, HVAC Systems, for fire safety, means of egress and general safety.

❖ The provisions of Chapter 34 are intended to evaluate the HVAC system's potential for resisting the movement and spread of fire, smoke or products of combustion. This section does not address HVAC systems that are used exclusively for smoke control in the building. The systems evaluated in this section are those that use either supply air, return air or exhaust air. For example, a typical building might have supply air ducts or shafts; return air ducts or shafts; toilet exhaust ducts or shafts and kitchen exhaust ducts or shafts, all of which are considered HVAC systems. All systems in the building are evaluated and the lowest score obtained by any of the systems is the score that must be assigned to the entire building.

These provisions include two other safety aspects that are also applicable for new construction: plenums and air movement in egress elements, such as exit access corridors and exit stairways. These factors can significantly affect the safety of the occupants of the existing building. In some cases, these safety aspects can be more important than just the number of stories connected by an HVAC system.

3412.6.7.1 Categories. The categories for HVAC systems are:

1. Category a—Plenums not in accordance with Section 602 of the *International Mechanical Code*. -10 points.
2. Category b—Air movement in egress elements not in accordance with Section 1018.5. -5 points.
3. Category c—Both categories a and b are applicable. -15 points.
4. Category d—Compliance of the HVAC system with Section 1018.5 and Section 602 of the *International Mechanical Code*. 0 points.
5. Category e—Systems serving one *story*; or a central boiler/chiller system without ductwork connecting two or more stories. 5 points.

❖ The five categories that must be used in the evaluation process are defined in this section, along with their applicable values. The values are to be entered in Table 3412.7 under Safety Parameter 3412.6.7, HVAC Systems. These values are not occupancy sensitive, since the spread of fire is dependent on the HVAC system present in the existing building, not on its occupancy classification.

Category a requires a value of -10 for existing buildings that contain plenums not in compliance with the requirements of Section 602 of the IMC. Locations of plenums within the existing buildings, the materials they are built of relative to the building's type of construction and the materials exposed to plenum air must all be evaluated.

Category b corresponds to existing buildings that have exit access corridors or exit stairways that are used to supply, return or exhaust air or for other ventilation purposes. That type of layout puts the existing building's occupants at greater risk and must, therefore, be penalized with a value of -5. An existing building complying with one of the exceptions of Section 1018.5 is not considered as Category b.

Category c is applicable when an existing building has both noncomplying plenums and corridors/stairways used for air movement. A value of -15 must be assigned to such a building because the movement of smoke and fire throughout would pose an even greater hazard to the occupants.

Category d represents the base value of zero. A newly constructed building is required to meet all the provisions of Section 1018.5 of the code and Section 602 of the IMC. An existing building that complies with these provisions is meeting this same minimum compliance level and, therefore, is neither penalized nor given benefit for that compliance.

Category e specifies a value of 5 points for HVAC systems that serve only one story of a building. The hazards of a fire spreading laterally through a story of a building via the HVAC system are minimal; therefore, the code assigns a positive value. A boiler/chiller system also does not lend itself to fire spread as long as there is no air movement in ducts.

3412.6.8 Automatic fire detection. Evaluate the smoke detection capability based on the location and operation of *automatic fire detectors* in accordance with Section 907 and the *International Mechanical Code*. Under the categories and occupancies in Table 3412.6.8, determine the appropriate value and enter that value into Table 3412.7 under Safety Parameter 3412.6.8, Automatic Fire Detection, for fire safety, means of egress and general safety.

❖ This section considers the use of smoke detectors in a building. To receive credit for the smoke detectors, they must be connected to audible alarms and installed in accordance with Section 907 of the code and Section 606 of the IMC.

TABLE 3412.6.8. See next column.

❖ Table 3412.6.8 assigns values for each occupancy. The use of detectors increases the safety in a building by providing early warning to occupants. The lack of detectors and the associated early warning are considered less than optimum for safety in occupancies with high population densities and high combustible loads. Large deficiency points are accrued if adequate detection systems are not provided.

TABLE 3412.6.8
AUTOMATIC FIRE DETECTION VALUES

OCCUPANCY	CATEGORIES				
	a	b	c	d	e
A-1, A-3, F, M, R, S-1	-10	-5	0	2	6
A-2	-25	-5	0	5	9
A-4, B, E, S-2	-4	-2	0	4	8

Example:

Assume a four-story Group R-1 motel with corridors and an elevator lobby on each story. Smoke detectors are installed throughout the corridors, closets, rooms and elevator lobbies. Only single-station detectors are installed in the guestrooms. There are smoke detectors in the HVAC return-air system and a fire alarm system is provided. An analysis of these building characteristics results in the selection of Category d. To be classified in Category e, guestrooms must have detectors connected to the building's emergency electrical system and annunciated by each room at a constantly attended location, such as the front desk. Additionally, the fire alarm system must be capable of being manually activated by the front desk when a smoke detector operates.

Category d value from Table 3412.6.8 = 2.

Enter 2 in Table 3412.7 under Safety Parameter 3412.6.8, Automatic Fire Detection.

3412.6.8.1 Categories. The categories for automatic fire detection are:

1. Category a—None.
2. Category b—Existing *smoke detectors* in HVAC systems and maintained in accordance with the *International Fire Code*.
3. Category c—*Smoke detectors* in HVAC systems.The detectors are installed in accordance with the requirements for new buildings in the *International Mechanical Code*.
4. Category d—*Smoke detectors* throughout all floor areas other than individual *sleeping units*, tenant spaces and *dwelling units*.
5. Category e—*Smoke detectors* installed throughout the floor area.

❖ The categories are based on the location and completeness of the smoke detectors. The categories represent a graduation in the levels of smoke detection that ranges from no detectors in Category a to full detection throughout all fire area spaces in Category e.

Category a is a facility that has no automatic smoke detection system.

Category b assumes that there are some smoke detectors in an existing building's HVAC system, but not to the extent required for new construction. If the limited HVAC system detectors are maintained in accordance with the IFC, the detection system qualifies as Category b.

Category c addresses existing buildings that have upgraded HVAC systems with duct detectors, as specified by the IMC. The detection values for Category c are zero, since this is the level of protection required for all new construction.

Category d is the classification for existing buildings that have full detection coverage throughout the public common spaces and secondary rooms and areas.

Category e has smoke detectors throughout floor areas including individual units.

3412.6.9 Fire alarm systems. Evaluate the capability of the *fire alarm system* in accordance with Section 907. Under the categories and occupancies in Table 3412.6.9, determine the appropriate value and enter that value into Table 3412.7 under Safety Parameter 3412.6.9, Fire Alarm Systems, for fire safety, means of egress and general safety.

❖ This section evaluates the capabilities of building fire alarm systems that are separate from the automatic fire detection system evaluated in Section 3412.6.8. A fire alarm system that is manually operated or activated by smoke detectors or sprinkler waterflow devices alerts the occupants to a fire situation. The fire alarm system will notify the occupants with visible or audible alarms so they may begin their egress and discharge from the building. Buildings of assembly, business, educational or residential occupancies can have large numbers of occupants in rooms with concentrated seating or people who are sleeping.

TABLE 3412.6.9
FIRE ALARM SYSTEM VALUES

OCCUPANCY	CATEGORIES			
	a	b[a]	c	d
A-1, A-2, A-3, A-4, B, E, R	-10	-5	0	5
F, M, S	0	5	10	15

a. For buildings equipped throughout with an *automatic sprinkler system*, add 2 points for activation by a sprinkler waterflow device.

❖ Table 3412.6.9 gives values for each occupancy and type of fire alarm system provided. It reflects the idea that the presence of an alarm system in a building usually creates a safer condition for the occupants when compared to a building without an occupant notification system.

Deficiency points are assigned to high population densities, such as Groups A-1, A-2, A-3, A-4 and E, which fit into Category a.

Example:

Assume a one-story assembly building has a complete manual fire alarm system, a voice/alarm communication system, a public address system and a fire command center that does not contain status indicators and controls for the air-handling system. The fire command center does not have emergency power or lighting system controls. It also does not have a fire department communication panel. As a result, the building is classified in Category c.

Category c value from Table 3412.6.9 = 0.

Enter 0 in Table 3412.7 under Safety Parameter 3412.6.9, Fire Alarm System.

3412.6.9.1 Categories. The categories for *fire alarm systems* are:

1. Category a—None.
2. Category b—*Fire alarm system* with *manual fire alarm boxes* in accordance with Section 907.4 and alarm notification appliances in accordance with Section 907.5.2.
3. Category c—*Fire alarm system* in accordance with Section 907.
4. Category d—Category c plus a required *emergency voice/alarm communications* system and a *fire command center* that conforms to Section 403.4.6 and contains the *emergency voice/alarm communications* system controls, fire department communication system controls and any other controls specified in Section 911 where those systems are provided.

❖ These categories are defined by the fire alarm system that is provided within the existing building.

Category a means that there is no fire alarm system in the building or it does not conform to all the requirements of Section 907. This includes a system that does not have a secondary power supply in accordance with Section 907 or one where the zones of a floor exceed 22,500 square feet (1858 m^2) as noted in Section 907.6.3.

Category b applies when the manual fire alarm boxes comply with Section 907.4.2 and NFPA 72. The alarm notification appliances, specifically the audible alarms, are in accordance with Section 907.5.2.1. Location, height and color of the manual fire alarm boxes must be specifically evaluated for compliance, along with the sound levels of the audible alarms when compared to the normal sound levels within the existing building.

Category c requires that the fire alarm system must comply with Section 907 and NFPA 72. This category indicates that a complete fire alarm system is present in the existing building and that the system complies with all the requirements for new construction.

Category d is applicable for a building that is provided with a fire command center for fire department operations that conforms to Section 911. Section 911 may be used when evaluating any building that complies with its performance requirements. This section specifies the fire command operations required, including items such as the following:

- Voice/alarm communication system panels.
- Fire department communication panels.
- Fire detection and alarm system annunciator panels.

- An annunciator for visually indicating floor locations of elevators and whether they are operational.
- Status indicators and controls for air-handling systems.
- Controls for unlocking all stairway doors simultaneously.
- Sprinkler valve and waterflow detector display panels.
- Emergency and standby power status indicators.
- Elevator fire recall switch in accordance with ASME A17.1.

3412.6.10 Smoke control. Evaluate the ability of a natural or mechanical venting, exhaust or pressurization system to control the movement of smoke from a fire. Under the categories and occupancies in Table 3412.6.10, determine the appropriate value and enter that value into Table 3412.7 under Safety Parameter 3412.6.10, Smoke Control, for means of egress and general safety.

❖ This section is used to evaluate characteristics that could limit smoke migration in the building, including operable windows, mechanical exhaust systems or pressurized stairway or smokeproof enclosures.

TABLE 3412.6.10
SMOKE CONTROL VALUES

OCCUPANCY	CATEGORIES					
	a	b	c	d	e	f
A-1, A-2, A-3	0	1	2	3	6	6
A-4, E	0	0	0	1	3	5
B, M, R	0	2[a]	3[a]	3[a]	3[a]	4[a]
F, S	0	2[a]	2[a]	3[a]	3[a]	3[a]

a. This value shall be 0 if compliance with Category d or e in Section 3412.6.8.1 has not been obtained.

❖ Table 3412.6.10 assigns values for each occupancy and category of smoke control in the building. The table indicates the relative risk to occupants if adequate smoke control methods are not provided in the building.

In occupancies having a higher population density, zero points are indicated if no smoke control or limited smoke control is provided. The table also indicates that smoke control is extremely important in the exit stairways of a building. Note a of the table assigns zero credit points for Groups B, M, R, F and S unless the building complies with Category d or e in Section 3412.6.8.1, which are the automatic fire detection system requirements.

Example 1:

Assume a one-story Group A-3 church sanctuary. It has no operable windows and no smoke control system.

Category a from Table 3412.6.10 = 0.

Enter 0 in Table 3412.7 under Safety Parameter 3412.6.10, Smoke Control, for only means of egress (ME) and general safety (GS) parameters. No entry is to be made under the fire safety (FS) parameter.

Example 2:

Assume a three-story Group E high school. It has operable windows throughout the building. The stairways are interior without windows or a pressurization system.

Category b from Table 3412.6.10 = 0.

Enter 0 in Table 3412.7 under Safety Parameter 3412.6.10, Smoke Control, for means of egress (ME) and general safety (GS) parameters.

Example 3:

Assume a two-story Group A-2 nightclub with a smoke control system that meets the requirements for Category e.

Category e from Table 3412.6.10 = 6.

Enter 6 in Table 3412.7 under Safety Parameter 3410.6.10, Smoke Control, for means of egress and general safety parameters.

Example 4:

Assume a six-story Group B office building with three stairways: one is a smokeproof enclosure conforming to Section 1022.10, one is pressurized in accordance with Section 909.20.5; and one has operable exterior windows. The building has smoke detectors throughout all floor spaces and fire areas.

Category f from Table 3412.6.10 = 4.

Enter 4 in Table 3412.7 under Safety Parameter 3412.6.10, Smoke Control, for means of egress (ME) and general safety (GS).

If detectors are omitted from some of the offices of the building, then a score of 0 must be entered in Table 3412.7. This is determined from Note a in Table 3412.6.10.

3412.6.10.1 Categories. The categories for smoke control are:

1. Category a—None.
2. Category b—The building is equipped throughout with an *automatic sprinkler system*. Openings are provided in exterior walls at the rate of 20 square feet (1.86 m^2) per 50 linear feet (15 240 mm) of *exterior wall* in each *story* and distributed around the building perimeter at intervals not exceeding 50 feet (15 240 mm). Such openings shall be readily openable from the inside without a key or separate tool and shall be provided with ready access thereto. In lieu of operable openings, clearly and permanently marked tempered glass panels shall be used.
3. Category c—One enclosed exit stairway, with ready access thereto, from each occupied floor of the build-

ing. The *stairway* has operable exterior windows and the building has openings in accordance with Category b.

4. Category d—One smokeproof enclosure and the building has openings in accordance with Category b.

5. Category e—The building is equipped throughout with an *automatic sprinkler system*. Each floor area is provided with a mechanical air-handling system designed to accomplish smoke containment. Return and exhaust air shall be moved directly to the outside without recirculation to other floor areas of the building under fire conditions. The system shall exhaust not less than six air changes per hour from the floor area. Supply air by mechanical means to the floor area is not required. Containment of smoke shall be considered as confining smoke to the floor area involved without migration to other floor areas. Any other tested and *approved* design which will adequately accomplish smoke containment is permitted.

6. Category f—Each *stairway* shall be one of the following: a smokeproof enclosure in accordance with Section 1022.9; pressurized in accordance with Section 909.20.5 or shall have operable exterior windows.

❖ The six categories to be evaluated are compared to the occupancies to determine the score to be entered in Table 3412.7 under Safety Parameter 3412.6.10, Smoke Control.

Category a means there is no method of controlling smoke in the building.

Category b means the existing building is sprinklered and there are exterior windows that can be readily opened without the use of keys or tools. This category recognizes the fire safety benefits of automatic sprinkler systems. Operable panels or windows in the exterior walls must be provided at the rate of 20 square feet per 50 lineal feet (1.85 m^2 per 15 240 mm) of exterior wall in each story. The openings must be distributed around the perimeter of the building at intervals no more than 50 feet (15 240 mm) between windows. A story with very high ceilings may have a strip of windows located with latches or controls that are not easily reachable. These openings would not meet Category b standards, even if they are the correct sizes.

Category c requires at least one enclosed exit stairway to have operable windows that open to the exterior. Additionally, the building must have operable windows complying with the size and spacing requirements of Category b.

Category d requires a minimum of one smokeproof enclosure and operable windows in the building. These operable windows are the same as those addressed in Category b. By definition, a "Smokeproof enclosure" refers to an enclosed interior exit stairway that conforms to Section 1009 as designated in Section 1022.9.

Section 1022.9 requires stairways to be protected from smoke by one of two methods: a smokeproof enclosure or stairway pressurization design (see Section 909.20.5).

Additional requirements that must be considered when evaluating a smokeproof enclosure are access (see Section 909.20.1), construction (see Section 909.20.2), ventilating equipment (see Section 909.20.6) and standby power (see Section 909.20.6.2). Category e recognizes the use of a mechanical smoke control system as described by this category. Each fire area within the building must have a mechanical smoke control system. Return and exhaust air from the system must be discharged directly to the exterior to achieve the necessary level of protection. A specific air change requirement is provided, independent of the fire area's volume or size. Note that a smoke control system designed in accordance with Section 909 is also considered acceptable.

Category f recognizes the merits of smokeproof enclosures, pressurized stairways and stairways with operable exterior windows. To be classified in this category, all stairways in the building must comply with the requirements of any one, or a combination of the three types of stairways. It is possible for a single building to have a smokeproof enclosure, a pressurized stairway and a stairway with operable exterior windows.

Both the categories and the occupancy of the building are used in determining the score that will be entered in Table 3412.7. This score is determined from Table 3412.6.10. Note a in Table 3412.6.10 can have a significant impact on the scores. The note states that, even if the building has some level of smoke control, it is to receive no credit if it does not have an automatic fire detection system complying with Category d or e in Section 3412.6.8.1.

3412.6.11 Means of egress capacity and number. Evaluate the *means of egress* capacity and the number of exits available to the building occupants. In applying this section, the *means of egress* are required to conform to the following sections of this code: 1003.7, 1004, 1005, 1014.2, 1014.3, 1015.2, 1021, 1024.1, 1027.2, 1027.5, 1028.2, 1028.3, 1028.4 and 1029. The number of exits credited is the number that is available to each occupant of the area being evaluated. Existing fire escapes shall be accepted as a component in the *means of egress* when conforming to Section 3406.

Under the categories and occupancies in Table 3412.6.11, determine the appropriate value and enter that value into Table 3412.7 under Safety Parameter 3412.6.11, Means of Egress Capacity, for means of egress and general safety.

❖ This section addresses the exit capacity and number of existing exits available to the building occupants. Before a building can be evaluated in this category, and Section 3412, in general, it must comply with the sections listed.

The means of egress is required to conform to the following sections for new construction before evaluation:

- Section 1003.7 establishes that elevators, escalators and moving walks cannot be considered as part of the means of egress.
- Section 1004 establishes the minimum number of occupants the exit facilities must accommodate.
- Section 1005.3 defines the capacity of the means of egress by identifying a minimum width of egress component per occupant. This is used to calculate the total capacity of the means of egress component.
- Section 1014.2 establishes requirements for egress through intervening spaces.
- Section 1014.3 establishes requirements for common path of egress travel.
- Section 1015.2 establishes requirements for the remoteness of exit doors and means of egress doors.
- Section 1021 establishes requirements for minimum number of exits.
- Sections 1027.1, 1027.2 and 1027.5 establish requirements related to the exit discharge.
- Sections 1028.2, 1028.3 and 1028.4 establish requirements related to exits and lobbies for assembly occupancies.
- Section 1029 establishes requirements for emergency escape and rescue.
- Section 3406.1.2 permits the continued use of existing fire escapes in existing buildings only.

Evaluation of the means of egress capacity involves all egress components, including exit access, exits and exit discharge. This evaluation is correlated to the requirements for the means of egress for new construction.

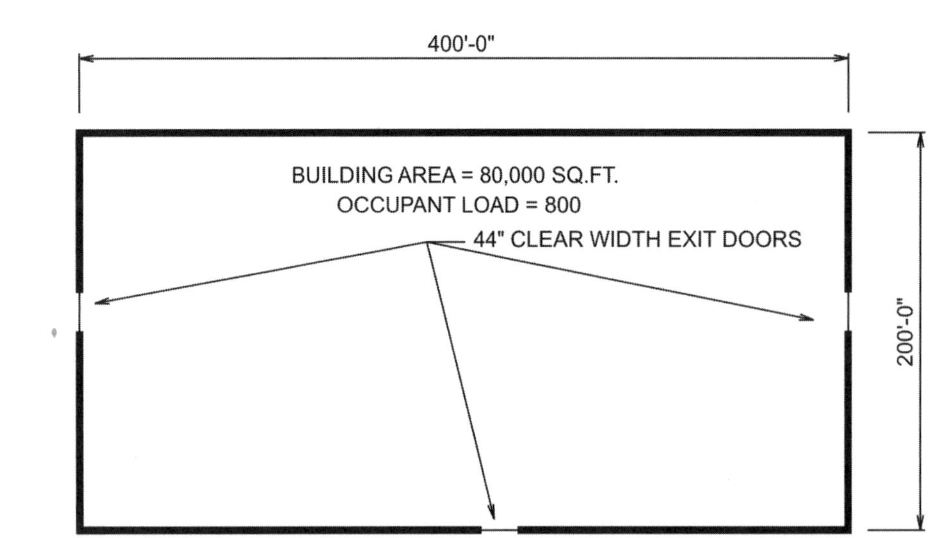

- OCCUPANCY GROUP F
- COMPLETE SPRINKLER SYSTEM
- EGRESS WIDTH PER SECTION 1005.3.2
- THE EXIT CAPACITY OF THE THREE 44" CLEAR WIDTH EXIT DOORS $= \frac{3 \times 44"}{0.15} = 880$ OCCUPANTS

- CLASSIFICATION = CATEGORY b. THE EXIT CAPACITY OF 880 EXCEEDS THE ALLOWABLE CAPACITY OF 800, COMPLYING WITH SECTION 1005
- SCORE = 0 FROM TABLE 3412.6.11
- ENTER 0 IN TABLE 3412.7

For SI: 1 foot = 304.8 mm,
1 square foot = 0.0929 m^2.

Figure 3412.6.11(1)
MEANS OF EGRESS VALUES—CATEGORY b

TABLE 3412.6.11
MEANS OF EGRESS VALUES

OCCUPANCY	CATEGORIES				
	a[a]	b	c	d	e
A-1, A-2, A-3, A-4, E	-10	0	2	8	10
M	-3	0	1	2	4
B, F, S	-1	0	0	0	0
R	-3	0	0	0	0

a. The values indicated are for buildings six stories or less in height. For buildings over six stories above grade plane, add an additional -10 points.

❖ Table 3412.6.11 assigns values for each occupancy and category from Section 3412.6.11.1. The Table gives credit for providing additional exits and exit capacities beyond the minimum. These positive values indicate a greater degree of safety than is required under current code provisions. This additional evaluation contributes to the overall assessment of the building's safety and may offset other safety deficiencies [see Figures 3412.6.11(1) and 3412.6.11(2)]. Note a of the Table provides a significant penalty for high-rise buildings that use fire escapes as part of the means of egress. In this case, if the building is seven stories or greater in height and uses fire escapes, -10 points must be added to the values already listed for Category a buildings.

3412.6.11.1 Categories. The categories for Means of Egress Capacity and number of exits are:

1. Category a—Compliance with the minimum required *means of egress* capacity or number of exits is achieved through the use of a fire escape in accordance with Section 3406.
2. Category b—Capacity of the *means of egress* complies with Section 1004 and the number of exits complies with the minimum number required by Section 1021.
3. Category c—Capacity of the *means of egress* is equal to or exceeds 125 percent of the required *means of egress* capacity, the *means of egress* complies with the minimum required width dimensions specified in the code and the number of exits complies with the minimum number required by Section 1021.
4. Category d—The number of exits provided exceeds the number of exits required by Section 1021. Exits shall be located a distance apart from each other equal to not less than that specified in Section 1015.2.

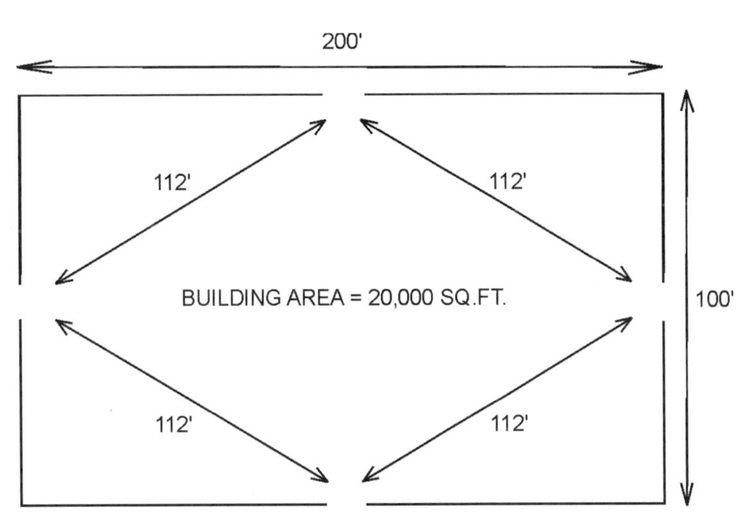

• OCCUPANCY GROUP A-3
• BUILDING FULLY SPRINKLERED
• NUMBER OF OCCUPANTS = 400
• TABLE 1021.2(2) AND SECTION 1021.3 REQUIRES TWO EXITS, FOUR EXTERIOR DOORS (EXITS) ARE PROVIDED
• THE $^1/_3$ DIAGONAL (Exception 2, Section 1015.2.1) = $\frac{224'}{3}$ = 75'
• EACH DOOR IS SPACED 112' APART
• CLASSIFICATION = CATEGORY d FROM TABLE 3412.6.11
• ENTER 8 IN TABLE 3412.7

For SI: 1 foot = 304.8 mm,
1 square foot = 0.0929 m^2.

Figure 3412.6.11(2)
MEANS OF EGRESS VALUES—CATEGORY d

5. Category e—The area being evaluated meets both Categories c and d.

❖ Five categories must be evaluated. These categories and the occupancy of the building will determine the score entered in Table 3412.7.

Category a is appropriate for buildings that comply with either the means of egress capacity or the number of exits, including the use of fire escapes. Although the code allows fire escapes to be used as an egress element in existing buildings, this is the least desirable option in any type of building. The code requires a building using fire escapes to use Category a and its corresponding negative points.

Category b is used for buildings that meet the minimum requirements of Section 1004 for capacity of means of egress and Section 1021 for minimum number of exits and continuity.

Category c is used for buildings that exceed the requirements for new construction. This category provides small positive points for existing buildings that meet all of the following requirements:

- The capacity of all the means of egress components is greater than or equal to 125 percent of the required capacity.
- All of the means of egress components comply with the minimum required widths [32 inch (813 mm) clear for doors, corridors that are 44 inches (1118 mm) wide and stairways].
- The minimum number of exits is provided based on the number of occupants on each floor level.

By providing oversized egress capacity and minimum egress width requirements, an existing building has added safety, which is rewarded with a small number of positive points.

Category d simply relates to buildings where more exits are provided than required by Section 1021, which contributes a positive factor to the building's exit capacity.

Before credit can be given to an existing building with a greater number of exits, the exits must be evaluated for their remoteness and independence from each other. All of the exits must be located at least one-half the length of the diagonal from each other before Category d can be used. If the building is equipped throughout with an automatic sprinkler system in accordance with NFPA 13 or 13R, the exits can be located one-third the length of the diagonal from each other.

Category e provides additional credit for existing buildings that have the characteristics of both Categories c and d. Such a building, which has oversized capacities in its egress elements, minimum required clear widths for the egress components and additional exits that are all remotely located from one another, is considered to have the highest degree of safety; however, this is true for only some of the applicable occupancies. In Group B, F, S and R occupancies, any additional capacity or increased number of exits is not a significant factor. No positive points are awarded for this life safety issue for these occupancies.

3412.6.12 Dead ends. In spaces required to be served by more than one *means of egress*, evaluate the length of the *exit* access travel path in which the building occupants are confined to a single path of travel. Under the categories and occupancies in Table 3412.6.12, determine the appropriate value and enter that value into Table 3412.7 under Safety Parameter 3412.6.12, Dead Ends, for means of egress and general safety.

❖ This section is used to evaluate dead-end exit access conditions within the building. This section uses the terminology "confined to a single path of travel." Another way to illustrate the meaning of this is "a single direction of travel to reach an exit." A typical corridor is a single path in which a building occupant can travel in two directions to reach exits. A dead end may be a single path, but the key feature of a dead end is that only one direction is available to reach an exit. When building occupants have only one direction of travel available, a potentially hazardous condition is created because they may become trapped if the direction of travel is blocked by fire or smoke.

This section addresses only dead ends that are a component of exit access travel, which is "that portion of a means of egress that leads to an entrance to an exit." Although a room containing only one door may restrict exiting from the room to only one direction, it is not considered a dead-end condition as described in Section 1018.4. This section notes that the dead-end path of travel is limited to 20 feet (6096 mm) for corridors that serve more than one exit; therefore, the single-door exit of a room enters a "corridor," which is subject to a dead-end limitation of 20 feet (6096 mm). The travel distance within the room to the exit access is limited by Section 1014.3 and Table 1015.1. The total travel distance from the remote point within a room to the exit must include the room travel length, plus corridor length (see commentary, Section 3412.6.13). Section 3412.6.12 addresses the condition of dead-end passageways and corridors (see commentary, Section 1018.4).

TABLE 3412.6.12
DEAD-END VALUES

OCCUPANCY	CATEGORIES[a]		
	a	b	c
A-1, A-3, A-4, B, E, F, M, R, S	-2	0	2
A-2, E	-2	0	2

a. For dead-end distances between categories, the dead-end value shall be obtained by linear interpolation.

❖ The table reflects the relative degree of hazard associated with dead-end passageways and corridors. This is shown by the deficiency points that apply where a dead-end exceeds 35 or 70 feet (10 668 or 21 336 mm) in an occupancy with a relatively high occupant load, such as an assembly occupancy. Note a allows for the interpolation for actual dead-end lengths between the distances specified in the categories. For example, if a building contains a corridor that has a dead-end length of 10 feet (3048 mm) at each end beyond the exits, the value of 1 is used; this is the midpoint between Categories b and c.

3412.6.12.1 Categories. The categories for dead ends are:

1. Category a—Dead end of 35 feet (10 670 mm) in nonsprinklered buildings or 70 feet (21 340 mm) in sprinklered buildings.
2. Category b—Dead end of 20 feet (6096 mm); or 50 feet (15 240 mm) in Group B in accordance with Section 1018.4, exception 2.
3. Category c—No dead ends; or ratio of length to width (l/w) is less than 2.5:1.

❖ This section defines the categories for dead-end exit access conditions. Category a allows dead-end conditions of up to 35 feet (10 668 mm) for existing buildings that are not fully sprinklered, and up to 70 feet (21 336 mm) for existing buildings that are fully sprinklered. These distances correspond to the dead-end lengths listed in the IFC. Since these lengths far exceed the allowable lengths permitted for new construction, negative values are associated with this category.

Category b is the classification for buildings that comply with the requirements for new construction. This zero-based category provides no extra credit for complying with new construction requirements.

Category c represents conditions that exceed the requirements for new construction. If there are no dead-end corridors or the "corridor" is more of a "space," Category c can be used. In Category c, the length-to-width ratio of 2.5 to define a space/corridor is the same as Exception 2 in Section 1018.4. Such a space allows the building user a more circular route and full view of the space; therefore, this does not constitute a dead-end situation.

These categories and occupancies of the building are used in Table 3412.6.12 to determine the score to be entered in Table 3412.7 under Safety Parameter 3412.6.12, Dead Ends [see Figures 3412.6.12.1(1), 3412.6.12.1(2) and 3412.6.12.1(3)].

Figure 3412.6.12.1(1)
TYPICAL DEAD-END CORRIDOR

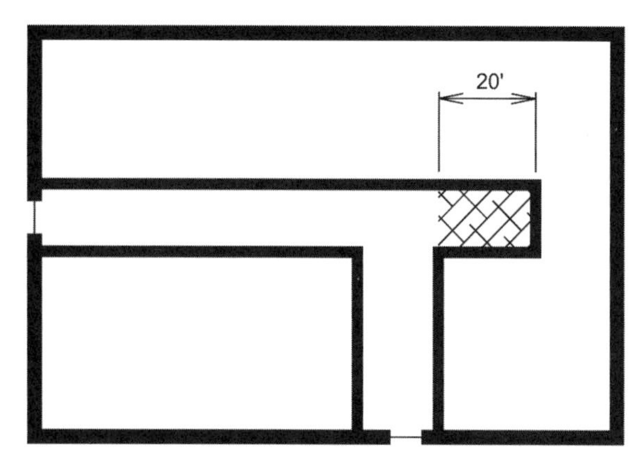

- OCCUPANCY GROUP B WITH MOVABLE PARTITIONS AND FURNITURE 6'-6" HIGH
- DEAD-END PASSAGEWAY IS 20'-0"
- CLASSIFICATION = CATEGORY b
- VALUE = 0 FROM TABLE 3412.6.12
- ENTER 0 IN TABLE 3412.7

For SI: 1 inch = 25.4 mm, 1 foot = 304.8 mm.

Figure 3412.6.12.1(2)
DEAD-END VALUES—CATEGORY b

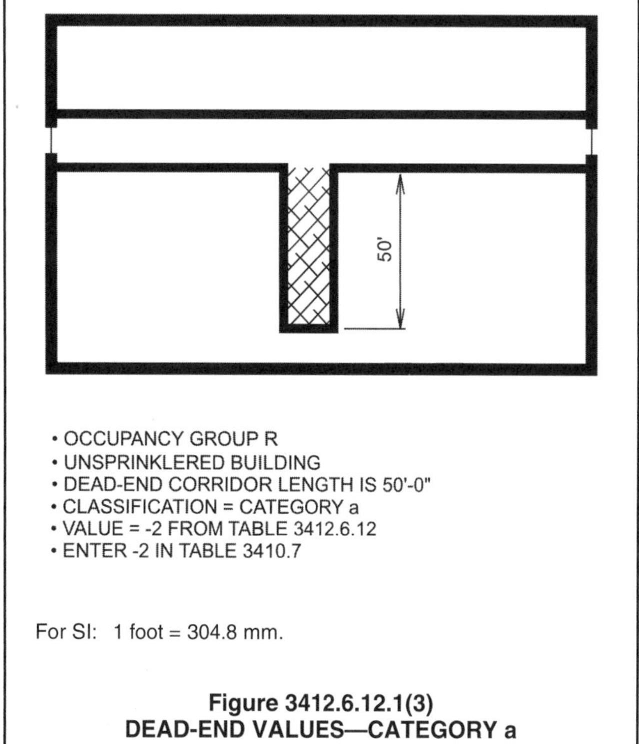

50'
• OCCUPANCY GROUP R
• UNSPRINKLERED BUILDING
• DEAD-END CORRIDOR LENGTH IS 50'-0"
• CLASSIFICATION = CATEGORY a
• VALUE = -2 FROM TABLE 3412.6.12
• ENTER -2 IN TABLE 3410.7

For SI: 1 foot = 304.8 mm.

Figure 3412.6.12.1(3)
DEAD-END VALUES—CATEGORY a

3412.6.13 Maximum exit access travel distance. Evaluate the length of *exit access* travel to an *approved exit*. Determine the appropriate points in accordance with the following equation and enter that value into Table 3412.7 under Safety Parameter 3412.6.13, Maximum Exit Access Travel Distance, for means of egress and general safety. The maximum allowable *exit access* travel distance shall be determined in accordance with Section 1016.1.

$$\text{Points} = 20 \times \frac{\text{Maximum allowable travel distance} - \text{Maximum actual travel distance}}{\text{Max. allowable travel distance}}$$

(Equation 34-6)

❖ The length of exit access travel distance is evaluated in comparison to the travel distance allowed by Table 1016.1 for a particular occupancy. The exit access travel distance is measured from the most remote point in the building to the nearest exit. Total exit access travel length of Table 1016.1 includes travel distance within a room or space plus exit access corridor travel to the exit. Some modifications to the Table 1016.1 requirements are listed for certain occupancies and buildings as listed in the table's notes.

To determine the value to be assigned for maximum travel distances to an exit, the specified equation must be used. This equation allows a graduated scale to be used to evaluate compliance of an existing situation with new construction travel distance requirements. With the equation, a more definite evaluation can occur with a broader range of scores. Any existing building having overall travel distances less than those specified in Table 1016.1 will achieve a positive credit. Existing buildings with travel distances greater than those allowed for new construction will be assigned negative points based on the extent the travel distance exceeds the allowable limit.

For example, an existing Group B business building has a travel distance of 150 feet (45 720 mm). The distance is measured from the most remote corner of a partitioned office, down a corridor and to an exit door. Table 1016.1 permits an unsprinklered Group B building to have a travel distance of 200 feet (60 960 mm). The equation yields the following:

$$\text{Points} = 20 \times \left(\frac{200\text{ feet} - 150\text{ feet}}{200\text{ feet}}\right) = 5.0$$

See Figures 3412.6.13(1), 3412.6.13(2) and 3412.6.13(3) for other examples.

3412.6.14 Elevator control. Evaluate the passenger elevator equipment and controls that are available to the fire department to reach all occupied floors. Emergency recall and in-car operation of elevator recall controls shall be provided in accordance with the *International Fire Code*. Under the categories and occupancies in Table 3412.6.14, determine the appropriate value and enter that value into Table 3412.7 under Safety Parameter 3412.6.14, Elevator Control, for fire safety, means of egress and general safety. The values shall be zero for a single-story building.

❖ The availability of elevators in the building and their incapability in an emergency are evaluated. This section addresses four different categories of elevators in existing buildings. This section does not address the requirements of Chapter 30. Access must be provided to all occupied floors by passenger elevators for the purposes of this evaluation. Freight elevators cannot be considered, since they may be in locations not readily accessible for fire-department use. Emergency elevator operation must comply with the IFC for either a Phase I Emergency Recall or II Emergency In-car Operation category. One-story buildings, including those with mezzanines served by an elevator, are awarded zero value.

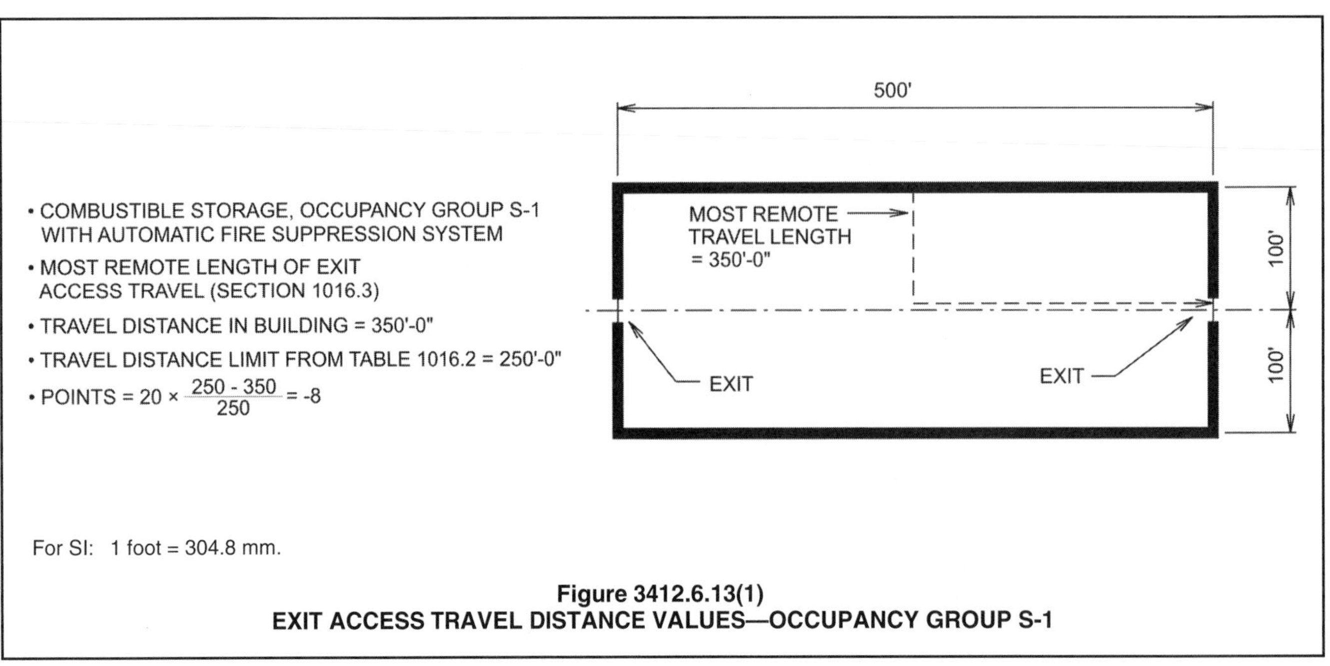

• COMBUSTIBLE STORAGE, OCCUPANCY GROUP S-1 WITH AUTOMATIC FIRE SUPPRESSION SYSTEM
• MOST REMOTE LENGTH OF EXIT ACCESS TRAVEL (SECTION 1016.3)
• TRAVEL DISTANCE IN BUILDING = 350'-0"
• TRAVEL DISTANCE LIMIT FROM TABLE 1016.2 = 250'-0"
• POINTS = 20 × (250 - 350)/250 = -8
500'
MOST REMOTE TRAVEL LENGTH = 350'-0"
100'
100'
EXIT
EXIT
For SI: 1 foot = 304.8 mm.

Figure 3412.6.13(1)
EXIT ACCESS TRAVEL DISTANCE VALUES—OCCUPANCY GROUP S-1

TABLE 3412.6.14. See below.

❖ The table assigns values based on the types of controls the elevators have and the distance the elevators must travel to reach the floors they serve. The 25-foot (7620 mm) threshold of elevator travel is based on the IFC requirements and ASME A17.3 for existing elevators. Elevator travel distance is based on the level where the elevator is accessed by the fire department. Usually, this is the ground floor or grade-level floor and corresponds to the level where fire command stations or fire alarm system control units are located. Any elevator that travels 25 feet (7620 mm) or more must always be provided with Phase I Emergency Recall and Phase II Emergency In-car Operation Capabilities.

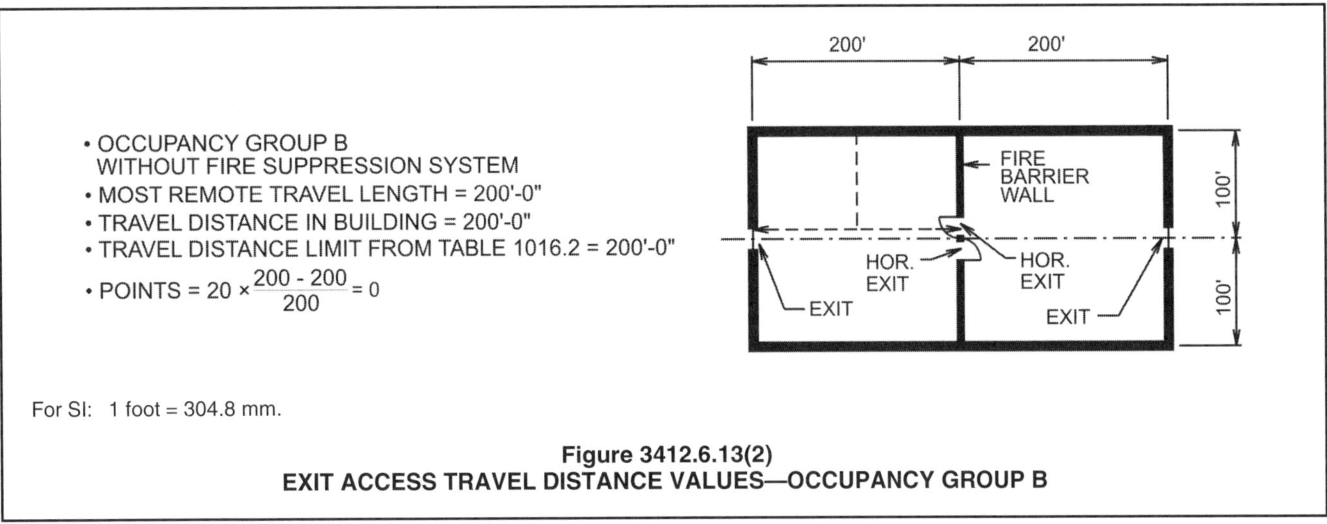

Figure 3412.6.13(2)
EXIT ACCESS TRAVEL DISTANCE VALUES—OCCUPANCY GROUP B

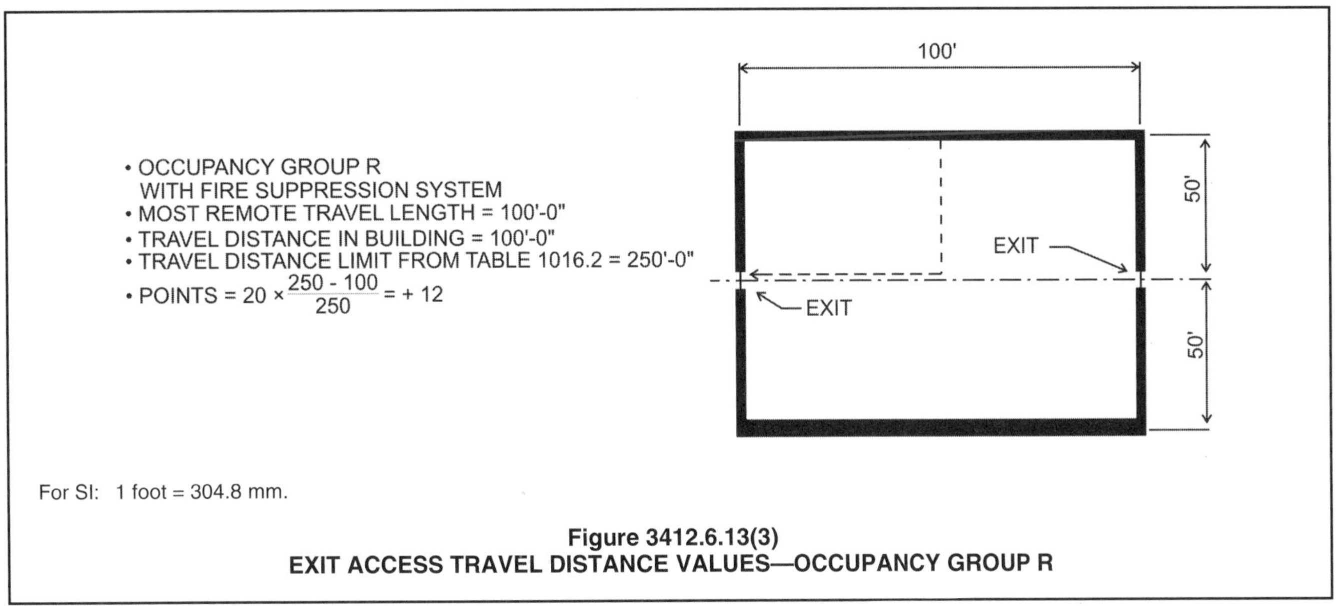

Figure 3412.6.13(3)
EXIT ACCESS TRAVEL DISTANCE VALUES—OCCUPANCY GROUP R

TABLE 3412.6.14
ELEVATOR CONTROL VALUES

ELEVATOR TRAVEL	CATEGORIES			
	a	b	c	d
Less than 25 feet of travel above or below the primary level of elevator access for emergency fire-fighting or rescue personnel	-2	0	0	+2
Travel of 25 feet or more above or below the primary level of elevator access for emergency fire-fighting or rescue personnel	-4	NP	0	+4

For SI: 1 foot = 304.8 mm.
NP = Not permitted

Example:

Assume a Group B business occupancy, three stories in height. The elevator lobbies are equipped with smoke detectors that recall the elevator to the main floor or to an alternate floor if the main floor detector is activated. The elevator travels more than 25 feet (7620 mm) above the main floor.

This building must be placed in Category c. Because the elevator travels more than 25 feet (7620 mm) above the main floor, it must be brought into compliance with the IFC, which requires Phase I Emergency Recall and Phase II Emergency In-car Operation. When the elevators are brought into compliance with the IFC, they are automatically placed in Category c. The value from Table 3412.6.14 is 0; therefore, 0 must be entered in Table 3412.7 under Safety Parameter 3412.6.14, Elevator Control.

3412.6.14.1 Categories. The categories for elevator controls are:

1. Category a — No elevator.
2. Category b—Any elevator without Phase I emergency recall operation and Phase II emergency in-car operation.
3. Category c — All elevators with Phase I emergency recall operation and Phase II emergency in-car operation as required by the *International Fire Code*.
4. Category d—All meet Category c; or Category b where permitted to be without Phase I emergency recall operation and Phase II emergency in-car operation; and at least one elevator that complies with new construction requirements serves all occupied floors.

❖ The categories that a building may be placed into range from buildings with no elevators to buildings with elevators complying with new construction requirements and total elevator recall capability.

Category a is for buildings with no elevators. Without an elevator, fire department personnel are required to use the stairways for rescuing people and accessing fire floors. Category a is assigned a negative value.

Category b is for buildings in which elevators are present but have no ability to be recalled or operated by fire fighters. Without recall or fire department control, the elevators are allowed vertical travel distances of less than 25 feet (7620 mm). This approach is consistent with the IFC, which requires controls for elevators reaching 25 feet (7620 mm) or more.

Category c is used for buildings with elevators that have automatic recall when required by the IFC.

Category d is used for buildings when the elevator controls comply with either Category b or c and at least one of the elevators in the existing building serves all occupied floor levels. The controls must also comply with all the requirements for new construction. This category recognizes the benefits of having all elevators with Phase I Emergency Recall and II Emergency In-car Operation controls, as well as having a newly constructed elevator with all the features to facilitate fire-fighting and rescue operations.

Based on the controls provided, or the lack of elevators, the appropriate category is determined. This category and the distance the elevator travels are then used along with Table 3412.6.14 to determine the value score the building will receive for this item. The value is then entered in Table 3412.7.

3412.6.15 Means of egress emergency lighting. Evaluate the presence of and reliability of *means of egress* emergency lighting. Under the categories and occupancies in Table 3412.6.15, determine the appropriate value and enter that value into Table 3412.7 under Safety Parameter 3412.6.15, Means of Egress Emergency Lighting, for means of egress and general safety.

❖ Lighting throughout the entire means of egress is evaluated in this section. Illumination of the means of egress is essential in a building during normal occupancy. During a fire condition, illumination becomes even more important, since visibility will more than likely be reduced by the buildup of smoke. Failure to provide adequate lighting in a fire condition has led to numerous deaths because occupants become disoriented and are unable to follow the means of egress to safety. It is appropriate to include egress lighting as one of the major building components that must be evaluated. The importance of this lighting is shown by the minimal values assigned by Table 3412.6.15 when egress lighting is not provided.

TABLE 3412.6.15
MEANS OF EGRESS EMERGENCY LIGHTING VALUES

NUMBER OF EXITS REQUIRED BY SECTION 1015	CATEGORIES		
	a	b	c
Two or more exits	NP	0	4
Minimum of one exit	0	1	1

❖ This table is used to assess the relative risk to the occupants of the building when egress lighting is not adequate.The table reinforces the idea that when a building is required to have two or more exits, lighting has more significance. The lighting values are less critical in smaller buildings where only one exit is required because the hazard of a substantial occupant load locating and using the means of egress is minimal.

Example:

A Group A church building with a number of classrooms that will accommodate up to 60 people has at least 1 footcandle (11 lux) of illumination at the floor level throughout the entire means of egress. The means of egress lighting in the sanctuary, corridors and stairways is provided with battery back-up power that is sized to provide lighting for 1 hour. The classroom lighting is wired to the main switch panel in the building and has no emergency power source.

The lighting must be placed in Category a because the classrooms are part of the means of egress (exit access) and the lack of emergency power does not comply with Section 1006.3. The classroom will accommodate up to 60 occupants; therefore, in accordance with Section 1015.1 and Table 1015.1, the rooms are required to have two means of egress. The lighting must be provided with an emergency power source in accordance with Section 1006.3. The value determined from Table 3412.6.15 is "NP." Emergency power must be added to the lighting in the classrooms. The resulting Category b has a value of 0. Enter 0 in Table 3412.7.

3412.6.15.1 Categories. The categories for means of egress emergency lighting are:

1. Category a—*Means of egress* lighting and *exit* signs not provided with emergency power in accordance with Chapter 27.
2. Category b—*Means of egress* lighting and *exit* signs provided with emergency power in accordance with Chapter 27.
3. Category c—Emergency power provided to *means of egress* lighting and exit signs which provides protection in the event of power failure to the site or building.

❖ There are three categories of egress lighting. These categories are consistent with the requirements established by the IFC.

Category a is used when there is no egress lighting in any occupied room or space within the building or when the level of illumination is less than 1 foot-candle (11 lux) at the floor level at any location in the means of egress, other than the exempted locations indicated in Section 1006 (see commentary, Section 1006). If means of egress lighting and exit signs are provided but are not connected to emergency power, Category a is also applicable. Without emergency power providing a 90-minute duration after primary power loss, the means of egress lighting and exit signs would not provide as reliable a level of safety to the building occupants.

Category b includes buildings where the means of egress lighting and exit signs comply with the requirements of Sections 1006 and 1011, respectively, including emergency power requirements. Section 1006.3 includes requirements for emergency power to maintain continuous illumination of the means of egress during a power failure. This category represents the zero-based criteria, which is the level required for new construction.

Category c is used when emergency power is provided for means of egress lighting and exit signs in excess of the minimum requirements for new construction. If the emergency power provides full protection to the site or building during power failure, Category c is applicable. Campus-type complexes or buildings that require extra security may have a power plant available to provide complete back-up power for indefinite periods of time. This will qualify for the positive points of this category.

These categories, along with the number of required exits, are used to determine the appropriate value from Table 3412.6.15. The value determined from Table 3412.6.15 is then entered in Table 3412.7.

3412.6.16 Mixed occupancies. Where a building has two or more occupancies that are not in the same occupancy classification, the separation between the mixed occupancies shall be evaluated in accordance with this section. Where there is no separation between the mixed occupancies or the separation between mixed occupancies does not qualify for any of the categories indicated in Section 3412.6.16.1, the building shall be evaluated as indicated in Section 3412.6 and the value for mixed occupancies shall be zero. Under the categories and occupancies in Table 3412.6.16, determine the appropriate value and enter that value into Table 3412.7 under Safety Parameter 3412.6.16, Mixed Occupancies, for fire safety and general safety. For buildings without mixed occupancies, the value shall be zero.

❖ These provisions are used to evaluate mixed occupancies within an existing building and whether the method of separating mixed occupancies conforms to the requirements of Section 508.4. If Section 508.4 is not completely understood, see the associated commentary before proceeding with the evaluation of the existing building in this section (see also commentary, Section 3412.6).

This section is applicable only to separated mixed occupancies. If a building is a single occupancy, the applicable value for this section is zero. If an existing mixed occupancy building has no fire-resistance-rated separation between the different uses, or the separation is fire-resistance rated for less than 1 hour, the applicable value is also zero. The building must also be evaluated in accordance with Section 3412.6. The zero-based category for this section is equivalent to full compliance with Section 508.4, which is the requirement for new construction.

TABLE 3412.6.16
MIXED OCCUPANCY VALUES[a]

OCCUPANCY	CATEGORIES		
	a	b	c
A-1, A-2, R	-10	0	10
A-3, A-4, B, E, F, M, S	-5	0	5

a. For fire-resistance ratings between categories, the value shall be obtained by linear interpolation.

❖ This table addresses the relative risk of a building in or close to compliance with the provisions for separated mixed occupancies. When mixed occupancies are not separated from each other, the risk from hazards is greater in high-density occupancies, such as Groups A-1 and A-2. This risk is also greater in residential occupancies, because occupants may be sleeping and not fully alert. For this reason, inadequate separation is given greater negative values. In buildings with lower occupant loads, and where the occupants are alert, the risks are relatively lower.

Note a permits linear interpolation of the corresponding values in each category. For example, an unsprinklered, Group B/Group F-2 mixed occupancy building is separated with a $1^1/_2$-hour fire-resistance-rated assembly. From Table 508.4, the required fire-resistance rating of the separation between a Group B and a Group F-2 occupancy for Category b is 2 hours. Since the rating that is provided is only $1^1/_2$ hours, an interpolation halfway between Categories a and b can be used. The resulting points are -2.5.

Example 1:

Assume a three-story building. The first floor is a Group M mercantile occupancy and the upper two stories are Group R residential occupancies. The building is Type VB construction, 9,000 square feet (836 m^2) per floor. It is fully sprinklered and only 25 percent of the perimeter can be reached for fire-fighting purposes. Although Type VB construction is not required to be protected, the Group M occupancy on the first floor is separated from the Group R occupancy on the second floor with a 2-hour fire-resistance-rated floor/ceiling assembly. In this case, Table 508.4 requires only 1-hour separation because the building is equipped throughout with an automatic sprinkler system in accordance with Section 903.3.1.1. The provided 2-hour rating is twice that required; therefore, this building qualifies for Category c. A value of 10 points is assigned to the residential portion of the building, and a value of 5 points is assigned to the mercantile portion (see Section 3412.6 for commentary about mixed uses).

Example 2:

Assume a four-story building of Type IIB construction, fully sprinklered in accordance with NFPA 13. It has 15,000 square feet (13 934 m^2) per floor with 25-percent open perimeter. The building is to be a Group R-1 hotel, except for a portion of the first floor that will be used as a Group A-2 nightclub, occupying 1,000 square feet (93 m^2). Although the building is unprotected Type IIB construction, the floor/ceiling assemblies of the first and second floors are 3 hours and the fire barrier assembly at the first floor between the Group R-1 and A-2 occupancies is also 3 hours. Considering these building characteristics, the following determinations are made:

Because of the presence of an automatic sprinkler system, Table 508.4 requires a 1-hour separation. Although this building is required to have 1-hour separation, it has a 3-hour separation, which is more than double the minimum. This building is classified as Category c. A value of 10 points is awarded to both occupancies.

3412.6.16.1 Categories. The categories for mixed occupancies are:

1. Category a—Occupancies separated by minimum 1-hour *fire barriers* or minimum 1-hour *horizontal assemblies*, or both.
2. Category b—Separations between occupancies in accordance with Section 508.4.
3. Category c—Separations between occupancies having a *fire-resistance rating* of not less than twice that required by Section 508.4.4.

❖ This section addresses three different conditions:

Category a is used when the separation of mixed occupancies does not conform to the fire-resistance ratings specified in Section 508.4. If the ratings between mixed occupancies are less than specified in Section 508.4, but at least a 1-hour rating is provided, Category a is applicable.

Category b is used when the separation of mixed occupancies conforms to Section 508.4 for fire-resistance ratings.

Category c is used when the fire-resistance ratings that separate occupancies in an existing building are no less than twice the requirements in Section 508.4. Category c gives bonus points or increased credits to existing buildings that have higher fire-resistance ratings.

Section 3412 does not require compliance with Section 508.4. Section 3412.6.16 evaluates whether the existing building's compliance with Section 508.4 provides relative safety within the building. If the building does comply with Section 508.4, Section 3412.6.16 acknowledges that there is no basis for assigning negative points, nor is there any basis for assigning positive points for safety; therefore, a neutral zero value is assigned. Whatever the value, it is entered in Table 3412.7 under Safety Parameter 3412.6.16, Mixed Occupancies.

3412.6.17 Automatic sprinklers. Evaluate the ability to suppress a fire based on the installation of an *automatic sprinkler system* in accordance with Section 903.3.1.1. "Required sprinklers" shall be based on the requirements of this code. Under the categories and occupancies in Table 3412.6.17, determine the appropriate value and enter that value into Table 3412.7 under Safety Parameter 3412.6.17, Automatic Sprinklers, for fire safety, means of egress divided by 2 and general safety.

❖ These provisions are used to determine the amount of credit that can be applied to the evaluation for the installation of an automatic sprinkler system in an existing building.

Sprinkler safety parameters appear to provide a double credit for automatic sprinkler systems. Basic credits are provided for: building height, Section 3412.6.1; building area, Section 3412.6.2 and maximum travel distance to an exit, Section 3412.6.13. Throughout the code, additional credits are offered for the use of an automatic sprinkler system.

The evaluation of sprinklers in an existing building is based on whether an automatic sprinkler system is both required and installed. The criteria used to determine when an automatic sprinkler system is required are tied to the same requirements for new construction in Section 903. The thresholds listed in Sections

903.2 through 903.2.12 must be used to evaluate whether those characteristics and occupancies are present and whether a sprinkler system is needed.

The exceptions to Section 903; the requirements for stories (including basements) without openings of Section 903.2.11.1 and the suppression requirements for covered mall buildings, high-rise buildings, public garages and unlimited area buildings must also be evaluated. The determination of whether sprinklers are required in a building or a portion of a building must be done to correctly determine the category applicable to the existing building.

These guidelines for sprinklers allow for a more equitable evaluation of the contribution of that feature to overall building safety. This factor encourages the installation of automatic sprinkler systems in existing buildings by providing substantial negative and positive points.

The values for sprinklers are given in Table 3412.6.17. The appropriate credit values from Table 3412.6.17 are entered in Table 3412.7 under Safety Parameter 3412.6.17, Automatic Sprinklers, for both fire safety (FS) and general safety (GS), but only one-half the value is entered under the means of egress (ME). The one-half credit for egress is allowed because some credits for sprinklers are incorporated into the parameters for means of egress capacity (see Section 3412.6.11), dead-ends (see Section 3412.6.12) and maximum travel distance to an exit (see Section 3412.6.13).

TABLE 3412.6.17
SPRINKLER SYSTEM VALUES

OCCUPANCY	CATEGORIES					
	a	b	c	d	e	f
A-1, A-3, F, M, R, S-1	-6	-3	0	2	4	6
A-2	-4	-2	0	1	2	4
A-4, B, E, S-2	-12	-6	0	3	6	12

❖ This table lists the credit values for the respective categories of Section 3412.6.17.1, based on the occupancy being evaluated in the existing building. The assembly occupancies containing large combustible fuel loads with large occupant loads are included with those occupancies containing large fuel loads. These occupancies represent buildings where experience has shown that adequate sprinkler systems save lives and where an on-site sprinkler system is necessary to supplement the local fire department capabilities.

Group A-2 buildings are contained in a separate line in the table for determining sprinkler system values. This occupancy with its densely packed high occupant loads, ill-defined seating and aisle arrangements and facilities services must be separately evaluated to adequately match its hazards with sprinkler system requirements. Sports arenas and schools, which have high occupant/low-fuel loads, are combined with other low-fuel load occupancies in the last line item in the table. Category c is the zero-based category. The categories to the left of this column contain negative values that define buildings and occupancies that are required to be, but are not, sprinklered or are provided with inadequate sprinkler systems. The three categories to the right of this column contain positive values that represent buildings and occupancies with sprinkler systems that are deemed adequate or comply with current standards and requirements.

3412.6.17.1 Categories. The categories for *automatic sprinkler system* protection are:

1. Category a—Sprinklers are required throughout; sprinkler protection is not provided or the sprinkler system design is not adequate for the hazard protected in accordance with Section 903.
2. Category b—Sprinklers are required in a portion of the building; sprinkler protection is not provided or the sprinkler system design is not adequate for the hazard protected in accordance with Section 903.
3. Category c—Sprinklers are not required; none are provided.
4. Category d—Sprinklers are required in a portion of the building; sprinklers are provided in such portion; the system is one which complied with the code at the time of installation and is maintained and supervised in accordance with Section 903.
5. Category e—Sprinklers are required throughout; sprinklers are provided throughout in accordance with Chapter 9.
6. Category f—Sprinklers are not required throughout; sprinklers are provided throughout in accordance with Chapter 9.

❖ Six categories are defined in this section for evaluating the automatic sprinkler system in an existing building. These categories address all aspects, from existing buildings that are unsprinklered but are required to be sprinklered if new construction, to existing buildings that are sprinklered and are not required to be sprinklered if new construction. These categories reduce the impact of providing sprinkler protection required by Chapter 9 to be sprinklered in new construction. This greater range of points increases the flexibility in the use of this evaluation method.

Category a includes buildings or occupancies that exceed the requirements of Section 903 and are required to be sprinklered throughout the building. Category a buildings are provided with either no sprinkler system, or one that is inadequate and does not provide the required level of protection. To evaluate the existing sprinkler system, a trained fire protection engineer should evaluate the system's design against the applicable referenced standards. This is not the same type of consideration needed in Category d, where the existing system actually meets all of the requirements of any earlier edition of the applicable referenced standards. This category is considered the lowest acceptable level of compliance and is therefore associated with the largest negative values in Table 3412.6.17.

Category b is applicable when only a portion of the existing building is required to be sprinklered by the provisions of Section 903. For example, a multistory building may have two of its stories qualify as windowless stories that require sprinklers, or a mercantile building may have one fire area exceeding 12,000 square feet (1114.8 m^2) and require sprinklers. In both cases, the buildings are without an automatic sprinkler system, or the sprinkler systems that are in place are inadequate and do not comply with the technical provisions of the code. Since only portions of the buildings are required to be sprinklered, the negative sprinkler system values for this category in Table 3412.6.17 are exactly half of the Category a values.

Category c is the zero-based category for this safety parameter. The existing building is not required by Section 903 to be sprinklered and no sprinkler system is provided. Such buildings are neither penalized nor rewarded by Table 3412.6.17.

Category d is similar to Category b because only a portion of the existing building is required by Section 903 to be sprinklered. In this case, there is a sprinkler system in place in that portion of the building and the system was designed at the time of its installation to comply with the requirements of an earlier edition of the applicable standards in Section 903. An example of this is a sprinkler system designed to comply with the hydraulic design criteria of the 1989 edition of NFPA 13 for an ordinary Group 2 hazard. As long as this existing sprinkler system is still properly maintained and supervised, the building qualifies for the small positive values listed in Table 3412.6.17.

Category e addresses existing buildings that are required by the provisions of Chapter 9 to be sprinklered throughout, and are protected throughout by a properly designed, installed and supervised sprinkler system. Although this category more closely represents the requirements for new construction, moderate positive values are awarded by Table 3412.6.17. The evaluation rewards existing buildings that are sprinklered with higher points.

Category f is the highest level of protection and is rewarded with maximum positive points in the accompanying table. Existing buildings, which are not required by Section 903 to be sprinklered but are voluntarily provided with a fully designed, installed and supervised system, provide an added level of protection that justifies substantial bonus points that may ultimately determine whether an existing building meets its mandatory safety scores.

3412.6.18 Standpipes. Evaluate the ability to initiate attack on a fire by making a supply of water available readily through the installation of standpipes in accordance with Section 905. Required standpipes shall be based on the requirements of this code. Under the categories and occupancies in Table 3412.6.18, determine the appropriate value and enter that value into Table 3412.7 under Safety Parameter 3412.6.18, Standpipes, for fire safety, means of egress and general safety.

❖ These provisions are used to determine the amount of credit that can be applied to the evaluation for the installation of a standpipe system in an existing building. Standpipes are provided as a tool for use by fire fighters to aid their fire-fighting operations. The general consensus is that they are not recommended or intended for use by untrained building occupants.

The evaluation of standpipe systems in an existing building, much like that for automatic sprinkler systems, is based on whether a standpipe system is both required and installed. The criteria used to determine when a standpipe system is required are tied to the same requirements for new construction in Section 905.3. The thresholds listed in Sections 905.3.1 through 905.3.8 must be used to evaluate whether those characteristics and occupancies are present and whether a standpipe system would be required for new construction. These parameters for standpipes encourage the installation of standpipe systems in existing buildings by providing substantial negative and positive points.

The values for standpipe systems are given in Table 3412.6.18. The appropriate values from Table 3412.6.18 are entered in Table 3412.7 under Safety Parameter 3412.6.18, Standpipes, for fire safety (FS), means of egress (ME) and general safety (GS).

TABLE 3412.6.18
STANDPIPE SYSTEM VALUES

OCCUPANCY	CATEGORIES			
	a[a]	b	c	d
A-1, A-3, F, M, R, S-1	-6	0	4	6
A-2	-4	0	2	4
A-4, B, E, S-2	-12	0	6	12

a. This option cannot be taken if Category a or b in Section 3412.6.17 is used.

❖ This table lists the credit values for the respective categories of Section 3412.6.18.1, based on the occupancy being evaluated in the existing building. The grouping of occupancies is the same as provided for the sprinkler system parameter, since the need for and value of a standpipe system, like sprinkler systems, is a function of the likelihood and potential severity of a fire that the various occupancies present.

Category b is the zero-based category. Category a to the left of this column contains negative values that define buildings and occupancies that would be required to have a standpipe system as new construction but are not provided with one, or are provided with a standpipe system that is not designed and installed in accordance with the design and installation requirements of Section 905 for new construction. The three categories to the right of this column contain positive values that represent buildings and occupancies with a standpipe system that are deemed adequate or comply with current design standards and installation requirements.

3412.6.18.1 Standpipe. The categories for standpipe systems are:

1. Category a—Standpipes are required; standpipe is not provided or the standpipe system design is not in compliance with Section 905.3.
2. Category b—Standpipes are not required; none are provided.
3. Category c—Standpipes are required; standpipes are provided in accordance with Section 905.
4. Category d—Standpipes are not required; standpipes are provided in accordance with Section 905.

❖ Four categories are defined in this section for evaluating the standpipe system in an existing building. These categories address all aspects from existing buildings that do not have standpipes but would be required to have standpipes if new construction, to existing buildings that have standpipes and would not be required to have standpipes if new construction.

Category a is applicable to buildings or occupancies that would be required by the new construction criteria to be provided with a Class I, II or III standpipe system. Category a buildings are provided with either no standpipe system, or one that is inadequate and does not comply with the applicable design and installation criteria of Section 905. To evaluate the existing standpipe system, a trained fire protection engineer should evaluate the system's design against the applicable referenced standards. This category is considered the lowest acceptable level of compliance and is, therefore, associated with the largest negative values in Table 3412.6.18. Category a is not an option in unsprinklered or inadequately sprinklered buildings that would be required to be partially or fully sprinklered as new construction. This is reflected in Note a to Table 3412.6.18, which indicates that Category a cannot be used if Category a or b is used in the automatic sprinkler parameter in Section 3412.6.17. The effect of this provision is that the absence of both sprinklers and standpipes is unacceptable in buildings that would be required to provide them. One or the other system must be provided in order to complete this evaluation method. If only one of the required systems is provided, negative points will be accrued in the parameter for the other system, but it is still possible to have an overall passing score for the building. This reflects the concept that when both systems would be required for new construction, it is permissible to only have one or the other system in an existing building because the absence of the other system will have other positive attributes or features that offset the negative points accrued because of the absence of the other system.

Category b is the zero-based category for this safety parameter. The existing building is not required by Section 905 to have a standpipe system and none is provided. Such buildings are neither penalized nor rewarded by Table 3412.6.18.

Category c is applicable to buildings that are required by Section 905 to have a standpipe system, and are provided with the class standpipe system that would be required for new construction, designed and installed in accordance with Section 905.2. Although this category more closely represents the requirements for new construction, moderate positive values are awarded by Table 3412.6.18. The evaluation rewards existing buildings that have a contemporary, reliable standpipe system with higher points.

Category d is the highest level of protection and is rewarded with maximum positive points in the accompanying table. Existing buildings, which are not required by Section 905 to have a standpipe system but are voluntarily provided with a fully designed and installed system, provide an added level of protection that justifies substantial bonus points.

3412.6.19 Incidental uses. Evaluate the protection of incidental uses in accordance with Section 509.4.2. Do not include those where this code requires automatic sprinkler systems throughout the buildings, including *covered or open mall buildings*, *high-rise buildings*, public garages and unlimited area buildings. Assign the lowest score from Table 3412.6.19 for the building or floor area being evaluated and enter that value into Table 3412.7 under safety Parameter 3412.6.19, Incidental Use Area, for fire safety, means of egress and general safety. If there are no specific occupancy areas in the building or floor area being evaluated, the value shall be zero.

❖ This section includes an evaluation system for the separation/protection requirements indicated for incidental uses in an existing building. This evaluation is based on Table 509.4.2. If the building designer chooses to separate/protect the small rooms or areas in an existing building in accordance with this table, the building is classified according to its main use. Because some existing buildings may have been designed in this fashion, an evaluation procedure has been added to account for the level of protection provided. The lowest score must be assigned to the building or fire area for the specific use areas. For example, an existing Group E high school has three laboratory classrooms. If two of the classrooms are protected with limited area sprinkler systems and separated with construction capable of resisting the passage of smoke, and the third room is not protected in the same manner, the lowest score for all the classrooms is determined by the single unprotected classroom. If the building designer chooses to treat the specific occupancy areas as a separate occupancy, the provisions of Section 3412.9.1 are applicable and a zero is inserted in Table 3412.7 under Safety Parameter 3412.6.19, Incidental Use Areas.

TABLE 3412.6.19. See below.

❖ This table provides a matrix of characteristics arranged in a format of rows and columns for determining values for incidental uses. The values are inserted into the building score sheet. The left-hand column of the table is arranged for the separation/protection specified by Table 509.4.2 for a particular occupancy room or area. Table 509.4.2 must first be consulted to determine the level of separation/protection required by the code for new construction. This same entry is then found in the left-hand column of Table 3412.6.19. The top row represents the actual level of protection that is provided in the existing building. The corresponding use area value is read and then inserted into Table 3412.7 under Safety Parameter 3412.6.19, Incidental Use Areas. Values of zero are assigned to all arrangements that represent compliance with the requirements for new construction. Negative values are assigned based on the degree of noncompliance with the requirements for new construction.

3412.7 Building score. After determining the appropriate data from Section 3412.6, enter those data in Table 3412.7 and total the building score.

❖ This section is the tally sheet for all of the 19 safety parameters evaluated in Sections 3412.6.1 through 3412.6.19, which determine the building's overall safety profile for fire safety (FS), means of egress (ME) and general safety (GS).

This section directs the data and values of the 19 safety parameters of Sections 3412.6.1 through 3412.6.19 to be entered in Table 3412.7 for totaling the building scores.

TABLE 3412.7. See page 34-55.

❖ Table 3412.7 is the summary sheet containing all of the relative attributes of the building. The summary sheet also contains a complete listing of the 19 safety parameters that have been evaluated. The upper portion of the summary sheet serves as a guide to the user to catalog and highlight existing building elements that relate to the 19 safety parameters and to the evaluation. The lower portion of the summary sheet is used to record the results of the 19 safety parameters that have been evaluated. These are added to produce the building score total values for fire safety (FS), means of egress (ME) and general safety (GS).

3412.8 Safety scores. The values in Table 3412.8 are the required mandatory safety scores for the evaluation process listed in Section 3412.6.

❖ This section lists the minimum scores for fire safety (FS), means of egress (ME) and general safety (GS) that must be obtained from the evaluation of the 19 safety parameters to be acceptable as a building meeting the code's objectives for public safety and health. This section summarizes the mandatory values of Table 3412.8 that must be met from the evaluation program.

TABLE 3412.8. See page 34-56.

❖ The table lists the minimum mandatory safety scores for the evaluation of various occupancies for the three major mandatory safety scores of fire safety (MFS), means of egress (MME) and general safety (MGS). The mandatory safety values are based on the scores considered to be in compliance with the code for new construction. This is the zero-based concept. The scores have been determined as representing one level of compliance higher than the code's minimum requirements for new construction. The mandatory safety scores are consistent with the idea of establishing an equivalent level of safety, even though the existing building is evaluated only for the 19 safety parameters.

TABLE 3412.6.19
INCIDENTAL USE AREA VALUES[a]

PROTECTION REQUIRED BY TABLE 508.2.5	PROTECTION PROVIDED						
	None	1 Hour	AS	AS with SP	1 Hour and AS	2 Hours	2 Hours and AS
2 Hours and AS	-4	-3	-2	-2	-1	-2	0
2 Hours, or 1 Hour and AS	-3	-2	-1	-1	0	0	0
1 Hour and AS	-3	-2	-1	-1	0	-1	0
1 Hour	-1	0	-1	0	0	0	0
1 Hour, or AS with SP	-1	0	-1	0	0	0	0
AS with SP	-1	-1	-1	0	0	-1	0
1 Hour or AS	-1	0	0	0	0	0	0

a. AS = Automatic sprinkler system; SP = Smoke partitions (See Section 508.2.5).

Note: For Table 3412.7, see next page.

TABLE 3412.7
SUMMARY SHEET—BUILDING CODE

Existing occupancy: ________________________________ Proposed occupancy: ____________________

Year building was constructed: ____________________________ Number of stories: ______ Height in feet: ______

Type of construction: ________________________________ Area per floor:____________________________

Percentage of open perimeter increase: ______%

Completely suppressed: Yes ____ No ____ Corridor wall rating: ____________________

Compartmentation: Yes ____ No ____ Required door closers: Yes ______ No ______

Fire-resistance rating of vertical opening enclosures:____

Type of HVAC system: ______________________________, serving number of floors: ____________________

Automatic fire detection: Yes _____ No _____ Type and location: ____________________

Fire alarm system: Yes _____ No _____ Type: ______________________________

Smoke control: Yes _____ No _____ Type: ______________________________

Adequate exit routes: Yes _____ No _____ Dead ends: __________ Yes ______ No ______

Maximum exit access travel distance: ________________________ Elevator controls: Yes ______ No ______

Means of egress emergency lighting: Yes ______ No ______ Mixed occupancies: Yes ______ No ______

SAFETY PARAMETERS	FIRE SAFETY (FS)	MEANS OF EGRESS (ME)	GENERAL SAFETY (GS)
3412.6.1 Building Height 3412.6.2 Building Area 3412.6.3 Compartmentation			
3412.6.4 Tenant and Dwelling Unit Separations 3412.6.5 Corridor Walls 3412.6.6 Vertical Openings			
3412.6.7 HVAC Systems 3412.6.8 Automatic Fire Detection 3412.6.9 Fire Alarm Systems			
3412.6.10 Smoke Control 3412.6.11 Means of Egress Capacity 3412.6.12 Dead Ends	* * * * * * * * * * * *		
3412.6.13 Maximum Exit Access Travel Distance 3412.6.14 Elevator Control 3412.6.15 Means of Egress Emergency Lighting	* * * * * * * *		
3412.6.16 Mixed Occupancies 3412.6.17 Automatic Sprinklers 3412.6.18 Standpipes 3412.6.19 Incidental Use		* * * * ÷ 2 =	
Building score — total value			

* * * *No applicable value to be inserted.

TABLE 3412.8
MANDATORY SAFETY SCORES[a]

OCCUPANCY	FIRE SAFETY (MFS)	MEANS OF EGRESS (MME)	GENERAL SAFETY (MGS)
A-1	20	31	31
A-2	21	32	32
A-3	22	33	33
A-4, E	29	40	40
B	30	40	40
F	24	34	34
M	23	40	40
R	21	38	38
S-1	19	29	29
S-2	29	39	39

a. MFS = Mandatory Fire Safety;
MME = Mandatory Means of Egress;
MGS = Mandatory General Safety.

3412.9 Evaluation of building safety. The mandatory safety score in Table 3412.8 shall be subtracted from the building score in Table 3412.7 for each category. Where the final score for any category equals zero or more, the building is in compliance with the requirements of this section for that category. Where the final score for any category is less than zero, the building is not in compliance with the requirements of this section.

❖ The sections and Table 3412.9 that follow are the final steps in the evaluation process. This section also discusses how mixed occupancies must to be treated during the final step of the evaluation process.

This section compares the three building scores from Table 3412.7 to the three mandatory safety scores from Table 3412.8. If the values in all three categories from Table 3412.7 exceed the corresponding mandatory safety scores in Table 3412.8, the building passes and is in compliance with the code. If the score in any one category is less than the mandatory safety score, the building is deemed to have failed and additional measures must be taken to bring the scores to a point that will at least equal the mandatory safety scores.

TABLE 3412.9. See below.

❖ Figure 3412.9 shows in simple equations whether a building passes the evaluation by subtracting the mandatory safety score from Table 3412.7. This is done for each of the three general categories of evaluation: fire safety, means of egress and general safety. If the difference for each category is zero or greater, the existing building passes and is considered to comply with the code objectives for public safety.

Example:

A Group M mercantile occupancy receives evaluations for the 19 safety parameters that result in total building scores in Table 3412.7 as follows:

Fire safety (FS) = 23

Means of egress (ME) = 41

General safety (GS) = 42

These scores are compared to the mandatory safety scores for a Group M occupancy from Table 3412.8 (see Figure 3412.9).

Mandatory fire safety (MFS) = 23

Mandatory means of egress (MME) = 40

Mandatory general safety (MGS) = 40

The mandatory score is subtracted from the building score.

TABLE 3412.9
EVALUATION FORMULAS[a]

FORMULA	T.3410.7			T.3410.8	SCORE	PASS	FAIL
FS-MFS ≥ 0	_____	(FS)	—	_____ (MFS) =	_____	_____	_____
ME-MME ≥ 0	_____	(ME)	—	_____ (MME) =	_____	_____	_____
GS-MGS ≥ 0	_____	(GS)	—	_____ (MGS) =	_____	_____	_____

a. FS = Fire Safety
ME = Means of Egress
GS = General Safety
MFS = Mandatory Fire Safety
MME = Mandatory Means of Egress
MGS = Mandatory General Safety

Table 3412.9
EVALUATION FORMULAS[a]

Formula	Table 3412.7	Table 3412.8	Score	Pass	Fail
FS-MFS ≥ 0	23 (FS)	- 23 (MFS) =	0	X	—
ME-MME ≥ 0	41 (ME)	- 40 (MME) =	1	X	—
GS-MGS ≥ 0	42 (GS)	- 40 (MGS) =	1	X	—

Note a.
FS = Fire Safety MFS = Mandatory Fire Safety
ME = Means of Egress MME = Mandatory Means of Egress
GS = General Safety MGS = Mandatory General Safety

Figure 3412.9
EXAMPLE OF EVALUATION FORMULAS

Conclusion:

The building is acceptable; all of the final safety scores are zero or greater.

3412.9.1 Mixed occupancies. For mixed occupancies, the following provisions shall apply:

1. Where the separation between mixed occupancies does not qualify for any category indicated in Section 3412.6.16, the mandatory safety scores for the occupancy with the lowest general safety score in Table 3412.8 shall be utilized (see Section 3412.6).
2. Where the separation between mixed occupancies qualifies for any category indicated in Section 3412.6.16, the mandatory safety scores for each occupancy shall be placed against the evaluation scores for the appropriate occupancy.

❖ This section explains how to determine whether a mixed occupancy building passes or fails the process of evaluation. It restates the information in Sections 3412.6 and 3410.6.16. The mixed occupancy evaluation is based on the requirements of Sections 508.3 and 508.4. The two procedures described are:

Procedure 1:

For nonseparated occupancies in accordance with Section 508.3, or for occupancies that are separated with a fire-resistance rating of less than 1 hour, mandatory safety scores for the occupancy with the lowest general safety score of Table 3412.8 apply.

Procedure 2:

For separated occupancies in accordance with Section 508.4, or one of the categories listed in Section 3412.6.16, mandatory safety scores for each occupancy are compared to the evaluation scores for the appropriate occupancy. The total building score of Table 3412.7 is computed for each appropriate occupancy.

This section does not include a category or condition for a third option for mixed occupancies, which is the separation of multiple occupancies within a single building or structure with one or more fire walls. Like wise, this option is not addressed in Section 508. A fire wall creates separate and independent buildings; therefore, a separate evaluation must be done for each area within a building that is separated by fire walls.

Example:

A 40-foot (12 192 mm), three-story building has an open perimeter of 25 percent. It is Type IIB unprotected construction, unsprinklered. The building has 9,000 square feet (836 m^2) per floor. A Group M mercantile is on the first floor with Group R residential occupancies on the second and third floors. Although the building is unprotected Type IIB construction, there is a 2-hour fire-resistance-rated floor/ceiling assembly separating the first and second floors. An analysis of this building will result in the following: "In accordance with Table 508.4, the mixed occupancies are separated by the required fire-rated assembly of 2 hours."

This building qualifies for Category b in Section 3412.6.16.1. A separate summary sheet for tabulating the building score in Table 3412.7 has to be completed for each of the two occupancies. The values of the building scores for fire safety (FS), means of egress and general safety (GE) for the Group M mercantile occupancy and the Group R residential occupancy must be calculated. The mercantile building scores are compared to the mandatory safety scores of 23, 40 and 40, and the residential building scores are compared to the mandatory safety scores of 21, 38 and 38. If any one of the three building scores from either of the two occupancies does not equal or exceed the applicable safety score, the entire building fails the evaluation. For the building to pass, the safety parameters that are less than the mandatory safety score must be upgraded until a passing score is achieved.

Bibliography

The following resource materials are referenced in this chapter or are relevant to the subject matter addressed in this chapter.

36 CFR, Parts 1190 and 1191 Draft, *The Americans with Disabilities Act Accessibility Guidelines* (ADAAG). Washington, DC: Architectural and Transportation Barriers Compliance Board, April 2, 2002.

42 USC 3601-88, *Fair Housing Amendments Act* (FHAA). Washington, DC: United States Code, 1988.

"Accessibility and Egress for People with Physical Disabilities." Falls Church, VA: CABO Board for the Coordination of the Model Codes Report, October 5, 1993.

ASME A17.1/CSA B44-07, *Safety Code for Elevators and Escalators—with A17.1a/CSA B44a-08 Addenda.* New York: American Society of Mechanical Engineers, 2007.

ASME A17.3-08, *Safety Code for Existing Elevators and Escalators*. New York: American Society of Mechanical Engineers, 2008.

ASME 18.1-08, *Safety Standard for Platform Lifts and Stairway Chairlifts*. New York: American Society of Mechanical Engineers, 2008.

DOJ 28 CFR, Part 36-91, *Americans with Disabilities Act* (ADA). Washington, DC: Department of Justice, 1991.

DOJ 28 CFR, Part 36-91(Appendix A), *ADA Accessibility Guidelines for Buildings and Facilities* (ADAAG). Washington, DC: Department of Justice, 1991.

FED-STD-795-88, *Uniform Federal Accessibility Standards*. Washington, DC: General Services Administration; Department of Defense; Department of Housing and Urban Development; U.S. Postal Service, 1988.

FEMA 311-07, *Residential Substantial Damage Estimators In Your Community*. Washington, DC: Federal Emergency Management Agency, 2007.

FEMA P-467-2, *Bulletin on Historic Subways*. Washington, DC: Federal Emergency Management Agency, 2008.

FEMA P-758, *Substantial Improvement/Substantial Damage Desk Reference*. Washington, DC: Federal Emergency Management Agency, 2009.

"Final Report, Recommendations for a New ADAAG." Washington DC: ADAAG Review Federal Advisory Committee, September 30, 1996.

ICC A117.1-09, *Accessible and Usable Buildings and Facilities*. Falls Church, VA: International Code Council, 2011.

IFC-12, *International Fire Code*. Washington, DC: International Code Council, 2011.

IMC-12, *International Mechanical Code*. Washington, DC: International Code Council, 2012.

IPMC-12, *International Property Maintenance Code*. Washington, DC: International Code Council, 2011.

IPSDC-12, *International Private Sewage Disposal Code*. Washington, DC: International Code Council, 2011.

IRC-12, *International Residential Code*. Washington, DC: International Code Council, 2012.

NFPA 13-2010, *Installation of Sprinkler Systems*. Quincy, MA: National Fire Protection Association, 2010.

NFPA 13D-2010, *Installation of Sprinkler Systems in One- and Two-family Dwellings and Manufactured Homes*. Quincy, MA: National Fire Protection Association, 2010.

NFPA 13R-2010, *Installation of Sprinkler Systems in Residential Occupancies Up to and Including Four Stories in Height*. Quincy, MA: National Fire Protection Association, 2010.

NFPA 70-2011, *National Electrical Code*. Quincy, MA: National Fire Protection Association, 2011.

NFPA 72-2010, *National Fire Alarm Code*. Quincy, MA: National Fire Protection Association, 2010.

NFPA 101-2012, *Alternative Approaches to Life Safety*. Quincy, MA: National Fire Protection Association, 2012.

Rehabilitation Guidelines/1980. Washington, DC: National Institute of Building Science for the U.S. Department of Housing and Urban Development, 1980.

1. Guideline for Setting and Adopting Standards for Building Rehabilitation;
2. Guideline for Approval of Building Rehabilitation;
3. Statutory Guideline for Building Rehabilitation;
4. Guideline for Managing Official Liability Associated with Building Rehabilitation;
5. Egress Guideline for Residential Rehabilitation;
6. Electrical Guideline for Residential Rehabilitation;
7. Plumbing DWV Guideline for Residential Rehabilitation; and
8. Guideline on Fire Ratings of Archaic Materials and Assemblies.

UL 555-06, *Fire Dampers*. Northbrook, IL: Underwriters Laboratories Inc., 2006.

Chapter 35: Referenced Standards

General Comments

Chapter 35 contains a comprehensive list of all standards that are referenced in the code. It is organized in a manner that makes it easy to locate specific document references.

This chapter lists the standards that are referenced in various sections of this document. The standards are listed herein by the promulgating agency of the standard, the standard identification, the date and title, and the section or sections of this document that reference the standard. The application of the referenced standards shall be as specified in Section 102.4.

It is important to understand that not every document related to building design and construction is qualified to be a "referenced standard." The International Code Council® (ICC®) has adopted a criterion that standards referenced in and intended for adoption into the *International Codes®* must meet in order to qualify as a referenced standard. The policy is summarized as follows:

- Code references: The scope and application of the standard must be clearly identified in the code text.
- Standard content: The standard must be written in mandatory language and appropriate for the subject covered. The standard shall not have the effect of requiring proprietary materials or prescribing a proprietary testing agency.
- Standard promulgation: The standard must be readily available, and developed and maintained in a consensus process, such as ASTM or ANSI.

It should be noted that the ICC Code Development Procedures, of which the standards policy is a part, are updated periodically. A copy of the latest version can be obtained from the ICC offices.

Once a standard is incorporated into the code through the code development process, it becomes an enforceable part of the code. When the code is adopted by a jurisdiction, the standard also is part of that jurisdiction's adopted code. It is for this reason that the criteria were developed. Compliance with this policy provides that documents incorporated into the code are, among others, developed through the use of a consensus process, written in mandatory language and do not mandate the use of proprietary materials or agencies. The requirement for a standard to be developed through a consensus process is vital, as it means that the standard will be representative of the most current body of available knowledge on the subject as determined by a broad spectrum of interested or affected parties without dominance by any single interest group. A true consensus process has many attributes, including but not limited to:

- An open process that has formal (published) procedures that allow for the consideration of all viewpoints;
- A definitive review period that allows for the standard to be updated or revised;
- A process of notification to all interested parties; and
- An appeals process.

Many available documents related to design, installation and construction, though useful, are not "standards" and are not appropriate for reference in the code. Often, these documents are developed or written with the intention of being used for regulatory purposes and are unsuitable for use as a regulation due to extensive use of recommendations, advisory comments and nonmandatory terms. Typical examples of such documents include installation instructions, guidelines and practices.

The objective of ICC's standards policy is to provide regulations that are clear, concise and enforceable; thus the requirement for standards to be written in mandatory language. This requirement is not intended to mean that a standard cannot contain informational or explanatory material that will aid the user of the standard in its application. When it is the desire of the standard's promulgating agency for such material to be included, however, the information must appear in a nonmandatory location, such as an annex or appendix, and be clearly identified as not being part of the standard.

Overall, standards referenced by the code must be authoritative, relevant, up-to-date and, most important, reasonable and enforceable. Standards that comply with ICC's standards policy fulfill these expectations.

Purpose

As a performance-oriented code, the code contains numerous references to documents that are used to regulate materials and methods of construction. The references to these documents within the code text consist of the promulgating agency's acronym and its publication designation (e.g., ASME A18.1) and a further indication that the document being referenced is the one that is listed in Chapter 35. Chapter 35 contains all of the information that is necessary to identify the specific referenced document. Included is the following information on a document's promulgating agency (see Figure 35):

- The promulgating agency (i.e., the agency's title);
- The promulgating agency's acronym; and
- The promulgating agency's address.

For example, a reference to an ASME standard within the code indicates that the document is promulgated by the American Society of Mechanical Engineers (ASME), which is located in New York City. This chapter lists the standards agencies alphabetically for ease of identification.

This chapter also includes the following information on the referenced document itself (see Figure 35):

- The document's publication designation;
- The document's edition year;
- The document's title;
- Any addenda or revisions to the document that are applicable; and
- Every section of the code in which the document is referenced.

For example, a reference to ASME A18.1 indicates that this document can be found in Chapter 35 under the heading ASME. The specific standards designation is A18.1. For convenience, these designations are listed in alphanumeric order. This chapter identifies that ASME A18.1 is titled *Safety Standard for Platform Lifts and Stairway Chairlifts*, the applicable edition (i.e., its year of publication) is 2005 and it is referenced in numerous sections of the code.

This chapter will also indicate when a document has been discontinued or replaced by its promulgating agency. When a document is replaced by a different one, a note will appear to tell the user the designation and title of the new document.

The key aspect of the manner in which standards are referenced by the code is that a specific edition of a specific standard is clearly identified. In this manner, the requirements necessary for compliance can be readily determined. The basis for code compliance is, therefore, established and available on an equal basis to the building official, contractor, designer and owner.

This chapter lists the standards that are referenced in various sections of this document. The standards are listed herein by the promulgating agency of the standard, the standard identification, the effective date and title and the section or sections of this document that reference the standard. The application of the referenced standards shall be as specified in Section 102.4.

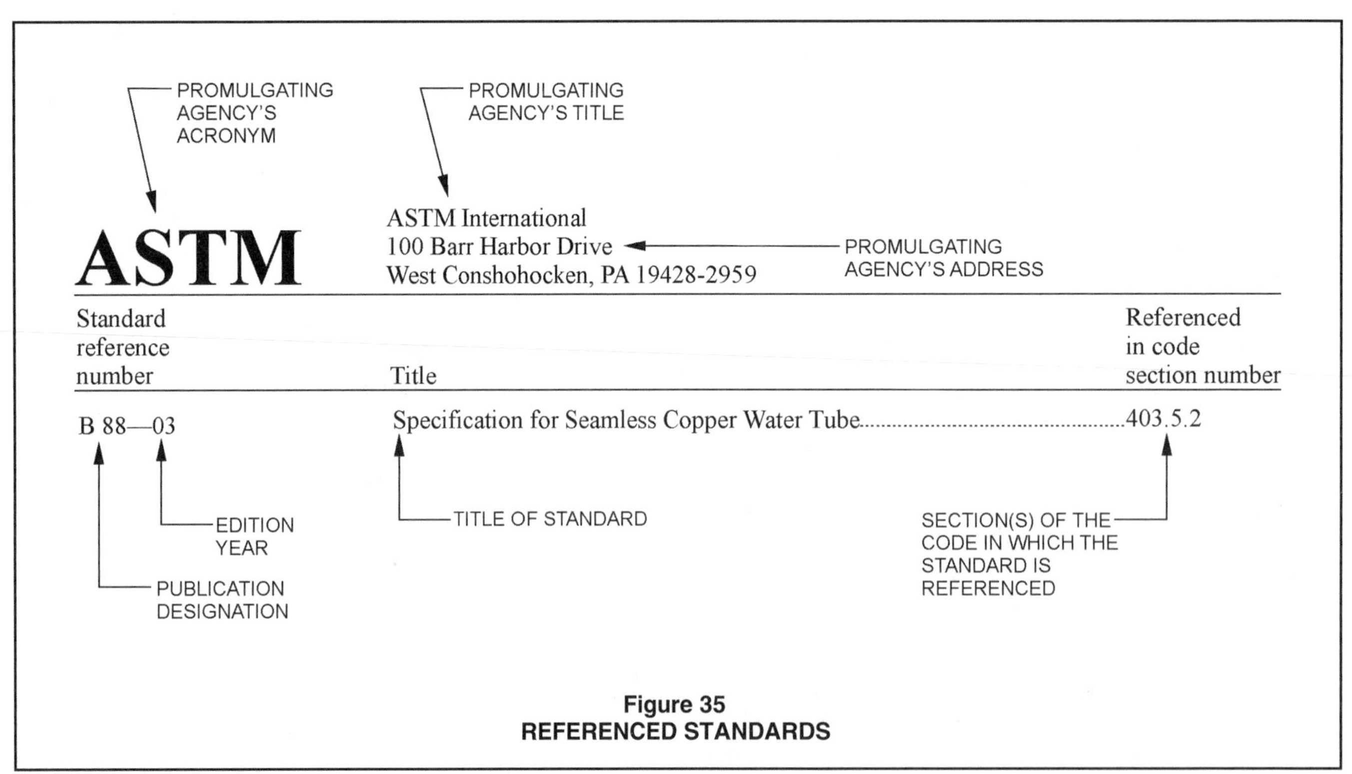

Figure 35
REFERENCED STANDARDS

This chapter lists the standards that are referenced in various sections of this document. The standards are listed herein by the promulgating agency of the standard, the standard identification, the effective date and title, and the section or sections of this document that reference the standard. The application of the referenced standards shall be as specified in Section 102.4.

AA

Aluminum Association
1525 Wilson Boulevard, Suite 600
Arlington, VA 22209

Standard reference number	Title	Referenced in code section number
ADM1—2010	Aluminum Design Manual: Part 1—A Specification for Aluminum Structures	1604.3.5, 2002.1
ASM 35—00	Aluminum Sheet Metal Work in Building Construction (Fourth Edition)	2002.1

AAMA

American Architectural Manufacturers Association
1827 Waldon Office Square, Suite 550
Schaumburg, IL 60173

Standard reference number	Title	Referenced in code section number
1402—86	Standard Specifications for Aluminum Siding, Soffit and Fascia	1404.5.1
AAMA/WDMA/CSA 101/I.S.2/A440—11	North American Fenestration Standard/Specifications for Windows, Doors and Skylights	1710.5.1, 2405.5

ACI

American Concrete Institute
38800 Country Club Drive
Farmington Hills, MI 48331

Standard reference number	Title	Referenced in code section number
216.1—07	Standard Method for Determining Fire Resistance of Concrete and Masonry Construction Assemblies	Table 721.1(2), 722.1
318—11	Building Code Requirements for Structural Concrete	1604.3.2, 1615.3.1, 1615.4.1, Table 1705.2.2, 1705.2.2.1.2, Table 1705.3, 1705.3.1, 1705.12.1, 1808.8.2, Table 1808.8.2, 1808.8.5, 1808.8.6, 1810.1.3, 1810.2.4.1, 1810.3.2.1.1, 1810.3.2.1.2, 1810.3.8.3.1, 1810.3.8.3.3, 1810.3.9.4.2.1, 1810.3.9.4.2.2, 1810.3.10.1, 1810.3.11.1, 1901.2, 1902.1, 1903.1, 1904.1, 1904.2, Table 1904.2, 1905.1, 1905.1.1, 1905.1.2, 1905.1.3, 1905.1.4, 1905.1.5, 1905.1.6, 1905.1.7, 1905.1.8, 1905.1.9, 1905.1.10, 1906.1, 1909.1, 2108.3, 2206.1
530—11	Building Code Requirements for Masonry Structures	1405.6, 1405.6.1, 1405.6.2, 1405.10, 1604.3.4, 1705.4, 1705.4.1, 1807.1.6.3, 1807.1.6.3.2, 1808.9 2101.2.2, 2101.2.3, 2101.2.4, 2101.2.5, 2101.2.6, 2103.9, 2103.12 2103.13, 2103.14, 2104.1, 2104.1.1, 2104.1.2, 2104.1.3, 2104.2, 2104.3, 2104.4, 2105.2.2.1, 2105.2.2.1.2, 2105.2.2.1.3, 2106.1, 2107.1, 2107.2, 2107.3, 2107.4, 2108.1, 2108.2, 2108.3, 2109.1, 2109.1.1, 2109.2, 2109.2.1, 2109.3, 2110.1
530.1—11	Specifications for Masonry Structures	1405.6.1, 1705.4, 1807.1.6.3, 2103.9, 2103.12, 2103.13, 2103.14, 2104.1, 2104.1.1, 2104.1.2, 2104.1.3, 2104.2, 2104.3, 2104.4, 2105.2.2.1.1, 2105.2.2.1.2, 2105.2.2.1.3

AF&PA

American Forest & Paper Association
1111 19th St, NW Suite 800
Washington, DC 20036

Standard reference number	Title	Referenced in code section number
WCD No. 4—2003	Wood Construction Data—Plank and Beam Framing for Residential Buildings	2306.1.2
ANSI/AF&PA WFCM—2012	Wood Frame Construction Manual for One- and Two-Family Dwellings	1609.1.1, 1609.1.1.1, 2301.2, 2308.1, 2308.2.1
NDS—2012	National Design Specification (NDS) for Wood Construction with 2012 Supplement	722.1, 722.6.3.2, 1711.1.1, 1711.1.2.1, 1809.12, 1810.3.2.4, Table 1810.3.2.6, 1905.1.9, 2302.1, 2304.12, 2306.1, 2306.2, Table 2306.2(1), Table 2306.2(2), Table 2306.3(1), Table 2306.3(2), 2307.1
AF&PA—2012	Span Tables for Joists and Rafters	202, 2306.1.1, 2308.8, 2308.10.2, 2308.10.3
ANSI/AF&PA PWF—2007	Permanent Wood Foundation Design Specification	1805.2, 1807.1.4, 2304.9.5.2
ANSI/AF&PA SDPWS—2008	Special Design Provisions for Wind and Seismic	202, 2305.1, 2305.2, 2305.3, 2306.2, 2306.3, Table 2306.3(1), Table 2306.3(3), 2307.1

AISC

American Institute of Steel
Construction One East Wacker Drive, Suite 700
Chicago, IL 60601-18021

Standard reference number	Title	Referenced in code section number
341—10	Seismic Provisions for Structural Steel Buildings	1613.4.1, 1705.11.1, 1705.12.2, 2205.2.1, 2205.2.2, 2206.2
360—10	Specification for Structural Steel Buildings	722.5.2.2.1, 1604.3.3, 1705.2.1, 2203.1, 2203.2, 2205.1, 2205.2.1, 2206.1

AISI

American Iron and Steel Institute
1140 Connecticut Avenue, 705
Suite 705
Washington, DC 20036

Standard reference number	Title	Referenced in code section number
AISI S100—07/SI—10	North American Specification for the Design of Cold-formed Steel Structural Members, with Supplement 1, dated 2010	1604.3.3, 1905.1.9, 2203.1, 2203.2, 2210.1, 2210.2, 2211.2, 2211.4, 2211.6
AISI S110—07	Standard for Seismic Design of Cold-formed Steel Structural Systems—Special Bolted Moment Frames, with Supplement 1, dated 2009	2210.2
S200—07	North American Standard for Cold-formed Steel Framing-General Provisions	2203.1, 2203.2, 2211.1
S210—07	North American Standard for Cold-formed Steel Framing-Floor and Roof System Design	2211.5
S211—07	North American Standard for Cold-formed Steel Framing-Wall Stud Design	2211.4
S212—07	North American Standard for Cold-formed Steel Framing-Header Design	2211.2
AISI S213—07/ SI—10	North American Standard for Cold-formed Steel Framing-Lateral Design, with Supplement 1, dated 2010	2211.6
S214—07	North American Standard for Cold-formed Steel Framing-Truss Design, with Supplement 2, dated 2008	2211.3, 2211.3.1, 2211.3.2
S230—07	Standard for Cold-formed Steel Framing-Prescriptive Method for One- and Two-family Dwellings, with Supplement 2, dated 2008	1609.1.1, 1609.1.1.1, 2211.7

AITC

American Institute of Timber Construction
Suite 140
7012 S. Revere Parkway
Englewood, CO 80112

Standard reference number	Title	Referenced in code section number
AITC Technical Note 7—96	Calculation of Fire Resistance of Glued Laminated Timbers	722.6.3.3
AITC 104—03	Typical Construction Details	2306.1
AITC 110—01	Standard Appearance Grades for Structural Glued Laminated Timber	2306.1
AITC 113—10	Standard for Dimensions of Structural Glued Laminated Timber	2306.1
AITC 117—10	Standard Specifications for Structural Glued Laminated Timber of Softwood Species	2306.1
AITC 119—96	Standard Specifications for Structural Glued Laminated Timber of Hardwood Species	2306.1
ANSI/AITC A 190.1—07	Structural Glued Laminated Timber	2303.1.3, 2306.1
AITC 200—09	Manufacturing Quality Control Systems Manual for Structural Glued Laminated Timber	2306.1

ALI

Automotive Lift Institute
P.O. Box 85
Courtland, NY 13045

Standard reference number	Title	Referenced in code section number
ALI ALCTV—2006	Standard for Automotive Lifts—Safety Requirements for Construction, Testing and Validation (ANSI)	3001.2

AMCA

Air Movement and Control Association International
30 West University Drive
Arlington Heights, IL 60004

Standard reference number	Title	Referenced in code section number
540—08	Test Method for Louvers Impacted by Wind Borne Debris	1609.1.2.1

ANSI

American National Standards Institute
25 West 43rd Street, Fourth Floor
New York, NY 10036

Standard reference number	Title	Referenced in code section number
A13.1—96 (Reaffirmed 2002)	Scheme for the Identification of Piping Systems	415.10.6.5
A108.1A—99	Installation of Ceramic Tile in the Wet-set Method, with Portland Cement Mortar	2103.11
A108.1B—99	Installation of Ceramic Tile, quarry Tile on a Cured Portland Cement Mortar Setting Bed with Dry-set or Latex-portland Mortar	2103.11
A108.4—99	Installation of Ceramic Tile with Organic Adhesives or Water-cleanable Tile-setting Epoxy Adhesive	2103.11.6
A108.5—99	Installation of Ceramic Tile with Dry-set Portland Cement Mortar or Latex-Portland Cement Mortar	2103.11.1, 2103.11.2
A108.6—99	Installation of Ceramic Tile with Chemical-resistant, Water Cleanable Tile-setting and -grouting Epoxy	2103.11.3
A108.8—99	Installation of Ceramic Tile with Chemical-resistant Furan Resin Mortar and Grout	2103.11.4
A108.9—99	Installation of Ceramic Tile with Modified Epoxy Emulsion Mortar/Grout	2103.11.5
A108.10—99	Installation of Grout in Tilework	2103.11.7
A118.1—99	American National Standard Specifications for Dry-set Portland Cement Mortar	2103.11.1

ANSI—continued

A118.3—99	American National Standard Specifications for Chemical-resistant, Water-cleanable Tile-setting and -grouting Epoxy and Water Cleanable Tile-setting Epoxy Adhesive	2103.11.3
A118.4—99	American National Standard Specifications for Latex-portland Cement Mortar	2103.11.2
A118.5—99	American National Standard Specifications for Chemical Resistant Furan Mortar and Grouts for Tile Installation	2103.11.4
A118.6—99	American National Standard Specifications for Cement Grouts for Tile Installation	2103.11.7
A118.8—99	American National Standard Specifications for Modified Epoxy Emulsion Mortar/Grout	2103.11.5
A136.1—99	American National Standard Specifications for Organic Adhesives for Installation of Ceramic Tile	2103.11.6
A137.1—08	American National Standard Specifications for Ceramic Tile	202, 2103.6
A208.1—09	Particleboard	2303.1.7, 2303.1.7.1
Z 97.1—09	Safety Glazing Materials Used in Buildings—Safety Performance Specifications and Methods of Test	2406.1.2, 2406.2, Table 2406.2(2), 2406.3.1, 2407.1, 2407.1.4.1, 2408.2.1, 2408.3, 2409.1, 2409.2, 2409.3.1

APA

APA - Engineered Wood Association
7011 South 19th
Tacoma, WA 98466

Standard reference number	Title	Referenced in code section number
ANSI/APA PRP 210—8	Standard for Performance-Rated Engineered Wood Siding	2303.1.4, 2304.6.2, 2306.3, Table 2306.3(1)
APA PDS—04	Panel Design Specification	2306.1
APA PDS Supplement 1—90	Design and Fabrication of Plywood Curved Panels (revised 1995)	2306.1
APA PDS Supplement 2—92	Design and Fabrication of Plywood-lumber Beams (revised 1998)	2306.1
APA PDS Supplement 3—90	Design and Fabrication of Plywood Stressed-skin Panels (revised 1996)	2306.1
APA PDS Supplement 4—90	Design and Fabrication of Plywood Sandwich Panels (revised 1993)	2306.1
APA PDS Supplement 5—08	Design and Fabrication of All-plywood Beams (revised 2008)	2306.1
EWS R540—02	Builders Tips: Proper Storage and Handling of Glulam Beams	2306.1
EWS S475—01	Glued Laminated Beam Design Tables	2306.1
EWS S560—03	Field Notching and Drilling of Glued Laminated Timber Beams	2306.1
EWS T300—05	Glulam Connection Details	2306.1
EWS X440—03	Product Guide-Glulam	2306.1
EWS X450—01	Glulam in Residential Construction-Western Edition	2306.1

APSP

The Association of Pool & Spa Professionals
2111 Eisenhower Avenue
Alexandria, VA 22314

Standard reference number	Title	Referenced in code section number
ANSI/APSP 7—06	Standard for Suction Entrapment Avoidance in Swimming Pools, Wading Pools, Spas, Hot Tubs and Catch Basins	3109.5

ASABE

American Society of Agricultural and Biological Engineers
2950 Niles Road
St. Joseph, MI 49085

Standard reference number	Title	Referenced in code section number
EP 484.2 June 1998 (R2008)	Diaphragm Design of Metal-clad, Wood-frame Rectangular Buildings	2306.1
EP 486.1 Dec 1999 (R2005)	Shallow-post Foundation Design	2306.1
EP 559 1997 Dec 1996 (R2008)	Design Requirements and Bending Properties for Mechanically Laminated Columns	2306.1

ASCE/SEI

American Society of Civil Engineers
Structural Engineering Institute
1801 Alexander Bell Drive
Reston, VA 20191-4400

Standard reference number	Title	Referenced in code section number
5—11	Building Code Requirements for Masonry Structures	1405.6, 1405.6.1, 1405.6.2, 1405.10, 1604.3.4, 1705.4, 1705.4.1, 1807.1.6.3, 1807.1.6.3.2, 1808.9, 2101.2.2, 2101.2.3, 2101.2.4, 2101.2.5, 2101.2.6, 2103.9, 2103.12, 2103.13, 2103.14, 2104.1, 2104.1.1, 2104.1.2, 2104.1.3, 2104.2, 2104.3, 2104.4, 2105.2.2.1, 2105.2.2.1.2, 2105.2.2.1.3, 2106.1, 2107.1, 2107.2, 2107.3, 2107.4, 2108.1, 2108.2, 2108.3, 2109.1, 2109.1.1, 2109.2, 2109.2.1, 2109.3, 2110.1
6—11	Specification for Masonry Structures	1405.6.1, 1705.4, 1807.1.6.3, 2103.9, 2103.12, 2103.13, 2103.14, 2104.1, 2104.1.1, 2104.1.2, 2104.1.3, 2104.2, 2104.3, 2104.4, 2105.2.2.1.1, 2105.2.2.1.2, 2105.2.2.1.3
7—10	Minimum Design Loads for Buildings and Other Structures	202, Table 1504.8, 1602.1, 1604.3, Table 1604.5, 1604.8.2, 1604.10, 1605.1, 1605.2.1, 1605.3.1, 1605.3.1.2, 1605.3.2, 1605.3.2.1, 1607.8.1, 1607.8.1.1, 1607.8.1.2, 1607.8.3, 1607.12.1, 1608.1, 1608.2, 1608.3, 1609.1.1, 1609.1.2, 1609.3, 1609.5.1, 1609.5.3, 1609.6, 1609.6.1, 1609.6.1.1, 1609.6.2, Table 1609.6.2, 1609.6.3, 1609.6.4.1, 1609.6.4.2, 1609.6.4.4.1, 1611.2, 1612.4, 1613.1, 1613.3.2, Table 1613.3.3(1), Table 1613.3.3(2), 1613.3.5, 1613.3.5.1, 1613.3.5.2, 1613.4, 1613.4.1, 1614.1, 1705.11, 1705.12, 1705.12.3, 1705.12.4, 1803.5.12, 1808.3.1, 1810.3.6.1, 1810.3.9.4, 1810.3.11.2, 1810.3.12, 1905.1.1, 1905.1.2, 1905.1.9, 2205.2.1, 2205.2.2, 2206.2, 2209.1, 2210.2, 2304.6.1, 2404.1, 2505.1, 2505.2, 2506.2.1, 3404.4, 3404.5
8—02	Standard Specification for the Design of Cold-formed Stainless Steel Structural Members	1604.3.3, 2210.1, 2210.2
19—09	Structural Applications of Steel Cables for Buildings	2208.1, 2208.2
24—05	Flood Resistant Design and Construction	1203.3.2, 1612.4, 1612.5, 3001.2, G103.1, G401.3, G401.4
29—05	Standard Calculation Methods for Structural Fire Protection	722.1
32—01	Design and Construction of Frost Protected Shallow Foundations	1809.5

ASME

American Society of Mechanical Engineers
Three Park Avenue
New York, NY 10016-5990

Standard reference number	Title	Referenced in code section number
ASME/A17.1 2007/CSA B44—07	Safety Code for Elevators and Escalators – with A17.1a/CSA B44a-08 Addenda	907.3.3, 911.1.5, 1007.4, 1607.9.1, 3001.2, 3001.4, 3002.5, 3003.2, 3007.1, 3007.2, 3008.2, 3008.2.1, 3008.7.6, 3008.8.1, 3411.8.2

ASME—continued

A18.1—2008	Safety Standard for Platform Lifts and Stairway Chairlifts	1109.8, 2702.2.6, 3411.8.3
A90.1—09	Safety Standard for Belt Manlifts	3001.2
B16.18—2001 (Reaffirmed 2005)	Cast Copper Alloy Solder Joint Pressure Fittings	909.13.1
B16.22—2001 (Reaffirmed 2005)	Wrought Copper and Copper Alloy Solder Joint Pressure Fittings	909.13.1
B20.1—2009	Safety Standard for Conveyors and Related Equipment	3005.3
B31.3—2004	Process Piping	415.10.6

ASTM

ASTM International
100 Barr Harbor Drive
West Conshohocken, PA 19428-2959

Standard reference number	Title	Referenced in code section number
A 36/A 36M—08	Specification for Carbon Structural Steel	1810.3.2.3
A 153/A 153M—05	Specification for Zinc Coating (Hot-dip) on Iron and Steel Hardware	2304.9.5
A 240/A 240M—09a	Standard Specification for Chromium and Chromium-nickel Stainless Steel Plate, Sheet and Strip for Pressure Vessels and for General Applications	Table 1507.4.3(1)
A 252—98 (2007)	Specification for Welded and Seamless Steel Pipe Piles	1810.3.2.3
A 283/A 283M—03(2007)	Specification for Low and Intermediate Tensile Strength Carbon Steel Plates	1810.3.2.3
A 307—07b	Specification for Carbon Steel Bolts and Studs, 60,000 psi Tensile Strength	1908.1
A 416/A 416M—06	Specification for Steel Strand, Uncoated Seven-wire for Prestressed Concrete	1810.3.2.2
A 463/A 463M—06	Standard Specification for Steel Sheet, Aluminum-coated, by the Hot-dip Process	Table 1507.4.3(2)
A 572/A 572M—07	Specification for High-strength Low-alloy Columbium-vanadium Structural Steel	1810.3.2.3
A 588/A 588M—05	Specification for High-strength Low-alloy Structural Steel with 50 ksi (345 MPa) Minimum Yield Point with Atmospheric Corrosion Resistance	1810.3.2.3
A 615/A 615M—09	Specification for Deformed and Plain Billet-steel Bars for Concrete Reinforcement	1705.12.1, 1810.3.10.2
A 653/A 653M—08	Specification for Steel Sheet, Zinc-coated Galvanized or Zinc-iron Alloy-coated Galvannealed by the Hot-dip Process	Table 1507.4.3(1), Table 1507.4.3(2), 2304.9.5.1
A 690/A 690M—07	Standard Specification for High-strength Low-alloy Nickel, Copper, Phosphorus Steel H-piles and Sheet Piling with Atmospheric Corrosion Resistance for Use in Marine Environments	1810.3.2.3
A 706/A 706M—09	Specification for Low-alloy Steel Deformed and Plain Bars for Concrete Reinforcement	Table 1705.2.2, 1705.3.1, 2107.4, 2108.3
A 722/A 722M—07	Specification for Uncoated High-strength Steel Bar for Prestressing Concrete	1810.3.10.2
A 755/A 755M—03(2008)	Specification for Steel Sheet, Metallic-coated by the Hot-dip Process and Prepainted by the Coil-coating Process for Exterior Exposed Building Products	Table 1507.4.3(1), Table 1507.4.3(2)
A 792/A 792M—08	Specification for Steel Sheet, 55% Aluminum-zinc Alloy-coated by the Hot-dip Process	Table 1507.4.3(1), Table 1507.4.3(2)
A 875/A 875M—06	Standard Specification for Steel Sheet Zinc-5 percent, Aluminum Alloy-coated by the Hot-dip Process	Table 1507.4.3(2)
A 913/A 913M—07	Specification for High-strength Low-alloy Steel Shapes of Structural Quality, Produced by Quenching and Self-tempering Process (QST)	1810.3.2.3
A 924/A 924M—08a	Standard Specification for General Requirements for Steel Sheet, Metallic-coated by the Hot-dip Process	Table 1507.4.3(1)
A 992/A 992M—06a	Standard Specification for Structural Shapes	1810.3.2.3
B 42—02e01	Specification for Seamless Copper Pipe, Standard Sizes	909.13.1
B 43—98(2004)	Specification for Seamless Red Brass Pipe, Standard Sizes	909.13.1
B 68—02	Specification for Seamless Copper Tube, Bright Annealed (Metric)	909.13.1
B 88—03	Specification for Seamless Copper Water Tube	909.13.1
B 101—07	Specification for Lead-coated Copper Sheet and Strip for Building Construction	1404.5.3, Table 1507.2.9.2, Table 1507.4.3(1)
B 209—07	Specification for Aluminum and Aluminum Alloy Steel and Plate	Table 1507.4.3(1)
B 251—02e01	Specification for General Requirements for Wrought Seamless Copper and Copper-alloy Tube	909.13.1
B 280—03	Specification for Seamless Copper Tube for Air Conditioning and Refrigeration Field Service	909.13.1
B 370—09	Specification for Cold-rolled Copper Sheet and Strip for Building Construction	1404.5.2, Table 1507.2.9.2, Table 1507.4.3(1)

ASTM—continued

Standard	Title	Referenced in code section number
B 695—04	Standard Specification for Coatings of Zinc Mechanically Deposited on Iron and Steel Strip for Building Construction	2304.9.5.1, 2304.9.5.3
C 5—03	Specification for Quicklime for Structural Purposes	Table 2507.2
C 22/C 22M—00 (2005) e01	Specification for Gypsum	Table 2506.2
C 27—98(2008)	Specification for Standard Classification of Fireclay and High-alumina Refractory Brick	2111.5
C 28/C 28M—00 (2005)	Specification for Gypsum Plasters	Table 2507.2
C 31/C 31M—08b	Practice for Making and Curing Concrete Test Specimens in the Field	Table 1705.3
C 33/C33M—08	Specification for Concrete Aggregates	722.3.1.4, 722.4.1.1.3
C 34—03	Specification for Structural Clay Load-bearing Wall Tile	2103.2
C 35—01(2005)	Specification for Inorganic Aggregates for Use in Gypsum Plaster	Table 2507.2
C 55—06e01	Specification for Concrete Building Brick	Table 722.3.2, 2103.1, 2105.2.2.1.2
C 56—05	Specification for Structural Clay Nonload Bearing Tile	2103.2
C 59/C 59M—00 (2006)	Specification for Gypsum Casting and Molding Plaster	Table 2507.2
C 61/C 61M—00 (2006)	Specification for Gypsum Keene's Cement	Table 2507.2
C 62—08	Specification for Building Brick (Solid Masonry Units Made from Clay or Shale)	1807.1.6.3, 2203.2, 2105.2.2.1.1
C 67—08	Test Methods of Sampling and Testing Brick and Structural Clay Tile	721.4.1.1.1, 2109.3.1.1
C 73—05	Specification for Calcium Silicate Face Brick (Sand-lime Brick)	Table 722.3.2, 2103.1
C 90—08	Specification for Loadbearing Concrete Masonry Units	Table 722.3.2, 1807.1.6.3, 2103.1
C 91—05	Specification for Masonry Cement	Table 2507.2
C 94/C 94M—09	Specification for Ready-mixed Concrete	110.3.1
C 126—99 (2005)	Specification for Ceramic Glazed Structural Clay Facing Tile, Facing Brick and Solid Masonry Units	2103.2
C 140—08a	Test Method Sampling and Testing Concrete Masonry Units and Related Units	722.3.1.2
C 150—07	Specification for Portland Cement	Table 2507.2
C 172—08	Practice for Sampling Freshly Mixed Concrete	Table 1705.3
C 199—84 (2005)	Test Method for Pier Test for Refractory Mortars	2111.5, 2111.8, 2113.12
C 206—03	Specification for Finishing Hydrated Lime	Table 2507.2
C 208—08a	Specification for Cellulosic Fiber Insulating Board	Table 1508.2, 2303.1.5
C 212—00 (2006)	Specification for Structural Clay Facing Tile	2103.2
C 216—07a	Specification for Facing Brick (Solid Masonry Units Made from Clay or Shale)	1807.1.6.3, 2103.2, 2105.2.2.1.1
C 270—08a	Specification for Mortar for Unit Masonry	2103.9
C 315—07	Specification for Clay Flue Liners and Chimney Pots	2111.8, 2113.11.1, Table 2113.16(1)
C 317/C 317M—00 (2005)	Specification for Gypsum Concrete	1911.1
C 330—05	Specification for Lightweight Aggregates for Structural Concrete	202
C 331—05	Specification for Lightweight Aggregates for Concrete Masonry Units	722.3.1.4, 722.4.1.1.3
C 406—06e01	Specification for Roofing Slate	1507.7.5
C 472—99 (2004)	Specification for Standard Test Methods for Physical Testing of Gypsum, Gypsum Plasters and Gypsum Concrete	Table 2506.2
C 473—07	Test Method for Physical Testing of Gypsum Panel Products	Table 2506.2
C 474—05	Test Methods for Joint Treatment Materials for Gypsum Board Construction	Table 2506.2
C 475/C 475M—02 (2007)	Specification for Joint Compound and Joint Tape for Finishing Gypsum Wallboard	Table 2506.2
C 503—08a	Specification for Marble Dimension Stone (Exterior)	2103.4
C 514—04	Specification for Nails for the Application of Gypsum Board	Table 721.1(2), Table 721.1(3), Table 2306.7, Table 2506.2
C 516—08a	Specifications for Vermiculite Loose Fill Thermal Insulation	722.3.1.4, 722.4.1.1.3
C 547—07e1	Specification for Mineral Fiber Pipe Insulation	Table 721.1(2), Table 721.1(3)
C 549—06	Specification for Perlite Loose Fill Insulation	722.3.1.4, 722.4.1.1.3
C 552—07	Standard Specification for Cellular Glass Thermal Insulation	Table 1508.2
C 557—03e01	Specification for Adhesives for Fastening Gypsum Wallboard to Wood Framing	Table 2506.2
C 568—08a	Specification for Limestone Dimension Stone	2103.4
C 578—08b	Standard Specification for Rigid, Cellular Polystyrene Thermal Insulation	Table 1508.2
C 587—04	Specification for Gypsum Veneer Plaster	Table 2507.2
C 595—08a	Specification for Blended Hydraulic Cements	Table 2507.2
C 615—03	Specification for Granite Dimension Stone	2103.4
C 616—08a	Specification for Quartz Dimension Stone	2103.4
C 629—08	Specification for Slate Dimension Stone	2103.4
C 631—09	Specification for Bonding Compounds for Interior Gypsum Plastering	Table 2507.2
C 635/C635M—07	Specification for the Manufacture, Performance and Testing of Metal Suspension Systems for Acoustical Tile and Lay-in Panel Ceilings	808.1.1, 2506.2.1, H107.1.1
C 636/C 636M—08	Practice for Installation of Metal Ceiling Suspension Systems for Acoustical Tile and Lay-in Panels	808.1.1.1
C 645—08a	Specification for Nonstructural Steel Framing Members	Table 2506.2, Table 2507.2

ASTM—continued

C 652—09 Specification for Hollow Brick (Hollow Masonry Units Made from Clay or Shale) 1807.1.6.3, 2103.2, 2105.2.2.1.1

C 726—05e1 Standard Specification for Mineral Fiber Roof Insulation Board. . . . Table 1508.2

C 728—05 Standard Specification for Perlite Thermal Insulation Board . . . . Table 1508.2

C 744—08 Specification for Prefaced Concrete and Calcium Silicate Masonry Units . . . . Table 722.3.2, 2103.1

C 754—08 Specification for Installation of Steel Framing Members to Receive Screw-attached Gypsum Panel Products . . . . Table 2508.1, Table 2511.1.1

C 836—06 Specification for High-solids Content, Cold Liquid-applied Elastomeric Waterproofing Membrane for Use with Separate Wearing Course 1507.15.2

C 840—08 Specification for Application and Finishing of Gypsum Board . . . . Table 2508.1, 2509.2

C 841—03 (2008) e1 Specification for Installation of Interior Lathing and Furring . . . . Table 2508.1, Table 2511.1.1

C 842—05 Specification for Application of Interior Gypsum Plaster . . . . Table 2511.1.1, 2511.3, 2511.4

C 843—99 (2006) Specification for Application of Gypsum Veneer Plaster . . . . Table 2511.1.1

C 844—04 Specification for Application of Gypsum Base to Receive Gypsum Veneer Plaster . . . . Table 2508.1

C 847—09 Specification for Metal Lath . . . . Table 2507.2

C 887—05 Specification for Packaged, Dry Combined Materials for Surface Bonding Mortar 1805.2.2, 2103.10

C 897—05 Specification for Aggregate for Job-Mixed Portland Cement-based Plaster . . . . Table 2507.2

C 920—08 Standard for Specification for Elastomeric Joint Sealants . . . . Table 2506.2

C 926—06 Specification for Application of Portland Cement-based Plaster . . . . 2109.3.4.8, 2510.3, Table 2511.1.1, 2511.3, 2511.4, 2512.1, 2512.1.2, 2512.2, 2512.6, 2512.8.2, 2512.9, 2513.7

C 932—06 Specification for Surface-applied Bonding Compounds Agents for Exterior Plastering . . . . Table 2507.2

C 933—07b Specification for Welded Wire Lath . . . . Table 2507.2

C 946—91 (2001) Specification for Practice for Construction of Dry-stacked, Surface-bonded Walls 2103.10, 2109.2.2

C 954—07 Specification for Steel Drill Screws for the Application of Gypsum Panel Products or Metal Plaster Bases to Steel Studs from 0.033 inch (0.84 mm) to 0.112 inch (2.84 mm) in Thickness . . . . Table 2506.2, Table 2507.2

C 955—09 Standard Specification for Load-bearing Transverse and Axial Steel Studs, Runners Tracks, and Bracing or Bridging, for Screw Application of Gypsum Panel Products and Metal Plaster Bases . . . . Table 2506.2, Table 2507.2

C 956—04 Specification for Installation of Cast-in-place Reinforced Gypsum Concrete 1911.1

C 957—06 Specification for High-solids Content, Cold Liquid-applied Elastomeric Waterproofing Membrane with Integral Wearing Surface 1507.15.2

C 1002—07 Specification for Steel Self-piercing Tapping Screws for the Application of Gypsum Panel Products or Metal Plaster Bases to Wood Studs or Steel Studs . . . Table 2506.2, Table 2507.2

C 1007—08a Specification for Installation of Load Bearing (Transverse and Axial) Steel Studs and Related Accessories . . . . Table 2508.1, Table 2511.1.1

C 1019—09 Test Method of Sampling and Testing Grout 2105.2.2.1.1, 2105.2.2.1.2, 2105.2.2.1.3

C 1029—08 Specification for Spray-applied Rigid Cellular Polyurethane Thermal Insulation 1507.14.2

C 1032—06 Specification for Woven Wire Plaster Base . . . . Table 2507.2

C 1047—09 Specification for Accessories for Gypsum Wallboard and Gypsum Veneer Base . . . . Table 2506.2, Table 2507.2

C 1063—08 Specification for Installation of Lathing and Furring to Receive Interior and Exterior Portland Cement-based Plaster . . . . 2109.3.4.8, 2510.3, Table 2511.1.1, 2512.1.1

C 1088—09 Specification for Thin Veneer Brick Units Made from Clay or Shale . . . . Table 721.1(2), 2103.2

C 1167—03 Specification for Clay Roof Tiles 1507.3.4

C 1177/C 1177M—08 Specification for Glass Mat Gypsum Substrate for Use as Sheathing . . . . Table 2506.2

C 1178/C 1178M—06 Specification for Coated Mat Water-resistant Gypsum Backing Panel . . . . Table 2506.2, 2509.2

C 1186—08 Specification for Flat Fiber Cement Sheets 1404.10, 1405.16.1, 1405.16.2

C 1261—07 Specification for Firebox Brick for Residential Fireplaces 2111.5, 2111.8

C 1278/C 1278M—07a Specification for Fiber-reinforced Gypsum Panels . . . . Table 2506.2

C 1280—09 Specification for Application of Gypsum Sheathing . . . . Table 2508.1, 2508.2

C 1283—07a Practice for Installing Clay Flue Lining 2113.9.1, 2113.12

C 1288—99 (2004) e1 Standard Specification for Discrete Nonasbestos Fiber-cement Interior Substrate Sheets 2509.2

C 1289—08 Standard Specification for Faced Rigid Cellular Polyisocyanurate Thermal Insulation Board . . . . Table 1508.2

C 1314—07 Test Method for Compressive Strength of Masonry Prisms 2105.2.2.2.2, 2105.3.1, 2105.3.2

C 1325—08b Standard Specification for Nonasbestos Fiber-mat Reinforced Cementitious Backer Units 2509.2

C 1328—05 Specification for Plastic (Stucco Cement) . . . . Table 2507.2

C 1364—07 Standard Specification for Architectural Cast Stone 2103.5

C 1386—07 Specification for Precast Autoclaved Aerated Concrete (AAC) Wall Construction Units 202, 2103.3, 2105.2.2.1.3

C 1396M/C1396M—06a Specification for Gypsum Board . . . . Figure 722.5.1(2), Figure 722.5.1(3)

C 1405—08 Standard Specification for Glazed Brick (Single Fired, Solid Brick Units) 2103.2

C 1492—03 Standard Specification for Concrete Roof Tile 1507.3.5

C 1629/C 1629M—06 Standard Classification for Abuse-resistant Nondecorated Interior Gypsum Panel Products and Fiber-reinforced Cement Panels 403.2.3.1, 403.2.3.2, 403.2.3.4

ASTM—continued

C 1658/C 1658M—06	Standard Specification for Glass Mat Gypsum Panels	Table 2506.2
D 25—99 (2005)	Specification for Round Timber Piles	1810.3.2.4, 2303.1.11
D 41—05	Specification for Asphalt Primer Used in Roofing, Dampproofing and Waterproofing	Table 1507.10.2
D 43—00 (2006)	Specification for Coal Tar Primer Used in Roofing, Dampproofing and Waterproofing	Table 1507.10.2
D 56—05	Test Method for Flash Point By Tag Closed Tester	202
D 86—09	Test Method for Distillation of Petroleum Products at Atmospheric Pressure	202
D 93—08	Test Method for Flash Point By Pensky-Martens Closed Cup Tester	202
D 225—07	Specification for Asphalt Shingles (Organic Felt) Surfaced with Mineral Granules	1507.2.5
D 226—06	Specification for Asphalt-saturated Organic Felt Used in Roofing and Waterproofing	1404.2, 1507.2.3, 1507.2.8.1, 1507.3.3, 1507.4.5 1507.5.3, 1507.6.3, 1507.6.3.1, 1507.7.3, Table 1507.8, 1507.8.3, 1507.9.3, 1507.9.5, Table 1507.10.2
D 227—03	Specification for Coal-tar-saturated Organic Felt Used in Roofing and Waterproofing	Table 1507.10.2
D 312—00 (2006)	Specification for Asphalt Used in Roofing	Table 1507.10.2
D 422—63 (2007)	Test Method for Particle-size Analysis of Soils	1803.5.3
D 448—08	Standard Classification for Sizes of Aggregate for Road and Bridge Construction	1507.12.3, 1507.13.3
D 450—07	Specification for Coal-tar Pitch Used in Roofing, Dampproofing and Waterproofing	Table 1507.10.2
D 635—06	Test Method for Rate of Burning and/or Extent and Time of Burning of Self-supporting Plastics in a Horizontal Position	2606.4, H107.1.1
D 1143/D 1143M—07e1	Test Method for Piles Under Static Axial Compressive Load	1810.3.3.1.2
D 1227—95 (2007)	Specification for Emulsified Asphalt Used as a Protective Coating for Roofing	Table 1507.10.2, 1507.15.2
D 1557—07	Test Method for Laboratory Compaction Characteristics of Soil Using Modified Effort [56,000 ft-lb/ft^3 (2,700 KN m/m^3)]	1705.6, 1804.5, J107.6
D 1761—06	Test Method for Mechanical Fasteners in Wood	1711.1.1, 1711.1.2, 1711.1.3
D 1863—05	Specification for Mineral Aggregate Used on Built-up Roofs	Table 1507.10.2
D 1929—96 (2001) e01	Test Method for Determining Ignition Properties of Plastics	402.6.4.4, 406.7.2, 1407.11.2.1, 1407.11.3.3, 1407.11.4.2, 2606.4
D 1970—09	Specification for Self-adhering Polymer Modified Bituminous Sheet Materials Used as Steep Roof Underlayment for Ice Dam Protection	1507.2.4, 1502.2.8.1, 1507.2.9.2, 1507.3.3.3 1507.3.9, 1507.4.5, 1507.5.3.1, 1507.5.7, 1507.6.3.1, 1507.7.3.1, 1507.8.3.1, 1507.8.8, 1507.9.3.1, 1507.9.9
D 2178—04	Specification for Asphalt Glass Felt Used in Roofing and Waterproofing	Table 1507.10.2
D 2487—06e1	Practice for Classification of Soils for Engineering Purposes (Unified Soil Classification System)	Table 1610.1, 1803.5.1
D 2626—04	Specification for Asphalt Saturated and Coated Organic Felt Base Sheet Used in Roofing	1507.3.3, Table 1507.10.2
D 2822—05	Specification for Asphalt Roof Cement, Asbestos Containing	Table 1507.10.2
D 2823—05	Specification for Asphalt Roof Coatings, Asbestos Containing	Table 1507.10.2
D 2824—06	Standard Specification for Aluminum-Pigmented Asphalt Roof Coating, Nonfibered, Asbestos Fibered and Fibered without Asbestos	Table 1507.10.2
D 2843—99 (2004) e01	Test for Density of Smoke from the Burning or Decomposition of Plastics	2606.4
D 2859—06	Standard Test Method for Ignition Characteristics of Finished Textile Floor Covering Materials	804.4.1
D 2898—04	Test Methods for Accelerated Weathering of Fire-retardant-treated Wood for Fire Testing	1505.1, 2303.2.4, 2303.2.6
D 3019—08	Specification for Lap Cement Used with Asphalt Roll Roofing, Nonfibered, Asbestos Fibered and Nonasbestos Fibered	Table 1507.10.2
D 3161—09	Test Method for a Wind Resistance of Asphalt Shingles (Fan Induced Method)	1507.2.7.1, Table 1507.2.7.1(2)
D 3200—74 (2005)	Standard Specification and Test Method for Establishing Recommended Design Stresses for Round Timber Construction Poles	2303.1.11
D 3201—08a	Test Method for Hygroscopic Properties of Fire-retardant-treated Wood and Wood-based Products	2303.2.7
D 3278—96 (2004)	Test Methods for Flash Point of Liquids by Small Scale Closed-cup Apparatus	202
D 3462—09	Specification for Asphalt Shingles Made from Glass Felt and Surfaced with Mineral Granules	1507.2.5
D 3468—99 (2006) e01	Specification for Liquid-applied Neoprene and Chlorosulfonated Polyethylene Used in Roofing and Waterproofing	1507.15.2
D 3679—09	Specification for Rigid Poly [Vinyl Chloride (PVC) Siding]	1404.9, 1405.14
D 3689—07	Test Methods for Deep Foundations Under Static Axial Tensile Load	1810.3.3.1.5
D 3737—08	Practice for Establishing Allowable Properties for Structural Glued Laminated Timber (Glulam)	2303.1.3
D 3746—85 (2008)	Test Method for Impact Resistance of Bituminous Roofing Systems	1504.7
D 3747—79 (2007)	Specification for Emulsified Asphalt Adhesive for Adhering Roof Insulation	Table 1507.10.2
D 3909—97b (2004)	Specification for Asphalt Roll Roofing (Glass Felt) Surfaced with Mineral Granules	1507.2.9.2, 1507.6.5, Table 1507.10.2

ASTM—continued

D 3957—06	Standard Practices for Establishing Stress Grades for Structural Members Used in Log Buildings	2303.1.10
D 4022—07	Specification for Coal Tar Roof Cement, Asbestos Containing	Table 1507.10.2
D 4272—08a	Test Method for Total Energy Impact of Plastic Films by Dart Drop	1504.7
D 4318—05	Test Methods for Liquid Limit, Plastic Limit and Plasticity Index of Soils	1803.5.3
D 4434/D4434M—09	Specification for Poly (Vinyl Chloride) Sheet Roofing	1507.13.2
D 4479—07	Specification for Asphalt Roof Coatings-Asbestos-free	Table 1507.10.2
D 4586—07	Specification for Asphalt Roof Cement-Asbestos-free	Table 1507.10.2
D 4601—04	Specification for Asphalt-coated Glass Fiber Base Sheet Used in Roofing	Table 1507.10.2
D 4637—08	Specification for EPDM Sheet Used in Single-ply Roof Membrane	1507.12.2
D 4829—08a	Test Method for Expansion Index of Soils	1803.5.3
D 4869—05e01	Specification for Asphalt-saturated (Organic Felt) Underlayment Used in Steep Slope Roofing	1507.2.3, 1507.2.8.1, 1507.4.5, 1507.5.3, 1507.5.3.1, 1507.6.3, 1507.7.3, 1507.7.3.1, 1507.8.3, 1507.8.3.1, 1507.9.3, 1507.9.3.1
D 4897—01	Specification for Asphalt-coated Glass Fiber Venting Base Sheet Used in Roofing	Table 1507.10.2
D 4945—08	Test Method for High-strain Dynamic Testing of Piles	1810.3.3.1.2
D 4990—97a (2005) e01	Specification for Coal Tar Glass Felt Used in Roofing and Waterproofing	Table 1507.10.2
D 5019—07a	Specification for Reinforced Nonvulcanized Polymeric Sheet Used in Roofing Membrane	1507.12.2
D 5055—09	Specification for Establishing and Monitoring Structural Capacities of Prefabricated Wood I-joists	2303.1.2
D 5456—09	Specification for Evaluation of Structural Composite Lumber Products	2303.1.9
D 5516—03	Test Method of Evaluating the Flexural Properties of Fire-retardant-treated Softwood Plywood Exposed to the Elevated Temperatures	2303.2.5.1
D 5643—06	Specification for Coal Tar Roof Cement, Asbestos-free	Table 1507.10.2
D 5664—08	Test Methods for Evaluating the Effects of Fire-retardant Treatment and Elevated Temperatures on Strength Properties of Fire-retardant-treated Lumber	2303.2.5.2
D 5665—99a (2006)	Specification for Thermoplastic Fabrics Used in Cold-applied Roofing and Waterproofing	Table 1507.10.2
D 5726—98 (2005)	Specification for Thermoplastic Fabrics Used in Hot-applied Roofing and Waterproofing	Table 1507.10.2
D 6083—05e01	Specification for Liquid Applied Acrylic Coating Used in Roofing	Table 1507.10.2, Table 1507.14.3, 1507.15.2
D 6162—00A (2008)	Specification for Styrene-butadiene-styrene (SBS) Modified Bituminous Sheet Materials Using a Combination of Polyester and Glass Fiber Reinforcements	1507.11.2
D 6163—00 (2008)	Specification for Styrene-butadiene-styrene (SBS) Modified Bituminous Sheet Materials Using Glass Fiber Reinforcements	1507.11.2
D 6164—05e1	Specification for Styrene-butadiene-styrene (SBS) Modified Bituminous Sheet Metal Materials Using Polyester Reinforcements	1507.11.2
D 6222—08	Specification for Atactic Polypropylene (APP) Modified Bituminous Sheet Materials Using Polyester Reinforcements	1507.11.2
D 6223—02	Specification for Atactic Polypropylene (APP) Modified Bituminous Sheet Materials Using a Combination of Polyester and Glass Fiber Reinforcements	1507.11.2
D 6298—05e1	Specification for Fiberglass Reinforced Styrene-butadiene-styrene (SBS) Modified Bituminous Sheets with a Factory Applied Metal Surface	1507.11.2
D 6305—08	Practice for Calculating Bending Strength Design Adjustment Factors for Fire-retardant-treated Plywood Roof Sheathing	2303.2.5.1
D 6380—03 (2009)	Standard Specification for Asphalt Roll Roofing (Organic) Felt	1507.2.9.2, 1507.3.3, 1507.6.5
D 6509/D6509M—09	Standard Specification for Atactic Polypropylene (APP) Modified Bituminous base Sheet Materials Using Glass Fiber Reinforcements	1507.11.2
D 6694—08	Standard Specification for Liquid-applied Silicone Coating Used in Spray Polyurethane Foam Roofing	Table 1507.14.3, 1507.15.2
D 6754—02	Standard Specification for Ketone Ethylene Ester Based Sheet Roofing	1507.13.2
D 6757—07	Standard Specification for Inorganic Underlayment for Use with Steep Slope Roofing Products	1507.2.3
D 6841—08	Standard Practice for Calculating Design Value Treatment Adjustment Factors for Fire-retardant-treated Lumber	2303.2.5.2
D 6878—08e1	Standard Specification for Thermoplastic Polyolefin Based Sheet Roofing	1507.13.2
D 6947—07	Standard Specification for Liquuid Applied Moisture Cured Polyurethane Coating Used in Spray Polyurethane Foam Roofing System	Table 1507.14.3, 1507.15.2
D 7158—08d	Standard Test Method for Wind Resistance of Sealed Asphalt Shingles (Uplift Force/Uplift Resistance Method)	1507.2.7.1, Table 1507.2.7.1(1)
D 7254—07	Standard Specification for polypropylene (PP) siding	1404.12
E 84—09	Test Methods for Surface Burning Characteristics of Building Materials	202, 402.6.4.4, 406.7.2, 703.5.2, 720.1, 720.4, 803.1.1, 803.1.4, 803.9, 803.13, 806.5, 1404.12.1, 1407.9, 1407.10.1, 1409.9, 1409.10.1, 1509.6.2, 1509.6.3, 2303.2, 2603.3, 2603.4.1.13, 2606.3.5.4, 2603.7, 2604.2.4, 2606.4, 2613.3, 3105.4

ASTM—continued

E 90—04	Test Method for Laboratory Measurement of Airborne Sound Transmission Loss of Building Partitions and Elements . . . 1207.2, 1207.2.1
E 96/E 96M—05	Test Method for Water Vapor Transmission of Materials . . . 202
E 108—07a	Test Methods for Fire Tests of Roof Coverings . . . 1505.1, 2603.6, 2610.2, 2610.3
E 119—08a	Test Methods for Fire Tests of Building Construction and Materials . . . 703.2, 703.2.1, 703.2.3, 703.3, 703.4, 703.6, 704.12, 705.7, 705.8.5, 711.3.2, 714.3.1, 714.4.1.1, 715.1, 716.2, Table 716.3, 716.5.6, 716.5.8.1.1, Table 716.6, 716.6.7.1, 717.5.2, 717.5.3, 717.6.1, 716.6.2.1, Table 721.1(1), 1409.10.2, 2103.2, 2603.5.1
E 136—09	Test Method for Behavior of Materials in a Vertical Tube Furnace at 750°C . . . 703.5.1
E 330—02	Test Method for Structural Performance of Exterior Windows, Curtain Walls and Doors by Uniform Static Air Pressure Difference . . . 1409.10.2, 1710.5.2
E 331—00 (2009)	Test Method for Water Penetration of Exterior Windows, Skylights, Doors and Curtain Walls by Uniform Static Air Pressure Difference . . . 1403.2
E 492—09	Test Method for Laboratory Measurement of Impact Sound Transmission Through Floor-ceiling Assemblies Using the Tapping Machine . . . 1207.3
E 605—93 (2006)	Test Method for Thickness and Density of Sprayed Fire-resistive Material (SFRM) Applied to Structural Members . . . 1705.13.4.1, 1705.13.4.2
E 681—04	Test Methods for Concentration Limits of Flammability of Chemical Vapors and Gases . . . 202
E 736—00 (2006)	Test Method for Cohesion/Adhesion of Sprayed Fire-resistive Materials Applied to Structural Members . . . 704.13.2, 1705.13.6
E 814—08b	Test Method of Fire Tests of Through-penetration Firestops . . . 202, 714.3.1.2, 714.3.2, 7143.4.1.1.2
E 970—08a	Test Method for Critical Radiant Flux of Exposed Attic Floor Insulation Using a Radiant Heat Energy Source . . . 720.3.1
E 1300—07e01	Practice for Determining Load Resistance of Glass in Buildings . . . 2404.1, 2404.2, 2404.3.1, 2404.3.2, 2404.3.3, 2404.3.4, 2404.3.5
E 1354—09	Standard Test Method for Heat and Visible Smoke Release Rates for Materials and Products Using an Oxygen Consumption Calorimeter . . . 424.2
E 1592—05	Test Method for Structural Performance of Sheet Metal Roof and Siding Systems by Uniform Static Air Pressure Difference . . . 1504.3.2
E 1602—03	Guide for Construction of Solid Fuel-burning Masonry Heaters . . . 2112.2
E 1886—05	Test Method for Performance of Exterior Windows, Curtain Walls, Doors and Storm Shutters Impacted by Missiles and Exposed to Cyclic Pressure Differentials . . . 1609.1.2
E 1966—07	Test Method for Fire-resistant Joint Systems . . . 202, 715.3
E 1996—09	Specification for Performance of Exterior Windows, Glazed Curtain Walls, Doors and Impact Protective Systems Impacted by Windborne Debris in Hurricanes . . . 1609.1.2, 1609.1.2.2
E 2072—04	Standard Specification for Photoluminescent (Phosphorescent) Safety Markings . . . 1024.4
E 2174—09	Standard Practice for On-Site Inspection of Installed Fire Stops . . . 1705.16.1
E 2273—03	Standard Test Method for Determining the Drainage Efficiency of Exterior Insulation and Finish Systems (EIFS) Clad Wall Assemblies . . . 1408.4.1
E 2307—04e01	Standard Test Method for Determining Fire Resistance of Perimeter Fire Barrier Systems Using Intermediate-scale, Multistory Test Apparatus . . . 715.4
E 2393—09	Standard Practice for On-Site Inspection of Installed Fire Resistive Joint Systems and Perimeter Fire Barrier . . . 1705.16.2
E 2404—08	Standard Practice for Specimen Preparation and Mounting of Textile, Paper or Vinyl Wall or Ceiling Coverings to Assess Surface Burning Characteristics . . . 803.1.4
E 2568—09e1	Standard Specification for PB Exterior Insulation and Finish Systems (EIFS) . . . 1408.2
E 2570—07	Standard Test Method for Evaluating Water-resistive Barrier (WRB) Coatings Used Under Exterior Insulation and Finish Systems (EIFS) for EIFS with Drainage . . . 1408.4.1.1, 1705.15.1
E 2573—07a	Standard Practice for Specimen Preparation and Mounting of Site-fabricated Stretch Systems to Assess Surface Burning Characteristics . . . 803.13
E 2599—09	Standard Practice for Specimen Preparation and Mounting of Reflective Insulation Materials and Radiant Barrier Materials for Building Applications to Assess Surface Burning Characteristics . . . 2613.3
E 2634—08	Standard Specification for Flat Wall Insulating Concrete Form (ICF) Systems . . . 1903.3
F 547—06	Terminology of Nails for Use with Wood and Wood-based Materials . . . Table 2506.2
F 1346—91 (2003)	Performance Specification for Safety Covers and Labeling Requirements for All Covers for Swimming Pools, Spas and Hot Tubs . . . 3109.4, 3109.4.1.8
F 1667—05	Specification for Driven Fasteners: Nails, Spikes and Staples . . . Table 721.1(2), Table 721.1(3), 1507.2.6, 2303.6, Table 2506.2
F 2006—00 (2005) 10	Standard/Safety Specification for Window Fall Prevention Devices for Nonemergency Escape (Egress) and Rescue (Ingress) Windows . . . 1013.8
F 2090—08	Specification for Window Fall Prevention Devices with Emergency Escape (Egress) Release Mechanisms . . . 1013.8, 1013.8.1
F 2200—05	Standard Specification for Automated Vehicular Gate Construction . . . 3110.3
G 152—06	Practice for Operating Open Flame Carbon Arc Light Apparatus for Exposure of Nonmetallic Materials . . . 1504.6

ASTM—continued

G 154—06	Practice for Operating Fluorescent Light Apparatus for UV Exposure of Nonmetallic Materials	1504.6
G 155—05a	Practice for Operating Xenon Arc Light Apparatus for Exposure of Nonmetallic Materials	1504.6

AWCI

Association of the Wall and Ceiling Industry
513 West Broad Street, Suite 210
Falls Church, VA 22046

Standard reference number	Title	Referenced in code section number
12-B—98	Technical Manual 12-B Standard Practice for the Testing and Inspection of Field Applied Thin Film Intumescent Fire-resistive Materials; an Annotated Guide, First Edition	1705.14

AWPA

American Wood Protection Association
P.O. Box 361784
Birmingham, AL 35236-1784

Standard reference number	Title	Referenced in code section number
C1—03	All Timber Products-Preservative Treatment by Pressure Processes	1505.6
M4—08	Standard for the Care of Preservative-treated Wood Products	1810.3.2.4.1, 2303.1.8
U1—11	USE CATEGORY SYSTEM: User Specification for Treated Wood Except Section 6, Commodity Specification H	1403.6, Table 1507.9.6, 1807.1.4, 1807.3.1, 1809.12, 1810.3.2.4.1, 2303.1.8, 2303.1.8.1, 2304.11.2, 2304.11.4, 2304.11.6, 2304.11.7

AWS

American Welding Society
550 N.W. LeJeune Road
Miami, FL 33126

Standard reference number	Title	Referenced in code section number
D1.3—98	Structural Welding Code-Sheet Steel	Table 1705.2.2, 1705.2.2.1.1
D1.4—98	Structural Welding Code-Reinforcing Steel	Table 1705.2.2, 1705.2.2.1.2, Table 1705.3, 2107.4

BHMA

Builders Hardware Manufacturers' Association
355 Lexington Avenue, 17th Floor
New York, NY 10017-6603

Standard reference number	Title	Referenced in code section number
A 156.10—2011	Power Operated Pedestrian Doors	1008.1.4.2
A 156.19—2007	Standard for Power Assist and Low Energy Operated Doors	1008.1.4.2

CGSB

Canadian General Standards Board
Place du Portage 111, 6B1
11 Laurier Street
Gatineau, Quebec, Canada KIA 1G6

Standard reference number	Title	Referenced in code section number
37-GP-52M (1984)	Roofing and Waterproofing Membrane, Sheet Applied, Elastomeric	1504.7, 1507.12.2
37-GP-56M (1980)	Membrane, Modified, Bituminous, Prefabricated and Reinforced for Roofing—with December 1985 Amendment	1507.11.2
CAN/CGSB 37.54—95	Polyvinyl Chloride Roofing and Waterproofing Membrane	1507.13.2

CPA

Composite Panel Association
19465 Deerfield Avenue, Suite 306
Leesburg, VA 20176

Standard reference number	Title	Referenced in code section number
ANSI A135.4—2004	Basic Hardboard	1404.3.1, 2303.1.6
ANSI A135.5—2004	Prefinished Hardboard Paneling	2303.1.6, 2304.6.2
ANSI A135.6—2006	Hardboard Siding	1404.3.2, 2303.1.6

CPSC

Consumer Product Safety Commission
4330 East West Highway
Bethesda, MD 20814-4408

Standard reference number	Title	Referenced in code section number
16 CFR Part 1201 (2002)	Safety Standard for Architectural Glazing Material	2406.2, Table 2406.2(1), 2406.3.1, 2407.1, 2407.1.4.1, 2408.2.1, 2408.3, 2409.1, 2409.2, 2409.3.1
16 CFR Part 1209 (2002)	Interim Safety Standard for Cellulose Insulation	720.6
16 CFR Part 1404 (2002)	Cellulose Insulation	720.6
16 CFR Part 1500 (2009)	Hazardous Substances and Articles; Administration and Enforcement Regulations	202
16 CFR Part 1500.44 (2009)	Method for Determining Extremely Flammable and Flammable Solids	202
16 CFR Part 1507 (2002)	Fireworks Devices	202
16 CFR Part 1630 (2007)	Standard for the Surface Flammability of Carpets and Rugs	804.4.1

CSA

Canadian Standards Association
5060 Spectrum Way
Mississauga, Ontario Canada L4W 5N6

Standard reference number	Title	Referenced in code section number
AAMA/WDMA/CSA 101/I.S.2/A440—11	Specifications for Windows, Doors and Unit Skylights	1710.5.1, 2405.5

CSSB

Cedar Shake and Shingle Bureau
P. O. Box 1178
Sumas, WA 98295-1178

Standard reference number	Title	Referenced in code section number
CSSB—97	Grading and Packing Rules for Western Red Cedar Shakes and Western Red Shingles of the Cedar Shake and Shingle Bureau	Table 1507.8.5, Table 1507.9.6

DASMA

Door and Access Systems Manufacturers Association International
1300 Summer Avenue
Cleveland, OH 44115-2851

Standard reference number	Title	Referenced in code section number
ANSI/DASMA 107—1997 (R2004)	Room Fire Test Standard for Garage Doors Using Foam Plastic Insulation	2603.4.1.9
108—05	Standard Method for Testing Sectional Garage Doors and Rolling Doors: Determination of Structural Performance Under Uniform Static Air Pressure Difference	1710.5.2
115—05	Standard Method for Testing Sectional Garage Doors and Rolling Doors: Determination of Structural Performance Under Missile Impact and Cyclic Wind Pressure	1609.1.2.3

DOC

U.S. Department of Commerce
National Institute of Standards and Technology
1401 Constitution Avenue NW
Washington, DC 20230

Standard reference number	Title	Referenced in code section number
PS-1—09	Structural Plywood	2303.1.4, 2304.6.2, Table 2304.7(4), Table 2304.7(5), Table 2306.2(1), Table 2306.2(2)
PS-2—10	Performance Standard for Wood-based Structural-use Panels	2303.1.4, 2304.6.2, Table 2304.7(5), Table 2306.2(1), Table 2306.2(2)
PS 20—05	American Softwood Lumber Standard	202, 1810.3.2.4, 2303.1.1

DOJ

U.S. Department of Justice
950 Pennsylvania Avenue, NW
Civil Rights Division, Disability Rights Section-NYA
Washington, DC 20530

Standard reference number	Title	Referenced in code section number
DOJ 36 CFR Part 1192	American with Disabilities Act (ADA) Accessibility Guidelines for Transportation Vehicles (ADAAG) Department of Justice, 1991	E109.2.4

DOL

U.S. Department of Labor
c/o Superintendent of Documents
U.S. Government Printing Office
Washington, DC 20402-9325

Standard reference number	Title	Referenced in code section number
29 CFR Part 1910.1000 (2009)	Air Contaminants	202

DOTn

U.S. Department of Transportation
c/o Superintendent of Documents
1200 New Jersey Avenue, SE
Washington, DC 20402-9325

Standard reference number	Title	Referenced in code section number
49 CFR Parts 100—185 2005	Hazardous Materials Regulations	202
49 CFR Parts 173.137 (2009)	Shippers—General Requirements for Shipments and Packaging—Class 8—Assignment of Packing Group	202
49 CFR—1998	Specification of Transportation of Explosive and Other Dangerous Articles, UN 0335, UN 0336 Shipping Containers	202

EN

European Committee for Standardization (EN)
Central Secretariat
Rue de Stassart 36
B-10 50 Brussels

Standard reference number	Title	Referenced in code section number
EN 1081—98	Resilient Floor Coverings—Determination of the Electrical Resistance	406.7.1

FEMA

Federal Emergency Management Agency
Federal Center Plaza
500 C Street S.W.
Washington, DC 20472

Standard reference number	Title	Referenced in code section number
FIA-TB-11—01	Crawlspace Construction for Buildings Located in Special Flood Hazard Areas	1805.1.2.1
P646—08	Guidelines for Design for Structures for Vertical Evacuation from Tsunamis	M101.4

FM

Factory Mutual Global Research
Standards Laboratories Department
1301 Atwood Avenue, P.O. Box 7500
Johnston, RI 02919

Standard reference number	Title	Referenced in code section number
4450 (1989)	Approval Standard for Class 1 Insulated Steel Deck Roofs—with Supplements through July 1992	1508.1, 2603.3, 2603.4.1.5
4470 (1992)	Approval Standard for Class 1 Roof Covers	1504.7

FM—continued

4474 (04)	Evaluating the Simulated Wind Uplift Resistance of Roof Assemblies Using Static Positive and/or Negative Differential Pressures	1504.3.1
4880 (2005)	American National Standard for Evaluating Insulated Wall or Wall and Roof/ Ceiling Assemblies, Plastic Interior Finish Materials, Plastic Exterior Building Panels, Wall/Ceiling Coating Systems,Interior and Exterior Finish Systems	2603.4, 2603.10

GA

Gypsum Association
810 First Street N.E. #510
Washington, DC 20002-4268

Standard reference number	Title	Referenced in code section number
GA 216—07	Application and Finishing of Gypsum Panel Products	Table 2508.1, 2509.2
GA 600—09	Fire-Resistance Design Manual, 18th Edition	Table 721.1(1), Table 721.1(2), Table 721.1(3)

HPVA

Hardwood Plywood Veneer Association
1825 Michael Faraday Drive
Reston, VA 20190

Standard reference number	Title	Referenced in code section number
HP-1—2009	Standard for Hardwood and Decorative Plywood	2303.3, 2304.6.2

HUD

U.S. Department of Housing and Urban Development
451 7th Street, SW
Washington, DC 20410

Standard reference number	Title	Referenced in code section number
HUD 24 CFR Part 3280 (2008)	Manufactured Home Construction and Safety Standards	G201

ICC

International Code Council, Inc.
500 New Jersey Ave, NW
6th Floor
Washington, DC 20001

Standard reference number	Title	Referenced in code section number
ICC A117.1—09	Accessible and Usable Buildings and Facilities	202, 907.5.2.3.4, 1007.9, 1010.1, 1010.7.5, 1010.10, 1011.4, 1022.9, 1101.2, 1107.2, 1109.1, 1109.2, 1109.5.1, 1109.5.2, 1110.3, 1110.4, 1110.4.2, 3008.7.7.1, 3008.7.7.2, 3411.8.2, 3411.8.3, E101.2, E104.2, E104.2.1, E104.3.4, E106.4.9, E107.3, E108.3, E108.4, E109.2.2.2, E109.2.2.3, E109.2.3, E109.2.5, E110.2
ICC 300—12	ICC Standard on Bleachers, Folding and Telescopic Seating and Grandstands	1028.1.1, 1028.14.2, Table 1607.1, 3401.1
ICC 400—12	Standard on Design and Construction of Log Structures	2301.2
ICC 500—08	ICC/NSSA Standard on the Design and Construction of Storm Shelters	202, 423.1
ICC 600—08	Standard for Residential Construction in High-wind Regions	1609.1.1, 1609.1.1.1, 2308.2.1
IEBC—12	International Existing Building Code®	3401.5
IECC—12	International Energy Conservation Code®	101.4.6, 201.3, 1203.1, 1203.3.2, 1301.1.1, 1405.3, 3401.3

ICC—continued

IFC—12	International Fire Code®	101.4.5, 102.6, 201.3, 202, 307.1, Table 307.1(1), Table 307.1(2), 307.1.1, 403.4.5, 404.2, 406.7, 406.8, 410.3.6, 411.1, 412.1, 412.6.1, 413.1, 414.1.1, 414.1.2, 414.1.2.1, 414.2, 414.2.5, Table 414.2.5(1), Table 414.2.5(2), 414.3, 414.5, 414.5.1, Table 414.5.1, 414.5.2, 414.5.3, 414.5.4, 414.6, 415.1, 415.5, 415.5.1, 415.5.1.1, 415.5.1.4, Table 415.5.2, 415.7.3, 415.8, 415.8.1, 415.8.1.4, 415.8.2, 415.8.2.3, 415.8.2.4, 415.8.2.6, 415.8.2.7, 415.8.2.8, 415.8.3, 415.8.4, 415.9, 415.10, 415.10.1.7, 415.10.4, 415.10.7.2, 415.10.9.3, 415.10.10.1, 416.1, 416.4, 421.1, 421.7, 507.3, 507.8.1.1.1, 507.8.1.1.2, 507.8.1.1.3, 705.8.1, 707.1, 901.2, 901.3, 901.5, 901.6.2, 901.6.3, 903.1.1, 903.2.7.1, 903.2.11.6, 903.2.12, 903.5, 904.2.1, 905.1, 905.3.6, 906.1, 907.1.8, 907.2.5, 907.2.13.2, 907.2.15, 907.2.16, 907.6.5, 907.8, 909.20, 910.2.2, 1001.3, 1001.4, 1008.1.9.6, 1203.4.2, 1203.5, 1507.16, 1511.1, Table 1604.5, 2603.4.1.12, 2702.1, 2702.2.9, 2702.2.11, 2702.2.12, 2702.2.13, 2702.3, 3003.3, 3008.1.2, 3102.1, 3103.1, 3111.1, 3302.3, 3303.7, 3309.2, 3401.3, 3403.5, 3404.6, 3412.3.2, 3412.6.8.1, 3412.6.14, 3412.6.14.1
IFGC—12	International Fuel Gas Code®	101.4.1, 201.3, Table 307.1(1), 415.8.3, 2113.11.1.2, 2113.15, 2801.1, 3401.3, A101.2
IMC—12	International Mechanical Code®	101.4.2, 201.3, 307.1, Table 307.1(1), 406.6.2, 406.8.2, 406.8.4, 409.3, 412.6.6, 414.1.2, 414.3, 415.8.1.4, 415.8.2, 415.8.2.7, 415.8.3, 415.8.4, 415.10.11 415.10.11.1, 416.2.2, 413.3, 416.3, 417.1, 419.8, 421.5, 603.1, 603.1.1, 603.1.2, 712.1.5, 717.2.2, 717.5.3, 717.5.4, 717.6.1, 717.6.2, 717.6.3, 718.5, 720.1, 720.7, 903.2.11.4, 904.2.1, 904.11, 907.3.1, 908.6, 909.1, 909.10.2, 909.13.1, 1015.5, 1018.5, 1203.1, 1203.2.1, 1203.4.2, 1203.4.2.1, 1203.5, 1209.3, 2304.5, 2801.1, 3004.3.1, 3401.3, 3412.6.7.1, 3412.6.8, 3412.6.8.1
IPC—12	International Plumbing Code®	101.4.3, 201.3, 415.8.4, 603.1.2, 718.5, 903.3.5, 912.5, 1206.3.3, 1503.4, 1503.4.1, 1805.4.3, 2901.1, Table 2902.1, 3305.1, 3401.3, A101.2
IPMC—12	International Property Maintenance Code®	101.4.4, 102.6, 103.3, 3401.3, 3412.3.2
IPSDC—12	International Private Sewage Disposal Code®	101.4.3, 2901.1, 3401.3
IRC—12	International Residential Code®	101.2, 305.2.3, 308.3.1, 308.4.1, 308.6.4, 310.1, 310.5.1, 2308.1, 3401.3
IWUIC—12	International Wildland-Urban Interface Code®	Table 1505.1
SBCCI SSTD 11—97	Test Standard for Determining Wind Resistance of Concrete or Clay Roof Tiles	1711.2.1, 1711.2.2

ISO

International Organization for Standardization
ISO Central Secretariat
1 ch, de la Voie-Creuse, Case Postale 56
CH-1211 Geneva 20, Switzerland

Standard reference number	Title	Referenced in code section number
ISO 8115—86	Cotton Bales—Dimensions and Density	Table 307.1(1), Table 415.10.1.1.1

NAAMM

National Association of Architectural Metal Manufacturers
800 Roosevelt Road, Bldg. C, Suite 312
Glen Ellyn, IL 60137

Standard reference number	Title	Referenced in code section number
FP 1001—07	Guide Specifications for Design of Metal Flag Poles	1609.1.1

NCMA

National Concrete Masonry Association
13750 Sunrise Valley
Herndon, VA 22071-4662

Standard reference number	Title	Referenced in code section number
TEK 5—84 (1996)	Details for Concrete Masonry Fire Walls	Table 721.1(2)

NFPA

National Fire Protection Association
1 Batterymarch Park
Quincy, MA 02169-7471

Standard reference number	Title	Referenced in code section number
10—10	Portable Fire Extinguishers	906.2, 906.3.2, 906.3.4, Table 906.3(1), Table 906.3(2)
11—10	Low Expansion Foam	904.7
12—11	Carbon Dioxide Extinguishing Systems	904.8, 904.11
12A—09 Halon 1301	Halon 1301 Fire Extinguishing Systems	904.9
13—10	Installation of Sprinkler Systems	708.2, 903.3.1.1, 903.3.2, 903.3.5.1.1, 903.3.5.2, 904.11, 905.3.4, 907.6.3, 1009.3
13D—10	Installation of Sprinkler Systems in One- and Two-family Dwellings and Manufactured Homes	903.3.1.3, 903.3.5.1.1
13R—10	Installation of Sprinkler Systems in Residential Occupancies Up to and Including Four Stories in Height	903.3.1.2, 903.3.5.1.1, 903.3.5.1.2, 903.4
14—10	Installation of Standpipe and Hose System	905.2, 905.3.4, 905.4.2, 905.6.2, 905.8
16—11	Installation of Foam-water Sprinkler and Foam-water Spray Systems	904.7, 904.11
17—09	Dry Chemical Extinguishing Systems	904.6, 904.11
17A—09	Wet Chemical Extinguishing Systems	904.5, 904.11
20—10	Installation of Stationary Pumps for Fire Protection	913.1, 913.2.1, 913.5
30—12	Flammable and Combustible Liquids Code	415.5, 507.8.1.1.1, 507.8.1.1.2
31—06	Installation of Oil-burning Equipment	2113.15
32—11	Dry Cleaning Plants	415.8.4
40—11	Storage and Handling of Cellulose Nitrate Film	409.1
58—11	Liquefied Petroleum Gas Code	415.8.3
61—08	Prevention of Fires and Dust Explosions in Agricultural and Food Product Facilities	415.8.1
70—11	National Electrical Code	108.3, 415.10.1.8, 904.3.1, 907.6.1, 909.12.1, 909.16.3, 1205.4.1, 2701.1, 3401.3, H106.1, H106.2, K101, K111.1
72—10	National Fire Alarm Code	901.6, 903.4.1, 904.3.5, 907.2, 907.2.5, 907.2.11, 907.2.13.2, 907.3, 907.3.3, 907.3.4, 907.5.2.1.2, 907.5.2.2, 907.6, 907.6.1, 907.6.5, 907.7, 907.7.1, 907.7.2, 907.2.9.2, 911.1.5, 3006.5, 3007.8
80—10	Fire Doors and Other Opening Protectives	410.3.5, 509.4.2, 716.5, 716.5.7, 716.5.8.1, 716.5.9.2, 716.6, 716.6.4, 1008.1.4.2, 1008.1.4.3
82—09	Standard for Incinerators and Waste and Linen Handling Systems and Equipment, 2009 Edition	713.13
85—11	Boiler and Combustion System Hazards Code (Note: NFPA 8503 has been incorporated into NFPA 85)	415.8.1
92B—09	Smoke Management Systems in Malls, Atria and Large Spaces	909.8
99—10	Standard for Health Care Facilities	407.10
101—12	Life Safety Code	1028.6.2
105—10	Standard for the Installation of Smoke Door Assemblies	405.4.2, 710.5.2.2, 716.5.3.1, 909.20.4.1
110—10	Emergency and Standby Power Systems	2702.1
111—10	Stored Electrical Energy Emergency and Standby Power Systems	2702.1
120—10	Coal Preparation Plants	415.8.1
170—09	Standard for Fire Safety and Emergency Symbols	1024.2.6.1
211—10	Chimneys, Fireplaces, Vents and Solid Fuel-burning Appliances	2112.5
221—09	Standard for High Challenge Fire Walls, Fire Walls, and Fire Barrier Walls, 2009 Edition	706.2
252—12	Standard Methods of Fire Tests of Door Assemblies	715.4.2, 715.4.3, 715.4.7.3.1, Table 716.3, 716.4, 716.5.1, 716.5.3, 716.5.8, 716.5.8.1.1, 716.5.8.3.1
253—11	Test for Critical Radiant Flux of Floor Covering Systems Using a Radiant Heat Energy Source	406.8.3, 424.2, 804.2, 804.3
257—12	Standard for Fire Test for Window and Glass Block Assemblies	Table 716.3, 716.4, 716.5.3.2, 716.6, 716.6.1, 716.6.2, 716.6.7.3
259—08	Test Method for Potential Heat of Building Materials	2603.4.1.10, 2603.5.3

NFPA—continued

265—11	Method of Fire Tests for Evaluating Room Fire Growth Contribution of Textile Wall Coverings on Full Height Panels and Walls	803.1.3, 803.1.3.1
268—12	Standard Test Method for Determining Ignitibility of Exterior Wall Assemblies Using a Radiant Heat Energy Source	1406.2.1.1, 1406.2.1.1.1, 1406.2.1.1.2, 2603.5.7, D105.1
275—09	Standard Method Of Fire Tests for the Evaluation of Thermal Barriers Used Over Foam Plastic Insulation	1407.10.2, 2603.4
285—11	Standard Method of Test for the Evaluation of Flammability Characteristics of Exterior Nonload-bearingWall Assemblies Containing Combustible Components	718.2.6, 1403.5, 1407.10.4, 1409.10.4, 1509.6.2, 2603.5.5
286—11	Standard Method of Fire Test for Evaluating Contribution of Wall and Ceiling Interior Finish to Room Fire Growth	402.6.4.4, 803.1.2, 803.1.2.1, 803.9, 2603.4, 2603.7, 2603.10, 2604.2.4, 2613.4
288—12	Standard Method of Fire Tests of Floor Fire Door Assemblies Installed Horizontally in Fire-resistance-rated Floor Systems	711.8
289—09	Standard Method of Fire Test for Individual Fuel Packages	402.6.2, 407.6.4.6, 424.2
409—10	Aircraft Hangars	412.4.6, Table 412.4.6, 412.4.6.1, 412.6.5
418—11	Standard for Heliports	412.7.4
484—12	Combustible Metals	415.8.1
654—11	Prevention of Fire & Dust Explosions from the Manufacturing, Processing and Handling of Combustible Particulate Solids	415.8.1
655—12	Prevention of Sulfur Fires and Explosions	415.8.1
664—12	Prevention of Fires and Explosions in Wood Processing and Woodworking Facilities	415.8.1
701—10	Standard Methods of Fire Tests for Flame-propagation of Textiles and Films	410.3.6, 424.2, 801.4, 806.1, 806.1.2, 806.2, 3102.3, 3102.3.1, 3102.6.1.1, 3105.4, D102.2.8, H106.1.1
704—12	Standard System for the Identification of the Hazards of Materials for Emergency Response	202, 414.7.2
720—09	Standard for the Installation of Carbon Monoxide (CO) Detection and Warning Equipment	908.7
1124—06	Manufacture, Transportation and Storage of Fireworks and Pyrotechnic Articles	415.5.1.1, 415.3.1
2001—08	Clean Agent Fire Extinguishing Systems	904.10

PCI

Precast Prestressed Concrete Institute
200 West Adams Street, Suite 2100
Chicago, IL 60606-5230

Standard reference number	Title	Referenced in code section number
MNL 124—89	Design for Fire Resistance of Precast Prestressed Concrete	722.2.3.1
MNL 128—01	Recommended Practice for Glass Fiber Reinforced Concrete Panels	1903.2

PTI

Post-Tensioning Institute
8601 North Black Canyon Highway, Suite 103
Phoenix, AZ 85021

Standard reference number	Title	Referenced in code section number
PTI—2007	Standard Requirements for Analysis of Shallow Concrete Foundations on Expansive Soils, Third Edition	1808.6.2
PTI—2007	Standard Requirements for Design of Shallow Post-tensioned Concrete Foundation on Expansive Soils, Second Edition	1808.6.2

RMI

Rack Manufacturers Institute
8720 Red Oak Boulevard, Suite 201
Charlotte, NC 28217

Standard reference number	Title	Referenced in code section number
ANSI/MH16.1—08	Specification for Design, Testing and Utilization of Industrial Steel Storage Racks	2209.1

SDI

Steel Deck Institute
P. O. Box 25
Fox River Grove, IL 60021

Standard reference number	Title	Referenced in code section number
ANSI/NC1.0—10	Standard for Noncomposite Steel Floor Deck	2210.1.1.1
ANSI/RD1.0—10	Standard for Steel Roof Deck	2210.1.1.2

SJI

Steel Joist Institute
1173B London Links Drive
Forest, VA 24551

Standard reference number	Title	Referenced in code section number
CJ—10	Standard Specification for Composite Steel Joists, CJ-series	1604.3.3, 2203.2, 2207.1
JG—10	Standard Specification for Joist Girders	1604.3.3, 2203.2, 2207.1
K—10	Standard Specification for Open Web Steel Joists, K-series	1604.3.3, 2203.2, 2207.1
LH/DLH—10	Standard Specification for Longspan Steel Joists, LH-series and Deep Longspan Steel Joists, DLH-series	1604.3.3, 2203.2, 2207.1

SPRI

Single-Ply Roofing Institute
411 Waverly Oaks Road, Suite 331B
Waltham, MA 02452

Standard reference number	Title	Referenced in code section number
ANSI/SPRI/ FM4435-ES-1—03	Wind Design Standard for Edge Systems Used with Low Slope Roofing Systems	1504.5
RP-4—08	Wind Design Guide for Ballasted Single-ply Roofing Systems	1504.4

TIA

Telecommunications Industry Association
2500 Wilson Boulevard
Arlington, VA 22201-3834

Standard reference number	Title	Referenced in code section number
222-G—05	Structural Standards for Antenna Supporting Structures and Antennas, including—Addendum 1, 222-G-1, Dated 2007 and Addendum 2, 222-G-2 Dated 2009	1609.1.1, 3108.1, 3108.2

TMS

The Masonry Society
3970 Broadway, Unit 201-D
Boulder, CO 80304-1135

Standard reference number	Title	Referenced in code section number
0216—97	Standard Method for Determining Fire Resistance of Concrete and Masonry Construction Assemblies	Table 721.1(2), 722.1
0302—07	Standard Method for Determining the Sound Transmission Class Rating for Masonry Walls	1207.2.1
402—11	Building Code Requirements for Masonry Structures	1405.6, 1405.6.1, 1405.6.2, 1405.10, 1604.3.4, 1705.4, 1705.4.1, 1807.1.6.3, 1807.1.6.3.2, 1808.9 2101.2.2, 2101.2.3, 2101.2.4, 2101.2.5, 2101.2.6, 2103.9, 2103.12 2103.13, 2103.14, 2104.1, 2104.1.1, 2104.1.2, 2104.1.3, 2104.2, 2104.3, 2104.4, 2105.2.2.1, 2105.2.2.1.2, 2105.2.2.1.3, 2106.1, 2107.1, 2107.2, 2107.3, 2107.4, 2108.1, 2108.2, 2108.3, 2109.1, 2109.1.1, 2109.2, 2109.2.1, 2109.3, 2110.1
403—10	Direct Design Handbook for Masonry Structures	2101.2.7
602—11	Specification for Masonry Structures	1405.6.1, 1705.4, 1807.1.6.3, 2103.9, 2103.12, 2103.13, 2103.14, 2104.1, 2104.1.1, 2104.1.2, 2104.1.3, 2104.2, 2104.3, 2104.4, 2105.2.2.1.1, 2105.2.2.1.2, 2105.2.2.1.3

TPI

Truss Plate Institute
218 N. Lee Street, Suite 312
Alexandria, VA 22314

Standard reference number	Title	Referenced in code section number
TPI 1—2007	National Design Standards for Metal-plate-connected Wood Truss Construction	2303.4.6, 2306.1

UL

Underwriters Laboratories, Inc.
333 Pfingsten Road
Northbrook, IL 60062-2096

Standard reference number	Title	Referenced in code section number
9—2009	Fire Tests of Window Assemblies—with Revisions through April 2005	715.5.2, 716.4, 716.5.3.2, 716.6, 716.6.1, 716.6.2, 716.6.8.1
10A—2009	Tin Clad Fire Doors	716.5
10B—2008	Fire Tests of Door Assemblies—with Revisions through April 2009	716.5.2
10C—2009	Positive Pressure Fire Tests of Door Assemblies	716.5.1, 716.5.3, 1008.1.10.1
14B—2008	Sliding Hardware for Standard Horizontally-mounted Tin Clad Fire Doors	716.5
14C—06	Swinging Hardware for Standard Tin Clad Fire Doors Mounted Singly and in Pairs—with revisions through December 2008	716.5
55A—04	Materials for Built-Up Roof Coverings	1507.10.2
103—01	Factory-built Chimneys, for Residential Type and Building Heating Appliances—with Revisions through *March 2010*	718.2.5.1
127—08	Factory-built Fireplaces—with Revisions through January 2010	718.2.5.1, 2111.11
199E—04	Outline of Investigation for Fire Testing of Sprinklers and Water Spray Nozzles for Protection of Deep Fat Fryers	904.11.4.1
217—06	Single and Multiple Station Smoke Alarms—with Revisions *through April 2010*	907.2.11
263—03	Standard for Fire Tests of Building Construction and Materials, *with revisions through October 2007*	703.2, 703.2.1, 703.2.3, 703.3, 703.4, 703.6, 704.12, 705.7, 705.8.5, 707.7, 711.3.2, 714.3.1, 714.4.1.1, 715.1, 716.2, Table 716.3, 716.5.6, 716.5.8.1.1, 716.7.1, 717.5.2, 717.5.3, 717.6.2.1, Table 721.1(1), 1407.10.2, 2103.2, 2603.4, 2603.5.1
268—06	Smoke Detectors for Fire Protective Signaling Systems—with Revisions through January 1999	407.8, 907.2.6.2
294—1999	Access Control System Units with revisions through 2009	1008.1.9.8

UL—continued

300—05	Fire Testing of Fire Extinguishing Systems for Protection of Commercial Cooking Equipment	904.11
305—97	Panic Hardware—with revisions through January 2007	1008.1.10.1
325—02	Door, Drapery, Gate, Louver and Window Operations and Systems—with Revisions through *February 2010*	405.3.5, 3110.4
555—2006	Fire Dampers—*with revisions through May 2010*	717.3
555C—2006	Ceiling Dampers—*with revisions through May 2010*	717.3
555S—99	Smoke Dampers—with Revisions *through May 2010*	717.3, 717.3.1
580—2006	Test for Uplift Resistance of Roof Assemblies—with Revisions through July 2009	1504.3.1, 1504.3.2
641—95	Type L Low-temperature Venting Systems—with Revisions through *July 2009*	2113.11.1.4
710B—04	Recirculating Systems—with Revisions through *December 2009*	904.11
723—2008	Standard for Test for Surface Burning Characteristics of Building Materials	202, 402.6.4.4, 406.7.2, 703.5.2, 720.1, 720.4, 803.1.1, 803.1.4, 803.9, 803.13, 806.5, 1404.12.1, 1407.9, 1407.10.1, 1409.9, 1409.10.1, 1509.6.2, 1509.6.3, 2303.2, 2603.3, 2603.4.1.13, 2606.3.5.4, 2603.7, 2604.2.4, 2606.4, 2613.3, 3105.4
790—04	Standard Test Methods for Fire Tests of Roof Coverings—with revisions through October 2008	1505.1, 2603.6, 2610.2, 2610.3
793—08	Standards for Automatically Operated Roof Vents for Smoke and Heat	406.8.5.1.1, 910.3.1
864—03	Standards for Control Units and Accessories for Fire Alarm Systems—with Revisions through *February 2010*	909.12
924—06	Standard for Safety Emergency Lighting and Power Equipment—with revisions through January 2009	1011.5
1040—96	Fire Test of Insulated Wall Construction—with Revisions through September 2007	1407.10.3, 1409.10.3, 2603.4, 2603.10
1256—02	Fire Test of Roof Deck Construction—with Revisions through January 2007	1508.1, 2603.3, 2603.4.1.5
1479—03	Fire Tests of Through-penetration Firestops—with Revisions through *March 2010*	202, 714.3.1.2, 714.3.2, 714.4.1.1.2, 714.5
1482—2010	Solid-Fuel-type Room Heater	2112.2, 2112.5
1703—02	Flat-Plate Photovoltaic Modules and Panels—with revisions through April 2008	1507.17.1, 1509.7.4
1715—97	Fire Test of Interior Finish Material—with Revisions through April 2008	1407.10.3, 1409.10.2, 1409.10.3, 2603.4, 2603.10, 2613.4
1777—2007	Chimney Liners—*with revisions through July 2009*	2113.11.1, 2113.19
1784—01	Air Leakage Tests of Door Assemblies—with Revisions through *July 2009*.	710.5.2.2, 713.14.1, 716.5.3.1, 716.5.7.1, 716.5.7.3, 3007.7.3, 3008.7.3
1897—04	Uplift Tests for Roof Covering Systems—with revisions through May 2008	1504.3.1
1975—06	Fire Test of Foamed Plastics Used for Decorative Purposes	402.6.2, 402.6.4.5, 424.2
1994—04	Luminous Egress Path Marking Systems—*with Revisions through April 2010*	411.7, 1024.2.1, 1024.2.3, 1024.2.4, 1024.4
2017—2008	Standards for General-purpose Signaling Devices and Systems—with Revisions through *October 2009*	406.8.5.1.1, 3109.4.1.8
2034—2008	Standard for Single- and Multiple Station Carbon Monoxide Alarm—with revision through February 2009	908.7
2079—04	Tests for Fire Resistance of Building Joint Systems—with Revisions through June 2008	202, 715.3, 715.6
2200—98	Stationary Engine Generator Assemblies—with Revisions through *December 2009*	2702.1.1

ULC

Underwriters Laboratories of Canada
7 Underwriters Road
Toronto, Ontario, Canada M1R3B4

Standard reference number	Title	Referenced in code section number
CAN/ULC S 102.2—1988	Standard Method of Test for Surface Burning Characteristics of Flooring, Floor Coverings and Miscellaneous Materials and Assemblies—with 2000 Revisions	720.4

USC

United States Code
c/o Superintendent of Documents
U.S. Government Printing Office
Washington, DC 20402-9325

Standard reference number	Title	Referenced in code section number
18 USC Part 1, Ch.40	Importation, Manufacture, Distribution and Storage of Explosive Materials	202

WDMA

Window and Door Manufacturers Association
1400 East Touhy Avenue #470
Des Plaines, IL 60018

Standard reference number	Title	Referenced in code section number
AAMA/WDMA/CSA 101/I.S.2/A440—11	Specifications for Windows, Doors and Unit Skylights	1710.5.1, 2405.5

WRI

Wire Reinforcement Institute, Inc.
942 Main Street, Suite 300
Hartford, CT 06103

Standard reference number	Title	Referenced in code section number
WRI/CRSI—81	Design of Slab-on-ground Foundations—with 1996 Update	1808.6.2

Appendix A: Employee Qualifications

The provisions contained in this appendix are not mandatory unless specifically referenced in the adopting ordinance.

General Comments

When adopted (see Section 101.2.1), this appendix provides jurisdictions with training, experience and certification requirements for those employees of the department of building safety responsible for enforcing the provisions of the building code, including the codes referenced in Section 101.4. The authorization to create these positions is found in Section 103 of the code. Note that jurisdictions may also be mandated by applicable state laws to employ only persons licensed by the state to perform certain duties, and may or may not be able to impose additional requirements.

Purpose

The purpose of this appendix is to provide optional criteria for qualifications for employees who enforce the code. A jurisdiction that wants to make this appendix a mandatory part of the code needs to specifically list this appendix in its adoption ordinance (see page xxi of the code for a sample legislation for adoption).

SECTION A101 BUILDING OFFICIAL QUALIFICATIONS

A101.1 Building official. The *building official* shall have at least 10 years' experience or equivalent as an architect, engineer, inspector, contractor or superintendent of construction, or any combination of these, five years of which shall have been supervisory experience. The *building official* should be certified as a *building official* through a recognized certification program. The building official shall be appointed or hired by the applicable governing authority.

❖ As the executive official in charge of administering the code enforcement program of the jurisdiction, the building official is required to possess significant experience in the construction industry, including at least five years as a supervisor. Although not mandated, it is recommended that the person hold certification as a building official, such as the Certified Building Official® (C.B.O.®) issued by the International Code Council® (ICC®). People with this certification have passed examinations that measure knowledge of legal and management principles, as well as code technology. This position may have a different title, as described in the commentary to Section 103.1.

A101.2 Chief inspector. The *building official* can designate supervisors to administer the provisions of the *International Building*, *Mechanical* and *Plumbing Codes* and *International Fuel Gas Code*. Each supervisor shall have at least 10 years' experience or equivalent as an architect, engineer, inspector, contractor or superintendent of construction, or any combination of these, five years of which shall have been in a supervisory capacity. They shall be certified through a recognized certification program for the appropriate trade.

❖ People who supervise inspectors and plans examiners that review construction for conformance to the building and referenced codes are expected to have considerable experience in their area of expertise, since they are responsible for overseeing the day-to-day operation of their divisions within the department. Large departments may have a separate division for each of the codes, while those of lesser size may group two or more together. Because of the responsibility placed on these persons, it is required that they have demonstrated knowledge of the codes that they administer by being certified in the appropriate category. ICC provides certification programs that cover the range of categories involved.

A101.3 Inspector and plans examiner. The *building official* shall appoint or hire such number of officers, inspectors, assistants and other employees as shall be authorized by the jurisdiction. A person shall not be appointed or hired as inspector of construction or plans examiner who has not had at least 5 years' experience as a contractor, engineer, architect, or as a superintendent, foreman or competent mechanic in charge of construction. The inspector or plans examiner shall be certified through a recognized certification program for the appropriate trade.

❖ Departments typically have inspectors and plans examiners who specialize in codes that regulate the various aspects of building construction. Additionally, there will be employees who handle clerical and various administrative duties. The building official will hire the number of employees authorized by the jurisdiction's elected officials. Newly hired inspectors and plans examiners are not necessarily expected to have had previous experience in code enforcement, but should be familiar with the construction industry, with at least five years experience in one or more of the fields listed. They also must possess certification in the job category in which they are employed. ICC provides certification programs that cover the range of categories involved.

A101.4 Termination of employment. Employees in the position of *building official*, chief inspector or inspector shall not be removed from office except for cause after full opportunity has been given to be heard on specific charges before such applicable governing authority.

❖ This section establishes that once employed, the building official and other technical employees of the department cannot be removed, except for cause, subject to a due process review of the specific charges.

SECTION A102 REFERENCED STANDARDS

IBC-12	International Building Code	A101.2
IFGC-12	International Fuel Gas Code	A101.2
IMC-2	International Mechanical Code	A101.2
IPC-12	International Plumbing Code	A101.2

Appendix B: Board of Appeals

The provisions contained in this appendix are not mandatory unless specifically referenced in the adopting ordinance.

General Comments

When adopted (see Section 101.2.1), this appendix provides jurisdictions with detailed appeals board member qualifications and administrative procedures to supplement the basic requirements found in Section 113.

Purpose

The purpose of this appendix is to provide optional criteria for administrative procedures of the board of appeals and board member qualifications. A jurisdiction that wants to make this appendix a mandatory part of the code needs to specifically list this appendix in its adoption ordinance (see page xxi of the code for sample legislation for adoption).

SECTION B101 GENERAL

B101.1 Application. The application for appeal shall be filed on a form obtained from the *building official* within 20 days after the notice was served.

❖ A party aggrieved by a decision of the building official and desiring to appeal that decision must file a written request for a hearing in a timely manner, in this case 20 days from the date of the decision. The building official's decision also needs to be in writing, as specified in Sections 105.3.1 for a permit application, 110.6 for an inspection and 114.2 for a violation, or in a letter written regarding a request for an approval, depending on the action being appealed. The request must be on the jurisdiction's standard appeals form, which requires the appellant to provide the information needed by the board members to make a decision. This is often supplemented by whatever additional information the appellant wants to provide to support his or her case.

B101.2 Membership of board. The board of appeals shall consist of persons appointed by the chief appointing authority as follows:

1. One for five years; one for four years; one for three years; one for two years; and one for one year.
2. Thereafter, each new member shall serve for five years or until a successor has been appointed.

The *building official* shall be an ex officio member of said board but shall have no vote on any matter before the board.

❖ The board of appeals is to consist of five members appointed by the "chief appointing authority," typically the mayor or city manager. One member must be appointed for five years, one for four, one for three, one for two and one for one year. This method of appointment allows for a smooth transition of board of appeals members, providing continuity of action over the years. The building official also sits as a member of the board, but cannot vote on matters regarding his or her own decisions.

B101.2.1 Alternate members. The chief appointing authority shall appoint two alternate members who shall be called by the board chairperson to hear appeals during the absence or disqualification of a member. Alternate members shall possess the qualifications required for board membership and shall be appointed for five years, or until a successor has been appointed.

❖ This section authorizes the chief appointing authority to appoint two alternate members who must be available if the principal members of the board are absent or disqualified. Alternate members must possess the same qualifications as the principal members and are appointed for a term of five years, or until such time that a successor is appointed.

B101.2.2 Qualifications. The board of appeals shall consist of five individuals, one from each of the following professions or disciplines:

1. Registered design professional with architectural experience or a builder or superintendent of building construction with at least ten years' experience, five of which shall have been in responsible charge of work.
2. Registered design professional with structural engineering experience.
3. Registered design professional with mechanical and plumbing engineering experience or a mechanical contractor with at least ten years' experience, five of which shall have been in responsible charge of work.
4. Registered design professional with electrical engineering experience or an electrical contractor with at least ten years' experience, five of which shall have been in responsible charge of work.

5. Registered design professional with fire protection engineering experience or a fire protection contractor with at least ten years' experience, five of which shall have been in responsible charge of work.

❖ The board of appeals consists of five persons with the qualifications and experience indicated in this section. One must be a registered design professional (see Item 2) with structural experience. The others must be registered design professionals, construction superintendents or contractors with experience in the various areas of building construction. This is necessary because the board will be hearing appeals regarding the codes referenced in Section 101.4 as well as the building code. These requirements are important in that technical people rule on technical matters. The board of appeals is not the place for policy or political deliberations. It is intended that these matters be decided purely on their technical merits, with due regard for state-of-the-art construction technology.

B101.2.3 Rules and procedures. The board is authorized to establish policies and procedures necessary to carry out its duties.

❖ The board can set forth the detailed procedures by which it operates, supplementing those required by Section B101.3.2.

B101.2.4 Chairperson. The board shall annually select one of its members to serve as chairperson.

❖ It is customary to determine a chairperson annually so that a regular opportunity is available to evaluate and either reappoint the current chairperson or appoint a new one.

B101.2.5 Disqualification of member. A member shall not hear an appeal in which that member has a personal, professional or financial interest.

❖ All members must disqualify themselves regarding any appeal in which they have a personal, professional or financial interest.

B101.2.6 Secretary. The chief administrative officer shall designate a qualified clerk to serve as secretary to the board. The secretary shall file a detailed record of all proceedings in the office of the chief administrative officer.

❖ The chief administrative officer is to designate a qualified clerk, typically an employee in the department, to serve as secretary to the board. The secretary is required to keep detailed records of the proceedings. These may be needed for any future review of the board's decision.

B101.2.7 Compensation of members. Compensation of members shall be determined by law.

❖ Members of the board of appeals are not required to be compensated unless required by state or local law.

B101.3 Notice of meeting. The board shall meet upon notice from the chairperson, within 10 days of the filing of an appeal or at stated periodic meetings.

❖ In order that an appellant's request be heard in a timely manner, the board must meet within 10 days of the filing of an appeal. In large jurisdictions, where there are likely to be more appeals, the board will often set a regular schedule of meeting dates, such as monthly, and the 10-day rule would not apply.

B101.3.1 Open hearing. All hearings before the board shall be open to the public. The appellant, the appellant's representative, the building official and any person whose interests are affected shall be given an opportunity to be heard.

❖ All hearings before the board must be open to the public. The appellant, the appellant's representative, the building official and any person whose interests are affected must be heard.

B101.3.2 Procedure. The board shall adopt and make available to the public through the secretary procedures under which a hearing will be conducted. The procedures shall not require compliance with strict rules of evidence, but shall mandate that only relevant information be received.

❖ The board is required to establish and make available to the public written procedures detailing how hearings are to be conducted. Additionally, this section provides that although strict rules of evidence are not applicable, the information presented must be deemed relevant so the hearings are not unnecessarily long.

B101.3.3 Postponed hearing. When five members are not present to hear an appeal, either the appellant or the appellant's representative shall have the right to request a postponement of the hearing.

❖ When all five members of the board are not present, either the appellant or the appellant's representative may request that the hearing be postponed. Note that this is not a requirement, and that the appellant can elect to have the case heard before the smaller board.

B101.4 Board decision. The board shall modify or reverse the decision of the *building official* by a concurring vote of two-thirds of its members.

❖ A concurring vote of two-thirds of the board is needed to modify or reverse the decision of the building official. This means that at least four members must vote to modify or reverse the building official, even if less than the full board is present in accordance with Section B101.3.3.

B101.4.1 Resolution. The decision of the board shall be by resolution. Certified copies shall be furnished to the appellant and to the *building official*.

❖ The building official is to be bound by the action of the board of appeals, unless it is the opinion of the building official that the board of appeals has acted improperly. In such cases, relief through the court having jurisdiction may be sought by the jurisdiction's counsel.

B101.4.2 Administration. The *building official* shall take immediate action in accordance with the decision of the board.

❖ To avoid any undue hindrance in the progress of construction, the building official is required to act without delay based on the board's decision. This action may be to enforce the decision or to seek judicial relief if the board's action can be demonstrated to be inappropriate.

Appendix C: Group U—Agricultural Buildings

The provisions contained in this appendix are not mandatory unless specifically referenced in the adopting ordinance.

General Comments

The provisions of this appendix are supplemental to the remainder of the code's requirements for agricultural buildings and are only applicable when specifically adopted, as stated in Section 101.2.1. This appendix contains provisions that may alter requirements found elsewhere in the code; however, the general requirements of the code still apply unless modified within this appendix. For example, the height and area limitations established in Chapter 5 apply to all Group U buildings and structures unless this appendix is specifically adopted by the jurisdiction. In such a case, the allowable height and area limitations in Table C102.1 would supersede those in Table 503.

As with all of the appendices to the code, the provisions of this appendix are not applicable unless the appendix is explicitly included in an adopting ordinance by the jurisdiction. Section 101.2.1 establishes the relationship of all appendices (see the commentary to that section for further discussion).

Section 312.1 of the code already contains a use and occupancy classification for buildings and structures considered as utility or miscellaneous in nature. Included in the Group U classification are agricultural buildings, barns, grain silos, greenhouses, livestock shelters, sheds and stables. Complete and detailed requirements are contained in the code for Group U buildings and structures; however, in most cases these provisions are more restrictive than a Group S-1 occupancy containing moderate-hazard contents that are likely to burn with moderate rapidity. This appendix provides alternative provisions that can be adopted by a jurisdiction to realistically match safeguards with the actual hazards associated with buildings and structures that are minimally occupied by persons and pose a limited fire risk.

Purpose

This appendix contains the requirements for protecting agricultural buildings. The provisions in this appendix reflect those occupancies and uses that deserve special consideration based on their limited occupant load. The code's foundation is the equivalent risk theory. Type of construction, fire-resistance ratings and means of egress requirements are balanced with the risks associated with agricultural buildings. This appendix contains general requirements, allowable heights and areas, mixed use provisions and exit requirements. The purpose of these provisions is to modify or supersede the basic code requirements for agricultural buildings. This approach allows the jurisdiction an alternative method that is more liberal and to address just the requirements for agricultural buildings that are included in this appendix.

SECTION C101 GENERAL

C101.1 Scope. The provisions of this appendix shall apply exclusively to agricultural buildings. Such buildings shall be classified as Group U and shall include the following uses:

1. Livestock shelters or buildings, including shade structures and milking barns.
2. Poultry buildings or shelters.
3. Barns.
4. Storage of equipment and machinery used exclusively in agriculture.
5. Horticultural structures, including detached production greenhouses and crop protection shelters.
6. Sheds.
7. Grain silos.
8. Stables.

❖ This section contains the scope and applicability requirements for agricultural buildings. A specific list is provided to identify which agricultural buildings are intended to be controlled by the alternative requirements of the appendix. This list expands and clarifies agricultural buildings as stated in Section 312.1.

This section of the code contains an explicit list of those types of agricultural buildings that are eligible for the alternative requirements of this appendix. Section 312.1 contains a similar list of utility and miscellaneous uses classified as Group U. Livestock shelters, barns, greenhouses, sheds, grain silos and stables are contained in both lists. Additionally, the term "livestock shelters" has been expanded to include weather protection structures and those used as milking barns. All of these buildings and structures are united in that they are used exclusively for agricultural purposes. As such, the occupant load for a Group U building is minimal. Although some of these structures may contain a considerable fuel load associated with their contents and uses, they would pose little fire hazard to persons or other occupiable buildings. The intent of the code is to match minimal requirements with these buildings based on the minimal fire and life hazards they pose.

SECTION C102 ALLOWABLE HEIGHT AND AREA

C102.1 General. Buildings classified as Group U Agricultural shall not exceed the area or height limits specified in Table C102.1.

❖ This section provides general building height and area limitations for Group U occupancies, just like Chapter 5 of the code. The height limitations, in terms of number of stories, and the area limitations from Table 503 for Group U buildings are in all cases the same or significantly less than those established for a Group S-1 structure. Those types of buildings are characterized by significant fuel loads that usually mandate the presence of an on-site automatic fire sprinkler system to abate the hazards to acceptable levels. A Group U building typically contains a lesser fire hazard and therefore, the allowable heights and areas should be greater than what is permitted for Group S-1 occupancies.

Revised entries for Table 503 are provided in this section for Group U buildings. Additionally, there are two modifications provided that supplement the requirements of Sections 507.1 and 507.3 for unlimited area buildings.

The provisions for governing the heights and areas of agricultural buildings are established in this section and are based on the Group U occupancy classification and the building's type of construction. Table C102.1 is referenced here as the primary determinant for the minimum type of construction required for Group U buildings.

TABLE C102.1. See below.

❖ This table is configured similar to Table 503. The nine types of construction are listed across the top of the table. Included are separate lines for the allowable area, allowable height in terms of the number of stories and allowable height in terms of feet. The values listed for height in terms of feet are the same values listed in Table 503. Only the allowable areas and heights in terms of number of stories have been increased. Although not specifically stated here, the rest of the provisions of Chapter 5 are still applicable for determining any modifications (see Sections 504 and 506).

C102.2 One-story unlimited area. The area of a one-story Group U agricultural building shall not be limited if the building is surrounded and adjoined by *public ways* or yards not less than 60 feet (18 288 mm) in width.

❖ Just as additional allowances are made for unlimited area buildings in Section 507, there are similar provisions for agricultural buildings. A single-story building without a basement level containing agricultural uses represents a low fire hazard due to the limited number of occupants that may be present. If such a building is further isolated with sufficient clear open space around it, the building area is not limited. No other structures may be located within the 60 feet (18 288 mm) of clear open space required around the building. The open space must be located on the same lot as the agricultural building to allow access for firefighting operations. The type of construction is not restricted with this option.

C102.3 Two-story unlimited area. The area of a two-story Group U agricultural building shall not be limited if the building is surrounded and adjoined by *public ways* or *yards* not less than 60 feet (18 288 mm) in width and is provided with an approved automatic sprinkler system throughout in accordance with Section 903.3.1.1.

❖ This provision of the appendix, in effect, expands the allowance of Section 507.4. A two-story unlimited area agricultural building is permitted with the same open space requirement of Section C102.2 when it is equipped throughout with an automatic NFPA 13 fire sprinkler system. An agricultural building represents a

TABLE C102.1
BASIC ALLOWABLE AREA FOR A GROUP U, ONE STORY IN HEIGHT AND MAXIMUM HEIGHT OF SUCH OCCUPANCY

I		II		III and IV		V	
A	B	A	B	III A and IV	III B	A	B
ALLOWABLE AREA (square feet)[a]							
Unlimited	60,000	27,100	18,000	27,100	18,000	21,100	12,000
MAXIMUM HEIGHT IN STORIES							
Unlimited	12	4	2	4	2	3	2
MAXIMUM HEIGHT IN FEET							
Unlimited	160	65	55	65	55	50	40

For SI: 1 square foot = 0.0929 m^2.

a. See Section C102 for unlimited area under certain conditions.

lesser hazard than a Group B, F, M or S building. Therefore, a two-story fully suppressed agricultural building can also be of unlimited area as long as it is isolated. Once again, the type of construction is not restricted in these buildings.

SECTION C103 MIXED OCCUPANCIES

C103.1 Mixed occupancies. Mixed occupancies shall be protected in accordance with Section 508.

❖ The appendix includes a reference for any mixed occupancies that may exist within the agricultural building; for example, a large milking barn with integral offices and lab areas used for the transaction of farm business and testing of milk products. Since these areas are occupiable spaces, they are not classified as part of the Group U portions. Most likely these areas are classified as Group B. As such, the building must be protected in accordance with Chapter 3. Another example would be a barn with sleeping accommodations for migrant farm workers. The sleeping rooms would be considered as guestrooms and would need to be classified as Group R.

Any of the options listed in Section 508 are available to the building designer to address the mixed uses and occupancies that may be present within an agricultural building. This would include accessory use areas in Section 508, nonseparated uses in Section 508.3 or separated uses in Section 508.4.

SECTION C104 EXITS

C104.1 Exit facilities. Exits shall be provided in accordance with Chapters 10 and 11.

Exceptions:

1. The maximum travel distance from any point in the building to an approved exit shall not exceed 300 feet (91 440 mm).
2. One exit is required for each 15,000 square feet (1393.5 m^2) of area or fraction thereof.

❖ The last section of this appendix provides a reminder to the permit applicant that the means of egress requirements of the code are still applicable to agricultural buildings. Although these buildings have a minimal number of occupants, provisions must still be made for their egress. This section includes two modifications to the Chapter 10 requirements that acknowledge the low occupant load and the limited hazards the occupants are exposed to.

The means of egress for agricultural buildings must comply with the applicable provisions of Chapter 10. For example, Table 1004.1.1 specifies an occupant load for agricultural buildings based on a rate of one person for every 300 square feet (28 m^2) of gross floor area. Typically, Group U buildings are exempt from the requirements of Chapter 11 for accessibility to buildings by physically disabled persons. Section 1103.2.5 only mandates that access is required to paved work areas and those areas within a building or structure that are open to the general public.

The two exceptions modify or reinforce certain requirements of Chapter 10. Table 1016 limits the exit access travel distance for Group U buildings to 300 feet (91 440 mm). Exception 1 restates this 300-foot (91 440 mm) limit.

Exception 2 refines Table 1021.1 in determining the number of exits required for agricultural buildings. An exit is required for every 15,000 square feet (1393 m2) of floor area. This works out to the same 50 occupants specified in Table 1021.1; however, the single-story height limitation and 75-foot (22 860 mm) travel distance are deleted with Exception 2.

Appendix D: Fire Districts

The provisions contained in this appendix are not mandatory unless specifically referenced in the adopting ordinance.

General Comments

The provisions contained in this appendix are not mandatory unless specifically referenced in the adopting ordinance, as stated in Section 101.2.1.

Purpose

Fire districts, in one form or another, have been in the model codes for years. Over time, they have been revised, and in some cases deleted, as the codes have evolved to address the conflagration hazards that the fire district concept was intended to abate. However, there are jurisdictions that continue to utilize the provisions for fire districts. It is for this reason, in order to transition from the previously adopted model code to the *International Building Code*® (IBC®), that the provisions have been incorporated into the appendix.

SECTION D101 GENERAL

D101.1 Scope. The fire district shall include such territory or portion as outlined in an ordinance or law entitled "An Ordinance (Resolution) Creating and Establishing a Fire District." Wherever, in such ordinance creating and establishing a fire district, reference is made to the fire district, it shall be construed to mean the fire district designated and referred to in this appendix.

❖ The establishment of fire districts is among the oldest recognized means of controlling the threat of urban conflagrations. (The term "conflagration" is used to describe a fire that involves three or more buildings and spreads across a natural or constructed barrier, such as a stream, street or other open space.) In fact, the use of fire district ordinances by local governments could be considered an early form of land use control similar to zoning as it is now practiced. Establishment of fire districts became popular at the beginning of the 20th century as a means of preventing the destruction of entire cities by fire. Massive urban conflagrations struck Chicago (1871), Baltimore (1872 and 1904) and San Francisco (1906), destroying all or most of the urbanized areas as they then existed just as these cities were experiencing perhaps their greatest development. Fire districts usually include areas, designated by ordinance or resolution by the local governing body, which are considered susceptible to mass fires involving multiple buildings or groups of buildings. While it may be true that fire district ordinances and building code adoption have significantly reduced the incidence of conflagrations in densely built or populated urbanized areas, continuing development of suburban and rural communities in areas where nearby wildlands are preserved or protected has spurred the increased incidence and severity of fires in urban/wildland interface and intermix areas, which often lie outside designated fire districts. This appendix is currently not designed to address this hazard. This appendix establishes regulatory guidelines for controlling construction, use and occupancy of buildings and conduct of dangerous operations or processes in urbanized areas that may be especially prone to the danger of fires spreading from one building to another. This appendix may be particularly useful in managing fire danger in areas built prior to the adoption and enforcement of a building code but, as stated, may not be suitable for managing the threat of urban/wildland interface and intermix area fire hazards. See the *International Wildland-Urban Interface Code*® (IWUIC®) for guidance on managing fire hazards in urban/wildland interface and intermix areas.

D101.1.1 Mapping. The fire district complying with the provisions of Section D101.1 shall be shown on a map that shall be available to the public.

❖ The requirement to prepare a map facilitates communication to the public of the additional controls found in these guidelines. In effect, establishing a fire district is a form of land use regulation, and many communities include fire district controls in their zoning ordinance or establish fire districts through overlay zoning. Zoning maps are frequently prepared and used to communicate land use controls to the public and to illustrate the relationships between land uses, infrastructure and other features of the community.

D101.2 Establishment of area. For the purpose of this code, the fire district shall include that territory or area as described in Sections D101.2.1 through D101.2.3.

❖ The following sections describe the conditions typical of fire districts established by ordinances that reference this appendix. Note that these sections are not customarily intended to include an entire jurisdiction. In fact, this is contrary to the fundamental purpose of a fire district—protection from catastrophic fires that could destroy an entire town. By dividing all jurisdictions into at least two separate fire districts, a city presumably reduces its risk to destruction of just a portion of the city or town. In newer communities developed largely after the adoption of local land use regulations and building codes, fire districts provide little added protection against conflagrations.

If a community is so compact and densely developed that designation of separate areas is not possible or practical, an entire jurisdiction could be designated a fire district as a means of controlling hazardous occupancies and regulating combustible construction.

When adopted for use in areas developed prior to the adoption of a building code, fire districts may serve as an effective means of abating the hazard posed when no single building constitutes a public nuisance, but taken in aggregate, the hazard of groups of similarly noncode-conforming buildings is considered unreasonable. By preventing new construction and new occupancies from contributing to the nonconformity of existing hazards and isolating these nonconforming districts from new development, a city may gain a measure of security from a mass fire.

D101.2.1 Adjoining blocks. Two or more adjoining blocks, exclusive of intervening streets, where at least 50 percent of the ground area is built upon and more than 50 percent of the built-on area is devoted to hotels and motels of Group R-1; Group B occupancies; theaters, nightclubs, restaurants of Group A-1 and A-2 occupancies; garages, express and freight depots, warehouses and storage buildings used for the storage of finished products (not located with and forming a part of a manufactured or industrial plant); or Group S occupancy. Where the average height of a building is two and one-half *stories* or more, a block should be considered if the ground area built upon is at least 40 percent.

❖ The fire district should include areas densely developed with light and moderate fire hazard occupancies that have either significant fuel loads or significant transient occupant loads (see Figure D101.2.1). Nightclubs; theaters; restaurants; hotels and motels; certain types of Group B occupancies and warehouses not associated with industrial operations have been the sites of some of the nation's largest conflagrations. Intentionally set fires and delayed detection or notification of fire-fighting forces have contributed significantly to these losses. Occupancies and occupants typical of those listed in this section may be particularly prone to both of these factors.

Although Group F and M occupancies and Group S warehouses containing raw or waste materials are not listed, they are typically found in these areas as well. One- and two-family dwellings; multiple-family housing; schools or day care centers and institutional occupancies are not included in the description because they are typically found in less densely developed areas.

Local officials should recognize that redevelopment of industrial and warehouse buildings in existing fire districts for new residential and mercantile uses is increasingly common (see Section D103.2). These activities typically do not pose the same degree of mass fire danger in such buildings, despite the greater frequency of fires in residential occupancies. This may be due to reduced fuel loading and increased occupant activity, which aids in early detection and notification of a fire.

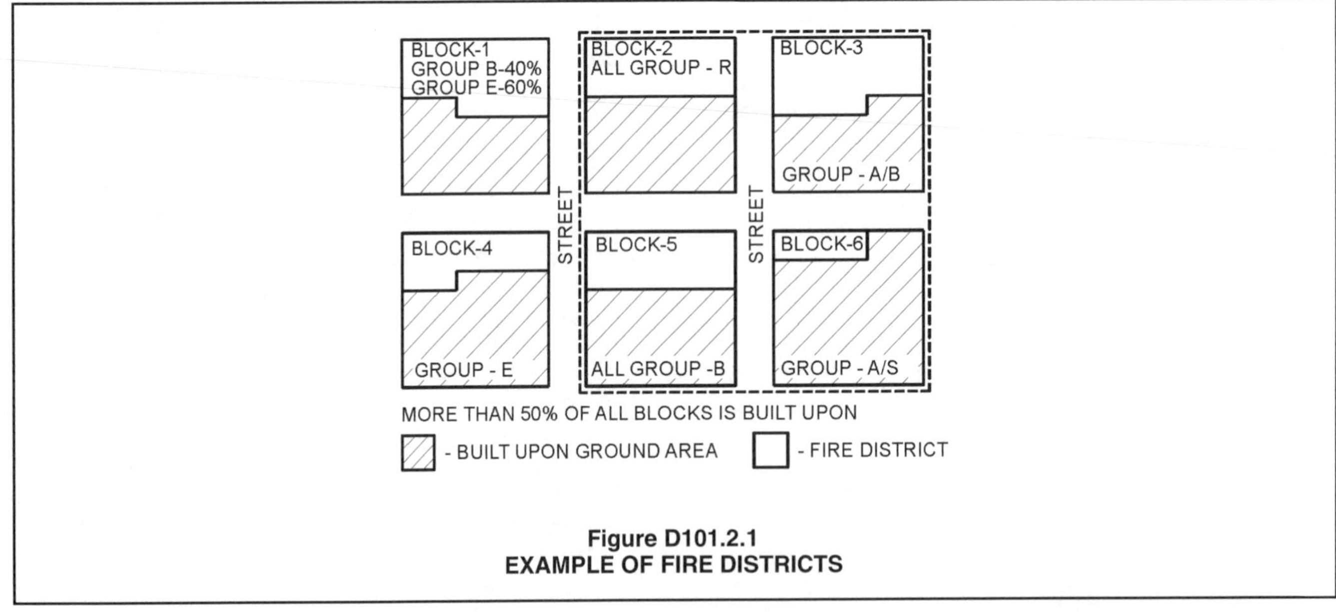

Figure D101.2.1
EXAMPLE OF FIRE DISTRICTS

D101.2.2 Buffer zone. Where four contiguous blocks or more comprise a fire district, there shall be a buffer zone of 200 feet (60 960 mm) around the perimeter of such district. Streets, rights-of-way and other open spaces not subject to building construction can be included in the 200-foot (60 960 mm) buffer zone.

❖ The buffer areas described by this section serve as fire or (preferably) fuel breaks by prohibiting the construction of buildings or structures that could become involved, allowing the fire to spread (see Figure D101.2.2). The green space or buffer zones may include the width of streets, streams or other dedicated rights-of-way that ensure segregation between adjacent fire district zones.

D101.2.3 Developed blocks. Where blocks adjacent to the fire district have developed to the extent that at least 25 percent of the ground area is built upon and 40 percent or more of the built-on area is devoted to the occupancies specified in Section D101.2.1, they can be considered for inclusion in the fire district, and can form all or a portion of the 200-foot (60 960 mm) buffer zone required in Section D101.2.2.

❖ Adjacent blocks may be added to the fire district as development of buffer spaces progresses (see Figure D101.2.3). However, care should be taken to ensure that the densities of buffer spaces do not increase to such an extent as to provide an avenue of fire spread between adjacent fire districts. If building densities remain between 25 and 50 percent of the land area, danger of fire spread between buildings in separate fire districts remains relatively low, as long as building separations required by Section D102.2 for the construction type are provided and maintained and combustible building features, such as roofing, sheathing, trim, canopies and signs, are similarly controlled.

SECTION D102
BUILDING RESTRICTIONS

D102.1 Types of construction permitted. Within the fire district every building hereafter erected shall be either Type I, II, III or IV, except as permitted in Section D104.

❖ The construction of Type V buildings is prohibited within a fire district. All other construction types are permitted, including unprotected Type IIB and IIIB structures with the further restrictions of Section D102.2 below.

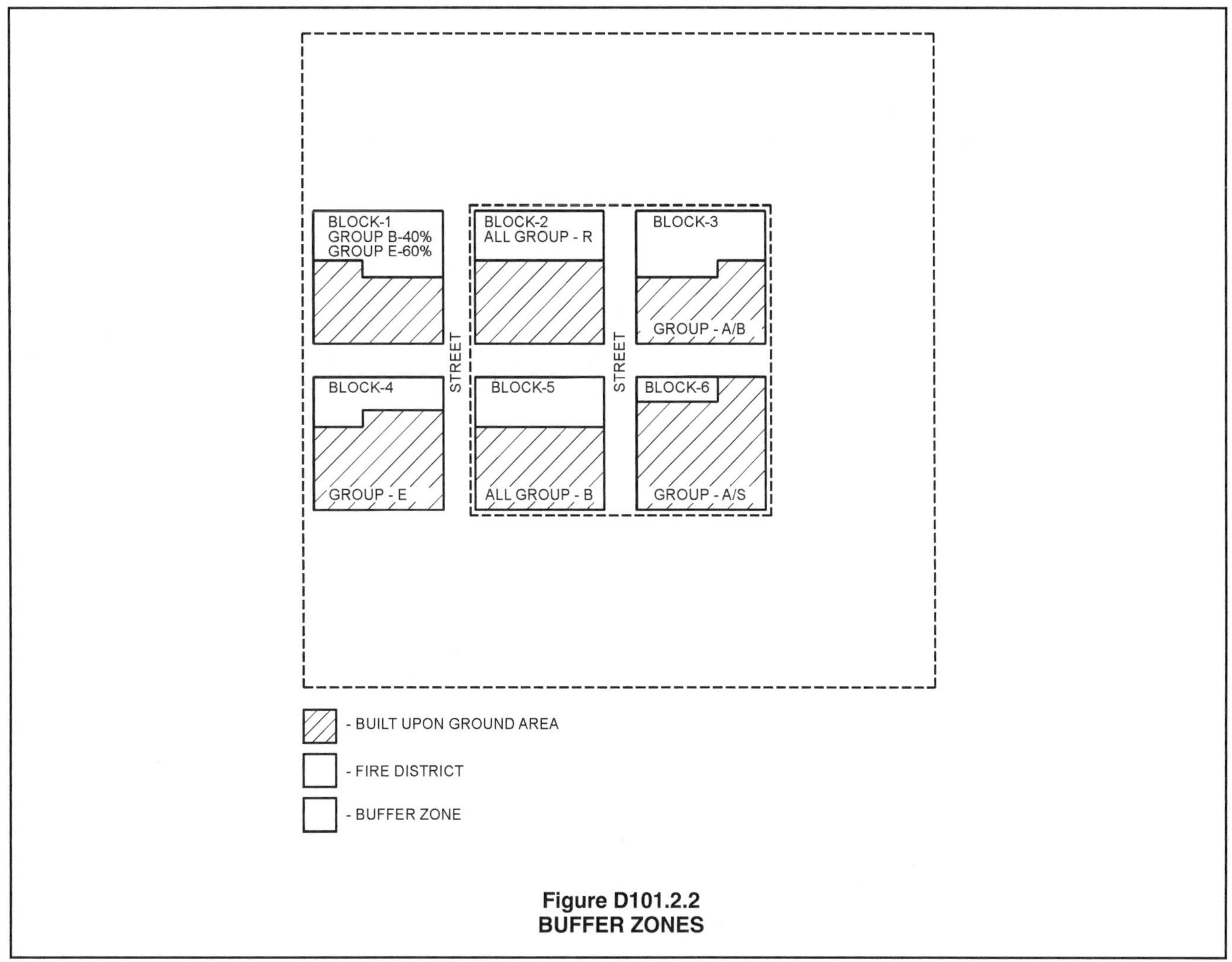

Figure D101.2.2
BUFFER ZONES

D102.2 Other specific requirements.

D102.2.1 Exterior walls. Exterior walls of buildings located in the fire district shall comply with the requirements in Table 601 except as required in Section D102.2.6.

❖ Fire-resistant separations between buildings, either by distance or barriers, as described by Table 601, provide protection against fire spread by radiant and convective heat transfer and, to a limited extent, from small flying brands.

D102.2.2 Group H prohibited. Group H occupancies shall be prohibited from location within the fire district.

❖ Group H-1, H-2, H-3 and H-5 hazardous occupancies are prohibited within the fire district because of the fire hazard posed by these occupancies themselves. Group H-4 occupancies are prohibited due to the inherent danger of release and contamination posed by fire should occupancies containing highly toxic, toxic, radioactive or other health-hazard materials be exposed.

D102.2.3 Construction type. Every building shall be constructed as required based on the type of construction indicated in Chapter 6.

❖ This section reinforces the need for new buildings to be constructed as required based on height and area and type of construction considerations in the body of the code.

D102.2.4 Roof covering. Roof covering in the fire district shall conform to the requirements of Class A or B roof coverings as defined in Section 1505.

❖ Class C and wood-shake shingle roofs are prohibited within the fire district due to the danger of fire spread and involvement through production of or exposure to burning brands.

D102.2.5 Structural fire rating. Walls, floors, roofs and their supporting structural members shall be a minimum of 1-hour fire-resistance-rated construction.

Exceptions:

1. Buildings of Type IV construction.

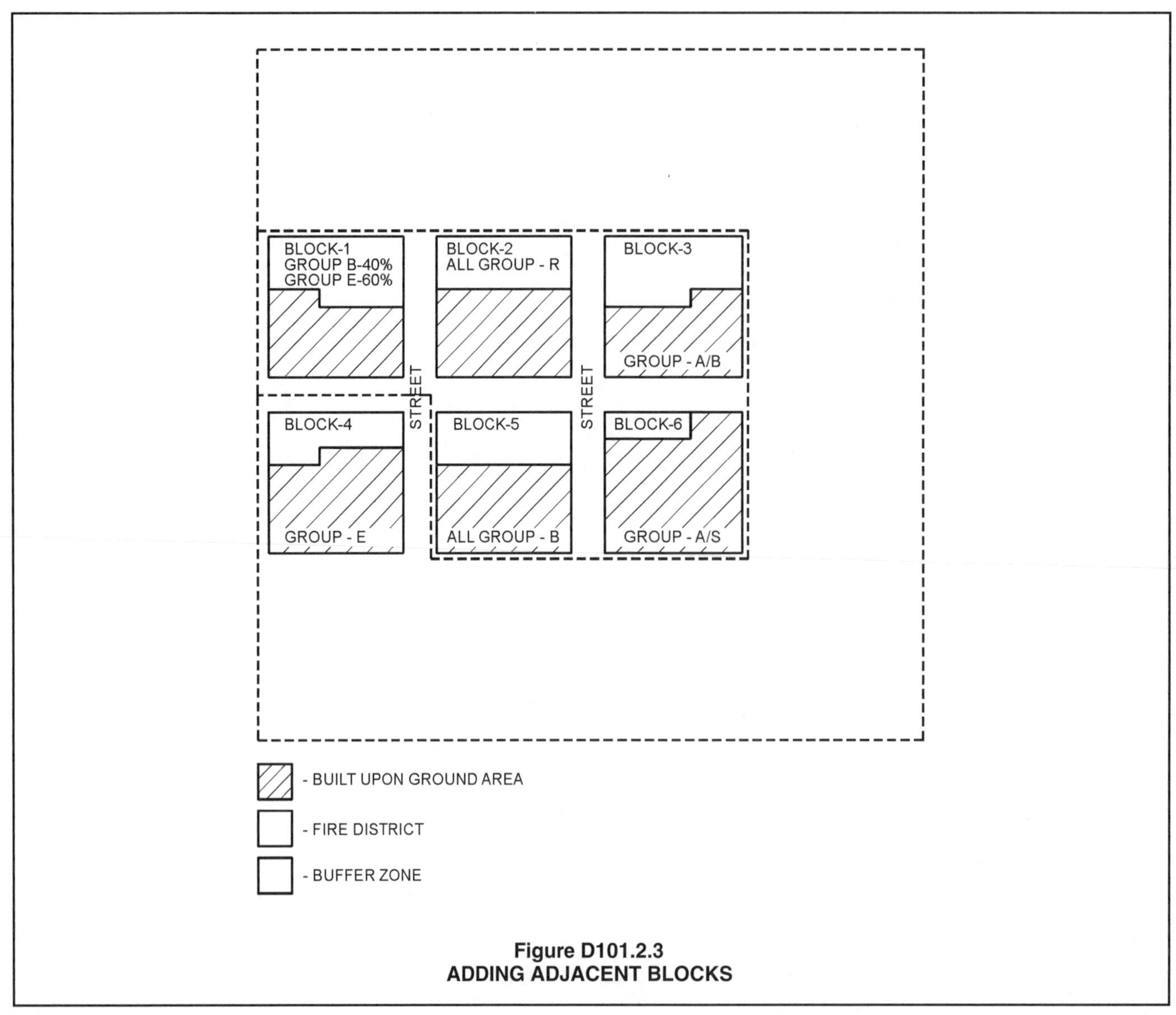

Figure D101.2.3
ADDING ADJACENT BLOCKS

2. Buildings equipped throughout with an *automatic sprinkler system* in accordance with Section 903.3.1.1.

3. Automobile parking structures.

4. Buildings surrounded on all sides by a permanently open space of not less than 30 feet (9144 mm).

5. Partitions complying with Section 603.1, Item 10.

❖ Unprotected Type IIB and IIIB construction is permitted within the fire district only for fully sprinklered buildings, automobile parking garages and buildings with 30 feet (9144 mm) of open space on all sides. All other buildings, except those of Type IV construction, must be of fire-resistance-rated construction.

D102.2.6 Exterior walls. Exterior load-bearing walls of Type II buildings shall have a *fire-resistance rating* of 2 hours or more where such walls are located within 30 feet (9144 mm) of a common property line or an assumed property line. Exterior nonload-bearing walls of Type II buildings located within 30 feet (9144 mm) of a common property line or an assumed property line shall have fireresistance ratings as required by Table 601, but not less than 1 hour. Exterior walls located more than 30 feet (9144 mm) from a common property line or an assumed property line shall comply with Table 601.

Exception: In the case of one-story buildings that are 2,000 square feet (186 m^2) or less in area, exterior walls located more than 15 feet (4572 mm) from a common property line or an assumed property line need only comply with Table 601.

❖ Although Type II buildings may be capable of resisting ignition because their components are generally noncombustible, their structural integrity could be threatened or contents ignited by radiant heat exposure from a fire in an adjacent building. Moreover, since Section 603.1, Item 1, permits the use of fire-retardant-treated wood for nonbearing partitions and roof systems of Type II buildings, these structures should not be assumed to be of noncombustible construction.

Small accessory buildings [less than 2,000 square feet (186 m^2)] are exempt from the exterior wall fireresistance requirements of this section, provided they comply with Table 601 and are located at least 15 feet (4572 mm) from an adjacent or assumed property line. Bear in mind that, except as provided in Section D105, such structures may not be of Type V construction. Such buildings present little danger of contributing to a mass fire, provided they comply with the requirements of the code.

D102.2.7 Architectural trim. Architectural *trim* on buildings located in the fire district shall be constructed of *approved* noncombustible materials or *fire-retardant-treated wood*.

❖ The ignition of architectural trim on the exterior of buildings from radiant heat transfer may pilot the ignition of other building components or contents of the building, providing an avenue for further fire spread. Many types of architectural trim may also catch flying brands from a fire, sparking the ignition of the exposed building.

D102.2.8 Permanent canopies. Permanent canopies are permitted to extend over adjacent open spaces provided all of the following are met:

1. The canopy and its supports shall be of noncombustible material, *fire-retardant-treated wood*, Type IV construction or of 1-hour fire-resistance-rated construction.

   Exception: Any textile covering for the canopy shall be flame resistant as determined by tests conducted in accordance with NFPA 701 after both accelerated water leaching and accelerated weathering.

2. Any canopy covering, other than textiles, shall have a *flame spread index* not greater than 25 when tested in accordance with ASTM E 84 or UL 723 in the form intended for use.

3. The canopy shall have at least one long side open.

4. The maximum horizontal width of the canopy shall not exceed 15 feet (4572 mm).

5. The *fire resistance* of *exterior walls* shall not be reduced.

❖ Although combustible exterior canopies may become involved in fire from radiant heat transfer, their relatively low mass and method of support and attachment to the building should minimize the risk that they will pilot the ignition of exterior walls or building contents. The requirements of this section restrict certain features of exterior canopies to ensure that their construction contributes little to fire growth or spread in the fire district.

D102.2.9 Roof structures. Structures, except aerial supports 12 feet (3658 mm) high or less, flagpoles, water tanks and cooling towers, placed above the roof of any building within the fire district shall be of noncombustible material and shall be supported by construction of noncombustible material.

❖ Vertical structures with little combustible mass, or which will be cooled by their contents, are permitted, provided their failure in a fire does not pose a significant secondary hazard. Of course, rooftop water tanks supplying fire protection systems should not be of combustible construction because their failure may undermine the effectiveness of required fire protection systems.

D102.2.10 Plastic signs. The use of plastics complying with Section 2611 for signs is permitted provided the structure of the sign in which the plastic is mounted or installed is noncombustible.

❖ Free-standing and projecting signs are unlikely to contribute to fire spread between buildings, provided their structural supports are noncombustible and structural collapse of the sign supports will not directly expose a building or its contents. Although rooftop, marquee and wall-mounted signs may present a more direct hazard to buildings, Sections

2611.2 and 2611.3 limit the area of any plastic components and require all plastic materials to meet specified combustibility criteria (see Section 2606.4).

D102.2.11 Plastic veneer. Exterior plastic veneer is not permitted in the fire district.

❖ All plastic veneers, including vinyl siding, are prohibited in the fire district. Most plastics deform under high radiant heat conditions, some melt and drip and all will burn. Plastics represent a wide range of fire hazard conditions, generally all of which are unfavorable.

This prohibition is consistent with restrictions on other combustible exterior elements, such as roof and architectural trim. Although this appendix does not explicitly prohibit other combustible exterior veneers, Section D105.1, Item 11, excludes wood veneer from these provisions, and the use of such combustible cladding materials remains severely restricted.

SECTION D103 CHANGES TO BUILDINGS

D103.1 Existing buildings within the fire district. An existing building shall not hereafter be increased in height or area unless it is of a type of construction permitted for new buildings within the fire district or is altered to comply with the requirements for such type of construction. Nor shall any existing building be hereafter extended on any side, nor square footage or floors added within the existing building unless such modifications are of a type of construction permitted for new buildings within the fire district.

❖ Bearing in mind that the principal purpose of Section D103 is to control the fire hazard posed by existing nonconforming construction by isolating areas where such buildings are located, it would be imprudent to allow any increase in the hazard within the fire district once it is established. Therefore, any new construction must comply with the requirements of the code, including limits on construction, area and height and fire protection (see Section D102.2.3).

D103.2 Other alterations. Nothing in Section D103.1 shall prohibit other alterations within the fire district provided there is no change of occupancy that is otherwise prohibited and the fire hazard is not increased by such *alteration*.

❖ All construction and any alteration of existing buildings must comply with the provisions of the code (see Section 101.2). Assuming all new work complies with the relevant requirements of the code and all applicable referenced standards, the hazard of any building in the fire district should not be increased.

Change of occupancy from one group to another is allowed within the fire district, provided that the new occupancy is permitted by the code for the type of construction, fire protection, means of egress and other requirements and is not a hazardous (Group H) occupancy (see Section D102.2.2).

D103.3 Moving buildings. Buildings shall not hereafter be moved into the fire district or to another lot in the fire district unless the building is of a type of construction permitted in the fire district.

❖ Relocated buildings, like new construction, must be of a type of construction permitted within the fire district. The type of construction will be determined by the area, height and occupancy of the building after its relocation. However, certain structural components and systems may require higher fire-resistance ratings than specified in Table 601, in accordance with Section D102.2.5.

SECTION D104 BUILDINGS LOCATED PARTIALLY IN THE FIRE DISTRICT

D104.1 General. Any building located partially in the fire district shall be of a type of construction required for the fire district, unless the major portion of such building lies outside of the fire district and no part is more than 10 feet (3048 mm) inside the boundaries of the fire district.

❖ Buildings that lie principally outside the fire district and project no more than 10 feet (3048 mm) within a fire district boundary are exempt from the fire district requirements and should be considered part of the buffer zone described in Sections D101.2.2 and D101.2.3.

In Figure D104(A), the building does not need to meet the provisions of Appendix D, since the majority of the building is outside and no more than 10 feet (3048 mm) of the building lies within the boundaries of the fire district.

In Figure D104(B), the building shall comply with the provisions of Appendix D, since more than 10 feet (3048 mm) of the building is within the boundaries of the fire district.

SECTION D105 EXCEPTIONS TO RESTRICTIONS IN FIRE DISTRICT

D105.1 General. The preceding provisions of this appendix shall not apply in the following instances:

1. Temporary buildings used in connection with duly authorized construction.
2. A private garage used exclusively as such, not more than one *story* in height, nor more than 650 square feet (60 m^2) in area, located on the same lot with a *dwelling*.
3. Fences not over 8 feet (2438 mm) high.
4. Coal tipples, material bins and trestles of Type IV construction.
5. Water tanks and cooling towers conforming to Sections 1509.3 and 1509.4.
6. Greenhouses less than 15 feet (4572 mm) high.

7. Porches on dwellings not over one *story* in height, and not over 10 feet (3048 mm) wide from the face of the building, provided such porch does not come within 5 feet (1524 mm) of any property line.
8. Sheds open on a long side not over 15 feet (4572 mm) high and 500 square feet (46 m^2) in area.
9. One- and two-family *dwellings* where of a type of construction not permitted in the fire district can be extended 25 percent of the floor area existing at the time of inclusion in the fire district by any type of construction permitted by this code.
10. Wood decks less than 600 square feet (56 m^2) where constructed of 2-inch (51 mm) nominal wood, pressure treated for exterior use.
11. Wood veneers on *exterior walls* conforming to Section 1405.5.
12. Exterior plastic veneer complying with Section 2605.2 where installed on exterior walls required to have a *fire-resistance rating* not less than 1 hour, provided the exterior plastic veneer does not exhibit sustained flaming as defined in NFPA 268.

❖ With the exception of Items 11 and 12, excluded structures are generally accessory to other buildings and pose little hazard of contributing to fire spreading from one building to another or between buildings in adjacent fire districts. Excluding these structures from the provisions of this appendix does not exempt any new structures of the types or uses listed from complying with the requirements of the code.

Water tanks for fire protection located within the fire district may not be governed by the requirements of this appendix.

Wood veneer applied to the exterior of structures must comply with the restrictions of Section 1405.5. The referenced provisions restrict the use of wood veneer to installation on noncombustible exterior walls of the fire resistance required by Table 601, and not exceeding two stories in height.

SECTION D106 REFERENCED STANDARDS

ASTM E 84-04	Test Method for Surface Burning Characteristics of Building Materials	D102.2.8
NFPA 268-01	Test Method for Determining Ignitability of Exterior Wall Assemblies Using a Radiant Heat Energy Source	D105.1
NFPA 701-99	Methods of Fire Tests for Flame-Propagation of Textiles and Films	D102.2.8
UL 723-03	Standard for Test for Surface Burning Characteristics of Building Materials, with Revisions through May 2005	D102.2.8

Bibliography

The following resource materials are referenced in this chapter or are relevant to the subject matter addressed in this chapter.

IWUIC-12, *International Wildland-Urban Interface Code*. Washington, DC: International Code Council, 2011.

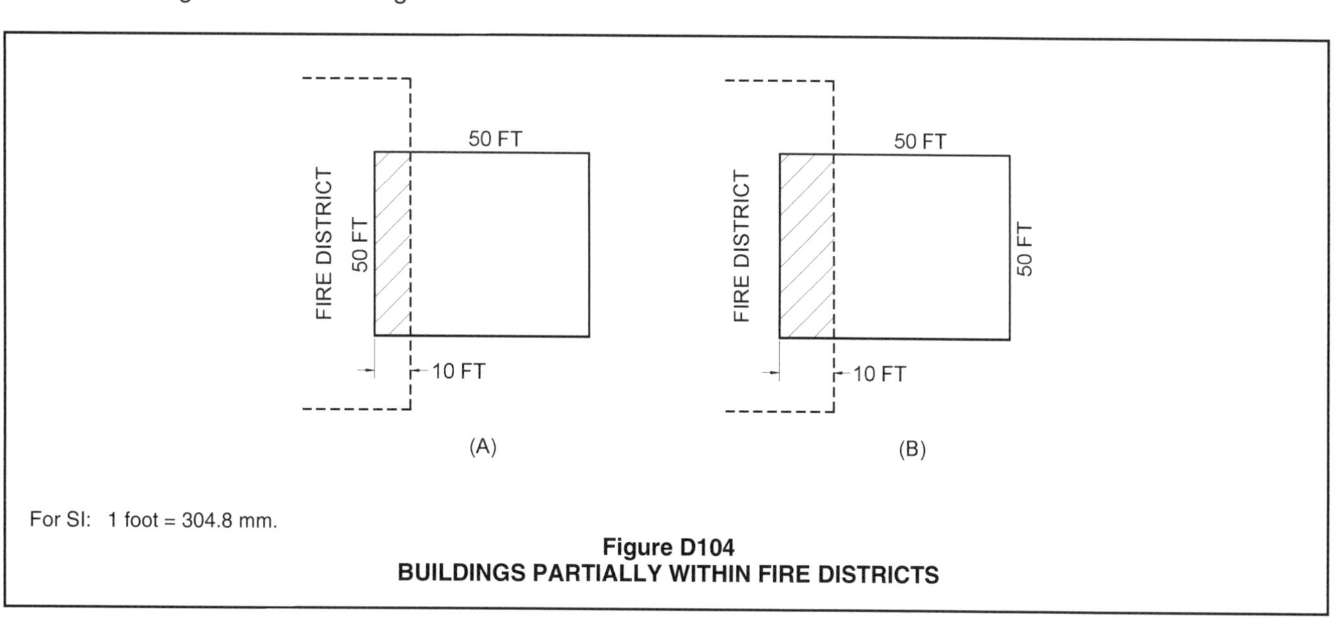

For SI: 1 foot = 304.8 mm.

Figure D104
BUILDINGS PARTIALLY WITHIN FIRE DISTRICTS

Appendix E: Supplementary Accessibility Requirements

The provisions contained in this appendix are not mandatory unless specifically referenced in the adopting ordinance.

General Comments

As stated in Section 101.2.1, the provisions of the appendices do not apply unless specifically adopted.

Chapter 11 sets forth requirements for accessibility to buildings and their associated sites and facilities for people with physical disabilities. This appendix was added to address accessibility in construction for items that were not typically enforceable through the traditional code enforcement process but were in the drafts for the 2004 Americans with Disabilities Act/Architectural Barriers Act (ADA/ABA) *Accessibility Guidelines.*

Section E101 contains the broad scope statement of the appendix and identifies the baseline criteria for accessibility as being in compliance with this appendix and ICC A117.1-2009, *Accessible and Usable Buildings and Facilities* (ICC A117.1). ICC A117.1 is the consensus national standard that sets forth the details, dimensions and construction specifications for accessibility.

Section El02 contains definitions of terms that are associated with accessibility that are not found in Chapter 2.

Section E103 contains additional requirements for interior accessible routes. An accessible route is a key component of the built environment that provides a person with a disability access to spaces, elements, facilities and buildings.

Section E104 contains various accessibility requirements that are unique to specific occupancies and are applicable in addition to other general requirements of this chapter. Specific provisions unique to transient lodging and Group I-3 facilities are included.

Section E105 contains various requirements that are applicable to special equipment provided for public use, including requirements for portable toilet and bathing rooms; laundry equipment; vending machines; mailboxes; automatic teller machines (ATMs) and fare machines and two-way communication systems.

Section E106 contains special criteria for public telephones, including TTY requirements.

Section E107 sets forth requirements for signage identifying permanent designations, such as room numbers in a hospital, directional and informational signs, and signage specifically related to the transportation facilities identified in Sections E108, E109 and E110.

Section E108 lists criteria specific to bus stops and terminals.

Section E109 lists criteria specific to fixed transportation facilities and stations, such as train stations.

Section E110 lists criteria specific to airports.

Section E111 enumerates the standards referenced in this appendix.

Purpose

The Architectural and Transportation Barriers Compliance Board (Access Board) has revised and updated its accessibility guidelines for buildings and facilities covered by the Americans with Disabilities Act of 1990 (ADA) and the Architectural Barriers Act of 1968 (ABA). The final ADA/ABA Guidelines, published by the Access Board in September 2004, will serve as the basis for the minimum standards when adopted by other federal agencies responsible for issuing enforceable standards. The guidelines have been adopted by the Government Service Administration (GSA), the Post Office, the Department of Defense, the Department of Transportation and the Department of Justice (DOJ). When the DOJ officially adopted the guidelines they renamed them the *2010 ADA Standard for Accessible Design* (2010 ADA Standard). The plan is to eventually use this new document in place of the Uniform Federal Accessibility Standard (UFAS) and the Americans with Disabilities Act *Accessibility Guidelines* (ADAAG). Refer to the Access Board web site (www.access-board.gov) for the current status of this adoption process. The 2010 ADA Standards will be mandatory as of March 15, 2013 for all alternations and new construction.

Appendix E includes scoping requirements from the 2010 ADA Standard that are not in Chapter 11 nor otherwise mentioned or mainstreamed throughout the code. Items in the appendix deal with topics not typically addressed in building codes. For example, items in Sections E102 through E107 are not part of the "built" environment, as they can be added, modified or removed without a building permit and without notifying the building official. Sections E108 through E110 deal specifically with transportation facilities. While there has been an attempt to include all provisions in the 2010 ADA Standard that are not in the *International Building Code®* (IBC®), at this time, the efforts are ongoing. Code users should refer to the DOJ web site for the revised ADA regulations at www.ada.gov.

This appendix is being included in the code for the convenience of building owners and designers who are required to comply with the 2010 ADA Standard and for the benefit of jurisdictions that may wish to go beyond the traditional boundary of the code and formally adopt these additional 2010 ADA Standard requirements. The requirements within the code indicate what is required and enforceable by the building official. The appendix includes items that are outside the scope of code enforcement, but must be complied with in addition to the code requirements to satisfy the 2010 ADA Standard.

SECTION E101
GENERAL

E101.1 Scope. The provisions of this appendix shall control the supplementary requirements for the design and construction of facilities for *accessibility* to physically disabled persons.

❖ This section establishes the scope of the appendix as providing for design and construction of facilities for accessibility to persons with disabilities other than items covered under Chapter 11. Items in the appendix are not enforceable under the typical code review process because these items can be altered or removed without notification to the building official. Items in this appendix are included in the 2010 ADA Standard.

E101.2 Design. Technical requirements for items herein shall comply with this code and ICC A117.1.

❖ This section establishes the primary and fundamental relationship of ICC A117.1 to the code. The code text is intended to "scope" or provide thresholds for application of required accessibility features. The referenced standard contains technical provisions indicating how compliance with the code is achieved. In short, this appendix specifies what, when and how many accessible features are required; the referenced standard ICC A117.1 indicates how to make these features accessible. Compliance with both the scoping and technical standard is necessary for accessibility.

SECTION E102
DEFINITIONS

E102.1 General. The following words and terms shall, for the purposes of this appendix, have the meanings shown herein. Refer to Chapter 2 of the *International Building Code* for general definitions.

❖ This section contains definitions of terms that are associated with the subject matter of this appendix.

Definitions of terms can help in the understanding and application of the code requirements. The purpose for including these definitions within this appendix is because appendices are not considered part of the code requirements unless specifically referenced in the adopting ordinance.

CLOSED-CIRCUIT TELEPHONE. A telephone with a dedicated line such as a house phone, courtesy phone or phone that must be used to gain entrance to a facility.

❖ Telephones described in this definition are utilized in hotels, apartments, airports, hospitals and similar facilities. Access to these systems is important to provide communication throughout a facility.

MAILBOXES. Receptacles for the receipt of documents, packages or other deliverable matter. Mailboxes include, but are not limited to, post office boxes and receptacles provided by commercial mail-receiving agencies, apartment houses and schools.

❖ The definition for "Mailboxes" clarifies that the scope of this provision is meant to address all types of facilities where a person could receive correspondence, packages or written notifications. The description includes post office boxes, mailboxes in the lobby of an apartment building or mailboxes in the lobby of a college dorm.

TRANSIENT LODGING. A building, facility or portion thereof, excluding inpatient medical care facilities and long-term care facilities, that contains one or more dwelling units or sleeping units. Examples of transient lodging include, but are not limited to, resorts, group homes, hotels, motels, dormitories, homeless shelters, halfway houses and social service lodging.

❖ Transient lodging facilities, such as hotels, typically provide short-term stays; however, this definition also includes facilities that provide longer term stays, such as dormitories. The 30-day or less stay typically associated with transient lodging is not applicable in this definition. The key is whether sleeping or dwelling units are provided. The exclusions are for inpatient and long-term care facilities, such as hospitals and nursing homes. Additionally, this would not be applicable to permanent lodging facilities with dwelling units, such as apartments, condominiums or townhouses.

SECTION E103
ACCESSIBLE ROUTE

E103.1 Raised platforms. In banquet rooms or spaces where a head table or speaker's lectern is located on a raised platform, an *accessible* route shall be provided to the platform.

❖ An accessible route is required to a temporary raised platform utilized for a speaker or head table. An accessible route would not be required to a raised platform used for other purposes, such as a band. Access to permanently raised platforms or stages is addressed in Chapter 11.

SECTION E104
SPECIAL OCCUPANCIES

E104.1 General. Transient lodging facilities shall be provided with *accessible* features in accordance with Sections E104.2 and E104.3. Group I-3 occupancies shall be provided with *accessible* features in accordance with Sections E104.3 and E104.4.

❖ In the following sections there are bed and communication feature requirements for transient lodging facilities. Also included are communication and visitation booth requirements for jails (Group I-3).

E104.2 Accessible beds. In rooms or spaces having more than 25 beds, 5 percent of the beds shall have a clear floor space complying with ICC A117.1.

❖ If dormitory-type transient lodging facilities have 25 or more beds in a room, 5 percent of the beds must have adequate clearance to be accessible by a person using a wheelchair. If separate-sex facilities are provided, accessible beds must be provided for both sexes. Clear floor space requirements are found in Section 305 of ICC A117.1.

E104.2.1 Sleeping areas. A clear floor space complying with ICC A117.1 shall be provided on both sides of the accessible bed. The clear floor space shall be positioned for parallel approach to the side of the bed.

Exception: This requirement shall not apply where a single clear floor space complying with ICC A117.1 positioned for parallel approach is provided between two beds.

❖ To be considered accessible, the bed must be approachable from either side in order to accommodate wheelchair users who may need to transfer from a particular side due to their mobility impairment. The alternative applies when a bed is provided on both sides of a single-approach location. Clear floor space requirements are found in Section 305 of ICC A117.1.

Some people require a type of lift system to get them into bed. Some hotels have their beds on platforms. Beds on platforms do not allow for the supports for these types of lifts to move under the beds. Allowing for these types of lifts to be used is the intent of the open-bedframe requirements in Section 1002.15.2 of the 2009 ICC A117.1.

E104.3 Communication features. Accessible communication features shall be provided in accordance with Sections E104.3.1 through E104.3.4.

❖ Examples of communication features are telephones, doorbells and peepholes. In units required to have accessible communication features, the designer should consider compatibility with different types of adaptive equipment used by people with hearing impairments. Telephone interface jacks should be compatible with both digital and analog signal use. If an audio headphone jack is provided on a speaker phone, a cutoff switch can be included in the jack so that insertion of the jack cuts off the speaker. If a telephone-like handset is used, the external speakers can be turned off when the handset is removed from the cradle. For headset or external amplification system compatibility, a standard subminiature jack installed in the telephone will provide the most flexibility. Peepholes in doors should be installed at heights for both persons who are standing and persons using a wheelchair.

E104.3.1 Transient lodging. In transient lodging facilities, sleeping units with accessible communication features shall be provided in accordance with Table E104.3.1. Units required to comply with Table E104.3.1 shall be dispersed among the various classes of units.

❖ In transient lodging facilities, an alternative notification system for persons with hearing impairments must be offered in the number of dwelling and sleeping units specified in Table E104.3.1 (see the definition for "Dwelling units" and "Sleeping units" in Chapter 2). The number of dwelling or sleeping units that are required to provide accessible communication features depends on the number of units provided. The number of accessible communication features is based on the number of units, not the number of beds. The rooms with accessible communication features are not required to be the same rooms that contain accessible beds. Requirements for communication features within dwelling and sleeping units are found in Section 1006 of ICC A117.1.

TABLE E104.3.1. See below.

❖ This table specifies the number of dwelling and sleeping units in transient lodging facilities that are required to provide accessible communication features. In accordance with the definition of "Sleeping unit," bedrooms in dwelling units are not considered sleeping units.

TABLE E104.3.1
DWELLING OR SLEEPING UNITS WITH ACCESSIBLE COMMUNICATION FEATURES

TOTAL NUMBER OF DWELLING OR SLEEPING UNITS PROVIDED	MINIMUM REQUIRED NUMBER OF DWELLING OR SLEEPING UNITS WITH ACCESSIBLE COMMUNICATION FEATURES
1	1
2 to 25	2
26 to 50	4
51 to 75	7
76 to 100	9
101 to 150	12
151 to 200	14
201 to 300	17
301 to 400	20
401 to 500	22
501 to 1,000	5% of total
1,001 and over	50 plus 3 for each 100 over 1,000

E104.3.2 Group I-3. In Group I-3 occupancies at least 2 percent, but no fewer than one of the total number of general holding cells and general housing cells equipped with audible emergency alarm systems and permanently installed telephones within the cell, shall comply with Section E104.3.4.

❖ Jails are required to have a fire alarm system in accordance with Section 907.2.6. When inmates or detainees are allowed independent means of egress, a minimum of 2 percent of the cells must have visible of as well as audible alarm notification. The code and Section 702.1of ICC A117.1 both reference NFPA 72 for installation requirements. If cells also have permanently installed telephones, 2 percent of the telephones shall have a visible notification device as well as volume control and a place to plug in a TTY. The visual notification device for the phone cannot be connected to the visual alarms for the fire alarm system.

E104.3.3 Dwelling units and sleeping units. Where *dwelling units* and *sleeping units* are altered or added, the requirements of Section E104.3 shall apply only to the units being altered or added until the number of units with accessible communication features complies with the minimum number required for new construction.

❖ When transient lodging dwelling and sleeping units are altered or added, only these units are counted to determine the number of units that will require accessible communication features. Once the entire building meets the number required for new construction, no additional accessible communication features are required. This is consistent with the intent of Sections 3411.3 and 3411.8.7.

E104.3.4 Notification devices. Visual notification devices shall be provided to alert room occupants of incoming telephone calls and a door knock or bell. Notification devices shall not be connected to visual alarm signal appliances. Permanently installed telephones shall have volume controls and an electrical outlet complying with ICC A117.1 located within 48 inches (1219 mm) of the telephone to facilitate the use of a TTY.

❖ A type of visual notification, such as a flashing light, must be provided in some sleeping accommodations. The purpose of the visual notification is to alert a person with a hearing impairment to incoming phone calls or someone knocking at the door. The notification cannot be connected to any type of visual alarm system. If phones are provided within the dwelling or sleeping unit, they must be equipped with a volume control and be located within 48 inches (1219 mm) of an electrical outlet to accommodate a TTY. Specific requirements for telephones are provided in Section 704 of ICC A117.1.

E104.4 Partitions. Solid partitions or security glazing that separates visitors from detainees in Group I-3 occupancies shall provide a method to facilitate voice communication. Such methods are permitted to include, but are not limited to, grilles, slats, talk-through baffles, intercoms or telephone handset devices. The method of communication shall be accessible to individuals who use wheelchairs and individuals who have difficulty bending or stooping. Hand-operable communication devices, if provided, shall comply with Section E106.3.

❖ Section 1109.11 requires 5 percent of cubicles or counters in jail (Group I-3) visiting areas to be accessible in accordance with the built-in counter/work surface requirements. This section deals with the visiting situation where solid partitions of security glazing separates visitors from detainees. To facilitate communication, the system may include some type of handset, headphones or microphone. The system must consider in its design that either user may use a wheelchair or have difficulty bending or stooping. Any type of telephone or headphone system must have a volume control in accordance with Section E106.3.

SECTION E105
OTHER FEATURES AND FACILITIES

E105.1 Portable toilets and bathing rooms. Where multiple single-user portable toilet or bathing units are clustered at a single location, at least 5 percent, but not less than one toilet unit or bathing unit at each cluster, shall be accessible. Signs containing the International Symbol of Accessibility shall identify *accessible* portable toilets and bathing units.

> **Exception:** Portable toilet units provided for use exclusively by construction personnel on a construction site.

❖ When portable toilets are provided for special events, at least 5 percent must be accessible. A "cluster" is a group of toilets adjacent or in close proximity to one another. This is consistent with the requirements for sinks in Section 1109.3. Appropriate signage is also required. If portable toilets are provided for personnel on a construction site, they are not required to be accessible. This is consistent with Section 1103.2.6. Technical requirements for bathroom and signage are found in Sections 602 and 703 of ICC A117.1, respectively.

E105.2 Laundry equipment. Where provided in spaces required to be *accessible*, washing machines and clothes dryers shall comply with this section.

❖ When laundry facilities are provided that are open to the general public, or are associated with Accessible, Type A or B units, the washers and dryers must be accessible in accordance with Sections E105.2.1 and E105.2.2. This is consistent with Section 1107.3. The requirements are based on the washers or dryers provided in each space, not the total within the building. For example, a dorm with a single washer and dryer on each floor of a building should provide an accessible washer and dryer in each location.

E105.2.1 Washing machines. Where three or fewer washing machines are provided, at least one shall be accessible.

Where more than three washing machines are provided, at least two shall be accessible.

❖ When three or fewer washing machines are provided in a laundry room, at least one must meet the requirements for accessibility in Section 611 in ICC A117.1 for clear floor space, operable parts and height. In larger facilities, at least two washers must be accessible. Requirements include criteria for clear floor space to access the unit; operable parts that meet height, graspability and force requirements and the location of doors to access the interior of the appliance.

E105.2.2 Clothes dryers. Where three or fewer clothes dryers are provided, at least one shall be accessible. Where more than three clothes dryers are provided, at least two shall be accessible.

❖ The same criteria for washers is expected for dryers (see commentary, Section E105.2.1).

E105.3 Depositories, vending machines, change machines and similar equipment. Where provided, at least one of each type of depository, vending machine, change machine and similar equipment shall be accessible.

> **Exception:** Drive-up-only depositories are not required to comply with this section.

❖ All types of self-service machinery must be accessible. Examples include the book return at a library, a change machine at a laundry mat or a soda machine. A drive-up-only depository, such as a drive-up night deposit at a bank, is not required to be accessible. To be accessible, the equipment must meet the provisions for clear floor space, height and have operable parts within reach ranges in accordance with Section 309 of ICC A117.1.

E105.4 Mailboxes. Where mailboxes are provided in an interior location, at least 5 percent, but not less than one, of each type shall be accessible. In residential and institutional facilities, where mailboxes are provided for each *dwelling unit* or *sleeping unit*, accessible mailboxes shall be provided for each unit required to be an *Accessible unit*.

❖ Whenever mailboxes are provided, such as in an office building lobby, at least 5 percent are required to be accessible. In residential and institutional facilities, if mailboxes are provided for each unit, the mailboxes for the Accessible units must have adequate clear floor space and meet operable parts provisions. Examples of such types of facilities would be dorms, hotels, assisted living facilities or nursing homes. The number of Accessible units depends on the type of facility (see Section 1107 for requirements). To be accessible, the mailbox must meet the provisions for clear floor space, height and have operable parts within reach ranges in accordance with Section 309 of ICC A117.1.

E105.5 Automatic teller machines and fare machines. Where automatic teller machines or self-service fare vending, collection or adjustment machines are provided, at least one machine of each type at each location where such machines are provided shall be *accessible*. Where bins are provided for envelopes, wastepaper or other purposes, at least one of each type shall be accessible.

❖ When such machines are provided, such as at banks and in bus or train stations, at least one of each type of machine must be accessible. If a site provides both interior and exterior ATMs, each such installation is considered a separate location. Accessible ATMs, including those with speech and those that are within reach of people who use wheelchairs, must provide all the functions provided to customers at that location at all times. For example, it is unacceptable for the accessible ATM only to provide cash withdrawals while inaccessible ATMs in that area also sell theater tickets. Specific criteria is provided in Section 707 of ICC A117.1.

E105.6 Two-way communication systems. Where two-way communication systems are provided to gain admittance to a building or facility or to restricted areas within a building or facility, the system shall be accessible.

❖ Apartments or condominiums with secured entrances often have some type of two-way communication system at the entry so the residents can screen persons entering the building. Specific criteria for systems connected with dwelling or sleeping units are provided in Sections 1006.6 and 1006.7 of ICC A117.1. Similar provisions would be applicable to two-way communication systems provided at other locations, such as office buildings, courthouses and other facilities where access to the building or restricted spaces are dependent on two-way communication systems.

SECTION E106
TELEPHONES

E106.1 General. Where coin-operated public pay telephones, coinless public pay telephones, public closed-circuit telephones, courtesy phones or other types of public telephones are provided, *accessible* public telephones shall be provided in accordance with Sections E106.2 through E106.5 for each type of public telephone provided. For purposes of this section, a bank of telephones shall be considered two or more adjacent telephones.

❖ Many types of public phones are provided, such as pay telephones, courtesy phones throughout airports, house phones in hotels, entrance phones in apartment complexes, etc. Wherever such phones are provided, at least one of each type must comply with Sections E106.2 through E106.5. A bank of telephones is defined as two or more telephones in one location.

E106.2 Wheelchair-accessible telephones. Where public telephones are provided, *wheelchair-accessible* telephones shall be provided in accordance with Table E106.2.

Exception: Drive-up-only public telephones are not required to be *accessible.*

❖ For each type of public telephone provided, the number of phones specified in Table E106.2 must be accessible. Specific criteria are provided in Section 704.2 of ICC A117.1.

TABLE E106.2
WHEELCHAIR-ACCESSIBLE TELEPHONES

NUMBER OF TELEPHONES PROVIDED ON A FLOOR, LEVEL OR EXTERIOR SITE	MINIMUM REQUIRED NUMBER OF WHEELCHAIR-ACCESSIBLE TELEPHONES
1 or more single unit	1 per floor, level and exterior site
1 bank	1 per floor, level and exterior site
2 or more banks	1 per bank

❖ If single telephones are provided in several locations or one bank of telephones is provided on a floor level or over a site, then at least one of each type must be accessible. The accessible telephone can be accessed using a parallel approach or a forward approach. If two or more banks are provided on a floor level or over a site, at least one telephone at each location must be accessible.

E106.3 Volume controls. All public telephones provided shall have accessible volume control.

❖ Every public telephone must have volume controls and be identified with appropriate signage. Amplifiers on pay phones are located in the base of the handset or are built into the telephone. Most are operated by pressing a button or key. If the microphone in the handset is not being used, a mute button that temporarily turns off the microphone can also reduce the amount of background noise that the person can hear in the earpiece. Consider compatibility issues when matching an amplified handset with a phone or phone system. Amplified handsets that can be switched with pay telephone handsets are available. Specific criteria is found in Section 704.3 of ICC A117.1.

E106.4 TTYs. TTYs shall be provided in accordance with Sections E106.4.1 through E106.4.9.

❖ TTYs are defined in ICC A117.1 as "machinery or equipment that employs interactive, graphic communications through the transmission of coded signals across the standard telephone network." The term "TTY" also refers to devices known as text telephones and TDDs. Text telephones enable a person with a hearing impairment to utilize a telephone to send and receive text messages. Consideration should be given to phone systems that can accommodate both digital and analog transmissions for compatibility with digital and analog TTYs. Sections E106.4.1 through E106.4.4 indicate when TTYs are required within a bank of telephones, on a floor, within a building or on a site. More than one section could apply. Sections E106.4.5 through E106.4.8 are special requirements to specific types of sites. Section E106.4.9 deals with appropriate signage.

E106.4.1 Bank requirement. Where four or more public pay telephones are provided at a bank of telephones, at least one public TTY shall be provided at that bank.

Exception: TTYs are not required at banks of telephones located within 200 feet (60 960 mm) of, and on the same floor as, a bank containing a public TTY.

❖ A TTY is required wherever four or more telephones are located together. If another bank is located both on the same floor and within 200 feet (60 960 mm) of the TTY, that bank is not required to contain a TTY.

E106.4.2 Floor requirement. Where four or more public pay telephones are provided on a floor of a privately owned building, at least one public TTY shall be provided on that floor. Where at least one public pay telephone is provided on a floor of a publicly owned building, at least one public TTY shall be provided on that floor.

❖ This section is similar to Section E106.4.1; however, the number of phones on the floor is counted, instead of within a bank. The term "publicly owned building" is intended to apply to facilities such as government buildings, courthouses, libraries, schools, police stations, fire stations, park district buildings, etc.

E106.4.3 Building requirement. Where four or more public pay telephones are provided in a privately owned building, at least one public TTY shall be provided in the building. Where at least one public pay telephone is provided in a publicly owned building, at least one public TTY shall be provided in the building.

❖ This section is similar to Sections E106.4.1 and E106.4.2; however, the number of phones in a building is counted, instead of within a bank or on a floor. In some situations, one TTY can satisfy more than one requirement. For example, the TTY required for a bank of phones could also satisfy the requirement for the TTY required for the building. The term "publicly owned building" is intended to apply to facilities such as government buildings, courthouses, libraries, schools, police stations, fire stations, park district buildings, etc.

E106.4.4 Site requirement. Where four or more public pay telephones are provided on a site, at least one public TTY shall be provided on the site.

❖ This section is similar to Sections E106.4.1, E106.4.2 and E106.4.3; however, the number of phones on a site is counted. In some situations, one TTY can satisfy more than one requirement. For example, the TTY required for a bank of phones could also satisfy the requirement for the TTY required for the building. However, the requirement for a TTY on an exterior site cannot be met by installing a TTY inside the building due to the phones in the building typically not being available 24 hours a day.

E106.4.5 Rest stops, emergency road stops, and service plazas. Where a public pay telephone is provided at a public rest stop, emergency road stop or service plaza, at least one public TTY shall be provided.

❖ A TTY is required at these types of facilities regardless of the number of phones provided. The possibility of emergencies or a need for assistance at these locations increases the need for someone to be able to make a telephone call.

E106.4.6 Hospitals. Where a public pay telephone is provided in or adjacent to a hospital emergency room, hospital recovery room or hospital waiting room, at least one public TTY shall be provided at each such location.

❖ In hospitals, when phones are located for public use in an emergency room, recovery room or waiting room, a TTY is required at each location. The possibility of emergencies or a need for assistance at these locations increases the need for someone to be able to make a telephone call.

E106.4.7 Transportation facilities. Transportation facilities shall be provided with TTYs in accordance with Sections E109.2.5 and E110.2 in addition to the TTYs required by Sections E106.4.1 through E106.4.4.

❖ Transportation facilities, such as bus stations, train stations and airports, have specific requirements as stated in Sections E109.2.5 and E110.2. Additionally, they must comply with the requirements based on the number of telephones in a bank, on a floor, within a building or on a site.

E106.4.8 Detention and correctional facilities. In detention and correctional facilities, where a public pay telephone is provided in a secured area used only by detainees or inmates and security personnel, then at least one TTY shall be provided in at least one secured area.

❖ Specific criteria is provided for the detainee side of detention and correctional facilities. Phones available on the public side should comply with Sections E106.4.1 through E106.4.4.

E106.4.9 Signs. Public TTYs shall be identified by the International Symbol of TTY complying with ICC A117.1. Directional signs indicating the location of the nearest public TTY shall be provided at banks of public pay telephones not containing a public TTY. Additionally, where signs provide direction to public pay telephones, they shall also provide direction to public TTYs. Such signs shall comply with visual signage requirements in ICC A117.1 and shall include the International Symbol of TTY.

❖ Signage must be provided indicating the location of TTYs. Technical requirements for signage can be found in Sections 703.2 and 703.6 of ICC A117.1.

E106.5 Shelves for portable TTYs. Where a bank of telephones in the interior of a building consists of three or more public pay telephones, at least one public pay telephone at the bank shall be provided with a shelf and an electrical outlet.

Exceptions:

1. In secured areas of detention and correctional facilities, if shelves and outlets are prohibited for purposes of security or safety shelves and outlets for TTYs are not required to be provided.
2. The shelf and electrical outlet shall not be required at a bank of telephones with a TTY.

❖ If a bank of telephones has three or more public telephones, a shelf and electrical outlet must be provided to allow a person with a hearing impairment to utilize his or her own portable TTY. In a bank of four or more phones a public TTY is required (see Section E106.4.1); therefore, a shelf and outlet for a portable TTY are not required within the same bank. Due to security concerns, there may be situations in jails where the shelf and outlet for a portable TTY is not required. Technical requirements for the shelf can be found in Section 704.6 of ICC A117.1.

SECTION E107 SIGNAGE

E107.1 Signs. Required *accessible* portable toilets and bathing facilities shall be identified by the International Symbol of Accessibility.

❖ Accessible portable toilets must be identified (see Section E105.2).

E107.2 Designations. Interior and exterior signs identifying permanent rooms and spaces shall be raised characters and Braille. Where pictograms are provided as designations of interior rooms and spaces, the pictograms shall have raised characters and Braille text descriptors.

Exceptions:

1. Exterior signs that are not located at the door to the space they serve are not required to comply.
2. Building directories, menus, seat and row designations in assembly areas, occupant names, building addresses and company names and logos are not required to comply.
3. Signs in parking facilities are not required to comply.
4. Temporary (seven days or less) signs are not required to comply.
5. In detention and correctional facilities, signs not located in public areas are not required to comply.

❖ This section applies to signs that provide designations, labels or names for interior rooms or spaces where the sign is not likely to change over time. Examples include room numbers in hospitals or hotels, restrooms, room names in conference centers, etc. Where permanent signage is provided, signage consisting of both raised letters and Braille must also be provided. When pictograms are utilized to identify a space, such as the women's restroom, they must also have Braille. Pictograms that provide information about a space, such as "no smoking" or the International Symbol of Accessibility, are not required to have text descriptors. For signage requirements, reference Section 703 of ICC A117.1. Exceptions are provided for types of signs where it may not be effective or practical to provide signage that needs to be felt to be read. Examples would be the remote sign at the street indicating the name or occupant of the building, the menu above the take-out counter, temporary signage or building directories.

E107.3 Directional and informational signs. Signs that provide direction to, or information about, permanent interior spaces of the site and facilities shall contain visual characters complying with ICC A117.1.

> **Exception:** Building directories, personnel names, company or occupant names and logos, menus and temporary (seven days or less) signs are not required to comply with ICC A117.1.

❖ Directional information must meet requirements for font style, location, finish and contrast in order for signage to be readable by persons with limited vision. The characters should contrast as much as possible with their background. Additional factors affecting the readability of signage include lighting, surface glare and text uniformity. When making color choices, remember that a percentage of the population is color blind; therefore, some color combinations may not have adequate contrast for these individuals (for example, red and blue). Color blindness is typically a genetic condition, and it is much more common in men than in women. Approximately one in 12 men has at least some color perception problems. Less common are acquired deficiencies that stem from injury, disease or the aging process. Also, although not called "color blindness," when people age, their corneas typically turn yellowish, severely hampering their ability to see violet and blue colors. The exception is for signage that may be changeable, such as building directories, personnel names, etc. For requirements for this signage, reference Section 703.2 of ICC A117.1.

E107.4 Other signs. Signage indicating special accessibility provisions shall be provided as follows:

1. At bus stops and terminals, signage must be provided in accordance with Section E108.4.
2. At fixed facilities and stations, signage must be provided in accordance with Sections E109.2.2 through E109.2.2.3.
3. At airports, terminal information systems must be provided in accordance with Section E110.3.

❖ This section is a reference to additional signage requirements found in subsequent sections. This was done in an effort to include all signage requirements in one section.

SECTION E108 BUS STOPS

E108.1 General. Bus stops shall comply with Sections E108.2 through E108.5.

❖ Bus stops and terminals provide access to public transportation and are considered an essential service; therefore, access in these facilities is of special concern. Technical requirements for transportation facilities are found in Section 805 of ICC A117.1.

E108.2 Bus boarding and alighting areas. Bus boarding and alighting areas shall comply with Sections E108.2.1 through E108.2.4.

❖ "Kneeling" buses (those with a lift) deploy a ramp to allow access by persons with physical disabilities. When new bus stop pads are constructed to allow use of these buses, the requirements in Sections E108.2.1 through E108.2.4 are applicable. At bus stops where a shelter is provided, the bus stop pad can be located either within or outside the shelter.

E108.2.1 Surface. Bus boarding and alighting areas shall have a firm, stable surface.

❖ To allow proper deployment of the ramp and approach by a person in a wheelchair, a firm, stable surface must be used for the pad. Typically, this will be some form of concrete or asphalt.

E108.2.2 Dimensions. Bus boarding and alighting areas shall have a clear length of 96 inches (2440 mm) minimum, measured perpendicular to the curb or vehicle roadway edge, and a clear width of 60 inches (1525 mm) minimum, measured parallel to the vehicle roadway.

❖ To allow maneuvering for both the ramp and the person using the wheelchair, a bus stop pad should be a minimum of 96 inches (2438 mm) deep and 60 inches (1524 mm) wide.

E108.2.3 Connection. Bus boarding and alighting areas shall be connected to streets, sidewalks or pedestrian paths by an accessible route complying with Section 1104.

❖ All bus stop pads must have an accessible route leading from public access points, such as the sidewalk, to the bus stop pad. A pad allowing for kneeling buses is not beneficial if the pad itself is not accessible.

E108.2.4 Slope. Parallel to the roadway, the slope of the bus boarding and alighting area shall be the same as the roadway, to the maximum extent practicable. For water drainage, a maximum slope of 1:48 perpendicular to the roadway is allowed.

❖ The bus stop pad should be as flat as possible to allow a wheelchair user to wait at this location comfortably. A maximum slope of one unit vertical in 48 units horizontal (1:48) is allowed for drainage of rainwater to the curb and gutter drainage system provided along the edge of the road.

E108.3 Bus shelters. Where provided, new or replaced bus shelters shall provide a minimum clear floor or ground space complying with ICC A117.1, Section 305, entirely within the shelter. Such shelters shall be connected by an accessible route to the boarding area required by Section E108.2.

❖ A wheelchair parking space [30 inches by 48 inches (762 mm by 1219 mm)] must be provided within the bus shelter. If the shelter is small enough to meet the alcove requirements in Section 305.7 of ICC A117.1, additional space is required for maneuvering. In addition, an accessible route is required from the bus shelter to the bus stop (see Section E108.2).

E108.4 Signs. New bus route identification signs shall have finish and contrast complying with ICC A117.1. Additionally, to the maximum extent practicable, new bus route identification signs shall provide visual characters complying with ICC A117.1.

Exception: Bus schedules, timetables and maps that are posted at the bus stop or bus bay are not required to meet this requirement.

❖ The bus routes must be identified with visual characters that have light/dark contrast and a nonglare surface so they are easily readable. As much as practicable, the signage must also meet the requirements in Section 703.2 of ICC A117.1, for character form and spacing, line spacing and mounting height. There is an exception for schedules, timetables and maps due to the amount of information provided.

E108.5 Bus stop siting. Bus stop sites shall be chosen such that, to the maximum extent practicable, the areas where lifts or ramps are to be deployed comply with Sections E108.2 and E108.3.

❖ Whenever possible, bus stop sites should be chosen based on efforts to comply with the previous requirements for bus stop pads and bus shelters. Site constraints, such as a steeply inclined street, may not always make this practical.

SECTION E109 TRANSPORTATION FACILITIES AND STATIONS

E109.1 General. Fixed transportation facilities and stations shall comply with the applicable provisions of Section E109.2.

❖ Fixed transportation facilities and stations, such as train or light rail stations, provide access to public transportation and are considered an essential service; therefore, access in these facilities is of special concern. Technical requirements for transportation facilities can be found in Section 805 of ICC A117.1.

E109.2 New construction. New stations in rapid rail, light rail, commuter rail, intercity rail, high speed rail and other fixed guideway systems shall comply with Sections E109.2.1 through E109.2.8.

❖ All new construction involved with fixed guideway systems, including stations and terminals, is required to comply with this section.

E109.2.1 Station entrances. Where different entrances to a station serve different transportation fixed routes or groups of fixed routes, at least one entrance serving each group or route shall comply with Section 1104.

❖ A fixed guideway system, such as a commuter train system, may have separate platforms or entrances for a group of routes. When this is the case, at least one accessible route leading from public access points (e.g., public sidewalk, public parking, etc.) to the transit platform utilized for each group must be provided. When transit systems are underground, an accessible route must provide access to each subway station. As much as possible, the accessible entrance should be the same as the main entrance to the facility.

E109.2.2 Signs. Signage in fixed transportation facilities and stations shall comply with Sections E109.2.2.1 through E109.2.2.3.

❖ Specific requirements are listed for raised character and Braille identification and informational signs.

E109.2.2.1 Raised character and Braille signs. Where signs are provided at entrances to stations identifying the station or the entrance, or both, at least one sign at each entrance shall be raised characters and Braille. A minimum of one raised character and Braille sign identifying the specific station shall be provided on each platform or boarding area. Such signs shall be placed in uniform locations at entrances and on platforms or boarding areas within the transit system to the maximum extent practicable.

Exceptions:

1. Where the station has no defined entrance but signs are provided, the raised characters and Braille signs shall be placed in a central location.
2. Signs are not required to be raised characters and Braille where audible signs are remotely transmitted to hand-held receivers, or are user or proximity actuated.

❖ When identification signs are provided at a facility entrance, a sign with raised letters and Braille for the visually impaired must also be provided. A raised letter and Braille sign must also be provided at each platform or boarding area. The location of such signage should be consistent throughout the rail system. Signage must include both raised letters and Braille and meet the requirements of Section 703 of ICC A117.1. Raised letter and Braille signage is not required when information is provided by alternative audible means. The purpose of Exception 2 is to allow for emerging technologies such as an audible sign systems using infrared transmitters and receivers. These new systems may provide greater accessibility in the transit environment than traditional Braille and raised letter signs. The transmitters are placed on or next to print signs and transmit their information to an infrared receiver that is held by a person. By scanning an area, the person will hear the sign. This means that signs can be placed well out of reach of Braille readers, even on parapet walls and on walls beyond barriers. Additionally, such signs can be used to provide wayfinding information that cannot be efficiently conveyed on Braille signs.

E109.2.2.2 Identification signs. Stations covered by this section shall have identification signs containing visual characters complying with ICC A117.1. Signs shall be clearly visible and within the sightlines of a standing or sitting passenger from within the train on both sides when not obstructed by another train.

❖ When identification signs are provided so that persons on the trains can identify the station, the signs must meet the requirements of Section 703.2 of ICC A117.1 for visual characters including font style, finish and contrast.

E109.2.2.3 Informational signs. Lists of stations, routes and destinations served by the station which are located on boarding areas, platforms or mezzanines shall provide visual characters complying with ICC A117.1 Signs covered by this provision shall, to the maximum extent practicable, be placed in uniform locations within the transit system.

❖ Signs indicating stations, routes and destinations served by the train system accessed from that platform must comply with Section 703.2 of ICC A117.1 for visual characters including font style, finish and contrast. The location of such signage should be consistent throughout the system. Route maps are not required to comply with the informational sign requirements.

E109.2.3 Fare machines. Self-service fare vending, collection and adjustment machines shall comply with ICC A117.1, Section 707. Where self-service fare vending, collection or adjustment machines are provided for the use of the general public, at least one accessible machine of each type provided shall be provided at each accessible point of entry and exit.

❖ Self-service fare machines must be accessible in accordance with Section 707 of ICC A117.1. The requirements are the same as for ATM machines. If self-service machines are provided, at least one accessible machine must be provided along the accessible route between the entrance and the platform.

E109.2.4 Rail-to-platform height. Station platforms shall be positioned to coordinate with vehicles in accordance with the applicable provisions of 36 CFR, Part 1192. Low-level platforms shall be 8 inches (250 mm) minimum above top of rail.

Exception: Where vehicles are boarded from sidewalks or street level, low-level platforms shall be permitted to be less than 8 inches (250 mm).

❖ The height and position of a platform must be coordinated with the floor of the vehicles it serves to minimize the vertical and horizontal gaps, in accordance with the *ADA Accessibility Guidelines for Transportation Vehicles* (36 CFR Part 1192). The vehicle guidelines, categorized by bus, van, light rail, rapid rail, commuter rail and intercity rail are available at www.access-board.gov. The preferred alignment is a high platform, level with the vehicle floor. In some cases, the vehicle guidelines permit use of a low platform in conjunction with a lift or ramp. Most such low platforms must have a minimum height of 8 inches (203 mm) above the top of the rail. Some vehicles are designed to be boarded from a street or the sidewalk along the street and the exception permits such boarding areas to be less than 8 inches (203 mm) high.

E109.2.5 TTYs. Where a public pay telephone is provided in a transit facility (as defined by the Department of Transportation) at least one public TTY complying with ICC A117.1, Section 704.4, shall be provided in the station. In addition, where one or more public pay telephones serve a particular entrance to a transportation facility, at least one TTY telephone complying with ICC A117.1, Section 704.4, shall be provided to serve that entrance.

❖ Due to the essential nature of the facility, at least one public telephone in a transit facility is required to have a permanent TTY in accordance with Section 704.4 of ICC A117.1. When a bank of four or more is provided at an entrance, at least one is required to have a permanent TTY in accordance with Section 704.4 of ICC A117.1 (see also commentary, Section E106.4.7).

E109.2.6 Track crossings. Where a circulation path serving boarding platforms crosses tracks, an accessible route shall be provided.

Exception: Openings for wheel flanges shall be permitted to be $2^1/_2$ inches (64 mm) maximum.

❖ If people utilizing the train station must cross tracks to access the platform, it is necessary to provide an accessible route that takes into account the changes in elevation and gaps associated with such a route. Front wheels of wheelchairs are small, and if a person in a wheelchair got stuck while crossing the tracks, the consequences could be severe. If it is not possible to meet the criteria specified, an alternative accessible route must be provided. This requirement is not a requirement for track crossings remote from the boarding platform.

E109.2.7 Public address systems. Where public address systems convey audible information to the public, the same or equivalent information shall be provided in a visual format.

❖ If a public address system is utilized within a facility to provide information to the passengers, there must be an alternative means for providing the same information to persons with a hearing impairment.

E109.2.8 Clocks. Where clocks are provided for use by the general public, the clock face shall be uncluttered so that its elements are clearly visible. Hands, numerals and digits shall contrast with the background either light-on-dark or dark-on-light. Where clocks are mounted overhead, numerals and digits shall comply with visual character requirements.

❖ Clocks provided in transportation facilities must be readily distinguishable. When clocks are provided overhead, the numerals must comply with Section 703.2 of ICC A117.1 for visual characters.

SECTION E110 AIRPORTS

E110.1 New construction. New construction of airports shall comply with Sections E110.2 through E110.4.

❖ Airports provide access to public transportation and are considered an essential service; therefore, access in these facilities is of special concern. Technical requirements for transportation facilities can be found in Section 805 of ICC A117.1.

E110.2 TTYs. Where public pay telephones are provided, at least one TTY shall be provided in compliance with ICC A117.1, Section 704.4. Additionally, if four or more public pay telephones are located in a main terminal outside the security areas, a concourse within the security areas or a baggage claim area in a terminal, at least one public TTY complying with ICC A117.1, Section 704.4, shall also be provided in each such location.

❖ Due to the essential nature of the facility, at least one public telephone in an airport facility is required to have a permanent TTY in accordance with Section 704.4 of ICC A117.1. Given the restricted areas in an airport, when a bank of four or more phones is provided at either side of security checkpoints or in a baggage claim area, at least one phone in each bank is required to have a permanent TTY in accordance with Section 704.4 of ICC A117.1 (see also commentary, Section E106.4.7).

E110.3 Terminal information systems. Where terminal information systems convey audible information to the public, the same or equivalent information shall be provided in a visual format.

❖ If a public address system is utilized within a facility to provide information to passengers, there must be an alternative means for providing the same information to persons with a hearing impairment.

E110.4 Clocks. Where clocks are provided for use by the general public, the clock face shall be uncluttered so that its elements are clearly visible. Hands, numerals and digits shall contrast with the background either light-on-dark or dark-on-light. Where clocks are mounted overhead, numerals and digits shall comply with visual character requirements.

❖ Clocks provided in transportation facilities must be readily distinguishable. When clocks are provided overhead, the numerals must comply with Section 703.2 of ICC A117.1 for visual characters.

SECTION E111 REFERENCED STANDARDS

DOJ 36 CFR Part 1192	Americans with Disabilities Act (ADA) Accessibility Guidelines for Transportation Vehicles (ADAAG). Washington, DC: Department of Justice, 1991	E109.2.4
ICC A117.1-09	Accessible and Usable Buildings and Facilities	E101.2, E104.2, E104.2.1, E104.3 E104.3.4, E105.1, E105.2.1, E105.2.2, E105.3, E105.4, E105.6, E106.2, E106.3, E106.4, E106.4.9, E106.5, E107.2, E107.3, E108.3, E108.4, E109.2.1, E109.2.2.1, E109.2.2.2, E109.2.2.3, E109.2.3

❖ This section lists the standards that are referenced in Appendix E. They are listed here and not in Chapter 35 of the code, since appendices are not considered part of the code unless specifically referenced in the adopting ordinance.

Bibliography

The following resource materials are referenced in this chapter or are relevant to the subject matter addressed in this chapter.

2010 *ADA Standard for Accessible Design*. Washington, DC: Department of Justice, Sept. 15, 2010.

"Accessibility and Egress for People with Physical Disabilities." CABO Board for the Coordination of the Model Codes Report: October 5, 1993.

Architectural and Transportation Barriers Compliance Board, 36 CFR, Parts 1190 and 1191 Final Rule, *The Americans with Disabilities Act (ADA) Accessibility Guidelines)*; *Architectural Barriers Act (ABA) Accessibility Guidelines*. Washington, DC: July 23, 2004.

DOJ 28 CFR, Part 36-91, *Americans with Disabilities Act* (ADA). Washington, DC: Department of Justice, 1991.

DOJ 28 CFR, Part 36-91 (Appendix A), *ADA Accessibility Guidelines for Buildings and Facilities* (ADAAG). Washington, DC: Department of Justice, 1991.

DOJ 36 CFR, Part 1192/DOT 49 CFR, Part 38, *ADA Accessibility Guidelines for Transportation Vehicles*

(ADAAG). Washington, DC: Department of Justice, 1991.

DOT 49 CFR, Part 37, *Transportation Services for Individuals with Disabilities* (ADA). Washington, DC: Department of Transportation, 1999.

FED-STD-795-88, *Uniform Federal Accessibility Standards*. Washington, DC: General Services Administration; Department of Defense; Department of Housing and Urban Development; U.S. Postal Service, 1988.

ICC A117.1-2009, *Accessible and Usable Buildings and Facilities*. Washington, DC: International Code Council, 2003.

NFPA 72, *National Fire Alarm Code*. Quincy, MA: National Fire Protection Association, 2007.

Appendix F: Rodentproofing

The provisions contained in this appendix are not mandatory unless specifically referenced in the adopting ordinance.

General Comments

The requirements of this appendix apply only where it has been specifically adopted by the governing jurisdiction, as stated in Section 101.2.1.

The provisions of this appendix are minimum mechanical methods to prevent the entry of rodents. If a food supply is available, a rodent problem cannot be eliminated solely by these methods. However, removal of the food supply and general cleanliness and maintenance of the premises will have a significant impact.

Purpose

The purpose of this appendix is to provide regulations for rodentproofing for health purposes.

SECTION F101 GENERAL

F101.1 General. Buildings or structures and the walls enclosing habitable or occupiable rooms and spaces in which persons live, sleep or work, or in which feed, food or foodstuffs are stored, prepared, processed, served or sold, shall be constructed in accordance with the provisions of this section.

❖ In an effort to protect the public health and welfare and to prevent the spread of disease, this section requires all rooms and spaces used for living, sleeping or working and all spaces used for the storage, preparation or serving of food to be protected from rodents in the manner described in the code. Note that a building used for the storage of materials other than foodstuffs is not required to comply with the requirements of this section.

F101.2 Foundation wall ventilation openings. Foundation wall ventilator openings shall be covered for their height and width with perforated sheet metal plates no less than 0.070 inch (1.8 mm) thick, expanded sheet metal plates not less than 0.047 inch (1.2 mm) thick, cast-iron grills or grating, extruded aluminum load-bearing vents or with hardware cloth of 0.035 inch (0.89 mm) wire or heavier. The openings therein shall not exceed $^1/_4$ inch (6.4 mm).

❖ The purpose of this section is to protect ventilation openings in the foundation walls from the entry of rodents. The thickness of the metal shields and the maximum openings that are specified in this section are necessary for the shields to be effective.

F101.3 Foundation and exterior wall sealing. Annular spaces around pipes, electric cables, conduits, or other openings in the walls shall be protected against the passage of rodents by closing such openings with cement mortar, concrete masonry or noncorrosive metal.

❖ During construction, the openings in the foundation and exterior walls that are made for pipes, electrical cables, conduits and the like are significantly larger than the component that passes through the wall. These openings around the component are required to be closed for rodentproofing.

F101.4 Doors. Doors on which metal protection has been applied shall be hinged so as to be free swinging. When closed, the maximum clearance between any door, door jambs and sills shall not be greater than $^3/_8$ inch (9.5 mm).

❖ This section addresses wood doors located at ground or basement floor levels. Solid sheet metal protection is needed to stop rodents from gnawing their way through the wood into areas where food is located. The limitation on the clearance between the door and the adjacent door jamb and sill is to deter rodent entry.

F101.5 Windows and other openings. Windows and other openings for the purpose of light or ventilation located in exterior walls within 2 feet (610 mm) above the existing ground level immediately below such opening shall be covered for their entire height and width, including frame, with hardware cloth of at least 0.035-inch (0.89 mm) wire or heavier.

❖ The requirement for wire screens or "hardware cloth" for rodent protection applies only to openings 2 feet (610 mm) or less from the ground. This is a minimum requirement and will not keep all rodents out of the building.

F101.5.1 Rodent-accessible openings. Windows and other openings for the purpose of light and ventilation in the exterior walls not covered in this chapter, accessible to rodents by way of exposed pipes, wires, conduits and other appurtenances, shall be covered with wire cloth of at least 0.035-inch (0.89 mm) wire. In lieu of wire cloth covering, said pipes, wires, conduits and other appurtenances shall be blocked from rodent usage by installing solid sheet metal guards 0.024 inch (0.61 mm) thick or heavier. Guards shall be fitted around pipes, wires, conduits or other appurtenances. In addition, they shall be fastened securely to and shall extend perpendicularly from the exterior wall for a minimum distance of 12 inches (305 mm) beyond and on either side of pipes, wires, conduits or appurtenances.

❖ This section addresses screen requirements for windows or other openings that are higher than 2 feet (610 mm) from grade and are made rodent accessible by pipes, wires, conduits and the like. The window or other opening is required to be screened or rodent guards are to be installed on the pipes, wires or conduits so that the window or other opening is not accessible to rodents.

F101.6 Pier and wood construction.

❖ The requirements of Sections F101.6.1 and F101.6.2 apply only if the building does not have a continuous foundation wall of either treated wood, concrete or coated metal. An example of such a building is a post and beam wood-frame building with a slab on grade. Also note that these requirements apply if the building contains habitable or occupiable rooms in which persons live, sleep or work, or in which foodstuffs are stored, prepared, processed, served or sold. Thus, a storage or utility building used for storage of materials other than foodstuffs is not required to meet this section, even though it does not have a continuous foundation wall.

F101.6.1 Sill less than 12 inches above ground. Buildings not provided with a continuous foundation shall be provided with protection against rodents at grade by providing either an apron in accordance with Section F101.6.1.1 or a floor slab in accordance with Section F101.6.1.2.

❖ See the commentary to Section F101.6 for an explanation of when the requirements of this section apply. This section cites two alternative methods that meet the requirements: an apron or a concrete grade floor. Either of these methods provides effective rodent protection.

F101.6.1.1 Apron. Where an apron is provided, the apron shall not be less than 8 inches (203 mm) above, nor less than 24 inches (610 mm) below, grade. The apron shall not terminate below the lower edge of the siding material. The apron shall be constructed of an approved nondecayable, water-resistant rodentproofing material of required strength and shall be installed around the entire perimeter of the building. Where constructed of masonry or concrete materials, the apron shall not be less than 4 inches (102 mm) in thickness.

❖ The apron is usually a concrete, masonry or treated wood wall. Each material meets the requirement of this section. The extension into the ground is to prevent rodents from burrowing underground into the building. Note that a concrete floor slab is not required where the apron is used for rodent protection in accordance with Section F101.6.2.

F101.6.1.2 Grade floors. Where continuous concrete grade floor slabs are provided, open spaces shall not be left between the slab and walls, and openings in the slab shall be protected.

❖ This is the second option for rodent protection in Section 101.6.1. When a grade floor that meets the requirements of this section is installed, the apron described in Section F101.6.1.2 is not required.

The grade floor is required to have any openings protected from rodents. For example, if the floor slab has an opening where drain pipes go through it, the area around the pipe is to be sealed with material that will keep rodents out of the building.

F101.6.2 Sill at or above 12 inches above ground. Buildings not provided with a continuous foundation and which have sills 12 or more inches (305 mm) above the ground level shall be provided with protection against rodents at grade in accordance with any of the following:

1. Section F101.6.1.1 or F101.6.1.2;
2. By installing solid sheet metal collars at least 0.024 inch (0.6 mm) thick at the top of each pier or pile and around each pipe, cable, conduit, wire or other item which provides a continuous pathway from the ground to the floor; or
3. By encasing the pipes, cables, conduits or wires in an enclosure constructed in accordance with Section F101.6.1.1.

❖ See the commentary to Section F101.6 for an explanation of when the requirements of this section apply. Where the sill height is 12 inches (305 mm) or more above the ground, this section provides two additional acceptable methods for rodent protection versus methods for instances where the sill height is less than 12 inches (305 mm).

Methods 2 and 3 described in this section provide the rodent barrier directly around the pipes, cables, conduits or wires that would otherwise be a continuous pathway from the ground to the raised floor.

Appendix G: Flood-Resistant Construction

The provisions contained in this appendix are not mandatory unless specifically referenced in the adopting ordinance.

General Comments

The provisions contained in this appendix are not mandatory unless specifically referenced in the adopting ordinance, as stated in Section 101.2.1. This appendix is intended to fulfill the flood-plain management and administrative requirements of the National Flood Insurance Program (NFIP) that are not included in the code. Communities that adopt the code and this appendix without modification will meet the minimum requirements of NFIP as set forth in Title 44 of the Code of Federal Regulations (CFR). This appendix includes administrative requirements of NFIP and requirements concerning modifications to watercourses, permits for flood hazard area development, conditions for the issuance of variances from flood-plain management requirements and site improvements, subdivision planning and installation of manufactured homes, recreation vehicles and tanks in flood hazard areas. It is important to note that many states and communities regulate flood-plain development to higher standards than the minimum requirements of the code and this appendix. Prior to adopting the code or this appendix, communities are advised to consult their state NFIP coordinator or Federal Emergency Management Agency (FEMA) regional office to determine additional actions that may be necessary to provide for continued participation in NFIP.

Purpose

The purpose of this appendix is to provide optional criteria for flood-resistant construction. A jurisdiction that wants to make this appendix mandatory needs to include it in its adoption ordinance. See page xv of the code for a sample ordinance for adoption.

SECTION G101 ADMINISTRATION

G101.1 Purpose. The purpose of this appendix is to promote the public health, safety and general welfare and to minimize public and private losses due to *flood* conditions in specific *flood hazard* areas through the establishment of comprehensive regulations for management of *flood hazard areas* designed to:

1. Prevent unnecessary disruption of commerce, access and public service during times of *flooding*;
2. Manage the alteration of natural *flood* plains, stream channels and shorelines;
3. Manage filling, grading, dredging and other development which may increase *flood* damage or erosion potential;
4. Prevent or regulate the construction of *flood* barriers which will divert floodwaters or which can increase flood hazards; and
5. Contribute to improved construction techniques in the flood plain.

❖ This section sets forth the purposes of this appendix. Communities that administer flood-plain management regulations such as in this appendix achieve multiple objectives beyond reducing physical damage to buildings.

G101.2 Objectives. The objectives of this appendix are to protect human life, minimize the expenditure of public money for *flood* control projects, minimize the need for rescue and relief efforts associated with *flooding*, minimize prolonged business interruption, minimize damage to public facilities and utilities, help maintain a stable tax base by providing for the sound use and development of flood-prone areas, contribute to improved construction techniques in the flood plain and ensure that potential owners and occupants are notified that property is within flood hazard areas.

❖ This appendix, combined with the provisions in the code, achieves several objectives intended to reduce the impacts of flooding. Management of flood hazard areas in ways that reduce exposure to damage also protects health and safety. Additionally, the provisions of this appendix promote sustainable development in communities subject to flooding. Flooding has both short-term and long-term impacts. Some impacts are obvious, such as damaged homes and businesses. Others are less apparent, including increases in flood levels due to changes in the flood plain; scour and erosion; impaired public and private water and sewage systems and reductions in the natural and beneficial functions of flood plains, including wetland areas.

G101.3 Scope. The provisions of this appendix shall apply to all proposed development in a flood hazard area established in Section 1612 of this code, including certain building work exempt from permit under Section 105.2.

❖ In communities that participate in NFIP, its minimum requirements must be adopted and applied to all development in flood hazard areas. The definition of "Development," as stated in Section G201.2, is inclusive. This broad definition captures all activities that take place in flood hazard areas that are subject to the requirements of the code or this appendix. Because NFIP requires regulation of all development within special flood hazard areas that could have an impact on flooding, a code that applies only to buildings or structures does not fulfill the minimum requirements for participation in the program. Work that is exempt from the requirement to obtain a permit, listed in Section 105.2, is subject to the requirements of this appendix in order to satisfy the NFIP definition of "Development," described in the commentary for Sections G102.1 and G201.2.

G101.4 Violations. Any violation of a provision of this appendix, or failure to comply with a permit or variance issued pursuant to this appendix or any requirement of this appendix, shall be handled in accordance with Section 114.

❖ Violations of the provisions of this appendix are to be corrected or adjudicated under the provisions of Section 114. Communities participating in NFIP enter into an agreement with FEMA: in exchange for the availability of federally backed flood insurance and many forms of federal disaster assistance, communities agree to adopt and effectively enforce the flood-plain management and administrative provisions of NFIP. A significant part of community responsibility takes the form of enforcing flood-plain management and other flood damage reduction ordinances and laws. The procedures set forth in Section 114 are to be followed if activities take place in a flood hazard area without a permit, or if a permit applicant fails to comply with a condition of a permit or variance. If reasonable enforcement actions do not result in compliance, the community is encouraged to contact its state NFIP coordinator or appropriate FEMA regional office for further guidance.

SECTION G102 APPLICABILITY

G102.1 General. This appendix, in conjunction with the *International Building Code*, provides minimum requirements for development located in flood hazard areas, including the subdivision of land; installation of utilities; placement and replacement of manufactured homes; new construction and repair, reconstruction, rehabilitation or additions to new construction; substantial improvement of existing buildings and structures, including restoration after damage, temporary structures, and temporary or permanent storage, utility and miscellaneous Group U buildings and structures, and certain building work exempt from permit under Section 105.2.

❖ The definition of "Development" in Section G201.2 encompasses a wide array of activities that are to be regulated if they are proposed to take place within a designated flood hazard area. Some activities have the potential to increase flooding on other properties. Other activities, if planned in a way that recognizes flood hazards, have the potential to allow reasonable use of land without placing people and site improvements at significant risk.

G102.2 Establishment of flood hazard areas. *Flood hazard areas* are established in Section 1612.3 of the *International Building Code*, adopted by the applicable governing authority on **[INSERT DATE]**.

❖ The flood hazard areas established in Section 1612.3 are subject to the additional provisions of this appendix. The flood hazard map to be adopted is, at a minimum, the flood insurance rate map (FIRM), supported by the flood insurance study prepared by FEMA. A community participating in NFIP may specify another map, provided it designates flood hazard areas that are the same or more extensive and regulatory flood elevations that are the same or higher than those shown on the FIRM. Regardless of which map is specified, flood elevations are the design flood elevations. From time to time, FEMA revises and republishes FIRMs. A formal review process is followed, including community and public comment. When new FIRMs are published, communities are required to use them. The flood hazard areas shown on maps prepared by FEMA are determined using the base flood, which is defined as having a 1-percent chance (one chance in 100) of occurring in any given year. The maps are not intended to show the worst-case flood, or the "flood of record," which usually refers to the most severe flood in the history of the community. Although a 1-percent chance seems fairly remote, larger floods occur regularly throughout the United States. Application of the provisions of this appendix is not intended to prevent or eliminate all future flood damage. These provisions are intended to represent a reasonable balance of the knowledge and awareness of flood hazards, methods to guide development to less hazard-prone locations, methods of design and construction intended to resist flood damage and each community's and landowner's reasonable expectations to use the land. Some coastal communities have FIRMs that show areas that are designated as units of the Coastal Barrier Resource System (CBRS) established by the Coastal Barrier Resource Act (CoBRA) of 1982 and subsequent amendments. Within these designated areas, NFIP is prohibited from offering flood insurance on new or substantially improved buildings. The community is responsible for application of the flood-resistant provisions of the code and this appendix in all designated flood hazard areas, whether or not flood insurance is made available.

SECTION G103 POWERS AND DUTIES

G103.1 Permit applications. The *building official* shall review all *permit* applications to determine whether proposed development sites will be reasonably safe from flooding. If a proposed development site is in a flood hazard area, all site development activities (including grading, filling, utility installation and drainage modification), all new construction and substantial improvements (including the placement of prefabricated buildings and manufactured homes) and certain building work exempt from *permit* under Section 105.2 shall be designed and constructed with methods, practices and materials that minimize flood damage and that are in accordance with this code and ASCE 24.

❖ The building official is empowered to examine all permit applications to determine if the proposed activities will take place in designated flood hazard areas on the community's flood hazard map. Such activities are to be designed and constructed in accordance with both the code and this appendix, as applicable. For additional guidance for development activities that involve the placement of fill in a flood hazard area, refer to FEMA FIA-TB #10, *Ensuring that Structures Built on Fill In or Near Special Flood Hazard Areas are Reasonably Safe from Flooding.*

G103.2 Other permits. It shall be the responsibility of the *building official* to assure that approval of a proposed development shall not be given until proof that necessary permits have been granted by federal or state agencies having jurisdiction over such development.

❖ Other federal, state and local regulatory authorities may have jurisdiction over activities within designated flood hazard areas. Examples at the federal level include permitting under Section 404 of the Clean Water Act of 1977 and Section 10 of the Rivers and Harbors Act of 1899 and consultation or permitting under the Endangered Species Act of 1973. State and regional agencies may also regulate activities in flood hazard areas, including activities that impact wetlands, forestry resources, dunes, the coastal zone, subaquatic vegetation, threatened and endangered species, navigation and waterways. The intent is to provide coordination among various levels of government and to avoid added costs to applicants in situations where multiple permits are required. Building officials may satisfy this requirement by either withholding the permit until other permits have been obtained, or by issuing the permit contingent on the applicant obtaining other specified permits.

G103.3 Determination of design flood elevations. If design flood elevations are not specified, the *building official* is authorized to require the applicant to:

1. Obtain, review and reasonably utilize data available from a federal, state or other source, or
2. Determine the *design flood elevation* in accordance with accepted hydrologic and hydraulic engineering techniques. Such analyses shall be performed and sealed by a *registered design professional.* Studies, analyses and computations shall be submitted in sufficient detail to allow review and approval by the *building official.* The accuracy of data submitted for such determination shall be the responsibility of the applicant.

❖ Many FIRMs show flood hazard areas without specifying the base flood elevations (BFE). These areas, often referred to as "unnumbered A or V Zones," are subject to the flood-plain management requirements of the code and the appendix, although the minimum height to which buildings and structures are to be elevated has not been determined by FEMA. An important step in regulating these flood hazard areas is determination of the design flood elevation (DFE). As defined in the code, at a minimum the DFE is the BFE shown on a community's flood insurance rate map. In some instances, flood elevation information may have been developed by sources other than FEMA, including other federal or state agencies. The building official is to obtain the information or require the permit applicant to do so. Some communities develop flood hazard information and provide it to applicants so that all development in an area is conducted based on the same level of risk. If flood elevation information is not available, the building official may require the applicant to develop the DFE in accordance with accepted engineering practices. Local officials unfamiliar with establishing DFEs in unnumbered A and V Zones are encouraged to contact the state NFIP coordinator or appropriate FEMA regional office. For additional guidance, refer to FEMA 265, *Managing Floodplain Development in Approximate Zone A Areas: A Guide for Obtaining and Developing Base (100-Year) Flood Elevations.*

G103.4 Activities in riverine flood hazard areas. In riverine *flood hazard areas* where *design flood elevations* are specified but *floodways* have not been designated, the *building official* shall not permit any new construction, substantial improvement or other development, including fill, unless the applicant demonstrates that the cumulative effect of the proposed development, when combined with all other existing and anticipated flood hazard area encroachment, will not increase the design flood elevation more than 1 foot (305 mm) at any point within the community.

❖ Although FEMA has provided floodways along many rivers and streams shown on a community's FIRM, many other riverine flood hazard areas have BFEs but not designated floodways. In these areas the potential effects that flood-plain activities may have on flood elevations have not been evaluated. If FEMA has not designated a regulatory floodway on a community's FIRM, the community is responsible for regulating development so as not to increase flood elevations by more than 1 foot (305 mm) at any point in the community. In effect, this means a community must either prepare a hydraulic analysis for proposed activities or require permit applicants to do so. Several states have more restrictive requirements, which can be determined by contacting the state NFIP coordinator.

G103.5 Floodway encroachment. Prior to issuing a *permit* for any *floodway* encroachment, including fill, new construction, substantial improvements and other development or land-disturbing activity, the *building official* shall require submission of a certification, along with supporting technical data, that demonstrates that such development will not cause any increase of the level of the base *flood*.

❖ The floodway is that part of the riverine flood plain that must be reserved in order to convey the base flood without cumulatively increasing the water surface elevation more than a designated height specified in the flood insurance study provided by FEMA. Generally, this designated height is 1 foot (305 mm), although some states and communities have a more restrictive requirement. Development in the floodway could obstruct flood flows and increase flood levels, causing additional damage. Usually, the floodway is where the water will be deepest and move the fastest. As required in Section G103.5, communities are to prohibit any floodway encroachment, including fill, new construction and substantial improvement, if such activities will cause any increase in flood levels. Permit applications are to include a certification supported with engineering analyses to demonstrate the anticipated impacts. To be acceptable, the certification must demonstrate that the proposed development will not impact BFEs, floodway elevations or floodway widths and is to be signed and sealed by a registered professional engineer. The building official is to review the certification and determine that it meets accepted engineering practices. If necessary to make this determination, the building official is advised to seek technical assistance from the community's engineer, the state NFIP coordinator or the FEMA regional office. In limited situations, a community may decide to permit development in the floodway that causes an increase in flood elevations. This may be appropriate for dams or other water resource structures and for bridges and culverts where the most cost-effective alternative results in an increase in flood elevations. Section 103.5 allows for these activities, provided the community or applicant requests and obtains a conditional letter of map revision (CLOMR) and a floodway revision from FEMA. The proposed development may not increase flooding of existing buildings and all impacted property owners must be notified. A permit for the development may be issued after the CLOMR is approved by FEMA and after the community adopts the increased flood elevations. When the project is completed, the community is required to submit as-built certifications to FEMA so that a final map revision can be issued. For guidance on preparing and certifying revisions to NFIP map products, see FEMA Form 81-89 Series, *Revisions to National Flood Insurance Rate Maps: Application/Certification Forms and Instructions for Conditional Letters of Map Revision, Letters of Map Revision and Physical Map Revisions*. Although each floodway proposal is to be reviewed carefully and some states have specific and more restrictive regulations that apply within floodways, there are some activities and uses that may be permitted without extensive engineering analysis. The most important factor is that the fill and grading do not change the shape of the land. Possible uses include:

- Agricultural uses not involving buildings or structures;
- At-grade uses such as parking, loading areas and small airport landing strips;
- Passive recreational uses, such as hiking, biking, horse trails, wildlife and nature preserves and hunting and fishing areas;
- Active recreational uses (where improvements are anchored to prevent flotation), such as picnic and playground areas, ball fields, boat launch ramps, swimming areas, target shooting ranges and similar uses; and
- Uses incidental to residential buildings, such as lawns, gardens, parking areas, playgrounds and tot lots.

G103.5.1 Floodway revisions. A *floodway* encroachment that increases the level of the base *flood* is authorized if the applicant has applied for a conditional Flood Insurance Rate Map (FIRM) revision and has received the approval of the Federal Emergency Management Agency (FEMA).

❖ Under certain circumstances, a permit may be issued for a proposed floodway encroachment that, based on analysis and supporting technical data, will cause an increase in the level of the base flood. The community is to require that the analysis be submitted to FEMA in accordance with the guidance described in Section G103.5 (FEMA Form 81-89). The permit may be issued if FEMA approves the proposal by issuance of a conditional revision to the flood hazard area map. Prior to issuance of a final FIRM revision, FEMA will require that the applicant submit post-construction documentation to demonstrate that the work was conducted as proposed.

G103.6 Watercourse alteration. Prior to issuing a permit for any alteration or relocation of any watercourse, the *building official* shall require the applicant to provide notification of the proposal to the appropriate authorities of all affected adjacent government jurisdictions, as well as appropriate state agencies. A copy of the notification shall be maintained in the permit records and submitted to FEMA.

❖ Relocation or alteration of the channel of a river or stream can significantly and adversely impact the flood plain. Because such impact may extend a significant distance from a proposed relocation or alteration, NFIP requires notification of adjacent communities, appropriate state agencies and FEMA.

G103.6.1 Engineering analysis. The *building official* shall require submission of an engineering analysis which demonstrates that the flood-carrying capacity of the altered or relocated portion of the watercourse will not be decreased. Such watercourses shall be maintained in a manner which preserves the channel's flood-carrying capacity.

❖ Prior to issuing a permit that allows modification of a waterway, the applicant is to submit a hydraulic analysis demonstrating that the ability of the altered channel and flood plain to carry flood discharges is not diminished. An important condition of a permit, if issued, is that the applicant or other designated entity is to maintain the altered channel in order to preserve its flood-carrying capacity and to avoid increasing future flood risks.

G103.7 Alterations in coastal areas. Prior to issuing a permit for any alteration of sand dunes and mangrove stands in flood hazard areas subject to high velocity wave action, the *building official* shall require submission of an engineering analysis which demonstrates that the proposed alteration will not increase the potential for flood damage.

❖ As part of the review of proposed site work in coastal areas, a community is to examine proposals for possible impacts to sand dunes or mangrove stands, which mitigate the effects of coastal flooding. If a proposal is to alter a sand dune or mangrove stand, an engineering analysis of the effect on flood damages is required as part of the documentation. If engineering analysis indicates that the potential for flood damage will be increased due to the proposed alteration, the building official is to require the permit applicant to make necessary modifications to the proposed site work to avoid such increases.

G103.8 Records. The *building official* shall maintain a permanent record of all *permits* issued in *flood hazard areas*, including copies of inspection reports and certifications required in Section 1612.

❖ The building official's responsibility to maintain department records is established in Section 104.7. This section extends that responsibility to require that records for permits in flood hazard areas be maintained permanently. Specifically identified for retention are the inspection reports and certifications required to be submitted in Section 1612. As a condition for participation in the NFIP, a community agrees that FEMA or its designee may examine records relating to administration of the community's flood management regulations.

SECTION G104 PERMITS

G104.1 Required. Any person, owner or authorized agent who intends to conduct any development in a flood hazard area shall first make application to the *building official* and shall obtain the required *permit*.

❖ The definition of "Development" in Section G201 encompasses a wide array of human activities that may be proposed within designated flood hazard areas. Communities are to require that all such activities are authorized by permits issued in compliance with the applicable flood-plain management requirements of the code and this appendix.

G104.2 Application for permit. The applicant shall file an application in writing on a form furnished by the *building official*. Such application shall:

1. Identify and describe the development to be covered by the permit.
2. Describe the land on which the proposed development is to be conducted by legal description, street address or similar description that will readily identify and definitely locate the site.
3. Include a site plan showing the delineation of flood hazard areas, floodway boundaries, flood zones, design flood elevations, ground elevations, proposed fill and excavation and drainage patterns and facilities.
4. Indicate the use and occupancy for which the proposed development is intended.
5. Be accompanied by construction documents, grading and filling plans and other information deemed appropriate by the building official.
6. State the valuation of the proposed work.
7. Be signed by the applicant or the applicant's authorized agent.

❖ This section details the information to be shown on or included in an application for a permit to conduct activities within a flood hazard area. The site plan is to show sufficient detail and information about the designated flood hazard area, including floodway, to allow for a complete review of the proposed activities. The horizontal boundary of the designated flood hazard area shown on the community's flood hazard map is an approximate boundary based on topographic information that was available when the map was prepared by FEMA. The determination of the actual boundary of the designated flood hazard is to be based on the best topographic information that is available at the time of the application. All land area below the DFE is subject to regulation even if it is not shown on the community's flood hazard map as being within the designated flood hazard area. Along with necessary flood hazard information, the storm drainage conveyance requirements are to be shown on the site plan.

G104.3 Validity of permit. The issuance of a *permit* under this appendix shall not be construed to be a permit for, or approval of, any violation of this appendix or any other ordinance of the jurisdiction. The issuance of a *permit* based on submitted documents and information shall not prevent the *building official* from requiring the correction of errors. The *building official* is authorized to prevent occupancy or use of a structure or site which is in violation of this appendix or other ordinances of this jurisdiction.

❖ This section clarifies that it is the owner's responsibility to comply with the code. Permits are approved based on a review of the materials submitted and a determination that the proposed activities are in accordance with the provisions of the code. If errors or deviations are made during construction, or it is discovered that the information and documentation submitted to obtain the permit were in error, the code official has the authority to require correction and compliance.

G104.4 Expiration. A *permit* shall become invalid if the proposed development is not commenced within 180 days after its issuance, or if the work authorized is suspended or abandoned for a period of 180 days after the work commences. Extensions shall be requested in writing and justifiable cause demonstrated. The *building official* is authorized to grant, in writing, one or more extensions of time, for periods not more than 180 days each.

❖ Permits issued for activities in a flood hazard area are to expire after 180 days if work either has not commenced during that period of time or is abandoned for any 180-day period after it commences. The building official may grant extensions to the permit for up to 180 days at a time. Section 1612.2 defines "Start of construction" as "the date of permit issuance for new construction and substantial improvements to existing structures, provided the actual start of construction, repair, reconstruction, rehabilitation, addition placement or other improvement is within 180 days after the date of issuance. The actual start of construction means the first placement of permanent construction of a building (including a manufactured home) on a site, such as the pouring of a slab or footings, installation of pilings or construction of columns."

Building officials may not grant permit extensions in situations where construction has not begun during the 180-day period if FEMA has issued a revised FIRM that shows increases in flood hazard areas or DFEs at the location of the permitted activity. A new permit is required if the revised FIRM changes the flood-risk information at the site of the permit. The new application is to be reviewed based on the revised flood-risk information.

G104.5 Suspension or revocation. The building official is authorized to suspend or revoke a *permit* issued under this appendix wherever the permit is issued in error or on the basis of incorrect, inaccurate or incomplete information, or in violation of any ordinance or code of this jurisdiction.

❖ A permit issued in error or based on incorrect, inaccurate or incomplete information is to be suspended or revoked. This provision does not preclude issuance of a new or revised permit that is based on accurate information. Violation of a permit or permit condition may also prompt suspension or revocation.

SECTION G105 VARIANCES

G105.1 General. The *board of appeals* established pursuant to Section 112 shall hear and decide requests for variances. The *board of appeals* shall base its determination on technical justifications, and has the right to attach such conditions to variances as it deems necessary to further the purposes and objectives of this appendix and Section 1612.

❖ Section 112 empowers the board of appeals to hear and decide requests for variances. Variances from the flood plain management requirements of the code and this appendix may place people and property at significant risk. Therefore, communities are cautioned to carefully evaluate the impacts of issuing a variance from the flood-resistant construction provisions of the code and this appendix, particularly the requirements to elevate or floodproof buildings to the DFE. The impacts to be evaluated include impacts on the site, the permit applicant and other parties that may be affected, such as adjacent property owners and the community. Flood-plain development that is not undertaken in accordance with the flood-resistant construction provisions of the code and this appendix will be exposed to increased flood damages. As a consequence, flood insurance premium rates will be significantly higher. Variance decisions made by the board of appeals are to be based solely on technical justifications outlined in this section, not on the personal circumstances of an owner or applicant.

Applicants sometimes request variances to the minimum elevation requirements for the lowest floor of buildings in flood hazard areas. Such requests may be based on the need to improve access for the disabled and elderly. Generally, variances of this nature are not to be granted since these are personal circumstances that will change as the property changes ownership. Not only would persons of limited mobility be at risk from flooding, but the building would continue to be exposed to flood damage long after the personal need for a variance changes. More appropriate alternatives are to be considered to serve the needs of disabled or elderly persons, such as varying setbacks to allow construction on less flood-prone portions of sites or installing personal elevators.

G105.2 Records. The building official shall maintain a permanent record of all variance actions, including justification for their issuance.

❖ The permanent records of the community that document official actions on permit applications and variance requests concerning flood-plain management requirements are to include the justifications on which variance determinations are based. These records will be examined periodically by FEMA or its designee in an effort to determine if the community is effectively administering its flood management provisions consistent with the minimum requirements of NFIP.

G105.3 Historic structures. A variance is authorized to be issued for the repair or rehabilitation of a historic structure upon a determination that the proposed repair or rehabilitation will not preclude the structure's continued designation as a historic structure, and the variance is the minimum necessary to preserve the historic character and design of the structure.

Exception: Within *flood hazard areas*, *historic structures* that are not:

1. Listed or preliminarily determined to be eligible for listing in the National Register of Historic Places; or
2. Determined by the Secretary of the U.S. Department of Interior as contributing to the historical significance of a registered historic district or a district preliminarily determined to qualify as an historic district; or
3. Designated as *historic* under a state or local historic preservation program that is approved by the Department of Interior.

❖ This provision recognizes the importance and value of historic buildings and structures. If an owner proposes repair or restoration of a historic building or structure that would otherwise be found to be a substantial improvement, the board of appeals may grant a variance to perform the work without requiring full compliance with the flood-resistant provisions of the code and this appendix. This is not to say that work towards flood-resistant provisions should not be performed. It is intended that the historic building should be brought into compliance as much as possible without adversely affecting the historic significance. To qualify for this treatment, the building or structure must be individually listed or eligible to be listed as a historic structure, or be certified as contributing to the historic significance of a historic district. Merely being located in a historic district is insufficient justification for a variance.

A variance may be issued, provided the proposed work does not change the historic designation of the building or structure. The board of appeals or the building official may require that the permit applicant obtain a written review and determination to that effect from the appropriate state or local organization responsible for the historic designation.

All variances are to be the minimum necessary to afford relief [see Section G105.7(4)]. There are a number of actions that may be taken to reduce future flood damage to substantially damaged or substantially improved historic buildings or structures, while maintaining their historic and cultural value. For example, if elevation of a historic building or structure would negatively affect its character or cause a loss of its historic designation, a variance from the elevation requirement may be appropriate. However, it may still be possible to require that building utility systems be elevated or otherwise protected from flood damage to the maximum extent possible. A condition of a variance may require that building materials, including interior finishes, be resistant to flood damage (refer to FEMA FIA-TB #2, *Flood Resistant Materials Requirements for Buildings Located in Special Flood Hazard Areas*). Another condition may require that building contents be elevated to the DFE, or the maximum extent possible. For further information, contact the state NFIP coordinator, state historic preservation officer or appropriate FEMA regional office.

G105.4 Functionally dependent facilities. A variance is authorized to be issued for the construction or substantial improvement of a functionally dependent facility provided the criteria in Section 1612.1 are met and the variance is the minimum necessary to allow the construction or substantial improvement, and that all due consideration has been given to methods and materials that minimize flood damages during the design flood and create no additional threats to public safety.

❖ Functionally dependent facilities include those that must be located in close proximity to water to fulfill their intended purpose, such as shipbuilding, ship or boat repair, as well as docks and port facilities for loading or unloading cargo or passengers. Some facilities that often are located close to water are specifically excluded from the definition, including long-term storage and manufacturing, sales or service facilities that are associated with water activities. Functionally dependent facilities represent a special case recognized by NFIP as warranting treatment under the variance provisions. Under some circumstances it may impossible for a facility to perform its function unless one or more of the flood-resistant construction provisions of the code or this appendix are varied. For example, temporary storage facilities may need to be at dock level or a building in a designated flood hazard area subject to high-velocity wave action may need to be elevated on armored fill rather than on piles or columns.

At a minimum, functionally dependent facilities are to be wet floodproofed. This type of flood protection reduces flood damage while allowing floodwaters to enter the building or structure. The keys to successful wet floodproofing are to design the building or structure, to the maximum extent possible, to resist flood and wind loads anticipated during conditions of the design flood; to use flood damage-resistant materials and to elevate or use damage-resistant building utility systems as required by the flood-resistant construction provisions of the code and this appendix. For further guidance, refer to FEMA FIA-TB #1, *Openings in Foundation Walls for Buildings Located in Special Flood Hazard Areas*; FEMA FIA-TB #2, *Flood Resistant Material Requirements for Buildings Located in Special Flood Hazard Areas* and FEMA FIA-TB #7, *Wet Floodproofing Requirements for Structures Located in Special Flood Hazard Areas.*

Within functionally dependent facilities, only those functions that must be carried out below the DFE are to be included within the scope of the variance. For example, a boat repair facility may have areas that are used for a variety of purposes. The actual boat repair activities may need to be conducted at an elevation below the DFE. Other areas that are not functionally dependent, such as employee locker rooms, bathrooms and offices, must either be elevated or dry floodproofed (A Zones only) to or above the DFE as required by the code.

G105.5 Restrictions. The *board of appeals* shall not issue a variance for any proposed development in a floodway if any increase in flood levels would result during the base flood discharge.

❖ The board of appeals may not issue variances to the requirements in Section G103.5, which addresses activities in floodways that may increase flood elevations. The commentary to Section G103.5 describes submission of engineering analyses and approval by FEMA prior to approving activities that increase flood elevations in designated floodways.

G105.6 Considerations. In reviewing applications for variances, the *board of appeals* shall consider all technical evaluations, all relevant factors, all other portions of this appendix and the following:

1. The danger that materials and debris may be swept onto other lands resulting in further injury or damage;
2. The danger to life and property due to flooding or erosion damage;
3. The susceptibility of the proposed development, including contents, to flood damage and the effect of such damage on current and future owners;
4. The importance of the services provided by the proposed development to the community;
5. The availability of alternate locations for the proposed development that are not subject to flooding or erosion;
6. The compatibility of the proposed development with existing and anticipated development;
7. The relationship of the proposed development to the comprehensive plan and flood plain management program for that area;
8. The safety of access to the property in times of flood for ordinary and emergency vehicles;
9. The expected heights, velocity, duration, rate of rise and debris and sediment transport of the floodwaters and the effects of wave action, if applicable, expected at the site; and
10. The costs of providing governmental services during and after flood conditions including maintenance and repair of public utilities and facilities such as sewer, gas, electrical and water systems, streets and bridges.

❖ This section provides factors to be taken into account by the board of appeals when it reviews requests for variances from the flood-resistant construction and flood-plain management provisions of the code and this appendix. Section G105.2 requires that documentation be maintained as part of the permanent record to indicate how the board of appeals addressed each consideration.

G105.7 Conditions for issuance. Variances shall only be issued by the *board of appeals* upon:

1. A technical showing of good and sufficient cause that the unique characteristics of the size, configuration or topography of the site renders the elevation standards inappropriate;
2. A determination that failure to grant the variance would result in exceptional hardship by rendering the lot undevelopable;
3. A determination that the granting of a variance will not result in increased flood heights, additional threats to public safety, extraordinary public expense, nor create nuisances, cause fraud on or victimization of the public or conflict with existing local laws or ordinances;
4. A determination that the variance is the minimum necessary, considering the flood hazard, to afford relief; and
5. Notification to the applicant in writing over the signature of the building official that the issuance of a variance to construct a structure below the base flood level will result in increased premium rates for flood insurance up to amounts as high as $25 for $100 of insurance coverage, and that such construction below the base flood level increases risks to life and property.

❖ Each of the listed conditions for issuance of variances is to be addressed by the board of appeals, especially the requirement that it determine that failure to grant the variance would result in exceptional hardship by rendering the lot undevelopable. By itself, this determination may be insufficient to result in exceptional hardship if other conditions for issuance of a variance cannot be met. The determination of hardship is to be based on the unique characteristics of the site and not the personal circumstances of the applicant.

In guidance materials, FEMA cautions that economic hardship alone is not to be considered an exceptional hardship. Building officials and boards of appeals are cautioned that granting a variance does not affect how the building will be rated for the purposes of NFIP flood insurance. Even if circumstances justify granting a variance to build a floor that is below the design flood elevation, the rate used to calculate the cost of a flood insurance policy will be based on the risk to the building. Flood insurance, required by certain mortgage lenders, may be extremely expensive, costing as much as $25 for each $100 of insurance coverage. Although the applicant may not be required to purchase flood insurance, the requirement may be imposed on subsequent owners. The building official is to provide the applicant a written notice to this effect, along with the other conditions listed in this section.

SECTION G201 DEFINITIONS

G201.1 General. The following words and terms shall, for the purposes of this appendix, have the meanings shown herein. Refer to Chapter 2 of the *International Building Code* for general definitions.

❖ Definitions of terms can help in the understanding and application of the code requirements. Terms specific to this appendix are defined herein. The use and application of the general definitions of the code are found in Section 202.

G201.2 Definitions.

DEVELOPMENT. Any manmade change to improved or unimproved real estate, including but not limited to, buildings or other structures, temporary structures, temporary or permanent storage of materials, mining, dredging, filling, grading, paving, excavations, operations and other land-disturbing activities.

❖ Any activity within a flood hazard area has the potential to damage and thereby alter the characteristics of the area. Some alterations may increase flood frequency or flood levels that in turn may adversely impact adjacent property owners. This term is broadly defined so that communities examine and regulate all activities for potential flood hazard area impacts. Land development activities that are not buildings and structures are also subject to the provisions of this appendix.

FUNCTIONALLY DEPENDENT FACILITY. A facility which cannot be used for its intended purpose unless it is located or carried out in close proximity to water, such as a docking or port facility necessary for the loading or unloading of cargo or passengers, shipbuilding or ship repair. The term does not include long-term storage, manufacture, sales or service facilities.

❖ By the very nature of their use, certain buildings and structures would not be able to fulfill their intended functions if located away from bodies of water. In addition, their functions may be hampered if they are elevated in full compliance with the code. Very specific facilities are defined as functionally dependent facilities and other facilities that often are located near bodies of water are specifically excluded from the definition.

MANUFACTURED HOME. A structure that is transportable in one or more sections, built on a permanent chassis, designed for use with or without a permanent foundation when attached to the required utilities, and constructed to the Federal Mobile Home Construction and Safety Standards and rules and regulations promulgated by the U.S. Department of Housing and Urban Development. The term also includes mobile homes, park trailers, travel trailers and similar transportable structures that are placed on a site for 180 consecutive days or longer.

❖ Manufactured homes are defined separately from buildings and approvals for them are administered differently by some local jurisdictions. In flood hazard areas, manufactured homes that are not installed according to the provisions of the code will be more vulnerable to flood damage.

MANUFACTURED HOME PARK OR SUBDIVISION. A parcel (or contiguous parcels) of land divided into two or more manufactured home lots for rent or sale.

❖ Typically, manufactured home parks and subdivisions are developed by a single owner and, in most cases, the utilities, pads and foundations are installed before the lots are rented or sold. These development activities are included in this appendix (see Sections G301.1 and G301.2) so that manufactured homes that are placed or replaced will comply with the requirements for flood hazard areas.

RECREATIONAL VEHICLE. A vehicle that is built on a single chassis, 400 square feet (37.16 m^2) or less when measured at the largest horizontal projection, designed to be self-propelled or permanently towable by a light-duty truck, and designed primarily not for use as a permanent dwelling but as temporary living quarters for recreational, camping, travel or seasonal use. A recreational vehicle is ready for highway use if it is on its wheels or jacking system, is attached to the site only by quick disconnect-type utilities and security devices and has no permanently attached additions.

❖ Recreational vehicles are intended to be used temporarily and, therefore, typically are not placed on permanent foundations that are designed and constructed to resist flood damage. Due to the nature of their placement and susceptibility to movement, this appendix (see Section G601.1) provides that they not be placed in flood hazard areas subject to high-velocity wave action (V Zones) and floodways.

VARIANCE. A grant of relief from the requirements of this section which permits construction in a manner otherwise prohibited by this section where specific enforcement would result in unnecessary hardship.

❖ Within flood hazard areas, a variance, especially to build below the design flood elevation, exposes people and property to higher risks and increased potential for flood damage. For the purpose of handling requests for variances to the flood-resistant design and construction provisions of the code, specific restrictions, considerations and conditions are set forth in this appendix (see Section G105). Note that repair or rehabilitation of historic structures, as well as construction or substantial improvements of functionally dependent facilities, may be handled by variance.

VIOLATION. A development that is not fully compliant with this appendix or Section 1612, as applicable.

❖ Work that is approved by issuance of a permit is to be conducted in accordance with the code and conditions of the permit. If the actual work is not in compliance, then it is in violation. Violations are to be handled in accordance with the procedures set forth in the code (see Section 113) and this appendix (see Sections G101.4, G104.3 and G104.5).

SECTION G301 SUBDIVISIONS

G301.1 General. Any subdivision proposal, including proposals for manufactured home parks and subdivisions, or other proposed new development in a flood hazard area shall be reviewed to assure that:

1. All such proposals are consistent with the need to minimize flood damage;
2. All public utilities and facilities, such as sewer, gas, electric and water systems are located and constructed to minimize or eliminate flood damage; and
3. Adequate drainage is provided to reduce exposure to flood hazards.

❖ When land is subdivided, the opportunity arises to recognize flood hazard areas and design the lot layout to reduce future flood damage. It is during this process that many communities elect to guide development to less hazardous locations. At a minimum, public utilities are to be designed and located to avoid or minimize impairment and damage. For additional guidance, refer to APA PAS #473.

G301.2 Subdivision requirements. The following requirements shall apply in the case of any proposed subdivision, including proposals for manufactured home parks and subdivisions, any portion of which lies within a flood hazard area:

1. The flood hazard area, including floodways and areas subject to high velocity wave action, as appropriate, shall be delineated on tentative and final subdivision plats;
2. Design flood elevations shall be shown on tentative and final subdivision plats;
3. Residential building lots shall be provided with adequate buildable area outside the floodway; and
4. The design criteria for utilities and facilities set forth in this appendix and appropriate *International Codes* shall be met.

❖ To facilitate the review of subdivision proposals with respect to flood hazards, tentative and final subdivision plats are to include adequate information. Where the land to be subdivided includes areas that are not within the designated flood hazard area, lots may be laid out to avoid or minimize flood-plain impacts. Residential lots are to be laid out so that building footprints and fills or other obstructions are outside the floodway.

SECTION G401 SITE IMPROVEMENT

❖ The definition of "Development" in Section G201 includes site work necessary to place buildings or structures, as well as site improvements that do not involve buildings or structures. When buildings and structures cannot be located outside of the designated flood hazard area, they are to be elevated so that the lowest floor is at or above the DFE. The most common ways to elevate buildings or structures are on fill, on solid foundation walls surrounding crawl spaces or on posts, pilings or columns. Compacted fill, placed in accordance with the code, may be placed to raise a building pad above the DFE. Note that fill is not to be used to elevate buildings in flood hazard areas subject to high-velocity wave action (V Zones).

Before allowing fill in a designated flood hazard area, the building official is to determine that the fill will not increase flooding or cause drainage problems on neighboring properties. In areas with known drainage and flooding problems, the building official has the authority to require submission of an engineering analysis. Under Section G103.4, this analysis is required when a floodway has not been delineated and, under Sections G103.5 and G401.1, if fill will encroach into a designated floodway. Some states and communities require encroachment analyses for all large fill proposals. In areas that are already developed, increased flood levels resulting from flood hazard area activities may be considered threats to public safety and are to be mitigated to the maximum extent possible by the permit applicant.

Permit applicants may propose placing fill in designated flood hazard areas with the intent of later constructing buildings with excavated basements. The definition of "Basement" in Section 1612.2 is "the portion of a building having its floor subgrade (below ground level) on all sides." When excavated into fill, basements may be subject to damage, especially in designated flood hazard areas where waters remain high for more than a few hours. Fill materials can become saturated and provide inadequate support or water pressure can collapse below-grade walls. Basements below residential buildings are not to be constructed below the DFE, even if excavated into fill that is placed above the DFE. Basements beneath nonresidential buildings are similarly not to be built unless they are designed, constructed and certified to be dry floodproofed in accordance with flood-resistant construction provisions of the code and ASCE 24.

If elevated on individual fill pads in designated flood hazard areas, buildings will be surrounded by water during general conditions of flooding. To address emergency management concerns about evacuation and public safety, it may be appropriate to consult with local emergency personnel responsible for evacuations. This could be especially important when reviewing applications for large subdivisions or critical facilities such as hospitals and care facilities. Many states and communities have provisions that require uninterrupted access to all buildings during design flood conditions. These requirements may be administered by agencies responsible for permitting the construction of public rights-of-way.

G401.1 Development in floodways. Development or land disturbing activity shall not be authorized in the *floodway* unless it has been demonstrated through hydrologic and hydraulic analyses performed in accordance with standard engineering practice that the proposed encroachment will not result in any increase in the level of the base *flood*.

❖ See the commentary to Section G103.5 for more information.

G401.2 Flood hazard areas subject to high-velocity wave action. In *flood hazard areas* subject to high-velocity wave action:

1. New buildings and buildings that are substantially improved shall only be authorized landward of the reach of mean high tide.
2. The use of fill for structural support of buildings is prohibited.

❖ All new construction within flood hazard areas subject to high-velocity wave action (commonly called V Zones) is to be located landward of the reach of mean high tide. The use of fill to provide structural support is prohibited because areas subject to wave action are more likely to experience flood-related erosion, which may cause failure of foundations on fill. For additional guidance, refer to FEMA FIA-TB #5, *Free of Obstruction Requirements for Buildings Located in Coastal High Hazard Areas* and FEMA 55, *Coastal Construction Manual: Principles and Practices of Planning, Siting, Designing, Constructing, and Maintaining Residential Buildings in Coastal Areas.*

G401.3 Sewer facilities. All new or replaced sanitary sewer facilities, private sewage treatment plants (including all pumping stations and collector systems) and on-site waste disposal systems shall be designed in accordance with Chapter 7, ASCE 24, to minimize or eliminate infiltration of floodwaters into the facilities and discharge from the facilities into floodwaters, or impairment of the facilities and systems.

❖ New and replacement sewer facilities are to be designed in accordance with Chapter 7 of ASCE 24. Adhering to these requirements will minimize damage to such facilities, as well as the likelihood of contamination from such facilities.

G401.4 Water facilities. All new or replacement water facilities shall be designed in accordance with the provisions of Chapter 7, ASCE 24, to minimize or eliminate infiltration of floodwaters into the systems.

❖ New and replacement water supply facilities are to be designed in accordance with Chapter 7 of ASCE 24. Adhering to these requirements will minimize damage to water supply facilities and infiltration of contamination, which may be carried by floodwaters.

G401.5 Storm drainage. Storm drainage shall be designed to convey the flow of surface waters to minimize or eliminate damage to persons or property.

❖ In addition to designing a site to minimize damage from flood conditions, storm runoff is to be addressed to minimize or eliminate related damage. Storm drainage conveyance patterns and facilities are to be shown on the site plan described in Section G104.2.

G401.6 Streets and sidewalks. Streets and sidewalks shall be designed to minimize potential for increasing or aggravating flood levels.

❖ The layout of streets and sidewalks may impact both storm runoff and flood flows. The design and construction of streets and roadways, including bridges and culverts, are to reduce or eliminate damage from both storm runoff and flooding. In addition to removing flood storage from designated flood hazard areas, the placement of fill associated with the construction of roads and sidewalks has the potential to alter the flow of both storm runoff and floodwater. Bridges and culverts are to be sized to minimize their effects on flooding. Refer to Section G401.1 if such activities are proposed within floodways.

SECTION G501 MANUFACTURED HOMES

❖ This section requires that all new and substantially improved manufactured homes are to be elevated such that their lowest floors are at or above the DFE. Replacement of an existing manufactured home is considered "new construction" and must comply with the provisions of this section. In effect, manufactured homes are to meet the same flood-resistant construction requirements as other buildings and structures. In part, this provision exceeds the minimum requirements of NFIP, which allows, in some limited situations in existing manufactured home parks and subdivisions, replacement manufactured homes to be located below the DFE. Manufactured homes are to be anchored to permanent foundations to resist flotation, collapse and lateral movement. Foundations are to be reinforced and elevated so that the floor of the manufactured home is at or above the DFE. Foundations are to meet all applicable requirements to resist wind, seismic and flood loads prescribed in the code.

Manufactured homes may be placed either on prepared sites in manufactured home parks or subdivisions or on individually owned parcels of land. Manufactured home parks have a single owner, and the pads, foundations and utility hookups are rented to the manufactured home owners. As in a typical subdivision, a manufactured home subdivision normally is developed by a single developer and the lots are sold to individuals for use as manufactured home sites.

For additional guidance on installation, foundations, anchoring and ties for manufactured homes in flood hazard areas, refer to FEMA 85, *Manufactured Homes in Flood Hazard Areas.*

G501.1 Elevation. All new and replacement manufactured homes to be placed or substantially improved in a *flood hazard area* shall be elevated such that the lowest floor of the manufactured home is elevated to or above the design flood elevation.

❖ The elevation requirements set forth in Section 1612 and ASCE 24 apply to manufactured homes, thereby treating this type of construction the same as other buildings and structures. The certification of the lowest floor elevation, as required in Section 1612.5, also applies to manufactured homes that are placed in designated flood hazard areas.

G501.2 Foundations. All new and replacement manufactured homes, including substantial improvement of existing manufactured homes, shall be placed on a permanent, reinforced foundation that is designed in accordance with Section 1612.

❖ Flood loads experienced in most designated flood hazard areas include hydrostatic and hydrodynamic components. Impacts from floating debris and waves can impart additional loads. Additionally, in flood hazard areas subject to high-velocity wave action (V Zones), flood and wind loads acting simultaneously, as required in Section 1603, are to be addressed in foundation design. Permanent, reinforced foundations for manufactured homes are necessary to resist anticipated loads. A permanent foundation system for a manufactured home includes the following:

- A below-grade footing capable of providing resistance against overturning;
- Footing depth below the frost line;
- Reinforced piers, driven pilings, embedded posts or poured concrete or reinforced block foundation walls; and
- Anchors and connections (see Section G501.3).

G501.3 Anchoring. All new and replacement manufactured homes to be placed or substantially improved in a *flood hazard area* shall be installed using methods and practices which minimize flood damage. Manufactured homes shall be securely anchored to an adequately anchored foundation system to resist flotation, collapse and lateral movement. Methods of anchoring are authorized to include, but are not limited to, use of over-the-top or frame ties to ground anchors. This requirement is in addition to applicable state and local anchoring requirements for resisting wind forces.

❖ The manufactured home anchoring requirement has two elements:

1. Manufactured homes are to be anchored to a permanent, reinforced foundation to limit movement and resist uplift and overturning due to flood, wind and seismic forces; and
2. Manufactured homes are to have ties firmly attached to ground anchors to transfer loads to the ground. Manufacturer specifications for ground anchor depths typically do not include flood loads. Steel strapping, cable, chain or other approved material may be used for ties, which are to be fastened to ground anchors and drawn tight with turnbuckles or other adjustable tensioning devices. Manufacturer specifications for ties typically do not include flood loads. Ties may be over-the-top straps or frame ties, as appropriate to the manufactured home.

SECTION G601 RECREATIONAL VEHICLES

G601.1 Placement prohibited. The placement of recreational vehicles shall not be authorized in *flood hazard areas* subject to high-velocity wave action and in *floodways.*

❖ By their very nature, recreational vehicles are not normally elevated or permanently attached to a foundation and, therefore, may be very susceptible to flooding. This provision limits the permanent placement of recreational vehicles in designated flood hazard areas where velocities may be high and flood loads may be significant, such as floodways and flood hazard areas subject to high-velocity wave action.

G601.2 Temporary placement. Recreational vehicles in *flood hazard areas* shall be fully licensed and ready for highway use, and shall be placed on a site for less than 180 consecutive days.

❖ Unless otherwise prohibited under Section G601.1, recreational vehicles may be temporarily placed in designated flood hazard areas if they are fully licensed, remain highway ready and can be moved when flooding threatens. "Highway ready" means that a recreational vehicle is on its wheels or internal jacking system, is connected to the site only by quick disconnect-type utilities and security devices and has no permanently attached additions.

G601.3 Permanent placement. Recreational vehicles that are not fully licensed and ready for highway use, or that are to be placed on a site for more than 180 consecutive days, shall meet the requirements of Section G501 for manufactured homes.

❖ Recreational vehicles that do not meet the requirements of Section G601.2 are considered permanent buildings and are to be installed in accordance with the requirements for manufactured homes in Section G501.

SECTION G701 TANKS

❖ This section addresses requirements for tanks located in flood hazard areas. Physical damage as well as environment contamination and health risks result when above-ground and underground tanks are not adequately protected and restrained during conditions of flooding. Many hazardous materials, such as gas or oil, are lighter than water and, if released due to damage or unprotected vents and fill opening, contribute to contamination and risks to public health and safety. For additional guidance, reference FEMA 348 and ASCE 24.

G701.1 Underground tanks. Underground tanks in *flood hazard areas* shall be anchored to prevent flotation, collapse or lateral movement resulting from hydrostatic loads, including the effects of buoyancy, during conditions of the design *flood*.

❖ Most tanks will not be entirely filled at all times, and thus will contain at least some air that contributes to buoyancy. When the surrounding ground becomes saturated during conditions of flooding, underground tanks may be dislodged or shifted, causing damage and loss of product. Installation designs and methods are to account for buoyancy.

G701.2 Above-ground tanks. Above-ground tanks in flood hazard areas shall be elevated to or above the design *flood* elevation or shall be anchored or otherwise designed and constructed to prevent flotation, collapse or lateral movement resulting from hydrodynamic and hydrostatic loads, including the effects of buoyancy, during conditions of the design *flood*.

❖ Above-ground tanks are to be raised above the DFE, which may be accomplished by installation on platforms or other foundations that are designed to resist flood loads. Above-ground tanks that are not elevated are subject to flood loads imposed by moving water and buoyancy and are to be adequately anchored to resist flood damage. Tanks that are exposed to floodwaters may be impacted by floating debris or become dislodged, resulting in loss of product and contributing to contamination and becoming floating debris themselves.

G701.3 Tank inlets and vents. In *flood hazard areas*, tank inlets, fill openings, outlets and vents shall be:

1. At or above the design flood elevation or fitted with covers designed to prevent the inflow of floodwater or outflow of the contents of the tanks during conditions of the design *flood*.
2. Anchored to prevent lateral movement resulting from hydrodynamic and hydrostatic loads, including the effects of buoyancy, during conditions of the design *flood*.

❖ Tank inlets and vents are to be protected from the entry of floodwater either by raising them above the DEF or by protective covers or devices. In addition, inlets and vents, and the piping that serve them, are to be anchored and protected from flood damage.

SECTION G801 OTHER BUILDING WORK

❖ Certain activities that are specified in Section 105.2 are exempt from the requirement to obtain a permit, although they must not be conducted in a manner that violates the code. As explained in Section G101.3, this appendix addresses flood-resistant construction provisions that, in conjunction with provisions of the code that relate to flood-resistant design and construction, can be used to satisfy the requirements of the NFIP. Under the NFIP, all development is to meet the requirements for flood-resistant construction. In order to ensure that activities that are exempt from a permit meet specific flood-resistant provisions, those activities are specifically listed in this appendix.

G801.1 Detached accessory structures. Detached accessory structures shall be anchored to prevent flotation, collapse or lateral movement resulting from hydrostatic loads, including the effects of bouyancy, during conditions of the design *flood*. Fully enclosed accessory structures shall have flood openings to allow for the automatic entry and exit of *flood* waters.

❖ Detached accessory structures are not required to be elevated, provided they are constructed or installed to minimize damage to the structures themselves, and to avoid flotation that may adversely impact other structures or block the flow of floodwaters and drainage.

G801.2 Fences. Fences in floodways that may block the passage of floodwaters, such as stockade fences and wire mesh fences, shall meet the requirement of Section G103.5.

❖ Fences that are solid or that may trap floating leaves and debris alter the flow of water and may increase flooding on adjacent properties. Such fences in floodways are to be avoided or, if unavoidable, their impact on flooding and flood heights is to be examined as floodway encroachments

G801.3 Oil derricks. Oil derricks located in *flood hazard areas* shall be designed in conformance with the flood loads in Sections 1603.1.7 and 1612.

❖ As permanent structures, if located in flood hazard areas, oil derricks are to be designed to resist the flood loads. If located in floodways, analyses are required to determine the impact on flood levels. Attendant equipment and utilities are to meet the requirements in ASCE 24, a referenced standard in Section 1612 of the code.

G801.4 Retaining walls, sidewalks and driveways. Retaining walls, sidewalks and driveways shall meet the requirements of Section 1803.4.

❖ Retaining walls, sidewalks and driveways may involve grading and fill. When constructed in flood hazard areas they may be subject to saturation or the erosive action of moving water and waves. If placed in floodways, they may contribute to increased flood levels and flooding of adjacent properties. Section 1803.4 outlines requirements, including engineering analyses, which are applicable depending on the nature of the flood hazard area.

G801.5 Prefabricated swimming pools. Prefabricated swimming pools in *floodways* shall meet the requirements of Section G103.5.

❖ Prefabricated swimming pools are designed for ready placement in a variety of situations. However, if placed in a floodway, they would constitute an encroachment that could increase flood levels and flood damage on adjacent properties. The definition for "Swimming pool" is found in Section 3109.2.

SECTION G901 TEMPORARY STRUCTURES AND TEMPORARY STORAGE

G901.1 Temporary structures. Temporary structures shall be erected for a period of less than 180 days. Temporary structures shall be anchored to prevent flotation, collapse or lateral movement resulting from hydrostatic loads, including the effects of buoyancy, during conditions of the design *flood*. Fully enclosed temporary structures shall have flood openings to allow for the automatic entry and exit of floodwaters.

❖ To be consistent with the regulations of the NFIP, which includes a broad definition of "Development" (44 CFR §59.1), this appendix contains temporary structures and temporary storage of materials to the definition of "Development" in Appendix G. It also adds minimum requirements that apply to temporary structures and temporary storage of materials in flood hazard areas. Temporary structures are to be provided with flood openings and anchored to prevent flotation during the design flood so that they do not contribute to damage of downstream structures or blockage of bridges and culverts. Floodways are portions of riverine floodplains that are to be reserved to convey the base flood; placement of development in floodways may alter flood elevations and increase flood depths, contributing to increased damage. Prior to placement of temporary buildings or temporary storage of materials in floodways, the effect on floodways is to be considered as set forth in Section G103.5.

Temporary structures are addressed in the code (Sections 107 and 3103). Appendix G addresses flood-resistant provisions that, if adopted in conjunction with provisions of the code related to flood-resistant design and construction, offer an option for jurisdictions to satisfy the minimum requirements of the NFIP. Appendix G includes administrative provisions required by the NFIP and specific provisions for certain activities that are not within the scope of the code, including subdivisions, certain site development activities, manufactured homes, recreational vehicles and tanks.

G901.2 Temporary storage. Temporary storage includes storage of goods and materials for a period of less than 180 days. Stored materials shall not include hazardous materials.

❖ See the commentary to Section G901.1.

G901.3 Floodway encroachment. Temporary structures and temporary storage in floodways shall meet the requirements of G103.5.

❖ See the commentary to Section G901.1.

SECTION G1001 UTILITY AND MISCELLANEOUS GROUP U

G1001.1 Utility and miscellaneous Group U. Utility and miscellaneous Group U includes buildings that are accessory in character and miscellaneous structures not classified in any specific occupancy in the *International Building Code*, including, but not limited to, agricultural buildings, aircraft hangars (accessory to a one- or two-family residence), barns, carports, fences more than 6 feet (1829 mm) high, grain silos (accessory to a residential occupancy), greenhouses, livestock shelters, private garages, retaining walls, sheds, stables and towers.

❖ Utility and miscellaneous structures are now included in Appendix G because the flood-resistant provisions must be applied to all buildings and structures that are listed in Section 312 in order to minimize damage during conditions of the design flood. To be consistent with the regulations of the NFIP, which includes a broad definition of "Development" (44 CFR §59.1) and also includes all buildings and structures, this section outlines flood-resistant provisions for utility and miscellaneous Group U buildings and structures. Section 312 of the code requires that "buildings and structures of an accessory character and miscellaneous structures not classified in any specific occupancy shall be constructed, equipped and maintained to conform to the requirements of this code commensurate with the fire and life hazard incidental to their occupancy." Flood resistance is appropriate, in addition to fire and life safety, and is required for consistency with the minimum requirements of the NFIP. Appendix G addresses flood-resistant provisions that, if adopted in conjunction with provisions of the code related to flood-resistant design and construction, offer an option for jurisdictions to satisfy the minimum requirements of the NFIP. Appendix G includes administrative provisions required by the NFIP and specific provisions for certain activities that are not within the scope of the code, including subdivisions, certain site development activities, manufactured homes, recreational vehicles and tanks.

G1001.2 Flood loads. Utility and miscellaneous Group U buildings and structures, including substantial improvement of such buildings and structures, shall be anchored to prevent flotation, collapse or lateral movement resulting from flood loads, including the effects of buoyancy, during conditions of the design *flood*.

❖ See the commentary to Section G1001.1.

G1001.3 Elevation. Utility and miscellaneous Group U buildings and structures, including substantial improvement of such buildings and structures, shall be elevated such that the lowest floor, including basement, is elevated to or above the design *flood* elevation in accordance with Section 1612 of the *International Building Code*.

❖ See the commentary to Section G1001.1.

G1001.4 Enclosures below design flood elevation. Fully enclosed areas below the design flood elevation shall be at or above grade on all sides and conform to the following:

1. In *flood hazard areas* not subject to high-velocity wave action, enclosed areas shall have flood openings to allow for the automatic inflow and outflow of floodwaters.
2. In *flood hazard areas* subject to high-velocity wave action, enclosed areas shall have walls below the design flood elevation that are designed to break away or collapse from a water load less than that which would occur during the design flood, without causing collapse, displacement or other structural damage to the building or structure.

❖ See the commentary to Section G1001.1.

G1001.5 Flood-damage-resistant materials. Flood-damage-resistant materials shall be used below the design *flood* elevation.

❖ See the commentary to Section G1001.1.

G1001.6 Protection of mechanical, plumbing and electrical systems. Mechanical, plumbing and electrical systems, including plumbing fixtures, shall be elevated to or above the design *flood* elevation.

> **Exception:** Electrical systems, equipment and components, and heating, ventilating, air conditioning, and plumbing appliances, plumbing fixtures, duct systems and other service equipment shall be permitted to be located below the design *flood* elevation provided that they are designed and installed to prevent water from entering or accumulating within the components and to resist hydrostatic and hydrodynamic loads and stresses, including the effects of buoyancy, during the occurrence of flooding to the design flood elevation in compliance with the flood-resistant construction requirements of this code. Electrical wiring systems shall be permitted to be located below the design flood elevation provided they conform to the provisions of NFPA 70.

❖ See the commentary to Section G1001.1.

SECTION G1101
REFERENCED STANDARDS

ASCE 24—05	Flood Resistance Design and Construction	G103.1, G401.3, G401.4
HUD 24 CFR Part 3280 (1994)	Manufactured Home Construction and Safety Standards	G201
IBC—12	International Building Code	G102.2
NFPA 70—08	National Electrical Code	G1001.6

Bibliography

The following resource materials are referenced in this appendix or are relevant to the subject matter addressed in this appendix. See the commentary to Chapter 16 for information on obtaining additional related publications from FEMA.

APA PAS #473, *Subdivision Design in Flood Hazard Areas.* Washington, DC: American Planning Association, 1997.

ASCE 24—05, *Flood-resistant Design and Construction.* Reston, VA: American Society of Civil Engineers Structural Engineering Institute, 2005.

FEMA 44 CFR, Parts 59-73, *National Flood Insurance Program* (NFIP). Washington, DC: Federal Emergency Management Agency.

FEMA 55, *Coastal Construction Manual: Principles and Practices of Planning, Siting, Designing, Constructing, and Maintaining Residential Buildings in Coastal Areas.* Washington, DC: Federal Emergency Management Agency, 2000.

FEMA 85, *Manufactured Homes in Flood Hazard Areas.* Washington, DC: Federal Emergency Management Agency, 1985.

FEMA 265, *Managing Floodplain Development in Approximate Zone A Areas: A Guide for Obtaining and Developing Base (100-year) Flood Elevations.* Washington, DC: Federal Emergency Management Agency, 1995.

FEMA 348, *Protecting Building Utilities from Flood Damage: Principles and Practices for the Design and Construction of Flood-resistant Building Utility Systems.* Washington, DC: Federal Emergency Management Agency, 1999.

FEMA FIA-TB #1, *Openings in Foundation Walls for Buildings Located in Special Flood Hazard Areas.* Washington, DC: Federal Emergency Management Agency, 1993.

FEMA FIA-TB #2, *Flood-resistant Material Requirements for Buildings Located in Special Flood Hazard Areas.* Washington, DC: Federal Emergency Management Agency, 1993.

FEMA FIA-TB #3, *Nonresidential Floodproofing Requirements and Certification for Buildings Located in Special Flood Hazard Areas.* Washington, DC: Federal Emergency Management Agency,1993.

FEMA FIA-TB #4, *Elevator Installation for Buildings Located in Special Flood Hazard Areas.* Washington, DC: Federal Emergency Management Agency, 1993.

FEMA FIA-TB #5, *Free of Obstruction Requirements for Buildings Located in Coastal High-hazard Areas.* Washington, DC: Federal Emergency Management Agency, 1993.

FEMA FIA-TB #6, *Below Grade Parking Requirements for Buildings Located in Special Flood Hazard Areas.* Washington, DC: Federal Emergency Management Agency, 1993.

FEMA FIA-TB #7, *Wet Floodproofing Requirements for Structures Located in Special Flood Hazard Areas.* Washington, DC: Federal Emergency Management Agency, 1993.

FEMA FIA-TB #8, *Corrosion Protection for Metal Connectors in Coastal Areas for Structures Located in Special Flood Hazard Areas.* Washington, DC: Federal Emergency Management Agency,1996.

FEMA FIA-TB #9, *Design and Construction Guidance for Breakaway Walls Below Elevated Coastal Buildings.* Washington, DC: Federal Emergency Management Agency,1999.

FEMA FIA-TB #10, *Ensuring that Structures Built on Fill In or Near Special Flood Hazard Areas are Reasonably Safe from Flooding.* Washington, DC: Federal Emergency Management Agency, 2001.

FEMA FIA-TB #11, *Crawl Space Construction for Buildings Located in Special Flood Hazard Areas.* Washington, DC: Federal Emergency Management Agency, 2001.

FEMA Form 81-89 *Series, Revisions to National Flood Insurance Rate Maps: Application/Certification Forms and Instructions for Conditional Letters of Map Revision, Letters of Map Revision and Physical Map Revisions.* Washington, DC: Federal Emergency Management Agency, 2002.

FEMA (various dates). *NFIP Technical Bulletin Series.* Washington, DC: National Flood Insurance Program. [Online]. Available: http://www.fema.gov/MIT/ techbul.htm.

Appendix H: Signs

The provisions contained in this appendix are not mandatory unless specifically referenced in the adopting ordinance.

General Comments

The provisions contained in this appendix are not mandatory unless specifically referenced in the adopting ordinance, as stated in Section 101.2.1.

The regulations contained in this appendix were prompted by improper design, erection, placement, location and maintenance of outdoor signs. Improperly designed and installed signs pose a hazard to people and property. Improperly wired signs or improperly maintained electrical signs can cause fires. The criteria in this chapter are straightforward, which, in turn, facilitates straightforward enforcement. Section H102 contains definitions for nine distinct types of signs.

Purpose

The purpose of this appendix is to locate in one place the provisions that regulate construction or protection requirements for outdoor signs. This appendix also provides the owner, designer, installer and maintainer of outdoor signs with minimum safe practices presented in performance language whenever possible, thus allowing the widest possible application.

SECTION H101 GENERAL

H101.1 General. A sign shall not be erected in a manner that would confuse or obstruct the view of or interfere with exit signs required by Chapter 10 or with official traffic signs, signals or devices. Signs and sign support structures, together with their supports, braces, guys and anchors, shall be kept in repair and in proper state of preservation. The display surfaces of signs shall be kept neatly painted or posted at all times.

❖ The provisions of this appendix are confined to that which is displayed in any manner out of doors for recognized advertising purposes. An orderly procedure is to be followed in erecting and maintaining outdoor signs. Outdoor signs must not block the view of signs regulated in Section 1011 that identify the location and path of travel to exits or street signs regulated by the jurisdiction. Signs must be properly maintained in accordance with this section, including corrosion prevention on all parts of the sign and its supports. Corrosion of a structural member results in loss of cross section, which could render that member incapable of handling the load of the sign or loads (snow, wind, etc.) on the sign. The building official has the authority to order the removal of a sign that is not maintained properly, as it can represent a hazard to the public and, therefore, can be determined to be unsafe (see Section 115).

H101.2 Signs exempt from permits. The following signs are exempt from the requirements to obtain a permit before erection:

1. Painted nonilluminated signs.
2. Temporary signs announcing the sale or rent of property.
3. Signs erected by transportation authorities.
4. Projecting signs not exceeding 2.5 square feet (0.23 m^2).
5. The changing of moveable parts of an approved sign that is designed for such changes, or the repainting or repositioning of display matter shall not be deemed an alteration.

❖ The objective of this section is to describe when a permit is not required to regulate the erection, construction, alteration and maintenance of a new sign. Note that this section applies to new signs only, except for existing signs that are to be altered, and that the owner of an exempt sign is responsible for erecting and maintaining it in a safe manner (see Section H104.1 for a related issue).

A permit is not required for the erection of wall signs (commonly of limited area) without electrical service that, when not maintained, can be painted over or the surface repaired to its original state.

A sale or rental sign is temporary in nature, pertains to the premises on which it is placed and its installation is not likely to represent a hazard; for example, it is substantially fixed in place and without electrical service.

Regulation of street signs in the code would be redundant, since signs erected by the jurisdiction inevitably undergo one or more levels of regulatory review before installation. Directional (informational) signs in conjunction with transit lines, such as subways, trains, buses, ferries and airports, are exempt from permit requirements since it is generally not necessary for a jurisdiction to take out a permit for an activity it is performing.

Projecting signs that do not exceed an area of 2.5 square feet (0.23 m^2) do not require a permit, as the size limitation allows access for maintenance. The sign will likely be constructed so as not to create a hazard, for example, lack of headroom for pedestrians, protrusions into a walking surface and electrical supply disrepair.

When an existing sign is altered (i.e., enlarged, relocated, etc.), the alterations must comply with the requirements for new signs. A permit is required for such alterations. The exceptions to the requirements for alterations apply to a sign that has been designed and previously authorized (through the permit process) to have changeable or movable parts. Repainting or reposting of display matter (lettering) is not considered an alteration. If a sign is enlarged or relocated without a permit and, in turn, without a field inspection, it could add excessive loads to a roof, interfere with the exhaust or intake ventilation facilities from a building or create a potential hazard to the public.

SECTION H102
DEFINITIONS

H102.1 General. The following words and terms shall, for the purposes of this appendix, have the meanings shown herein. Refer to Chapter 2 of the *International Building Code* for general definitions.

❖ Definitions of terms that are associated with the content of this section are contained herein. These definitions can help in the understanding and application of the code requirements. The use and application of all defined terms are set forth in Section 202 of the code.

COMBINATION SIGN. A sign incorporating any combination of the features of pole, projecting and roof signs.

❖ The purpose of this definition is to identify which combination of features constitutes a combination sign. These signs are characterized by being simultaneously supported partly by one or more poles, partly by the wall, partly by the roof of a structure or any combination thereof.

DISPLAY SIGN. The area made available by the sign structure for the purpose of displaying the advertising message.

❖ A display sign constitutes the surface of a sign where the alphabetic or pictographic components of a message are arranged for display.

ELECTRIC SIGN. A sign containing electrical wiring, but not including signs illuminated by an exterior light source.

❖ This definition states that an electrical sign is any sign activated or illuminated by means of electrical energy. Electric signs are characterized by the use of artificial light projecting through, but not reflecting off, its surface(s).

GROUND SIGN. A billboard or similar type of sign which is supported by one or more uprights, poles or braces in or upon the ground other than a combination sign or pole sign, as defined by this code.

❖ This definition identifies those signs that are not part of other buildings or structures. The structural supports of these signs are such that the wind loading is directly transferred into its own foundation system rather than into other buildings or structures. Section H109.1 contains the specific material and height requirements of ground signs.

POLE SIGN. A sign wholly supported by a sign structure in the ground.

❖ Pole signs are those where the structural supports are constructed so that loads are directly transferred to the ground.

PORTABLE DISPLAY SURFACE. A display surface temporarily fixed to a standardized advertising structure which is regularly moved from structure to structure at periodic intervals.

❖ This definition identifies the surfaces available for displaying commercial or noncommercial advertising messages that are not permanently attached to a structure and are periodically placed at different locations by means of manual or remote input.

PROJECTING SIGN. A sign other than a wall sign, which projects from and is supported by a wall of a building or structure.

❖ A projecting sign is the opposite of a wall sign in that it is projecting from the wall that provides its support. Section H112 contains requirements for projecting signs to regulate their materials, maximum projections, clearances and structural capabilities.

ROOF SIGN. A sign erected upon or above a roof or parapet of a building or structure.

❖ This definition identifies that any exterior sign supported on the roof of a building or structure is considered a roof sign. Such a sign directly impacts the design of the roof because of increased wind and snow drifting loads. Section H110.1 contains explicit material and clearance requirements along with height limitations for the erection of roof signs.

SIGN. Any letter, figure, character, mark, plane, point, marquee sign, design, poster, pictorial, picture, stroke, stripe, line, trademark, reading matter or illuminated service, which shall be constructed, placed, attached, painted, erected, fastened or manufactured in any manner whatsoever, so that the same shall be used for the attraction of the public to any place, subject, person, firm, corporation, public performance, article, machine or merchandise, whatsoever, which is displayed in any manner outdoors. Every sign shall be classified and conform to the requirements of that classification as set forth in this chapter.

❖ This definition provides the basis for determining which building components are considered as signs, and thus must comply with the provisions of this appendix. For the purposes of the code, signs are always considered as exterior elements, since the code addresses the hazards associated with exterior exposure. Interior signs must meet the specific code requirements for the materials involved, along with the code provisions for interior finishes and trim (see Chapter 8). The code further subdivides signs into different groups, with specific requirements for each group.

SIGN STRUCTURE. Any structure which supports or is capable of supporting a sign as defined in this code. A sign structure is permitted to be a single pole and is not required to be an integral part of the building.

❖ Structures supporting signs, which may be affixed to the ground and include one or more columns, poles or braces placed in or upon the ground, are not required to be part of the building structural system.

WALL SIGN. Any sign attached to or erected against the wall of a building or structure, with the exposed face of the sign in a plane parallel to the plane of said wall.

❖ This definition details those signs that are part of the wall surfaces of a building or structure, or which are separate signs and are then fastened to the wall surface. Wall signs are separately fastened signs of limited extension from the wall surface; otherwise, they become projecting signs. The code allows certain-sized wall signs to be exempt from permit requirements (see Section H101.2). Section H111.1 specifies the material requirements and extension limitations of wall signs.

SECTION H103
LOCATION

H103.1 Location restrictions. Signs shall not be erected, constructed or maintained so as to obstruct any fire escape or any window or door or opening used as a *means of egress* or so as to prevent free passage from one part of a roof to any other part thereof. A sign shall not be attached in any form, shape or manner to a fire escape, nor be placed in such manner as to interfere with any opening required for ventilation.

❖ The intent of this section is to prohibit the installation of signs at locations that may interfere with either the evacuation process during an emergency situation or free movement throughout the roof of a building. Fire escapes and openings that are part of the means of egress or required for ventilation are to be free from obstructions that may affect their intended use.

SECTION H104
IDENTIFICATION

H104.1 Identification. Every outdoor advertising display sign hereafter erected, constructed or maintained, for which a permit is required shall be plainly marked with the name of the person, firm or corporation erecting and maintaining such sign and shall have affixed on the front thereof the permit number issued for said sign or other method of identification *approved* by the *building official*.

❖ All signs must bear the name of the owner to identify the party responsible for installation and maintenance. The construction documents on file must show where such identification is located on each sign. If the sign is erected improperly or becomes damaged, the field inspector will be able to identify the responsible person to whom any notification is to be directed.

SECTION H105
DESIGN AND CONSTRUCTION

H105.1 General requirements. Signs shall be designed and constructed to comply with the provisions of this code for use of materials, loads and stresses.

❖ This section addresses the general requirements pertaining to the construction, design loads and working stresses of signs. The design and construction of signs must be in accordance with all of the pertinent requirements of the code. In many instances, structural calculations and the resulting design must be submitted showing the type of materials used, their strength capabilities and the loads and resultant stresses placed on the sign and its supports (see Chapter 16).

H105.2 Permits, drawings and specifications. Where a permit is required, as provided in Chapter 1, construction documents shall be required. These documents shall show the dimensions, material and required details of construction, including loads, stresses and anchors.

❖ The intent of this section is to require construction documents when a permit is required for the erection of a sign. In order to grant a permit to the owner, the sign is to be delineated on construction documents, such as plans and specifications, in sufficient detail to allow the determination of code compliance. Section H104.1 affects this section in that it explains that the owner of a sign (new or altered) is not relieved from obtaining the necessary permit, even if he or she holds a building or other type of permit at the time the sign is under construction. The sign permit is usually a separate permit and may also require other related permits (e.g., electrical).

H105.3 Wind load. Signs shall be designed and constructed to withstand wind pressure as provided for in Chapter 16.

❖ A reference is made to Section 1609, which relates to wind loads. Although a reference is not made to snow loads (see Section 1608), consideration should also be given thereto, especially for signs with a significant horizontal surface that will have snow accumulation, such as marquee signs.

H105.4 Seismic load. Signs designed to withstand wind pressures shall be considered capable of withstanding earthquake loads, except as provided for in Chapter 16.

❖ The designer and plan reviewer are to consider the provisions set forth in Section 1613 before the sign and its construction can be considered in compliance with the code.

H105.5 Working stresses. In outdoor advertising display signs, the allowable working stresses shall conform to the requirements of Chapter 16. The working stresses of wire rope and its fastenings shall not exceed 25 percent of the ultimate strength of the rope or fasteners.

Exceptions:

1. The allowable working stresses for steel and wood shall be in accordance with the provisions of Chapters 22 and 23.
2. The working strength of chains, cables, guys or steel rods shall not exceed one-fifth of the ultimate strength of such chains, cables, guys or steel.

❖ Consideration is to be given to the imposed loads and resultant working stresses in the design of signs. The allowable working stresses are to be evaluated in accordance with the provisions of Chapter 16. Structural members are to be of the appropriate size to support the loads applied to the sign (snow, wind, etc.) or the load of the sign itself. All members and connections are to be sufficient to carry all design loads of Chapter 16 without exceeding the specified design values set forth in the referenced chapter.

An exception to this requirement applies only to steel and wood allowable working stresses, which are to comply with the provisions of Chapters 22 and 23. A limitation is placed on wire rope and fastening utilized as structural members supporting signs, where the working stresses are to be below 25 percent of the ultimate strength of the member. An exception to this limitation applies to chains, cables, guys or steel rods where the working stresses of such members may not exceed one-fifth of the ultimate strength of the members.

H105.6 Attachment. Signs attached to masonry, concrete or steel shall be safely and securely fastened by means of metal anchors, bolts or approved expansion screws of sufficient size and anchorage to safely support the loads applied.

❖ In accordance with this section, metal connectors are the prescriptive requirement for the support and transfer of loads when signs are bearing on walls, floors and roofs of masonry, concrete or steel construction. Metal connectors provide points of anchorage at intersecting components and are also used as bonding elements. Section 2103.11 contains the standards and material requirements for joint reinforcement, wire fabric, wire accessories, metal anchors, ties and accessories and the required corrosion protection of these components.

SECTION H106 ELECTRICAL

H106.1 Illumination. A sign shall not be illuminated by other than electrical means, and electrical devices and wiring shall be installed in accordance with the requirements of NFPA 70. Any open spark or flame shall not be used for display purposes unless specifically approved.

❖ The reference to NFPA 70, *National Electrical Code*, when dealing with electrical aspects is common throughout the *International Codes*® (see commentary, Chapter 27). Improperly installed lighting and wiring are a potential fire or electrocution hazard. Other illumination, utilizing such means as flame or sparks, must be approved by the building official so that such an arrangement does not induce a fire hazard for which commensurate safety precautions are not considered.

H106.1.1 Internally illuminated signs. Except as provided for in Sections 402.16 and 2611, where internally illuminated signs have facings of wood or approved plastic, the area of such facing section shall not be more than 120 square feet (11.16 m^2) and the wiring for electric lighting shall be entirely enclosed in the sign cabinet with a clearance of not less than 2 inches (51 mm) from the facing material. The dimensional limitation of 120 square feet (11.16 m^2) shall not apply to sign facing sections made from flame-resistant-coated fabric (ordinarily known as "flexible sign face plastic") that weighs less than 20 ounces per square yard (678 g/m^2) and that, when tested in accordance with NFPA 701, meets the fire propagation performance requirements of both Test 1 and Test 2 or that when tested in accordance with an approved test method, exhibits an average burn time of 2 seconds or less and a burning extent of 5.9 inches (150 mm) or less for 10 specimens.

❖ This section applies to outdoor signs that are internally illuminated. The exception is a reminder that the area limitations of this section are not applicable to internally illuminated signs located within a building. Wood and approved plastic are limited because of the need to control the amount of combustibles incorporated into the sign. Once the amount of combustibles is limited, maintenance of a 2-inch (51 mm) clearance is still necessary between this material and electrical wiring. Square-foot limitations are not imposed on flame-resistant controlled fabrics that weigh less than 20 ounces per square yard (678 g/m^2), provided that the material meets the specific criteria listed. This is because of the low fuel content of such fabrics and the resulting lower potential for fire propagation. One standard that is referenced, NFPA 701, deals with not only average flame time and average length of char, but also outdoor conditions and weathering.

H106.2 Electrical service. Signs that require electrical service shall comply with NFPA 70.

❖ This section is concerned with sign wiring being installed in a safe and protected manner so as not to create a potential tripping or electrical shock hazard. Electrical safety is as important as structural safety when dealing with illuminated signs. This section emphasizes the need for control by the building official so that the referenced NFPA 70 is utilized properly.

SECTION H107 COMBUSTIBLE MATERIALS

H107.1 Use of combustibles. Wood, approved plastic or plastic veneer panels as provided for in Chapter 26, or other materials of combustible characteristics similar to wood, used for moldings, cappings, nailing blocks, letters and latticing, shall comply with Section H109.1, and shall not be used for other ornamental features of signs, unless approved.

❖ The use of combustibles for ornamentation and sign facings is addressed in Sections H107.1.1 and H107.1.2. The control of combustible signs and portions of signs is necessary to limit the fuel load that the combustible features add to a structure. The use of the term "approved" means approved by the building official in accordance with the provisions of Chapter 26 and other applicable code sections. The reference to Section H109.1 controls the height at which combustibles may be used (see Section H109.1 for details).

H107.1.1 Plastic materials. Notwithstanding any other provisions of this code, plastic materials which burn at a rate no faster than 2.5 inches per minute (64 mm/s) when tested in accordance with ASTM D 635 shall be deemed approved plastics and can be used as the display surface material and for the letters, decorations and facings on signs and outdoor display structures.

❖ These provisions specify those plastic materials, mostly consisting of acrylics, that burn at a faster rate when tested in accordance with ASTM D 635 are to be considered approved plastics for the purpose of installation as display surface materials and alphabetic characters, ornaments and coverings on signs. ASTM D 635 measures the burning rate of a small sample held horizontally and exposed to a Bunsen burner flame to compare the relative linear rate of burning, extent and time of burning or both of plastics in the horizontal position.

H107.1.2 Electric sign faces. Individual plastic facings of electric signs shall not exceed 200 square feet (18.6 m^2) in area.

❖ Consistent with other provisions of the code to limit the spread of fire, this section places restrictions on the size of plastic facings utilized in signs activated by electrical means. This limitation is intended to control the amount of combustible materials that potentially soften or melt when subject to heat.

H107.1.3 Area limitation. If the area of a display surface exceeds 200 square feet (18.6 m^2), the area occupied or covered by approved plastics shall be limited to 200 square feet (18.6 m^2) plus 50 percent of the difference between 200 square feet (18.6 m^2) and the area of display surface. The area of plastic on a display surface shall not in any case exceed 1,100 square feet (102 m^2).

❖ This section provides requirements for how to determine the permissible amount of combustible materials when the area of a display surface exceeds the 200-square-foot (18.6 m^2) threshold. As an additional safeguard against exterior fire spread, the aggregate area of combustible materials covering a display surface should not exceed the sum of 200 square feet (18.6 m^2) and 50 percent of the difference between the limitation and the area of the display surface. However, at no time can the aggregate area of plastic in display surfaces occupy more than 1,100 square feet (102 m^2).

H107.1.4 Plastic appurtenances. Letters and decorations mounted on an approved plastic facing or display surface can be made of approved plastics.

❖ Under the provisions of this section, combustible lettering and ornaments are permitted to be mounted on display surfaces constructed of approved plastic materials.

SECTION H108 ANIMATED DEVICES

H108.1 Fail-safe device. Signs that contain moving sections or ornaments shall have fail-safe provisions to prevent the section or ornament from releasing and falling or shifting its center of gravity more than 15 inches (381 mm). The fail-safe device shall be in addition to the mechanism and the mechanism's housing which operate the movable section or ornament. The fail-safe device shall be capable of supporting the full dead weight of the section or ornament when the moving mechanism releases.

❖ Animated signs may be operated for many hours at a time, perhaps even continuously. Thus, it is imperative that if the operating mechanism malfunctions, the animated part of the sign will not release or fall, creating a hazard or damaging property.

SECTION H109 GROUND SIGNS

H109.1 Height restrictions. The structural frame of ground signs shall not be erected of combustible materials to a height of more than 35 feet (10668 mm) above the ground. Ground signs constructed entirely of noncombustible material shall not be erected to a height of greater than 100 feet (30 480 mm) above the ground. Greater heights are permitted where approved and located so as not to create a hazard or danger to the public.

❖ Combustible materials exceeding 35 feet (10 668 mm) in height are prohibited in a sign's structural frame. This correlates with the height limitation for structures of Type V construction in Table 503. The primary concern is the sign and its structure's potential involvement in a fire conflagration. The height of a noncombustible sign structure is limited to 100 feet (30 480 mm) so as to limit reasonably the area exposed to a sign or the sign's structural failure. The building official is permitted to approve signs that exceed 100 feet (30 480 mm) in height, provided that a public hazard is not created. An example of this is an area remote from other buildings and occupied areas, where structural failure would not jeopardize the public.

H109.2 Required clearance. The bottom coping of every ground sign shall be not less than 3 feet (914 mm) above the ground or street level, which space can be filled with platform decorative trim or light wooden construction.

❖ This section regulates the bottom clearance of ground signs. This clear space of 3 feet (914 mm) is permitted to be decorated with nonstructural or decorative trim or light wooden construction, provided this arrangement will not create a hazard.

H109.3 Wood anchors and supports. Where wood anchors or supports are embedded in the soil, the wood shall be pressure treated with an approved preservative.

❖ The intent of this section is to require wood support members in direct contact with the earth to be preservative treated. This provision is consistent with the requirements of Section 2304.11.4. Typically, wood species that are untreated are susceptible to decay. However, when preservative treated with chemicals in accordance with standardized procedures, wood becomes less susceptible to failure due to decay and rot.

SECTION H110 ROOF SIGNS

H110.1 General. Roof signs shall be constructed entirely of metal or other approved noncombustible material except as provided for in Sections H106.1.1 and H107.1. Provisions shall be made for electric grounding of metallic parts. Where combustible materials are permitted in letters or other ornamental features, wiring and tubing shall be kept free and insulated therefrom. Roof signs shall be so constructed as to leave a clear space of not less than 6 feet (1829 mm) between the roof level and the lowest part of the sign and shall have at least 5 feet (1524 mm) clearance between the vertical supports thereof. No portion of any roof sign structure shall project beyond an exterior wall.

Exception: Signs on flat roofs with every part of the roof accessible.

❖ Section H110.1 controls the materials, bottom clearances and arrangements of roof signs (refer to definition of "Sign" in the Section H102.1 for special sign descriptions). Noncombustible materials are required for roof signs. This arrangement will not add to the fuel load present on the building (see Sections H106.1.1 and H107.1 for exceptions that deal with sign facings, electrical wiring and electrical clearances). Grounding of metal signs for lightning control and preventing electrical shock and fire is mandatory. A 6-foot (1829 mm) clear space is to be maintained between the bottom of the sign and the roof. This arrangement will enable wind to travel under the sign, thereby minimizing the collection of snow and debris. It will also facilitate roof repair. An exception to these requirements applies to signs constructed on flat roofs where every portion of the roof is available for access.

H110.2 Bearing plates. The bearing plates of roof signs shall distribute the load directly to or upon masonry walls, steel roof girders, columns or beams. The building shall be designed to avoid overstress of these members.

❖ The provisions of this section require bearing plates of roof signs to transfer the loads directly to masonry or steel structural members. The intent is to minimize the possibility of imposing additional loads on bearing plates not prepared to carry them, which may affect the structural stability of the roof sign. The structural design is to consider the load-carrying capability of the bearing plates to avoid distress of these elements.

H110.3 Height of solid signs. A roof sign having a solid surface shall not exceed, at any point, a height of 24 feet (7315 mm) measured from the roof surface.

❖ This section specifies a height limitation on roof signs of solid construction. The 24-foot (7315 mm) threshold provides a conservative criteria where solid signs are to remain stable against the forces of wind.

H110.4 Height of open signs. Open roof signs in which the uniform open area is not less than 40 percent of total gross area shall not exceed a height of 75 feet (22 860 mm) on buildings of Type 1 or Type 2 construction. On buildings of other construction types, the height shall not exceed 40 feet (12 192 mm). Such signs shall be thoroughly secured to the building upon which they are installed, erected or constructed by iron, metal anchors, bolts, supports, chains, stranded cables, steel rods or braces and they shall be maintained in good condition.

❖ An open sign is the opposite of a closed sign (see Section H110.5) in that the membranes and edges are substantially open, thus allowing wind loads to pass through without any significant structural effects. The code allows increased heights for open signs located on the roofs of buildings based on the type of construction. On buildings of combustible construction, the height is limited to 40 feet (12.2 m), whereas on buildings of noncombustible construction, the height is permitted to be 75 feet (22.9 m). Consistent with other provisions of the code, the use of noncombustible materials will not add to the fuel load present in the building. Open signs are to be properly anchored to their supporting structure and structural supports are to be kept in proper state of preservation and free from rust to avoid failure of the members.

H110.5 Height of closed signs. A closed roof sign shall not be erected to a height greater than 50 feet (15 240 mm) above the roof of buildings of Type 1 or Type 2 construction, nor more than 35 feet (10 668 mm) above the roof of buildings of Type 3, 4 or 5 construction.

❖ As in the previous section, there are different height limitations for closed signs located on the roofs of buildings. A closed sign usually consists of membranes containing the lettering or design fastened to the structural framework with its edges also enclosed. The 50-foot (15 240 mm) threshold identifies the point at which wind loading on the sign becomes significant and a fire could go undetected for some time.

SECTION H111 WALL SIGNS

H111.1 Materials. Wall signs which have an area exceeding 40 square feet (3.72 m^2) shall be constructed of metal or other approved noncombustible material, except for nailing rails and as provided for in Sections H106.1.1 and H107.1.

❖ This section regulates the materials used and the arrangement of wall signs (see the definition of "Wall sign"). Signs in excess of 40 square feet (3.72 m^2) must be constructed of noncombustible materials. Nailing rails and the exceptions regarding sign facings, electrical wiring and clearances to electrical wires, as stipulated in Section H107.1, are permitted. Signs of less than 40 square feet (3.72 m^2) may be of combustible construction. To date, experience has shown that fire exposure from a combustible sign of this size is minimal.

H111.2 Exterior wall mounting details. Wall signs attached to *exterior walls* of solid masonry, concrete or stone shall be safely and securely attached by means of metal anchors, bolts or expansion screws of not less than $^3/_8$ inch (9.5 mm) diameter and shall be embedded at least 5 inches (127 mm). Wood blocks shall not be used for anchorage, except in the case of wall signs attached to buildings with walls of wood. A wall sign shall not be supported by anchorages secured to an unbraced parapet wall.

❖ This section indicates the size, type and location of masonry anchors intended to be used for the attachment of signs to exterior walls of masonry construction. Note that the anchorage required is a minimum $^3/_8$-inch-diameter (9.5 mm) metal anchor, bolt or expansion screw embedded a minimum of 5 inches (127 mm) in solid masonry, concrete or stone construction. The use of wood blocking for attachment of wall signs is prohibited except in buildings where the exterior walls are constructed of wood. The anchors of wall signs are not permitted to be supported by unbraced parapet walls that are not providing adequate structural support for load transfer.

H111.3 Extension. Wall signs shall not extend above the top of the wall, nor beyond the ends of the wall to which the signs are attached unless such signs conform to the requirements for roof signs, projecting signs or ground signs.

❖ Signs cannot extend above the top of the wall (access to the roof must be maintained for repair and fire fighting) or beyond the ends of the wall. The signs must be kept on the property, unless exceptions for other types of signs apply.

SECTION H112 PROJECTING SIGNS

H112.1 General. Projecting signs shall be constructed entirely of metal or other noncombustible material and securely attached to a building or structure by metal supports such as bolts, anchors, supports, chains, guys or steel rods. Staples or nails shall not be used to secure any projecting sign to any building or structure. The *dead load* of projecting signs not parallel to the building or structure and the load due to wind pressure shall be supported with chains, guys or steel rods having net cross-sectional dimension of not less than $^3/_8$ inch (9.5 mm) diameter. Such supports shall be erected or maintained at an angle of at least 45 percent (0.78 rad) with the horizontal to resist the *dead load* and at angle of 45 percent (0.78 rad) or more with the face of the sign to resist the specified wind pressure. If such projecting sign exceeds 30 square feet (2.8 m^2) in one facial area, there shall be provided at least two such supports on each side not more than 8 feet (2438 mm) apart to resist the wind pressure.

❖ This section establishes explicit provisions for the construction and anchorage of projecting signs (see the definition of "Projecting sign"). Projecting signs are required to be constructed of metal or approved noncombustible materials and are to be anchored to a building or structure only by metal bolts, anchors, supports, chains, guys or steel rods. Metal chains, guys or steel bolts with a minimum $^3/_8$-inch (9.5 mm) diameter are specifically prescribed to support the dead loads and wind pressures imposed on projecting signs perpendicular to the building or structure. These anchorage requirements are based on common construction techniques and are considered to provide a minimum resistance to pressures caused by wind loads. To resist the dead loads, these structural supports are to be located at a minimum angle of 45 degrees (0.78 rad). However, to resist wind pressures, these supports are to be at an angle of more than 45 degrees (0.78 rad) with the face of the sign. The use of staples or nails is prohibited for the attachment of projecting signs to buildings or structures. These limitations recognize that nails, when subjected to wind forces, may not provide adequate connection strength and structural resistance. A minimum of two supports is to be provided with a space between them of at least 8 feet (2438 mm) when the facial area of projecting signs exceeds 30 square feet (2.8 m^2). The intent is to provide a minimum number of structural elements so that adequate load transfer is provided.

H112.2 Attachment of supports. Supports shall be secured to a bolt or expansion screw that will develop the strength of the supporting chains, guys or steel rods, with a minimum $^5/_8$-inch (15.9 mm) bolt or lag screw, by an expansion shield. Turn buckles shall be placed in chains, guys or steel rods supporting projecting signs.

❖ This section stipulates that structural supports of projecting signs are to be secured to a bolt or lag screw with a diameter of at least $^3/_8$-inch (15.9 mm). The bolt or expansion screw is to develop the strength of the supporting members. Structural supports are to be provided with turnbuckles to adjust the tension of the members.

H112.3 Wall mounting details. Chains, cables, guys or steel rods used to support the live or dead load of projecting signs are permitted to be fastened to solid masonry walls with expansion bolts or by machine screws in iron supports, but such supports shall not be attached to an unbraced parapet wall. Where the supports must be fastened to walls made of wood, the supporting anchor bolts must go through the wall and be plated or fastened on the inside in a secure manner.

❖ This section specifies the use of expansion bolts or machine screws in iron supports to fasten tension members required to support the live or dead loads of projecting signs at all locations of solid masonry walls, except on unbraced parapets. Where walls constructed of wood are provided for fastening the supports, anchor bolts are required to penetrate the wall and be securely plated or fastened on the inside face of the wall.

H112.4 Height limitation. A projecting sign shall not be erected on the wall of any building so as to project above the roof or cornice wall or above the roof level where there is no cornice wall; except that a sign erected at a right angle to the building, the horizontal width of which sign is perpendicular to such a wall and does not exceed 18 inches (457 mm), is permitted to be erected to a height not exceeding 2 feet (610 mm) above the roof or cornice wall or above the roof level where there is no cornice wall. A sign attached to a corner of a building and parallel to the vertical line of such corner shall be deemed to be erected at a right angle to the building wall.

❖ Consistent with other provisions of the code, projecting signs cannot extend beyond the ends of the wall or above the roof or cornice wall of a building. An exception to this limitation is projecting signs constructed perpendicular to the building, with a horizontal width of 18 inches (457 mm) and a height not exceeding 2 feet (610 mm) above the top of the roof or cornice wall or roof level without a cornice wall. In accordance with this section, signs attached to a corner of a building and parallel to the corner's vertical line are considered projecting signs constructed at a right angle to the building wall. Thus, their extension beyond the end of the wall is permitted within the limitations prescribed in this section.

H112.5 Additional loads. Projecting sign structures which will be used to support an individual on a ladder or other servicing device, whether or not specifically designed for the servicing device, shall be capable of supporting the anticipated additional load, but not less than a 100-pound (445 N) concentrated horizontal load and a 300-pound (1334 N) concentrated vertical load applied at the point of assumed or most eccentric loading. The building component to which the projecting sign is attached shall also be designed to support the additional loads.

❖ The intent of this section is to require additional structural capability in projecting signs that may be subjected to loads during maintenance procedures as shown in Figure H112.5. The face of the sign and the

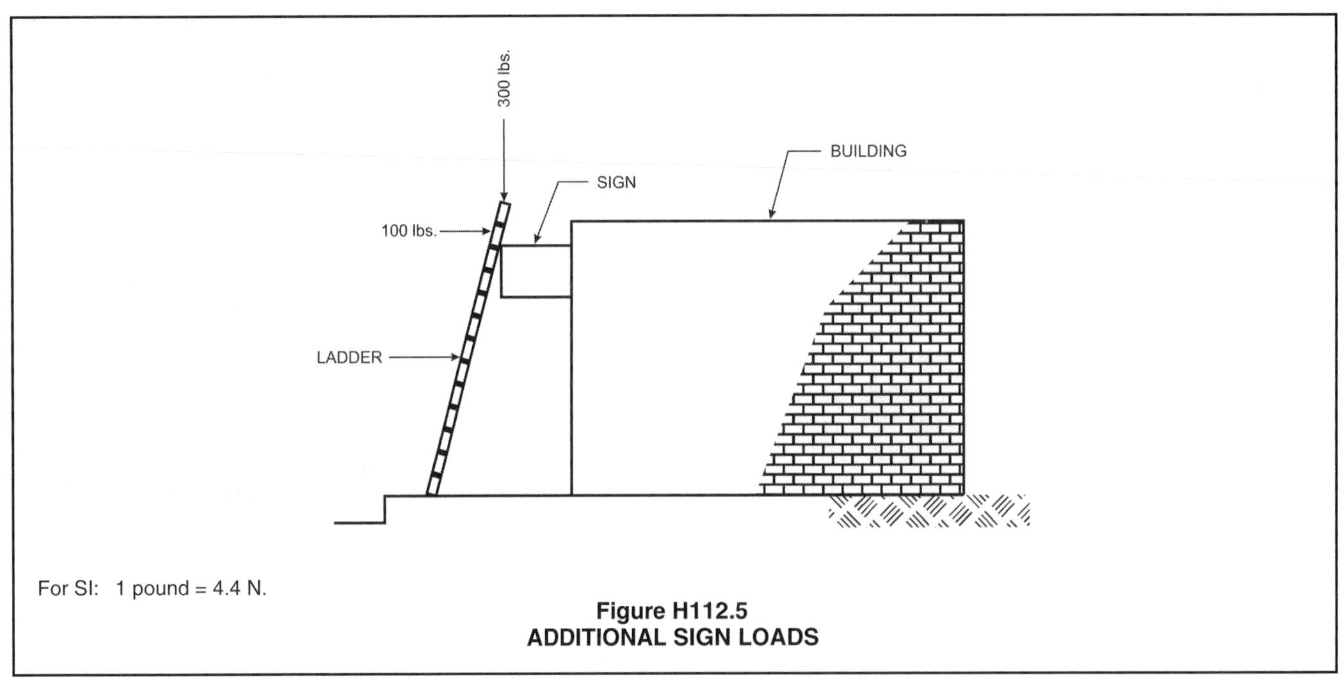

For SI: 1 pound = 4.4 N.

Figure H112.5
ADDITIONAL SIGN LOADS

supports, therefore, must be capable of withstanding not only an additional concentrated horizontal load anywhere on the face of the sign, but also a 300-pound (1334 N) concentrated load acting vertically down at the point most eccentric from the sign fasteners. The building or building component to which the sign is attached must be capable of withstanding the resultant load.

SECTION H113 MARQUEE SIGNS

H113.1 Materials. Marquee signs shall be constructed entirely of metal or other approved noncombustible material except as provided for in Sections H106.1.1 and H107.1.

❖ A marquee sign is always supported by the building or structure and projects from the building into surrounding spaces. Marquee signs are usually located on canopies or other extensions of the building or structure to obtain a greater visual effect. The code has specific provisions for marquee signs concerning allowable materials. Marquee signs must be of metal or approved noncombustible materials. Sections H113.2 through H113.4 regulate the attachment, dimensions and height limitations of a marquee sign. Section H107.1 describes exceptions regarding sign facings, electrical wiring and clearances to electrical wires.

H113.2 Attachment. Marquee signs shall be attached to approved marquees that are constructed in accordance with Section 3106.

❖ Marquee signs are attached to a structure called a "marquee." The marquee is to be designed and constructed in accordance with Section H113.1. A marquee sign should not be confused with a projecting sign (see Section H112.1).

H113.3 Dimensions. Marquee signs, whether on the front or side, shall not project beyond the perimeter of the marquee.

❖ Marquees accommodate a specially designed marquee sign arrangement. In other words, the sign fits into the marquee. A projecting sign cannot be placed on a marquee because the sign would project beyond the perimeter of the marquee.

H113.4 Height limitation. Marquee signs shall not extend more than 6 feet (1829 mm) above, nor 1 foot (305 mm) below such marquee, but under no circumstances shall the sign or signs have a vertical dimension greater than 8 feet (2438 mm).

❖ This section requires marquee signs to project less than 6 feet (1829 mm) above and not more than 1 foot (305 mm) below the marquee structure. The total vertical dimension of the marquee sign is limited to 8 feet (2438 mm). The intent of this section is to place a limitation on structures with a higher fuel load due to decorations incorporated on the sign that are subject to outdoor conditions and weathering.

SECTION H114 PORTABLE SIGNS

H114.1 General. Portable signs shall conform to requirements for ground, roof, projecting, flat and temporary signs where such signs are used in a similar capacity. The requirements of this section shall not be construed to require portable signs to have connections to surfaces, tie-downs or foundations where provisions are made by temporary means or configuration of the structure to provide stability for the expected duration of the installation.

❖ Portable signs must conform to the general regulations as set forth in this appendix for all signs, except that temporary support or stability methods may be utilized when approved by the building official. Portable signs must also conform to the requirements of this section commensurate to the manner in which the sign is used. This section establishes criteria for those types of signs that are neither mounted on buildings nor provided with a permanent foundation support. A typical example of a portable sign is a trailer-mounted illuminated sign. Such signs are used temporarily at locations and do not require any permanent support to resist wind overturning loads because of their limited size. The code contains electrical supply requirements for portable signs.

TABLE 4-A SIZE, THICKNESS AND TYPE OF GLASS PANELS IN SIGNS

MAXIMUM SIZE OF EXPOSED PANEL		MINIMUM THICKNESS OF GLASS (inches)	TYPE OF GLASS
Any dimension (inches)	Area (square inches)		
30	500	$^1/_8$	Plain, plate or wired
45	700	$^3/_{16}$	Plain, plate or wired
144	3,600	$^1/_4$	Plain, plate or wired
> 144	> 3,600	$^1/_4$	Wired glass

For SI: 1 inch = 25.4 mm, 1 square inch = 645.16 mm^2.

❖ Glazing materials installed in all types of signs are required to conform to the limitations of Table 4-A. This table details the maximum allowable size, thickness limitations and type of glass panels to be used in signs. Glass subject to unusual loading conditions may require special engineering analysis to determine its size.

TABLE 4-B
THICKNESS OF PROJECTION SIGN

PROJECTION (feet)	MAXIMUM THICKNESS (feet)
5	2
4	2.5
3	3
2	3.5
1	4

For SI: 1 foot = 304.8 mm.

❖ The requirements of Table 4-B are applicable only for projecting signs constructed in accordance with Section H112.

SECTION H115
REFERENCED STANDARDS

ASTM D 635-03	Test Method for Rate of Burning and/or Extent and Time of Burning of Self-Supporting Plastics in a Horizontal Position	H107.1.1
NFPA 70-08	National Electrical Code	H106.1, H106.2
NFPA 701-99	Methods of Fire Test for Flame Propagation of Textiles and Films	H106.1.1

Appendix I: Patio Covers

The provisions contained in this appendix are not mandatory unless specifically referenced in the adopting ordinance.

General Comments

These provisions are optional. They apply only if specifically adopted by the jurisdiction as stated in Section 101.2.1. The requirements are intended to apply only to patio covers associated with residential dwelling units.

Purpose

These provisions for patio covers are intended to simplify the requirements for residential patio installations.

SECTION I101 GENERAL

I101.1 General. Patio covers shall be permitted to be detached from or attached to *dwelling units*. Patio covers shall be used only for recreational, outdoor living purposes and not as carports, garages, storage rooms or habitable rooms.

❖ The provisions of this appendix are intended to be used in conjunction with Groups R-1, R-2, R-3 and U with attached or detached patio covers. Patio covers must be used for recreational or outdoor living purposes only and not for the uses generally ascribed to Group U and R occupancies. If used for storage or as a garage, an entire new fuel load is introduced, thereby qualifying the patio cover as a private garage governed by Sections 312 and 406.

SECTION I102 DEFINITIONS

I102.1 General. The following words and terms shall, for the purposes of this appendix, have the meanings shown herein. Refer to Chapter 2 of the *International Building Code* for general definitions.

❖ The definition of the term that is associated with the content of this appendix is provided. This definition can help in the understanding and application of the code requirements. It is important to emphasize that this term is not exclusively related to this appendix, but is applicable everywhere the term is used throughout the code. The purpose for including this definition within this appendix is to provide more convenient access to it rather than referring back to Chapter 2.

PATIO COVER. A structure with open or glazed walls which is used for recreational, outdoor living purposes associated with a dwelling unit.

❖ A patio cover is defined in order to make a distinction between carports, garages and covered awnings. The code does not consider a patio a hazardous condition with a high fuel load; therefore, special construction or consideration is not necessary. Patio covers are distinct from sunrooms, which are addressed in Chapter 12 as well as in the *International Energy Conservation Code*® (IECC®).

SECTION I103 EXTERIOR WALLS AND OPENINGS

I103.1 Enclosure walls. Enclosure walls shall be permitted to be of any configuration, provided the open or glazed area of the longer wall and one additional wall is equal to at least 65 percent of the area below a minimum of 6 feet 8 inches (2032 mm) of each wall, measured from the floor. Openings shall be permitted to be enclosed with insect screening, approved translucent or transparent plastic not more than 0.125 inch (3.2 mm) in thickness, glass conforming to the provisions of Chapter 24 or any combination of the foregoing.

❖ Despite its name, a patio cover is not limited to being simply a cover. The walls of the patio cover may be enclosed, provided a minimum of 65 percent of the longest wall and one other wall be of a material that will let light come into the area beneath the cover. The code allows simple screening, plastic materials or glass; however, the code doesn't mandate any of these. The patio cover can simply be open to the surrounding yard. Finally, the code clearly allows a mixture of materials for providing the enclosure.

I103.2 Light, ventilation and emergency egress. Exterior openings of the dwelling unit required for light and ventilation shall be permitted to open into a patio structure. However, the patio structure shall be unenclosed if such openings are serving as emergency egress or rescue openings from sleeping rooms. Where such exterior openings serve as an exit from the dwelling unit, the patio structure, unless unenclosed, shall be provided with exits conforming to the provision of Chapter 10.

❖ A patio is allowed to be placed over an opening that is to be used as a means of natural light or ventilation required by Chapter 12. Where openings of a building required for light and ventilation open to a patio cover, no further requirements are imposed on the patio cover other than the provisions of this appendix. However, if the patio covers an opening used for egress by the occupants of the building, the patio must provide an appropriate means of egress conforming with the provisions of Chapter 10.

SECTION I104 HEIGHT

I104.1 Height. Patio covers shall be limited to one-story structures not exceeding 12 feet (3657 mm) in height.

❖ The maximum height of a patio cover is 12 feet (3658 mm) (see Figure I102.1). A patio cover that is taller than 12 feet (3658 mm) should not be built under the provisions of this appendix, but would need to comply with provisions of the rest of the code.

SECTION I105 STRUCTURAL PROVISIONS

I105.1 Design loads. Patio covers shall be designed and constructed to sustain, within the stress limits of this code, all dead loads plus a minimum vertical live load of 10 pounds per square foot (0.48 kN/m^2) except that snow loads shall be used where such snow loads exceed this minimum. Such patio covers shall be designed to resist the minimum wind and seismic loads set forth in this code.

❖ This section provides a reduced roof live load from that stated in the code for typical roofs. The minimum value for the roof live load of the patio is 10 pounds per square foot (psf) (0.48 kN/m^2). Standard load combinations (as presented within the code) are still required to be considered when designing patio cover structures. Load combinations shall include dead, roof live, snow, wind and seismic forces.

I105.2 Footings. In areas with a frost depth of zero, a patio cover shall be permitted to be supported on a concrete slab on grade without footings, provided the slab conforms to the provisions of Chapter 19 of this code, is not less than $3^1/_2$ inches (89 mm) thick and further provided that the columns do not support loads in excess of 750 pounds (3.36 kN) per column.

❖ This section provides an allowance that permits foundations to be constructed without footings. The allowance takes into account the minor value of these structures and the lack of any substantive safety issues involved. This provision addresses the exemption of footings in regard to frost penetration; it does not address allowable load-bearing capacity of the soil. This provision addresses the required structural integrity of the slab, therefore eliminating the need to provide structural calculations for a slab complying with the provision when a column load is less than 750 pounds (3.36 kN).

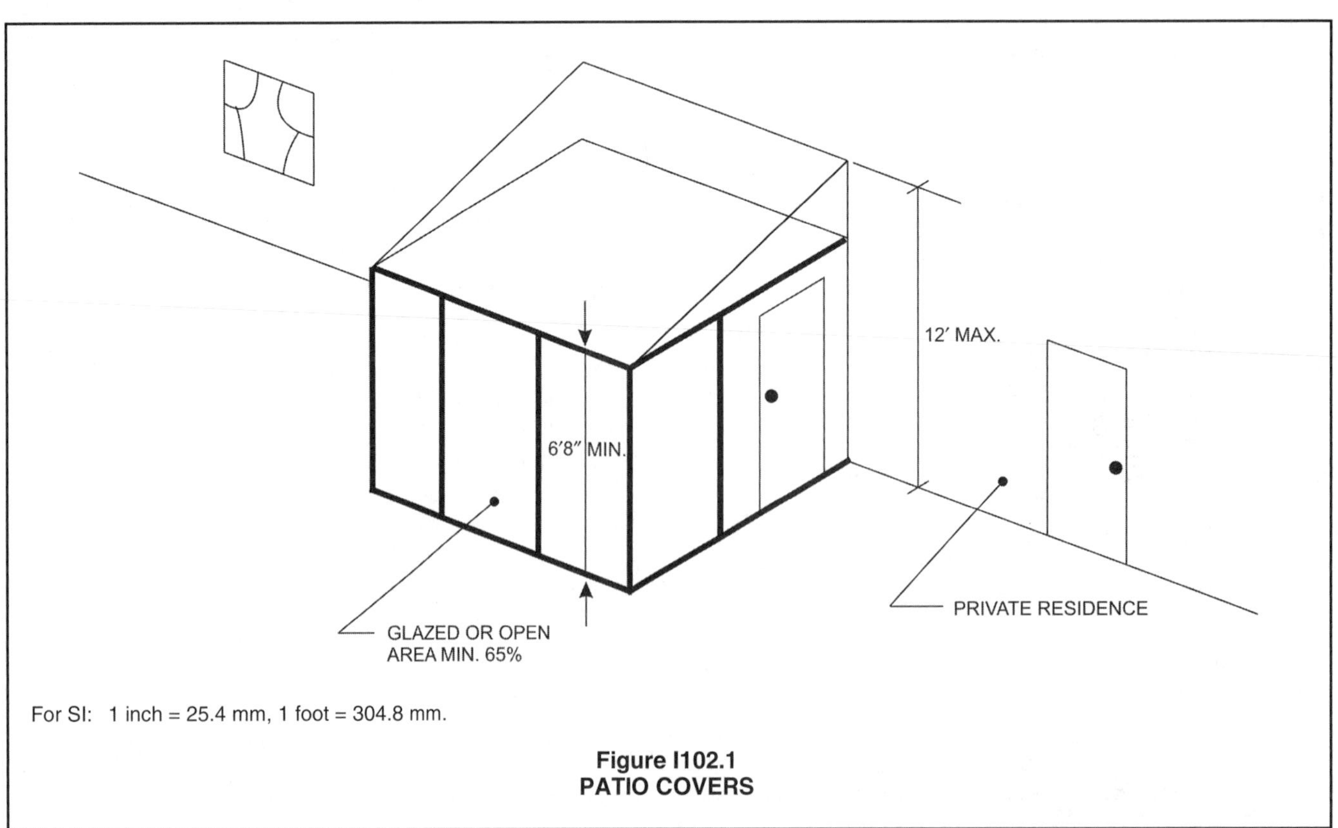

Figure I102.1
PATIO COVERS

Appendix J: Grading

The provisions contained in this appendix are not mandatory unless specifically referenced in the adopting ordinance.

General Comments

This appendix contains grading provisions that address soil-related hazards such as slope failure, landslides and erosion. Jurisdictions that include mountainous or hilly terrain often encounter grading work in connection with large commercial developments or the development of residential subdivisions that can pose grading problems that are not addressed by any other section of the code. Note that only excavation, grading and fill in connection with foundation construction is regulated by Section 1803.

In western states the *Uniform Building Code®* (UBC™) appendix chapter on excavating and grading has found very widespread usage in many areas where topography is an issue. While the need for this appendix is based on the extensive use of that UBC appendix, which was originally developed in the 1960s, suggestions from numerous individuals and groups involved in the code development process have been incorporated into this grading appendix to make it compatible for use with the code.

Purpose

This appendix provides a set of grading requirements that can be utilized in lieu of (or in addition to) developing local regulations. Where grading is an important consideration, a comprehensive code is available simply by adopting this appendix. It is intended to provide consistent and uniform code requirements anywhere grading is considered an issue.

SECTION J101 GENERAL

J101.1 Scope. The provisions of this chapter apply to grading, excavation and earthwork construction, including fills and embankments. Where conflicts occur between the technical requirements of this chapter and the geotechnical report, the geotechnical report shall govern.

❖ This section states the scope of the chapter and also states that the soils report, which is required by Section J104.3, would supercede the requirements of this chapter when a conflict exists. This gives the professional responsible for the report much latitude to provide recommendations that are appropriate for each individual site.

J101.2 Flood hazard areas. The provisions of this chapter shall not apply to grading, excavation and earthwork construction, including fills and embankments, in *floodways* within *flood hazard areas* established in Section 1612.3 or in *flood hazard areas* where design *flood* elevations are specified but floodways have not been designated, unless it has been demonstrated through hydrologic and hydraulic analyses performed in accordance with standard engineering practice that the proposed work will not result in any increase in the level of the base flood.

❖ As the definition in Section 1612.2 indicates, a floodway is that portion of a flood hazard area that is reserved for the discharge of the design flood event. The National Flood Insurance Program (NFIP) requires that the impact of development or encroachment on the floodway be considered. If the authority having jurisdiction has not established a floodway, then the earthwork construction, grading or excavation is prohibited in the flood hazard area, unless appropriate analysis has been done. The intent of this section is to permit grading in a floodway only if is demonstrated that this activity will not adversely affect surrounding areas by increasing the base flood elevation.

SECTION J102 DEFINITIONS

J102.1 Definitions. The following words and terms shall, for the purposes of this appendix, have the meanings shown herein. Refer to Chapter 2 of the *International Building Code* for general definitions.

BENCH. A relatively level step excavated into earth material on which fill is to be placed.

❖ This term is used to identify the steps that are cut into an existing slope prior to placing fill over that slope (see Figure J107.3).

COMPACTION. The densification of a fill by mechanical means.

❖ This definition aides the application of Section J107.5, which is applicable to all fill material.

CUT. See "Excavation."

❖ Earthwork is typically referred to as "cut" and "fill." The term "cut" is synonymous with excavation.

DOWN DRAIN. A device for collecting water from a swale or ditch located on or above a slope, and safely delivering it to an approved drainage facility.

❖ The inclusion of a definition for "Down drain" helps to clarify the term and make the requirements of Section J109.2 enforceable.

EROSION. The wearing away of the ground surface as a result of the movement of wind, water or ice.

❖ This definition describes what is a naturally occurring process of wearing away the ground surface. The provisions of this chapter regulate grading with the intent to minimize the susceptibility to erosion of man-made slopes primarily due to surface water.

EXCAVATION. The removal of earth material by artificial means, also referred to as a cut.

❖ Excavations are regulated by Section J106.

FILL. Deposition of earth materials by artificial means.

❖ Earthwork is typically referred to as "cut" and "fill." Fill is regulated by Section J107.

GRADE. The vertical location of the ground surface.

❖ The elevation of the existing ground surface is usually established by topographic surveys. This must be done in order to methodically evaluate a site to determine the need for regrading. The process of developing an appropriate grading plan leads to the required finished grade that must be achieved either by excavating where the existing grade is above the desired finished grade or by filling where the existing grade is below the desire finished grade.

GRADE, EXISTING. The grade prior to grading.

❖ See the definition of "Grade."

GRADE, FINISHED. The grade of the site at the conclusion of all grading efforts.

❖ See the definition of "Grade."

GRADING. An excavation or fill or combination thereof.

❖ This definition clarifies that the term "grading" refers to all earthwork.

KEY. A compacted fill placed in a trench excavated in earth material beneath the toe of a slope.

❖ This term refers to the level area that creates a notch at the bottom of a slope requiring benching in accordance with Section J107.3 (see Figure J107.3).

All slope references in the chapter have been modified to show the horizontal/vertical relationship.

SLOPE. An inclined surface, the inclination of which is expressed as a ratio of horizontal distance to vertical distance.

❖ This definition establishes that slopes in this chapter are expressed in the form of the horizontal-to-vertical distance ratio.

TERRACE. A relatively level step constructed in the face of a graded slope for drainage and maintenance purposes.

❖ This definition helps to clarify the term "terrace" and makes the requirements of Section J109.2 enforceable.

SECTION J103
PERMITS REQUIRED

J103.1 Permits required. Except as exempted in Section J103.2, no grading shall be performed without first having obtained a *permit* therefor from the *building official*. A grading *permit* does not include the construction of retaining walls or other structures.

❖ A grading permit does not include the construction of retaining walls or other structures, since that is covered by Section 1803.

J103.2 Exemptions. A grading *permit* shall not be required for the following:

1. Grading in an isolated, self-contained area, provided there is no danger to the public, and that such grading will not adversely affect adjoining properties.
2. Excavation for construction of a structure permitted under this code.
3. Cemetery graves.
4. Refuse disposal sites controlled by other regulations.
5. Excavations for wells, or trenches for utilities.
6. Mining, quarrying, excavating, processing or stockpiling rock, sand, gravel, aggregate or clay controlled by other regulations, provided such operations do not affect the lateral support of, or significantly increase stresses in, soil on adjoining properties.
7. Exploratory excavations performed under the direction of a registered design professional.

Exemption from the permit requirements of this appendix shall not be deemed to grant authorization for any work to be done in any manner in violation of the provisions of this code or any other laws or ordinances of this jurisdiction.

❖ The exemptions listed are for grading work that is relatively minor in nature (e.g., graves) or grading work that is normally regulated by other laws (e.g., utility trenching, mining, etc.).

SECTION J104
PERMIT APPLICATION AND SUBMITTALS

J104.1 Submittal requirements. In addition to the provisions of Section 105.3, the applicant shall state the estimated quantities of excavation and fill.

❖ Section 105.3 lists the information that is required for a building permit application, which also applies for a grading permit. Additionally, the quantities of earthwork, commonly referred to as "cut" and "fill," must be furnished.

J104.2 Site plan requirements. In addition to the provisions of Section 107, a grading plan shall show the existing grade and finished grade in contour intervals of sufficient clarity to indicate the nature and extent of the work and show in detail that it complies with the requirements of this code. The plans shall show the existing grade on adjoining properties in sufficient detail to identify how grade changes will conform to the requirements of this code.

❖ It is necessary for the permit applicant to furnish all applicable construction documents that are required in Section 106. Additionally, grading plans must provide sufficient grading information so that compliance with these requirements is verifiable.

J104.3 Geotechnical report. A geotechnical report prepared by a *registered design professional* shall be provided. The report shall contain at least the following:

1. The nature and distribution of existing soils;
2. Conclusions and recommendations for grading procedures;
3. Soil design criteria for any structures or embankments required to accomplish the proposed grading; and
4. Where necessary, slope stability studies, and recommendations and conclusions regarding site geology.

Exception: A geotechnical report is not required where the building code official determines that the nature of the work applied for is such that a report is not necessary.

❖ This section requires the applicant to furnish a geotechnical report unless the building official determines that it is not necessary because the proposed grading is of relatively minor consequence. The intent is that the appropriate professional provides a report that assesses the hazards of a particular site and includes recommendations for accomplishing the proposed grading. Note that Section J101.1 states that the soils report would supercede the requirements of this chapter when a conflict exists, giving great weight to the professional's recommendations.

J104.4 Liquefaction study. For sites with mapped maximum considered earthquake spectral response accelerations at short periods (S_s) greater than 0.5g as determined by Section 1613, a study of the liquefaction potential of the site shall be provided, and the recommendations incorporated in the plans.

Exception: A liquefaction study is not required where the building official determines from established local data that the liquefaction potential is low.

❖ This section intends to provide a guideline for the building official to determine if a liquefaction study is necessary. The exception provides the latitude to waive this requirement where it is evidenced that liquefaction is a low risk.

The need for a liquefaction study is triggered by the mapped spectral acceleration at short periods (see Section 1613.5.6) rather than seismic design category (see Section 1613.1), which is based on the nature of an occupancy in addition to the mapped spectral accelerations modified for the site soil classification. While seismic design category is an entirely appropriate criteria for such a requirement in connection with a building foundation under Chapter 18, the mapped spectral acceleration is more direct and appropriate for the application of this grading appendix since occupancy is most likely not applicable. While the mapped acceleration is used as the trigger for a study, it is ultimately the soil profile that determines the susceptibility to liquefaction.

SECTION J105 INSPECTIONS

J105.1 General. Inspections shall be governed by Section 109 of this code.

❖ A general reference is made to the inspection requirements in Section 109 of the code. The type and frequency of inspections specific to grading are not enumerated and are, therefore, left to the judgment of the building official.

J105.2 Special inspections. The special inspection requirements of Section 1704.7 shall apply to work performed under a grading permit where required by the *building official*.

❖ The special inspection requirements of Section 1704.7 that are referred to in this section concern site preparation and fill placement.

SECTION J106 EXCAVATIONS

J106.1 Maximum slope. The slope of cut surfaces shall be no steeper than is safe for the intended use, and shall be no steeper than two units horizontal to one unit vertical (50-percent slope) unless the owner or authorized agent furnishes a geotechnical report justifying a steeper slope.

Exceptions:

1. A cut surface shall be permitted to be at a slope of 1.5 units horizontal to one unit vertical (67-percent slope) provided that all of the following are met:
   - 1.1. It is not intended to support structures or surcharges.
   - 1.2. It is adequately protected against erosion.
   - 1.3. It is no more than 8 feet (2438 mm) in height.
   - 1.4. It is approved by the building code official.
   - 1.5. Ground water is not encountered.
2. A cut surface in bedrock shall be permitted to be at a slope of one unit horizontal to one unit vertical (100-percent slope).

❖ The code establishes a maximum two units horizontal to one unit vertical (50 percent) slope even though many professionals experienced in hillside urban development are likely to consider a one and one-half units horizontal to one unit vertical (67 percent) slope to be adequate for stability and erosion control. The two units horizontal to one unit vertical (50 percent) slope gradient is required to provide an extra margin of safety and easier maintenance capability, due to the fact that slope failures of cuts (as well as fills) with one and one-half units horizontal to one unit vertical (67 percent) slopes have been observed in the past. The cause of these failures was frequently traced to a lack of geotechnical inspection during grading or of proper maintenance after they were constructed.

The building official may allow steeper cut slopes if they are justified by the geotechnical report. Exception 1 allows the steeper cut surface slope (discussed above) of one and one-half units horizontal to one unit vertical (67 percent) where the listed conditions are met, indicating a lower slope failure hazard. A steeper cut surface slope in bedrock is permitted because it is inherently more stable.

SECTION J107
FILLS

J107.1 General. Unless otherwise recommended in the geotechnical report, fills shall comply with the provisions of this section.

❖ These provisions govern all fills. As is indicated in Section J101.1, the geotechnical report takes precedence in the event of conflicts with this chapter.

J107.2 Surface preparation. The ground surface shall be prepared to receive fill by removing vegetation, topsoil and other unsuitable materials, and scarifying the ground to provide a bond with the fill material.

❖ Fill must only be placed onto sound bedrock or other competent material to minimize subsidence or settlement; therefore, the removal of all unsuitable materials is necessary.

J107.3 Benching. Where existing grade is at a slope steeper than five units horizontal to one unit vertical (20-percent slope) and the depth of the fill exceeds 5 feet (1524 mm) benching shall be provided in accordance with Figure J107.3. A key shall be provided which is at least 10 feet (3048 mm) in width and 2 feet (610 mm) in depth.

❖ Benching requirements are illustrated in Figure J107.3. The fill over an existing slope is potentially unstable unless constructed properly. Benching creates a series of (temporary) level steps on which the fill can be reliably placed and compacted. It is required for fills over 5 feet (1524 mm) deep where the existing slope is steeper than five units horizontal to one unit vertical (20 percent). The toe-of-fill bench is referred to as a key and the code specifies a minimum depth and width. The area of ground surface left at the planned toe-of-fill slope should be graded to drain away from the fill (see Section J108.3).

This figure illustrates the intention of the benching requirements. It shows the key that is located at the toe- of-slope and the series of level benches that

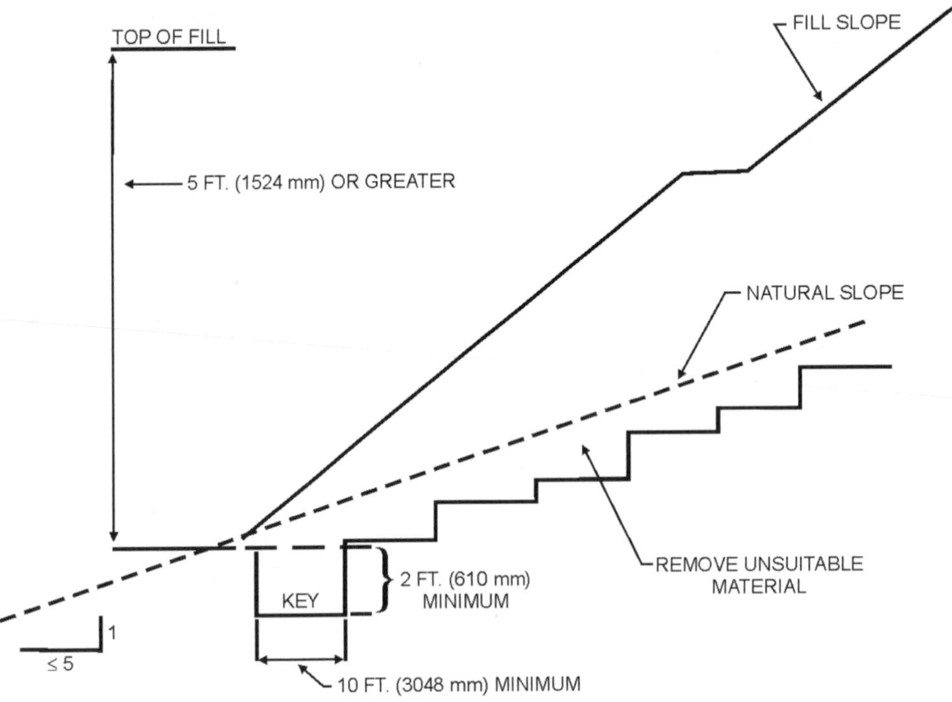

For SI: 1 foot = 304.8 mm.

FIGURE J107.3
BENCHING DETAILS

must be established in order to accomplish the proposed earthwork.

J107.4 Fill material. Fill material shall not include organic, frozen or other deleterious materials. No rock or similar irreducible material greater than 12 inches (305 mm) in any dimension shall be included in fills.

❖ Organic materials should not be allowed in fills because they decompose over time, causing settlement or creating planes of weakness. Rocks of limited size are permitted in fills. Where doing so, they should be placed to ensure that the fill material can be placed in the voids around the rocks and compacted as required.

J107.5 Compaction. All fill material shall be compacted to 90 percent of maximum density as determined by ASTM D 1557, Modified Proctor, in lifts not exceeding 12 inches (305 mm) in depth.

❖ This section specifies a minimum soil density to be achieved uniformly throughout the fill as well as the test to be used. This sets minimum acceptable levels of fill quality to achieve performance that is consistent with the purpose of this chapter.

J107.6 Maximum slope. The slope of fill surfaces shall be no steeper than is safe for the intended use. Fill slopes steeper than two units horizontal to one unit vertical (50-percent slope) shall be justified by a geotechnical report or engineering data.

❖ The maximum gradient of two units horizontal to one unit vertical (50 percent) for fill slopes minimizes the risk of slope failures (see commentary, Section J106.1). Compaction of fill can be accomplished more readily on slopes of two units horizontal to one unit vertical (50 percent) or less. Also, such slopes are more accessible, increasing the likelihood they will be maintained. Greater slopes are only permitted if substantiated by the geotechnical report or other engineering data.

SECTION J108 SETBACKS

J108.1 General. Cut and fill slopes shall be set back from the property lines in accordance with this section. Setback dimensions shall be measured perpendicular to the property line and shall be as shown in Figure J108.1, unless substantiating data is submitted justifying reduced setbacks.

❖ This section establishes minimum setbacks from property lines between the property being graded and adjacent properties. It is intended to apply at the exterior boundary of a tract and should not be applied, for instance, at the interior lot lines of a proposed development.

The figure specifies the setback requirements at the top and bottom of graded slopes.

J108.2 Top of slope. The setback at the top of a cut slope shall not be less than that shown in Figure J108.1, or than is required to accommodate any required interceptor drains, whichever is greater.

❖ The intention of the top-of-slope setback is to protect the adjacent property. The setback allows for some

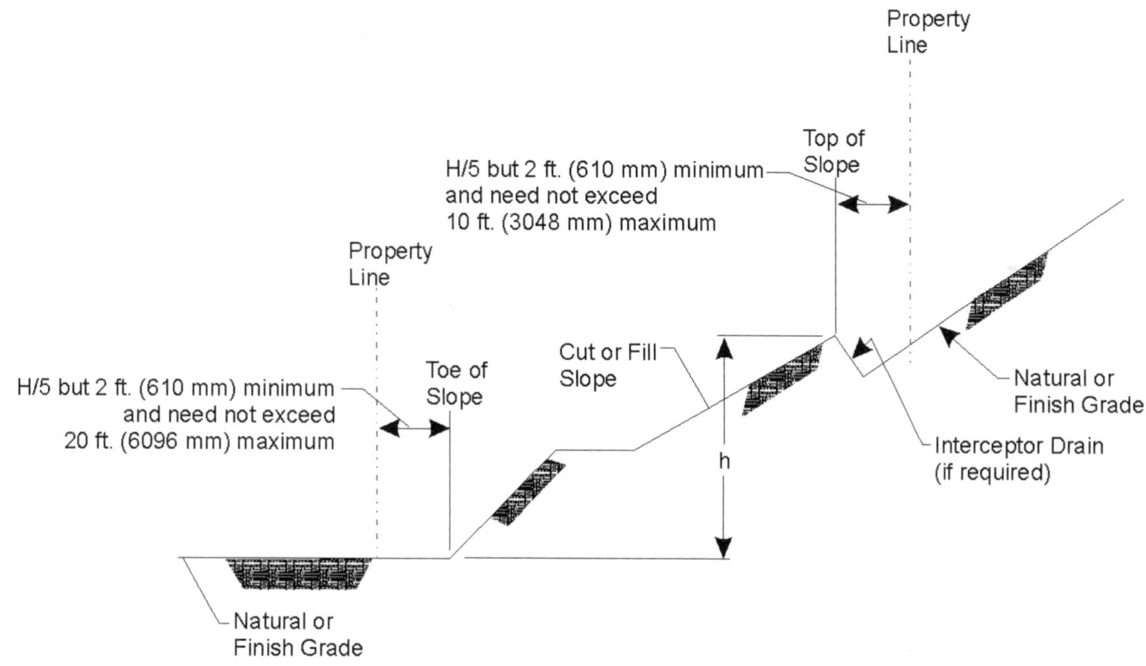

For SI: 1 foot = 304.8 mm.

FIGURE J108.1 DRAINAGE DIMENSIONS

soil erosion without losing lateral support of the adjacent property, thus protecting structures on that property. It also provides some space for an interceptor drain if needed (see Section J109.3). The top-of-slope setback requirements are illustrated in Figure J108.1.

J108.3 Slope protection. Where required to protect adjacent properties at the toe of a slope from adverse effects of the grading, additional protection, approved by the *building official*, shall be included. Such protection may include but shall not be limited to:

1. Setbacks greater than those required by Figure J108.1.
2. Provisions for retaining walls or similar construction.
3. Erosion protection of the fill slopes.
4. Provision for the control of surface waters.

❖ The minimum toe-of-fill setback from the adjacent property line is specified in Figure J108.1. Additional clearance must be provided as necessary, for example, to allow for the construction of retaining walls. For example, it may also be necessary to provide an area to collect slope runoff along the toe of slope and to construct a drain if needed.

SECTION J109 DRAINAGE AND TERRACING

J109.1 General. Unless otherwise recommended by a *registered design professional*, drainage facilities and terracing shall be provided in accordance with the requirements of this section.

Exception: Drainage facilities and terracing need not be provided where the ground slope is not steeper than 3 horizontal to 1 vertical (33 percent).

❖ This section establishes minimum requirements for drainage that are largely based on past experience. They are minimum standards and the design professional should use them as such, evaluating each site based on its own merits. Consideration also should be given to long-term performance in addition to the likelihood of inadequate maintenance of slopes in residential developments. Controlling drainage onto and off the proposed site is one of the most important considerations in grading design, since even a well-compacted fill or graded cut slope can be compromised by inadequate drainage.

J109.2 Terraces. Terraces at least 6 feet (1829 mm) in width shall be established at not more than 30-foot (9144 mm) vertical intervals on all cut or fill slopes to control surface drainage and debris. Suitable access shall be provided to allow for cleaning and maintenance.

Where more than two terraces are required, one terrace, located at approximately mid-height, shall be at least 12 feet (3658 mm) in width.

Swales or ditches shall be provided on terraces. They shall have a minimum gradient of 20 horizontal to 1 vertical (5 percent) and shall be paved with concrete not less than 3 inches (76 mm) in thickness, or with other materials suitable to the application. They shall have a minimum depth of 12 inches (305 mm) and a minimum width of 5 feet (1524 mm).

A single run of swale or ditch shall not collect runoff from a tributary area exceeding 13,500 square feet (1256 m^2) (projected) without discharging into a down drain.

❖ Terraces are required where a graded slope has a height over 30 feet (9144 mm). They provide level surfaces along the graded slope to allow access for slope maintenance and accommodate drainage.

The terrace ditch gradient of 5 percent helps make the drain more self-cleansing and recognizes a typical lack of maintenance, since most of the hillside dwellers will not climb up or down the slope to keep a drain clear of debris. Swales serving larger drainage areas must discharge into a down drain, which is defined in Section J102.1.

J109.3 Interceptor drains. Interceptor drains shall be installed along the top of cut slopes receiving drainage from a tributary width greater than 40 feet (12 192 mm), measured horizontally. They shall have a minimum depth of 1 foot (305 mm) and a minimum width of 3 feet (915 mm). The slope shall be approved by the *building official*, but shall not be less than 50 horizontal to 1 vertical (2 percent). The drain shall be paved with concrete not less than 3 inches (76 mm) in thickness, or by other materials suitable to the application. Discharge from the drain shall be accomplished in a manner to prevent erosion and shall be approved by the building official.

❖ Interceptor drains are important in protecting the face of cut slopes from excessive erosion and are, therefore, required under certain conditions. The discharge of an interceptor drain must not contribute to erosion and requires the building official's approval.

J109.4 Drainage across property lines. Drainage across property lines shall not exceed that which existed prior to grading. Excess or concentrated drainage shall be contained on site or directed to an approved drainage facility. Erosion of the ground in the area of discharge shall be prevented by installation of nonerosive down drains or other devices.

❖ This section protects adjacent properties from exposure to added runoff as a result of the grading as well as any concentrated runoff created by the grading. It requires collection of any excess or concentrated drainage and allows it to be held on site or discharged by approved methods.

SECTION J110 EROSION CONTROL

J110.1 General. The faces of cut and fill slopes shall be prepared and maintained to control erosion. This control shall be permitted to consist of effective planting.

Exception: Erosion control measures need not be provided on cut slopes not subject to erosion due to the erosion-resistant character of the materials.

Erosion control for the slopes shall be installed as soon as practicable and prior to calling for final inspection.

❖ Adequate provision should be made to prevent surface water from damaging the face of a graded slope. Effective planting is a lasting method of minimizing erosion. Other measures may include berms, interceptor drains and terrace drains.

J110.2 Other devices. Where necessary, check dams, cribbing, riprap or other devices or methods shall be employed to control erosion and provide safety.

❖ This requires use of the alternative means of providing erosion control where necessary.

SECTION J111
REFERENCED STANDARDS

ASTM D 1557-e01	Test Method for Laboratory Compaction Characteristics of Soil Using Modified Effort [56,000 ft-lb/ft^3 (2,700kN-m/m^3)].	J107.6

❖ This section lists the standards referenced by this appendix chapter. Since appendices are not considered part of the code unless specifically referred to in the adopting ordinance, the standards that are referenced in this chapter are listed here rather than in Chapter 35.

Bibliography

The following resource materials are referenced in this appendix or are relevant to the subject matter addressed in this appendix.

ASTM D 1557-e01, *Test Method for Laboratory Compaction Characteristics of Soil Using Modified Effort (56,000ft-lb/ft^2)*. West Conshohocken, PA: ASTM International, 2001.

Scullin, C. Michael. *Excavation and Grading Code Administration, Inspection and Enforcement*. Englewood Cliffs, NJ: Prentice-Hall, Inc., 1983.

UBC-97, *Uniform Building Code*. Whittier, CA: International Conference of Building Officials, 1997.

Appendix K: Administrative Provisions

The provisions contained in this appendix are not mandatory unless specifically referenced in the adopting ordinance.

With the exception of Section K111, this appendix contains only administrative provisions that are intended to be used by a jurisdiction to implement and enforce NFPA 70, the National Electrical Code. Annex H of NFPA 70 also contains administrative and enforcement provisions, and these provisions may or may not be completely compatible with or consistent with Chapter 1 of the IBC, whereas the provisions in IBC Appendix K are compatible and consistent with Chapter 1 of the IBC and other ICC codes. Section K111 contains technical provisions that are unique to this appendix and are in addition to those of NFPA 70.

The provisions of Appendix K are specific to what might be designated as an Electrical Department of Inspection and Code Enforcement and could be implemented where other such provisions are not adopted.

SECTION K101 GENERAL

K101.1 Purpose. A purpose of this code is to establish minimum requirements to safeguard public health, safety and general welfare by regulating and controlling the design, construction, installation, quality of materials, location, operation and maintenance or use of electrical systems and equipment.

K101.2 Scope. This code applies to the design, construction, installation, *alteration*, repairs, relocation, replacement, *addition* to, use or maintenance of electrical systems and equipment.

SECTION K102 APPLICABILITY

K102.1 General. The provisions of this code apply to all matters affecting or relating to structures and premises, as set forth in Section K101.

K102.2 Existing installations. Except as otherwise provided for in this chapter, a provision in this code shall not require the removal, *alteration* or abandonment of, nor prevent the continued utilization and maintenance of, existing electrical systems and equipment lawfully in existence at the time of the adoption of this code.

K102.3 Maintenance. Electrical systems, equipment, materials and appurtenances, both existing and new, and parts thereof shall be maintained in proper operating condition in accordance with the original design and in a safe, hazard-free condition. Devices or safeguards that are required by this code shall be maintained in compliance with the code edition under which installed. The owner or the owner's designated agent shall be responsible for the maintenance of the electrical systems and equipment. To determine compliance with this provision, the *building official* shall have the authority to require that the electrical systems and equipment be reinspected.

K102.4 Additions, alterations and repairs. Additions, alterations, renovations and repairs to electrical systems and equipment shall conform to that required for new electrical systems and equipment without requiring that the existing electrical systems or equipment comply with all of the requirements of this code. Additions, alterations and repairs shall not cause existing electrical systems or equipment to become unsafe, hazardous or overloaded.

Minor additions, alterations, renovations and repairs to existing electrical systems and equipment shall meet the provisions for new construction, except where such work is performed in the same manner and arrangement as was in the existing system, is not hazardous and is *approved*.

K102.5 Subjects not regulated by this code. Where no applicable standards or requirements are set forth in this code, or are contained within other laws, codes, regulations, ordinances or bylaws adopted by the jurisdiction, compliance with applicable standards of nationally recognized standards as are *approved* shall be deemed as prima facie evidence of compliance with the intent of this code. Nothing herein shall derogate from the authority of the *building official* to determine compliance with codes or standards for those activities or installations within the building official's jurisdiction or responsibility.

SECTION K103 PERMITS

K103.1 Types of permits. An owner, authorized agent or contractor who desires to construct, enlarge, alter, repair, move, demolish or change the occupancy of a building or structure, or to erect, install, enlarge, alter, repair, remove, convert or replace electrical systems or equipment, the installation of which is regulated by this code, or to cause such work to be done, shall first make application to the *building official* and obtain the required *permit* for the work.

> **Exception:** Where repair or replacement of electrical systems or equipment must be performed in an emergency situation, the *permit* application shall be submitted within the next working business day of the department of electrical inspection.

K103.2 Work exempt from permit. The following work shall be exempt from the requirement for a *permit*:

1. Listed cord- and plug-connected temporary decorative lighting.

2. Reinstallation of attachment plug receptacles, but not the outlets therefor.
3. Replacement of branch circuit overcurrent devices of the required capacity in the same location.
4. Temporary wiring for experimental purposes in suitable experimental laboratories.
5. Electrical wiring, devices, appliances, apparatus or equipment operating at less than 25 volts and not capable of supplying more than 50 watts of energy.

Exemption from the permit requirements of this code shall not be deemed to grant authorization for work to be done in violation of the provisions of this code or other laws or ordinances of this jurisdiction.

SECTION K104 CONSTRUCTION DOCUMENTS

K104.1 Information on construction documents. *Construction documents* shall be drawn to scale upon suitable material. Electronic media documents are permitted to be submitted where *approved* by the *building official. Construction documents* shall be of sufficient clarity to indicate the location, nature and extent of the work proposed and show in detail that such work will conform to the provisions of this code and relevant laws, ordinances, rules and regulations, as determined by the *building official*.

K104.2 Penetrations. *Construction documents* shall indicate where penetrations will be made for electrical systems and shall indicate the materials and methods for maintaining required structural safety, *fire-resistance rating* and *fireblocking*.

K104.3 Load calculations. Where an *addition* or *alteration* is made to an existing electrical system, an electrical load calculation shall be prepared to determine if the existing electrical service has the capacity to serve the added load.

SECTION K105 ALTERNATIVE ENGINEERED DESIGN

K105.1 General. The design, documentation, inspection, testing and approval of an alternative engineered design electrical system shall comply with this section.

K105.2 Design criteria. An alternative engineered design shall conform to the intent of the provisions of this code and shall provide an equivalent level of quality, strength, effectiveness, *fire-resistance*, durability and safety. Materials, equipment or components shall be designed and installed in accordance with the manufacturer's installation instructions.

K105.3 Submittal. The *registered design professional* shall indicate on the *permit* application that the electrical system is an alternative engineered design. The *permit* and permanent *permit* records shall indicate that an alternative engineered design was part of the *approved* installation.

K105.4 Technical data. The *registered design professional* shall submit sufficient technical data to substantiate the proposed alternative engineered design and to prove that the performance meets the intent of this code.

K105.5 Construction documents. The *registered design professional* shall submit to the *building official* two complete sets of signed and sealed *construction documents* for the alternative engineered design. The *construction documents* shall include floor plans and a diagram of the work.

K105.6 Design approval. Where the *building official* determines that the alternative engineered design conforms to the intent of this code, the electrical system shall be *approved*. If the alternative engineered design is not *approved*, the *building official* shall notify the *registered design professional* in writing, stating the reasons therefor.

K105.7 Inspection and testing. The alternative engineered design shall be tested and inspected in accordance with the requirements of this code.

SECTION K106 REQUIRED INSPECTIONS

K106.1 General. The *building official*, upon notification, shall make the inspections set forth in this section.

K106.2 Underground. Underground inspection shall be made after trenches or ditches are excavated and bedded, piping and conductors installed, and before backfill is put in place. Where excavated soil contains rocks, broken concrete, frozen chunks and other rubble that would damage or break the raceway, cable or conductors, or where corrosive action will occur, protection shall be provided in the form of granular or selected material, *approved* running boards, sleeves or other means.

K106.3 Rough-in. Rough-in inspection shall be made after the roof, framing, *fireblocking* and bracing are in place and all wiring and other components to be concealed are complete, and prior to the installation of wall or ceiling membranes.

K106.4 Contractors' responsibilities. It shall be the responsibility of every contractor who enters into contracts for the installation or repair of electrical systems for which a *permit* is required to comply with adopted state and local rules and regulations concerning licensing.

SECTION K107 PREFABRICATED CONSTRUCTION

K107.1 Prefabricated construction. Prefabricated construction is subject to Sections K107.2 through K107.5.

K107.2 Evaluation and follow-up inspection services. Prior to the approval of a prefabricated construction assembly having concealed electrical work and the issuance of an electrical *permit*, the *building official* shall require the submittal of an evaluation report on each prefabricated construction assembly, indicating the complete details of the electrical system, including a description of the system and its components, the basis upon which the system is being evaluated, test results and similar information, and other data as necessary for the *building official* to determine conformance to this code.

K107.3 Evaluation service. The *building official* shall designate the evaluation service of an *approved* agency as the evaluation agency, and review such agency's evaluation report for adequacy and conformance to this code.

K107.4 Follow-up inspection. Except where ready access is provided to electrical systems, service equipment and accessories for complete inspection at the site without disassembly or dismantling, the *building official* shall conduct the in-plant inspections as frequently as necessary to ensure conformance to the *approved* evaluation report or shall designate an independent, *approved* inspection agency to conduct such inspections. The inspection agency shall furnish the *building official* with the follow-up inspection manual and a report of inspections upon request, and the electrical system shall have an identifying label permanently affixed to the system indicating that factory inspections have been performed.

K107.5 Test and inspection records. Required test and inspection records shall be available to the *building official* at all times during the fabrication of the electrical system and the erection of the building; or such records as the *building official* designates shall be filed.

SECTION K108 TESTING

K108.1 Testing. Electrical work shall be tested as required in this code. Tests shall be performed by the *permit* holder and observed by the *building official*.

K108.1.1 Apparatus, material and labor for tests. Apparatus, material and labor required for testing an electrical system or part thereof shall be furnished by the *permit* holder.

K108.1.2 Reinspection and testing. Where any work or installation does not pass an initial test or inspection, the necessary corrections shall be made so as to achieve compliance with this code. The work or installation shall then be resubmitted to the *building official* for inspection and testing.

SECTION K109 RECONNECTION

K109.1 Connection after order to disconnect. A person shall not make utility service or energy source connections to systems regulated by this code, which have been disconnected or ordered to be disconnected by the *building official*, or the use of which has been ordered to be discontinued by the *building official* until the *building official* authorizes the reconnection and use of such systems.

SECTION K110 CONDEMNING ELECTRICAL SYSTEMS

K110.1 Authority to condemn electrical systems. Wherever the *building official* determines that any electrical system, or portion thereof, regulated by this code has become hazardous to life, health or property, the *building official* shall order in writing that such electrical systems either be removed or restored to a safe condition. A time limit for compliance with such order shall be specified in the written notice. A person shall not use or maintain a defective electrical system or equipment after receiving such notice.

Where such electrical system is to be disconnected, written notice as prescribed in this code shall be given. In cases of immediate danger to life or property, such disconnection shall be made immediately without such notice.

SECTION K111 ELECTRICAL PROVISIONS

K111.1 Adoption. Electrical systems and equipment shall be designed, constructed and installed in accordance with the *International Residential Code* or NFPA 70 as applicable, except as otherwise provided in this code.

[F] K111.2 Abatement of electrical hazards. All identified electrical hazards shall be abated. All identified hazardous electrical conditions in permanent wiring shall be brought to the attention of the *building official* responsible for enforcement of this code. Electrical wiring, devices, appliances and other equipment which is modified or damaged and constitutes an electrical shock or fire hazard shall not be used.

[F] K111.3 Appliance and fixture listing. Electrical appliances and fixtures shall be tested and *listed* in published reports of inspected electrical equipment by an *approved* agency and installed in accordance with all instructions included as part of such listing.

K111.4 Nonmetallic-sheathed cable. The use of Type NM, NMC and NMS (nonmetallic sheathed) cable wiring methods shall not be limited based on height, number of stories or construction type of the building or structure.

K111.5 Cutting, notching and boring. The cutting, notching and boring of wood and steel framing members, structural members and engineered wood products shall be in accordance with this code.

K111.6 Smoke alarm circuits. Single- and multiple-station smoke alarms required by this code and installed within *dwelling* units shall not be connected as the only load on a branch circuit. Such alarms shall be supplied by branch circuits having lighting loads consisting of lighting outlets in habitable spaces.

K111.7 Equipment and door labeling. Doors into electrical control panel rooms shall be marked with a plainly visible and legible sign stating ELECTRICAL ROOM or similar *approved* wording. The disconnecting means for each service, feeder or branch circuit originating on a switchboard or panelboard shall be legibly and durably marked to indicate its purpose unless such purpose is clearly evident.

Appendix L: Earthquake Recording Instrumentation

The provisions contained in this appendix are not mandatory unless specifically referenced in the adopting ordinance.

General Comments

Data collected from earthquake recording instrumentation provides fundamental information that is needed to better understand structural behavior and ultimately to improve the seismic performance of buildings. These provisions are based on Appendix Chapter 16, Division II of the 1997 *Uniform Building Code®* (UBC™). Because this is an optional appendix, it will only affect construction in jurisdictions that specifically adopt it. Where jurisdictions do so, the impact will further depend on whether similar provisions, such as those of the UBC, have previously been in place. Where that is the case, the cost impact should be negligible.

Purpose

The purpose of this appendix is to foster the collection of data, particularly from strong-motion earthquakes. When this data is synthesized, it can be useful in developing future improvements to the earthquake provisions of the code.

SECTION L101 GENERAL

L101.1 General. Every structure located where the 1-second spectral response acceleration, S_1, in accordance with Section 1613.3 is greater than 0.40 that either 1) exceeds six stories in height with an aggregate floor area of 60,000 square feet (5574 m^2) or more, or 2) exceeds ten stories in height regardless of floor area, shall be equipped with not less than three approved recording accelerographs. The accelerographs shall be interconnected for common start and common timing.

❖ The intent of this section is to require instrumentation in a sufficient number of multiple-story buildings, so that representative data on strong ground motion can be obtained during seismic events. As more and more seismic events are recorded and the data analyzed, the earthquake response of structures can be better understood and building code provisions can be improved accordingly. This section only requires instrumentation in newly constructed buildings. Because it only applies to very few structures in areas of higher seismic hazards, the overall cost impact on new construction is relatively small.

L101.2 Location. As a minimum, instruments shall be located at the lowest level, mid-height, and near the top of the structure. Each instrument shall be located so that access is maintained at all times and is unobstructed by room contents. A sign stating "MAINTAIN CLEAR ACCESS TO THIS INSTRUMENT" in 1-inch block letters shall be posted in a conspicuous location.

❖ This section requires at least three instruments to be located in a place where the recorded data covers the broadest range of structural response.

L101.3 Maintenance. Maintenance and service of the instrumentation shall be provided by the owner of the structure. Data produced by the instrument shall be made available to the building official on request.

Maintenance and service of the instruments shall be performed annually by an approved testing agency. The owner shall file with the building official a written report from an approved testing agency certifying that each instrument has been serviced and is in proper working condition. This report shall be submitted when the instruments are installed and annually thereafter. Each instrument shall have affixed to it an externally visible tag specifying the date of the last maintenance or service and the printed name and address of the testing agency.

❖ In order to assure that the earthquake instrumentation remains in working order, annual maintenance and service must be performed by an approved testing agency. Maintenance is the responsibility of the owner of the structure and he or she must file a written report certifying that each instrument has been serviced and is in proper working condition. The recorded data must be made available to the building official on request.

Appendix M: Tsunami-generated Flood Hazard

The provisions within this appendix are not mandatory unless referenced in the adopting ordinance.

General Comments

Tsunami is a Japanese word meaning "harbor" ("tsu") and "wave" ("nami"). It is a naturally occurring series of waves that can result when there is a rapid, large-scale disturbance of a body of water. The most common triggering events are earthquakes below or near the ocean floor, but a tsunami can also be created by volcanic activity, landslides, undersea slumps and impacts of extraterrestrial objects. In deep water, the waves; however, are gentle sea-surface slopes that can be unnoticeable; as the waves approach the shallower waters of the coast, the velocity decreases while the height increases. Upon reaching the shoreline the waves can have hazardous height and force, penetrating inland, damaging structures and flooding normally dry areas.

The combination of a great ocean seismic event with the right bathymetry can have devastating results, as was brought to the world's attention by the Indian Ocean tsunami of December 26, 2004. The tsunami created by the magnitude-9.3 underwater earthquake devastated coastal areas around the northern Indian Ocean. The tsunami took anywhere from 15 minutes to 7 hours to hit the various coastlines it affected. It is estimated that the tsunami took over 220,000 lives and displaced over 1.5 million people.

Wave propagation times from far-source-generated tsunamis can allow for advance warning to distant coastal communities. Near-source-generated tsunamis, however, can strike suddenly and with very little warning. The 1993 tsunami that hit Okushiri, Hokkaido, Japan, for example, reached the shoreline within 5 minutes after the earthquake and resulted in 202 fatalities as victims were trapped by debris from the earthquake and unable to flee toward higher ground and more secure places.

Although considered rare events, tsunamis occur on a regular basis around the world. Each year, on average, there are 20 tsunami-generating earthquake events, with five of these large enough to generate tsunami waves capable of causing damage and loss of life. In the period between 1990 and 1999 there were 82 tsunamis reported, 10 of which resulted in more than 4,000 fatalities. With the trend toward increased habitation of coastal areas, more populations will be exposed to tsunami hazards.

Alaska is considered to have the highest potential for tsunami-generating events in the United States. Earthquakes along the Alaska-Aleutian subduction zone, particularly in the vicinity of the Alaskan Peninsula, the Aleutian Islands and the Gulf of Alaska, have the capability of generating tsunamis that affect both local and distant sites. The 1964 earthquake in Prince William Sound resulted in 122 fatalities, including 12 in California and four in Oregon.

The Cascadia Subduction Zone along the Pacific Northwest coast poses a threat from northern California to British Columbia, Canada. An earthquake along the southern portion of the Cascadia Subduction Zone could create tsunami waves that would hit the coasts of Humboldt and Del Norte counties in California and Curry County in Oregon within a few minutes of the earthquake. Areas further north, along the Oregon and Washington coasts, could see tsunami waves within 20 to 40 minutes after a large earthquake. The last major earthquake on the Cascadia Subduction Zone occurred in January 1700 and generated a tsunami that struck Japan.

Communities along the entire United States Pacific coastline are at risk for both far-source-generated (trans-Pacific) tsunamis and locally triggered tsunamis. In southern California there is evidence that movement from local offshore strike-slip earthquakes and submarine landslides have generated tsunamis affecting areas extending from Santa Barbara to San Diego. The largest of these occurred in 1930, when a magnitude-5.2 earthquake reportedly generated a 20-foot-high (6096 mm) wave in Santa Monica, California.

Hawaii, located in the middle of the Pacific Ocean, has experienced both far-source-generated tsunamis and locally triggered tsunamis. The most recent damaging tsunami occurred in 1975, the result of a magnitude-7.2 earthquake off the southeast coast of the island of Hawaii. This earthquake resulted in tsunami wave heights more than 20 feet (6096 mm) and, in one area, more than 40 feet (12 192 mm). Two deaths and more than $1 million in property damage were attributed to this local Hawaiian tsunami.

Although the Atlantic and Gulf Coast regions of the United States are perceived to be at less risk, there are examples of deadly tsunamis that have occurred in the Atlantic Ocean. Since 1600, more than 40 tsunamis and tsunami-like waves have been cataloged in the eastern United States. In 1929, a tsunami generated in the Grand Banks region of Canada hit Nova Scotia, killing 51 people (Lockridge et al., 2002).

Puerto Rico and the U.S. Virgin Islands are at risk from earthquakes and underwater landslides that could occur in the Puerto Rico Trench subduction zone. Since 1530, more than 50 tsunamis of varying intensity have occurred in the Caribbean. In 1918, an earthquake in this zone generated a tsunami that caused an estimated

40 deaths in Puerto Rico. In 1867, an earthquake-generated tsunami caused damage and 12 deaths on the islands of St. Thomas and St. Croix. In 1692 a tsunami generated by massive landslides in the Puerto Rican Trench reached the coast of Jamaica, causing an estimated 2,000 deaths (Lander, 1999).

Information from historic tsunami events indicates that tsunami behaviors and characteristics are quite distinct from other coastal hazards, primarily because of the unique timescale associated with tsunami phenomena. Unlike typical wind-generated water waves with periods between 5 and 20 seconds, tsunamis can have wave periods ranging from a few minutes to over 1 hour (FEMA, 2005).

There is significant uncertainty in the prediction of characteristics of tsunamis because they are highly influenced by the tsunami waveform and the surrounding topography and bathymetry. Although there are exceptions, previous research and field surveys indicate that tsunamis have the following general characteristics:

- The magnitude of the triggering event determines the period of the resulting waves and, generally (but not always), the tsunami magnitude and damage potential (FEMA, 2005).
- A tsunami can propagate more than several thousand kilometers without losing energy.
- Tsunamis are highly reflective at the shore and capable of sustaining their motion for several hours without dissipating energy. Typically, several tsunami waves will attack a coastal area and the first wave is not necessarily the largest.
- Tsunami runup height varies significantly in neighboring areas. The configuration of the continental shelf and shoreline affect tsunami impacts at the shoreline. Variations in offshore bathymetry and shoreline irregularities can focus or disperse tsunami wave energy along certain shoreline reaches, increasing or decreasing tsunami impacts (FEMA, 2005).
- The majority of eyewitness accounts and visual records (videos and photos) indicate that an incident tsunami will break offshore, forming a bore or a series of bores as it approaches the shore. A "turbulent bore" is defined as a broken wave having a steep, violently foaming and turbulent wave front, propagating over still water of a finite depth. Such bore formations were often observed in video footage of the 2004 Indian Ocean tsunami.

Tsunamis are rare events often accompanied by advance warning. As such, strategies for protecting life safety from tsunami risk have generally involved evacuation to areas of naturally occurring high ground outside of the tsunami inundation zone. Most mitigation efforts to date have focused on the development of more effective warning systems, improved inundation maps for evacuation planning and greater tsunami awareness to improve evacuation efficiency. However, post-tsunami investigations have documented that the type of built environment in the tsunami inundation zone can play a significant role in the type and amount of losses.

Purpose

Tsunami-resistant construction is generally not practical at an individual building level, unless the building in question presents a significant risk in terms of occupant load; has a critical function in the communities' response and recovery or contains a high-hazard occupancy that can significantly add to the damage. Therefore, addressing tsunami risk for all types of structures through normal building code requirements would normally not be cost effective. However, the adoption and enforcement of code requirements regulating the presence of high-risk or high-hazard structures may be a viable alternative for at-risk communities.

SECTION M101 TSUNAMI-GENERATED FLOOD HAZARD

M101.1 General. The purpose of this appendix is to provide tsunami regulatory criteria for those communities that have a tsunami hazard and have elected to develop and adopt a map of their tsunami hazard inundation zone.

❖ While the tsunami hazard generally has a return frequency that is too low to justify code requirements for all classes of structures and design values that are too high for normal construction to withstand, it may still be in a community's best interest in controlling the presence of certain structures that present a significant risk to the population. The goal of this appendix is to provide a community with the ability to use a map of its tsunami inundation zone to regulate the new construction of structures that would present a higher than usual risk, either because of high occupancy loads, because the structure houses a critical facility necessary for the communities continued functionality or because the structure houses a hazardous function that could adversely impact the community if it was damaged or destroyed.

M101.2 Definitions. The following words and terms shall, for the purposes of this appendix, have the meanings shown herein. Refer to Chapter 2 of the *International Building Code*® (IBC®) for general definitions.

TSUNAMI HAZARD ZONE. The area vulnerable to being flooded or inundated by a design event tsunami as identified on a community's Tsunami Hazard Zone Map.

❖ A "Tsunami Hazard Zone" is defined as the geographic area that is designated on a community's Tsunami Hazard Map as being capable of being inundated in a design tsunami event. This zone is the geographic area where this appendix would apply and would be enforced by a community to restrict new construction.

TSUNAMI HAZARD ZONE MAP. A map adopted by the community that designates the extent of inundation by a design event tsunami. This map shall be based on the tsunami inundation map which is developed and provided to a community by either the applicable State agency or the National Atmospheric and Oceanic Administration (NOAA) under the National Tsunami Hazard Mitigation Program, but shall be permitted to utilize a different probability or hazard level.

❖ This appendix is based on the adoption and use of a Tsunami Hazard Map. The selection and definition of such a map is intentionally left up to the community. Such maps may be tsunami inundation maps prepared for specific communities for evacuation planning under the National Tsunami Hazard Mitigation Program (NTHMP); however, in some cases, evacuation maps may have been prepared using too restrictive criteria to justify its use for code purposes. In those cases, a tsunami hazard map may be developed by the community using its own criteria. Since the NTHMP is not a regulatory program, there are no federal or state requirements for the selection of a Tsunami Hazard Map under this appendix.

Under the NTHMP, the National Oceanic and Atmospheric Administration (NOAA) either develops community-specific tsunami inundation mapping or establishes criteria for modeling that is then used by states to prepare such mapping. Tsunami inundation mapping performed under the NTHMP uses credible worst-case scenarios. Credible worst-case scenario maps are based on a tsunami source and modeling that can be scientifically defended as a worst-case scenario for a particular community. The simulation output becomes the basis for maps that typically display maximum inundation depth and maximum current speed or velocity. For more information on these mapping criteria, see Chapter 2 of FEMA P646.

M101.3 Establishment of Tsunami Hazard Zone. Where applicable, if a community has adopted a Tsunami Hazard Zone Map, that map shall be used to establish a community's Tsunami Hazard Zone.

❖ This section contains the charging language that specifies and establishes that the Tsunami Hazard Zone, as designated on a community's adopted Tsunami Hazard Map, is the geographic area specified under this appendix.

M101.4 Construction within the Tsunami Hazard Zone. Construction of structures designated Risk Category III and IV as specified under Section 1604.5 shall be prohibited within a Tsunami Hazard Zone.

Exceptions:

1. A vertical evacuation tsunami refuge shall be permitted to be located in a Tsunami Hazard Zone provided it is constructed in accordance with FEMA P646.
2. Community critical facilities shall be permitted to be located within the Tsunami Hazard Zone when such a location is necessary to fulfill their function, providing suitable structural and emergency evacuation measures have been incorporated.

❖ This section contains the primary requirement for new construction. That requirement prohibits the construction of structures designated Risk Category III and IV under Section 1604.5 of the code within a community's Tsunami Hazard Zone. Risk Category III structures are high-occupancy buildings that represent a substantial hazard to human life in the event of failure, while Risk Category IV structures are designated as essential facilities, either because of their role in a community's response and recovery from a disaster or because they contain significant amounts of hazardous materials.

Exception 1 is for structures that have been designed and constructed to meet the vertical evacuation tsunami refuge criteria specified in FEMA P646. These structures provide a means to create areas of refuge for communities in which evacuation out of the inundation zone is not feasible. Such structures need to be designated by the community as a vertical evacuation tsunami refuge and be usable for that purpose. A community may determine and permit that such structures serve multiple uses in addition to that of a vertical evacuation tsunami refuge as described within FEMA P646. Such uses could include a parking garage, hotel or community center. When such structures are designed and constructed to the criteria specified within FEMA P646, they should be capable of withstanding the initial earthquake loads as well as any subsequent tsunami wave loads.

Exception 2 is for a community's critical facilities where such facilities are required by a community's code to be located in a Tsunami Hazard Zone. An example would be where a fire station is required by code to be located within a certain area to meet distance and response time requirements that also happens to place the station within a Tsunami Hazard Zone. In such cases, suitable structural and emergency evacuation requirements should be incorporated into the design.

SECTION M102
REFERENCED STANDARDS

FEMA P646-08	Guidelines for Design of Structures for Vertical Evacuation from Tsunamis	M101.4

❖ Vertical evacuation is a programmatic issue central to the National Tsunami Hazard Mitigation Program, driven by the fact that there are coastal communities along the West Coast of the United States that are vulnerable to tsunamis that could be generated within minutes of an earthquake on the Cascadia Subduction Zone. This guidance document includes the following information to assist in the planning and design of tsunami vertical evacuation structures: general information on the tsunami hazard and its history; guidance on determining the tsunami hazard, including tsunami depth and velocity; different options for tsunami vertical evacuation structures; guidance on siting, spacing, sizing and elevation considerations; determining tsunami and earthquake loads and related structural design criteria and structural design concepts and other considerations. FEMA P646 is available for download at www.fema.gov/library/viewRecord.do?id=3463 or may be ordered free of charge from the Federal Emergency management Agency (FEMA) by calling 1-800-480-2520.

Bibliography

The following resource materials are referenced in this chapter or are relevant to the subject matter addressed in this chapter.

FEMA P646, *Guidelines for Design of Structures for Vertical Evacuation from Tsunamis*. Washington, DC: Federal Emergency Management Agency, 2008.

INDEX

A

C

D

E

G

H

I

J

K

L

M

N

O

P

R

S

T

U

V

W

Y